D0202476

Basic Algebra II

Basic Algebra II

Second Edition

NATHAN JACOBSON
YALE UNIVERSITY

W. H. FREEMAN AND COMPANY
New York

To Mike and Polly

Library of Congress Cataloging-in-Publication Data

(Revised for vol. 2)

Jacobson, Nathan, 1910–
Basic algebra.

Includes bibliographical references and indexes.
1. Algebra. I. Title.
QA154.2.J32 1985 512.9 84-25836
ISBN 0-7167-1480-9 (v. 1)
ISBN 0-7167-1933-9 (v. 2)

Printed in the United States of America

1 2 3 4 5 6 7 8 9 0 VB 7 6 5 4 3 2 1 0 8 9

Contents

Contents of Basic Algebra I

Preface

The most extensive changes in this edition occur in the segment of the book devoted to commutative algebra, especially in Chapter 7, Commutative Ideal Theory: General Theory and Noetherian Rings; Chapter 8, Field Theory; and Chapter 9, Valuation Theory. In Chapter 7 we give an improved account of integral dependence, highlighting relations between a ring and its integral extensions ("lying over," "going-up," and "going-down" theorems). Section 7.7, Integrally Closed Domains, is new, as are three sections in Chapter 8: 8.13, Transcendency Bases for Domains; 8.18, Tensor Products of Fields; and 8.19, Free Composites of Fields. The latter two are taken from Volume III of our *Lectures in Abstract Algebra* (D. Van Nostrand 1964; Springer-Verlag, 1980). The most notable addition to Chapter 9 is Krasner's lemma, used to give an improved proof of a classical theorem of Kurschak's lemma (1913). We also give an improved proof of the theorem on extensions of absolute values to a finite dimensional extension of a field (Theorem 9.13) based on the concept of composite of a field considered in the new section 8.18.

In Chapter 4, Basic Structure Theory of Rings, we give improved accounts of the characterization of finite dimensional splitting fields of central simple algebras and of the fact that the Brauer classes of central simple algebras over a given field constitute a set—a fact which is needed to define the Brauer group $Br(F)$. In the chapter on homological algebra (Chapter 6), we give an improved proof of the existence of a projective resolution of a short exact sequence of modules.

A number of new exercises have been added and some defective ones have been deleted.

Some of the changes we have made were inspired by suggestions made by our colleagues, Walter Feit, George Seligman, and Tsuneo Tamagawa. They, as well as Ronnie Lee, Sidney Porter (a former graduate student), and the Chinese translators of this book, Professors Cao Xi-hua and Wang Jian-pan, pointed out some errors in the first edition which are now corrected. We are indeed grateful for their interest and their important inputs to the new edition. Our heartfelt thanks are due also to F. D. Jacobson, for reading the proofs of this text and especially for updating the index.

January 1989 *Nathan Jacobson*

Preface to the First Edition

This volume is a text for a second course in algebra that presupposes an introductory course covering the type of material contained in the Introduction and the first three or four chapters of *Basic Algebra* I. These chapters dealt with the rudiments of set theory, group theory, rings, modules, especially modules over a principal ideal domain, and Galois theory focused on the classical problems of solvability of equations by radicals and constructions with straight-edge and compass.

Basic Algebra II contains a good deal more material than can be covered in a year's course. Selection of chapters as well as setting limits within chapters will be essential in designing a realistic program for a year. We briefly indicate several alternatives for such a program: Chapter 1 with the addition of section 2.9 as a supplement to section 1.5, Chapters 3 and 4, Chapter 6 to section 6.11, Chapter 7 to section 7.13, sections 8.1–8.3, 8.6, 8.12, Chapter 9 to section 9.13. A slight modification of this program would be to trade off sections 4.6–4.8 for sections 5.1–5.5 and 5.9. For students who have had no Galois theory it will be desirable to supplement section 8.3 with some of the material of Chapter 4 of *Basic Algebra* I. If an important objective of a course in algebra is an understanding of the foundations of algebraic structures and the relation between algebra and mathematical logic, then all of Chapter 2 should be included in the course. This, of course, will necessitate thinning down other parts, e.g., homological algebra. There are many other possibilities for a one-year course based on this text.

The material in each chapter is treated to a depth that permits the use of the text also for specialized courses. For example, Chapters 3, 4, and 5 could constitute a one-semester course on representation theory of finite groups, and Chapter 7 and parts of Chapters 8, 9, and 10 could be used for a one-semester course in commutative algebras. Chapters 1, 3, and 6 could be used for an introductory course in homological algebra.

Chapter 11 on real fields is somewhat isolated from the remainder of the book. However, it constitutes a direct extension of Chapter 5 of *Basic Algebra* I and includes a solution of Hilbert's problem on positive semi-definite rational functions, based on a theorem of Tarski's that was proved in Chapter 5 of the first volume. Chapter 11 also includes Pfister's beautiful theory of quadratic forms that gives an answer to the question of the minimum number of squares required to express a sum of squares of rational functions of n real variables (see section 11.5).

Aside from its use as a text for a course, the book is designed for independent reading by students possessing the background indicated. A great deal of material is included. However, we believe that nearly all of this is of interest to mathematicians of diverse orientations and not just to specialists in algebra. We have kept in mind a general audience also in seeking to reduce to a minimum the technical terminology and in avoiding the creation of an overly elaborate machinery before presenting the interesting results. Occasionally we have had to pay a price for this in proofs that may appear a bit heavy to the specialist.

Many exercises have been included in the text. Some of these state interesting additional results, accompanied with sketches of proofs. Relegation of these to the exercises was motivated simply by the desire to reduce the size of the text somewhat. The reader would be well advised to work a substantial number of the exercises.

An extensive bibliography seemed inappropriate in a text of this type. In its place we have listed at the end of each chapter one or two specialized texts in which the reader can find extensive bibliographies on the subject of the chapter. Occasionally, we have included in our short list of references one or two papers of historical importance. None of this has been done in a systematic or comprehensive manner.

Again it is a pleasure for me to acknowledge the assistance of many friends in suggesting improvements of earlier versions of this text. I should mention first the students whose perceptions detected flaws in the exposition and sometimes suggested better proofs that they had seen elsewhere. Some of the students who have contributed in this way are Monica Barattieri, Ying Cheng, Daniel Corro, William Ellis, Craig Huneke, and Kenneth McKenna. Valuable suggestions have been communicated to me by Professors Kevin McCrimmon, James D. Reid, Robert L. Wilson, and Daniel Zelinsky. I have received such suggestions also from my colleagues Professors Walter Feit, George Seligman, and Tsuneo Tamagawa. The arduous task of proofreading was largely taken over by Ying Cheng, Professor Florence Jacobson, and James Reid. Florence Jacobson assisted in compiling the index. Finally we should mention the fine job of typing that was done by Joyce Harry and Donna Belli. I am greatly indebted to all of these individuals, and I take this opportunity to offer them my sincere thanks.

January 1980 *Nathan Jacobson*

Introduction

In the Introduction to *Basic Algebra* I (abbreviated throughout as "BAI") we gave an account of the set theoretic concepts that were needed for that volume. These included the power set $\mathscr{P}(S)$ of a set S, the Cartesian product $S_1 \times S_2$ of two sets S_1 and S_2, maps (=functions), and equivalence relations. In the first volume we generally gave preference to constructive arguments and avoided transfinite methods altogether.

The results that are presented in this volume require more powerful tools, particularly for the proofs of certain existence theorems. Many of these proofs will be based on a result, called Zorn's lemma, whose usefulness for proving such existence theorems was first noted by Max Zorn. We shall require also some results on the arithmetic of cardinal numbers. All of this fits into the framework of the Zermelo–Fraenkel axiomatization of set theory, including the axiom of choice (the so-called ZFC set theory). Two excellent texts that can be used to fill in the details omitted in our discussion are P. R. Halmos' *Naive Set Theory* and the more substantial *Set Theory and the Continuum Hypothesis* by P. J. Cohen.

Classical mathematics deals almost exclusively with structures based on sets. On the other hand, category theory—which will be introduced in Chapter 1—

deals with collections of sets, such as all groups, that need to be treated differently from sets. Such collections are called classes. A brief indication of a suitable foundation for category theory is given in the last section of this Introduction.

0.1 ZORN'S LEMMA

We shall now formulate a maximum principle of set theory called Zorn's lemma. We state this first for subsets of a given set. We recall that a set C of subsets of a set S (that is, a subset of the power set $\mathscr{P}(S)$) is called a *chain* if C is totally ordered by inclusion, that is, for any $A, B \in C$ either $A \subset B$ or $B \subset A$. A set T of subsets of S is called *inductive* if the union $\bigcup A_\alpha$ of any chain $C = \{A_\alpha\} \subset T$ is a member of T. We can now state

ZORN'S LEMMA (First formulation). *Let T be a non-vacuous set of subsets of a set S. Assume T is inductive. Then T contains a maximal element, that is, there exists an $M \in T$ such that no $A \in T$ properly contains M.*

There is another formulation of Zorn's lemma in terms of partially ordered sets (BAI, p. 456). Let $P, \geqslant$ be a partially ordered set. We call $P, \geqslant$ (*totally* or *linearly*) *ordered* if for every $a, b \in P$ either $a \geqslant b$ or $b \geqslant a$. We call P *inductive* if every non-vacuous subset C of P that is (totally) ordered by $\geqslant$ as defined in P has a least upper bound in P, that is, there exists a $u \in P$ such that $u \geqslant a$ for every $a \in C$ and if $v \geqslant a$ for every $a \in C$ then $v \geqslant u$. Then we have

ZORN'S LEMMA (Second formulation). *Let $P, \geqslant$ be a partially ordered set that is inductive. Then P contains maximal elements, that is, there exists $m \in P$ such that no $a \in P$ satisfies $m < a$.*

It is easily seen that the two formulations of Zorn's lemma are equivalent, so there is no harm in referring to either as "Zorn's lemma." It can be shown that Zorn's lemma is equivalent to the

AXIOM OF CHOICE. *Let S be a set, $\mathscr{P}(S)^*$ the set of non-vacuous subsets of S. Then there exists a map f (a "choice function") of $\mathscr{P}(S)^*$ into S such that $f(A) \in A$ for every $A \in \mathscr{P}(S)^*$.*

This is equivalent also to the following: If $\{A_\alpha\}$ is a set of non-vacuous sets A_α, then the Cartesian product $\prod A_\alpha \neq \varnothing$.

The statement that the axiom of choice implies Zorn's lemma can be proved by an argument that was used by E. Zermelo to prove that every set can be well ordered (see Halmos, pp. 62–65). A set S is *well ordered* by an order relation $\geqslant$ if every non-vacuous subset of S has a least element. The well-ordering theorem is also equivalent to Zorn's lemma and to the axiom of choice. We shall illustrate the use of Zorn's lemma in the next section.

0.2 ARITHMETIC OF CARDINAL NUMBERS

Following Halmos, we shall first state the main results on cardinal arithmetic without defining cardinal numbers. We say that the sets A and B have the *same cardinality* and indicate this by $|A| = |B|$ if there exists a bijective map of A onto B. We write $|A| \leqslant |B|$ if there is an injective map of A into B and $|A| < |B|$ if $|A| \leqslant |B|$ and $|A| \neq |B|$. Using these notations, the Schröder-Bernstein theorem (BAI, p. 25) can be stated as: $|A| \leqslant |B|$ and $|B| \leqslant |A|$ if and only if $|A| = |B|$. A set F is *finite* if $|F| = |N|$ for some $N = \{0, 1, \ldots, n-1\}$ and A is *countably infinite* if $|A| = |\omega|$ for $\omega = \{0, 1, 2, \ldots\}$. It follows from the axiom of choice that if A is infinite (=not finite), then $|\omega| \leqslant |A|$. We also have Cantor's theorem that for any A, $|A| < |\mathscr{P}(A)|$.

We write $C = A \dot{\cup} B$ for sets A, B, C if $C = A \cup B$ and $A \cap B = \varnothing$. It is clear that if $|A_1| \leqslant |A_2|$ and $|B_1| \leqslant |B_2|$, then $|A_1 \dot{\cup} B_1| \leqslant |A_2 \dot{\cup} B_2|$. Let $C = F \dot{\cup} \omega$ where F is finite, say, $F = \{x_0, \ldots, x_{n-1}\}$ where $x_i \neq x_j$ for $i \neq j$. Then the map of C into ω such that $x_i \rightsquigarrow i$, $k \rightsquigarrow k+n$ for $k \in \omega$ is bijective. Hence $|C| = |\omega|$. It follows from $|\omega| \leqslant |A|$ for any infinite A that if $C = F \dot{\cup} A$, then $|C| = |A|$. For we can write $A = D \dot{\cup} B$ where $|D| = |\omega|$. Then we have a bijective map of $F \cup D$ onto D and we can extend this by the identity on B to obtain a bijective map of C onto A.

We can use the preceding result and Zorn's lemma to prove the main result on "addition of cardinals," which can be stated as: If A is infinite and $C = A \cup B$ where $|B| = |A|$, then $|C| = |A|$. This is clear if A is countable from the decomposition $\omega = \{0, 2, 4, \ldots\} \dot{\cup} \{1, 3, 5, \ldots\}$. It is clear also that the result is equivalent to $|A \times 2| = |A|$ if $2 = \{0, 1\}$, since $|A \times 2| = |A \dot{\cup} B|$. We proceed to prove that $|A \times 2| = |A|$ for infinite A. Consider the set of pairs (X, f) where X is an infinite subset of A and f is a bijective map of $X \times 2$ onto X. This set is not vacuous, since A contains countably infinite subsets X and for such an X we have bijective maps of $X \times 2$ onto X. We order the set $\{(X, f)\}$ by $(X, f) \leqslant (X', f')$ if $X \subset X'$ and f' is an extension of f. It is clear that $\{(X, f)\}, \leqslant$ is an inductive partially ordered set. Hence, by Zorn's lemma, we have a maximal (Y, g) in $\{(X, f)\}$. We claim that $A - Y$ is finite. For, if $A - Y$ is infinite, then this contains a countably infinite subset D, and g can be extended to a bijective map of $(Y \dot{\cup} D) \times 2$ onto $Y \dot{\cup} D$ contrary to the maximality of

(Y, g). Thus $A - Y$ is finite. Then

$$|A \times 2| = |((Y \times 2) \cup (A - Y)) \times 2| = |Y \times 2| \\ = |Y| = |A|.$$

We can extend the last result slightly to

$$|A \dot{\cup} B| = |B| \tag{1}$$

if B is infinite and $|B| \geqslant |A|$. This follows from

$$|B| \leqslant |A \cup B| \leqslant |B \times 2| = |B|.$$

The reader is undoubtedly familiar with the fact that $|A \times A| = |A|$ if A is countably infinite, which is obtained by enumerating $\omega \times \omega$ as

$$(0,0), (0,1), (1,0), (0,2), (1,1), (2,0), \ldots .$$

More generally we have $|A \times A| = |A|$ for any infinite A. The proof is similar to the one for addition. We consider the set of pairs (X, f) where X is an infinite subset of A and f is a bijective map of $X \times X$ onto X and we order the set $\{(X, f)\}$ as before. By Zorn's lemma, we have a maximal (Y, g) in this set. The result we wish to prove will follow if $|Y| = |A|$. Hence suppose $|Y| < |A|$. Then the relation $A = Y \dot{\cup} (A - Y)$ and the result on addition imply that $|A| = |A - Y|$. Hence $|A - Y| > |Y|$. Then $A - Y$ contains a subset Z such that $|Z| = |Y|$. Let $W = Y \cup Z$, so $W = Y \dot{\cup} Z$ and $W \times W$ is the disjoint union of the sets $Y \times Y$, $Y \times Z$, $Z \times Y$, and $Z \times Z$. We have

$$|(Y \times Z) \cup (Z \times Y) \cup (Z \times Z)| = |Z \times Z| = |Z|.$$

Hence we can extend g to a bijective map of $W \times W$ onto W. This contradicts the maximality of (Y, g). Hence $|Y| = |A|$ and the proof is complete.

We also have the stronger result that if $A \neq \varnothing$ and B is infinite and $|B| \geqslant |A|$, then

$$|A \times B| = |B|. \tag{2}$$

This follows from

$$|B| \leqslant |A \times B| \leqslant |B \times B| = |B|.$$

0.3 ORDINAL AND CARDINAL NUMBERS

In axiomatic set theory no distinction is made between sets and elements. One writes $A \in B$ for "the set A is a member of the set B." This must be distinguished from $A \subset B$, which reads "A is a subset of B." (In the texts on set theory one finds $A \subseteq B$ for our $A \subset B$ and $A \subset B$ for our $A \subsetneq B$.) One defines $A \subset B$ to mean that $C \in A \Rightarrow C \in B$. One of the axioms of set theory is that given an arbitrary set C of sets, there is a set that is the union of the sets belonging to C, that is, for any set C there exists a set U such that $A \in U$ if and only if there exists a B such that $A \in B$ and $B \in C$. In particular, for any set A we can form the *successor* $A^+ = A \cup \{A\}$ where $\{A\}$ is the set having the single member A.

The process of forming successors gives a way of defining the set ω $(=\mathbb{N})$ of natural numbers. We define $0 = \varnothing$, $1 = \varnothing^+ = \{\varnothing\}$, $2 = 1^+, \ldots, n+1 = n^+, \ldots$ and we define ω to be the union of the set of natural numbers n. The natural number n and the set ω are ordinal numbers in a sense that we shall now define. First, we define a set S to be *transitive* if $A \in S$ and $B \in A \Rightarrow B \in S$. This is equivalent to saying that every member of S is a subset. We can now give

DEFINITION 0.1. *An* ordinal *is a set α that is well ordered by $\in$ and is transitive.*

The condition $A \in A$ is excluded by the axioms of set theory. We write $A \leqslant B$ for $A \in \alpha$, $B \in \alpha$ if $A \in B$ or $A = B$. It is readily seen that every natural number n is an ordinal and the set ω of natural numbers is an ordinal. Also $\omega^+, (\omega^+)^+, \ldots$ are ordinals. The union of these sets is also an ordinal. This is denoted as $\omega + \omega$ or $\omega \times 2$.

We shall now state without proofs some of the main properties of ordinal numbers.

Two partially ordered sets S_1, $\leqslant_1$ and S_2, $\leqslant_2$ are said to be *similar* if there exists an order-preserving bijective map of S_1 onto S_2. The ordinals constitute a set of representatives for the similarity classes of well-ordered sets. For we have the following theorem: If S, $\leqslant$ is well ordered, then there exists a unique ordinal α and a unique bijective order-preserving map of S onto α. If α and β are ordinals, either $\alpha = \beta$, $\alpha < \beta$, or $\beta < \alpha$. An ordinal α is called a *successor* if there exists an ordinal β such that $\alpha = \beta^+$. Otherwise, α is called a *limit ordinal.* Any non-vacuous set of ordinals has a least element.

DEFINITION 0.2. *A* cardinal number *is an ordinal α such that if β is any ordinal such that the cardinality $|\beta| = |\alpha|$, then $\alpha \leqslant \beta$.*

A cardinal number is either finite or it is a limit ordinal. On the other hand, not every limit ordinal is a cardinal. For example $\omega + \omega$ is not a cardinal. The smallest infinite cardinal is ω. Cardinals are often denoted by the Hebrew letter "aleph" $\aleph$ with a subscript. In this notation one writes $\aleph_0$ for ω.

Since any set can be well ordered, there exists a uniquely determined cardinal α such that $|\alpha| = |S|$ for any given set S. We shall now call α the *cardinal number* or *cardinality* of S and indicate this by $|S| = \alpha$. The results that we obtained in the previous section yield the following formulas for cardinalities

$$|A \cup B| = |B| \tag{3}$$

if B is infinite and $|B| \geqslant |A|$. Here, unlike in equation (1), $|S|$ denotes the cardinal number of the set S. Similarly we have

$$|A \times B| = |B| \tag{4}$$

if A is not vacuous and $|B|$ is infinite and $\geqslant |A|$.

0.4 SETS AND CLASSES

There is an axiomatization of set theory different from the ZF system that permits the discussion of collections of sets that may not themselves be sets. This is a system of axioms that is called the Gödel–Bernays (or GB) system. The primitive objects in this system are "classes" and "sets" or more precisely class variables and set variables together with a relation $\in$. A characteristic feature of this system is that classes that are members of other classes are sets, that is, we have the axiom: $Y \in X \Rightarrow Y$ is a set.

Intuitively classes may be thought of as collections of sets that are defined by certain properties. A part of the GB system is concerned with operations that can be performed on classes, corresponding to combinations of properties. A typical example is the intersection of classes, which is expressed by the following: For all X and Y there exists a Z such that $u \in Z$ if and only if $u \in X$ and $u \in Y$. We have given here the intuitive meaning of the axiom: $\forall X \forall Y \exists Z \forall u \, (u \in Z \Leftrightarrow u \in X$ and $u \in Y)$ where $\forall$ is read "for all" and $\exists$ is read "there exists." Another example is that for every X there exists a Y such that $u \in Y$ if and only if $u \notin X$ ($\forall X \exists Y \forall u \, (u \in Y \Leftrightarrow \sim u \in X)$, where $\sim \cdots$ is the negation of $\cdots$). Other class formations correspond to unions, etc. We refer to Cohen's book for a discussion of the ZF and the GB systems and their relations. We note here only that it can be shown that any theorem of ZF is a theorem of GB and every theorem of GB that refers only to sets is a theorem of ZF.

In the sequel we shall use classes in considering categories and in a few other places where we encounter classes and then show that they are sets by showing that they can be regarded as members of other classes. The familiar algebraic structures (groups, rings, fields, modules, etc.) will always be understood to be "small," that is, to be based on sets.

REFERENCES

Paul R. Halmos, *Naive Set Theory*, Springer, New York, 1960.

Paul J. Cohen, *Set Theory and the Continuum Hypothesis*, Addison-Wesley, Reading, Mass., 1966.

1

Categories

In this chapter and the next one on universal algebra we consider two unifying concepts that permit us to study simultaneously certain aspects of a large number of mathematical structures. The concept we shall study in this chapter is that of category, and the related notions of functor and natural transformation. These were introduced in 1945 by Eilenberg and MacLane to provide a precise meaning to the statement that certain isomorphisms are "natural." A typical example is the natural isomorphism between a finite-dimensional vector space V over a field and its double dual V^{**}, the space of linear functions on the space V^* of linear functions on V. The isomorphism of V onto V^{**} is the linear map associating with any vector $x \in V$ the evaluation function $f \rightsquigarrow f(x)$ defined for all $f \in V^*$. To describe the "naturality" of this isomorphism, Eilenberg and MacLane had to consider simultaneously all finite-dimensional vector spaces, the linear transformations between them, the double duals of the spaces, and the corresponding linear transformations between them. These considerations led to the concepts of categories and functors as preliminaries to defining natural transformation. We shall discuss a generalization of this example in detail in section 1.3.

The concept of a category is made up of two ingredients: a class of objects and a class of morphisms between them. Usually the objects are sets and the morphisms are certain maps between them, e.g., topological spaces and continuous maps. The definition places on an equal footing the objects and the morphisms. The adoption of the category point of view represents a shift in emphasis from the usual one in which objects are primary and morphisms secondary. One thereby gains precision by making explicit at the outset the morphisms that are allowed between the objects collected to form a category.

The language and elementary results of category theory have now pervaded a substantial part of mathematics. Besides the everyday use of these concepts and results, we should note that categorical notions are fundamental in some of the most striking new developments in mathematics. One of these is the extension of algebraic geometry, which originated as the study of solutions in the field of complex numbers of systems of polynomial equations with complex coefficients, to the study of such equations over an arbitrary commutative ring. The proper foundation of this study, due mainly to A. Grothendieck, is based on the categorical concept of a scheme. Another deep application of category theory is K. Morita's equivalence theory for modules, which gives a new insight into the classical Wedderburn–Artin structure theorem for simple rings and plays an important role in the extension of a substantial part of the structure theory of algebras over fields to algebras over commutative rings.

A typical example of a category is the category of groups. Here one considers "all" groups, and to avoid the paradoxes of set theory, the foundations need to be handled with greater care than is required in studying group theory itself. One way of avoiding the well-known difficulties is to adopt the Gödel–Bernays distinction between sets and classes. We shall follow this approach, a brief indication of which was given in the Introduction.

In this chapter we introduce the principal definitions of category theory—functors, natural transformations, products, coproducts, universals, and adjoints—and we illustrate these with many algebraic examples. This provides a review of a large number of algebraic concepts. We have included some nontrivial examples in order to add a bit of seasoning to a discussion that might otherwise appear too bland.

1.1 DEFINITION AND EXAMPLES OF CATEGORIES

DEFINITION 1.1. *A* category $\mathbf{C}$ *consists of*

1. *A class ob* $\mathbf{C}$ *of* objects (*usually denoted as* A, B, C, *etc.*).
2. *For each ordered pair of objects* (A, B), *a set* $hom_{\mathbf{C}}(A, B)$ (*or simply* $hom(A, B)$ *if* $\mathbf{C}$ *is clear*) *whose elements are called* morphisms *with* domain A *and* codomain B (*or* from A to B).

3. *For each ordered triple of objects* (A, B, C), *a map* $(f, g) \rightsquigarrow gf$ *of the product set* hom $(A, B) \times$ hom (B, C) *into* hom (A, C).

It is assumed that the objects and morphisms satisfy the following conditions:

C1. *If* $(A, B) \neq (C, D)$, *then* hom (A, B) *and* hom (C, D) *are disjoint.*

C2. (*Associativity*). *If* $f \in \text{hom}(A, B)$, $g \in \text{hom}(B, C)$, *and* $h \in \text{hom}(C, D)$, *then* $(hg)f = h(gf)$. (*As usual, we simplify this to* hgf.)

C3. (*Unit*). *For every object* A *we have an element* $1_A \in \text{hom}(A, A)$ *such that* $f1_A = f$ *for every* $f \in \text{hom}(A, B)$ *and* $1_A g = g$ *for every* $g \in \text{hom}(B, A)$. (1_A *is unique.*)

If $f \in \text{hom}(A, B)$, we write $f: A \to B$ or $A \xrightarrow{f} B$ (sometimes $A \underset{f}{\to} B$), and we call f an *arrow from* A *to* B. Note that gf is defined if and only if the domain of g coincides with the codomain of f and gf has the same domain as f and the same codomain as g.

The fact that $gf = h$ can be indicated by saying that

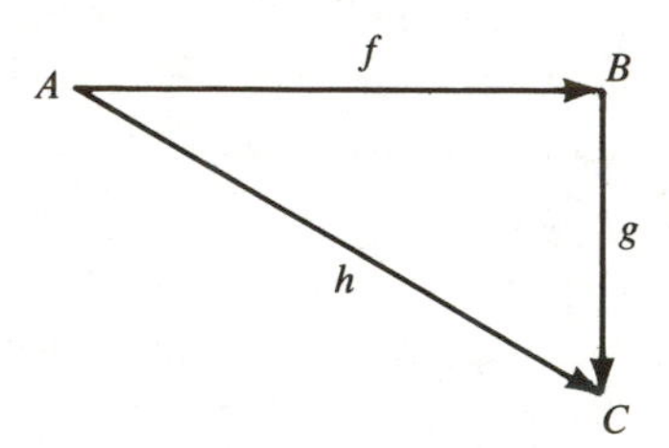

is a commutative diagram. The meaning of more complicated diagrams is the same as for maps of sets (BAI, pp. 7–8). For example, the commutativity of

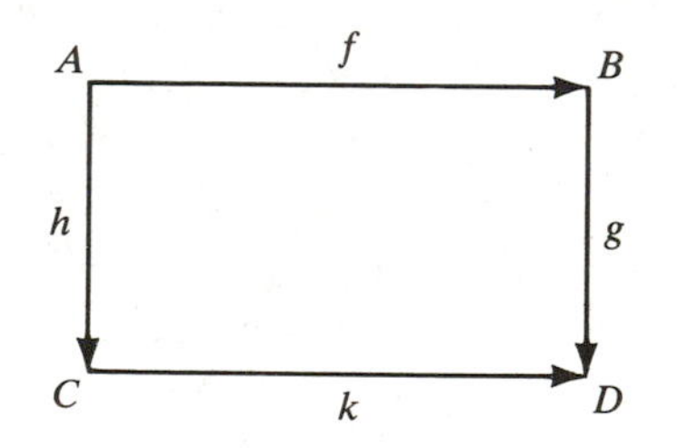

means that $gf = kh$, and the associativity condition $(hg)f = h(gf)$ is expressed by the commutativity of

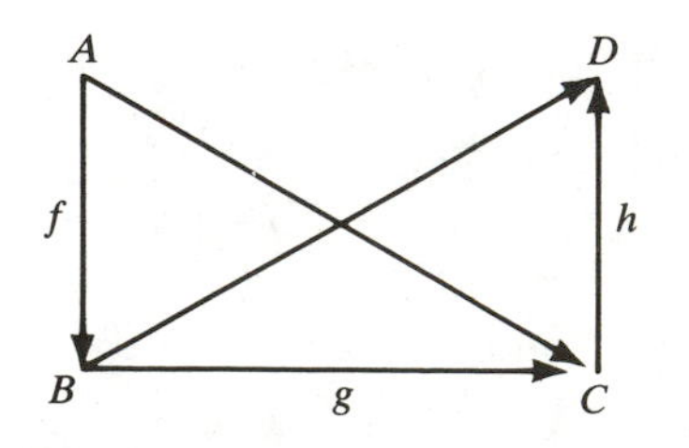

The condition that 1_A is the unit in hom (A, A) can be expressed by the commutativity of

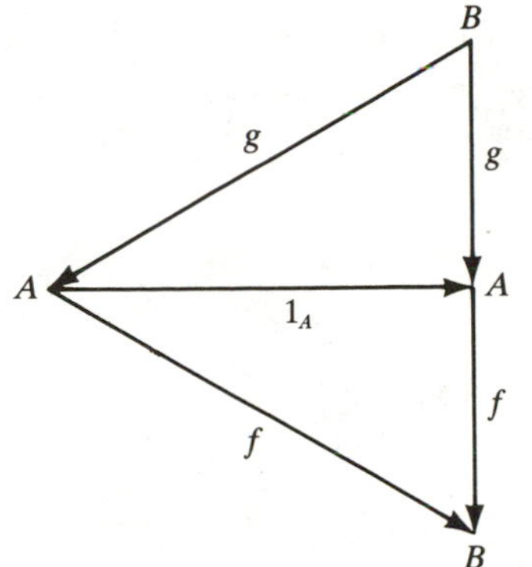

for all $f \in$ hom (A, B) and all $g \in$ hom (B, A).

We remark that in defining a category it is not necessary at the outset that the sets hom (A, B) and hom (C, D) be disjoint whenever $(A, B) \neq (C, D)$. This can always be arranged for a given class of sets hom (A, B) by replacing the given set hom (A, B) by the set of ordered triples (A, B, f) where $f \in$ hom (A, B). This will give us considerably greater flexibility in constructing examples of categories (see exercises 3–6 below).

We shall now give a long list of examples of categories.

EXAMPLES

1. **Set**, the category of sets. Here ob **Set** is the class of all sets. If A and B are sets, hom $(A, B) = B^A$, the set of maps from A to B. The product gf is the usual composite of maps and 1_A is the identity map on the set A. The validity of the axioms C1, C2, and C3 is clear.

2. **Mon**, the category of monoids. ob **Mon** is the class of monoids (BAI, p. 28), hom (M, N) for monoids M and N is the set of homomorphisms of M into N, gf is the composite of the homomorphisms g and f, and 1_M is the identity map on M (which is a homomorphism). The validity of the axioms is clear.

3. **Grp**, the category of groups. The definition is exactly like example 2, with groups replacing monoids.

4. **Ab**, the category of abelian groups. ob **Ab** is the class of abelian groups. Otherwise, everything is the same as in example 2.

A category **D** is called a *subcategory* of the category **C** if ob **D** is a subclass of ob **C** and for any $A, B \in$ ob **D**, $\hom_{\mathbf{D}}(A, B) \subset \hom_{\mathbf{C}}(A, B)$. It is required also (as part of the definition) that 1_A for $A \in$ ob **D** and the product of morphisms for **D** is the same as for **C**. The subcategory **D** is called *full* if $\hom_{\mathbf{D}}(A, B) = \hom_{\mathbf{C}}(A, B)$ for every $A, B \in \mathbf{D}$. It is clear that **Grp** and **Ab** are full subcategories of **Mon**. On the other hand, since a monoid is not just a set but a triple $(M, p, 1)$ where p is an associative binary composition in M and 1 is the unit, the category **Mon** is not a subcategory of **Set**. We shall give below an example of a subcategory that is not full (example 10).

We continue our list of examples.

5. Let M be a monoid. Then M defines a category $\mathbf{M}$ by specifying that $\mathrm{ob}\,\mathbf{M} = \{A\}$, a set with a single element A, and defining $\hom(A, A) = M$, 1_A the unit of M, and xy for $x, y \in \hom(A, A)$, the product of x and y as given in M. It is clear that $\mathbf{M}$ is a category with a single object. Conversely, let $\mathbf{M}$ be a category with a single object: $\mathrm{ob}\,\mathbf{M} = \{A\}$. Then $M = \hom(A, A)$ is a monoid. It is clear from this that monoids can be identified with categories whose object classes are single-element sets.

A category is called *small* if $\mathrm{ob}\,\mathbf{C}$ is a set. Example 5 is a small category; 1–4 are not.

An element $f \in \hom(A, B)$ is called an *isomorphism* if there exists a $g \in \hom(B, A)$ such that $fg = 1_B$ and $gf = 1_A$. It is clear that g is uniquely determined by f, so we can denote it as f^{-1}. This is also an isomorphism and $(f^{-1})^{-1} = f$. If f and h are isomorphisms and fh is defined, then fh is an isomorphism and $(fh)^{-1} = h^{-1}f^{-1}$. In **Set** the isomorphisms are the bijective maps, and in **Grp** they are the usual isomorphisms (=bijective homomorphisms).

6. Let G be a group and let this define a category $\mathbf{G}$ with a single object as in example 5. The characteristic property of this type of category is that it has a single object and all arrows (=morphisms) are isomorphisms.

7. A *groupoid* is a small category in which morphisms are isomorphisms.

8. A *discrete* category is a category in which $\hom(A, B) = \varnothing$ if $A \neq B$ and $\hom(A, A) = \{1_A\}$. Small discrete categories can be identified with their sets of objects.

9. **Ring**, the category of (associative) rings (with unit for the multiplication composition). ob **Ring** is the class of rings and the morphisms are homomorphisms (mapping 1 into 1).

10. **Rng**, the category of (associative) rings without unit (BAI, p. 155), homomorphisms as usual. **Ring** is a subcategory of **Rng** but is not a full subcategory, since there exist maps of rings with unit that preserve addition and multiplication but do not map 1 into 1. (Give an example.)

11. R-**mod**, the category of left modules for a fixed ring R. (We assume $1x = x$ for x in a left R-module M.) ob R-**mod** is the class of left modules for R and the morphisms are R-module homomorphisms. Products are composites of maps. If $R = \Delta$ is a division ring (in particular, a field), then R-**mod** is the category of (left) vector spaces over Δ. In a similar manner one defines **mod**-R as the category of right modules for the ring R.

12. Let S be a pre-ordered set, that is, a set S equipped with a binary relation $a \leqslant b$ such that $a \leqslant a$ and $a \leqslant b$ and $b \leqslant c$ imply $a \leqslant c$. S defines a category $\mathbf{S}$ in which $\text{ob}\,\mathbf{S} = S$ and for $a, b \in S$, hom (a, b) is vacuous or consists of a single element according as $a \nleqslant b$ or $a \leqslant b$. If $f \in \text{hom}\,(a, b)$ and $g \in \text{hom}\,(b, c)$, then gf is the unique element in hom (a, c). It is clear that the axioms for a category are satisfied. Conversely, any small category such that for any pair of objects A, B, hom (A, B) is either vacuous or a single element is the category of a pre-ordered set as just defined.

13. **Top**, the category of topological spaces. The objects are topological spaces and the morphisms are continuous maps. The axioms are readily verified.

We conclude this section by giving two general constructions of new categories from old ones. The first of these is analogous to the construction of the opposite of a given ring (BAI, p. 113). Suppose $\mathbf{C}$ is a category; then we define $\mathbf{C}^{\text{op}}$ by $\text{ob}\,\mathbf{C}^{\text{op}} = \text{ob}\,\mathbf{C}$; for $A, B \in \text{ob}\,\mathbf{C}^{\text{op}}$, $\text{hom}_{\mathbf{C}^{\text{op}}}(A, B) = \text{hom}_{\mathbf{C}}(B, A)$; if $f \in \text{hom}_{\mathbf{C}^{\text{op}}}(A, B)$ and $g \in \text{hom}_{\mathbf{C}^{\text{op}}}(B, C)$, then $g \cdot f$ (in $\mathbf{C}^{\text{op}}$) $= fg$ (as given in $\mathbf{C}$). 1_A is as in $\mathbf{C}$. It is clear that this defines a category. We call this the *dual category* of $\mathbf{C}$. Pictorially we have the following: If $A \xrightarrow{f} B$ in $\mathbf{C}$, then $A \xleftarrow{f} B$ in $\mathbf{C}^{\text{op}}$ and if

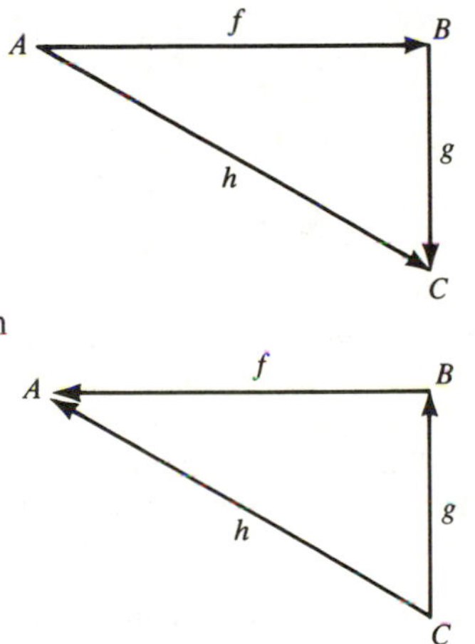

is commutative in $\mathbf{C}$, then

is commutative in $\mathbf{C}^{\text{op}}$. More generally, any commutative diagram in $\mathbf{C}$ gives rise to a commutative diagram in $\mathbf{C}^{\text{op}}$ by reversing all of the arrows.

Next let $\mathbf{C}$ and $\mathbf{D}$ be categories. Then we define the *product category* $\mathbf{C} \times \mathbf{D}$ by the following recipe: $\text{ob}\,\mathbf{C} \times \mathbf{D} = \text{ob}\,\mathbf{C} \times \text{ob}\,\mathbf{D}$; if $A, B \in \text{ob}\,\mathbf{C}$ and $A', B' \in \text{ob}\,\mathbf{D}$, then $\text{hom}_{\mathbf{C} \times \mathbf{D}}((A, A'), (B, B')) = \text{hom}_{\mathbf{C}}(A, B) \times \text{hom}_{\mathbf{D}}(A', B')$, and $1_{(A, A')} = (1_A, 1_{A'})$; if $f \in \text{hom}_{\mathbf{C}}(A, B)$, $g \in \text{hom}_{\mathbf{C}}(B, C)$, $f' \in \text{hom}_{\mathbf{D}}(A', B')$, and $g' \in \text{hom}_{\mathbf{D}}(B', C')$, then

$$(g, g')(f, f') = (gf, g'f').$$

The verification that this defines a category is immediate. This construction can be generalized to define the product of indexed sets of categories. We leave it to the reader to carry out this construction.

EXERCISES

1. Show that the following data define a category **Ring***: ob **Ring*** is the class of rings; if R and S are rings, then $\hom_{\mathbf{Ring}^*}(R,S)$ is the set of homomorphisms and anti-homomorphisms of R into S; gf for morphisms is the composite g following f for the maps f and g; and 1_R is the identity map on R.

2. By a *ring with involution* we mean a pair (R,j) where R is a ring (with unit) and j is an involution in R; that is, if $j:a \rightsquigarrow a^*$, then $(a+b)^* = a^*+b^*$, $(ab)^* = b^*a^*$, $1^* = 1$, $(a^*)^* = a$. (Give some examples.) By a *homomorphism* of a ring with involution (R,j) into a second one (S,k) we mean a map η of R into S such that η is a homomorphism of R into S (sending 1 into 1) such that $\eta(ja) = k(\eta a)$ for all $a \in R$. Show that the following data define a category **Rinv**: ob **Rinv** is the class of rings with involution; if (R,j) and (S,k) are rings with involution, then $\hom((R,j),(S,k))$ is the set of homomorphisms of (R,j) into (S,k); gf for morphisms is the composite of maps; and $1_{(R,j)} = 1_R$.

3. Let $\mathbf{C}$ be a category, A an object of $\mathbf{C}$. Let $\operatorname{ob}\mathbf{C}/A = \bigcup_{X\in \operatorname{ob}\mathbf{C}} \hom(X,A)$ so $\operatorname{ob}\mathbf{C}/A$ is the class of arrows in $\mathbf{C}$ ending at A. If $f \in \hom(B,A)$ and $g \in \hom(C,A)$, define $\hom(f,g)$ to be the set of $u:B \to C$ such that

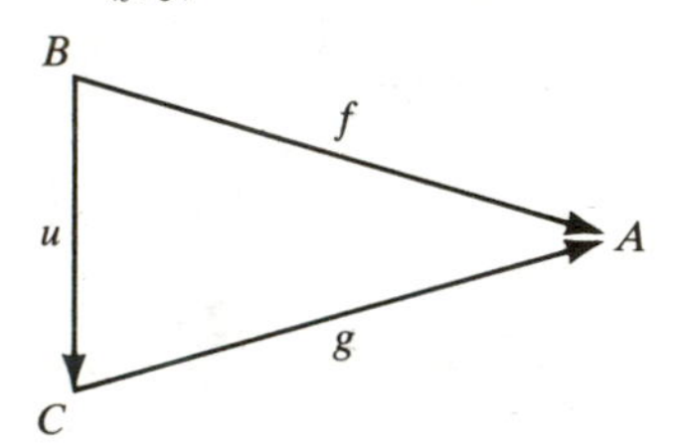

is commutative. Note that $\hom(f,g)$ and $\hom(f',g')$ may not be disjoint for $(f,g) \neq (f',g')$. If $u \in \hom(f,g)$ and $v \in \hom(g,h)$ for $h:D \to A$, then $vu \in \hom(f,h)$. Use this information to define a product from $\hom(f,g)$ and $\hom(g,h)$ to $\hom(f,h)$. Define $1_f = 1_B$ for $f:B \to A$. Show that these data and the remark on page 11 can be used to define a category $\mathbf{C}/A$ called the *category of objects of* $\mathbf{C}$ *over* A.

4. Use $\mathbf{C}^{\mathrm{op}}$ to dualize exercise 3. This defines the *category* $\mathbf{C}$ A *of objects of* $\mathbf{C}$ *below* A.

5. Let $\mathbf{C}$ be a category, $A_1, A_2 \in \operatorname{ob}\mathbf{C}$. Show that the following data define a category $\mathbf{C}/\{A_1, A_2\}$: The objects are the triples (B,f_1,f_2) where $f_i \in \hom_{\mathbf{C}}(B,A_i)$. A morphism $h:(B,f_1,f_2) \to (C,g_1,g_2)$ is a morphism $h:B \to C$ in $\mathbf{C}$ such that

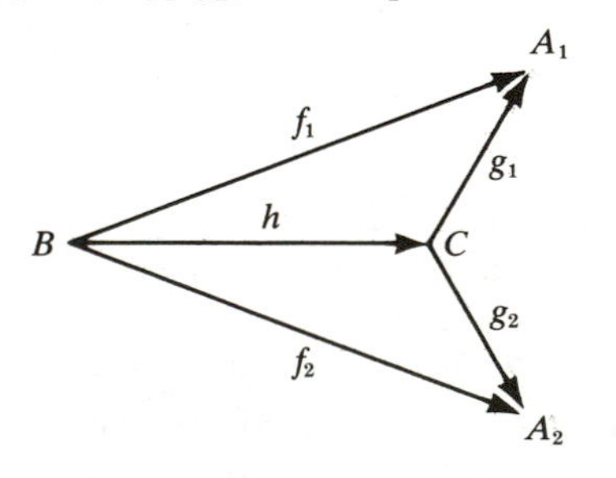

is commutative. Arrange to have the hom sets disjoint as before. Define $1_{(B,f_1,f_2)} = 1_B$ and the product of morphisms as in **C**. Verify the axioms C2 and C3 for a category.

6. Dualize exercise 5 by applying the construction to $\mathbf{C}^{\text{op}}$ and interpreting in **C**. The resulting category is denoted as $\mathbf{C}\backslash\{A_1, A_2\}$.

7. (Alternative definition of a groupoid.) Let **G** be a groupoid as defined in example 7 above and let $G = \bigcup_{A,B \in \text{ob}\,\mathbf{G}} \hom(A, B)$. Then G is a set equipped with a composition fg that is defined for some pairs of elements (f, g), $f, g \in G$, such that the following conditions hold:
   (i) For any $f \in G$ there exist a uniquely determined pair (u, v), $u, v \in G$ such that uf and fv are defined and $uf = f = fv$. These elements are called the *left* and *right units* respectively of f.
   (ii) If u is a unit (left and right for some $f \in G$), then u is its own left unit and hence its own right unit.
   (iii) The product fg is defined if and only if the right unit of f coincides with the left unit of g.
   (iv) If fg and gh are defined, then $(fg)h$ and $f(gh)$ are defined and $(fg)h = f(gh)$.
   (v) If f has left unit u and right unit v, then there exists an element g having right unit u and left unit v such that $fg = u$ and $gf = v$.

   Show that, conversely, if G is a set equipped with a partial composition satisfying conditions (i)–(v), then G defines a groupoid category **G** in which ob **G** is the set of units of G; for any objects u, v, $\hom(u, v)$ is the subset of G of elements having u as left unit and v as right unit; the product composition of $\hom(u,v) \times \hom(v,w)$ is that given in G.

8. Let G be as in exercise 7 and let G^* be the disjoint union of G and a set $\{0\}$. Extend the composition in G to G^* by the rules that $0a = 0 = a0$ for all $a \in G^*$ and $fg = 0$ if $f, g \in G$ and fg is not defined in G. Show that G^* is a semigroup (BAI, p. 29).

1.2 SOME BASIC CATEGORICAL CONCEPTS

We have defined a morphism f in a category **C** to be an isomorphism if $f: A \to B$ and there exists a $g: B \to A$ such that $fg = 1_B$ and $gf = 1_A$. If $f: A \to B$, $g: B \to A$, and $gf = 1_A$, then f is called a *section* of g and g is called a *retraction* of f. More interesting than these two concepts are the concepts of monic and epic that are defined by cancellation properties: A morphism $f: A \to B$ is called *monic* (*epic*) if it is left (right) cancellable in **C**; that is, if g_1 and $g_2 \in \hom(C, A)$ ($\hom(B, C)$) for any C and $fg_1 = fg_2$ ($g_1 f = g_2 f$), then $g_1 = g_2$. The following facts are immediate consequences of the definitions:

1. If $A \xrightarrow{f} B$ and $B \xrightarrow{g} C$ and f and g are monic (epic), then gf is monic (epic).

2. If $A \xrightarrow{f} B$ and $B \xrightarrow{g} C$ and gf is monic (epic), then f is monic (g is epic).
3. If f has a section then f is epic, and if f has a retraction then f is monic.

If f is a map of a set A into a set B, then it is readily seen that f is injective (that is, $f(a) \neq f(a')$ for $a \neq a'$ in A) if and only if for any set C and maps g_1, g_2 of C into A, $fg_1 = fg_2$ implies $g_1 = g_2$ (exercise 3, p. 10 of BAI). Thus $f \in \hom_{\mathbf{Set}}(A, B)$ is monic if and only if f is injective. Similarly, f is epic in **Set** if and only if f is surjective ($f(A) = B$). Similar results hold in the categories R-**mod** and **Grp**: We have

PROPOSITION 1.1. *A morphism f in R-**mod** or in **Grp** is monic (epic) if and only if the map of the underlying set is injective (surjective).*

Proof. Let $f: A \to B$ in R-**mod** or **Grp**. If f is injective (surjective) as a map of sets, then it is left (right) cancellable in **Set**. It follows that f is monic (epic) in R-**mod** or **Grp**. Now suppose the set map f is not injective. Then $C = \ker f \neq 0$ in the case of R-**mod** and $\neq 1$ in the case of **Grp**. Let g be the injection homomorphism of C into A (denoted by $C \hookrightarrow A$), so $g(x) = x$ for every $x \in C$. Then fg is the homomorphism of C into B, sending every $x \in C$ into the identity element of B. Next let h be the homomorphism of C into A, sending every element of C into the identity element of A. Then $h \neq g$ since $C \neq 0$ (or $\neq 1$), but $fg = fh$. Hence f is not monic.

Next suppose we are in the category R-**mod** and f is not surjective. The image $f(A)$ is a submodule of B and we can form the module $C = B/f(A)$, which is $\neq 0$ since $f(A) \neq B$. Let g be the canonical homomorphism of B onto C and h the homomorphism of B into C, sending every element of B into 0. Then $g \neq h$ but $gf = hf$. Hence f is not epic.

Finally, suppose we are in the category **Grp** and f is not surjective. The foregoing argument will apply if $C = f(A) \lhd B$ (C is a normal subgroup of B). This will generally not be the case, although it will be so if $[B:C] = 2$. Hence we assume $[B:C] > 2$ and we shall complete the proof by showing that in this case there exist distinct homomorphisms g and h of B into the group $\operatorname{Sym} B$ of permutations of B such that $gf = hf$. We let g be the homomorphism $b \rightsquigarrow b_L$ of B into $\operatorname{Sym} B$ where b_L is the left translation $x \rightsquigarrow bx$ in B. We shall take h to have the form kg where k is an inner automorphism of $\operatorname{Sym} B$. Thus k has the form $y \rightsquigarrow pyp^{-1}$ where $y \in \operatorname{Sym} B$ and p is a fixed element of $\operatorname{Sym} B$. Then $h = kg$ will have the form $b \rightsquigarrow pb_Lp^{-1}$ and we want this to be different from $g: b \rightsquigarrow b_L$. This requires that the permutation p does not commute with every b_L. Since the permutations commuting with all of the left translations are the right translations (exercise 1, p. 42 of BAI), we shall have $h = kg \neq g$ if p is not a right translation. Since translations $\neq 1$ have no fixed points, our condition will be satisfied if p is any permutation $\neq 1$ having a fixed point. On the other hand, to achieve the condition $gf = hf$ we require that p commutes with every c_L, $c \in C$. To construct a p satisfying all of our conditions, we choose a permutation π of the set $C \backslash B$ of right cosets Cb, $b \in B$, that is not the identity and has a fixed point. Since $|C \backslash B| > 2$, this can be done. Let I be a set of representatives of the right cosets Cb. Then every element of B can be written in one and only one way as a product cu, $c \in C$, $u \in I$. We now define the map p by $p(cu) = cu'$ where $\pi(Cu) = Cu'$. Then $p \in \operatorname{Sym} B$, $p \neq 1$, and p has fixed points since $\pi \neq 1$ and π has fixed points. It is clear that p commutes with every d_L, $d \in C$. Hence p satisfies all of our requirements and f is not epic. □

What can be said about monics and epics in the category **Ring**? In the first place, the proof of Proposition 1.1 shows that injective homomorphisms are monic and surjective ones are epic. The next step of the argument showing that monics are injective breaks down totally, since the kernel of a ring homomorphism is an ideal and this may not be a ring (with unit). Moreover, even if it were, the injection map of the kernel is most likely not a ring homomorphism. We shall now give a different argument, which we shall later generalize (see p. 82), to show that monics in **Ring** are injective.

Let f be a homomorphism of the ring A into the ring B that is not injective. Form the ring $A \oplus A$ of pairs (a_1, a_2), $a_i \in A$, with component-wise addition and multiplication and unit $1 = (1, 1)$. Let K be the subset of $A \oplus A$ of elements (a_1, a_2) such that $f(a_1) = f(a_2)$. It is clear that K is a subring of $A \oplus A$ and $K \supsetneq D = \{(a, a) | a \in A\}$. Let g_1 be the map $(a_1, a_2) \rightsquigarrow a_1$ and g_2 the map $(a_1, a_2) \rightsquigarrow a_2$ from K to A. These are ring homomorphisms and $fg_1 = fg_2$, by the definition of K. On the other hand, since $K \supsetneq D$, we have a pair $(a_1, a_2) \in K$ with $a_1 \neq a_2$. Then $g_1(a_1, a_2) = a_1 \neq a_2 = g_2(a_1, a_2)$. Hence $g_1 \neq g_2$, which shows that f is not monic.

Now we can show by an example that epics in **Ring** need not be surjective. For this purpose we consider the injection homomorphism of the ring $\mathbb{Z}$ of integers into the field $\mathbb{Q}$ of rationals. If g and h are homomorphisms of $\mathbb{Q}$ into a ring R, then $gf = hf$ if and only if the restrictions $g|\mathbb{Z} = h|\mathbb{Z}$. Since a homomorphism of $\mathbb{Q}$ is determined by its restriction to $\mathbb{Z}$, it follows that $gf = hf$ implies $g = h$. Thus f is epic and obviously f is not surjective.

We have proved

PROPOSITION 1.2. *A morphism in* **Ring** *is monic if and only if it is injective. However, there exist epics in* **Ring** *that are not surjective.*

The concept of a monic can be used to define subobjects of an object A of a category **C**. We consider the class of monics ending in A

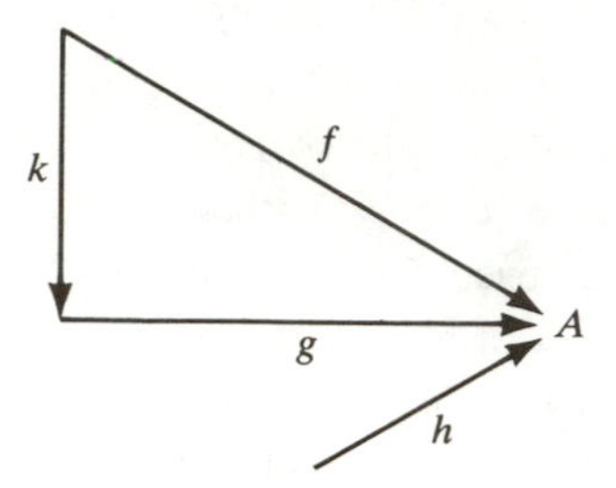

We introduce a preorder in the class of these monics by declaring that $f \leqslant g$ if there exists a k such that $f = gk$. It follows that k is monic. We write $f \equiv g$ if $f \leqslant g$ and $g \leqslant f$. In this case the element k is an isomorphism. The relation $\equiv$ is an equivalence and its equivalence classes are called the *subobjects* of A.

By duality we obtain the concept of a quotient object of A. Here we consider the epics issuing from A and define $f \geqslant g$ if there exists a k such that $f = kg$. We have an equivalence relation $f \equiv g$ defined by $f = kg$ where k is an isomorphism. The equivalence classes determined by this relation are called the *quotient objects* of A.

If the reader will consider the special case in which $\mathbf{C} = \mathbf{Grp}$, he will convince himself that the foregoing terminology of subobjects and quotient objects of the object A is reasonable. However, it should be observed that the quotient objects defined in **Ring** constitute a larger class than those provided by surjective homomorphisms.

EXERCISES

1. Give an example in **Top** of a morphism that is monic and epic but does not have a retraction.

2. Let G be a finite group, H a subgroup. Show that the number of permutations of G that commute with every h_L, $h \in H$ (acting in G), is $[G:H]!|H|^{[G:H]}$ where $[G:H]$ is the index of H in G.

1.3 FUNCTORS AND NATURAL TRANSFORMATIONS

In this section we introduce the concept of a functor or morphism from one category to another and the concept of maps between functors called natural transformations. Before proceeding to the definitions we consider an example.

Let R be a ring and let $U(R)$ denote the multiplicative group of units (=invertible elements) of R. The map $R \rightsquigarrow U(R)$ is a map of rings into groups, that is, a map of ob **Ring** into ob **Grp**. Moreover, if $f: R \to S$ is a homomorphism of rings, then the restriction $f|U(R)$ maps $U(R)$ into $U(S)$ and so may be regarded as a map of $U(R)$ into $U(S)$. Evidently this is a group homomorphism. It is clear also that if $g: S \to T$ is a ring homomorphism, then $(gf)|U(R) = (g|U(S))(f|U(R))$. Moreover, the restriction $1_R|U(R)$ is the identity map on $U(R)$.

The map $R \rightsquigarrow U(R)$ of rings into groups and $f \rightsquigarrow f|U(R)$ of ring homomorphisms into group homomorphisms constitute a functor from **Ring** to **Grp** in the sense of the following definition.

DEFINITION 1.2. *If* **C** *and* **D** *are categories, a* (covariant) functor F *from* **C** *to* **D** *consists of*

1. *A map* $A \rightsquigarrow FA$ *of* ob **C** *into* ob **D**.
2. *For every pair of objects* (A, B) *of* **C**, *a map* $f \rightsquigarrow F(f)$ *of* $\text{hom}_{\mathbf{C}}(A, B)$ *into* $\text{hom}_{\mathbf{D}}(FA, FB)$.

We require that these satisfy the following conditions:

F1. *If* gf *is defined in* **C**, *then* $F(gf) = F(g)F(f)$.

F2. $F(1_A) = 1_{FA}$.

The condition F1 states that any commutative triangle

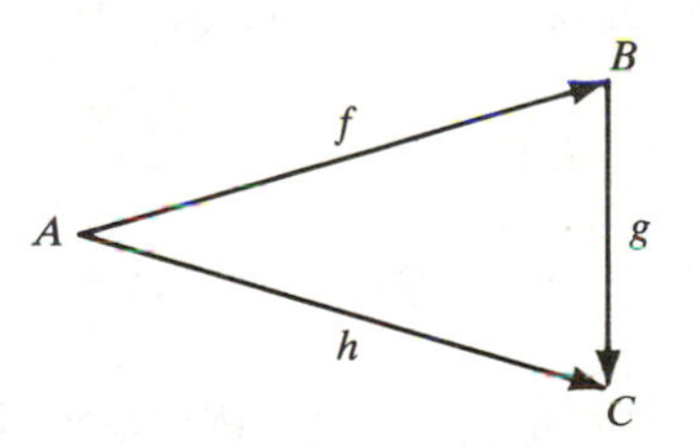

in **C** is mapped into a commutative triangle in **D**

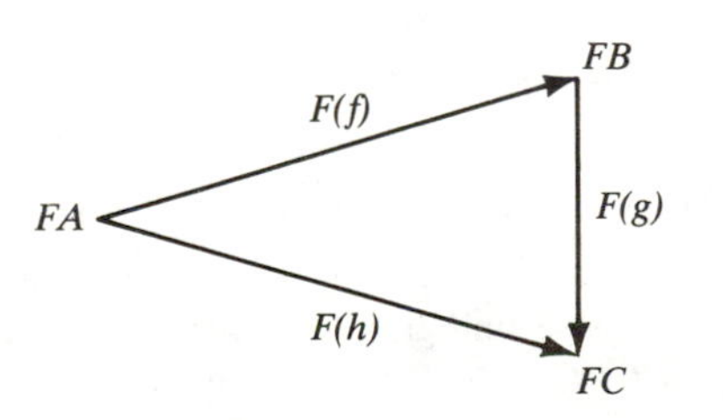

A *contravariant functor* from **C** to **D** is a functor from $\mathbf{C}^{\text{op}}$ to **D**. More directly, this is a map F of ob **C** into ob **D** and for each pair (A, B) of objects in

C, a map F of hom (A, B) into hom (FB, FA) such that $F(fg) = F(g)F(f)$ and $F(1_A) = 1_{FA}$. A functor from $\mathbf{B} \times \mathbf{C}$ to **D** is called a *bifunctor* from **B** and **C** into **D**. We can also combine bifunctors with contravariant functors to obtain functors from $\mathbf{B}^{\mathrm{op}} \times \mathbf{C}$ to **D** and from $\mathbf{B}^{\mathrm{op}} \times \mathbf{C}^{\mathrm{op}}$ to **D**. The first is called a bifunctor that is *contravariant in* **B** and *covariant in* **C** and the second is a bifunctor that is *contravariant in* **B** and **C**.

EXAMPLES

1. Let **D** be a subcategory of the category **C**. Then we have the *injection functor* of **D** into **C** that maps every object of **D** into the same object of **C** and maps any morphism in **D** into the same morphism in **C**. The special case in which $\mathbf{D} = \mathbf{C}$ is called the *identity functor* $1_{\mathbf{C}}$.

2. We obtain a functor from **Grp** to **Set** by mapping any group into the underlying set of the group and mapping any group homomorphism into the corresponding set map. The type of functor that discards some of the given structure is called a "forgetful" functor. Two other examples of forgetful functors are given in the next example.

3. Associated with any ring $(R, +, \cdot, 0, 1)$ we have the additive group $(R, +, 0)$ and the multiplicative monoid $(R, \cdot, 1)$. A ring homomorphism is in particular a homomorphism of the additive group and of the multiplicative monoid. These observations lead in an obvious manner to definitions of the forgetful functors from **Ring** to **Ab** and from **Ring** to **Mon**.

4. Let n be a positive integer. For any ring R we can form the ring $M_n(R)$ of $n \times n$ matrices with entries in R. A ring homomorphism $f: R \to S$ determines a homomorphism $(r_{ij}) \rightsquigarrow (f(r_{ij}))$ of $M_n(R)$ into $M_n(S)$. In this way we obtain a functor M_n of **Ring** into **Ring**.

5. Let n and R be as in example 4 and let $GL_n(R)$ denote the group of units of $M_n(R)$, that is, the group of $n \times n$ invertible matrices with entries in R. The maps $R \rightsquigarrow GL_n(R)$, f into $(r_{ij}) \rightsquigarrow (f(r_{ij}))$ define a functor GL_n from **Ring** to **Grp**.

6. We define the *power functor* $\mathscr{P}$ in **Set** by mapping any set A into its power set $\mathscr{P}(A)$ and any set map $f: A \to B$ into the induced map $f_{\mathscr{P}}$ of $\mathscr{P}(A)$ into $\mathscr{P}(B)$, which sends any subset A_1 of A into its image $f(A_1) \subset B$ $(\varnothing \rightsquigarrow \varnothing)$.

7. The *abelianizing functor* from **Grp** to **Ab**. Here we map any group G into the abelian group $G/(G, G)$ where (G, G) is the commutator group of G (BAI, p. 238). If f is a homomorphism of G into a second group H, f maps (G, G) into (H, H) and so induces a homomorphism $\bar{f}$ of $G/(G, G)$ into $H/(H, H)$. The map $f \rightsquigarrow \bar{f}$ completes the definition of the abelianizing functor.

8. Let **Poset** be the category of partially ordered sets. Its objects are partially ordered sets (BAI, p. 456) and the morphisms are order-preserving maps. We obtain a functor

from R-**mod** to **Poset** by mapping any R-module M into $L(M)$, the set of submodules of M ordered by inclusion. If $f: M \to N$ is a module homomorphism, f determines an order-preserving map of $L(M)$ into $L(N)$. These maps define a functor.

9. We define a *projection functor* of $\mathbf{C} \times \mathbf{D}$ into $\mathbf{C}$ by mapping any object (A, B) of $\mathbf{C} \times \mathbf{D}$ into the object A of $\mathbf{C}$ and mapping $(f, g) \in \hom((A, B), (A', B'))$ into $f \in \hom(A, A')$.

10. We define the *diagonal functor* $\mathbf{C} \to \mathbf{C} \times \mathbf{C}$ by mapping $A \rightsquigarrow (A, A)$ and $f: A \to B$ into $(A, A) \xrightarrow{(f,f)} (B, B)$.

11. Consider the categories R-**mod** and **mod**-R of left R-modules and right R-modules respectively for the ring R. We shall define a contravariant functor D from R-**mod** to **mod**-R as follows. If M is a left R-module, we consider the set $M^* = \hom_R(M, R)$ of homomorphisms of M into R regarded as left R-module in the usual way. Thus M^* is the set of maps of M into R such that

$$f(x+y) = f(x)+f(y)$$
$$f(rx) = rf(x)$$

for $x, y \in M$, $r \in R$. If $f, g \in M^*$ and $s \in R$, then $f+g$ defined by $(f+g)(x) = f(x)+g(x)$ and fs defined by $(fs)(x) = f(x)s$ are in M^*. In this way M^* becomes a right R-module and we have the map $M \rightsquigarrow M^*$ of $\mathrm{ob}\, R$-**mod** into $\mathrm{ob}\,$**mod**-R. Now let $L: M \to N$ be a homomorphism of the R-module M into the R-module N. We have the *transposed* map $L^*: N^* \to M^*$ defined as

$$L^*: g \rightsquigarrow gL,$$

the composite of g and L:

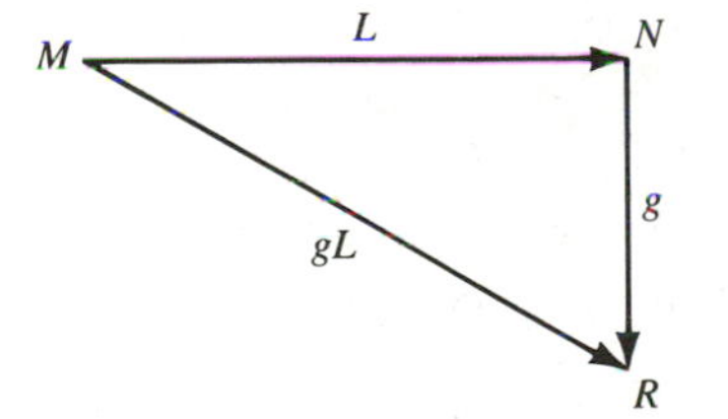

If $M_1 \xrightarrow{L_1} M_2 \xrightarrow{L_2} M_3$ in R-**mod** and $g \in M_3^*$, then $(L_2L_1)^*(g) = gL_2L_1 = (gL_2)L_1 = L_1^*L_2^*(g)$. Hence

$$(L_2L_1)^* = L_1^*L_2^*.$$

It is clear that $(1_M)^* = 1_{M^*}$. It follows that

$$D: M \rightsquigarrow M^*, \qquad L \rightsquigarrow L^*$$

defines a contravariant functor, the *duality functor* D from R-**mod** to **mod**-R. In a similar fashion one obtains the duality functor D from **mod**-R to R-**mod**.

It is clear that a functor maps an isomorphism into an isomorphism: If we have $fg = 1_B$, $gf = 1_A$, then $F(f)F(g) = 1_{FB}$ and $F(g)F(f) = 1_{FA}$. Similarly, sections are mapped into sections and retractions are mapped into retractions by functors. On the other hand, it is easy to give examples to show that monics (epics) need not be mapped into monics (epics) by functors (see exercise 3 below).

If F is a functor from **C** to **D** and G is a functor from **D** to **E**, we obtain a functor GF from **C** to **E** by defining $(GF)A = G(FA)$ for $A \in \text{ob}\,\mathbf{C}$ and $(GF)(f) = G(F(f))$ for $f \in \hom_{\mathbf{C}}(A, B)$. In a similar manner we can define composites of functors one or both of which are contravariant. Then FG is contravariant if one of F, G is contravariant and the other is covariant, and FG is covariant if both F and G are contravariant. Example 5 above can be described as the composite UM_n where M_n is the functor defined in example 4 and U is the functor from **Ring** to **Grp** defined at the beginning of the section. As we shall see in a moment, the double dual functor D^2 from R-**mod** to itself is a particularly interesting covariant functor.

A functor F is called *faithful* (*full*) if for every pair of objects A, B in **C** the map $f \rightsquigarrow F(f)$ of $\hom_{\mathbf{C}}(A, B)$ into $\hom_{\mathbf{D}}(FA, FB)$ is injective (surjective). In the foregoing list of examples, example 1 is faithful and is full if and only if **D** is a full subcategory of **C**; examples 2 and 3 are faithful but not full (why?); and example 9 is full but not faithful.

We shall define next the important concept of natural transformation between functors. However, before we proceed to the definition, it will be illuminating to examine in detail the example mentioned briefly in the introduction to this chapter. We shall consider the more general situation of modules. Accordingly, we begin with the category R-**mod** for a ring R and the double dual functor D^2 in this category. This maps a left R-module M into $M^{**} = (M^*)^*$ and a homomorphism $L: M \to N$ into $L^{**} = (L^*)^*: M^{**} \to N^{**}$. If $x \in M$, $g \in N^*$, then $L^*g \in M^*$ and $(L^*g)(x) = g(Lx)$. If $\varphi \in M^{**}$, $L^{**}\varphi \in N^{**}$ and $(L^{**}\varphi)(g) = \varphi(L^*g)$. We now consider the map

$$\eta_M(x): f \rightsquigarrow f(x)$$

of M^* into R. This is contained in $M^{**} = \hom_R(M^*, R)$ and the map $\eta_M: x \rightsquigarrow \eta_M(x)$ is an R-homomorphism of M into M^{**}. Now for any homomorphism $L: M \to N$, the diagram

(1)

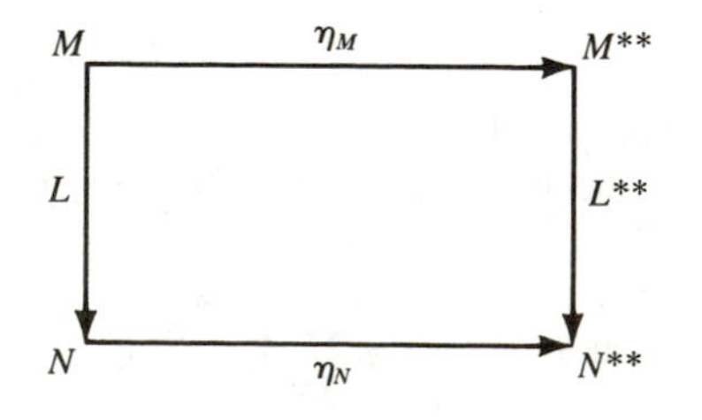

is commutative, because if $x \in M$, then $\eta_N(Lx)$ is the map $g \rightarrow g(Lx)$ of N^* into R and for $\varphi = \eta_M(x) \in M^{**}$, $(L^{**}\varphi)(g) = \varphi(L^*g)$. Hence $(L^{**}\eta_M(x))(g) = \eta_M(x)(L^*g) = (L^*g)(x) = g(Lx)$.

We now introduce the following definition of "naturality."

DEFINITION 1.3. *Let F and G be functors from* $\mathbf{C}$ *to* $\mathbf{D}$. *We define a* natural transformation *η from F to G to be a map that assigns to every object A in* $\mathbf{C}$ *a morphism $\eta_A \in \hom_{\mathbf{D}}(FA, GA)$ such that for any objects A, B of* $\mathbf{C}$ *and any $f \in \hom_{\mathbf{C}}(A, B)$ the rectangle in*

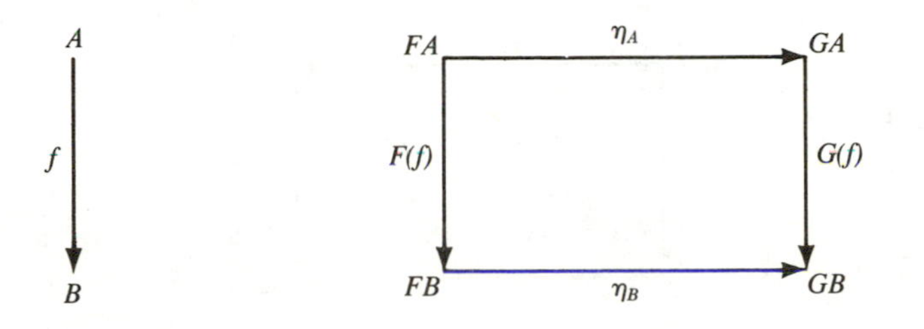

is commutative. Moreover, if every η_A is an isomorphism then η is called a natural isomorphism.

In the foregoing example we consider the identity functor $1_{R\text{-}\mathbf{mod}}$ and the double dual functor D^2 on the category of left R-modules. For each object M of R-**mod** we can associate the morphism η_M of M into M^{**}. The commutativity of (1) shows that $\eta : M \rightarrow \eta_M$ is a natural transformation of the identity functor into the double dual functor.

We can state a stronger result if we specialize to finite dimensional vector spaces over a division ring Δ. These form a subcategory of Δ-**mod**. If V is a finite dimensional vector space over Δ, we can choose a base $(e_1, e_2, \cdots, e_n)$ for V over Δ. Let e_i^* be the linear function on V such that $e_i^*(e_j) = \delta_{ij}$. Then $(e_1^*, e_2^*, \cdots, e_n^*)$ is a base for V^* as right vector space over Δ—the *dual* (or *complementary*) *base* to the base $(e_1, \cdots, e_n)$. This shows that V^* has the same dimensionality n as V. Hence V^{**} has the same dimensionality as V. Since any non-zero vector x can be taken as the element e_1 of the base $(e_1, e_2, \cdots, e_n)$, it is clear that for any $x \neq 0$ in V there exists a $g \in V^*$ such that $g(x) \neq 0$. It follows that for any $x \neq 0$ the map $\eta_V(x) : f \rightarrow f(x)$ is non-zero. Hence $\eta_V : x \rightarrow \eta_V(x)$ is an injective linear map of V into V^{**}. Since $\dim V^{**} = \dim V$, it follows that η_V is an isomorphism. Thus, in this case, η is a natural isomorphism of the identity functor on the category of finite dimensional vector spaces over Δ onto the double dual functor on this category.

We shall encounter many examples of natural transformations as we proceed in our discussion. For this reason it may be adequate to record at this point only two additional examples.

EXAMPLES

1. We define the functor $\oplus_n$ in R-**mod** by mapping any module M into $M^{(n)}$, the direct sum of n copies of M (BAI, p. 175), and any homomorphism $f: M \rightarrow N$ into

$$f^{(n)}: (a_1, \cdots, a_n) \rightsquigarrow (f(a_1), \cdots, f(a_n)).$$

For any M we define the diagonal homomorphism

$$\delta_M{}^{(n)}: a \rightsquigarrow (a, \ldots, a).$$

Then $\delta^{(n)}: M \rightsquigarrow \delta_M{}^{(n)}$ is a natural transformation of $1_{R\text{-mod}}$ into $\oplus_n$ since we have the commutative diagram

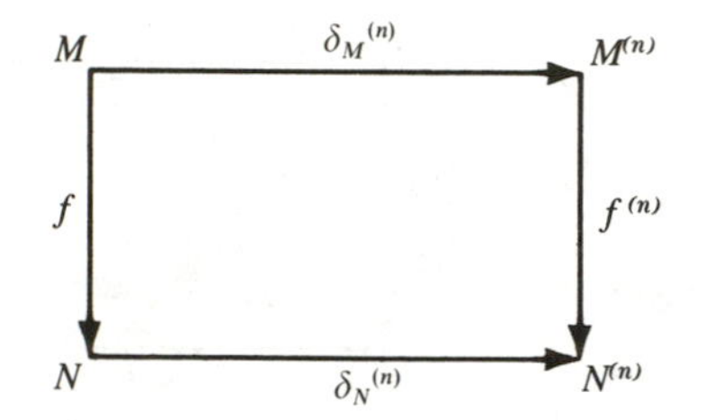

2. We consider the abelianizing functor as defined in example 7 above, but we now regard this as a functor from **Grp** to **Grp** rather than from **Grp** to **Ab**. (This is analogous to changing the codomain of a function.) Let ν_G denote the canonical homomorphism of G onto the factor group. Then we have the commutative diagram

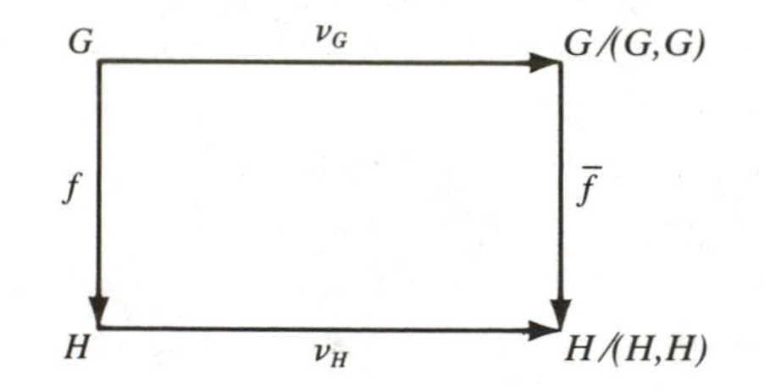

which shows that $\nu: G \rightsquigarrow \nu_G$ is a natural transformation of the identity functor of **Grp** to **Grp** to the abelianizing functor.

Let F, G, H be functors from **C** to **D**, η a natural transformation of F to G, and ζ a natural transformation from G to H. If $A \in ob\,\mathbf{C}$ then $\eta_A \in \hom_{\mathbf{D}}(FA, GA)$ and $\zeta_A \in \hom_{\mathbf{D}}(GA, HA)$. Hence $\zeta_A \eta_A \in \hom_{\mathbf{D}}(FA, HA)$. We have the commutativity of the two smaller rectangles in

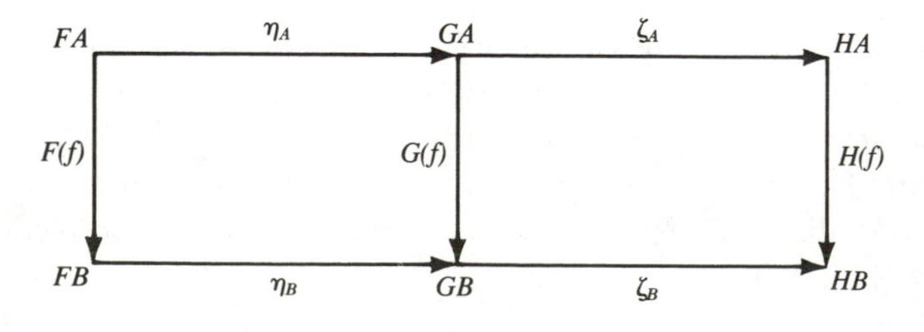

which implies the commutativity of the large rectangle with vertices FA, HA, FB, HB. This implies that $A \rightsquigarrow \zeta_A \eta_A$ is a natural transformation from F to H. We call this $\zeta\eta$, the *product* of ζ and η.

If F is a functor from $\mathbf{C}$ to $\mathbf{D}$, we obtain a natural transformation 1_F of F into itself by mapping any $A \in \text{ob}\,\mathbf{C}$ into $1_{FA} \in \hom_{\mathbf{D}}(FA, FA)$. If η is any natural transformation from F to a functor G from $\mathbf{C}$ to $\mathbf{D}$, then we evidently have $\eta 1_F = \eta = 1_G \eta$.

Let η be a natural isomorphism of F to G. Then η_A is an isomorphism $\eta_A : FA \to GA$ for every $A \in \text{ob}\,\mathbf{C}$. Hence we have the isomorphism $\eta_A{}^{-1} : GA \to FA$. The required commutativity is clear, so $A \to \eta_A{}^{-1}$ is a natural isomorphism of G to F. We call this the *inverse* η^{-1} of η ($\eta^{-1}{}_A = \eta_A{}^{-1}$). It is clear that we have $\eta^{-1}\eta = 1_F$ and $\eta\eta^{-1} = 1_G$. Conversely, if η is a natural transformation from F to G and η is one from G to F such that $\zeta\eta = 1_F$ and $\eta\zeta = 1_G$, then η is a natural isomorphism with $\eta^{-1} = \zeta$.

If η is a natural isomorphism of a functor E of $\mathbf{C}$ to $\mathbf{C}$ with the functor $1_{\mathbf{C}}$, then the commutativity of

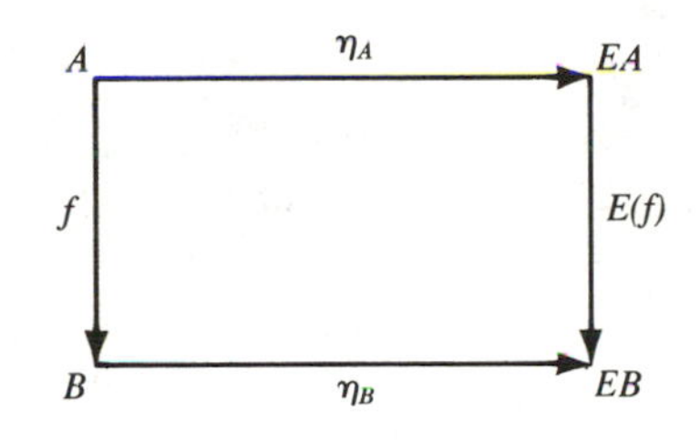

shows that $E(f) = \eta_B f \eta_A{}^{-1}$, which implies that the map $f \rightsquigarrow E(f)$ of $\hom(A, B)$ into $\hom(EA, EB)$ is bijective.

EXERCISES

1. Let F be a functor from $\mathbf{C}$ to $\mathbf{D}$ that is faithful and full and let $f \in \hom_{\mathbf{C}}(A, B)$. Show that any one of the following properties of $F(f)$ implies the same property for f: $F(f)$ is monic, is epic, has a section, has a retraction, is an isomorphism.

2. Let M and N be monoids regarded as categories with a single object as in example 5, p. 12. Show that in this identification, a functor is a homomorphism of M into N and that a natural transformation of a functor F to a functor G corresponds to an element $b \in N$ such that $b(Fx) = (Gx)b$, $x \in M$.

3. Use exercise 2 to construct a functor F and a monic (epic) f such that $F(f)$ is not monic (epic).

4. Let G be a group, $\mathbf{G}$ the one object category determined by G as in example 6 on p. 12. Show that a functor from $\mathbf{G}$ to **Set** is the same thing as a homomorphism of G into the group $\operatorname{Sym} S$ of permutations of a set S, or, equivalently, an action of G on S (BAI, p. 72). Show that two such functors are naturally isomorphic if and only if the actions of G are equivalent (BAI, p. 74).

5. Let $\mathbf{B}, \mathbf{C}, \mathbf{D}, \mathbf{E}$ be categories, F and G functors from $\mathbf{C}$ to $\mathbf{D}$, K a functor from $\mathbf{B}$ to $\mathbf{C}$, and H a functor from $\mathbf{D}$ to $\mathbf{E}$. Show that if η is a natural transformation from F to G, then $A \rightsquigarrow H\eta_A$ is a natural transformation from HF to HG for $A \in \operatorname{ob} \mathbf{C}$ and $B \rightsquigarrow \eta_{KB}$ is a natural transformation from FK to GK for $B \in \operatorname{ob} \mathbf{B}$.

6. Define the *center* of a category $\mathbf{C}$ to be the class of natural transformations of the identity functor $1_{\mathbf{C}}$ to $1_{\mathbf{C}}$. Let $\mathbf{C} = R$-**mod** and let c be an element of the center of R. For any $M \in \operatorname{ob} R$-**mod** let $\eta_M(c)$ denote the map $x \rightsquigarrow cx$, $x \in M$. Show that $\eta(c) : M \rightsquigarrow \eta_M(c)$ is in the center of $\mathbf{C}$ and every element of the center of $\mathbf{C}$ has this form. Show that $c \rightsquigarrow \eta(c)$ is a bijection and hence that the center of R-**mod** is a set.

1.4 EQUIVALENCE OF CATEGORIES

We say that the categories $\mathbf{C}$ and $\mathbf{D}$ are *isomorphic* if there exist functors $F : \mathbf{C} \to \mathbf{D}$ (from $\mathbf{C}$ to $\mathbf{D}$) and $G : \mathbf{D} \to \mathbf{C}$ such that $GF = 1_{\mathbf{C}}$ and $FG = 1_{\mathbf{D}}$. This condition is rather strong, so that in most cases in which it holds one tends to identify the isomorphic categories. Here is an example. Let $\mathbf{C} = \mathbf{Ab}$ and $\mathbf{D} = \mathbb{Z}$-**mod**. If M is an abelian group (written additively), M becomes a $\mathbb{Z}$-module by defining nx for $n \in \mathbb{Z}$, $x \in M$, as the nth multiple of x (BAI, p. 164). On the other hand, if M is a $\mathbb{Z}$-module, then the additive group of M is an abelian group. In this way we have maps of $\operatorname{ob} \mathbf{Ab}$ into $\operatorname{ob} \mathbb{Z}$-**mod** and of $\operatorname{ob} \mathbb{Z}$-**mod** into $\operatorname{ob} \mathbf{Ab}$ that are inverses. If f is a homomorphism of the abelian group M into the abelian group N, then $f(nx) = nf(x)$, $n \in \mathbb{Z}$, $x \in M$. Hence f is a homomorphism of M as $\mathbb{Z}$-module into N as $\mathbb{Z}$-module. Conversely, any $\mathbb{Z}$-homomorphism is a group homomorphism. It is clear from this that **Ab** and $\mathbb{Z}$-**mod** are isomorphic categories, and one usually identifies these two categories.

Another example of isomorphic categories are R-**mod** and **mod**-R^{op} for any ring R. If M is a left R-module, M becomes a right R^{op}-module by defining $xr = rx$ for $x \in M$, $r \in R^{\mathrm{op}} = R$ (as sets). Similarly, any right R^{op}-module becomes a left R-module by reversing this process. It is clear also that a homomorphism of ${}_RM$, M as left R-module, into ${}_RN$ is a homomorphism of $M_{R^{\mathrm{op}}}$, M as right R^{op}-module, into $N_{R^{\mathrm{op}}}$. We have the obvious functors F and G such that $GF = 1_{R\text{-}\mathbf{mod}}$ and $FG = 1_{\mathbf{mod}\text{-}R^{\mathrm{op}}}$. Hence the two categories are

The concept of isomorphism of categories is somewhat too restrictive; a considerably more interesting notion is obtained by broadening this in the following manner. We define **C** and **D** to be *equivalent* categories if there exist functors $F:\mathbf{C} \to \mathbf{D}$ and $G:\mathbf{D} \to \mathbf{C}$ such that $GF \cong 1_{\mathbf{C}}$ and $FG \cong 1_{\mathbf{D}}$ where $\cong$ denotes the natural isomorphism of functors. Evidently isomorphism of categories implies equivalence. It is clear also that the relation of equivalence between categories is what the name suggests: it is reflexive, symmetric, and transitive.

We note that the functor G in the definition of equivalence is not uniquely determined by F. It is therefore natural to shift the emphasis from the pair (F, G) to the functor F and to seek conditions on a functor $F:\mathbf{C} \to \mathbf{D}$ in order that there exists a $G:\mathbf{D} \to \mathbf{C}$ such that (F, G) gives an equivalence, that is, $GF \cong 1_{\mathbf{C}}$ and $FG \cong 1_{\mathbf{D}}$. We have seen that $GF \cong 1_{\mathbf{C}}$ implies that the map $f \rightsquigarrow GF(f)$ of $\hom_{\mathbf{C}}(A, B)$ onto $\hom_{\mathbf{C}}(GFA, GFB)$ is bijective. Similarly, $g \rightsquigarrow FG(g)$ is bijective of $\hom_{\mathbf{D}}(A', B')$ onto $\hom_{\mathbf{D}}(FGA', FGB')$. Now the injectivity of $f \rightsquigarrow GF(f)$ implies the injectivity of the map $f \rightsquigarrow F(f)$ of the set $\hom_{\mathbf{C}}(A, B)$ into the set $\hom_{\mathbf{D}}(FA, FB)$ and the surjectivity of $g \rightsquigarrow FG(g)$ implies the surjectivity of $f \rightsquigarrow F(f)$. Thus we see that the functor F is faithful and full. We note also that given any $A' \in \operatorname{ob}\mathbf{D}$, the natural isomorphism $FG \cong 1_{\mathbf{D}}$ gives an isomorphism $\eta_{A'} \in \hom_{\mathbf{D}}(A', FGA')$. Thus if we put $A = GA' \in \operatorname{ob}\mathbf{C}$, then there is an isomorphism contained in $\hom_{\mathbf{D}}(A', FA)$ or, equivalently, in $\hom_{\mathbf{D}}(FA, A')$.

We shall now show that the conditions we have sorted out are also sufficient and thus we have the following important criterion.

PROPOSITION 1.3. *Let F be a functor from* **C** *to* **D**. *Then there exists a functor $G:\mathbf{D} \to \mathbf{C}$ such that (F, G) is an equivalence if and only if F is faithful and full and for every object A' of* **D** *there exists an object A of* **C** *such that FA and A' are isomorphic in* **D**, *that is, there is an isomorphism contained in* $\hom_{\mathbf{D}}(FA, A')$.

Proof. It remains to prove the sufficiency of the conditions. Suppose these hold. Then for any $A' \in \operatorname{ob}\mathbf{D}$ we choose $A \in \operatorname{ob}\mathbf{C}$ such that FA and A' are isomorphic and we choose an isomorphism $\eta_{A'}:A' \to FA$. We define a map G of $\operatorname{ob}\mathbf{D}$ into $\operatorname{ob}\mathbf{C}$ by $A' \rightsquigarrow A$ where A is as just chosen. Then $\eta_{A'}:A' \to FGA'$. Let B' be a second object of **D** and let $f' \in \hom_{\mathbf{D}}(A', B')$. Consider the diagram

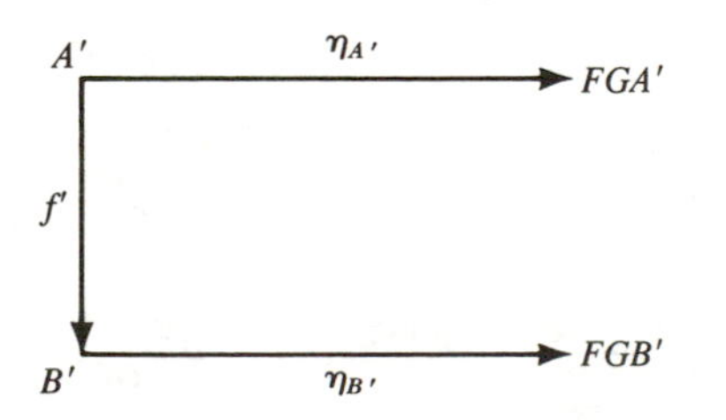

Since $\eta_{A'}$ is an isomorphism, we have a unique morphism $\eta_{B'}f'\eta_{A'}^{-1}$: $FGA' \to FGB'$, making a commutative rectangle. Since F is full and faithful, there is a unique $f: GA' \to GB'$ in $\mathbf{C}$ such that $F(f) = \eta_{B'}f'\eta_{A'}^{-1}$. We define the map G from $\hom_{\mathbf{D}}(A', B')$ to $\hom_{\mathbf{C}}(GA', GB')$ by $f' \rightsquigarrow f$. Then we have the commutative rectangle

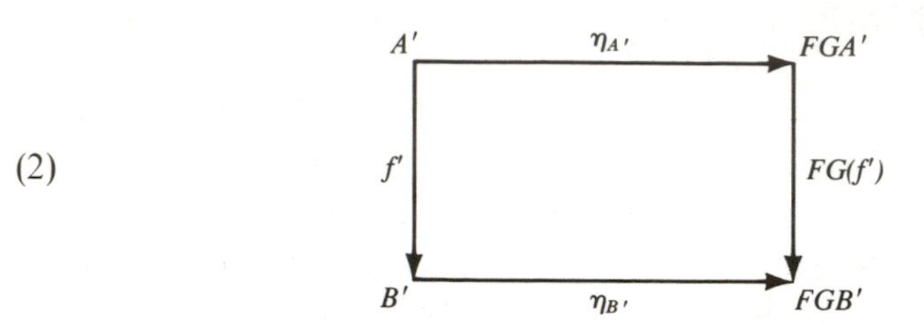

and $G(f')$ is the only morphism $GA' \to GB'$ such that (2) is commutative.

Now let $g' \in \hom_{\mathbf{D}}(B', C')$. Then we have the diagram

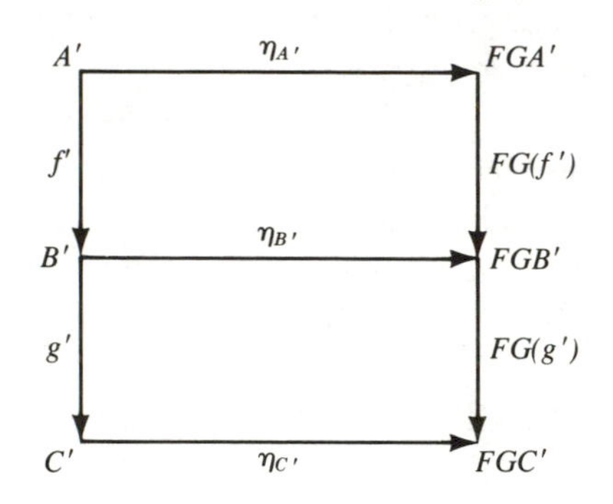

in which the two small rectangles and hence the large one are commutative. Since F is a functor, we have the commutative rectangle

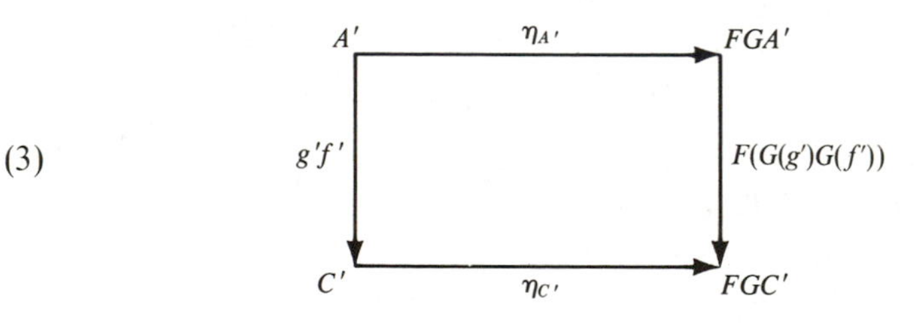

On the other hand, we have the commutative rectangle

(4)

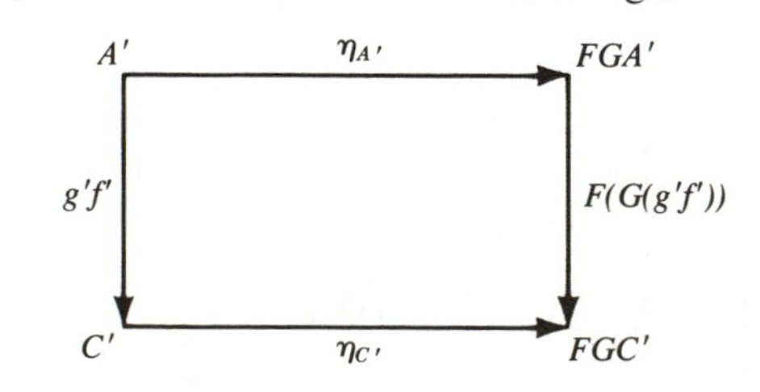

and $G(g'f'):GA' \to GC'$ is the only morphism for which (4) is commutative. Hence we have $G(g')G(f') = G(g'f')$. In a similar manner we see that $G(1_{A'}) = 1_{GA'}$. Thus the maps $G:A' \rightsquigarrow GA'$, $G:f' \rightsquigarrow G(f')$ for all of the hom sets $\hom_{\mathbf{D}}(A', B')$ constitute a functor from $\mathbf{D}$ to $\mathbf{C}$. Moreover, the commutativity of (2) shows that $\eta' : A' \rightsquigarrow \eta_{A'}$ is a natural isomorphism of $1_{\mathbf{D}}$ to FG.

We observe next that since F is faithful and full, if $A, B \in \operatorname{ob}\mathbf{C}$ and $f':FA \to FB$ is an isomorphism, then the morphism $f:A \to B$ such that $F(f) = f'$ is an isomorphism (exercise 1, p. 25). It follows that since $\eta_{FA}:FA \to FGFA$ is an isomorphism, there exists a unique isomorphism $\zeta_A:A \to GFA$ such that $F(\zeta_A) = \eta_{FA}$. The commutativity of (2) for $A' = FA$, $B' = FB$, and $f' = F(f)$ where $f:A \to B$ in $\mathbf{C}$ implies that

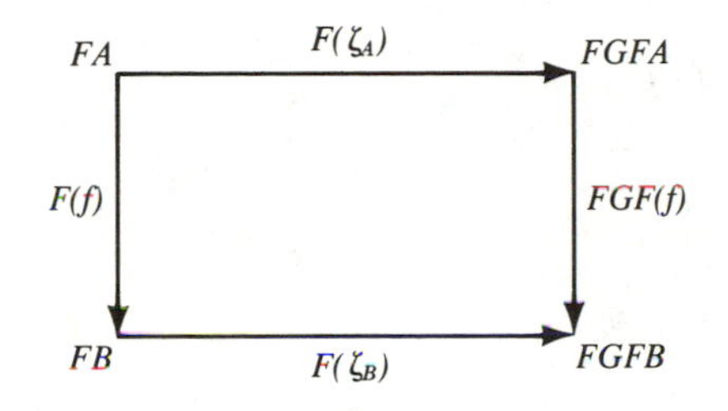

is commutative. Since F is faithful, this implies that

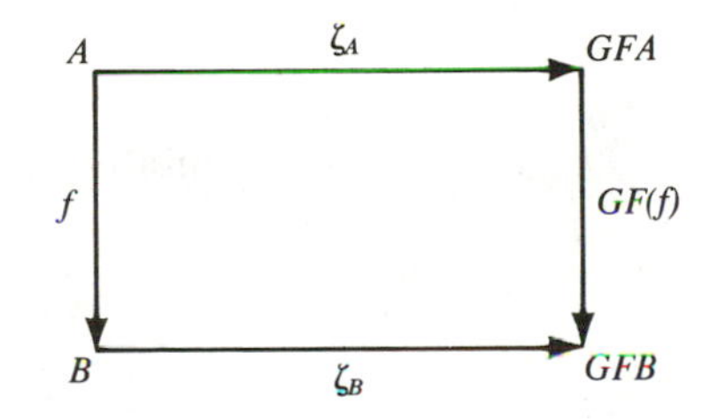

is commutative. Hence $\zeta : A \rightsquigarrow \zeta_A$ is a natural isomorphism of $1_{\mathbf{C}}$ into GF. □

As an illustration of this criterion we prove the following very interesting proposition.

PROPOSITION 1.4. *Let R be a ring, and $M_n(R)$ the ring of $n \times n$ matrices with entries in R. Then the categories* **mod**-R *and* **mod**-$M_n(R)$ *of right modules over R and $M_n(R)$ respectively are equivalent.*

Before proceeding to the proof we recall some elementary facts about matrix units in $M_n(R)$ (BAI, pp. 94–95). For $i, j = 1, \dots, n$, we define e_{ij} to be the matrix whose (i, j)-entry is 1 and other entries are 0, and for any $a \in R$ we let a' denote diag$\{a, \dots, a\}$, the diagonal matrix in which all diagonal entries are a.

Then we have the multiplication table

$$e_{ij}e_{kl} = \delta_{jk}e_{il} \tag{5}$$

and

$$\Sigma e_{ii} = 1. \tag{6}$$

Moreover,

$$a'e_{ij} = e_{ij}a' \tag{7}$$

and this matrix has a in the (i,j)-position and 0's elsewhere. Hence if

$$A = \begin{bmatrix} a_{11} & a_{12} & \cdot & \cdot & \cdot & a_{1n} \\ a_{21} & a_{22} & \cdot & \cdot & \cdot & a_{2n} \\ \cdot & \cdot & \cdot & \cdot & \cdot & \cdot \\ a_{n1} & a_{n2} & \cdot & \cdot & \cdot & a_{nn} \end{bmatrix}, \qquad a_{ij} \in R \tag{8}$$

then

$$A = \Sigma a'_{ij}e_{ij} = \Sigma e_{ij}a'_{ij}. \tag{9}$$

We can now give the

Proof of Proposition 1.4. Let M be a right R-module and let $M^{(n)}$ be a direct sum of n copies of M (BAI, p. 175). If $x = (x_1, x_2, \cdots, x_n) \in M^{(n)}$ and $A \in M_n(R)$ as in (8), we define xA to be the matrix product

$$(x_1, x_2, \cdots, x_n) \begin{bmatrix} a_{11} & a_{12} & \cdot & \cdot & \cdot & a_{1n} \\ a_{21} & a_{22} & \cdot & \cdot & \cdot & a_{2n} \\ \cdot & \cdot & \cdot & \cdot & \cdot & \cdot \\ a_{n1} & a_{n2} & \cdot & \cdot & \cdot & a_{nn} \end{bmatrix} = (y_1, y_2, \cdots, y_n) \tag{10}$$

where

$$y_i = \Sigma x_j a_{ji}, \qquad 1 \leqslant i \leqslant n \tag{11}$$

and the right-hand side is as calculated in M. Using the associativity of matrix multiplication (in this mixed case of multiplication of "vectors" by matrices) we can verify that $M^{(n)}$ is a right $M_n(R)$-module under the action we have defined. Thus we have a map $M \rightsquigarrow M^{(n)}$ of ob **mod**-R into ob **mod**-$M_n(R)$. If f is a module homomorphism of M into N, then the diagonal homomorphism $f^{(n)}: (x_1, x_2, \ldots, x_n) \rightarrow (f(x_1), f(x_2), \ldots, f(x_n))$ is a homomorphism of the right

$M_n(R)$-module $M^{(n)}$ into the right $M_n(R)$-module $N^{(n)}$. The maps $M \rightsquigarrow M^{(n)}, f \rightsquigarrow f^{(n)}$ constitute a functor F from **mod**-R to **mod**-$M_n(R)$. We shall verify that F satisfies the conditions of Proposition 1.3.

1. F is faithful. Clear.

2. F is full: Let g be an $M_n(R)$-homomorphism of $M^{(n)}$ into $N^{(n)}$ where M and N are right R-modules. Now $M^{(n)}e_{11}$ is the set of elements $(x,0,\ldots,0)$, $x \in M$, and $N^{(n)}e_{11}$ is the set of elements $(y,0,\ldots,0)$, $y \in N$. Since g is an $M_n(R)$-homomorphism, $g(M^{(n)}e_{11}) \subset N^{(n)}e_{11}$. Hence $g(x,0,\ldots,0) = (f(x),0,\ldots,0)$. It is clear that f is additive, and $f(xa) = f(x)a$ for $a \in R$ follows from $g((x,0,\ldots,0)a') = (g(x,0,\ldots,0))a'$. Hence f is an R-homomorphism of M into N. Now $(x,0,\ldots,0)e_{1i} = (0,\ldots,0,\overset{i}{x},0,\ldots,0)$, so $g((x,0,\ldots,0)e_{1i}) = (g(x,0,\ldots,0))e_{1i}$ implies that $g(0,\ldots,0,\overset{i}{x},0,\ldots,0) = (0,\ldots,0,\overset{i}{f(x)},0,\ldots,0)$. Then $g = f^{(n)}$ and F is full.

3. Any right $M_n(R)$-module M' is isomorphic to a module FM, M a right R-module: The map $a \rightsquigarrow a'$ is a homomorphism of R into $M_n(R)$. Combining this with the action of $M_n(R)$ on M', we make M' a right R-module in which $x'a = x'a'$, $x' \in M'$. Then $M = M'e_{11}$ is an R-submodule of M' since $e_{11}a' = a'e_{11}$, $a \in R$. Moreover, $x'e_{i1} = x'e_{i1}e_{11} \in M$ for any i. We define a map $\eta_{M'}: M' \to FM = M^{(n)}$ by

$$x' \rightsquigarrow (x'e_{11}, x'e_{21}, \ldots, x'e_{n1}). \tag{12}$$

Direct verification using (9) and the definition of $x'a$ shows that $\eta_{M'}$ is an $M_n(R)$-homomorphism. If $x'e_{i1} = 0$ for $1 \leqslant i \leqslant n$, then $x' = \sum x'e_{ii} = \sum x'e_{i1}e_{1i} = 0$. Hence $\eta_{M'}$ is injective. Moreover, $\eta_{M'}$ is surjective: If $(x_1, x_2, \ldots, x_n) \in M^{(n)}$, then $x_i = x_i'e_{11} = (x_i'e_{1i})e_{i1}$ and

$$\begin{aligned}(x_1, x_2, \ldots, x_n) = {} & ((x_1'e_{11})e_{11}, (x_1'e_{11})e_{21}, \ldots, (x_1'e_{11})e_{n1}) \\ & + ((x_2'e_{12})e_{11}, (x_2'e_{12})e_{21}, \ldots, (x_2'e_{12})e_{n1}) + \cdots.\end{aligned}$$

Thus $\eta_{M'}$ is an isomorphism. $\square$

EXERCISES

1. Let (F, G) be an equivalence of $\mathbf{C}$ into $\mathbf{D}$ and let $f \in \hom_{\mathbf{C}}(A, B)$. Show that any one of the following properties of f implies the same property for $F(f)$: f is monic, is epic, has a section, has a retraction, is an isomorphism.

2. Are **mod**-R and **mod**-$M_n(R)$ isomorphic for $n > 1$?

1.5 PRODUCTS AND COPRODUCTS

There are many basic constructions in mathematics and especially in algebra—such as the (Cartesian) product of sets, the direct product of groups, the disjoint union of sets, and the free product of groups—that have simple characterizations by means of properties of maps. This fact, which has been known for some time, can be incorporated into category theory in the definition and examples of products and coproducts in categories. We shall begin with an example: the direct product of two groups.

Let $G = G_1 \times G_2$, the direct product of the groups G_1 and $G_2 : G$ is the group of pairs (g_1, g_2), $g_i \in G_i$, with the multiplication defined by

$$(g_1, g_2)(h_1, h_2) = (g_1 h_1, g_2 h_2),$$

the unit $1 = (1_1, 1_2)$ where 1_i is the unit of G_i, and $(g_1, g_2)^{-1} = (g_1{}^{-1}, g_2{}^{-1})$. We have the projections $p_i : G \to G_i$ defined by

$$p_1 : (g_1, g_2) \rightarrow g_1, \qquad p_2 : (g_1, g_2) \rightarrow g_2. \tag{13}$$

These are homomorphisms, since $p_1((g_1, g_2)(h_1, h_2)) = p_1(g_1 h_1, g_2 h_2) = g_1 h_1 = (p_1(g_1, g_2))(p_1(h_1, h_2))$ and $p_1(1) = 1_1$. Similar relations hold for p_2.

Now let H be another group and let $f_i : H \to G_i$ be a homomorphism of H into G_i. Then we define a map f of H into $G = G_1 \times G_2$ by

$$h \rightarrow (f_1(h), f_2(h)).$$

It is clear that this is a homomorphism and $p_i f(h) = f_i(h)$. Hence we have the commutativity of

(14)

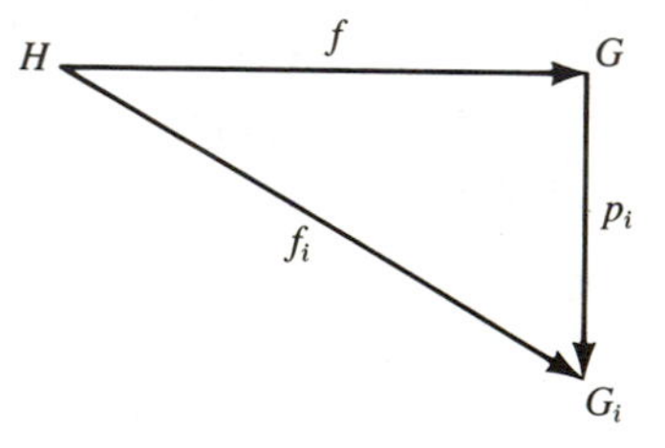

Next let f' be any homomorphism of H into G such that $p_i f' = f_i$, $i = 1, 2$. Then $f'(h) = (f_1(h), f_2(h)) = f(h)$ so $f' = f$. Thus f is the only homomorphism $H \to G$ making (14) commutative.

We now formulate the following

DEFINITION 1.4. *Let A_1 and A_2 be objects of a category* $\mathbf{C}$. *A* product *of A_1 and A_2 in $\mathbf{C}$ is a triple (A, p_1, p_2) where $A \in \operatorname{ob} \mathbf{C}$ and $p_i \in \hom_{\mathbf{C}}(A, A_i)$ such that if B is any object in $\mathbf{C}$ and $f_i \in \hom_{\mathbf{C}}(B, A_i)$, $i = 1, 2$, then there exists a unique*

$f \in \hom_{\mathbf{C}}(B, A)$ such that the diagrams

(15)

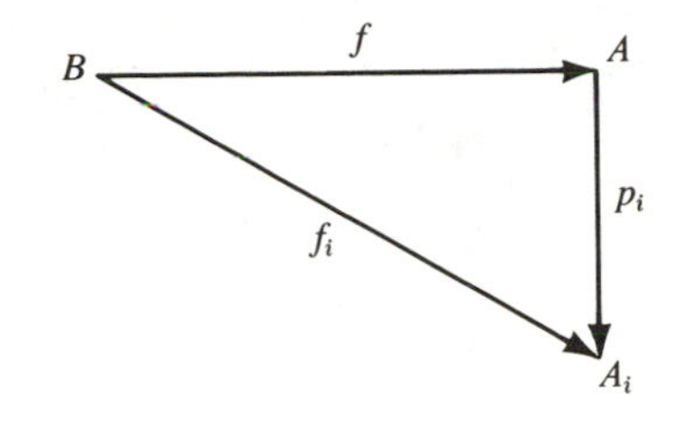

are commutative.

It is clear from our discussion that if G_1 and G_2 are groups, then $(G = G_1 \times G_2, p_1, p_2)$ is a product of G_1 and G_2 in the category of groups. The fact that $(G_1 \times G_2, p_1, p_2)$ is a product of G_1 and G_2 in **Grp** constitutes a characterization of the direct product $G_1 \times G_2$ and the projections p_i. This is a special case of

PROPOSITION 1.5. *Let (A, p_1, p_2) and (A', p_1', p_2') be products of A_1 and A_2 in* **C**. *Then there exists a unique isomorphism $h: A \to A'$ such that $p_i = p_i'h$, $i = 1, 2$.*

Proof. If we use the fact that (A', p_1', p_2') is a product of A_1 and A_2, we obtain a unique homomorphism $h: A \to A'$ such that $p_i = p_i'h$, $i = 1, 2$. Reversing the roles of (A, p_1, p_2) and (A', p_1', p_2'), we obtain a unique homomorphism $h': A' \to A$ such that $p_i' = p_i h'$. We now have $p_i = p_i h'h$ and $p_i' = p_i hh'$. On the other hand, if we apply Definition 1.4 to $B = A$ and $f_i = p_i$, we see that 1_A is the only homomorphism $A \to A$ such that $p_i = p_i 1_A$. Hence $h'h = 1_A$ and similarly $hh' = 1_{A'}$. Thus h is an isomorphism and $h' = h^{-1}$. □

Because of the essential uniqueness of the product we shall denote any product of A_1 and A_2 in the category **C** by $A_1 \Pi A_2$. The concept of product in a category can be generalized to more than two objects.

DEFINITION 1.4′. *Let $\{A_\alpha | \alpha \in I\}$ be an indexed set of objects in a category* **C**. *We define a product $\prod A_\alpha$ of the A_α to be a set $\{A, p_\alpha | \alpha \in I\}$ where $A \in \operatorname{ob} \mathbf{C}$, $p_\alpha \in \hom_{\mathbf{C}}(A, A_\alpha)$ such that if $B \in \operatorname{ob} \mathbf{C}$ and $f_\alpha \in \hom_{\mathbf{C}}(B, A_\alpha)$, $\alpha \in I$, then there exists a unique $f \in \hom_{\mathbf{C}}(B, A)$ such that every diagram*

(15′)

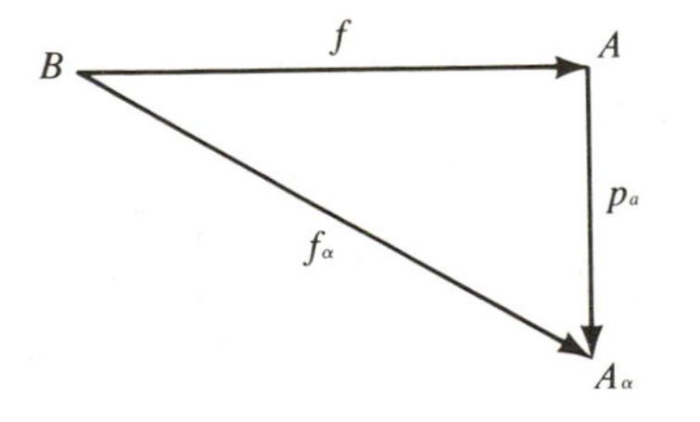

is commutative.

We do not assume that the map $\alpha \rightsquigarrow A_\alpha$ is injective. In fact, we may have all the A_α equal. Also, if a product exists, then it is unique in the sense defined in Proposition 1.5. The proof of Proposition 1.5 carries over to the general case without change. We now consider some examples of products in categories.

EXAMPLES

1. Let $\{A_\alpha | \alpha \in I\}$ be an indexed set of sets. We have the product set $A = \Pi A_\alpha$, which is the set of maps $a: I \to \bigcup A_\alpha$ such that for every $\alpha \in I$, $a(\alpha) \in A_\alpha$. For each α we have the projection $p_\alpha : a \quad a(\alpha)$. We claim that $\{A, p_\alpha\}$ is a product of the A_α in the category **Set**. To see this, let B be a set and for each $\alpha \in I$ let f_α be a map: $B \to A_\alpha$. Then we have the map $f: B \to A$ such that $f(b)$ is $\alpha \rightsquigarrow f_\alpha(b)$. Then $p_\alpha f = f_\alpha$ and it is clear that f is the only map: $B \to A$ satisfying this condition. Hence $\{A, p_\alpha\}$ satisfies the condition for a product of the A_α.

2. Let $\{G_\alpha | \alpha \in I\}$ be an indexed set of groups. We define a product in $G = \Pi G_\alpha$ by $gg'(\alpha) = g(\alpha)g'(\alpha)$ for $g, g' \in G$ and we let $1 \in G$ be defined by $1(\alpha) = 1_\alpha$, the unit of G_α for $\alpha \in I$. It is easy to verify that this defines a group structure on G and it is clear that the projections as defined in **Set** are homomorphisms. As in example 1, $\{G, p_\alpha)$ is a product of the G_α in the category of groups.

3. The argument expressed in example 2 applies also in the category of rings. If $\{R_\alpha | \alpha \in I\}$ is an indexed set of rings, we can endow ΠR_α with a ring structure such that the projections p_α become ring homomorphisms. Then $\{\Pi R_\alpha, p_\alpha\}$ is a product of the R_α in the category **Ring**.

4. In a similar manner we can define products of indexed sets of modules in R-**mod** for any ring R.

We now consider the dual of the concept of a product. This is given in the following

DEFINITION 1.5. *Let $\{A_\alpha | \alpha \in I\}$ be an indexed set of objects of a category* **C**. *We define a coproduct* $\quad A_\alpha$ *to be a set $\{A, i_\alpha | \alpha \in I\}$ where $A \in \text{ob}\,\mathbf{C}$ and $i_\alpha \in \hom(A_\alpha, A)$ such that if $B \in \text{ob}\,\mathbf{C}$ and $g_\alpha \in \hom_{\mathbf{C}}(A_\alpha, B)$, $\alpha \in I$, then there exists a unique $g \in \hom_{\mathbf{C}}(A, B)$ such that every diagram*

(16)

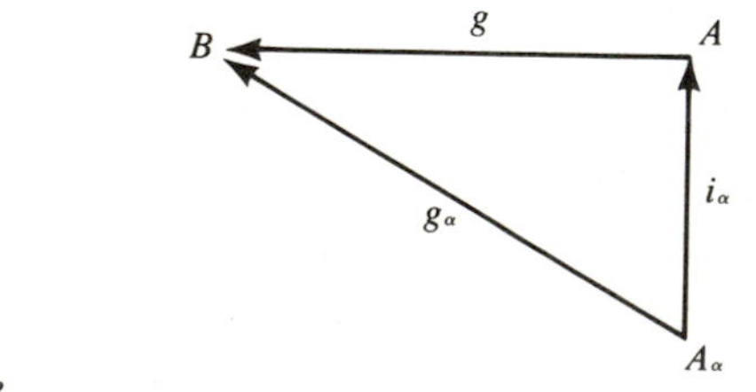

is commutative.

It is readily seen that if $\{A, i_\alpha | \alpha \in I\}$ and $\{A', i'_\alpha | \alpha \in I\}$ are coproducts in **C** of $\{A_\alpha | \alpha \in I\}$, then there exists a unique isomorphism $k: A' \to A$ such that $i_\alpha = ki'_\alpha$ for all $\alpha \in I$.

If $\{A_\alpha | \alpha \in I\}$ is an indexed set of sets, then there exists a set $\dot{\bigcup} A_\alpha$ that is a disjoint union of the sets A_α. Let i_α denote the injection map of A_α into $\dot{\bigcup} A_\alpha$. Let B be a set and suppose for each α we have a map g_α of A_α into B. Then there exists a unique map g of $\dot{\bigcup} A_\alpha$ into B such that the restriction $g|A_\alpha = g_\alpha$, $\alpha \in I$. It follows that $\{\dot{\bigcup} A_\alpha, i_\alpha\}$ is a coproduct of the A_α in **Set**.

We shall show in the next chapter that coproducts exist for any indexed set of objects in **Grp** or in **Ring** (see p. 84). In the case of R-**mod** this is easy to see: Let $\{M_\alpha | \alpha \in I\}$ be an indexed set of left R-modules for the ring R and let $\prod M_\alpha$ be the product set of the M_α endowed with the left R-module structure in which for $x, y \in \prod M_\alpha$ and $r \in R$, $(x+y)(\alpha) = x(\alpha) + y(\alpha)$, $(rx)(\alpha) = r(x(\alpha))$. Let $\oplus M_\alpha$ be the subset of $\prod M_\alpha$ consisting of the x such that $x(\alpha) = 0$ for all but a finite number of the $\alpha \in I$. Clearly $\oplus M_\alpha$ is a submodule of $\prod M_\alpha$. If $x_\alpha \in M_\alpha$, we let $i_\alpha x_\alpha$ be the element of $\prod M_\alpha$ that has the value x_α at α and the value 0 at every $\beta \neq \alpha, \beta \in I$. The map $i_\alpha : x_\alpha \rightsquigarrow i_\alpha x_\alpha$ is a module homomorphism of M_α into $\oplus M_\alpha$. Now let $N \in R$-**mod** and suppose for every $\alpha \in I$ we have a homomorphism $g_\alpha : M_\alpha \to N$. Let $x \in \oplus M_\alpha$. Then $x(\alpha) = 0$ for all but a finite number of the α; hence $\sum g_\alpha x(\alpha)$ is well defined. We define g as the map $x \rightsquigarrow \sum g_\alpha x(\alpha)$ of $\oplus M_\alpha$ into N. It is readily verified that this is a homomorphism of $\oplus M_\alpha$ into N and it is the unique homomorphism of $\oplus M_\alpha$ into N such that $gi_\alpha = g_\alpha$. Thus $\{\oplus M_\alpha, i_\alpha\}$ is a coproduct in R-**mod** of the M_α. We call $\oplus M_\alpha$ the *direct sum* of the modules M_α.

Since the category **Ab** is isomorphic to $\mathbb{Z}$-**mod**, coproducts of arbitrary indexed sets of objects in **Ab** exist.

EXERCISES

1. Let S be a partially ordered set and **S** the associated category as defined in example 12 on p. 13. Let $\{a_\alpha | \alpha \in I\}$ be an indexed set of elements of S. Give a condition on $\{a_\alpha\}$ that the corresponding set of objects in **S** has a product (coproduct). Use this to construct an example of a category in which every finite subset of objects has a product (coproduct) but in which there are infinite sets of objects that do not have a product (coproduct).

2. A category **C** is called a category *with a product* (*coproduct*) if any pair of objects in **C** has a product (coproduct) in **C**. Show that if **C** is a category with a product

(coproduct), then any finitely indexed set of objects in **C** has a product (coproduct).

3. An object A of a category **C** is called *initial* (*terminal*) if for every object X of **C**, $\hom_{\mathbf{C}}(A, X)$ $(\hom_{\mathbf{C}}(X, A))$ consists of a single element. An object that is both initial and terminal is called a *zero* of **C**. Show that if A and A' are initial (terminal), then there exists a unique isomorphism h in $\hom_{\mathbf{C}}(A, A')$.

4. Let A_1 and A_2 be objects of a category **C** and let $\mathbf{C}/\{A_1, A_2\}$ be the category defined in exercise 5 of p. 14. Show that A_1 and A_2 have a product in **C** if and only if $\mathbf{C}/\{A_1, A_2\}$ has a terminal object. Note that this and exercise 3 give an alternative proof of Proposition 1.5. Generalize to indexed sets of objects in **C**.

5. Use exercise 6 on p. 15 to give an alternative definition of a coproduct of objects of a category.

6. Let $f_i : A_i \to B$, $i = 1, 2$, in a category **C**. Define a *pullback diagram* of $\{f_1, f_2\}$ to be a commutative diagram

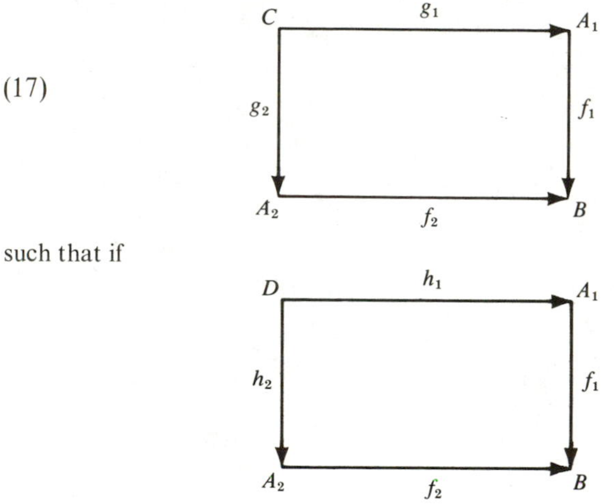

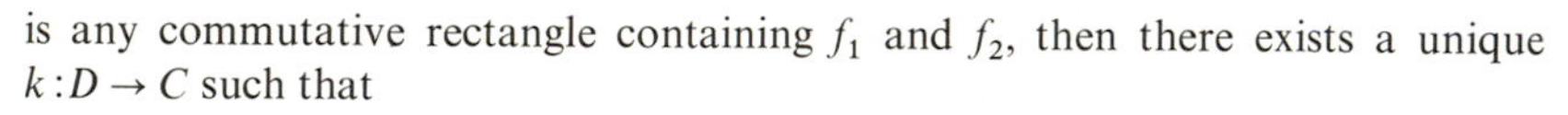

is any commutative rectangle containing f_1 and f_2, then there exists a unique $k : D \to C$ such that

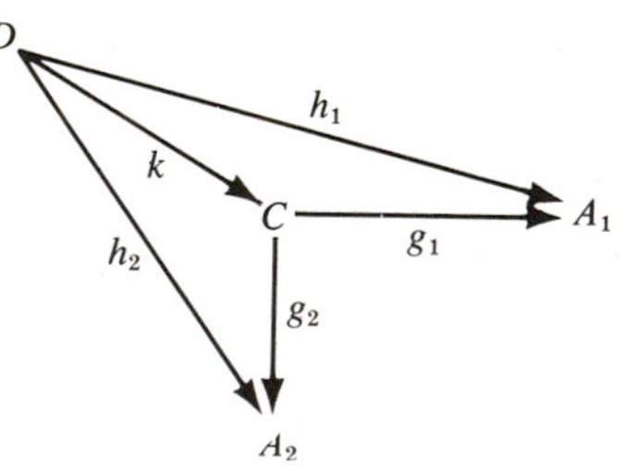

is commutative. Show that if (C, g_1, g_2) and (C', g'_1, g'_2) determine pullbacks for f_1 and f_2 as in (17), then there exists a unique isomorphism $k : C' \to C$ such that $g'_i = g_i k$, $i = 1, 2$.

7. Let $f_i: G_i \to H$ in **Grp**. Form $G_1 \times G_2$ and let M be the subset of $G_1 \times G_2$ of elements (a_1, a_2) such that $f_1(a_1) = f_2(a_2)$. This is a subgroup. Let $m_i = p_i|M$ where p_i is the projection of $G_1 \times G_2$ on G_i. Show that $\{m_1, m_2\}$ defines a pullback diagram of f_1 and f_2.

8. Dualize exercise 6 to define a *pushout diagram* determined by $f_i: B \to A_i$, $i = 1, 2$, in **C**. Let $f_i: B \to A_i$, $i = 1, 2$, in R-**mod**. Form $A_1 \oplus A_2$. Define the map $f: b \rightsquigarrow (-f_1(b), f_2(b))$ of B into $A_1 \oplus A_2$. Let $I = \operatorname{Im} f$ and put $N = (A_1 \oplus A_2)/I$. Define $n_i: A_i \to N$ by $n_1 a_1 = (a_1, 0) + I$, $n_2 a_2 = (0, a_2) + I$. Verify that $\{n_1, n_2\}$ defines a pushout diagram for f_1 and f_2.

1.6 THE HOM FUNCTORS. REPRESENTABLE FUNCTORS

We shall now define certain important functors from a category **C** and the related categories $\mathbf{C}^{\text{op}}$ and $\mathbf{C}^{\text{op}} \times \mathbf{C}$ to the category of sets. We consider first the functor hom from $\mathbf{C}^{\text{op}} \times \mathbf{C}$. We recall that the objects of $\mathbf{C}^{\text{op}} \times \mathbf{C}$ are the pairs (A, B), $A, B \in \operatorname{ob} \mathbf{C}$, and a morphism in this category from (A, B) to (A', B') is a pair (f, g) where $f: A' \to A$ and $g: B \to B'$. If (f', g') is a morphism in $\mathbf{C}^{\text{op}} \times \mathbf{C}$ from (A', B') to (A'', B''), so $f': A'' \to A'$, $g': B' \to B''$, then $(f', g')(f, g) = (ff', g'g)$. Also $1_{(A,B)} = (1_A, 1_B)$.

We now define the functor hom from $\mathbf{C}^{\text{op}} \times \mathbf{C}$ to **Set** by specifying that this maps the object (A, B) into the set $\hom(A, B)$ (which is an object of **Set**) and the morphism $(f, g): (A, B) \to (A'. B')$ into the map of $\hom(A, B)$ into $\hom(A', B')$ defined by

$$\hom(f, g): k \rightsquigarrow gkf. \tag{18}$$

This makes sense since $f: A' \to A$, $g: B \to B'$, $k: A \to B$, so $gkf: A' \to B'$. These rules define a functor, since if $(f', g'): (A', B') \to (A'', B'')$, then $(f', g')(f, g) = (ff', g'g)$ and

$$\begin{aligned}\hom(ff', g'g)(k) &= (g'g)k(ff') \\ &= g'(gkf)f' \\ &= g' \hom(f, g)(k) f' \\ &= \hom(f', g')(\hom(f, g)(k)).\end{aligned}$$

Thus $\hom((f', g')(f, g)) = \hom(f', g') \hom(f, g)$. Moreover, if $f = 1_A$ and $g = 1_B$, then (18) shows that $\hom(1_A, 1_B)$ is the identity map on the set $\hom(A, B)$. Thus the defining conditions for a functor from $\mathbf{C}^{\text{op}} \times \mathbf{C}$ to **Set** are satisfied.

We now fix an object A in **C** and we define a functor $\hom(A, -)$ from **C** to **Set** by the rules

$$\hom(A, -)B = \hom(A, B)$$

$$\hom(A, -)(g) \text{ for } g: B \to B' \text{ is the map}$$

$$\hom(A, g): k \rightsquigarrow gk \tag{19}$$

of $\hom(A, B)$ into $\hom(A, B')$. It is clear that this defines a functor. We call this functor the *(covariant) hom functor determined by the object A* in **C**.

In a similar manner we define the *contravariant hom functor* $\hom(-, B)$ *determined by* $B \in \mathrm{ob}\,\mathbf{C}$ by

$$\hom(-, B)A = \hom(A, B)$$

$$\hom(-, B)(f) \text{ for } f: A' \to A \text{ is the map}$$

$$\hom(f, B): k \rightsquigarrow kf \tag{20}$$

of $\hom(A, B)$ into $\hom(A', B)$. Now let $f: A' \to A$, $g: B \to B'$, $k: A \to B$. Then

$$\hom(f, B')\hom(A, g)(k) = (gk)f$$
$$\hom(A', g)\hom(f, B)(k) = g(kf)$$

and $(gk)f = g(kf) = \hom(f, g)(k)$. Hence

(21)

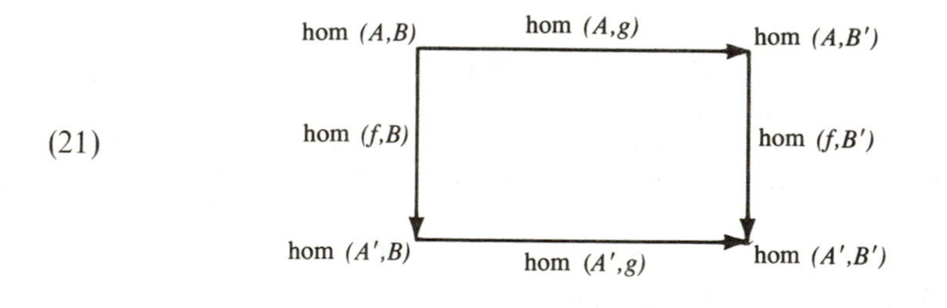

is commutative. We can deduce two natural transformations from this commutativity. First, fix $g: B \to B'$ and consider the map $A \rightsquigarrow \hom(A, g) \in \hom_{\mathbf{Set}}(\hom(A, B), \hom(A, B'))$. The commutativity of the foregoing diagram states that $A \rightsquigarrow \hom(A, g)$ is a natural transformation of the contravariant functor $\hom(-, B)$ into the contravariant functor $\hom(-, B')$. Similarly, the commutativity of (21) can be interpreted as saying that for $f: A' \to A$ the map $B \rightsquigarrow \hom(f, B)$ is a natural transformation of $\hom(A, -)$ into $\hom(A', -)$.

In the applications one is often interested in "representing" a given functor by a hom functor. Before giving the precise meaning of representation we shall determine the natural transformations from a functor $\hom_{\mathbf{C}}(A, -)$, which is a functor from **C** to **Set**, to any functor F from **C** to **Set**. Let a be any element of the set FA and let $B \in \operatorname{ob} \mathbf{C}$, $k \in \hom_{\mathbf{C}}(A, B)$. Then $F(k)$ is a map of the set FA into the set FB and its evaluation at a, $F(k)(a) \in FB$. Thus we have a map

$$a_B : k \rightsquigarrow F(k)(a) \tag{22}$$

of $\hom_{\mathbf{C}}(A, B)$ into FB. We now have the important

YONEDA'S LEMMA. *Let F be a functor from* **C** *to* **Set**, *A an object of* **C**, *a an element of the set FA. For any $B \in \operatorname{ob} \mathbf{C}$ let a_B be the map of $\hom_{\mathbf{C}}(A, B)$ into FB such that $k \rightsquigarrow F(k)(a)$. Then $B \rightsquigarrow a_B$ is a natural transformation $\eta(a)$ of $\hom_{\mathbf{C}}(A, -)$ into F. Moreover, $a \rightsquigarrow \eta(a)$ is a bijection of the set FA onto the class of natural transformations of $\hom_{\mathbf{C}}(A, -)$ to F. The inverse of $a \rightsquigarrow \eta(a)$ is the map $\eta \rightsquigarrow \eta_A(1_A) \in FA$.*

Proof. We have observed that (22) is a map of $\hom_{\mathbf{C}}(A, B)$ into FB. Now let $g : B \to C$. Then

$$F(g)a_B(k) = F(g)F(k)(a) = F(gk)(a)$$

$$a_C \hom(A, g)(k) = a_C(gk) = F(gk)(a).$$

Hence we have the commutativity of

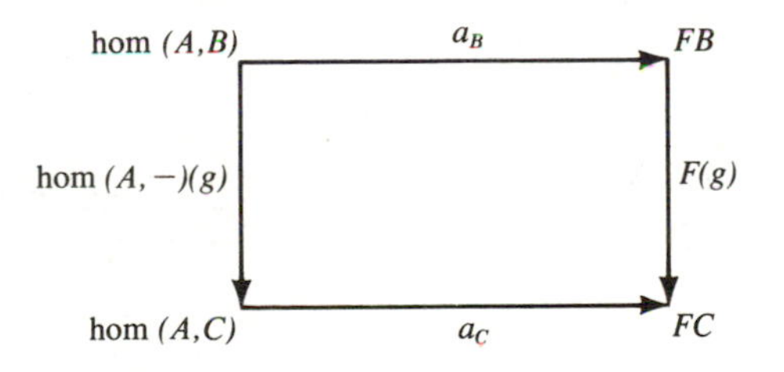

Then $\eta(a) : B \rightsquigarrow a_B$ is a natural transformation of $\hom(A, -)$ into F. Moreover $\eta(a)_A(1_A) = a_A(1_A) = F(1_A)(a) = a$.

Next let η be any natural transformation of $\hom(A, -)$ into F. Suppose $f \in \hom_{\mathbf{C}}(A, B)$. The commutativity of

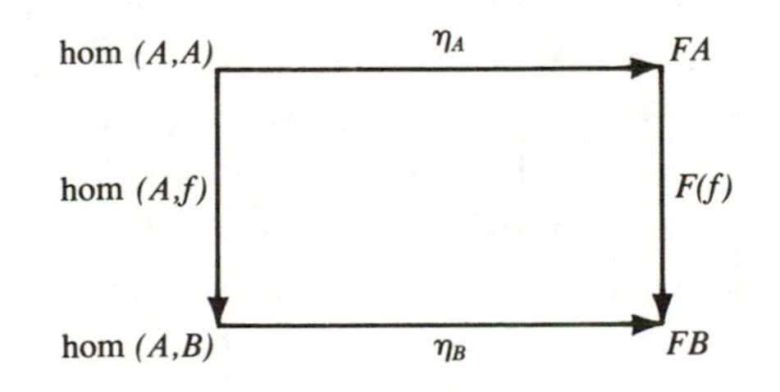

implies that $\eta_B(f) = \eta_B(f1_A) = \eta_B(\hom(A,f)(1_A)) = F(f)\eta_A(1_A) = F(f)(a)$ where $a = \eta_A(1_A) \in FA$. This shows that $\eta = \eta(a)$ as defined before.

The foregoing pair of results proves the lemma. □

We shall call a functor F from **C** to **Set** *representable* if there exists a natural isomorphism of F with a functor $\hom(A, -)$ for some $A \in \text{ob}\,\mathbf{C}$. If η is this natural isomorphism then, by Yoneda's lemma, η is determined by A and the element $a = \eta_A(1_A)$ of FA. The pair (A, a) is called a *representative* of the representable functor F.

EXERCISES

1. Apply Yoneda's lemma to obtain a bijection of the class of natural transformations of $\hom(A, -)$ to $\hom(A', -)$ with the set $\hom_{\mathbf{C}}(A', A)$.
2. Show that $f: B \to B'$ is monic in **C** if and only if $\hom(A, f)$ is injective for every $A \in \text{ob}\,\mathbf{C}$.
3. Dualize Yoneda's lemma to show that if F is a contravariant functor from **C** to **Set** and $A \in \text{ob}\,\mathbf{C}$ then any natural transformation of $\hom_{\mathbf{C}}(-, A)$ to F has the form $B \rightsquigarrow a_B$ where a_B is a map of $\hom_{\mathbf{C}}(B, A)$ into FB determined by an element $a \in FA$ as

$$a_B : k \rightsquigarrow F(k)a.$$

Show that we obtain in this way a bijection of the set FA with the class of natural transformations of $\hom_{\mathbf{C}}(-, A)$ to F.

1.7 UNIVERSALS

Two of the earliest instances of the concept of universals are those of a free group determined by a set X and the universal (associative) enveloping algebra of a Lie algebra. We have considered the first for a finite set X in BAI, pp. 68–69, where we constructed for a set X, of cardinality $r < \infty$, a group $FG^{(r)}$ and a map $i: x \to \bar{x}$ of X into $FG^{(r)}$ such that if G is any group and g is a map of X into G then there exists a unique homomorphism $\tilde{g}: FG^{(r)} \to G$,

making the diagram

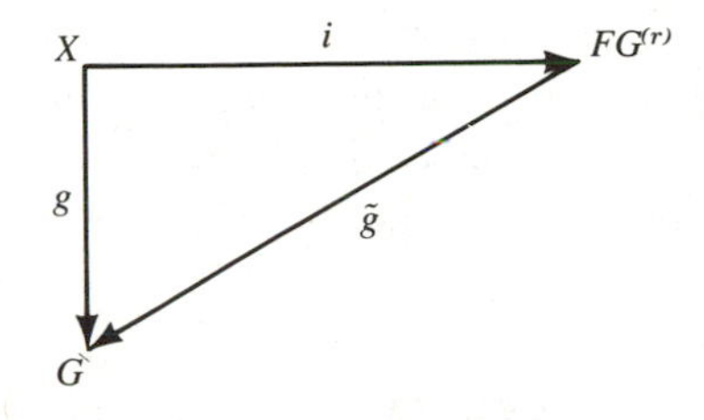

commutative. Here $\tilde{g}$ is regarded as a map of sets.

We recall that a Lie algebra L over a field is a vector space equipped with a bilinear product $[xy]$ such that $[xx] = 0$ and $[[xy]z]+[[yz]x]+[[zx]y] = 0$. If A is an associative algebra, A defines a Lie algebra A^- in which the composition is the Lie product (or additive commutator) $[xy] = xy-yx$ where xy is the given associative product in A (BAI, pp. 431 and 434). It is clear that if A and B are associative algebras and f is a homomorphism of A into B, then f is also a Lie algebra homomorphism of A^- into B^-.

If L is a Lie algebra, a universal enveloping algebra of L is a pair $(U(L), u)$ where $U(L)$ is an associative algebra and u is a homomorphism of L into the Lie algebra $U(L)^-$ such that if g is any homomorphism of L into a Lie algebra A^- obtained from an associative algebra A, then there exists a unique homomorphism $\tilde{g}$ of the associative algebra $U(L)$ into A such that

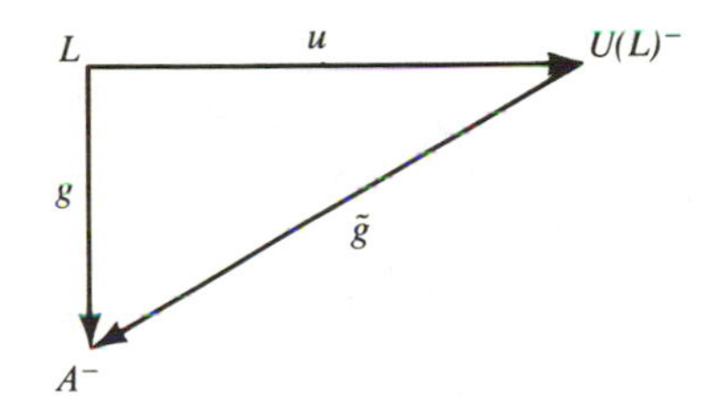

is a commutative diagram of Lie algebra homomorphisms. We shall give a construction of $(U(L), u)$ in Chapter 3 (p. 142).

Both of these examples can be formulated in terms of categories and functors. In the first we consider the categories **Grp** and **Set** and let F be the forgetful functor from **Grp** to **Set** that maps a group into the underlying set and maps a group homomorphism into the corresponding set map. Given a set X, a free group determined by X is a pair (U, u) where U is a group and u is a map of X into U such that if G is any group and g is a map of X into G, then there exists a unique homomorphism $\tilde{g}$ of U into G such that $\tilde{g}u = g$ holds for the set maps.

For the second example we consider the category **Alg** of associative algebras and the category **Lie** of Lie algebras over a given field. We have the functor F

from **Alg** to **Lie** defined by $FA = A^-$ for an associative algebra A, and if $f: A \to B$ for associative algebras, then $F(f) = f: A^- \to B^-$ for the corresponding Lie algebras. For a given Lie algebra L, a universal envelope is a pair $(U(L), u)$ where $U(L)$ is an associative algebra and u is a Lie algebra homomorphism of L into $U(L)^-$ such that if g is any homomorphism of L into a Lie algebra A^-, A associative, then there exists a unique homomorphism $\tilde{g}$ of $U(L)$ into A such that $\tilde{g}u = g$.

We now give the following general definition of universals.

DEFINITION 1.6. *Let* **C** *and* **D** *be categories, F a functor from* **C** *to* **D**. *Let B be an object in* **D**. *A* universal from B to the functor F *is a pair* (U, u) *where U is an object of* **C** *and u is a morphism from B to FU such that if g is any morphism from B to FA, then there exists a unique morphism* $\tilde{g}$ *of U into A in* **C** *such that*

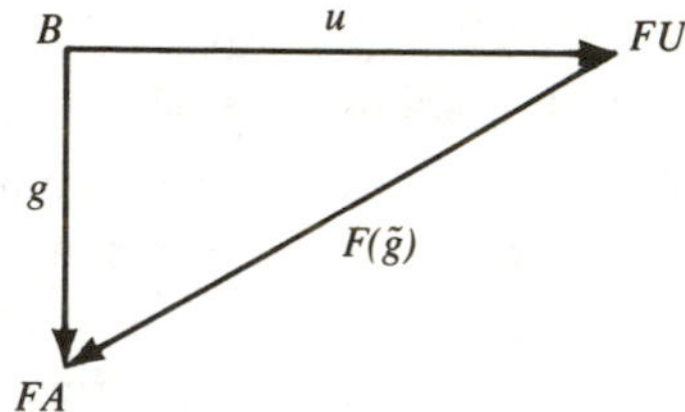

is commutative. U is called a universal **C**-object *for B and u the corresponding* universal map.

It is clear that the two examples we considered are special cases of this definition. Here are some others.

EXAMPLES

1. *Field of fractions of a commutative domain.* Let **Dom** denote the subcategory of the category **Ring** whose objects are commutative domains (=commutative rings without zero divisors $\neq 0$) with monomorphisms as morphisms. Evidently this defines a subcategory of **Ring**. Moreover, **Dom** has the full subcategory **Field** whose objects are fields and morphisms are monomorphisms. If D is a commutative domain, D has a field of fractions F (see BAI, p. 115). The important property of F is that we have the monomorphism $u: a \rightsquigarrow a/1$ of D into F, and if g is any monomorphism of D into a field F', then there exists a unique monomorphism $\tilde{g}$ of F into F' such that $g = \tilde{g}u$. We can identify D with the set of fractions $d/1$ and thereby take u to be the injection map. When this is done, the result we stated means that any monomorphism of D ($\subset F$) has a unique extension to a monomorphism of F into F' (see Theorem 2.9, p. 117 of BAI).

To put this result into the mold of the definition of universals, we consider the injection functor of the subcategory **Field** into the category **Dom** (see example 1 on p. 20). If D is a commutative domain, hence an object of **Dom**, we take the universal object for D in **Field** to be the field F of fractions of D and we take the universal map u

to be the injection of D into F. Then (F, u) is a universal from D to the injection functor as defined in Definition 1.6.

2. *Free modules*. Let R be a ring, X a non-vacuous set. We define the *free* (*left*) *R-module* $\oplus_X R$ to be the direct sum of X copies of R, that is, $\oplus_X R = \oplus M_a$, $a \in X$, where every $M_a = R$. Thus $\oplus_X R$ is the set of maps f of X into R such that $f(x) = 0$ for all but a finite number of the x's. Addition and the action of R are the usual ones: $(f+g)(x) = f(x)+g(x)$, $(rf)(x) = rf(x)$. We have the map $u: x \rightarrow \bar{x}$ of X into $\oplus_X R$ where $\bar{x}$ is defined by

$$\bar{x}(x) = 1, \qquad \bar{x}(y) = 0 \quad \text{if} \quad y \neq x.$$

If $f \in \oplus_X R$ and $\{x_1, \ldots, x_n\}$ is a subset of X such that $f(y) = 0$ for $y \notin \{x_1, \ldots, x_n\}$, then $f = \sum r_i \bar{x}_i$ where $f(x_i) = r_i$. Moreover, it is clear that for distinct $\bar{x}_i$, $\sum r_i \bar{x}_i = 0$ implies every $r_i = 0$. Hence the set $\bar{X} = \{\bar{x} | x \in X\}$ is a *base* for $\oplus_X R$ in the sense that every element of this module is a sum $\sum_{x \in X} r_x \bar{x}$, which is finite in the sense that only a finite number of the r_x are $\neq 0$, and $\sum r_x \bar{x} = 0$ for such a sum implies that every $r_x = 0$.

Now suppose M is any (left) R-module and $\varphi: x \rightarrow m_x$ is a map of X into M. Then

$$\tilde{\varphi}: \sum r_x \bar{x} \rightarrow \sum r_x m_x$$

is a well-defined map of $\oplus_X R$ into M. It is clear that this is a module homomorphism. Moreover, $\tilde{\varphi}(\bar{x}) = m_x$ so we have the commutativity of

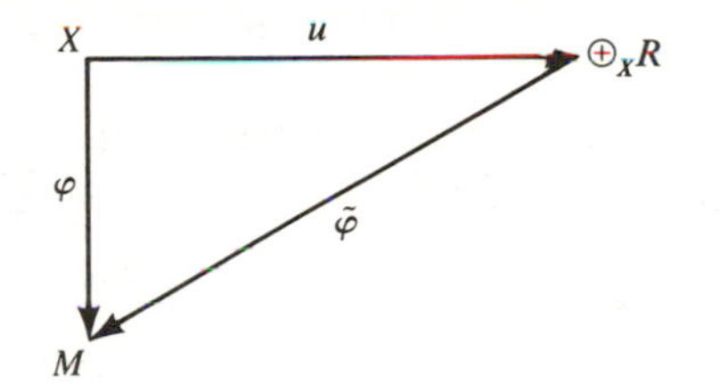

Since a module homomorphism is determined by its restriction to a set of generators, it is clear that $\tilde{\varphi}$ is the only homomorphism of the free module $\oplus_X R$ into M, making the foregoing diagram commutative.

Now consider the forgetful functor F from R-**mod** to **Set** that maps an R-module into its underlying set and maps any module homomorphism into the corresponding set map. Let X be a non-vacuous set. Then the results mean that $(\oplus_X R, u)$ is a universal from X to the functor F.

3. *Free algebras and polynomial algebras*. If K is a commutative ring, we define an (*associative*) *algebra over* K as a pair consisting of a ring $(A, +, \cdot, 0, 1)$ and a K-module A such that the underlying sets and additions are the same in the ring and in the module (equivalently, the ring and module have the same additive group), and

$$a(xy) = (ax)y = x(ay)$$

for $a \in K$, $x, y \in A$ (cf. BAI, p. 407). The algebra A is said to be *commutative* if its multiplication is commutative. *Homomorphisms* of K-algebras are K-module homomorphisms such that $1 \rightarrow 1$, and if $x \rightarrow x'$ and $y \rightarrow y'$ then $xy \rightarrow x'y'$. We have a category K-**alg** of K-algebras and a category K-**comalg** of commutative K-algebras. In the first, the objects are K-algebras and the morphisms are K-algebra homomorphisms. K-**comalg** is the full subcategory of commutative K-algebras.

We have the forgetful functors from K-**alg** and K-**comalg** to **Set**. If X is a non-vacuous set, then a universal from X to the forgetful functor has the form $(K\{X\},u)$ where $K\{X\}$ is a K-algebra and u is a map of X into $K\{X\}$ such that if g is any map of X into a K-algebra A, then there is a unique K-algebra homomorphism $\tilde{g}$ of $K\{X\}$ into A such that $\tilde{g}u = g$ (as set maps). If $(K\{X\},u)$ exists, then $K\{X\}$ is called the *free K-algebra* determined by X. In a similar manner we can replace K-**alg** by K-**comalg**. A universal from X to the forgetful functor to **Set** is denoted as $(K[X],u)$. If $X = \{x_1, x_2, \ldots, x_n\}$, we let $K[X_1, \ldots, X_n]$ be the polynomial ring in the indeterminates X_i with coefficients in K. If g is a map of X into a commutative K-algebra A, then there is a unique homomorphism of $K[X_1, \ldots, X_n]$ into A as K-algebras such that $X_i \rightsquigarrow g(x_i)$, $1 \leqslant i \leqslant n$ (BAI, p. 124). Hence $K[X_1, \ldots, X_n]$ and the map $x_i \rightsquigarrow X_i$ constitute a universal from X to the forgetful functor from K-**comalg** to **Set**.

4. *Coproducts*. Let **C** be a category. We have the diagonal functor Δ of **C** to $\mathbf{C}\times\mathbf{C}$ that maps an object A of **C** into the object (A,A) of $\mathbf{C}\times\mathbf{C}$ and a morphism $f: A \to B$ into the morphism $(f,f):(A,A) \to (B,B)$. A universal from (A_1, A_2) to Δ is a pair (U,u) where U is an object of **C** and $u = (u_1, u_2)$, $u_i: A_i \to U$ such that if $C \in \mathrm{ob}\,\mathbf{C}$ and $g_i: A_i \to C$, then there is a unique $\tilde{g}: U \to C$ such that $g_i = \tilde{g}u_i$, $i = 1, 2$. This is equivalent to saying that (U, u_1, u_2) is a coproduct in **C** of A_1 and A_2. This has an immediate generalization to coproducts of indexed sets of objects of **C**. If the index set is $I = \{\alpha\}$, then a coproduct $\coprod C_\alpha$ is a universal from (A_α) to the diagonal functor from **C** to the product category $\mathbf{C}^I$ where $\mathbf{C}^I$ is the product of I-copies of **C** (cf. p. 21).

One can often construct universals in several different ways. It is immaterial which determination of a universal we use, as we see in the following strong uniqueness property.

PROPOSITION 1.6. *If (U,u) and (U',u') are universals from an object B to a functor F, then there exists a unique isomorphism $h: U \to U'$ such that $u' = F(h)u$.*

We leave the proof to the reader. We remark that this will follow also from exercise 4 below by showing that (U,u) is the initial element of a certain category.

As one might expect, the concept of a universal from an object to a functor has a dual. This is given in

DEFINITION 1.7. *Let **C** and **D** be categories, G a functor from **D** to **C**. Let $A \in \mathrm{ob}\,\mathbf{C}$. A universal from G to A is a pair (V,v) where $V \in \mathrm{ob}\,\mathbf{D}$ and $v \in \hom_{\mathbf{C}}(GV, A)$ such that if $B \in \mathrm{ob}\,\mathbf{D}$ and $g \in \hom_{\mathbf{C}}(GB. A)$, then there exists a unique $\bar{g}: B \to V$ such that $vG(\bar{g}) = g$.*

As an illustration of this definition, we take $\mathbf{C} = R$-**mod**, $\mathbf{D} = $ **Set** as in example 2 above. If X is a non-vacuous set, we let $GX = \oplus_X R$, the free module determined by X. It is convenient to identify the element $x \in X$ with the corresponding element $\bar{x}$ of $\oplus_X R$ and we shall do this from now on. Then $\oplus_X R$ contains X and X is a base for the free module. The basic property of X is that any map φ of X into a module M has a unique extension to a homomorphism of $\oplus_X R$ into M. If $X = \varnothing$, we define $\oplus_X R = 0$. If X and Y are sets and φ is a map of X into Y, then φ has a unique extension to a homomorphism $\bar{\varphi}$ of $\oplus_X R$ into $\oplus_Y R$. We obtain a functor G from **Set** to R-**mod** by putting $G(X) = \oplus_X R$ and $G(\varphi) = \bar{\varphi}$.

Now let M be a (left) R-module, FM the underlying set so $GFM = \oplus_{FM} R$. Let v be the homomorphism of GFM into M, extending the identity map on FM. Let X be a set and g a homomorphism of $\oplus_X R$ into M, and put $\bar{g} = g|X$. Then $G(\bar{g})$ is the homomorphism of $\oplus_X R$ into $\oplus_{FM} R$ such that $G(\bar{g})(x) = g(x)$, $x \in X$, and $vG(\bar{g})(x) = g(x)$. Hence $vG(\bar{g}) = g$. Moreover, $\bar{g}$ is the only map of X into FM satisfying this condition. Hence (FM, v) is a universal from the functor G to the module M.

EXERCISES

1. Let $\mathbf{D}$ be a category, Δ the diagonal functor from $\mathbf{D}$ to $\mathbf{D} \times \mathbf{D}$. Show that (V, v), $v = (v_1, v_2)$, is a universal from Δ to (A_1, A_2) if and only if (V, v_1, v_2) is a product of A_1 and A_2 in $\mathbf{D}$.

2. Let **Rng** be the category of rings without unit, F the functor from **Ring** to **Rng** that forgets the unit. Show that any object in **Rng** possesses a universal from it to F. (See BAI, p. 155.)

3. Let F_1 be a functor from $\mathbf{C}_1$ to $\mathbf{C}_2$, F_2 a functor from $\mathbf{C}_2$ to $\mathbf{C}_3$. Let (U_2, u_2) be a universal from B to F_2, (U_1, u_1) a universal from U_2 to F_1. Show that $(U_1, F_2(u_1)u_2)$ is a universal from B to F_2F_1.

4. Let F be a functor from $\mathbf{C}$ to $\mathbf{D}$, B an object of $\mathbf{D}$. Verify that the following data define a category $\mathbf{D}(B, F)$: The objects are the pairs (A, g) where $A \in \operatorname{ob}\mathbf{C}$ and $g \in \hom_{\mathbf{D}}(B, FA)$. Define $\hom((A, g), (A', g'))$ as the subset of $\hom_{\mathbf{C}}(A, A')$ of h such that $g' = F(h)g$ and arrange to make these hom sets non-overlapping. Define multiplication of morphisms as in $\mathbf{C}$ and $1_{(A,g)} = 1_A$. Show that (U, u) is a universal from B to F if and only if (U, u) is an initial element of $\mathbf{D}(B, F)$. Dualize.

1.8 ADJOINTS

We shall now analyze the situation in which we are given a functor F from $\mathbf{C}$ to $\mathbf{D}$ such that for every object B of $\mathbf{D}$ there exists a universal (U, u) from B to

F. The examples of the previous section are all of this sort. First, we need to consider some alternative definitions of universals.

Let F be a functor from $\mathbf{C}$ to $\mathbf{D}$, B an object of $\mathbf{D}$, (U,u) a pair such that $U \in \mathrm{ob}\,\mathbf{C}$ and $u \in \hom_{\mathbf{D}}(B, FU)$. Then if $f \in \hom_{\mathbf{C}}(U,A)$, $F(f) \in \hom_{\mathbf{D}}(FU, FA)$ and $F(f)u \in \hom_{\mathbf{D}}(B, FA)$. Accordingly, we have the map

$$\eta_A : f \rightsquigarrow F(f)u \tag{23}$$

of the set $\hom_{\mathbf{C}}(U,A)$ into the set $\hom_{\mathbf{D}}(B, FA)$. By definition, (U,u) is a universal from B to F if for any object A of $\mathbf{C}$ and any morphism $g: B \to FA$ in $\mathbf{D}$ there is one and only one $\tilde{g}: U \to A$ such that $g = F(\tilde{g})u$. Evidently this means that (U,u) is a universal from B to F if and only if for every $A \in \mathrm{ob}\,\mathbf{C}$ the map η_A from $\hom_{\mathbf{C}}(U,A)$ to $\hom_{\mathbf{D}}(B, FA)$ given in (23) is bijective. If this is the case and $h: A \to A'$ in $\mathbf{C}$, then the diagram

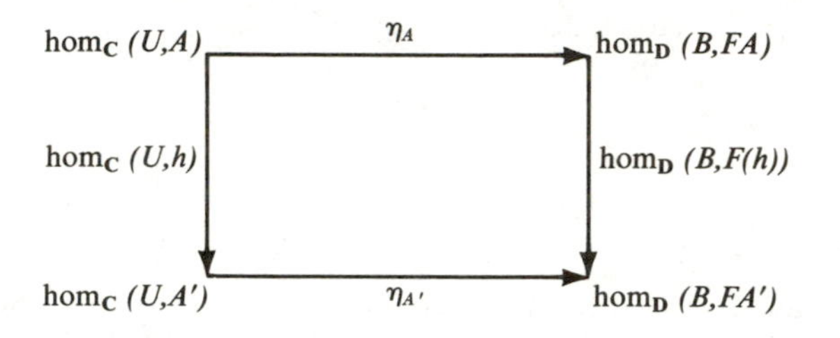

is commutative, since one of the paths from an $f \in \hom_{\mathbf{C}}(U,A)$ gives $F(hf)u$ and the other gives $F(h)F(f)u$, which are equal since $F(hf) = F(h)F(f)$. It follows that $\eta: A \rightsquigarrow \eta_A$ is a natural isomorphism of the functor $\hom_{\mathbf{C}}(U, -)$ to the functor $\hom_{\mathbf{D}}(B, F-)$ that is obtained by composing F with the functor $\hom_{\mathbf{D}}(B, -)$. Note that both of these functors are from $\mathbf{C}$ to $\mathbf{Set}$, and since $\eta_U(1_U) = F(1_U)u = 1_{FU}u = u$, the result is that $\hom_{\mathbf{D}}(B, F-)$ is representable with (U,u) as representative.

Conversely, suppose $\hom_{\mathbf{D}}(B, F-)$ is representable with (U,u) as representative. Then, by Yoneda's lemma, for any object A in $\mathbf{C}$, the map η_A of (23) is a bijection of $\hom_{\mathbf{C}}(U,A)$ onto $\hom_{\mathbf{D}}(B, FA)$. Consequently (U,u) is a universal from B to F.

Similar considerations apply to the other kind of universal. Let G be a functor from $\mathbf{D}$ to $\mathbf{C}$, A an object of $\mathbf{C}$, (V,v) a pair such that $V \in \mathrm{ob}\,\mathbf{D}$ and $v \in \hom_{\mathbf{C}}(GV, A)$. Then for any $B \in \mathrm{ob}\,\mathbf{D}$ we have the map

$$\zeta_B : g \rightsquigarrow vG(g) \tag{24}$$

of $\hom_{\mathbf{D}}(B, V)$ into $\hom_{\mathbf{C}}(GB, A)$, and (V,v) is a universal from G to A if and only if ζ_B is a bijection for every $B \in \mathrm{ob}\,\mathbf{D}$. Moreover, this is the case if and only if $\zeta: B \rightsquigarrow \zeta_B$ is a natural isomorphism of the contravariant functor $\hom_{\mathbf{D}}(-, V)$

with the contravariant functor $\hom_{\mathbf{C}}(G-, A)$ obtained by composing G with $\hom_{\mathbf{C}}(-, A)$.

We now assume that for every $B \in \mathrm{ob}\,\mathbf{D}$ we have a universal from B to the functor F from $\mathbf{C}$ to $\mathbf{D}$. For each B we choose a universal that we denote as (GB, u_B). Then for any $A \in \mathrm{ob}\,\mathbf{C}$ we have the bijective map

$$\eta_{B,A} : f \rightsquigarrow F(f)u_B \tag{25}$$

of $\hom_{\mathbf{C}}(GB, A)$ onto $\hom_{\mathbf{D}}(B, FA)$, and for fixed B, $A \rightsquigarrow \eta_{B,A}$ is a natural isomorphism of the functor $\hom_{\mathbf{C}}(GB, -)$ to the functor $\hom_{\mathbf{D}}(B, F-)$.

Let B' be a second object in $\mathbf{D}$ and let $h: B \to B'$. We have the diagram

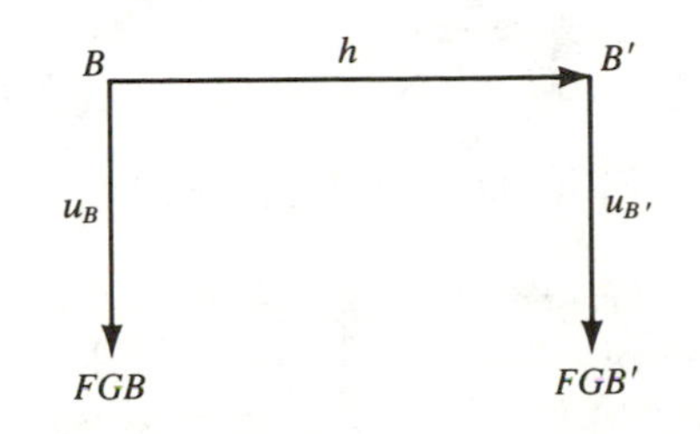

so $u_{B'}h \in \hom_{\mathbf{D}}(B, FGB')$, and since (GB, u_B) is a universal from B to F, there is a unique morphism $G(h): GB \to GB'$ such that the foregoing diagram becomes commutative by filling in the horizontal $FG(h)$:

$$FG(h)u_B = u_{B'}h. \tag{26}$$

The commutativity of the diagram

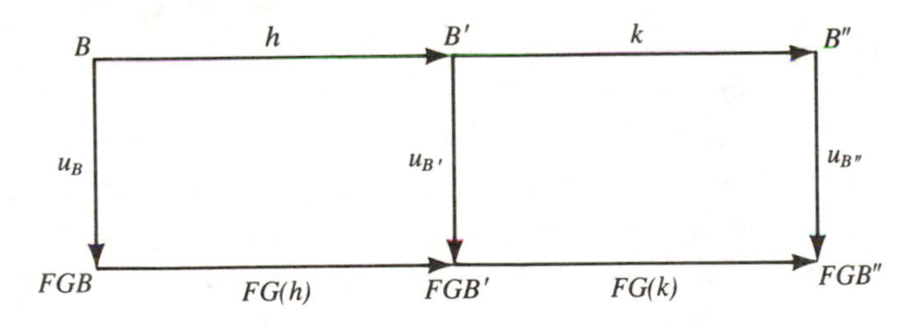

and the functorial property $(FG(k))(FG(h)) = F(G(k)G(h))$ imply that

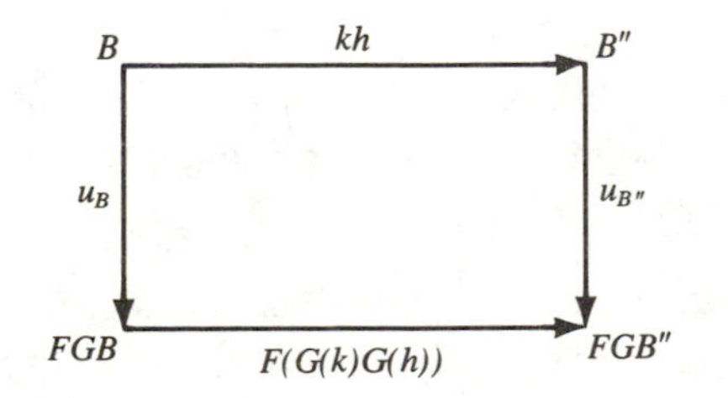

is commutative. Since $G(kh)$ is the only morphism $GB \to GB''$ such that $FG(kh)u_B = u_{B''}kh$, it follows that $G(kh) = G(k)G(h)$. In a similar manner we have $G(1_B) = 1_{GB}$. Thus G is a functor from **D** to **C**.

We wish to study the relations between the two functors F and G. Let $A \in \text{ob}\,\mathbf{C}$. Then $FA \in \text{ob}\,\mathbf{D}$, so applying the definition of a universal from FA to F to the map $g = 1_{FA}$, we obtain a unique $v_A: GFA \to A$ such that

$$F(v_A)u_{FA} = 1_{FA}. \tag{27}$$

Then for any $B \in \text{ob}\,\mathbf{D}$ we have the map

$$\zeta_{A,B}: g \rightsquigarrow v_A G(g) \tag{28}$$

of $\hom_{\mathbf{D}}(B, FA)$ into $\hom_{\mathbf{C}}(GB, A)$. Now let $g \in \hom_{\mathbf{D}}(B, FA)$. Then

$$\eta_{B,A}\zeta_{A,B}(g) = F(V_A G(g))u_B = F(V_A)(FG(g))u_B = F(v_A)u_{FA}g = g$$

(by (26) and (27)). Since $\eta_{B,A}$ is bijective, it follows that $\zeta_{A,B} = \eta_{B,A}{}^{-1}$ and this is a bijective map of $\hom_{\mathbf{D}}(B, FA)$ onto $\hom_{\mathbf{C}}(GB, A)$.

The fact that for every $B \in \text{ob}\,\mathbf{D}$, $\zeta_{A,B}: g \rightsquigarrow v_A G(g)$ is a bijective map of $\hom_{\mathbf{D}}(B, FA)$ onto $\hom_{\mathbf{C}}(GB, A)$ implies that (FA, v_A) is a universal from G to A, and this holds for every $A \in \text{ob}\,\mathbf{C}$. Moreover, for fixed A, $B \rightsquigarrow \zeta_{A,B}$ is a natural isomorphism of the contravariant functor $\hom_{\mathbf{D}}(-, FA)$ to the contravariant functor $\hom_{\mathbf{C}}(G-, A)$. Consequently, $B \rightsquigarrow \eta_{B,A} = \zeta_{A,B}{}^{-1}$ is a natural isomorphism of $\hom_{\mathbf{C}}(G-, A)$ to $\hom_{\mathbf{D}}(-, FA)$.

We summarize the results obtained thus far in

PROPOSITION 1.7. *Let F be a functor from* **C** *to* **D** *such that for every $B \in \text{ob}\,\mathbf{D}$ there is a universal from B to F. For each B choose one and call it (GB, u_B). If $h: B \to B'$ in* **D**, *define $G(h): GB \to GB'$ to be the unique element in* $\hom_{\mathbf{C}}(GB, GB')$ *such that $F(G(h))u_B = u_{B'}h$. Then G is a functor from* **D** *to* **C**. *If $A \in \text{ob}\,\mathbf{C}$, there is a unique $v_A: GFA \to A$ such that $F(v_A)u_{FA} = 1_{FA}$. Then (FA, v_A) is a universal from G to A. If $\eta_{B,A}$ is the map $f \rightsquigarrow F(f)u_B$ of* $\hom_{\mathbf{C}}(GB, A)$ *into* $\hom_{\mathbf{D}}(B, FA)$, *then for fixed B, $A \rightsquigarrow \eta_{B,A}$ is a natural isomorphism of* $\hom_{\mathbf{C}}(GB, -)$ *to* $\hom_{\mathbf{D}}(B, F-)$ *and for fixed A, B $\eta_{B,A}$ is a natural isomorphism of* $\hom_{\mathbf{C}}(G-, A)$ *to* $\hom_{\mathbf{D}}(-, FA)$.

The last statement of this proposition can be formulated in terms of the important concept of adjoint functors that is due to D. M. Kan.

DEFINITION 1.8. *Let F be a functor from* **C** *to* **D**, *G a functor from* **D** *to* **C**. *Then F is called a* right adjoint *of G and G a* left adjoint *of F if for every (B, A),*

$B \in \text{ob}\,\mathbf{D}$, $A \in \text{ob}\,\mathbf{C}$, *we have a bijective map* $\eta_{B,A}$ *of* $\hom_{\mathbf{C}}(GB, A)$ *onto* $\hom_{\mathbf{D}}(B, FA)$ *that is natural in A and B in the sense that for every B,* $A \rightsquigarrow \eta_{B,A}$ *is a natural isomorphism of* $\hom_{\mathbf{C}}(GB, -)$ *to* $\hom_{\mathbf{D}}(B, F-)$ *and for every A,* $B \quad \eta_{B,A}$ *is a natural isomorphism of* $\hom_{\mathbf{C}}(G-, A)$ *to* $\hom_{\mathbf{D}}(-, FA)$. *The map* $\eta{:}(B, A) \rightsquigarrow \eta_{B,A}$ *is called an adjugant from G to F, and the triple* (F, G, η) *is an* adjunction.

Evidently the last statement of Proposition 1.7 implies that (F, G, η) is an adjunction. We shall now show that, conversely, any adjunction (F, G, η) determines universals so that the adjunction can be obtained from the universals as in Proposition 1.7.

Thus suppose (F, G, η) is an adjunction. Let $B \in \text{ob}\,\mathbf{D}$ and put $u_B = \eta_{B,GB}(1_{GB}) \in \hom_{\mathbf{D}}(B, FGB)$. Keeping B fixed, we have the natural isomorphism $A \rightsquigarrow \eta_{B,A}$ of $\hom_{\mathbf{C}}(GB, -)$ to $\hom_{\mathbf{D}}(B, F-)$. By Yoneda's lemma, if $f \in \hom_{\mathbf{C}}(GB, A)$ then

$$\eta_{B,A}(f) = \hom_{\mathbf{D}}(B, F(f))u_B = F(f)u_B.$$

Then $f \rightsquigarrow F(f)u_B$ is a bijective map of $\hom_{\mathbf{C}}(GB, A)$ onto $\hom_{\mathbf{D}}(B, FA)$. The fact that this holds for all $A \in \text{ob}\,\mathbf{C}$ implies that (GB, u_B) is a universal from B to F.

Let $k \in \hom_{\mathbf{C}}(A, A')$, $f \in \hom_{\mathbf{C}}(GB, A)$, $g \in \hom_{\mathbf{C}}(GB', A)$, $h \in \hom_{\mathbf{D}}(B, B')$. The natural isomorphism $A \rightsquigarrow \eta_{B,A}$ of $\hom_{\mathbf{C}}(GB, -)$ to $\hom_{\mathbf{D}}(B, F-)$ gives the relation

$$\eta_{B,A'}(kf) = F(k)\eta_{B,A}(f) \tag{29}$$

and the natural isomorphism $B \rightsquigarrow \eta_{B,A}$ of $\hom_{\mathbf{C}}(G-, A)$ to $\hom_{\mathbf{D}}(-, FA)$ gives

$$\eta_{B,A}(gG(h)) = \eta_{B',A}(g)h. \tag{30}$$

(Draw the diagrams.) These imply that

$$\begin{aligned} F(G(h))u_B &= F(G(h))\eta_{B,GB}(1_{GB}) = \eta_{B,GB'}(G(h)1_{GB}) \\ &\qquad \text{(by (29) with } A = GB,\ A' = GB',\ k = G(h), f = 1_{GB}) \\ &= \eta_{B,GB'}(1_{GB'}G(h)) = \eta_{B',GB'}(1_{GB'})h = u_{B'}h \\ &\qquad \text{(by (30) with } A = GB',\ g = 1_{GB'}). \end{aligned}$$

The relation $F(G(h))u_B = u_{B'}h$, which is the same as that in (26), shows that the given functor G is the one determined by the choice of the universal (GB, u_B) for every $B \in \text{ob}\,\mathbf{D}$. We have therefore completed the circle back to the situation we considered at the beginning.

An immediate consequence of the connection between adjoints and universals is the following

PROPOSITION 1.8. *Any two left adjoints G and G' of a functor F from* **C** *to* **D** *are naturally isomorphic.*

Proof. Let η and η' be adjugants from G and G' respectively to F. For $B \in \text{ob}\,\mathbf{D}$, (GB, u_B) and $(G'B, u'_B)$, where u_B and u'_B are determined as above, are universals from B to F. Hence there exists a unique isomorphism $\lambda_B : GB \to G'B$ such that $u'_B = F(\lambda_B)u_B$ (Proposition 1.5). We shall be able to conclude that $\lambda : B \rightsquigarrow \lambda_B$ is a natural isomorphism of G to G' if we can show that for any $h : B \to B'$ the diagram

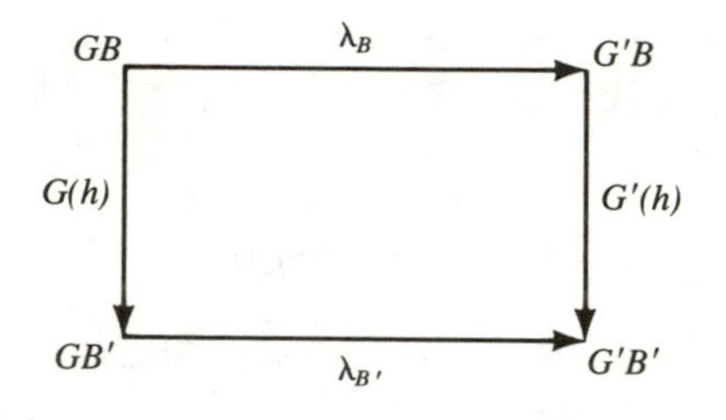

is commutative. To see this we apply $\eta_{B,GB'}$ to $G'(h)\lambda_B$ and $\lambda_{B'}G(h)$. This gives

$$F(G'(h))F(\lambda_B)u_B = F(G'(h))u'_B = u'_{B'}h \quad \text{(by (26))}$$
$$F(\lambda_{B'})F(G(h))u_B = F(\lambda_{B'})u_{B'}h = u'_B h \quad \text{(by (26))}.$$

Since $\eta_{B,GB'}$ is an isomorphism, it follows that $G'(h)\lambda_B = \lambda_{B'}G(h)$ and so the required commutativity holds.

Everything we have done dualizes to universals from a functor to an object. We leave it to the reader to verify this.

EXERCISES

1. Determine left adjoints for the functors defined in the examples on pp. 42–44.

2. Let (F, G, η) be an adjunction. Put $u_B = \eta_{B,GB}(1_{GB}) \in \text{hom}_{\mathbf{D}}(B, FGB)$. Verify that $u : B \rightsquigarrow u_B$ is a natural transformation of $1_{\mathbf{D}}$ to FG. u is called the *unit* of the adjugant η. Similarly, $v : A \rightsquigarrow v_A = \eta_{FA,A}{}^{-1}(1_{FA})$ is a natural transformation of GF to $1_{\mathbf{C}}$. This is called the *co-unit* of η.

3. Let (G_1, F_1, η_1) be an adjunction where $F_1 : \mathbf{C}_1 \to \mathbf{C}_2$, $G_1 : \mathbf{C}_2 \to \mathbf{C}_1$, and let (G_2, F_2, η_2) be an adjunction where $F_2 : \mathbf{C}_2 \to \mathbf{C}_3$, $G_2 : \mathbf{C}_3 \to \mathbf{C}_2$. Show that $G_1 G_2$ is a left adjoint of $F_2 F_1$ and determine the adjugant.

4. Let (G, F, η) be an adjunction. Show that $F : \mathbf{C} \to \mathbf{D}$ preserves products; that is, if $A = \prod A_\alpha$ with maps $p_\alpha : A \to A_\alpha$, then $FA = \prod FA_\alpha$ with maps $F(p_\alpha)$. Dualize.

REFERENCES

S. Eilenberg and S. MacLane, General theory of natural equivalences, *Trans. Amer. Math. Soc.*, vol. 58 (1945), 231–294.

S. MacLane, *Categories for the Working Mathematician*, New York, Springer, 1971.

2

Universal Algebra

The idea of "universal" algebra as a comparative study of algebraic structures appears to have originated with the philosopher-mathematician Alfred North Whitehead. In his book *A Treatise on Universal Algebra with Applications*, which appeared in 1898, Whitehead singled out the following topics as particularly worthy of such a comparative study: "Hamilton's Quaternions, Grassmann's Calculus of Extensions and Boole's Symbolic Logic." Perhaps the time was not yet ripe for the type of study that Whitehead undertook. At any rate, the first significant results on universal algebra were not obtained until considerably later—in the 1930's and 1940's, by G. Birkhoff, by Tarski, by Jònsson and Tarski, and others.

The basic concept we have to deal with is that of an "Ω-algebra," which, roughly speaking, is a non-vacuous set equipped with a set of finitary compositions. Associated with this concept is an appropriate notion of homomorphism. We adopt the point of view that we have a set Ω of operator symbols (for example, $+$, $\cdot$, $\wedge$) that is given à priori, and a class of non-vacuous sets, the carriers, in which the operator symbols are realized as compositions in such a way that a given symbol ω is realized in every carrier as a composition with a fixed "arity," that is, is always a binary composition, or a ternary composition, etc. The class of Ω-algebras for a fixed Ω constitutes a category with homomorphisms of Ω-algebras as the morphisms. Categorical ideas will play an important role in our account.

The concept of Ω-algebras can be broadened to encompass relations as well as operations. In this way one obtains the basic notion of a relation structure that serves as the vehicle for applying mathematical logic to algebra. We shall not consider the more general concept here.

We shall develop first some general results on homomorphisms and isomorphisms of Ω-algebras that the reader has already encountered in special cases. Beyond this, we introduce the concept of a subdirect product of algebras and the important constructions of direct limits, inverse limits, and ultraproducts of Ω-algebras. We shall be particularly interested in varieties (or equational classes) of Ω-algebras, free algebras, and free products in varieties, and we shall prove an important theorem due to G. Birkhoff giving an internal characterization of varieties. The important special case of free products of groups and of free groups will be treated in a more detailed manner.

2.1 Ω-ALGEBRAS

We review briefly some set theoretic notions that will be useful in the study of general (or universal) algebras.

If A and B are sets, then one defines a *correspondence from* A to B to be any subset of $A \times B$. A map $f: A \to B$ is a correspondence such that for every $a \in A$ there is a b in B such that $(a, b) \in f$, and if (a, b) and $(a, b') \in f$, then $b = b'$. The uniquely determined b such that $(a, b) \in f$ is denoted as $f(a)$. This gives the connection between the notation for correspondences and the customary function or map notation. If Φ is a correspondence from A to B, the *inverse correspondence* Φ^{-1} from B to A is the set of pairs (b, a) such that $(a, b) \in \Phi$. If Φ is a correspondence from A to B and Ψ is a correspondence from B to C, then $\Psi\Phi$ is defined to be the correspondence from A to C consisting of the pairs (a, c), $a \in A$, $c \in C$, for which there exists a $b \in B$ such that $(a, b) \in \Phi$ and $(b, c) \in \Psi$. If f is a map from A to B and g is a map from B to C, then the product of the correspondence gf is the usual composite of g following f: $(gf)(a) = g(f(a))$. As usual, we denote the identity map on the set A by 1_A. This is the correspondence from A to A consisting of the elements (a, a), $a \in A$. It is also called the *diagonal* on the set A. If Φ is a correspondence from A to B, Ψ a correspondence from B to C, Θ a correspondence from C to D, then the following set relations are readily checked:

$$(\Theta\Psi)\Phi = \Theta(\Psi\Phi). \tag{1}$$

$$(\Psi\Phi)^{-1} = \Phi^{-1}\Psi^{-1}. \tag{2}$$

$$(\Phi^{-1})^{-1} = \Phi. \tag{3}$$

$$\Phi 1_A = \Phi = 1_B\Phi. \tag{4}$$

The set of maps from A to B is denoted as B^A. If $A = \{1, 2, \ldots, n\}$ we write $B^{(n)}$ for B^A and call its elements *n-tuples* (or sequences of n elements) in the set B. We identify $B^{(1)}$ with B. We define an *n-ary relation R* on the set A to be a subset R of $A^{(n)}$. Thus a *binary* ($n = 2$) relation on A is a correspondence from A to A. An alternative notation for $(a, b) \in R$ is aRb. In this case we say that *a is in the relation R to b*.

An *equivalence relation E* on A is a binary relation that is reflexive: aEa for every $a \in A$; symmetric: if aEb then bEa; and transitive: if aEb and bEc then aEc. These conditions can be expressed more concisely in the following way:

$$1_A \subset E \qquad \text{(reflexivity)}.$$

$$E = E^{-1} \qquad \text{(symmetry)}.$$

$$EE \subset E \qquad \text{(transitivity)}.$$

The element of the power set $\mathscr{P}(A)$ (set of subsets of A) consisting of the elements b such that bEa is denoted as $\bar{a}_E$ or $\bar{a}$ if E is clear. The set of these $\bar{a}$ constitutes a partition of the set A called the *quotient set A/E* of A with respect to the equivalence relation E (see, for example, BAI, p. 11). By a partition of A we mean a set of non-vacuous subsets of A such that A is their disjoint union, that is, A is their union and the intersection of distinct subsets is vacuous. There is a 1–1 correspondence between partitions of A and equivalence relations on A (BAI, pp. 11–12).

Another important type of relation on a set A is that of a partial order $\geqslant$. This is defined by the following properties: $a \geqslant a$ (reflexivity); if $a \geqslant b$ and $b \geqslant a$, then $a = b$ (anti-symmetry); and if $a \geqslant b$ and $b \geqslant c$, then $a \geqslant c$ (transitivity). If we write O for the subset of $A \times A$ of (a, b) such that $a \geqslant b$, then these conditions are respectively:

$$O \supset 1_A.$$

$$O \cap O^{-1} \subset 1_A.$$

$$OO \subset O.$$

If n is a positive integer, we define an *n-ary product* on A to be a map of $A^{(n)}$ to A. Thus the n-ary products are just the elements of the set $A^{A^{(n)}}$. If $n = 1, 2, 3$, etc., we have products that are *unary*, *binary*, *ternary*, etc. It is convenient to introduce also *nullary* products, which are simply distinguished elements of A (e.g., the element 0 of a ring). We can extend our notation $A^{A^{(n)}}$ to include $A^{A^{(0)}}$, which is understood to be A, the set of nullary products on A. Thus we have n-ary products for any $n \in \mathbb{N} = \{0, 1, 2, 3, \ldots\}$. If $n \geqslant 1$ and $\omega \in A^{A^{(n)}}$, so ω is a map $(a_1, a_2, \ldots, a_n) \rightarrow \omega(a_1, \ldots, a_n) \in A$, then the set of elements of the form

$$(a_1, a_2, \ldots, a_n, \omega(a_1, \ldots, a_n))$$

is a subset of $A^{(n+1)}$ and this is an $(n+1)$-ary relation on A. Since the nullary products are just the elements of A, we see that for any $n \in \mathbb{N}$ the notion of $(n+1)$-ary relation includes that of n-ary product on A.

We are now ready to define the concept of a "general" algebra or Ω-algebra. Roughly speaking, this is just a non-vacuous set A equipped with a set Ω of n-ary products, $n = 0, 1, 2, \ldots$. In order to compare different algebras—more precisely, to define homomorphisms—it is useful to regard Ω as having an existence apart from A and to let the elements of Ω determine products in different sets $A, B, \ldots$ in such a way that the products determined by a given $\omega \in \Omega$ in $A, B, \ldots$ all have the same *arity*, that is, are n-ary with a fixed n. We therefore begin with a set Ω together with a given decomposition of Ω as a disjoint union of subsets $\Omega(n)$, $n = 0, 1, 2, \ldots$. The elements of $\Omega(n)$ are called *n-ary product* (or *operator*) *symbols*. For the given Ω and decomposition $\Omega = \bigcup \Omega(n)$ we introduce the following

DEFINITION 2.1. *An* Ω-algebra *is a non-vacuous set A together with a map of Ω into products on A such that if $\omega \in \Omega(n)$, then the corresponding product is n-ary on A. The underlying set A is called the* carrier *of the algebra, and we shall usually denote the algebra by the same symbol as its carrier.*

If ω is nullary, so $\omega \in \Omega(0)$, the corresponding distinguished element of A, is denoted as ω_A or, if there is no danger of confusion, as ω (e.g., the element 1 in every group G rather than 1_G). If $n \geqslant 1$ and $\omega \in \Omega(n)$, the corresponding product in A is

$$(a_1, a_2, \ldots, a_n) \rightarrow \omega(a_1, a_2, \ldots, a_n),$$

$a_i \in A$. We shall now abbreviate the right-hand side as

$$\omega a_1 a_2 \cdots a_n$$

thus dropping the parentheses as well as the commas. The observation that this can be done without creating ambiguities even when more than one operator symbol occurs is due to Lukasiewicz. For example, if ω is 5-ary, φ ternary, ψ unary, and λ nullary, then

$$\varphi(a_1, a_2, \omega(a_3, \lambda, a_4, a_5, \psi(a_6)))$$

becomes $\varphi a_1 a_2 \omega a_3 \lambda a_4 a_5 \psi a_6$ in Lukasiewicz's notation, and it is easily seen that if we *know* that φ is ternary, ω is 5-ary, λ is nullary, and ψ is unary, then the displayed element is the only meaning that can be assigned to $\varphi a_1 a_2 \omega a_3 \lambda a_4 a_5 \psi a_6$. The reader should make a few more experiments with this notation. We shall return to this later in our discussion of free Ω-algebras. We remark also that we do not exclude the possibility that $\omega \neq \omega'$ in $\Omega(n)$, but the associated n-ary products in A are identical.

EXAMPLES

1. A monoid is an Ω-algebra with $\Omega = \{p, 1\}$ where p is binary, 1 is nullary, satisfying the following identities (or laws): $ppabc = papbc$ (the associative law for p) and $pa1 = a = p1a$.

2. A group is an Ω-algebra with $\Omega = \{p, 1, i\}$ where p is binary, 1 is nullary, and i is unary. Corresponding to the group axioms we have the following identities: $ppabc = papbc$, $pa1 = a = p1a$, $paia = 1$, $piaa = 1$. Thus ia is the usual a^{-1}. It is necessary to introduce this unary operation to insure that the general theory of Ω-algebras has a satisfactory specialization to the usual group theory.

3. A ring is an Ω-algebra with $\Omega = \{s, p, 0, 1, -\}$ where s and p are binary, 0 and 1 are nullary, and $-$ is unary. Here s gives the sum, p is the product, 0 and 1 have their usual significance, and $-a$ is the negative of a. We leave it to the reader to formulate the identities that are required to complete the definition of a ring. For example, one of the distributive laws reads $pasbc = spabpac$.

4. Groups with operators. This concept is designed to study a group relative to a given set of endomorphisms. Examples are the sets of inner automorphisms, all automorphisms, all endomorphisms. From the point of view of Ω-algebras, we have a set $\Omega = \Lambda \cup \{p, 1, i\}$ where Λ is a set of unary operator symbols distinct from i and $p, 1, i$ are as in the definition of groups. Besides the group conditions on $p, 1, i$ we assume that if $\lambda \in \Lambda$, then

$$\lambda pab = p\lambda a \lambda b \tag{5}$$

or, in the usual notation,

$$\lambda(ab) = (\lambda a)(\lambda b). \tag{5'}$$

Thus $a \rightsquigarrow \lambda a$ is an endomorphism of the group. This type of algebra is called a *group with operators* or a Λ-*group*.

Any module M (left or right) for a ring R can be regarded as an abelian group with R as operator set.

5. Lattices. Here Ω consists of two binary product symbols. If L is the carrier, we denote the result of applying these to $a, b \in L$ in the usual way as $a \vee b$ and $a \wedge b$, which are called the *join* and *meet* of a and b respectively. The lattice axioms are

$$a \vee b = b \vee a, \qquad a \wedge b = b \wedge a.$$

$$(a \vee b) \vee c = a \vee (b \vee c), \qquad (a \wedge b) \wedge c = a \wedge (b \wedge c).$$

$$a \vee a = a, \qquad a \wedge a = a.$$

$$(a \vee b) \wedge a = a, \qquad (a \wedge b) \vee a = a.$$

There is another way of defining a lattice—as a certain type of partially ordered set. One begins with a set L with a partial order $\geqslant$. If S is a subset, an *upper bound* u (*lower bound* l) of S is an element of L such that $u \geqslant s$ ($l \leqslant s$) for every $s \in S$. A *least upper bound* or *sup* (*greatest lower bound* or *inf*) of S is an upper bound B (lower bound b) such that $B \leqslant u$ ($b \geqslant l$) for every upper bound u (lower bound l) of S. If a sup or inf exists for a set then it is unique. This is clear from the definition. One can define a lattice as a partially ordered set in which every pair of elements has a sup $a \vee b$ and an inf $a \wedge b$. It is easy to see that this definition is equivalent to the one given above. The equivalence is established by showing that the sup and inf satisfy the relations listed in example 5 and showing that conversely if one has a lattice in the algebraic sense, then one obtains a partial order by defining $a \geqslant b$ if $a \wedge b = b$ (or, equivalently, $a \vee b = a$). Then the given $a \vee b$ and $a \wedge b$ in L are sup and inf in this partially ordered set. The details are easily carried out and are given in full in BAI, pp. 459–460.

A lattice (viewed as a partially ordered set) is called *complete* if every non-vacuous subset of L has a sup and an inf. Then the element $1 = \sup\{a | a \in L\}$ satisfies $1 \geqslant a$ for every $a \in L$ and $0 = \inf\{a | a \in L\}$ satisfies $0 \leqslant a$ for every a. These are called respectively the greatest and least elements of the lattice. A very useful result is the following theorem:

THEOREM 2.1. *A partially ordered set* L *is a complete lattice if and only if* L *contains a greatest element* 1 ($\geqslant a$ *for every* a) *and every non-vacuous subset of* L *has an inf.*

The proof is obtained by showing that if S is a non-vacuous subset, the set U of upper bounds of S is not vacuous and its inf is a sup for S (BAI, p. 458).

The preceding examples show that the study of Ω-algebras has relevance for the study of many important algebraic structures. It should be noted, however, that some algebraic structures, e.g., fields, are not Ω-algebras, since one of the operations in a field, the inverse under multiplication, is not defined everywhere. Also the results on Ω-algebras have only a limited application to module theory since the passage to the Ω-algebra point of view totally ignores the ring structure of the operator set.

EXERCISES

1. Let Φ and Φ' be correspondences from A to B, Ψ and Ψ' correspondences from B to C. Show that if $\Phi \subset \Phi'$ and $\Psi \subset \Psi'$, then $\Psi\Phi \subset \Psi'\Phi'$.

2. Let E and F be equivalence relations on a set A. Show that EF is an equivalence relation if and only if $EF = FE$.

3. Let H be a subgroup of a group G and define $a \equiv_H b$ if $b^{-1}a \in H$. This is an equivalence relation. Show that if K is a second subgroup, then $\equiv_H$ and $\equiv_K$ commute if and only if $HK = KH$. Hence show that if $H \lhd G$ (H is normal in G), then $\equiv_H \equiv_K$ is an equivalence relation.

4. List all of the partitions of $\{1, 2, \ldots, n\}$ for $n = 1, 2, 3$, and 4.

5. Let $E(n)$ be the set of equivalence relations on a set of n elements, $|E(n)|$ its cardinality. Prove the following recursion formula for $|E(n)|$:

$$|E(n)| = |E(n-1)| + \binom{n-1}{1}|E(n-2)| + \binom{n-1}{2}|E(n-3)| + \cdots + \binom{n-1}{n-2}|E(1)| + 1.$$

6. Let A be an Ω-algebra. Let $\omega_j^{(i)}$ be an i-ary operator symbol. Write

$$\omega_1^{(2)}\omega_2^{(3)}\omega_3^{(2)}\omega_4^{(0)}\omega_5^{(4)}a_1a_2a_3a_4a_5a_6\omega_6^{(3)}a_7a_8a_9$$

with parentheses and commas.

2.2 SUBALGEBRAS AND PRODUCTS

Let A be an arbitrary Ω-algebra. A non-vacuous subset B of A is called a *subalgebra* of A if for any $\omega \in \Omega(n)$ and $(b_1, b_2, \ldots, b_n)$, $b_i \in B$, $\omega b_1 b_2 \cdots b_n \in B$. In particular, if $n = 0$, then B contains the distinguished element ω_A associated with ω. For example, if A is a group, a subset B is a subgroup if and only if B contains bc, 1, and b^{-1} for every $b, c \in B$. This coincides with the usual definition of a subgroup. Observe that this would not have been the case if we had omitted the unary product symbol i in the Ω-algebra definition of a group. If A is a Λ-group, a subalgebra is a subgroup B such that $\lambda b \in B$ for $b \in B$, $\lambda \in \Lambda$. This is called a Λ-*subgroup* of A.

If B is a subalgebra of A, B becomes an Ω-algebra by defining the action of every $\omega \in \Omega$ by the restriction to B of its action in A. It is clear that if B is a subalgebra of A and C is a subalgebra of B, then C is a subalgebra of A. It is clear also that if $\{B_\alpha | \alpha \in I\}$ is a set of subalgebras, then $\bigcap B_\alpha$ is a subalgebra or is vacuous. It is convenient to adjoin $\varnothing$ to the set of subalgebras of A. If we partially order the resulting set by inclusion, then it is clear that Theorem 2.1 can be invoked to conclude that this set is a complete lattice. Moreover, the proof we sketched shows how to obtáin the sup of a given set $\{B_\alpha\}$ of subalgebras: take the intersection of the set of subalgebras C containing every B_α. Generally this is not the union $\bigcup B_\alpha$ of the sets B_α, which is the sup of the B_α in the partially ordered set $\mathscr{P}(A)$, since $\bigcup B_\alpha$ need not be a subalgebra.

However, there is an important case in which it is $\bigcup B_\alpha$, namely, if $\{B_\alpha\}$ is *directed* by inclusion in the sense that for any B_β, B_γ in $\{B_\alpha\}$ there is a B_δ in this set such that $B_\delta \supset B_\beta$ and $B_\delta \supset B_\gamma$. For example, this is the case if $\{B_\alpha\}$ is a chain (or is totally ordered). If $\{B_\alpha\}$ is directed and $\{B_{\beta_1}, \ldots, B_{\beta_n}\}$ is any finite subset of $\{B_\alpha\}$, then there exists a $B_\delta \in \{B_\alpha\}$ such that $B_\delta \supset B_{\beta_i}$, $1 \leqslant i \leqslant n$.

THEOREM 2.2 *If $\{B_\alpha\}$ is a directed set of subalgebras, then $\bigcup B_\alpha$ is a subalgebra.*

Proof. Let $\omega \in \Omega(n)$. If $n = 0$ it is clear that $\omega_A \in B = \bigcup B_\alpha$. Now let $n \geqslant 1$, $(b_1, \ldots, b_n) \in B^{(n)}$. Then $b_i \in B_{\beta_i} \in \{B_\alpha\}$ and so every $b_i \in B_\delta$ for a suitable B_δ in $\{B_\alpha\}$. Then $\omega b_1 \cdots b_n \in B_\delta \subset \bigcup B_\alpha$. Hence $\bigcup B_\alpha$ is a subalgebra. □

It is clear that if $\bigcup B_\alpha$ is a subalgebra, then this is the sup of the B_α.

Let X be a non-vacuous subset of the Ω-algebra A. Let $\{B_\alpha\}$ be the set of subalgebras of A containing X and put $[X] = \bigcap B_\alpha$. Then $[X]$ is a subalgebra containing X and contained in every subalgebra of A containing X. Clearly $[X]$ is uniquely determined by these properties. We call $[X]$ the *subalgebra of A generated by the set X*. We can also define this subalgebra constructively as follows: Put $X_0 = X \cup U$ where U is the set of distinguished elements ω_A, $\omega \in \Omega(0)$. For $k \geqslant 0$ let $X_{k+1} = X_k \cup \{y \mid y = \omega x_1 \cdots x_n, x_i \in X_k, \omega \in \Omega(n), n \geqslant 1\}$. Evidently $X_0 \subset X_1 \subset \cdots$ and it is clear that $\bigcup X_k$ is a subalgebra containing X. Moreover, if B is any subalgebra containing X, then induction on k shows that B contains every X_k. Hence $B \supset \bigcup X_k$ and $[X] = \bigcup X_k$.

Let $\{A_\alpha \mid \alpha \in I\}$ be a family of Ω-algebras indexed by a set I, that is, we have a map $\alpha \rightsquigarrow A_\alpha$ of I into $\{A_\alpha\}$. Moreover, we allow $A_\alpha = A_\beta$ for $\alpha \neq \beta$. In particular, we may have all of the $A_\alpha = A$. It is convenient to assume that all of the A_α are subsets of the same set A. We recall that the product set $\prod_I A_\alpha$ is the set of maps $a: \alpha \rightsquigarrow a_\alpha$ of I into A such that for every α, $a_\alpha \in A_\alpha$. We now define an Ω-algebra structure on $\prod A_\alpha$ by defining the products component-wise: If $\omega \in \Omega(0)$, we define the corresponding element of $\prod A_\alpha$ to be the map $\alpha \rightsquigarrow \omega_{A_\alpha}$ where ω_{A_α} is the element of A_α singled out by ω. If $\omega \in \Omega(n)$ for $n \geqslant 1$ and $a^{(1)}, \ldots, a^{(n)} \in \prod A_\alpha$, then we define $\omega a^{(1)} \cdots a^{(n)}$ to be the map

$$\alpha \qquad \omega a_\alpha^{(1)} a_\alpha^{(2)} \cdots a_\alpha^{(n)},$$

which is evidently an element of $\prod A_\alpha$. If we do this for all ω, we obtain an Ω-algebra structure on $\prod A_\alpha$. $\prod A_\alpha$ endowed with this structure is called the *product of the indexed family of Ω-algebras $\{A_\alpha \mid \alpha \in I\}$*.

If $I = \{1, 2, \ldots, r\}$, we write $A_1 \times A_2 \times \cdots \times A_r$ for $\prod_I A_\alpha$. The elements of this algebra are the r-tuples $(a_1, a_2, \ldots, a_r)$, $a_i \in A_i$. If ω is nullary, the associated element in $A_1 \times \cdots \times A_r$ is $(\omega_{A_1}, \omega_{A_2}, \ldots, \omega_{A_r})$, and if $n \geqslant 1$ and $a^{(i)} = (a_1{}^{(i)}, a_2{}^{(i)}, \ldots, a_r{}^{(i)})$, $1 \leqslant i \leqslant n$, then $\omega a^{(1)} \cdots a^{(n)}$ is the element of $A_1 \times \cdots \times A_r$ whose ith component is $\omega a_i{}^{(1)} a_i{}^{(2)} \cdots a_i{}^{(n)}$.

EXERCISES

1. Let G be a group and adjoin to the set of product symbols defining G the set G whose action is $g(x) = g^{-1}xg$. Verify that G is a group with operator set G. What are the G-subgroups?
2. Let X be a set of generators for an Ω-algebra A (that is, $A = [X]$). Show that A is the union of its subalgebras $[F]$, F a finite subset of X.
3. Show that an Ω-algebra A is finitely generated if and only if it has the following property: The union of any directed set of proper ($\neq A$) subalgebras is proper.
4. Give examples of the following: (i) an Ω-algebra containing two subalgebras having vacuous intersection, and (ii) an Ω-algebra in which the intersection of any two subalgebras is a subalgebra but there exist infinite sets of subalgebras with vacuous intersection.

2.3 HOMOMORPHISMS AND CONGRUENCES

DEFINITION 2.2. *If A and B are Ω-algebras, a* homomorphism from A to B *is a map f of A into B such that for any $\omega \in \Omega(n)$, $n = 0, 1, 2, \cdots$, and every $(a_1, \ldots, a_n) \in A^{(n)}$ we have*

$$f(\omega a_1 a_2 \cdots a_n) = \omega f(a_1) f(a_2) \cdots f(a_n). \tag{6}$$

In the case $n = 0$ it is understood that if ω_A is the element of A corresponding to ω then $f(\omega_A) = \omega_B$.

It is clear that the composite gf of the homomorphisms $f: A \rightarrow B$ and $g: B \rightarrow C$ is a homomorphism from A to C and that 1_A is a homomorphism from A to A. It follows that we obtain a category, **Ω-alg**, whose objects are the Ω-algebras and morphisms are the homomorphisms. It is interesting to see how the important category ideas apply to **Ω-alg** and its important subcategories. We observe now that the product construction of Ω-algebras provides a product in the sense of categories for **Ω-alg**. Let $\{A_\alpha | \alpha \in I\}$ be an indexed family of algebras and $\prod_I A_\alpha$ the product of these algebras as defined in section 2.2. For each α we have the projection map $p_\alpha : a \quad a_\alpha$ of $P = \prod A_\alpha$ into A_α. It is clear from the definition of the action of the ω's in A that p_α is a homomorphism of P into A_α. We claim that $\{P, p_\alpha\}$ is a product in **Ω-alg** of the set $\{A_\alpha | \alpha \in I\}$. To verify this, let B be an Ω-algebra, $f_\alpha : B \to A_\alpha$ a homomorphism of B into A_α for every $\alpha \in I$. It is immediate that if we map any $b \in B$ into the element of P whose "α-component" is $f_\alpha(b)$, we obtain a homomorphism f of B into P. Moreover, $p_\alpha f = f_\alpha$ and f is the only homomorphism of B into P satisfying this condition for all $\alpha \in I$. Hence $\{P, p_\alpha\}$ is indeed a product of the A_α in **Ω-alg**.

DEFINITION 2.3. *A* congruence *on an Ω-algebra A is an equivalence relation on A, which is a subalgebra of $A \times A$.*

If Φ is an equivalence relation on the Ω-algebra A and $\omega \in \Omega(0)$, then $(\omega_A, \omega_A) \in \Phi$ and this is the element $\omega_{A \times A}$ corresponding to the nullary symbol ω in the algebra $A \times A$. Hence an equivalence relation Φ is a congruence if and only if for every $\omega \in \Omega(n)$, $n \geqslant 1$, and $(a_i, a_i') \in \Phi$, $i = 1, 2, \ldots, n$, we have $\omega(a_1, a_1')(a_2, a_2') \cdots (a_n, a_n') \in \Phi$. Since $\omega(a_1, a_1')(a_2, a_2') \cdots (a_n, a_n') = (\omega a_1 a_2 \cdots a_n, \omega a_1' a_2' \cdots a_n')$, the requirement is that $a_i \Phi a_i'$, $1 \leqslant i \leqslant n$, implies $\omega a_1 \cdots a_n \Phi \omega a_1' \cdots a_n'$. Thus the condition that an equivalence relation be a congruence is that this last property holds for every n-ary operator symbol with $n \geqslant 1$. Let A/Φ be the quotient set of A determined by the congruence relation Φ and let $\bar{a}$ be the element of A/Φ corresponding to the element a of A. Since $a_i \Phi a_i'$, $1 \leqslant i \leqslant n$, implies $\omega a_1 \cdots a_n \Phi \omega a_1' \cdots a_n'$, we see that we have a well-defined map of $(A/\Phi)^{(n)}$ into A/Φ such that

$$(\bar{a}_1, \ldots, \bar{a}_n) \rightsquigarrow \overline{\omega a_1 \cdots a_n}.$$

We can use this to define

$$\omega \bar{a}_1 \cdots \bar{a}_n = \overline{\omega a_1 \cdots a_n} \tag{7}$$

and we do this for every $\omega \in \Omega(n)$, $n \geqslant 1$. Also for $\omega \in \Omega(0)$ we define

$$\omega_{A/\Phi} = \overline{\omega_A}. \tag{8}$$

In this way we endow A/Φ with an Ω-algebra structure. We call A/Φ with this structure the *quotient algebra of A relative to the congruence* Φ. In terms of the natural map $v : a \rightsquigarrow \bar{a}$ of A into A/Φ, equations (7) and (8) read $\omega v a_1 \cdots v a_n = v \omega a_1 \cdots a_n$, $\omega_{A/\Phi} = v \omega_A$. Evidently this means that the natural map v is a homomorphism of A into A/Φ.

Let f be a map of a set A into a set B. Then $\Phi = f^{-1} f$ is a relation on A.

Since f is the subset of $A \times B$ consisting of the pairs $(a, f(a))$, $a \in A$, and f^{-1} is the subset of $B \times A$ of pairs $(f(a), a)$, $f^{-1}f$ is the set of pairs (a, a') in $A \times A$ such that $f(a) = f(a')$. Thus $a\Phi a'$ if and only if $f(a) = f(a')$ and it is clear from this that Φ is an equivalence relation on the set A (BAI, p. 17). We shall now call this equivalence relation the *kernel* of the map f. Since $a\Phi a'$ implies $f(a) = f(a')$, it is clear that $\bar{f}(\bar{a}) = f(a)$ defines a map $\bar{f}$ of A/Φ into B such that we have commutativity of

(9)

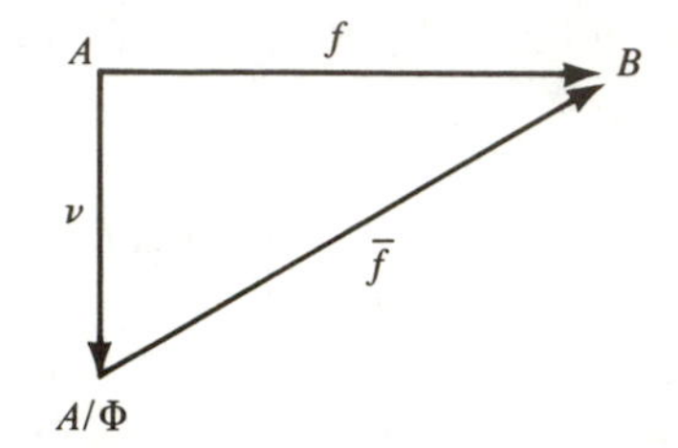

Here $\nu = \nu_\Phi$ is the map $a \rightsquigarrow \bar{a} = \bar{a}_\Phi$. Moreover, $\bar{f}$ is uniquely determined by this property and this map is injective. Clearly ν is surjective. We now extend this to Ω-algebras in the so-called

FUNDAMENTAL THEOREM OF HOMOMORPHISMS OF Ω-ALGEBRAS. *Let A and B be Ω-algebras, f a homomorphism of A into B. Then $\Phi = f^{-1}f$ is a congruence on A and the image $f(A)$ is a subalgebra of B. Moreover, we have a unique homomorphism $\bar{f}$ of A/Φ into B such that $f = \bar{f}\nu$ where ν is the homomorphism $a \rightsquigarrow \bar{a}$ of A into A/Φ. The homomorphism $\bar{f}$ is injective and ν is surjective.*

Proof. Let $\omega \in \Omega(n)$, $n \geqslant 1$, $(a_i, a_i') \in \Phi$ for $1 \leqslant i \leqslant n$. Then $f(a_i) = f(a_i')$ and $f(\omega a_1 \cdots a_n) = \omega f(a_1) \cdots f(a_n) = \omega f(a_1') \cdots f(a_n') = f(\omega a_1' \cdots a_n')$. Hence $(\omega a_1 \cdots a_n, \omega a_1' \cdots a_n') \in \Phi$, so $\omega(a_1, a_1') \cdots (a_n, a_n') \in \Phi$. Since Φ is an equivalence relation, this shows that Φ is a congruence on the algebra A. It is immediate that $f(A)$ is a subalgebra of B. Now consider the map $\bar{f}$. If $\omega \in \Omega(0)$, then $\bar{f}(\omega_{A/\Phi}) = \bar{f}(\overline{\omega}_A) = f(\omega_A) = \omega_B$ and if $\omega \in \Omega(n)$, $n \geqslant 1$, and $a_i \in A$, $1 \leqslant i \leqslant n$, then

$$\begin{aligned}
\bar{f}(\omega \bar{a}_1 \ldots \bar{a}_n) &= \bar{f}(\overline{\omega a_1 \ldots a_n}) \\
&= f(\omega a_1 \ldots a_n) \\
&= \omega f(a_1) \ldots f(a_n) \\
&= \omega \bar{f}(\bar{a}_1) \ldots \bar{f}(\bar{a}_n).
\end{aligned}$$

Hence $\bar{f}$ is a homomorphism. The remaining assertions are clear from the results on maps that we noted before. □

We shall consider next the extension to Ω-algebras of the circle of ideas centering around the correspondence between the subgroups of a group and those of a homomorphic image (BAI, pp. 64–67). Let f be a surjective homomorphism of an Ω-algebra A onto an Ω-algebra B. If A_1 is a subalgebra of A, the restriction $f|A_1$ is a homomorphism of A_1; hence $f(A_1)$ is a subalgebra of B. Next let B_1 be a subalgebra of B and let $A_1 = f^{-1}(B_1)$, the inverse image of B_1. Then $\omega_A \in A_1$ for every nullary ω, and if $a_i \in A_1$ and $\omega \in \Omega(n)$, $n \geqslant 1$, then $f(\omega a_1 \ldots a_n) = \omega f(a_1) \ldots f(a_n) \in B_1$ so $\omega a_1 \ldots a_n \in A_1$. Hence $A_1 = f^{-1}(B_1)$ is a subalgebra of A. Moreover, $f(A_1) = B_1$ and A_1 is a *saturated* subalgebra of A in the sense that if $a_1 \in A_1$, then every a_1' such that $f(a_1') = f(a_1)$ is contained in A_1. It is clear also that if A_1 is any saturated subalgebra of A, then $A_1 = f^{-1}(f(A_1))$. It now follows that the map

$$A_1 \rightsquigarrow f(A_1)$$

of the set of saturated subalgebras of A into the set of subalgebras of B is a bijection with inverse $B_1 \rightsquigarrow f^{-1}(B_1)$.

This applies in particular if we have a congruence Φ on A and we take $B = A/\Phi, f = v$, the natural homomorphism $a \rightsquigarrow \bar{a} = \bar{a}_\Phi$ of A onto A/Φ. In this case, if A_1 is any subalgebra of A, $v^{-1}(v(A_1))$ is the union of the congruence classes of A (determined by Φ) that meet A_1, that is, have an element in common with A_1. The map $v_1 = v|A_1$ is a homomorphism of A_1 into A/Φ. Now one sees that the image is A_1'/Φ_1' where $A_1' = v^{-1}(v(A_1))$ and Φ_1' is the congruence $\Phi \cap (A_1' \times A_1')$ on A_1'. On the other hand, the kernel of v_1 is $\Phi \cap (A_1 \times A_1)$. We may therefore apply the fundamental theorem of homomorphisms to conclude that

$$\bar{a}_{1\Phi_1} \rightsquigarrow \bar{a}_{1\Phi_1'} \tag{10}$$

for $a_1 \in A_1$ is an isomorphism of A_1/Φ_1 onto A_1'/Φ_1'. We state this result as the

FIRST ISOMORPHISM THEOREM. *Let Φ be a congruence on an Ω-algebra A, A_1 a subalgebra of A. Let A_1' be the union of the Φ equivalence classes that meet A_1. Then A_1' is a subalgebra of A containing A_1, $\Phi_1' = \Phi \cap (A_1' \times A_1')$ and $\Phi_1 = \Phi \cap (A_1 \times A_1)$ are congruences on A_1' and A_1 respectively, and* (10) *is an isomorphism of A_1/Φ_1 onto A_1'/Φ_1'.*

We shall consider next the congruences on the quotient algebra A/Φ, Φ a given congruence on A. We note first that if Θ is a second congruence, then we have the correspondence

$$v_{\Theta/\Phi} = \{(\bar{a}_\Phi, \bar{a}_\Theta) | a \in A\} \tag{11}$$

from A/Φ to A/Θ. This is a map of A/Φ to A/Θ if and only if $a\Phi b$ implies $a\Theta b$, which means if and only if $\Phi \subset \Theta$. In this case $\nu_{\Theta/\Phi}: \bar{a}_\Phi \rightsquigarrow \bar{a}_\Theta$ is the unique map of A/Φ into A/Θ such that

(12)

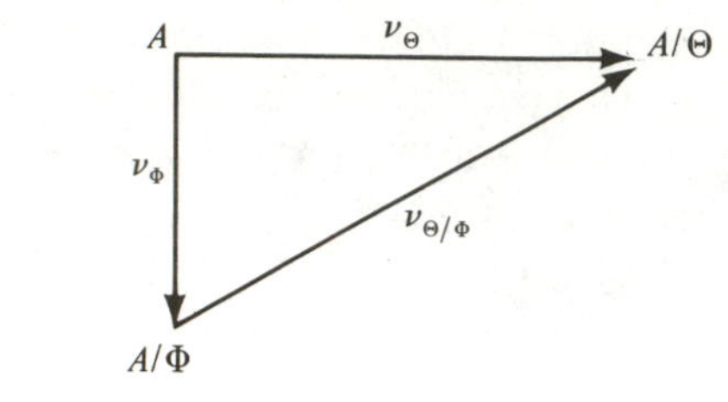

is commutative. Direct verification, which we leave to the reader, shows that $\nu_{\Theta/\Phi}$ is an algebra homomorphism. It is clear that $\nu_{\Theta/\Phi}$ is surjective and it is injective if and only if $\Theta = \Phi$.

We now denote the kernel of the homomorphism $\nu_{\Theta/\Phi}$ of A/Φ into A/Θ by Θ/Φ. This is a congruence on A/Φ consisting of the pairs $(\bar{a}_\Phi, \bar{b}_\Phi)$ such that $\bar{a}_\Theta = \bar{b}_\Theta$. It follows from this that if Θ_1 and Θ_2 are two congruences on A such that $\Theta_i \supset \Phi$, then $\Theta_1 \supset \Theta_2$, if and only if $\Theta_1/\Phi \supset \Theta_2/\Phi$. In particular, $\Theta_1/\Phi = \Theta_2/\Phi$ implies $\Theta_1 = \Theta_2$.

We shall show next that any congruence $\bar{\Theta}$ on A/Φ has the form Θ/Φ where Θ is a congruence on A containing Φ. Let $\bar{\nu}$ be the canonical homomorphism of A/Φ onto $(A/\Phi)/\bar{\Theta}$. Then $\bar{\nu}\nu_\Phi$ is a surjective homomorphism of A onto $(A/\Phi)/\bar{\Theta}$. If Θ is the kernel, by the fundamental theorem of homomorphisms, we have a unique isomorphism $\nu': A/\Theta \to (A/\Phi)/\bar{\Theta}$ such that

(13)

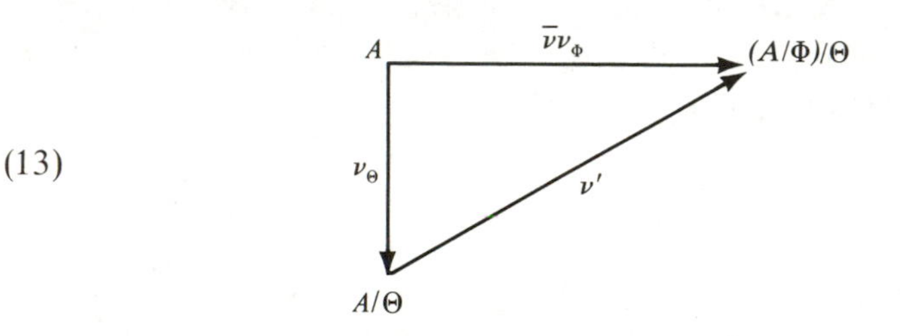

is commutative. Then $\nu_\Theta = (\nu')^{-1}\bar{\nu}\nu_\Phi$. This implies that $\Theta \supset \Phi$ and that we have the homomorphism $\nu_{\Theta/\Phi}$ as in the commutative diagram (12). Since ν_Φ is surjective on the algebra A/Φ and $\nu_\Theta = (\nu')^{-1}\bar{\nu}\nu_\Phi = \nu_{\Theta/\Phi}\nu_\Phi$, we have the equality of the two homomorphisms $(\nu')^{-1}\bar{\nu}$ and $\nu_{\Theta/\Phi}$ of the algebra A/Φ. Since ν' is an isomorphism, the kernel of $(\nu')^{-1}\bar{\nu}$ is the same as that of $\bar{\nu}$. Hence the homomorphisms $\bar{\nu}$ and $\nu_{\Theta/\Phi}$ of A/Φ have the same kernel. For $\bar{\nu}$ this is $\bar{\Theta}$ and for $\nu_{\Theta/\Phi}$ it is Θ/Φ. Hence $\bar{\Theta} = \Theta/\Phi$ as we claimed.

The results we derived can be stated in the following way:

THEOREM 2.3. *Let A be an Ω-algebra, Φ a congruence on A, A/Φ the corresponding quotient algebra. Let Θ be a congruence on A such that $\Theta \supset \Phi$.*

Then there exists a unique homomorphism $v_{\Theta/\Phi}: A/\Phi \to A/\Theta$ *such that* (12) *is commutative, and if* Θ/Φ *denotes the kernel of* $v_{\Theta/\Phi}$ *then* Θ/Φ *is a congruence on* A/Φ. *The map* $\Theta \rightsquigarrow \Theta/\Phi$ *is a bijective map of the set of congruences on* A *containing* Φ *onto the set of congruences on* A/Φ. *Moreover,* $\Theta_1 \supset \Theta_2$ *for two congruences on* A *containing* Φ *if and only if* $\Theta_1/\Phi \supset \Theta_2/\Phi$.

We also have the

SECOND ISOMORPHISM THEOREM. *Let* Θ *and* Φ *be congruences on the* Ω*-algebra* A *such that* $\Theta \supset \Phi$ *and let* Θ/Φ *be the corresponding congruence of* A/Φ *given in Theorem* 2.3. *Then*

$$(\overline{\bar{a}_\Phi})_{\Theta/\Phi} \rightsquigarrow \bar{a}_\Theta, \qquad a \in A, \tag{14}$$

is an isomorphism of $(A/\Phi)/(\Theta/\Phi)$ *onto* A/Θ.

Proof. The homomorphism $v_{\Theta/\Phi}$ of A/Φ onto A/Θ maps $\bar{a}_\Phi \rightsquigarrow \bar{a}_\Theta$ and has kernel Θ/Φ. Accordingly, by the fundamental theorem, (14) is an isomorphism of $(A/\Phi)/(\Theta/\Phi)$ onto A/Θ. □

The results on homomorphisms of Ω-algebras specialize to familiar ones in the case of the category **Grp** (see BAI, pp. 61–66). If f is a homomorphism of a group G into a group H, then $K = f^{-1}(1)$ is a normal subgroup of G. The kernel $\Phi = f^{-1}f$ is the set of pairs (a, b), $a, b \in G$, such that $a^{-1}b \in K$. It follows that $\bar{a}_\Phi = aK = Ka$ and G/Φ is the usual factor group G/K. The fundamental theorem on homomorphisms, as we stated it, is equivalent in the case of groups to the standard theorem having this name in group theory. The other results we obtained also have familiar specializations (see exercise 1, below). Similarly, if R and S are rings and f is a homomorphism of R into S, then $K = f^{-1}(0)$ is an ideal in R and $R/\Phi = R/K$. Our results specialize to the classical ones for rings (see exercise 2, below).

EXERCISES

1. Let G and H be groups, f a homomorphism of G into H, and $K = f^{-1}(1)$. Then in the standard terminology, K is called the kernel of f. The standard formulation of the *fundamental theorem of homomorphism for groups* states that $\bar{f}: aK \rightsquigarrow f(a)$ is a monomorphism of G/K into H and we have the factorization $f = \bar{f}v$ where v is

the epimorphism $a \rightsquigarrow aK$. The *first isomorphism theorem for groups* states that if H is a subgroup of a group G and K is a normal subgroup of G, then the subgroup generated by H and K is $HK = \{hk | h \in H, k \in K\}$, K is normal in HK, $K \cap H$ is normal in H, and we have the isomorphism

$$hK \rightsquigarrow h(H \cap K) \tag{15}$$

of HK/K onto $H/(H \cap K)$. The *second isomorphism theorem* states that if $G \rhd H$, K and $H \supset K$, then H/K is normal in G/K and

$$gH \rightsquigarrow (gK)(H/K) \tag{16}$$

is an isomorphism of G/H onto $(G/K)/(H/K)$. Derive the fundamental theorem of homomorphisms and the isomorphisms (15) and (16) from the results on congruences on Ω-algebras.

State the corresponding results for Λ-groups (groups with operators). Do the same for modules for a ring R.

2. Derive the results on ring homomorphisms analogous to the preceding ones on groups from the theorems on congruences on Ω-algebras.

3. Let A be an Ω-algebra, X a set of generators for A (that is, the subalgebra $[X] = A$). Show that if f and g are homomorphisms of A into a second Ω-algebra B such that $f | X = g | X$, then $f = g$.

2.4 THE LATTICE OF CONGRUENCES. SUBDIRECT PRODUCTS

In this section we shall consider first some additional properties of the set of congruences on an Ω-algebra A. Then we shall apply these to study the important notion of subdirect products of algebras.

We investigate first the set $\Gamma(A)$ of congruences on A as a partially ordered set in which the ordering $\geqslant$ is the usual inclusion $\Phi \supset \Theta$. This, of course, means that $a\Theta b \Rightarrow a\Phi b$ or, equivalently, $\bar{a}_\Theta = \bar{b}_\Theta \Rightarrow \bar{a}_\Phi = \bar{b}_\Phi$.

THEOREM 2.4 *If $\{\Phi_\alpha\}$ is a set of congruences on the Ω-algebra A, then $\bigcap \Phi_\alpha$ is a congruence. Moreover, if $\{\Phi_\alpha\}$ is directed, then $\bigcup \Phi_\alpha$ is a congruence.*

Proof. We note first that if $\{\Phi_\alpha\}$ is a set of equivalence relations on A then $\bigcap \Phi_\alpha$ is an equivalence relation, and if $\{\Phi_\alpha\}$ is directed then $\bigcup \Phi_\alpha$ is an equivalence relation: Since every $\Phi_\alpha \supset 1_A$, $\bigcap \Phi_\alpha \supset 1_A$. If $(x, y) \in \Phi_\alpha$ then $(y, x) \in \Phi_\alpha$; hence, if $(x, y) \in \bigcap \Phi_\alpha$ then $(y, x) \in \bigcap \Phi_\alpha$. Now let (x, y) and $(y, z) \in \Phi_\alpha$; then $(x, z) \in \Phi_\alpha$. Thus if (x, y) and $(y, z) \in \bigcap \Phi_\alpha$ so does (x, z). Hence $\bigcap \Phi_\alpha$ is an equivalence relation. Next assume $\{\Phi_\alpha\}$ is directed. It is clear that $\bigcup \Phi_\alpha \supset 1_A$ and $(\bigcup \Phi_\alpha)^{-1} = \bigcup \Phi_\alpha$. Now let (x, y) and $(y, z) \in \bigcup \Phi_\alpha$. Since $\{\Phi_\alpha\}$ is directed, there exists a $\Phi_\delta \in \{\Phi_\alpha\}$ such that (x, y) and $(y, z) \in \Phi_\delta$. Then $(x, z) \in \Phi_\delta$ so $(x, z) \in \bigcup \Phi_\alpha$. Hence $\bigcup \Phi_\alpha$ is an equivalence relation.

Now let the Φ_α be congruences so these are subalgebras of $A \times A$ as well as equivalence relations. Then $\bigcap \Phi_\alpha$ is an equivalence relation and so this is not vacuous. Hence $\bigcap \Phi_\alpha$ is also a subalgebra and thus a congruence. Similarly if $\{\Phi_\alpha\}$ is directed then $\bigcup \Phi_\alpha$ is a subalgebra. □

Clearly $\bigcap \Phi_\alpha$ is an inf of the set of congruences $\{\Phi_\alpha\}$. It is evident also that the subset $A \times A$ of $A \times A$ is a congruence and this is the greatest element in the partially ordered set of congruences. The corresponding quotient algebra $A/(A \times A)$ is the *trivial algebra*, that is, a one-element algebra (in which the action of every $\omega \in \Omega$ is uniquely determined). It is clear also that the diagonal 1_A is a congruence, and $A/1_A$ can be identified with A since the homomorphism ν_{1_A} is an isomorphism. The fact that the set $\Gamma(A)$ of congruences on A has a greatest element and has the property that infs exist for arbitrary non-vacuous subsets of $\Gamma(A)$ implies, via Theorem 2.1 (p. 57), that $\Gamma(A)$ is a complete lattice. In particular, any two congruences Φ and Θ have a sup $\Phi \vee \Theta$ in $\Gamma(A)$. We shall indicate an alternative, more constructive proof of this fact at the end of this section, in exercise 3.

Given a binary relation R on the algebra A we can define the *congruence* $[R]$ *generated* by R to be the inf (or intersection) of all the congruences on A containing R. This is characterized by the usual properties: (1) $[R]$ is a congruence containing R and (2) $[R]$ is contained in every congruence containing R.

We recall that *a maximal element* of a subset S of a partially ordered set A is an element $m \in S$ such that there exists no s in S such that $s \neq m$ and $s \geq m$. A basic existence theorem for maximal congruences is

THEOREM 2.5. *Let a and b be distinct elements of an Ω-algebra A and let $D(a,b)$ be the set of congruences Φ on A such that $\Phi \not\ni (a,b)$. Then $D(a,b)$ is not vacuous and contains a maximal element.*

Proof. Evidently $1_A \in D(a,b)$ so $D(a,b) \neq \varnothing$. Now let $\{\Phi_\alpha\}$ be a totally ordered subset of $D(a,b)$, that is, for any $\Phi_\beta, \Phi_\gamma \in \{\Phi_\alpha\}$ either $\Phi_\beta \supset \Phi_\gamma$ or $\Phi_\gamma \supset \Phi_\beta$. Then $\{\Phi_\alpha\}$ is directed. Hence $\bigcup \Phi_\alpha$ is a congruence. Evidently $\bigcup \Phi_\alpha \in D(a,b)$ and $\bigcup \Phi_\alpha \supset \Phi_\alpha$. Thus every totally ordered subset of $D(a,b)$ has an upper bound in $D(a,b)$. We can therefore apply Zorn's lemma to conclude that $D(a,b)$ contains maximal elements. □

The following is a useful extension of Theorem 2.5.

COROLLARY 1. *Let Θ be a congruence on A and a and b elements such that $\bar{a}_\Theta \neq \bar{b}_\Theta$. Then the set of congruences Φ such that $\Phi \supset \Theta$ and $\bar{a}_\Phi \neq \bar{b}_\Phi$ contains a maximal element.*

Proof. We apply Theorem 2.5 to the algebra A/Θ and the pair of elements $(\bar{a}_\Theta, \bar{b}_\Theta)$ of this algebra. We have a congruence in this algebra that is maximal among the congruences not containing $(\bar{a}_\Theta, \bar{b}_\Theta)$. By Theorem 2.3, such a congruence has the form Φ/Θ where Φ is a congruence in A containing Θ. Since $\Phi/\Theta \not\ni (\bar{a}_\Theta, \bar{b}_\Theta)$, $v_{\Phi/\Theta}\bar{a}_\Theta \neq v_{\Phi/\Theta}\bar{b}_\Theta$, which means that $\bar{a}_\Phi \neq \bar{b}_\Phi$. Moreover, Φ is maximal among the congruences containing Θ having this property. □

There are many important special cases of this result. One of these, which we now consider, concerns the existence of maximal ideals of a ring. An ideal B (left, right ideal) of a ring R is called *maximal* if $B \neq R$ and there exists no ideal (left, right ideal) B' such that $R \supsetneq B' \supsetneq B$. Observe that an ideal (left, right ideal) B is proper ($B \neq R$) if and only if $1 \notin B$. We now have the following result.

COROLLARY 2. *Any proper ideal (one-sided ideal) in a ring $R \neq 0$ is contained in a maximal ideal (one-sided ideal).*

Proof. To obtain the result on ideals we use the fact that any congruence in a ring R is determined by an ideal B as the set of pairs (a, a'), $a, a' \in R$, such that $a - a' \in B$. Conversely, any ideal is determined by a congruence in this way, and if Φ_B is the congruence associated with the ideal B, then $\Phi_B \supset \Phi_{B'}$ for the ideal B' if and only if $B \supset B'$. Now if B is proper, then Φ_B does not contain $(1, 0)$. Then Corollary 1 gives a congruence Φ_M, M an ideal, such that $\Phi_M \supset \Phi_B$, $\Phi_M \not\ni (1, 0)$, and Φ_M is maximal among such congruences. Then $M \supset B$, $M \not\ni 1$, so M is proper, and M is maximal in the set of ideals having this property. It follows that M is a maximal ideal in the sense defined above. The proof for one-sided ideals is obtained in the same way by regarding R as an Ω-algebra that is a group (the additive group) with operators, the set R of operators acting by left multiplications for the case of left ideals and by right multiplications for right ideals. □

Of course, these results can also be obtained by applying Zorn's lemma directly to the sets of ideals (see exercise 1 below). However, we thought it would be instructive to deduce the results on ideals from the general theorem on congruences.

We consider again an arbitrary Ω-algebra A and let $\{\Theta_\alpha | \alpha \in I\}$ be an indexed set of congruences on A such that $\bigcap \Theta_\alpha = 1_A$. For each $\alpha \in I$ we have the quotient $A_\alpha = A/\Theta_\alpha$ and we can form the product algebra $P = \prod_{\alpha \in I} A_\alpha$. We have the homomorphism $v_\alpha : a \rightarrow \bar{a}_{\Theta_\alpha}$ of A onto A_α and the homomorphism $v : a \rightarrow \bar{a}$ where $\bar{a}_\alpha = \bar{a}_{\Theta_\alpha}$ of A into P. The kernel of v is the set of pairs (a, b) such that $\bar{a}_\Theta = \bar{b}_\Theta$ for all $\alpha \in I$. This is the intersection of the kernels Θ_α of the

v_α. Since $\bigcap\Theta_\alpha = 1_A$, we see that the kernel of v is 1_A and so v is a monomorphism. It is clear also from the definition that for any α the image $p_\alpha v(A)$ of $v(A)$ under the projection p_α of P onto A_α is A_α.

If $\{A_\alpha | \alpha \in I\}$ is an arbitrary indexed set of algebras and $P = \prod A_\alpha$, then we call an algebra A a *subdirect product of the* A_α if there exists a monomorphism i of A into P such that for every α, $i_\alpha = p_\alpha i$ is surjective on A_α. Then we have shown that if an algebra A has a set of congruences $\{\Theta_\alpha | \alpha \in I\}$ such that $\bigcap\Theta_\alpha = 1_A$, then A is a subdirect product of the algebras $A_\alpha = A/\Theta_\alpha$. Conversely, let A be a subdirect product of the A_α via the monomorphism i. Then if Θ_α is the kernel of $i_\alpha = p_\alpha i$, it is immediate that $\bigcap\Theta_\alpha = 1_A$ and the image $i_\alpha(A) \cong A/\Theta_\alpha$.

DEFINITION 2.4. *An* Ω*-algebra* A *is* subdirectly irreducible *if the intersection of all of the congruences of* $A \neq 1_A$ *is* $\neq 1_A$.

In view of the results we have noted we see that this means that if A is subdirectly irreducible and A is a subdirect product of algebras A_α, then one of the homomorphisms $i_\alpha = p_\alpha i$ is an isomorphism. If $a \neq b$ in an algebra A, then we have proved in Theorem 2.5 the existence of a congruence $\Phi_{a,b}$ that is maximal in the set of congruences not containing (a, b). We claim that $A/\Phi_{a,b}$ is subdirectly irreducible. By Theorem 2.3, any congruence in $A/\Phi_{a,b}$ has the form $\Theta_\alpha/\Phi_{a,b}$ where Θ_α is a congruence in A containing $\Phi_{a,b}$. If this is not the diagonal then $\Theta_\alpha \supsetneqq \Phi_{a,b}$ and so by the maximality of $\Phi_{a,b}$, $(a, b) \in \Theta_\alpha$. Then $(a_{\Phi_{a,b}}, b_{\Phi_{a,b}}) \in \Theta_\alpha/\Phi_{a,b}$ for every congruence $\Theta_\alpha/\Phi_{a,b}$ on $A/\Phi_{a,b}$ different from the diagonal. Evidently this means that $A/\Phi_{a,b}$ is subdirectly irreducible. We use this to prove

THEOREM 2.6 (Birkhoff). *Every* Ω*-algebra* A *is a subdirect product of subdirectly irreducible algebras.*

Proof. We may assume that A contains more than one element. For every (a, b), $a \neq b$, let $\Phi_{a,b}$ be a maximal congruence such that $(a, b) \notin \Phi_{a,b}$. Then $A/\Phi_{a,b}$ is subdirectly irreducible. Also $\bigcap\Phi_{a,b} = 1_A$. Otherwise, let $(c, d) \in \bigcap\Phi_{a,b}$, $c \neq d$. Then $(c, d) \in \Phi_{c,d}$, a contradiction. Hence $\bigcap\Phi_{a,b} = 1_A$ and consequently A is a subdirect product of the subdirectly irreducible algebras $A/\Phi_{a,b}$. □

EXERCISES

1. Give a direct proof of Corollary 2 to Theorem 2.5.
2. Show that $\mathbb{Z}$ is a subdirect product of the fields $\mathbb{Z}/(p)$, p prime.
3. Let Θ and Φ be congruences on A. Show that $\Theta\Phi \subset (\Theta\Phi)^2 \subset (\Theta\Phi)^3 \subset \cdots$ and $\bigcup_k (\Theta\Phi)^k$ is a congruence that is a sup of Θ and Φ. It follows from this that any finite set of congruences has a sup. Use this and the second statement of Theorem 2.4 to show that any set of congruences has a sup.
4. Let $\{\Theta_\alpha\}$ be an arbitrary set of congruences in an algebra A and let $\Theta = \bigcap \Theta_\alpha$. Show that A/Θ is a subdirect product of the algebras A/Θ_α.

2.5 DIRECT AND INVERSE LIMITS

There are two constructions of Ω-algebras that we shall now consider: direct limits and inverse limits. The definitions can be given for arbitrary categories, and it is of interest to place them in this setting, since there are other important instances of these concepts besides the one of Ω-algebras that is our present concern. We begin with the definition of direct limit in a category $\mathbf{C}$.

Let I be a pre-ordered set (the index set) and let $\mathbf{I}$ be the category defined by I as in example 12, p. 13: $\text{ob}\,\mathbf{I} = I$ and for $\alpha, \beta \in I$, $\hom(\alpha, \beta)$ is vacuous unless $\alpha \leqslant \beta$, in which case $\hom(\alpha, \beta)$ has a single element. If $\mathbf{C}$ is a category, a functor F from $\mathbf{I}$ to $\mathbf{C}$ consists of a map $\alpha \rightsquigarrow A_\alpha$ of I into $\text{ob}\,\mathbf{C}$, and a map $(\alpha, \beta) \rightsquigarrow \varphi_{\alpha\beta} \in \hom_{\mathbf{C}}(A_\alpha, A_\beta)$ defined for all pairs (α, β) with $\alpha \leqslant \beta$ such that

1. $\varphi_{\alpha\gamma} = \varphi_{\beta\gamma}\varphi_{\alpha\beta}$ if $\alpha \leqslant \beta \leqslant \gamma$.
2. $\varphi_{\alpha\alpha} = 1_{A_\alpha}$.

Given these data we define a *direct* (or *inductive*) *limit* $\lim (A_\alpha, \varphi_{\alpha\beta})$ as a set $(A, \{\eta_\alpha\})$ where $A \in \text{ob}\,\mathbf{C}$, $\eta_\alpha : A_\alpha \to A$, $\alpha \in I$, satisfies $\eta_\alpha = \eta_\beta \varphi_{\alpha\beta}$ and $(A, \{\eta_\alpha\})$ is universal for this property in the sense that if $(B, \{\zeta_\alpha\})$ is another such set, then there exists a unique morphism $\theta : A \to B$ such that $\zeta_\alpha = \theta\eta_\alpha$, $\alpha \in I$. (This can be interpreted as a universal from an object to a functor for suitably defined categories. See exercise 1 at the end of this section.) It is clear that if a direct limit exists, then it is unique up to isomorphism in the sense that if $(A', \{\eta'_\alpha\})$ is a second one, then there exists a unique isomorphism $\theta : A \to A'$ such that $\eta'_\alpha = \theta\eta_\alpha$, $\alpha \in I$.

A simple example of direct limits can be obtained by considering the finitely generated subalgebras of an Ω-algebra. Let A be any Ω-algebra and let I be the set of finite subsets of A ordered by inclusion. If $F \in I$, let A_F be the subalgebra of A generated by F and if $F \subset G$, let φ_{FG} denote the injection homomorphism of A_F into $A_G \supset A_F$. Evidently $\varphi_{FF} = 1_F$ and $\varphi_{GH}\varphi_{FG} = \varphi_{FH}$ if $F \subset G \subset H$. Moreover, if η_F denotes the injection homomorphism of A_F into A, then $\eta_G\varphi_{FG} = \eta_F$. Now let B be an Ω-algebra and $\{\zeta_F | F \in I\}$ a set of homomorphisms such that $\zeta_F : A_F \to B$, satisfies $\zeta_G\varphi_{FG} = \zeta_F$ for $F \subset G$. Since φ_{FG} is the injection of A_F into A_G, this means that the homomorphism ζ_G on A_G is an extension of ζ_F on A_F. Since $\bigcup A_F = A$, there is a unique homomorphism θ of A into B such that $\zeta_F = \theta\eta_F$, $F \in I$. It follows that A and $\{\eta_F\}$ constitute $\varinjlim (A_F, \varphi_{FG})$. The result we have proved can be stated in a slightly imprecise form in the following way:

THEOREM 2.7. *Any Ω-algebra is isomorphic to a direct limit of finitely generated algebras.*

We shall now show that if the index set I is *directed* in the sense that for any $\alpha, \beta \in I$ there exists a $\gamma \in I$ such that $\gamma \geqslant \alpha$ and $\gamma \geqslant \beta$, then the direct limit exists in the category $\mathbf{C} = \mathbf{\Omega\text{-alg}}$ for any functor F from $\mathbf{I}$ to $\mathbf{C}$.

Suppose the notations are as above where the A_α are Ω-algebras and $\varphi_{\alpha\beta}$ for $\alpha \leqslant \beta$ is an algebra homomorphism of A_α into A_b. To construct a direct limit we consider the disjoint union $\dot{\bigcup} A_\alpha$ of the sets A_α. We introduce a relation $\sim$ in $\dot{\bigcup} A_\alpha$ by defining $a \sim b$ for $a \in A_\alpha$, $b \in A_\beta$ if there exists a $\rho \geqslant \alpha, \beta$ such that $\varphi_{\alpha_\rho}(a) = \varphi_{\beta\rho}(b)$. (We allow $\alpha = \beta$). This relation is evidently reflexive and symmetric. Also if $\varphi_{\alpha\rho}(a) = \varphi_{\beta\rho}(b)$ and $\varphi_{\beta\sigma}(b) = \varphi_{\gamma\sigma}(c)$ for $\rho \geqslant \alpha, \beta$ and $\sigma \geqslant \beta, \gamma$, then there exists a $\tau \geqslant \rho, \sigma$. Then

$$\begin{aligned}\varphi_{\alpha\tau}(a) &= \varphi_{\text{oq}}\varphi_{\alpha\rho}(a) = \varphi_{\rho\tau}\varphi_{\beta\rho}(b) = \varphi_{\beta\tau}(b)\\ &= \varphi_{\sigma\tau}\varphi_{\beta\sigma}(b) = \varphi_{\sigma\tau}\varphi_{\gamma\sigma}(c) = \varphi_{\gamma\tau}(c),\end{aligned}$$

which shows that $a \sim b$ and $b \sim c$ imply $a \sim c$. Hence our relation is an equivalence.

Let A be the corresponding quotient set of equivalence classes $\bar{a}$, $a \in \dot{\bigcup} A_\alpha$. If $\omega \in \Omega(0)$, then $\omega_{A_\alpha} = \omega_{A_\beta}$ since there exists a $\rho \geqslant \alpha, \beta$ and $\varphi_{\alpha\rho}(\omega_{A_\alpha}) = \omega_{A_\rho} = \varphi_{\beta\rho}(\omega_{A_\beta})$. We define $\omega_A = \omega_{A_\alpha}$. Now let $n \geqslant 1$ and let $1 \leqslant i \leqslant n$. Let $a_i \in A_{\alpha_i}$. Choose $\rho \geqslant \alpha_i$, $1 \leqslant i \leqslant n$, which can be done since I is directed. Now define

$$\bar{\omega}_1\bar{a}_2\cdots\bar{a}_n = \omega\varphi_{\alpha_1\rho}(a_1)\varphi_{\alpha_2\rho}(a_2)\cdots\varphi_{\alpha_n\rho}(a_n). \tag{17}$$

This is clearly independent of the choice of the a_i n $\bar{a}_i$. We also have to show that this is independent of the choice of ρ. Hence let $\sigma \geqslant \alpha_i$, $1 \leqslant i \leqslant n$, and choose $\tau \geqslant \rho, \sigma$. Then

$$\begin{aligned}\varphi_{\rho\tau}\omega\varphi_{\alpha_1\rho}(a_1)\cdots\varphi_{\alpha_n\rho}(a_n) &= \omega\varphi_{\rho\tau}\varphi_{\alpha_1\rho}(a_1)\ldots\varphi_{\rho\tau}\varphi_{\alpha_n\rho}(a_n)\\ &= \omega\varphi_{\alpha_1\tau}(a_1)\ldots\varphi_{\alpha_n\tau}(a_n)\end{aligned}$$

and similarly, $\varphi_{\sigma\tau}\omega\varphi_{\alpha_1\sigma}(a_1)\ldots\varphi_{\alpha_n\sigma}(a_n) = \omega\varphi_{\alpha_1\tau}(a_1)\ldots\varphi_{\alpha_n\tau}(a_n)$ and so (17) is independent of the choice of $\rho \geqslant \alpha_i$, $1 \leqslant i \leqslant n$. This definition of the action of every $\omega \in \Omega(n)$, $n \geqslant 1$, together with the previous one of the nullary operations in A gives an Ω-algebra structure on A.

Now fix α and consider the map $\eta_\alpha : a_\alpha \rightsquigarrow \bar{a}_\alpha$ of A_α into A. If $\omega \in \Omega(0)$, then $\eta_\alpha \omega_{A_\alpha} = \omega_{A_\alpha} = \omega_A$ and if $n \geqslant 1$ and $a_\alpha^{(1)}, \ldots, a_\alpha^{(n)} \in A_\alpha$, then taking $\rho = \alpha$ in (17) gives

$$\omega \bar{a}_\alpha^{(1)} \ldots \bar{a}_\alpha^{(n)} = \overline{\omega a_\alpha^{(1)} \ldots a_\alpha^{(n)}}.$$

Hence η_α is a homomorphism. We have for $\alpha \leqslant \beta$, $\overline{\varphi_{\alpha\beta}(a_\alpha)} = \bar{a}_\alpha$ since $\varphi_{\alpha\beta}(a_\alpha) = \varphi_{\beta\beta}\varphi_{\alpha\beta}(a_\alpha)$, which shows that $\eta_\beta\varphi_{\alpha\beta}(a_\alpha) = \eta_\alpha(a_\alpha)$. Thus $\eta_\beta\varphi_{\alpha\beta} = \eta_\alpha$ if $\alpha \leqslant \beta$.

Next suppose we are given an Ω-algebra B and homomorphisms $\zeta_\alpha : A_\alpha \to B$ satisfying $\zeta_\beta\varphi_{\alpha\beta} = \varphi_\alpha$ for $\alpha \leqslant \beta$ in I. Suppose $\bar{a}_\alpha = \bar{a}_\beta$ so we have a $\rho \geqslant \alpha, \beta$ such that $\varphi_{\alpha\rho}(a_\alpha) = \varphi_{\beta\rho}(a_\beta)$. Then

$$\zeta_\alpha(a_\alpha) = \zeta_\rho\varphi_{\alpha\rho}(a_\alpha) = \zeta_\rho\varphi_{\beta\rho}(a_\beta) = \zeta_\beta(a_\beta).$$

Hence we have the map $\theta : \bar{a}_\alpha \rightsquigarrow \zeta_\alpha(a_\alpha)$ with domain $A = \bigcup_\alpha \eta_\alpha(A_\alpha)$ and codomain B. This satisfies $\zeta_\alpha = \theta\eta_\alpha$ since $\theta\eta_\alpha(a_\alpha) = \theta(\bar{a}_\alpha) = \zeta_\alpha(a_\alpha)$ for all $a_\alpha \in A_\alpha$, $\alpha \in I$. Since A is the union of the $\eta_\alpha(A_\alpha)$, it is clear that θ is determined by this condition. Also since $\varphi_{\alpha\beta}a_\alpha \sim a_\alpha$ if $\beta \geqslant \alpha$, any finite set of equivalence classes relative to $\sim$ has representatives in the same A_ρ. It follows from this that θ is an Ω-algebra homomorphism. Thus we have the universality of $(A, \{\eta_\alpha\})$, which we require in the definition of direct limit. We can therefore state the following

THEOREM 2.8. *Direct limits exist in the category of Ω-algebras for every directed set of indices I.*

The notion of inverse limit is dual to that of direct limit. Again we begin with a pre-ordered set $I = \{\alpha\}$ and a category **C**, but now we let F be a contravariant functor from the category **I** defined by I to **C**. Thus we have a map $\alpha \rightsquigarrow A_\alpha \in \text{ob}\,\mathbf{C}$, $\alpha \in I$, and for every α, β with $\alpha \leqslant \beta$ we have a morphism $\varphi_{\beta\alpha} : A_\beta \to A_\alpha$ such that

1′. $\varphi_{\gamma\alpha} = \varphi_{\beta\alpha}\varphi_{\gamma\beta}$ if $\alpha \leqslant \beta \leqslant \gamma$.

2′. $\varphi_{\alpha\alpha} = 1_{A_\alpha}$.

Then an *inverse* (or *projective*) *limit* is a set $(A, \{\eta_\alpha\})$ where $A \in \text{ob}\,\mathbf{C}$, $\eta_\alpha : A \to A_\alpha$ such that $\varphi_{\beta\alpha}\eta_\beta = \eta_\alpha$ if $\alpha \leqslant \beta$, and if $(B, \{\zeta_\alpha\})$ is a set such that $B \in \text{ob}\,\mathbf{C}$,

$\zeta_\alpha: B \to A_\alpha$ and $\varphi_{\beta\alpha}\zeta_\beta = \zeta_\alpha$ for $\alpha \leqslant \beta$, then thhere exists a unique morphism $\theta: B \to A$ such that $\eta_\alpha\theta = \zeta_\alpha$, $\alpha \in I$. We denote $(A, \{\eta_\alpha\})$ by $\lim(A_\alpha, \varphi_{\alpha\beta})$. We have the usual uniqueness up to isomorphism: If $(A', \{\eta'_\alpha\})$ is a second inverse limit, then there exists a unique isomorphism $\theta: A' \to A$ such that $\eta_\alpha\theta = \eta'_\alpha$.

Now suppose $\mathbf{C} = \mathbf{\Omega}$**-alg** so the A_α are $\mathbf{\Omega}$-algebras and for $\alpha \leqslant \beta$, $\varphi_{\beta\alpha}$ is a homomorphism of A_β into A_α. To attempt to construct an inverse limit algebra, we begin with the product algebra $\prod A_\alpha$ and the projections $p_\alpha: \prod A_\alpha \to A_\alpha$ defined by $p_\alpha(a) = a_\alpha$ where a is α $a_\alpha \in A_\alpha$, We now define

$$A = \{a \in \prod A_\alpha \,|\, p_\alpha(a) = \varphi_{\beta\alpha}p_\beta(a), \alpha \leqslant \beta\}, \tag{18}$$

$\alpha, \beta \in I$. It may happen, as will be seen in exercise 2 below, that A is vacuous. In this case, no solution of our problem exists. For, if B is any algebra and $\zeta_\alpha: B \to A_\alpha$ is a homomorphism for $\alpha \in \mathrm{I}$ such that $\zeta_\alpha = \varphi_{\beta\alpha}\zeta_\beta$, then for any $b \in B$, any element $a \in \prod A_\alpha$ such that $a_\alpha = \zeta_\alpha(b)$ satisfies the condition in (18).

Now suppose A defined by (18) is not vacuous. Then we claim that A is a subalgebra of $\prod A_\alpha$. If $\omega \in \Omega(0)$, then $p_\alpha(\omega_{\prod A_\alpha}) = \omega_{A_\alpha}$ so if $\alpha \leqslant \beta$, then $\varphi_{\beta\alpha}(\omega_{A_\beta}) = \omega_{A_\alpha}$ gives $p_\alpha(\omega_{\prod A_\alpha}) = \varphi_{\beta\alpha}p_\beta(\omega_{\prod A_\alpha})$. Hence $\omega_{\prod A_\alpha} \in A$. (Observe that this shows that $A \neq \varnothing$ if Ω contains nullary operator symbols.) Next let $\omega \in \Omega(n)$, $n \geqslant 1$, and let $a^{(1)}, \ldots, a^{(n)} \in A$. Then

$$\begin{aligned} p_\alpha(\omega a^{(1)} \cdots a^{(n)}) &= \omega a_\alpha{}^{(1)} \ldots a_\alpha{}^{(n)} = \omega p_\alpha(a^{(1)}) \ldots p_\alpha(a^{(n)}) \\ &= \omega \varphi_{\beta\alpha}p_\beta(a^{(1)}) \ldots \varphi_{\beta\alpha}p_\beta(a^{(n)}) \\ &= \varphi_{\beta\alpha}p_\beta(\omega a^{(1)} \ldots a^{(n)}). \end{aligned}$$

Hence $\omega a^{(1)} \ldots a^{(n)} \in A$ and A is a subalgebra. If A is not vacuous, we claim that this subalgebra of $\prod A_\alpha$ together with the homomorphisms $\eta_\alpha = p_\alpha | A$ is an inverse limit $\lim(A_\alpha, \varphi_{\alpha\beta})$ in $\mathbf{\Omega}$**-alg**. By definition of A we have $\eta_\alpha = \varphi_{\beta\alpha}\eta_\beta$ if $\alpha \leqslant \beta$. Now suppose we have an Ω-algebra B and homomorphisms $\zeta_\alpha: B \to A_\alpha$, $\alpha \in I$, satisfying $\zeta_\alpha = \varphi_{\beta\alpha}\zeta_\beta$ if $\alpha \leqslant \beta$. We have a unique homomorphism $\theta': B \to \prod A_\alpha$ such that $b \rightsquigarrow a$ where $a_\alpha = \zeta_\alpha(b)$. Then, as we saw before, $a \in A$, so θ' defines a homomorphism θ of B into A such that $\theta(b) = \theta'(b)$. Now $\eta_\alpha\theta(b) = \zeta_\alpha(b)$ and θ is uniquely determined by this property. This completes the verification that $(A, \{\eta_\alpha\}) = \lim(A_\alpha, \varphi_{\alpha\beta})$.

An important special case of inverse limits is obtained when we are given a set of congruences $\{\mathbf{\Phi}_\alpha\}$ on an algebra B. We pre-order the set $I = \{\alpha\}$ by agreeing that $\alpha \leqslant \beta$ if $\mathbf{\Phi}_\alpha \supset \mathbf{\Phi}_\beta$. Put $A_\alpha = B/\mathbf{\Phi}_\alpha$ and for $\alpha \leqslant \beta$ let $\varphi_{\beta\alpha}$ be the homomorphism $\bar{b}_{\Phi_\beta} \rightsquigarrow \bar{b}\mathrm{b}_{\Phi_\alpha}$ of A_β into A_α. Then $\varphi_{\alpha\alpha} = 1_{A_\alpha}$, $\varphi_{\beta\alpha}\varphi_{\gamma\beta} = \varphi_{\gamma\alpha}$ if $\alpha \leqslant \beta \leqslant \gamma$. Let ν_α denote the homomorphism b $\bar{b}_{\Phi_\alpha}$ of B onto A_α. Then $\varphi_{\beta\alpha}\nu_\beta = \nu_\alpha$ if $\alpha \leqslant \beta$. This implies that $\lim(A_\alpha, \varphi_{\alpha\beta}) = (A, \{\eta_\alpha\})$ exists. Moreover, we have the homomorphism $\theta: B \to A$ such that $\eta_\alpha\theta = \nu_\alpha$. Now suppose

$\bigcap\Phi_\alpha = 1_B$. Let $b, c \in B$ satisfy $\theta b = \theta c$. Then $v_\alpha b = v_\alpha c$ or $\bar{b}_{\Phi_\alpha} = \bar{c}_{\Phi_\alpha}$ for all α. Since $\bigcap\Phi_\alpha = 1_B$, we obtain $b = c$. Thus in this case the homomorphism θ of B into A is a monomorphism.

EXAMPLE

Ring of p-adic integers. As an example of the last construction we shall define the important ring of p-adic integers for any prime p as an inverse limit. It should be remarked that this construction is essentially the original one due to Hensel, which preceded the valuation theoretic approach that is now the standard one (see Chapter 9, p. 548).

Let p be a prime in $\mathbb{Z}$ and let (p^k) be the principal ideal of multiples of p^k for $k = 1, 2, \ldots$. Let Φ_k be the congruence determined by $p^k : a\Phi_k b$ means $a \equiv b \pmod{p^k}$. Then $\mathbb{Z}/\Phi_k = \mathbb{Z}/(p^k)$, the ring of residues modulo p^k. We have $\Phi_l \subset \Phi_k$ if $l \geqslant k$ and $\bigcap\Phi_k = 1_{\mathbb{Z}}$. We can form the inverse limit of the finite rings $\mathbb{Z}/(p^k)$, which we call the *ring* $\mathbb{Z}_p$ *of p-adic integers.* An element of $\mathbb{Z}_p$ is a sequence of residue classes (or cosets) $(a_1 + (p), a_2 + (p^2), a_3 + (p^3), \ldots)$ where the a_i are integers and for $l \geqslant k$, $a_k \equiv a_l \pmod{p^k}$. We can represent this element by the sequence of integers $(a_1, a_2, \ldots)$ where $a_k \equiv a_l \pmod{p^k}$ for $k \leqslant l$. Then two such sequences $(a_1, a_2, \ldots)$ and $(b_1, b_2, \ldots)$ represent the same element if and only if $a_k \equiv b_k \pmod{p^k}$, $k = 1, 2, \ldots$. Addition and multiplication of such sequences is component-wise. If $a \in \mathbb{Z}$, we can write $a = r_0 + r_1 p + \cdots + r_n p^n$ where $0 \leqslant r_i < p$. Then we can replace the representative $(a_1, a_2, \ldots)$ in which $a_k \equiv a_l \pmod{p^k}$ if $k \leqslant l$ by a representative of the form $(r_0, r_0 + r_1 p, r_0 + r_1 p + r_2 p^2, \ldots)$ where $0 \leqslant r_i < p$. In this way we can associate with any element of $\mathbb{Z}_p$ a uniquely determined *p-adic number*

$$r_0 + r_1 p + r_2 p^2 + \cdots \tag{19}$$

where $0 \leqslant r_i < p$. Addition and multiplication of these series corresponding to these compositions in $\mathbb{Z}_p$ are obtained by applying these compositions on the r_i and "carrying." For example, if $p = 5$, we have

$$(1 + 2.5 + 3.5^2 + \cdots) + (3 + 3.5 + 2.5^2 + \cdots) = 4 + 0.5 + 1.5^2 + \cdots$$
$$(1 + 2.5 + 3.5^2 + \cdots)(3 + 3.5 + 2.5^2 + \cdots) = 3 + 4.5 + 3.5^2 + \cdots$$

EXERCISES

1. Let I be a pre-ordered set, $\mathbf{I}$ the associated category defined as usual. Let $\alpha \rightsquigarrow A_\alpha$, $\varphi_{\alpha\beta} : A_\alpha \to A_\beta$ for $\alpha \leqslant \beta$ and $\alpha \rightsquigarrow B_\alpha$, $\psi_{\alpha\beta} : B_\alpha \to B_\beta$ be functors from $\mathbf{I}$ to a category $\mathbf{C}$ (as at the beginning of the section). Call these F and G respectively. A natural transformation λ of F into G is a map $\alpha \rightsquigarrow \lambda_\alpha$ where $\lambda_\alpha : A_\alpha \to B_\alpha$ such that for $\alpha \leqslant \beta$, $\psi_{\alpha\beta}\lambda_\alpha = \lambda_\beta\varphi_{\alpha\beta}$. Verify that one obtains a category $\mathbf{C}^{\mathbf{I}}$ by specifying that the

objects are the functors from **I** to **C**, and if F and G are two such functors, define hom(F,G) to be the class of natural transformations of F to G (which is a set). Moreover, define 1_F as on p. 25 and the product of natural transformations as on p. 25. (This completes the definition of $\mathbf{C}^\mathbf{I}$.) If $A \in \operatorname{ob} \mathbf{C}$, let $F_A \in \mathbf{C}^\mathbf{I}$ be defined by the following: $\alpha \rightsquigarrow A$, $\varphi_{\alpha\beta} = 1_A$ for $\alpha \leqslant \beta$. Define a functor from **C** to $\mathbf{C}^\mathbf{I}$ by $A \rightsquigarrow F_A$ and if $f: A \to B$, then f is mapped onto the natural transformation of the function F_A into the functor F_B, which is the morphism f from $A_\alpha = A$ to $B_\alpha = B$ for every $\alpha \in I$. Show that a direct limit $\varinjlim (A_\alpha, \varphi_{\alpha\beta})$ can be defined as a universal from F (as object of $\mathbf{C}^\mathbf{I}$) to the functor from **C** to $\mathbf{C}^\mathbf{I}$ that we have just defined.

2. Let I be the ordered set of positive integers (with the natural order) and for $k \in I$, let $A_k = I$ be regarded as a semigroup under addition. For $k \leqslant l$ let φ_{lk} be the map $x \rightsquigarrow 2^{l-k}x$ of A_l to A_k. Note that this is a homomorphism of semigroups and $\varphi_{mk} = \varphi_{lk}\varphi_{ml}$, $\varphi_{ll} = 1_{A_l}$. Show that an inverse limit does not exist in the category of semigroups for this functor from **I** to the category.

3. Let $\mathbb{Z}_p$ be the ring of p-adic integers and represent the elements of $\mathbb{Z}_p$ as sequences $a = (a_1, a_2, a_3, \ldots)$, $a_k \in \mathbb{Z}$, $a_k \equiv a_l \pmod{p^k}$ if $k \leqslant l$ (as above). Show that $a \quad (a, a, a, \ldots)$ is a monomorphism of $\mathbb{Z}$ in $\mathbb{Z}_p$. In this way $\mathbb{Z}$ is imbedded in $\mathbb{Z}_p$ and we may identify a with $(a, a, a, \ldots)$. Show that $\mathbb{Z}_p$ has no zero divisors. Hence this has a field of fractions Q_p. This is called the *field of p-adic numbers*.

4. Assume $p \neq 2$ and let a be an integer (element of $\mathbb{Z}$) such that $a \not\equiv 0 \pmod{p}$ and the congruence $x^2 \equiv a \pmod{p}$ is solvable. Show that $x^2 = a$ has a solution in $\mathbb{Z}_p$.

5. Show that the units of $\mathbb{Z}_p$ are the elements that are represented by sequences $(a_1, a_2, \ldots)$ with $a_k \equiv a_l \pmod{p^k}$, if $k \leqslant l$, such that $a_1 \not\equiv 0 \pmod{p}$.

6. Let $a \in \mathbb{Z}$, $a \not\equiv 0 \pmod{p}$, and put $u = (a, a^p, a^{p^2}, \ldots)$. Show that this represents an element of $\mathbb{Z}_p$ and $u^{p-1} = 1$ in $\mathbb{Z}_p$. Hence prove that $\mathbb{Z}_p$ contains $p-1$ distinct roots of 1.

2.6 ULTRAPRODUCTS

The concept of ultraproducts of algebras and of more general structures that involve relations as well as compositions plays an important role in mathematical logic, where it was first introduced. It has turned out to be a useful tool in algebra as well. For the sake of simplicity, we confine our attention to ultraproducts of Ω-algebras. We need to recall some results on filters in Boolean algebras that were given in BAI. Although these will be required only for the Boolean algebra of subsets of a given set, it seems worthwhile to discuss the results on filters for arbitrary Boolean algebras, especially since it is conceptually simpler to treat the general case.

We recall that a Boolean algebra is a lattice with least and greatest elements 0 and 1 that is distributive and complemented. The first of these conditions is that $a \wedge (b \vee c) = (a \wedge b) \vee (a \wedge c)$ and the second is that for any a there exists an element a' in the lattice such that $a \vee a' = 1$ and $a \wedge a' = 0$. We refer the reader to BAI, pp. 474–480, for the results on Boolean algebras that we shall require. In the Boolean algebra $\mathscr{P}(S)$ of subsets of a set S, $A \vee B$ and $A \wedge B$ are the union and intersection respectively of the subsets A and B of S, $1 = S$, $0 = \varnothing$ and for any A, A' is the complementary set of A in S. We recall that a *filter* in a Boolean algebra B is a subset F that is closed under $\wedge$ and contains every $b \geqslant$ any $u \in F$. A filter is *proper* if and only if $F \neq B$, which is equivalent to $0 \notin F$. An *ultrafilter* F is a proper filter that is maximal in the sense that it is not properly contained in any other proper filter of B. A filter F is an ultrafilter if and only if it is proper and for any $a \in B$ either a or its complement a' (which is unique) is contained in F.

A Boolean algebra can be made into a ring in two ways. In the first of these one defines $a+b = (a \wedge b') \vee (a' \wedge b)$, $ab = a \wedge b$, and takes 0 and 1 to be the given 0 and 1 (BAI, p. 478). In the second, one dualizes and takes the addition composition to be $a+'b = (a \vee b') \wedge (a' \vee b)$ and the multiplication $a \cdot' b = a \vee b$. The zero element is $0' = 1$ and the unit is $1' = 0$. If one does this, it turns out that a filter in B is simply an ideal in the Boolean ring $(B, +', \cdot', 0', 1')$. As a consequence of this we have the following special case of Corollary 2 to Theorem 2.5.

COROLLARY 3 (to Theorem 2.5). *Any proper filter F in a Boolean algebra can be imbedded in an ultrafilter.*

Now let $\{A_\alpha\}$ be a set of Ω-algebras indexed by a set I and let F be a filter in the power set $\mathscr{P}(I)$: If $S_1, S_2 \in F$, then $S_1 \cap S_2 \in F$ and if $S \in F$ and $T \supset S$, then $T \in F$. Consider $\prod A_\alpha$. In this algebra we define a relation $\sim_F$ by $a \sim_F$ if the set of indices

$$I_{a,b} = \{\alpha \in I \mid a_\alpha = b_\alpha\} \in F. \tag{20}$$

We claim that this is a congruence on $\prod A_\alpha$. It is clear that this relation is symmetric, and it is reflexive since $I \in F$. Now suppose $a \sim_F b\, b$ and $b \sim_F c$ so $I_{a,b}$ and $I_{b,c} \in F$. Since $I_{a,b} \cap I_{b,c} \subset I_{a,c}$ and F is a filter, it follows that $I_{a,c} \in F$. Hence $a \sim_F c$. Thus $\sim_F$ is an equivalence relation on $\prod A_\alpha$. To show that it is a congruence, we must show that if $\omega \in \Omega(n)$, $n \geqslant 1$, and $a^{(1)}, \dots, a^{(n)}$, $b^{(1)}, \dots, b^{(n)} \in \prod A_\alpha$ satisfy $a^{(i)} \sim_F b^{(i)}$, $1 \leqslant i \leqslant n$, then $\omega a^{(1)} \cdots a^{(n)} \sim_F \omega b^{(1)} \cdots b^{(n)}$. By assumption, $I_{a^{(i)},b^{(i)}} \in F$ for $1 \leqslant i \leqslant n$. Then $\bigcap I_{a^{(i)},b^{(i)}} \in F$ and if α is in this set, then

$$\omega a_\alpha{}^{(1)} \cdots a_\alpha{}^{(n)} = \omega b_\alpha{}^{(1)} \cdots b_\alpha{}^{(n)}.$$

Thus $\bigcap I_{a^{(i)},b^{(i)}} \subset I_{\omega a^{(1)} \cdots a^{(n)}, \omega b^{(1)} \cdots b^{(n)}}$. Hence the latter set is in F and consequently $\mathscr{F} = \sim_F$ is a congruence on the algebra $\prod A_\alpha$. We now define the quotient algebra $\prod A_\alpha/\mathscr{F}$ to be the *F-reduced product* of the Ω-algebras A_α. If F is an ultrafilter, then $\prod A_\alpha/\mathscr{F}$ is called an *ultraproduct* of the A_α and if every $A_\alpha = A$, then we speak of an *ultrapower* of A.

Ultraproducts owe their importance to the fact that any "elementary" statement valid for all of the A_α is valid for every ultraproduct of the A_α. We refer the reader to books on model theory for a precise statement and proof of this result. An excellent reference is the chapter "Ultraproducts for Algebraists" by Paul Eklof in *Handbook of Mathematical Logic*, ed. by Jon Barwise, North Holland Company, 1977. An illustration of the result is

PROPOSITION 2.1. *Any ultraproduct of division rings is a division ring.*

Proof. The statement that D is a division ring is that D is a ring in which $1 \neq 0$ and every $a \neq 0$ in D has a (two-sided) multiplicative inverse. (This is an example of an elementary statement.) Suppose the D_α, $\alpha \in I$, are division rings. We wish to show that any ultraproduct $\prod D_\alpha/\mathscr{F}$ determined by an ultrafilter F is a division ring. Let a be an element of $\prod D_\alpha$ such that $\bar{a} = \bar{a}_{\mathscr{F}} \neq 0$. Then the set $I_a \equiv I_{a,0} = \{\alpha \in I \mid a_\alpha = 0_{D_\alpha}\} \notin F$. Since F is an ultrafilter, the complementary set $I'_a = \{\alpha \in I \mid a_\alpha \neq 0_{D_\alpha}\}$ is in F. Define b by $b_\alpha = a_\alpha{}^{-1}$ if $\alpha \in I'_\alpha$, $b_\alpha = 0$ otherwise. Then the set of α such that $a_\alpha b_\alpha = 1_{D_\alpha} = b_\alpha a_\alpha$ is I'_a and so belongs to F. Hence $\overline{ab} = 1 = \overline{ba}$ in $\prod D_\alpha/\mathscr{F}$ and D is a division ring. $\square$

The same argument shows that any ultraproduct of fields is a field. It should be noted that the proof makes strong use of the fact that F is an ultrafilter and not just a filter. In fact, the argument shows that $\prod D_\alpha/\mathscr{F}$ for F a filter is a division ring if and only if the complement of any set S of indices that is not contained in F is contained in F, hence, if and only if F is an ultrafilter.

Another result whose proof can be based on ultraproducts is

PROPOSITION 2.2 (A. Robinson). *Let R be a ring without zero divisors that is a subring of a direct product $\prod D_\alpha$ of division rings D_α. Then R can be imbedded in a division ring.*

Proof (M. Rabin). Let I be the index set. For each $a \in R \subset \prod D_\alpha$ let $I'_a = \{\alpha \in I \mid a_\alpha \neq 0_{D_\alpha}\}$. Then $I'_a \neq \varnothing$ if $a \neq 0$. Let $B = \{I'_a \mid a \neq 0\}$. If $a_1, \ldots, a_n$ are non-zero elements of R, then $a_1 a_2 \cdots a_n \neq 0$ so $I'_{a_1 \cdots a_n} \in B$. Evidently, $I'_{a_1 \cdots a_n} \subset \bigcap_1^n I'_{a_i}$. Thus the set B is a *filter base* in $\mathscr{P}(I)$ in the sense that $\varnothing \notin B$, and the intersection of any finite subset of B contains an element of B. Then the set G of subsets of I, which contain the sets I'_a of B, is a filter in $\mathscr{P}(I)$ not containing $\varnothing$. By Corollary 3 to Theorem 2.5, G can be imbedded in an ultrafilter F of $\mathscr{P}(I)$. Put $D = \prod D_\alpha/\mathscr{F}$. By Proposition 2.1 this is a division ring. We have the nomomorphism $a \rightsquigarrow \bar{a}_{\mathscr{F}}$ of R into D. We claim that this is a monomorphism. Otherwise, we have an $a \neq 0$ in R such that the set $I_a = \{\alpha \in I \mid a_\alpha = 0_{D_\alpha}\} \in F$. On the other hand, $I'_a = \{\alpha \in I \mid a_\alpha \neq 0_{D_\alpha}\} \in B \subset F$ and I'_a is the complement of I_α in I. Since F is a filter, it contains $\varnothing = I_\alpha \bigcap I'_a$. This contradicts the fact that F is an ultrafilter. Hence $a \rightsquigarrow \bar{a}_{\mathscr{F}}$ is a monomorphism and this provides an imbedding of R in a division ring. □

EXERCISES

1. Use exercise 2, p. 70 and Proposition 2.2 to prove that $\mathbb{Z}$ can be imbedded in a field.
2. If a is an element of a Boolean algebra, the subset $F_a = \{x \mid x \geqslant a\}$ is a filter called the *principal filter determined by a*. Let I be a set, S a subset, F_S the principal filter determined by S. Show that if $\mathscr{F}_S$ is the congruence in $\prod_{\alpha \in I} A_\alpha$ determined by F_S, then $\prod A_\alpha/\mathscr{F}_S \cong \prod_{\beta \in S} A_\beta$.
3. Show that an ultrafilter in $\mathscr{P}(I)$, I an infinite set, is not principal if and only if it contains the filter consisting of the complements of finite subsets of I.

2.7 FREE Ω-ALGEBRAS

Let $\Omega = \bigcup_{n=0}^{\infty} \Omega(n)$ be a given set of operator symbols where $\Omega(n)$ is the set of n-ary symbols, and let X be a non-vacuous set. Now let Y be the disjoint union of the sets Ω and X and form the disjoint union $W(\Omega, X)$ of the sets $Y^{(m)}$, $m \geqslant 1$, where $Y^{(m)}$ is the set of m-tuples $(y_1, y_2, \ldots, y_m)$, $y_i \in Y$. To simplify the writing, we write $y_1 y_2 \cdots y_m$ for $(y_1, y_2, \ldots, y_m)$. This suggests calling the elements of $W(\Omega, X)$ *words* in the *alphabet* Y. We define the *degree* of the word $w = y_1 y_2 \cdots y_m$ to be m. It is useful also to introduce a notion of *valence* of a word according to the following rule: The valence $v(x) = 1$ if $x \in X$, $v(\omega) = 1 - n$ if $\omega \in \Omega(n)$, and

$$v(y_1 \cdots y_m) = \sum_1^m v(y_i)$$

for $y_i \in Y$. For example, if $w = \omega_1{}^{(1)} \omega_2{}^{(1)} x_1 \omega_3{}^{(2)} x_2 x_3 \omega_4{}^{(5)}$ where the superscripts on the symbols indicate their arities, then w has degree 7 and valence -2.

We now introduce the juxtaposition multiplication in $W(\Omega, X)$ by putting

$$(y_1 \cdots y_m)(y'_1 \cdots y'_r) = y_1 \cdots y_m y'_1 \cdots y'_r.$$

It is clear that this product is associative, so no parentheses are required to indicate products of more than two words.

Next we shall use this associative product to define an Ω-algebra structure on the set $W(\Omega, X)$. If $\omega \in \Omega(0)$, we define an element of Y to be the element corresponding to ω in $W(\Omega, X)$. Now let $\omega \in \Omega(n)$, $n \geqslant 1$, and let $w_1, w_2, \ldots, w_n \in W(\Omega, X)$. Then we define the action of ω on the n-tuple $(w_1, w_2, \ldots, w_n)$ as the word $\omega w_1 w_2 \cdots w_n$. Evidently, these definitions make $W(\Omega, X)$ an Ω-algebra. We now let $F(\Omega, X)$ be the subalgebra of the Ω-algebra $W(\Omega, X)$ generated by the subset X and we shall show that this is a free Ω-algebra generated by X in the sense that any map of X into an Ω-algebra A has a unique extension to a homomorphism of $F(\Omega, X)$ into A. We shall establish first the following criterion for an element to belong to $F(\Omega, X)$ and a unique factorization property.

LEMMA. *A word $w \in F(\Omega, X)$ if and only if its valence is* 1 *and the valence of any right factor w' of w ($w = w''w'$ or $w' = w$) is positive. The subset of $F(\Omega, X)$ of elements of degree* 1 *is $X \cup \Omega(0)$. If $w \in F(\Omega, X)$ has degree >1, then w begins with an $\omega \in \Omega(n)$, $n \geqslant 1$, and w can be written in one and only one way as $\omega w_1 w_2 \cdots w_n$ where $w_i \in F(\Omega, X)$.*

Proof. The statement on the elements of degree 1 in $F(\Omega, X)$ is clear. It is clear also that $v(w) = 1$ for these elements and w is the only right factor of w of degree 1 so the conditions on valence hold for w. Now let $w \in F(\Omega, X)$ have degree >1. It is clear from the inductive procedure given on p. 59 for producing the elements of the subalgebra of an Ω-algebra generated by a subset that w has the form $w = \omega w_1 w_2 \cdots w_n$ where $\omega \in \Omega(n)$ and every $w_i \in F(\Omega, X)$. Using induction on the degree, we may assume that every w_i satisfies the stated conditions on the valences. It then follows that $w = \omega w_1 w_2 \cdots w_n$ satisfies the valence conditions. Conversely, let w be any word that satisfies the valence conditions. If the degree of w is 1, then either w is an $x \in X$ or $w = \omega \in \Omega(0)$. In either case, $w \in F(\Omega, X)$. Now suppose the degree of w is greater than 1. Then the valence conditions imply that w begins with an $\omega \in \Omega(n)$, $n \geqslant 1$. Thus $w = \omega y_1 y_2 \cdots y_m$ where $\omega \in \Omega(n)$, $n \geqslant 1$, and the $y_i \in Y$. For any i, $1 \leqslant i \leqslant n$, let u_i be the right factor of $u = y_1 y_2 \cdots y_m$ of maximum degree having valence i. For example, let

$$w = \omega_1{}^{(4)} x_1 x_2 x_3 \omega_2{}^{(2)} x_4 \omega_3{}^{(4)} x_5 x_6 \omega_4{}^{(0)} \omega_5{}^{(0)}$$

where the superscripts give the arities of the operator symbols to which they are attached. Then $u = x_1 x_2 x_3 \omega_2{}^{(2)} x_4 \omega_3{}^{(4)} x_5 x_6 \omega_4{}^{(0)} \omega_5{}^{(0)}$ and if we write the valence of the right-hand factor, which begins with a symbol above this symbol, then we have

$$u = \overset{4}{x_1}\overset{3}{x_2}\overset{2}{x_3}\overset{1}{\omega_2}{}^{(2)}\overset{2}{x_4}\overset{1}{\omega_3}{}^{(4)}\overset{4}{x_5}\overset{3}{x_6}\overset{2}{\omega_4}{}^{(0)}\overset{1}{\omega_5}{}^{(0)}.$$

Hence, in this case, $u_1 = \omega_2{}^{(2)}x_4\omega_3{}^{(4)}x_5x_6\omega_4{}^{(0)}\omega_5{}^{(0)}$, $u_2 = x_3u_1$, $u_3 = x_2u_2$, $u = u_4 = x_1u_3$. In general, we have $u_n = u = y_1y_2\cdots y_m$ and u_{i-1} is a right factor of u_i for $2 \leqslant i \leqslant n$. Then we can write $u_1 = w_n$, $u_2 = w_{n-1}u_1$, $u_3 = w_{n-2}u_2, \ldots$ where the w_i satisfy the valence conditions. Thus we have $w = \omega y_1y_2\cdots y_m = \omega w_1w_2\cdots w_n$. Since the degree of every w_i is less than that of w, using induction on the degree, we may assume that $w_i \in F(\Omega, X)$. Then $w = \omega w_1w_2\cdots w_n \in F(\Omega, X)$. This proves that the valence conditions are sufficient to insure that an element is in $F(\Omega, X)$. Now suppose we have a second factorization of w as $w = \omega w'_1w'_2\cdots w'_n$ where $w'_i \in F(\Omega, X)$. Then the valence of w'_n is 1, of $w'_{n-1}w'_n$ is 2, of $w'_{n-2}w'_{n-1}w'_n$ is 3, etc. Hence the definition of the u_i implies that the degree $\deg w'_iw'_{i+1}\cdots w'_n \leqslant \deg w_iw_{i+1}\cdots w_n$, $1 \leqslant i \leqslant n$. If equality holds for all i, then $w'_1 = w_1$, $w'_2 = w_2, \ldots, w'_n = w_n$. Hence assume $\deg w'_i\cdots w'_n < \deg w_i\cdots w_n$ for some i and let i be minimal for this relation. Evidently $i > 1$ and $\deg w'_{i-1}\cdots w'_n = \deg w_{i-1}\cdots w_n$ so $w'_{i-1}\cdots w'_n = w_{i-1}\cdots w_n$. Then $\deg w'_1\cdots w'_{i-1} > \deg w_1\cdots w_{i-1}$ and $w'_{i-1} = w_{i-1}z$ where z is a word. Since $v(w'_{i-1}) = 1 = v(w_{i-1})$, we have $v(z) = 0$, which contradicts the valence conditions that must be satisfied by $w'_{i-1} \in F(\Omega, X)$. □

We can use this to prove the "freeness" property of $F(\Omega, X)$:

THEOREM 2.9. *Let f be a map of X into an Ω-algebra A. Then f has a unique extension to a homomorphism $\bar{f}$ of $F(\Omega, X)$ into A.*

Proof. If $x \in X$, we put $\bar{f}(x) = f(x)$ and if $\omega \in \Omega(0)$ so $\omega \in F(\Omega, X)$, then we let $\bar{f}(\omega) = \omega_A$. The remaining elements of $F(\Omega, X)$ have the form $\omega w_1w_2\cdots w_n$, where the $w_i \in F(\Omega, X)$, $\omega \in \Omega(n)$, $n \geqslant 1$. Suppose we have already defined $\bar{f}(w')$ for all w' of degree $< \deg \omega w_1\cdots w_n$. Then $\bar{f}(w_i)$ is defined and we extend the definition to $\omega w_1\cdots w_n$ by $\bar{f}(\omega w_1\cdots w_n) = \omega\bar{f}(w_1)\cdots\bar{f}(w_n)$. This inductive procedure defines $\bar{f}$ for all $F(\Omega, X)$. It is clear also that $\bar{f}$ is a homomorphism. Since X generates $F(\Omega, X)$, $\bar{f}$ is unique (see exercise 3, p. 66). □

We can express the result of Theorem 2.9 in categorical terms. We have the forgetful functor F from the category **Ω-alg** to **Set** mapping any Ω-algebra into its carrier (underlying set) and algebra homomorphisms into the corresponding set maps. Then it is clear from the definition of universals (p. 42) that if X is a non-vacuous set, then the pair $(F(\Omega, X), i)$ where i is the injection of X into $F(\Omega, X)$ constitutes a universal from X to the functor F. The fact that $(F(\Omega, X), i)$ is a universal implies its uniqueness in the usual sense.

2.8 VARIETIES

The concept of a free Ω-algebra $F(\Omega, X)$ permits us to define identities for Ω-algebras. Let (w_1, w_2) be a pair of elements in $F(\Omega, X)$ for some set X. Then we say that an Ω-algebra A *satisfies the identity* $w_1 = w_2$ (or $w_1 = w_2$ *is a law in* A) if $f(w_1) = f(w_2)$ for every homomorphism f of $F(\Omega, X)$ into A. If S is a subset of $F(\Omega, X) \times F(\Omega, X)$, then the class of Ω-algebras satisfying the identities $w_1 = w_2$ for all $(w_1, w_2) \in S$ is called *the variety of* Ω*-algebras* $V(S)$ *defined by* S (or *the equational class defined by* S). For example, let $\omega^{(2)}, \omega^{(1)}, \omega^{(0)}$ be respectively binary, unary, and nullary operator symbols, $\Omega = \{\omega^{(2)}, \omega^{(1)}, \omega^{(0)}\}$, $X = \{x_1, x_2, x_3\}$; then the class of groups is defined by the following subset of $F(\Omega, X) \times F(\Omega, X)$:

$$(\omega^{(2)}\omega^{(2)}x_1x_2x_3, \omega^{(2)}x_1\omega^{(2)}x_2x_3)$$

$$(\omega^{(2)}\omega^{(0)}x_1, x_1)$$

$$(\omega^{(2)}x_1\omega^{(0)}, x_1)$$

$$(\omega^{(2)}x_1\omega^{(1)}x_1, \omega^{(0)})$$

$$(\omega^{(2)}\omega^{(1)}x_1x_1, \omega^{(0)}).$$

For, the statement that A is an Ω-algebra for the indicated Ω is that A is equipped with a binary, a unary, and a nullary composition. If we denote the effect of these in A by ab, a^{-1}, and 1 respectively, then the statement that A is in $V(S)$ for S, the set of five elements we have listed, amounts to the following laws in A: $(ab)c = a(bc)$, $1a = a$, $a1 = a$, $aa^{-1} = 1$, $a^{-1}a = 1$. Evidently, this means simply that A is a group, so $V(S)$ is the variety of groups. In a similar manner one sees that the classes of monoids, of lattices, of rings, of commutative rings, and of Boolean algebras are varieties.

It is easily seen that in defining varieties there is no loss in generality in assuming that X is a countably infinite set $X_0 = \{x_1, x_2, \ldots\}$. At any rate, we shall do this from now on. Let $V(S)$ be the variety of Ω-algebras defined by the subset S of $F(\Omega, X_0) \times F(\Omega, X_0)$. It is clear that if $A \in V(S)$, then any subalgebra of A is contained in $V(S)$. Moreover, every homomorphic image $\bar{A}$ of A is contained in $V(S)$. To see this we observe that any homomorphism $\bar{f}$ of $F(\Omega, X_0)$ into $\bar{A}$ can be lifted to a homomorphism f of $F(\Omega, X_0)$ into A in the sense that if λ is a surjective homomorphism of A onto $\bar{A}$, then there exists a homomorphism $f: F(\Omega, X_0) \to A$ such that $\bar{f} = \lambda f$. To define such an f we choose for each $i = 1, 2, \ldots$ an element $a_i \in A$ such that $\lambda(a_i) = \bar{f}(x_i)$ and we let f be the homomorphism of $F(\Omega, X_0)$ into A extending the map $x_i \quad a_i$ of X_0 into A. Then $\lambda f(x_i) = \lambda(a_i) = \bar{f}(x_i)$. Since X_0 generates $F(\Omega, X_0)$, we have $\lambda f = \bar{f}$ for the homomorphisms λf and $\bar{f}$ of $F(\Omega, X_0)$ into $\bar{A}$. Now suppose $(w_1, w_2) \in S$ and let $\bar{f}$ be any homomorphism of $F(\Omega, X_0)$ into $\bar{A}$, f a homomorphism of $F(\Omega, X_0)$ into A such that $\lambda f = \bar{f}$. Then $\bar{f}(w_1) = \lambda f(w_1) = \lambda f(w_2) = \bar{f}(w_2)$. Thus $\bar{A}$ satisfies every identity $w_1 = w_2$ for $(w_1, w_2) \in S$ and hence $\bar{A} \in V(S)$.

We note next that if $\{A_\alpha | \alpha \in I\}$ is an indexed set of algebras contained in $V(S)$, then $\prod A_\alpha \in V(S)$. This is clear since if p_α denotes the projection of $A = \prod A_\alpha$ into A_α and f is a homomorphism of $F(\Omega, X_0)$ into A, then $p_\alpha f(w_1) = p_\alpha f(w_2)$ for every $(w_1, w_2) \in S$. Then $f(w_1) = f(w_2)$ and hence $A \in V(S)$.

Any variety $V(S)$ defines a full subcategory $\mathbf{V}(S)$ of Ω-**alg**. We shall now consider some of the important properties of such categories and we prove first

PROPOSITION 2.3. *A morphism in $\mathbf{V}(S)$ is monic if and only if it is an injective map.*

Proof. The proof is an immediate generalization of the one given on p. 17 in the special case of the category of rings. If f is an injective homomorphism of A into B, $A, B \in \mathbf{V}(S)$, then f is monic. Now suppose $f: A \to B$ is not injective. We have the subalgebra $\Phi = f^{-1}f$ of $A \times A$ (the kernel of f) and $\Phi \in V(S)$. Let p and p' be the restrictions to Φ of the projections $(a, a') \rightsquigarrow a$ and $(a, a') \rightsquigarrow a'$. Since $(a, a') \in \Phi$ is equivalent to $f(a) = f(a')$, we have $fp = fp'$. On the other hand, since f is not injective, we have a pair (a, a') such that $f(a) = f(a')$ and $a \neq a'$. Then $(a, a') \in \Phi$ and $p(a, a') \neq p'(a, a')$. Since $fp = fp'$, this shows that f is not monic in $\mathbf{V}(S)$. □

We recall that in the category of rings, epics need not be surjective. Hence the analogue for epics of Proposition 2.3 is not valid.

If **C** is a subcategory of Ω-**alg** and X is a set, we call a universal from X to the forgetful functor from **C** to **Set** a *free algebra* for **C** determined by X. We have the following extension of Theorem 2.9 to varieties:

THEOREM 2.10. *Let $\mathbf{V}(S)$ be the category defined by a variety $V(S)$ and let X be a non-vacuous set. Then there exists a free algebra for $\mathbf{V}(S)$ determined by X.*

Proof. Consider the free algebras $F(\Omega, X)$ and $F(\Omega, X_0)$ where $X_0 = \{x_1, x_2, \ldots\}$. Then S is a subset of $F(\Omega, X_0) \times F(\Omega, X_0)$. If φ is a homomorphism of $F(\Omega, X_0)$ into $F(\Omega, X)$ and $(w_1, w_2) \in S$, then $(\varphi(w_1), \varphi(w_2)) \in F(\Omega, X) \times F(\Omega, X)$. Let $\Phi(S)$ denote the congruence on $F(\Omega, X)$ generated by all the elements $(\varphi(w_1), \varphi(w_2))$ obtained from all the homomorphisms φ and all of the pairs $(w_1, w_2) \in S$. Put $F_{\mathbf{V}(S)}(\Omega, X) = F(\Omega, X)/\Phi(S)$ and let ν be the canonical homomorphism $a \rightsquigarrow \bar{a}_{\Phi(S)}$ of $F(\Omega, X)$ onto $F_{\mathbf{V}(S)}(\Omega, X)$.

We claim that $F_{\mathbf{V}(S)}(\Omega, X)$ together with the restriction i of y to X constitutes a free $\mathbf{V}(S)$-algebra for the set X. We show first that $F_{\mathbf{V}(S)}(\Omega, X) \in V(S)$. Let $\bar{\varphi}$ be a homomorphism of $F(\Omega, X_0)$ into $F_{\mathbf{V}(S)}(\Omega, X)$. Then as we showed above, there exists a homomorphism $\varphi: F(\Omega, X_0) \to F(\Omega, X)$ such that $\bar{\varphi} = \nu\varphi$. If $(w_1, w_2) \in S$, $(\varphi(w_1), \varphi(w_2)) \in \Phi(S)$, so $\nu\varphi(w_1) = \nu\varphi(w_2)$. Thus $\bar{\varphi}(w_1) = \bar{\varphi}(w_2)$. Since this holds for every $(w_1, w_2) \in S$ and all homomorphisms $\bar{\varphi}$ of $F(\Omega, X_0)$ into $F_{\mathbf{V}(S)}(\Omega, X)$, we see that $F_{\mathbf{V}(S)}(\Omega, X) \in V(S)$. Next suppose g is a map of X into an algebra $A \in V(S)$. Then g can be extended to a homomorphism $\bar{g}$ of $F(\Omega, X)$ into A. If φ is a homomorphism of $F(\Omega, X_0)$ into $F(\Omega, X)$, then $\bar{g}\varphi$ is a homomorphism of $F(\Omega, X_0)$ into A. Hence, if $(w_1, w_2) \in S$, then $\bar{g}\varphi(w_1) = \bar{g}\varphi(w_2)$. Then $(\varphi(w_1), \varphi(w_2))$ is in the kernel $\ker \bar{g}$. Since this holds for all $(w_1, w_2) \in S$ and all homomorphisms φ of $F(\Omega, X_0)$ into $F(\Omega, X)$, the congruence $\Phi(S) \subset \ker \bar{g}$. Hence we have a unique homomorphism f of $F_{\mathbf{V}(S)}(\Omega, X) = F(\Omega, X)/\Phi(S)$ into A such that

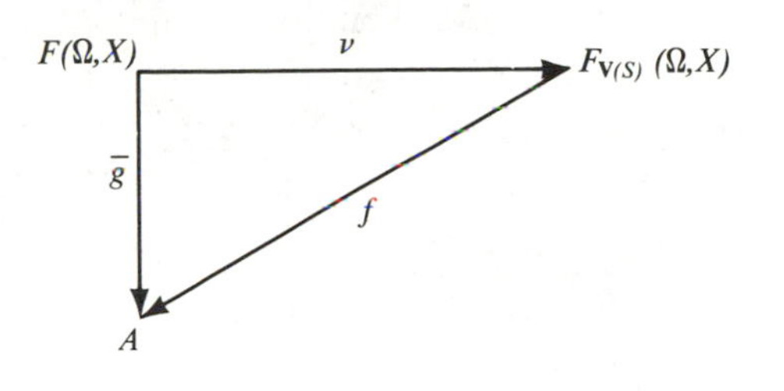

is commutative. Then for $i = \nu|X$ we have the commutative diagram

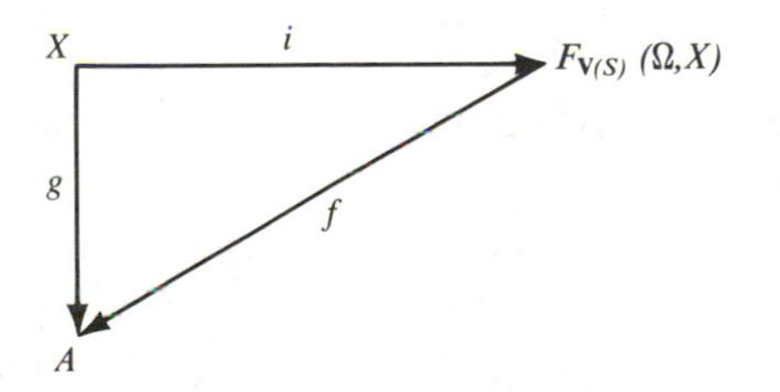

Since X generates $F(\Omega, X)$, $i(X)$ generates $F_{\mathbf{V}(S)}(\Omega, X)$. Hence f is unique and $(F_{\mathbf{V}(S)}(\Omega, X), i)$ is a free algebra for $\mathbf{V}(S)$ determined by X. □

COROLLARY. *Any algebra in a variety is a homomorphic image of a free algebra.*

Proof. Let X be a set of generators for the given algebra $A \in V(S)$. For example, we may take $X = A$. Then we have a homomorphism of $F_{\mathbf{V}(S)}(\Omega, X)$ into A whose image includes X. Since X generates A, the image is A. Thus we have a surjective homomorphism of $F_{\mathbf{V}(S)}(\Omega, X)$ onto A. □

The result we noted before—that if $\{A_\alpha | \alpha \in I\}$ is an indexed set of algebras contained in a variety $V(S)$ then $\prod A_\alpha \in V(S)$—implies that products exist in $\mathbf{V}(S)$ for any indexed set of objects in this category. This holds also for coproducts:

THEOREM 2.11. *Coproducts exist in* $\mathbf{V}(S)$ *for any indexed set* $\{A_\alpha | \alpha \in I\}$ *of algebras in* $V(S)$.

Proof. We reduce the proof to the case of free algebras in $V(S)$ by showing that if a coproduct exists for a set of algebras $\{A_\alpha | \alpha \in I\}$, $A_\alpha \in V(S)$, then it exists for every set $\{\bar{A}_\alpha | \alpha \in I\}$ where $\bar{A}_\alpha$ is a homomorphic image of A_α. Let $\{A, i_\alpha\} = \coprod A_\alpha$ in $\mathbf{V}(S)$ and let η_α be a surjective homomorphism of A_α onto $\bar{A}_\alpha$, $\alpha \in I$. Let Θ_α be the kernel of η_α and let $i_\alpha \Theta_\alpha$ denote the subset of $A \times A$ of elements $(i_\alpha(a_\alpha), i_\alpha(b_\alpha))$ for $(a_\alpha, b_\alpha) \in \Theta_\alpha$. Next, let Θ be the congruence on A generated by $\bigcup_\alpha i_\alpha \Theta_\alpha$ and put $\bar{A} = A/\Theta$, $i'_\alpha = \nu i_\alpha$ where ν is the canonical homomorphism of A onto $\bar{A} = A/\Theta$. Then i'_α is a homomorphism of A_α into $\bar{A}$ and if $(a_\alpha, b_\alpha) \in \Theta_\alpha$, $(i_\alpha(a_\alpha), i_\alpha(b_\alpha)) \in \Theta$, so $\nu i_\alpha(a_\alpha) = \nu i_\alpha(b_\alpha)$. Thus Θ_α is contained in the kernel of $i'_\alpha = \nu i_\alpha$; consequently we have a unique homomorphism $\bar{i}_\alpha : \bar{A}_\alpha \to \bar{A}$, making the diagram

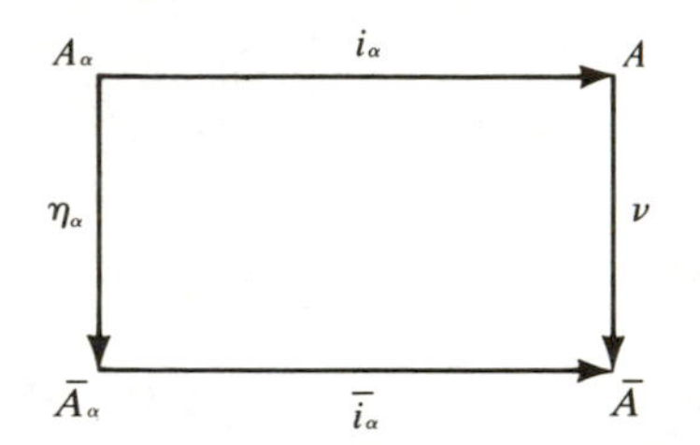

commutative. Now $\bar{A} \in V(S)$ since it is a homomorphic image of $A \in V(S)$.

We claim that $\{\bar{A}, \bar{i}_\alpha\} = \coprod \bar{A}_\alpha$. Let $B \in V(S)$ and for each $\alpha \in I$ let $\bar{f}_\alpha$ be a homomorphism of $\bar{A}_\alpha$ into B. Then $f_\alpha = \bar{f}_\alpha \eta_\alpha$ is a homomorphism of A_α into B whose kernel contains Θ_α. Since $\{A, i_\alpha\} = \coprod A_\alpha$, we have a unique homomorphism $f: A \to B$ such that $f i_\alpha = f_\alpha$, $\alpha \in I$. If $(a_\alpha, b_\alpha) \in \Theta_\alpha$, then $\eta_\alpha(a_\alpha) = \eta_\alpha(b_\alpha)$ and $f_\alpha(a_\alpha) = \bar{f}_\alpha \eta_\alpha(a_\alpha) = \bar{f}_\alpha \eta_\alpha(b_\alpha) = f_\alpha(b_\alpha)$. Thus $f i_\alpha(a_\alpha) = f i_\alpha(b_\alpha)$, so $i_\alpha \Theta_\alpha$ is contained in the kernel of f. Since this holds for all α and since Θ is the congruence on A generated by $\bigcup_\alpha i_\alpha \Theta_\alpha$, we see that Θ is contained in the kernel of f. Hence we have a unique homomorphism $\bar{f}: \bar{A} \to B$ such that

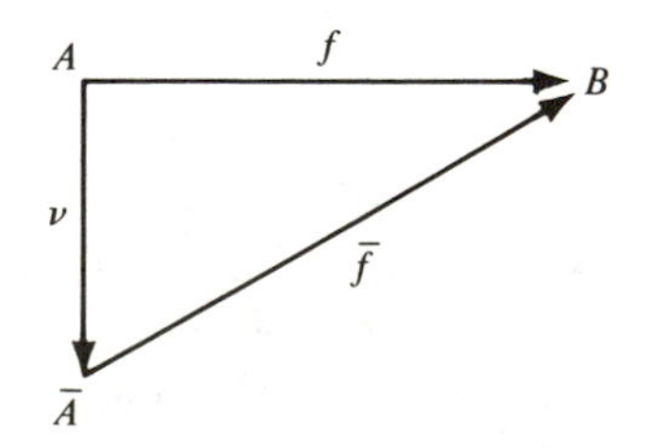

is commutative. Now we have the diagram

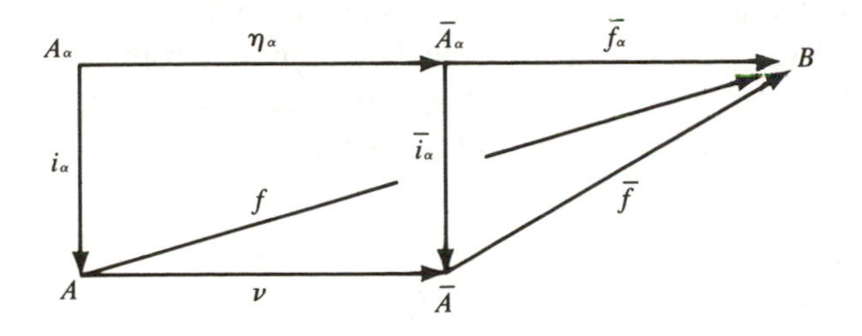

in which the rectangle and the triangles $A\bar{A}B$ and $A_\alpha AB$ are commutative. Now the "diagram chase" $\bar{f}\bar{i}_\alpha\eta_\alpha = \bar{f}\nu i_\alpha = fi_\alpha = \bar{f}_\alpha\eta_\alpha$ and the fact that η_α is surjective, hence epic, imply that $\bar{f}\bar{i}_\alpha = \bar{f}_\alpha$, $\alpha \in I$. To prove the uniqueness of $\bar{f}$ we observe that if $\bar{f}: \bar{A} \to B$ satisfies $\bar{f}_\alpha = \bar{f}\bar{i}_\alpha$, $\alpha \in I$, and we define $f = \bar{f}\nu$, then the rectangle and the triangles $A\bar{A}B$ and $\bar{A}_\alpha\bar{A}B$ in the foregoing diagram commute. Then the triangle $A_\alpha AB$ is commutative, since $f_\alpha = \bar{f}_\alpha\eta_\alpha = \bar{f}\bar{i}_\alpha\eta_\alpha = \bar{f}\nu i_\alpha = fi_\alpha$. Since $\{A, i_\alpha\} = \quad A_\alpha$, this implies that f is unique. Since ν is surjective, $\bar{f}$ is unique. Hence $\{\bar{A}, \bar{i}_\alpha\} = \quad \bar{A}_\alpha$.

The result just proved and the fact that every $A_\alpha \in V(S)$ is a homomorphic image of a free algebra in $V(S)$ imply that it suffices to prove the theorem for free algebras $A_\alpha = F_{\mathbf{V}(S)}(\Omega, X_\alpha)$. For this purpose we form $A = F_{\mathbf{V}(S)}(\Omega, X)$ where $X = \dot{\bigcup} X_\alpha$; the disjoint union of the X_α. We denote the maps $X \to A$, $X_\alpha \to A_\alpha$ given in the definition of the free algebras in $\mathbf{V}(S)$ indiscriminately by i and let i_α be the homomorphism $A_\alpha \to A$ corresponding to the injection of X_α into X. Hence $i_\alpha i(x_\alpha) = i(x_\alpha)$. Now suppose we have homomorphisms $f_\alpha: A_\alpha \to B$ where $B \in \mathbf{V}(S)$. Let g be the map of X into B such that $g(x_\alpha) = f_\alpha i(x_\alpha)$, $\alpha \in I$. Then we have a homomorphism f of A into B such that $fi(x) = g(x)$, $x \in X$. Then $f_i(x_\alpha) = g(x_\alpha) = f_\alpha i(x_\alpha) = fi_\alpha i(x_\alpha)$. Since the elements $i(x_\alpha)$ generate A_α, we have $fi_\alpha = f_\alpha$, $\alpha \in I$. Since the elements $i(x_\alpha)$, $x_\alpha \in X_\alpha$, $\alpha \in I$, generate A, f is unique. Hence $\{A, i_\alpha\} = \quad A_\alpha$. $\square$

The preceding construction of $\coprod A_\alpha = \{A, i_\alpha\}$ in a variety shows that A is generated by the images $i_\alpha(A_\alpha)$. This can also be seen directly from the definition of a coproduct of algebras. Of particular interest is the situation in which the i_α are injective. In this case we shall call $\coprod A_\alpha$ the *free product*. A useful sufficient condition for the coproduct to be a free product is given in

PROPOSITION 2.4. *Let $\coprod A_\alpha$ be a coproduct of Ω-algebras in a category* $\mathbf{C}$. *Then $\coprod A_\alpha$ is a free product if every A_α has a one-element subalgebra $\{\varepsilon_\alpha\}$.*

Note that this is applicable in the category of monoids and of groups, but not in the category of rings.

Proof. If A is any Ω-algebra, the map of A into A_β sending every element of A into ε_β is a homomorphism. Let $f_{\alpha\beta}$ be this homomorphism for $A = A_\alpha$ into A_β if $\alpha \neq \beta$ and let $f_{\beta\beta} = 1_{A_\beta}$. Applying the definition of a coproduct, we obtain a homomorphism $f_\beta : A = \quad A_\alpha \to A_\beta$ such that $f_{\alpha\beta} = f_\beta i_\alpha$, $\alpha \in I$. Since $f_{\beta\beta}$ is injective, i_β is injective. Thus every i_β is injective and $\quad A_\alpha$ is a free product. □

EXERCISES

1. Give an example in the category of rings in which the coproduct is not a free product.

2. Let $F_{\mathbf{C}}(\Omega, X)$ be a free algebra for a category $\mathbf{C}$ of Ω-algebras determined by a set X. Show that if λ is a surjective homomorphism of A into $\bar{A}$ where $A \in \operatorname{ob} \mathbf{C}$ and $\bar{f}$ is a homomorphism of $F_{\mathbf{C}}(\Omega, X)$ into $\bar{A}$, then there exists a homomorphism f of $F_{\mathbf{C}}(\Omega, X)$ into A such that $\bar{f} = \lambda f$.

3. Let $F_{\mathbf{C}}(\Omega, X)$ be as in exercise 2. Prove the following: (i) $F_{\mathbf{C}}(\Omega, X)$ is generated by $i(X)$ (i the canonical map of X into $F_{\mathbf{C}}(\Omega, X)$). (ii) If Y is a non-vacuous subset of X, then the subalgebra of $F_{\mathbf{C}}(\Omega, X)$ generated by $i(Y)$ together with the restriction of i to Y regarded as a map into the indicated subalgebra constitutes a free $\mathbf{C}$ algebra determined by Y. (iii) Let X' be a second set, $F_{\mathbf{C}}(\Omega, X')$ a free $\mathbf{C}$ algebra for X. Show that if $|X| = |X'|$, then $F_{\mathbf{C}}(\Omega, X)$ and $F_{\mathbf{C}}(\Omega, X')$ are isomorphic.

4. Show that if $V(S)$ contains an algebra A with $|A| > 1$, then it contains algebras of cardinality exceeding any given cardinal. Use this to prove that for any X the map i of X into $F_{\mathbf{V}(S)}(\Omega, X)$ is injective if $V(S)$ contains non-trivial algebras.

5. Show that if the $A_\alpha \in V(S)$ for $\alpha \in I$, a pre-ordered set, and $\varprojlim A_\alpha$ exists, then $\varprojlim A_\alpha \in V(S)$. Show also that if I is directed, then $\varinjlim A_\alpha \in V(S)$. Show that any ultraproduct of $A_\alpha \in V(S)$ is contained in $V(S)$.

6. Let $FM^{(r)}$ be the free monoid defined by $X = \{x_1, x_2, \ldots, x_r\}$ as in BAI, p. 67. $FM^{(r)}$ consists of 1 and the monomials (or sequences) $x_{i_1} x_{i_2} \cdots x_{i_k}$ where $1 \leqslant i_j \leqslant r$, and the product is defined by juxtaposition and the condition that 1 is a unit. Form the free $\mathbb{Z}$-module with $FM^{(r)}$ as base and define a product by

$$(\sum r_i w_i)(\sum r'_j w'_j) = \sum r_i r'_j w_i w'_j$$

for $r_i, r'_j \in \mathbb{Z}$, $w_i, w'_j \in FM^{(r)}$. This defines a ring $\mathbb{Z}\{X\} = \mathbb{Z}[FM^{(r)}]$ (exercise 8, p. 127 of BAI). Show that $\mathbb{Z}\{X\}$ together with the injection map of X into $\mathbb{Z}\{X\}$ constitutes a free ring determined by X.

2.9 FREE PRODUCTS OF GROUPS

We shall now have a closer look at the coproduct of an indexed set $\{G_\alpha | \text{Å} \in I\}$ of groups G_α. By Proposition 2.4, $\Omega\, G_\alpha$ is a free product. In group theory it is customary to denote this group as $\prod^* G_\alpha$ and to identify G_α with its image $i_\alpha(G_\alpha)$ in $\prod^* G_\alpha$. When this is done we have the following situation: $\prod^* G_\alpha$ contains the G_α as subgroups and is generated by $\bigcup G_\alpha$. Moreover, if H is any group and f_α is a homomorphism of G_α into H for every $\alpha \in I$, then there is a unique homomorphism f of $\prod^* G_\alpha$ into H such that $f | G_\alpha = f_\alpha$, $\alpha \in I$.

If G is any group containing subgroups G_α, $\alpha \in I$, and generated by $\bigcup G_\alpha$, then every element of G is a product of elements of the G_α. This is clear, since the set G' of these products contains $\bigcup G_\alpha$ and thus contains 1. Moreover, G' is closed under multiplication and taking inverses. Hence G' is a subgroup, so $G' = G$. It is clear that if we take a product of elements of the G_α, then we may suppress the factors $= 1$ and we may replace any product of elements that are in the same G_α by a single element in G_α. In this way we may replace any product by one that either is 1 or has the form $x_1 x_2 \cdots x_n$ where $x_i \neq 1$ and if $x_i \in G_\alpha$, then $x_{i+1} \notin G_\alpha$. Products of this form are called *reduced*. We shall now show that in the free product $\prod^* G_\alpha$ any two reduced products that look different are different. For this purpose we shall give an alternative, more direct construction of $\prod^* G_\alpha$. There are a number of ways of doing this. The one we have chosen is a particularly simple one due to van der Waerden.

We start from scratch. Given a set of groups $\{G_\alpha | \alpha \in I\}$, let G'_α be the subset of G_α of elements $\neq 1$. Form the disjoint union S of the G'_α and $\{1\}$. In this set we can identify G_α with $G'_\alpha \cup \{1\}$. Then we shall have a set that is the union of the G_α; moreover, $G_\alpha \cap G_\beta = 1$ if $\alpha \neq \beta$. Now let G be the set of *reduced words* based on the G_α. By this we mean either 1 or the words $x_1 x_2 \cdots x_n$ where $x_i \neq 1$ and if $x_i \in G_\alpha$, then $x_{i+1} \notin G_\alpha$, $1 \leqslant i \leqslant n-1$. We shall now define an action of G_α on G(cf. BAI, pp. 71–73) by the following rules:

$$1w = w \quad \text{for any } w \in G. \tag{21}$$

If $x \neq 1$ in G_α, then $x1 = x$ and if $x_1 \cdots x_n$ is a reduced word $\neq 1$, then

$$x x_1 \cdots x_n = \begin{cases} x x_1 \cdots x_n & \text{if } x_1 \notin G_\alpha \\ (x x_1) x_2 \cdots x_n & \text{if } x_1 \in G_\alpha \text{ and } x x_1 \neq 1 \\ x_2 \cdots x_n & \text{if } x_1 \in G_\alpha \text{ and } x x_1 = 1. \end{cases} \tag{22}$$

In the last formula, it is understood that if $n = 1$, so that an empty symbol results, then this is taken to be 1.

Our definitions give $xw \in G$ for $x \in G_\alpha$, $w \in G$. We shall now verify the axioms for an action: $1w = w$ and $(xy)w = x(yw)$ for $x, y \in G_\alpha$ and $w \in G$. The first is the definition (21) and the second is clear if either x, y, or w is 1 or if $x_1 \notin G_\alpha$. It remains to consider the situation in which $x, y, x_1 \in G_\alpha$, $x \neq 1$, $y \neq 1$. We distinguish three cases: (α) $yx_1 \neq 1$, $xyx_1 \neq 1$; (β) $yx_1 \neq 1$, $xyx_1 = 1$; (γ) $yx_1 = 1$, $xyx_1 \neq 1$. Note that $yx_1 = 1$, $xyx_1 = 1$ is ruled out, since $x = 1$ in this case.

(α) Here $(xy)x_1 \cdots x_n = (xyx_1)x_2 \cdots x_n$ and $x(yx_1 \cdots x_n) = x(yx_1)x_2 \cdots x_n = (xyx_1)x_2 \cdots x_n$, so $(xy)x_1 \cdots x_n = x(yx_1 \cdots x_n)$.

(β) Here $(xy)x_1 \cdots x_n = x_2 \cdots x_n$ and $x(yx_1 \cdots x_n) = x(yx_1)x_2 \cdots x_n = x_2 \cdots x_n$, so $(xy)x_1 \cdots x_n = x(yx_1 \cdots x_n)$.

(γ) Here $(xy)x_1 \cdots x_n = (xyx_1)x_2 \cdots x_n = xx_2 \cdots x_n$ and $x(yx_1 \cdots x_n) = x(x_2 \cdots x_n) = xx_2 \cdots x_n$, so again $(xy)x_1 \cdots x_n = x(yx_1 \cdots x_n)$.

If T_x denotes the map $w \quad xw$ for $w \in G$ and $x \in G_\alpha$, then we have $T_x T_y = T_{xy}$ and $T_1 = 1$. Hence T_x is bijective with inverse $T_{x^{-1}}$ and $x \rightsquigarrow T_x$ is a homomorphism of G_α into the group Sym G of bijective transformations of the set G. Since $T_x 1 = x, x \rightsquigarrow T_x$ is injective, so G_α is isomorphic to its image $T(G_\alpha)$, which is a subgroup of Sym G. Let G^* denote the subgroup of Sym G generated by $\bigcup T_\alpha(G)$. Let $T_{x_1} \cdots T_{x_n}$ be a reduced product of elements of $\bigcup T_\alpha(G)$ where no $x_i = 1$. The formulas (21) and (22) give $T_{x_1} \cdots T_{x_n} 1 = x_1 \cdots x_n$. Then $T_{x_1} \cdots T_{x_n} \neq 1$ and if $T_{y_1} \cdots T_{y_m}$ is a second such product, then $T_{x_1} \cdots T_{x_n} = T_{y_1} \cdots T_{y_m}$ implies that $m = n$ and $x_i = y_i$, $1 \leqslant i \leqslant n$. We can identify the $T(G_\alpha)$ with G_α. Then we have proved the first conclusion in the following.

THEOREM 2.12. *Let $\{G_\alpha | \alpha \in I\}$ be a set of groups. Then there exists a group G^* containing the G_α as subgroups that is generated by $\bigcup G_\alpha$ and has the property that if $x_1 \cdots x_n$ and $y_1 \cdots y_m$, $x_i, y_j \in \bigcup G_\alpha$, are reduced products relative to the G_α, then $x_1 \cdots x_n = y_1 \cdots y_m$ implies that $m = n$ and $x_i = y_i$, $1 \leqslant i \leqslant n$. Moreover, if G^* is any group having this property and i_α is the injection of G_α in G, then $\{G^*, i_\alpha\}$ is a free product of the G_α.*

Proof. To prove the last statement we invoke the existence of the free product $\prod^* G_\alpha$, which is a consequence of Theorem 2.11, and Proposition 2.4. By the defining property of $\prod^* G_\alpha$ we have a homomorphism η of $\prod^* G_\alpha$ into G^*, which is the identity map on every G_α. We claim η is an isomorphism, which will imply that $\{G^*, i_\alpha\}$ is a free product of the G_α. Since the G_α generate G^*, η is surjective. Now let a be any element $\neq 1$ in $\prod^* G_\alpha$. Then we can write a in the reduced form $a = x_1 \cdots x_n$ where the ${}_i \neq 1$ and successive x_i are in different G_α. Then $\eta(a) = x_1 \cdots x_n \neq 1$ in G^*. Hence $\ker \eta$ (in the usual sense) is 1 so η is injective. □

The free product can be used to give an alternative construction of the free group $FG(X)$ determined by a non-vacuous set X, which is more explicit than the ones we have considered before (in BAI or in the previous section). With each $x \in X$ we associate the infinite cyclic group $\langle x \rangle$ of powers x^k, $k = 0, \pm 1, \pm 2, \ldots$, and we construct the free product $\prod_{x\in X}^{*}\langle x \rangle$ of the groups $\langle x \rangle$ as above. This contains the subgroups $\langle x \rangle$ and has the properties stated for the G_α in Theorem 2.12. Thus $\prod_{x\in X}^{*}\langle x \rangle$ consists of 1 and the products ${x_1}^{k_1}{x_2}^{k_2}\cdots{x_n}^{k_n}$ where the $x_i \in X$, $k_i = \pm 1, \pm 2, \ldots$, and $x_i \neq x_{i+1}$ for $1 \leqslant i \leqslant n-1$. Moreover, these products are all $\neq 1$ and two of them are equal only if they look alike. Evidently $\prod_{x\in X}^{*}\langle x \rangle$ contains the subset of elements x^1, which we can identify with X. Now suppose f is a map of X into a group H. Then for each $x \in X$ we have the homomorphism f_x of $\langle x \rangle$ into H sending $x \rightsquigarrow f(x)$. Hence, by the property of $\prod_{x\in X}^{*}\langle x \rangle$ as free product of its subgroups $\langle x \rangle$, we have a unique homomorphism $\bar{f}$ of $\prod_{x\in X}^{*}\langle x \rangle$ into H such that for every x the restriction of $\bar{f}$ to $\langle x \rangle$ is f_x. Evidently, this means that $\bar{f}(x) = f(x)$ and so we have a homomorphism of $\prod_{x\in X}^{*}\langle x \rangle$ into H, which extends the given map f of X into H. It is clear also that there is only one such homomorphism $\bar{f}$. Hence $\prod_{x\in X}^{*}\langle x \rangle$ is a free group determined by the set X. The uniqueness of free groups in the category sense permits us to state the result we have obtained from this construction in the following way:

THEOREM 2.13. *Let X be a non-vacuous set, $(FG(X), i)$ a free group determined by X where i denotes the given map of X into $FG(X)$. Then i is injective, so we may identify X with $i(X)$. If we do this then we can write any element $\neq 1$ of $FG(X)$ in the reduced form ${x_1}^{k_1}{x_2}^{k_2}\cdots{x_n}^{k_n}$ where the $k_i = \pm 1, \pm 2, \ldots$ and consecutive x_i are different. Moreover, any element of the form indicated is $\neq 1$ and if ${y_1}^{l_1}{y_2}^{l_2}\cdots{y_m}^{l_m}$ is a second element in reduced form, then ${x_1}^{k_1}{x_2}^{k_2}\cdots{x_n}^{k_n} = {y_m}^{l_m}$ implies $m = n$, $y_i = x_i$, $k_i = l_i$, $1 \leqslant i \leqslant n$.*

The usefulness of the normal forms of elements of free products and free groups given in Theorems 2.12 and 2.13 will be illustrated in the exercises.

EXERCISES

1. Show that if $G_\alpha \neq 1$, $\alpha \in I$, and $|I| > 1$, then the center of $\prod^{*}G_\alpha$ is 1.
2. Show that the free group $FG(X)$ has no elements of finite order.
3. Let $\prod^{*}G_\alpha$ be as in exercise 1. Determine the elements of finite order in $\prod^{*}G_\alpha$.

The next two exercises are designed to prove that the group of transformations of the form $z \rightsquigarrow \dfrac{az+b}{cz+d}$, $a, b, c, d \in \mathbb{Z}$, $ad-bc = 1$, of the complex plane plus ∞ is a free product of a cyclic group of order two and a cyclic group of order three. We note a fact, easily verified, that this group is isomorphic to $SL_2(\mathbb{Z})/\{1, -1\}$ where $SL_2(\mathbb{Z})$ is the group of 2×2 integral matrices of determinant 1 and $1 = \begin{pmatrix} 1 & 0 \\ 0 & 1 \end{pmatrix}$, $-1 = \begin{pmatrix} -1 & 0 \\ 0 & -1 \end{pmatrix}$.

4. Show that $SL_2(\mathbb{Z})$ is generated by the matrices $\begin{pmatrix} 1 & 1 \\ 0 & 1 \end{pmatrix}$ and $\begin{pmatrix} 1 & 0 \\ 1 & 1 \end{pmatrix}$. Let $S: z \rightsquigarrow -\dfrac{1}{z}$, $T: z \rightsquigarrow \dfrac{-1}{z+1}$. Verify that $S^2 = 1$, $T^3 = 1$, $ST: z \rightsquigarrow z+1$, $ST^2: z \rightsquigarrow \dfrac{z}{z+1}$. Hence prove that the group G of transformations $z \rightsquigarrow \dfrac{az+b}{cz+d}$, $a, b, c, d \in \mathbb{Z}$, $ad-bc = 1$ of $\mathbb{C} \,\dot\cup\, \{\infty\}$ is generated by S and T.

5. Show that all products of the form

$$(ST)^{k_1}(ST^2)^{k_2}(ST)^{k_3}\cdots \quad \text{or} \quad (ST^2)^{k_1}(ST)^{k_2}(ST^2)^{k_3}\cdots$$

with $k_i > 0$ can be written as $z \rightsquigarrow \dfrac{az+b}{cz+d}$ where $a, b, c, d \geqslant 0$ and at most one of these is 0. Hence show that no such product is 1 or S. Use this to prove that G is a free product of a cyclic group of order two and a cyclic group of order three.

6. Let $\mathbb{Z}\{X\}$ be the free ring generated by $X = \{x_1, x_2, \ldots, x_n\}$ as in exercise 6, p. 86, and let B be the ideal in $\mathbb{Z}\{X\}$ generated by the elements $x_1^2, x_2^2, \ldots, x_n^2$. Put $A = \mathbb{Z}\{X\}/B$. Call a word in the x's *standard* if no x_i^2 occurs in it. Show that the cosets of 1 and of the standard words form a $\mathbb{Z}$-base for A. Identify 1 and the standard words with their cosets. Then A has a $\mathbb{Z}$-base consisting of 1 and the standard words with the obvious multiplication, which results in either 1, a standard word, or 0 (e.g., $x_i^2 = 0$). Put $y_i = 1 + x_i$, $1 \leqslant i \leqslant n$. Note that y_i has the inverse $1 - x_i$, so the y_i generate a group under the multiplication in A. Show that this is the free group $FG(Y)$, $Y = \{y_1, y_2, \ldots, y_n\}$.

7. Let $FG(Y)$ be as in exercise 6 and let $FG(Y)_k$, $k = 1, 2, 3, \ldots$, be the subset of $FG(Y)$ of $a \equiv 1 \pmod{(X)^k}$ where X is the ideal in A generated by $x_1, \ldots, x_n$. Show that $FG(Y)_k$ is a normal subgroup of $FG(Y)$ and $\bigcap_{k=1}^{\infty} FG(Y)_k = 1$. If G is any group, define $G^{(k)}$ inductively by $G^{(1)} = G$ and $G^{(r)}$ is the subgroup generated by the commutators $xyx^{-1}y^{-1}$ where $x \in G^{(r-1)}$ and $y \in G$. One has $G^{(1)} \supset G^{(2)} \supset G^{(3)} \supset \cdots$ and this is called the *lower central series* for G. Prove that $F(Y)^{(k)} \subset F(Y)_k$. Hence prove *Magnus' theorem*: $\bigcap FG(Y)^{(k)} = 1$.

8. Let B, A_α, $\alpha \in I$, be algebras in a variety $V(S)$, and f_α a homomorphism of $B \to A_\alpha$, $\alpha \in I$. Show that there exists an *amalgamated sum* (or *pushout*) *of the* f_α in the following sense: an algebra $A \in V(S)$ and homomorphisms $i_\alpha : A_\alpha \to A$, $\alpha \in I$, such that $i_\alpha f_\alpha = i_\beta f_\beta$ for all α, β, and if $C \in V(S)$ and $g_\alpha : A_\alpha \to C$ satisfies $g_\alpha f_\alpha = g_\beta f_\beta$ for all α, β, then there is a unique homomorphism $g : A \to C$ such that $gi_\alpha = g_\alpha$, $\alpha \in I$.

9. Let $\{G_\alpha | \alpha \in I\}$ be a set of groups all containing the same subgroup H. Construct a group G with the following properties: (i) G contains every G_α as a subgroup so

that $G_\alpha \cap G_\beta = H$ if $\alpha \neq \beta$. (ii) If X_α is a set of representatives of the right cosets of H in G_α, $\alpha \in I$, and $X = \bigcup X_\alpha$, then any element of G has a unique representation as a product $hx_1x_2\cdots x_n$ where $h \in H$, $x_i \in X$ and if $x_i \in X_\alpha$ then $x_{i+1} \notin X_\alpha$, $1 \leqslant i \leqslant n-1$. (We allow $n = 0$, in which case it is understood that the element is in H.) G is called the *free product of the G with amalgamated subgroup H* and is denoted as $\prod_H^* G_\alpha$.

2.10 INTERNAL CHARACTERIZATION OF VARIETIES

Let $C = V(S)$ be a variety of Ω-algebras defined by a set of identities S. Here S is a subset of $F(\Omega, x_0) \times F(\Omega, X_0)$ where $X_0 = \{x_1, x_2, x_3, \ldots\}$ and $F(\Omega, X_0)$ is the free Ω-algebra determined by X_0. We have seen that C has the following closure properties:

1. If $A \in C$, then every subalgebra of A is in C.
2. If $A \in C$, then every homomorphic image of A is in C.
3. If $\{A_\alpha | \alpha \in I\} \subset C$, then $\prod A_\alpha \in C$.

Our principal goal in this section is to prove the converse: If C is any class of Ω-algebras satisfying 1, 2, and 3, then C is a variety.

As a preliminary to the proof we consider the identities satisfied by a given Ω-algebra A. For our purposes, we need to consider identities in any set of elements and not just the standard set X_0. Accordingly, let X be any non-vacuous set, and $F(\Omega, X)$ the free Ω-algebra determined by X. If η is a homomorphism of $F(\Omega, X)$ into A, then the kernel $\ker \eta$ is a congruence on $F(\Omega, X)$ and $F(\Omega, X)/\ker \eta$ is isomorphic to a subalgebra of A. Now put $\mathrm{Id}(X, A) = \bigcap \ker \eta$ where the intersection is taken over all the homomorphisms η of $F(\Omega, X)$ into A. Then $\mathrm{Id}(X, A)$ is the set of elements (w_1, w_2), $w_i \in F(\Omega, X)$, such that $\eta(w_1) = \eta(w_2)$ for every homomorphism η of $F(\Omega, X)$ into A. Thus $\mathrm{Id}(X, A)$ is the subset of $F(\Omega, X) \times F(\Omega, X)$ of identities in the set X satisfied by the algebra A. Evidently, $\mathrm{Id}(X, A)$ is a congruence on $F(\Omega, X)$. We call this the *congruence of identities in X of the algebra A*. We note that $F(\Omega, X)/\mathrm{Id}(X, A)$ is a subdirect product of the algebras $F(\Omega, X)/\ker \eta$ (exercise 4, p. 70); hence $F(\Omega, X)/\mathrm{Id}(X, A)$ is a subdirect product of subalgebras of A.

Next let C be any class of Ω-algebras. We wish to consider the set of identities in X satisfied by every $A \in C$. Evidently this set is $\mathrm{Id}(X, C) \equiv \bigcap_{A \in C} \mathrm{Id}(X, A)$. $\mathrm{Id}(X, C)$ is a congruence that we shall call the *congruence of identities in X for the class C*. It is clear that $F(\Omega, X)/\mathrm{Id}(X, C)$ is a subdirect product of algebras that are subalgebras of algebras in the class C.

Now suppose C has the closure properties 1, 2, and 3 above. Then it is clear that any subdirect product of algebras in the class C is an algebra in C. It follows that $F(\Omega, X)/\mathrm{Id}(X, C) \in C$. Now let S be the congruence of identities in X_0 for the class $C: S = \mathrm{Id}(X_0, C)$, X_0 the standard set, and let $V(S)$ be the variety defined by S. We shall show that $C = V(S)$. This will be an immediate consequence of

THEOREM 2.14. *If X is infinite, then $F(\Omega, X)/\mathrm{Id}(X, C)$ together with the canonical map $x \rightsquigarrow \bar{x}_{\mathrm{Id}(X,C)}$ constitutes a free algebra for $V(S)$ determined by X.*

Proof. We recall the construction given in section 2.8 for a free algebra $F_{V(S)}(\Omega, X)$ for the variety $V(S)$ and the set X: $F_{V(S)}(\Omega, X) = F(\Omega, X)/\Phi(S)$ where $\Phi(S)$ is the congruence on $F(\Omega, X)$ generated by all pairs $(\varphi(w_1), \varphi(w_2))$ where $(w_1, w_2) \in S$ and φ is a homomorphism of $F(\Omega, X_0)$ into $F(\Omega, X)$. Our result will follow if we can show that $\Phi(S) = \mathrm{Id}(X, C)$. We have seen that the closure properties of C imply that $F(\Omega, X)/\mathrm{Id}(X, C) \in C$. Hence we have the homomorphism of $F(\Omega, X)/\Phi(S)$ into $F(\Omega, X)/\mathrm{Id}(X, C)$ sending $\bar{x}_{\Phi(S)}$ into $\bar{x}_{\mathrm{Id}(X,C)}$. Consequently, $\Phi(S) \subset \mathrm{Id}(X, C)$. Now let $(z_1, z_2) \in \mathrm{Id}(X, C)$. Then z_1 and z_2 are contained in a subalgebra $F(\Omega, X')$ of $F(\Omega, X)$ generated by a countable subset X' of X. We have maps $\zeta: X_0 \to X$, $\lambda: X \to X_0$ such that $\zeta(X_0) = X'$, $\lambda\zeta = 1_{X_0}$. These have unique extensions to homomorphisms $\bar{\zeta}: F(\Omega, X_0) \to F(\Omega, X)$, $\bar{\lambda}: F(\Omega, X) \to F(\Omega, X_0)$ such that $\bar{\zeta}(F(\Omega, X_0)) = F(\Omega, X')$ and $\bar{\lambda}\bar{\xi} = 1_{F(\Omega, X_0)}$. Choose $w_i \in F(\Omega, X_0)$ so that $\bar{\xi}(w_i) = z_i$, $i = 1, 2$. Let f be a homomorphism of $F(\Omega, X_0)$ into an $A \in C$. Then $f = f1_{F(\Omega, X_0)} = f\bar{\lambda}\bar{\zeta} = \bar{f}\bar{\zeta}$ where $\bar{f} = f\bar{\lambda}$ is a homomorphism of $F(\Omega, X)$ into A. Then $\bar{f}(z_1) = \bar{f}(z_2)$ since $(z_1, z_2) \in \mathrm{Id}(X, C)$ and hence $f(w_1) = \bar{f}\bar{\zeta}(w_1) = \bar{f}(z_1) = \bar{f}(z_2) = f(w_2)$. f *Thu* $f(w_1) = f(w_2)$ for every homomorphism f of $F(\Omega, X_0)$ into an algebra $A \in C$. This means that $(w_1, w_2) \in S = \mathrm{Id}(X_0, C)$. Then $(z_1, z_2) = \bar{\zeta}(w_1), \bar{\zeta}(w_2)) \in \Phi(S)$; hence, $\mathrm{Id}(X, C) \subset \Phi(S)$ and so $\mathrm{Id}(X, C) = \Phi(S)$. □

We can now prove the main result.

THEOREM 2.15 (Birkhoff). *A class C of Ω-algebras is a variety if and only if it has the closure properties* 1, 2, *and* 3 *listed above.*

Proof. It is clear that $C \subset V(S)$ where S is the set of identities in X_0 satisfied by every $A \in C$. On the other hand, if $A \in V(S)$, A is a homomorphic image of an algebra $F(\Omega, X)/\mathrm{Id}(X, C)$ for some infinite set X. Since $F(\Omega, X)/\mathrm{Id}(X, C) \in C$, it follows that $A \in C$. Hence $V(S) \subset C$ and so $V(S) = C$. □

REFERENCES

P. M. Cohn, *Universal Algebra*, Harper and Row, New York, 1965, revised ed., D. Redei, 1981.

G. Grätzer, *Universal Algebra*, Van Nostrand-Reinhold, New York, 1967, 2nd ed., Springer-Verlag, 1979.

3

Modules

In this chapter we resume the study of modules which we initiated in BAI, Chapter 3. Our earlier discussion focused on the structure theory of finitely generated modules over a principal ideal domain and its applications to the structure theory of finitely generated abelian groups and the theory of a single linear transformation in a finite-dimensional vector space. The present chapter does not have such a singleness of purpose or immediacy of objective. It is devoted to the study of modules for their own sake and with a view of applications that occur later.

The theory of modules is of central importance in several areas of algebra, notably structure theory of rings, representation theory of rings and of groups, and homological algebra. These will be developed in the three chapters that follow.

We shall begin our study by noting the special features of the categories R-**mod** and **mod**-R of left and right modules respectively over the ring R. The most important of these is that for any pair of modules (M, N) the hom set $\hom_R(M, N)$ is an abelian group. This has led to the definition of an abelian category. A substantial part of the theory of modules can be absorbed into the more general theory of abelian categories. However, this is not the case for some important parts of module theory. For this reason and for the added gain of concreteness, we shall stick to the case of modules in this chapter.

The topics we shall consider are basic properties of categories R-**mod** and **mod**-R (section 1), structure theory of modules (sections 2–5), tensor products of modules and bimodules (sections 6–8), projective and injective modules (sections 9–10), and Morita theory (sections 11–14). The concepts of algebras and coalgebras over commutative rings will be defined by means of tensor products. In this connection, we shall single out some important examples: tensor algebras, symmetric algebras, and exterior algebras. As an application of the Morita theory, we shall give a proof of the Wedderburn-Artin structure theorem on simple rings. We shall not attempt to spell out in greater detail the topics that will be discussed. The section titles will perhaps be sufficiently indicative of these.

3.1 THE CATEGORIES R-mod AND mod-R

We begin our systematic study of modules by adopting the category point of view, that is, we shall consider first the special features of the categories R-**mod** and **mod**-R of left and right modules for a given ring R. We have seen that we may pass freely from left to right modules and vice versa by changing R to its opposite ring R^{op}. Hence, until we have to consider relations between left and right modules for the same ring R, we may confine our attention either to left or to right modules. Since there is a slight notational advantage in having the ring R and R-homomorphisms act on opposite sides, we shall give preference to right modules and to the category **mod**-R in this chapter.

Perhaps the most important fact about **mod**-R (which we have hitherto ignored) is that the hom sets for this category have a natural abelian group structure. If M and $N \in \text{ob}\,$**mod**-R, then there is a natural way of introducing a group structure in the set $\hom_R(M, N)$. If f and $g \in \hom(M, N)$ $(= \hom_R(M, N))$, then we define $f+g$ by

$$(f+g)x = fx + gx, \qquad x \in M.$$

Here we have abbreviated $f(x)$ to fx, etc. We define the map 0 from M to N by $0x = 0$ and $-f$ by $(-f)x = -fx$, $x \in M$. Direct verification shows that $f+g$, 0, and $-f \in \hom(M, N)$, and $(\hom(M, N), +\cdot 0)$ is an abelian group (BAI, pp. 168–169).

We observe next that the product of morphisms in our category is distributive on both sides with respect to addition. Let P be a third module. Then for $f, g \in \hom(M, N)$ and $h, k \in \hom(N, P)$ we have

$$h(f+g) = hf + hg, \qquad (h+k)f = hf + kf.$$

In particular, if $M = N = P$, then we have the addition $+$ and the multiplication, the composite of maps, in End M = hom (M, M). Moreover, the identity $1_M \in \text{End}\, M$ and this acts as unit with respect to multiplication. Thus (End M, $+, \cdot, 0, 1$) is a ring, the *ring of endomorphisms of the module M*.

An important module concept (which originated in algebraic topology) is that of an exact sequence of modules and homomorphisms. To begin with, we call a diagram of module homomorphisms

$$M \xrightarrow{f} N \xrightarrow{g} P$$

exact if $\operatorname{im} f\,(=f(M)) = \ker g$, that is, $gy = 0$ for $y \in N$ if and only if there exists an $x \in M$ such that $fx = y$. More generally, a sequence of modules and homomorphisms

$$\cdots \to M_1 \xrightarrow{f_1} M_2 \xrightarrow{f_2} M_3 \to \cdots$$

that may be finite or run to infinity in either direction is called *exact* if for any three consecutive terms the subsequence $M_i \to M_{i+1} \to M_{i+2}$ is exact. The exactness of

$$0 \to M \xrightarrow{f} N$$

means that $\ker f = 0$, which is equivalent to: f is injective. It is customary to write $M \overset{f}{\rightarrowtail} N$ for "$0 \to M \xrightarrow{f} N$ is exact." Similarly "$M \xrightarrow{f} N \to 0$ is exact" is equivalent to: f is surjective. This can also be indicated by $M \overset{f}{\twoheadrightarrow} N$.

An exact sequence of the form

$$0 \to M' \xrightarrow{f} M \xrightarrow{g} M'' \to 0$$

is called a *short exact sequence*. This means that f is a monomorphism, g is an epimorphism, and $\ker g = \operatorname{im} f$. A special case of this is obtained by taking a submodule N of a module M and the quotient M/N. Then we have the injection i of N into M and the canonical homomorphism v of M into M/N. The first is injective, the second surjective, and $\ker v = N = \operatorname{im} i$. Hence we have the short exact sequence

$$0 \to N \overset{i}{\hookrightarrow} M \xrightarrow{v} M/N \to 0.$$

If $f: M \to N$, we define the *cokernel*, coker f, as $N/\operatorname{im} f$. Evidently, f is surjective if and only if $\operatorname{coker} f = 0$. In any case we have

$$\ker f \overset{i}{\rightarrowtail} M \xrightarrow{f} N \twoheadrightarrow \operatorname{coker} f,$$

that is, $0 \to \ker f \hookrightarrow M \xrightarrow{f} N \to \operatorname{coker} f \to 0$ is exact.

We have seen in Chapter 1, pp. 35–36, that products and coproducts exist for arbitrary indexed sets of modules. If $\{M_\alpha|\alpha\in I\}$ is an indexed set of right R-modules, then the product in **mod**-R is $\{\prod M_\alpha, p_\alpha\}$ where p_α is the projection homomorphism of $\prod M_\alpha$ onto M_α. Moreover, if $\oplus M_\alpha$ is the submodule of $\prod M_\alpha$ of elements (x_α) having only a finite number of the $x_\alpha \neq 0$, and if i_α is the homomorphism of M_α into $\oplus M_\alpha$ sending the element $x_\alpha \in M_\alpha$ into the element of $\oplus M_\alpha$ whose value at α is x_α and whose remaining components are 0, then $\{\oplus M_\alpha, i_\alpha\} = \coprod M_\alpha$, the coproduct of the M_α.

Now let I be finite: $I = \{1, 2, \ldots, n\}$. Then clearly $M \equiv \coprod_1^n M_i = \oplus_1^n M_i$. Hence this module and the maps $i_1, \ldots, i_n$ constitute a coproduct of the M_i, and M and the maps $p_1, \ldots, p_n$ constitute a product of the M_i. It is immediate from the definition of the i_j and the p_j that we have the following relations on these homomorphisms:

$$p_j i_j = 1_{M_j}, \qquad p_k i_j = 0 \quad \text{if} \quad j \neq k$$
$$i_1 p_1 + i_2 p_2 + \cdots + i_n p_n = 1_M. \tag{1}$$

An important observation is that these conditions on a set of homomorphisms characterize the module M up to isomorphism. Precisely, suppose we have a module M' and homomorphisms $p'_j : M' \to M_j$ and $i'_j : M_j \to M'$, $1 \leqslant j \leqslant n$, such that

$$p'_j i'_j = 1_{M_j}, \qquad p'_k i'_j = 0 \quad \text{if} \quad j \neq k$$
$$\sum_1^n i'_j p'_j = 1_{M'}. \tag{2}$$

Then

$$\theta' = \sum_1^n i_j p'_j, \qquad \theta = \sum_1^n i'_j p_j \tag{3}$$

are homomorphisms $M' \to M$ and $M \to M'$ respectively. By (1) and (2) we have $\theta'\theta = 1_M$, $\theta\theta' = 1_{M'}$ so θ and θ' are isomorphisms and $\theta' = \theta^{-1}$. Moreover, $\theta i_j = i'_j : M_j \to M'$. It follows from the definition of a coproduct that M' and the i'_j constitute a coproduct of the M_j. Similarly, $p_j\theta' = p'_j$, and M' and the p'_j constitute a product of the M_j in **mod**-R.

In dealing with functors between categories of modules (for possibly different rings) we are primarily interested in the functors that are *additive* in the sense that for any pair of modules (M, N) the map $f \to F(f)$ of hom(M, N) into hom(FM, FN) is a group homomorphism. From now on, unless the contrary is stated explicitly, *functors between categories of modules will always be assumed to be additive.*

The foregoing characterization of the product and coproduct of $M_1, M_2, \ldots, M_n$ by the relations (1) on the i_j and p_j implies that any functor F from a category **mod**-R to a category **mod**-S respects finite products and coproducts. Let M and the i_j and p_j be as before. Then if we apply F to the relations (1) and use the multiplicative and additive properties of F, we obtain

$$F(p_j)F(i_j) = 1_{FM_j}, \qquad F(p_k)F(i_j) = F(0) = 0, \qquad j \neq k,$$
$$\sum F(i_j)F(p_j) = 1_{FM}.$$

Hence $\{FM, F(i_j)\} = \coprod FM_j$ and $\{FM, F(p_j)\} = \prod FM_j$.

A functor F from **mod**-R to **mod**-S is called *exact* if it maps short exact sequences into short exact sequences; that is, if $0 \to M' \xrightarrow{f} M \xrightarrow{g} M'' \to 0$ is exact in **mod**-R, then $0 \to FM' \xrightarrow{F(f)} FM \xrightarrow{F(g)} FM'' \to 0$ is exact in **mod**-S. This happens rarely. More common are functors that are *left exact* or *right exact*, where the former is defined by the condition that if $0 \to M' \xrightarrow{f} M \xrightarrow{g} M''$ is exact, then $0 \to FM' \xrightarrow{F(f)} FM \xrightarrow{F(g)} FM''$ is exact and the latter means that if $M' \xrightarrow{f} M \xrightarrow{g} M'' \to 0$ is exact, then $FM' \xrightarrow{F(f)} FM \xrightarrow{F(g)} FM'' \to 0$ is exact. Similar definitions apply to contravariant functors between categories of modules. These are assumed to be additive in the sense that for every (M, N) the map of hom (M, N) into hom (FN, FM) is a group homomorphism. Then F is *left exact* if the exactness of $M' \xrightarrow{f} M \xrightarrow{g} M'' \to 0$ implies that of $0 \to FM'' \xrightarrow{F(g)} FM \xrightarrow{F(f)} FM'$ and F is *right exact* if the exactness of $0 \to M' \xrightarrow{f} M \xrightarrow{g} M''$ implies that of $FM'' \xrightarrow{F(g)} FM \xrightarrow{F(f)} FM' \to 0$.

We have defined the covariant and contravariant hom functors from an arbitrary category **C** to **Set** (p. 38). If **C** = **mod**-R, it is more natural to regard these as functors to **mod**-$\mathbb{Z}$. We recall that if $A \in \text{ob}\,\mathbf{C}$, then the (covariant) hom functor hom $(A, -)$ from **C** to **Set** is defined as the map $B \rightsquigarrow \text{hom}\,(A, B)$ on the objects, and hom $(A, -)\,(f)$ for $f: B \to B'$ is the map of hom (A, B) into hom (A, B'), which multiplies the elements of the former on the left by f to obtain the corresponding elements in the latter. Now let **C** = **mod**-R and let $M, N \in \text{ob}\,\mathbf{mod}\text{-}R$. Then hom (M, N) is an abelian group, hence a $\mathbb{Z}$-module and if $f: N \to N'$ and $g, h \in \text{hom}\,(M, N)$, then $f(g+h) = fg + fh$. Thus hom $(M, -)\,(f)$ is a homomorphism of the $\mathbb{Z}$-module hom (M, N) into the $\mathbb{Z}$-module hom (M, N'). Moreover, if f' is a second element of hom (N, N'), then $(f+f')g = fg + f'g$; so $\text{hom}\,(M, -)(f+f') = \text{hom}\,(M, -)(f) + \text{hom}\,(M, -)(f')$. Thus hom $(M, -)$ may be regarded as an (additive) functor from **mod**-R to **mod**-$\mathbb{Z}$. This will be our point of view from now on.

A basic property of the hom functors is left exactness:

THEOREM 3.1. *The hom functor* hom $(M, -)$ *from* **mod**-R *to* **mod**-$\mathbb{Z}$ *is left exact.*

Proof. To prove this we have to show that if $0 \to N' \xrightarrow{f} N \xrightarrow{g} N''$ is exact, then

$$(4) \qquad 0 \to \hom(M, N') \xrightarrow{\hom(M,-)(f)} \hom(M, N) \xrightarrow{\hom(M,-)(g)} \hom(M, N'')$$

is exact. We are given that f is a monomorphism and $\operatorname{im} f = \ker g$ and we have to show that the same thing holds for $\hom(M, -)(f)$ and $\hom(M, -)(g)$. Suppose $\varphi \in \hom(M, N')$ and $(\hom(M, -)(f))(\varphi) = f\varphi = 0$. Then if $x \in M$, $f\varphi x = 0$ and since f is injective, this implies that $\varphi x = 0$. Thus $\hom(M, -)(f)(\varphi) = 0$ implies that $\varphi = 0$. Since $\hom(M, -)(f)$ is a group homomorphism, this implies that this is a monomorphism. Next we note that $gf\varphi = 0$ since $gf = 0$, so $gf\varphi(x) = 0$ for all $x \in M$. Thus $\hom(M, -)(g)\hom(M, -)(f)(\varphi) = 0$ for all $\varphi: M \to N'$, so we have $\hom(M, -)(g)\hom(M, -)(f) = 0$. Then

$$\operatorname{im} \hom(M, -)(f) \subset \ker \hom(M, -)(g).$$

To prove the reverse inclusion let $\psi \in \ker \hom(M, -)(g)$, so $\psi \in \hom(M, N)$ and $g\psi = 0$ or $g\psi x = 0$ for all $x \in M$. Then if $x \in M$, $g\psi x = 0$ shows that ψx, which is an element of N, is in $\ker g$. Then the hypothesis implies that there is a $y \in N'$ such that $fy = \psi x$. Moreover, since f is a monomorphism, y is uniquely determined. We now have a map $\varphi: x \rightsquigarrow y$ of M into N'. It follows directly from the definition that $\varphi \in \hom(M, N')$ and $f\varphi = \psi$. Thus $\ker \hom(M, -)(g) \subset \operatorname{im} \hom(M, -)(f)$ and hence we have the equality of these two sets. This completes the proof. □

In a similar manner, the contravariant hom functor $\hom(-, M)$ determined by $M \in \operatorname{ob}$ **mod**-R can be regarded as a contravariant functor from **mod**-R to **mod**-$\mathbb{Z}$. Here $\hom(-, M)N = \hom(N, M)$ and if $f: N \to N'$, then $\hom(-, M)(f)$ is the right multiplication of the elements of $\hom(N', M)$ by f to yield elements of $\hom(N, M)$. We have

THEOREM 3.1′. *The contravariant hom functor* $\hom(-, M)$ *is left exact.*

We leave the proof to the reader.

EXERCISES

1. Regard R as right R-module in the usual way (the action of R on R is the right multiplication as given in the ring R). Show that the right module M and $\hom_R(R, M)$ are isomorphic as abelian groups. More precisely, verify that for any $x \in M$, the map $\mu_x: a \rightsquigarrow xa$, $a \in R$, is in $\hom_R(R, M)$ and $x \rightsquigarrow \mu_x$ is an isomorphism of M as abelian group onto $\hom_R(R, M)$. Use this to prove that the hom functor

hom $(R, -)$ from **mod**-R to **mod**-$\mathbb{Z}$ and the forgetful functor from **mod**-R to **mod**-$\mathbb{Z}$, which maps a module into its underlying abelian group, etc. are naturally isomorphic.

2. Let $M = \mathbb{Z}$, $N = \mathbb{Z}/(m)$, $m > 1$, and let v be the canonical homomorphism of M *onto* N. Show that 1_N cannot be written as vf for any homomorphism $f: N \to M$. Hence show that the image of the exact sequence $M \to N \to 0$ under hom $(N, -)$ is not exact.

3. Let $M = \mathbb{Z}$, $N = m\mathbb{Z}$, $m > 1$, and let i be the injection of N into M. Show that the homomorphism $1/m : mx \rightsquigarrow x$ of N into M cannot be written as fi for any $f: M \to M$. Hence show that the image of the exact sequence $0 \to N \to M$ under hom $(-, M)$ is not exact.

4. (The "short five lemma.") Assume that

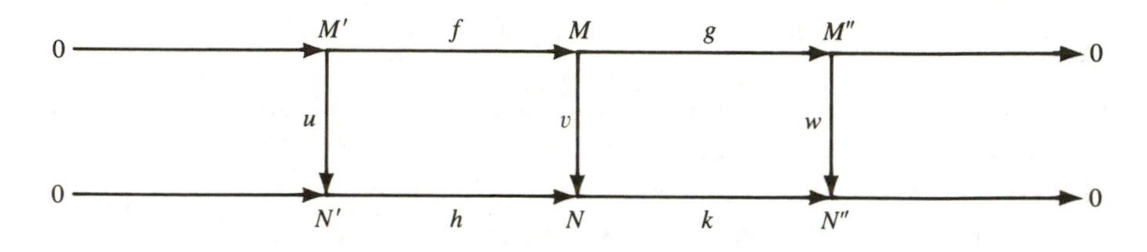

has exact rows and is commutative. Show that if any two of u, v, and w are isomorphisms, then so is the third.

3.2 ARTINIAN AND NOETHERIAN MODULES

After the aerial view of module theory that we experienced in the previous section, we shall now come down to earth to study modules as individuals. We shall begin with some aspects of the structure theory of modules. Categorical ideas will play only a minor role in the next six sections; they will come to the fore again in section 3.9 and thereafter.

We shall first collect some tool results on modules, which are special cases of results we proved for Ω-algebras (pp. 63–65), namely, the correspondence between the set of subalgebras of a quotient of an Ω-algebra with the set of saturated subalgebras of the algebra, and the two isomorphism theorems. To obtain the corresponding results for modules one can regard these as groups with operators, the set of operators being the given ring. The results then become the following results on modules: If M is a module and P is a submodule, we have a bijection of the set of submodules of the quotient $\bar{M} = M/P$ and the set of submodules of M containing P. If N is a submodule of M containing P, the corresponding submodule of $\bar{M} = M/P$ is $\bar{N} = N/P = \{y + P \mid y \in N\}$ and if $\bar{N}$ is a submodule of $\bar{M}$, then we put $N = \{y \in M \mid y + P \in \bar{N}\}$. This is a submodule of M containing P and $\bar{N} = N/P$.

This result can also be verified directly as in the case of groups (BAI, p. 64).

We have the following special cases of the isomorphism theorems for Ω-algebras.

FIRST ISOMORPHISM THEOREM FOR MODULES. *If N_1 and N_2 are submodules of M, then*

$$(5) \qquad (y_1 + y_2) + N_2 \quad y_1 + (N_1 \cap N_2), \qquad y_i \in N_i$$

is an isomorphism of $(N_1 + N_2)/N_2$ onto $N_1/(N_1 \cap N_2)$.

SECOND ISOMORPHISM THEOREM FOR MODULES. *If M is a module, N and P submodules such that $N \supset P$, then*

$$(6) \qquad (x + P) + N/P \quad x + N, \qquad x \in M$$

is an isomorphism of $(M/P)/(N/P)$ onto M/N.

These two results can also be established directly as with groups (BAI, pp. 64–65). In fact, the group results extend immediately to groups with operators and the case of modules is a special case of these. The fact that the results are valid for groups with operators will be needed in the next section.

A module M is called *noetherian* (*artinian*) if it satisfies the

Ascending (*descending*) *chain condition.* There exists no infinite properly ascending (descending) sequence of modules $M_1 \quad M_2 \quad M_3 \quad \cdots$) in M.

Another way of putting this is that if we have an ascending sequence of submodules $M_1 \subset M_2 \subset M_3 \subset \cdots$, then there exists an n such that $M_n = M_{n+1} = \cdots$. One has a similar formulation for the descending chain condition. It is easily seen that the foregoing condition is equivalent to the

Maximum (*minimum*) *condition.* Every non-vacuous set of S of submodules of M contains a maximal (minimal) submodule in the set, that is, a module $P \in S$ such that if $N \in S$ and $N \supset P$ ($N \subset P$), then $N = P$.

EXAMPLES

1. The ring $\mathbb{Z}$ regarded as $\mathbb{Z}$-module satisfies the ascending but not the descending chain condition. Recall that every ideal in $\mathbb{Z}$ is a principal ideal (BAI, p. 143) and the submodules of $\mathbb{Z}$ are ideals. If $I_1 \subset I_2 \subset \cdots$ is an ascending sequence of ideals in $\mathbb{Z}$, then $I = \bigcup I_j$ is an ideal. This is principal: $I = (m)$ where $m \in I$, so $m \in I_n$ for some n. Then $I_n = I_{n+1} = \cdots$. (The same argument shows that the ascending chain condition holds for every commutative principal ideal domain (p.i.d.), D regarded as a D-module. This

is proved in BAI, p. 147.) On the other hand, if $m \in \mathbb{Z}$ and $m \neq 0, \pm 1$, then $(m) \supsetneqq (m^2) \supsetneqq (m^3) \supsetneqq \cdots$ so the descending chain condition fails in $\mathbb{Z}$.

2. Let P be the additive group of rationals whose denominators are powers of a fixed prime $p: m/p^k$, $m \in \mathbb{Z}$, $k = 0, 1, 2, \ldots$. We regard P as $\mathbb{Z}$-module. We have the ascending chain of submodules

$$\mathbb{Z} \subsetneqq \mathbb{Z}(1/p) \subsetneqq \mathbb{Z}(1/p^2) \subsetneqq \mathbb{Z}(1/p^3) \subsetneqq \cdots.$$

It is easily seen that the submodules of this chain are the only submodules of P containing $\mathbb{Z}$. It follows that $M = P/\mathbb{Z}$ is artinian but not noetherian. On the other hand, since $\mathbb{Z}$ is a submodule of P, it is clear that P is neither artinian nor noetherian.

3. Any finite abelian group is both artinian and noetherian as $\mathbb{Z}$-module.

It is clear that if M is artinian or noetherian, then so is every submodule of M. The same is true of every quotient M/N, hence of every homomorphic image, since the submodule correspondence $P \rightsquigarrow \bar{P} = P/N$ has the property that $P_1 \supset P_2$ if and only if $\bar{P}_1 \supset \bar{P}_2$. We shall now show that conversely if M contains a submodule N such that N and M/N are artinian (noetherian), then M is artinian (noetherian). For this we need the

LEMMA. *Let N, P_1, P_2 be submodules of M such that*

$$P_1 \supset P_2, \qquad N + P_1 = N + P_2, \qquad N \cap P_1 = N \cap P_2. \tag{7}$$

Then $P_1 = P_2$.

Proof. Let $z_1 \in P_1$. Then $z_1 \in N + P_1 = N + P_2$, so $z_1 = y + z_2$, $y \in N$, $z_2 \in P_2$. Then $y = z_1 - z_2 \in P_1$, so $y \in P_1 \cap N = P_2 \cap N$. Thus $y \in P_2$ and so $z_1 = y + z_2 \in P_2$. Hence $P_1 \subset P_2$ and $P_1 = P_2$. □

THEOREM 3.2. *If M contains a submodule N such that N and M/N are artinian (noetherian), then M is artinian (noetherian).*

Proof. Let $P_1 \supset P_2 \supset P_3 \supset \cdots$ be a descending chain of submodules of M. Then $(N \cap P_1) \supset (N \cap P_2) \supset (N \cap P_3) \supset \cdots$ is a descending chain of submodules of M. Hence there exists a k such that $N \cap P_k = N \cap P_{k+1} = \cdots$. Also $(N + P_1)/N \supset (N + P_2)/N \supset \cdots$ is a descending chain of submodules of M/N. Hence there exists an l such that $(N + P_l)/N = (N + P_{l+1})/N = \cdots$. Then $N + P_l = N + P_{l+1} = \cdots$. Taking $n = \max(k, l)$ and applying the lemma, we conclude that $P_n = P_{n+1} = \cdots$. The proof in the noetherian case is obtained by replacing $\supset$ by $\subset$ everywhere. □

THEOREM 3.3. *Let $M = N + P$ where N and P are artinian (noetherian) submodules of M. Then M is artinian (noetherian).*

Proof. We have $M/N = (N+P)/N \cong P/(N \cap P)$, which is a homomorphic image of the artinian (noetherian) module P. Hence M/N and N are artinian (noetherian). Then M is artinian (noetherian) by Theorem 3.2. □

A ring R is called *right* (*left*) *noetherian* or *artinian* if R as right (left) R-module is noetherian or artinian respectively. A module M is called *finitely generated* if it is generated by a finite number of elements. In the case of right modules this means that M contains elements $x_1, x_2, \ldots, x_m$ such that $M = x_1R + x_2R + \cdots + x_mR$.

THEOREM 3.4. *If R is right noetherian (artinian), then every finitely generated right R-module M is noetherian (artinian).*

Proof. We have $M = x_1R + \cdots + x_mR$ and the epimorphism $a \rightsquigarrow x_ia$ of R onto x_iR, $1 \leqslant i \leqslant m$. Hence x_iR is noetherian (artinian). Hence M is noetherian (artinian) by Theorem 3.3 and indication. □

If R is a division ring, the only left or right ideals of R are 0 and R. Hence R is both left and right artinian and noetherian. Hence we have the

COROLLARY. *Let M be a vector space over a division ring. Assume M is finitely generated. Then M is artinian and noetherian.*

EXERCISES

1. Prove that M is noetherian if and only if every submodule of M is finitely generated.

2. Show that if an endomorphism of a noetherian (artinian) module is an epimorphism (monomorphism), then it is an isomorphism.

3. Let V be a vector space over a field F, T a linear transformation in V over F, and let $F[\lambda]$ be the polynomial ring over F in an indeterminate λ. Then V becomes an $F[\lambda]$-module if we define

$$(a_0 + a_1\lambda + a_2\lambda^2 + \cdots + a_m\lambda^m)x = a_0x + a_1(Tx) + a_2(T^2x) + \cdots + a_m(T^mx)$$

(BAI, p. 165). Let V have the countable base $(x_1, x_2, \ldots)$ and let T be the linear transformation such that $Tx_1 = 0$, $Tx_{i+1} = x_i$, $i = 1, 2, 3, \ldots$. Show that V as $F[\lambda]$-module defined by T is artinian. Let T' be the linear transformation in V such that $T'x_i = x_{i+1}$, $i = 1, 2, \ldots$. Show that V as $F[\lambda]$-module defined by T' is noetherian.

3.3 SCHREIER REFINEMENT THEOREM. JORDAN-HÖLDER THEOREM

The results that we will derive next are important also for non-commutative groups. In order to encompass both the cases of groups and modules we consider groups with operators. We remark that the results we shall obtain can also be generalized to Ω-algebras—we refer the reader to Cohn's *Universal Algebra*, pp. 91–93, for these (see References at the end of Chapter 2).

We now consider the class of groups having a fixed operator set Λ; that is, Λ is a fixed set of unary operator symbols: for each $\lambda \in \Lambda$ and each G in our class, $x \rightsquigarrow \lambda x$ is a map of G into itself. Moreover, we require that

$$\lambda(xy) = (\lambda x)(\lambda y). \tag{8}$$

Evidently left modules are a special case of this in which $G = M$, a commutative group written additively, and $\Lambda = R$, a given ring. We observe that even in this case the shift from the module to the group with operator point of view amounts to a generalization, since it means that we drop the module axioms involving the ring structure. We shall refer to groups with the operator set Λ as Λ-*groups*. We also speak of Λ-*subgroups* and *normal* Λ-*subgroups* meaning subgroups and normal subgroups closed under the action of Λ, that is, if h is in the subgroup and $\lambda \in \Lambda$, then λh is in the subgroup. We write $G \rhd H$ or $H \lhd G$ to indicate that H is a normal Λ-subgroup of G. A *homomorphism* f of a Λ-group G into a Λ-group H is a group homomorphism of G into H such that

$$f(\lambda x) = \lambda f(x) \tag{9}$$

for every $x \in G$, $\lambda \in \Lambda$. The image $f(G)$ is a Λ-subgroup of H and the kernel $f^{-1}(1) = \{k \in G \mid f(k) = 1\}$ is a normal Λ-subgroup. If $H \lhd G$ we define the Λ-*factor group* G/H whose elements are the cosets $xH = Hx$ with the usual group structure and with the action of Λ defined by $\lambda(xH) = (\lambda x)H$, $\lambda \in \Lambda$. The fact that H is a Λ-subgroup assures that $\lambda(xH)$ is well defined. Moreover, the defining property (8) for G/H is an immediate consequence of this property for G.

We recall also that if A and B are subsets of a group G, then $AB = \{ab \mid a \in A, b \in B\}$. If $K \lhd G$ and H is a subgroup, then $HK = KH$ since $HK = \bigcup_{h \in H} hK = \bigcup_{h \in H} Kh = KH$. This is a subgroup of G and is a Λ-subgroup if G is a Λ-group and H and K are Λ-subgroups. In this case $(H \cap K) \lhd H$ and we have the isomorphism of Λ-groups of $H/H \cap K$ onto HK/K given by

$$h(H \cap K) \rightsquigarrow hK, \qquad h \in H. \tag{10}$$

The proof for groups without operators carries over without change (BAI, p. 65). In a similar manner the other basic results such as those given in BAI, pp. 64–65, carry over. For convenience we collect here the results we shall need.

We have the bijective map $H \rightsquigarrow H/K$ of the set of Λ-subgroups of G containing $K \lhd G$ with the set of Λ-subgroups of G/K. Moreover, $H \lhd G$ if and only if $H/K \lhd G/K$. In this case we have the isomorphism of $(G/K)/(H/K)$ with G/H given by

$$(gK)H/K \rightsquigarrow gH, \qquad g \in G. \tag{11}$$

More generally, suppose we have an epimorphism f of the Λ-group G onto the Λ-group G' and H is a normal Λ-subgroup of G containing the kernel K of f. Then we have the isomorphism of Λ-groups of G/H onto $G'/f(H) = f(G)/f(H)$ given by

$$gH \rightsquigarrow f(g)f(H). \tag{12}$$

The concept of Λ-group achieves more than just an amalgamation of the group and the module cases. It also provides a method for dealing simultaneously with several interesting classes of subgroups of a group. Given a group G we let Λ be the set of inner automorphisms of G. Then the Λ-subgroups of G are the normal subgroups and the study of G as Λ-group focuses on these, discarding the subgroups that are not normal. Similarly, if we take Λ to be all automorphisms of G, then the Λ-subgroups are the subgroups that are mapped into themselves by all automorphisms. These are called *characteristic subgroups*. If we take Λ to be all endomorphisms of G, we obtain the *fully invariant subgroups*, that is, the subgroups mapped into themselves by every endomorphism of G.

We define a *normal series* for the Λ-group G to be a descending chain of Λ-subgroups

$$G = G_1 \rhd G_2 \rhd G_3 \rhd \cdots \rhd G_{s+1} = 1, \tag{13}$$

that is, every G_{i+1} is a Λ-subgroup normal in G_i (although not necessarily normal in G). The corresponding sequence of Λ-factor groups

$$G_1/G_2, G_2/G_3, \ldots, G_s/G_{s+1} = G_s \tag{14}$$

is called the *sequence of factors* of the normal series (13). The integer s is called the *length* of (13). The normal series

$$G = H_1 \rhd H_2 \rhd H_3 \rhd \cdots \rhd H_{t+1} = 1 \tag{15}$$

is *equivalent* to (13) if $t = s$ and there exists a permutation $i \rightsquigarrow i'$ of $\{1, 2, \ldots, s\}$ such that $G_i/G_{i+1} \cong H_{i'}/H_{i'+1}$ (isomorphism as Λ-groups). The normal series

(15) is a *refinement* of (13) if the sequence of G_i is a subsequence of the H_j. We wish to prove the

SCHREIER REFINEMENT THEOREM. *Any two normal series for a Λ-group have equivalent refinements.*

The proof of the theorem will be based on

ZASSENHAUS' LEMMA. *Let G_i, G_i', $i = 1, 2$, be Λ-subgroups of a group G such that $G_i' \lhd G_i$. Then*

$$(G_1 \cap G_2')G_1' \lhd (G_1 \cap G_2)G_1' \tag{16}$$

$$(G_1' \cap G_2)G_2' \lhd (G_1 \cap G_2)G_2' \tag{17}$$

and

$$\frac{(G_1 \cap G_2)G_1'}{(G_1 \cap G_2')G_1'} \cong \frac{(G_1 \cap G_2)G_2'}{(G_1' \cap G_2)G_2'}. \tag{18}$$

Proof. We shall prove (16),

$$(G_1' \cap G_2)(G_1 \cap G_2') = (G_1 \cap G_2')(G_1' \cap G_2) \lhd (G_1 \cap G_2) \tag{19}$$

and

$$\frac{(G_1 \cap G_2)G_1'}{(G_1 \cap G_2')G_1'} \cong \frac{G_1 \cap G_2}{(G_1 \cap G_2')(G_1' \cap G_2)} = \frac{G_1 \cap G_2}{(G_1' \cap G_2)(G_1 \cap G_2')}. \tag{20}$$

A diagram that will help visualize this is the following:

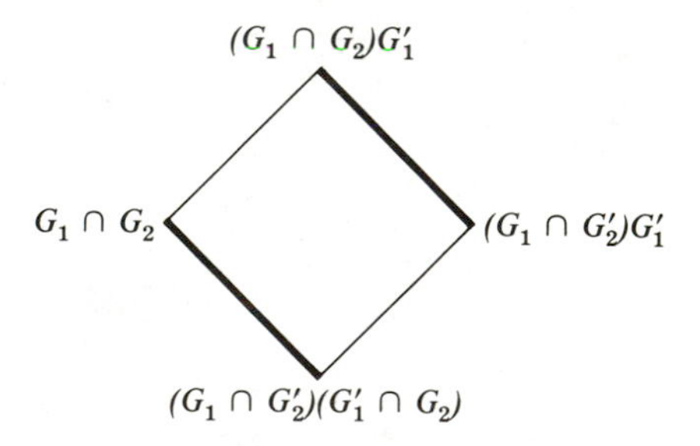

where the factor groups represented by the heavy lines are the ones whose isomorphism we wish to establish. Once we have proved the indicated results, then, by symmetry, we shall have (17) and

$$(G_1 \cap G_2)G_2'/(G_1' \cap G_2)G_2' \cong (G_1 \cap G_2)/(G_1 \cap G_2')(G_1' \cap G_2),$$

which together with (20) will give the desired isomorphism (18).

Now consider the Λ-subgroups $G_1 \cap G_2$ and G_1' of G_1. Since $G_1' \lhd G_1$, we have $G_1' \cap G_2 = G_1' \cap (G_1 \cap G_2) \lhd G_1 \cap G_2$, $(G_1 \cap G_2)G_1'$ is a Λ-subgroup, and $G_1' \lhd (G_1 \cap G_2)G_1'$. Similarly, $G_1 \cap G_2' \lhd G_1 \cap G_2$. Then we have (19). Now form the Λ-factor group $(G_1 \cap G_2)G_1'/G_1'$ and consider the homomorphism

$$f: x \to xG_1'$$

of $G_1 \cap G_2$ into $(G_1 \cap G_2)G_1'/G_1'$. This is an epimorphism and

$$\begin{aligned} f((G_1 \cap G_2')(G_1' \cap G_2)) &= (G_1 \cap G_2')(G_1' \cap G_2)G_1'/G_1' \\ &= (G_1 \cap G_2')G_1'/G_1'. \end{aligned}$$

Since $(G_1 \cap G_2')(G_1' \cap G_2)$ is normal in $G_1 \cap G_2$ and f is an epimorphism,

$$(G_1 \cap G_2')G_1'/G_1' \lhd (G_1 \cap G_2)G_1'/G_1'.$$

Then $(G_1 \cap G_2')G_1' \lhd (G_1 \cap G_2)G_1'$, which is (16). Moreover,

$$\begin{aligned} (G_1 \cap G_2)/(G_1 \cap G_2')(G_1' \cap G_2) &\cong f(G_1 \cap G_2)/f((G_1 \cap G_2')(G_1' \cap G_2)) \\ &= \frac{(G_1 \cap G_2)G_1'}{G_1'} \quad \frac{(G_1 \cap G_2')G_1'}{G_1'} \cong \frac{(G_1 \cap G_2)G_1'}{(G_1 \cap G_2')G_1'} \end{aligned}$$

which gives (20). $\square$

We can now give the

Proof of the Schreier refinement theorem. Suppose the given normal series are (13) and (15). Put

$$G_{ik} = (G_i \cap H_k)G_{i+1}, \qquad H_{ki} = (H_k \cap G_i)H_{k+1}. \tag{21}$$

Then $G_{i1} = G_i$, $G_{i,t+1} = G_{i+1}$, $H_{k1} = H_k$, $H_{k,s+1} = H_{k+1}$. By Zassenhaus' lemma, $G_{i,k+1} \lhd G_{i,k}$, $H_{k,i+1} \lhd H_{ki}$ and

$$G_{ik}/G_{i,k+1} \cong H_{ki}/H_{k,i+1}. \tag{22}$$

We have the normal series

$$\begin{aligned} (G=)\quad G_1 &\supset G_{12} \supset \cdots \supset G_{1t} \\ &\supset G_2 \supset G_{22} \supset \cdots \supset G_{2t} \\ &\quad . \quad . \quad . \quad \cdots \quad . \\ &\supset G_s \supset G_{s2} \supset \cdots G_{st} \supset G_{s,t+1} \quad (=1) \end{aligned} \tag{23}$$

and

$$
(24)\qquad
\begin{aligned}
(G=)\quad & H_1 \supset H_{12} \supset \cdots \supset H_{1s} \\
& \supset H_2 \supset H_{22} \supset \cdots \supset H_{2s} \\
& \quad . \quad . \quad . \ \cdots \ . \\
& \supset H_t \supset H_{t2} \supset \cdots \supset H_{ts} \supset H_{t,s+1} \quad (=1).
\end{aligned}
$$

The corresponding factors are

$$
(25)\qquad
\begin{aligned}
& G_{11}/G_{12}, G_{12}/G_{13}, \ldots, G_{1t}/G_{1,t+1}, \\
& G_{21}/G_{22}, G_{22}/G_{23}, \ldots, G_{2t}/G_{2,t+1}, \\
& \quad . \qquad . \quad \cdots \quad . \\
& G_{s1}/G_{s2}, G_{s2}/G_{s3}, \ldots, G_{st}/G_{s,t+1}
\end{aligned}
$$

$$
(26)\qquad
\begin{aligned}
& H_{11}/H_{12}, H_{12}/H_{13}, \ldots, H_{1s}/H_{1,s+1}, \\
& H_{21}/H_{22}, H_{22}/H_{23}, \ldots, H_{2s}/H_{2,s+1}, \\
& \quad . \qquad . \quad \cdots \quad . \\
& H_{t1}/H_{t2}, H_{t2}/H_{t3}, \ldots, H_{ts}/H_{t,s+1}.
\end{aligned}
$$

The factor in the ith row and kth column of (25) is isomorphic as Λ-group to the one in the kth row and ith column of (26). Hence (23) and (24) are equivalent. Clearly (23) is a refinement of (13) and (24) is a refinement of (15), so the theorem is proved. □

A normal series (13) will be called *proper* if every inclusion is proper: $G_i \supsetneqq G_{i+1}$. A normal series is a *composition* series if it is proper and it has no proper refinement. This means that we have $G_i \supsetneqq G_{i+1}$ and there is no G' such that $G_i \supsetneqq G' \supsetneqq G_{i+1}$ where $G' \lhd G_i$ and $G_{i+1} \lhd G'$. Equivalently, G_i/G_{i+1} is *simple* as Λ-group in the sense that it is $\neq 1$ and it contains no proper normal Λ-subgroup $\neq 1$. A Λ-group need not have a composition series (for example, an infinite cyclic group has none). However, if it does then we have the

JORDAN-HÖLDER THEOREM. *Any two composition series for a Λ-group are equivalent.*

Proof. The factors of a composition series are groups $\neq 1$ and the factors of a refinement of a composition series are the factors of the series augmented by factors $=1$. Now if we have two composition series, they have equivalent refinements. The equivalence matches the factors $=1$, so it matches those $\neq 1$, that is, those which are factors of the given composition series. Hence these are equivalent. □

If we take Λ to be the inner automorphisms, then the terms of a composition series are normal subgroups. Such a series is called a *chief series*. The Jordan-Hölder theorem shows that any two of these are equivalent. The result is applicable also to *characteristic series* defined to be composition series obtained by taking Λ to be all automorphisms and to *fully invariant series* obtained by taking Λ to be all endomorphisms.

All of this applies also to modules. In this case we have an important criterion for the existence of composition series.

THEOREM 3.5. *A module $M \neq 0$ has a composition series if and only if it is both artinian and noetherian.*

Proof. Assume M has a composition series $M = M_1 \supset M_2 \supset \cdots \supset M_{s+1} = 0$ (necessarily $M_i \rhd M_{i+1}$ since the groups are abelian). Let $N_1 \supsetneqq N_2 \supsetneqq \cdots \supsetneqq N_t$ be submodules. Then the given composition series and the normal series $M \supset N_1 \supset N_2 \supset \cdots \supset N_t \supset N_{t+1} = 0$ have equivalent refinements. It follows that $t \leqslant s+1$. This implies that M is artinian and noetherian. Now assume $M \neq 0$ has these properties. Consider the set of proper submodules of $M_1 = M$. This has a maximal element M_2 since M is noetherian. If $M_2 \neq 0$, we can repeat the process and obtain a maximal proper submodule M_3 of M_2. Continuing this we obtain the descending sequence $M_1 \supsetneqq M_2 \supsetneqq M_3 \supsetneqq \cdots$, which breaks off by the descending chain condition with $M_{t+1} = 0$. Then $M = M_1 \supset M_2 \supset \cdots \supset M_{t+1} = 0$ is a composition series. □

EXERCISES

1. Let G be a group with operator set Λ. Show that G has a composition series if and only if G satisfies the following two conditions: (i) If $G = G_1 \rhd G_2 \rhd \cdots$ then there exists an n such that $G_n = G_{n+1} = \cdots$. (ii) If H is a term of a normal series of G and $H_1 \subset H_2 \subset \cdots$ is an ascending sequence of normal subgroups of H, then there exists an n such that $H_n = H_{n+1}$.

2. Let G be a Λ-group having a composition series and let $H \lhd G$. Show that there exists a composition series in which one term is H. Hence show that H and G/H have composition series.

3. Let G and H be as in exercise 2. Call the number of composition factors the *length* $l(G)$. Show that $l(G) = l(H) + l(G/H)$. Show also that if $H_i \lhd G$, $i = 1, 2$, then

$$l(H_1 H_2) = l(H_1) + l(H_2) - l(H_1 \cap H_2).$$

3.4 THE KRULL-SCHMIDT THEOREM

The results we shall give in this section are valid for groups with operators and are of interest for these also. In fact, these results were obtained first for groups (without operators). However, the proofs are simpler in the module case. For this reason we shall confine our attention to modules in the text and indicate the extension of the theory to groups with operators in the exercises.

We recall that if $M = \oplus_1^n M_i$, then we have the homomorphisms $i_j : M_j \to M$, $p_j : M \to M_j$ such that the relations (1) hold. Now put

$$e_j = i_j p_j. \tag{27}$$

Then $e_j \in \operatorname{End} M$ $(= \hom_R(M, M))$ and (1) gives the relations

$$\begin{gathered} e_j^2 = e_j, \qquad e_j e_k = 0 \quad \text{if} \quad j \neq k \\ e_1 + e_2 + \cdots + e_n = 1. \end{gathered} \tag{28}$$

We recall that an element e of a ring is called idempotent if $e^2 = e$. Two idempotents e and f are said to be *orthogonal* if $ef = 0 = fe$. Thus the e_j are pair-wise orthogonal idempotents in $\operatorname{End} M$ and $\sum_1^n e_j = 1$. Put $M'_j = i_j(M_j) = i_j p_j(M) = e_j(M)$. If $x \in M$, $x = 1x = \sum e_j x$ and $e_j x \in M'_j$. Thus $M = M'_1 + M'_2 + \cdots + M'_n$. Moreover, if $x_j \in M'_j$ then $x_j = e_j x$ for some $x \in M$. Then $e_j x_j = e_j^2 x_j = e_j x = x_j$ and $e_k x_j = e_k e_j x = 0$ if $j \neq k$. This implies that $M'_j \cap (M'_1 + \cdots + M'_{j-1} + M'_{j+1} + \cdots + M'_n) = 0$ for every j.

Conversely, suppose a module M contains submodules $M_1, M_2, \ldots, M_n$ satisfying the two conditions

$$M = M_1 + M_2 + \cdots + M_n \tag{29}$$

$$M_j \quad (M_1 + \cdots + M_{j-1} + M_{j+1} + \cdots + M_n) = 0, \qquad 1 \leqslant j \leqslant n. \tag{30}$$

Then we have the injection homomorphism i_j of M_j into M and if $x \in M$, we have a unique way of writing $x = x_1 + \cdots + x_n$, $x_j \in M_j$, so we obtain a map $p_j : M \to M_j$ defined by $p_j x = x_j$. It is clear that the conditions (1) for the i_j and p_j are satisfied. Then, as before, we have an isomorphism of M onto $\oplus M_j$. In view of this result, we shall say that a module M is an (*internal*) *direct sum* of the submodules $M_1, M_2, \ldots, M_n$ if these satisfy (29) and (30). We shall write

$$M = M_1 \oplus M_2 \oplus \cdots \oplus M_n \tag{31}$$

to indicate this. The endomorphisms $e_j = i_j p_j$ are called the *projections determined by the direct decomposition* (31).

It is useful to express the conditions (29) and (30) in element form: Any element of M can be written in one and only one way in the form $x_1 + x_2 + \cdots + x_n$, $x_j \in M_j$. Using this, one sees easily that if M is a direct sum of submodules M_i and each M_i is a direct sum of submodules M_{ij}, then M is a direct sum of the submodules M_{ij}. There are other simple results of this sort that we shall use when needed (cf. BAI, pp. 176–177).

A module M is *decomposable* if $M = M_1 \oplus M_2$ where $M_i \neq 0$. Otherwise, M is called *indecomposable*.

PROPOSITION 3.1. *A module $M \neq 0$ is indecomposable if and only if* End M *contains no idempotent* $\neq 0, 1$.

Proof. If $M = M_1 \oplus \cdots \oplus M_n$ where the M_j are $\neq 0$ and $n > 1$, then the projections e_j are idempotents and $e_j \neq 0$. If $e_j = 1$, then $1 = \sum_1^n e_k$ gives $e_l = e_l 1 = e_l e_j = 0$ for every $l \neq j$. Hence $e_j \neq 1$. Conversely, suppose End M contains an idempotent $e \neq 0, 1$. Put $e_1 = e$, $e_2 = 1 - e$. Then $e_2{}^2 = (1-e)^2 = 1 - 2e + e^2 = 1 - 2e + e = 1 - e = e_2$ and $e_1 e_2 = e(1-e) = 0 = e_2 e_1$. Hence the e_j are non-zero orthogonal idempotents. Then we have $M = M_1 \oplus M_2$ where $M_j = e_j(M) \neq 0$. □

Can every module be written as a direct sum of a finite number of indecomposable submodules? It is easy to see that the answer to this is no (for example, take M to be an infinite dimensional vector space). Suppose M is a direct sum of a finite number of indecomposable submodules. Are the components unique? Again, it is easy to see that the answer is no (take a vector space of finite dimensionality > 1). Are the indecomposable components determined up to isomorphism? It can be shown that this need not be the case either. However, there is a simple condition to assure this. We shall first give the condition and later show that it holds for modules that are both artinian and noetherian, or equivalently, have composition series. The condition is based on a ring concept that is of considerable importance in the theory of rings.

DEFINITION 3.1. *A ring R is called* local *if the set I of non-units of R is an ideal.*

If R is local, it contains no idempotents except 0 and 1. For, if e is an idempotent $\neq 0, 1$, then e is not a unit since $e(1-e) = 0$. Also $1-e$ is not a unit. Hence if R is local, then e and $1-e$ are contained in the ideal I of non-units. Then $1 \in I$, which is absurd. It follows that if End M for a module $M \neq 0$ is local, then M is indecomposable. We shall now call a module M *strongly indecomposable* if $M \neq 0$ and End M is local. The main uniqueness theorem for direct decompositions into indecomposable modules, which we shall prove, is

THEOREM 3.6. *Let*

$$M = M_1 \oplus M_2 \oplus \cdots \oplus M_n \tag{32}$$

$$N = N_1 \oplus N_2 \oplus \cdots \oplus N_m \tag{33}$$

where the M_i are strongly indecomposable and the N_i are indecomposable and suppose $M \cong N$. Then $m = n$ and there is a permutation $j \rightsquigarrow j'$ such that $M_j \cong N_{j'}$, $1 \leqslant j \leqslant n$.

We shall first separate off a

LEMMA. *Let M and N be modules such that N is indecomposable and $M \neq 0$. Let f and g be homomorphisms $f: M \to N$, $g: N \to M$ such that gf is an automorphism of M. Then f and g are isomorphisms.*

Proof. Let k be the inverse of gf so $kgf = 1_M$. Put $l = kg: N \to M$. Then $lf = 1_M$ and if we put $e = fl: N \to N$, then $e^2 = flfl = f1_M l = fl = e$. Since N is indecomposable, either $e = 1$ or $e = 0$ and the latter is ruled out since it would imply that $1_M = 1_M{}^2 = lflf = lef = 0$, contrary to $M \neq 0$. Hence $fl = e = 1_N$. Then f is an isomorphism and $g = k^{-1}f^{-1}$ is an isomorphism. $\square$

Proof of Theorem 3.6. We shall prove the theorem by induction on n. The case $n = 1$ is clear, so we may assume $n > 1$. Let $e_1, \ldots, e_n$ be the projections determined by the decomposition (32) of M, and let $f_1, \ldots, f_m$ be those determined by the decomposition (33) of N. Let g be an isomorphism of M into N and put

$$h_j = f_j g e_1, \qquad k_j = e_1 g^{-1} f_j, \qquad 1 \leqslant j \leqslant m. \tag{34}$$

Then $\sum_1^m k_j h_j = \sum_1^m e_1 g^{-1} f_j g e_1 = e_1 g^{-1} \sum f_j g e_1 = e_1 g^{-1} 1_N g e_1 = e_1$. Now the restrictions of e_1 and $k_j h_j$ to M_1 map M_1 into M_1, so these may be regarded as endomorphisms e_1', $(k_j h_j)'$ respectively of M_1. Since $M_1 = e_1(M)$ and $e_1{}^2 = e_1$, $e_1' = 1_{M_1}$. Hence we have $1_{M_1} = \sum (k_j h_j)'$. Since End M_1 is local, this implies that one of the $(k_jh_j)'$ is a unit, therefore an automorphism of M_1. By reordering the N_j we may assume $j = 1$, so $(k_1h_1)'$ is an automorphism of M_1. The restriction of h_1 to M_1 may be regarded as a homomorphism h_1' of M_1 into N_1. Similarly, we obtain the homomorphism k_1' of N_1 into M_1 by restricting k_1 to N_1 and we have $k_1'h_1' = (k_1h_1)'$ is an automorphism. Then, by the lemma, $h_1' = (f_1ge_1)': M_1 \to N_1$ and $k_1' = (e_1g^{-1}f_1)': N_1 \to M_1$ are isomorphisms.

Next we prove

$$M = g^{-1}N_1 \oplus (M_2 + \cdots + M_n). \tag{35}$$

Let $x \in g^{-1}N_1 \cap (M_2 + \cdots + M_n)$. Then $e_1x = 0$ since $x \in M_2 + \cdots + M_n$ and $x = g^{-1}y$, $y \in N_1$. Then $0 = e_1x = e_1g^{-1}y = e_1g^{-1}f_1y = k_1y = k_1'y$. Then $y = 0$ and $x = 0$; hence, $g^{-1}N \cap (M_2 + \cdots + M_n) = 0$. Next let $x \in g^{-1}N_1 \subset M' \equiv g^{-1}N_1 + M_2 + \cdots + M_n$. Then $x, e_2x, \ldots, e_nx \in M'$ and hence $e_1x \in M'$. Thus $M' \supset e_1g^{-1}N_1 = e_1g^{-1}f_1N_1 = k_1N_1 = k_1'N_1 = M_1$. Then $M' \supset M_j$, $1 \leqslant j \leqslant n$, and so $M' = M$. We therefore have (35).

We now observe that the isomorphism g of M onto N maps $g^{-1}N_1$ onto N_1. Hence this induces an isomorphism of $M/g^{-1}N_1$ onto N/N_1. By (35), $M/g^{-1}N_1 \cong M_2 + \cdots + M_n = M_2 \oplus \cdots \oplus M_n$. Since $N/N_1 \cong N_2 \oplus \cdots \oplus N_m$, we have an isomorphism of $M_2 \oplus \cdots \oplus M_n$ onto $N_2 \oplus \cdots \oplus N_m$. The theorem now follows by induction. $\square$

We shall now show that these results apply to modules that have composition series.

If f is an endomorphism of a module M, we define $f^\infty M = \bigcap_{n=0}^\infty f^n(M)$ and $f^{-\infty}0 = \bigcup_{n=0}^\infty \ker f^n$. Thus $x \in f^\infty M$ if and only if for every $n = 1, 2, \ldots$ there exists a y_n such that $f^n y_n = x$ and $x \in f^{-\infty}0$ if and only if $f^n x = 0$ for some $n = 1, 2, \ldots$. The $f^n(M)$ form a descending chain

$$M \supset f(M) \supset f^2(M) \supset \cdots \tag{36}$$

and since $f^n x = 0$ implies that $f^{n+1}x = 0$, we have the ascending chain

$$0 \subset \ker f \subset \ker f^2 \subset \cdots. \tag{37}$$

Since the terms of both chains are submodules stabilized by f, $f^\infty M$ and $f^{-\infty}0$ are submodules stabilized by f. We have the following important result.

FITTING'S LEMMA. *Let f be an endomorphism of a module M that is both artinian and noetherian (equivalently, M has a composition series). Then we have* *the Fitting decomposition*

$$M = f^{\infty}M \oplus f^{-\infty}0. \tag{38}$$

Moreover, the restriction of f to $f^{\infty}M$ is an automorphism and the restriction of f to $f^{-\infty}M$ is nilpotent.

Proof. Since M is artinian, there is an integer s such that $f^s(M) = f^{s+1}(M) = \cdots = f^{\infty}M$. Since M is noetherian, we have a t such that $\ker f^t = \ker f^{t+1} = \cdots = f^{-\infty}0$. Let $r = \max(s,t)$, so $f^{\infty}M = f^r(M)$ and $f^{-\infty}0 = \ker f^r$. Let $z \in f^{\infty}M \cap f^{-\infty}0$, so $z = f^r y$ for $y \in M$. Then $0 = f^r z = f^{2r}y$ and $y \in \ker f^{2r} = \ker f^r$. Hence $z = f^r y = 0$. Thus $f^{\infty}M \cap f^{-\infty}0 = 0$. Now let $x \in M$. Then $f^r x \in f^r(M) = f^{2r}(M)$ so $f^r x = f^{2r}y$, $y \in M$. Then $f^r(x - f^r(y)) = 0$ and so $z = x - f^r(y) \in f^{-\infty}0$. Then $x = f^r y + z$ and $f^r y \in f^{\infty}M$. Thus $M = f^{\infty}M + f^{-\infty}0$ and we have the decomposition (38). Since $f^{-\infty}0 = \ker f^r$, the restriction of f to $f^{-\infty}0$ is nilpotent. The restriction of f to $f^{\infty}M = f^r(M) = f^{r+1}(M)$ is surjective. Also it is injective, since $f^{\infty}M \cap f^{-\infty}0 = 0$ implies that $f^{\infty}M \cap \ker f = 0$. □

An immediate consequence of the lemma is

THEOREM 3.7. *Let M be an indecomposable module satisfying both chain conditions. Then any endomorphism f of M is either nilpotent or an automorphism. Moreover, M is strongly indecomposable.*

Proof. By Fitting's lemma we have either $M = f^{\infty}M$ or $M = f^{-\infty}0$. In the first case f is an automorphism and in the second, f is nilpotent. To prove M strongly indecomposable, that is, End M is local, we have to show that the set I of endomorphisms of M that are not automorphisms is an ideal. It suffices to show that if $f \in I$ and g is arbitrary, then fg and $gf \in I$ for any $g \in \text{End}\, M$ and if $f_1, f_2 \in I$, then $f_1 + f_2 \in I$. The first one of these follows from the result that if $f \in I$, then f is nilpotent. Hence f is neither surjective nor injective. Then fg and gf are not automorphisms for any endomorphism g. Now assume $f_1 + f_2 \notin I$, so $f_1 + f_2$ is a unit in End M. Multiplying by $(f_1 + f_2)^{-1}$ we obtain $h_1 + h_2 = 1$ where $h_i = f_i(f_1 + f_2)^{-1} \in I$. Then $h_1 = 1 - h_2$ is invertible since $h_2{}^n = 0$ for some n, and so $(1 - h_2)(1 + h_2 + \cdots + h_2{}^{n-1}) = 1 = (1 + h_2 + \cdots + h_2{}^{n-1})(1 - h_2)$. This contradicts $h_1 \in I$. □

We prove next the existence of direct decompositions into indecomposable modules for modules satisfying both chain conditions.

THEOREM 3.8. *If $M \neq 0$ is a module that is both artinian and noetherian, then M contains indecomposable submodules M_i, $1 \leqslant i \leqslant n$, such that $M = M_1 \oplus M_2 \oplus \cdots \oplus M_n$.*

Proof. We define the length of M to be the number of composition factors in a composition series for M (exercise 3, p. 110). If N is a proper submodule, then the normal series $M \supsetneqq N \supset 0$ has a refinement that is a composition series. This follows from the Schreier refinement theorem. It follows that the length of $N <$ length of M. If M is indecomposable, the result holds. Otherwise, $M = M_1 \oplus M_2$ where the $M_i \neq 0$. Then the length $l(M_i) < l(M)$, so applying induction on length we may assume that $M_1 = M_{11} \oplus M_{12} \oplus \cdots \oplus M_{1n_1}$, $M_2 = M_{21} \oplus M_{22} \oplus \cdots \oplus M_{2n_2}$ where the M_{ij} are indecomposable. The element criterion for direct decomposition then implies that $M = M_{11} \oplus M_{12} \oplus \cdots \oplus M_{1n_1} \oplus M_{21} \oplus \cdots \oplus M_{2n_2}$.

The uniqueness up to isomorphism of the indecomposable components of M is an immediate consequence of Theorems 3.6 and 3.7. More precisely, we have the following theorem, which is generally called the

KRULL-SCHMIDT THEOREM. *Let M be a module that is both artinian and noetherian and let $M = M_1 \oplus M_2 \oplus \cdots \oplus M_n = N_1 \oplus N_2 \oplus \cdots \oplus N_m$ where the M_i and N_j are indecomposable. Then $m = n$ and there is a permutation $i \rightarrow i'$ such that $M_i \simeq N_{i'}$, $1 \leqslant i \leqslant n$.*

This theorem was first formulated for finite groups by J. H. M. Wedderburn in 1909. His proof contained a gap that was filled by R. Remak in 1911. The extension to abelian groups with operators, hence to modules, was given by W. Krull in 1925 and to arbitrary groups with operators by O. Schmidt in 1928. For a while the theorem was called the "Wedderburn-Remak-Krull-Schmidt theorem." This was shortened to the "Krull-Schmidt theorem."

EXERCISES

1. Let $p \in \mathbb{Z}$ be a prime and let R be the subset of $\mathbb{Q}$ of rational numbers a/b with $(p, b) = 1$. Show that this is a subring of $\mathbb{Q}$, which is local.
2. Show that the ring $\mathbb{Z}_p$ of p-adic numbers (p. 74) is local.
3. Let A be an algebra over a field that is generated by a single nilpotent element ($z^n = 0$). Then A is local.

4. Let F be a field and R the set of triangular matrices of the form

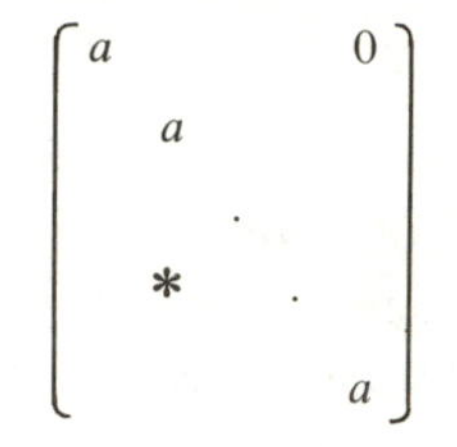

in $M_n(F)$. Show that R is a local ring.

5. Let M and f be as in Fitting's lemma and assume $M = M_0 \oplus M_1$ where the M_i are submodules stabilized by f such that $f|M_0$ is nilpotent and $f|M_1$ is an automorphism. Show that $M_0 = f^{-\infty}0$ and $M_1 = f^{\infty}M$.

6. Use Theorem 3.6 to prove that if $R^{(1)}$ and $R^{(2)}$ are local rings and n_1 and n_2 are positive integers such that $M_{n_1}(R^{(1)}) \cong M_{n_2}(R^{(2)})$, then $n_1 = n_2$ and $R^{(1)} \cong R^{(2)}$.

7. Let M be a finite dimensional vector space over a field F, T a linear transformation in M, and introduce an $F[\lambda]$-module structure of M via T, as in BAI, p. 165. Here λ is an indeterminate. Use the Krull-Schmidt theorem to prove the uniqueness of the elementary divisors of T (see BAI, p. 193).

8. Let the hypotheses be as in Theorem 3.6. Prove the following *exchange property*: For a suitable ordering of the N_j we have $N_j \cong M_j$ and
$$M = g^{-1}(N_1) \oplus \cdots \oplus g^{-1}(N_k) \oplus M_{k+1} \oplus \cdots \oplus M_n,$$
$1 \leqslant k \leqslant n$.

The remainder of the exercises deal with the extension of the results of this section to Λ-groups (groups with an operator set Λ) and further results for these groups.

9. Let G and H be Λ-groups and let $\hom(G,H)$ denote the set of homomorphisms of G into H (as Λ-groups). If $f, g \in \hom(G,H)$, define $f+g$ by $(f+g)(x) = f(x)g(x)$. Give an example to show that this need not be a homomorphism. Show that if K is a third Λ-group and $h, k \in \hom(H,K)$, then $hf \in \hom(G,K)$ and $h(f+g) = hf+hg$, $(h+k)f = hf+kf$. Define the 0 map from G to H by $x \rightsquigarrow 1$. Note that this is a homomorphism.

10. Let G be a Λ-group, $G_1, G_2, \ldots, G_n$ normal Λ-subgroups such that
$$G = G_1 G_2 \cdots G_n \tag{39}$$
$$G_i \cap G_1 \cdots G_{i-1} G_{i+1} \cdots G_n = 1 \tag{40}$$
for $1 \leqslant i \leqslant n$. Show that if $x_i \in G_i$, $x_j \in G_j$ for $i \neq j$, then $x_i x_j = x_j x_i$ and every element of G can be written in one and only way in the form $x_1 x_2 \cdots x_n$, $x_i \in G_i$. We say that G is an (*internal*) *direct product* of the G_i if conditions (39) and (40) hold and indicate this by $G = G_1 \times G_2 \times \cdots \times G_n$. Let e_i denote the map $x_1 x_2 \cdots x_n \quad x_i$. Verify that $e_i \in \operatorname{End} G = \hom(G,G)$ and e_i is *normal* in the sense that it commutes with every inner automorphism of G. Verify that $e_i^2 = e_i$, $e_i e_j = 0$ if $i \neq j$, $e_i + e_j = e_j + e_i$ for any i and j, and $e_1 + e_2 + \cdots + e_n = 1$. (Note

that we have associativity of $+$ so no parentheses are necessary.) Show that for distinct $i_1,\ldots,i_r$ in $\{1,2,\ldots,n\}$, $e_{i_1}+e_{i_2}+\cdots+e_{i_r}\in \text{END}\, G$.

11. Call G *indecomposable* if $G \neq G_1 \times G_2$ for any two normal Λ-subgroups $\neq G, 1$. Call G *strongly indecomposable* if $G \neq 1$ and the following condition on $\text{End}\, G$ holds: If f and g are normal endomorphisms of G such that $f+g$ is an automorphism, then either f or g is an automorphism. Show that this implies indecomposability. Use this concept to extend Theorem 3.6 to groups with operators.
12. Extend Fitting's lemma to normal endomorphisms of a Λ-group satisfying both chain conditions on normal Λ-subgroups.
13. Extend the Krull-Schmidt theorem to Λ-groups satisfying the condition given in exercise 12.
14. Show that if f is a surjective normal endomorphism of G, then $f = 1_G + g$ where $g(G) \subset C$, the center of G.
15. Prove that a finite group G has only one decomposition as a direct product of indecomposable normal subgroups if either one of the following conditions holds: (i) $C = 1$, (ii) $G' = G$ for G' the derived group (BAI, p. 245).

3.5 COMPLETELY REDUCIBLE MODULES

From the structural point of view the simplest type of module is an irreducible one. A module M is called *irreducible* if $M \neq 0$ and M contains no submodule N such that $M \supsetneq N \supsetneq 0$. This is a special case of the concept of a simple Λ-group that we encountered in discussing the Jordan-Hölder theorem (p. 108). We have the following characterizations, assuming $R \neq 0$.

THEOREM 3.9. *The following conditions on a module are equivalent*: (1) M *is irreducible*, (2) $M \neq 0$ *and M is generated by any* $x \neq 0$ *in* M, (3) $M \cong R/I$ *where I is a maximal right ideal in R.*

Proof. (1)$\Leftrightarrow$(2). Let M be irreducible. Then $M \neq 0$ and if $x \neq 0$ in M, then the cyclic submodule $xR \neq 0$. Hence $xR = M$. Conversely, suppose $M \neq 0$ and $M = xR$ for any $x \neq 0$ in M. Let N be a submodule $\neq 0$ in M and let $x \in N$, $x \neq 0$. Then $N \supset xR = M$. Hence M is irreducible.

(1)$\Leftrightarrow$(3). Observe first that M is cyclic ($M = xR$) if and only if $M \cong R/I$ where I is a right ideal in R. If $M = xR$, then we have the surjective module homomorphism $a \rightsquigarrow xa$ of R into M. The kernel is the annihilator $I = \text{ann}\, x$ of x in R (BAI, p. 163). This is a right ideal and $M \cong R/I$. Conversely, R/I is cyclic with generator $1+I$ and if $M \cong R/I$, then M is cyclic. We observe next that the submodules of R/I have the form I'/I where I' is a right ideal of R

containing I. Hence R/I is irreducible if and only if I is a maximal right ideal in R. Now if M is irreducible then M is cyclic, so $M \cong R/I$ and I is a maximal right ideal. The converse is clear. Hence $(1) \Leftrightarrow (3)$. □

Perhaps the most important fact about irreducible modules is the following basic result.

SCHUR'S LEMMA. *Let M and N be irreducible modules. Then any homomorphism of M into N is either* 0 *or an isomorphism. Moreover, the ring of endomorphisms* End M *is a division ring.*

Proof. Let f be a homomorphism of M into N. Then $\ker f$ is a submodule of M and $\operatorname{im} f$ is a submodule of N, so the irreducibility of M and N imply that $\ker f = M$ or 0 and $\operatorname{im} f = N$ or 0. If $f \neq 0$, $\ker f \neq M$ and $\operatorname{im} f \neq 0$. Then $\ker f = 0$ and $\operatorname{im} f = N$, which means that f is an isomorphism. In the case $N = M$ the result is that any endomorphism $f \neq 0$ is an automorphism. Then $f^{-1} \in \operatorname{End} M$ and this ring is a division ring. □

The modules that we shall now consider constitute a direct generalization of irreducible modules and of vector spaces over a division ring. As we shall see, there are several equivalent ways of defining the more general class. Perhaps the most natural one is in terms of direct sums of arbitrary (not necessarily finite) sets of submodules. We proceed to define this concept.

Let $S = \{M_\alpha\}$ be a set of submodules of M. Then the submodule $\sum M_\alpha$ generated by the M_α is the set of sums

$$x_{\alpha_1} + x_{\alpha_2} + \cdots + x_{\alpha_k}, \qquad x_{\alpha_i} \in \bigcup M_\alpha.$$

The set S is called *independent* if for every $M_\alpha \in S$ we have $M_\alpha \cap \sum_{M_\beta \neq M_\alpha} M_\beta = 0$. Otherwise, S is *dependent*. If N is a submodule, then $\{N\}$ is independent. If we look at the meaning of the definition of dependence in terms of elements, we see that a non-vacuous dependent set of modules contains finite dependent subsets. This property implies, via Zorn's lemma, that if S is a non-vacuous set of submodules and T is an independent subset of S, then T can be imbedded in a maximal independent subset of S. If $M = \sum M_\alpha$ and $S = \{M_\alpha\}$ is independent, then we say that M is a (*internal*) *direct sum* of the submodules M_α and we write $M = \oplus M_\alpha$. This is a direct generalization of the concept for finite sets of submodules that we considered in section 3.4.

We shall require the following

LEMMA 1. *Let $S = \{M_\alpha\}$ be an independent set of submodules of M, N a submodule such that $N \cap \sum M_\alpha = 0$. Then $S \cup \{N\}$ is independent.*

Proof. If not, we have an $x_\alpha \in M_\alpha$ $(\in S)$ such that $x_\alpha = y + x_{\alpha_1} + \cdots + x_{\alpha_k} \neq 0$ where $y \in N$, $x_{\alpha_i} \in M_{\alpha_i} \in S$ and $M_\alpha \neq M_{\alpha_i}$, $1 \leqslant i \leqslant k$. If $y = 0$ we have $x_\alpha = x_{\alpha_1} + \cdots + x_{\alpha_k}$ contrary to the independence of the set S. Hence $y \neq 0$. Then $y = x_\alpha - x_{\alpha_1} - \cdots - x_{\alpha_k} \in \sum M_\alpha$ contrary to $N \cap \sum M_\alpha \neq 0$. □

We can now give the following

DEFINITION 3.2. *A module M is* completely reducible *if M is a direct sum of irreducible submodules.*

Evidently any irreducible module is completely reducible. Now let M be a right vector space over a division ring Δ. It is clear that if $x \neq 0$ is in M, then $x\Delta$ is an irreducible submodule of M and $M = \sum_{x \neq 0} x\Delta$, so M is a sum of irreducible submodules. Now M is completely reducible. This will follow from the following

LEMMA 2. *Let $M = \sum M_\alpha$, M_α irreducible, and let N be a submodule of M. Then there exists a subset $\{M_\beta\}$ of $\{M_\alpha\}$ such that $\{N\} \cup \{M_\beta\}$ is independent and $M = N + \sum M_\beta$.*

Proof. Consider the set of subsets of the set of submodules of $\{M_\alpha\} \cup \{N\}$ containing N and let $\{M_\beta\} \cup \{N\}$ be a maximal independent subset among these. Put $M' = N + \sum M_\beta$. If $M' \subsetneq M$, then there exists an M_α such that $M_\alpha \not\subset M'$. Since M_α is irreducible and $M_\alpha \cap M'$ is a submodule of M_α, we must have $M_\alpha \cap M' = 0$. Then, by Lemma 1, $\{M_\beta\} \cup \{N\} \cup \{M_\alpha\}$ is independent contrary to the maximality of $\{M_\beta\} \cup \{N\}$. Hence $M = N + \sum M_\beta$. □

This lemma has two important consequences. The first is

COROLLARY 1. *If M is a sum of irreducible modules, say, $M = \sum M_\alpha$, M_α irreducible, then $M = \oplus M_\beta$ for a subset $\{M_\beta\}$ of $\{M_\alpha\}$.*

This is obtained by taking $N = 0$ in Lemma 2.

If M is a right vector space over Δ, then $M = \sum_{x \neq 0} x\Delta$ and every $x\Delta$ is irreducible. Hence M is completely reducible by Corollary 1.

We have also

COROLLARY 2. *If M is completely reducible and N is a submodule, then there exists a submodule N' of M such that $M = N \oplus N'$.*

Proof. Let $M = \sum M_\alpha$ where the M are irreducible. Then, by Lemma 2, there exists a subset $\{M_\beta\}$ of $\{M_\alpha\}$ such that $M = \sum M_\beta + N$ and $\{M_\beta\} \cup \{N\}$ is independent. If we put $N' = \sum M_\beta$ we have $M = N \oplus N'$. □

The property stated in Corollary 2 is that the lattice $L(M)$ of submodules of a completely reducible module M is complemented (BAI, p. 469). We shall show that this property characterizes completely reducible modules. We prove first

LEMMA 3. *Let M be a module such that the lattice $L(M)$ of submodules of M is complemented. Then $L(N)$ and $L(\bar{M})$ are complemented for any submodule N of M and any homomorphic image $\bar{M}$ of M.*

Proof. Let P be a submodule of N. Then $M = P \oplus P'$ where P' is a submodule of M. Put $P'' = P' \cap N$. Then $N = N \cap M = N \cap (P + P') = P + P''$, by modularity of $L(M)$ (BAI, p. 463). Hence $N = P \oplus P''$. This proves the first statement. To prove the second we may assume $\bar{M} = M/P$ where P is a submodule of M. Then $M = P \oplus P'$ where P' is a submodule and $\bar{M} \cong P'$. Since $L(P')$ is complemented by the first result, $L(\bar{M})$ is complemented. □

The key result for proving that $L(M)$ complemented implies complete reducibility is

LEMMA 4. *Let M be a non-zero module such that $L(M)$ is complemented. Then M contains irreducible submodules.*

Proof. Let $x \neq 0$ be in M and let $\{N\}$ be the set of submodules of M such that $x \notin N$. This contains 0 and so it is not vacuous. Hence by Zorn's lemma there exists a maximal element P in $\{N\}$. Evidently $P \neq M$, but every submodule $P_1 \supsetneqq P$ contains x. Hence if P_1 and P_2 are submodules such that $P_1 \supsetneqq P$ and $P_2 \supsetneqq P$, then $P_1 \cap P_2 \supsetneqq P$. It follows that the intersection of any two non-zero submodules of M/P is non-zero. Hence if P' is a submodule of M such that $M = P \oplus P'$, then $P' \cong M/P$ has the property that the intersection of any two non-zero submodules of P' is non-zero. Then P' is irreducible. For, $P' \neq 0$ since $M \supsetneqq P$ and if P'_1 is a submodule of P' such that $P'_1 \neq P', 0$, then by Lemma 3, we have a submodule P'_2 of P' such that $P' = P'_1 \oplus P'_2$. Then $P'_1 \cap P'_2 = 0$, but $P'_1 \neq 0$, $P'_2 \neq 0$ contrary to the property we established before for P'. □

We are now in a position to establish the following characterizations of completely reducible modules.

THEOREM 3.10. *The following conditions in a module are equivalent*: (1) $M = \sum M_\alpha$ *where the* M_α *are irreducible*, (2) M *is completely reducible*, (3) $M \neq 0$ *and the lattice* $L(M)$ *of submodules of* M *is complemented.*

Proof. The implication (1) ⇒ (2) has been proved in Corollary 1, and (2) ⇒ (3) has been proved in Corollary 2. Now assume condition (3). By Lemma 4, M contains irreducible submodules. Let $\{M_\alpha\}$ be the set of these and put $M' = \sum M_\alpha$. Then $M = M' \oplus M''$ where M'' is a submodule. By Lemma 3, $L(M'')$ is complemented. Hence if $M'' \neq 0$, then M'' contains one of the M_α. This contradicts $M'' \cap M' = M'' \cap \sum M_\alpha = 0$. Thus $M'' = 0$ and $M = M' = \sum M_\alpha$. Hence (3) ⇒ (1). □

Let N be an irreducible submodule of the completely reducible module M. We define the *homogeneous component* H_N of M *determined* by N to be $\sum N'$ where the sum is taken over all of the submodules $N' \cong N$. We shall show that M is a direct sum of the homogeneous components. In fact, we have the following stronger result.

THEOREM 3.11. *Let* $M = \oplus M_{\alpha\beta}$ *where the* $M_{\alpha\beta}$ *are irreducible and the indices are chosen so that* $M_{\alpha\beta} \cong M_{\alpha'\beta'}$ *if and only if* $\alpha = \alpha'$. *Then* $H_\alpha = \sum_\beta M_{\alpha\beta}$ *is a homogeneous component,* $M = \oplus H_\alpha$, *and every homogeneous component coincides with one of the* H_α.

Proof. Let N be an irreducible submodule of M. Then N is cyclic and hence $N \subset M' = M_1 \oplus M_2 \oplus \cdots \oplus M_n$ where $M_i \in \{M_{\alpha\beta}\}$. If we apply to N the projections on the M_i determined by the decomposition of M', we obtain homomorphisms of N into the M_i. By Schur's lemma, these are either 0 or isomorphisms. It follows that the non-zero ones are isomorphisms to M_i contained in a subset of $\{M_{\alpha\beta}\}$, all having the same first index. Then $N \subset H_\alpha$ for some α. If N' is any submodule isomorphic to N, we must also have $N' \subset H_\alpha$. Hence the homogeneous component $H_N \subset H_\alpha$. Since $H_\alpha \subset H_N$ is clear, we have $H_N = H_\alpha$. Since N was arbitrary irreducible, we account for every homogeneous component in this way. Also the fact that $M = \oplus H_\alpha$ is clear, since $M = \oplus M_{\alpha\beta}$ and $H_\alpha = \sum M_{\alpha\beta}$. □

EXERCISES

1. Let M be a finite dimensional (left) vector space over a field F. Show that M regarded as a right module for $R = \text{End}_F M$ in the natural way is irreducible. Does this hold if F is replaced by a division ring? Does it hold if the condition of finiteness of dimensionality is dropped?

2. Let M be the vector space with base (e_1, e_2) over a field F and let R' be the set of linear transformations a of M/F such that $ae_1 = \alpha e_1$, $ae_2 = \beta e_1 + \gamma e_2$ where $\alpha, \beta, \gamma \in F$. Show that R' is a ring of endomorphisms and that if M is regarded in the natural way as right R'-module, then M is not irreducible. Show that $\text{End}_{R'} M$ is the field consisting of the scalar multiplications $x \rightsquigarrow \alpha x$. (This example shows that the converse of Schur's lemma does not hold.)

3. Exercise 2, p. 193 of BAI.

4. Let V be the $F[\lambda]$-module defined by a linear transformation T as in exercise 3, p. 103. Show that V is completely reducible if and only if the minimal polynomial $m(\lambda)$ of T is a product of distinct irreducible factors in $F[\lambda]$.

5. Show that $\oplus M_\alpha$ together with the injections $i_\beta : M_\beta \to \oplus M_\alpha$ constitute a coproduct of the M_α.

6. Suppose $\{M_\alpha\}$ is a set of irreducible modules. Is $\prod M_\alpha$ completely reducible?

7. Show that a completely reducible module is artinian if and only if it is noetherian.

8. Let M be completely reducible. Show that if H is any homogeneous component of M, then H is an $\text{End}_R M$ submodule under the natural action.

3.6 ABSTRACT DEPENDENCE RELATIONS. INVARIANCE OF DIMENSIONALITY

A well-known result for finite dimensional vector spaces is that any two bases have the same number of elements. We shall now prove an extensive generalization of this result, namely if M is a completely reducible module and $M = \oplus M_\alpha = \oplus M'_\beta$ where the M_α and M'_β are irreducible, then the cardinality $|\{M_\alpha\}| = |\{M'_\beta\}|$. This will be proved by developing some general results on abstract dependence relations. The advantage of this approach is that the results we shall obtain will be applicable in a number of other situations of interest (e.g., algebraic dependence in fields).

We consider a non-vacuous set X and a correspondence Δ from X to the power set $\mathscr{P}(X)$. We write $x \prec S$ if $(x, S) \in \Delta$. We shall call Δ a *dependence relation* in X if the following conditions are satisfied:

1. If $x \in S$, then $x \prec S$.
2. If $x \prec S$, then $x \prec F$ for some finite subset F of S.

3. If $x \prec S$ and every $y \in S$ satisfies $y \prec T$, then $x \prec T$.
4. If $x \prec S$ but $x \not\prec S-\{y\}$ (complementary set of $\{y\}$ in S), then $y \prec (S-\{y\}) \cup \{x\}$. This is called the *Steinitz exchange axiom.*

The case of immediate concern is that in which X is the set of irreducible submodules of a completely reducible module M and we define $M_\alpha \prec S$ for $M_\alpha \in X$ and $S \subset X$ to mean that $M_\alpha \subset \sum_{M_\beta \in S} M_\beta$. Property 1 is clear. Now suppose $M_\alpha \subset \sum_{M_\beta \in S} M_\beta$ and let $x_\alpha \neq 0$ be in M_α. Then $x_\alpha \in \sum_{M_\gamma \in F} M_\gamma$ for some finite subset F of S. Then $M_\alpha = x_\alpha R \subset \sum M_\gamma$. Hence 2 holds. Property 3 is clear. Now let $M_\alpha \subset \sum_{M_\beta \in S} M_\beta$ and $M_\alpha \not\subset \sum_{M_\gamma \neq M_\beta} M_\gamma$. As before, let $x_\alpha \neq 0$ in M_α and write

$$x_\alpha = x_{\beta_1} + x_{\beta_2} + \cdots + x_{\beta_m}$$

where $x_{\beta_i} \in M_{\beta_i} \in S$. Then $M_\alpha \subset M_{\beta_1} + M_{\beta_2} + \cdots + M_{\beta_m}$. We may assume that the $x_{\beta_i} \neq 0$ and $M_{\beta_i} \neq M_{\beta_j}$ if $i \neq j$. Moreover, the condition $M_\alpha \not\subset \sum_{M_\gamma \neq M_\beta} M_\gamma$ implies that one of the $M_{\beta_i} = M_\beta$. We may assume $i = 1$. Then $0 \neq x_\beta = x_\alpha - x_{\beta_2} - \cdots - x_{\beta_m}$, which implies that $M_\beta = x_\beta R \subset M_\alpha + M_{\beta_2} + \cdots + M_{\beta_m}$. Then $M_\beta \prec \{M_\alpha, M_\gamma | M_\gamma \neq M_\beta\}$. This proves the exchange axiom.

We now consider an arbitrary dependence relation on a set X. We call a subset S of X *independent* (relative to Δ) if no $x \in S$ is dependent on $S-\{x\}$. Otherwise, S is called a *dependent set*. We now prove the analogue of Lemma 1 of section 3.5.

LEMMA 1. *If S is independent and $x \prec S$, then $S \cup \{x\}$ is independent.*

Proof. We have $x \prec S$ so $x \notin S$. Then if $S \cup \{x\}$ is not independent, we have a $y \in S$ such that $y \prec (S-\{y\}) \cup \{x\}$. Since S is independent, $y \prec S-\{y\}$. Hence by the exchange axiom, $x \prec (S-\{y\}) \cup \{y\} = S$ contrary to hypothesis. Hence $S \cup \{x\}$ is independent. □

DEFINITION 3.3. *A subset B of X is called a* base *for X relative to the dependence relation Δ if* (i) *B is independent,* (ii) *every $x \in X$ is dependent on B. (Note $B = \varnothing$ is allowed.)*

THEOREM 3.12. *There exists a base for X and any two bases have the same cardinal number.*

Proof. The "finiteness" property 2 of a dependence relation permits us to apply Zorn's lemma to prove that X contains a maximal independent subset, for it implies that the union of any totally ordered set of independent subsets of X is independent. Then Zorn's lemma is applicable to give the desired conclusion. Now let B be a maximal independent subset of X and let $x \in X$. Then $x \prec B$; otherwise, Lemma 1 shows that $B \cup \{x\}$ is independent contrary to the maximality of B. Hence every $x \in X$ is dependent on B, so B is a base.

Now let B and C be bases. If $B = \varnothing$, it is clear that $C = \varnothing$ so we assume $B \neq \varnothing$, $C \neq \varnothing$. Suppose first that B is finite, say, $B = \{x_1, x_2, \ldots, x_n\}$. We prove by induction on k that for $1 \leqslant k \leqslant n+1$ there exist $y_i \in C$ such that $\{y_1, \ldots, y_{k-1}, x_k, \ldots, x_n\}$ is a base. This holds for $k = 1$ since B is a base. Now assume it for some k. We claim that there is a y in C such that $y \not\prec \{y_1, \ldots, y_{k-1}, x_{k+1}, \ldots, x_n\}$. Otherwise, every $y \in C$ is dependent on $D = \{y_1, \ldots, y_{k-1}, x_{k+1}, \ldots, x_n\}$. Then $x \prec D$ for every $x \in B$ contrary to the independence of $\{y_1, \ldots, y_{k-1}, x_k, \ldots, x_n\}$. Now choose $y = y_k$ so that $y_k \not\prec D$. Then $\{y_1, \ldots, y_k, x_{k+1}, \ldots, x_n\}$ is independent, by Lemma 1. Moreover, since $y_k \prec \{y_1, \ldots, y_{k-1}, x_k, \ldots, x_n\}$, it follows from the exchange axiom that $x_k \prec \{y_1, \ldots, y_{k-1}, y_k, x_{k+1}, \ldots, x_n\}$. Then all of the elements of the base $\{y_1, \ldots, y_{k-1}, x_k, \ldots, x_n\}$ are dependent on $\{y_1, \ldots, y_k, x_{k+1}, \ldots, x_n\}$ and hence every $x \in X$ is dependent on $\{y_1, \ldots, y_k, x_{k+1}, \ldots, x_n\}$. Then $\{y_1, \ldots, y_k, x_{k+1}, \ldots, x_n\}$ is a base. We have therefore proved that for every k we have a base of the form $\{y_1, \ldots, y_k, x_{k+1}, \ldots, x_n\}$ with the y's in C. Taking $k = n$ we obtain a base $\{y_1, \ldots, y_n\}$, $y_i \in C$. It follows that $C = \{y_1, \ldots, y_n\}$. Thus $|C| = n = |B|$. The same argument applies if $|C|$ is finite, so it remains to consider the case in which both $|B|$ and $|C|$ are infinite. To prove the result in this case we shall use a counting argument that is due to H. Löwig. For each $y \in C$ we choose a finite subset F_y of B such that $y \prec F_y$. Then we have a map $y \rightarrow F_y$, which implies that $|C| \geqslant |\{F_y\}|$. Since every F_y is finite, this implies that $\aleph_0 |C| \geqslant |\bigcup F_y|$. Since $|C|$ is infinite, we have $\aleph_0 |C| = |C|$. Thus $|C| \geqslant |\bigcup F_y|$. Now $\bigcup F_y = B$. Otherwise, every $y \in C$ is dependent on a proper subset B' of B and since C is a base, every $x \in B$ is dependent on B'. This contradicts the independence of the set B. Hence $|C| \geqslant |\bigcup F_y| = |B|$. By symmetry $|B| \geqslant |C|$. Hence $|B| = |C|$. □

We remark that the first part of the preceding proof shows that any maximal independent subset of X is a base. The converse is evident. We mention also two other useful supplements to the result on existence of a base:

(i) If S is a subset such that every element is dependent on S then S contains a base.

(ii) If S is an independent subset then there exists a base containing S.

We now apply Theorem 3.12 to completely reducible modules by taking X to be the set of irreducible submodules of a completely reducible module M and defining $M_\alpha \prec \{M_\beta\}$ if $M_\alpha \subset \Sigma M_\beta$. We have seen that this is a dependence relation. Moreover, independence of a set of irreducible modules in the sense of $\prec$ is the independence we defined before for submodules of a module. A set $B = \{M_\beta\}$ of irreducible modules is a base for X if it is independent and every irreducible module $M_\alpha \subset \sum M_\beta$. Since M is a sum of irreducible submodules, this condition is equivalent to $M = \sum M_\beta$. Hence $B = \{M_\beta\}$ is a base if and only if $M = \oplus M_\beta$. The theorem on bases for a set X relative to a dependence relation now gives the following

THEOREM 3.13. *Let* $M = \oplus M_\beta = \oplus N_\gamma$ *where the* M_β *and* N_γ *are irreducible. Then* $|\{M_\beta\}| = |\{N_\gamma\}|$.

We shall call $|\{M_\beta\}|$ the *dimensionality* of the completely reducible module M. If M is a right vector space over Δ with base (x_β), then $M = \oplus x_\beta \Delta$ and every $x_\beta \Delta$ is irreducible. Hence the dimensionality of M is the cardinality of (x_β). Thus any two bases have the same cardinality. This is the usual "invariance of dimensionality theorem."

We can also apply Theorem 3.13 and the decomposition of a completely reducible module into homogeneous components to obtain a Krull-Schmidt theorem for completely reducible modules.

THEOREM 3.14. *Let* $M = \oplus M_\beta = \oplus N_\gamma$ *where the* M_β *and* N_γ *are irreducible. Then we have a bijection between the sets* $\{M_\beta\}$ $^{\alpha\pi\delta}$ $\{\notin_\gamma\}$ *such that the corresponding modules are isomorphic.*

This is clear, since for each homogeneous component the cardinality of the set of M_β in this component is the same as that of the set of N_γ in this component.

EXERCISE

1. Show that if M is a completely reducible module that has a composition series, then the length of M as defined in exercise 3 (p. 110) coincides with the dimensionality.

3.7 TENSOR PRODUCTS OF MODULES

The concept of tensor product of modules, which we shall now introduce, is a generalization of an old concept for vector spaces that has played an important role for quite a long time in several branches of mathematics, notably, differential geometry and representation theory. In the case of vector spaces, this was originally defined in terms of bases. The first base-free definition seems to have been given by H. Whitney. This was subsequently improved to take the standard form that we shall present. Before giving this, we look at an example that is somewhat in the spirit of the older point of view.

EXAMPLE

Let R be a ring. We wish to consider simultaneously free left and free right modules for R. To distinguish these we use the notation $R^{(n)}$ for the free right module with base of n elements and ${}^{(n)}R$ for the free left module having a base of n elements. For our purpose, it is convenient to write the elements of ${}^{(n)}R$ as rows: $(x_1, \ldots, x_n)$, $x_i \in R$, and those of $R^{(n)}$ as columns:

$$x = \begin{pmatrix} x_1 \\ \cdot \\ \cdot \\ \cdot \\ x_n \end{pmatrix}. \tag{41}$$

In the first case the module action is by left multiplication and in the second by right multiplication by elements of R. We now consider $R^{(m)}$, ${}^{(n)}R$ and $M_{m,n}(R)$ the additive group of $m \times n$ matrices with entries in R. If $x \in R^{(m)}$ as in (41) and $y = (y_1, \ldots, y_n) \in {}^{(n)}R$, then we can form the matrix product

$$xy = \begin{matrix} x_1 \\ \cdot \\ \cdot \\ \cdot \\ x_m \end{matrix} \quad (y_1, \ldots, y_n) = \begin{matrix} x_1y_1 & x_1y_2 & \cdot & \cdot & \cdot & x_1y_n \\ x_2y_1 & x_2y_2 & \cdot & \cdot & \cdot & x_2y_n \\ \cdot & \cdot & \cdot & \cdot & \cdot & \cdot \\ \cdot & \cdot & \cdot & \cdot & \cdot & \cdot \\ x_my_1 & x_my_2 & \cdot & \cdot & \cdot & x_my_n \end{matrix}. \tag{42}$$

This product is distributive on both sides and satisfies an associativity condition $(xa)y = x(ay)$, $a \in R$. Hence it is an instance of a balanced product in the sense that we shall now define for an arbitrary pair consisting of a right module and a left module.

Let M be a right module, N a left module for the ring R. We indicate this situation by writing M_R and ${}_RN$ in the respective cases. We define a *balanced product* of M and N to be an abelian group P (written additively) together with a map f of the product set $M \times N$ into P satisfying the following conditions:

B1. $f(x+x', y) = f(x, y) + f(x', y)$.

B2. $f(x, y+y') = f(x, y) + f(x, y')$.

B3. $f(xa, y) = f(x, ay)$.

Here $x, x' \in M_R$, $y, y' \in {}_RN$, $a \in R$. Note that these conditions imply as in the case of addition and multiplication in a ring that $f(0, y) = 0 = f(x, 0)$ and $f(-x, y) = -f(x, y) = f(x, -y)$. We denote the balanced product by (P, f). If (Q, g) is a second one, then we define a *morphism from* (P, f) *to* (Q, g) to be a homomorphism η of the additive group P into the additive group Q such that for any $x, y \in M \times N$ we have

$$g(x, y) = \eta f(x, y). \tag{43}$$

We can now introduce the following

DEFINITION 3.4. *A tensor product of M_R and ${}_RN$ is a balanced product $(M\otimes_R N, \otimes)$ such that if (P,f) is any balanced product of M_R and ${}_RN$, then there exists a unique morphism of $(M\otimes_R N, \otimes)$ to (P,f). In other words, there is a unique homomorphism of the abelian group $M\otimes_R N$ into the abelian group P sending every $x\otimes y$, $x\in M$, $y\in N$ into $f(x,y)\in P$.*

It is clear from the definition that if $((M\otimes_R N)_1, \otimes_1)$ and $((M\otimes_R N)_2, \otimes_2)$ are tensor products, then there exists a unique isomorphism of $(M\otimes_R N)_1$ onto $(M\otimes_R N)_2$ such that $x\otimes_1 y \rightarrow x\otimes_2 y$. It is also clear that if $(M\otimes_R N, \otimes)$ is a tensor product, then the group $M\otimes_R N$ is generated by the products $x\otimes y$. In fact, since $-(x\otimes y) = (-x)\otimes y$, it is clear that every element of $M\otimes_R N$ has the form $\sum x_i\otimes y_i$, $x_i\in M$, $y_i\in N$.

We proceed to construct a tensor product for a given pair $(M_R, {}_RN)$. For this purpose we begin with the free abelian group F having the product set $M\times N$ as base. The elements of F have the form

$$n_1(x_1, y_1) + n_2(x_2, y_2) + \cdots + n_r(x_r, y_r) \tag{44}$$

where the $n_i\in\mathbb{Z}$, $x_i\in M$, $y_i\in N$. Addition is the obvious one, and if $(x_i, y_i) \neq (x_j, y_j)$ for $i\neq j$, then (44) is 0 if and only if every $n_i = 0$. Let G be the subgroup of F generated by all of the elements

$$\begin{gathered}(x + x', y) - (x, y) - (x', y)\\ (x, y + y') - (x, y) - (x, y')\\ (xa, y) - (x, ay)\end{gathered} \tag{45}$$

where $x, x'\in M$, $y, y'\in N$, $a\in R$. Now define

$$M\otimes_R N = F/G, \qquad x\otimes y = (x, y) + G\in M\otimes_R N. \tag{46}$$

We claim that $(M\otimes_R N, \otimes)$ is a tensor product of M and N. First, we have

$$\begin{aligned}(x + x')\otimes y - x\otimes y - x'\otimes y &= ((x + x', y) + G) - ((x, y) + G) - ((x', y) + G)\\ &= ((x + x', y) - (x, y) - (x', y)) + G\\ &= G = 0.\end{aligned}$$

Similarly, $x\otimes(y + y') = x\otimes y + x\otimes y'$ and $xa\otimes y = x\otimes ay$. Hence $(M\otimes_R N, \otimes)$ is a balanced product. Now let (P,f) be any balanced product. Since F is the free abelian group with base $M\times N$, we have a (unique) homomorphism of F into P sending $(x, y)\rightarrow f(x, y)$. Let K denote the kernel of this homomorphism.

The conditions B1, B2, B3 imply that

$$(x+x', y)-(x, y)-(x', y)$$
$$(x, y+y')-(x, y)-(x, y')$$
$$(xa, y)-(x, ay)$$

$x, x' \in M$, $y, y' \in N$, $a \in R$, are in K. Then $G \subset K$ and so we have a homomorphism of $M \otimes_R N = F/G$ into P sending $x \otimes y = (x, y)+G \rightsquigarrow f(x, y)$. This is unique, since the cosets $x \otimes y$ generate $M \otimes_R N$. Hence we have verified that $(M \otimes_R N, \otimes)$ is a tensor product of M_R and ${}_RN$.

We shall now assume that somehow we have made a determination of $(M \otimes_R N, \otimes)$. We simplify the notation to $M \otimes N$ or $M \otimes_R N$ if we need to specify the ring R, and we speak of *the* tensor product of M and N. Now suppose we have module homomorphisms $f: M \rightarrow M'$ and $g: N \rightarrow N'$. Then we have the map

$$(x, y) \rightsquigarrow fx \otimes gy$$

of $M \times N$ into $M' \otimes N'$, which satisfies the conditions for a balanced product of M and N since

$$f(x+x') \otimes gy = fx \otimes gy + fx' \otimes gy$$
$$fx \otimes g(y+y') = fx \otimes gy + fx \otimes gy'$$
$$f(xa) \otimes gy = fx \otimes g(ay).$$

Hence we have a unique homomorphism of $M \otimes N$ into $M' \otimes N'$ such that

$$x \otimes y \rightsquigarrow fx \otimes gy.$$

We denote this as $f \otimes g$, so, by definition,

$$(f \otimes g)(x \otimes y) = fx \otimes gy \tag{47}$$

for all $x \in M$, $y \in N$. Suppose next that we have homomorphisms $f': M' \rightarrow M''$, $g': N' \rightarrow N''$. Then $f'f: M \rightarrow M''$, $g'g: N \rightarrow N''$ and

$$\begin{aligned}(f'f \otimes g'g)(x \otimes y) &= f'fx \otimes g'gy = (f' \otimes g')(fx \otimes gy)\\ &= (f' \otimes g')((f \otimes g)(x \otimes y))\\ &= ((f' \otimes g')(f \otimes g))(x \otimes y).\end{aligned}$$

Since the elements $x \otimes y$ generate $M \otimes N$, this implies that

$$(f' \otimes g')(f \otimes g) = f'f \otimes g'g. \tag{48}$$

Since $(1_M \otimes 1_N)(x \otimes y) = x \otimes y$, we have

$$1_M \otimes 1_N = 1_{M \otimes N}. \tag{49}$$

These results amount to saying that the maps

$$(M, N) \rightsquigarrow M \otimes N, \qquad (f, g) \rightsquigarrow f \otimes g$$

define a functor from the category **mod**-$R \times R$-**mod** to the category **Ab** (or $\mathbb{Z}$-**mod**). We shall denote this functor as $\otimes_R$.

Again let $f: M \to M'$ in **mod**-R and $g: N \to N'$ in R-**mod**. Then we have the commutativity of the diagram

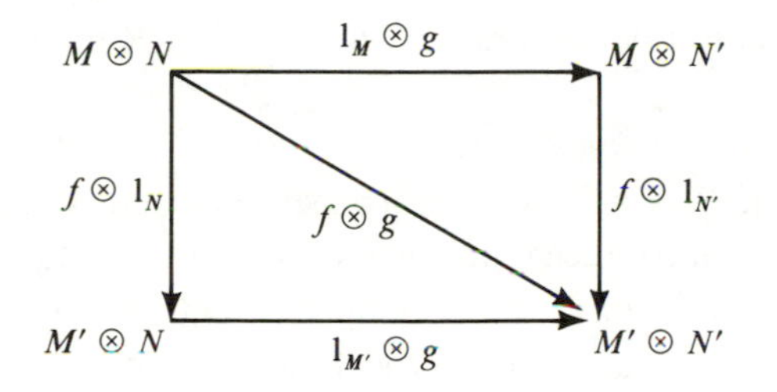

that is, we have

$$(f \otimes 1_{N'})(1_M \otimes g) = f \otimes g = (1_{M'} \otimes g)(f \otimes 1_N). \tag{50}$$

This is an immediate consequence of (48). We also have that the following distributive laws

$$\begin{aligned} (f_1 + f_2) \otimes g &= f_1 \otimes g + f_2 \otimes g \\ f \otimes (g_1 + g_2) &= f \otimes g_1 + f \otimes g_2 \end{aligned} \tag{51}$$

for $f_i: M \to M'$, $g_i: N \to N'$ will follow directly by applying the two sides to $x \otimes y$, $x \in M$, $y \in N$.

We shall now fix one of the arguments in the functor $\otimes_R$ and study the resulting functors of one variable. Let M be a fixed right R-module. Then we define the functor $M \otimes_R$ (or $M \otimes$) by specifying that for any left R-module N and any homomorphism $g: N \to N'$ of left R-modules, we have

$$(M \otimes_R) N = M \otimes_R N, \qquad (M \otimes_R)(g) = 1_M \otimes g.$$

We have $1_M \otimes g'g = (1_M \otimes g')(1_M \otimes g)$ for $g': N' \to N''$ and $1_M \otimes 1_N = 1_{M \otimes N}$. Moreover, we have the additivity property since $1_M \otimes (g_1 + g_2) = 1_M \otimes g_1 + 1_M \otimes g_2$, by (51). Hence $M \otimes_R$ is indeed a functor from R-**mod** to $\mathbb{Z}$-**mod** in the sense that we have adopted in this chapter. In a similar manner, any left module defines a functor $\otimes_R N$ (or $\otimes N$) from **mod**-R to **mod**-$\mathbb{Z}$ by

$$(\otimes_R N)M = M \otimes_R N, \qquad (\otimes_R N)(f) = f \otimes 1_N$$

for $f: M \to M'$ in **mod**-R.

We now specialize $M = R = R_R$, R regarded as right R-module. Then we have

PROPOSITION 3.2. *Let N be a left R-module. Then the map*

$$\eta_N : y \rightsquigarrow 1 \otimes y \tag{52}$$

*is an isomorphism of N regarded as a $\mathbb{Z}$-module with $R \otimes_R N$. Moreover, $N \rightsquigarrow \eta_N$ is a natural isomorphism of the forgetful functor from R-**mod** to $\mathbb{Z}$-**mod** (or **Ab**) with the functor $R \otimes_R$.*

Proof. Evidently (52) is a $\mathbb{Z}$-homomorphism. On the other hand, for $r \in R$, $y \in N$, $f(r, y) = ry$ is a balanced product of R and N. Hence we have a $\mathbb{Z}$-homomorphism ζ_N of $R \otimes_R N$ into N such that $\zeta_N(r \otimes y) = ry$. It is clear that $\zeta_N \eta_N = 1_N$ and $\eta_N \zeta_N = 1_{R \otimes N}$. Hence the first assertion is valid. The second assertion is an immediate consequence of the definitions. □

In a similar manner, if we define $\eta'_M : M \to M \otimes_R R$ by $x \rightsquigarrow x \otimes 1$, we see that this is an isomorphism and $M \rightsquigarrow \eta'_M$ is a natural isomorphism of the forgetful functor from **mod**-R to **mod**-$\mathbb{Z}$ with the functor $\otimes_R R$. The reader should note that a similar result for the hom functor was given in exercise 1, p. 99.

The general result we noted in section 3.1 (p. 98) that functors on categories of modules respect finite coproducts can be applied to the functors $M \otimes_R$ and $\otimes_R N$. In this way we obtain canonical isomorphisms of $M \otimes (\oplus N_j)$ with $\oplus (M \otimes N_j)$ and $(\oplus M_i) \otimes N$ with $\oplus (M_i \otimes N)$. Combining these two we obtain an isomorphism of $(\oplus_1^m M_i) \otimes (\oplus_1^n N_j)$ with $\oplus_{(1,1)}^{(m,n)} M_i \otimes N_j$.

Proposition 3.2 shows that we have an isomorphism of $R \otimes_R R$ onto the additive group of R such that $r \otimes s \rightsquigarrow rs$. With this result and the isomorphism of $(\oplus M_i) \otimes (\oplus N_j)$ onto $\oplus (M_i \otimes N_j)$ it is easy to see that the example given at the beginning of this section of the balanced product of $R^{(m)}$ and ${}^{(n)}R$ into $M_{m,n}(R)$ is the tensor product.

The result on the isomorphism of a finite coproduct $\oplus (M_i \otimes N)$ with $(\oplus M_i) \otimes N$ can be extended to arbitrary coproducts. Let $\{M_\alpha | \alpha \in I\}$ be an indexed set of right R-modules. Then $\oplus M_\alpha$ is the set of functions $(x_\alpha): \alpha \quad x_\alpha \in M_\alpha$ such that $x_\beta = 0$ for all but a finite number of β, and if N is a left R-module, $(\oplus M_\alpha) \otimes N$ is the set of sums of elements $(x_\alpha) \otimes y$, $y \in N$. Similarly, $\oplus (M_\alpha \otimes N)$ is the set of functions (z_α), $\alpha \quad z_\alpha \in M_\alpha \otimes N$ such that

$z_\beta = 0$ for all but a finite number of β. Since every element of $M_\alpha \otimes N$ is a sum of elements $x_\alpha \otimes y$, $x_\alpha \in M_\alpha$, $y \in N$, it follows that every element of $\oplus(M_\alpha \otimes N)$ is a sum of elements $(x_\alpha \otimes y)$ where this denotes the function $\alpha \quad x_\alpha \otimes y$. We shall now prove

PROPOSITION 3.3 *If the M_α, $\alpha \in I$, are right R-modules and N is a left R-module, then we have an isomorphism*

$$\eta_N : (\oplus M_\alpha) \otimes N \to \oplus(M_\alpha \otimes N)$$

such that

$$(x_\alpha) \otimes y \quad (x_\alpha \otimes y). \tag{53}$$

Proof. Let $(x_\alpha) \in M_\alpha$, $y \in N$. Then $(x_\alpha \otimes y) \in \oplus(M_\alpha \otimes N)$ and $((x_\alpha), y) \quad (x_\alpha \otimes y)$ defines a balanced product of $\oplus M_\alpha$ and N into $\oplus(M_\alpha \otimes N)$. Hence we have a homomorphism of $(\oplus M_\alpha) \otimes N$ into $\oplus(M_\alpha \otimes N)$ such that $(x_\alpha) \otimes y \quad (x_\alpha \otimes y)$. On the other hand, if i_α is the canonical homomorphism of M_α into $\oplus M_\alpha$ (sending x_α into the function whose value at α is x_α and whose value at every $\beta \neq \alpha$ is 0), we have the homomorphism $i_\alpha \otimes 1_N$ of $M_\alpha \otimes N$ into $(\oplus M_\alpha) \otimes N$. By the defining coproduct property of $\oplus(M_\alpha \otimes N)$ we obtain a hoomomorphism $\zeta_N : \oplus(M_\alpha \otimes N) \to (\oplus M_\alpha) \otimes N$ such that

$$(x_\alpha \otimes y) \quad (x_\alpha) \otimes y.$$

Checking on generators, we see that η_N and ζ_N are inverses. Hence (53) is an isomorphism. □

Now let N' be a second left R-module and let $f : N' \to N$. We have the homomorphism $1_{M_\alpha} \otimes f$ of $M_\alpha \otimes N' \to M_\alpha \otimes N$, which defines the homomorphism f^* of $\oplus(M_\alpha \otimes N') \to \oplus(M_\alpha \otimes N)$ such that $(x_\alpha \otimes y') \quad (x_\alpha \otimes fy')$ for $x_\alpha \in M_\alpha$, $y' \in N$. We also have the homomorphism $1 \otimes f$ of $(\oplus M_\alpha) \otimes N'$ into $(\oplus M_\alpha) \otimes N$ where $1 = 1_{\oplus M_\alpha}$. Now we have the naturality of the isomorphism η_N given in Proposition 3.3, that is,

(54)

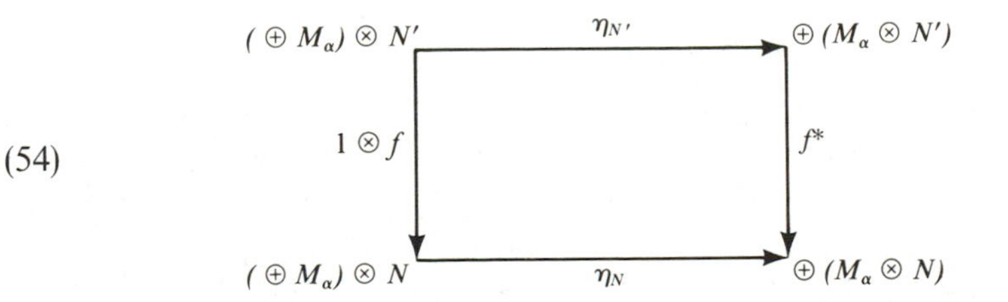

is commutative. This follows directly by checking on the generators $(x_\alpha) \otimes y'$.

We shall derive next the fundamental exactness property of the tensor functors.

THEOREM 3.15. *The functors $M\otimes_R$ and $\otimes_R N$ are right exact.*

Proof. Suppose $N' \xrightarrow{f} N \xrightarrow{g} N''$. Since g is surjective, every element of $M\otimes_R N''$ has the form $\sum x_i \otimes g y_i$, $x_i \in M$, $y_i \in N$. Thus $1\otimes g$ is an epimorphism of $M\otimes N$ onto $M\otimes N''$. Hence to prove that $M\otimes N' \xrightarrow{1\otimes f} M\otimes N \xrightarrow{1\otimes g} M\otimes N''$, it remains to show that $\ker(1\otimes g) = \operatorname{im}(1\otimes f)$. Since $gf = 0$, we have $(1\otimes g)(1\otimes f) = 0$ and $\operatorname{im}(1\otimes f) \subset \ker(1\otimes g)$. Hence we have a group homomorphism Θ of $(M\otimes N)/\operatorname{im}(1\otimes f)$ into $M\otimes N''$ such that $x\otimes y + \operatorname{im}(1\otimes f) \rightsquigarrow x\otimes gy$. This is an isomorphism if and only if $\operatorname{im}(1\otimes f) = \ker(1\otimes g)$. Thus it suffices to show that Θ is an isomorphism. Let $x \in M$, $y'' \in N''$ and choose a $y \in N$ such that $gy = y''$. We claim that the coset $x\otimes y + \operatorname{im}(1\otimes f)$ in $(M\otimes N)/\operatorname{im}(1\otimes f)$ is independent of the choice of y. To see this, let $y_1, y_2 \in N$ satisfy $gy_1 = y'' = gy_2$. Then $g(y_1 - y_2) = 0$ and $y_1 - y_2 = fy'$ for $y' \in N'$. Then $x\otimes y_1 + \operatorname{im}(1\otimes f) = x\otimes y_2 + \operatorname{im}(1\otimes f)$. We now have a map of $M\times N''$ into $(M\otimes N)/\operatorname{im}(1\otimes f)$ such that $(x, y'') \rightsquigarrow x\otimes y + \operatorname{im}(1\otimes f)$ where $gy = y''$. It is clear that this is a balanced product. Hence we have a homomorphism θ' of $M\otimes N''$ into $(M\otimes N)/\operatorname{im}(1\otimes f)$ such that $\theta'(x\otimes y'') = x\otimes y + \operatorname{im}(1\otimes f)$ *where* $gy = y''$. Checking on generators we see that $\theta\theta' = 1_{M\otimes N''}$ and $\theta'\theta = 1_{(M\otimes N)/\operatorname{im}(1\otimes f)}$. Thus θ is an isomorphism. This proves the first assertion. The second is obtained in a similar fashion. □

EXERCISES

1. Let R^{op} be the opposite ring of R and regard a right (left) R-module as a left (right) R^{op}-module in the usual way. Show that there is a group isomorphism of $M\otimes_R N$ onto $N\otimes_{R^{\mathrm{op}}} M$ mapping $x\otimes y$ into $y\otimes x$, $x\in M$, $y \in N$. Note that if R is commutative, this applies to $M\otimes_R N$ and $N\otimes_R M$.

2. Show that if m is a positive integer, then $\mathbb{Z}/(m)\otimes m\mathbb{Z} \cong \mathbb{Z}/(m)$ ($\otimes = \otimes_{\mathbb{Z}}$). Hence show that $0 \to m\mathbb{Z}\otimes(\mathbb{Z}/(m)) \xrightarrow{i\otimes 1} \mathbb{Z}\otimes(\mathbb{Z}/m)$ is not exact for i the injection of $m\mathbb{Z}$ in $\mathbb{Z}$. Note that this shows that $M\otimes$ and $\otimes M$ need not be exact functors.

3. Use Proposition 3.2 and Theorem 3.15 to show that if m and n are positive integers, then $(\mathbb{Z}/(m))\otimes(\mathbb{Z}/(n)) \cong \mathbb{Z}/(d)$ where $d = (m, n)$ the g.c.d. of m and n.

4. Generalize exercise 3 to R a p.i.d. Use this and the structure theorem for finitely

generated abelian groups (BAI, p. 195) to determine the structure of the tensor product of two finitely generated modules over a p.i.d.

5. Let $i_\alpha, j_\beta, k_{\alpha,\beta}$ be the canonical monomorphisms of M_α into $\oplus M_\alpha$, of N_β into $\oplus N_\beta$ and of $M_\alpha \otimes N_\beta$ into $\oplus M_\alpha \otimes N_\beta$. Show that there is an isomorphism of $(\oplus M_\alpha)\otimes(\oplus N_\beta)$ onto $\oplus M_\alpha \otimes N_\beta$ sending $i_\alpha x_\alpha \otimes j_\beta y_\beta$ into $k_{\alpha\beta}(x_\alpha \otimes y_\beta)$.

3.8 BIMODULES

Let M be a right R-module and let $R' = \text{End}\, M$. Then M can be regarded as a left R'-module if we define fx for $f \in R'$, $x \in M$, to be the image of x under the map f. Since f is a homomorphism of the additive group of M, we have $f(x+y) = fx+fy$, and by definition of the sum and product of homomorphisms, we have $(f+g)x = fx+gx$ and $(fg)x = f(gx)$ if $f, g \in R'$. Clearly also $1x = x$ so M is a left R'-module. Also the definition of module homomorphism gives the relation

$$f(xa) = (fx)a \tag{55}$$

for $x \in M$, $a \in R$, $f \in R'$. This associativity connects the given right R-module structure with the left R'-module structure of M. More generally we now introduce

DEFINITION 3.5. *If R and S are rings, an S-R-*bimodule *is a commutative group M (written additively) together with actions of S and R on M such that M with the action of S (that is, the product sx, $s \in S$, $x \in M$) is a left S-module, M with the action of R is a right R-module, and we have the associativity condition*

$$s(xr) = (sx)r \tag{56}$$

for all $s \in S$, $x \in M$, $r \in R$.

The foregoing considerations show that if M is a right R-module, then M can be regarded in a natural way as an R'-R-bimodule for $R' = \text{End}\, M$.

EXAMPLES

1. Let R be commutative. Then any right R-module M can be regarded also as a left R-module by putting $ax = xa$, $a \in R$, $x \in M$. It is often advantageous to regard M as an R-R-bimodule with $ax = xa$. The associativity condition $a(xb) = (ax)b$ is an immediate consequence of the commutativity of R.

2. Again let R be commutative and let η be an automorphism of R. If M is a right R-module, M becomes a left R-module by defining $ax = x\eta(a)$. This action together with the given right action makes M an R-R-bimodule.

3. Any right R-module can be regarded as a $\mathbb{Z}$-R-bimodule by defining mx for $m \in \mathbb{Z}$ in the usual way. Similarly, any left R-module becomes an R-$\mathbb{Z}$-bimodule. In this way the theory of one-sided modules can be subsumed in that of bimodules.

4. If R is a ring, we have the left module ${}_RR$ and the right module R_R. Since we have the associative law $(ax)b = a(xb)$, we can put the two module structures together to obtain a bimodule.

We write $M = {}_SM_R$ to indicate that M is an S-R-bimodule. We define a *homomorphism* of ${}_SM_R$ into ${}_SN_R$ to be a map of M into N that is simultaneously an S-homomorphism and an R-homomorphism of M into N. It is clear from the associativity condition (54) that for any s, x $\quad sx$ is an endomorphism of M_R (M as right R-module) and $x \quad xr$ is an endomorphism of ${}_SM$ if $r \in R$. Given ${}_SM_R$ and ${}_TN_R$ one is often interested in the homomorphisms of M into N as right R-modules. The set of these is denoted as $\hom_R({}_SM_R, {}_TN_R)$, which is the same thing as $\hom_R(M, N)$. This is an abelian group. Since for $s \in S$, $x \quad sx$ is an endomorphism of M_R, the composite map f_s:

$$(57) \qquad (fs)x = f(sx)$$

is a homomorphism of M_R into N_R if $f \in \hom_R(M, N)$. Similarly, if $t \in T$, then if we follow f by the map $y \quad ty$, $y \in N$, we obtain a homomorphism of M_R into N_R. Thus tf defined by

$$(58) \qquad (tf)x = t(fx)$$

is in $\hom_R(M_R, N_R)$. It is clear from the associative and distributive laws for homomorphisms of modules that we have $t(f_1+f_2) = tf_1+tf_2$, $(t_1+t_2)f = t_1f+t_2f$, $(t_1t_2)f = t_1(t_2f)$, $(f_1+f_2)s = f_1s+f_2s$, $f(s_1+s_2) = fs_1+fs_2$, $f(s_1s_2) = (fs_1)s_2$ if $f_i \in \hom_R(M_R, N_R)$, $s, s_i \in S$, $t, t_i \in T$. Also we have $1f = f = f1$ and $(tf)s = t(fs)$. We therefore have the following

PROPOSITION 3.4. *The abelian group* $\hom_R({}_SM_R, {}_TN_R)$ *becomes a* T-S-*bimodule if we define* tf *and* fs *for* $f \in \hom_R({}_SM_R, {}_TN_R)$, $s \in S$, $t \in T$, *by*

$$(tf)x = t(fx), \qquad (fs)x = f(sx),$$

$x \in M$.

In a similar manner, in the situation ${}_RM_S$, ${}_RN_T$ we obtain the following

PROPOSITION 3.5. *The abelian group* $\hom_R({}_RM_S, {}_RN_T)$ *becomes an* S-T-*bimodule if we define*

$$(59) \qquad (sf)x = f(xs), \qquad ft(x) = f(x)t.$$

We leave the proof to the reader.

An important special case of Proposition 3.4 is obtained by taking the dual module $M^* = \hom(M_R, R_R)$ of a right R-module M. We showed at the beginning of our discussion that the right-module M can be considered in a natural way as an R'-R-bimodule for $R' = \operatorname{End} M$. Also $R = {}_RR_R$. Hence, by Proposition 3.4, M^* is an R-R'-bimodule if we define

$$(60) \qquad (ry^*)x = r(y^*x), \qquad (y^*r')x = y^*(r'x)$$

for $r \in R$, $r' \in R'$, $y^* \in M^*$, $x \in M$.

We now consider tensor products for bimodules. Given ${}_SM_R$ and ${}_RN_T$ we can form the tensor product with respect to R of the right module M and the left module N. Then we have

PROPOSITION 3.6. *The tensor product* ${}_SM_R \otimes {}_RN_T$ *is an* S-T-*bimodule if we define* $sz = (s \otimes 1)z$ *and* $zt = z(1 \otimes t)$ *for* $z \in {}_SM_R \otimes {}_RN_T$. *Here* s *is the endomorphism* $x \rightsquigarrow sx$ *of* M_R *and* t *is the endomorphism* $y \rightsquigarrow yt$ *of* ${}_RN$.

The verification is immediate and is left to the reader.

Propositions 3.4, 3.5, and 3.6 show one advantage in dealing with bimodules rather than with one-sided modules: Tensoring or "homing" of bimodules having a common ring in the right place yields bimodules. It is interesting to see what happens when one iterates these processes. A first result of this sort is an associativity of tensor products, which we give in

PROPOSITION 3.7. *We have an* R-U-*bimodule isomorphism of* $({}_RM_S \otimes {}_SN_T) \otimes {}_TP_U$ *onto* ${}_RM_S \otimes ({}_SN_T \otimes {}_TP_U)$ *such that*

$$(x \otimes y) \otimes z \rightsquigarrow x \otimes (y \otimes z)$$

for $x \in M$, $y \in N$, $z \in P$.

Proof. For $s \in S$ we have

$$xs \otimes (y \otimes z) = x \otimes s(y \otimes z) = x \otimes (s \otimes 1)(y \otimes z) = x \otimes (sy \otimes z).$$

Hence for fixed z, $f_z(x, y) = x \otimes (y \otimes z)$ is a balanced product of M and N. Hence we have a group homomorphism of $M \otimes N$ into $M \otimes (N \otimes P)$ sending $x \otimes y \rightsquigarrow x \otimes (y \otimes z)$. This maps $\sum x_i \otimes y_i \rightsquigarrow \sum x_i \otimes (y_i \otimes z)$. Now define $f(\sum x_i \otimes y_i, z) = \sum x_i \otimes (y \otimes z)$. This defines a balanced product of $M \otimes N$ as right T-module and P as left T-module. Then we have a group homomorphism of $(M \otimes N) \otimes P$ into $M \otimes (N \otimes P)$ such that $(x \otimes y) \otimes z \rightsquigarrow x \otimes (y \otimes z)$. It is clear that this is, in fact, an R-U-bimodule homomorphism. In a similar manner, we can show that we have a bimodule homomorphism of $M \otimes (N \otimes P)$ into $(M \otimes N) \otimes P$ such that $x \otimes (y \otimes z) \rightarrow (x \otimes y) \otimes z$. It is clear that composites in both orders of the two homomorphisms we have defined are identity maps. Hence both are isomorphisms. □

We now consider the bimodules $M = {}_RM_S$, $N = {}_SN_T$, $P = {}_UP_T$. Then $M \otimes N$ is an R-T-bimodule and since P is a U-T-bimodule, $\hom_T(M \otimes N, P)$ is a U-R-bimodule by Proposition 3.4. Next we note that $\hom_T(N, P)$ is a U-S-bimodule and hence $\hom_S(M, \hom_T(N, P))$ is a U-R-bimodule, again by Proposition 3.4. Now let $f \in \hom_T(M \otimes N, P)$. Then $f_x : y \quad f(x \otimes y) \in \hom_T(N, P)$ and $x \quad f_x \in \hom_S(M, \hom_T(N, P))$. We have

PROPOSITION 3.8. *If* $M = {}_RM_S$, $N = {}_SN_T$, $P = {}_UP_T$, *then the map* $\varphi : f \quad \varphi(f)$ *of* $\hom_T(M \otimes N, P)$ *such that* $\varphi(f)$ *is* $x \quad f_x$, $x \in M$, *where* f_x *is* $y \quad f(x \otimes y)$, $y \in N$ *is an isomorphism of* $\hom_T(M \otimes N, P)$ *onto* $\hom_S(M, \hom_T(N, P))$ *as* U-R-*bimodules.*

Proof. Direct verification shows that φ is a U-R-bimodule homomorphism. In the other direction, let $g \in \hom_S(M, \hom_T(N, P))$. Then if $x \in M$ and $y \in M$, $g(x)(y) \in P$. If $s \in S$, we have $g(xs)(y) = (g(x)s)(y) = g(x)(sy)$. Since $g(x)(y)$ is additive in x and y, this defines a balanced product of M_S and ${}_SN$. Hence we have a group homorphism f of $M \otimes N$ into P such that $f(x \otimes y) = g(x)y$. Replacing y by yt and using the uniqueness of the corresponding f we see that f is a T-homomorphism. Put $\psi(g) = f$. Then the definitions show that $\varphi\psi(g) = g$ and $\psi\varphi(f) = f$. Hence φ is an isomorphism. □

EXERCISES

1. Let M be an R-R-bimodule and let $S = R \times M$ endowed with the abelian group structure given by these structures in R and M: $(r, x) + (r', x') = (r + r', x + x')$. Define a product in S by

$$(r, x)(r', x') = (rr', rx' + xr').$$

Verify that this defines a ring with unit $1 = (1, 0)$ containing the subring of elements of the form $(r, 0)$ and the ideal of elements of the form $(0, x)$. These can be identified with R and M respectively. Then $S = R \oplus M$ and $M^2 = 0$.

2. Let M be an R-S-bimodule for the rings R and S. Form the ring $T = R \oplus S$. Show that M becomes a T-T-bimodule by defining for $(r, s) \in R \oplus S$ and $x \in M$

$$(r, s)x = rx, \qquad x(r, s) = xs.$$

Apply the construction in exercise 1 to define a ring determined by R, S and M as a set of triples (r, s, x).

3. Specialize Proposition 3.8 by taking $R = T = U = \mathbb{Z}$ to obtain an isomorphism of $\hom_{\mathbb{Z}}(M \otimes N, P)$ onto $\hom_S(M, \hom_{\mathbb{Z}}(N, P))$ for $M = {}_{\mathbb{Z}}M_S$, $N = {}_SN_{\mathbb{Z}}$, $P = {}_{\mathbb{Z}}P_{\mathbb{Z}}$. Use this to show that the functor $M \otimes_R$ from R-**mod** to $\mathbb{Z}$-**mod** (or **Ab**) is a left adjoint of the functor $\hom_{\mathbb{Z}}(M, -)$ from $\mathbb{Z}$-**mod** to R-**mod**.

4. Let S be a subring of a ring R. Then any right R-module becomes a right S-module by restricting the action to S. Homomorphisms of right R-modules are homomorphisms of these modules as right S-modules. In this way one obtains a functor F from **mod**-R to **mod**-S ("restricting the scalars to S"). Show that F has both a left and a right adjoint. (*Hint*: Given a right S-module M, consider $M \otimes_S R$ where R is regarded as left S-module in the obvious way. Consider also $\hom_S(R, M)$.)

3.9 ALGEBRAS AND COALGEBRAS

We shall now specialize the theory of tensor products of modules to the case in which the ring is a commutative ring K. Considerably more can be said in this case. In particular, tensor products of modules over K provide an alternative definition of an algebra that in many ways is preferable to the one we gave originally (p. 44). For one thing, the new definition suggests the dual concept of a coalgebra. Moreover, it facilitates the definition of tensor products of algebras. Both of these concepts are of considerable importance. We shall give a number of important constructions of algebras, notably, tensor algebras, exterior algebras, and symmetric algebras defined by K-modules.

If K is a commutative ring, then K is isomorphic to its opposite ring K^{op}. Hence any left (right) K-module can be regarded as a right (left) K-module by putting $kx = xk$, $k \in K$. Thus we need not distinguish between left and right modules; we shall simply speak of modules over K. Another special feature of modules over a commutative ring is that for a fixed $l \in K$, the scalar multiplication $x \rightarrow lx$ is an endomorphism. This is an immediate consequence of the commutative law.

It is also sometimes useful to regard a module over K as a K-bimodule in which the two actions of any $k \in K$ on the module M coincide: $kx = xk$. Again, the commutativity of K insures that the conditions for a bimodule are satisfied.

If M and N are K-modules, we can regard these as K-K-bimodules and form $M \otimes_K N$, which is again a K-K-bimodule. If $k \in K$, $x \in M$, $y \in N$, then

$$k(x \otimes y) = kx \otimes y = x \otimes ky = (x \otimes y)k.$$

In view of this relation, we may as well regard $M \otimes_K N$ simply as a K-module (left or right). The map $(x, y) \rightarrow x \otimes y$ of $M \times N$ into $M \otimes N$ is K-*bilinear* in the sense that for fixed x, $y \rightarrow x \otimes y$ is a K-homomorphism and for fixed y, $x \rightarrow x \otimes y$ is a K-homomorphism (cf. BAI, p. 344).

Now let $f:(x, y) \rightarrow f(x, y)$ be any K-bilinear map of $M \times N$ into a K-module P. Then we have

B1. $f(x_1 + x_2, y) = f(x_1, y) + f(x_2, y)$

B2. $f(x, y_1 + y_2) = f(x, y_1) + f(x, y_2)$

B3. $f(kx, y) = kf(x, y) = f(x, ky)$

for $k \in K$, $x \in M$, $y \in N$. Since $kx = xk$, $k \in K$, $x \in M$, B1, B2, and B3 imply that (P, f) is a balanced product of M and N. Hence we have the additive group homomorphism of $M \otimes_K N$ into P such that $x \otimes y \rightarrow f(x, y)$. Evidently, by B3, this is a K-module homomorphism. Conversely, if π is a K-homomorphism of $M \otimes_K N$ into a K-module P, then $f(x, y) \equiv \pi(x \otimes y)$ is a K-bilinear map of $M \times N$ into P.

In particular, $f(x, y) \equiv y \otimes x$ defines a K-bilinear map of $M \times N$ into $N \otimes M$. Hence we have a K-homomorphism σ of $M \otimes N$ into $N \otimes M$ such that $x \otimes y \rightarrow y \otimes x$. Similarly we have a K-homomorphism σ' of $N \otimes M$ into $M \otimes N$ such that $y \otimes x \rightarrow x \otimes y$. Then $\sigma' = \sigma^{-1}$, so σ is an isomorphism. Taking $N = M$ we obtain the automorphism of $M \otimes M$ such that $x \otimes y \rightarrow y \otimes x$.

We recall also that we have the associativity of tensor multiplication in the sense that if M, N, and P are K-modules, then we have a K-isomorphism of $(M \otimes N) \otimes P$ into $M \otimes (N \otimes P)$ such that $(x \otimes y) \otimes z \rightarrow x \otimes (y \otimes z)$ for $x \in M$, $y \in N$, $z \in P$ (Proposition 3.6, p. 135).

We can now give the alternative definition of an algebra over K:

DEFINITION 3.6. *If K is a commutative ring, an* (associative) algebra over K *(or a K-*algebra*) is a triple (A, π, ε) where A is a K-module, π is a K-homomorphism of $A \otimes A$ into A, and ε is a K-homomorphism of K into A such that the following diagrams are commutative*

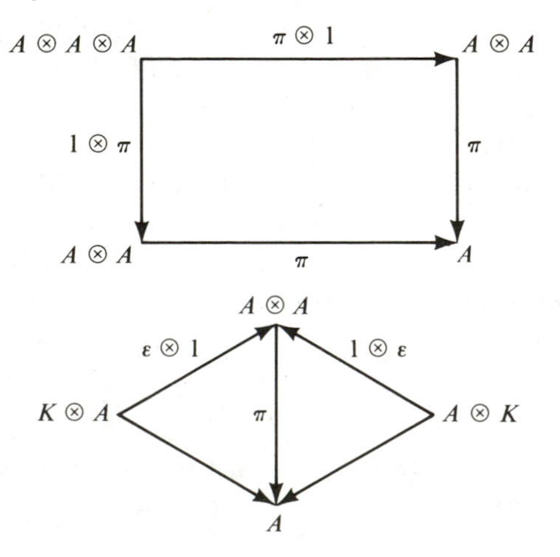

Here we have written $A\otimes A\otimes A$ for either $(A\otimes A)\otimes A$ or $A\otimes(A\otimes A)$, which we identify via the isomorphism mapping $(x\otimes y)\otimes z \rightsquigarrow x\otimes(y\otimes z)$, $x, y, z\in A$, and $\pi\otimes 1$ is the K-homomorphism applied to $(A\otimes A)\otimes A$ while $1\otimes\pi$ is applied to $A\otimes(A\otimes A)$. The unlabelled arrows are the canonical isomorphisms sending $k\otimes x$, $k\in K$, $x\in A$, into kx and $x\otimes k$ into $xk = kx$.

Suppose we have an algebra according to this definition. We introduce a product

$$xy = \pi(x\otimes y) \tag{61}$$

in A and we define 1 in A by

$$\varepsilon 1 = 1. \tag{62}$$

Let $x, y, z\in A$ and apply the commutativity of the first diagram to $(x\otimes y)\otimes z = x\otimes(y\otimes z)$ in $A\otimes A\otimes A$. This gives the associative law $(xy)z = x(yz)$. If we take the element $k\otimes x$ in $K\otimes A$ and apply the commutativity of the left-hand triangle in the second diagram, we obtain $kx = \pi(\varepsilon\otimes 1)(k\otimes x) = \pi(k1\otimes x) = (k1)x$. Similarly, the right-hand triangle gives $x(k1) = kx$. It follows that $1x = x = x1$ and we have the algebra condition $k(xy) = (kx)y = x(ky)$ by the associative law applied to $k1$, x, and y.

Conversely, suppose we have an algebra in the first sense. Then the given product xy is K-bilinear, so we have a K-homomorphism π of $A\otimes A$ into A such that $\pi(x\otimes y) = xy$. We define $\varepsilon: K\rightarrow A$ by $k \rightsquigarrow k1$. Then we have the commutativity of the first diagram by the associative law and of the second by $kx = (k1)x = x(k1)$, which follows from $k(xy) = (kx)y = x(ky)$.

Thus the two definitions are equivalent.

We have given a number of examples of algebras over fields in BAI, pp. 406–409 and pp. 411–414. We shall now give some examples of algebras over an arbitrary commutative ring K.

EXAMPLES

1. $\mathrm{End}_K M$. Let M be a K-module, $A = \mathrm{End}\, M$. We know that this has a ring structure with the natural definitions of addition, zero, negatives, product, and 1. Also A is a K-module and $k(fg) = (kf)g = f(kg)$ for $k\in K$, $f, g\in A$, since applying these to any $x\in M$ gives $k((fg)x)$, $((kf)g)x$, $(f(kg))x$. All of these are equal by the definition of $(kf)x = f(kx) = k(fx)$. Hence we have an algebra over K in the old sense.

2. *Tensor algebras*. Before giving our next construction, we need to consider a general associativity property of tensor products of K-modules. Let $M_1, M_2, \ldots, M_n$ be K-modules. We define $M_1\otimes\cdots\otimes M_n$ inductively by $M_1\otimes\cdots\otimes M_i = (M_1\otimes\cdots\otimes M_{i-1})\otimes M_i$. Also if $x_j\in M_j$, we define $x_1\otimes\cdots\otimes x_n$ inductively by $x_1\otimes\cdots\otimes x_i = (x_1\otimes\cdots\otimes x_{i-1})\otimes x_i$. Now we claim that if $1\leqslant m < n$, we

have a unique isomorphism $\pi_{m,n}$ of $(M_1\otimes\cdots\otimes M_m)\otimes(M_{m+1}\otimes\cdots\otimes M_n)$ onto $M_1\otimes\cdots\otimes M_n$ such that

$$\pi_{m,n}((x_1\otimes\cdots\otimes x_m)\otimes(x_{m+1}\otimes\cdots\otimes x_n)) = x_1\otimes\cdots\otimes x_m\otimes_{m+1}\otimes\cdots\otimes x_n. \tag{63}$$

This is immediate by induction on $n-m$ using the isomorphism $(M\otimes N)\otimes P \cong M\otimes(N\otimes P)$ given in Proposition 3.7.

Now suppose all of the $M_i = M$, a given K-module, and put $M^{(i)} = M\otimes\cdots\otimes M$, i times, $i = 1,2,3,\ldots$. Also put $M^{(0)} = K$ and define $\pi_{0,n}: K\otimes M^{(n)} \to M^{(n)}$, $n = 0,1,\ldots$, to be the canonical isomorphism mapping $k\otimes x$ into kx. Define $\pi_{n,n}$ as the canonical isomorphism sending $x\otimes k$ into $kx = xk$. Now form the K-module

$$T(M) = \bigoplus_0^\infty M^{(i)} = K\oplus M^{(1)}\oplus M^{(2)}\oplus\cdots \tag{64}$$

We can modify the definition of $\pi_{m,n}$, $0 \leqslant m \leqslant n$, $n = 0,1,2,\ldots$ so as to regard this as a K-homomorphism into $T(M)$. Using the fact that $\oplus(M^{(i)}\otimes M^{(j)})$ is the coproduct of the modules $M^{(i)}\otimes M^{(j)}$ we obtain a homomorphism π' of $\oplus(M^{(i)}\otimes M^{(j)})$ into $T(M)$, which coincides with $\pi_{m,n}$ on $M^{(m)}\otimes M^{(n-m)}$ (identified with their images in $\oplus(M^{(i)}\otimes M^{(j)})$). We also have an isomorphism of $T(M)\otimes T(M) = (\oplus M^{(i)})\otimes(\oplus M^{(j)})$ onto $\oplus(M^{(i)}\otimes M^{(j)})$ (exercise 5, p. 133). Hence we obtain a homomorphism π of $T(M)\otimes T(M)$ into $T(M)$, which is found by following this isomorphism with the homomorphism π'. Now let ε be the injection of K into $T(M) = \oplus M^{(i)}$. We claim that $(T(M),\pi,\varepsilon)$ is an algebra in the sense of Definition 3.6, or equivalently, if we put $xy = \pi(x\otimes y)$, we have the associative law and $1x = x = x1$. The second of these is clear. To prove the first, we note that any element of $T(M)$ is a sum of terms of the form $k1$ and elements of the form $x_1\otimes\cdots\otimes x_m$, so it suffices to prove associativity of multiplication of terms of these forms. Then the general associativity will follow by the distributive law. Associativity of products, one of which is $k1$, is clear since the definition of the product gives $(k1)x = kx = x(k1)$. Now consider $x = x_1\otimes\cdots\otimes x_m$, $y = y_1\otimes\cdots\otimes y_n$, $z = z_1\otimes\cdots\otimes z_p$, $m,n,p > 0$. The definition of π gives

$$\begin{aligned}(xy)z &= ((x_1\otimes\cdots\otimes x_m)(y_1\otimes\cdots\otimes y_n))(z_1\otimes\cdots\otimes z_p)\\ &= (x_1\otimes\cdots\otimes x_m\otimes y_1\otimes\cdots\otimes y_n)\otimes(z_1\otimes\cdots\otimes z_p)\\ &= x_1\otimes\cdots\otimes x_m\otimes y_1\otimes\cdots\otimes y_n\otimes z_1\otimes\cdots\otimes z_p.\end{aligned}$$

In a similar fashion, one obtains the same result for $x(yz)$. Hence our definitions do give a K-algebra. This is called the *tensor algebra* $T(M)$ defined by the given K-module M.

We now consider the forgetful functor F from the category **K-alg** to the category **K-mod**. We claim that $(T(M),i)$ where i is the injection of M in $T(M)$ constitutes a universal from M to F. To see this, let f be a K-module homomorphism of M into a K-algebra A. We define a K-homomorphism $f^{(n)}$ of $M^{(n)}$ into A, $n = 1,2,\ldots$ by $f^{(1)} = f$, and $f^{(i)}$ is the homomorphism of $M^{(i)} = M^{(i-1)}\otimes M$ obtained by composing $f^{(i-1)}\otimes f$ with the homomorphism of $A\otimes A$ into A, sending $x\otimes y$ into xy. Then for $x_i \in M$ we have

$$f^{(n)}(x_1\otimes\cdots\otimes x_n) = f(x_1)\cdots f(x_n). \tag{65}$$

We define $f^{(0)}: K \to A$ as $k \to k1$ and we let f^* be the K-module homomorphism of $T(M)$ into A that coincides with $f^{(n)}$ on $M^{(n)}$, $n = 0,1,2,\ldots$. It is immediate from (65) that f^* is a K-algebra homomorphism. Now it is clear from the definition of $T(M)$ that $T(M)$ is generated by M. Hence f^* is the only homomorphism of $T(M)$ into A that coincides with f on M. Thus $(T(M), i)$ is a universal from M to the forgetful functor from K-**alg** to K-**mod**.

An algebra A is called *graded* (by the natural numbers $0, 1, \ldots$) if $A = \oplus_0^\infty A_i$ where A_i is a submodule of A and $A_i A_j \subset A_{i+j}$. The submodule A_i is called the *homogeneous part of degree* i of A. An example of this sort that should be familiar to the reader is obtained by taking $A = K[\lambda_1, \ldots, \lambda_r]$, the polynomial algebra over K in indeterminates $\lambda_1, \ldots, \lambda_r$ (BAI, pp. 124–126). Let A_i be the set of homogeneous polynomials of (total) degree i (BAI, p. 138). Then these provide a grading for $A = K[\lambda_1, \ldots, \lambda_r]$. The tensor algebra $T(M)$ is graded by its submodules $M^{(i)}$ since we have $T(M) = \oplus M^{(i)}$ and $M^{(i)}M^{(j)} = M^{(i)} \otimes M^{(j)} \subset M^{(i+j)}$.

The tensor algebra is the starting point for defining several other algebras. Perhaps the most important of these are exterior algebras and symmetric algebras, which we shall now define.

3. *Exterior algebras.* We define the *exterior algebra* $E(M)$ of the K-module M as

$$E(M) = T(M)/B \tag{66}$$

where B is the ideal in $T(M)$ generated by the elements

$$x \otimes x, \tag{67}$$

$x \in M$. It is clear from the definition that the elements in the ideal B are contained in $\sum_{i \geq 2} M^{(i)}$. Hence $B \cap M = 0$ and the restriction to M of the canonical homomorphism of $T(M)$ onto $E(M) = T(M)/B$ is injective. Then we can identify M with its image and so regard M as a subset of $E(M)$. It is clear that M generates $E(M)$. It is clear also that since the ideal B is generated by the homogeneous elements $x \otimes x$, B is homogeneous in the sense that $B = \sum B^{(i)}$ where $B^{(i)} = B \cap M^{(i)}$. It follows that $E(M)$ is graded by the subsets $E^{(i)} = (M^{(i)} + B)/B$, that is, we have $E(M) = \oplus E^{(i)}$ and $E^{(i)}E^{(j)} \subset E^{(i+j)}$.

Now suppose A is a K-algebra and f is a K-module homomorphism of M into A such that $f(x)^2 = 0$ in A for every $x \in M$. The universality property of $T(M)$ gives a homomorphism f^* of $T(M)$ into A, extending the given map f of M. Since $f^*(x)^2 = f(x)^2 = 0$, $x \in M$, the kernel of f^* contains every $x \otimes x$, $x \in M$. Hence it contains the ideal B defining $E(M)$. Then we have a homomorphism $\bar{f}$ of $E(M)$ into A sending x (as element of $E(M)$) into $f(x)$. In other words, any K-module homomorphism f of M into an algebra A satisfying $f(x)^2 = 0$, $x \in M$, can be extended to a K-algebra homomorphism of $E(M)$ into A. Since $E(M)$ is generated by M, it is clear that this homomorphism is unique. Thus $E(M)$ has the universality property that any K-module homomorphism f of M into a K-algebra such that $f(x)^2 = 0$ for all $x \in M$ has a unique extension to an algebra homomorphism of $E(M)$ into A. This property implies that in the case in which M is a finite dimensional vector space over a field, the exterior algebra we have defined here is the same as the algebra we constructed in BAI, pp. 411–414, using bases.

4. *Symmetric algebras.* We start again with $T(M)$ and we now factor out the ideal C generated by the elements of the form

$$x \otimes y - y \otimes x \tag{68}$$

$x, y \in M$. The resulting algebra $S(M) = T(M)/C$ is called the *symmetric algebra* of the K-module M. The arguments used for the exterior algebra show that we may regard M as a subset of $S(M)$ which generates $S(M)$ and that $S(M)$ is graded by the submodules

$S^{(i)} = (M^{(i)} + C)/C$. We have $xy = yx$ in $S(M)$ for the generators $x, y \in M$. This implies that $S(M)$ is a commutative algebra. Moreover, it is easily seen that $S(M)$ together with the injection of M into $S(M)$ constitutes a universal from M to the forgetful functor from the category **Comalg** of commutative algebras to the category K-**mod**.

5. *Universal enveloping algebra of a Lie algebra.* We recall that a Lie algebra L over a commutative ring K is a K-module equipped with a bilinear composition $(x, y) \rightsquigarrow [xy]$ that satisfies the two conditions

$$[xx] = 0$$
$$[[xy]z] + [[yz]x] + [[zx]y] = 0.$$

If A is an associative algebra, A defines a Lie algebra A^- whose K-module structure is the same as that of A and whose Lie product is $[xy] \equiv xy - yx$ (BAI, p. 432). If A_1 and A_2 are associative algebras, a homomorphism of A_1 into A_2 is also a homomorphism of $A_1{}^-$ into $A_2{}^-$. In this way we obtain a functor F from the category K-**alg** of associative algebras over K to the category K-**Lie** of Lie algebras over K. We proceed to show that given any Lie algebra L there exists a universal from L to the functor F. To construct this, we form the tensor algebra $T(L)$ for the K-module $L: T(L) = K \oplus L^{(1)} \oplus L^{(2)} \oplus \cdots$ and we let B be the ideal in $T(L)$ generated by all of the elements of the form

$$[xy] - x \otimes y + y \otimes x, \qquad x, y \in L \quad (= L^{(1)}).$$

Put $U(L) = T(L)/B$ and let u be the restriction to L of the canonical homomorphism of $T(L)$ onto $U(L) = T(L)/B$. Then u is a homomorphism of the Lie algebra L into $U(L)^-$. We call $U(L)$ the (*universal*) *enveloping algebra* of the Lie algebra L. We claim that $(U(L), u)$ constitutes a universal from L to the functor F.

We have to show that if A is any associative algebra and g is a homomorphism of L into A^-, then there exists a unique homomorphism $\tilde{g}$ of associative algebras, $\tilde{g}: U(L) \to A$, such that the following diagram of Lie algebra homomorphisms is commutative

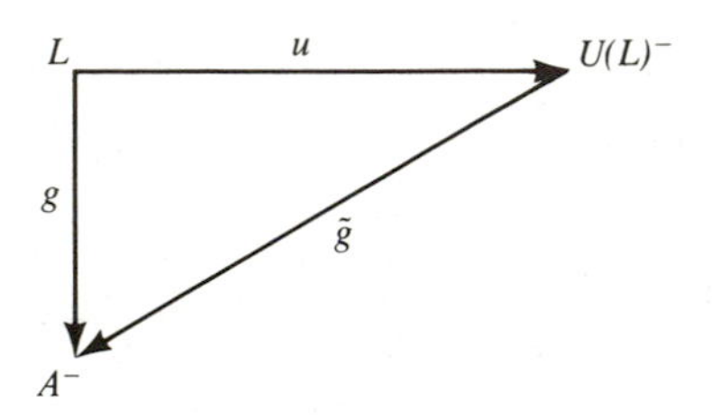

We have seen that we have a unique homomorphism g^* of $T(L)$ into A extending the K-homomorphism g of L into A. Now $g([xy]) = g(x)g(y) - g(y)g(x)$ for $x, y \in L$, since g is a homomorphism of L into A^-. Hence

$$g^*([xy] - x \otimes y + y \otimes x) = g([xy]) - g(x)g(y) + g(y)g(x) = 0.$$

which implies that $B \subset \ker g^*$. Hence we have a homomorphism $\tilde{g}$ of $U(L) = T(L)/B$ into A such that $\tilde{g}(x + B) = g(x)$ or $\tilde{g}u(x) = g(x)$. Thus $\tilde{g}$ makes the preceding diagram commutative. Since L generates $T(L)$, $u(L)$ generates $U(L)$. Hence $\tilde{g}$ is unique.

We remark that the result we have established: existence of a universal to F for every Lie algebra is equivalent to the statement that the functor F from K-**alg** to K-**Lie** has a left adjoint from K-**Lie** to K-**alg**. A similar remark can be made for symmetric algebras.

We observe also that a symmetric algebra of a K-module M can be regarded as the enveloping algebra of the Lie algebra M in which the composition is the trivial one: $[xy] \equiv 0$.

We consider next the tensor products of associative algebras. Let $(A_i, \pi_i, \varepsilon_i)$, $i = 1, 2$, be algebras in the sense of Definition 3.6. Then $\pi_i : A_i \otimes A_i \to A_i$, $\varepsilon_i : K \to A_i$ and we have the commutativity of the diagrams given in the definition. Now consider the K-modules $A_1 \otimes A_2$ and $(A_1 \otimes A_2) \otimes (A_1 \otimes A_2)$. Using the fact that we have an isomorphism of $A_2 \otimes A_1$ onto $A_1 \otimes A_2$ sending every $x_2 \otimes x_1 \rightsquigarrow x_1 \otimes x_2$, $x_i \in A_i$ and the associativity of tensor products, we obtain an isomorphism of $(A_1 \otimes A_2) \otimes (A_1 \otimes A_2)$ onto $(A_1 \otimes A_1) \otimes (A_2 \otimes A_2)$ such that

$$(x_1 \otimes x_2) \otimes (y_1 \otimes y_2) \rightsquigarrow (x_1 \otimes y_1) \otimes (x_2 \otimes y_2).$$

Following this with $\pi_1 \otimes \pi_2$ we obtain a K-homomorphism π of $(A_1 \otimes A_2) \otimes (A_1 \otimes A_2)$ into $A_1 \otimes A_2$ such that

$$(x_1 \otimes x_2) \otimes (y_1 \otimes y_2) \rightsquigarrow \pi_1(x_1 \otimes y_1) \otimes \pi_2(x_2 \otimes y_2).$$

In other words, we have a product composition in $A_1 \otimes A_2$ such that

$$(x_1 \otimes x_2)(y_1 \otimes y_2) = x_1 y_1 \otimes x_2 y_2 \tag{69}$$

in terms of the products in A_1 and A_2. Also put

$$1 = 1_1 \otimes 1_2. \tag{70}$$

Then for $x_i, y_i, z_i \in A_i$ we have

$$\begin{aligned} ((x_1 \otimes x_2)(y_1 \otimes y_2))(z_1 \otimes z_2) &= (x_1 y_1) z_1 \otimes (x_2 y_2) z_2 \\ (x_1 \otimes x_2)((y_1 \otimes y_2)(z_1 \otimes z_2)) &= x_1 (y_1 z_1) \otimes x_2 (y_2 z_2). \end{aligned}$$

This implies the associativity of the multiplication in $A = A_1 \otimes A_2$. We have

$$1(x_1 \otimes x_2) = x_1 \otimes x_2 = (x_1 \otimes x_2)1,$$

which implies that 1 is the unit under multiplication. Hence A is a K-algebra. We call $A = A_1 \otimes A_2$ the *tensor product* of A_1 and A_2.

The tensor product of algebras is characterized by a universal property, which we proceed to describe. We note first that the maps $x_1 \rightsquigarrow x_1 \otimes 1_2$, $x_2 \rightsquigarrow 1_1 \otimes x_2$ are homomorphisms of A_1 and A_2 respectively into $A_1 \otimes A_2$. (These need not be injective.) We have

$$(x_1 \otimes 1_2)(1_1 \otimes x_2) = x_1 \otimes x_2 = (1_1 \otimes x_2)(x_1 \otimes 1_2), \tag{71}$$

that is, if $e_1 : x_1 \to x_1 \otimes 1_2$ and $e_2 : x_2 \to 1_1 \otimes x_2$, then $e_1(x_1)e_2(x_2) = e_2(x_2)e_1(x_1)$ for all $x_1 \in A_1$, $x_2 \in A_2$. Now suppose f_i is a homomorphism of A_i, $i = 1, 2$, into an algebra B such that $f_1(x_1)f_2(x_2) = f_2(x_2)f_1(x_1)$, $x_i \in A_i$. Then we claim that we have a unique algebra homomorphism f of $A_1 \otimes A_2$ into B such that $fe_i = f_i$. To prove this we define a map $(x_1, x_2) \quad f_1(x_1)f_2(x_2)$ of $A_1 \times A_2$ into B. This is K-bilinear, so we have a K-module homomorphism f of $A_1 \otimes A_2$ into B such that $f(x_1 \otimes x_2) = f_1(x_1)f_2(x_2)$. Since $f(1) = f(1_1 \otimes 1_2) = f_1(1_1)f_2(1_2) = 11 = 1$ and

$$
\begin{aligned}
f((x_1 \otimes x_2)(y_1 \otimes y_2)) &= f(x_1 y_1 \otimes x_2 y_2) \\
&= f_1(x_1 y_1)f_2(x_2 y_2) = f_1(x_1)f_1(y_1)f_2(x_2)f_2(y_2) \\
&= f_1(x_1)f_2(x_2)f_1(y_1)f_2(y_2) = f(x_1 \otimes x_2)f(y_1 \otimes y_2),
\end{aligned}
$$

f is a K-algebra homomorphism. We have $fe_1(x_1) = f(x_1 \otimes 1_2) = f_1(x_1)f_2(1_2) = f_1(x_1)$ and $fe_2(x_2) = f_2(x_2)$, so $fe_i = f_i$. The relation (71) and the fact that every element of $A_1 \otimes A_2$ is a sum $\sum_j x_{1j} \otimes x_{2j}$, $x_{ij} \in A_i$, imply that f is the only homomorphism of $A_1 \otimes A_2$ satisfying $fe_i = f_i$.

Tensor multiplication of algebras is associative and commutative in the sense of the following

THEOREM 3.16. *We have an isomorphism of $A_1 \otimes A_2$ onto $A_2 \otimes A_1$ mapping $x_1 \otimes x_2$ into $x_2 \otimes x_1$ and an isomorphism of $(A_1 \otimes A_2) \otimes A_3$ onto $A_1 \otimes (A_2 \otimes A_3)$ mapping $(x_1 \otimes x_2) \otimes x_3$ into $x_1 \otimes (x_2 \otimes x_3)$, $x_i \in A_i$.*

Proof. We have seen that we have module isomorphisms of the form indicated. Direct verification shows that these are algebra maps. □

We have observed several instances in which the use of Definition 3.6 for an algebra is advantageous. Perhaps its most important advantage is that this definition suggests a dual concept. We present this as

DEFINITION 3.7. *A* coalgebra over K (*or* K-coalgebra) *is a triple* (C, δ, α) *such that C is a K-module, δ is a K-homomorphism $C \to C \otimes_K C$, and α is a K-homomorphism $C \to K$ such that the following diagrams are commutative:*

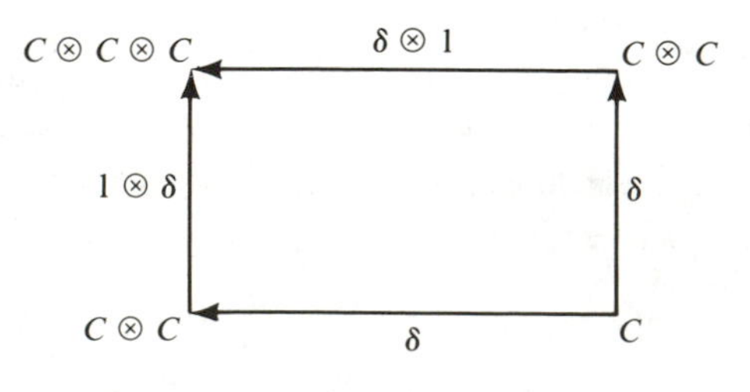

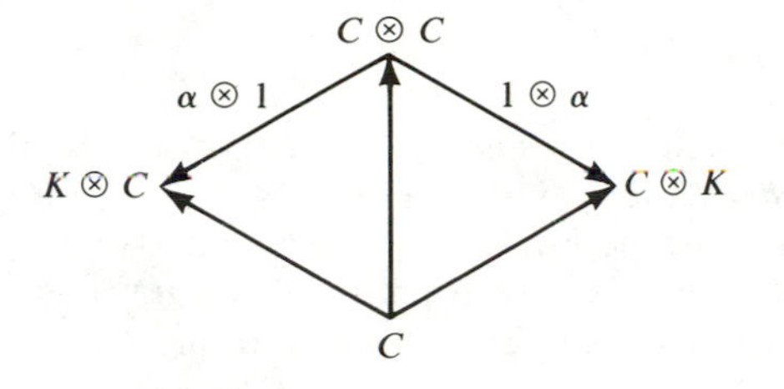

where the maps $C \to C \otimes K$ *and* $C \to K \otimes C$ *are* $x \rightsquigarrow x \otimes 1$ *and* $x \rightsquigarrow 1 \otimes x$. *The map* δ *is called a* diagonalization *and* α *is called an* augmentation.

It is nice to have a pretty definition, but it is even nicer to have some pretty examples. Here are a couple for coalgebras.

EXAMPLES

1. Let G be a group and let $K[G]$ be the group algebra over K of G (see BAI, p. 127 (exercise 8) and p. 408). This has G as module base over K and the product in $K[G]$ is given by the formula $(\sum k_i s_i)(\sum l_j t_j) = \sum k_i l_j s_i t_j$ where $k_i, l_j \in K$, $s_i, t_j \in G$. As usual, we identify the element $1s$, $s \in G$, with s and so regard G as imbedded in $K[G]$. The fundamental property of $K[G]$ is that if g is a homomorphism of G into the multiplicative group of invertible elements of an algebra A, then g has a unique extension to a homomorphism of $K[G]$ into A. Now take $A = K[G] \otimes_K K[G]$. It is clear that the map $s \rightsquigarrow s \otimes s$ of G into A is a group homomorphism, so this has an extension to an algebra homomorphism δ of $K[G]$ into $K[G] \otimes K[G]$. We also have a unique algebra homomorphism α of $K[G]$ into K extending the homomorphism $s \rightsquigarrow 1$ of G into the multiplicative group of K. Then $(K[G], \delta, \alpha)$ is a coalgebra. We leave it to the reader to verify the conditions in the definition.

2. Let $U(L)$ be the enveloping algebra of a Lie algebra L. If u is the map of L into $U(L)$, then we have $u([xy]) = [u(x), u(y)]$ $(= u(x)u(y) - u(y)u(x))$ for $x, y \in L$. Put $u_2(x) = u(x) \otimes 1 + 1 \otimes u(x) \in U(L) \otimes U(L)$. It follows from the commutativity of any $a \otimes 1$ and $1 \otimes b$, $a, b \in U(L)$ that

$$[u_2(x), u_2(y)] = u_2([xy])$$

so u_2 is a homomorphism of L into $U(L) \otimes U(L)$. This has a unique extension to a homomorphism δ of $U(L)$ into $U(L) \otimes U(L)$. Similarly, we have a homomorphism α of $U(L)$ into K such that $u(x) \rightsquigarrow 0$ for all $x \in L$. We leave it to the reader to verify that $(U(L), \delta, \alpha)$ is a coalgebra.

Both of the preceding examples are algebras as well as coalgebras. Moreover, the maps δ and α are algebra homomorphisms. In this situation, the composite system $(C, \pi, \varepsilon, \delta, \alpha)$ is called a *bialgebra*. Bialgebras of a special type, now called Hopf algebras, were introduced by Heinz Hopf in his studies of the topology of Lie groups.

EXERCISES

1. Let M and N be vector spaces over a field K, M^* and N^* the dual spaces. Verify that if $f \in M^*$, $g \in N^*$, then $(x, y) \to f(x)g(y)$ is a bilinear map of $M \times N$ into K. Hence there is a unique $h \in (M \otimes N)^*$ such that $h(x \otimes y) = f(x)g(y)$. Show that there is a vector space monomorphism of $M^* \otimes N^*$ into $(M \otimes N)^*$ such that $f \otimes g \to h$ and that this is an isomorphism if M and N are finite dimensional.

2. Let M and N be finite dimensional vector spaces over a field K. Show that there is an isomorphism of $\text{End}_K M \otimes_K \text{End}_K N$ onto $\text{End}_K(M \otimes_K N)$ such that $f \otimes g$ for $f \in \text{End}_K M$ and $g \in \text{End}_K N$ is mapped into the linear transformation of $M \otimes N$ such that $x \otimes y \to f(x) \otimes g(y)$. Does this hold if M or N is infinite dimensional?

3. Let $(A, \pi_A, \varepsilon_A)$ and $(B, \pi_B, \varepsilon_B)$ be algebras over K. Show that a K-homomorphism $\theta: A \to B$ is an algebra homomorphism if and only if the diagrams

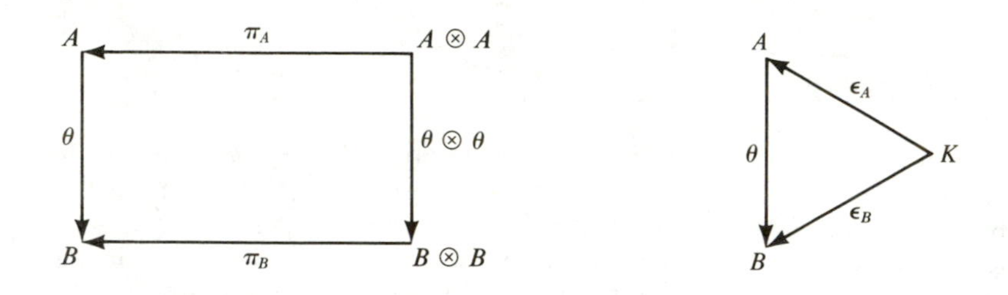

commute.

Dualize these diagrams to obtain a definition of homomorphism of coalgebras.

4. Let (C, π, ε) be an algebra and (C, δ, α) be a coalgebra. Show that $(C, \pi, \varepsilon, \delta, \alpha)$ is a bialgebra if and only if π and ε are coalgebra homomorphisms.

5. Let (C, δ, α) be a coalgebra over a field K and let ρ denote the monomorphism of $C^* \otimes C^*$ into $(C \otimes C)^*$ given in exercise 1. Put $\pi = \delta^* \rho$ where $\delta^*: (C \otimes C)^* \to C^*$ is the transpose of the map $\delta: C \to C \otimes C$ (p. 21). Let $\varepsilon = \alpha^* i$ where i is the canonical isomorphism of K onto $K^* = \hom(K, K)$ and α^* is the transpose of α. Verify that (C^*, π, ε) is an algebra. This is called the *dual algebra* of (C, δ, α).

6. Let (A, π, ε) be a finite dimensional algebra over a field K. Let τ be the isomorphism of $(A \otimes A)^*$ onto $A^* \otimes A^*$ given in exercise 1. Put $\delta = \tau \pi^*$ where $\pi^*: A^* \to (A \otimes A)^*$ is the transpose of π and $\alpha = i^{-1} \varepsilon^*$ where i is as in exercise 5 and ε^* is the transpose of ε. Verify that (A^*, δ, α) is a coalgebra. This is called the *dual coalgebra* of A.

7. Let $C(n)$ have the base $\{x_{ij} \mid 1 \leqslant i, j \leqslant n\}$ over a field K. Let δ and α be the linear maps such that $\delta(x_{ij}) = \sum_{k=1}^{n} x_{ik} \otimes x_{kj}$ and $\alpha(x_{ij}) = \delta_{ij}$. Verify that $(C(n), \delta, \alpha)$ is the dual coalgebra of $M_n(K)$.

8. If (A, π, ε) is an algebra, a left A-module is defined as a K-module M and a map $\mu: A \otimes_K M \to M$ such that

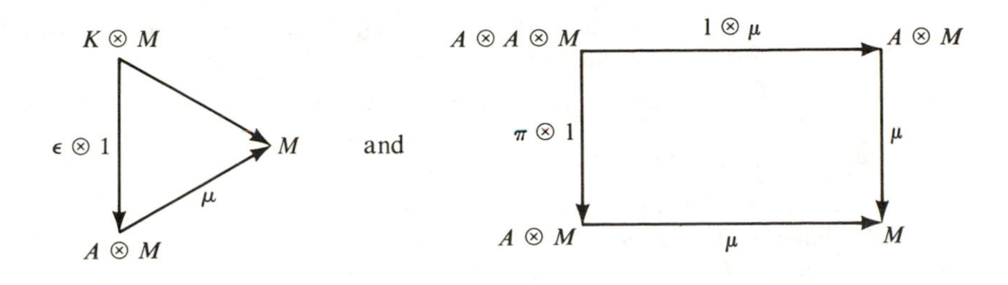

commute.

Dualize to define comodules for coalgebras.

9. An algebra A is called *filtered* if for every $i = 0, 1, 2, \ldots$ there is defined a subspace $A^{(i)}$ such that (i) $A^{(i)} \subset A^{(j)}$ if $i \leqslant j$, (ii) $\bigcup A^{(i)} = A$, (iii) $A^{(i)}A^{(j)} \subset A^{(i+j)}$. Show that if X is a set of generators of A, then A is filtered by defining $A^{(i)} = K1 + KX + KX^2 + \cdots + KX^i$ where KY is the K-submodule spanned by the set Y and X^j is the set of products of j elements in X.

10. Let A be a filtered algebra and define $G(A) = \oplus_0^\infty \bar{A}_i$ where $\bar{A}_i = A^{(i)}/A^{(i-1)}$ for $i > 0$, $\bar{A}_0 = A^{(0)}$. Show that $G(A)$ can be given the structure of a graded algebra in which $(a_i + A^{(i-1)})(a_j + A^{(j-1)}) = a_i a_j + A^{(i+j-1)}$. $G(A)$ is called the *associated graded algebra* of the filtered algebra A.

11. If K is a commutative ring, let $\Delta = K[\lambda]/(\lambda^2)$, λ an indeterminate. Show that 1 and the coset $d = \lambda + (\lambda^2)$ form a K-base for Δ and $d^2 = 0$. Δ is called the algebra of *dual numbers* over K. Let A be a K-algebra and form the algebra $\Delta \otimes_K A$. Note that this is a right A-module in which $(b \otimes x)y = b \otimes xy$ for $b \in \Delta$, $x, y \in A$. Show that 1 $(= 1 \otimes 1)$ and $d \equiv d \otimes 1$ form a base for $\Delta \otimes A$ over A and we have

$$(x_1 + dy_1)(x_2 + dy_2) = x_1x_2 + d(x_1y_2 + y_1x_2)$$

where $x_i = 1 \cdot x_i$ and $x_i, y_i \in A$. Show that a K-endomorphism D of A is a derivation, that is,

$$D(xy) = (Dx)y + x(Dy)$$

for $x, y \in A$ if and only if the Δ-endomorphism of $\Delta \otimes_K A$ such that $x \rightsquigarrow x + d(Dx)$, $x \in A$, is an automorphism of $\Delta \otimes_K A$. Show that if $a \in A$, then $1 + da$ is invertible in $\Delta \otimes_K A$ with inverse $1 - da$ and $(1 + da)x(1 - da) = x + d(ax - xa)$. Hence conclude that $x \rightsquigarrow [a, x] = ax - xa$ is a derivation. This is called the *inner derivation* determined by a.

12. Generalize exercise 11 by replacing Δ by $\Delta^{(n)} = K[\lambda]/(\lambda^n)$, $n \geqslant 2$, to obtain $\Delta^{(n)} \otimes_K A$, which is a right A-module with base $(1, d, d^2, \ldots, d^{n-1})$ where $d = \lambda + (\lambda^n)$ and d is identified with the element $d \otimes 1$ of $\Delta^{(n)} \otimes A$. Call a sequence $D \equiv (D_0 = 1, D_1, \ldots, D_{n-1})$ of K-endomorphisms of A a *higher derivation of order* n if

$$D_i(xy) = \sum_{j+k=i} (D_j x)(D_k y)$$

for $x, y \in A$, $0 \leqslant i \leqslant n-1$. Show that $D = (1, D_1, \ldots, D_{n-1})$ is a higher derivation of order n if and only if the Δ-endomorphism of $\Delta \otimes_K A$ such that

$$x \rightsquigarrow x + d(D_1 x) + d^2(D_2 x) + \cdots + d^{n-1}(D_{n-1} x)$$

is an algebra automorphism.

13. Let M and N be modules over a commutative ring K, A and B algebras over K, K' a commutative algebra over K, K'' a commutative algebra over K'. Write $M_{K'} = K' \otimes_K M$ etc. Note that K'' is an algebra over K in which $kk'' = (k1)k''$ for $k \in K$, $k'' \in K''$ and 1 is the unit of K'. Prove the following isomorphisms:
    (i) $M_{K'} \otimes_{K'} N_{K'} \cong (M \otimes_K N)_{K'}$ as K'-modules.
    (ii) $A_{K'} \otimes_{K'} B_{K'} \cong (A \otimes_K B)_{K'}$ as K'-algebras.
    (iii) $(M_{K'})_{K''} \cong M_{K''}$ as K''-modules.
    (iv) $(A_{K'})_{K''} \cong A_{K''}$ as K''-algebras.

3.10 PROJECTIVE MODULES

In this section we shall introduce a class of modules that are more general than free modules and that, from the categorical point of view, constitute a more natural class. We recall a basic homomorphism lifting property of free Ω-algebras: If F is a free Ω-algebra and λ is a surjective homomorphism of the Ω-algebra A onto the Ω-algebra $\bar{A}$, then any homomorphism $\bar{f}$ of F into $\bar{A}$ can be "lifted" to a homomorphism f of F into A, that is, there exists a homomorphism f of F into A such that $\bar{f} = \lambda f$ (p. 80). A similar result holds for free modules. If we have a diagram

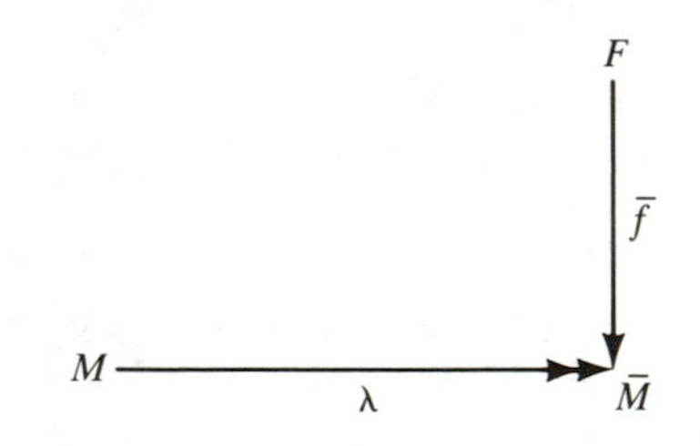

where F is free, we can complete it by an $f: F \to M$ to obtain a commutative diagram. The proof is the same as for Ω-algebras: Let X be a base for F. For each $x \in X$ choose an element $u \in M$ such that $\lambda u = \bar{f} x$ and let f be the homomorphism of F into M such that $fx = u$. Then $\lambda f x = \bar{f} x$ for all $x \in X$ so $\lambda f = \bar{f}$, as required.

We single out this property of free modules as the basis for the following generalization.

DEFINITION 3.8. *A module P is called* projective *if given any diagram*

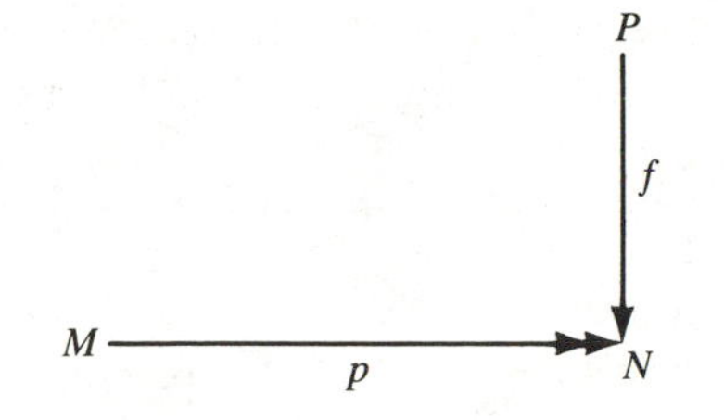

there exists a homomorphism $g: P \to M$ *such that*

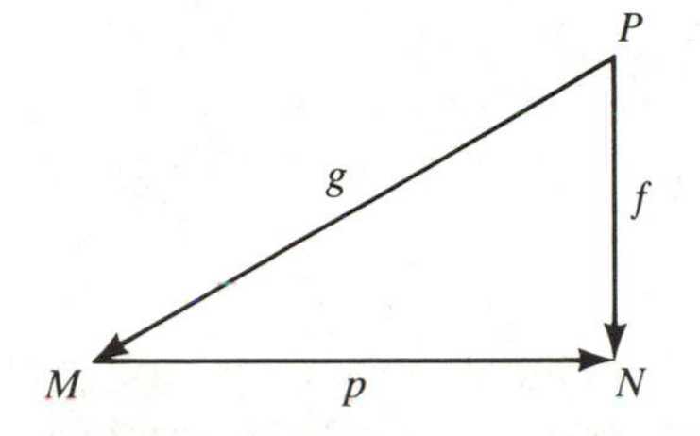

is commutative. In other words, given an epimorphism $p: M \to N$*, then any homomorphism* $f: P \to N$ *can be factored as* $f = pg$ *for some* $g: P \to M$.

We recall that for any module M, the functor $\hom(M, -)$ is left exact. Hence if

$$0 \to N' \xrightarrow{i} N \xrightarrow{p} N'' \to 0 \tag{72}$$

is exact, then

$$0 \to \hom(M, N') \xrightarrow{\hom(M,i)} \hom(M, N) \xrightarrow{\hom(M,p)} \hom(M, N'')$$

is exact. Now suppose $M = P$ is projective. Then given $f \in \hom(P, N'')$ there exists a $g \in \hom(P, N)$ such that $\hom(P, p)(g) = pg = f$. Thus in this case, $\hom(P, p)$ is surjective and so we actually have the exactness of

$$0 \to \hom(P, N') \xrightarrow{\hom(P,i)} \hom(P, N) \xrightarrow{\hom(P,p)} \hom(P, N'') \to 0$$

as a consequence of the exactness of (72). Hence if P is projective, then $\hom(P, -)$ is an exact functor from **mod**-R to **mod**-$\mathbb{Z}$.

The converse holds also. Suppose $\hom(P, -)$ is exact and suppose $M \overset{p}{\twoheadrightarrow} N$. Let $K = \ker p$. Then we have the exact sequence $0 \to K \xrightarrow{i} M \xrightarrow{p} N \to 0$ where i is the injection map. Applying the exactness of $\hom(P, -)$, we obtain the property stated in Definition 3.8. We have therefore established the following categorical characterization of projective modules.

PROPOSITION 3.9. *A module* P *is projective if and only if* $\hom(P, -)$ *is an exact functor.*

How close are projective modules to being free? We shall give some answers to this question. First, we need a simple result on short exact sequences

$$0 \to M' \xrightarrow{i} M \xrightarrow{p} M'' \to 0. \tag{73}$$

Suppose that for such a sequence there exists a homomorphism $i': M'' \to M$ such that $pi' = 1_{M''}$. Then $p(1_M - i'p) = p - p = 0$. This implies that there exists a $p': M \to M'$ such that $1_M - i'p = ip'$. For, if $x \in M$ then $p(1_M - i'p)x = 0$, so $(1_M - i'p)x$ is in the kernel of p, hence, in the image of i. Then we have an $x' \in M'$ such that $ix' = (1_M - i'p)x$ and x' is unique since i is a monomorphism. We now have a map $p': x \rightsquigarrow x'$ of M into M'. It follows directly that p' is a module homomorphism and $ix' = ip'x$ so $ip' = 1_M - i'p$ or

$$i'p + ip' = 1_M. \tag{74}$$

Also we have $i'pi'p = i'1_{M''}p = i'p$. Hence multiplying (74) on the right by $i'p$ gives $ip'i'p = 0$. Since p is surjective and i is injective, this implies that $p'i' = 0$ (on M''). If $x' \in M'$, then $ip'ix' = (1_M - i'p)ix' = ix'$ and since i is injective, we have $p'i = 1_{M'}$. Since we had at the outset $pi = 0$ and $pi' = 1_{M''}$, we have the four relations

$$\begin{aligned} pi' &= 1_{M''}, \qquad & p'i' &= 0 \\ p'i &= 1_{M'}, \qquad & pi &= 0. \end{aligned} \tag{75}$$

These together with (74) imply that M is canonically isomorphic to $M' \oplus M''$ (see p. 97).

In a similar manner, suppose there exists a $p': M \to M'$ such that $p'i = 1_{M'}$. Then $(1_M - ip')i = i - i1_{M'} = i - i = 0$. This implies that $1_M - ip' = i'p$ for a homomorphism $i': M'' \to M$. For, if f is any homomorphism $M \to M$ such that $fi = 0$, then $\ker p = \operatorname{im} i \subset \ker f$. Hence $px \rightsquigarrow fx$ is a well-defined map g of $pM = M''$ into M. Direct verification shows that this is a module homomorphism. Evidently, we have $gp = f$. Applying this to $f = 1_M - ip'$, we obtain $i': M'' \to M$ such that $1_M = ip' + i'p$. Since $ip'ip' = i1_{M'}p' = ip'$, we obtain $ip'i'p = 0$, which implies $p'i' = 0$. Since $pi'p = p(1_M - ip') = p - pip' = p$ (by $pi = 0$), $pi'p = p$. Since p is surjective, this implies that $pi' = 1_{M''}$. Again we have the relations (74) and (75), so $M \cong M' \oplus M''$ canonically.

We shall now say that the exact sequence (73) *splits* if there exists an $i': M'' \to M$ such that $pi' = 1_{M''}$ or, equivalently, there exists a $p': M \to M'$ such that $p'i = 1_{M'}$.

We can now give two important characterizations of projective modules.

PROPOSITION 3.10. *The following properties of a module P are equivalent*:

(1) *P is projective.*

(2) *Any short exact sequence $0 \to M \to N \to P \to 0$ splits.*

(3) *P is a direct summand of a free module (that is, there exists a free module F isomorphic to $P \oplus P'$ for some P').*

Proof. (1) ⇒ (2). Let $0 \to M \xrightarrow{f} N \xrightarrow{g} P \to 0$ be exact and consider the diagram

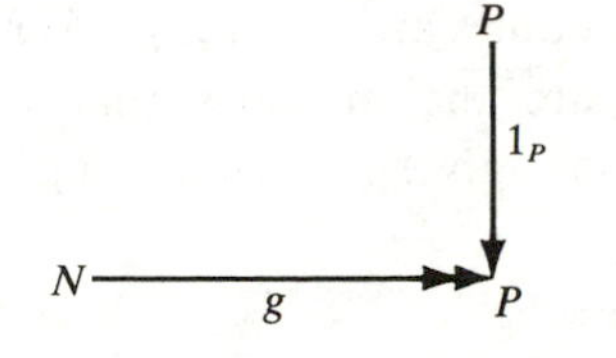

By hypothesis we can fill this in with $g': P \to N$ to obtain a commutative diagram. Then $gg' = 1_P$ and the given short exact sequence splits.

(2) ⇒ (3). Since any module is a homomorphic image of a free module, we have a short exact sequence $0 \to P' \xrightarrow{i} F \xrightarrow{p} P \to 0$ where F is a free module. If P satisfies property 2, then $0 \to P' \xrightarrow{i} F \xrightarrow{p} P \to 0$ splits and hence $F \cong P \oplus P'$.

(3) ⇒ (1). We are given that there exists a split exact sequence $0 \to P' \xrightarrow{i} F \xrightarrow{p} P \to 0$ with F free. Now suppose we have a diagram

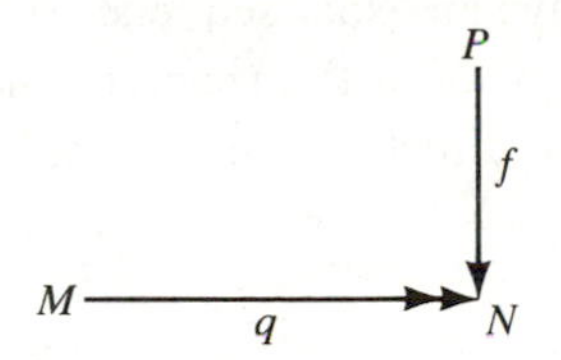

Combining the two diagrams, we obtain

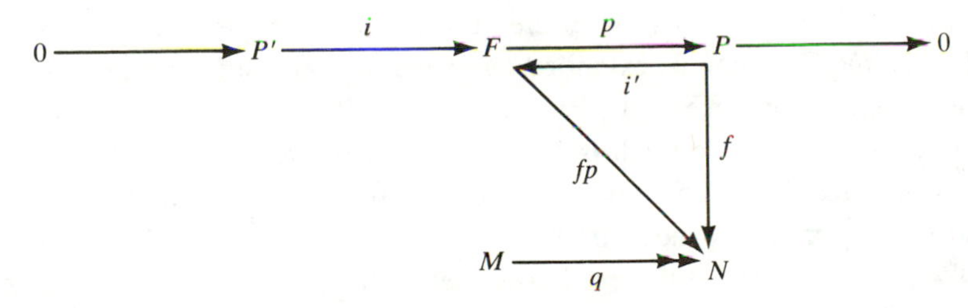

where $pi' = 1_P$ (since the top line splits). Since F is free, hence projective, we can fill in $g: F \to M$ to obtain $fp = qg$. Then $f = f1_P = fpi' = qgi'$ and $gi': P \to M$ makes

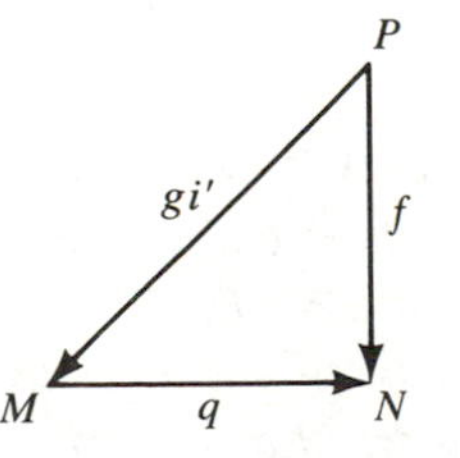

commutative. Hence P is projective. □

As a first consequence of Proposition 3.10, we note that there exist projective modules that are not free. The simplest example is perhaps the following one. Let $F = \mathbb{Z}/(6)$ regarded as a module for $R = \mathbb{Z}/(6)$, so F is free. Now $F \cong \mathbb{Z}/(2) \oplus \mathbb{Z}/(3)$. Hence $\mathbb{Z}/(2)$ and $\mathbb{Z}/(3)$ are projective by Proposition 3.10 and evidently these modules are not free $\mathbb{Z}/(6)$-modules.

Of particular interest are the modules that are finitely generated and projective. The proposition gives the following characterization of these modules.

COROLLARY. *A module P is finitely generated and projective if and only if P is a direct summand of a free module with a finite base.*

Proof. If P is a direct summand of a free module F with finite base, then P is projective. Moreover, P is a homomorphic image of F, so P has a finite set of generators (the images of the base under an epimorphism of F onto P). Conversely, suppose P is finitely generated and projective. Then the first condition implies that we have an exact sequence $0 \to P' \to F \to P \to 0$ where F is free with finite base. The proof of the theorem shows that if P is projective, then $F \cong P \oplus P'$, so P is a direct summand of a free module with finite base. □

Some rings have the property that all of their projective modules are free. This is the case for p.i.d. (commutative principal ideal domains). We showed in BAI, pp. 179–180, that any submodule of a free module with a finite base over a p.i.d. D is free. This can be extended to prove that any submodule of any free module over D is free. By Proposition 3.10.3, this implies that any projective module over a p.i.d. is free. We shall prove later (p. 416) that any finitely generated projective module over a local ring is free. Another result of this sort, proved independently by D. Quillen and A. Suslin, is that if $R = D[\lambda_1, \dots, \lambda_r]$ where D is a p.i.d., then every finitely generated module over R is free. Their work settled a question that had been raised in 1955 by J.-P. Serre and that had been unanswered for more than twenty years.

We shall give next an "elementary" criterion for projectivity that is extremely useful.

PROPOSITION 3.11. *A (right) R-module P is projective if and only if there exists a set $\{x_\alpha | \alpha \in I\}$ of elements P and elements $\{x_\alpha^* | \alpha \in I\}$ of the dual module $P^* = \hom(P, R)$ such that for any $x \in P$, $x_\alpha^*(x) = 0$ for all but a finite number of the x_α^*, and $x = \sum_{\alpha \in I} x_\alpha x_\alpha^*(x)$ (with the obvious meaning of this sum). If the condition holds, then $\{x_\alpha\}$ is a set of generators and if P is projective, then $\{x_\alpha\}$ can be taken to be any set of generators for P.*

Proof. Suppose first that P is projective and $\{x_\alpha | \alpha \in I\}$ is a set of generators for P. We have a free module F with base $\{X_\alpha | \alpha \in I\}$ and an epimorphism $p: F \to P$ such that $p(X_\alpha) = x_\alpha$. Since P is projective, we have a homomorphism $i': P \to F$ such that $pi' = 1_P$. Since $\{X_\alpha\}$ is a base for F, any X can be written as a sum $X_{\alpha_1}a_{\alpha_1} + X_{\alpha_2}a_{\alpha_2} + \cdots + X_{\alpha_r}a_{\alpha_r}$ where the $a_{\alpha_j} \in R$. We may write also $X = \sum_{\alpha \in I} X_\alpha a_\alpha$ where only a finite number of the a_α are $\neq 0$ and the sum is taken over the α_i for which $a_{\alpha_i} \neq 0$. Since the X_α form a base, the a_α (0 or not) are uniquely determined by X. Hence we have the maps $X_\alpha^*: X \rightsquigarrow a_\alpha$ and for a particular X we have $X_\alpha^*(X) = 0$ for all but a finite number of the α. It is immediate from the definition of X_α^* that $X_\alpha^* \in F^* = \hom(F, R)$. Then $x_\alpha^* = X_\alpha^* i' \in P^*$ and for any $x \in P$, $x_\alpha^*(x) = X_\alpha^*(i'(x)) = 0$ for all but a finite set of α. If $x \in P$ we have $i'(x) = X \in F$ and $X = \sum X_\alpha X_\alpha^*(X)$. Since $x_\alpha = p(X_\alpha)$, we have $x = pi'(x) = p(X) = \sum p(X_\alpha) X_\alpha^*(X) = \sum x_\alpha X_\alpha^*(X) = \sum x_\alpha X_\alpha^* i'(x) = \sum x_\alpha x_\alpha^*(x)$. Thus $\{x_\alpha\}$ and $\{x_\alpha^*\}$ have the stated properties. Conversely, assume that for a module P we have $\{x_\alpha\}$, $\{x_\alpha^*\}$ with the stated properties. Again let F be the free module with base $\{X_\alpha\}$ and let p be the homomorphism $F \to P$ such that $p(X_\alpha) = x_\alpha$. For each α we have the homomorphism $x \rightsquigarrow X_\alpha x_\alpha^*(x)$ of P into F. Since for a given x, $x_\alpha^*(x) = 0$ for all but a finite number of α, we have a homomorphism $i': P \to F$ such that $i'(x) = \sum X_\alpha x_\alpha^*(x)$. Then $pi'(x) = \sum p(X_\alpha) x_\alpha^*(x) = \sum x_\alpha x_\alpha^*(x) = x$. Thus $pi' = 1_P$, so we have a split exact sequence $0 \to P' \xrightarrow{i} F \xrightarrow{p} P \to 0$. Then P is a direct summand of F and hence P is projective. □

The result proved in Proposition 3.11 is often called the "dual basis lemma" for projective modules. It should be noted, however, that the x_α in the statement need not be a base for P.

The dual basis lemma is usually used to characterize finitely generated projective modules. Proposition 3.11 and its proof have the following immediate consequence.

COROLLARY. *A module P is finitely generated and projective if and only if there exists a finite set of pairs (x_i, x_i^*) where $x_i \in P$ and $x_i^* \in P^*$ such that for any $x \in P$ we have $x = \sum x_i x_i^*(x)$.*

There are a number of important properties of free modules that carry over to projective modules. One of these is flatness, which we define in

DEFINITION 3.9. *A right module $M = M_R$ is called* flat *if for every monomorphism $N' \xrightarrow{f} N$ of left R-modules we have the monomorphism $M \otimes N' \xrightarrow{1 \otimes f} M \otimes N$ of $\mathbb{Z}$-modules.*

We recall that the functor $M \otimes_R$ from **R-mod** to **$\mathbb{Z}$-mod** is right exact (Theorem 3.15, p. 132). As in the proof of Proposition 3.9, we have the following immediate consequence

PROPOSITION 3.12. *A right module M is flat if and only if the tensor functor $M \otimes_R$ is exact.*

The main result on flatness that we shall establish in this section is that projectives are flat. The proof will follow quite readily from the following

PROPOSITION 3.13. *$M = \otimes M_\alpha$ is flat if and only if every M_α is flat.*

Proof. Suppose $N' \; ^{f} \; N$ for left modules N' and N. Our result will follow if we can show that $M \otimes N' \qquad M \otimes N$ if and only if $M_\alpha \otimes N' \qquad M_\alpha \otimes N$ for every α. If we use the isomorphisms η_N and $\eta_{N'}$ given in Proposition 3.3 and the naturality given by (54), we see that it suffices to show that we have $\otimes(M_\alpha \otimes N') \; ^{f} \; \otimes(M_\alpha \otimes N)$ if and only if $M_\alpha \otimes N' \quad M_\alpha \otimes N$ for every α. This is clear from the definition of f^*:f^* is injective if and only if every $1_{M_\alpha} \otimes f$ is injection. □

We can now prove

THEOREM 3.17. *Projective modules are flat.*

Proof. Observe first that $R = R_R$ is flat, since we have an isomorphism of $R \otimes N$ into N sending $a \otimes y \rightarrow ay$. Hence $N' \xrightarrow{f} N$ is injective if and only if $R \otimes N' \qquad R \otimes N$ is injective. Both the flatness of R and Proposition 3.13 imply that any free R-module is flat. Since any projective module is a direct summand of a free module, another application of Proposition 3.13 gives the result that any projective module is flat. □

The technique of reduction to R by means of a result on direct sums is a standard one for establishing properties of projective modules. As another illustration of this method, we sketch a proof of

PROPOSITION 3.14. *Let K be a commutative ring, P a finitely generated projective K-module, N an arbitrary K-module. Then the homomorphism of* $\mathrm{End}_K P \otimes \mathrm{End}_K N$ *into* $\mathrm{End}_K(P \otimes N)$ *sending $f \otimes g$ for $f \in \mathrm{End}\, P$, $g \in \mathrm{End}\, N$, into the endomorphism $f \otimes g$ of $P \otimes N$ such that $(f \otimes g)\,(x \otimes y) = f(x) \otimes g(y)$ is an isomorphism (as K-algebras).*

Proof. It is clear that the homomorphism is an algebra homomorphism. Moreover, it is an isomorphism if $P = K$. Now suppose $M = \oplus_1^n M_j$. Then we have a canonical isomorphism of $\mathrm{End}\, M = \hom(M, M)$ onto $\oplus_{j,k} \hom(M_j, M_k)$ and of $\mathrm{End}\, M \otimes \mathrm{End}\, N$ onto $\oplus(\hom(M_j, M_k) \otimes \mathrm{End}\, N)$. Also we have a canonical isomorphism of $\mathrm{End}\,(M \otimes N) = \hom(M \otimes N, M \otimes N)$

onto $\oplus(\hom(M_j \otimes N), (M_k \otimes N))$. Using these isomorphisms, we see that the homomorphism of End $M \otimes$ End N into End $(M \otimes N)$ sending $f \otimes g$ into $f \otimes g$, as in the statement of the proposition, is an isomorphism if and only if for every (j, k) the same map of $\hom(M_j, M_k) \otimes \text{End } N$ into $\hom(M_j \otimes N, M_k \otimes N)$ is an isomorphism. Taking the $M_j = K$, we see that the result stated holds for P, a free module with a finite base. Then the result follows for finitely generated projective P, since such a P is a direct summand of a free module with a finite base. □

We leave it to the reader to fill in the details of this proof.

EXERCISES

1. Show that if we have a diagram

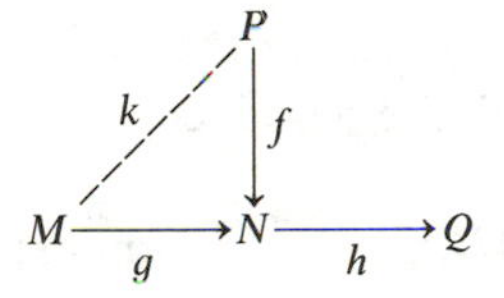

in which the row is exact, P is projective, and $hf = 0$ then there exists a $k: P \to M$ such that $f = gk$.

2. Show that if P_α, $\alpha \in I$, is projective, then $\oplus P_\alpha$ is projective.

3. Show that if e is an idempotent in a ring R, then eR is a projective right module and Re is a projective left module.

4. Let $\{e_{ij}\}$ be the usual matrix units in $M_n(\Delta)$, Δ a division ring. Show that $e_{11}M_n(\Delta)$ is irreducible as right $M_n(\Delta)$-module. Hence conclude that $e_{11}M_n(\Delta)$ is projective but not free if $n > 1$.

5. Prove that any submodule of a free module over a p.i.d. D is free. (*Hint*: Extend the argument of BAI, p. 179, by transfinite induction or by Zorn's lemma.)

6. Show that the additive group of $\mathbb{Q}$ regarded as $\mathbb{Z}$-module is flat.

7. Let R and S be rings. Let P be a finitely generated projective left R-module, M an R-S-bimodule, N a left S-module. Show that there is a group isomorphism

$$\eta : \hom_R(P, M) \otimes_S N \to \hom_R(P, M \otimes_S N)$$

such that for $f \in \hom_R(P, M)$ and $y \in N$, $\eta(f \otimes y)$ is the homomorphism $x \rightsquigarrow f(x) \otimes y$ of P into $M \otimes_S N$.

8. (Schanuel's lemma.) Suppose we have a short exact sequence $0 \to N_i \to P_i \to M \to 0$ where P_i is projective and $i = 1, 2$. Show that $P_1 \oplus N_2 \cong P_2 \oplus N_1$. (*Hint*: Consider the pullback of g_1 and g_2 as constructed in exercise 6, p. 36.)

3.11 INJECTIVE MODULES. INJECTIVE HULL

The concept of a projective module has a dual obtained by reversing the arrows in the definition. This yields the following

DEFINITION 3.10. *A module Q is called* injective *if given any diagram of homomorphisms*

(76)

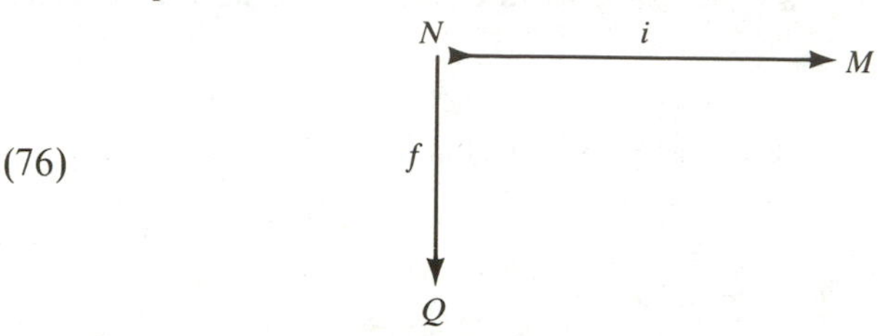

there exists a homomorphism $g: M \to Q$ such that the diagram obtained by filling in g is commutative. In other words, given $f: N \to Q$ and a monomorphism $i: N \to M$ there exists a $g: M \to Q$ such that $f = gi$.

With a slight change of notation, the definition amounts to this: Given an exact sequence $0 \to N' \xrightarrow{i} N$, the sequence

$$\hom(N, Q) \xrightarrow{\hom(i,Q)} \hom(N', Q) \to 0$$

is exact. Since we know that exactness of $N' \xrightarrow{i} N \xrightarrow{p} N'' \to 0$ implies exactness of

$$0 \to \hom(N'', M) \xrightarrow{\hom(p,M)} \hom(N, M) \xrightarrow{\hom(\iota,M)} \hom(N', M)$$

(Theorem 3.1, p. 99), it is clear that Q is injective if and only if the contravariant hom functor $\hom(-, Q)$ is exact in the sense that it maps any short exact sequence $0 \to N' \to N \to N'' \to 0$ into a short exact sequence

$$0 \to \hom(N'', Q) \to \hom(N, Q) \to \hom(N', Q) \to 0.$$

It is easily seen also that the definition of injective is equivalent to the following: If N is a submodule of a module M, then any homomorphism of N into Q can be extended to a homomorphism of M into Q. Another result, which is easily established by dualizing the proof of the analogous result on projectives (Proposition 3.9, p. 149), is that if Q is injective, then any short exact sequence $0 \to Q \to M \to N \to 0$ splits. The converse of this holds also. However, the proof requires the dual of the easy result that any module is a homomorphic image of a projective module (in fact, a free module). The dual statement is that any module can be imbedded in an injective one. We shall see that this is the case, but the proof will turn out to be fairly difficult.

The concept of an injective module was introduced in 1940 by Reinhold Baer before projective modules were thought of. Most of the results presented here are due to him; in particular, this is the case with the following criterion.

PROPOSITION 3.15. *A module Q is injective if and only if any homomorphism of a right ideal I of R into Q can be extended to a homomorphism of R into Q.*

Proof. Obviously, the condition is necessary. Now suppose it holds and suppose M is a module and f is a homomorphism of a submodule N of M into Q. Consider the set $\{(g, M')\}$ where M' is a submodule of M containing N and g is a homomorphism of M' into Q such that $g|N = f$. We define a partial order in the set $\{(g, M')\}$ by declaring that $(g_1, M'_1) \geqslant (g_2, M'_2)$ if $M'_1 \supset M'_2$ and $g_1|M'_2 = g_2$. It is clear that any totally ordered subset has an upper bound in this set. Hence, by Zorn's lemma, there exists a maximal (g, M'); that is, we have an extension of f to a homomorphism g of $M' \supset N$ which is maximal in the sense that if g_1 is a homomorphism of an $M'_1 \supset M'$ such that $g_1|M' = g$, then necessarily $M'_1 = M'$. We claim that $M' = M$. Otherwise, there is an $x \in M, \notin M'$ and so $xR + M'$ is a submodule of M properly containing M'. Now let

$$I = \{s \in R | xs \in M'\}. \tag{77}$$

Then $I = \text{ann}\,(x + M')$ in M/M', so I is a right ideal of R. If $s \in I$ then $xs \in M'$, so $g(xs) \in Q$. It is immediate that the map $h: s \rightsquigarrow g(xs)$ is a module homomorphism of I into Q. Hence, by hypothesis, h can be extended to a homomorphism k of R into Q. We shall use this to obtain an extension of g to a homomorphism of $xR + M'$ to Q. The elements of $xR + M'$ have the form $xr + y$, $r \in R$, $y \in M'$. If we have a relation $xs + y' = 0$, $s \in R$, $y' \in M'$, then the definition (77) shows that $s \in I$. Then

$$k(s) = h(s) = g(xs) = -g(y').$$

Thus $xs + y' = 0$ for $s \in R$, $y' \in M'$, implies that $k(s) + g(y') = 0$. It follows that

$$xr + y \rightsquigarrow k(r) + g(y), \tag{78}$$

$r \in R$, $y \in M'$, is a well-defined map. For, if $xr_1 + y_1 = xr_2 + y_2$, $r_i \in R$, $y_i \in M'$, then $xs + y' = 0$ for $s = r_1 - r_2$, $y' = y_1 - y_2$. Then $k(s) + g(y') = 0$ and $k(r_1 - r_2) + g(y_1 - y_2) = 0$. Since k and g are homomorphisms, this implies that $k(r_1) + g(y_1) = k(r_2) + g(y_2)$. Thus, (78) is single-valued. It is immediate that the map $rx + y \rightsquigarrow k(r) + g(y)$ is a module homomorphism of $xR + M'$ into Q extending the homomorphism g of M'. This contradicts the maximality of

(g, M'). Hence $M' = M$ and we have proved that if f is a homomorphism of a submodule N of M into Q, then f can be extended to a homomorphism of M into Q. Hence Q is injective. $\square$

For certain "nice" rings, the concept of injectivity of modules is closely related to the simpler notion of divisibility, which we proceed to define. If $a \in R$, then the module M is said to be *divisible by* a if the map $x \rightsquigarrow xa$ of M into M is surjective. A module is called *divisible* if it is divisible by every $a \neq 0$. It is clear that if M is divisible by a or if M is divisible, then any homomorphic image of M has the same property. In some sense injectivity is a generalization of divisibility, for we have

PROPOSITION 3.16. *If R has no zero divisors $\neq 0$, then any injective R-module is divisible. If R is a ring such that every right ideal of R is principal ($= aR$ for some $a \in R$), then any divisible R-module is injective.*

Proof. Suppose R has no zero-divisors $\neq 0$ and let Q be an injective R-module. Let $x \in Q$, $r \in R$, $r \neq 0$. If $a, b \in R$ and $ra = rb$, then $a = b$. Hence we have a well-defined map $ra \rightsquigarrow xa$, $a \in R$, of the right ideal rR into Q. Clearly this is a module homomorphism. Since Q is injective, the map $ra \rightsquigarrow xa$ can be extended to a homomorphism of R into Q. If $1 \rightsquigarrow y$ under this extension, then $r = 1r \rightsquigarrow yr$. Since $r = r1 \rightsquigarrow x1 = x$, we have $x = yr$. Since x was arbitrary in Q and r was any non-zero element of R, this shows that Q is divisible. Now suppose R is a ring in which every right ideal is principal. Let M be a divisible R-module and let f be a homomorphism of the right ideal rR into M. If $r = 0$, then f is the 0 map and this can be extended to the 0 map of R. If $r \neq 0$ and $f(r) = x \in M$, then there exists a y in M such that $x = yr$. Then $a \rightsquigarrow ya$ is a module homomorphism of R into M and since $rb \rightsquigarrow yrb = xb = f(r)b = f(rb)$, $a \rightsquigarrow ya$ is an extension of f. Thus any module homomorphism of a right ideal of R into M can be extended to a homomorphism of R. Hence M is injective by Baer's criterion. $\square$

If R satisfies both conditions stated in the proposition, then an R-module is injective if and only if it is divisible. In particular, this holds if R is a p.i.d. We can use this to construct some examples of injective modules.

EXAMPLES

1. Let R be a subring of a field F and regard F as an R-module in the natural way. Evidently F is a divisible R-module. Hence if K is any R-submodule of F, then F/K is a divisible R-module.

2. Let D be a p.i.d., F its field of fractions. If $r \in D$, then $F/(r)$ ($(r) = rD$) is divisible and hence is injective by Proposition 3.16.

Our next objective is to prove that any module can be imbedded in an injective module, that is, given any M there exists an exact sequence $0 \to M \xrightarrow{i} Q$ with Q injective. The first step in the proof we shall give is

LEMMA 1. *Any abelian group can be imbedded in a divisible group* ($=\mathbb{Z}$-*module*).

Proof. First let F be a free abelian group with base $\{x_\alpha\}$ and let F' be the vector space over $\mathbb{Q}$ with $\{x_\alpha\}$ as base. Then F is imbedded in F' and it is clear that F' is divisible. Now let M be an arbitrary abelian group. Then M is isomorphic to a factor group F/K of a free abelian group F. Then F'/K is a divisible group and $F/K \cong M$ is a subgroup. □

An immediate consequence of this and Proposition 3.16 is the

COROLLARY. *Any $\mathbb{Z}$-module can be imbedded in an injective $\mathbb{Z}$-module.*

We now consider an arbitrary R-module M. We have the isomorphism of M onto $\hom_R(R, M)$ which maps an element $x \in M$ into the homomorphism f_x such that $1 \to x$. This is an R-isomorphism if we make $\hom_R(R, M)$ into a right R-module as in Proposition 3.4 by defining fa, $a \in R$, by $(fa)(b) = f(ab)$. Also $\hom_{\mathbb{Z}}(R, M)$ is a right R-module using this definition of fa. Clearly, $\hom_R(R, M)$ is a submodule of $\hom_{\mathbb{Z}}(R, M)$. Since M is isomorphic to $\hom_R(R, M)$, we have an imbedding of M in $\hom_{\mathbb{Z}}(R, M)$. Now imbed M in an injective $\mathbb{Z}$-module Q, which can be done by the foregoing corollary. Then we have an imbedding of $\hom_{\mathbb{Z}}(R, M)$ into $\hom_{\mathbb{Z}}(R, Q)$ and hence of M in $\hom_{\mathbb{Z}}(R, Q)$ as R-modules. Now this gives an imbedding of M in an injective R-module, since we have the following

LEMMA 2. *If Q is a $\mathbb{Z}$-injective $\mathbb{Z}$-module, then* $\hom_{\mathbb{Z}}(R, Q)$ *is an injective R-module.*

Proof. We have to show that if $0 \to N' \xrightarrow{f} N$ is an exact sequence of R-modules, then

$$\hom_R(N, \hom_{\mathbb{Z}}(R, Q)) \xrightarrow{f^*} \hom_R(N', \hom_{\mathbb{Z}}(R, Q)) \to 0 \tag{79}$$

is exact where $f^* = \hom_R(f, \hom_{\mathbb{Z}}(R, Q))$. By Proposition 3.8 we have an isomorphism

$$\varphi_N : \hom_{\mathbb{Z}}(N \otimes_R R, Q) \to \hom_R(N, \hom_{\mathbb{Z}}(R, Q))$$

and the definition shows that this is natural in N. Since the isomorphism of $N \otimes_R R$ onto N such that $y \otimes 1 \rightarrow y$ is natural in N, we have an isomorphism

$$\psi_N : \hom_{\mathbb{Z}}(N, Q) \rightarrow \hom_R(N, \hom_{\mathbb{Z}}(R, Q))$$

which is natural in N, that is, we have the commutativity of

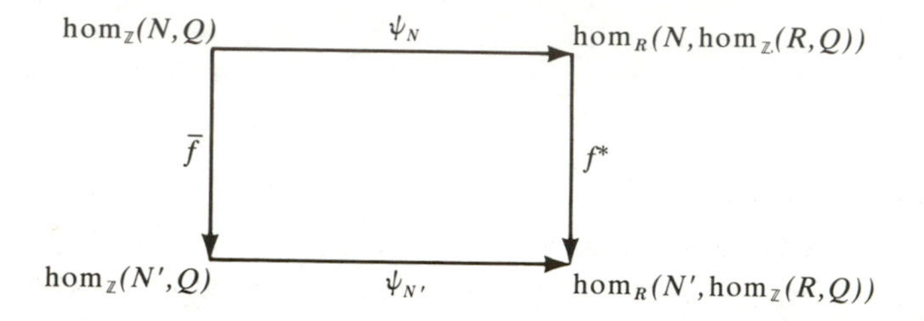

where $\bar{f} = \hom(f, Q)$. Now $\bar{f}$ is surjective since $\mathbb{Q}$ is $\mathbb{Z}$-injective. Since ψ_N and $\psi_{N'}$ are isomorphisms, this implies that f^* is surjective. Thus (79) is exact. □

The foregoing lemma completes the proof of the imbedding theorem.

THEOREM 3.18. *Any module can be imbedded in an injective module.*

The proof we have given is due to B. Eckmann and A. Schöpf.

We can apply the theorem to complete the following characterization of injectives, which we indicated earlier.

PROPOSITION 3.17. *A module Q is injective if and only if every short exact sequence $0 \rightarrow Q \xrightarrow{i} M \xrightarrow{p} N \rightarrow 0$ splits. Equivalently, Q is injective if and only if it is a direct summand of every module containing it as a submodule.*

Proof. We have seen that if Q is injective then every exact sequence $0 \rightarrow Q \rightarrow M \rightarrow N \rightarrow 0$ splits (p. 156). Conversely, suppose Q has this property. By the imbedding theorem we have an exact sequence $0 \rightarrow Q \xrightarrow{i} M$ where M is injective. Then we have the short exact sequence $0 \rightarrow Q \xrightarrow{i} M \xrightarrow{p} M/Q \rightarrow 0$ where p is the canonical homomorphism of M onto M/Q. By hypothesis, we can find a $p' : M \rightarrow Q$ such that $p'i = 1_Q$. Now suppose we have a diagram

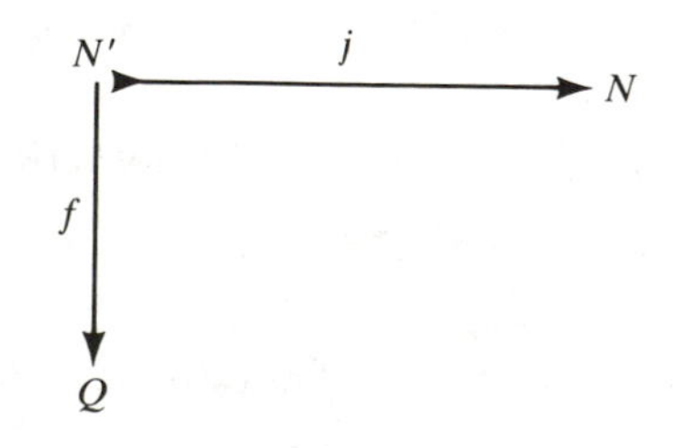

Since M is injective, we can enlarge this to a commutative diagram

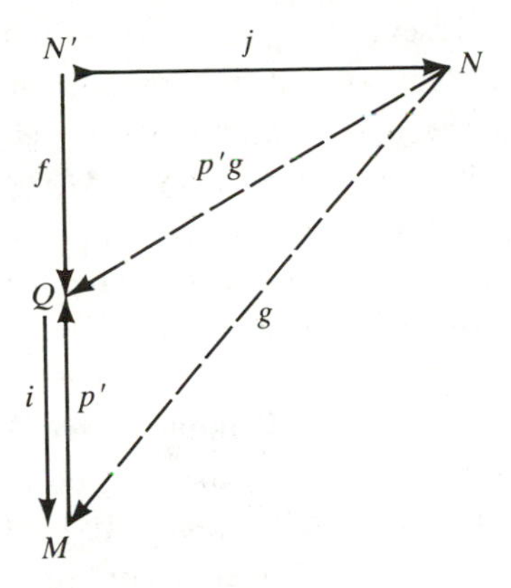

This means that by the injectivity of M we have $g: N \to M$ such that $if = gj$. Then $f = 1_Q f = p'if = (p'g)j$. Hence Q is injective. □

The imbedding theorem of modules in injective modules has an important refinement, namely, there exists such an imbedding, which is in a sense minimal, and any two minimal injective extensions are isomorphic. The key concepts for developing these results are those of a large submodule of a module and of an essential monomorphism. A submodule M of a module N is called *large in* N if $N' \cap M \neq 0$ for every submodule $N' \neq 0$ of N. A monomorphism $i: M \to N$ is called *essential* if $i(M)$ is a large submodule of N. In this case, N is called an *essential extension* of M.

It is clear that if M is a large submodule of N and N is a large submodule of Q, then M is large in Q. Let M be any submodule of a module N and let $E = \{S\}$ be the collection of submodules of N containing M as large submodule. Evidently, $E \neq \varnothing$ since $M \in E$. Moreover, if $\{S_\alpha\}$ is a totally ordered subset of E, then $S' = \bigcup S_\alpha$ is a submodule and if $N' \neq 0$ is a submodule of S', then $N' \cap S_\alpha \neq 0$ for some S_α. Since M is large in S_α, $(N' \cap S_\alpha) \cap M \neq 0$; hence, $N' \cap M \neq 0$. Thus M is large in $S' = \bigcup S_\alpha$. It follows from Zorn's lemma that E contains a maximal element S, that is, we have a submodule S of N containing M as a large submodule and such that for any submodule $T \supsetneqq S$, M is not large in T.

We can give a characterization of injective modules in terms of essential homomorphisms:

PROPOSITION 3.18. *A module Q is injective if and only if every essential monomorphism of Q is an isomorphism.*

Proof. Suppose Q is injective and let $Q \xrightarrow{i} M$ be an essential monomorphism of Q. We have an exact sequence $0 \to Q \xrightarrow{i} M \to M/Q \to 0$ that splits by

Proposition 3.17. Then $M = i(Q) \oplus Q'$ where Q' is a submodule. Since $Q' \cap i(Q) = 0$ and i is essential, $Q' = 0$. Hence $M = i(Q)$ and i is an isomorphism. Conversely, suppose Q is a module such that every essential monomorphism of Q is an isomorphism and let $0 \to Q \xrightarrow{i} M \xrightarrow{p} N \to 0$ be exact. By Zorn's lemma, there exists a submodule S of M such that $S \cap i(Q) = 0$ and S is maximal among the submodules of M having this property. Consider the module M/S. Since $S \cap i(Q) = 0$, we have the monomorphism $j: x \rightsquigarrow i(x) + S$ of Q into M/S whose image is $(i(Q)+S)/S$. If T/S, $T \supset S$, is a submodule of M/S such that $T/S \cap (i(Q)+S)/S = S/S$, then $T \cap (i(Q)+S) = S$. Hence $T \cap i(Q) \subset S \cap i(Q) = 0$. Then $T = S$ by the maximality of S. This shows that j is an essential monomorphism. Hence this is an isomorphism. Then $M/S = (i(Q)+S)/S$ and $M = i(Q)+S$. Since $S \cap i(Q) = 0$ we have $M = S \oplus i(Q)$, which implies that $0 \to Q \to M \to N \to 0$ splits and proves that Q is injective by Proposition 3.17. □

The main result on imbedding of modules in injective modules is

THEOREM 3.19. *If M is a module, there exists an essential monomorphism i of M into an injective module Q. If i and i' are essential monomorphisms of M into injective modules Q and Q' respectively, then there exists an isomorphism $l: Q \to Q'$ such that $i' = li$.*

For the proof, we require the following

LEMMA. *Suppose we have a diagram of monomorphisms*

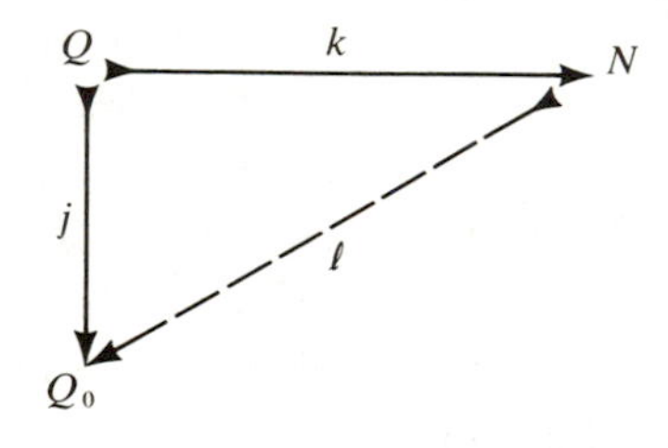

such that k is essential and Q_0 is injective. Then there exists a monomorphism $l: N \to Q_0$ to make a commutative triangle.

Proof. Since Q_0 is injective, we have a homomorphism $l: N \to Q_0$ to make a commutative triangle. Let $y \in \ker l \cap \operatorname{im} k$. Then $y = kx$, $x \in Q$, and $0 = ly = lkx = jx$. Then $x = 0$ and $y = 0$. Thus $\ker l \cap \operatorname{im} k = 0$. Since k is essential, we have $\ker l = 0$. Hence l is a monomorphism. □

We can now give the

Proof of Theorem 3.19. Let $i_0 : M \to Q_0$ be a monomorphism of M into an injective module Q_0. The existence of i_0 and Q_0 were proved in Theorem 3.18. We showed also that there exists a submodule Q of Q_0 in which $i_0(M)$ is large and that is maximal for this property. Let i denote the monomorphism of M into Q obtained by restricting the codomain of i_0 to Q. Let j denote the injection of Q in Q_0 and let k be an essential monomorphism of Q into a module N. Then we have the diagram and conditions of the preceding lemma. Hence we have a monomorphism $l : N \to Q_0$ such that $lk = j$. Then $l(N)$ is a submodule of Q_0 containing $l(k(Q)) = j(Q) = Q$. Since k is essential, $k(Q)$ is large in N and since l is a monomorphism, $Q = l(k(Q))$ is large in $l(N)$. Since $i(M)$ is large in Q, $i(M)$ is large in $l(N) \supset Q$. It follows from the maximality of Q that $l(N) = Q$. As is readily seen, this implies that $k(Q) = N$ and so k is an isomorphism. We have therefore shown that any essential monomorphism k of Q is an isomorphism. Then Q is injective by Proposition 3.18. Now let i' be any essential monomorphism of M into an injective module Q'. Applying the lemma to

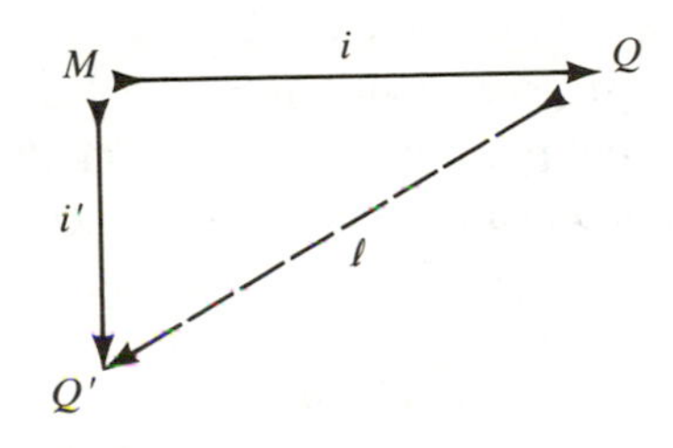

we obtain the monomorphism $l : Q \to Q'$ such that $li = i'$. Since $l(Q)$ is an injective submodule of Q', we have a submodule Q'' of Q' such that $Q' = l(Q) \oplus Q''$. Since $l(Q) \supset i'(M) = l(i(M))$ and $l(Q) \cap Q'' = 0$, $i'(M) \cap Q'' = 0$. Since i' is an essential monomorphism, $i'(M)$ is large in Q'. Hence $Q'' = 0$. Then $Q' = l(Q)$ and l is an isomorphism of Q onto Q' such that $li = i'$. □

An injective module Q such that there exists an essential monomorphism i of M into Q is called an *injective hull* (or *injective envelope*) of M. Its existence and uniqueness is given in Theorem 3.19. The argument shows that if Q' is any injective module and i' is a monomorphism of M into Q, then there exists a monomorphism l of Q into Q' such that $i' = li$. In this sense an injective hull provides a minimal imbedding of M into an injective module.

EXERCISES

1. Let D be a domain (commutative ring with no zero divisors $\neq 0$), F the field of fractions of D ($F \supset D$). Show that if M is a D-module, then $M_F = F \otimes_D M$ is a divisible D-module and $x \rightsquigarrow 1 \otimes x$ is an essential monomorphism of M into M_F if M is torsion-free.

2. Show that $\prod Q_\alpha$ is injective if and only if every Q_α is injective.

3. Show that a submodule M is large in the module N if and only if for any $x \neq 0$ in N there exists an $r \in R$ such that $xr \neq 0$ and $xr \in M$.

3.12 MORITA CONTEXTS

The remainder of this chapter is devoted to the Morita theory of equivalence of categories of modules. The principal questions considered in this theory are when are two categories **mod**-R and **mod**-S equivalent, how are such equivalences realized, and what are the auto-equivalences of **mod**-R? We showed in Chapter 1, pp. 29–31, that for any ring R and any positive integer n, **mod**-R and **mod**-$M_n(R)$ are equivalent categories. We shall see (section 3.14) that this is almost the most general situation in which equivalence of **mod**-R and **mod**-S occurs. From the point of view of the applications, the machinery used to develop the results on equivalence of modules is more important than the results themselves. The central concept of this machinery is that of a Morita context (or "set of pre-equivalence data"). We shall begin with the Morita context, considering first an example.

We consider a foursome consisting of 1) a ring R, 2) a right R-module M, 3) the dual $M^* = \hom(M, R)$, and 4) $R' = \operatorname{End} M$. We have seen that we can regard M in a natural way as an R'-R-bimodule (p. 133). Here, if $r' \in R'$ and $x \in M$, $r'x$ is the image of x under r'. We have also seen that M^* becomes an R-R'-bimodule if we define

$$(ry^*)x = r(y^*x), \qquad (y^*r')x = y^*(r'x) \tag{80}$$

for $r \in R$, $r' \in R$, $y^* \in M^*$, $x \in M$ (p. 135). It is a good idea to treat M and M^* in a symmetric fashion. To facilitate this, we write (y^*, x) for $y^*x = y^*(x)$ and we consider the map of $M^* \times M$ sending the pair (y^*, x) into the element $(y^*, x) \in R$. We list the properties of this map:

$$(y^*, x_1 + x_2) = (y^*, x_1) + (y^*, x_2)$$

$$(y^*, xr) = (y^*, x)r$$

$$(y_1^* + y_2^*, x) = (y_1^*, x) + (y_2^*, x) \tag{81}$$

$$(ry^*, x) = r(y^*, x)$$

$$(y^*r', x) = (y^*, r'x).$$

The first two of these amount to the definition of a homomorphism from M_R to R_R and the third to the definition of the sum of two homomorphisms; the last two are the definitions (80). The first, third, and fifth of (81) show that $(y^*, x) \rightarrow (y^*, x)$ is an R'-balanced product of M^* and M. Hence we have a homomorphism τ of $M^* \otimes_{R'} M$ into R such that $\tau(y^* \otimes x) = (y^*, x)$. The second and fourth equations now read $\tau(y^* \otimes xr) = (\tau(y^* \otimes x))r$ and $\tau(ry^* \otimes x) = r\tau(y^* \otimes x)$. Hence if we regard $M^* = {}_R M^*_{R'}$ and $M = {}_{R'} M_R$ and, consequently, $M^* \otimes_{R'} M$ as an R-R-bimodule, then τ is a bimodule homomorphism of $M^* \otimes_{R'} M$ into R.

We define next a bimodule homomorphism μ of $M \otimes_R M^*$ into R'. Let (x, y^*) be an element of $M \times M^*$. Then this defines a map

$$[x, y^*] : y \rightarrow x(y^*, y) \tag{82}$$

of M into M. Evidently, $[x, y^*](y_1 + y_2) = [x, y^*]y_1 + [x, y^*]y_2$ and $[x, y^*](yr) = ([x, y^*]y)r$ so $[x, y^*] \in R' = \text{End}\, M$. Moreover, we have

$$\begin{aligned} [x_1 + x_2, y^*] &= [x_1, y^*] + [x_2, y^*] \\ [x, y_1^* + y_2^*] &= [x, y_1^*] + [x, y_2^*] \\ [xr, y^*] &= [x, ry^*] \\ [r'x, y^*] &= r'[x, y^*] \\ [x, y^*r'] &= [x, y^*]r', \end{aligned} \tag{83}$$

which follow directly from the definition (82) and from (81). It follows from (83) that we have an R'-R'-bimodule homomorphism μ of $M \otimes_R M^*$ into R' such that $\mu(x \otimes y^*) = [x, y^*]$. The definition (82) can be rewritten as

$$[x, y^*]y = x(y^*, y) \tag{84}$$

and we have

$$x^*[x, y^*] = (x^*, x)y^*, \tag{85}$$

since

$$(x^*[x, y^*])y = x^*([x, y^*]y) = x^*(x(y^*, y)) = (x^*x)(y^*, y) = (x^*, x)(y^*, y)$$

and

$$((x^*, x)y^*)y = (x^*, x)(y^*y) = (x^*, x)(y^*, y).$$

We shall now abstract the essential elements from this situation by formulating the following definition.

DEFINITION 3.11. *A* Morita context *is a set* $(R, R', M, M', \tau, \mu)$ *where* R *and* R' *are rings,* $M = {}_{R'}M_R$ *is an* R'-R-*bimodule,* $M' = {}_RM'_{R'}$ *is an* R-R'-*bimodule,* τ *is an* R-R-*homomorphism of* $M' \otimes_{R'} M$ *into* R, *and* μ *is an* R'-R'-*homomorphism of* $M \otimes_R M'$ *into* R' *such that if we put* $\tau(x' \otimes x) = (x', x)$ *and* $\mu(x \otimes x') = [x, x']$, *then*

(i) $[x, y']y = x(y', y)$.
(ii) $x'[x, y'] = (x', x)y'$.

We remark that conditions (i) and (ii) are equivalent respectively to the commutativity of the following diagrams:

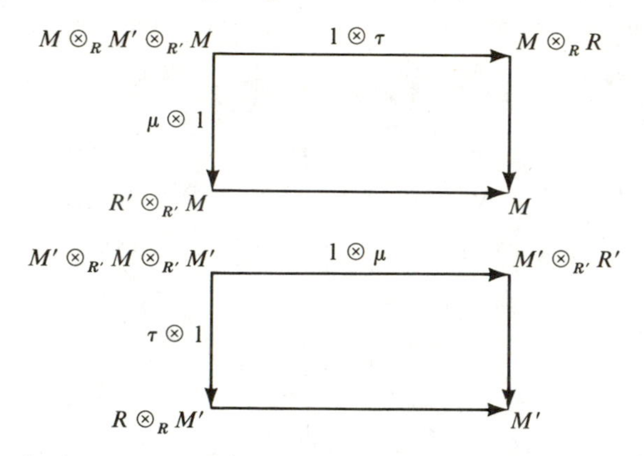

Here the unmarked arrows are the canonical isomorphisms of $M \otimes_R R$ with M, of $R' \otimes_{R'} M$ with M, etc.

The foregoing considerations amount to the definition of a Morita context $(R, R' = \operatorname{End} M_R, M, M^*, \tau, \mu)$ from a given right R-module M. We shall call this the *Morita context defined by* M_R. We now give a simpler

EXAMPLE

Consider the free left module ${}^{(n)}R$ and the free right module $R^{(n)}$ as on p. 126. Put $M = R^{(n)}$, $M' = {}^{(n)}R$ and denote the elements of M as columns

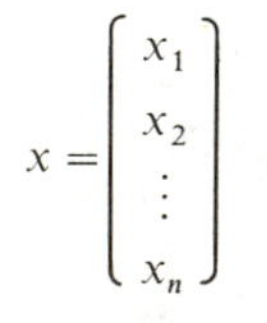

$$x = \begin{pmatrix} x_1 \\ x_2 \\ \vdots \\ x_n \end{pmatrix}$$

and those of M' as rows $x' = (x_1, x_2, \ldots, x_n)$. Let $R' = M_n(R)$, the ring of $n \times n$ matrices with entries in R. Then M is a left R'-module if we define $r'x$ for $r' \in M_n(R)$ to be the matrix product. Then M is an R'-R-bimodule. Similarly, M' is an R-R'-bimodule if we defined the action of r' on x' as the matrix product $x'r'$. Also we define $[x, y'] = xy'$, the matrix product, as on p. 126. Then we have the homomorphism μ of $M \otimes_R M'$ into $R' = M_n(R)$ such that $\mu(x \otimes y') = [x, y']$, and this is an R'-R'-bimodule homomorphism. We define $(y', x) = y'x$. This is a 1×1 matrix, so it may be regarded as an element of R. Then we have an R-R-bimodule homomorphism τ of $M' \otimes_{R'} M$ into R such that $\tau(y' \otimes x) = (y', x)$. The relations (i) and (ii) follow from the associative law of multiplication of matrices. Hence $(R, R' = M_n(R), R^{(n)}, {}^{(n)}R, \tau, \mu)$ is a Morita context.

If M is a right R-module and $M^* = \hom(M, R)$, then we put

$$(86) \qquad T(M) = \left\{ \sum_{x,y^*} (y^*, x) \right\} \subset R.$$

Evidently this is the subgroup of the additive group R generated by the values $(y^*, x) = y^*(x)$, $y^* \in M^*$, $x \in M. T(M)$, which is called the *trace* of the module M, is an ideal in R, since if $r \in R$ then

$$(y^*, x)r = (y^*, xr) \in T(M)$$
$$r(y^*, x) = (ry^*, x) \in T(M).$$

We are now ready to prove the first main theorem on Morita contexts. This draws a large number of important conclusions from the hypotheses that τ and μ are surjective. The results are implicit in K. Morita's basic paper on duality for modules. Subsequently the formulation was improved substantially by H. Bass and others, who called the result "Morita I." To shorten the formulation we shall now call a module a *progenerator* if it is finitely generated projective and its trace ideal is the whole ring. The appropriateness of this terminology will appear in section 3.13.

MORITA I. *Let $(R, R', M, M', \tau, \mu)$ be a Morita context in which τ and μ are surjective. Then*

(1) *M_R, ${}_{R'}M$, $M'_{R'}$, ${}_RM'$ are progenerators.*
(2) *τ and μ are isomorphisms.*
(3) *The map $l: x' \rightarrow l(x')$ where $l(x')$ is the map $y \rightarrow (x', y)$ of M into R is a bimodule isomorphism of ${}_RM'_{R'}$ onto ${}_RM^*_{R'}$ where $M^* = \hom(M_R, R_R)$. We have similar bimodule isomorphisms of ${}_RM'_{R'}$ with $\hom({}_{R'}M, {}_{R'}R')$, of ${}_{R'}M_R$ with $\hom({}_RM', {}_RR)$, and of ${}_{R'}M_R$ with $\hom(M'_{R'}, R'_{R'})$.*
(4) *The map $\lambda: r' \rightarrow \lambda(r')$ where $\lambda(r')$ is the map $y \rightarrow r'y$ is a ring isomorphism of R' onto $\operatorname{End} M_R = \hom(M_R, M_R)$. Similarly, $\rho: r \rightarrow \rho(r)$ where $\rho(r)$ is $y \rightarrow yr$ is a ring anti-isomorphism of R onto $\operatorname{End}_{R'}M$ and we have a similar isomorphism λ' of R onto $\operatorname{End} M'_{R'}$ and anti-isomorphism ρ' of R' onto $\operatorname{End}_R M'$.*

(5) *The pair of functors* $\otimes_R M'$ *and* $\otimes_{R'} M$ *define an equivalence of the categories* **mod**-R *and* **mod**-R'. *Similarly,* $M\otimes_R$ *and* $M'\otimes_{R'}$ *define an equivalence of* R-**mod** *and* R'-**mod**.

(6) *If* I *is a right ideal of* R *put*

$$\eta(I) = IM' = \{\sum b_i x_i' | b_i \in I, x_i' \in M'\} \tag{87}$$

and if N' *is a submodule of* $M'_{R'}$ *put*

$$\xi(N') = (N', M) = \{\sum (y_i', x_i) | y_i' \in N', x_i \in M\}. \tag{88}$$

Then η *and* ξ *are inverses and are lattice isomorphisms between the lattice of right ideals of* R *and the lattice of submodules of* $M'_{R'}$. *Moreover, these induce lattice isomorphisms between the lattice of (two-sided) ideals of* R *and the lattice of submodules of* ${}_R M'_{R'}$. *Similar statements hold for the lattices of left ideals of* R, *of left ideals of* R', *and of right ideals of* R'. *These imply that* R *and* R' *have isomorphic lattices of ideals.*

(7) *The centers of* R *and* R' *are isomorphic.*

Proof. (1) Since μ is surjective, we have $u_i \in M$, $v_i' \in M'$, $1 \leqslant i \leqslant m$, such that $\sum[u_i, v_i'] = 1$ (the unit of R'). Then if $x \in M$, $x = 1x = \sum[u_i, v_i']x = \sum u_i(v_i', x)$. Now the map $l(v_i'): x \rightsquigarrow (v_i', x)$ is in $\hom(M_R, R)$. Hence it follows from the dual basis lemma (Proposition 3.11, p. 152) that M_R is finitely generated projective. The hypothesis that τ is surjective means that any $r \in R$ can be written as $r = \sum (y_i', x_i)$, $y_i' \in M'$, $x_i \in M$. Then $r = \sum l(y_i')\,(x_i) \in T(M)$ and $T(M) = R$. Thus M_R is a progenerator. The other three statements follow by symmetry.

(2) Suppose $\sum x_i' \otimes y_i \in \ker \tau$, so $\sum (x_i', y_i) = 0$. Since τ is surjective, we have $w_j \in M$, $z_j' \in M'$ such that $\sum (z_j', w_j) = 1$ (in R). Then

$$\begin{aligned}\sum x_i' \otimes y_i &= \sum_{i,j} x_i' \otimes y_i(z_j', w_j) = \sum_{i,j} x_i' \otimes [y_i, z_j'] w_j \\ &= \sum_{i,j} x_i'[y_i, z_j'] \otimes w_j = \sum_{i,j} (x_i', y_i) z_j' \otimes w_j = 0.\end{aligned}$$

Thus $\ker \tau = 0$ and τ is an isomorphism. Similarly, μ is an isomorphism.

(3) We have noted that $l(y'): x \rightsquigarrow (y', x)$ is in $\hom(M_R, R)$. Direct verification, using the definition of the actions of R and R' in $\hom(M_R, R)$, shows that $y' \rightsquigarrow l(y')$ is an R-R'-bimodule homomorphism of M' into $\hom(M_R, R)$. Suppose $l(y') = 0$, so $(y', x) = 0$ for all x. As before, we can write $1 = \sum [u_i, v_i']$, $u_i \in M$, $v_i' \in M'$. Then $y' = \sum y'[u_i, v_i'] = \sum (y', u_i) v_i' = 0$. Hence $l: y' \rightsquigarrow l(y')$ is injective. Now let $x^* \in \hom(M_R, R)$ and put $y' = \sum (x^* u_i) v_i'$ where $\sum [u_i, v_i'] = 1$.

Then

$$(y', x) = \sum_i ((x^*u_i)v_i', x) = \sum_i (x^*u_i)(v_i', x)$$
$$= x^*(\sum u_i(v_i', x)) = x^*(\sum [u_i, v_i']x) = x^*(x).$$

Hence $x^* = l(y')$ and l is surjective. The other three assertions follow by symmetry.

(4) Since M is an R'-R-bimodule, $\lambda(r'): x \rightsquigarrow r'x$ is an endomorphism of M_R. It is clear that λ is a ring homomorphism of R' into End M_R. Now suppose $\lambda(r') = 0$. Then $r'x = 0$ for all $x \in M$; hence, $r' = r'1 = r'\sum[u_i, v_i']$ where $\sum[u_i, v_i'] = 1$ (in R'), and $r'\sum[u_i, v_i'] = \sum r'[u_i, v_i'] = \sum[r'u_i, v_i] = 0$. Thus $r' = 0$ and so λ is injective. Now let $f \in \text{End}\, M_R$. Then $r' = \sum[fu_i, v_i'] \in R'$ and

$$r'x = \sum[fu_i, v_i']x = \sum fu_i(v_i', x) = f(\sum u_i(v_i', x))$$
$$= f(\sum[u_i, v_i']x) = f(x).$$

Hence $f = \lambda(r')$ so λ is surjective. Thus λ is an isomorphism of rings. The other cases follow in a similar fashion.

(5) If N is a right R-module, then $N \otimes_R M'$ is a right R'-module and if N' is a right R'-module, then $N' \otimes_{R'} M$ is a right R-module. Hence we obtain the functors $\otimes_R M'$ and $\otimes_{R'} M$ from **mod**-R to **mod**-R' and from **mod**-R' to **mod**-R respectively. Iteration gives the functors $\otimes_R M' \otimes_{R'} M$ and $\otimes_{R'} M \otimes_R M'$ from **mod**-R to itself and from **mod**-R' to itself. Now we have the associativity isomorphism of $(N \otimes_R M') \otimes_{R'} M$ with $N \otimes_R (M' \otimes_R M)$. Following this with the isomorphism τ of $M' \otimes_{R'} M$ onto R we obtain an isomorphism of $N \otimes_R (M' \otimes_{R'} M)$ onto $N \otimes_R R$. Applying the canonical isomorphism of $N \otimes_R R$ to N and combining all of these we obtain a right R-module isomorphism

$$(N \otimes_R M') \otimes_{R'} M \to N. \tag{89}$$

Since all of the intermediate isomorphisms that we defined are natural in N, (89) is natural in N. Hence $\otimes_R M' \otimes_{R'} M$ is naturally isomorphic to the identity functor $1_{\textbf{mod}\text{-}R}$. Similarly, $\otimes_{R'} M \otimes_R M'$ is naturally isomorphic to $1_{\textbf{mod}\text{-}R'}$. Hence $\otimes_R M'$, $\otimes_{R'} M$ provide a natural equivalence between the categories **mod**-R and **mod**-R'. The other equivalence asserted in (5) follows by symmetry.

(6) It is clear from the definition that if I is a right ideal of R, then $\eta(I)$ is a submodule of $M'_{R'}$. Moreover, if I is an ideal, then $\eta(I)$ is a submodule of ${}_R M'_{R'}$. Similarly, if N' is a submodule of $M'_{R'}$, then $\zeta(N')$ is a right ideal of R and if N' is a submodule of ${}_R M'_{R'}$, then $\zeta(N')$ is an ideal of R. If I is a right ideal of R, then $\zeta(\eta(I))$ is the set of sums of elements of the form $(by', x) = b(y', x)$ where

$b \in I$, $y' \in M'$, $x \in M$. Since τ is surjective, this set is I. Thus, $\zeta\eta$ is the identity map on the lattice of right ideals. If N' is a submodule of ${}_RM'_{R'}$ then $\eta(\zeta(N'))$ is the set of sums of elements of the form $(y',x)x'$, $y' \in N'$, $x \in M$, $x' \in M'$. Since $(y',x)x' = y'[x,x']$ and μ is surjective, we have $\eta(\zeta(N')) = N'$. It is clear also that η and ζ are order-preserving for the order given by inclusion. Hence these are isomorphisms between the lattices of right ideals of R and of submodules of $M'_{R'}$ (see BAI, p. 460). Evidently, these isomorphisms induce isomorphisms between the lattices of ideals of R and of submodules of ${}_RM'_{R'}$. The other statements follow by symmetry. In particular, we have a lattice isomorphism $I' \rightsquigarrow M'I' = \{\sum x_i'b_i' | b_i' \in I', x_i' \in M'\}$ of the lattice of ideals I' of R' with the lattice of submodules of the bimodule ${}_RM'_{R'}$. Combining isomorphisms, we see that given any ideal I of R there is a unique ideal I' of R' such that $IM' = M'I'$ and $I \rightsquigarrow I'$ is an isomorphism of the lattice of ideals of R onto the lattice of ideals of R'.

(7) To establish an isomorphism between the center $C(R)$ and the center $C(R')$, we consider the two rings of endomorphisms $\mathrm{End}\, M_R$ and $\mathrm{End}_{R'} M$. Both of these are subrings of $\mathrm{End}\, M$, the ring of endomorphisms of M regarded just as an abelian group. Now each of the rings $\mathrm{End}\, M_R$, $\mathrm{End}_{R'} M$ is the centralizer of the other in the ring $\mathrm{End}\, M$: By (4), $\mathrm{End}\, M_R$ is the set of maps $y \rightsquigarrow r'y$, $r' \in R'$, and $\mathrm{End}_{R'} M$ is the set of maps $y \rightsquigarrow yr$. On the other hand, the definition of $\mathrm{End}\, M_R$ shows that this subring of $\mathrm{End}\, M$ is the centralizer of the subring consisting of the maps $y \rightsquigarrow yr$. Similarly, $\mathrm{End}_{R'} M$ is the centralizer in $\mathrm{End}\, M$ of the set of maps $y \rightsquigarrow r'y$. Hence each of $\mathrm{End}\, M_R$, $\mathrm{End}_{R'} M$ is the centralizer of the other in $\mathrm{End}\, M$. Clearly this implies that the center

$$C(\mathrm{End}\, M_R) = \mathrm{End}\, M_R \cap \mathrm{End}_{R'} M = C(\mathrm{End}_{R'} M).$$

Since we have an isomorphism of $\mathrm{End}\, M_R$ with R' and an anti-isomorphism of $\mathrm{End}_{R'} M$ with R, this gives an isomorphism of $C(R)$ with $C(R')$. $\square$

We shall now show that Morita I is applicable to any Morita context $(R, R' = \mathrm{End}\, P_R, P, P^* = \hom(P_R, R), \tau, \mu)$ determined by a progenerator $P = P_R$. Here τ and μ are as defined at the beginning of the section. The hypothesis that P is a progenerator includes the condition $T(P) = R$ and this means that τ is surjective. Since P is finitely generated projective, by the dual basis lemma, we have $x_i \in P$, $x_i^* \in P^*$, $1 \leqslant i \leqslant m$, such that $x = \sum x_i(x_i^* x)$ for any x in P. Then $x = \Sigma x_i(x_i^*, x) = \Sigma[x_i, x_i^*]\, x = (\Sigma[x_i, x_i^*])x$. Thus $\Sigma[x_i, x_i^*] = 1_P$, which is the unit element of $R' = \mathrm{End}\, P_R$. It follows that any $r' \in \mathrm{End}\, P_R$ has the form $r' = r'1_P = \Sigma[r'x_i, x_i^*]$ and so μ is surjective on R'. We therefore have

THEOREM 3.20. *If P is a progenerator in* **mod R**, *then μ and τ are surjective for the Morita context $(R, R' = \mathrm{End}\, P_R, P, P^* = \hom(P_R, R), \tau, \mu)$. Hence Morita I is applicable.*

EXERCISES

1. Show that Morita I is applicable in the example given on pp. 166–167. Interpret the results of the theorem in this case as theorems on matrices. In particular, use the correspondence between the lattice of ideals of R and of R' given in the proof of Morita I (6) to show that the map $B \rightsquigarrow M_n(B)$ is an isomorphism of the lattice of ideals of R onto the lattice of ideals of $M_n(R)$. (This was exercise 8, p. 103 of BAI.)

2. Let $(R, R', M, M', \tau, \mu)$ be a Morita context. Let $\begin{pmatrix} R & M' \\ M & R' \end{pmatrix}$ be the set of matrices $\begin{pmatrix} a & x' \\ y & b' \end{pmatrix}$ where $a \in R$, $b' \in R'$, $x' \in M'$, $y \in M$. Define addition component-wise (usual matrix addition) and multiplication by
$$\begin{pmatrix} a_1 & x'_1 \\ y_1 & b'_1 \end{pmatrix}\begin{pmatrix} a_2 & x'_2 \\ y_2 & b'_2 \end{pmatrix} = \begin{pmatrix} a_1a_2 + (x'_1, y_2) & a_1x'_2 + x'_1b'_2 \\ y_1a_2 + b'_1y_2 & [y_1, x'_2] + b'_1b'_2 \end{pmatrix}.$$
Verify that this addition and multiplication together with the obvious 0 and 1 constitute a ring. We shall call this the *ring of the Morita context* $(R, R', M, M', \tau, \mu)$. Let B be an ideal in R, B' an ideal in R'. Verify that the set of matrices $\begin{pmatrix} B & BM' + M'B' \\ B'M + MB & B' \end{pmatrix}$ is an ideal in the foregoing ring.

3. Consider the special case of exercise 2 in which $M' = 0$, $\tau = 0$, $\mu = 0$. Show that the resulting ring is isomorphic to the ring defined as in exercise 1, p. 136, regarding M as a T-T-bimodule relative to $T = R \times R'$ as in exercise 2, p. 136.

3.13 THE WEDDERBURN-ARTIN THEOREM FOR SIMPLE RINGS

We shall now apply Morita I to derive a classical structure theorem for simple rings—the Wedderburn-Artin theorem. This was proved for finite dimensional algebras over a field by Wedderburn in 1908 and for rings with descending chain condition on one-sided ideals by Artin in 1928. In the next chapter we shall integrate this theorem into the general structure theory of rings. Here we treat it somewhat in isolation and we formulate the result in the following way.

WEDDERBURN-ARTIN THEOREM FOR SIMPLE RINGS. *The following conditions on a ring R are equivalent:*

(1) *R is simple and contains a minimal right (left) ideal.*
(2) *R is isomorphic to the ring of linear transformations of a finite dimensional vector space over a division ring.*
(3) *R is simple, left and right artinian, and left and right noetherian.*

Proof. (1) $\Rightarrow$ (2). Let I be a minimal right ideal of R, so $I = I_R$ is an irreducible right R-module. We show first that I_R is finitely generated projective. Consider $RI = \sum_{a \in R} aI$. This is an ideal in R containing I. Hence $RI = R$. Then $1 = \sum_1^m a_i b_i$ where $a_i \in R$, $b_i \in I$. This implies that $R = \sum_1^m a_i I$. Observe next that since for $a \in R$, the map $x \rightsquigarrow ax$, $x \in I$, is a homomorphism of I as right module, by Schur's lemma either $aI = 0$ or $aI \cong I$ as right module. Then $R = \sum a_i I$ is a sum of irreducible right modules. Then by Thoerem 3.9 (p. 117), $R = I \oplus I'$ where I' is a second right ideal. This implies that I is projective and can be generated by a single element. By the dual basis lemma, the trace ideal $T(I) \neq 0$. Since R is simple, we have $T(I) = R$ and I is a progenerator. Hence Morita I is applicable to the Morita context $(R, R' = \text{End}\, I_R, I, I^* = \hom(I_R, R), \tau, \mu)$. By Schur's lemma, R' is a division ring. By Morita I, I^* is a finitely generated right module over the division ring R' and R is isomorphic to $\text{End}\, I^*_{R'}$. If we replace R' in the usual way by its opposite $\Delta = R'^{\text{op}}$, then I^* becomes a finite dimensional (left) vector space over Δ and R is isomorphic to the ring of linear transformations of this vector space. The proof for I, a minimal left ideal, is similar.

(2) $\Rightarrow$ (3). Let V be a finite dimensional vector space over a division ring Δ and let L be its ring of linear transformations. Put $R = \Delta^{\text{op}}$ and regard V as right module over R. Then V_R is free with finite base so it is finitely generated projective. Moreover, if $x \neq 0$ in V, then there is a linear function f such that $f(x) = 1$. This implies that the trace ideal $T(V_R) = R$ and so V_R is a progenerator. Then Morita I applies to $(R, L = \text{End}\, V_R = \text{End}_\Delta V, V, V^* = \hom(V_R, R), \tau, \mu)$. Then Morita I (6) shows that L and the division ring R have isomorphic lattices of ideals. Hence L is simple. Also this result shows that the lattice of left ideals of L is isomorphic to the lattice of submodules of V^* and the lattice of right ideals is isomorphic to the lattice of submodules of V. Since finite dimensional vector spaces satisfy the descending and ascending chain conditions for subspaces, it follows that L is left and right artinian and noetherian.

(3) $\Rightarrow$ (1) is clear. $\square$

EXERCISES

1. Use Morita I (6) to show that if L is the ring of linear transformations of a finite dimensional vector space V over a division ring, then L acts irreducibly on V (that is, V regarded in the natural way as left L-module is irreducible).

2. Let V be a finite dimensional vector space over a division ring Δ, V^* the right vector space of linear functions on V. If $x \in V$ and $f \in V^*$, write $\langle x, f \rangle = f(x)$. If U (U^*) is a subspace of V (V^*), put

$$U^\perp = \{f \in V^* \mid \langle y, f \rangle = 0, y \in U\} \; (U^{*\perp} = \{y \in V \mid \langle y, g \rangle = 0, g \in U^*\}).$$

Show that the maps $U \rightsquigarrow U^{\perp}, U^* \rightsquigarrow U^{*\perp}$ are inverses and hence are anti-isomorphisms between the lattices of subspaces of V and V^*.

3. Let V and L be as in exercise 1. Use Morita I (6) and exercise 2 to show that the map

$$\alpha : U \rightsquigarrow \alpha(U) = \{l \in L \mid l(U) = 0\}$$

is an anti-isomorphism of the lattice of subspaces of V onto the lattice of left ideals of L.

4. Show that the map

$$I \rightsquigarrow \mathrm{ann}_L I = \{l \in L \mid lI = 0\}$$

is an anti-isomorphism of the lattice of right ideals of L onto the lattice of left ideals of L.

5. Show that any left (right) ideal of L is a principal left (right) ideal generated by an idempotent (that is, has the form Le, $e^2 = e$ or eL, $e^2 = e$).

6. Show that if $l \in L$, then there exists a $u \in L$ such that $lul = l$.

3.14 GENERATORS AND PROGENERATORS

If R is a ring and M is a right R-module, any $x \in M$ is contained in the submodule xR. Hence $M = \sum_{x \in M} xR$ and xR is a homomorphic image of $R = R_R$. We shall now call a right R-module X a *generator* of the category **mod**-R if any module M is a sum of submodules all of which are homomorphic images of X. Thus R is a generator for **mod**-R. Evidently, if X is a generator and X is a homomorphic image of Y, then Y is a generator. Since a homomorphic image of $X^{(n)}$ is a sum of homomorphic images of X, it is clear also that if $X^{(n)}$ is a generator, then X is a generator. The concept of generator will play a central role in the study of equivalences between categories of modules. The following theorem gives a number of important characterizations of generators.

THEOREM 3.21. *The following conditions on a module X are equivalent*:

(1) *X is a generator.*

(2) *The functor* $\hom(X, -)$ *is faithful* (see p. 22).

(3) *The trace ideal $T(X) = R$.*

(4) *There exists an n such that R is a homomorphic image of $X^{(n)}$.*

Proof. (1) $\Rightarrow$ (2). We have to show that for any two right R-modules M and N, the map $f \rightsquigarrow \hom(X, f)$ of $\hom(M_R, N_R)$ into the set of homomorphisms of $\hom(X, M)$ into $\hom(X, N)$ is injective. Here $\hom(X, f)$ is the map $g \rightsquigarrow fg$ of hom (X, M) into hom(X, N). Since hom(X,f) is a homomorphism, it suffices to show that if $f \neq 0$ then the map hom(X,f) of hom(X, M) into hom(X, N) is $\neq 0$. This means that we have to show that for a given $f \neq 0$, $f: M \to N$, there exists a $g \in \hom(X, M)$ such that $fg \neq 0$. Suppose this is not the case. Then $fg = 0$ for every $g \in \hom(X, M)$. Since X is a generator $M = \sum gX$ where the summation is taken over all homomorphisms g of X into M. Then $fM = f(\sum gX) = \sum fgX = 0$, contrary to $f \neq 0$.

(2) $\Rightarrow$ (3). Let $M = R$, $N = R/T(X)$, and let v be the canonical homomorphism of M into N. Given any $g \in X^* = \hom(X, M)$, then $g(x) \in T(X)$ for every $x \in X$. Hence $vg(x) = 0$. Thus $vg = 0$ for every $g \in \hom(X, M)$. By hypothesis, this implies $v = 0$. Then $N = 0$ and $R = T(X)$.

(3) $\Rightarrow$ (4). If 3 holds, we have $1 = \sum_1^m f_i x_i$ for $x_i \in X$, $f_i \in X^*$. Now consider $X^{(m)}$ and the map $(y_1, \ldots, y_m) \rightsquigarrow \sum_1^m f_i y_i$ of $X^{(m)}$ into R. This is a homomorphism of right R-modules, so its image is a right ideal of R. Since $1 = \sum f_i x_i$ is in the image, the image is all of R. Thus we have an epimorphism of $X^{(n)}$ onto R.

(4) $\Rightarrow$ (1). Since R is a generator and R is a homomorphic image of $X^{(n)}$, it follows that $X^{(n)}$ is a generator. Then X is a generator. □

The characterization 3 of generators shows that a module X is a progenerator in the sense defined in section 3.12 if and only if X is finitely generated projective and X is a generator of **mod**-R.

We shall call a module M *faithful* if the only $a \in R$ such that $Ma = 0$ is $a = 0$. For any module M we define $\mathrm{ann}_R M = \{b \in R \mid Mb = 0\}$. It is clear that this is an ideal in R and that M is faithful if and only if $\mathrm{ann}_R M = 0$. It is clear also that R regarded as the module R_R is faithful and it follows from Theorem 3.21.4 that any generator X of **mod**-R is faithful. For, if R is a homomorphic image of some $X^{(n)}$, then $\mathrm{ann}_R X \subset \mathrm{ann}_R R = 0$ so X is faithful.

In the important special case in which the ring R is commutative, condition (3) of Theorem 3.21 can be replaced by the simpler condition that X is faithful. To prove this we require the following

LEMMA. *Let R be a commutative ring, M a finitely generated R-module. An ideal I of R satisfies $MI = M$ if and only if $I + \mathrm{ann}_R M = R$.*

Proof. (1) Suppose $I + \mathrm{ann}_R M = R$. Then $1 = b + c$ where $b \in I$, $c \in \mathrm{ann}_R M$. Any $x \in M$ can be written as $x = x1 = xb + xc = xb \in MI$. Thus $MI = M$. (Note that finite generation is not needed for this part.) (2) Let $x_1, \ldots, x_n$ generate M. The condition $MI = M$ implies that any $x \in M$ has the form $\sum_1^n x_i b_i$, $b_i \in I$. In particular, $x_i = \sum_{j=1}^n x_j b_{ji}$, $1 \leqslant i \leqslant n$, or $\sum_{j=1}^n x_j(\delta_{ji} - b_{ji}) = 0$. These equations imply $x_j \det(1 - B) = 0$, $1 \leqslant j \leqslant n$, $B = (b_{ij})$. Evidently

$c = \det(1-B) = 1-b$, $b \in I$. Since R is commutative, $x_jc = 0$, $1 \leqslant j \leqslant n$, implies that $xc = 0$ for all $x \in M$ so $c \in \text{ann}_R M$. Then $1 = b+c \in I + \text{ann}_R M$ and hence $R = I + \text{ann}_R M$. □

We can now prove

THEOREM 3.22. *Any faithful finitely generated projective module over a commutative ring is a generator (hence a progenerator).*

Proof. If M is finitely generated projective, by the dual basis lemma, we have elements $x_1, \dots, x_n \in M$ and elements $f_1, \dots, f_n \in \hom_R(M, R)$ such that $x = \sum x_i f_i(x)$. Then $f_i(x) \in T(M)$, so this shows that $M = MT(M)$. Hence, by the lemma, $R = \text{ann}_R M + T(M)$ and so if M is faithful, then $R = T(M)$. Hence M is a generator by Theorem 3.21. □

In considering equivalences between **mod**-R and **mod**-R' for a second ring R' we assume, of course, that the pair of functors (F, G) defining the equivalence are additive. We recall that F is faithful and full (Proposition 1.3, p. 27). Hence for any R-modules M and N, the map F of $\hom_R(M, N)$ into $\hom_{R'}(FM, FN)$ is an isomorphism. Moreover, F respects the composition of homomorphisms. It follows that properties of an R-module or an R-homomorphism that can be expressed in categorical terms carry over from **mod**-R to the equivalent **mod**-R'. For example, $f: M \to N$ is injective (surjective) if and only if f is monic (epic) in **mod**-R. Hence f is injective (surjective) if and only if $F(f): FM \to FN$ is injective (surjective).

The concept of a subobject of an object in a category (p. 18) provides a categorical way of dealing with the submodules of a given module N. In **mod**-R, a subobject of the module N is an equivalence class $[f]$ of monics $f: M \to N$ where the equivalence relation is defined by $f \sim f': M' \to N$ if there exists an isomorphism $g: M' \to M$ such that $f' = fg$. In this case, $f'M' = fM$, so all of the f in $[f]$ have the same image in N and this is a submodule. Moreover, if M is any submodule, then we have the injection $i: M \to N$, which is a monic, and $iM = M$. Thus we have a bijection of the set of submodules of N with the set of subobjects of N. This is order-preserving if submodules are ordered in the usual way by inclusion, and we define $[f'] \leqslant [f]$ for subobjects of N to mean $f' = fg$ for a monomorphism g. If $\{N_\alpha\}$ is a directed set of submodules of N, then $\bigcup N_\alpha$ is a submodule and this is a sup for the set of N_α in the partial ordering by inclusion. It follows that any directed set of subobjects of N has a sup in the ordering of subobjects. If (F, G) is an equivalence of **mod**-R to **mod**-R' and $\{[f_\alpha]\}$ is a directed set of subobjects of N with sup $[f]$, then it is clear that $\{[F(f_\alpha)]\}$ is a directed set of subobjects of FN with sup $[F(f)]$. A subobject $[f]$ is called *proper* if f is not an isomorphism or, equivalently, if fM for $f: M \to N$ is a proper submodule of N. If $[f]$ is proper then $[F(f)]$ is proper.

We wish to show that if X is a progenerator of **mod**-R, then FX is a progenerator of **mod**-R'. This will follow from

PROPOSITION 3.19. *Let (F, G) be a pair of functors giving an equivalence of* **mod**-R *and* **mod**-R'. *Then* (1) *If X is a generator of* **mod**-R *then FX is a generator of* **mod**-R'. (2) *If X is projective then FX is projective.* (3) *If X is finitely generated then FX is finitely generated.*

Proof. (1) By Theorem 3.21.2, X is a generator if and only if for $f: M \to N$, $f \neq 0$, there exists a $g: X \to M$ such that $fg \neq 0$. Now consider FX and let $f': M' \to N'$, $f' \neq 0$, in **mod**-R'. Then $G(f') \neq 0$ so there exists a $g: X \to GM'$ such that $G(f')g \neq 0$. Then $FG(f')F(g) \neq 0$. The fact that $FG \cong 1_{\textbf{mod-}R'}$ implies the existence of a $g': FX \to M'$ such that $f'g' \neq 0$ (draw a diagram). Hence FX is a generator in **mod**-R'.

(2) Since surjectivity of a homomorphism f is equivalent to the condition that f is epic in **mod**-R, the statement that X is projective is equivalent to the following: given an epic $p: M \to N$, the map $g \rightsquigarrow pg$ of $\hom_R(X, M)$ into $\hom_R(X, N)$ is surjective. From this it follows easily that X projective implies FX is projective. We leave it to the reader to carry out the details of the proof.

(3) We note first that X is finitely generated if and only if it cannot be expressed as a union of a directed set of proper submodules (exercise 3, p. 60). The categorical form of this condition is that the sup of any directed set of proper subobjects of X is a proper subobject. Because of the symmetry of equivalence, it suffices to show that if $\{[f_\alpha]\}$ is a directed set of proper subobjects of an R-module N such that $[f] = \sup\{[f_\alpha]\}$ is not proper, then the same thing holds for $\{[F(f_\alpha)]\}$ as subobjects of FN. This is clear from the remark made above. □

EXERCISES

1. Show that Proposition 3.19 holds for the property of injectivity of a module.
2. Let (F, G) be as in Proposition 3.19. Show that if M is noetherian, then FM is noetherian. What if N is artinian?

3. A module X is called a *cogenerator* of **mod**-R if for any module M there exists a monomorphism of M into a product $\prod_I X$ of copies of X. Show that X is a cogenerator if and only if the functor hom $(-, X)$ is faithful.

4. Determine a cogenerator for **mod**-$\mathbb{Z}$.

3.15 EQUIVALENCE OF CATEGORIES OF MODULES

Before proceeding to the next main result, which will give a condition for equivalence of the categories **mod**-R and **mod**-R' for two rings R and R' and the form of such equivalences, we shall establish a natural isomorphism between two functors that arise in the situation of Morita I.

We recall that if $P = {}_{R'}P_R$ and $M = M_R$ (hence $= {}_{\mathbb{Z}}M_R$) then, as in Proposition 3.4, hom (P_R, M_R) can be regarded as a right R'-module by defining fr' for $f \in \hom(P_R, M_R)$, $r' \in R'$ by $(fr')x = f(r'x)$ (see p. 134). This defines a functor hom $(P_R, -)$ from **mod**-R to **mod**-R'. We shall now show that if we have a Morita context $(R, R', P, P', \tau, \mu)$ in which τ and μ are surjective, as in Morita I, then hom $(P_R, -)$ and $\otimes_R P'$ are naturally isomorphic functors. Let $u \in M$, $z' \in P'$. Then we define the map

$$\{u, z'\}: z \rightsquigarrow u(z', z), \qquad z \in P \tag{90}$$

of P into M. It is clear that $\{u, z'\} \in \hom(P_R, M_R)$ and the map $(u, z') \rightsquigarrow \{u, z'\}$ defines a balanced product of M_R and ${}_RP'$. Hence we have the homomorphism of $M \otimes_R P'$ into hom (P_R, M_R) such that $u \otimes z' \rightsquigarrow \{u, z'\}$. If $r' \in R'$, then $\{u, z'\}r'$ is defined by $(\{u, z'\}r')z = \{u, z'\}r'z = u(z', r'z) = u(z'r', z)$. Thus $\{u, z'\}r' = \{u, z'r'\}$ so the homomorphism such that $u \otimes z' \rightsquigarrow \{u, z'\}$ is a right R'-module homomorphism. We claim that this is an isomorphism. As in the proof of Morita I, we write $1 = \sum [u_j, v_j']$ for $u_j \in P$, $v_j' \in P'$. Then if $f \in \hom(P_R, M_R)$ and $z \in P$, we have $fz = f(1z) = f(\sum(\&[u_j, v_j']z) = f(\sum u_j(v_j', z)) = \sum f(u_j)(v_j', z) = \sum \{fu_j, v_j'\}z$. Thus $f = \sum \{fu_j, v_j'\}$. Hence the homomorphism of $M \otimes_R P'$ into hom (P_R, M_R) is surjective. Now suppose $\sum \{w_i, w_i'\} = 0$ for $w_i \in M$, $w_i' \in P'$. Then

$$\begin{aligned}
\sum w_i \otimes w_i' &= \sum w_i \otimes w_i' 1 = \sum_{i,j} w_i \otimes w_i'[u_j, v_j'] \\
&= \sum_{i,j} w_i \otimes (w_i', u_j)v_j' = \sum_{i,j} w_i(w_i', u_j) \otimes v_j' \\
&= \sum_{i,j} \{w_i, w_i'\}u_j \otimes v_j' = 0.
\end{aligned}$$

Hence our homomorphism is injective. Thus we have the right R'-module isomorphism

(91) $$\eta_M : M \otimes_R P' \to \hom(P_R, M_R)$$

such that $\eta_M(u \otimes z') = \{u, z'\}$. This is natural in M, that is, if $f \in \hom(M_R, N_R)$ then

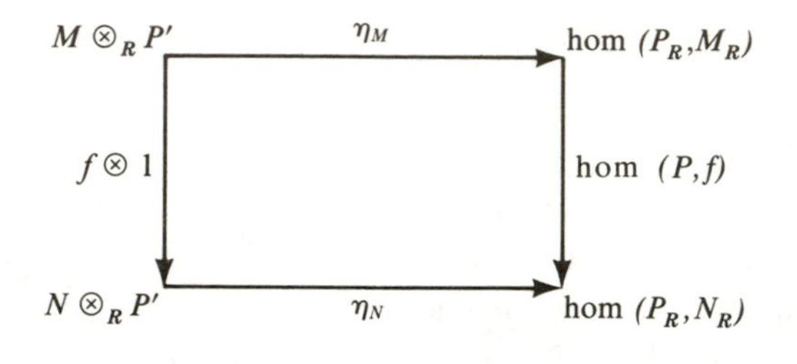

is commutative. To verify this let $u \otimes z' \in M \otimes_R P'$. Then $\hom(P, f)\eta_M(u \otimes z') = f\{u, z'\} = \{fu, z'\}$ (by the definition (90)) and

$$\eta_N(f \otimes 1)(u \otimes z') = \{fu, z'\}.$$

Hence the diagram is commutative. We have therefore shown that the functors $\otimes_R P'$ and $\hom(P_R, -)$ are naturally isomorphic.

We can now prove the second main result of the Morita theory.

MORITA II. *Let R and R' be rings such that the categories* **mod**-R *and* **mod**-R' *are equivalent. Then*

(1) *There exist bimodules ${}_{R'}P_R$, ${}_RP'_{R'}$ and a Morita context $(R, R', P, P', \tau, \mu)$ for which τ and μ are surjective so Morita* I *holds. In particular, R' is isomorphic to* End P_R *for a progenerator P of* **mod**-R *and R is isomorphic to* End $P'_{R'}$, *P' a progenerator of* **mod**-R'.

(2) *If (F, G) is a pair of functors giving an equivalence of* **mod**-R *and* **mod**-R', *then F is naturally isomorphic to $\otimes_R P'$ and G to $\otimes_{R'} P$ where P and P' are as in* (1).

Proof. (1) Let (F, G) be as in (2) and put $P = GR'$, $P = FR$. Since R is a progenerator of **mod**-R, P' is a progenerator of **mod**-R' by Proposition 3.19. Similarly, P is a progenerator of *mod-R*. Since R' is a ring, the map $\lambda : r' \rightsquigarrow \lambda(r')$ where $\lambda(r')$ is the left multiplication $x' \rightsquigarrow r'x'$ is an isomorphism of R' onto End $R'_{R'}$. Since G applied to End $R'_{R'} = \hom(R'_{R'}, R'_{R'})$ is an isomorphism of this ring onto $\hom(P_R, P_R) = \text{End } P_R$, we have the ring isomorphism $r' \rightsquigarrow G\lambda(r')$ of R' onto End P_R. It is clear that $r'x = G\lambda(r')x$ makes P an R'-R-bimodule. Similarly, $r \rightsquigarrow F\lambda(r)$ is an isomorphism of R onto End $P'_{R'}$ and P' is an R-R'-bimodule with $rx' = F\lambda(r)x'$. We have the chain of group isomorphisms

$$(92) \qquad \begin{gathered} P'_{R'} \cong \hom(R', P'_{R'}) = \hom(R', FR) \cong \\ \hom(P_{R'}, GFR) \cong \hom(P_R, R) = P^* \end{gathered}$$

where the first is the canonical one (exercise 1, p. 99), the second is G, and the third is $g \to \zeta_{GFR}g$ where ζ is the given natural isomorphism of GF to $1_{\mathbf{mod}\text{-}R}$. All of the modules R', $P'_{R'}$, P_R, R and GFR in (92) except GFR are bimodules and GFR becomes an R-R-bimodule if we define $rx = GF\lambda(r)x$ for $r \in R$, $x \in GFR$. With these definitions, all of the homs in (92) have R-R'-bimodule structures given by Proposition 3.4 (p. 134). Now it is immediate that the canonical isomorphism of $P'_{R'}$ and $\hom(R', P'_{R'})$ is an R-R'-bimodule isomorphism. The left R-action on $\hom(R', FR)$ is $rf = F\lambda(r)f$ and the right R'-action is $fr' = f\lambda(r')$. Applying G we obtain $G(rf) = GF\lambda(r)G(f)$ and $G(fr') = G(f)G\lambda(r')$. Taking into account the R and R' actions on $\hom(P_R, GFR)$, we see that the second isomorphism in (92) is an R-R'-bimodule isomorphism. Similarly, if we use the naturality condition $\zeta_{GFR}GF\lambda(r) = \lambda(r)\zeta_{GFR}$, we can show that the last isomorphism in (92) is R-R'. Thus we obtain an R-R'-bimodule isomorphism of ${}_RP'_{R'}$ with ${}_RP^*_{R'}$. If we apply this to the Morita context $(R, R', P, P^*, -\ -)$ determined by P, we obtain a Morita context $(R, R', P, P', \tau, \mu)$ in which τ and μ are surjective. Then Morita I and the result noted at the beginning of the section are valid.

(2) If M is a right R-module, we have $FM \cong \hom(R'_{R'}, FM) \cong \hom(GR'_{R'}, GFM)$ (by G) $\cong \hom(P, M)$ (by the natural isomorphism of GF and $1_{\mathbf{mod}\text{-}R}$) $\cong M \otimes_R P'$ (as above), and all of the isomorphisms are natural in M. Hence the given functor F is naturally isomorphic to $\otimes_R P'$. Similarly, G is naturally isomorphic to $\otimes_{R'} P$. $\square$

The existence of a Morita context $(R, R', P, P', \tau, \mu)$ with τ and μ surjective implies the equivalence of **mod**-R and **mod**-R' and the equivalence of R-**mod** and R'-**mod** (Morita I (5)). A consequence of this result and Morita II is that R and R' have equivalent categories of right modules **mod**-R and **mod**-R' if and only if the categories R-**mod** and R'-**mod** are equivalent. If this is the case, we say that R and R' are (Morita) *similar*. Morita II and the characterization of generators in Theorem 3.21 (p. 173) permit a precise identification of the rings R', which are similar to a given ring R, namely, we have

THEOREM 3.23. *For a given ring R let e be an idempotent in a matrix ring $M_n(R)$, $n \geqslant 1$, such that the ideal $M_n(R)eM_n(R)$ generated by e is $M_n(R)$. Then $R' = eM_n(R)e$ is similar to R. Moreover, any ring similar to R is isomorphic to a ring of this form.*

We shall require the following lemma, which collects a good deal of useful information.

LEMMA. (1) *If e is an idempotent in R and N is a right R-module, then* $\hom_R(eR, N)$ *is the set of maps* $ea \rightsquigarrow uea$, $u \in N$. *The map* $f \rightsquigarrow f(e)$ *is a group isomorphism of* $\hom_R(eR, N)$ *onto* Ne *and this is a ring isomorphism of* $\mathrm{End}_R eR = \hom_R(eR, eR)$ *onto* eRe *if* $N = eR$. (2) *The trace ideal* $T(eR) = ReR$. (3) *A right R-module M is a cyclic progenerator if and only if* $M \cong eR$ *where e is an idempotent in R such that* $ReR = R$. (4) *If* $M = M_1 \oplus M_2$ *for right R-modules and e_i are the projections determined by this direct sum, then* $\mathrm{End}_R M_i \cong e_i(\mathrm{End}_R M)e_i$.

Proof. (1) Since e is a generator of eR, a homomorphism f of eR into N is determined by the image $f(e) \in N$. Then $f(e) = f(e^2) = f(e)e$, so $f(e) = ue \in Ne$ and $f(ea) = uea$. Conversely, for any $u \in N$, the map $ea \rightsquigarrow uea$ is a module homomorphism of eR into N. Direct verification shows that $f \rightsquigarrow f(e) = ue$ is a group isomorphism of $\hom(eR, N)$ onto Ne and in the special case in which $N = eR$, this is a ring isomorphism (eRe is a subring of R and has e as unit).

(2) The determination of $\hom_R(eR, N)$ for the case $N = R$ shows that $(eR)^* = \hom_R(eR, R)$ is the set of maps $ea \rightsquigarrow bea$, $a, b \in R$. Hence $T(eR)$ is the set of sums $\sum b_i e a_i$, $a_i, b_i \in R$. Thus $T(eR) = ReR$.

(3) It is clear that M is a cyclic projective right R-module if and only if M is isomorphic to a direct summand of R, hence, if and only if $M \cong eR$ where $e^2 = e \in R$. By Theorem 3.21, eR is a generator if and only if $T(eR) = R$. Hence this holds if and only if $ReR = R$. Combining these results we see that M is a cyclic progenerator if and only if $M \cong eR$ where e is an idempotent in R and $ReR = R$.

(4) Let $M = M_1 \oplus M_2$ and let e_i, $i = 1, 2$, be the projections determined by this decomposition. Then $e_i \in \mathrm{End}_R M$, $e_1 + e_2 = 1$, $e_i^2 = e_i$, and $e_1 e_2 = 0 = e_2 e_1$. $e_1(\mathrm{End}_R M)e_1$ is a ring with the unit e_1. If $\eta \in \mathrm{End}_R M$, $e_1 \eta e_1$ maps M_1 into itself and hence $e_1 \eta e_1 | M_1 \in \mathrm{End}_R M_1$. Since $(e_1 \eta e_1) M_2 = 0$, the ring homomorphism $e_1 \eta e_1 \rightsquigarrow (e_1 \eta e_1)|M_1$ is a monomorphism. Moreover, this map is surjective, since if $\zeta \in \mathrm{End}_R M_1$ then $e_1 \zeta e_1 \in \mathrm{End}_R M$ and $e_1 \zeta e_1 | M_1 = \zeta$. Thus $e_1 \eta e_1 \rightsquigarrow e_1 \eta e_1 | M_1$ is an isomorphism of $e_1(\mathrm{End}_R M)e_1$ onto $\mathrm{End}_R M_1$. Similarly, we have an isomorphism of $e_2(\mathrm{End}_R M)e_2$ onto $\mathrm{End}_R M_2$. □

We can now give the

Proof of Theorem 3.23. We gave a direct proof of the equivalence of **mod**-R and **mod**-$M_n(R)$ in Proposition 1.4 (p. 29). A better approach to this is obtained by applying Morita I to the example of the left module $M' = {}^{(n)}R$ of rows of n elements of R and the right module $R^{(n)}$ of columns of n elements of R as in the example on p. 166. It is readily seen that the pairings μ and τ are surjective, so Morita I is applicable. Statement (5) of this theorem gives an equivalence of R and $M_n(R)$. Hence R and $M_n(R)$ are similar. If e is an idempotent in $M_n(R)$ such that $M_n(R)eM_n(R) = M_n(R)$ then, by (3) in the lemma, $eM_n(R)$ is a progenerator of **mod**-$M_n(R)$. Then, by Theorem 3.20, $M_n(R)$ and $\mathrm{End}_{M_n(R)}(eM_n(R))$ are similar. By statement (1) of the lemma, the latter ring is isomorphic to $eM_n(R)e$. Thus $M_n(R)$ and $eM_n(R)e$ are similar and hence R and $eM_n(R)e$ are similar if $M_n(R)eM_n(R) = M_n(R)$.

Conversely, let R' be any ring similar to R. By Morita II, $R' \cong \mathrm{End}_R P$ where P is a progenerator for **mod**-R. Then P is a direct summand of $R^{(n)}$ for some n and $T(P) = R$. Thus $P = eR^{(n)}$ where $e^2 = e \in \mathrm{End}_R R^{(n)}$. Hence, by (4) of the lemma, $\mathrm{End}\, P_R \cong e(\mathrm{End}\, R^{(n)})e$. Also the application of Morita I to the example $(R, M_n(R), R^{(n)}, {}^{(n)}R, \tau, \mu)$ shows that $\mathrm{End}\, R^{(n)}$ is the set of left multiplications of the elements of $R^{(n)}$ by $n \times n$ matrices with entries in R and $\mathrm{End}\, R^{(n)} \cong M_n(R)$. Identifying e with the corresponding matrix, we see that $R' \cong eM_n(R)e$. Since P is a generator, we have $T(P) = R$ and since P is a direct summand of $R^{(n)}$, the elements of $P^* = \hom(P, R)$ are the restrictions to P of the elements of $\hom(R^{(n)}, R)$. The latter are the set of left multiplications by rows $(a_1, a_2, \ldots, a_n)$, $a_i \in R$. Since $P = eR^{(n)}$, it follows that $T(P)$ is the set of sums of elements of the form

$$(a_1, \ldots, a_n)e \begin{pmatrix} b_1 \\ \vdots \\ b_n \end{pmatrix}$$

$a_i, b_i \in R$. Hence 1 can be expressed as such a sum. It follows that $M_n(R)eM_n(R)$ contains the matrix e_{11} whose (1–1)-entry is 1 and other entries are 0. Since $M_n(R)e_{11}M_n(R) = M_n(R)$, we have $M_n(R)eM_n(R) = M_n(R)$. Thus $R' \cong eM_n(R)e$ where $e^2 = e$ and $M_n(R)eM_n(R) = M_n(R)$. □

Having settled the question of what the rings similar to a given ring look like, we now study more closely the equivalences of **mod**-R and **mod**-R' (and R-**mod** and R'-**mod**). By Morita II, these are given up to natural isomorphism by tensoring by an R'-R-bimodule P in a Morita context $(R, R', P, P', \tau, \mu)$ in which τ and μ are surjective, hence, isomorphisms. Thus μ is an R'-R'-bimodule isomorphism of $P \otimes_R P'$ onto R' and τ is an R-R-bimodule isomorphism of $P' \otimes_{R'} P$ onto R. Conversely, suppose we have such isomorphisms for an R'-R-bimodule P and an R-R'-bimodule P' for any pair of rings R and R'. Then, as in the proof of Morita I (5), $\otimes_R P'$ and $\otimes_{R'} P$ are functors giving an equivalence of **mod**-R and **mod**-R' and $P\otimes_R$ and $P'\otimes_{R'}$ give an equivalence of R'-**mod** and R-**mod**. Hence R and R' are similar. Also since an equivalence of module categories sends progenerators into progenerators and $R\otimes_R P' \cong P'$, it is clear that P' is a progenerator for **mod**-R'. Similarly, it is a progenerator for R-**mod** and P is one for **mod**-R and for R'-**mod**. We shall now call an R'-R-bimodule P *invertible* if there exists an R-R'-bimodule P' such that $P'\otimes_{R'} P \cong R$ as R-R-bimodule and $P\otimes_R P' \cong R'$ as R'-R'-bimodule. One can tensor multiply such modules. More precisely, if R'' is a third ring and Q is an invertible R''-R'-bimodule (so R'' and R' are similar), then the associative law for tensor products shows that $Q\otimes_{R'} P$ is invertible. We can now relate this to isomorphism classes of functors giving equivalences between the categories of modules of two rings. This is the content of

MORITA III. *Let $R, R', R'', \ldots$ be similar rings. Then the map $P \rightsquigarrow \otimes_{R'} P$ defines a bijection of the class of isomorphism classes of invertible R'-R-bimodules and the class of natural isomorphism classes of functors giving equivalences of* **mod**-*R' and* **mod**-*R. In this correspondence, composition of equivalences corresponds to tensor products of invertible bimodules.*

Proof. The first statement amounts to this: If P is an invertible R'-R-bimodule, then $\otimes_{R'} P$ gives an equivalence of **mod**-R' and **mod**-R and every such equivalence is naturally isomorphic to one of this form. Moreover, the functors $\otimes_{R'} P_1$ for $\otimes_{R'} P_2$ for invertible R'-R-bimodules P_1 and P_2 are naturally isomorphic functors if and only if P_1 and P_2 are isomorphic as bimodules. The first assertion has been proved in the first two Morita theorems. Now suppose P_1 and P_2 are isomorphic invertible R'-R-bimodules. Then for any $N' = N'_{R'}$, $M'\otimes_{R'} P_1$ and $N'\otimes_{R'} P_2$ are isomorphic under an isomorphism that is natural in N'. Hence $\otimes_{R'} P_1$ and $\otimes_{R'} P_2$ are naturally isomorphic. Conversely, suppose $\otimes_{R'} P_1$ and $\otimes_{R'} P_2$ are naturally isomorphic. Then we have an R-isomorphism η of $R'\otimes_{R'} P_1$ onto $R'\otimes_{R'} P_2$ such that for any $a' \in R'$

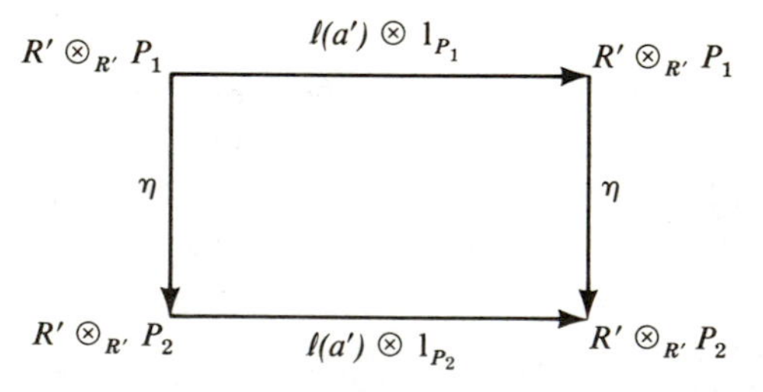

is commutative. This means that $R'\otimes_{R'} P_1$ and $R'\otimes_{R'} P_2$ are isomorphic as R'-R-bimodules. Then P_1 and P_2 are isomorphic as R'-R-bimodules. Now suppose Q is an invertible R''-R'-bimodule. Then we have the functor $\otimes_{R''}Q$ from **mod**-R'' to **mod**-R' and the functor $\otimes_{R'}P$ from **mod**-R' to **mod**-R. The composite is the functor $(\otimes_{R''}Q)\otimes_{R'}P$ from **mod**-R'' to **mod**-R. By associativity, this is the functor $\otimes_{R''}(Q\otimes_{R'}P)$. This is the meaning of the last statement of the theorem. □

In the special case in which $R' = R$, the preceding theorem relates isomorphism classes of invertible R-R-bimodules and natural isomorphism classes of auto-equivalences of **mod**-R. It is readily seen that the first of these classes is a set. (Prove this.) Hence it constitutes a group in which the multiplication is defined by the tensor product. The theorem gives an isomorphism of this group with the group of natural isomorphism classes of auto-equivalences of **mod**-R. The first group is denoted as $\mathrm{Pic}\,R$. Of particular interest is the case in which R is commutative. We shall consider this group later (sections 7.8 and 10.6).

EXERCISES

1. Let e be an idempotent in a ring such that $ReR = R$. Show that the map $B \rightarrow eBe = B \cap eRe$ is a lattice isomorphism of the lattice of ideals of R onto the lattice of ideals of eRe. Does this hold without the assumption that $ReR = R$?
2. Define a relation $\sim$ among rings by $R \sim R'$ if $R' \cong eM_n(R)e$ for some n and an idempotent e such that $M_n(R)eM_n(R) = M_n(R)$. By Theorem 3.23 this is an equivalence relation. Can you prove this directly without using the Morita theory?
3. Show that if R is a p.i.d., then any ring similar to R is isomorphic to a matrix ring $M_n(R)$.

REFERENCES

K. Morita, Duality for modules and its application to the theory of rings with minimum condition, *Science Reports of the Tokyo Kyoiku Daigaku*, vol. 6, Sec. A (1958), pp. 83–142.

H. Bass, *The Morita Theorems*, University of Oregon lecture notes, 1962, 21 pages.

C. Faith, *Algebra: Rings, Modules and Categories* I, New York, Springer, 1973.

4

Basic Structure Theory of Rings

The structure theory of rings, which we shall consider in this chapter, is an outgrowth of the structure theory of finite-dimensional algebras over a field that was developed during the latter part of the nineteenth century and the early part of this century. The discovery of quaternions by Hamilton in 1843 led to the construction of other "hypercomplex number systems," that is, finite dimensional algebras over $\mathbb{R}$ and $\mathbb{C}$, including some non-associative ones (e.g., the octonions of Cayley and Graves). The problem of classifying such systems was studied intensively during the second half of the nineteenth century by a number of mathematicians: Molien, Frobenius, Cartan, and others. Their results constituted a very satisfactory theory for algebras over $\mathbb{R}$ and $\mathbb{C}$, which included a complete classification of an important subclass, the semi-simple algebras. This structure theory was extended to finite dimensional algebras over an arbitrary field by Wedderburn in 1908. In 1927, Artin, stimulated by Emmy Noether's earlier strikingly successful simplification of ideal theory of commutative rings by the introduction of chain conditions, and motivated by the needs of arithmetic in finite dimensional algebras, extended the Wedderburn theory to rings satisfying chain conditions for one-sided ideals. In 1945 this author developed a structure theory of rings without finiteness assumptions. The principal ingredients of this theory were a new definition of a radical, the concepts of primitivity and semi-primitivity (formerly called semi-simplicity), and a density theorem (due independently to Chevalley) that constituted an extensive generalization of a classical theorem of Burnside.

In this chapter we shall reverse the historical order of development by first giving an account of the general theory and then specializing successively to Artin's theory and to Wedderburn's theory of finite-dimensional algebras over fields. The Wedderburn structure theorem for simple algebras reduces the classification of these algebras to division algebras. The proper vehicle for studying these is the Brauer group of algebra classes with composition defined by tensor products. We shall give an introduction to the study of these groups and related results on central simple algebras. At the end of the chapter we shall apply the results to the study of Clifford algebras. These are needed to round out the structure theory of orthogonal groups that was presented in Chapter 6 of BAI. Further applications of the theory to representation theory of finite groups will be given in the next chapter.

4.1 PRIMITIVITY AND SEMI-PRIMITIVITY

In the development of ring theory that we shall give, rings of endomorphisms of abelian groups and concretizations of rings as rings of endomorphisms play a predominant role. We recall that if M is an abelian group (written additively), the set of endomorphisms of M has a natural structure of a ring. Thus we obtain $\operatorname{End} M$ (or $\operatorname{End}_{\mathbb{Z}} M$), the ring of endomorphisms of M. By a *ring of endomorphisms* we mean a subring of $\operatorname{End} M$ for some abelian group M. We define a *representation* ρ of a ring R to be a homomorphism of R into a ring $\operatorname{End} M$ of endomorphisms of an additive abelian group M. A representation ρ of R *acting on* M (that is, with codomain $\operatorname{End} M$) defines a left R-module structure on M by specifying that the action of R on M is given through ρ:

$$ax = \rho(a)x \tag{1}$$

for $a \in R$, $x \in M$. Equation (1) defines a left module action since $\rho(a)$ is an endomorphism and $a \rightarrow \rho(a)$ is a ring homomorphism. Conversely, if M is a left module for the ring R, then M is an abelian group and M affords a representation $\rho = \rho_M$ of R by defining a_M for $a \in R$ to be the endomorphism

$$a_M : x \rightarrow ax.$$

Then $\rho : a \rightarrow a_M$ is a homomorphism of R into $\operatorname{End} M$, hence a representation of R.

By an *irreducible representation* of a ring we shall mean a representation ρ for which the associated module M (as just defined) is irreducible: $M \neq 0$ and M and 0 are the only submodules of M. The structure theory that we shall develop is based on two articles of faith: irreducible representations are the best kinds of representations and the best behaved rings are those that have enough irreducible representations to distinguish between elements of the ring, that is, given a pair of distinct elements a, b of R, there exists an irreducible representation ρ such that $\rho(a) \neq \rho(b)$. Evidently this is equivalent to the simpler condition that given any $a \neq 0$ in R there exists an irreducible representation ρ of R such that $\rho(a) \neq 0$.

Before we proceed to the main definitions, it will be useful to collect some elementary facts on representations and modules. Because of the connection between representations and left modules that we have noted, we give preference in this chapter to left modules. Accordingly, in this chapter the unadorned term "module" will always mean *left module*.

If ρ is a representation of the ring R acting in M, the kernel of ρ is evidently the ideal

$$\text{ann}_R M = \{b \in R \mid bM = 0\}, \tag{2}$$

$(bx = \rho(b)x)$. We shall call the representation ρ (and occasionally the module M) *faithful* if $\ker \rho = 0$. For any $x \in M$ we have the order ideal or annihilator of x, $\text{ann}_R x = \{b \in R \mid bx = 0\}$. This is a left ideal and the cyclic submodule Rx generated by x is isomorphic to $R/\text{ann}_R x$ where R is regarded as left R-module in the natural way. It is evident that

$$\ker \rho = \text{ann}_R M = \bigcap_{x \in M} \text{ann}_R x. \tag{3}$$

If f is a homomorphism of a ring S into R and ρ is a representation of R acting on M, then ρf is a representation of S acting on M. The corresponding module action of S on M is given by

$$bx = f(b)x \tag{4}$$

for $b \in S$. As in the last chapter, the notations ${}_S M$ and ${}_R M$ serve to distinguish M as S-module from M as R-module (as well as to indicate that the action is on the left). In most cases, it will be clear from the context which of these is intended and "M" will be an adequate notation. It is clear that

$$\text{ann}_S M = f^{-1}(\text{ann}_R M) \tag{5}$$

where, of course, $f^{-1}(\ )$ denotes the inverse image under f. If the representation ρ of R is faithful, then $\text{ann}_S M = f^{-1}(0) = \ker f$.

Let B be an ideal of R contained in $\mathrm{ann}_R M$. Then M becomes an $\bar{R} = R/B$ module by defining the action of the coset $a+B$ on $x \in M$ by

$$(a+B)x = ax. \tag{6}$$

Since $bx = 0$ for every $b \in B$, $x \in M$, it is clear that (6) is a well-defined action. It is immediate also that this is a module action of $\bar{R}$ on M and it is clear that a subset N of M is an R-submodule if and only if it is an $\bar{R}$-submodule. In other words, ${}_RM$ and ${}_{\bar{R}}M$ have the same lattice of submodules. The relation between the kernels of the representations of R and of $\bar{R}$ is given by the formula

$$\mathrm{ann}_{\bar{R}} M = (\mathrm{ann}_R M)/B. \tag{7}$$

We now proceed to introduce two of our main definitions.

DEFINITION 4.1. *A ring R is called* primitive *if it has a faithful irreducible representation. R is called* semi-primitive *if for any $a \neq 0$ in R there exists an irreducible representation ρ such that $\rho(a) \neq 0$.*

We shall be interested also in representations that are *completely reducible* in the sense that the corresponding module M is completely reducible, that is, $M = \sum M_\alpha$ where the M_α are irreducible submodules. We establish first two characterizations of semi-primitivity in the following

PROPOSITION 4.1. *The following conditions on a ring R are equivalent*: (1) *R is semi-primitive.* (2) *R has a faithful completely reducible representation.* (3) *R is a subdirect product of primitive rings* (*see p. 69*).

Proof. (1) $\Rightarrow$ (2). For each $a \neq 0$ in R let M_a be an irreducible module such that for the representation ρ_{M_a} we have $\rho_{M_a}(a) \neq 0$. Form $M = \oplus_{a\neq 0} M_a$. This is a completely reducible module and it is clear that

$$\mathrm{ann}_R M = \bigcap_{a\neq 0} \mathrm{ann}_R M_a = 0. \tag{8}$$

Hence ρ_M is a faithful completely reducible representation for R.

(2) $\Rightarrow$ (3). Suppose ρ is a faithful completely reducible representation for R so $M = \sum M_\alpha$, M_α irreducible, for the corresponding module M. We have $0 = \mathrm{ann}_R M = \bigcap_\alpha \mathrm{ann}_R M_\alpha$, and $\mathrm{ann}_R M_\alpha$ is an ideal in R. Hence R is a subdirect product of the rings $R_\alpha = R/\mathrm{ann}_R M_\alpha$. On the other hand, since $(\mathrm{ann}_R M_\alpha)M_\alpha = 0$, M_α can be regarded as an irreducible module for R and the representation of R_α defined by M_α is faithful. Hence R_α is primitive and R is a subdirect product of primitive rings.

$(3) \Rightarrow (1)$. Let R be a subdirect product of the primitive rings R_α. For each R_α let ρ_α be a faithful irreducible representation of R_α. We have the canonical homomorphism π_α of R onto R_α and $\bigcap_\alpha \ker \pi_\alpha = 0$. We have the irreducible representation $\rho_\alpha \pi_\alpha$ of R whose kernel is $\ker \pi_\alpha$. Accordingly, we have a family of irreducible representations $\{\rho_\alpha \pi_\alpha\}$ of R such that $\bigcap_\alpha \ker \rho_\alpha \pi_\alpha = 0$. Then R is semi-primitive. □

The definitions we have given thus far involve objects (representations and modules) that are external to the ring R. It is easy and useful to replace the definitions by internal characterizations. Let I be a left ideal of the ring R. We define

$$(I:R) = \{b \in R \mid bR \subset I\}. \tag{9}$$

It is clear that if we put $M = R/I$ and regard this as a left R-module then

$$(I:R) = \operatorname{ann}_R R/I. \tag{10}$$

It follows from this or directly from (9) that $(I:R)$ is an ideal. Moreover, by (9), $(I:R) \subset I$ and $(I:R)$ contains every ideal of R contained in I. In other words, $(I:R)$ is the (unique) largest ideal of R contained in I. We can now give the following characterization of primitivity and semi-primitivity.

PROPOSITION 4.2. *A ring R is primitive if and only if R contains a maximal left ideal I that contains no non-zero ideal of R. A ring R is semi-primitive if and only if $R \neq 0$ and $\bigcap_I (I:R) = 0$ where the intersection is taken over all maximal left ideals I of R.*

Proof. If R is primitive, we have an irreducible module M such that $\operatorname{ann}_R M = 0$. Now $M \cong R/I$ for a maximal left ideal I (p. 117). Then $(I:R) = \operatorname{ann}_R R/I = \operatorname{ann}_R M = 0$. Thus I is a maximal left ideal containing no non-zero ideal of R. Conversely, if the condition holds, we have a maximal left ideal I such that $(I:R) = 0$. Then $M = R/I$ provides a faithful irreducible representation of R, so R is primitive.

Now let R be semi-primitive and let $\{\rho_\alpha\}$ be a set of irreducible representations of R that is adequate for distinguishing the elements of R. Then $\bigcap \ker \rho_\alpha = 0$. If M_α is the module for ρ_α then $M_\alpha \cong R/I_\alpha$, I_α a maximal left ideal in R. Hence $(I_\alpha : R) = \operatorname{ann}_R M_\alpha = \ker \rho_\alpha$ and $\bigcap_\alpha (I_\alpha : R) = 0$. A fortiori, $\bigcap (I:R)$ taken over all of the maximal left ideals of R is 0. The converse follows by retracing the steps. □

The internal characterizations of primitivity and semi-primitivity have some important consequences that we shall now record. The first of these is

COROLLARY 1. *Any simple ring* ($\neq 0$) *is primitive.*

Proof. Any ring $R \neq 0$ contains a maximal left ideal I (p. 68). If R is simple, $(I:R) = 0$. Hence R is primitive. □

Next we obtain characterizations of primitive and semi-primitive commutative rings.

COROLLARY 2. *Let R be a commutative ring. Then R is primitive if and only if R is a field and R is semi-primitive if and only if it is a subdirect product of fields.*

Proof. If R is a commutative primitive ring, R contains a maximal left ideal I containing no ideal $\neq 0$ of R. Since R is commutative, I is an ideal. Hence $I = 0$. Then 0 is a maximal ideal in R, which means that $R \neq 0$ and 0 and R are the only ideals in the commutative ring R. Then R is a field. Conversely, it is clear that any field satisfies the first condition of Proposition 4.2. Hence any field is primitive. It now follows from Proposition 4.1.3 that R is semi-primitive if and only if it is a subdirect product of fields. □

We shall now give some examples of primitive and semi-primitive rings. A convenient way of constructing examples of rings is as rings of endomorphisms of abelian groups. If R is a ring of endomorphisms of an abelian group M, then M is an R-module in the natural way in which the action ax, $a \in R$, $x \in M$ is the image of x under a. Obviously, the corresponding representation is the injection of R into End M and so this is faithful. Hence if R acts irreducibly on $M \neq 0$ in the sense that there is no subgroup of M other than 0 and M that is stabilized by R, then R is a primitive ring. Several of the following examples are of this type.

EXAMPLES

1. The ring L of linear transformations of a finite dimensional vector space over a division ring is simple (p. 171). Hence L is primitive.

2. Let V be a vector space over a division ring Δ that need not be finite dimensional. We claim that the ring L of linear transformations in V over Δ acts irreducibly on V and hence is a primitive ring. To prove irreducibility, we have to show that if x is any non-zero vector and y is any vector, there exists an $l \in L$ such that $lx = y$. Now this is clear, since we can take x to be an element in a base (e_α) for V over Δ, and given any base (e_α) and corresponding elements f_α for every e_α, then there exists a linear transformation l such that $le_\alpha = f_\alpha$. If V is infinite dimensional, the subset F of

transformations l with finite dimensional range ($l(V)$ finite dimensional) is a non-zero ideal in L. Since $1 \notin F$, $F \neq L$. Hence L is not simple, so this is an example of a primitive ring that is not simple. If the dimensionality of V is countably infinite, it is easy to see that F is the only proper non-zero ideal of L. Hence L/F is simple and thus primitive.

3. If $(x_1, x_2, x_3, \ldots)$ is a base for a vector space V over a division ring Δ, then a linear transformation l is determined by its action on the base. We can write

$$lx_i = \sum_j \lambda_{ij} x_j \tag{11}$$

where the $\lambda_{ij} \in \Delta$ and for a given i only a finite number of the λ_{ij} are $\neq 0$. Thus l can be described by a matrix $\Lambda = (\lambda_{ij})$, which is *row finite* in the sense that there are only a finite number of non-zero entries in each row. The set of row-finite matrices with entries in Δ is a ring under the usual matrix addition and multiplication, and the map $l \to \Lambda$ where Λ is the matrix of l relative to the base (x_i) is an anti-isomorphism. Now let R be the set of linear transformations whose matrices relative to the given base have the form

$$\begin{pmatrix} \Phi & & & & 0 \\ & \lambda & & & \\ & & \lambda & & \\ & & & \cdot & \\ 0 & & & & \cdot \end{pmatrix} \tag{12}$$

where Φ is a finite square block and $\lambda \in \Delta$. The set of matrices of the form (12) is a subring of the ring of row-finite matrices. Hence R is a subring of the ring of linear transformations L. R acts irreducibly on V. For, if x is a non-zero vector and y is any vector, then we can write $x = \xi_1 x_1 + \cdots + \xi_n x_n$, $y = \eta_1 x_1 + \cdots + \eta_n x_n$ for some n where the $\xi_i, \eta_i \in \Delta$. Thus x and y are contained in a finite dimensional subspace V' of V. There exists a linear transformation l' of V' into itself such that $l'x = y$ and l' can be extended to a linear transformation l of V contained in R (for example, by specifying that $lx_i = 0$ if $i > n$). Then $lx = y$. Hence V is irreducible as R-module and so R is primitive.

4. Let V be as in example 3 with $\Delta = F$, a field, and let p and q be the linear transformations in V over F such that

$$\begin{aligned} px_1 = 0, \qquad px_i &= x_{i-1}, \qquad i > 1 \\ qx_i &= x_{i+1}, \qquad i = 1, 2, 3, \ldots. \end{aligned} \tag{13}$$

Then $pqx_i = x_i$ for all i and $qpx_1 = 0$, $qpx_i = x_i$ for $i > 1$. Hence

$$pq = 1, \qquad qp \neq 1. \tag{14}$$

Let R be the set of linear transformations

$$\sum_{i,j=0}^{n} \alpha_{ij} q^i p^j, \qquad n = 0, 1, 2, \ldots \tag{15}$$

where the $\alpha_{ij} \in F$. The first relation in (14) implies that R is closed under multiplication, so R is a ring of endomorphisms of V. Now put

$$e_{ij} = q^{i-1} p^{j-1} - q^i p^j, \qquad i, j = 1, 2, \ldots. \tag{16}$$

The matrix of e_{ij} has a 1 in the (j, i)-position, 0's elsewhere. It follows that the matrix ring corresponding to R includes all the matrices (12). Thus R contains the ring given in example 3 as a subring. Hence the present R acts irreducibly on V, so R is primitive.

5. For any ring R we have the *regular representation* ρ_R whose module is the additive group of R on which the ring R acts by left multiplication (see BAI, pp. 422–426). This representation is faithful and the submodules of R are the left ideals. Now let A be an algebra over the field F. Then the regular representation ρ_A of A provides a monomorphism of A into the algebra L of linear transformations of the vector space A/F. Thus any algebra over a field can be imbedded in a primitive ring. We shall now show that A is also a homomorphic image of a primitive ring. Since A is an arbitrary algebra over a field, this will show that there is not much that can be said about homomorphic images of primitive rings. For our construction we take $V = \oplus A$, a direct sum of a countably infinite number of copies of the vector space A. Let A act on V in the obvious way and identify A with the corresponding algebra of linear transformations in V. For $n = 1, 2, 3, \ldots$ let V_n be the subspace of V of elements of the form $(x_1, x_2, \ldots, x_n, 0, 0, \ldots)$, $x_i \in A$, and let L_n be the set of linear transformations that map V_n into itself and annihilate all of the elements $(0, \ldots, 0, x_{n+1}, x_{n+2}, \ldots)$. Then $L_1 \subset L_2 \subset \cdots$. Put $L = \bigcup L_i$ and let R be the ring of linear transformations generated by A and L. Then R acts irreducibly on V, so R is primitive. Moreover, $R = A + L$, $A \cap L = 0$, and L is an ideal in R. Then $R/L \cong A$, so the given algebra A is a homomorphic image of the primitive algebra R.

6. The ring $\mathbb{Z}$ is semi-primitive since $\bigcap_{p \text{ prime}} (p) = 0$ and $\mathbb{Z}/(p)$ is a field. Hence $\mathbb{Z}$ is a subdirect product of fields, so $\mathbb{Z}$ is semi-primitive. Similarly any p.i.d. with an infinite number of primes is semi-primitive.

EXERCISES

1. Let $F\{x, y\}$ be the free algebra generated by x and y, and let K be the ideal in $F\{x, y\}$ generated by $xy - 1$. Show that $F\{x, y\}/K$ is isomorphic to the algebra R defined in example 4 above.

2. (Samuel.) Let V, F be as in example 4 and let S be the algebra of linear transformations generated by p as in example 4 and the linear transformation r such that $rx_i = x_{i^2+1}$. Show that S is primitive and S is isomorphic to the free algebra $F\{x, y\}$.

3. Let $F\{x, y\}$ be as in exercise 1 and let K be the ideal generated by $xy - yx - x$. Assume that F is of characteristic 0. Show that $R = F\{x, y\}/K$ is primitive.

4. Let R be a ring containing an ideal $N \neq 0$ that is *nilpotent* in the sense that there exists an integer m such that $N^m = 0$ (that is, the product of any m elements of N is 0). Show that R is not semi-primitive.

5. Show that if R is primitive then the matrix ring $M_n(R)$ is primitive.

6. Show that if R is primitive and e is an idempotent element $\neq 0$ in R, then the ring eRe (with unit e) is primitive.

7. Show that if R and R' are Morita similar (p. 179), then R is primitive if and only if R' is primitive.

8. Let F be a field, $P = F[t_1,\dots,t_n]$, the ring of polynomials in n indeterminates with coefficients in F. Show that for any $i = 1,2,\dots,n$, there exists a unique derivation D_i of P such that

$$D_i t_j = \delta_{ij}$$

(see exercise 11, p. 147). For $f \in P$ let $\bar{f}$ denote the multiplication $g \rightsquigarrow fg$ in P. Show that if D is a derivation in P and $f \in P$, then $\bar{f}D$ is a derivation in P. Show that the derivations of P into P are the maps $\sum_1^n \bar{f}_i D_i$ where D_i is as above and $f_i \in P$. Let R be the ring of endomorphisms of the additive group of P generated by the derivations and the multiplications of P. Show that if F is of characteristic 0, then R is an irreducible ring of endomorphisms and hence R is primitive.

9. Let F be a field of characteristic 0 and $F\{x_1,\dots,x_n,y_1,\dots,y_n\}$ the free algebra over F determined by the $2n$ elements $x_1,\dots,x_n,y_1,\dots,y_n$. Let I be the ideal in $F\{x_1,\dots,x_n,y_1,\dots,y_n\}$ generated by the elements

$$[x_i,x_j], \qquad [y_i,y_j], \qquad [y_i,x_j]-\delta_{ij}1,$$

$1 \leqslant i,j \leqslant n$. The algebra $W_n = F\{x_1,\dots,x_n,y_1,\dots,y_n\}/I$ is called a *Weyl algebra*. Show that if P and R are as in exercise 8, then the homomorphism η of $F\{x_1,\dots,x_n,y_1,\dots,y_n\}$ into R such that $x_i \rightsquigarrow \bar{t}_i$, $y_i \rightsquigarrow D_i$, $1 \leqslant i \leqslant n$, is surjective and $\ker \eta = I$. Hence conclude that W_n is primitive.

10. Show that W_n is simple. (*Suggestion*: First treat the case in which $n = 1$.)

4.2 THE RADICAL OF A RING

From the point of view that we took in the previous section, the purpose of defining the radical of a ring is to isolate that part of the ring whose elements are mapped into 0 by every irreducible representation of the ring. Accordingly, we introduce the following

DEFINITION 4.2. *The* (Jacobson) radical *of a ring is the intersection of the kernels of the irreducible representations of the ring.*

Evidently, the radical, rad R, of the ring R is an ideal in R.

We shall call an ideal P of R *primitive* (*in* R) if R/P is a primitive ring. Let I be a maximal left ideal of R and let $P = (I:R)$. Then $M = R/I$ is an irreducible module for R whose annihilator is P. Hence M can be regarded as an $\bar{R} = R/P$-module by (6) and M is irreducible as $\bar{R}$-module. By (7), $\operatorname{ann}_{\bar{R}} M = \operatorname{ann}_R M/P = P/P = 0$. Hence M is a faithful irreducible module for $\bar{R}$ and so

$\bar{R}$ is primitive. Then, by definition, P is a primitive ideal in R. Conversely, let P be a primitive ideal in R. Then we have an irreducible module M for $\bar{R} = R/P$ such that the associated representation is faithful. Regarding M as R-module via (4) we see that M is an irreducible R-module such that $\text{ann}_R M = P$ (see (5)). Now $M \cong R/I$ where I is a maximal left ideal of R and hence $\text{ann}_R R/I = P$. Since $\text{ann}_R R/I = (I:R)$, we have $P = (I:R)$. We therefore have the

LEMMA 1. *An ideal P of R is primitive in R if and only if $P = (I:R)$ for some maximal left ideal I of R.*

We can now give our first internal characterization of the radical.

PROPOSITION 4.3. (1) rad R *is the intersection of the primitive ideals of R.* (2) rad R *is the intersection of the maximal left ideals of R.*

Proof. (1) By definition,

$$\text{rad}\, R = \bigcap \text{ann}_R M \tag{17}$$

where $\{M\}$ is the class of irreducible modules of R. Since $M \cong R/I$ and $\text{ann}\, R/I = (I:R)$, we have

$$\text{rad}\, R = \bigcap (I:R) \tag{18}$$

where I runs over the maximal left ideals of R. By the lemma, this can be written also as

$$\text{rad}\, R = \bigcap P \tag{18'}$$

where P runs over the primitive ideals in R. This proves (1). To prove (2), we recall that for any module M we have $\text{ann}_R M = \bigcap_{x \in M} \text{ann}_R x = \bigcap_{x \neq 0} \text{ann}_R x$. If M is irreducible, this expresses $\text{ann}_R M$ as an intersection of maximal left ideals. Hence, by (17), rad R is an intersection of maximal left ideals, so $\text{rad}\, R \supset \bigcap I$ where I ranges over the maximal left ideals of R. On the other hand, since $I \supset (I:R)$, $\bigcap I \supset \bigcap (I:R) = \text{rad}\, R$ so

$$\text{rad}\, R = \bigcap I \tag{19}$$

where the intersection is taken over the set of maximal left ideals of R. □

We prove next

PROPOSITION 4.4. (1) *R is semi-primitive if and only if* rad $R = 0$. (2) *If R*

$\neq 0$, *then* $R/\text{rad}\, R$ *is semi-primitive and* $\text{rad}\, R$ *is contained in every ideal* B *of* R *such that* R/B *is semi-primitive.*

Proof. (1) If $\text{rad}\, R = 0$, then $\bigcap P = 0$ for the primitive ideals of R. Then R is a subdirect product of the primitive rings R/P and R is semi-primitive by Proposition 4.1. Conversely, if R is semi-primitive, R is a subdirect product of primitive rings and so there exists a set of primitive ideals whose intersection is 0. Then $\text{rad}\, R = \bigcap P = 0$ if P runs over the set of primitive ideals of R.

(2) If B is an ideal of R, then any ideal of $\bar{R} = R/B$ has the form P/B where P is an ideal of R containing B. Since $R/P \cong \bar{R}/\bar{P}$ where $\bar{P} = P/B$, it is clear that $\bar{P}$ is primitive in $\bar{R}$ if and only if P is primitive in R. Since $\bar{R}$ is semi-primitive if and only if $\bigcap \bar{P} = 0$, $\bar{P}$ primitive in $\bar{R}$, it follows that $\bar{R}$ is semi-primitive if and only if B is the intersection of all of the primitive ideals of R containing B. Then $B \supset \text{rad}\, R$, by (18′). If $B = \text{rad}\, R$, (18′) implies that B is an intersection of primitive ideals. Then $R/B = R/\text{rad}\, R$ is semi-primitive. This proves (2). □

We shall give next an important element characterization of the radical. For this purpose, we introduce a number of definitions. First, we call an element z of a ring R *left* (*right*) *quasi-regular* in R if $1-z$ has a left (right) inverse in R. Evidently, left (right) quasi-regularity is equivalent to the following: the principal left (right) ideal $R(1-z)\,((1-z)R) = R$. If z is both left and right quasi-regular, then z is called *quasi-regular*. A left (right) ideal I is called *quasi-regular* if all of its elements are left (right) quasi-regular.

If z is left quasi-regular, $1-z$ has a left inverse that we can write as $1-z'$. Then $(1-z')(1-z) = 1$ gives the relation

$$z' \circ z \equiv z + z' - z'z = 0. \tag{20}$$

The binary product $z' \circ z = z + z' - z'z$ defines the *circle composition* $\circ$ in the ring R. Let σ denote the map $x \rightarrow 1-x$ in R. Then $\sigma^2 = 1$, so σ is bijective. We have

$$\begin{aligned} z' \circ z &= 1-(1-z')(1-z) \\ &= \sigma((\sigma(z')\sigma(z)) \\ &= \sigma^{-1}(\sigma(z')\sigma(z)). \end{aligned}$$

This implies that the circle composition is associative. Also since $\sigma(0) = 1$, $0 \circ z = z = z \circ 0$. Thus $(R, \circ, 0)$ is a monoid.

LEMMA 2. (1) *If* I *is a quasi-regular left* (*right*) *ideal, then* I *is a subgroup of* $(R, \circ, 0)$. (2) *Nilpotent elements are quasi-regular.*

Proof. (1) Let $z \in I$. Then we have a $z' \in R$ such that $z' \circ z = 0$. Then $z + z' - z'z = 0$ and $z' = z'z - z \in I$. Hence there exists a z'' such that $z'' \circ z' = 0$. Since $(R, \circ, 0)$ is a monoid, the equations $z' \circ z = 0 = z'' \circ z'$ imply that $z'' = z$, so z' is the inverse of z in $(R, \circ, 0)$ and $z' \in I$. Since I is closed under $\circ$ and contains 0, statement (1) is clear.

(2) If $z^n = 0$, then $(1-z)(1+z+z^2+\cdots+z^{n-1}) = 1 = (1+z+z^2+\cdots+z^{n-1})(1-z)$. Hence z is quasi-regular. □

An ideal (left, right ideal) is called *nil* if all of its elements are nilpotent. The second part of Lemma 2 shows that any nil left (right) ideal is quasi-regular.

We can now give the following element characterizations of rad R.

PROPOSITION 4.5. (1) rad R *is a quasi-regular left ideal that contains every quasi-regular left ideal of* R. (2) rad R *is the set of elements* z *such that* az *is left quasi-regular for every* $a \in R$.

Proof. (1) Suppose $z \in \text{rad}\, R$ is not left quasi-regular. Then $R(1-z) \neq R$, so this left ideal can be imbedded in a maximal left ideal I. Since rad R is the intersection of the maximal left ideals of R, $z \in I$. But $1 - z \in R(1-z) \subset I$, so $1 = 1 - z + z \in I$ contrary to $I \subsetneqq R$. This contradiction shows that every $z \in \text{rad}\, R$ is left quasi-regular, so it proves that rad R is a quasi-regular left ideal. Now let Z be any quasi-regular left ideal of R. If $Z \not\subset \text{rad}\, R$, then there exists a maximal left ideal I such that $Z \not\subset I$. Then $I + Z \supsetneqq I$ and so $I + Z = R$. Then $1 = b + z$ for some $b \in I$, $z \in Z$, and $1 = (1-z)^{-1} b \in I$ contrary to $I \subsetneqq R$. Thus $Z \subset R$ and part (1) is proved.

(2) If $z \in \text{rad}\, R$, then $az \in \text{rad}\, R$ for any a in R, so az is left quasi-regular. Conversely, suppose z satisfies this condition. Then Rz is a quasi-regular left ideal. Hence $Rz \subset \text{rad}\, R$ by (1) and $z \in \text{rad}\, R$. □

The concept of primitivity of a ring and of an ideal in a ring is not left-right symmetric. For, there exist primitive rings that are not *right primitive* in the sense that they have no irreducible right modules M' such that the associated *anti-representation* $\rho' : a \rightsquigarrow \rho'(a)$, where $\rho'(a)x = xa$, is monic. The first examples of such rings were given by George Bergman. In spite of this lack of symmetry in the concept of primitivity, the concepts of the radical and semi-primitivity are symmetric. Evidently, everything we have done can be carried over to anti-representations, right modules, and right ideals. Suppose we denote the corresponding right radical as rad$'\, R$. Then we have the analogue of Proposition 4.5 for rad$'\, R$. On the other hand, by Lemma 2, the elements of rad R and rad$'\, R$ are quasi-invertible. Hence, by Proposition 4.5 and its analogue for right ideals, we have rad$'\, R \subset \text{rad}\, R$ and rad $R \subset \text{rad}'\, R$. Thus rad R = rad$'$ R. This permits us to give a number of additional characterizations of rad R. We give these and summarize the ones we obtained before in

THEOREM 4.1. *We have the following characterizations of the radical of a ring*: rad R *is*

1. *the intersection of the primitive ideals of* R,
2. *the intersection of the maximal left ideals of* R,
3. *the intersection of the right primitive ideals of* R,
4. *the intersection of the maximal right ideals of* R,
5. *a quasi-regular left ideal containing every quasi-regular left ideal,*
6. *the set of elements* z *such that* az *is left quasi-regular for every* $a \in R$,
7. *a quasi-regular right ideal containing every quasi-regular right ideal,*
8. *the set of elements* z *such that* za *is right quasi-regular for every* $a \in R$.

It is clear also that rad R contains every nil right ideal as well as every nil left ideal of R. Moreover, since right semi-primitivity (defined in the obvious way) is equivalent to rad$'$ $R = 0$, it is clear that a ring is semi-primitive if and only if it has a faithful completely reducible anti-representation and if and only if it is a subdirect product of right primitive rings. Another way of putting the left-right symmetry is that if we regard R and R^{op} as the same sets, then rad R = rad R^{op} and if R is semi-primitive, then so is R^{op}.

EXERCISES

1. Show that a ring R is a local ring (p. 111) if and only if $R/\text{rad } R$ is a division ring and that if this is the case, then rad R is the ideal of non-units of R.
2. Determine the radical of $\mathbb{Z}/(n)$, $n \geqslant 0$.
3. (McCrimmon.) Show that $z \in \text{rad } R$ if and only if for every $a \in R$ there exists a w such that $z + w = waz = zaw$.
4. Show that ab is quasi-regular if and only if ba is quasi-regular.
5. Call an element a of a ring R *von Neumann regular* if there exists a b such that $aba = a$. Show that the only element in rad R having this property is 0. (A ring is called *von Neumann regular* if all of its elements are von Neumann regular. Hence this exercise implies that such rings are semi-primitive.)
6. Let R be a ring such that for any $a \in R$ there is an integer $n(a) > 1$ such that $a^{n(a)} = a$. Show that R is semi-primitive.
7. Let e be a non-zero idempotent in a ring R. Show that rad $eRe = e(\text{rad } R)e = eRe \cap \text{rad } R$. (*Sketch of proof*: Show that every element of $e(\text{rad } R)e$ is quasi-regular in eRe, so $e(\text{rad } R)e \subset \text{rad } eRe$. Next let $z \in \text{rad } eRe$ and let $a \in R$. Write

$a = eae + ea(1-e) + (1-e)ae + (1-e)a(1-e)$ (the *two-sided Peirce decomposition*). Then $za = zeae + zea(1-e)$ and $zeae$ has a quasi-inverse z' in eRe. Then $za \circ z' = zea(1-e)$, which is nilpotent, thus quasi-regular. Hence za is right quasi-regular. Hence $z \in \text{rad}\, R \cap eRe$ and $\text{rad}\, eRe \subset e(\text{rad}\, R)e$.)

8. Show that for any ring R, $\text{rad}\, M_n(R) = M_n(\text{rad}\, R)$. (*Sketch of proof*: We have a 1–1 correspondence $B \rightsquigarrow M_n(B)$ of the set of ideals of R onto the set of ideals of $M_n(R)$ (see exercise 1, p. 171). Let $\{e_{ij} | 1 \leqslant i,j \leqslant n\}$ be a set of matrix units for $M_n(R)$. We have the isomorphism $r \rightsquigarrow (r1)e_{11}$ of R onto $e_{11}M_n(R)e_{11}$. Under this the image of $\text{rad}\, R$ is $\text{rad}\, e_{11}M_n(R)e_{11} = e_{11}\, \text{rad}\, M_n(R)e_{11}$ (by exercise 7) $= e_{11}M_n(B)e_{11}$. It follows that $B = \text{rad}\, R$.)

9. Show that if R and R' are (Morita) similar, then in any correspondence between the ideals of R and R' given by Morita I, the radicals of R and R' are paired.

10. Prove that any ultraproduct of semi-primitive (primitive) rings is semi-primitive (primitive).

4.3 DENSITY THEOREMS

If V is a vector space over a division ring Δ, then a set S of linear transformations in V is called *dense* if for any given finite set of linearly independent vectors $x_1, x_2, \ldots, x_n$ and corresponding vectors $y_1, y_2, \ldots, y_n$ there exists an $l \in S$ such that $lx_i = y_i$, $1 \leqslant i \leqslant n$. If V is finite dimensional, then we can take the x_i to be a base and $y_i = ax_i$ for any given linear transformation a. Then we have an $l \in S$ such that $lx_i = ax_i$, $1 \leqslant i \leqslant n$, from which it follows that $l = a$. Hence the only dense set of linear transformations in a finite dimensional vector space is the complete set of linear transformations.

In this section we shall prove a basic theorem that permits the identification of any primitive ring with a dense ring of linear transformations. We consider first a more general situation in which we have an arbitrary ring R and a completely reducible (left) module M for R. Let $R' = \text{End}_R M$, $R'' = \text{End}_{R'} M$ where M is regarded in the natural way as a left R'-module. We shall prove first the

DENSITY THEOREM FOR COMPLETELY REDUCIBLE MODULES. *Let M be a completely reducible module for a ring R. Let $R' = \text{End}_R M, R'' = \text{End}_{R'} M$. Let $\{x_1, \ldots, x_n\}$ be a finite subset of M, a'' an element of R''. Then there exists an $a \in R$ such that $ax_i = a''x_i$, $1 \leqslant i \leqslant n$.*

The proof we shall give is due to Bourbaki and is based on the following pair of lemmas.

LEMMA 1. *Any R-submodule N of M is an R''-submodule.*

Proof. Since M is completely reducible as R-module, we can write $M = N \oplus P$ where P is a second submodule (p. 121). Let e be the projection on N determined by this decomposition. Then $e \in R'$ and $N = e(M)$. Hence if $a'' \in R''$, then $a''(N) = a''e(M) = ea''(M) \subset N$. Thus N is a submodule of ${}_{R''}M$. □

LEMMA 2. *Let M be a module,* $M^{(n)}$ *the direct sum of n copies of M,* $n = 1, 2, \ldots$ *Then* $\mathrm{End}_R M^{(n)}$ *is the set of maps* $(u_1, u_2, \ldots, u_n) \to (v_1, v_2, \ldots, v_n)$ *where* $v_i = \sum a'_{ij} u_j$, $a'_{ij} \in R' = \mathrm{End}_R M$. *Moreover, for any* $a'' \in R'' = \mathrm{End}_{R'} M$, *the map* $(u_1, u_2, \ldots, u_n) \to (a''u_1, a''u_2, \ldots, a''u_n)$ *is contained in the ring of endomorphisms of* $M^{(n)}$ *regarded as a left* $\mathrm{End}_R M^{(n)}$*-module.*

Proof. Let $l \in \mathrm{End}_R M^{(n)}$, and consider the action of l on the element $(0, \ldots, 0, u_i, 0, \ldots, 0)$ where u_i is in the ith place. We have $l(0, \ldots, 0, u_i, 0, \ldots, 0) = (u_{1i}, u_{2i}, \ldots, u_{ni})$, which defines the map $u_i \to u_{ji}$, $1 \leqslant j \leqslant n$. This is contained in $\mathrm{End}_R M$. Denoting this map as a'_{ji}, we obtain $l(u_1, \ldots, u_n) = (\sum a'_{1i} u_i, \sum a'_{2i} u_i, \ldots, \sum a'_{ni} u_i)$. It is clear also that any map of this form is contained in $\mathrm{End}_R M^{(n)}$. This proves the first assertion. The second follows, since for any $a'' \in R''$ the map $(u_1, u_2, \ldots, u_n) \to (a''u_1, a''u_2, \ldots, a''u_n)$ commutes with every map $(u_1, u_2, \ldots, u_n) \to (\sum a'_{1i} u_i, \sum a'_{2i} u_i, \ldots, \sum a'_{ni} u_i)$, $a'_{ji} \in R'$. □

We can now give the

Proof of the theorem. Suppose first that $n = 1$. Then $N = Rx_1$ is an R-submodule, hence an R''-submodule. Since $x_1 \in N$, $a''x_1 \in N = Rx_1$. Hence we have an $a \in R$ such that $ax_1 = a''x_1$. Now suppose n is arbitrary. Consider $M^{(n)}$ the direct sum of n copies of the R-module M. The complete reducibility of M implies that $M^{(n)}$ is completely reducible. By Lemma 2, if $a'' \in R''$, the map $(u_1, u_2, \ldots, u_n) \to (a''u_1, a''u_2, \ldots, a''u_n)$ is in the ring of endomorphisms of $M^{(n)}$, regarded as left $\mathrm{End}_R M^{(n)}$ module. We apply the result established for $n = 1$ to this endomorphism and the element $x = (x_1, \ldots, x_n)$ of $M^{(n)}$. This gives an element $a \in R$ such that $ax = (a''x_1, \ldots, a''x_n)$. Then $ax_i = a''x_i$, $1 \leqslant i \leqslant n$, as required. □

There is a good deal more that can be done at the level of completely reducible modules (see the exercises below). However, for the sake of simplicity, we shall now specialize to the case that is of primary interest for the structure theory: a faithful irreducible representation of a ring. In this case we have the

DENSITY THEOREM FOR PRIMITIVE RINGS. *A ring R is primitive if and only if it is isomorphic to a dense ring of linear transformations in a vector space over a division ring.*

Proof. Suppose R is a primitive ring, so R has a faithful irreducible representation ρ acting in M. By Schur's lemma, $\Delta = \text{End}_R M$ is a division ring. Since ρ is faithful, R is isomorphic to $\rho(R)$, which is a ring of linear transformations in M regarded as a vector space over Δ. Now let $x_1, \ldots, x_n$ be linearly independent vectors in M, $y_1, \ldots, y_n$ arbitrary vectors. Since the x's can be taken to be part of a base, there exists a linear transformation l such that $lx_i = y_i$, $1 \leqslant i \leqslant n$. Since $l \in \text{End}_\Delta M$ and $\Delta = \text{End}_R M$, it follows from the previous theorem that there exists an $a \in R$ such that $ax_i = lx_i = y_i$, $1 \leqslant i \leqslant n$. Then $\rho(R)$ is a dense ring of linear transformations in M over Δ. Conversely, suppose R is isomorphic to a dense ring of linear transformations in a vector space M over a division ring Δ. If ρ is the given isomorphism, then putting $ax = \rho(a)x$ for $a \in R$, $x \in M$, makes M an R-module. Moreover, this is irreducible since if $x \neq 0$ and y is arbitrary, we have an $a \in R$ such that $ax = y$. Hence ρ is a faithful irreducible representation of R and so R is primitive. □

For some applications it is important to have an internal description of the division ring Δ in the foregoing theorem. This can be obtained by identifying an irreducible R-module with an isomorphic one of the form R/I where I is a maximal left ideal of R. We need to determine $\text{End}_R(R/I)$. We shall do this for any left ideal I or, equivalently, for any cyclic module. We define the *idealizer of I in R* by

$$B = \{b \in R \mid Ib \subset I\}. \tag{21}$$

It is clear that this is a subring of R containing I as an ideal and, in fact, B contains every subring of R having this property. We have

PROPOSITION 4.6. *If I is a left ideal in R, then* $\text{End}_R(R/I)$ *is anti-isomorphic to B/I where B is the idealizer of I in R.*

Proof. Since $1+I$ is a generator of R/I, $f \in \text{End}_R(R/I)$ is determined by $f(1+I) = b+I = b(1+I)$ as the map

$$a+I = a(1+I) \rightsquigarrow a(b+I) = ab+I. \tag{22}$$

If $a \in I$, $a+I = I = 0$ in R/I and hence its image $ab+I = I$. Then $ab \in I$. Thus $b \in$ the idealizer B of I. Conversely, let $b \in B$ and consider the homomorphism of the free R-module R into R/I sending $1 \rightsquigarrow b+I$. Any $d \in I$ is mapped into db

$+I = I$ by this homomorphism. Hence we have an induced homomorphism of R/I into R/I sending $1+I \to b+I$. Thus any $b \in B$ determines an endomorphism of R/I given by (22). Denote the correspondence between elements of B and endomorphisms of R/I determined in this way by g. Evidently g is additive and $g(1) = 1$. Also if $b_1, b_2 \in B$, we have $g(b_1 b_2)(1+I) = b_1 b_2 + I$ and $g(b_2)g(b_1)(1+I) = g(b_2)(b_1+I) = b_1 b_2 + I$. Hence g is an anti-homomorphism of B onto $\mathrm{End}_R(R/I)$. The ideal in B of elements mapped into 0 by this map is I. Hence B/I and $\mathrm{End}_R(R/I)$ are anti-isomorphic. □

The reader should be warned that in spite of this nice determination of the division ring Δ in the density theorem, given by Proposition 4.6, this division ring is not an invariant of the primitive ring. A given primitive ring may have non-isomorphic irreducible modules giving faithful representations and even some irreducible modules whose endomorphism division rings are not isomorphic. There are important cases in which the division ring is an invariant of the primitive ring. For example, as we shall show in the next section, this is the case for artinian primitive rings. Our next theorem gives the structure of these rings.

THEOREM 4.2. *The following conditions on a ring R are equivalent*:

(1) *R is simple, left artinian, and non-zero.*

(2) *R is primitive and left artinian.*

(3) *R is isomorphic to a ring $\mathrm{End}_\Delta M$ where M is a finite dimensional vector space over a division ring Δ.*

Proof. (1) ⇒ (2) is clear.

(2) ⇒ (3). Let M be a faithful irreducible module for R, $\Delta = \mathrm{End}_R M$. We claim that M is finite dimensional over Δ. Otherwise, we have an infinite linearly independent set of elements $\{x_i | i = 1, 2, 3, \ldots\}$ in M. Let $I_j = \mathrm{ann}_R x_j$ so I_j is a left ideal. Evidently, $I_1 \cap \cdots \cap I_n$ is the subset of R of elements annihilating $x_1, \ldots, x_n$. By the density theorem there exists an $a \in R$ such that $ax_1 = \cdots = ax_n = 0$, $ax_{n+1} \neq 0$. Hence $I_1 \cap \cdots \cap I_n \supsetneq I_1 \cap \cdots \cap I_{n+1}$. Then

$$I_1 \supsetneq I_1 \cap I_2 \supsetneq I_1 \cap I_2 \cap I_3 \supsetneq \cdots$$

is an infinite properly descending sequence of left ideals in R contrary to the artinian property of R. Thus M is finite dimensional over Δ and if we let $(x_1, x_2, \ldots, x_n)$ be a base for M over Δ, the density property implies that for any $y_1, \ldots, y_n$ there exists an $a \in R$ such that $ax_i = y_i$, $1 \leqslant i \leqslant n$. Hence if ρ is the representation determined by M, $\rho(R) = \mathrm{End}_\Delta M$ so R is isomorphic to the ring of linear transformations in the finite dimensional vector space M over Δ.

(3) ⇒ (1). This has been proved in proving the Wedderburn-Artin theorem (p. 171). It can also be seen directly by using the anti-isomorphism of $\mathrm{End}_\Delta M$ with $M_n(\Delta)$ and recalling that if R is any ring, the map $B \rightsquigarrow M_n(B)$ is an isomorphism of the lattice of ideals of R with the lattice of ideals of $M_n(R)$ (exercise 8, p. 103 of BAI; or exercise 1, p. 171). The result then follows from the simplicity of Δ. □

EXERCISES

1. Let R be a primitive ring, M a faithful irreducible module for R, $\Delta = \mathrm{End}_R M$. Show that either $R \cong M_n(\Delta)$ for some n or for any $k = 1, 2, 3, \ldots, R$ contains a subring having $M_k(\Delta)$ as a homomorphic image.

2. Use exercise 1 to show that if R is a primitive ring such that for any $a \in R$ there exists an integer $n(a) > 1$ such that $a^{n(a)} = a$, then R is a division ring.

3. Call a set S of linear transformations *k-fold transitive* if for any $l \leqslant k$ linearly independent vectors $x_1, \ldots, x_l$ and arbitrary vectors $y_1, \ldots, y_l$ there exists an $a \in S$ such that $ax_i = y_i$, $1 \leqslant i \leqslant l$. Show that a ring of linear transformations that is two-fold transitive is dense.

In the next three exercises, $M = {}_RM$ is completely reducible, $R' = \mathrm{End}_R M$, $R'' = \mathrm{End}_{R'} M$.

4. Show that any homomorphism of a submodule ${}_RN$ of ${}_RM$ into ${}_RM$ can be extended to an endomorphism of ${}_RM$ (an element of R') and that if ${}_RN$ is irreducible, any non-zero homomorphism of ${}_RN$ into ${}_RM$ can be extended to an automorphism of ${}_RM$.

5. Show that if $x \neq 0$ is an element of an irreducible submodule ${}_RN$ of ${}_RM$, then $R'x$ is an irreducible submodule of ${}_{R'}M$. (*Hint*: Let $y = a'x \neq 0$, $a' \in R'$, and consider the map $u \rightsquigarrow a'u$ of ${}_RN = Rx$ into ${}_RM$.)

6. Show that ${}_{R'}M$ is completely reducible and that its dimensionality (defined on p. 125) is finite if R is left artinian. Show also that in this case $R'' = \rho(R)$ where ρ is the representation defined by M.

7. Let M be a completely reducible R-module, B an ideal in R. Show that the following conditions are equivalent: (i) $BM = M$. (ii) If $x \in M$ satisfies $Bx = 0$, then $x = 0$. (iii) For any x, $x \in Bx$.

8. Prove the following extension of the density theorem for completely reducible modules: If R, M, and B are as in exercise 7 and B satisfies (i), (ii), (iii), then for any $a'' \in R''$ and $x_1, \ldots, x_n \in M$ there exists a $b \in B$ such that $bx_i = a''x_i$, $1 \leqslant i \leqslant n$.

9. Use exercise 8 to prove the simplicity of the ring of linear transformations in a finite dimensional vector space over a division ring.

4.4 ARTINIAN RINGS

In this section we shall derive the classical results on structure and representations of (left) artinian rings. We shall show first that the radical of such a ring is nilpotent. We recall that if B and C are ideals of a ring R, then $BC = \{\sum b_i c_i | b_i \in B, c_i \in C\}$. This is an ideal in R. Also, if D is another ideal, then $(BC)D = B(CD)$. We define B^k inductively by $B^1 = B$, $B^i = B^{i-1}B$ and B is called *nilpotent* if $B^k = 0$ for some k. This is equivalent to the following: the product of any k elements of B is 0. In particular, $b^k = 0$ for every B. Hence B nilpotent implies that B is a nil ideal. On the other hand, it is easy to give examples of nil ideals that are not nilpotent (exercise 1, at the end of this section).

THEOREM 4.3. *The radical of a (left) artinian ring is nilpotent.*

Proof. Let $N = \operatorname{rad} R$, R artinian. By induction, $N \supset N^2 \supset N^3 \supset \cdots$. Since these are left ideals, there is a k such that $P \equiv N^k = N^{k+1} = \cdots$. Then $P = P^2 = N^{2k}$. N is nilpotent if and only if $P = 0$. Suppose this is not the case and consider the set S of left ideals I of R such that (1) $I \subset P$ and (2) $PI \neq 0$. Since P has these properties, S is not vacuous, and since R is artinian, S contains a minimal element I. Evidently, I contains an element b such that $Pb \neq 0$. Then Pb is a left ideal contained in I and in P and $P(Pb) = P^2 b = Pb \neq 0$. Hence $Pb = I$ by the minimality of I in S. Since $b \in I$, we have a $z \in P$ such that $zb = b$ or $(1-z)b = 0$. Since $z \in P \subset \operatorname{rad} R$, z is quasi-regular. Hence $(1-z)^{-1}$ exists and $b = (1-z)^{-1}(1-z)b = 0$. This contradicts $Pb \neq 0$ and proves $(\operatorname{rad} R)^k = P = 0$. $\square$

We recall that the radical of any ring contains all quasi-regular one-sided ideals, hence all nil one-sided ideals of the ring. The foregoing result implies that all such ideals in an artinian ring are nilpotent. We shall now derive the main result on the structure of semi-primitive artinian rings. For the proof we shall need the

LEMMA. *Let M be an artinian module that is a subdirect product of irreducible modules. Then M is a direct sum of a finite number of irreducible submodules.*

Proof. The second hypothesis is equivalent to the following: M contains a set of submodules $\{N_\alpha\}$ such that $\bigcap N_\alpha = 0$ and every $M_\alpha = M/N_\alpha$ is irreducible. Consider the set of finite intersections $N_{\alpha_1} \cap \cdots \cap N_{\alpha_k}$ of $N_{\alpha_i} \in \{N_\alpha\}$. The artinian condition on M implies that there exists a minimal element in this set, say, $N_1 \cap \cdots \cap N_m$. Then for any N_α, $N_\alpha \cap (N_1 \cap \cdots \cap N_m) = N_1 \cap \cdots \cap N_m$ so $N_\alpha \supset N_1 \cap \cdots \cap N_m$. Since $\bigcap N_\alpha = 0$, we have $N_1 \cap \cdots \cap N_m = 0$. Then we have a monomorphism of M into $\prod M_i = \oplus M_i$, $M_i = M/N_i$. Thus M is isomorphic to a submodule of the completely reducible module $\oplus M_i$. Then M is completely reducible and is a direct sum of irreducible modules isomorphic to some of the M_i. □

We shall now call a ring *semi-simple* if it is a subdirect product of simple rings. We have the following easy consequence of the preceding lemma.

PROPOSITION 4.7. *If R is semi-simple and R satisfies the minimum condition for two-sided ideals, then R is a direct sum $\oplus_1^n R_i$ of simple rings R_i.*

Proof. Let $M(R)$ be the ring of endomorphisms of the additive group of R generated by the left and the right multiplications $x \rightsquigarrow ax$ and $x \rightsquigarrow xa$. This ring is called the *multiplication ring* of R. The additive group R can be regarded in the natural way as $M(R)$-module and when this is done, the submodules of ${}_{M(R)}R$ are the ideals of the ring R. The hypothesis that R is semi-simple is equivalent to the following: R as $M(R)$-module is a subdirect product of irreducible $M(R)$-modules. The other hypothesis of the lemma is also fulfilled. The conclusion then follows from the lemma. □

We can now prove the main

STRUCTURE THEOREM FOR SEMI-PRIMITIVE ARTINIAN RINGS. *The following conditions on a ring R are equivalent*:

(1) *R is artinian and contains no nilpotent ideals $\neq 0$.*
(2) *R is semi-primitive and artinian.*
(3) *R is semi-simple and artinian.*
(4) *${}_RR$ is a completely reducible R-module.*
(5) *R is a direct sum of a finite number of rings R_i each of which is isomorphic to the ring of linear transformations of a finite dimensional vector space over a division ring.*

The implication (1) ⇒ (5) is called the *Wedderburn-Artin theorem.*

Proof. (1) ⇔ (2). This is clear, since the radical of an artinian ring is a nilpotent ideal containing all nilpotent ideals and $\operatorname{rad} R = 0$ is equivalent to semi-primitivity.

(2) ⇔ (3) is clear from Theorem 4.2, which establishes the equivalence of simplicity and primitivity for artinian rings.

$(2) \Rightarrow (4)$. Since rad R is the intersection of the maximal left ideals, semi-primitivity of R implies that ${}_RR$ is a subdirect product of irreducible modules. Then (4) follows from the lemma.

$(4) \Rightarrow (2)$. The semi-primitivity is clear, since the representation determined by ${}_RR$ is faithful and completely reducible. Also we have $R = \sum I_\alpha$, I_α a minimal left ideal. Then $1 = e_1 + \cdots + e_k$, $e_j \in I_j \in \{I_\alpha\}$. Then for $a \in R$, we have $a = a1 = ae_1 + \cdots + ae_k \in I_1 + \cdots + I_k$. Thus $R = I_1 + \cdots + I_k$ and hence $R = I_1 \oplus I_2 \oplus \cdots \oplus I_l$ for a subset of the I_j. Then it is clear that

$$R \supset I_2 + \cdots + I_l \supset I_3 + \cdots + I_l \supset \cdots \supset I_l \supset 0$$

is a composition series for ${}_RR$. The existence of such a series implies that R is artinian and noetherian. Then (2) holds.

$(3) \Rightarrow (5)$. Evidently the artinian property implies that R satisfies the minimum condition for two-sided ideals. Hence, by Proposition 4.7, R is a direct sum of simple rings. Thus $R = R_1 \oplus R_2 \oplus \cdots \oplus R_s$ where the R_i are ideals and are simple rings. Then for $i \neq j$, $R_iR_j \subset R_i \cap R_j = 0$. It follows that any left ideal of R_i is a left ideal of R; hence, each R_i is artinian. Then R_i is isomorphic to the ring of linear transformations of a finite dimensional vector space over a division ring. Hence (5) holds.

$(5) \Rightarrow (2)$. If $R = R_1 \oplus R_2 \oplus \cdots \oplus R_s$ where the R_i are ideals and R_i is isomorphic to $\mathrm{End}_{\Delta_i} M_i$, M_i a finite dimensional vector space over the division ring Δ_i, then R_i is primitive artinian by Theorem 4.2, and R is a subdirect product of primitive rings, so R is semi-primitive. Since R_i is a submodule of ${}_RR$ and the R-submodules of R_i are left ideals, ${}_RR_i$ is artinian. Hence R is artinian. Thus (2) holds. This completes the proof. □

Let R be any ring and suppose $R = R_1 \oplus \cdots \oplus R_s$ where the R_i are ideals that are indecomposable in the sense that if $R_i = R_i' \oplus R_i''$, R_i', R_i'' ideals, then either $R_i' = 0$ or $R_i'' = 0$. Now let B be any ideal in R. Then $B = B1 = BR = BR_1 + \cdots + BR_s$ and since $BR_i \subset R_i$, we have $B = BR_1 \oplus \cdots \oplus BR_s$. Also BR_i is an ideal of R, so if B is indecomposable then we have $BR_i = B$ for some i. Similarly, we have $B = R_jB$ for some j and since $R_i \cap R_j = 0$ if $i \neq j$, we have $B = R_iB$. It follows that if $R = R_1' \oplus \cdots \oplus R_t'$ is a second decomposition of R as a direct sum of indecomposible ideals, then $s = t$ and the R_j' are, apart from order, the same as the R_i. This applies in particular to the decomposition of a semi-simple artinian ring as direct sum of ideals that are simple rings. We see that there is only one such decomposition. Accordingly, we call the terms R_i of the decomposition *the simple components* of R.

The structure theorem for semi-primitive artinian rings (or Theorem 4.2) shows that if R is simple artinian, then $R \cong \mathrm{End}_\Delta M$, M a finite dimensional vector space over the division ring Δ. We now consider the question of uniqueness of M and Δ. The problem can be posed in the following way: Suppose $\mathrm{End}_{\Delta_1}M_1$ and $\mathrm{End}_{\Delta_2}M_2$ are isomorphic for two finite dimensional vector spaces M_i over Δ_i; what does this imply about the relations between the two spaces? We shall see that the vector spaces are semi-linearly isomorphic. We recall that if σ is an isomorphism of Δ_1 onto Δ_2, then a map $s: M_1 \to M_2$ is called a *σ-semi-linear map* if

$$s(x+y) = sx+sy, \qquad s(\delta x) = \sigma(\delta)(sx) \tag{23}$$

for $x, y \in M_1$, $\delta \in \Delta_1$ (BAI, p. 469). It is immediately apparent that if s is bijective, then s^{-1} is a σ^{-1}-semi-linear map of M_2 onto M_1. We shall say that M_1 and M_2 are *semi-linearly isomorphic* if there exists a semi-linear isomorphism (=bijective semi-linear map) of M_1 onto M_2. This assumes tacitly that the underlying division rings Δ_1 and Δ_2 are isomorphic. Moreover, if $\{x_\alpha\}$ is a base for M_1 over Δ_1 and s is a semi-linear isomorphism of M_1 onto M_2, then $\{sx_\alpha\}$ is a base for M_2 over Δ_2. Hence the spaces have the same dimensionality. Conversely, let M_1 over Δ_1 and M_2 over Δ_2 have the same dimensionality and assume that Δ_1 and Δ_2 are isomorphic. Let $\{x_\alpha\}$, $\{y_\alpha\}$ be bases for M_1 and M_2 respectively, indexed by the same set, and let σ be an isomorphism of Δ_1 onto Δ_2. Define the map s of M_1 into M_2 by

$$\sum \delta_\alpha x_\alpha \rightsquigarrow \sum \sigma(\delta_\alpha) y_\alpha \qquad \text{(finite sums).} \tag{24}$$

One checks directly that s is σ-semi-linear and bijective. Hence we see that M_1 over Δ_1 and M_2 over Δ_2 are semi-linearly isomorphic if and only if they have the same dimensionality and Δ_1 and Δ_2 are isomorphic.

The key result for the isomorphism theorem (and for representation theory as well) is the following

LEMMA 1. *Any two irreducible modules for a simple artinian ring are isomorphic.*

Proof. We choose a minimal left ideal I of R, so I is an irreducible R-module. The result will follow by showing that any irreducible module M for R is isomorphic to I. Since R is simple, the representation of R given by M is faithful. Hence, since $I \neq 0$, there exists an $x \in M$ such that $Ix \neq 0$. We have the homomorphism $b \rightsquigarrow bx$ of I into M. Since $Ix \neq 0$, the homomorphism is not 0. By Schur's lemma, it is an isomorphism $\square$

We shall require also the following

LEMMA 2. *Let M be a vector space over a division ring Δ, $R = \mathrm{End}_\Delta M$ the* *ring of linear transformations of M over Δ. Then the centralizer of R in the ring of endomorphisms* End M *of M as abelian group is the set of maps $\delta' : u \rightsquigarrow \delta u$, $\delta \in \Delta$. Moreover, $\delta \rightsquigarrow \delta'$ is an isomorphism of Δ onto this centralizer.*

Proof. Since M is a completely reducible Δ-module, the first statement is a special case of the density theorem. We can also give a simple direct proof of this result: Let d be an endomorphism of $(M, +, 0)$ that centralizes $\mathrm{End}_\Delta M$. Let $x \neq 0$ in M. Then $dx \in \Delta x$. Otherwise, x and dx are linearly independent, and hence there is an $l \in \mathrm{End}_\Delta M$ such that $lx = 0$ and $l(dx) \neq 0$. Since $l(dx) = d(lx) = 0$, this is impossible. Hence for any $x \neq 0$ we have a $\delta_x \in \Delta$ such that $dx = \delta_x x$. Let y be a second non-zero element of M and choose $l \in \mathrm{End}_\Delta M$ such that $lx = y$. Then $\delta_y y = dy = dlx = ldx = l\delta_x x = \delta_x lx = \delta_x y$. Hence $\delta_x = \delta_y$ for all non-zero x and y, which implies that d has the form $\delta' : u \rightsquigarrow \delta u$. It is clear that $\delta \rightsquigarrow \delta'$ is an isomorphism. □

We can now prove the

ISOMORPHISM THEOREM FOR SIMPLE ARTINIAN RINGS. *Let M_i be a finite dimensional vector space over a division ring Δ_i, $i = 1, 2$, and let g be an isomorphism of $R_1 = \mathrm{End}_{\Delta_1} M_1$ onto $R_2 = \mathrm{End}_{\Delta_2} M_2$. Then there exists a semi-linear isomorphism s of M_1 onto M_2 such that*

$$g(a) = sas^{-1} \tag{25}$$

for all $a \in R_1$.

Proof. We consider M_1 as $R_1 = \mathrm{End}_{\Delta_1} M_1$ module in the natural way and regard M_2 as R_1-module by defining $ay = g(a)y$ for $a \in R_1$, $y \in M_2$. These two modules for the simple artinian ring R_1 are irreducible. Hence they are isomorphic. Accordingly, we have a group isomorphism s of M_1 onto M_2 such that for any $x \in M_1$, $a \in R_1$, we have $s(ax) = a(sx)$. Since $ay = g(a)y$, $y \in M_2$, this gives the operator relation $g(a) = sas^{-1}$, which is (25).

Since s is an isomorphism of M_1 onto M_2, the map $b \rightsquigarrow sbs^{-1}$, $b \in \mathrm{End}\, M_1$, is an isomorphism of the endomorphism ring End M_1 of the abelian group M_1 onto the endomorphism ring End M_2 of the abelian group M_2. Since this maps $\mathrm{End}_{\Delta_1} M_1$ onto $\mathrm{End}_{\Delta_2} M_2$, its restriction to the centralizer of $\mathrm{End}_{\Delta_1} M_1$ in End M_1 is an isomorphism of this centralizer onto the centralizer of $\mathrm{End}_{\Delta_2} M_2$ in End M_2. By Lemma 2, the two centralizers are isomorphic respectively to Δ_1 and Δ_2 under the maps sending $\delta_i \in \Delta_i$ into the endomorphism $\delta_i' : u_i \rightsquigarrow \delta_i u_i$. Hence we have an isomorphism σ of Δ_1 onto Δ_2 that is the composite of $\delta_1 \rightsquigarrow \delta_1' \rightsquigarrow s\delta_1' s^{-1}$ with the inverse of $\delta_2 \rightsquigarrow \delta_2'$. Thus we have $s\delta_1' s^{-1} = (\sigma\delta_1)'$

and hence for any $x \in M_1$ we have $s(\delta_1 x) = s\delta'_1 x = s\delta'_1 s^{-1}(sx) = (\sigma\delta_1)'(sx) = (\sigma\delta_1)sx$. This shows that s is a σ-semi-linear isomorphism of M_1 onto M_2. □

An immediate consequence of this theorem is that if $\mathrm{End}_{\Delta_1} M_1$ and $\mathrm{End}_{\Delta_2} M_2$ are isomorphic, then M_1 and M_2 are semi-linearly isomorphic. Conversely, if s is a semi-linear isomorphism of M_1 onto M_2 then $sas^{-1} \in \mathrm{End}_{\Delta_2} M_2$ for any $a \in \mathrm{End}_{\Delta_1} M_1$ and it is immediate that $a \rightsquigarrow sas^{-1}$ is an isomorphism of $\mathrm{End}_{\Delta_1} M_1$ onto $\mathrm{End}_{\Delta_2} M_2$. We have seen that M_1 and M_2 are semi-linearly isomorphic if and only if M_1 and M_2 have the same dimensionality and Δ_1 and Δ_2 are isomorphic. This shows that a simple artinian ring is determined up to isomorphism by an integer n, the dimensionality of M, and the isomorphism class of a division ring Δ.

The isomorphism theorem for simple artinian rings also gives a determination of the automorphism group of such a ring. If we take $R = \mathrm{End}_\Delta M$ where M is a finite dimensional vector space over the division ring Δ, then the isomorphism theorem applied to $\mathrm{End}_\Delta M$ and $\mathrm{End}_\Delta M$ states that the automorphisms of R are the maps

$$a \rightsquigarrow sas^{-1} \tag{26}$$

where $s \in G$, the group of bijective semi-linear transformations of the vector space M. Denoting the map in (26) as I_s, we have the epimorphism $s \rightsquigarrow I_s$ of G onto $\mathrm{Aut}\, R$, the group of automorphisms of R. The kernel of this epimorphism is the set of bijective semi-linear transformations s such that $sas^{-1} = a$ for every $a \in R = \mathrm{End}_\Delta M$. By Lemma 2, any such map has the form $\delta' : x \rightsquigarrow \delta x$, $\delta \in \Delta$. We note also that if $\delta \neq 0$, then δ' is a semi-linear automorphism of M whose associated automorphism of Δ is the inner automorphism $i_\delta : \alpha \rightsquigarrow \delta\alpha\delta^{-1}$. This follows from

$$\delta'(\alpha x) = \delta(\alpha x) = (\delta\alpha\delta^{-1})\delta x = (\delta\alpha\delta^{-1})\delta' x.$$

Thus the kernel of $s \rightsquigarrow I_s$ is the group $\Delta^{*\prime}$ of multiplications $x \rightsquigarrow \delta x$ determined by the non-zero elements δ of Δ. Evidently, our results on automorphisms can be stated in the following way.

COROLLARY. *We have an exact sequence*

$$1 \to \Delta^* \xrightarrow{i} G \xrightarrow{I} \mathrm{Aut}\, R \to 1 \tag{27}$$

where Δ^* is the multiplicative group of non-zero elements of Δ, G is the multiplicative group of bijective semi-linear transformations of the vector space M, Aut *R is the group of automorphisms of $R = \mathrm{End}_\Delta M$, i is the map $\delta \rightsquigarrow \delta'$, and I is the map $s \rightsquigarrow I_s$.*

Of course, this implies that $\mathrm{Aut}\, R \cong G/\Delta^{*\prime}$.

A reader who is familiar with the fundamental theorem of projective geometry (see BAI, pp. 468–473) should compare the results obtained here on isomorphisms and automorphisms of simple artinian rings with the fundamental theorem and its consequences. We remark that the present results could be derived from the geometric theorem. However, it would be somewhat roundabout to do so.

We shall derive next the main result on representations of artinian rings. This is the following

THEOREM 4.4. *Any module for a semi-simple artinian ring R is completely reducible and there is a 1–1 correspondence between the isomorphism classes of irreducible modules for R and the simple components of the ring. More precisely, if $R = R_1 \oplus \cdots \oplus R_s$ where the R_j are the simple components and I_j is a minimal left ideal in R_j, then $\{I_1, \ldots, I_s\}$ is a set of representatives of the isomorphism classes of irreducible R-modules.*

Proof. Write each simple component as a sum of minimal left ideals of R. This gives $R = \sum I_k$ where I_k is a minimal left ideal of R contained in a simple component. Now let M be an R-module and let x be any non-zero element of M. Then $x = 1x \in Rx = \sum I_k x$. By Schur's lemma, either $I_k x = 0$ or $I_k x$ is an irreducible module isomorphic to I_k. Now $M = \sum_{x \in M} Rx = \sum_{k,x} I_k x$ and this shows that M is a sum of irreducible submodules that are isomorphic to minimal left ideals of the simple components of R. Hence M is completely reducible and if M is irreducible, it is isomorphic to a minimal left ideal I of a simple component of R. To complete the proof, we need to show that if I is a minimal left ideal in R_j and I' is a minimal left ideal in $R_{j'}$, then $I \cong I'$ as R-modules if and only if $j = j'$. Suppose first that $j = j'$. Then I and I' are R_j-isomorphic by Lemma 1 on p. 205. Since $R_k I = 0 = R_k I'$ for $k \neq j$, it follows that I and I' are R-isomorphic. Next assume $j \neq j'$. Then $R_j I = I$ and $R_j I' = 0$; hence I and I' are not isomorphic as R-modules. □

We also have a definitive result on irreducible representations for arbitrary (left) artinian rings. This is

THEOREM 4.5. *Let R be (left) artinian, $N = \mathrm{rad}\, R$, and $\bar{R} = R/N = \bar{R}_1 \oplus \cdots \oplus \bar{R}_s$ where the $\bar{R}_i$ are the simple components. For each j, $1 \leqslant j \leqslant s$, let $\bar{I}_j$ be a minimal left ideal of $\bar{R}_j$. Then $\bar{I}_j$ is an irreducible module for R, $\bar{I}_j \ncong \bar{I}_{j'}$ (as R-modules) if $j \neq j'$, and any irreducible R-module is isomorphic to one of the $\bar{I}_j$.*

Proof. Any irreducible R-module M is annihilated by N, so it can be regarded as an irreducible $\bar{R}$-module. Conversely, any irreducible $\bar{R}$-module is an irreducible R-module. It is clear also that irreducible R-modules are isomorphic if and only if they are isomorphic as $\bar{R}$-modules. Hence the theorem is an immediate consequence of Theorem 4.4. □

EXERCISES

1. If z is a nilpotent element of a ring, then the *index of nilpotency* is the smallest t such that $z^t = 0$, $z^{t-1} \neq 0$. A similar definition applies to nilpotent ideals. Note that if N is a nilpotent ideal, then the indices of nilpotency of the elements of N are bounded. Use this to construct an example of a nil ideal that is not nilpotent.

2. Prove that the center of any simple ring is a field.

3. Show that if $R = R_1 \oplus \cdots \oplus R_s$ where the R_i are simple, then the center C of R is $C = C_1 \oplus \cdots \oplus C_s$ where $C_i = C \cap R_i$ is the center of R_i. Hence C is a commutative semi-simple artinian ring. Note that the R_i are determined by the simple components C_i of C since $R_i = C_i R$.

4. Let R be left artinian, $N = \operatorname{rad} R$, and assume that $N^t = 0$, $N^{t-1} \neq 0$. Show that R/N and N^i/N^{i+1}, $1 \leqslant i \leqslant t-1$, are completely reducible modules for R of finite dimensionality. Use this to prove that ${}_R R$ has a composition series and hence that R is left noetherian.

The next four exercises are designed to give conditions for anti-isomorphism of simple artinian rings and to determine the anti-isomorphisms and involutions of such rings.

5. Let V be a vector space over a division ring Δ that has an anti-automorphism $j: \delta \rightarrow \bar{\delta}$. A *sesquilinear form on* V (*relative to* j) is a map f of $V \times V$ into Δ such that

$$f(x_1 + x_2, y) = f(x_1, y) + f(x_2, y)$$

$$f(x, y_1 + y_2) = f(x, y_1) + f(x, y_2)$$

$$f(\delta x, y) = \delta f(x, y), \qquad f(x, \delta y) = f(x, y)\bar{\delta}$$

for $x, y \in V$, $\delta \in \Delta$. Call f *non-degenerate* if $f(z, y) = 0$ for all y implies $z = 0$ and $f(x, z) = 0$ for all x implies $z = 0$. Show that if V is finite dimensional and f is non-degenerate, then for a linear transformation l in V over Δ there exists a (left) *adjoint relative to* f defined to be a linear transformation l' such that $f(lx, y) = f(x, l'y)$ for all x, y. Show that $l \rightarrow l'$ is an anti-isomorphism and that this is an involution if f is *hermitian* in the sense that $f(y, x) = \overline{f(x, y)}$ for all x, y or *anti-hermitian*: $f(y, x) = -\overline{f(x, y)}$.

6. Let V and Δ be as in exercise 5 and let V^* be the dual space $\hom_\Delta(V, \Delta)$. This is a right vector space over Δ, hence a left vector space over Δ^{op}. If l is a linear

transformation of V over Δ, let l^* be the transposed linear transformation in V^*. Show that $l \rightsquigarrow l^*$ is an anti-isomorphism of $\mathrm{End}_\Delta V$ onto $\mathrm{End}\, V^*_\Delta$. Hence show that if V_i, $i = 1, 2$, is a finite dimensional vector space over a division ring Δ_i, then $\mathrm{End}_{\Delta_1} V_1$ and $\mathrm{End}_{\Delta_2} V_2$ are anti-isomorphic if and only if Δ_1 and Δ_2 are anti-isomorphic and V_1 and V_2 have the same dimensionality.

7. Let V be a finite dimensional vector space over a division ring Δ and assume $\mathrm{End}_\Delta V$ has an anti-automorphism J. Note that $J(l) \rightsquigarrow l^*$ is an isomorphism of $\mathrm{End}_\Delta V$ onto $\mathrm{End}_{\Delta^{\mathrm{op}}} V^*$ where V^* is regarded in the usual way as left vector space over Δ^{op} $(\alpha x^* = x^*\alpha)$. Hence there exists a semi-linear isomorphism s of V over Δ onto V^* over Δ^{op} such that
$$l^* = sJ(l)s^{-1}$$
for $l \in \mathrm{End}_\Delta V$. If $x, y \in V$, define
$$f(x, y) = (sy)(x).$$
Show that f is a non-degenerate sesquilinear form on V corresponding to the anti-automorphism σ that is the isomorphism of Δ onto Δ^{op} associated with s and that J is the adjoint map relative to f.

8. (Continuation of exercise 7.) Show that if J is an involution ($J^2 = 1$), then either f is skew hermitian or f can be replaced by ρf, $\rho \neq 0$, in Δ so that ρf is hermitian and J is the adjoint map relative to ρf. (*Sketch*: Note that for any $v \in V$, the map $x \rightsquigarrow \sigma^{-1} f(v, x)$ is a linear function, so there exists a $v' \in V$ such that $f(v, x) = \overline{f(x, v')}$, $x \in V$. Show that for any $u \in V$ the map $l : x \rightsquigarrow f(x, u)v$ is in $\mathrm{End}_\Delta V$ and its adjoint relative to f is $y \rightsquigarrow \sigma^{-1} f(v, y)u = \overline{f(y, v')}u$. Conclude from $J^2(l) = l$ that there exists a $\delta \neq 0$ in Δ such that $f(x, y) = \delta f(y, x)$ for all $x, y \in V$. Show that $\bar{\delta} = \delta^{-1}$ and if $\delta \neq -1$, then ρf for $\rho = \delta + 1$ is hermitian and J is the adjoint map relative to ρf.)

9. Note that the Wedderburn-Artin theorem for simple rings has the following matrix form: Any simple artinian ring is isomorphic to a matrix ring $M_n(\Delta')$, Δ' a division ring ($\Delta' \cong \Delta^{\mathrm{op}}$ if Δ is as in the statement of the theorem). Note that the isomorphism theorem implies that if Δ_1' and Δ_2' are division rings, then $M_{n_1}(\Delta_1') \cong M_{n_2}(\Delta_2')$ implies that $n_1 = n_2$ and $\Delta_1' \cong \Delta_2'$. Show that the isomorphism theorem implies also that the automorphisms of $M_n(\Delta')$ have the form $A \rightsquigarrow S({}^\sigma A)S^{-1}$ where $S \in M_n(\Delta')$, σ is an automorphism of Δ, and ${}^\sigma A = (\sigma(\alpha_{ij}))$ for $A = (\alpha_{ij})$.

10. Formulate the results of exercises 6–8 in matrix form.

4.5 STRUCTURE THEORY OF ALGEBRAS

In this section we shall extend the structure theory of rings that we developed in sections 1–4 to algebras over commutative rings. We shall then specialize to the case of finite dimensional algebras over a field.

We recall that an algebra A over a commutative ring K is a (left) K-module and a ring such that the module A and the ring A have the same additive group $(A, +, 0)$ and the following relations connecting the K-action and ring multiplication hold:

$$k(ab) = (ka)b = a(kb) \tag{28}$$

for $a, b \in A$, $k \in K$ (p. 44). We have the ring homomorphism

$$\varepsilon : k \rightsquigarrow k1 \tag{29}$$

of K into A (p. 138). The image under ε is $K1 = \{k1 \mid k \in K\}$. This is contained in the center of A, and ka for $k \in K$, $a \in A$, coincides with $(k1)a = a(k1)$. Conversely, given a ring R and a subring K of the center of R, R becomes a K-algebra by defining ka for $k \in K$, $a \in R$, to be the ring product ka. We remark also that rings are $\mathbb{Z}$-algebras in which na for $n \in \mathbb{Z}$, $a \in R$, is the nth multiple of a. Hence the theory of rings is a special case of that of algebras.

The prime examples of K-algebras are the algebras $\text{End}_K M$ where M is a K-module for the commutative ring K. Here kf for $k \in K$, $f \in \text{End}_K M$, is defined by $(kf)(x) = kf(x) = f(kx)$, $x \in M$ (p. 139). If A is an arbitrary K-algebra, then we define a *representation* of A to be a (K-algebra) homomorphism of A into an algebra $\text{End}_K M$. If ρ is a representation of A acting in the K-module M, then we define $ax = \rho(a)x$, $a \in A$, $x \in M$, to make M an A-module as well as a K-module. Since $\rho(ka) = k\rho(a)$, $k \in K$, and $\rho(a)$ is a K-endomorphism, we have

$$k(ax) = (ka)x = a(kx) \tag{30}$$

for $k \in K$, $a \in A$, $x \in M$. We now give the following

DEFINITION 4.3. *If A is an algebra over the commutative ring K, a* (left) A-module *is an abelian group written additively that is both a (left) K-module and a (left) A-module such that* (30) *holds for all $k \in K$, $a \in A$, $x \in M$.*

If M is a module for A, we obtain a representation ρ_M of A by defining $\rho_M(a)$ for $a \in A$ to be the map $x \rightsquigarrow ax$. This is contained in $\text{End}_K M$, and $a \rightsquigarrow \rho_M(a)$ is an algebra homomorphism. It follows from (30) that $kx = (k1)x$. Hence, if M and N are modules for the algebra A, then any homomorphism f of M into N regarded as modules for the ring A is also a K-homomorphism. It is clear also that any A-submodule is also a K-submodule and this gives rise to a factor module M/N for the algebra A in the sense of Definition 4.3. We remark also that if M is a ring A-module, then M is a K-module in which $kx = (k1)x$, and (30) holds; hence, this is an algebra A-module. The algebra A itself is an A-module in which the action of A is left multiplication and the action of K is the one given in the definition of A. The submodules are the left ideals. These are the same as the left ideals of A as ring. If $k \in K$ and $f \in \hom_A(M,N)$ where M and N are modules for A, then kf is defined by $(kf)(x) = kf(x)$, $x \in M$. This is contained in $\hom_A(M,N)$, and this action of K on $\hom_A(M,N)$ is a module action. It is immediate that $\text{End}_A M$ is an algebra over K.

We can now carry over the definitions and results on the radical, primitivity, and semi-primitivity to algebras. An algebra is called primitive if it has a faithful irreducible representation, semi-primitive if there are enough irreducible representations to distinguish elements. The radical is defined to be the intersection of the kernels of the irreducible representations of the algebra. The results we proved for rings carry over word for word to algebras. We have observed that if A is an algebra over K, then the left ideals of A as ring are also left ideals of A as algebra and, of course, the converse holds. Since the radical of A as algebra or as ring is the intersection of the maximal left ideals, it is clear that $\text{rad}\, A$ is the same set with the same addition and multiplication whether A is regarded as ring or as algebra. Similarly, A is primitive (semi-primitive) as algebra if and only if it is primitive (semi-primitive) as ring.

The structure and representation theory of artinian rings carries over to algebras. In particular, the theory is applicable to algebras that are finite dimensional over fields, since any left ideal in such an algebra is a subspace and the descending chain condition for subspaces is satisfied. For the remainder of this section, except in some exercises, we consider this classical case: A a finite dimensional algebra over a field F. Then $\text{rad}\, A$ is a nilpotent ideal containing every nil one-sided ideal of A. If $\text{rad}\, A = 0$, then $A = A_1 \oplus A_2 \oplus \cdots \oplus A_s$ where the A_i are ideals that are simple algebras. If A is simple, then A is isomorphic to the algebra of linear transformations in a finite dimensional vector space M over a division algebra Δ. Here M is any irreducible module for A and Δ is the division algebra $\text{End}_A M$ given by Schur's lemma. We recall also that M can be taken to be A/I where I is a maximal left ideal of A and Δ is anti-isomorphic to B/I where B is the idealizer of I (Proposition 4.7, p. 203). Hence Δ is a finite dimensional division algebra over F. Since $\text{End}_\Delta M$ for M n-dimensional over Δ is anti-isomorphic to $M_n(\Delta)$ and $M_n(\Delta)$ is anti-isomorphic to $M_n(\Delta^{\text{op}})$, it is clear that A is a finite dimensional simple algebra over F if and only if it is isomorphic to an algebra $M_n(\Delta')$, Δ' a finite dimensional division algebra. These results were proved by Wedderburn in 1908.

In BAI, pp. 451–454, we determined the finite dimensional division algebras over algebraically closed, real closed, and finite fields. We showed that if F is algebraically closed, then F is the only finite dimensional division algebra over F and if F is real closed, then there are three possibilities for finite dimensional algebras over F: (1) F, (2) $F(\sqrt{-1})$, and (3) the quaternion algebra over F with base $(1, i, j, k)$ such that $i^2 = -1 = j^2$ and $ij = k = -ji$. If F is finite, then one has Wedderburn's theorem that the finite dimensional algebras over F are fields. There is one of these for every finite dimensionality and any two of the same (finite) dimensionality are isomorphic (see BAI, pp. 287–290). Combining these results with the structure theorems, one obtains a complete determination of the finite dimensional semi-simple algebras over algebraically closed, real closed, or finite fields.

The results on representations of artinian rings are also applicable to finite dimensional algebras over fields. Thus if A over F is semi-simple, then any representation of A is completely reducible and we have a 1–1 correspondence between the isomorphism classes of irreducible modules and the simple components of the algebra.

One can apply the theory of algebras to the study of arbitrary sets of linear transformations in a finite dimensional vector space V over a field F. If S is such a set, we let $\operatorname{Env} S$ denote the subalgebra of $\operatorname{End}_F V$ generated by S. We call this the *enveloping algebra* of S. Evidently, $\operatorname{Env} S$ consists of the linear combinations of 1 and the products $s_1 s_2 \cdots s_r$, $s_i \in S$, and the dimensionality $\dim \operatorname{Env} S \leqslant n^2$ where $n = \dim V$. The injection map into $\operatorname{End}_F V$ is a representation of $\operatorname{Env} S$. The theory of algebras is applicable to the study of S via this representation. We shall now apply this method and the density theorem to obtain a classical theorem of the representation theory of groups. This is

BURNSIDE'S THEOREM. *Let G be a monoid of linear transformations in a finite dimensional vector space V over an algebraically closed field F that acts irreducibly on V in the sense that the only subspaces stabilized by G are V and 0. Then G contains a base for* $\operatorname{End}_F V$ *over F.*

Proof. The hypothesis is that G contains 1 and G is closed under multiplication. Then $\operatorname{Env} G = FG$, the set of F-linear combinations of the elements of G. Then V is an irreducible module for $A = \operatorname{Env} G$ and $A' = \operatorname{End}_A V$ is a division algebra by Schur's lemma. Evidently, A' is the centralizer of A in $\operatorname{End}_F V$, so A' is a finite dimensional algebra over F. Since F is algebraically closed, we have $A' = F1$. By the density theorem, $\operatorname{End}_{A'} V = A$. Hence $A = FG = \operatorname{End}_F V$. Evidently this implies that G contains a base for $\operatorname{End}_F V$. □

If S is an arbitrary set of linear transformations in V over F that acts irreducibly, then Burnside's theorem can be applied to the submonoid G generated by S. The result one obtains is that there exists a base for $\operatorname{End}_F V$ consisting of 1 and certain products of the elements of S.

EXERCISES

In these exercises A is an algebra over a field F. In exercises 1–5, A may be infinite dimensional; thereafter all vector spaces are finite dimensional.

1. Let $a \in A$ and let $\alpha_1, \ldots, \alpha_m$ be elements of F such that $a - \alpha_i 1$ is invertible in A. Show that either a is algebraic over F or the elements $(a - \alpha_i 1)^{-1}$, $1 \leqslant i \leqslant m$, are linearly independent over F.

2. Let A be a division algebra with a countable base over an uncountable algebraically closed field F (e.g., $F = \mathbb{C}$). Show that $A = F$.

3. Let A be finitely generated over an uncountable algebraically closed field F and let M be an irreducible A-module. Show that $\mathrm{End}_A M = F$.

4. Show that the elements of rad A are either nilpotent or transcendental over F.

5. Let A be finitely generated over an uncountable field F. Show that rad A is a nil ideal.

6. Assume F contains n distinct nth roots of 1 and let ε be a primitive one of these. Let s and t be the linear transformations of the vector space V/F with base $(x_1, x_2, \ldots, x_n)$ whose matrices relative to this base are respectively

$$\sigma = \begin{pmatrix} 1 & & & & 0 \\ & \varepsilon & & & \\ & & \varepsilon^2 & & \\ & & & \ddots & \\ 0 & & & & \varepsilon^{n-1} \end{pmatrix}, \qquad \tau = \begin{pmatrix} 0 & 1 & & & \\ & & \ddots & & \\ & & & \ddots & \\ & & & \cdot & 1 \\ 1 & \cdot & \cdot & \cdot & 0 \end{pmatrix}.$$

   Verify that

$$s^n = 1 = t^n, \qquad st = \varepsilon ts,$$

   and that $\mathrm{Env}\,\{s,t\} = \mathrm{End}_F V$. Let u and v be linear transformations of a vector space W/F such that $u^n = 1 = v^n$, $uv = \varepsilon vu$. Apply the representation theory of simple algebras to show that $\dim W = r \dim V$ and there exists a base for W over F such that the matrices of u and v relative to this base are respectively $\mathrm{diag}\,\{\sigma, \sigma, \ldots, \sigma\}$ and $\mathrm{diag}\,\{\tau, \tau, \ldots, \tau\}$.

7. If a is a linear transformation in a vector space V over F with base $(x_1, \ldots, x_n)$ and $ax_i = \sum \alpha_{ij} x_j$, then the trace of a, $\mathrm{tr}\, a = \sum \alpha_{ii}$, is independent of the choice of the base (BAI, p. 196). Moreover, $\mathrm{tr}\, a$ is the sum of the characteristic roots of a. Let

$$t(a, b) = \mathrm{tr}\, ab.$$

   Show that $t(a, b)$ is a non-degenerate symmetric bilinear form on $\mathrm{End}_F V$ (BAI, p. 346).

8. Let G, V, F be as in Burnside's theorem. Suppose $\operatorname{tr} G = \{\operatorname{tr} a | a \in G\}$ is a finite set. Show that G is finite and $|G| \leqslant |\operatorname{tr} G|^{n^2}$, $n = \dim V$. (*Hint*: G contains a base $(a_1, \ldots, a_{n^2})$ for $\operatorname{End}_F V$ by Burnside's theorem. Consider the map $a \rightsquigarrow (t(a, a_1), t(a, a_2), \ldots, t(a, a_{n^2}))$ where $t(a, b) = \operatorname{tr} ab$ as in exercise 7.)

9. A linear transformation is called *unipotent* if it has the form $1 + z$ where z is nilpotent. Let G be a monoid of unipotent linear transformations in a finite dimensional vector space V over an algebraically closed field F. Prove *Kolchin's theorem* that V has a base relative to which the matrix of every $a \in G$ has the form

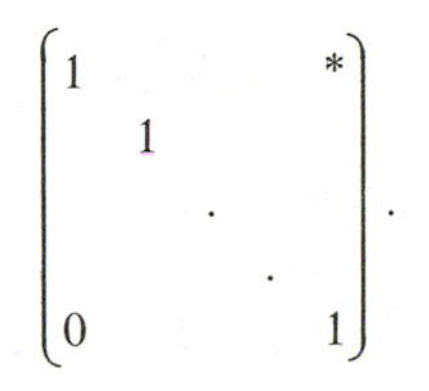

(*Hint*: Let $V = V_1 \supset V_2 \supset \cdots \supset V_{s+1} = 0$ be a composition series for V as $A = FG$-module and apply exercise 7 to the induced transformations in the irreducible modules V_i/V_{i+1}.) Extend the theorem to arbitrary base fields F.

10. (Burnside.) Let G be a group of linear transformations in a finite dimensional vector space V over an algebraically closed field F such that there exists an integer m not divisible by the characteristic of F such that $a^m = 1$ for all $a \in G$. Prove that G is finite. (*Hint*: If G acts irreducibly, the theorem follows from exercise 8. Otherwise, we have a subspace $U \neq 0$, $\neq V$ stabilized by G. Use induction on the dimensionality to conclude that the induced groups of transformations in U and in V/U are finite. Let K_1 and K_2 be the kernels of the homomorphisms of G onto the induced groups in U and V/U. Then K_1 and K_2 and hence $K_1 \cap K_2$ are of finite index in G. Finally, show that $K_1 \cap K_2 = 1$.)

4.6 FINITE DIMENSIONAL CENTRAL SIMPLE ALGEBRAS

An algebra A over a field F is called *central* if its center $C = F1 = \{\alpha 1 | \alpha \in F\}$. If A is a finite dimensional simple algebra over F, then $A \cong M_n(\Delta)$ where Δ is a finite dimensional division algebra over Δ. The center C of A is isomorphic to the center of Δ and hence C is a field (cf. exercise 2, p. 209). Now A can be regarded as an algebra over C. When this is done, A becomes finite dimensional central simple over C. As we shall see in this section and the next, the class of finite dimensional central simple algebras has some remarkable properties. To facilitate their derivation it will be useful to collect first some simple results on tensor products of modules and algebras over fields that will be needed in the main part of our discussion.

We note first that since any F-module (=vector space over F) is free, it is projective and thus flat (p. 154). Hence if we have an injective linear map $i: V' \to V$ of F-modules, then $i \otimes 1$ and $1 \otimes i$ are injective maps of $V' \otimes U \to V \otimes U$ and of $U \otimes V'$ into $U \otimes V$ respectively for any F-module U. If $\{x_\alpha | \alpha \in I\}$ is a base for V/F, then any element of $U \otimes V$ can be written in one and only one way as a finite sum $\sum a_{\alpha_i} \otimes x_{\alpha_i}$, $a_{\alpha_i} \in U$. The fact that it is such a sum is clear since any element of $U \otimes V$ has the form $\sum a_j \otimes b_j$, $a_j \in U$, $b_j \in V$. The uniqueness follows from the isomorphism of $U \otimes V = U \otimes \sum Fx \cong \oplus U \otimes Fx_\alpha$ (Proposition 3.3, p. 131), which implies that if $\sum a_{\alpha_i} \otimes x_{\alpha_i} = 0$, then every $a_{\alpha_i} \otimes x_{\alpha_i} = 0$. Then the isomorphisms $U \otimes Fx_\alpha \cong U \otimes F \cong U$ imply that $a_\alpha \otimes x_\alpha = 0$ if and only if $a_\alpha = 0$. Similarly, if $\{y_\beta | \beta \in J\}$ is a base for U, then every element of $U \otimes V$ can be written in one and only one way as $\sum y_{\beta_j} \otimes b_j$, $b_j \in V$. It is clear also that if $\{x_\alpha | \alpha \in I\}$ is a base for V and $\{y_\beta | \beta \in J\}$ is a base for U, then $\{y_\beta \otimes x_\alpha | \beta \in J, \alpha \in I\}$ is a base for $U \otimes V$. Hence the dimensionality $[U \otimes V : F] = [U:F][V:F]$.

If A and B are algebras over the field F, then $A \otimes B$ contains the subalgebras $1 \otimes B = \{1 \otimes b | b \in B\}$ and $A \otimes 1 = \{a \otimes 1 | a \in A\}$. These are isomorphic to B and A respectively under the maps $b \rightsquigarrow 1 \otimes b$ and $a \rightsquigarrow a \otimes 1$. Hence we can identify B with the subalgebra $1 \otimes B$ and A with the subalgebra $A \otimes 1$ of $A \otimes B$. If we write a for $a \otimes 1$ and b for $1 \otimes b$, then we have $ab = ba$ and $A \otimes B = AB = BA$.

We shall now prove the following result, which in effect gives internal characterizations of tensor products of algebras over a field.

PROPOSITION 4.8. *If A and B are subalgebras of an algebra D over a field F, then $D \cong A \otimes_F B$ if the following conditions are satisfied:*

(1) *$ab = ba$ for all $a \in A$, $b \in B$.*

(2) *There exists a base (x_α) for A such that every element of D can be written in one and only one way as $\sum b_i x_{\alpha_i}$, $b_i \in B$.*

If D is finite dimensional, then condition (2) can be replaced by

(2′) *$D = AB$ and $[D:F] = [A:F][B:F]$.*

Proof. The first condition implies that we have an algebra homomorphism of $A \otimes B$ into D such that $a \otimes b \rightsquigarrow ab$, $a \in A$, $b \in B$ (p. 144). The second condition insures that this is bijective, hence, an isomorphism. It is immediate that (2′) $\Rightarrow$ (2) if D is finite dimensional. □

As a first application of this criterion, we prove

PROPOSITION 4.9. *If B is an algebra over F, then $M_n(B) \cong M_n(F) \otimes_F B$.*

Proof. This follows by applying the criterion to the subalgebras $M_n(F)$ and $B1 = \{b1 | b \in B\}$ where

$$b1 = \begin{pmatrix} b & & & & 0 \\ & b & & & \\ & & \cdot & & \\ & & & \cdot & \\ 0 & & & & b \end{pmatrix}.$$

If $\{e_{ij}|1 \leqslant i,j \leqslant n\}$ is the usual base of matrix units for $M_n(F)$, then $(b1)e_{ij} = e_{ij}(b1)$. Hence condition (1) holds. Moreover, $(b1)e_{ij}$ is the matrix that has b in the (i,j)-position and 0's elsewhere. This implies condition (2). □

Next we prove

PROPOSITION 4.10. $M_m(F) \otimes M_n(F) \simeq M_{mn}(F)$.

Proof. This follows by applying the criterion to the subalgebras consisting of the matrices

$$\begin{pmatrix} b & & & & 0 \\ & b & & & \\ & & \cdot & & \\ & & & \cdot & \\ 0 & & & & b \end{pmatrix}$$

where $b \in M_m(F)$, and of the matrices

$$\begin{pmatrix} \alpha_{11}1_m & \cdot & \cdot & \cdot & \alpha_{1n}1_m \\ \cdot & \cdot & \cdot & \cdot & \cdot \\ \alpha_{n1}1_m & \cdot & \cdot & \cdot & \alpha_{nn}1_m \end{pmatrix}, \qquad \alpha_{ij} \in F,$$

respectively. We leave the details to the reader. □

If A is any algebra, we define $A^e = A \otimes_F A^{\text{op}}$ and we call this the *enveloping algebra* of A. If A is a subalgebra of another algebra B, then we have a natural module action of A^e on B defined by

$$(31) \qquad (\Sigma a_i \otimes b_i)y = \Sigma a_i y b_i$$

where the $a_i \in A$, $b_i \in A^{\text{op}}$, $y \in B$. It is clear from the universal property of the tensor product that we have a homomorphism of A^e into B such that

$\sum a_i \otimes b_i \rightsquigarrow \sum a_i y b_i$. Hence (31) is well-defined. Direct verification shows that it is a module action. In particular, we have a module action of A^e on A. The submodules of A as A^e-module are the ideals of A. Hence if A is simple, then A is A^e-irreducible. We recall that the endomorphisms of A regarded as left (right) A-module in the natural way are the right (left) multiplications $x \rightsquigarrow xa$ ($x \rightsquigarrow ax$). It follows that $\mathrm{End}_{A^e} A$ is the set of maps that are both left and right multiplications $x \rightsquigarrow cx = xd$. Putting $x = 1$, we obtain $d = c$; hence $\mathrm{End}_{A^e} A$ is the set of maps $x \rightsquigarrow cx$, c in the center. If A is central, this is the set of maps $x \rightsquigarrow (\alpha 1)x = \alpha x$, $\alpha \in F$. We shall now apply this to prove our first main theorem on finite dimensional central simple algebras.

THEOREM 4.6. *If A is a finite dimensional central simple algebra over a field F, then $A^e = A \otimes_F A^{\mathrm{op}} \cong M_n(F)$ where $n = \dim A$.*

Proof. We regard A as A^e-module as above. Then A is irreducible and $\mathrm{End}_{A^e} A = F$. Also A is finite dimensional over F. Hence by the density theorem A^e maps onto $\mathrm{End}_F A$. Since both A^e and $\mathrm{End}_F A$ have dimensionality n^2, we have an isomorphism of A^e onto $\mathrm{End}_F A$. Since $\mathrm{End}_F A \cong M_n(F)$, the result follows. □

We consider next how a finite dimensional central simple algebra A sits in any algebra B (not necessarily finite dimensional) containing A as subalgebra. The result is

THEOREM 4.7. *Let A be a finite dimensional central simple subalgebra of an algebra B. Then $B \cong A \otimes_F C$ where C is the centralizer of A in B. The map $I \rightsquigarrow AI$ is a bijection of the set of ideals of C with the set of ideals of B. Moreover, the center of B coincides with the center of C.*

Proof. We regard B as A^e-module as above. By the preceding theorem, A^e is simple. Hence B is a direct sum of A^e-irreducible modules all of which are isomorphic to A (since A is A^e-irreducible and any two irreducible A^e-modules are isomorphic). Now observe that the generator 1 of A as A^e-module satisfies $(a \otimes 1)1 = a1 = 1a = (1 \otimes a)1$ and $(a \otimes 1)1 = 0$ implies $a = 0$. Hence in any irreducible A^e-module, we can choose a c such that $(a \otimes 1)c = (1 \otimes a)c$ and $(a \otimes 1)c = 0$ implies $a = 0$. Applying this to B as A^e-module, we see that we can write $B = \oplus A c_\alpha$ where $ac_\alpha = c_\alpha a$ for all $a \in A$ and $ac_\alpha = 0$ implies $a = 0$. Then $c_\alpha \in C$ and any element of B can be written in one and only one way as a finite sum $\sum a_\alpha c_\alpha$, $a_\alpha \in A$. If $c \in C$, we have $c = \sum a_\alpha c_\alpha$ and $ac = ca$ implies $aa_\alpha = a_\alpha a$. It follows that every $a_\alpha \in F1$ and $c \in \sum F c_\alpha$. This implies that $C = \sum F c_\alpha$ and $\{c_\alpha\}$ is a base for C. It is now clear from Proposition 4.8 that $A \otimes_F C \cong B$.

Now let I be an ideal in C. Then AI is an ideal in $B = AC$. Moreover, $AI \cap C = I$. For, if $(x_1 = 1, x_2, \ldots, x_n)$ is a base for A, then the canonical isomorphism of $A \otimes_F C$ onto B (p. 144) implies that any element of B can be written in one and only one way as $\sum_1^n c_i x_i$, $c_i \in C$. Then the elements of AI have the form $\sum d_i x_i$, $d_i \in I$, and if such an element is in C, it also has the form $c_1 x_1$, $c_1 \in C$. It follows that $C \cap AI$ is the set of elements $d_1 x_1 = d_1 \in I$. Thus $C \cap AI = I$, which implies that the map $I \rightsquigarrow AI$ of the ideals of C into ideals of B is injective. To see that it is surjective, let I' be any ideal of B. Then I' is a submodule of B as A^e-module. Hence $I' = \sum A d_\beta$ where $d_\beta \in I = I' \cap C$, which implies that $I' = AI$. Thus the map $I \rightsquigarrow AI$ is also surjective.

It remains to prove that the center of B is the center of C. Since the center of B is contained in C, it is contained in the center of C. On the other hand, if c is contained in the center of C, it commutes with every element of $B = AC$ and so it is in the center of B. □

It follows from the preceding theorem that B is simple if and only if C is simple and B is central if and only if C is central. The theorem can be applied to tensor products in which one of the factors is finite dimensional central simple. Thus let $B = A \otimes_F C$ where A is finite dimensional central simple. We identify A and C with the corresponding subalgebras $A \otimes 1$ and $1 \otimes C$ of $A \otimes C$. Now we claim that C is the centralizer of A in B. For, let $\{y_\beta\}$ be a base for C over F. Then any element of B can be written in one and only one way as $b = \sum a_\beta y_\beta$, $a_\beta \in A$, and $ab = ba$ for $a \in A$ is equivalent to $aa_\beta = a_\beta a$ for every a_β. This implies that b commutes with every $a \in A$ if and only if $b = \sum \alpha_\beta y_\beta$, $\alpha_\beta \in F$, that is, if and only if $b \in C$. We can now apply Theorem 4.7 to obtain the

COROLLARY 1. *Let A be a finite dimensional central simple algebra over F, C an arbitrary algebra over F. Then the map $I \rightsquigarrow A \otimes_F I$ is a bijection of the set of ideals of C with the set of ideals of $B = A \otimes_F C$. Moreover, the center of B coincides with the center of C (identified with the corresponding subalgebra of B).*

An important special case of this result is

COROLLARY 2. *Let A be a finite dimensional central simple algebra over F, C any algebra over F. Then $A \otimes_F C$ is simple if C is simple and $A \otimes_F C$ is central if C is central.*

Iteration of Corollary 2 has the following consequence.

COROLLARY 3. *The tensor product $A_1 \otimes_F A_2 \otimes_F \cdots \otimes_F A_r$ of a finite number of finite dimensional central simple algebras is finite dimensional central simple.*

Of particular interest are tensor products in which one of the factors is an extension field E of the base field F. If A is any algebra over F, then $A \otimes_F E$ can be regarded as an algebra over E. This algebra over E is denoted as A_E and is called *the algebra obtained from A by extending the base field to E* (see exercise 13, p. 148). If (x_α) is a base for A over F, then by identifying A with $A \otimes 1$ in A_E, we can say that (x_α) is a base for A_E over E. Hence the dimensionality of A over F is the same as that of A_E over E. It follows also that if K is an extension field of E, then $(A_E)_K \cong A_K$. If B is a second algebra over F with base (y_β), then $A \otimes_F B$ has the base $(x_\alpha y_\beta)$ $(=(x_\alpha \otimes y_\beta))$ over F and this is also a base for $(A \otimes_F B)_E$ over E and $A_E \otimes_E B_E$ over E. It follows that $(A \otimes_F B)_E \cong A_E \otimes_E B_E$. (More general results are indicated in exercise 13, p. 148.)

If A is finite dimensional central simple over F, then Corollary 2 to Theorem 4.7 shows that A_E is finite dimensional central simple over E. We recall also that $M_n(F)$ is finite dimensional central simple over F. These are the simplest examples of central simple algebras. We now give

DEFINITION 4.4. *The matrix algebra $M_n(F)$ is called a* split *central simple algebra over F. If A is finite dimensional central simple over F, then an extension field E of F is called a* splitting field *for A if A_E is split over E ($A_E \cong M_n(E)$ for some n).*

Evidently any extension field of F is a splitting field of $M_n(F)$ and if E is a splitting field for A and B, then E is a splitting field for the central simple algebra $A \otimes_F B$. This follows since $A_E \cong M_n(E)$ and $B_E \cong M_m(E)$ imply $(A \otimes_F B)_E \cong A_E \otimes_E B_E \cong M_n(E) \otimes_E M_m(E) \cong M_{nm}(E)$. We note also that if E is a splitting field for A, then it is a splitting field for A^{op}. For, $(A^{\mathrm{op}})_E = (A_E)^{\mathrm{op}} \cong M_n(E)^{\mathrm{op}}$, and since $M_n(E)$ has an anti-automorphism, $M_n(E)^{\mathrm{op}} \cong M_n(E)$. If $A \cong M_r(B)$ where B is central simple over F then E is a splitting field for A if and only if it is a splitting field for B. We have $A_E \cong M_r(B_E)$ since $M_r(B) \cong M_r(F) \otimes_F B$ and $M_r(B)_E \cong M_r(F)_E \otimes_E B_E \cong M_r(B_E)$. Now $B_E \cong M_s(\Delta)$ where Δ is a central division algebra over E. Then $A_E \cong M_{rs}(\Delta)$. If E splits A (that is, is a splitting field for A) then $A_E \cong M_n(E)$ and hence $M_n(E) \cong M_{rs}(\Delta)$ which implies that $rs = n$ and $\Delta = E$. Then $B_E \cong M_s(E)$ and E splits B. The converse is clear from Proposition 4.10. In particular, if $A \cong M_r(\Delta)$ where Δ is a central division algebra over F then E splits A if and only if E splits Δ. Thus it suffices to consider splitting fields for central division algebras. We remark finally that any extension field K of a splitting field is a splitting field. This is clear since $(A_E)_K \cong A_K$ and $M_n(E)_K \cong M_n(K)$.

If E is an algebraically closed extension field of F, then the only finite dimensional simple algebras over E are the algebras $M_n(E)$ (p. 213). Accordingly, any algebraically closed extension field E of F is a splitting field for every finite dimensional central simple algebra over F. We shall see later that every field can be imbedded in an algebraically closed field (section 8.1). When this result is available, it proves the existence of splitting fields for finite dimensional central simple algebras.

We shall now prove the existence of finite dimensional splitting fields for any finite dimensional central simple algebra A. Writing $A = M_n(\Delta)$ where Δ is a finite dimensional central division algebra, we shall show that any maximal subfield of Δ is a splitting field. This will follow from the following result (which we shall improve in Theorem 4.12).

THEOREM 4.8. *If Δ is a finite dimensional central division algebra over F, then a finite dimensional extension field E of F is a splitting field for Δ if and only if E is a subfield of an algebra $A = M_r(\Delta)$ such that the centralizer $C_A(E) = E$.*

Proof. Suppose the condition holds: E is a subfield of $A = M_r(\Delta)$ such that $C_A(E) = E$. We can identify A with $\mathrm{End}_{\Delta'}V$ where V is an r-dimensional vector space over $\Delta' = \Delta^{op}$. Then V is a $\Delta' \otimes_F E$ module such that $(d \otimes e)x = dex = edx$ for $d \in \Delta'$, $e \in E \subset A$. Since $\Delta' \otimes_F E$ is simple, the representation of $\Delta' \otimes_F E$ in V is faithful and we can identify $\Delta' \otimes_F E$ with the corresponding ring of endomorphisms in V. Since $\Delta' \otimes_F E$ is finite dimensional simple over F, V is completely reducible as $\Delta' \otimes_F E$ module and so we can apply the density theorem. Now $\mathrm{End}_{\Delta'} V = A$ so $\mathrm{End}_{\Delta' \otimes E} V$ is the centralizer of E in A. By hypothesis, this is E. Now V is finite dimensional over F since it is finite dimensional over Δ' and Δ' is finite dimensional over F. Thus V is finite dimensional over E. Hence, by the density theorem, $\Delta' \otimes_F E$ is isomorphic to the complete algebra of linear transformations in V over E. If the dimensionality $[V:E] = n$, then $\Delta' \otimes_F E \cong M_n(E)$. Thus E is a splitting field for Δ' over F and hence of Δ over F.

Conversely, suppose $\Delta \otimes_F E \cong M_n(E)$. Then also $\Delta' \otimes_F E \cong M_n(E)$. Let V be an irreducible module for $\Delta' \otimes_F E$. Since $\Delta' \otimes_F E \cong M_n(E)$, V is an n-dimensional vector space over E and $\Delta' \otimes_F E$ can be identified with $\mathrm{End}_E V$. Also V is a vector space over Δ' and if $[V:\Delta'] = r$, then since E centralizes Δ', $E \subset \mathrm{End}_{\Delta'} V$. The centralizer of E in $\mathrm{End}_{\Delta'} V$ is contained in the centralizer of $\Delta' \otimes E$ in $\mathrm{End}_E V$. Since $\Delta' \otimes E = \mathrm{End}_E V$, this is E. Hence $C_{\mathrm{End}_{\Delta'} V}(E) = E$ and $C_{M_n(\Delta)}(E) = E$. □

The existence of a maximal subfield of Δ is clear since $[\Delta:F] < \infty$. We now have the following corollary, which proves the existence of finite dimensional splitting fields for any finite dimensional central simple algebra.

COROLLARY. *Let A be finite dimensional central simple over F and let $A \cong M_r(\Delta)$ where Δ is a division algebra. Then any maximal subfield E of Δ is a splitting field for A.*

Proof. It suffices to show that E is a splitting field for Δ. Let $E' = C_\Delta(E)$. Then $E' \supset E$ and if $E' \neq E$, we can choose $c \in E' \notin E$. The division algebra generated by E and c is commutative, hence this is a subfield of Δ properly containing E, contrary to the maximality of E. Thus $C_\Delta(E) = E$. Hence Theorem 4.8 with $A = \Delta$ shows that E is a splitting field. □

If E is a splitting field for A so that $A_E \cong M_n(E)$, then $[A:F] = [A_E:E] = [M_n(E):E] = n^2$. Thus we see that the dimensionality of any central simple algebra is a square. The square root n of this dimensionality is called the *degree* of A.

We shall prove next an important theorem on extension of homomorphisms:

THEOREM 4.9. *Let A be a simple subalgebra of a finite dimensional central simple algebra B. Then any homomorphism of A into B can be extended to an inner automorphism of B.*

Proof. We form the algebra $E = A \otimes_F B^{\text{op}}$, which is finite dimensional and simple. Any module for E is completely reducible and any two irreducible E-modules are isomorphic. It follows that any two E-modules of the same finite dimensionality over F are isomorphic. We now make B into an E-module in two ways. In the first, the action is $(\sum a_i \otimes b_i)x = \sum a_i x b_i$ and in the second it is $(\sum a_i \otimes b_i)x = \sum f(a_i)xb_i$ where f is the given homomorphism of A into B. Clearly these are module actions and the two modules are isomorphic. Hence there exists a bijective linear transformation l of B over F such that

$$\sum a_i l(x) b_i = l(\sum f(a_i) x b_i) \tag{32}$$

for all $a_i \in A$, $x, b_i \in B$. In particular, $l(x)b = l(xb)$ for all $x, b \in B$. It follows that l has the form $x \rightsquigarrow dx$ where d is an invertible element of B. We have also $al(x) = l(f(a)x)$ for all $a \in A$, $x \in B$. Then $adx = df(a)x$. Putting $x = 1$ we obtain $f(a) = d^{-1}ad$. Hence f can be extended to the inner automorphism $b \rightsquigarrow d^{-1}bd$ in B. □

If we take $A = F1$ in Theorem 4.9 we obtain the important

COROLLARY (Skolem-Noether). *Any automorphism of a finite dimensional central simple algebra is inner.*

We prove next an important double centralizer theorem for central simple algebras.

THEOREM 4.10. *Let A be a semi-simple subalgebra of a finite dimensional central simple algebra B. Then the double centralizer $C_B(C_B(A)) = A$.*

Proof. Consider the algebra $B \otimes B^{\mathrm{op}}$ and its subalgebra $A \otimes B^{\mathrm{op}}$. If we identify B with $B \otimes 1$, then it is easily seen as in the proof of Theorem 4.7 that $B \cap (A \otimes B^{\mathrm{op}}) = A$. Now regard B as $B \otimes B^{\mathrm{op}}$-module in the usual way. By the proof of Theorem 4.6, the algebra $B \otimes B^{\mathrm{op}}$ is isomorphic to the algebra of linear transformations of B over F of the form $x \rightsquigarrow \sum b_i x b_i'$ where $b_i, b_i' \in B$. Then the result that $B \cap (A \otimes B^{\mathrm{op}}) = A$ implies that if a linear transformation of B is simultaneously of the form $x \rightsquigarrow \sum a_i x b_i$ where $a_i \in A$, $b_i \in B$, and of the form $x \rightsquigarrow bx$, $b \in B$, then $b = a \in A$. We now consider the algebra $A \otimes B^{\mathrm{op}}$. We claim that this is semi-simple. To see this we write $A = A_1 \oplus \cdots \oplus A_s$ where the A_i are simple ideals. Then $A \otimes B^{\mathrm{op}} = (A_1 \otimes B^{\mathrm{op}}) \oplus \cdots \oplus (A_s \otimes B^{\mathrm{op}})$ and $A_i \otimes B^{\mathrm{op}}$ is an ideal in $A \otimes B^{\mathrm{op}}$. Since A_i is simple and B^{op} is finite dimensional central simple, $A_i \otimes B^{\mathrm{op}}$ is simple by Corollary 2, p. 219. Thus $A \otimes B^{\mathrm{op}}$ is a direct sum of ideals that are simple algebras. Hence $A \otimes B^{\mathrm{op}}$ is semi-simple. Now regard B as $A \otimes B^{\mathrm{op}}$-module by restricting the action of $B \otimes B^{\mathrm{op}}$ to the subalgebra $A \otimes B^{\mathrm{op}}$. Since $A \otimes B^{\mathrm{op}}$ is semi-simple, B is completely reducible as $E = A \otimes B^{\mathrm{op}}$-module. Now consider $E' = \mathrm{End}_E B$. If we use the fact that the endomorphisms of B as right B-module are the left multiplications, we see that E' is the set of maps $x \rightsquigarrow cx$, $c \in C_B(A)$. Since B is finite dimensional over F and B is completely reducible as $A \otimes B^{\mathrm{op}}$-module, it follows from the density theorem that $\mathrm{End}_{E'} B$ is the set of maps $x \rightsquigarrow \sum a_i x b_i$, $a_i \in A$, $b_i \in B$. Now let $c' \in C_B(C_B(A))$. Then $x \rightsquigarrow c'x$ commutes with $x \rightsquigarrow cx$ for $c \in C_B(A)$ and with $x \rightsquigarrow xb$ for $b \in B$. It follows that $x \rightsquigarrow c'x$ is in $\mathrm{End}_{E'} B$. Hence $x \rightsquigarrow c'x$ also has the form $x \rightsquigarrow \sum a_i x b_i$, $a_i \in A$, $b_i \in B$. We have seen that this implies that $c' \in A$. Thus we have proved that $C_B(C_B(A)) \subset A$. Since $A \subset C_B(C_B(A))$ is clear, we have $C_B(C_B(A)) = A$. □

The foregoing result does not give us any information on the structures of the various algebras involved in the proof or on their relations. We shall now look at this question and for the sake of simplicity, we confine our attention to the most interesting case in which A is a simple subalgebra of the central simple algebra B. The extension to A semi-simple is readily made and will be indicated in the exercises.

Assume A simple. Then $E = A \otimes B^{\mathrm{op}}$ is simple, so $E \cong M_r(\Delta)$ where Δ is a division algebra. Let $d = [\Delta : F]$. Then we have

$$[E:F] = [A:F][B:F] = r^2 d. \tag{33}$$

If $\{e_{ij} | 1 \leqslant i, j \leqslant r\}$ is the usual set of matrix units for $M_r(\Delta)$, then it is clear that $\Delta \cong e_{11} M_r(\Delta) e_{11}$, which is just the set of matrices with $(1,1)$-entry in Δ and other entries 0. Also, $M_r(\Delta)$ is a direct sum of the r minimal left ideals $M_r(\Delta)e_{ii}$ and these are isomorphic modules for the algebra $M_r(\Delta)$. Each of these irreducible modules has dimensionality $r^2 d/r = rd$ over F. Now consider B as $E = A \otimes B^{\mathrm{op}}$-module. Since E is simple, B as E-module is a direct sum of, say, s irreducible submodules all of which are isomorphic to minimal left ideals of E. Since $E \cong M_r(\Delta)$, the dimensionality over F of any irreducible E-module is rd. Hence we have

$$[B:F] = srd. \tag{34}$$

By (33) and (34) we see that $s|r$ and

$$[A:F] = r/s. \tag{35}$$

Now consider $C_B(A) \cong E' = \mathrm{End}_E B$. Since B is a direct sum of s irreducible E-modules and these are isomorphic to minimal left ideals Ee, $e^2 = e$, of E, we can determine the structure of $\mathrm{End}_E B$, hence of $C_B(A)$, by using the following

LEMMA. *Let e be a non-zero idempotent in an algebra E, $I = Ee$, $I^{(s)}$ the direct sum of s copies of I. Then* $\mathrm{End}_E I^{(s)}$ *is isomorphic to* $M_s((eEe)^{\mathrm{op}})$.

Proof. By Lemma 2, p. 198, $\mathrm{End}_E I^{(s)}$ consists of the maps $(u_1,\dots,u_s) \rightsquigarrow (v_1,\dots,v_s)$ where $v_i = \sum_j a'_{ij}u_j$ and $a'_{ij} \in \mathrm{End}_E I$. This gives a map $(a'_{ij}) \rightsquigarrow \eta(a'_{ij})$ of $M_s(\mathrm{End}_E I)$ into $\mathrm{End}_E I^{(s)}$. Direct verification shows that this is an isomorphism. Next we consider $\mathrm{End}_E I = \mathrm{End}_E Ee$. As in the proof of Lemma (1), p. 180, $\mathrm{End}_E Ee$ is the set of maps $ae \rightsquigarrow aebe$, $a, b \in E$. In this way we obtain a map of eEe into $\mathrm{End}_E Ee$ such that $ebe \rightsquigarrow \eta(ebe): ae \rightsquigarrow aebe$. Direct verification shows that this is an anti-isomorphism. Hence $\mathrm{End}_E I \cong (eEe)^{\mathrm{op}}$ and $\mathrm{End}_E I^{(s)} \cong M_s((eEe)^{\mathrm{op}})$. □

Applying this lemma to B as E-module, we see that $\mathrm{End}_E B \cong M_s((e_{11}M_r(\Delta)e_{11})^{\mathrm{op}}) \cong M_s(\Delta^{\mathrm{op}})$. Hence $C_B(A) \cong M_s(\Delta^{\mathrm{op}})$ and

$$[C_B(A):F] = s^2 d. \tag{36}$$

Then

$$[A:F][C_B(A):F] = (r/s)s^2 d = rsd = [B:F]. \tag{37}$$

We summarize these results in

THEOREM 4.11. *Let A be a simple subalgebra of a finite dimensional central simple algebra B. Then $[B:F] = [A:F][C_B(A):F]$ and $A \otimes B^{\mathrm{op}} \cong M_r(\Delta)$, Δ a division algebra, and $C_B(A) \cong M_s(\Delta^{\mathrm{op}})$. Moreover, $s|r$, $[B:F] = rsd$ where $d = [\Delta:F]$ and $[A:F] = r/s$.*

The relation $[B:F] = [A:F][C_B(A):F]$ shows that if A is a field, then $C_B(A) = A \Leftrightarrow [B:F] = [A:F]^2$. This gives the following precise determination of the finite dimensional splitting fields of a central division algebra.

THEOREM 4.12. *A finite dimensional extension field E of F is a splitting field for a central division algebra Δ if and only if $[E:F] = nr$ where n is the degree of Δ, and E is isomorphic to a subfield of $M_r(\Delta)$.*

EXERCISES

1. Let Δ be a finite dimensional central division algebra over F, E a maximal subfield of Δ. Show that $[\Delta:F] = [E:F]^2$.

2. (Proof of Wedderburn's theorem on finite division rings.) Let Δ be a finite division ring, F the center of Δ so Δ is a central division algebra over F. Show that if E_1 and E_2 are maximal subfields of Δ, then these are conjugates in the sense that there exists an $s \in \Delta$ such that $E_2 = sE_1s^{-1}$. Conclude from this that the multiplicative group Δ^* of non-zero elements of Δ is a union of conjugates of the subgroup E_1^*. Hence conclude that $\Delta = E_1 = F$ is commutative (see exercise 7, p. 77 of BAI).

3. (Proof of Frobenius' theorem on real division algebras.) Let Δ be a finite dimensional division algebra over $\mathbb{R}$. If Δ is commutative, then either $\Delta = \mathbb{R}$ or $\Delta = \mathbb{C}$. If Δ is not commutative, its center is either $\mathbb{R}$ or $\mathbb{C}$; $\mathbb{C}$ is ruled out since it is algebraically closed and so has no finite dimensional division algebras over it. Now suppose Δ is central over $\mathbb{R}$ and $\Delta \supsetneq \mathbb{R}$. Then Δ contains an element i such that $i^2 = -1$ and since $\mathbb{C} = \mathbb{R}(i)$ has an automorphism such that $i \rightsquigarrow -i$, there exists a j in Δ such that $ji = -ij$. Show that the subalgebra generated by i and j is Hamilton's quaternion algebra $\mathbb{H}$. Use exercise 1 to prove that $[\Delta:\mathbb{R}] = 4$ and that $\Delta = \mathbb{H}$.

4. Let E be a cyclic field over F, $[E:F] = n$, $\operatorname{Gal} E/F$, the Galois group of $E/F, = \langle s \rangle$. Let (E, s, γ) be the subalgebra of $M_n(E)$ generated by all of the diagonal matrices

$$\bar{b} = \operatorname{diag}\{b, sb, \ldots, s^{n-1}b\}, \tag{38}$$

$b \in E$, and the matrix

$$z = \begin{pmatrix} 0 & 1 & 0 & . & . & . \\ 0 & 0 & 1 & 0 & . & . \\ . & . & . & . & . & . \\ . & . & . & . & 0 & 1 \\ \gamma & 0 & . & . & . & 0 \end{pmatrix} \tag{39}$$

where γ is a non-zero element of F. Note that $b \rightsquigarrow \bar{b}$ is a monomorphism and identify b with $\bar{b}$. Then we have the relations

$$zb = s(b)z, \qquad z^n = \gamma \tag{40}$$

and every element of (E, s, γ) can be written in one and only one way in the form

$$b_0 + b_1z + \cdots + b_{n-1}z^{n-1} \tag{41}$$

where the $b_i \in E$. Prove that (E, s, γ) is central simple and $[(E, s, \gamma):F] = n^2$.

5. Let E, F, G, s be as in exercise 4. Note that $s \in \operatorname{End}_F E$ and for any $b \in E$, $x \rightsquigarrow bx$ is in $\operatorname{End}_F E$. Identifying this with b, we have $sb = s(b)s$. Show that the algebra L of linear transformations in E over F generated by the multiplications $b \in E$ and s is isomorphic to the algebra $(E, s, 1)$ of exercise 4. Hence conclude that $L = \operatorname{End}_F E$ and $(E, s, 1) \cong M_n(F)$.

6. Prove that $(E, s, \gamma) \cong M_n(F)$ if and only if γ is the norm of an element of E. (*Hint*: Use Theorem 4.9, p. 222). Hence show that if $n = p$ is a prime and γ is not a norm in E, then (E, s, γ) is a division algebra.

7. Let $E = \mathbb{Q}(r)$ where r is a root of $x^3 + x^2 - 2x - 1$. Show that $E/\mathbb{Q}$ is cyclic (see BAI, p. 236, exercise 1). Show that if γ is an integer not divisible by 8, then γ is not a norm of any element of E. Hence conclude from exercise 6 that (E, s, γ) is a division algebra.

8. Let A be an algebra over F with base $(u_1, \ldots, u_n)$ and let $a \rightsquigarrow \rho(a)$ be the regular matrix representation determined by this base. Put $N(a) = \det \rho(a)$ (see BAI, p. 403). Show that A is a division algebra if and only if $N(a) \neq 0$ for every $a \neq 0$ in A. Use this criterion to show that if t is an indeterminate, then A is a division algebra if and only if $A_{F(t)}$ is a division algebra. Generalize to show that if $t_1, \ldots, t_r$ are indeterminates, then $A_{F(t_1, \ldots, t_r)}$ is a division algebra if and only if A is a division algebra.

9. Show that if Δ_1 and Δ_2 are finite dimensional division algebras over F and Δ_1 is central, then $\Delta_1 \otimes_F \Delta_2 = M_r(E)$ where E is a division algebra and $r | [\Delta_i : F]$, $i = 1, 2$. Hence show that $\Delta_1 \otimes \Delta_2$ is a division algebra if $([\Delta_1 : F], [\Delta_2 : F]) = 1$.

10. If A is a subalgebra of an algebra B, a linear map D of A into B is called a *derivation of A into B* if

$$D(a_1 a_2) = a_1 D(a_2) + D(a_1) a_2 \tag{42}$$

for $a_i \in A$. Form $B^{(2)}$ and define

$$a(x, y) = (ax + D(a)y, ay) \tag{43}$$

for $a \in A$, $(x, y) \in B^{(2)}$. Verify that $B^{(2)}$ is an A-module with this action. Show that $B^{(2)}$ is an $A \otimes B^{\mathrm{op}}$-module in which

$$a \otimes b(x, y) = (axb + D(a)yb, ayb).$$

11. A derivation of B into B is called a *derivation in B*. If $d \in B$, the map $x \rightsquigarrow [d, x] \equiv dx - xd$ is a derivation in B called the *inner derivation* determined by d. Prove that if A is a semi-simple subalgebra of a finite dimensional central simple algebra B, then any derivation of A into B can be extended to an inner derivation of B.

12. Let A be a semi-simple subalgebra of a finite dimensional central simple algebra B. Show that $C_B(A)$ is semi-simple and investigate the relation between the structure of A, $A \otimes B^{\mathrm{op}}$ and $C_B(A)$.

4.7 THE BRAUER GROUP

The results on tensor products of finite dimensional central simple algebras given in the previous section lead to the introduction of an important group that was first defined by R. Brauer in 1929. The elements of this group are similarity classes of finite dimensional central simple algebras. If A and B are such algebras, we say that A is *similar* to B ($A \sim B$) if there exist positive integers m and n such that $M_m(A) \cong M_n(B)$ or, equivalently, $M_m(F) \otimes A \cong M_n(F) \otimes B$. The relation of similarity is evidently reflexive and symmetric. It is also transitive since if

$$M_m(F) \otimes A \cong M_n(F) \otimes B$$

and

$$M_r(F) \otimes B \cong M_s(F) \otimes C,$$

then

$$\begin{aligned} M_{mr}(F) \otimes A &\cong M_r(F) \otimes M_m(F) \otimes A \cong M_r(F) \otimes M_n(F) \otimes B \\ &\cong M_n(F) \otimes M_r(F) \otimes B \cong M_n(F) \otimes M_s(F) \otimes C \cong M_{ns}(F) \otimes C. \end{aligned}$$

Note that we have used associativity and commutativity of tensor multiplication as well as the formula $M_n(F) \otimes M_m(F) \cong M_{mn}(F)$, which was established in Proposition 4.10, p. 217. We now see that $\sim$ is an equivalence relation. Let $[A]$ denote the similarity class of A, that is, the equivalence class of finite dimensional central simple algebras similar to A. Suppose $A \sim A'$ and $B \sim B'$, so we have positive integers m, m', n, n' such that $A \otimes M_n(F) \cong A' \otimes M_{n'}(F)$ and $B \otimes M_m(F) \cong B' \otimes M_{m'}(F)$. Then $A \otimes B \otimes M_{nm}(F) \cong A' \otimes B' \otimes M_{n'm'}(F)$. Hence $A \otimes B \sim A' \otimes B'$. Thus we have a well-defined binary composition of similarity classes given by

$$[A][B] = [A \otimes B]. \tag{44}$$

Evidently this is associative and commutative. Moreover, the set of matrix algebras $M_n(F)$ constitutes a similarity class 1 and this acts as unit since $A \sim A \otimes M_n(F)$. We have seen in Theorem 4.6, p. 218, that $A \otimes A^{\text{op}} \sim 1$. Hence $[A][A^{\text{op}}] = 1 = [A^{\text{op}}][A]$.

If A is finite dimensional central simple over F, we can write $A \cong M_n(F) \otimes \Delta$ for Δ, a finite dimensional central division algebra. Conversely, if Δ is as indicated, then $M_n(F) \otimes \Delta$ is finite dimensional central simple over F. We recall that the isomorphism theorem for simple artinian rings (p. 206) implies that if $M_n(\Delta) \cong M_m(\Delta')$ for division algebras Δ and Δ', then $m = n$ and $\Delta \cong \Delta'$. It follows that the division algebra Δ in the formula $A \cong M_n(F) \otimes \Delta$ is determined up to isomorphism by A. Hence a similarity class $[A]$ contains a single isomorphism class of finite dimensional central division algebras and distinct similarity classes are associated with non-isomorphic division algebras. Now the class of subalgebras of the matrix algebras $M_n(F)$, $n = 1, 2, 3, \ldots$ is a set and every algebra over F is isomorphic to a member of this set. Hence the isomorphism classes of algebras over F constitute a set. This set has as subset the isomorphism classes of central division algebras over F. Hence the similarity classes of central simple algebras over F is a set. Our results show that this set is a group under the composition defined by (44). This group is denoted as $Br(F)$ and is called the *Brauer group of the field* F. It is clear that the determination of $Br(F)$ for a field F is equivalent to a complete classification of the finite dimensional central division algebras over F.

If F is algebraically closed or is a finite field, then $Br(F) = 1$. If $F = \mathbb{R}$, then $Br(F)$ is a cyclic group of order two by Frobenius' theorem. We shall determine $Br(F)$ for F a p-adic field in Chapter 9 (section 9.14). One of the most important achievements of algebra and number theory in the 1930's was the determination of $Br(\mathbb{Q})$ and, more generally, $Br(F)$ for F a number field (that is, a finite dimensional extension field of $\mathbb{Q}$). This involves some deep arithmetic results.

Let E be an extension field of F. If A is finite dimensional central simple over F, then A_E is finite dimensional central simple over E. We have $(A \otimes_F B)_E \cong A_E \otimes_E B_E$ and $M_n(F)_E \cong M_n(E)$. This implies that we have a homomorphism of $Br(F)$ into $Br(E)$ sending $[A]$ into $[A_E]$. The kernel, which we shall denote as $Br(E/F)$, is the set of classes $[A]$ such that $A_E \sim 1$ (1 for F is customary here), that is, the $[A]$ such that A is split by E. We shall consider this group in Chapter 8 for the case in which E is a finite dimensional Galois extension field of F.

EXERCISE

1. Use Theorem 4.11 (p. 224) to show that if A is finite dimensional central simple over F and E is a subfield of A/F, then $C_A(E) \sim A_E$ (in Br(E)).

4.8 CLIFFORD ALGEBRAS

In this section we apply some of the results on central simple algebras to the study of certain algebras—the Clifford algebras—defined by quadratic forms. These algebras play an important role in the study of quadratic forms and orthogonal groups. The results on these matters that we require can be found in BAI, Chapter 6.

Let V be a finite dimensional vector space over a field F equipped with a quadratic form Q. We recall the definition: Q is a map of V into the base field F such that

1. $Q(\alpha x) = \alpha^2 Q(x)$, $\alpha \in F$, $x \in V$.
2. $B(x, y) \equiv Q(x+y) - Q(x) - Q(y)$ is bilinear.

Evidently, $B(x, y)$ is symmetric, that is, $B(y, x) = B(x, y)$ and $B(x, x) = 2Q(x)$. We use Q to define an algebra in the following manner.

DEFINITION 4.5. *Let V be a vector space over a field F, Q a quadratic form on V. Let $T(V) = F \oplus V \oplus (V \otimes V) \oplus (V \otimes V \otimes V) \oplus \cdots$ be the tensor algebra defined by V (p. 140) and let K_Q be the ideal in $T(V)$ generated by all of the elements of the form*

$$x \otimes x - Q(x)1, \qquad x \in V. \tag{45}$$

Then we define the Clifford algebra of the quadratic form Q *to be the algebra*

$$C(V, Q) = T(V)/K_Q.$$

If $a \in T(V)$, we write $\bar{a} = a + K_Q \in C(V, Q)$ and we have the map $i: x \rightsquigarrow \bar{x}$ of V into $C(V, Q)$. Since V generates $T(V)$, $i(V)$ generates the Clifford algebra. We have $\bar{x}^2 = Q(x)\bar{1}$. Moreover, we claim that we have the following universality of the map i: If f is a linear map of V into an algebra A such that

$$f(x)^2 = Q(x)1, \qquad x \in V,$$

then there exists a unique algebra homomorphism g of $C(V, Q)$ into A such that

(46)

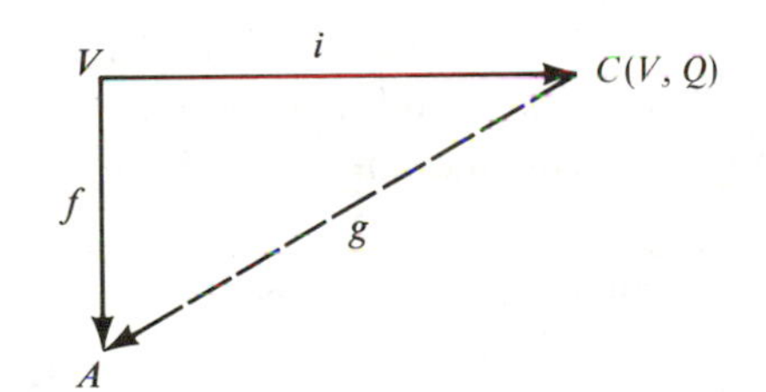

is commutative. To see this, we recall the basic property of the tensor algebra that any linear map f of V into an algebra has a unique extension to an algebra homomorphism f' of $T(V)$ into A (see p. 140). Now for the given f, the kernel, $\ker f'$, contains every element $x \otimes x - Q(x)1$, $x \in V$, since $f'(x \otimes x - Q(x)1) = f(x)^2 - Q(x)1 = 0$. Hence the ideal $K_Q \subset \ker f'$ and so we have the induced homomorphism $g: \bar{a} + K_Q \rightsquigarrow f'(a)$. In particular, $g(\bar{x}) = f'(x) = f(x)$, which is the commutativity of (46). The uniqueness of g is clear since the $\bar{x}$ generate $C(V, Q)$.

The ideal K_Q defining $C(V, Q)$ contains every

$$(x+y) \otimes (x+y) - Q(x+y)1 = x \otimes x + y \otimes x + x \otimes y + y \otimes y$$
$$- Q(x)1 - B(x, y)1 - Q(y)1, \qquad x, y \in V.$$

Since $x \otimes x - Q(x)1$ and $y \otimes y - Q(y)1 \in K_Q$, we see that

$$x \otimes y + y \otimes x - B(x, y)1 \in K_Q, \qquad x, y \in V. \tag{47}$$

Equivalently, we have the relations

$$\bar{x}\bar{y} + \bar{y}\bar{x} = B(x, y)\bar{1} \tag{47'}$$

in $C(V, Q)$ as well as $\bar{x}^2 = Q(x)\bar{1}$. We can use these to prove

LEMMA 1. *If the elements $u_1, u_2, \ldots, u_n$ span the vector space V over F, then the elements*

$$1 = \bar{1}, \qquad \bar{u}_{i_1}\bar{u}_{i_2}\cdots\bar{u}_{i_r}, \qquad i_1 < i_2 < \cdots < i_r, \qquad 1 \leqslant r \leqslant n \tag{48}$$

span the vector space $C(V, Q)$ over F.

Proof. Since $T(V)$ is generated by V and the u_i span V, it is clear that the u_i generate $T(V)$, so every element of $T(V)$ is a linear combination of 1 and monomials in the u_i of positive degree. We now call 1 and the monomials $u_{i_1}u_{i_2}\cdots u_{i_r}$ with $i_1 < i_2 < \cdots < i_r$, $1 \leqslant r \leqslant n$, *standard*. Let S be the set of these. We proceed to prove by induction on i and on the degree of $u \in S$ that $u_i u$ is congruent modulo the ideal K_Q to a linear combination of standard monomials of degree $\leqslant \deg u + 1$. This is clear if $\deg u = 0$. Now suppose $u = u_{i_1}u_{i_2}\cdots u_{i_r}$, $r \geqslant 1$. If $i = 1$, the result is clear if $i_1 > 1$ and if $i_1 = 1$, then $u_1 u = {u_1}^2 u_{i_2}\cdots u_{i_r} \equiv Q(u_1)u_{i_2}\cdots u_{i_r} \pmod{K_Q}$. Now let $i > 1$. If $i \leqslant i_1$ the result follows as in the case $i = 1$. Hence assume $i > i_1$. Then, by (47)

$$u_i u_{i_1} u_{i_2}\cdots u_{i_r} \equiv -u_{i_1}u_i u_{i_2}\cdots u_{i_r} + B(u_i, u_{i_1})u_{i_2}\cdots u_{i_r} \pmod{K_Q}.$$

By the degree induction, $u_i u_{i_2}\cdots u_{i_r}$ is congruent modulo K_Q to a linear combination of standard monomials of degree $\leqslant r$. Then induction on the subscript i implies that $u_{i_1}u_i u_{i_2}\cdots u_{i_r}$ and hence $u_i u_{i_1}u_{i_2}\cdots u_{i_r}$ is congruent modulo K_Q to a linear combination of standard monomials of degree $\leqslant r+1$. The result we have proved implies that if C' is the subspace of $C(V, Q)$ spanned by the elements $\bar{1}, \bar{u}_{i_1}\bar{u}_{i_2}\cdots\bar{u}_{i_r}$, $i_1 < i_2 < \cdots < i_r$, then $\bar{u}_i C' \subset C'$. This implies that $C'C' \subset C'$ so C' is a subalgebra of $C(V, Q)$. Since C' contains $\bar{V} = \{\bar{x} \mid x \in V\}$ and $\bar{V}$ generates $C(V, Q)$, we have $C' = C(V, Q)$. □

Evidently, the lemma implies that if $\dim V = n$, then $\dim C(V, Q) \leqslant 2^n$, which is the number of standard monomials in the u_i.

For the sake of simplicity, we assume in the remainder of the text of this section that $\operatorname{char} F \neq 2$. The extension of the results to the characteristic two case will be indicated in the exercises. We assume first that B is non-degenerate and we separate off the lowest dimensional cases $n = 1, 2$.

$n = 1$. We have $\dim C(V,Q) \leqslant 2$ and if $V = Fu$, then $\bar{u}^2 = Q(u)1$ and $Q(u) \neq 0$, since B is non-degenerate. Let $A = F[t]/(t^2 - Q(u)1)$ where t is an indeterminate. Then A has the base $(1, \bar{t})$ where $\bar{t} = t + (t^2 - Q(u)1)$. Then $\bar{t}^2 = Q(u)1$, which implies that the linear map f of V into A such that $f(u) = \bar{t}$ satisfies $f(x)^2 = Q(x)\bar{1}$. Hence we have a homomorphism of $C(V,Q)$ into A such that $\bar{u} \rightsquigarrow \bar{t}$. This is surjective. Since $\dim A = 2$ and $\dim C(V,Q) \leqslant 2$, we have $\dim C(V,Q) = 2$ and our homomorphism is an isomorphism. If $Q(u)$ is not a square in F, then $t^2 - Q(u)1$ is irreducible in $F[t]$ and $C(V,Q)$ is a field. If $Q(u) = \beta^2$, $\beta \in F$, then $e' = \bar{u} - \beta\bar{1}$ satisfies

$$\begin{aligned} e'^2 &= \bar{u}^2 - 2\beta\bar{u} + \beta^2\bar{1} = 2Q(u)\bar{1} - 2\beta\bar{u} \\ &= -2\beta(\bar{u} - \beta\bar{1}) = -2\beta e'. \end{aligned}$$

Hence $e = -(2\beta)^{-1}e'$ is an idempotent $\neq 0$, 1 in $C(V,Q)$. Then $C(V,Q) = Fe \oplus F(1-e)$, a direct sum of two copies of F. Hence we have the following lemma.

LEMMA 2. *If $B(x,y)$ is non-degenerate and $n = 1$, then* $\dim C(V,Q) = 2$ *and if $V = Fu$, then $C(V,Q)$ is a field or a direct sum of two copies of F according as $Q(u)$ is not or is a square in F.*

$n = 2$. Choose an orthogonal base (u, v) for V. Then $Q(u)Q(v) \neq 0$. Let C' be the Clifford algebra of Fu relative to the restriction Q' of Q to Fu. Then C' has a base $(1, \bar{u})$ where $\bar{u}^2 = Q(u)1$. Now consider the matrices

$$u' = \begin{pmatrix} \bar{u} & 0 \\ 0 & -\bar{u} \end{pmatrix}, \qquad v' = \begin{pmatrix} 0 & 1 \\ Q(v)1 & 0 \end{pmatrix}$$

in $M_2(C')$. We have

$$u'v' = \begin{pmatrix} 0 & \bar{u} \\ -Q(v)\bar{u} & 0 \end{pmatrix}, \qquad v'u' = \begin{pmatrix} 0 & -\bar{u} \\ Q(v)\bar{u} & 0 \end{pmatrix}$$

so $u'v' + v'u' = 0$. Also $u'^2 = Q(u)1$, $v'^2 = Q(v)1$. It follows that $A = F1 + Fu' + Fv' + Fu'v'$ is a subalgebra of $M_2(C')$. It is clear from the form of the matrices that 1, u', v', $u'v'$ are linearly independent. Hence $\dim A = 4$. The relations on u' and v' imply that the linear map f of V into A such that $u \rightsquigarrow u'$, $v \rightsquigarrow v'$ satisfies $f(x)^2 = Q(x)1$, $x \in V$. Hence we have a homomorphism g of $C(V,Q)$ into A such that $g(\bar{u}) = u'$, $g(\bar{v}) = v'$. Then g is surjective and since $\dim C(V,Q) \leqslant 4$ and $\dim A = 4$, it follows that g is an isomorphism and $\dim C(V,Q) = 4$.

An algebra A over a field F of characteristic $\neq 2$ is called a (*generalized*) *quaternion algebra* if A is generated by two elements i and j such that

$$i^2 = \alpha 1 \neq 0, \qquad j^2 = \beta 1 \neq 0, \qquad ij = -ji \tag{49}$$

where $\alpha, \beta \in F$. We denote this algebra as (α, β) or $(\alpha, \beta)/F$ if we need to call attention to the base field F. We now prove

LEMMA 3. *Any quaternion algebra is four-dimensional central simple over F.*

Proof. The relations (49) imply that any element of A is a linear combination of the elements 1, i, j, ij. Now suppose $\lambda 1 + \mu i + \nu j + \rho ij = 0$ for $\lambda, \mu, \nu, \rho \in F$. If we multiply this relation on the left by i and on the right by $i^{-1} = \alpha^{-1} i$, we obtain $\lambda 1 + \mu i - \nu j - \rho ij = 0$. Then $\lambda 1 + \mu i = 0$ and $\nu j + \rho ij = 0$. Multiplication of these on the left by j and on the right by $j^{-1} = \beta^{-1} j$ then gives $\lambda 1 - \mu i = 0$ and $\nu j - \rho ij = 0$. Then $\lambda 1 = \mu i = \nu j = \rho ij = 0$ and $\lambda = \mu = \nu = \rho = 0$. Hence the elements 1, i, j, ij are linearly independent and so these constitute a base for A. Hence $\dim A = 4$. Now let I be an ideal $\neq A$ in A and let $\bar{1} = 1 + I$, $\bar{i} = i + I$, $\bar{j} = j + I$. Then $\bar{i}^2 = \alpha \bar{1} \neq 0$, $\bar{j}^2 = \beta \bar{1} \neq 0$, $\bar{i}\bar{j} = -\bar{j}\bar{i}$. Hence A/I is a quaternion algebra and so $\dim A/I = 4$. Then $I = 0$ and hence A is simple. It remains to show that the center of A is $F1$. Let $c = \lambda 1 + \mu i + \nu j + \rho ij \in$ center of A. Then $ci = ic$ implies that $\nu = \rho = 0$. The fact that $\lambda 1 + \mu i$ commutes with j implies that $\mu = 0$. Hence $c = \lambda 1$ and the center is $F1$. □

The result we obtained before on $C(V, Q)$ in the case $n = 2$ can now be stated in the following way.

LEMMA 4. *If $B(x, y)$ is non-degenerate and $n = 2$, then $C(V, Q)$ is a quaternion algebra.*

Proof. We had an isomorphism of $C(V, Q)$ with the algebra A generated by u' and v' such that $u'^2 = Q(u)1 \neq 0$, $v'^2 = Q(v)1 \neq 0$, and $u'v' = -v'u'$. Then A and, hence, $C(V, Q)$ are quaternion algebras. □

We recall that if $B(x, y)$ is a symmetric bilinear form on a vector space V and $(u_1, u_2, \dots, u_n)$ is a base for V, then $\delta = \det(B(u_i, u_j))$ is called a discriminant of B. A change of base replaces δ by $\delta\beta^2$ where $\beta \neq 0$ is the determinant of the matrix giving the change of base. B is non-degenerate if and only if $\delta \neq 0$. We recall also that if U is a subspace of V on which the restriction of B to U is non-degenerate, then $V = U \oplus U^{\perp}$. Moreover, the restriction of B to $U^{\perp}$ is non-degenerate.

We shall now prove the following factorization property.

LEMMA 5. *Let $B(x,y)$ be non-degenerate and* $\dim V \geqslant 3$. *Let U be a two-dimensional subspace of V on which the restriction of B to U is non-degenerate. Write $V = U \oplus U^{\perp}$ and let Q' and Q'' denote the restrictions of Q to U and $U^{\perp}$ respectively. Then*

$$C(V,Q) \cong C(U,Q') \otimes C(U^{\perp}, -\delta' Q'') \tag{50}$$

where δ' is a discriminant of the restriction of B to U.

Proof. We shall prove (50) by using two universal map properties to produce inverse isomorphisms between $C(V,Q)$ and the right-hand side of (50). We denote the canonical maps of U into $C(U,Q')$ and of $U^{\perp}$ into $C(U^{\perp}, -\delta' Q'')$ as $i' : y \rightsquigarrow y'$ and $i'' : z \rightsquigarrow z''$ respectively. Let (u,v) be an orthogonal base for U and put $\bar{d} = 2\bar{u}\bar{v}$ (in $C(V,Q)$). We have $\bar{u}^2 = Q(u)1$, $\bar{v}^2 = Q(v)1$, $\bar{u}\bar{v} = -\bar{v}\bar{u}$, so $\bar{d}^2 = -4Q(u)Q(v)1 = -B(u,u)B(v,v)1$ (since $B(x,x) = 2Q(x)$) $= -\delta'1$ where δ' is the discriminant of the restriction of B to U defined by the base (u,v)). Also if $y \in U$ and $z \in U^{\perp}$, then $\bar{y}\bar{d} = -\bar{d}\bar{y}$, $\bar{y}\bar{z} = -\bar{z}\bar{y}$, and $\bar{d}\bar{z} = \bar{z}\bar{d}$; hence

$$\bar{y}(\bar{d}\bar{z}) = (\bar{d}\bar{z})\bar{y}, \qquad (\bar{d}\bar{z})^2 = -\delta' Q(z)1. \tag{51}$$

Since $\bar{y}^2 = Q(y)1$ and $(\bar{d}\bar{z})^2 = -\delta' Q(z)1$, the universal map property of Clifford algebras implies that we have homomorphisms of $C(U,Q')$ and $C(U^{\perp}, -\delta' Q'')$ into $C(V,Q)$ sending $y' \rightsquigarrow \bar{y}$, $z'' \rightsquigarrow \bar{d}\bar{z}$ respectively. The elements $\bar{y}$ and $\bar{d}\bar{z}$ generate the images under the homomorphisms and, by (51), these elements commute. Hence the images under our homomorphisms centralize each other, so by the universal map property of tensor products, we have a homomorphism h of $C(U,Q') \otimes C(U^{\perp}, -\delta' Q'')$ into $C(V,Q)$ such that

$$y' \rightsquigarrow \bar{y}, \qquad z'' \rightsquigarrow \bar{d}\bar{z}, \qquad y \in U, z \in U^{\perp}. \tag{52}$$

Now consider the element $d' = 2u'v'$ in $C(U,Q')$. The calculations made before show that $d'y' = -y'd'$ and $d'^2 = -\delta'1$, so d' is invertible in $C(U,Q')$. Next consider the element $y' + d'^{-1}z''$ of $C(U,Q') \otimes C(U^{\perp}, -\delta' Q'')$. We have

$$\begin{aligned}(y' + d'^{-1}z'')^2 &= y'^2 + (y'd'^{-1} + d'^{-1}y')z'' + z''^2 \\ &= y'^2 + z''^2 = Q(y)1 + Q(z)1 = Q(y+z)1.\end{aligned}$$

Hence by the universal map property of $C(V,Q)$ we have a homomorphism g of $C(V,Q)$ into $C(U,Q') \otimes C(U^{\perp}, -\delta' Q'')$ such that

$$\bar{y} + \bar{z} \rightsquigarrow y' + d'^{-1}z''. \tag{53}$$

Checking on generators, by (52) and (53), we see that $gh = 1$ on $C(U,Q') \otimes C(U^{\perp}, -\delta' Q'')$ and $hg = 1$ on $C(V,Q)$. Hence $C(V,Q) \cong C(U,Q') \otimes C(U^{\perp}, -\delta' Q'')$. □

We are now ready to prove the main theorem on the structure of Clifford algebras.

THEOREM 4.13. *Let Q be a quadratic form of an n-dimensional vector space over a field F of characteristic $\neq 2$. Then*

(1) $\dim C(V,Q) = 2^n$ *and if $(u_1, u_2, \ldots, u_n)$ is a base for V, the elements*

$$1 = \bar{1}, \bar{u}_{i_1}\bar{u}_{i_2}\cdots\bar{u}_{i_r}, i_1 < i_2 < \cdots < i_r, 1 \leqslant r \leqslant n \tag{54}$$

constitute a base for $C(V,Q)$ over F.

(2) *The canonical map $i: x \rightsquigarrow \bar{x}$ of V into $C(V,Q)$ is injective.*

(3) *If the bilinear form B associated with Q is non-degenerate, then $C(V,Q)$ is central simple if n is even and if $n = 2\nu+1$, $\nu \in \mathbb{Z}$, then $C(V,Q)$ is either simple with a two-dimensional field as center or is a direct sum of two isomorphic central simple algebras according as $(-1)^\nu 2\delta$, δ a discriminant of B, is not or is a square in F.*

Proof. We note first that the second statement in (1) and statement (2) are consequences of the dimensionality relation $\dim C(V,Q) = 2^n$. For, by Lemma 1, the elements in (54) span $C(V,Q)$. Since the number of these elements is $\leqslant 2^n$, if $\dim C(V,Q) = 2^n$, their number is 2^n (that is, they are distinct) and they form a base. This implies also that the $\bar{u}_i$, $1 \leqslant i \leqslant n$, are linearly independent and hence the linear map $i: x \rightsquigarrow \bar{x}$ is injective.

Next we prove the dimensionality relation and (3) in the non-degenerate case. If $n = 1$, the results follow from Lemma 2 and if $n = 2$, they follow from Lemmas 3 and 4. Now assume $n > 2$. Then we can pick a two-dimensional subspace U on which the restriction of B is non-degenerate. Then $V = U \oplus U^\perp$ and the restriction of B to $U^\perp$ is non-degenerate. By Lemma 5, $C(V,Q) \cong C(U,Q') \otimes C(U^\perp, -\delta' Q'')$ where δ' is a discriminant of the restriction of B to U. Moreover, $C(U,Q')$ is a quaternion algebra, hence, is four-dimensional central simple. Using induction on the dimensionality, we may assume the results for the quadratic form $-\delta' Q''$ on $U^\perp$. Then $\dim C(U^\perp, -\delta' Q'') = 2^{n-2}$ and this algebra is central simple if $n-2$ is even. Hence $\dim C(V,Q) = 2^2 2^{n-2} = 2^n$, and by Corollary 2 to Theorem 4.7, p. 219, this algebra is central simple if n is even. If $n = 2\nu+1$, $n-2 = 2(\nu-1)+1$ and the induction hypothesis implies that if δ'' is a discriminant of the restriction of B to $U^\perp$, then $C(U^\perp, -\delta' Q'')$ is simple with two-dimensional center or is a direct sum of two isomorphic central simple algebras according as $(-1)^{\nu-1} 2(-\delta')^{n-2}\delta''$ is not or is a square in F. Accordingly, $C(V,Q)$ is simple with two-dimensional center or is a direct sum of two isomorphic central simple algebras according as $(-1)^{\nu-1} 2(-\delta')^{n-2}\delta''$ is not or is a square. Now

$$\begin{aligned}(-1)^{\nu-1}2(-\delta')^{n-2}\delta'' &= (-1)^{\nu}(\delta')^{n-3}2\delta'\delta'' \\ &= (-1)^{\nu}(\delta')^{n-3}2\delta\end{aligned}$$

and since $n-3$ is even this is a square if and only if $(-1)^{\nu}2\delta$ is a square. This proves (3).

It remains to prove the dimensionality formula in the degenerate case. For this purpose, we shall imbed V in a finite dimensional space W with a quadratic form that has a non-degenerate associated bilinear form and is an extension of Q. To do this we write $V = V^{\perp} \oplus U$ for some subspace U. Then the restriction of B to U is non-degenerate. Now put $W = V^{\perp} \oplus U \oplus (V^{\perp})^*$ where $(V^{\perp})^*$ is the space of linear functions on $V^{\perp}$. Let $x = z+y+f$ where $z \in V^{\perp}$, $y \in U$, and $f \in (V^{\perp})^*$ and define

$$\tilde{Q}(x) = Q(z)+Q(y)+f(z). \tag{55}$$

It is readily seen that $\tilde{Q}$ is a quadratic form on W whose associated symmetric bilinear form $\tilde{B}$ is non-degenerate. Let $x \rightsquigarrow \tilde{x}$ be the canonical map of W into $C(W,\tilde{Q})$. It follows from the universal property of $C(V,Q)$ that we have a homomorphism of $C(V,Q)$ into $C(W,\tilde{Q})$ such that $\bar{x} \rightsquigarrow \tilde{x}$ for $x \in V$. Let $(u_1,\dots,u_n,u_{n+1},\dots,u_q)$ be a base for W such that $(u_1,\dots,u_n)$ is a base for V. Since $\tilde{B}$ is non-degenerate, the elements $\tilde{u}_{j_1}\cdots\tilde{u}_{j_s}$, $j_1 < \cdots < j_s$, $1 \leqslant s \leqslant q$, are distinct and linearly independent. Then this holds also for the elements $\tilde{u}_{i_1}\cdots\tilde{u}_{i_r}$, $i_1 < \cdots < i_r$, $1 \leqslant r \leqslant n$. Since the homomorphism of $C(V,Q)$ into $C(W,\tilde{Q})$ maps $\bar{u}_{i_1}\cdots\bar{u}_{i_r}$ into $\tilde{u}_{i_1}\cdots\tilde{u}_{i_r}$, the elements $\bar{u}_{i_1}\cdots\bar{u}_{i_r}$, $i_1 < \cdots < i_r$, $1 \leqslant r \leqslant n$, are linearly independent. Hence $\dim C(V,Q) = 2^n$. □

Since the map $i: x \rightsquigarrow \bar{x}$ of V into $C(V,Q)$ is injective, we can identify V with the corresponding subspace of $C(V,Q)$. Hence from now on we assume $V \subset C(V,Q)$. If U is a subspace of V, then the subalgebra of $C(V,Q)$ generated by U can be identified with $C(U,Q')$ where Q' is the restriction of Q to U. This is clear from the last part of the proof of Theorem 4.13. For, if $(u_1,\dots,u_m)$ is a base for U over F, then the argument shows that the elements 1, $u_{i_1}\cdots u_{i_r}$, $i_1 < i_2 < \cdots < i_r$, $1 \leqslant r \leqslant m$, are linearly independent and this implies that the canonical homomorphism of $C(U,Q')$ into $C(V,Q)$ is a monomorphism.

It is clear from the definitions that if $Q = 0$, then $C(V,Q)$ is the exterior algebra $E(V)$ defined by V (p. 141). The results (1) and (2) of Theorem 4.13 give another proof of properties of $E(V)$ that were derived in BAI, pp. 411–414.

We remark finally that the proof of statement (3) in Theorem 4.13 yields a stronger result than we stated in this theorem. We state this as the following

COROLLARY. *Let Q be a quadratic form on an n-dimensional vector space V over a field F of characteristic not two such that the associated bilinear form is* *non-degenerate. Then $C(V,Q)$ is a tensor product of quaternion algebras if n is even and is a tensor product of quaternion algebras and its center if n is odd. Moreover, the center C is two-dimensional of the form $F(c)$ where $c^2 = (-1)^\nu 2^{-n}\delta 1$, δ a discriminant, and C is a field or a direct sum of two copies of F according as $(-1)^\nu(2\delta)$ is not or is a square in F.*

Proof. The first statement follows by induction on the dimensionality and the factorization lemma (Lemma 5). To determine the center in the odd dimensional case, we choose an orthogonal base $(u_1, u_2, \ldots, u_n)$ where $n = 2\nu+1$. Then $u_iu_j = -u_ju_i$ for $i \neq j$, which implies that the element $c = u_1u_2\cdots u_n$ commutes with every u_i. Hence c is in the center and since $c \notin F1$ and the center is two-dimensional, the center is $F[c]$. We have

$$c^2 = u_1u_2\cdots u_nu_1u_2\cdots u_n = (-1)^{n(n-1)/2}u_1{}^2u_2{}^2\cdots u_n{}^2 \tag{56}$$
$$= (-1)^\nu \prod_1^n Q(u_i)1 = (-1)^\nu 2^{-n}\delta 1$$

where δ is the discriminant determined by the base $(u_1, u_2, \ldots, u_n)$. Then $F[c]$ is a field or a direct sum of two copies of F according as $(-1)^\nu 2^{-n}\delta$ is not or is a square. Since $n = 2\nu+1$, this holds if and only if $(-1)^\nu 2\delta$ is not or is a square. $\square$

In the remainder of this section, we shall give a brief indication of some applications of Clifford algebras to the study of orthogonal groups. For this purpose, we need to introduce the *even* (or *second*) *Clifford algebra* $C^+(V,Q)$ defined to be the subalgebra of $C(V,Q)$ generated by all of the products uv, $u, v \in V$ (that is, by V^2). We recall that a vector u is called non-isotropic if $Q(u) \neq 0$. If u_1 is non-isotropic, then

$$(u_1u)(u_1v) = u_1(-u_1u + B(u,u_1)1)v = -Q(u_1)uv + B(u,u_1)u_1v.$$

Hence

$$uv = Q(u_1)^{-1}(B(u,u_1)u_1v - (u_1u)(u_1v)),$$

which shows that $C^+ = C^+(V,Q)$ is generated by the elements u_1u, $u \in V$. Now we can write $V = Fu_1 + (Fu_1)^\perp$ and $u = \alpha u_1 + v$ where $\alpha \in F$ and $v \perp u_1$. Then $u_1u = \alpha Q(u_1)1 + u_1v$. It follows that C^+ is generated by the $n-1$ dimensional subspace $V_1 = u_1(Fu_1)^\perp$. We have

$$(u_1v)^2 = -u_1{}^2v^2 = -Q(u_1)Q(v)1 \tag{57}$$

and the restriction of $-Q(u_1)Q$ to $(Fu_1)^\perp$ is a quadratic form Q_1 with non-degenerate bilinear form B_1. Hence we have a surjective homomorphism of

$C((Fu_1)^{\perp}, Q_1)$ onto C^+. On the other hand, if $(u_1, \ldots, u_n)$ is a base for V, then 1, $u_{i_1} \cdots u_{i_r}$, $i_1 < \cdots < i_r$ is a base for $C(V, Q)$. Then the elements 1 and $u_{i_1} \cdots u_{i_r}$ with even r are contained in C^+ and there are 2^{n-1} of these. Thus $\dim C^+ \geqslant 2^{n-1}$, while $\dim C((Fu_1)^{\perp}, Q_1) = 2^{n-1}$. It follows that $C^+ \cong C((Fu_1)^{\perp}, Q_1)$. This proves the first statement in

THEOREM 4.14. *Let B be non-degenerate and* $\operatorname{char} F \neq 2$. *Then the even Clifford algebras $C^+(V, Q) \cong C((Fu_1)^{\perp}, Q_1)$ where u_1 is any non-isotropic vector and Q_1 is the restriction of $-Q(u_1)Q$ to $F(u_1)^{\perp}$. $C^+(V, Q)$ is central simple if the dimensionality n of V is odd and is a tensor product of a central simple algebra and a two-dimensional algebra D, which is either a field or a direct sum of two copies of F if $n = 2\nu$. The two alternatives for D correspond respectively to the following: $(-1)^{\nu}\delta$ is not or is a square in F, where δ is a discriminant of B_1.*

Proof. The second assertion is an immediate consequence of the first and Theorem 4.13. Now assume n is even and let $c = u_1 u_2 \cdots u_n$ where $(u_1, u_2, \ldots, u_n)$ is an orthogonal base for V. Then $c \in C^+$ and $u_i c = -c u_i$ and $u_i u_j c = c u_i u_j$. Hence c is in the center of C^+ and $c \notin F1$. Hence the center of C^+ is $F[c]$. As in (56) we have

$$c^2 = (-1)^{\nu} \beta^2 \delta 1$$

where $\beta = 2^{-\nu}$ and $F[c]$ is a field or a direct sum of two copies of F according as $(-1)^{\nu}\beta^2\delta$ and hence $(-1)^{\nu}\delta$ is not or is a square. $\square$

In both the even and the odd dimensional case, the subspace Fc, where $c = u_1 u_2 \cdots u_n$ and $(u_1, u_2, \ldots, u_n)$ is an orthogonal base, is independent of the choice of this base. For $F[c]$ is either the center of $C(V, Q)$ or the center of $C(V, Q)^+$. Moreover, $c \notin F1$ and $c^2 \in F1$. It is clear that these conditions characterize the set of non-zero elements of Fc.

Now let f be an orthogonal transformation in V. Then $f(x)^2 = Q(f(x))1 = Q(x)1$. Hence the universal map property of $C(V, Q)$ implies that f has a unique extension to an automorphism g of $C(V, Q)$. Now g stabilizes $C(V, Q)^+$ and it stabilizes the center of $C(V, Q)$ and of $C(V, Q)^+$. Since one of these centers is $F[c]$, g stabilizes $F[c]$. It follows from the characterization we have given of the set of non-zero elements of Fc that $g(c) = \alpha c$, $\alpha \neq 0$ in F. Since $g(c)^2 = g(c^2)$, we have $\alpha = \pm 1$ so $g(c) = \pm c$. Now we can write $g(u_i) = f(u_i) = \sum \alpha_{ij} u_j$, $\alpha_{ij} \in F$, where the matrix (α_{ij}) is orthogonal. Then

$$g(c) = f(u_1) f(u_2) \cdots f(u_n) = \sum \alpha_{1j_1} \alpha_{2j_2} \cdots \alpha_{nj_n} u_{j_1} u_{j_2} \cdots u_{j_n}.$$

Since $u_iu_j = -u_ju_i$ if $i \neq j$, and $u_i^2 = Q(u_i)1$, it is clear from the definition of determinants that the sum can be written as $\det(\alpha_{ij})u_1u_2\cdots u_n +$ a linear combination of elements $u_{i_1}u_{i_2}\cdots u_{i_r}$, $i_1 < i_2 < \cdots < i_r$ with $r < n$ (cf. BAI, p. 416). Since these elements together with $u_1u_2\cdots u_n$ constitute a base and since $g(c) = \pm c$, we have $g(c) = \det(\alpha_{ij})c$. Hence $g(c) = c$ if f is a rotation and $g(c) = -c$ otherwise.

We now observe that any automorphism g of a finite dimensional semi-simple algebra A that fixes the elements of the center C of A is inner. For, if $A = A_1 \oplus A_2 \oplus \cdots \oplus A_s$ where the A_i are the simple components of A, then $1 = 1_1 + 1_2 + \cdots + 1_s$ where 1_i is the unit of A_i and $A = A1_1 \oplus A1_2 \oplus \cdots \oplus A1_s$ and $C = C1_1 \oplus C1_2 \oplus \cdots \oplus C1_s$. Hence $g(1_i) = 1_i$ so g stabilizes every A_i and g fixes the elements of the center $C1_i$ of A_i. By the Skolem-Noether theorem, there exists an element u_i invertible in A_i such that the restriction of g to A_i is the inner automorphism determined by u_i. Then g is the inner automorphism determined by $u = \sum u_i$.

We can apply this to the foregoing situation. Then we see that if the given orthogonal transformation f is a rotation, there exists an invertible element $u \in C(V, Q)$ such that

$$f(x) = uxu^{-1}, \qquad x \in V. \tag{58}$$

These considerations lead to the introduction of the following groups.

DEFINITION 4.6. *The* Clifford group $\Gamma(V, Q)$ *is the subgroup of invertible elements* $u \in C(V, Q)$ *such that* $uxu^{-1} \in V$ *for all* $x \in V$. *Clearly this is a subgroup of the multiplicative group of invertible elements of* $C(V, Q)$. *The* even Clifford group *is* $\Gamma^+(V, Q) = \Gamma(V, Q) \cap C^+(V, Q)$.

If $x \in V$ and $u \in \Gamma = \Gamma(V, Q)$, then $uxu^{-1} \in V$ and $(uxu^{-1})^2 = ux^2u^{-1} = Q(x)1$. Hence the linear transformation $x \rightsquigarrow uxu^{-1}$ of V is in the orthogonal group $O(V, Q)$. The map χ, where $\chi(u)$ is $x \rightsquigarrow uxu^{-1}, x \in V$, is a homomorphism of $\Gamma(V, Q)$ into $O(V, Q)$ called the *vector representation* of the Clifford group.

Let $v \in V$ be non-isotropic. Then v is invertible in $C(V, Q)$ and for $x \in V$ we have

$$2vxv = v(vx + xv) + (xv + vx)v - v^2x - xv^2 = 2B(v, x)v - 2Q(v)x.$$

Since $vxv^{-1} = vxvv^{-2} = Q(v)^{-1}vxv$, this gives

$$vxv^{-1} = -x + Q(v)^{-1}B(v, x)v. \tag{59}$$

Thus $v \in \Gamma$. We recall that a map of the form

$$x \rightarrow x - Q(v)^{-1}B(v,x)v \tag{60}$$

is orthogonal and is called the symmetry S_v associated with the non-isotropic v (BAI, p. 363). Evidently $S_{\alpha v} = S_v$ if $\alpha \neq 0$ in F. We recall also that any orthogonal transformation is a product of symmetries and any rotation is a product of an even number of symmetries. The formula (59) now reads

$$\chi(v) = -S_v. \tag{61}$$

Now it is clear that if the v_i are non-isotropic, then $v_1 \cdots v_r \in \Gamma(V,Q)$ and this element is in $\Gamma^+(V,Q)$ if r is even. It is clear also that $\Gamma^+(V,Q)$ contains the group F^* of non-zero elements of F. We can now prove

THEOREM 4.15. *The even Clifford group Γ^+ coincides with the set of products $v_1 \cdots v_{2r}$, v_i non-isotropic in V. We have the exact sequence*

$$1 \rightarrow F^* \rightarrow \Gamma^+(V,Q) \xrightarrow{\chi} O^+(V,Q) \rightarrow 1 \tag{62}$$

where the second map is the injection of F^*.

Proof. Let $u \in \Gamma^+$. Then the automorphism $a \rightarrow uau^{-1}$ of $C(V,Q)$ fixes the element $c = u_1u_2 \cdots u_n$ where $(u_1, u_2, \ldots, u_n)$ is an orthogonal base for V. Then, as above, $\chi(u) \in O^+(V,Q)$. On the other hand, let $f \in O^+(V,Q)$ and write $f = S_{v_1} \cdots S_{v_{2r}}$ where the v_i are non-isotropic vectors in V. Then $v_1 \cdots v_{2r} \in \Gamma^+$ and $\chi(v_1 \cdots v_{2r}) = S_{v_1} \cdots S_{v_{2r}}$, by (61). Thus $\chi(\Gamma^+) = O^+$. The kernel of χ restricted to Γ^+ is the intersection of Γ^+ with the center of $C(V,Q)$. Since either $C(V,Q)$ or $C(V,Q)^+$ is central simple, it is clear that this intersection is F^*. This completes the proof of the exactness of (62). Moreover, if $u \in \Gamma^+$ then $\chi(u) \in O^+$, so there exists an element $v_1 \cdots v_{2r}$ such that $\chi(v_1 \cdots v_{2r}) = \chi(u)$. Then $u = \alpha v_1 \cdots v_{2r}$, $\alpha \in F^*$, and hence $u = (\alpha v_1)v_2 \cdots v_{2r}$. This proves the first statement of the theorem. □

Let C^{op} be the opposite algebra of $C = C(V,Q)$. We have $x^2 = Q(x)1$ for $x \in V \subset C^{\text{op}}$. Hence we have a unique homomorphism of C into C^{op} sending $x \rightarrow x$. This means that we have an anti-homomorphism ι of C into C such that $x \rightarrow x$. Then ι^2 is a homomorphism fixing every $x \in V$, so $\iota^2 = 1$. Thus ι is an involution in C, that is, an anti-automorphism satisfying $\iota^2 = 1$. This can be characterized as *the* involution of C fixing all of the elements of V. We call ι the *main involution* of C. Evidently C^+ is stabilized by ι, so the restriction $\iota | C^+$ is an involution in C^+.

Now let $u \in \Gamma^+$ and write $u = v_1 \cdots v_{2r}$. Then

$$N(u) \equiv i(u)u = v_{2r} \cdots v_1 v_1 \cdots v_{2r} = \prod_1^{2r} Q(v_i) \in F^*. \tag{63}$$

If $u' \in \Gamma^+$, we have

$$N(uu') = \imath(uu')uu' = \imath(u')\imath(u)uu' = N(u)\imath(u')u' = N(u)N(u'). \tag{64}$$

Hence $u \rightarrow N(u)$ is a homomorphism of Γ^+ into F^*. It is clear from (63) that $N(\Gamma^+)$ is the set of products $\prod_1^{2r} Q(v_i)$ where the v_i are non-isotropic in V and this contains F^{*2}, the set of squares of elements of F^*. If f is any rotation, there exists a $u \in \Gamma^+$ such that $\chi(u) = f$ and u is determined up to a factor in F^*. Hence the coset $N(u)F^{*2}$ in the group F^*/F^{*2} is determined by f. We now give the following definition, which ties together these concepts.

DEFINITION 4.7. *The kernel of the homomorphism N of Γ^+ into F^* is called the* spin group Spin (V, Q). *Its image $O'(V, Q)$ under the vector representation χ is called the* reduced orthogonal group. *If f is any rotation and $\chi(u) = f$, then $N(u)F^{*2}$ is called the* spinorial norm *of f.*

The spinorial norm map is also a homomorphism (of $O^+(V, Q)$ into F^*/F^{*2}). The spinorial norm of a rotation can be defined directly, without the intervention of the Clifford algebra. If f is a given rotation, then we can write $f = S_{v_1} \cdots S_{v_{2r}}$ and we can simply define the spinorial norm of f to be the coset $\prod_1^{2r} Q(v_i)F^{*2}$. Since $\chi(v_1 \cdots v_{2r}) = f$ and $N(v_1 \cdots v_{2r}) = \prod_1^{2r} Q(v_i)$, this is the same element of F^*/F^{*2}, which we have called the spinorial norm. The difficulty with the direct definition is that it is not apparent that the spinorial norm is well defined, since there are many ways of writing a rotation as product of symmetries. The definition using the Clifford algebra shows that we get the same elements of F^*/F^{*2} no matter what factorization of f as product of symmetries is used.

Now the reduced orthogonal group can also be defined as the kernel of the spinorial norm map. For, if $f \in O' = O'(V, Q)$, then there exists a $u \in$ Spin (V, Q) such that $\chi(u) = f$. Then the spinorial norm of f is $N(u)F^{*2} = F^{*2}$. Conversely, if the spinorial norm of the rotation f is F^{*2}, then $f = S_{v_1} \cdots S_{v_{2r}}$ and $\prod_1^{2r} Q(v_i) = \beta^2$, $\beta \neq 0$. Replacing v_1 by $\beta^{-1} v_1$ we may assume $\prod_1^{2r} Q(v_i) = 1$. Then $\chi(v_1 \cdots v_{2r}) = f$ and $N(v_1 \cdots v_{2r}) = 1$, so the spinorial norm of f is F^{*2}.

The reduced orthogonal group contains the commutator subgroup Ω of O, since any commutator can be written in the form

$$(S_{v_1}\cdots S_{v_r})(S_{v_{r+1}}\cdots S_{v_s})(S_{v_1}\cdots S_{v_r})^{-1}(S_{v_{r+1}}\cdots S_{v_s})^{-1}$$
$$= S_{v_1}\cdots S_{v_r}S_{v_{r+1}}\cdots S_{v_s}S_{v_r}\cdots S_{v_1}S_{v_s}\cdots S_{v_{r+1}}$$

from which it is clear that the spinorial norm is F^{*2}. Thus we have the following inclusions among the various subgroups of $O = O(V,Q)$ that we have defined:

(65) $$O \supset O^+ \supset O' \supset \Omega$$

and $O' \lhd O^+$ such that O^+/O' is isomorphic to the subgroup of F^*/F^{*2} of cosets of the form βF^{*2} where β has the form $\prod_1^{2r} Q(v_i)$, v_i non-isotropic. We can say considerably more in the case in which Q is of positive Witt index (BAI, p. 369).

THEOREM 4.16. *Let Q be of positive Witt index. Then the reduced orthogonal group $O'(V,Q)$ coincides with the commutator subgroup Ω of $O(V,Q)$ and $O^+(V,Q)/O'(V,Q) \cong F^*/F^{*2}$.*

Proof. Since Q is of positive Witt index, there exists a subspace U of V that is a hyperbolic plane, so we have $V = U \oplus U^\perp$ and U has a base (u,v) such that $Q(u) = 0 = Q(v)$ and $B(u,v) = 1$. The orthogonal transformations that stabilize U and act as the identity map on $U^\perp$ form a subgroup O_1 isomorphic to the orthogonal group in U. $O_1^+ = O_1 \cap O^+(V,Q)$ is the set of linear maps f_α such that $u \to \alpha u$, $v \to \alpha^{-1}v$, $w \to w$ for $w \in U^\perp$ where $\alpha \in F^*$ (BAI, pp. 365–366). We have $f_\alpha = S_{u-v}S_{u-\alpha v}$, so that the spinorial norm of f_α is $Q(u-v)Q(u-\alpha v) = (-1)(-\alpha) = \alpha$. Since α can be taken to be any element of F^*, this proves that $O^+(V,Q)/O'(V,Q) \cong F^*/F^{*2}$. Next let y be any non-isotropic vector in V and let $Q(y) = \alpha$. Then $Q(u+\alpha v) = \alpha = Q(y)$. Hence by Witt's theorem there is an orthogonal transformation g such that $g(u+\alpha v) = y$ (BAI, p. 351). Then $S_y = gS_{u+\alpha v}g^{-1}$. Let $f \in O'$ and write $f = S_{v_1}\cdots S_{v_{2r}}$ where $\alpha_i = Q(v_i)$ and hence $\prod_1^{2r}\alpha_i \in F^{*2}$. Then we have orthogonal transformations g_i such that $S_{v_i} = g_i S_{u+\alpha_i v} g_i^{-1}$, $1 \leq i \leq 2r$. Then $f = g_1 S_{u+\alpha_1 v} g_1^{-1} \cdots g_{2r} S_{u+\alpha_{2r} v} g_{2r}^{-1}$. Since $O(V,Q)/\Omega$ is abelian,

$$f \equiv h = S_{u+\alpha_1 v}\cdots S_{u+\alpha_{2r} v} \qquad (\text{mod}\ \Omega)$$

so to prove that $f \in \Omega$, it suffices to prove that $h \in \Omega$. Now $f \in O'$ and since $\Omega \subset O'$, $h \in O'$. Hence $h \in O_1^+$, so $h = f_\alpha = S_{u-c}S_{u-\alpha v}$, as above, and $\alpha = \beta^2$. Then $h = f_\alpha = S_{u-v}f_\beta^{-1}S_{u-v}f_\beta \in \Omega$. □

The main structure theorem on orthogonal groups, which we derived in BAI, states that if Q has positive Witt index and $n \geqslant 3$, then $\Omega/(\Omega \cap \{1, -1\})$ is simple except in the cases $n = 4$, Witt index 2 and $n = 3$, $|F| = 3$. An interesting question is, when does $-1 \in \Omega$? This can happen only for even n. In this case we have

PROPOSITION 4.11. *Let n be even and let Q be of positive Witt index. Then $-1 \in \Omega$ if and only if the discriminant is a square.*

Proof. Let $(u_1, u_2, \ldots, u_n)$ be an orthogonal base. Then the discriminant obtained from this base is $2^{-n}\prod_1^n Q(u_i)$ and this is a square if and only if $\prod_1^n Q(u_i)$ is a square. On the other hand $-1 = S_{u_1}S_{u_2}\cdots S_{u_n}$, so the spinorial norm of -1 is $\prod_1^n Q(u_i)$. Hence by Theorem 4.15, $-1 \in \Omega$ if and only if $\prod_1^n Q(u_i)$ is a square. □

EXERCISES

1. Show that any central simple algebra of degree two (p. 222) over a field F of characteristic $\neq 2$ is a quaternion algebra as defined on p. 232 and that any such algebra is isomorphic to a Clifford algebra $C(V, Q)$ where $\dim V = 2$.

In exercises 2–9, Q is a quadratic form with non-degenerate bilinear form B on an n-dimensional vector space V over F of characteristic $\neq 2$.

2. Let n be even and $(u_1, u_2, \ldots, u_n)$ an orthogonal base for V over F so $Q(u_i) = \gamma_i \neq 0$. Obtain an explicit formula

$$C(V, Q) \cong (\alpha_1, \beta_1) \otimes (\alpha_2, \beta_2) \otimes \cdots \otimes (\alpha_\nu, \beta_\nu) \tag{66}$$

where $\nu = n/2$ and (α, β) denotes the algebra with base $(1, i, j, k)$ such that $i^2 = \alpha 1$, $j^2 = \beta 1$, $ij = k = -ji$.

3. Use exercise 2 to show that $C(V, Q) \sim 1$ if Q has maximal Witt index ($= \nu$, see p. 370 of BAI).

4. Let n be odd, but otherwise let the notations be as in 2. Obtain a formula like (66) tensored with the center for $C(V, Q)$ and a formula like (66) for $C^+(V, Q)$.

5. Apply exercise 4 to obtain the structure of $C(V, Q)$ and $C^+(V, Q)$ if n is odd and Q has maximal Witt index.

6. Let $n = 4$, $F = \mathbb{R}$. Obtain the structure of $C(V, Q)$ for the following cases in which (u_1, u_2, u_3, u_4) is an orthogonal base for V and the matrix $\operatorname{diag}\{Q(u_1), Q(u_2), Q(u_3), Q(u_4)\}$ is respectively

I. $\operatorname{diag}\{1,1,1,1\}$
II. $\operatorname{diag}\{1,1,1-1\}$
III. $\operatorname{diag}\{1,1,-1,-1\}$
IV. $\operatorname{diag}\{1,-1,-1,-1\}$
V. $\operatorname{diag}\{-1,-1,-1,-1\}$.

7. Note that $a \to {}^t a$ (the transpose of a) is an involution in $M_n(F)$. Use the Skolem-Noether theorem to show that any involution in $M_n(F)$ has the form

$$J_s : a \to s^t a s^{-1}$$

where ${}^t s = \pm s$. Let $\operatorname{Sym} J_s$ be the subspace of $M_n(F)$ of J_s-symmetric elements, that is, satisfying $J_s(a) = a$. Show that $\dim J_s = n(n+1)/2$ if ${}^t s = s$ and $\dim J_s = n(n-1)/2$ if ${}^t s = -s$.

8. Let A be a finite dimensional central simple algebra over F, E a splitting field for A. Show that if A is viewed as contained in $A_E \cong M_n(E)$, then any involution J in A has a unique extension to an involution J_E of A_E over E. Then $J_E = J_s$ for some $s \in M_n(E)$ such that ${}^t s = \pm s$. Call J of *orthogonal* or *symplectic type* according as s is symmetric or skew. Show that J is of orthogonal (symplectic) type if and only if $\dim \operatorname{Sym} J = n(n+1)/2 \quad (n(n-1)/2)$.

9. Determine the type of the main involution in $C(V,Q)$ for n even and the type of the involution induced in $C^+(V,Q)$ by the main involution if n is odd.

If J and K are involutions in A and B respectively then we write $(A,J) \cong (B,K)$ if there exists an isomorphism η to A onto B such that $\eta J = K\eta$.

10. Prove the following extension of Lemma 5: If U is a 2ν-dimensional subspace of V on which the restriction of B is non-degenerate, then $C(V,Q) \cong C(U,Q') \otimes C(U,-\delta' Q'')$ and $(C(V,Q),\iota) \cong (C(U,Q') \otimes C(U,-\delta' Q''), \iota \otimes \iota)$ if ν is even. (Here ι is the main involution.)

11. Let c denote the conjugation $a \to \bar{a}$ in $\mathbb{C}$ and s the standard involution $a \quad \bar{a}$ in $\mathbb{H}$. Show that $(\mathbb{C} \otimes \mathbb{H}, c \otimes s) \cong (M_2(\mathbb{C}), *)$, $(a_{ij})^* = t_{(\bar{a}_{ij})}$, and $(\mathbb{H} \otimes \mathbb{H}, s \otimes s) \cong (M_4(\mathbb{R}), t)$.

The next three exercises concern isomorphisms for $(C(V,Q),\iota)$ where V is a finite dimensional vector space over $\mathbb{R}$ and Q is a positive definite quadratic form on V.

12. Show that

   (i) $(C(V,Q),\iota) \cong (\mathbb{R} \oplus \mathbb{R}, 1_{\mathbb{R}\oplus\mathbb{R}})$ if $[V:\mathbb{R}] = 1$.
   (ii) $(C(V,Q),\iota) \cong (M_2(\mathbb{R}), t)$ if $[V:\mathbb{R}] = 2$.
   (iii) $(C(V,Q),\iota) \cong (M_2(\mathbb{C}), *)$ where $* : a \to {}^t\bar{a}$ if $[V:\mathbb{R}] = 3$.
   (iv) $(C(V,Q),\iota) \cong (M_2(\mathbb{H}), *)$, $* : a \to {}^t\bar{a}$, if $[V:\mathbb{R}] = 4$.

13. Show that if $n = 2\nu$, then

$$(C(V,Q),\iota) \cong (M_{2^{2\nu}}(\mathbb{R}), t)) \text{ if } \nu \equiv 0, 1 \pmod 4$$

$$(C(V,Q),\iota) \cong (M_{2^{2\nu-1}}(\mathbb{H}), *) \text{ if } \nu \equiv 2, 3 \pmod 4$$

(* as in Exercise 12.)

14. Show that if $n = 2\nu - 1$ then

$$(C(V,Q),\iota) \cong (M_{2^{\nu-1}}(\mathbb{R}),t) \oplus (M_{2^{2\nu-1}}(\mathbb{R}),t) \text{ if } \nu \equiv 1$$

or 3 (mod 4)

$$(C(V,Q),\iota) \cong (M_{2^{\nu-1}}(\mathbb{C}),*) \text{ if } \nu \equiv 0 \text{ or } 2 \pmod 4.$$

Exercises 15–18 sketch a derivation of the main structure theorem for Clifford algebras for arbitrary fields including characteristic two.

15. Define a quaternion algebra over an arbitrary field F to be an algebra generated by two elements i, j such that

$$(67) \qquad i^2 = i + \alpha 1, \qquad 4\alpha + 1 \neq 0, \qquad j^2 = \beta 1 \neq 0, \qquad ji = (1-i)j.$$

Show that any such algebra is four-dimensional central simple. Show that this definition is equivalent to the one given in the text if char $F \neq 2$.

Now let V be an n-dimensional vector space equipped with a quadratic form Q with non-degenerate bilinear form. Note that if char $F = 2$, then B is an alternate form and hence n is even.

16. Show that if $n = 2$, then $C(V,Q)$ is a quaternion algebra.

17. Show that Lemma 5 is valid for arbitrary F. (*Hint*: Let (u,v) be any base for U and let $\bar{d} = \bar{u}\bar{v} - \bar{v}\bar{u}$. Use this $\bar{d}$ in place of the one used in the proof given in the text to extend the proof to arbitrary F.)

18. Prove Theorem 4.13 for B non-degenerate and F arbitrary. Note that the proof of $\dim C(V,Q) = 2^n$ given in the text by reduction to the non-degenerate case carries over to arbitrary F.

In exercises 19–21 we assume that char $F = 2$.

19. Define $C^+ = C^+(V,Q)$ as for char $F \neq 2$: the subalgebra generated by all products uv, $u, v \in V$. Show that $\dim C^+ = 2^{n-1}$ and C^+ is generated by the elements $u_1 v$ for any non-isotropic u_1. Show that the subalgebra C' of C^+ generated by the $u_1 v$ such that $Q(u_1) \neq 0$ and $v \perp u_1$ is isomorphic to a Clifford algebra determined by an $(n-2)$-dimensional vector space and that this algebra is central simple of dimension 2^{n-2}. Let $(u_1, v_1, \dots, u_\nu, v_\nu)$ be a symplectic base (that is, $B(u_i, v_j) = \delta_{ij}, B(u_i, u_j) = 0 = B(v_i, v_j)$). Show that $c = \sum_1^\nu u_i v_i$ is in the center of C^+, that $c \notin F1$, and that $c^2 + c + \sum_1^\nu Q(u_i)Q(v_i) = 0$. Hence conclude that $C^+ \cong F[c] \otimes C'$ and $F[c]$ is the center of C^+. Show that $F[c]$ is a field or a direct sum of two copies of F according as $\sum_1^\nu Q(u_i)Q(v_i)$ is not or is of the form $\beta^2 + \beta$, $\beta \in F$. Thus conclude that C^+ is simple or a direct sum of two isomorphic central simple algebras according as $\sum_1^\nu Q(u_i)Q(v_i) \neq \beta^2 + \beta$ or $= \beta^2 + \beta$, $\beta \in F$.

20. Let $F^{[2]}$ denote the set of elements of the form $\beta^2 + \beta$, $\beta \in F$. Show that $F^{[2]}$ is a subgroup of the additive group of F and put $G = F/F^{[2]}$. Let $(u_1, v_1, \dots, u_\nu, v_\nu)$ be a symplectic base for V and define

$$(68) \qquad \operatorname{Arf} Q = \sum_1^\nu Q(u_i)Q(v_i) + F^{[2]}$$

in G. Show that this is independent of the choice of the symplectic base. Arf Q is called the *Arf invariant* of Q. Note that the last result in exercise 14 can be stated as the following: $C^+(V,Q)$ is simple if and only if $\operatorname{Arf} Q \neq 0$.

21. Show that there exists a unique derivation D in $C = C(V,Q)$ such that $Dx = x$ for $x \in V$. Show that $C^+(V,Q)$ is the subalgebra of D-constants (=elements a such that $Da = 0$).

Historical Note

Clifford algebras defined by means of generators and relations were introduced by W. K. Clifford in a paper published in 1878 in the first volume of the *American Journal of Mathematics*. In this paper Clifford gave a tensor factorization of his algebras into quaternion algebras and the center. The first application of Clifford algebras to orthogonal groups was given by R. Lipschitz in 1884. Clifford algebras were rediscovered in the case $n = 4$ by the physicist P. A. M. Dirac, who used these in his theory of electron spin. This explains the terminology *spin group* and *spinorial norm*. The spin group for orthogonal groups over $\mathbb{R}$ are simply connected covering groups for the proper orthogonal groups. As in the theory of functions of a complex variable, multiple-valued representations of the orthogonal group become single-valued for the spin group. Such representations occurred in Dirac's theory. This was taken up in more or less general form by R. Brauer and H. Weyl (1935) and by E. Cartan (1938) and in complete generality by C. Chevalley (1954).

REFERENCES

N. Jacobson, *Structure of Rings*, American Mathematical Society Colloquium Publication XXXVII, Providence 1956, rev. ed., 1964.

I. Herstein, *Noncommutative Rings*, Carus Mathematical Monograph, No. 15, Mathematical Association of America, John Wiley, New York, 1968.

C. Faith, *Algebra* II. *Ring Theory*, Springer, New York, 1976.

For Clifford algebras:
C. Chevalley, *The Algebraic Theory of Spinors*, Columbia University Press, New York, 1954.

5

Classical Representation Theory of Finite Groups

One of the most powerful tools for the study of finite groups is the theory of representations and particularly the theory of characters. These subjects as generalized to locally compact groups constitute a major area of modern analysis that generalizes classical Fourier analysis. The subject of representation theory of finite groups is almost wholly the creation of Frobenius. Notable improvements and simplifications of the theory are due to Schur. During the past fifty years, very deep results have been added to the theory by Brauer, and representation theory has played an important role in the explosive growth of the structure theory of finite groups, which began with the Feit-Thompson proof of a hundred-year-old conjecture by Burnside that every finite group of odd order is solvable.

In this chapter, we are concerned with the classical theory of representations of finite groups acting on finite dimensional vector spaces over the field $\mathbb{C}$ of complex numbers or more generally over fields of characteristics not dividing the group order. This restriction on the characteristic implies complete reducibility of the representation or, equivalently, semi-simplicity of the group algebras. This permits the application of the structure and representation theory of finite dimensional semi-simple algebras to the representation theory of finite groups. However, there is considerably more to the story than this, namely, the theory of characters, much of which can be developed without recourse to the theory of algebras.

We shall derive the classical results of Frobenius, Schur, and Burnside as well as Brauer's results on induced characters and splitting fields. One of the most important contributions of Brauer is his modular theory, which deals with representations over fields whose characteristics do divide the group order and the relation between the representations over such fields and representations over $\mathbb{C}$. We shall not consider this theory in our account. A number of applications of character theory to structural results on finite groups will be given.

5.1 REPRESENTATIONS AND MATRIX REPRESENTATIONS OF GROUPS

DEFINITION 5.1. *By a* representation ρ *of a group* G *we shall mean a homomorphism of* G *into the group* $GL(V)$ *of bijective linear transformations of a finite dimensional vector space* V *over a field* F.

We shall say that ρ is a *representation of* G *over* F and that it *acts on* V/F. The dimensionality of V is called the *degree* of the representation. The defining conditions for a representation are that ρ is a map of G into $GL(V)$ such that

$$\rho(g_1g_2) = \rho(g_1)\rho(g_2) \tag{1}$$

for $g_i \in G$. These have the immediate consequences that $\rho(1) = 1_V$, $\rho(g^{-1}) = \rho(g)^{-1}$, and in general, $\rho(g^m) = \rho(g)^m$ for $m \in \mathbb{Z}$. We remark also that the conditions that $\rho(g) \in GL(V)$ and (1) holds can be replaced by the following: $\rho(g) \in \operatorname{End}_F V$, (1) holds, and $\rho(1) = 1$. These immediately imply that $\rho(g) \in GL(V)$, so ρ is a representation.

Let $B = (u_1, u_2, \ldots, u_n)$ be a base for V/F. If $a \in \operatorname{End}_F V$, we write

$$au_i = \sum_{j=1}^{n} \alpha_{ji} u_j, \qquad 1 \leqslant i \leqslant n \tag{2}$$

and obtain the matrix (α) whose (i,j)-entry is α_{ij}. The map $a \rightsquigarrow (\alpha)$ is an isomorphism of $\operatorname{End}_F V$ onto $M_n(F)$. Hence if ρ is a representation of G acting on V, then the map

$$g \rightsquigarrow \rho_B(g), \tag{3}$$

where $\rho_B(g)$ denotes the matrix of $\rho(g)$ relative to the base B, is a homomorphism of G into the group $GL_n(F)$ of invertible $n \times n$ matrices with entries in F. Such a homomorphism is called a *matrix representation* of G of *degree* n. A change of the base B to $C = (v_1, \ldots, v_n)$ where $v_i = \sum \mu_{ji} u_j$ and $(\mu) \in GL_n(F)$ replaces ρ_B by ρ_C where

$$\rho_C(g) = (\mu)^{-1} \rho_B(g)(\mu). \tag{4}$$

This matrix representation is said to be *similar* to ρ_B. It is clear that any matrix representation can be obtained from a representation in the manner indicated above.

A homomorphism of G into the symmetric group S_n or, equivalently, an action of G on the finite set $\{1, 2, \ldots, n\}$ (BAI, p. 71), gives rise to a representation. Let the action be described by

$$gi = \pi(g)i \tag{5}$$

where π is the homomorphism of G into S_n. Let V be the vector space with base $(u_1, u_2, \ldots, u_n)$ and let $\rho(g)$ be the linear transformation of V such that

$$\rho(g)u_i = u_{\pi(g)i}. \tag{6}$$

Then $\rho(g_1 g_2) = \rho(g_1)\rho(g_2)$ and $\rho(1) = 1$, so ρ is a representation of G. Representations of this type are called *permutation* representations. They are characterized by the property that they are representations that stabilize some base B of V/F. Of particular interest is the permutation representation obtained from the action of G on itself by left translations (BAI, p. 72). The corresponding representation of G is called the *regular* representation.

EXAMPLES

1. Let $G = \langle g \rangle$, the cyclic group generated by an element g of order n. We have the homomorphism of G into S_n mapping g into the cycle $(12 \cdots n)$. This action is equivalent to the action of G on itself by left translations. The associated permutation representation maps g into the linear transformation $\rho(g)$ such that

$$\rho(g)u_i = u_{i+1}, \qquad 1 \leqslant i \leqslant n-1, \qquad \rho(g)u_n = u_1. \tag{7}$$

This gives a matrix representation in which g is represented by the matrix

$$\begin{pmatrix} 0 & \cdot & \cdot & \cdot & 0 & 1 \\ 1 & 0 & \cdot & \cdot & \cdot & 0 \\ 0 & 1 & \cdot & \cdot & \cdot & \cdot \\ \cdot & \cdot & \cdot & \cdot & \cdot & \cdot \\ \cdot & \cdot & \cdot & \cdot & \cdot & \cdot \\ 0 & \cdot & \cdot & \cdot & 1 & 0 \end{pmatrix}. \tag{8}$$

This is obtained by specializing (2) to (6).

2. Let $G = S_3$ whose elements are

$$1, (12), (13), (23), (123), (132).$$

The identity map is an isomorphism of G onto itself. This gives rise to a permutation representation and an associated matrix representation in which

$$1 \rightsquigarrow \begin{pmatrix} 1 & 0 & 0 \\ 0 & 1 & 0 \\ 0 & 0 & 1 \end{pmatrix} \qquad (12) \rightsquigarrow \begin{pmatrix} 0 & 1 & 0 \\ 1 & 0 & 0 \\ 0 & 0 & 1 \end{pmatrix}$$

$$(13) \rightsquigarrow \begin{pmatrix} 0 & 0 & 1 \\ 0 & 1 & 0 \\ 1 & 0 & 0 \end{pmatrix} \qquad (23) \rightsquigarrow \begin{pmatrix} 1 & 0 & 0 \\ 0 & 0 & 1 \\ 0 & 1 & 0 \end{pmatrix}$$

$$(123) \rightsquigarrow \begin{pmatrix} 0 & 0 & 1 \\ 1 & 0 & 0 \\ 0 & 1 & 0 \end{pmatrix} \qquad (132) \rightsquigarrow \begin{pmatrix} 0 & 1 & 0 \\ 0 & 0 & 1 \\ 1 & 0 & 0 \end{pmatrix}.$$

For a given group G and given field F we now consider the class $\sum(G, F)$ of representations of G acting on vector spaces over F as base field. There is a rich algebraic structure that can be defined on $\sum(G, F)$. First, let ρ_1 and ρ_2 be representations of G acting on the vector spaces V_1/F and V_2/F respectively. Form the vector space $V_1 \otimes_F V_2$ and define $\rho_1 \otimes \rho_2$ by

$$(\rho_1 \otimes \rho_2)(g) = \rho_1(g) \otimes \rho_2(g) \tag{9}$$

where, as usual, $a_1 \otimes a_2$ for the linear transformations a_i of V_i is the linear transformation of $V_1 \otimes V_2$ such that

$$(a_1 \otimes a_2)(x_1 \otimes x_2) = a_1 x_1 \otimes a_2 x_2. \tag{10}$$

If b_i, $i = 1, 2$, is a second linear transformation of V_i, then we have $(a_1 \otimes a_2)(b_1 \otimes b_2) = a_1 b_1 \otimes a_2 b_2$. It follows that if $g_1, g_2 \in G$, then

$$\begin{aligned} (\rho_1 \otimes \rho_2)(g_1 g_2) &= \rho_1(g_1 g_2) \otimes \rho_2(g_1 g_2) \\ &= \rho_1(g_1)\rho_1(g_2) \otimes \rho_2(g_1)\rho_2(g_2) \\ &= (\rho_1(g_1) \otimes \rho_2(g_1))(\rho_1(g_2) \otimes \rho_2(g_2)) \\ &= ((\rho_1 \otimes \rho_2)(g_1))((\rho_1 \otimes \rho_2)(g_2)). \end{aligned}$$

Hence $\rho_1 \otimes \rho_2$ is a representation. We call this the *tensor product* of the given representations ρ_1 and ρ_2.

If $(u_1, u_2, \ldots, u_n)$ is a base for V_1/F and $(v_1, v_2, \ldots, v_m)$ is a base for V_2/F, then the mn vectors $u_i \otimes v_j$ constitute a base for $(V_1 \otimes V_2)/F$. We order these lexicographically:

$$(u_1 \otimes v_1, \ldots, u_1 \otimes v_m, u_2 \otimes v_1, \ldots, u_2 \otimes v_m, \ldots, u_n \otimes v_m). \tag{11}$$

Then if $a_1 \in \mathrm{End}_F V_1$ has the matrix $(\alpha^{(1)})$ relative to the base $(u_1, \ldots, u_n)$, and $a_2 \in \mathrm{End}_F V_2$ has the matrix $(\alpha^{(2)})$ relative to $(v_1, \ldots, v_m)$, we have

$$a_1 u_i = \sum_j \alpha_{ji}^{(1)} u_j, \qquad a_2 v_k = \sum_l \alpha_{lk}^{(2)} v_l$$

and

$$(a_1 \otimes a_2)(u_i \otimes v_k) = \sum_{j,l} \alpha_{ji}^{(1)} \alpha_{lk}^{(2)} (u_j \otimes v_l).$$

Hence the matrix of $a_1 \otimes a_2$ relative to the base (11) is

$$\begin{bmatrix} \alpha_{11}^{(1)}(\alpha^{(2)}) & \alpha_{12}^{(1)}(\alpha^{(2)}) & \cdots & \alpha_{1n}^{(1)}(\alpha^{(2)}) \\ \alpha_{21}^{(1)}(\alpha^{(2)}) & \alpha_{22}^{(1)}(\alpha^{(2)}) & \cdots & \alpha_{2n}^{(1)}(\alpha^{(2)}) \\ \cdot & \cdot & \cdots & \cdot \\ \alpha_{n1}^{(1)}(\alpha^{(2)}) & \alpha_{n2}^{(1)}(\alpha^{(2)}) & \cdots & \alpha_{nn}^{(1)}(\alpha^{(2)}) \end{bmatrix}. \tag{12}$$

We denote this matrix as $((\alpha^{(1)}) \otimes (\alpha^{(2)}))$. In particular, we see that if $\rho_{1B_1}(g)$ is the matrix of $\rho_1(g)$ relative to $B_1 = (u_1, \ldots, u_n)$ and $\rho_{2B_2}(g)$ is the matrix of $\rho_2(g)$ relative to $B_2 = (v_1, \ldots, v_m)$, then the matrix of $(\rho_1 \otimes \rho_2)(g)$ relative to the base (11) is $\rho_{1B_1}(g) \otimes \rho_{2B_2}(g)$.

As usual, we denote the dual space of linear functions on V/F by V^*. If $(u_1, u_2, \ldots, u_n)$ is a base for V/F, we have the dual (or complementary) base $(u_1^*, u_2^*, \ldots, u_n^*)$ of V^*/F where u_i^* is the linear function on V such that $u_i^*(u_j) = \delta_{ij}$, $1 \leqslant j \leqslant n$. If a is a linear transformation in V/F, we have the transposed transformation a^* in V^* such that

$$a^* x^*(y) = x^*(ay) \tag{13}$$

for $y \in V$ and $x^* \in V^*$. If $au_i = \sum_j \alpha_{ji} u_j$ and $a^* u_k^* = \sum \beta_{lk} u_l^*$, then $a^* u_k^*(u_i) = \sum \beta_{lk} u_l^*(u_i) = \beta_{ik}$ and $u_k^*(au_i) = u_k^*(\sum_j \alpha_{ji} u_j) = \alpha_{ki}$. Hence $\beta_{ik} = \alpha_{ki}$ and so the matrix of a^* relative to the dual base $(u_1^*, \ldots, u_n^*)$ of $(u_1, \ldots, u_n)$ is the transpose of the matrix of a relative to $(u_1, \ldots, u_n)$. The map $a \rightarrow a^*$ is an anti-isomorphism of the algebra $\mathrm{End}_F V$ onto $\mathrm{End}_F V^*$ and hence $a \rightarrow (a^*)^{-1}$ is an isomorphism of the group $GL(V)$ of bijective linear transformations in V into $GL(V^*)$.

Now let ρ be a representation of G acting on V. We compose this with the isomorphism $a \rightsquigarrow (a^*)^{-1}$ of $GL(V)$ onto $GL(V^*)$. This gives a representation $g \rightsquigarrow (\rho(g)^*)^{-1}$ of G acting on V^*. We call this the *contragredient representation* of ρ and denote it as ρ^*. Evidently it has the same degree as ρ. Moreover, if $B = (u_1, \ldots, u_n)$ and B^* is the dual base for V^*, then we have the following relation for the matrix representations $g \rightsquigarrow \rho_B(g)$ and $g \rightsquigarrow \rho^*_{B^*}(g)$: $\rho^*_{B^*}(g) = ({}^t\rho_B(g))^{-1}$.

The map $\rho \rightsquigarrow \rho^*$ may be regarded as a unary composition in $\sum(G, F)$. We also have an important nullary composition. This is the *unit representation* 1 or 1_G for which V is one-dimensional and $1(g) = 1_V$ for all g. The corresponding matrix representation (determined by any base) is $g \rightsquigarrow (1)$ where (1) is the 1×1 matrix with entry 1.

Let ρ_i, $i = 1, 2$, be a representation of G acting on V_i/F. Then we say that ρ_1 and ρ_2 are *equivalent* if there exists a bijective linear map η of V_1 onto V_2 such that

$$\rho_2(g) = \eta \rho_1(g) \eta^{-1}, \qquad g \in G. \tag{14}$$

EXERCISES

1. Let ρ be the representation of the cyclic group $\langle g \rangle$ given in example 1. Let $\mathbb{C}$ be the base field. Show that ρ is equivalent to the representation ρ' such that $\rho'(g)u_i = \zeta^i u_i$ where $\zeta = e^{2\pi i/n}$.

2. Let ρ_i, $i = 1, 2$, be a representation of G acting on V_i/F and put $V = \hom_F(V_1, V_2)$. If $l \in V$, define $\rho(g)l = \rho_2(g) l \rho_1(g)^{-1}$. Verify that ρ is a representation and show that ρ is equivalent to $\rho_1^* \otimes \rho_2$.

3. Let $\imath$ denote the identity map of $GL(V)$. This is a representation of $GL(V)$ acting on V. Consider the representation $\imath^* \otimes \imath$ acting on $V^* \otimes V$. Show that the set of vectors $c \in V^* \otimes V$ such that $(\imath^* \otimes \imath)(a)c = c$ for all $a \in GL(V)$ is a one-dimensional subspace of $V^* \otimes V$. Find a non-zero vector in this space.

4. Show that if ρ_i, $i = 1, 2$, is a representation acting on V_i, then $\rho_1 \otimes \rho_2$ and $\rho_2 \otimes \rho_1$ are equivalent.

5.2 COMPLETE REDUCIBILITY

The study of the representations of a group G can be reduced to the study of the representations of the group algebra of G. Let G be a group, F a field, then the group algebra $F[G]$ is the algebra over F having the set G as base and multiplication defined by

$$\left(\sum_{g\in G} \alpha_g g\right)\left(\sum_{h\in G} \beta_h h\right) = \sum_{g,h} \alpha_g \beta_h gh \tag{15}$$

for α_g, $\beta_h \in F$. Let ρ be a representation of G acting on the vector space V/F. Then the group homomorphism ρ has a unique extension to the algebra homomorphism

$$\sum \alpha_g g \rightsquigarrow \sum \alpha_g \rho(g) \tag{16}$$

of $F[G]$ into $\text{End}_F V$. Conversely, given a homomorphism ρ of $F[G]$ into $\text{End}_F V$, where V is a finite dimensional vector space over F, the restriction of ρ to G is a representation since $\rho(1) = 1$, which implies that every $\rho(g) \in GL(V)$. Now we recall that a representation of $F[G]$ (=homomorphism of $F[G]$ into $\text{End}_F V$) can be used to make V into an $F[G]$-module. One simply defines the action of $F[G]$ on V by

$$(\sum \alpha_g g)x = \sum \alpha_g \rho(g)x \tag{17}$$

for $x \in V$. Again, this can be turned around: Given a module V for $F[G]$, this is a module for F, hence, a vector space over F and, assuming finite dimensionality of V/F, we obtain a representation ρ of G where $\rho(g)$ is the map $x \rightsquigarrow gx$, which is a linear transformation of V/F. Thus representations of G acting on (finite dimensional) vector spaces over a field F are equivalent to $F[G]$-modules, which as F-modules are finite dimensional.

The standard concepts of module theory can be carried over to representations of groups via the group algebra. If ρ is a representation of G acting on V, a submodule U of V as $F[G]$-module is the same thing as a subspace of V that is $\rho(G)$-*invariant* in the sense that it is stabilized by every $\rho(g)$, $g \in G$. Then we obtain a representation $\rho|U$ of G acting on U in which $(\rho|U)(g)$ is the restriction of $\rho(g)$ to U. We shall call this a *subrepresentation* of ρ. We also have the module V/U. The associated representation of G is $\rho|V/U$ where $(\rho|V/U)(g)$ is the linear transformation $x+U \rightsquigarrow \rho(g)x+U$. This will be called a *factor representation* of ρ.

Let $B = (u_1, \ldots, u_n)$ be a base for V such that $(u_1, \ldots, u_r)$ is a base for the $\rho(G)$-invariant subspace U. Consider the matrix representation ρ_B determined by B. Since U is stabilized by every $\rho(g)$, $\rho(g)u_i$ for $1 \leqslant i \leqslant r$ is a linear combination of the vectors $(u_1, \ldots, u_r)$. Hence every matrix $\rho_B(g)$, $g \in G$, has the "reduced" form

$$\left[\begin{array}{ccc|ccc} \beta_{11} & \cdots & \beta_{1r} & \beta_{1r+1} & \cdots & \beta_{1n} \\ \cdot & \cdots & \cdot & \cdot & \cdots & \cdot \\ \beta_{r1} & \cdots & \beta_{rr} & \beta_{r,r+1} & \cdots & \beta_{rn} \\ \hline & & & \beta_{r+1,r+1} & \cdots & \beta_{r+1,n} \\ & 0 & & \cdot & \cdots & \cdot \\ & & & \beta_{n,r+1} & \cdots & \beta_{nn} \end{array}\right]. \tag{18}$$

The matrices in the upper left-hand corner are those of the matrix representation of G acting on U associated with the base $(u_1,\ldots,u_r)$ and those in the lower right-hand corner are the matrices of the representation of G on V/U relative to the base $(u_{r+1}+U,\ldots,u_n+U)$. It is clear that conversely if there exists a base $(u_1,\ldots,u_n)$ such that the matrices of the $\rho(g)$ relative to this base all have the form (18), then $U = \sum_1^r Fu_j$ is a $\rho(G)$-invariant subspace, hence, an $F[G]$-submodule of V.

If $V = U \oplus U'$ where U and U' are submodules, then we can choose a base $B = (u_1,\ldots,u_n)$ for V such that $(u_1,\ldots,u_r)$ is a base for U and $(u_{r+1},\ldots,u_n)$ is a base for U'. Then the corresponding matrices $\rho_B(g)$ have the form (18) in which the $r \times (n-r)$ blocks in the upper right-hand corner are all 0. If $\rho_1 = \rho|U$ and $\rho_2 = \rho|U'$, then we say that the representation ρ is a *direct sum of the subrepresentations* ρ_1 and ρ_2 and we write $\rho = \rho_1 \oplus \rho_2$. Let p be the projection on U determined by the decomposition $V = U \oplus U'$. Then if we write any $x \in V$ as $x = y+y'$, $y \in U$, $y' \in U'$, we have $\rho(g)x = \rho(g)y+\rho(g)y'$ with $\rho(g)y \in U$, $\rho(g)y' \in U'$ for all $g \in G$. Since p is the map $x \rightsquigarrow y$, we have $p\rho(g)x = \rho(g)y = \rho(g)px$. Thus p commutes with every $\rho(g)$. Conversely, suppose U is any $\rho(G)$-invariant subspace and there exists a projection p of V on U that commutes with every $\rho(g)$. We can write $V = pV \oplus (1-p)V$ and $pV = U$. Also, $\rho(g)(1-p)V = (1-p)\rho(g)V = (1-p)V$. Hence $U' = (1-p)V$ is $\rho(G)$-invariant and we have the decomposition $V = U \oplus U'$.

We shall call a representation ρ of G *irreducible* (*completely reducible*) if the corresponding $F[G]$-module is irreducible (completely reducible). We recall that a module V is completely reducible if and only if it satisfies either one of the following conditions: (1) $V = \sum V_i$ where the V_i are irreducible submodules, (2) $V \neq 0$ and for every submodule U there exists a submodule U' such that $V = U \oplus U'$ (Theorem 3.10, p. 121). We shall use these conditions to prove two theorems giving sufficient conditions for complete reducibility. The first of these is a fundamental theorem in the representation theory of finite groups. This is

MASCHKE'S THEOREM. *Every representation ρ of a finite group G acting on a vector space V/F such that the characteristic* char $F \nmid |G|$ *is completely reducible.*

Proof. Let U be a $\rho(G)$-invariant subspace of V and write $V = U \oplus U_0$ where U_0 is a second subspace (not necessarily invariant). Let p_0 be the projection on U determined by this decomposition. We shall now construct by an averaging process a projection on U that commutes with every $\rho(g)$, $g \in G$. We put

$$p = \frac{1}{|G|} \sum_{g \in G} \rho(g)^{-1} p_0 \rho(g). \tag{19}$$

Since char $F \nmid |G|$, $|G|^{-1}$ exists in F and p is well defined. If $g' \in G$, then

$$\begin{aligned} \rho(g')^{-1} p \rho(g') &= \frac{1}{|G|} \sum_{g \in G} \rho(g')^{-1} \rho(g)^{-1} p_0 \rho(g) \rho(g') \\ &= \frac{1}{|G|} \sum_{g \in G} \rho(gg')^{-1} p_0 \rho(gg') \\ &= \frac{1}{|G|} \sum_{g} \rho(g)^{-1} p_0 \rho(g) \\ &= p. \end{aligned}$$

Hence $\rho(g')p = p\rho(g')$ for $g' \in G$. Evidently, $p \in \operatorname{End}_F V$. If $y \in U$, then $p_0 y = y$ and since $\rho(g)y \in U$, $p_0 \rho(g) y = \rho(g) y$. Hence $\rho(g)^{-1} p_0 \rho(g) y = y$ and

$$py = \frac{1}{|G|} \sum_{g} \rho(g)^{-1} p_0 \rho(g) y = \frac{1}{|G|} |G| y = y. \tag{20}$$

If $x \in V$, then $p_0 x \in U$ and $\rho(g)^{-1} p_0 \rho(g) x \in U$. Hence $px \in U$. The two conditions on $p \in \operatorname{End}_F V$, $py = y$ for $y \in U$ and $px \in U$ for $x \in V$, imply that p is a projection on U. Then $V = U \oplus U'$ where $U' = (1-p)V$ and since p commutes with every $\rho(g)$, U' is $\rho(G)$-invariant. □

This result has a formulation in terms of matrix representations that should be obvious from the discussion above. The result in its matrix form was proved by H. Maschke for the case in which $F = \mathbb{C}$. The validity of the result for any F with char $F \nmid |G|$ was first noted by L. E. Dickson.

If ρ is a completely reducible representation of G acting on V, then the $F[G]$-module V decomposes as $V = V_1^{(1)} \oplus \cdots \oplus V_1^{(m_1)} \oplus V_2^{(1)} \oplus \cdots \oplus V_2^{(m_2)} \oplus \cdots \oplus V_r^{(1)} \oplus \cdots \oplus V_r^{(m_r)}$ where the $V_i^{(k)}$ are irreducible and $V_i^{(k)} \cong V_i^{(l)}$ for any k, l but $V_i^{(k)} \not\cong V_j^{(l)}$ if $i \neq j$. By Theorem 3.12 (p. 123), the submodules $W_i = \sum_k V_i^{(k)}$ are the homogeneous components of V. If ρ_i is any irreducible representation of G equivalent to the subrepresentation determined by the $V_i^{(k)}$, then we write

$$\rho \sim m_1 \rho_1 \oplus m_2 \rho_2 \oplus \cdots \oplus m_r \rho_r. \tag{21}$$

We call ρ_i an *irreducible constituent* of ρ and m_i its *multiplicity*. By Theorem 3.14, the equivalence classes of the irreducible constituents and their multiplicities are uniquely determined. We remark also that the multiplicity $m_i = q_i/n_i$ where $q_i = \dim W_i$, $n_i = \dim V_i^{(k)}$, which shows also that m_i is independent of the particular decomposition of V as a direct sum of irreducible submodules.

If H is a subgroup of G and ρ is a representation of G acting on V, then the restriction of ρ to H is a representation of H that we shall denote as ρ_H. We shall now show that if $H \lhd G$ and ρ is completely reducible, then ρ_H is completely reducible. It suffices to prove this for ρ irreducible. Moreover, in this case we can say considerably more on the relation between the irreducible constituents and their multiplicities. First, we need to give a definition of conjugacy of representations of a normal subgroup of a group.

Let $H \lhd G$ and let σ be a representation of H acting on U. For any $g \in G$ we can define the map

$$h \rightsquigarrow \sigma(ghg^{-1}). \tag{22}$$

This is the composite of the automorphism $h \rightsquigarrow ghg^{-1}$ of H with the homomorphism σ. Hence (22) is a representation of H acting on U, which we shall denote as ${}^g\sigma$. Any representation equivalent to ${}^g\sigma$ will be called a conjugate of σ or, more precisely, a *g-conjugate* of σ. Evidently, any $\sigma(H)$-invariant subspace is ${}^g\sigma(H)$-invariant and since $\sigma = {}^{g^{-1}}({}^g\sigma)$, it is clear that σ and ${}^g\sigma$ have the same lattices of $F[H]$-submodules. In particular, if σ is irreducible, then any conjugate of σ is irreducible. Now suppose σ_1 and σ_2 are equivalent representations of H acting on U_1 and U_2 respectively. Then we have a bijective linear map η of U_1 onto U_2 such that $\sigma_2(h) = \eta\sigma_1(h)\eta^{-1}$, $h \in H$. Then also, $\sigma_2(ghg^{-1}) = \eta\sigma_1(ghg^{-1})\eta^{-1}$, $h \in H$, so ${}^g\sigma_1$ and ${}^g\sigma_2$ are equivalent.

We can now prove

(A. H.) CLIFFORD'S THEOREM. *Let $H \lhd G$ and let ρ be an irreducible representation of G. Then ρ_H is completely reducible and all of its irreducible constituents are conjugate and have the same multiplicity.*

Proof. Let ρ act on V and let U be an irreducible $F[H]$-submodule of V. Evidently, $\sum_{g \in G}\rho(g)U$ is a $\rho(G)$-invariant subspace of V containing U. Hence $V = \sum\rho(g)U$. Let $h \in H$, $y \in U$. Then

$$\rho(h)\rho(g)y = \rho(hg)y = \rho(gg^{-1}hg)y = \rho(g)\rho(g^{-1}hg)y \in \rho(g)U. \tag{23}$$

Thus $\rho(g)U$ is $\rho(H)$-invariant. Let σ denote the representation of H acting on

U and σ' the representation of H acting on $\rho(g)U$. Now by (23), $\sigma'(h)\rho(g)y = \rho(g)\sigma^{g^{-1}}(h)y$. Since $y \rightarrow \rho(g)y$ is a bijective linear map of U onto $\rho(g)U$, we see that σ' is a conjugate of σ. Hence $\rho(g)U$ is an irreducible $F[H]$-module and $V = \sum \rho(g)U$ is a decomposition of V as sum of irreducible $F[H]$-modules such that the corresponding representations are conjugates. Then V is a direct sum of certain ones of the $\rho(g)U$. Hence ρ_H is completely reducible and its irreducible constituents are conjugate. If U_1 and U_2 are isomorphic irreducible $F[H]$-submodules, then $\rho(g)U_1$ and $\rho(g)U_2$ are isomorphic $F[H]$-submodules. It follows that every $\rho(g)$ permutes the homogeneous components of V as $F[H]$-module. Thus we have an action of G through $\rho(G)$ on the set S of homogeneous components of V as $F[H]$-module. Moreover, this action is transitive since if U is any irreducible $F[H]$-submodule, then any other irreducible $F[H]$-submodule is isomorphic to a $\rho(g)U$ for some $g \in G$. This implies that all of the irreducible constituents have the same multiplicity. □

EXERCISES

1. Prove the following extension of Maschke's theorem. Let ρ be a representation of a group G that is not necessarily finite. Suppose G contains a normal subgroup H of finite index $[G:H]$ not divisible by char F. Show that if $\rho|H$ is completely reducible, then ρ is completely reducible.

2. Let $F = \mathbb{C}$ and let $H(x, y)$ be a hermitian form on $V/\mathbb{C}$ that is positive definite in the sense that $H(x, x) > 0$ for all $x \neq 0$ (BAI, pp. 381–384). Let ρ be a representation of a group G acting on V such that every $\rho(g)$ is unitary with respect to H, $(H(\rho(g)x, \rho(g)y) = H(x, y), x, y \in V)$. Such a representation is called *unitary*. Show that if U is a $\rho(G)$-invariant subspace, then $U^{\perp}$ is $\rho(G)$-invariant. Use this to prove that ρ is completely reducible.

3. Same notations as in exercise 2. Let G be finite, ρ a representation of G acting on V. Let $H_0(x, y)$ be any positive definite hermitian form on $V/\mathbb{C}$. Put $H(x, y) = \sum_{g \in G} H_0(\rho(g)x, \rho(g)y)$. Show that $H(x, y)$ is positive definite hermitian and that every $\rho(g)$ is unitary relative to H. Use this and exercise 2 to prove Maschke's theorem for $F = \mathbb{C}$.

4. Define the *projective linear group* $PGL(V)$ of a finite dimensional vector space V/F as $GL(V)/F^*1$ where F^* is the multiplicative group of non-zero elements of F. A homomorphism ρ of a group G into $PGL(V)$ is called a *projective representation* of G. For each $g \in G$ let $\mu(g)$ denote a representative in $GL(V)$ of the coset of $\rho(g) \in PGL(V)$. Then $\mu(g_1 g_2) = \gamma_{g_1, g_2} \mu(g_1) \mu(g_2)$ where $\gamma_{g_1, g_2} \in F^*$. Define $\rho(G)$-invariant subspaces and complete reducibility as for ordinary representations. Prove the analogue of Maschke's theorem for projective representations.

5. Show that ρ is irreducible (completely reducible) if and only if ρ^* is irreducible (completely reducible).

6. Let ψ be a representation of G acting on W, V a $\psi(G)$-invariant subspace. Put $\rho = \psi|V$. Show that $\rho^* \otimes \psi$ has a subrepresentation equivalent to the unit representation.

7. Let ψ and ρ be representations of G acting on W and V respectively. Assume ρ is irreducible and $\rho^* \otimes \psi$ has a subrepresentation equivalent to the unit representation. Show that ψ has a subrepresentation equivalent to ρ.

8. Show that the following are irreducible groups of linear transformations of the finite dimensional vector space V/F: (1) $GL(V)$, (2) $O(V, Q)$ the group of orthogonal linear transformations relative to a non-degenerate quadratic form Q, (3) $O^+(V, Q)$ the rotation subgroup of $O(V, Q)$. (See Chapter 6 of BAI.)

5.3 APPLICATION OF THE REPRESENTATION THEORY OF ALGEBRAS

There are two main methods for developing the representation theory of finite groups: the structure and representation theory of finite dimensional algebras applied to group algebras and the theory of characters. Many results can be obtained by both methods, and the choice of the method is often a matter of taste. The first method is introduced in this section and the theory of characters appears in section 5. Our primary concern will be with representations of finite groups over fields whose characteristics do not divide the order of the group. However, the first two theorems that are given do not require these restrictions.

We begin by defining certain algebras of linear transformations associated with a representation ρ of a group G acting on a vector space V/F. First, we have the enveloping algebra $\operatorname{Env} \rho(G)$, which is the set of linear transformations of the form $\sum_{g \in G} \alpha_g \rho(g)$, $\alpha_g \in F$. If we extend ρ to a homomorphism ρ of $A = F[G]$ as in (16), then $\operatorname{Env} \rho(G) = \rho(A)$. Next we have the algebra $A' = \operatorname{End}_A V$. Evidently, this is the set of linear transformations of V that commute with every $a \in \rho(A)$, so $\operatorname{End}_A V$ is the centralizer in $\operatorname{End}_F V$ of $\rho(A)$. If ρ is irreducible, then by Schur's lemma, A' is a division algebra. Next we have $A'' = \operatorname{End}_{A'} V$, which is the algebra of linear transformations that commute with every $a' \in A'$. Evidently, $A'' \supset \operatorname{Env} \rho(G) = \rho(A)$. We remark that if we define $A''' = \operatorname{End}_{A''} V$, then it is trivial to see that $A''' = A'$. Hence the process of creating algebras by taking endomorphism algebras breaks off with A''. We call this the *double centralizer* of $\rho(A)$ and we say that ρ has the *double centralizer property* if $A'' = \rho(A)$. We have the following general result on completely reducible representations.

THEOREM 5.1. *Any completely reducible representation ρ of a group G has the double centralizer property.*

Proof. It is easily seen that this is a special case of Theorem 4.10 (p. 222). However, it is somewhat simpler to base the proof directly on the density theorem: Let $(u_1, u_2, \ldots, u_n)$ be a base for V/F and let l be an element of the double centralizer of $\rho(A)$. By the density theorem, there exists an $a \in \rho(A)$ such that $au_i = lu_i$, $1 \leqslant i \leqslant n$. Hence $l = a \in \rho(A)$ and $A'' = \rho(A)$. □

We state next

THEOREM 5.2. *A finite group has only a finite number of inequivalent irreducible representations.*

This follows by applying Theorem 4.5 (p. 208) to the artinian ring $A = F[G]$. We omit the details, since this result will not play any role in the sequel. We shall now prove

THEOREM 5.3. *The group algebra $A = F[G]$ of a finite group G over a field F is semi-simple if and only if* char $F \nmid |G|$.

Proof. Suppose first that char $F \nmid |G|$. Then every representation of G is completely reducible. Hence A regarded as (left) A-module is completely reducible. Then A is semi-simple by the structure theorem for semi-primitive artinian rings (p. 203). Now suppose char $F \mid |G|$. Consider the element

(24) $$z = \sum_{g \in G} g$$

of A. This is non-zero and we evidently have

(25) $$g'z = z = zg'$$

for all $g' \in G$. Hence Fz is an ideal in A. From (25) we obtain $z^2 = \sum_{g' \in G} g'z = |G|z = 0$, since char $F \mid |G|$. Hence Fz is a non-zero nilpotent ideal in A and A is not semi-simple. □

For the remainder of this section, we assume that G is finite and that char $F \nmid |G|$. We have

(26) $$A = F[G] = A_1 \oplus A_2 \oplus \cdots \oplus A_s$$

where the A_i are the simple components. By the matrix form of Wedderburn's theorem,

(27) $$A_i \cong M_{n_i}(\Delta_i)$$

where Δ_i is a division algebra over F. If I_i is a minimal left ideal of A_i, it is a minimal left ideal of A and hence gives an irreducible representation of G. Moreover, $I_i \not\cong I_j$ if $i \neq j$ and every irreducible representation of G is equivalent to one obtained from one of the I_i (Theorem 4.4, p. 208). Hence there are s equivalence classes of irreducible representations of G. We determine next the degree of the irreducible representations provided by I_i. Let $\{e_{ij} | 1 \leqslant i,j \leqslant n\}$ be the usual set of matrix units in $M_n(\Delta)$, Δ a division algebra. Then

(28) $$M_n(\Delta) = M_n(\Delta)e_{11} \oplus \cdots \oplus M_n(\Delta)e_{nn}$$

and $M_n(\Delta)e_{ii}$ is a minimal left ideal and every minimal left ideal of $M_n(\Delta)$ is isomorphic as $M_n(\Delta)$-left module to $M_n(\Delta)e_{ii}$. We have the subalgebra $\Delta 1$ of matrices $d1 = \text{diag}\{d, d, \ldots, d\}$ and we can regard $M_n(\Delta)$ as (left) vector space over Δ by defining $d(a)$ for the matrix (a) to be $(d1)(a)$. Then $M_n(\Delta)$ has the base $\{e_{ij}\}$ over Δ, so its dimensionality over Δ is n^2. Similarly, $M_n(\Delta)e_{ii} = \Delta e_{1i} \oplus \cdots \oplus \Delta e_{ni}$, so the dimensionality of $M_n(\Delta)e_{ii}$ over Δ is n. Then the dimensionality of $M_n(\Delta)e_{ii}$ over F is

(29) $$[M_n(\Delta)e_{ii} : \Delta][\Delta : F] = nd$$

where $d = [\Delta : F]$. It now follows that $[I_i : F] = n_i d_i$ where $d_i = [\Delta_i : F]$ and this is the degree of the irreducible representation provided by I_i. We can summarize these results as

THEOREM 5.4. *Let $A = F[G]$ be the group algebra of a finite group G over a field F such that* char $F \nmid |G|$ *and let $A = A_1 \oplus \cdots \oplus A_s$ where the A_i are the simple components of the semi-simple algebra A. Assume that $A_i \cong M_{n_i}(\Delta_i)$ where Δ_i is a division algebra and let $[\Delta_i : F] = d_i$. If I_i is a minimal left ideal of A_i, then I_i provides an irreducible representation of G and the irreducible representations obtained from the I_i for $1 \leqslant i \leqslant s$ form a set of representatives of the equivalence classes of irreducible representations of G. Moreover, the degree of the irreducible representation provided by I_i is $n_i d_i$.*

Let $\{\rho_i | 1 \leqslant i \leqslant s\}$ be a set of representatives of the equivalence classes of irreducible representations of G. As above, we may take ρ_i to be the irreducible representation provided by a minimal left ideal I_i of A_i. At any rate we may assume that ρ_i is equivalent to the irreducible representation determined by I_i.

If ρ is any representation of G acting on V, then $V = V_1 \oplus \cdots \oplus V_r$ where the V_j are $\rho(G)$-invariant and $\rho|V_j$ is irreducible. Then $\rho|V_j \cong \rho_i$ for some i and the number m_i of j such that $\rho|V_j \cong \rho_i$ for a fixed i is independent of the decomposition of V as $V_1 \oplus \cdots \oplus V_r$. As on p. 254, we can write $\rho \sim m_1\rho_1 \oplus m_2\rho_2 \oplus \cdots \oplus m_s\rho_s$ and call m_i the *multiplicity* of ρ_i in the representation ρ.

Now consider A as (left) A-module. Let ρ denote the representation of G determined by this module. The base G of A is stabilized by every $\rho(g): x \rightarrow gx$. It is clear that ρ is the regular representation of G (p. 248). The decompositions (26) and (28) show that the multiplicity of the irreducible representation ρ_i determined by I_i in the regular representation is n_i. We therefore have the following result on the regular representation of G.

THEOREM 5.5. *Let the notation be as in Theorem 5.4. Then the multiplicity of ρ_i in the regular representation is n_i.*

The number of equivalence classes of irreducible representations of G is the number s of simple components A_i of $A = F[G]$. Let cent A denote the center of A. Then

$$\text{cent}\, A = \text{cent}\, A_1 \oplus \cdots \oplus \text{cent}\, A_s \tag{30}$$

where the center, cent A_i, is isomorphic to the center of $M_{n_i}(\Delta_i)$ and hence to the center of Δ_i. Hence cent A_i is a field and is a simple component of the semi-simple commutative algebra cent A. Then s is the number of simple components of cent A. We now determine a base for this algebra.

PROPOSITION 5.1. *Let $G = C_1(=\{1\}) \cup C_2 \cup \cdots \cup C_r$ be the decomposition of G into conjugacy classes (BAI, p. 74). Put*

$$c_i = \sum_{g_i \in C_i} g_i. \tag{31}$$

Then $(c_1, c_2, \ldots, c_r)$ is a base for cent $F[G]$.

Proof. If $g \in G$, then $g^{-1}c_i g = \sum_{g_i \in C_i} g^{-1}g_i g = \sum g_i = c_i$ since the map $x \rightarrow g^{-1}xg$ permutes the elements of the conjugacy class C_i. Hence c_i commutes with every $g \in G$ and with every $\sum \alpha_g g \in F[G]$. Thus $c_i \in \text{cent}\, F[G]$. Now let $c = \sum \gamma_g g \in \text{cent}\, F[G]$. If $h \in G$,

$$h^{-1}ch = \sum_g \gamma_g h^{-1}gh = \sum_g \gamma_{hgh^{-1}} g$$

so the condition $h^{-1}ch = c$ gives $\gamma_{hgh^{-1}} = \gamma_g$, $h \in G$. Thus any two elements g

and g' in the same conjugacy class have the same coefficient in the expression $c = \sum \gamma_g g$ and hence c is a linear combination of the c_i. It is clear that the c_i are linearly independent and so $(c_1, c_2, \ldots, c_r)$ is a base for cent $F[G]$. $\square$

Both the foregoing result and the fact that the number s of equivalence classes of irreducible representations of G is the number of simple components show that $s \leqslant r$, the number of conjugacy classes. More precisely, we have $r = \sum_1^s [\text{cent } A_i : F]$. This shows that $s = r$ if and only if cent $A_i = F$ for all i, that is, the A_i are central simple. Since cent A_i is a finite dimensional field extension of F, we have cent $A_i = F$ for all i if F is algebraically closed. Hence in this important case the number of equivalence classes of irreducible representations is the number of conjugacy classes. In the algebraically closed case we have the following important result.

THEOREM 5.6. *Let G be a finite group, F an algebraically closed field such that* char $F \nmid |G|$. *Let s be the number of conjugacy classes of G. Then the number of equivalence classes of irreducible representations over F (acting on vector spaces V/F) is s and if* $\rho_1, \ldots, \rho_s$ *are representatives of these classes and* n_i *is the degree of* ρ_i, *then*

$$|G| = \sum_1^s n_i^2. \tag{32}$$

Proof. The first statement has been proved. To see the second, we use (26) and (27) and the fact that since F is algebraically closed, the only finite dimensional division algebra over F is F itself. Then $A = F[G] = A_1 \oplus \cdots \oplus A_s$ and $A_i \cong M_{n_i}(F)$. Then $[A:F] = |G| = \sum_1^s n_i^2$ and Theorem 5.4 shows that n_i is the degree of the irreducible representation ρ_i associated with A_i. $\square$

EXAMPLES

1. Let $G = \langle z \rangle$, the cyclic group of order n generated by z and let $F = \mathbb{C}$. Then $A = \mathbb{C}[G]$ is a direct sum of n copies of $\mathbb{C}$. Hence we have n inequivalent irreducible representations, all of degree 1. It is clear that in the corresponding matrix representations we have $z \rightsquigarrow (e^{2\pi i r/n})$ where ( ) is a 1×1 matrix and $r = 1, 2, \ldots, n$.

2. Let $G = D_n$, the dihedral group of order $2n$ generated by two elements r, s such that $r^n = 1$, $s^2 = 1$, $srs^{-1} = r^{-1}$ (BAI, pp. 34, 70). The elements of D_n are r^k, $r^k s$, $0 \leqslant k \leqslant n-1$, and we have the relation $sr^k = r^{-k}s$. Hence $(r^k s)^2 = 1$, so the n elements $r^k s$ are of period two. Using the multiplication table: $r^k r^l = r^{k+l}$, $r^k(r^l s) = r^{k+l}s$, $(r^l s)r^k = r^{l-k}s$, $(r^k s)(r^l s) = r^{k-l}$, it is readily seen that if n is odd $= 2v+1$, $v \geqslant 1$, then there are $v+2$ conjugacy classes with representatives: 1, r^k, $1 \leqslant k \leqslant v$, s, and if $n = 2v$, $v \geqslant 1$, then there are $v+3$ conjugacy classes with representatives 1, r^k, $1 \leqslant k \leqslant v$, s, rs.

On the other hand, we can list these numbers of inequivalent irreducible matrix representations over C as follows:

$n = 2v+1$, $v \geqslant 1$.
ρ_1, the unit representation.
ρ_2, the matrix representation of degree 1 such that $r \rightsquigarrow (1)$, $s \rightsquigarrow (-1)$.
σ_l, $1 \leqslant l \leqslant v$, the matrix representation of degree 2 such that

$$r \rightsquigarrow \begin{pmatrix} \omega^l & 0 \\ 0 & \omega^{-l} \end{pmatrix}, \qquad s \rightsquigarrow \begin{pmatrix} 0 & 1 \\ 1 & 0 \end{pmatrix}$$

where $\omega = e^{2\pi i/n}$.

$n = 2v$, $v \geqslant 1$.
ρ_1, the unit representation.
ρ_2, the matrix representation of degree 1 such that $r \rightsquigarrow (1)$, $s \rightsquigarrow (-1)$.
ρ_3, the matrix representation of degree 1 such that $r \rightsquigarrow (-1)$, $s \rightsquigarrow (1)$.
ρ_4, the matrix representation of degree 1 such that $r \rightsquigarrow (-1)$, $s \rightsquigarrow (-1)$.
The representations σ_l, $1 \leqslant l \leqslant v-1$, as above.

It is easy to verify that the representations listed are irreducible and inequivalent. Hence they constitute a set of representatives of the equivalence classes of irreducible representations of D_n. As a check, we can verify the degree relation (32) in the two cases:

$$n = 2v+1, \qquad 1+1+4(v) = 2(2v+1) = 2n.$$
$$n = 2v, \qquad 4(1)+4(v-1) = 4v = 2n.$$

We shall consider next a process of extension of the base field of a representation. We need to recall some simple facts about extension of the base field of a vector space. For our purpose, it suffices to restrict our attention to finite dimensional vector spaces. Thus let V/F be an n-dimensional vector space over the field F and let K be an extension field of F. We can form the tensor product $V_K = K \otimes_F V$, which can be regarded as a vector space over K. The injective map $v \rightsquigarrow 1 \otimes v$ permits us to regard V as contained in V_K as an F-subspace such that $KV = V_K$. Moreover, F-independent elements of V are K-independent and any base for V/F is a base for V_K/K (see p. 220). A linear transformation l of V/F has a unique extension to a linear transformation of V_K/K. We denote the extension by l also. These extensions span $\text{End}_K V_K$ as vector space over K, that is, $\text{End}_K V_K = K\,\text{End}_F V$. Moreover, $\text{End}_K V_K \cong K \otimes_F \text{End}_F V$.

Now suppose we have a representation ρ of G acting on V/F. Then the defining relations $\rho(g_1g_2) = \rho(g_1)\rho(g_2)$, $\rho(1) = 1_V$ give the relations $\rho(g_1g_2) = \rho(g_1)\rho(g_2)$, $\rho(1) = 1_{V_K}$ for the extensions. Hence $\rho_K : g \rightsquigarrow \rho(g)$ (in V_K) is a representation of G acting on V_K. We call this the *representation obtained from ρ by extending the base field to K*. It is clear that if $A = F[G]$, then $K[G] \cong A_K$ and $\rho_K(K[G]) = K\rho(A) \cong \rho(A)_K$. We also have the following useful result.

PROPOSITION 5.2. $\mathrm{End}_{K[G]} V_K = K\,\mathrm{End}_{F[G]} V \cong (\mathrm{End}_{F[G]} V)_K$.

Proof. The elements of $\mathrm{End}_{F[G]} V$ ($\mathrm{End}_{K[G]} V_K$) are the linear transformations of V (V_K) that commute with every $\rho(g)$, $g \in G$. Hence it is clear that $K\,\mathrm{End}_{F[G]} V \subset \mathrm{End}_{K[G]} V_K$. To prove the reverse containment, we choose a base (λ_i) for K/F. Then any element of $\mathrm{End}_K V_K$ can be written in one and only one way in the form $\sum \lambda_i l_i$ where $l_i \in \mathrm{End}_F V$. The conditions that this commutes with every $\rho(g)$ imply that the l_i commute with every $\rho(g)$, hence, that $l_i \in \mathrm{End}_{F[G]} V$. Thus $\sum \lambda_i l_i \in \mathrm{End}_{K[G]} V_K$ and $\mathrm{End}_{K[G]} V_K = K\,\mathrm{End}_{F[G]} V$. We have noted that $\mathrm{End}_K V_K \cong K \otimes \mathrm{End}_F V$. This implies that if E is any subspace of $\mathrm{End}_F V/F$, then $KE \cong K \otimes_F E$. In particular, we have $K\,\mathrm{End}_{F[G]} V \cong K \otimes_F \mathrm{End}_{F[G]} V = (\mathrm{End}_{F[G]} V)_K$. $\square$

It is clear that if U is a $\rho(G)$-invariant subspace of V, then $KU \cong U_K$ is a $\rho(G)$-invariant subspace of V_K. Hence if ρ_K is irreducible, then ρ is irreducible. The converse need not hold, as the following examples show.

EXAMPLES

1. Let $G = \langle g \rangle$, the cyclic group of order 4 and let $F = \mathbb{R}$. We have the representation of degree two over $\mathbb{R}$ in which $\rho(g)$ is the linear transformation with matrix

$$\begin{pmatrix} 0 & 1 \\ -1 & 0 \end{pmatrix}$$

relative to the base (u, v) of $V/\mathbb{R}$. Since the characteristic polynomial of this matrix is $\lambda^2 + 1$, $\rho(g)$ acts irreducibly on V. Now consider $\rho_{\mathbb{C}}$. We have the base ($z = u + iv$, $w = u - iv$) for $V_{\mathbb{C}}/\mathbb{C}$ and

$$\rho(g)_{\mathbb{C}} z = \rho(g)u + i\rho(g)v = -v + iu = iz$$
$$\rho(g)_{\mathbb{C}} w = \rho(g)u - i\rho(g)v = -v - iu = -iw.$$

Hence $\mathbb{C}z$ and $\mathbb{C}w$ are $\rho_{\mathbb{C}}(G)$-invariant subspaces. The irreducible representations $\rho_{\mathbb{C}} | \mathbb{C}z$ and $\rho_{\mathbb{C}} | \mathbb{C}w$ are inequivalent.

2. Let G be the quaternion group $\{\pm 1, \pm i, \pm j, \pm k\}$, which is a subgroup of the multiplicative group of Hamilton's quaternion algebra $\mathbb{H}$ over $\mathbb{R}$. Let ρ be the representation of G such that $\rho(g)$ is the left multiplication $x \rightsquigarrow gx$, $x \in \mathbb{H}$, $g \in G$. ρ is irreducible since $\mathbb{H}$ is a division algebra. On the other hand, $\mathbb{H}_{\mathbb{C}} \cong M_2(\mathbb{C})$ since $\mathbb{C}$ is a splitting field for $\mathbb{H}$ (p. 228). Since $M_2(\mathbb{C})$ is a direct sum of two minimal left ideals, it follows that $\mathbb{H}_{\mathbb{C}}$ is a direct sum of two $\rho_{\mathbb{C}}(G)$-invariant subspaces I_1 and I_2. The representations $\rho_{\mathbb{C}} | I_1$ and $\rho_{\mathbb{C}} | I_2$ are irreducible and equivalent.

A representation ρ is called *absolutely irreducible* if ρ_K is irreducible for every extension field K of the base field F. We have the following criterion for irreducibility and absolute irreducibility.

THEOREM 5.7. *Let G be a finite group, ρ a representation of G acting on V/F where* $\operatorname{char} F \nmid |G|$, $A = F[G]$. *Then ρ is irreducible if and only if $A' = \operatorname{End}_A V$ is a division algebra and ρ is absolutely irreducible if and only if $A' = F1$.*

Proof. If ρ is irreducible, then A' is a division algebra by Schur's lemma and if ρ is absolutely irreducible, then $A'_K \cong \operatorname{End}_{K[G]} V_K$ is a division algebra for every extension field K of F. This implies that $A' = F1$. For, suppose that $A' \neq F1$ and let $c \in A', \notin F1$. Then the minimum polynomial $f(\lambda)$ of c over F has degree >1 and this is irreducible since A' is a division algebra. Put $K = F[\lambda]/(f(\lambda))$ and consider A'_K. The minimum polynomial of $c \in A'_K$ is $f(\lambda)$ and this is reducible in $K[\lambda]$. Hence A'_K is not a division algebra, contrary to what we had proved. Hence $A' = F1$.

Next assume ρ is reducible, so we have a $\rho(G)$-invariant subspace $U \neq V, 0$. By Maschke's theorem, there exists a projection p on U that commutes with every $\rho(g)$. Then p is an idempotent $\neq 0, 1$ in A' and A' is not a division algebra. Thus if A' is a division algebra, then ρ is irreducible. Since $A' = F1$ implies $A'_K = K1$, it follows that if $A' = F1$, then ρ is absolutely irreducible. □

A field F is called a *splitting field* for the group G if every irreducible representation of G over F is absolutely irreducible. We have the following

THEOREM 5.8. *Let G be a finite group, F a field with* $\operatorname{char} F \nmid |G|$. *Then F is a splitting field for G if and only if $F[G]$ is a direct sum of matrix algebras $M_n(F)$.*

Proof. Let $A = F[G] = A_1 \oplus \cdots \oplus A_s$ where the A_i are the simple components of A and let I_i be a minimal left ideal of A_i. The representation ρ_i of G acting on I_i is irreducible and every irreducible representation of G over F is equivalent to one of the ρ_i. Hence F is a splitting field for G if and only if every ρ_i is absolutely irreducible. By Theorem 5.7, this is the case if and only if $\operatorname{End}_A I_i = F1$ for $1 \leqslant i \leqslant s$. Now $\operatorname{End}_A I_i = \Delta'_i$ is a division algebra and by Theorem 5.1, $\rho_i(A) = \operatorname{End}_{\Delta'_i} I_i \cong M_{n_i}(\Delta_i)$ where $\Delta_i = \Delta_i'^{\mathrm{op}}$ and n_i is the dimensionality of I_i as vector space over Δ'_i. On the other hand, since the A_j, $j \neq i$, annihilate I_i and A_i is simple, $\rho_i(A) \cong A_i$. Thus $A_i \cong M_{n_i}(\Delta_i)$. Now suppose F is a splitting field. Then $\Delta'_i = F$ and $A_i \cong M_{n_i}(F)$. Conversely, suppose $A_i \cong M_{n'_i}(F)$ for some n'_i. Then $M_{n'_i}(F) \cong M_{n_i}(\Delta_i)$. By the isomorphism theorem for simple artinian algebras, this implies that $n'_i = n_i$ and $\Delta'_i \cong F$. Then $\Delta_i = F1$, $\operatorname{End}_A I_i = \Delta'_i = F1$, and F is a splitting field. □

EXERCISES

1. Let G be finite, ρ an irreducible representation of G over F where char $F \nmid |G|$. Show that $\sum_{g\in G}\rho(g) = |G|1$ if ρ is the unit representation and $\sum_{g\in G}\rho(g) = 0$ otherwise.

2. Let G be a finite group, $F = \mathbb{Q}$ or $\mathbb{Z}/(p)$ where $p \nmid |G|$. Show that there exists a finite dimensional extension field of F that is a splitting field for G.

3. Let ρ_1 and ρ_2 be representations of G over an infinite field F and let K be an extension field. Show that if ρ_{1K} and ρ_{2K} are equivalent, then ρ_1 and ρ_2 are equivalent. Does this hold for F finite?

5.4 IRREDUCIBLE REPRESENTATIONS OF S_n

It is generally a difficult problem to determine the structure of the group algebra $F[G]$ for a given finite group G. As we saw in the previous section, if char $F \nmid |G|$, this amounts to determining a set of representatives for the equivalence classes of irreducible representations of G over F, or, as we shall now say more briefly, determining the irreducible representations of G over F. In this section, we give one important example for which this can be done: $G = S_n$. We shall determine a set of idempotents generating minimal left ideals of $F[S_n]$ (char $F \nmid n!$) that give the irreducible representations of S_n over F. These results are due to Frobenius and to A. Young; our exposition follows one due to J. von Neumann as presented in van der Waerden's *Algebra* vol. 2, p. 246.

We recall that the number of conjugacy classes of S_n is $p(n)$, the number of (unordered) partitions of n (BAI, p. 75). If $\{r_1, r_2, \dots, r_h\}$ is such a partition (that is, $r_i \geqslant 1$ and $\sum_1^h r_i = n$), then the permutations that are products of disjoint cycles of lengths $r_1, r_2, \dots, r_h$ form a conjugacy class and every conjugacy class is obtained in this way. For the partition $\alpha = \{r_1, r_2, \dots, r_h\}$ we assume $r_1 \geqslant r_2 \geqslant \cdots \geqslant r_h$ and we use this order to order the set of partitions lexicographically. Thus if $\beta = \{s_1, s_2, \dots, s_k\}$ with $s_1 \geqslant s_2 \geqslant \cdots \geqslant s_k$, then $\alpha > \beta$ if $r_i > s_i$ at the first place where $r_i \neq s_i$. Each partition $\alpha = \{r_1, r_2, \dots, r_h\}$ defines a (Young) *tableau*

(33)

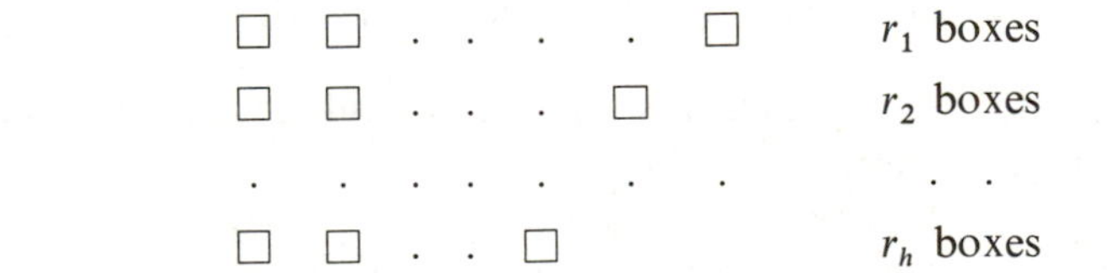

and each tableau defines a set of *diagrams* $D_\alpha, E_\alpha, \dots$ obtained by distributing the numbers $1, 2, \ldots, n$ in the boxes so that no two numbers appear in the same box. For example, if $\alpha = \{3, 2, 2\}$, then one of the D is

$$\begin{array}{ccc} \boxed{1} & \boxed{3} & \boxed{7} \\ \boxed{2} & \boxed{6} & \\ \boxed{4} & \boxed{5}. & \end{array} \tag{34}$$

If D $(=D)$ is a diagram, we define the *group* $R(D)$ of *row permutations* of D to be the subgroup of S_n of permutations stabilizing the subsets filled in the rows of D. Thus, for D as in (34), $R(D)$ is the set of products of cycles all of whose numbers appear in one of the rows $\{1,3,7\}$, $\{2,6\}$, or $\{4,5\}$ (so $R(D) = \{1, (13), (17), (37), (137), (173), (26), (45), (13), (26),$ etc.$\}$). Similarly, we define the subgroup $C(D)$ of column permutations of D to be the subgroup of S_n of permutations stabilizing the columns of D.

If D_α is a diagram and $\sigma \in S_n$, then σD_α is the diagram obtained from D_α by applying σ to its entries. Evidently this is an E_α and every E_α is a σD_α for some $\sigma \in S_n$. It is clear that $R(\sigma D_\alpha) = \sigma R(D_\alpha)\sigma^{-1}$ and $C(\sigma D_\alpha) = \sigma C(D_\alpha)\sigma^{-1}$.

With each diagram D $(=D_\alpha)$ we shall associate elements of the group algebra $F[S_n]$ as follows: Put

$$\begin{aligned} S_D &= \sum_{\sigma \in R(D)} \sigma, \qquad A_D = \sum_{\tau \in C(D)} (\operatorname{sg} \tau)\tau, \\ F_D &= A_D S_D = \sum_{\substack{\sigma \in R(D) \\ \tau \in C(D)}} (\operatorname{sg} \tau)\tau\sigma \end{aligned} \tag{35}$$

where $\operatorname{sg} \tau$ denotes the sign of the permutation τ. Evidently, $S_D \neq 0$ and $A_D \neq 0$ in $F[S_n]$. But also $F_D \neq 0$. To see this, we observe that $R(D) \cap C(D) = 1$, since an element common to these subgroups stabilizes every row and every column and hence fixes every element in $\{1, 2, \ldots, n\}$. It now follows that if $\sigma_1, \sigma_2 \in R(D)$ and $\tau_1, \tau_2 \in C(D)$, then $\tau_1\sigma_1 = \tau_2\sigma_2$ implies $\sigma_2\sigma_1^{-1} = \tau_2^{-1}\tau_1$ and $\sigma_1 = \sigma_2$ and $\tau_1 = \tau_2$. Thus the products $\tau\sigma$ appearing in F_D are distinct and hence $F_D \neq 0$.

Since $R(\rho D) = \rho R(D)\rho^{-1}$ and $C(\rho D) = \rho C(D)\rho^{-1}$ for $\rho \in S_n$ and $\operatorname{sg} \rho\tau\rho^{-1} = \operatorname{sg} \tau$, it follows that $S_{\rho D} = \rho S_D \rho^{-1}$, $A_{\rho D} = \rho A_D \rho^{-1}$, and $F_{\rho D} = \rho F_D \rho^{-1}$. Also if $\sigma \in R(D)$ and $\tau \in C(D)$, then the definition (35) gives

$$\sigma S_D = S_D = S_D \sigma, \qquad \tau A_D = (\operatorname{sg} \tau) A_D = A_D \tau. \tag{36}$$

The main properties of the elements S_D, A_D, F_D will be derived from the following combinatorial result.

LEMMA 1. *Let α and β be partitions such that $\alpha \geqslant \beta$ and let D_α and E_β be associated diagrams. Suppose no two numbers appearing in the same row in D_α*

are in the same column of E_β. *Then* $\alpha = \beta$ *and* $E_\alpha = \sigma\tau D_\alpha$ *for some* $\sigma \in R(D_\alpha)$ *and* $\tau \in C(D_\alpha)$.

Proof. The number of entries in the first row of D_α is the same as or exceeds that in the first row of E_β. If greater, then since the number of columns of E_β is the number of entries in the first row of E_β, two entries of the first row of D_α occur in the same column in E_β, contrary to hypothesis. Hence both diagrams have the same number of entries in the first row. Also we have a column permutation τ_1' of E_β so that the first row of $\tau_1' E_\beta$ has the same entries as the first row of D_α. Next we note that the entries in the second row of D_α occur in distinct columns of $\tau_1' E_\beta$ and in rows after the first. It follows that D_α and $\tau_1' E_\beta$ and hence D_α and E_β have the same number of entries in the second row and a column permutation τ_2' of $\tau_1' E_\beta$ (and of E_β) brings these into the second row. Continuing in this way, we see that $\beta = \alpha$ and there exists a $\tau' \in C(E_\alpha)$ such that the entries of each row of $\tau' E_\alpha$ and of D_α are the same. Hence there is a $\sigma \in R(D_\alpha)$ such that $\sigma D_\alpha = \tau' E_\alpha$. Now $\tau' \in C(E_\alpha) = C(\tau' E_\alpha) = C(\sigma D_\alpha) = \sigma C(D_\alpha)\sigma^{-1}$. Hence $\tau' = \sigma\tau^{-1}\sigma^{-1}$, $\tau^{-1} \in C(D_\alpha)$ and $\sigma\tau^{-1}\sigma^{-1}E_\alpha = \sigma D_\alpha$. Then $E_\alpha = \sigma\tau D_\alpha$. □

Now assume $\alpha > \beta$. Then Lemma 1 implies that there exist i, j, $i \neq j$, in a row of D_α and in a column of E_β. If $\pi = (ij)$ then $\pi \in R(D_\alpha)$. Then, by (36), $\pi S_{D_\alpha} = S_{D_\alpha} = S_{D_\alpha}\pi$ and $\pi A_{E_\beta} = -A_{E_\beta}\pi$. Hence

$$\begin{aligned} S_{D_\alpha}A_{E_\beta} &= (S_{D_\alpha}\pi)A_{E_\beta} = S_{D_\alpha}(\pi A_{E_\beta}) = -S_{D_\alpha}A_{E_\beta} \\ A_{E_\beta}S_{D_\alpha} &= A_{E_\beta}(\pi S_{D_\alpha}) = (A_{E_\beta}\pi)S_{D_\alpha} = -A_{E_\beta}S_{D_\alpha}. \end{aligned}$$

Hence $S_{D_\alpha}A_{E_\beta} = 0 = A_{E_\beta}S_{D_\alpha}$. If ρ is any element of S_n, then $S_{\rho D_\alpha} = \rho S_{D_\alpha}\rho^{-1}$ and since $S_{\rho D_\alpha}A_{E_\beta} = 0 = A_{E_\beta}S_{\rho D_\alpha}$ we have $S_{D_\alpha}\rho^{-1}A_{E_\beta} = 0$ and $A_{E_\beta}\rho S_{D_\alpha} = 0$. Thus

$$S_{D_\alpha}F[S_n]A_{E_\beta} = 0 = A_{E_\beta}F[S_n]S_{D_\alpha} \quad \text{if} \quad \alpha > \beta. \tag{37}$$

LEMMA 2. *An element* $a \in F[S_n]$ *satisfies the equations* $\tau a \sigma = (\operatorname{sg}\tau)a$ *for all* $\sigma \in R(D)$ *and* $\tau \in C(D)$ *if and only if* $a = \gamma F_D$, $\gamma \in F$, $F_D = A_D S_D$.

Proof. We have $F_D\sigma = A_D S_D \sigma = A_D S_D = F_D$ and $\tau F_D = \tau A_D S_D = (\operatorname{sg}\tau)A_D S_D = (\operatorname{sg}\tau)F_D$. Hence any γF_D, $\gamma \in F$, satisfies the conditions. Next let $a = \sum_{\rho \in S_n}\gamma_\rho \rho$ satisfy the given conditions. Then

$$\gamma_\rho = (\operatorname{sg}\tau)\gamma_{\tau\rho\sigma} \tag{38}$$

for $\sigma \in R(D)$ and $\tau \in C(D)$. In particular, $\gamma_{\tau\sigma} = \gamma_1 \operatorname{sg}\tau$, so if we can show that

$\gamma_\rho = 0$ if ρ is not of the form $\tau\sigma$, $\tau \in C(D)$, $\sigma \in R(D)$, then we shall have $a = \gamma_1 F_D$ by (35). Hence suppose $\rho \neq \tau\sigma$ for $\tau \in C(D)$, $\sigma \in R(D)$. Then $\rho^{-1} \neq \sigma\tau$, $\sigma \in R(D)$, $\tau \in C(D)$. Then $\rho^{-1}D \neq \sigma\tau D$ and Lemma 1 implies that there exists a transposition $\pi \in R(D)$, $\pi \in C(\rho^{-1}D) = \rho^{-1}C(D)\rho$. Then $\pi = \rho^{-1}\pi'\rho$ where π' is a transposition contained in $C(D)$. Then $\rho = \pi'\rho\pi$ and $\gamma_{\pi'\rho\pi} = \gamma_\rho = -\gamma_{\pi'\rho\pi}$ by (38). Hence $\gamma_\rho = 0$ and the proof is complete. □

Now let $x \in F[S_n]$ and consider the element $F_D x F_D = A_D S_D x A_D S_D$. By (36) we have $\tau F_D x F_D \sigma = \operatorname{sg} \tau F_D x F_D$ for $\sigma \in R_D$, $\tau \in C_D$. Hence, by Lemma 2, $F_D x F_D = \gamma F_D$ for $\gamma \in F$. In particular, $F_D{}^2 = \gamma F_D$. We proceed to show that $\gamma \neq 0$. For this purpose, we consider the map $x \rightsquigarrow F_D x$ of $F[S_n]$ into itself. If $\gamma = 0$, $F_D{}^2 = 0$ and the map $x \rightsquigarrow F_D x$ is nilpotent and hence has trace 0. On the other hand, if $\rho \in S_n$, then if we look at the matrix of $x \rightsquigarrow \rho x$ relative to a base, we see that the trace of this map is 0 if $\rho \neq 1$ and is $n!$ if $\rho = 1$. Since the formula (35) for F_D shows that the coefficient of 1 in the expression for F_D is 1, the trace of $x \rightsquigarrow F_D x$ is $n! \neq 0$ (since char $F \nmid n!$).

We now put $e_D = \gamma^{-1} F_D$. Then $e_D{}^2 = e_D \neq 0$ and $e_D F[S_n] e_D = F e_D$. Also, if α and β are distinct partitions, then $e_{D_\alpha} F[S_n] e_{E_\beta} = 0$ follows from (37) if D_α is a diagram associated with α and E_β is one associated with β.

We recall that if e and f are idempotents of a ring A, then the additive groups $\hom_A(Ae, Af)$ and eAf are isomorphic and the rings $\operatorname{End}_A Ae$ and eAe are anti-isomorphic (p. 180). If we apply this and Theorem 5.7 to the representations of G acting on $F[S_n]e_{D_\alpha}$, we see that this representation is absolutely irreducible and that if $\alpha \neq \beta$, then the representations provided by $F[S_n]e_{D_\alpha}$ and $F[S_n]e_{E_\beta}$ are inequivalent. Since the number of conjugacy classes of S_n is $p(n)$, we obtain in this way a full set of representatives of the equivalence classes of irreducible representations. In terms of the group algebra $F[S_n]$ the result we have proved is

THEOREM 5.9. *If F is a field of characteristic* 0 *or of prime characteristic exceeding n, then*

$$F[S_n] \cong M_{n_1}(F) \oplus M_{n_2}(F) \oplus \cdots \oplus M_{n_{p(n)}}(F).$$

The method of proof is constructive and in theory it can be used to carry out the decomposition of $F[S_n]$ into simple components. The determination of the n_i, which are the degrees of the irreducible representations, can be made by calculating the characters of S_n as defined in the next section for an arbitrary group. There is an extensive literature on the characters of S_n. We shall not consider any of this here. Evidently, Theorem 5.9 has the following consequence.

COROLLARY. *The field* $\mathbb{Q}$ *and any field* $\mathbb{Z}/(p)$ *with* $p > n$ *is a splitting field for* S_n.

5.5 CHARACTERS. ORTHOGONALITY RELATIONS

DEFINITION 5.2. *If* ρ *is a representation of a group* G *acting on a vector space* V/F, *then the* F*-valued function on* G *defined by*

$$\chi_\rho : g \rightsquigarrow \operatorname{tr} \rho(g),$$

where $\operatorname{tr} \rho(g)$ *is the trace of the linear transformation* $\rho(g)$, *is called the* character of G afforded by ρ. *If* ρ *is irreducible, then* χ_ρ *is called* irreducible *and if* $F = \mathbb{C}$, *then* χ_ρ *is called a* complex *character. The degree of* ρ *is called the* degree *of* χ_ρ.

As we shall see in a moment, two representations of G over $\mathbb{C}$ are equivalent if and only if they have the same character. Moreover, a great deal of experience has shown that the characters encapsulate precisely the information on the representations that is useable for the applications. For these reasons, it is fair to say that the central problem of representation theory is that of determining the complex irreducible characters of a given group, or, more precisely, of developing methods for this purpose.

We begin by listing some simple facts about characters.

1. Equivalent representations have the same character. If ρ_1 on V_1 is equivalent to ρ_2 on V_2, then there exists a bijective linear map η of V_1 onto V_2 such that $\rho_2(g) = \eta\rho_1(g)\eta^{-1}$. This implies that $\operatorname{tr} \rho_2(g) = \operatorname{tr} \rho_1(g)$ and $\chi_{\rho_2} = \chi_{\rho_1}$.

2. Any character is a *class function*, that is, it is constant on every conjugacy class and hence it defines a map of the set of conjugacy classes into the base field. Let $g, h \in G$. Then $\operatorname{tr} \rho(hgh^{-1}) = \operatorname{tr} \rho(h)\rho(g)\rho(h)^{-1} = \operatorname{tr} \rho(g)$. Hence $\chi_\rho(hgh^{-1}) = \chi_\rho(g)$.

3. If $\operatorname{char} F = 0$, then the degree of ρ is $\chi_\rho(1)$. This is clear since $\rho(1) = 1_V$, so $\chi_\rho(1) = \operatorname{tr} \rho(1) = \dim V$.

4. Let U be a $\rho(G)$-invariant subspace of the space V on which ρ acts and let $\rho|U$ and $\rho|V/U$ be the corresponding subrepresentation and factor representation. Then

$$\chi_\rho(g) = \chi_{\rho|U}(g) + \chi_{\rho|V/U}(g).$$

This follows by choosing a base $(u_1, \ldots, u_n)$ for V such that $(u_1, \ldots, u_r)$ is a base for U (and hence $(u_{r+1} + U, \ldots, u_n + U)$ is a base for V/U). Then the matrices of the $\rho(g)$ all have the reduced form (18). The result follows from this.

5. If ρ_1 and ρ_2 are representations of G, then

$$\chi_{\rho_1 \otimes \rho_2} = \chi_{\rho_1}\chi_{\rho_2}$$

that is, for every $g \in G$, $\chi_{\rho_1 \otimes \rho_2}(g) = \chi_{\rho_1}(g)\chi_{\rho_2}(g)$. To see this, we refer to the matrix (12) for the linear transformation $a_1 \otimes a_2$ in $V_1 \otimes V_2$ where a_i is a linear transformation in V_i. It is clear from this that $\operatorname{tr}(a_1 \otimes a_2) = (\operatorname{tr} a_1)(\operatorname{tr} a_2)$. Hence $\chi_{\rho_1 \otimes \rho_2} = \chi_{\rho_1}\chi_{\rho_2}$.

For the applications we are interested primarily in complex characters. We shall now note some of their properties.

6. If G is finite, then any complex character of G is a sum of mth roots of unity where m is the *exponent* of G, defined to be the least common multiple of the orders of the elements of G. If $g \in G$, then $g^m = 1$ and hence $\rho(g)^m = 1$. Hence the minimum polynomial of $\rho(g)$ is a factor of $\lambda^m - 1$, and so it has distinct roots that are mth roots of unity. It follows that $\rho(g)$ has a matrix of the form

$$\operatorname{diag}\{\omega_1, \omega_2, \ldots, \omega_n\} \tag{39}$$

where the ω_i are mth roots of unity. Then $\chi_\rho(g) = \sum \omega_i$.

There are several useful consequences of this result. First we have the following

PROPOSITION 5.3. *Let ρ be a complex representation of degree n of a finite group G. Then for any $g \in G$*

$$|\chi_\rho(g)| \leqslant n = \deg \chi_\rho \tag{40}$$

and

$$|\chi_\rho(g)| = n \tag{41}$$

if and only if $\rho(g) = \omega 1$ where ω is an mth root of unity, m the exponent of G. In particular, if

$$\chi_\rho(g) = n \tag{42}$$

then $\rho(g) = 1$.

Proof. We have $\chi_\rho(g) = \sum_1^n \omega_i$, ω_i an mth root of unity. Then $|\chi_\rho(g)| = |\sum_1^n \omega_i| \leqslant \sum_1^n |\omega_i| = n$. Moreover, equality holds if and only if all the ω_i are on the same ray through the origin. Since they are on the unit circle, this holds if and only if they are equal. Hence $|\chi_\rho(g)| = n$ if and only if $\rho(g) = \omega 1$. The last statement is an immediate consequence of this. □

The fact that $\chi_\rho(g) = n$ implies $\rho(g) = 1$, and the obvious converse of this leads us to define the *kernel of the character* χ_ρ as the set of g such that $\chi_\rho(g) = n$ (the degree of ρ). Then $\ker \chi_\rho = \ker \rho$ is a normal subgroup of G. Also we define

$$Z(\chi_\rho) = \{g \in G \mid |\chi_\rho(g)| = n\}.$$

Then we have shown that $Z(\chi_\rho)$ is the set of g such that $\rho(g) = \omega 1$. It is clear that these form a normal subgroup of G containing $\ker \chi_\rho$ and $Z(\chi_\rho)/\ker \rho$ is isomorphic to a subgroup of the multiplicative group of mth roots of unity. Hence $Z(\chi_\rho)/\ker \rho$ is cyclic.

Another important consequence of the proof of property 6 is

7. Let ρ be a complex representation of a finite group G, ρ^* the contragredient representation. Then

$$\chi_{\rho^*} = \bar{\chi}_\rho \tag{43}$$

(that is, $\chi_{\rho^*}(g) = \overline{\chi_\rho(g)}$, $g \in G$). To see this, we suppose $g \in G$ and we choose a base B in V such that the matrix of $\rho(g)$ relative to this base is (39). Then the matrix of $\rho^*(g)$ relative to the dual base B^* (p. 250) is ${}^t(\mathrm{diag}\{\omega_1,\ldots,\omega_n\})^{-1} = \mathrm{diag}\{\omega_1^{-1},\ldots,\omega_n^{-1}\} = \mathrm{diag}\{\bar{\omega}_1,\ldots,\bar{\omega}_n\}$. Hence $\chi_{\rho^*}(g) = \sum\bar{\omega}_i = \overline{\chi_\rho(g)}$.

EXAMPLES

1. Let 1 be the unit representation: V is one-dimensional and $1(g) = 1_V$, $g \in G$. The character afforded by this is the *unit* character $\chi_1 : g \rightsquigarrow 1 \in F$.

2. Let G be finite, ρ the regular representation of G. To determine χ_ρ we use the base $G = \{g_1 = 1, g_2, \ldots, g_n\}$ for $F[G]$. We have $\chi_\rho(1) = n$. On the other hand, if $i > 1$, then all of the diagonal elements of the matrix of $\rho(g_i)$ relative to the base G are 0 since $g_i g_j \neq g_j$. Hence, $\chi_\rho(g_i) = 0$. Thus the character of the regular representation is given by

$$\chi_\rho(1) = |G|1, \qquad \chi_\rho(g) = 0 \quad \text{if} \quad g \neq 1. \tag{44}$$

3. Let $G = D_n$, the dihedral group of order $2n$ generated by r, s such that $r^n = 1$, $s^2 = 1$, $srs^{-1} = r^{-1}$. If we refer to the results given in example 2, p. 261, we obtain the following character tables:

$$n = 2v+1, \qquad v \geqslant 1$$

	1	s	r^k
χ_1	1	1	1
χ_{ρ_2}	1	-1	1
χ_{σ_l}	2	0	$2\cos 2kl\pi/n$

$$n = 2v, \qquad v \geqslant 2$$

	1	s	rs	r^k
χ_1	1	1	1	1
χ_{ρ_2}	1	-1	-1	1
χ_{ρ_3}	1	1	-1	$(-1)^k$
χ_{ρ_4}	1	-1	1	$(-1)^k$
χ_{σ_l}	2	0	0	$2\cos 2kl\pi/n$

In these tables, the representatives of the conjugacy classes are listed in the top row and the rows correspond to the irreducible representations given before. In both cases, $1 \leqslant k \leqslant v$, and $1 \leqslant l \leqslant v$ if $n = 2v+1$, $1 \leqslant l \leqslant v-1$ if $n = 2v$.

We shall derive next some fundamental orthogonality relations connecting the irreducible complex characters of a finite group. We consider first a more general result in the situation in which G is a finite group, F a splitting field for G with char $F \nmid |G|$. Let $\{\rho_1, \ldots, \rho_s\}$ be a set of representatives of the (absolutely) irreducible representations of G over F and suppose ρ_i acts on V_i, $1 \leqslant i \leqslant s$. We assume also that ρ_1 is the unit representation. Let $1 \leqslant i,j \leqslant s$ and consider $\hom_F(V_i, V_j)$. We have a representation ρ_{ij} of G acting on $\hom_F(V_i, V_j)$ obtained by defining

$$\rho_{ij}(g)l = \rho_j(g)l\rho_i(g)^{-1} \tag{45}$$

for $l \in \hom_F(V_i, V_j)$. It is clear that this gives a representation of G (exercise 2, p. 251). Now let l be any element of $\hom_F(V_i, V_j)$ and form the element

$$\eta(l) = \sum_g \rho_{ij}(g)l = \sum_g \rho_j(g)l\rho_i(g)^{-1}. \tag{46}$$

We have

$$\rho_j(h)\eta(l) = \sum_g \rho_j(hg)l\rho_i(g)^{-1} = \sum_g \rho_j(g)l\rho_i(h^{-1}g)^{-1}$$
$$= \sum_g \rho_j(g)l\rho_i(g^{-1}h) = \eta(l)\rho_i(h).$$

This implies that $\eta(l)$ is a homomorphism of V_i regarded as $F[G]$-module into V_j regarded as $F[G]$-module, that is, $\eta(l) \in \hom_{F[G]}(V_i, V_j)$. Now if $i \neq j$, then V_i and V_j are irreducible and non-isomorphic $F[G]$-modules. Hence by Schur's lemma,

$$\sum_g \rho_j(g)l\rho_i(g)^{-1} = 0 \tag{47}$$

if $i \neq j$ and l is any element of $\hom_F(V_i, V_j)$. Next suppose $i = j$. Then $\eta(l) \in \operatorname{End}_{F[G]} V_i = F1$, so we have

(48) $$\sum_g \rho_i(g) l \rho_i(g)^{-1} \in F1$$

for any $l \in \operatorname{End}_F V_i$. We can use (47) and (48) to derive the following result, which is due to Schur.

THEOREM 5.10. *Let G be a finite group, F a splitting field for G with* $\operatorname{char} F \nmid |G|$. *Let $\{\rho_1, \ldots, \rho_s\}$ be a set of representatives of the equivalence classes of irreducible representations of G over F and for each i let $\rho^{(i)}$ be a matrix representation given by ρ_i. Then* $\operatorname{char} F \nmid \deg \rho_i$ *and we have the following relations*:

(49) $$\sum_g \rho_{kl}^{(j)}(g) \rho_{rt}^{(i)}(g^{-1}) = 0 \quad \text{if} \quad i \neq j.$$
$$\sum_g \rho_{kl}^{(i)}(g) \rho_{rt}^{(i)}(g^{-1}) = \delta_{kt} \delta_{lr} |G| / \deg \rho_i.$$

$(\rho^{(i)}(g) = (\rho_{rt}^{(i)}(g)))$.

These relations are called the *Schur relations.*

Proof. Let $(u_1^{(i)}, \ldots, u_{r_i}^{(i)})$ be a base for the space V_i on which ρ_i acts so that $\rho_i(g) u_t^{(i)} = \sum_r \rho_{rt}^{(i)}(g) u_r^{(i)}$ and let f_{lr} be the element of $\hom_F(V_i, V_j)$ such that $f_{lr} u_t^{(i)} = \delta_{rt} u_l^{(j)}$. Then $\rho_j(g) f_{lr} \rho_i(g^{-1}) u_t^{(i)} = \sum_k \rho_{rt}^{(i)}(g^{-1}) \rho_{kl}^{(j)}(g) u_k^{(j)}$. Then (47) implies the first set of relations (49). Also for $j = i$, the foregoing relations and (48) give

$$\sum_g \sum_k \rho_{rt}^{(i)}(g^{-1}) \rho_{kl}^{(i)}(g) u_k^{(i)} = \lambda_{lr} u_t^{(i)}, \qquad \lambda_{lr} \in F.$$

Hence

(50) $$\sum_g \rho_{rt}^{(i)}(g^{-1}) \rho_{kl}^{(i)}(g) = \delta_{kt} \lambda_{lr}.$$

Put $t = k$ in these equations and sum on k. This gives

$$(\deg \rho_i) \lambda_{lr} = \delta_{rl} |G|$$

which shows that $\operatorname{char} F \nmid \deg \rho_i$ and $\lambda_{lr} = \delta_{rl} |G| / \deg \rho_i$. Substituting this in (50) gives the second set of Schur's relations. □

We have $\chi_{\rho_i}(g) = \sum_k \rho_{kk}^{(i)}(g)$. Hence if we put $l = k$ and $t = r$ in (49) and sum on k and r, we obtain

$$\sum_g \chi_{\rho_j}(g)\chi_{\rho_i}(g^{-1}) = 0 \quad \text{if} \quad i \neq j$$

(51)

$$\sum_g \chi_{\rho_i}(g)\chi_{\rho_i}(g^{-1}) = |G|.$$

Now suppose $F = \mathbb{C}$. Then it is clear from the fact that $\rho_i(g)$ has a matrix of the form $\operatorname{diag}\{\omega_1, \ldots, \omega_{n_i}\}$ where the ω's are roots of unity (see property 6 above) that $\chi_{\rho_i}(g^{-1}) = \overline{\chi_{\rho_i}(g)}$. Hence we obtain from (51) the basic *orthogonality relations* for irreducible complex characters:

$$\sum_g \chi_{\rho_i}(g)\overline{\chi_{\rho_j}(g)} = 0 \quad \text{if} \quad i \neq j$$

(52)

$$\sum_g |\chi_{\rho_i}(g)|^2 = |G|.$$

We now consider the complex vector space $\mathbb{C}^G$ of complex valued functions on G. We have the usual definitions of addition and multiplication by complex numbers: If $\varphi, \psi \in \mathbb{C}^G$, then $(\varphi + \psi)(g) = \varphi(g) + \psi(g)$ and $(a\varphi)(g) = a\varphi(g)$ for $a \in \mathbb{C}$. We define a hermitian form on $\mathbb{C}^G$ by

$$(53) \qquad (\varphi|\psi) = \frac{1}{|G|}\sum_g \overline{\varphi(g)}\psi(g).$$

Then

$$(\varphi|\varphi) = \frac{1}{|G|}\sum |\varphi(g)|^2 \geqslant 0$$

and equality holds if and only if $\varphi = 0$. Hence $(\varphi|\psi)$ is positive definite. We shall now write χ_i for χ_{ρ_i}. Then the relations (52) are equivalent to

$$(54) \qquad (\chi_i|\chi_j) = \delta_{ij}.$$

These state that $\{\chi_1, \ldots, \chi_s\}$ is an orthonormal set of vectors in $\mathbb{C}^G$, that is, they are mutually orthogonal and all have length one. It is clear from this that the irreducible complex characters are linearly independent over $\mathbb{C}$. We have observed that characters are class functions. The set of class functions forms a subspace cf (G) of $\mathbb{C}^G$ whose dimensionality is the number of conjugacy classes. We have seen that this number is the same as the number s of irreducible representations over $\mathbb{C}$. Hence, it is clear that the irreducible characters constitute a base for cf (G).

Now let ρ be an arbitrary complex representation of G, χ its character. We have defined the multiplicity m_i of ρ_i in ρ on p. 255. It is clear from the definition and from the fact that the representations are completely reducible that two representations are equivalent if and only if for every $i = 1, 2, \ldots, s$, the multiplicity of ρ_i in the two representations is the same. It is clear also that we have the formula

(55) $$\chi = m_1\chi_1 + m_2\chi_2 + \cdots + m_s\chi_s$$

where m_i is the multiplicity of ρ_i in ρ. Now, by (54),

(56) $$m_i = (\chi_i|\chi).$$

Hence the m_i are determined by χ and consequently complex representations are equivalent if and only if they have the same character. Also, by (55), we have

(57) $$(\chi|\chi) = \sum m_i^2$$

and this has the value 1 if and only if one of the m_i is 1 and the rest are 0. This shows that a character is irreducible if and only if it has length 1 in the hermitian metric for $\mathbb{C}^G$.

Let $C_1 = \{1\}, C_2, \ldots, C_s$ be the conjugacy classes of G and let $h_k = |C_k|$. We now write χ_{ik} for $\chi_i(g)$, $g \in C_k$. Then the orthogonality relations give

$$\begin{aligned} |G|\delta_{ij} &= \sum_g \overline{\chi_i(g)}\chi_j(g) \\ &= \sum_{k=1}^{s} \sum_{g \in C_k} \overline{\chi_i(g)}\chi_j(g) \\ &= \sum_{k=1}^{s} \bar{\chi}_{ik} h_k \chi_{jk}. \end{aligned}$$

Thus we have

(58) $$\sum_{k=1}^{s} \bar{\chi}_{ik} h_k \chi_{jk} = |G|\delta_{ij}.$$

Let $X = (\chi_{ij}) \in M_s(\mathbb{C})$ and let $H = \operatorname{diag}\{h_1, \ldots, h_s\}$. Then the foregoing relations amount to the matrix equation

(59) $$\bar{X}H{}^tX = |G|1.$$

From this, one deduces ${}^tX\bar{X} = |G|H^{-1}$, which is equivalent to

(60) $$\sum_{i=1}^{s} \chi_{ij}\bar{\chi}_{ik} = \frac{|G|}{h_k}\delta_{jk}.$$

We shall call (58) and (60) the *row* and the *column orthogonality relations* respectively for the characters. We remark that if we take $j = k = 1$ in (60) we obtain $\sum n_i^s = |G|$ for $n_i = \chi_i(1) = \deg \chi_i$. This is the relation (32) that we had obtained in a more general situation by using the theory of algebras.

If ρ is a representation of a group G acting on V/F, then ρ has a unique extension to a homomorphism of $F[G]$ into $\mathrm{End}_F V$. This maps the element $\sum_g \gamma_g g$ of the group algebra into $\sum \gamma_g \rho(g)$. We shall denote this extension by ρ also. Similarly, the character χ_ρ defined on G can be extended uniquely to a linear map χ_ρ of $F[G]$ into F by putting $\chi_\rho(\sum \gamma_g g) = \mathrm{tr}\,\rho(\sum \gamma_g g) = \sum \gamma_g \,\mathrm{tr}\,\rho(g) = \sum \gamma_g \chi_\rho(g)$. If $c \in \mathrm{cent}\, F[G]$, the center of the group algebra, then $\rho(c)$ is in the centralizer of ρ, so if ρ is absolutely irreducible, then $\rho(c) = \omega(c)1$ where $\omega(c) \in F$. If the degree of ρ is n, then $\chi_\rho(c) = n\omega(c)$ so

$$\rho(c) = \frac{1}{n}\chi_\rho(c)1. \tag{61}$$

Since the map $c \rightsquigarrow \rho(c)$ is an algebra homomorphism of $\mathrm{cent}\, F[G]$ into $F1$, it is clear that $c \rightsquigarrow (1/n)\chi_\rho(c)$ is an algebra homomorphism.

We have seen that if we put $c_j = \sum_{g \in C_j} g$, then $(c_1, \ldots, c_s)$ is a base for $\mathrm{cent}\, F[G]$. Evidently, $c_j c_k = \sum_g n_{jkg} g$ where the $n_{jkg} \in \mathbb{N}$, the set of non-negative integers. Since $c_j c_k \in \mathrm{cent}\, F[G]$, we have $h^{-1} c_j c_k h = c_j c_k$, $h \in G$. This gives $n_{jkg} = n_{jk(hgh^{-1})}$, $g, h \in G$. It follows that we have a multiplication table of the form

$$c_j c_k = \sum_{l=1}^{s} m_{jkl} c_l, \qquad 1 \leqslant j, k \leqslant s \tag{62}$$

where $m_{jkl} \in \mathbb{N}$ for the base $(c_1, \ldots, c_s)$ of $\mathrm{cent}\, F[G]$ over F. It is readily seen that m_{jkl} is the number of pairs (x, y), $x \in C_j$, $y \in C_k$, such that xy is a given z in C_l.

Now suppose $F = \mathbb{C}$ and let the notation be as before. We apply the representation ρ_i to (62) to obtain $\rho_i(c_j)\rho_i(c_k) = \sum m_{jkl} \rho_i(c_l)$. If $n_i = \deg \rho_i = \chi_i(1)$, then we have $\rho_i(c_j) = (\chi_i(c_j)/n_i)1$. Hence we have the character relation

$$(\chi_i(c_j)/n_i)\,(\chi_i(c_k)/n_i) = \sum m_{jkl} \chi_i(c_l)/n_i. \tag{63}$$

If we use the definition of $c_j = \sum_{g \in C_j} g$ and the fact that characters are class functions, we obtain $\chi_i(c_j) = h_j \chi_i(g_j)$ for any $g_j \in C_j$, so $\chi_i(c_j) = h_j \chi_{ij}$. Hence we have the character relation

$$(h_j \chi_{ij}/n_i)\,(h_k \chi_{ik}/n_i) = \sum_{l=1}^{s} m_{jkl}(h_l \chi_{il}/n_i) \tag{64}$$

or

$$(64') \qquad (h_j\chi_{ij})(h_k\chi_{ik}) = \sum_{l=1}^{s} m_{jkl} h_l n_i \chi_{il}.$$

By a *character table* for a group G, we mean a table giving the values χ_{ij} for the irreducible complex characters. A fundamental problem for the study of a given group G is the computation of its character table. As an illustration of this, we consider the following

EXAMPLE

We wish to determine a character table for S_4. The following is a list of representatives for the conjugacy classes: (1, (12), (123), (1234), (12) (34)). We denote the corresponding conjugacy classes as $(C_1, C_2, C_3, C_4, C_5)$. Their cardinalities are respectively $(h_1, h_2, h_3, h_4, h_5) = (1, 6, 8, 6, 3)$. The alternating group A_4 is a normal subgroup of index 2 in S_4, and $V = \{1, (12)(34), (13)(24), (14)(23)\}$ is a normal subgroup of index 6 in S_4 (BAI, p. 261, exercise 4). If we let S_3 denote the subgroup of S_4 fixing 4, then S_4 is a semi-direct product of V and S_3: Any element of S_4 can be written in one and only one way as a product sv where $s \in S_3$, $v \in V$. If $s_i \in S_3$ and $v_i \in V$, then $(s_1v_1)(s_2v_2) = (s_1s_2)(s_2^{-1}v_1s_2v_2)$ and $s_2^{-1}v_1s_2 \in V$. Hence $sv \rightsquigarrow s$ is a homomorphism η of S_4 onto S_3 with kernel V. If ρ is a representation of S_3, then $\rho\eta$ is a representation of S_4 whose character is $\chi_\rho\eta$.

Now $S_3 \cong D_3$ under a map such that $(12) \rightsquigarrow s$, $(123) \rightsquigarrow r$. Hence the character table for D_3 gives the following character table for S_3:

	1	(12)	(123)
χ_1	1	1	1
χ_2	1	-1	1
χ_3	2	0	-1

We denote the characters obtained by composing this with η again by χ_1, χ_2, χ_3. Since (12) (13) (24) = (1324), the part of the character table obtained from these characters is

	1	(12)	(123)	(1234)	(12) (34)
χ_1	1	1	1	1	1
χ_2	1	-1	1	-1	1
χ_3	2	0	-1	0	2

If n_i is the degree of χ_i, $1 \leqslant i \leqslant 5$, then $\sum n_i^2 = 24$ gives $n_4^2 + n_5^2 = 18$. Hence $n_4 = n_5 = 3$, so the missing two irreducible characters are of degree three. Hence the last two rows of the character table have the form $(3, \alpha, \beta, \gamma, \delta)$ and $(3, \alpha', \beta', \gamma', \delta')$. If we use the relation (60) with $j = 1$ and $k = 2, 3, 4, 5$, we obtain

$$\alpha + \alpha' = 0, \qquad \beta + \beta' = 0, \qquad \gamma + \gamma' = 0, \qquad \delta + \delta' = -2.$$

Hence the last row is $(3, -\alpha, -\beta, -\gamma, -2-\delta)$. If we use (58) with $i = 4$ and $j = 1,2,3$, we obtain the relations

$$\begin{aligned} 3+6\alpha+8\beta+6\gamma+3\delta &= 0 \\ 3-6\alpha+8\beta-6\gamma+3\delta &= 0 \\ 6 \qquad -8\beta \qquad +6\delta &= 0. \end{aligned}$$

These equations give $\beta = 0$, $\delta = -1$, $\gamma = -\alpha$. The orthogonality relation (60) for $j = 2$, $k = 4$ gives $\alpha\bar{\gamma} = -1$. On the other hand, $\alpha = \chi_4(12)$ is a sum of square roots of 1, hence real. Thus $\alpha\bar{\gamma} = \alpha(-\bar{\alpha}) = -\alpha^2 = -1$ and $\alpha^2 = 1$, so $\alpha = \pm 1$, $\gamma = \mp 1$. Thus the last two rows are either $(3,1,0,-1,-1)$ and $(3,-1,0,1,-1)$ or $(3,-1,0,1,-1)$ and $(3,1,0,-1,-1)$. Both determinations give the same table except for the order of the last two rows.

The foregoing example illustrates how the orthogonality relations plus other information that one can get a hold of can be used to calculate a character table. Further results that we shall derive presently will supply additional information useful for calculating character tables. We should note also that there is a substantial literature on the characters of S_n, beginning with a classical paper published by Frobenius in 1900 that gives formulas for the characters of any S_n. See, for example, Weyl's *Classical Groups*, p. 213.

EXERCISES

1. Determine a character table for A_4.

2. Determine a character table for the quaternion group.

3. Let G be a subgroup of S_n, ρ the corresponding permutation representation of G over $\mathbb{C}$, χ its character. Show that $\sum_{g\in G}\chi(g) = r|G|$ where r is the multiplicity of the unit representation ρ_1 in ρ. Show that $\sum_{g\in G}\chi(g)$ is also the total number of fixed points in $\{1,2,\ldots,n\}$ for all $g \in G$.

4. Let the notations be as in exercise 3. Show that the number of orbits of G in $\{1,2,\ldots,n\}$ is the multiplicity r of ρ_1 in ρ.

5. A permutation group G is called *k-fold transitive* for $k = 1,2,\ldots$ if given any two k-tuples $(i_1,i_2,\ldots,i_k)$ and $(j_1,j_2,\ldots,j_k)$ of distinct i's and j's, there exists a $g \in G$ such that $gi_l = j_l$, $1 \leqslant l \leqslant k$. Let G be doubly (=2-fold) transitive, ρ the corresponding permutation representation. Show that $\rho \sim \rho_1 \oplus \rho_i$ where ρ_i is an irreducible representation $\neq \rho_1$.

6. Let $\chi_1,\ldots,\chi_s$ be the irreducible complex characters of G and $C_1 = \{1\}$, $C_2,\ldots,C_s$ be conjugacy classes of G, $h_i = |C_i|$. Put $c_i = \sum_{g\in C_i} g$ and $e_i = \sum_j (\chi_i(1)/|G|)\bar{\chi}_{ij}c_j$ where $\chi_{ij} = \chi_i(g)$ for some $g \in C_j$. Show that the e_i are orthogonal idempotents in the center of $\mathbb{C}[G]$ and $\sum e_i = 1$.

7. Show that an element g in a finite group G is conjugate to its inverse g^{-1} if and only if $\chi(g) \in \mathbb{R}$ for every irreducible complex character χ.

5.6 DIRECT PRODUCTS OF GROUPS. CHARACTERS OF ABELIAN GROUPS

It is easy to see that if G_1 and G_2 are finite groups, then $F[G_1 \times G_2] \cong F[G_1] \otimes_F F[G_2]$. If char $F \nmid |G_i|$ and F is a splitting field for G_1 and G_2, this can be used to reduce the study of the representations over F of $G_1 \times G_2$ to that of the components G_i. In the most important case, in which $F = \mathbb{C}$, we can obtain the results also by using characters. We shall follow this approach.

Let ρ_i, $i = 1, 2$, be a representation of G_i acting on V_i over any F. If we compose the projection $(g_1, g_2) \rightsquigarrow g_1$ with ρ_1, we obtain a representation ρ'_1 of $G_1 \times G_2$. Evidently $G_2 \subset \ker \rho'_1$. Similarly, we obtain a representation ρ'_2 of $G_1 \times G_2$ by composing the projection $(g_1, g_2) \rightsquigarrow g_2$ with the representation ρ_2 of G_2. We now form $\rho'_1 \otimes \rho'_2$, which we denote as $\rho_1 \# \rho_2$. Then

$$(\rho_1 \# \rho_2)(g_1, g_2) = \rho_1(g_1) \otimes \rho_2(g_2) \tag{65}$$

and hence

$$\chi_{\rho_1 \# \rho_2}(g_1, g_2) = \chi_{\rho_1}(g_1)\chi_{\rho_2}(g_2). \tag{66}$$

We have the canonical imbeddings $g_1 \rightsquigarrow (g_1, 1_2)$ and $g_2 \rightsquigarrow (1_1, g_2)$ of G_1 and G_2 in $G_1 \times G_2$. Then

$$\chi_{\rho_1 \# \rho_2}(g_1, 1_2) = \chi_{\rho_1}(g_1)\chi_{\rho_2}(1_2) = (\deg \rho_2)\chi_{\rho_1}(g_1). \tag{67}$$

Similarly,

$$\chi_{\rho_1 \# \rho_2}(1_1, g_2) = (\deg \rho_1)\chi_{\rho_2}(g_2). \tag{68}$$

All of this has an immediate extension to direct products of more than two factors.

Now suppose the G_i are finite, $F = \mathbb{C}$, and ρ_i is an irreducible representation of G_i. Then we have

$$(\chi_{\rho_1 \# \rho_2} | \chi_{\rho_1 \# \rho_2}) = \frac{1}{|G_1 \times G_2|} \sum_{(g_1, g_2)} |\chi_{\rho_1 \# \rho_2}(g_1, g_2)|^2$$

$$= \frac{1}{|G_1||G_2|} \sum_{(g_1,g_2)} |\chi_{\rho_1}(g_1)\chi_{\rho_2}(g_2)|^2$$
$$= \frac{1}{|G_1||G_2|} \sum_{(g_1,g_2)} |\chi_{\rho_1}(g_1)|^2|\chi_{\rho_2}(g_2)|^2$$
$$= \frac{1}{|G_1|}\left(\sum_{g_1} |\chi_{\rho_1}(g_1)|^2\right)\frac{1}{|G_2|}\left(\sum_{g_2} |\chi_{\rho_2}(g_2)|^2\right)$$
$$= (\chi_{\rho_1}|\chi_{\rho_1})(\chi_{\rho_2}|\chi_{\rho_2}) = 1.$$

Hence $\rho_1 \# \rho_2$ is an irreducible representation of $G_1 \times G_2$. Next suppose the irreducible characters of G_i are $\chi_1{}^{(i)}, \ldots, \chi_{s_i}{}^{(i)}$ and let $\rho_k{}^{(i)}$ be a representation affording $\chi_k{}^{(i)}$. The representations $\rho_k{}^{(1)} \# \rho_l{}^{(2)}$ of $G_1 \times G_2$ are irreducible and (67) and (68) imply that $\rho_k{}^{(1)} \# \rho_l{}^{(2)}$ and $\rho_{k'}{}^{(1)} \# \rho_{l'}{}^{(2)}$ have distinct characters and hence are inequivalent if $(k', l') \neq (k, l)$. Hence we obtain in this way $s_1 s_2$ inequivalent irreducible representations. Since the degree of $\rho_k{}^{(1)} \# \rho_l{}^{(2)}$ is the product of the degree of $\rho_k{}^{(1)}$ and the degree of $\rho_l{}^{(2)}$, the sum of the squares of the degrees of the $\rho_k{}^{(1)} \# \rho_l{}^{(2)}$ is the product of the sum of the squares of the degrees of the irreducible representations of G_1 and the sum of the squares of the degrees of the irreducible representations of G_2. This is $|G_1||G_2| = |G_1 \times G_2|$. It follows that the set of representations $\{\rho_k{}^{(1)} \# \rho_l{}^{(2)}\}$ is a set of representatives of the equivalence classes of the irreducible representations of $G_1 \times G_2$. This proves

THEOREM 5.11. *Let G_1 and G_2 be finite groups, $\{\rho_1{}^{(i)}, \ldots, \rho_{s_i}{}^{(i)}\}$ a set of representatives of the equivalence classes of irreducible representations over $\mathbb{C}$ of G_i. Then every $\rho_k{}^{(1)} \# \rho_l{}^{(2)}$ is an irreducible representation of $G_1 \times G_2$ and $\{\rho_k{}^{(1)} \# \rho_l{}^{(2)}\}$ is a set of representatives of the equivalence classes of irreducible representations of $G_1 \times G_2$.*

A character χ of degree one is called *linear*. Evidently, such a character is irreducible and may be identified with the representation affording it. Thus χ is a homomorphism of the given group into the multiplicative group of a field. Conversely, any homomorphism of a group G into the multiplicative group F^* of a field F is a linear character of G. We recall that these characters have played an important role in the Galois theory of fields (BAI, p. 291). If χ and χ' are linear characters of G, then $\chi_1\chi_2$ defined by $(\chi\chi')(g) = \chi(g)\chi'(g)$ is a linear character. In this way the set of linear characters of G form a group with the unit character: $g \rightsquigarrow 1$ as unit and the inverse of χ as $g \rightsquigarrow \chi(g)^{-1}$.

If G is an abelian group, every irreducible complex character of G is linear. For, if ρ is an irreducible representation of G acting on V over $\mathbb{C}$, then $\mathrm{End}_{\mathbb{C}[G]}V = \mathbb{C}1$ and since G is abelian, $\mathrm{Env}\,\rho \subset \mathrm{End}_{\mathbb{C}[G]}V$ so $\mathrm{Env}\,\rho = \mathbb{C}1$.

Then any subspace is $\rho(G)$-invariant and since ρ is irreducible, V is one-dimensional. Hence χ_ρ is linear. Since G is abelian, every conjugacy class consists of a single element; hence the number of these is $|G|$. Then G has $|G|$ irreducible complex characters.

The last result can also be seen without using representation theory. In fact, we can easily determine the structure of the group of irreducible (= linear) complex characters of any finite abelian group. If G is any abelian group, the group of irreducible complex characters is called the *character group* of G. Now let G be finite abelian. Then G is a direct product of cyclic groups (BAI, p. 195): $G = G_1 \times G_2 \times \cdots \times G_r$ where $G_i = \langle g_i \rangle$ and g_i has order e_i. We may assume the G_i are subgroups and every element of G can be written in one and only one way as $g = g_1^{k_1} g_2^{k_2} \cdots g_r^{k_r}$ where $0 \leqslant k_i < e_i$. Let $\chi \in \hat{G}$, the character group of G. Then $\chi(g_i)^{e_i} = \chi(g_i^{e_i}) = \chi(1) = 1$ so $\chi(g_i)$ is an e_ith root of unity. The set of these is a cyclic subgroup Z_{e_i} of the multiplicative group $\mathbb{C}^*$. We now define a map of $\hat{G}$ into $Z_{e_1} \times Z_{e_2} \times \cdots \times Z_{e_r}$ by

(69) $$\chi \rightsquigarrow (\chi(g_1), \ldots, \chi(g_r)).$$

It is clear that this is a homomorphism. The kernel is the set of χ such that $\chi(g_i) = 1$, $1 \leqslant i \leqslant r$. Then $\chi(g_1^{k_1} \cdots g_r^{k_r}) = \chi(g_1)^{k_1} \cdots \chi(g_r)^{k_r} = 1$ and $\chi = 1$. Hence (69) is a monomorphism. Moreover, the map is surjective. For, if $(u_1, \ldots, u_r) \in Z_{e_1} \times \cdots \times Z_{e_r}$, then $u_i^{e_i} = 1$ and hence we have a homomorphism of $\langle g_i \rangle$ into Z_{e_i} sending $g_i \to u_i$. Then we have a homomorphism χ of $G = G_1 \times \cdots \times G_r$ into $\mathbb{C}^*$ sending $g_i \to u_i$, $1 \leqslant i \leqslant r$. Evidently, χ is a character that is mapped into $(u_1, \ldots, u_r)$ by (69). Hence we have an isomorphism of G with its character group $\hat{G}$.

EXERCISES

1. Show that the number of complex linear characters of a finite group is the index $[G:G']$, G' the commutator group of G.

2. Let G be a finite group, ρ the regular representation of G over the field F. Form the field $F(x_{g_1}, \ldots, x_{g_m})$ in m-indeterminates where $m = |G|$ and $g_i \rightsquigarrow x_{g_i}$ is 1–1. The determinant

(70) $$\det \sum_{g \in G} x_g \rho(g)$$

is called the *group determinant* of G over F. (The study of such determinants was one of the chief motivations for Frobenius' introduction of the theory of characters.) Show that if $F = \mathbb{C}$ and G is abelian, then

$$\det\left(\sum_{g\in G} x_g\rho(g)\right) = \prod_{\chi\in\hat{G}}\left(\sum_{g\in G} x_g\chi(g)\right). \tag{71}$$

This result is due to R. Dedekind.

3. Let **Fab** denote the category of finite abelian groups with homomorphisms as morphisms. We have a contravariant functor D from **Fab** to itself such that $A \rightsquigarrow \hat{A}$ and if $f: A \to B$, then $\hat{f}: \hat{B} \to \hat{A}$ is defined by $\psi \rightsquigarrow \chi$ where $\chi(x) = \psi(f(x))$, $x \in A$. Show that $D^2: A \rightsquigarrow \hat{\hat{A}}, f \rightsquigarrow f$ is naturally equivalent to the identity functor (cf. p. 22).

4. Show that if H is a subgroup of a finite abelian group G, then the subgroup of $\hat{G}$ of χ such that $\chi_H = 1_H$ can be identified with $\widehat{G/H}$ and hence its order is $|G/H|$. Use this to show that the map $\chi \rightsquigarrow \chi_H$ of $\hat{G}$ into $\hat{H}$ is surjective. Note that this implies that any linear character of H can be extended to a linear character of G.

5.7 SOME ARITHMETICAL CONSIDERATIONS

In this section, G will be finite and all representations and characters are complex. We shall apply some elementary results on integral complex numbers that were given in BAI, pp. 279–281, to obtain important results on the degrees of the irreducible representations and on the characters of S_n.

We recall that $a \in \mathbb{C}$ is called algebraic if a is algebraic over the subfield $\mathbb{Q}$, that is, a is a root of a non-zero polynomial with coefficients in $\mathbb{Q}$. The complex number a is called integral (or an integer) if a is a root of a monic polynomial with integer coefficients. A useful criterion to prove integrality is that a is integral if and only if there exists a finitely generated $\mathbb{Z}$-submodule M of $\mathbb{C}$ such that $1 \in M$ and $aM \subset M$. The subset A of $\mathbb{C}$ of algebraic numbers is a subfield and the subset I of integral complex numbers is a subring. Moreover, if $a \in \mathbb{C}$ is a root of a monic polynomial in $A[\lambda]$ (in $I[\lambda]$), then $a \in A$ (I). We showed also that a is integral if and only if a is algebraic and its minimum polynomial over $\mathbb{Q}$ has integer coefficients, and that the only rational numbers that are integral in $\mathbb{C}$ are the elements of $\mathbb{Z}$.

We have seen that the characters of G are sums of roots of unity (property 6 on p. 270). Since roots of unity are integral, it follows that $\chi(g)$ is integral for any character χ and any $g \in G$. We recall the following notations that were introduced in section 5.5.

1. $\chi_1, \dots, \chi_s$ are the irreducible characters. χ_1 is the unit character, that is, the character of the unit representation ρ_1.
2. $C_1 = \{1\}, C_2, \dots, C_s$ are the conjugacy classes.

3. $\chi_{ij} = \chi_i(g_j)$ where $g_j \in C_j$.
4. $n_i = \chi_{i1}$. This is the degree of the representation ρ_i such that $\chi_{\rho_i} = \chi_i$.
5. $h_j = |C_j|$.

We recall also the following character relations (p. 276):

$$(h_j \chi_{ij}/n_i)(h_k \chi_{ik}/n_i) = \sum_{l=1}^{s} m_{jkl}(h_l \chi_{il}/n_i) \tag{64}$$

where the $m_{jkl} \in \mathbb{N}$. We shall deduce from these equations

PROPOSITION 5.4. *The complex numbers $h_j \chi_{ij}/n_i$ are integral.*

Proof. Fix i and put $u_k = h_k \chi_{ik}/n_i$, $1 \leqslant k \leqslant s$, $u_0 = 1$. Let $M = \sum_0^s \mathbb{Z} u_j$. Then (64) shows that $u_k M \subset M$. Hence u_k is integral by the criterion we noted above. □

We can now prove

THEOREM 5.12. $n_i | \, |G|$.

Proof. We use the formula (58) for $i = j$ to obtain $|G|/n_i = \sum_k \bar{\chi}_{ik} \chi_{ik} h_k/n_i$. Since the $\bar{\chi}_{ik}$ and $\chi_{ik} h_k/n_i$ are integral, so is $|G|/n_i$. Since this is rational, it is contained in $\mathbb{Z}$. Hence $n_i | \, |G|$. □

The foregoing result is due to Frobenius. We also have the theorem that every $n_i = 1$ if G is abelian. Both of these results are special cases of the following more general theorem, which is due to Schur.

THEOREM 5.13. $n_i | [G:Z]$, *Z the center of G.*

Proof (Tate). Let m be a positive integer and let $G_m = G \times G \times \cdots \times G$, m times. Let ρ_i be an irreducible representation of G over $\mathbb{C}$ affording the character χ_i, V the vector space on which ρ_i acts. We have the representation ρ of G_m acting in $V_m = V \otimes \cdots \otimes V$, m times, such that $\rho(g_1, \ldots, g_m) = \rho_i(g_1) \otimes \cdots \otimes \rho_i(g_m)$ (see section 5.6). By iterating the result of Theorem 5.11, we see that ρ is irreducible. If $c \in Z$, the irreducibility of ρ_i implies that $\rho_i(c) = \gamma_i(c) 1_V$, $\gamma_i(c) \in \mathbb{C}$. Evidently γ_i is a homomorphism of Z into $\mathbb{C}^*$. By the definition of ρ we have

$$\rho(1 \times \cdots \times 1 \times c \times 1 \times \cdots \times 1) = \gamma_i(c) 1_{V_m}$$

and hence for $c_j \in Z$,

$$\rho(c_1 \times \cdots \times c_m) = \gamma_i(c_1) \cdots \gamma_i(c_m) 1_{V_m} = \gamma_i(c_1 \cdots c_m) 1_{V_m}.$$

It follows that the subset D of elements of G_m of the form $(c_1, c_2, \ldots, c_m)$, $c_i \in Z$, $\prod_1^m c_i = 1$ is in the kernel of ρ. Evidently D is a normal subgroup of G_m and ρ defines a representation $\bar{\rho}$ of G_m/D that is irreducible. We now apply Theorem 5.12 to $\bar{\rho}$ to obtain $n_i^m \mid |G_m/D| = |G|^m/|Z|^{m-1}$. This implies that $(|G|/n_i|Z|)^m \in \mathbb{Z}|Z|^{-1}$ and this holds for all m. Put $u = |G|/n_i|Z|$. Then the relation shows that the $\mathbb{Z}$-submodule $M = \sum_{k=0}^{\infty} \mathbb{Z}u^k$ of $\mathbb{Q}$ is contained in $\mathbb{Z}|Z|^{-1}$. Since $\mathbb{Z}$ is noetherian, so is $\mathbb{Z}|Z|^{-1}$ and hence so is M. Thus M is finitely generated as $\mathbb{Z}$-module. Since $1 \in M$ and $uM \subset M$, it follows that $u \in \mathbb{Z}$. This proves that $(n_i|Z|) \mid |G|$. Then $n_i \mid [G:Z]$ as required. □

Let F be a subfield of $\mathbb{C}$. Then $I \cap F$ is a subring of F called the *ring of algebraic integers of* F. The study of the arithmetic of such rings constitutes the theory of algebraic numbers. We shall give an introduction to this theory in Chapter 10. Now suppose F is a splitting field for the finite group G. Then $F[G] = M_{n_1}(F) \oplus \cdots \oplus M_{n_s}(F)$ and s is the number of conjugacy classes of G (see p. 280). We have s inequivalent irreducible representations ρ_i over F and these remain irreducible on extension of F to $\mathbb{C}$. Thus $\{\rho_{1\mathbb{C}}, \ldots, \rho_{s\mathbb{C}}\}$ is a set of representatives of the classes of the irreducible complex representations. If ρ_i acts on V_i/F, then to compute $\chi_{\rho_i}(g)$, $g \in G$, we choose a base for V_i/F and take the trace of the matrix of $\chi_{\rho_i}(g)$ relative to this base. Nothing is changed if we pass to $\mathbb{C}$. Thus we see that $\chi_{\rho_{i\mathbb{C}}}(g) = \chi_{\rho_i}(g)$. This shows that the irreducible complex characters have values in F. We know also that these values are contained in I. Since any character is an integral combination of irreducible characters, this gives the following

THEOREM 5.14. *Let F be a subfield of $\mathbb{C}$ that is a splitting field for the finite group G. Then any complex character of G has values that are integral algebraic numbers of F.*

In section 4, we showed that $\mathbb{Q}$ is a splitting field for S_n. Hence we have the

COROLLARY. *The complex characters of S_n have values in $\mathbb{Z}$.*

5.8 BURNSIDE'S $p^a q^b$ THEOREM

One of the earliest applications of the theory of characters was to the proof of the following beautiful theorem due to Burnside.

THEOREM 5.15. *If p and q are primes, then any group of order $p^a q^b$ is solvable.*

Quite recently, John Thompson succeeded in giving a proof of this theorem that does not use representation theory. However, this is considerably more complicated than the original proof with characters, so the original proof—which we shall give here—remains a good illustration of the use of representation theory to obtain results on the structure of finite groups.

We prove first the following

LEMMA. *Let χ be an irreducible complex character of a finite group G, ρ a representation affording χ. Suppose C is a conjugacy class of G such that $(|C|, \chi(1)) = 1$. Then for any $g \in C$, either $\chi(g) = 0$ or $\rho(g) \in \mathbb{C}1$.*

Proof. Since $(|C|, \chi(1)) = 1$, there exist integers l and m such that $l|C| + m\chi(1) = 1$. Then

$$\frac{\chi(g)}{\chi(1)} = l|C|\frac{\chi(g)}{\chi(1)} + m\chi(g). \tag{72}$$

Now $\chi(g)$ is an algebraic integer, and by Proposition 5.4 (p. 283), $|C|\chi(g)/\chi(1) = |C|\chi(g)/n$, where $n = \deg \rho = \chi(1)$, is an algebraic integer. Hence $\chi(g)/n$ is an algebraic integer. We recall also that $|\chi(g)| \leqslant n$ and this was proved by showing that $\chi(g)$ is a sum of n roots of unity (p. 270). Thus $\chi(g) = \omega_1 + \cdots + \omega_n$ where the $\omega_i \in W$, a cyclotomic field of complex roots of unity (BAI, p. 252). Let $H = \mathrm{Gal}\, W/\mathbb{Q}$ and let $s \in H$. Then s maps roots of unity into roots of unity. Hence $s\chi(g)$ is a sum of n roots of unity. Hence $|s\chi(g)| \leqslant n$ and $|s\chi(g)/\chi(1)| \leqslant 1$. It is clear that since $a = \chi(g)/\chi(1)$ is an algebraic integer, so is $sa = s\chi(g)/\chi(1)$ and $|sa| \leqslant 1$. Then the norm $N_{W/\mathbb{Q}}(a) = \prod_{s \in H} sa$ satisfies $|N_{W/\mathbb{Q}}(a)| \leqslant 1$. Since this is an algebraic integer and a rational number, it follows that $|N_{W/\mathbb{Q}}(a)| \in \mathbb{Z}^+$. Hence either $N_{W/\mathbb{Q}}(a) = 0$, in which case $a = 0$ and $\chi(g) = 0$, or $|N_{W/\mathbb{Q}}(a)| = 1$. In the latter case, $|a| = 1$, $|\chi(g)| = n$, and Proposition 5.3 shows that $\rho(g) = \omega 1$, ω a root of unity. □

We recall that the only abelian simple groups are the cyclic groups of prime order. Since this class of simple groups is rather trivial, one generally excludes it from the study of simple groups. We follow this convention in the following

THEOREM 5.16. *Let G be a finite (non-abelian) simple group. Then no conjugacy class of G has cardinality of the form p^a, p a prime, $a > 0$.*

Proof. Suppose C is a conjugacy class of G such that $|C| = p^a$, p prime, $a > 0$. Let $\rho_1, \ldots, \rho_s$ be the irreducible representations of G, $\chi_1, \ldots, \chi_s$ the corresponding characters. We assume ρ_1 is the unit representation, so $\chi_1(g) = 1$ for all g.

Let $n_i = \chi_i(1)$, the degree of ρ_i. If $p \nmid n_i$, then the foregoing lemma shows that either $\chi_i(g) = 0$ or $\rho_i(g) \in \mathbb{C}1$ for every $g \in C$. Now the elements $g \in G$ such that $\rho_i(g) \in \mathbb{C}1$ form a normal subgroup G_i of G. Hence if $\chi_i(g) \neq 0$ for some $g \in C$, then $G_i \neq 1$. Then $G_i = G$ since G is simple. Also since G is simple and ρ_i is not the unit representation for $i > 1$, it follows that $G \cong \rho_i(G)$ if $i > 1$. Since $\rho_i(G)$ is abelian, this is excluded. Hence if $p \nmid n_i$ and $i > 1$, then $\chi_i(g) = 0$ for all $g \in C$. We now use the orthogonality relation (60), which gives

$$\sum_1^s \chi_i(g)\overline{\chi_i(1)} = \sum_i n_i \chi_i(g) = 0 \tag{73}$$

for $g \in C$. Since $n_1 = 1$ and $\chi_1(g) = 1$, we have some n_i, $i > 1$, divisible by p. Let $n_2, \ldots, n_t$ be the n_i, $i > 1$, divisible by p. Then (73) gives the relation

$$1 + \sum_2^t n_j \chi_j(g) = 0, \qquad g \in C.$$

Since the n_j are divisible by p and the $\chi_j(g)$ are algebraic integers, this implies that $1/p$ is an algebraic integer, contrary to the fact that the only rational numbers that are algebraic integers are the elements of $\mathbb{Z}$. □

We can now give the

Proof of Theorem 5.15. Let $|G| = p^a q^b$ where p and q are primes. Let P be a Sylow p-subgroup of G. Since P is of prime power order, its center $Z \neq 1$. If $z \neq 1$ is in Z, then the centralizer $C(z)$ of z contains P, so $[G:C(z)]$ is a power of q. Now $[G:C(z)]$ is the cardinality of the conjugacy class C containing z (BAI, p. 75). Hence if $[G:C(z)] > 1$, then G is not simple non-abelian by Theorem 5.16. On the other hand, if $[G:C(z)] = 1$, then z is in the center of G. Then the center of G is not trivial, so again G is not simple non-abelian. It is now clear that unless G is cyclic of prime order, it contains a normal subgroup $H \neq 1, \neq G$. Using induction on $|G|$ we can conclude that H and G/H are solvable. Then G is solvable (BAI, p. 247). □

5.9 INDUCED MODULES

Let G be a group, H a subgroup of finite index in G, σ a representation of H acting on the vector space U/F. There is an important process, introduced by Frobenius, for "extending" σ to a representation σ^G of G. We recall that we have an action of G on the set G/H of left cosets of H in G. If $G/H = \{H_1 = H, H_2, \ldots, H_r\}$ and $g \in G$, then $gH_i = H_{\pi(g)i}$ and $\pi(g)$ is a permutation of $\{1, \ldots, r\}$. The map $\pi: g \to \pi(g)$ is a homomorphism of G into the symmetric group S_r (BAI, p. 72). Now put $U^G \equiv U^{(r)}$, the direct sum of r copies of U. Let $\{s_1, s_2, \ldots, s_r\}$ be a set of representatives of the left cosets of H, say, $H_i = s_i H$, $1 \leqslant i \leqslant r$. From now on we shall call such a set a *(left) cross section of G relative to H*. If $g \in G$, then

$$gs_i = s_{\pi(g)i}\mu_i(g) \tag{74}$$

where $\mu_i(g) \in H$. We can define an action of G on $U^G = U^{(r)}$ by

$$g(u_1, \ldots, u_r) = (\sigma(\mu_{\pi(g)^{-1}1}(g))u_{\pi(g)^{-1}1}, \ldots, \sigma(\mu_{\pi(g)^{-1}r}(g))u_{\pi(g)^{-1}r}) \tag{75}$$

where the $u_j \in U$. Using the fact that π is a homomorphism and that H acts on U, we can verify that (75) does indeed define an action of G on U^G. Since this action is by linear transformations in a finite dimensional vector space, we have a representation σ^G of G acting on U^G. We can verify also that another choice of cross section of G relative to H defines an equivalent representation. All of this will become apparent without calculations by giving an alternative, conceptual definition of σ^G, as we shall now do.

Let $B = F[H]$, the group algebra of H, and $A = F[G]$, the group algebra of G. To be given a representation σ of H amounts to being given a module U for B that is finite dimensional as vector space over F. Now B is a subalgebra of A and hence A is a B-B-bimodule in which the actions of B on A are left and right multiplications by elements of B. We can form $A \otimes_B U$, which is a left A-module in which for $a_1, a_2 \in A$ and $u \in U$ we have

$$a_1(a_2 \otimes u) = a_1 a_2 \otimes u. \tag{76}$$

We shall now show that $[A \otimes_B U : F] < \infty$, so the $A = F[G]$-module $A \otimes_B U$ defines a representation of G. Moreover, we shall see that this representation is equivalent to σ^G as defined above.

As before, let $\{s_1, s_2, \ldots, s_r\}$ be a cross section of G with respect to H. Then any element of G can be written in one and only one way in the form $s_i h$, $h \in H$. Since the elements of G form a base for A, it follows that A regarded as right B-module is free with $(s_1, \ldots, s_r)$ as base. It follows that every element of $A \otimes_B U$ can be written in one and only one way in the form

$$s_1 \otimes u_1 + s_2 \otimes u_2 + \cdots + s_r \otimes u_r, \tag{77}$$

$u_i \in U$. Hence if $(u^{(1)}, \ldots, u^{(n)})$ is a base for U/F, then

$$(s_1 \otimes u^{(1)}, \ldots, s_1 \otimes u^{(n)}; \cdots ; s_r \otimes u^{(1)}, \ldots, s_r \otimes u^{(n)}) \tag{78}$$

is a base for $A \otimes_B U$, so $[A \otimes_B U : F] = rn < \infty$. Thus the module $A \otimes_B U$ defines a representation of G.

Now let $g \in G$. Then, by (76) and (74), we have

$$
\begin{aligned}
g(\sum s_i \otimes u_i) &= \sum g s_i \otimes u_i \\
&= \sum s_{\pi(g)i} \mu_i(g) \otimes u_i \\
&= \sum s_{\pi(g)i} \otimes \mu_i(g) u_i \\
&= \sum s_i \otimes \mu_{\pi(g)^{-1}i}(g) u_{\pi(g)^{-1}i} \\
&= \sum s_i \otimes \sigma(\mu_{\pi(g)^{-1}i}(g)) u_{\pi(g)^{-1}i}.
\end{aligned} \tag{79}
$$

Comparison of this with (75) shows that the map η: $(u_1, \ldots, u_r) \rightarrow \sum_1^r s_i \otimes u_i$ is an equivalence of the induced representation of G as defined first with the representative of G provided by $A \otimes_B U$.

The module U^G (or $A \otimes_B U$) is called the *induced G-module U* and the associated representation σ^G is called the *induced representation* of G. As we have seen, if $(u^{(1)}, \ldots, u^{(n)})$ is a base for U/F, then (78) is a base for the induced module. Now suppose

$$hu^{(i)} = \sum_j \alpha_{ji}(h) u^{(j)} \tag{80}$$

for $h \in H$, so $h \rightarrow \alpha(h) = (\alpha_{ij}(h))$ is a matrix representation of H determined by σ and

$$\chi_\sigma(h) = \sum \alpha_{ii}(h) \tag{81}$$

is the character of σ. By (79), the matrix of $\sigma^G(g)$ relative to the base (78) is obtained by applying the permutation $\pi(g)$ to the r rows of $n \times n$ blocks in the matrix

$$\operatorname{diag}\{\alpha(\mu_1(g)), \ldots, \alpha(\mu_r(g))\}. \tag{82}$$

Only the non-zero diagonal blocks of the matrix thus obtained contribute to the trace. These blocks occur in the ith row (of blocks) if and only if $\pi(g)i = i$, which is equivalent to $gs_iH = s_iH$ and to $s_i^{-1} g s_i \in H$. It follows that the character

$$\chi_{\sigma^G}(g) = \sum{}' \chi_\sigma(s_i^{-1} g s_i) \tag{83}$$

where the summation $\sum'$ is taken over all i such that $s_i^{-1} g s_i \in H$. This can be put in a better form by extending the definition of the function χ_σ from H to F to a function $\dot{\chi}_\sigma$ on G defined by $\dot{\chi}_\sigma(g) = 0$ if $g \in G - H$. Using this, we obtain the formula

$$\chi_{\sigma^G}(g) = \sum_{i=1}^{r} \dot{\chi}_\sigma(s_i^{-1} g s_i). \tag{84}$$

This can be written in another form if G is a finite group, by noting that $\dot{\chi}_\sigma(h^{-1}gh) = \dot{\chi}_\sigma(g)$. This is clear if $g \in H$ since χ_σ is a class function and if $g \notin H$, then $h^{-1}gh \notin H$ and $\dot{\chi}_\sigma(h^{-1}gh) = 0 = \dot{\chi}_\sigma(g)$. Hence

$$\dot{\chi}_\sigma(s_i{}^{-1}gs_i) = \frac{1}{|H|} \sum_{h \in H} \dot{\chi}_\sigma((s_i h)^{-1} g s_i h).$$

Then (84) gives the formula

$$\chi_{\sigma^G}(g) = \frac{1}{|H|} \sum_{a \in G} \dot{\chi}_\sigma(a^{-1}ga). \tag{85}$$

EXAMPLES

1. Let G be the dihedral group D_n of order $2n$ generated by r, s, such that $r^n = 1$, $s^2 = 1$, $srs^{-1} = r^{-1}$. Let $H = \langle r \rangle$, σ a representation of degree 1 of H such that $r \rightsquigarrow \omega 1$ where $\omega^n = 1$. We may take $s_1 = 1$, $s_2 = s$ as representatives of the cosets of H. Then $\pi(r) = 1$, $\pi(s) = (12)$, and

$$\begin{aligned} rs_1 &= s_1 r, & rs_2 &= s_2 r^{-1} \\ ss_1 &= s_2 1, & ss_2 &= s_1 1. \end{aligned}$$

It follows that a matrix representation determined by σ^G maps

$$r \rightsquigarrow \begin{pmatrix} \omega & 0 \\ 0 & \omega^{-1} \end{pmatrix}, \qquad s \rightsquigarrow \begin{pmatrix} 0 & 1 \\ 1 & 0 \end{pmatrix} \quad \text{(cf. p. 262).}$$

2. Let G be as in example 1, but now let $H = \langle s \rangle$. Let σ be the representation of degree 1 such that $s \rightsquigarrow -1$. We have $G = H \cup rH \cup \cdots \cup r^{n-1}H$, so we may take $\{s_i\}$ where $s_i = r^{i-1}$, $1 \leqslant i \leqslant n$, as a set of coset representatives. We have

$$\begin{aligned} rs_i &= s_{i+1}, & 1 \leqslant i &\leqslant n-1, & rs_n &= s_1 \\ ss_1 &= s_1 s, & ss_i &= s_{n+2-i} s & \text{if} \quad i &> 1. \end{aligned}$$

It follows that we have a matrix representation determined by σ^G such that

$$r \rightsquigarrow \begin{bmatrix} 0 & . & . & . & 0 & 1 \\ 1 & 0 & . & . & . & 0 \\ 0 & 1 & 0 & . & . & 0 \\ . & . & . & . & . & . \\ . & . & . & . & . & . \\ 0 & . & . & . & 1 & 0 \end{bmatrix}$$

$$s \rightsquigarrow \begin{bmatrix} -1 & 0 & . & . & . & 0 \\ 0 & 0 & . & . & 0 & -1 \\ 0 & 0 & . & . & -1 & 0 \\ . & . & . & . & . & . \\ . & . & . & . & . & . \\ 0 & -1 & 0 & . & . & 0 \end{bmatrix}.$$

3. An interesting special case of induction is obtained by beginning with the unit representation $1 = 1_H$ of the subgroup H acting on a one-dimensional space $U_1 = Fu$. Let $\{s_1, \ldots, s_r\}$ be a cross section of G relative to H and suppose $gs_i = s_{\pi(g)i}\mu_i(g)$ where $\mu_i(g) \in H$. The set of elements $s_i \otimes u$, $1 \leqslant i \leqslant r$, is a base for $U_1{}^G$ and, by definition of the induced representation, the action of g on $U_1{}^G$ is given by $g(s_i \otimes u) = s_{\pi(g)i} \otimes u$. On the other hand, we have the action of G on the set G/H of left cosets $\{s_iH | 1 \leqslant i \leqslant r\}$ given by $g(s_iH) = s_{\pi(g)i}H$. It is clear from this that 1^G is the permutation representation obtained from the action of G on G/H.

We can determine the character χ_{1^G} for the representation 1^G. It is clear from the formula $g(s_i \otimes u) = s_{\pi(g)i} \otimes u$ that the trace of the matrix of $1^G(g)$ relative to the base $(s_1 \otimes u, \ldots, s_n \otimes u)$ is the number of fixed points of $\pi(g)$. This is the number of cosets aH such that $g(aH) = aH$. Hence we have the formula

$$\chi_1{}^G(g) = |\{aH | gaH = aH\}|. \tag{86}$$

We note also that the condition $gaH = aH$ is equivalent to ${}^{a^{-1}}g \equiv a^{-1}ga \in H$.

We recall that the action of G in G/H is transitive and any transitive action of G on a finite set $N = \{1, 2, \ldots\}$ is equivalent to the action of G on a set of left cosets relative to a subgroup H of index n in G (BAI, p. 75). In fact, if G acts transitively on N, then we can take H to be the stabilizer of any element of N. It is clear from this that the study of induced representations of the form $1_H{}^G$, where 1_H is the unit representation of a subgroup H of finite index in G, is essentially the same thing as the study of transitive actions of G on finite sets.

We shall now give two useful characterizations of induced modules in the general case in which H is a subgroup of G of finite index in G, U a module for H. Let $A = F[G]$, $B = F[H]$, $\{s_1, \ldots, s_r\}$ a cross section of G relative to H, and consider the induced module U^G of G. We have $U^G = A \otimes_B U = \oplus (s_i \otimes U)$ and $[s_i \otimes U : F] = [U : F] = n$. Hence the map $x \rightsquigarrow s_i \otimes x$ of U into $s_i \otimes U$ is a linear isomorphism. Since we may take one of the $s_i = 1$, we see that the map $x \rightsquigarrow 1 \otimes x$ is a linear isomorphism of U onto the subspace $1 \otimes U = \{1 \otimes x | x \in U\}$ of U^G. If $b \in B$, then $b(1 \otimes x) = b \otimes x = b1 \otimes x = 1 \otimes bx$. Hence $x \rightsquigarrow 1 \otimes x$ is a B-isomorphism of U onto $1 \otimes U$, which is a B-submodule of U^G. We shall now identify U with its image $1 \otimes U$ and so regard U as a B-submodule of U^G. It is clear also that we have $U^G = s_1U \oplus \cdots \oplus s_rU$, a direct sum of the subspaces s_iU. The properties we have noted give a useful internal characterization of U^G that we state as

PROPOSITION 5.5. *Let V be an A-module. Then V is isomorphic to U^G for some B-module U if and only if V contains U (strictly speaking a B-submodule isomorphic to U) such that $V = s_1U \oplus \cdots \oplus s_rU$ for some cross section $\{s_1, \ldots, s_r\}$ of G relative to H.*

Proof. We have shown that U^G has the stated properties for the B-submodule U $(= 1 \otimes U)$ and any cross section of G relative to H. Conversely, let V be an

A-module having the stated properties. Then $V = AU$. Since $s_i \in G$, s_i is invertible in A and hence the map $x \rightarrow s_i x$ is a bijection. It follows that $[s_i U : F] = [U : F]$ and since $V = s_1 U \oplus \cdots \oplus s_r U$, $[V : F] = [G : H][U : F]$. We have the map $(a, x) \rightarrow ax$ of $A \times U$ into V, which is F-bilinear. Since $(ab)x = a(bx)$ for $b \in B$, it follows that we have an F-linear map η of $A \otimes_B U$ into V such that $\eta(a \otimes x) = ax$. Evidently, η is an A-homomorphism. Since $V = AU$, η is surjective and since $[V : F] = [A \otimes_B U : F] < \infty$, η is injective. Hence V and U^G are isomorphic A-modules. □

The second characterization of U^G that we shall give is a categorical one that applies to any ring A and subring B. We consider the categories of modules A-**mod** and B-**mod**. If M is an A-module, we can regard M as B-module. Evidently, homomorphisms of A-modules are homomorphisms of B-modules. In this way we obtain a functor R from the category A-**mod** to B-**mod** sending an A-module M into M regarded as B-module and A-homomorphisms into the same maps regarded as B-homomorphisms. We call R the *restriction of scalars functor* from A-**mod** to B-**mod**. Now let N be a given B-module. Then we can form $A \otimes_B N$ and regard this as an A-module in the usual way. We have a B-homomorphism u of N into $A \otimes_B N$ sending $y \in N$ into $1 \otimes y$. We claim that the pair $(A \otimes N, u)$ constitutes a universal from N to the functor R (p. 42). This means that if M is a (left) A-module and η is a B-homomorphism of N into $RM = M$, then there exists a unique homomorphism $\tilde{\eta}$ of $A \otimes_B N$ into M (as A-module) such that

(87)

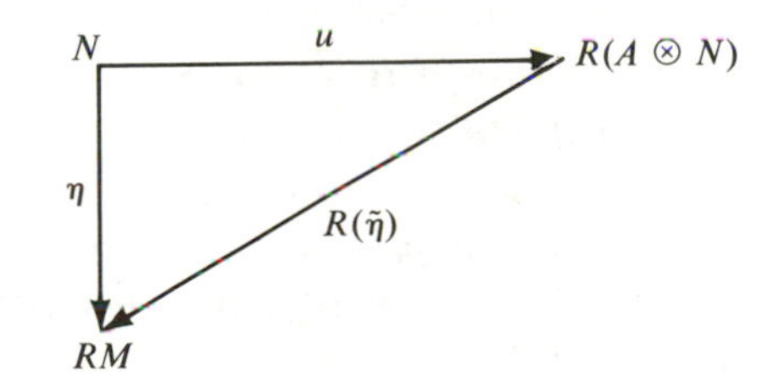

is commutative. To see this, we consider the map $(a, y) \rightarrow a(\eta y)$ of the product set $A \times N$ into M. Evidently this is additive in a and y and if $b \in B$, then $(ab)(\eta y) = a(b(\eta y)) = a(\eta(by))$. Hence $(a, y) \rightarrow a(\eta y)$ defines a balanced product from $A \times N$ into M (p. 126). Hence we have a group homomorphism $\tilde{\eta}$ of $A \otimes N$ into M such that $\tilde{\eta}(a \otimes y) = a(\eta y)$. It is clear that $\tilde{\eta}$ is a homomorphism of A-modules and if $y \in N$, then $R(\tilde{\eta})(uy) = R(\tilde{\eta})(1 \otimes y) = 1(\eta y) = \eta y$. Hence (87) is commutative. Now let η' be any A-homomorphism of $A \otimes_B N$ into M such that $\eta y = \eta' u y = \eta'(1 \otimes y)$. Then $\eta'(a \otimes y) = \eta'(a(1 \otimes y)) = a\eta'(1 \otimes y) = a(\eta y)$. Hence $\eta' = \tilde{\eta}$, so $\tilde{\eta}$ is unique.

The fact that for any B-module N, $(A \otimes_B N, u)$ is a universal from N to the functor R implies that we have a functor from B-**mod** to A-**mod** sending any B-

module N into the A-module $A \otimes_B N$ and mapping any homomorphism η of B-modules into $1 \otimes \eta$, which is a homomorphism of A-modules. This functor is a left adjoint of the functor R (p. 49).

This is applicable in particular to $A = F[G]$ and $B = F[H]$ where H is a subgroup of finite index in G. In this case for a given B-module U such that $[U:F] < \infty$, we obtain the universal object $U^G = A \otimes_B U$, which is finite dimensional over F and hence provides a representation of G.

EXERCISE

1. Let G be a finite group, H a subgroup, and let σ be the regular representation of H. Show that σ^G is the regular representation of G.

5.10 PROPERTIES OF INDUCTION. FROBENIUS RECIPROCITY THEOREM

We shall now derive the basic properties of induced representations. We prove first

THEOREM 5.17. *Let K be a subgroup of finite index in G, H a subgroup of finite index in K, σ a representation of H acting on U, ρ a representation of G acting on V, ρ_H the restriction of ρ to H. Then*

(1) *H is of finite index in G and σ^G and $(\sigma^K)^G$ are equivalent.*

(2) *If W is a $\sigma(H)$-invariant subspace of U, then W^G is a $\sigma^G(G)$-invariant subspace of U^G. Moreover, if $U = W_1 \oplus W_2$ where W_i is σ-invariant, then $U^G = {W_1}^G \oplus {W_2}^G$.*

(3) *$\sigma^G \otimes \rho$ and $(\sigma \otimes \rho_H)^G$ are equivalent.*

Proof. (1) Write $A = F[G]$, $B = F[H]$, $C = F[K]$. It is clear that H is of finite index in G. (In fact, $[G:H] = [G:K][K:H]$.) Now σ^K acts on $C \otimes_B U$, $(\sigma^K)^G$ acts on $A \otimes_C (C \otimes_B U)$, and σ^G acts on $A \otimes_B U$. The A-modules $A \otimes_C (C \otimes_B U)$ and $(A \otimes_C C) \otimes_B U$ are isomorphic (Proposition 3.7, p. 135) and $A \otimes_C C$ and A are isomorphic as left A-modules (Proposition 3.2, p. 130). Hence $A \otimes_C (C \otimes_B U)$ and $A \otimes_B U$ are isomorphic as A-modules, which means that $(\sigma^K)^G$ and σ^G are equivalent.

(2) If W is a $\sigma(H)$-invariant subspace of U, then this is a B-submodule of U. Since A is free as right module over B, $W^G = A \otimes_B W$ can be identified with

its image in $U^G = A\otimes_B U$. Then $(\sigma|W)^G$ is a subrepresentation of σ^G. The second statement follows in the same way.

(3) The two A-modules we have to consider are $(A\otimes_B U)\otimes_F V$ and $A\otimes_B(U\otimes_F V)$. In the first case the action of $g\in G$ is $g((a\otimes_B x)\otimes_F y) = (ga\otimes_B x)\otimes_F gy$ and in the second case we have $g(a\otimes_B(x\otimes_F y)) = ga\otimes_B(x\otimes_F y)$. Hence the F-isomorphism of $(A\otimes_B U)\otimes_F V$ onto $A\otimes_B(U\otimes_F V)$ sending $(a\otimes_B x)\otimes_F y \rightarrow a\otimes_B(x\otimes_F y)$ is not an A-isomorphism. We shall define an A-isomorphism of $A\otimes_B(U\otimes_F V)$ onto $(A\otimes_B U)\otimes_F V$ such that $1\otimes_B(x\otimes_F y)\rightarrow(1\otimes_B x)\otimes_F y$ and hence $g\otimes_B(x\otimes_F y)\rightarrow(g\otimes_B x)\otimes_F gy$. For a fixed $g\in G$ we consider the map $(x,y)\rightarrow(g\otimes_B x)\otimes_F gy$ of $U\times V$ into $(A\otimes_B U)\otimes_F V$. Since this is F-bilinear, we have an F-homomorphism τ_g of $U\otimes_F V$ into $(A\otimes_B U)\otimes_F V$ such that $x\otimes_F y\rightarrow(g\otimes_B x)\otimes_F gy$. Since G is a base for A/F, for any $a=\sum\alpha_g g$, $\alpha_g\in F$, we can define an F-homomorphism τ_a of $U\otimes_F V$ into $(A\otimes_B U)\otimes_F V$ by $\tau_a = \sum\alpha_g\tau_g$. Then $\tau_a(x\otimes_F y) = \sum\alpha_g(g\otimes_B x)\otimes_F gy$. We now have a map

$$(a,\textstyle\sum x_i\otimes_F y_i)\rightarrow\tau_a(\sum x_i\otimes_F y_i)$$

of $A\times(U\otimes_F V)$ into $(A\otimes_B U)\otimes_F V$, which is F-bilinear. Moreover, if $h\in H$, then

$$\begin{aligned}\tau_{gh}(x\otimes_F y)-\tau_g(hx\otimes hy) &= (gh\otimes_B x)\otimes_F ghy-(g\otimes_B hx)\otimes_F ghy\\ &= (g\otimes_B hx)\otimes_F ghy-(g\otimes_B hx)\otimes_F ghy = 0.\end{aligned}$$

It follows that we have a balanced product of A as right B-module and $U\otimes_F V$ as left B-module. Hence we have a homomorphism of $A\otimes_B(U\otimes_F V)$ into $(A\otimes_B U)\otimes_F V$ sending $g\otimes_B(x\otimes_F y)\rightarrow(g\otimes_B x)\otimes_F gy$. This is an F-homomorphism and if $g'\in G$, then $g'(g\otimes_B(x\otimes_F y)) = g'g\otimes_B(x\otimes_F y)$ and $(g'g\otimes_B x)\otimes_F g'gy = g'((g\otimes_B x)\otimes_F gy)$. Hence our map is an A-homomorphism. It is clear also that the map is surjective and since the two modules have the same dimensionality over F, the map is an isomorphism. □

For our next result on contragredience and induction we shall make use of an involution in the group algebra (BAI, p. 112). We note first that if $g,h\in G$, then $(gh)^{-1} = h^{-1}g^{-1}$ and $(g^{-1})^{-1} = g$. Hence if $a\rightarrow\tilde{a}$ is the linear transformation of $A = F[G]$ into itself such that $\tilde{g} = g^{-1}$, then $\widetilde{ab} = \tilde{b}\tilde{a}$ and $\tilde{\tilde{a}} = a$. Thus $a\rightarrow\tilde{a}$ is an involution in A. We shall call this the *main involution* of A and denote it as j. If H is a subgroup of G, then the main involution in $B = F[H]$ is the restriction to B of the main involution in A. Now let H have finite index and let $\{s_1,\dots,s_r\}$ be a cross section. Any element of A can be written in one and only one way as $a = \sum_1^r s_i b_i$, $b_i\in B$. If $c = \sum_1^r s_i c_i$, $c_i\in B$, we define

$$(a|c) = \sum\tilde{b}_i c_i\in B \tag{88}$$

and we have the map $(a,c) \rightsquigarrow (a|c)$ of $A \times A$ into B. We list its important properties:

(i) $(a|c)$ is independent of the choice of the representatives s_i, since any other choice has the form $s_i h_i$, $h_i \in H$. Using these replaces b_i by $h_i^{-1}b_i$, c_i by $h_i^{-1}c_i$. Then $(\widetilde{h_i^{-1}b_i})(h_i^{-1}c_i) = \bar{b}_i h_i h_i^{-1} c_i = \bar{b}_i c_i$ and (88) is unchanged.

(ii) $(a|c)$ is F-bilinear. Clear.

(iii) If $b \in B$, then

$$(ab|c) = \bar{b}(a|c), \qquad (a|cb) = (a|c)b.$$

This is clear from the definition.

(iv) If $e \in A$, then

$$(ea|c) = (a|\tilde{e}c).$$

It suffices to prove this for $e = g \in G$. Then $gs_i = s_{\pi(g)i}\mu_i(g)$ where $\pi(g) \in S_r$ and $\mu_i(g) \in H$. Then $ga = g(\sum s_i b_i) = \sum s_{\pi(g)i}\mu_i(g)b_i$ and $g^{-1}s_{\pi(g)i} = s_i\mu_i(g)^{-1}$, so $g^{-1}c = \sum s_i\mu_i(g)^{-1}c_{\pi(g)i}$. Hence

$$(ga|c) = \sum_i \bar{b}_i\mu_i(g)^{-1}c_{\pi(g)i}$$

$$(a|g^{-1}c) = \sum_i \bar{b}_i\mu_i(g)^{-1}c_{\pi(g)i}.$$

Thus $(ga|c) = (a|g^{-1}c)$ holds for all $g \in G$.

(v) $(s_i|s_j) = \delta_{ij}1$.
This is clear from the definition.

We shall now establish commutativity of induction and contragredience. This is given in

THEOREM 5.18 *If H is a subgroup of finite index and σ is a representation of H, then $(\sigma^G)^*$ and $(\sigma^*)^G$ are equivalent.*

Proof. Let U be the vector space on which σ acts and let U^* be the dual space of linear functions on U. If $x \in U$ and $y^* \in U^*$, we write $\langle y^*|x\rangle = y^*(x)$. Then $\langle y^*|x\rangle$ is F-bilinear. The space U^* is a left B-module defined by the contragredient representation σ^* of H and we can make this a right B-module by defining $y^*b = \bar{b}y^*$, $b \in B$. Since $\langle hy^*|x\rangle = \langle y^*|h^{-1}x\rangle$ for $h \in H$, by definition of σ^*, we have $\langle y^*h|x\rangle = \langle y^*|hx\rangle$, which implies that

$$\langle y^*b|x\rangle = \langle y^*|bx\rangle \tag{89}$$

for $x \in U$, $y^* \in U^*$, $b \in B$. We now introduce

$$\langle y^*(a|c)|x\rangle \in F. \tag{90}$$

This is defined for $x \in U$, $a, c \in A$, $y^* \in U^*$, and is F-bilinear in c and x for fixed a and y^*. Moreover, if $b \in B$, then by property (iii) above,

$$\langle y^*(a|cb)|x\rangle = \langle y^*(a|c)b|x\rangle = \langle y^*(a|c)|bx\rangle.$$

By definition of the tensor product, this implies that we have an F-linear map of $A \otimes_B U$ into F sending $c \otimes x \rightsquigarrow \langle y^*(a|c)|x\rangle$. Thus we have an element φ_{a,y^*} of the dual space $(A \otimes_B U)^*$ such that

$$\varphi_{a,y^*}(c \otimes x) = \langle y^*(a|c)|x\rangle. \tag{91}$$

Now the map $(a, y^*) \rightsquigarrow \varphi_{a,y^*}$ is F-bilinear and since

$$\langle y^*(ab|c)|x\rangle = \langle y^*\tilde{b}(a|c)|x\rangle = \langle (by^*)\,(a|c)|x\rangle,$$

$\varphi_{ab,y^*} = \varphi_{a,by^*}$. Hence we have a linear map η of $A \otimes_B U^*$ into $(A \otimes_B U)^*$ sending $a \otimes y^*$ into φ_{a,y^*}. If $g \in G$, $g\varphi_{a,y^*}(c \otimes x) = \varphi_{a,y^*}(g^{-1}(c \otimes x)) = \varphi_{a,y^*}(g^{-1}c \otimes x) = \langle y^*(a|g^{-1}c)|x\rangle$. On the other hand, $\varphi_{ga,y^*}(c \otimes x) = \langle y^*(ga|c)|x\rangle$. Hence by (iv), $g\varphi_{a,y^*}(c \otimes x) = \varphi_{ga,y^*}(c \otimes x)$, which implies that η is an A-homomorphism. To conclude the proof, we shall show that η is injective. Since $A \otimes_B U^*$ and $(A \otimes_B U)^*$ have the same finite dimensionality over F, this will imply that η is an isomorphism. Now suppose $\eta(\sum a_j \otimes y_j^*) = 0$. We can write $\sum a_j \otimes y_j^* = \sum_1^r s_i \otimes z_i^*$ and the condition $\eta(\sum s_i \otimes z_i^*) = 0$ implies that $\sum \langle z_i^*(s_i|c)|x\rangle = 0$ for all $c \in A$, $x \in U$ and hence $\sum z_i^*(s_i|c) = 0$ for all c. If we take $c = s_j$ and use (v), we obtain $z_j^* = 0$ and hence $\sum s_i \otimes z_i^* = 0$. This concludes the proof. □

Our next main objective is to derive an important reciprocity theorem due to Frobenius, which in its original form states that if ρ is an irreducible complex representation of a finite group and σ is an irreducible complex representation of a subgroup H, then the multiplicity of ρ in σ^G is the same as the multiplicity of σ in the representation ρ_H of H. As we shall show below, this can be proved by a simple calculation with characters. We shall first give a conceptual proof of an extension of Frobenius' theorem. Our proof will be based on the important concept of intertwining numbers for representations and on the following proposition.

PROPOSITION 5.6. *If H is a subgroup of finite index in G, U a module for $B = F[H]$, and V a module for $A = F[G]$, then* $\hom_A(U^G, V)$ *and* $\hom_B(U, V_H)$ *are isomorphic as vector spaces over F.*

Proof. We remark that this result can be deduced from a general result on adjoint functors applied to the restriction functor from $F[G]$-**mod** to $F[H]$-**mod** and its left adjoint. However, we shall give an independent proof. As before, let u be the map of U into U^G sending $x \to 1 \otimes x$. This is a B-homomorphism. Hence if $\zeta \in \hom_A(U^G, V)$, then $\zeta u \in \hom_B(U, V_H)$. The map $\zeta \to \zeta u$ is an F-linear map of $\hom_A(U^G, V)$ into $\hom_B(U, V_H)$. The result we proved in diagram (87) is that the map $\zeta \to \zeta u$ is surjective. It is also injective. For, if $\zeta u = 0$, then $\zeta(1 \otimes x) = 0$ for all $x \in U$. Since ζ is an A-homomorphism, $\zeta(a \otimes x) = \zeta(a(1 \otimes x)) = a\zeta(1 \otimes x) = 0$ for all $a \in A$, $x \in U$. Hence $\zeta = 0$. Thus $\zeta \to \zeta u$ is an isomorphism of $\hom_A(U^G, V)$ and $\hom_B(U, V_H)$ as vector spaces over F. $\square$

If ρ and τ are representations of G acting on V and W respectively, then the dimensionality

$$[\hom_A(V, W) : F]$$

is called the *intertwining number* $\iota(\rho, \tau)$ of ρ and τ (or of V and W as $F[G]$-modules). If $V = V_1 \oplus \cdots \oplus V_m$ where the V_i are $\rho(G)$-invariant subspaces, then we write $\rho = \rho_1 \oplus \cdots \oplus \rho_m$ where $\rho_i = \rho | V_i$. Similarly, let $\tau = \tau_1 \oplus \cdots \oplus \tau_l$ where $\tau_j = \tau | W_j$ and W_j is τ-invariant. We have the vector space decomposition

$$\hom_A(V, W) = \bigoplus_{i,j} \hom_A(V_i, W_j). \tag{92}$$

This implies that

$$\iota(\rho, \tau) = \sum_{i,j} \iota(\rho_i, \tau_j). \tag{93}$$

If ρ and τ are completely reducible and the ρ_i and τ_j are irreducible, then this formula reduces the calculation of $\iota(\rho, \tau)$ to that of the $\iota(\rho_i, \tau_j)$ where ρ_i and τ_j are irreducible. If ρ and τ are irreducible and inequivalent, then $\iota(\rho, \tau) = 0$ by Schur's lemma. It is clear also that $\iota(\rho, \tau)$ is unchanged if we replace ρ or τ by an equivalent representation. Hence if ρ and τ are equivalent, then $\iota(\rho, \tau) = \iota(\rho, \rho) = [\operatorname{End}_A(V, V) : F]$. Hence if ρ and τ are irreducible and F is algebraically closed, then

$$\iota(\rho, \tau) = \begin{cases} 0 & \text{if } \rho \text{ and } \tau \text{ are inequivalent,} \\ 1 & \text{if } \rho \text{ and } \tau \text{ are equivalent.} \end{cases} \tag{94}$$

An immediate consequence of (93) and (94) is

PROPOSITION 5.7. *If F is algebraically closed, ρ is irreducible, and τ is completely reducible, then $\iota(\rho, \tau) = \iota(\tau, \rho) =$ multiplicity of ρ in τ (that is, the number of irreducible components of τ isomorphic to ρ).*

We can now prove the fundamental

FROBENIUS RECIPROCITY THEOREM. *Let F be an algebraically closed field, ρ an irreducible representation of G over F, σ an irreducible representation of a subgroup H of finite index in G. Assume that σ^G and ρ_H are completely reducible. Then the multiplicity of ρ in σ^G is the same as the multiplicity of σ in ρ_H.*

Proof. By Proposition 5.6, $\imath(\sigma^G,\rho) = \imath(\sigma,\rho_H)$. By Proposition 5.7, $\imath(\sigma^G,\rho)$ is the multiplicity of ρ in σ^G and $i(\sigma,\rho_H)$ is the multiplicity of σ in ρ_H. Hence these multiplicities coincide. □

The Frobenius reciprocity theorem is usually stated for finite groups and algebraically closed fields whose characteristics do not divide the group order. In this situation the hypothesis that H is of finite index and that σ^G and ρ_H are completely reducible is automatically satisfied. The proof of this result and some of the others we gave on induced representations of finite groups can be made by calculations with characters. As an illustration of this method, we shall now prove the Frobenius reciprocity theorem for G finite and $F = \mathbb{C}$ by using characters. Let the notation be as above and let χ_ρ, χ_σ, etc. denote the characters of ρ, σ, etc. We recall that if ρ is irreducible and τ is arbitrary, then $(\chi_\rho|\chi_\tau) = (\chi_\tau|\chi_\rho)$ is the multiplicity of ρ in τ. The Frobenius reciprocity theorem is equivalent to

$$(\chi_{\sigma^G}|\chi_\rho)_G = (\chi_\sigma|\chi_{\rho_H})_H. \tag{95}$$

We use the formula (85) for χ_{σ^G}. Then we obtain

$$\begin{aligned}(\chi_{\sigma^G}|\chi_\rho)_G &= \sum_{a,g\in G} \frac{1}{|G|}\frac{1}{|H|}\overline{\dot{\chi}_\sigma(a^{-1}ga)\chi_\rho(g)}\\ &= \sum_{a,g\in G} \frac{1}{|G|}\frac{1}{|H|}\overline{\dot{\chi}_\sigma(g)}\chi_\rho(aga^{-1})\\ &= \sum_{\substack{a\in G\\ h\in H}} \frac{1}{|G|}\frac{1}{|H|}\overline{\chi_\sigma(h)}\chi_\rho(aha^{-1}).\end{aligned}$$

On the other hand,

$$\begin{aligned}(\chi_\sigma|\chi_{\rho_H})_H &= \sum_{h\in H} \frac{1}{|H|}\overline{\chi_\sigma(h)}\chi_\rho(h)\\ &= \sum_{\substack{h\in H\\ a\in G}} \frac{1}{|G|}\frac{1}{|H|}\overline{\chi_\sigma(h)}\chi_\rho(aha^{-1})\end{aligned}$$

since χ_ρ is a class function on G. Comparison of the two formulas gives (95).

EXERCISES

1. Construct a character table for $G = A_5$. *Sketch*: There are five conjugacy classes in G with representatives 1, (12)(34), (123), (12345), $(13524) = (12345)^2$ with respective cardinalities 1, 15, 20, 12, 12. Let χ_1 be the character of the unit representation, v the permutation representation obtained from the natural action of G on $\{1, 2, 3, 4, 5\}$. We have the following table

	1	(12)(34)	(123)	(12345)	(13524)
χ_1	1	1	1	1	1
χ_v	5	1	2	0	0

Calculation of $(\chi_v|\chi_v)$ and $(\chi_v|\chi_1)$ gives $\chi_v = \chi_1 + \chi_2$ where χ_2 is irreducible. This and the above table give χ_2. To obtain the three missing irreducible characters, we look at the induced characters obtained from linear (degree 1) characters of the subgroup

$$H = \{1, u = (12345), u^2 = (13524), u^3 = (14253),$$
$$u^4 = (54321), (13)(54), (24)(15), (35)(21), (14)(23), (25)(34)\}.$$

We find that $\langle u \rangle$ has index two in H, so we have the linear characters λ_1 and λ_2 on H where $\lambda_1 \equiv 1$ and $\lambda_2(u^k) = 1$, $\lambda_2(v) = -1$ if $v \notin \langle u \rangle$. The set

$$C = \{1, w = (13245), w^2 = (12534), w^3 = (14352), w^4 = (15423), (12)(34)\}$$

is a cross section of G relative to H. Using this we can compute $\lambda_1{}^G$ and $\lambda_2{}^G$ as

	1	(12)(34)	(123)	(12345)	(13524)
$\lambda_1{}^G$	6	2	0	1	1
$\lambda_2{}^G$	6	-2	0	1	1

Using $(\lambda_i{}^G|\lambda_i{}^G)$, $i = 1, 2$, and $(\lambda_1{}^G|\chi_1)$, we obtain $\lambda_1 = \chi_1 + \chi_3$ where χ_3 is irreducible. Also we have $(\lambda_2{}^G|\chi_1) = (\lambda_2{}^G|\chi_2) = (\lambda_2{}^G|\lambda_1{}^G) = 0$, which implies that $\lambda_2{}^G = \chi_4 + \chi_5$ where χ_4 and χ_5 are irreducible. The values of $\chi_i(1)$, $1 \leqslant i \leqslant 3$, and $\sum_{i=1}^5 \chi_i{}^2(1) = 60$ give $\chi_4(1) = 3 = \chi_5(1)$. It follows that we have the table

	1	(12)(34)	(123)	(12345)	(13524)
χ_1	1	1	1	1	1
χ_2	4	0	1	-1	-1
χ_3	5	1	-1	0	0
χ_4	3	a	b	c	d
χ_5	3	a'	b'	c'	d'

Since $(54321) = (12345)^{-1}$ is conjugate to (12345), χ_4 and χ_5 are real (exercise 7, p. 279). Using $\chi_4 + \chi_5 = \lambda_2{}^G$, we obtain $a + a' = -2$, $b + b' = 0$, $c + c' = 1 = d + d'$. Using $(\chi_4|\chi_i) = \delta_{i4}$ for $i = 1, 2, 3, 4$, we obtain a system of equations for a, b, c, d whose solutions are $a = -1$, $b = 0$, $c = (1 \pm \sqrt{5})/2$, $d = 1 - c$. Either determination can be used to complete the table.

2. Prove Theorems 5.17 and 5.18 for complex representations of finite groups by character calculations.

3. Let B be a subring of a ring A, M a (left) B-module. Then $\hom_B(A, M)$ becomes a left A-module if we define $(af)x = f(xa)$ for $f \in \hom_B(A, M), a, x \in A$ (Proposition 3.4, p. 134). This is called the *produced* module of A determined by M. Note that the map $v: f \rightarrow f1$ of $\hom_B(A, M)$ into M is a B-homomorphism. Show that the pair $(\hom_B(A, M), v)$ constitutes a universal from the restriction of scalars functor R to M (p. 137).

4. Let G be a group, H a subgroup of finite index, $A = F[G]$, $B = F[H]$, and let U be the B-module determined by a representation σ of H. Then $\hom_B(A, U)$ is finite dimensional over F and defines the *produced representation* ${}^G\sigma$ of G. Show that ${}^G(\sigma^*)$ is equivalent to $(\sigma^G)^*$. (This shows that in the theory of finite dimensional representations, produced modules are superfluous.)

5. Let G be finite, and let F be an algebraically closed field whose characteristic does not divide $|G|$. Let ρ be the permutation representation of G determined by the action of G on G/H where H is a subgroup. Use Frobenius reciprocity to show that the dimensionality of the space $\operatorname{Inv} \rho = \{x | gx = x, g \in G\}$ is one.

6. (Addendum to Clifford's theorem.) Let ρ be an irreducible representation of G acting on V/F and let $H \lhd G$. Let W be a homogeneous component of V as $F[H]$-module. (Recall that ρ_H is completely reducible by Clifford's theorem.) Let T be the subgroup of G stabilizing W. Show that (i) $H \subset T \subset G$ and $[G:T] = m$, the number of homogeneous components of V as $F[H]$-module, and (ii) if ψ is the representation of T acting on W, then $\rho = \psi^G$ and ψ is irreducible.

7. (Mackey.) Let H be a subgroup of finite index in G and let U be an $F[H]$-module. Let V be the set of maps $f: G \rightarrow U$ such that $f(hg) = hf(g)$. Using the usual vector space structure on V and an action of G defined by $(gf)(x) = f(xg)$, $f \in V, g, x \in G$. Show that V becomes an $F[G]$-module isomorphic to U^G.

5.11 FURTHER RESULTS ON INDUCED MODULES

One of the main problems that we shall consider in this section is that of obtaining useable criteria for an induced representation (or character) to be irreducible. To this end we shall need to study the following problem: Suppose that H and K are subgroups of a group G, H of finite index in G, and σ is a representation of H acting on the vector space U/F. What can be said about the structure of ${\sigma^G}_K \equiv (\sigma^G)_K$, the restriction to K of the induced representation σ^G of G? If V is an $F[G]$-module, we denote V regarded as an $F[K]$-module by V_K. Thus we are interested in the structure of ${U^G}_K \equiv (U^G)_K$. We identify U with the $F[H]$-submodule $1 \otimes U$ of U^G. If $g \in G$, then gU is stabilized by the action of the subgroup ${}^gH = \{ghg^{-1} | h \in H\}$ since if $x \in U$, then $(ghg^{-1})gx = ghx \in gU$. Hence we have the submodules $(gU)_{{}^gH}$ and $(gU)_{K \cap {}^gH}$ of ${U^G}_{{}^gH}$ and ${U^G}_{K \cap {}^gH}$ respectively. We remark that since ${}^gH$ is of finite index in

$G, K \cap {}^gH$ is of finite index in K. Hence we can form the induced $F[K]$-module $gU_{K \cap {}^gH}{}^K \equiv (gU_{K \cap {}^gH})^K$. We shall obtain a decomposition of $U^G{}_K$ into submodules of the form $gU_{K \cap {}^gH}{}^K$. To describe this, we require some simple remarks on double cosets.

We recall that G is a disjoint union of the double cosets KgH (BAI,p. 53). Moreover, KgH is a union of left cosets kgH, $k \in K$, and since H is of finite index in G, the number of left cosets of H contained in KgH is finite. We now note that if $k_1, k_2 \in K$, then $k_1 gH = k_2 gH$ if and only if $k_1(K \cap {}^gH) = k_2(K \cap {}^gH)$. For, $k_1 gH = k_2 gH$ if and only if $k_2 g = k_1 gh$, $h \in H$, and this holds if and only if $k_1{}^{-1} k_2 = ghg^{-1}$, $h \in H$. This last condition is equivalent to $k_1(K \cap {}^gH) = k_2(K \cap {}^gH)$. It now follows that $\{k_i\}$ is a cross section of K with respect to $K \cap {}^gH$ if and only if $\{k_i g\}$ is a set of representatives of the left cosets of H contained in KgH.

We can use this result to prove the following important theorem:

MACKEY'S DECOMPOSITION THEOREM. *Let H be a subgroup of finite index in G, K an arbitrary subgroup of G, and let U be an $F[H]$-module. Identify U with the $F[H]$-submodule $1 \otimes U$ of U^G and let Δ be the set of double cosets $D = KgH$. Then*

$$U^G{}_K = \bigoplus_{D \in \Delta} DU$$

where $DU \equiv \sum_{g \in D} gU$ is an $F[K]$-submodule of $U^G{}_K$ and $DU \cong ((gU)_{K \cap {}^gH})^K$ for any $g \in D$.

Proof. Let $\{s_1, \ldots, s_r\}$ be a cross section of G with respect to H. Then $U^G = s_1 U \oplus \cdots \oplus s_r U$ (Proposition 5.5, p. 290). If $g \in G$, then $D = KgH = s_{i_1} H \cup \cdots \cup s_{i_k} H$ where the s_{i_j} are distinct. Then $DU = KgHU = s_{i_1} U \oplus \cdots \oplus s_{i_k} U$. It follows that $U^G{}_K = \oplus_{D \in \Delta} DU$. Now $DU = KgU$ is an $F[K]$-submodule of $U^G{}_K$ and DU contains gU, which is an $F[K \cap {}^gH]$-submodule. We have seen that if $\{k_i | 1 \leqslant i \leqslant q\}$ is a cross section of K with respect to $K \cap {}^gH$, then $\{k_i g\}$ is a set of representatives of the left cosets of H contained in $D = KgH$. Thus we may assume that $\{k_i g\} = \{s_{i_1}, \ldots, s_{i_k}\}$, so $q = k$ and

$$DU = k_1 gU \oplus \cdots \oplus k_q gU.$$

By Proposition 5.5, this implies that DU as $F[K]$-module is isomorphic to $((gU)_{{}^gH \cap K})^K$. $\square$

We shall derive next a theorem on the structure of the tensor product of two induced representations. This is

THEOREM 5.19 (Mackey). *Let H_i, $i = 1, 2$, be a subgroup of finite index in G, σ_i a representation of H_i acting on U_i/F, and let R be a set of representatives of the double cosets H_1gH_2. Put $H^{(1,g)} = H_1 \cap {}^gH_2$. Then the $F[G]$ module*

$$U_1{}^G \otimes U_2{}^G \cong \bigoplus_{g \in R} ((U_1)_{H^{(1,g)}} \otimes (gU_2)_{H^{(1,g)}})^G.$$

Proof. By Theorem 5.17.3, $\sigma_1{}^G \otimes \sigma_2{}^G$ is equivalent to $(\sigma_1 \otimes (\sigma_2{}^G)_{H_1})^G$, and by Mackey's decomposition, $(U_2{}^G)_{H_1}$ is a direct sum of the $F[H_1]$-submodules isomorphic to the modules $((gU_2)_{H^{(1,g)}})^{H_1}$, $g \in R$. Hence $U_1{}^G \otimes U_2{}^G$ is $F[G]$ isomorphic to

$$\bigoplus_{g \in R} (U_1 \otimes ((gU_2)_{H^{(1,g)}})^{H_1})^G.$$

By Theorem 5.17.3,

$$U_1 \otimes ((gU_2)_{H^{(1,g)}})^{H_1} \cong ((U_1)_{H^{(1,g)}} \otimes (gU_2)_{H^{(1,g)}})^{H_1}.$$

Hence, by the transitivity of induction (Theorem 5.17.1),

$$(U_1 \otimes (gU_2)_{H^{(1,g)}})^G \cong ((U_1)_{H^{(1,g)}} \otimes (gU_2)_{H^{(1,g)}})^G.$$

Substituting this in the formula for $U_1{}^G \otimes U_2{}^G$ proves the theorem. □

The foregoing result can be used to obtain a criterion for an induced module to be irreducible. We shall require also some general results on modules, some of which have been indicated previously in exercises (p. 251). We note first that if V_i, $i = 1, 2$, is a finite dimensional vector space over a field F, then we have a canonical isomorphism between $V_1{}^* \otimes_F V_2$, $V_1{}^*$ the dual space of linear functions on V_1, and $\hom_F(V_1, V_2)$. This maps $u^* \otimes v$ for $u^* \in V_1{}^*$, $v \in V_2$, into the linear map $[u^*, v]$ such that $[u^*, v](x) = u^*(x)v$ for $x \in V_1$ (p. 165). The fact that we have such an isomorphism can be seen by noting that $u^* \otimes v$ is bilinear. Hence we have a linear map η of $V_1{}^* \otimes_F V_2$ into $\hom_F(V_1, V_2)$ such that $\eta(u^* \otimes v) = [u^*, v]$. It is readily seen that η is surjective and since the two spaces $V_1{}^* \otimes_F V_2$ and $\hom_F(V_1, V_2)$ have the same dimensionality $[V_1 : F][V_2 : F]$, it follows that η is a linear isomorphism.

Now suppose ρ_i, $i = 1, 2$, is a representation of a group G acting on V_i/F. As on p. 272, we obtain a representation ρ of G acting on $\hom_F(V_1, V_2)$ if we define $\rho(g)l = \rho_2(g)l\rho_1(g)^{-1}$, $g \in G$, $l \in \hom_F(V_1, V_2)$. The isomorphism η of $V_1{}^* \otimes_F V_2$ with $\hom_F(V_1, V_2)$ is an equivalence of the representations $\rho_1{}^* \otimes \rho_2$

and ρ. To see this, let $u^* \in V_1{}^*$, $v \in V_2$. Then $\eta(\rho_1{}^*(g)u^* \otimes \rho_2(g)v)$ is the map

$$\begin{aligned} x \rightsquigarrow (\rho_1{}^*(g)u^*)(x)\rho_2(g)v &= u^*(\rho_1(g)^{-1}x)\rho_2(g)v \\ &= \rho_2(g)(u^*(\rho_1(g)^{-1}x)v) \\ &= \rho_2(g)\eta(u^* \otimes v)\rho_1(g)^{-1}x. \end{aligned}$$

Hence $\eta(\rho_1{}^*(g)u^* \otimes \rho_2(g)v) = \rho(g)\eta(u^* \otimes v)$, which implies that η is an $F[G]$-isomorphism of $V_1{}^* \otimes V_2$ onto $\hom_F(V_1, V_2)$.

We now consider the vector space $\hom_{F[G]}(V_1, V_2)$. Evidently, this is a subspace of $\hom_F(V_1, V_2)$ and we can identify this subspace with the set of $l \in \hom_F(V_1, V_2)$ such that $\rho(g)l = l$, $g \in G$. For this condition is equivalent to $\rho_2(g)l\rho_1(g)^{-1} = l$ and to $\rho_2(g)l = l\rho_1(g)$, which is the condition that $l \in \hom_{F[G]}(V_1, V_2)$. In general, if ρ is a representation of G acting on V, then the set of $x \in V$ such that $\rho(g)x = x$ for all $g \in G$ is denoted as $\operatorname{Inv} \rho$. Evidently this is a subspace. It is clear also that $x \neq 0$ is in $\operatorname{Inv} \rho$ if and only if Fx is a one-dimensional $\rho(G)$-invariant subspace and the restriction of ρ to Fx is the unit representation. Hence if ρ is a completely reducible representation of G, then $[\operatorname{Inv} \rho : F]$ is the multiplicity of the unit representation in ρ.

If ρ_1 and ρ_2 are completely reducible representations, then we shall say that ρ_1 and ρ_2 are *disjoint* if they have no common irreducible constituents, that is, if there exists no irreducible representation that is isomorphic to a subrepresentation of both ρ_1 and ρ_2.

We have the following results, which are easy consequences of our definitions and of the $F[G]$-isomorphism of $V_1{}^* \otimes_F V_2$ and $\hom_F(V_1, V_2)$.

PROPOSITION 5.8. (i) *Let ρ_i, $i = 1, 2$, be a representation of a group G acting in V_i/F. Then the intertwining number $\iota(\rho_1, \rho_2) = [\operatorname{Inv} \rho_1{}^* \otimes \rho_2 : F]$. Moreover, if $\rho_1{}^* \otimes \rho_2$ is completely reducible, then $\iota(\rho_1, \rho_2)$ is the multiplicity of the unit representation in $\rho_1{}^* \otimes \rho_2$.* (ii) *If ρ_1, ρ_2, and $\rho_1{}^* \otimes \rho_2$ are completely reducible, then ρ_1 and ρ_2 are disjoint if and only if the unit representation is not a component of $\rho_1{}^* \otimes \rho_2$.* (iii) *Let F be algebraically closed and let ρ be a completely reducible representation of G acting in V/F such that $\rho^* \otimes \rho$ is also completely reducible. Then ρ is irreducible if and only if the unit representation has multiplicity one in $\rho^* \otimes \rho$.*

Proof. (i) By definition, $\iota(\rho_1, \rho_2) = [\hom_{F[G]}(V_1, V_2) : F]$. We have seen also that $\hom_{F[G]}(V_1, V_2) = \operatorname{Inv} \rho$ where ρ is the representation of G acting in $\hom_F(V_1, V_2)$ such that $\rho(g)l = \rho_2(g)l\rho_1(g)^{-1}$. Since ρ is equivalent to $\rho_1{}^* \otimes \rho_2$, $[\operatorname{Inv} \rho : F] = [\operatorname{Inv} \rho_1{}^* \otimes \rho_2 : F]$. Hence $\iota(\rho_1, \rho_2) = [\hom_{F[G]}(V_1, V_2) : F] =$

$[\mathrm{Inv}\,\rho : F] = [\mathrm{Inv}\,\rho_1{}^* \otimes \rho_2 : F]$. This proves the first statement in (i). The second statement is an immediate consequence of this and the fact that for ρ completely reducible, $[\mathrm{Inv}\,\rho : F]$ is the multiplicity of the unit representation in ρ. (ii) If ρ_1 and ρ_2 are completely reducible, then (94) and Schur's lemma show that ρ_1 and ρ_2 are disjoint if and only if $\imath(\rho_1, \rho_2) = 0$. By (i), this holds if and only if $\mathrm{Inv}\,\rho_1{}^* \otimes \rho_2 = 0$. If $\rho_1{}^* \otimes \rho_2$ is completely reducible, this holds if and only if the unit representation is not a component of $\rho_1{}^* \otimes \rho_2$. (iii) By (94) and (95), if F is algebraically closed and ρ is completely reducible, then ρ is irreducible if and only if $\imath(\rho, \rho) = 1$. By (i), this holds if and only if $[\mathrm{Inv}\,\rho^* \otimes \rho : F] = 1$. If $\rho^* \otimes \rho$ is completely reducible, this is the case if and only if the unit representation has multiplicity one in $\rho^* \otimes \rho$. □

The awkward hypotheses on complete reducibility in this proposition can be removed if G is finite and $\mathrm{char}\, F \nmid |G|$. In this case we have the following criterion for irreducibility of induced representations.

THEOREM 5.20. *Let G be a finite group, H a subgroup, F an algebraically closed field of characteristic not a divisor of $|G|$, and let σ be a representation of H acting on U/F. Then σ^G is irreducible if and only if* (1) *σ is irreducible and* (2) *for every $g \notin H$, the representations of $H \cap {}^gH$ on U and on gU are disjoint.*

Proof. Evidently (1) is a necessary condition for the irreducibility of σ^G. Now assume that σ is irreducible. By Proposition 5.8 (iii), σ^G is irreducible if and only if the unit representation has multiplicity 1 in $(\sigma^G)^* \otimes \sigma^G$. Since $(\sigma^G)^*$ and $(\sigma^*)^G$ are equivalent (Theorem 5.18, p. 294), we can replace $(\sigma^G)^*$ by $(\sigma^*)^G$. By Theorem 5.19, $U^{*G} \otimes U^G \cong \oplus (U^*_{H \cap {}^gH} \otimes (g_i U)_{H \cap {}^gH})^G$ where g runs over a set of representatives of the set Δ of double cosets HgH. Now $U^*_{H \cap {}^gH}$ and $(U_{H \cap {}^gH})^*$ mean the same thing: U^* as module for $F[H \cap {}^gH]$ determined by the contragredient action. Hence we have to consider the multiplicity in every $((U_{H \cap {}^{g_i}H})^* \otimes (g_i U)_{H \cap {}^{g_i}H})^G$ of the trivial module (giving the unit representation). We now note that if K is any subgroup of G and τ is a representation of K on V/F, then the multiplicity of the unit representation τ_1 of K in τ is the same as the multiplicity of the unit representation ρ_1 of G in τ^G. Since we have complete reducibility, it suffices to show that if τ is irreducible, then ρ_1 has multiplicity 0 in τ^G unless $\tau = \tau_1$, in which case, the multiplicity is 1. This follows immediately from Frobenius reciprocity. We now apply this to $K = H \cap {}^gH$, $g \in G$, and $V = (U_{H \cap {}^gH})^* \otimes (gU)_{H \cap {}^gH}$. If $g \in H$, then these become H and $U^* \otimes U$. Since σ is irreducible, the multiplicity of the unit representation in $\sigma^* \otimes \sigma$ is 1; hence, the component $(U^* \otimes U)^G$ contributes the multiplicity 1 for the unit representation of G on $U^{*G} \otimes U^G$. Hence σ^G is irreducible if and only if the multiplicity of the unit representation of $H \cap {}^gH$

on $(U_{H\cap {}^gH})^*\otimes(gU)_{H\cap {}^gH}$ is 0 for every $g\notin H$. By Proposition 5.8, this is the case if and only if the representations of $H\cap {}^gH$ on U and on gU are disjoint for every $g\notin H$. This completes the proof. □

An interesting special case of Theorem 5.20 is obtained by taking σ to be a representation of degree 1 of H. Then the representation σ^G is called a *monomial representation*. It is clear from the form of the matrix representation of an induced representation given on page 288 that a monomial representation has a matrix representation in which all of the matrices have a single non-zero entry in every row and column and this occurs in the same position for all of the matrices. Theorem 5.20 reduces to the following criterion for irreducibility of monomial representations.

COROLLARY 1 (K. Shoda). *Let ρ be a monomial representation of a finite group G obtained by inducing on a degree one representation σ of a subgroup H of G. Assume that the base field is algebraically closed of characteristic not a divisor of $|G|$. Then ρ is irreducible if and only if for every $g\in G-H$ there exists an $h\in H\cap {}^gH$ such that $\sigma(h)\neq\sigma(ghg^{-1})$.*

Another interesting special case of Theorem 5.20 is that in which $H\lhd G$. Then ${}^gH=H$ for every $g\in G$ and the representation of H on gU is the conjugate representation ${}^g\sigma$ (p. 255). Evidently the irreducibility criterion of Theorem 5.20 gives the following

COROLLARY 2. *Let G, H, F, and σ be as in Theorem 5.20 and assume that $H\lhd G$. Then σ^G is irreducible if and only if σ is irreducible and for every $g\in G-H$, ${}^g\sigma$ and σ are inequivalent.*

There is an important extension of this corollary, which we shall now derive. Again, let $H\lhd G$ and let σ be an irreducible representation of H acting in U/F. Let

$$T(\sigma)=\{g\in G\,|\,{}^g\sigma \text{ is equivalent to } \sigma\}.$$

Then $T(\sigma)$ is a subgroup of G containing H called the *inertial group* of σ. Then we have

COROLLARY 3. *Same hypothesis as Theorem 5.20, with $H\lhd G$. Let ψ be an irreducible representation of the inertial group $T(\sigma)$ such that ψ_H has σ as an irreducible component. Then ψ^G is irreducible.*

Proof. Let V be the $F[T]$-module on which ψ acts. By Clifford's theorem applied to $T = T(\sigma)$ and $H \lhd T$, σ is the only irreducible component of ψ_H. Now consider V^G, which we may assume contains V as a submodule, and consider the subspace gV of V^G as module for $F[T \cap {}^gT]$. Since $H \lhd G$, $H \subset T \cap {}^gT$ and gV is a sum of irreducible $F[H]$-submodules isomorphic to gU. The representation of H on gU is ${}^g\sigma$ and since $g \notin T$, this is not equivalent to σ. It follows that the representations of H on V and on gV are disjoint. Hence this is the case for the representations of $T \cap {}^gT$ on V and on gV. Then ψ^G is irreducible by Theorem 5.20. □

A consequence of Corollary 3 that will be useful in the sequel is

COROLLARY 4. *Let G be a finite group, $H \lhd G$, ρ an irreducible representation of G over an algebraically closed field F of characteristic not a divisor of $|G|$. Let σ be an irreducible component of ρ_H, T the inertial group of σ. Then $\rho = \psi^G$ for some irreducible representation ψ of T.*

Proof. Since σ is an irreducible component of ρ_H, there exists an irreducible component ψ of ρ_T such that σ is an irreducible component of ψ_H. Then ψ^G is irreducible by Corollary 3. Since ψ is an irreducible component of ρ_T, ρ is an irreducible component of ψ^G by Frobenius reciprocity. Hence $\rho = \psi^G$. □

The importance of this result is that if $T \subsetneq G$, then it gives a formula for ρ as an induced representation from a proper subgroup. This gives a way of establishing properties of irreducible representations of G by induction on $|G|$.

5.12 BRAUER'S THEOREM ON INDUCED CHARACTERS

For the remainder of this chapter, G will be finite and the base field F will be a subfield of $\mathbb{C}$—usually $\mathbb{C}$ itself. In this section, we shall prove a fundamental theorem of Brauer's on induced characters. To state this, we need the following

DEFINITION 5.3. *A group G is called* p-elementary *if $G = Z \times P$ where Z is cyclic of order prime to p and P is a p-group (that is, a group of order p^n). G is called* elementary *if it is p-elementary for some prime p.*

With this definition we can state

BRAUER'S THEOREM ON INDUCED CHARACTERS. *Any complex character of a group G is an integral linear combination of characters induced from linear characters of elementary subgroups of G.*

The proof we shall give of this theorem is due to D. Goldschmidt and M. Isaacs. Brauer's first proof was quite complicated. A considerably simpler one was given independently by Brauer and Tate and by K. Asano. These proofs as well as the Goldschmidt-Isaacs proof begin with the following simple observations.

Let ch(G) denote the set of $\mathbb{Z}$-linear combinations of the complex irreducible characters $\chi_1, \dots, \chi_s$ of the group G where χ_1 is the unit character. The elements of ch(G) are called *generalized characters*. Any complex character is a generalized character and since $\chi_i\chi_j$ is the character of the tensor product of the irreducible representations affording χ_i and χ_j, $\chi_i\chi_j \in \mathrm{ch}(G)$. This implies that ch($G$) is a ring, more precisely, a subring of the ring $\mathbb{C}^G$ of $\mathbb{C}$-valued functions on G with the usual component-wise addition and multiplication. χ_1 is the unit of ch(G). If H is a subgroup of G and ψ is a class function on H, then we define the *induced class function* ψ^G on G by

(96) $$\psi^G(g) = \frac{1}{|H|} \sum_{a \in G} \dot{\psi}(a^{-1}ga)$$

where $\dot{\psi}(h) = \psi(h)$ for $h \in H$ and $\dot{\psi}(g) = 0$ for $g \in G - H$. This is a class function on G and we have seen in (85) that if ψ is the character of a representation σ of H, then ψ^G is the character of σ^G. If χ is the character of a representation ρ of G, then we have

(97) $$\psi^G\chi = (\psi\chi_H)^G$$

where χ_H is the restriction of χ to H, since $\sigma^G \otimes \rho \cong (\sigma \otimes \rho_H)^G$ (Theorem 5.17.3, p. 292). Also transitivity of induction of representations implies the following transitivity formula

(98) $$(\psi^K)^G = \psi^G$$

if K is a subgroup of G containing H.

Now let $\mathscr{F}$ be any family of subgroups of G and let $\mathrm{ch}_{\mathscr{F}}(G)$ denote the set of $\mathbb{Z}$-linear combinations of characters of the form ψ^G where ψ is a complex character of a subgroup $H \in \mathscr{F}$. It is clear from (97) that $\mathrm{ch}_{\mathscr{F}}(G)$ is an ideal in ch(G). Hence $\mathrm{ch}_{\mathscr{F}}(G) = \mathrm{ch}(G)$ if and only if $\chi_1 \in \mathrm{ch}_{\mathscr{F}}(G)$.

The Goldschmidt-Isaacs proof of Brauer's theorem is made in two stages: first, the proof that $\chi_1 \in \mathrm{ch}_{\mathscr{F}}(G)$ for a family of subgroups, the so-called quasi-elementary subgroups, which is larger than the family of elementary subgroups and second, the proof of Brauer's theorem for quasi-elementary subgroups. For the proof of the second part we shall prove first an extension by Brauer of a theorem of H. Blichfeldt that any irreducible representation of a quasi-elementary group is monomial. Then we shall use this to prove Brauer's theorem for quasi-elementary groups.

A group G is called *p-quasi-elementary* for the prime p if G contains a normal cyclic subgroup Z such that $p \nmid |Z|$ and G/Z is a p-group. Evidently, if G is p-elementary, it is p-quasi-elementary. A group G is called *quasi-elementary* if it is p-quasi-elementary for some prime p.

Let G be p-quasi-elementary, Z a normal cyclic subgroup such that $p \nmid |Z|$ and G/Z is a p-group. Any subgroup H of G is p-quasi-elementary. For, $H \cap Z$ is cyclic of order prime to p and $H/(H \cap Z) \cong HZ/Z$ a subgroup of G/Z. Hence $H/(H \cap Z)$ is a p-group. The argument shows that if $p \nmid |H|$, then $|H/(H \cap Z)| = 1$ so $H \subset Z$. It follows that the subgroup Z specified in the definition of p-quasi-elementary is unique and hence this is a characteristic subgroup of G (p. 109). One can give a useful alternative definition of p-quasi-elementary, namely, $G = AP$ where A is cyclic and normal in G and P is a p-group. For, if this is the case, then we write $A = Z \times W$ where $p \nmid |Z|$ and $|W| = p^k$. Then Z and W are unique, hence characteristic in A and hence these are normal subgroups of G. Evidently $G/Z \cong WP$, which is a p-group. Conversely, suppose G is p-quasi-elementary and Z is as in the definition. Let P be a Sylow p-subgroup of G. Then $P \cap Z = 1$ and $PZ = ZP$ is a subgroup such that $|PZ| = |G|$. Hence $G = ZP$ where Z is cyclic and normal in G and P is a p-group. It is readily seen that a p-quasi-elementary group G is elementary if and only if the subgroup Z is contained in the center and if and only if a Sylow p-subgroup is normal. It follows that a subgroup of a p-elementary group is p-elementary.

The first part of the proof of Brauer's theorem is based on

LEMMA 1 (B. Banaschewski). *Let S be a non-vacuous finite set, R a subrng (rng = ring without unit, BAI, p. 155) of $\mathbb{Z}^S$. Then either R contains 1_S (and hence is a subring) or there exists an $x \in S$ and a prime p such that $f(x)$ is divisible by p for every $f \in R$.*

Proof. For any $x \in S$ let $I_x = \{f(x) | f \in R\}$. This is a subgroup of the additive group of $\mathbb{Z}$ and hence either $I_x = \mathbb{Z}$ or there exists a prime p such that $p | f(x)$ for every $f \in R$. If we have this for some x, then we have the second alternative. Now assume $I_x = \mathbb{Z}$ for every $x \in S$. Then for each x, we can choose an $f_x \in R$ such that $f_x(x) = 1$ and hence $(f_x - 1_S)(x) = 0$. Then $\prod_{x \in S}(f_x - 1_S) = 0$. Expanding this gives an expression for 1_S as a polynomial in the f_x with integer coefficients. Thus $1_S \in R$. $\square$

We shall also require

LEMMA 2. *For any $g \in G$ and any prime p there exists a p-quasi-elementary subgroup H of G such that $\chi_{\sigma_1^G}(g)$ is not divisible by p (σ_1 the unit representation of H).*

Proof. We can write $\langle g\rangle = Z \times W$ where $p \nmid |Z|$ and $|W| = p^k$. Let N be the normalizer of Z in G and let $\bar{H}$ be a Sylow p-subgroup of $\bar{N} = N/Z$ containing the p-group $\langle g\rangle/Z$ (BAI, p. 81). Then $\bar{H} = H/Z$ where H is a subgroup of N containing $\langle g\rangle$. Since $Z \lhd H$ and $p \nmid |Z|$ and $|H/Z|$ is a power of p, H is p-quasi-elementary. We wish to show that $\chi_{\sigma_1^G}(g)$ is not divisible by p. By (86), this is equivalent to showing that the number of cosets aH such that $g(aH) = aH$ is not divisible by p. If aH satisfies this condition, then $a^{-1}ga \in H$ and hence $a^{-1}Za \subset H$. Since H is p-quasi-elementary, Z is the only subgroup of H of order $|Z|$. Hence we have $a^{-1}Za = Z$ and $a \in N$. Hence we have to count the number of cosets aH of H in N such that $g(aH) = aH$. Now consider the action of $\langle g\rangle$ on N/H. Since $Z \lhd N$ and $Z \subset H$, $z(aH) = aH$ if $z \in Z$ and $a \in N$. Hence Z is contained in the kernel of the action of $\langle g\rangle$ on N/H. Hence we have the action of $\langle g\rangle/Z$ on N/H in which $(gZ)(aH) = gaH$. Since $|\langle g\rangle/Z| = p^k$, every orbit of the action of $\langle g\rangle/Z$ and hence of $\langle g\rangle$ on N/H has cardinality a power of p (BAI, p. 76). It follows that the number of non-fixed cosets aH under the action of g is divisible by p. Hence the number of fixed ones, $\chi_{\sigma_1^G}(g) \equiv [N:H] \pmod{p}$. Since H contains a Sylow p-subgroup of N, $[N:H]$ is not divisible by p. Hence $\chi_{\sigma_1^G}(g)$ is not divisible by p. □

We can now complete the first part of the proof of Brauer's theorem by proving

THEOREM 5.21. *Any complex character of G is an integral linear combination of characters induced from quasi-elementary subgroups.*

Proof. In the notation we introduced, this means that if Q denotes the family of quasi-elementary subgroups of G, then $\mathrm{ch}_Q(G) = \mathrm{ch}(G)$. Since $\mathrm{ch}_Q(G)$ is an ideal in $\mathrm{ch}(G)$, it suffices to show that $\chi_1 \in \mathrm{ch}_Q(G)$. Now let R be the subrng of $\mathrm{ch}_Q(G)$ generated by the induced characters ${\sigma_1}^G$ of unit characters σ_1 of quasi-elementary subgroups of G. It is clear that ${\sigma_1}^G \in \mathbb{Z}^G$; hence, $R \subset \mathbb{Z}^G$. Hence, by Banaschewski's lemma, if $\chi_1 \notin \mathrm{ch}_Q(G)$, then $\chi_1 \notin R$ and so for some $g \in G$ there exists a prime p such that $\chi(g) \equiv 0 \pmod{p}$ for every $\chi \in R$. This contradicts Lemma 2, which provides for any $g \in G$ and any prime p, a $\chi \in R$ such that $\chi(g)$ is not divisible by p. Hence $\mathrm{ch}_Q(G) = \mathrm{ch}(G)$. □

We prove next that any irreducible complex representation of a p-quasi-elementary group is monomial and has degree a power of p. In terms of characters this has the following form

THEOREM 5.22 (Blichfeldt-Brauer). *Let χ be an irreducible character of a p-quasi-elementary group G. Then*

(1) *the degree* $\chi(1)$ *is a power of* p, *and*
(2) $\chi = \lambda^G$ *for some linear character* λ *of a subgroup of* G.

Proof. We have $G = ZP$ where $Z \lhd G$, Z is cyclic with $p \nmid |Z|$, and P is a p-group. Let ρ be a representation of G affording χ.

(1) Let σ be an irreducible component of ρ_Z, and T the inertial group of σ. Then $Z \lhd T$ and the index of T in G is a power of p. By Corollary 4 to Theorem 5.20, $\rho = \psi^G$ for an irreducible representation ψ of T. Hence if $T \neq G$, the result follows by induction on $|G|$. Thus we may assume $T = G$. Since Z is abelian, σ is of degree 1, and since $Z \lhd G = T$ and ρ is irreducible, it follows from Clifford's theorem that every $a \in Z$ acts as a scalar in the space V on which ρ acts. Evidently this implies that ρ_P is irreducible. Since P is a p-group, the degree of ρ, which is a factor of $|P|$, is a power of p.

(2) Let the degree $\chi(1) = p^n$. We shall use induction on n. The result is clear if $n = 0$ since in this case χ is linear. Hence we assume $n > 0$. If λ is a linear character of G, then $\chi\lambda$ is a character whose degree $\chi\lambda(1) = \chi(1)$. Hence either $\chi = \chi\lambda$ or χ is not a component of $\chi\lambda$. The condition $\chi = \chi\lambda$ is equivalent to $(\chi|\chi\lambda) = 1$ and since

$$(\chi|\chi\lambda) = \frac{1}{|G|}\sum_g \bar{\chi}(g)\chi(g)\lambda(g) = (\chi\bar{\chi}|\lambda)$$

it is equivalent also to $(\chi\bar{\chi}|\lambda) = 1$. Now $\chi\bar{\chi}$ is the character of $\rho \otimes \rho^*$ and $(\chi\bar{\chi}|\lambda) = 1$ is equivalent to the fact that the multiplicity of λ in $\chi\bar{\chi}$ is 1. Let Λ be the set of linear characters λ of G such that $\chi\lambda = \chi$. It is clear that Λ is a group under multiplication. We have

$$\chi\bar{\chi} = \sum_{\lambda \in \Lambda} \lambda + \sum \chi' \tag{99}$$

where the χ' are non-linear characters. Since $\chi(1)$ and the $\chi'(1)$ are divisible by p and $\lambda(1) = 1$, it follows from (99) that $|\Lambda| \equiv 0 \pmod p$. It follows that there exists a $\lambda_1 \in \Lambda$ such that $\lambda_1 \neq \chi_1$ (the unit character) and $\lambda_1{}^p = \chi_1$. Since λ_1 is linear, it is a homomorphism of G into the multiplicative group of complex numbers, and since $\lambda_1{}^p = \chi_1$ and $\lambda_1 \neq \chi_1$, the image is the group of pth roots of unity. Hence, if $K = \ker \lambda_1$ then $K \lhd G$ and $[G:K] = p$. By (99), we have

$$\chi_K\bar{\chi}_K = \sum_{\lambda \in \Lambda} \lambda_K + \sum \chi'_K \tag{100}$$

and $\lambda_{1K} = \chi_{1K}$. This shows that the multiplicity of the unit representation in $\rho_K \otimes \rho_K^*$ is at least two and hence if $(\,|\,)_K$ denotes the scalar product on K, then $(\chi_K\bar{\chi}_K|\chi_{1K})_K \geqslant 2$. Then $(\chi_K|\chi_K)_K \geqslant 2$, which implies that ρ_K is reducible. Let σ

be an irreducible component of ρ_K, ψ the character of σ. Since $\chi(1)$ and $\psi(1)$ are powers of p and $\psi(1) < \chi(1)$, we have $\chi(1) \geqslant p\psi(1)$. Since σ is an irreducible component of ρ_K, ρ is an irreducible component of σ^G by Frobenius reciprocity. Since $\psi^G(1) = p\psi(1)$ and $\chi(1) \geqslant p\psi(1)$, it follows that $\chi = \psi^G$. The result now follows by induction. □

We shall now make the passage from quasi-elementary groups to elementary ones. As before, let $G = ZP$ where Z is a cyclic normal subgroup of G of order prime to p and P is a p-group. Let $W = C_Z(P) \equiv Z \cap C_G(P)$, $H = WP$. Then H is a p-elementary subgroup of G with P as normal Sylow p-subgroup. We shall need the following

LEMMA. *Let λ be a linear character of G such that $H \subset \ker \lambda$. Then $\lambda = \chi_1$, the unit character of G.*

Proof. It suffices to show that $Z \subset \ker \lambda$. Let $K = Z \cap \ker \lambda$. Let $b \in Z$, $d \in P$. Then $\lambda(b^{-1}dbd^{-1}) = 1$ since λ is a homomorphism. Hence $d(bK)d^{-1} = bK$ and so every coset bK of K in Z is stabilized under the conjugations by the elements of P. Thus we have an action of P by conjugations on the coset bK. Since $|P|$ is a power of p and $|bK| = |K|$ is prime to p, we have a fixed point under this action. Thus $bK \cap C_Z(P) \neq \varnothing$, which implies that $\lambda(b) = 1$. Since b was arbitrary in Z, we have $Z \subset \ker \lambda$. □

We can now prove

THEOREM 5.23. *Any irreducible character of a quasi-elementary group is an integral linear combination of characters induced from linear characters of elementary subgroups.*

Proof. Let χ be a complex irreducible character of a p-quasi-elementary group G. If $\chi(1) > 1$, the result follows from Theorem 5.22 by induction on $|G|$. Hence assume $\chi(1) = 1$, that is, χ is linear. Let H be the p-elementary subgroup WP defined above and let $\eta = \chi_H$. By the Frobenius reciprocity theorem the multiplicity of χ in η^G is 1. Now let χ' be a linear component of η^G. Again, by Frobenius reciprocity, η is a constituent of χ'_H. Hence $\chi'_H = \eta = \chi_H$. Then $\chi'' = \chi'\chi^{-1}$ is a linear character on G such that $H \subset \ker \chi''$. Hence by the lemma, χ'' is the unit character on G. Then $\chi' = \chi$. Thus $\eta^G = \chi + \theta$ where θ is a sum of characters of degree > 1. Since the induction on $|G|$ implies that θ is an integral linear combination of characters induced from linear characters of elementary subgroups and since η is linear on the elementary subgroup H, the required result follows for $\chi = \eta^G - \theta$. □

Evidently Brauer's theorem is an immediate consequence of Theorems 5.22 and 5.23 and transitivity of induction.

We shall now derive an important consequence of the theorem: a characterization of generalized characters among the class functions. Let cf(G) denote the set of complex class functions on G. Evidently cf(G) is a subring of $\mathbb{C}^G$ containing ch(G). We know also that cf(G) is the $\mathbb{C}$-vector space spanned by the irreducible characters $\chi_1, \ldots, \chi_s$ and we have defined ψ^G for a class function on a subgroup H by (96). Since the maps $\chi \rightsquigarrow \chi_H$ and $\psi \rightsquigarrow \psi^G$ are linear, it is clear that formula (97), $\psi^G\chi = (\psi\chi_H)^G$, is valid also for class functions. We can use this to establish

BRAUER'S CHARACTERIZATION OF GENERALIZED CHARACTERS. *A class function φ on G is a generalized character if and only if φ_H is a generalized character for every elementary subgroup H of G.*

Proof. Let ch$(G)'$ be the set of class functions φ of G satisfying the stated condition. It is clear that ch$(G)'$ is a subring of cf(G) containing ch(G). Now let ψ be a character on some elementary subgroup H of G and let $\chi \in \text{ch}(G)'$. Then $\chi_H \in \text{ch}(H)$, so $\psi\chi_H \in \text{ch}(H)$ and $(\psi\chi_H)^G \in \text{ch}(G)$. Then $\psi^G\chi = (\psi\chi_H)^G \in \text{ch}(G)$. Since any element of ch(G) is an integral linear combination of characters of the form ψ^G, ψ a character on some elementary subgroup, this implies that ch(G) is an ideal in ch$(G)'$. Since $\chi_1 \in \text{ch}(G)$, we have ch$(G)' = $ ch(G), which is equivalent to the statement of the criterion. □

We remark that Brauer's characterization also gives a characterization of the irreducible characters, since $\chi \in \text{ch}(G)$ is an irreducible character if and only if $(\chi|\chi) = 1$ and $\chi(1) > 0$. This is clear since $\chi \in \text{ch}(G)$ if and only if $\chi = \sum_1^s n_i\chi_i$, $n_i \in \mathbb{Z}$, and $(\chi|\chi) = \sum n_i^2 = 1$ implies that all $n_i = 0$ except one that has the value ± 1. The condition $\chi(1) > 0$ then gives $\chi = \chi_j$ for some j.

EXERCISES

1. A group G is called an *M-group* if every irreducible complex representation is monomial. Note that Theorem 5.22.2 states that every quasi-elementary group is an M-group. Show that the direct product of M-groups is an M-group. Show that any nilpotent group is an M-group (see BAI, pp. 250–251, exercises 4–11).

2. (Taketa.) Show that every M-group is solvable. (*Sketch of proof*: If the result is false, then there is a minimal counterexample: a non-solvable M-group G with $|G|$

minimal. Since any homomorphic image of an M-group is an M-group, if $A \lhd G$ and $A \neq G$, then G/A is solvable. Let A and B be minimal normal subgroups of G. If $A \neq B$, then $A \cap B = 1$ and G is isomorphic to a subgroup of $G/A \times G/B$ and so is solvable by the minimality of G. Hence $A = B$ and there exists a unique minimal normal subgroup A of G. There exist irreducible representations ρ of G such that $\ker \rho \not\supset A$. Let ρ be one of minimal degree and let $\rho = \sigma^G$ where σ is a representation of degree 1 of a subgroup H. Put $\rho' = {\sigma_1}^G$ where σ_1 is the unit representation of H. Show that $\ker \rho'$ is an abelian normal subgroup $\neq 1$ of G. Then G is solvable contrary to hypothesis.)

The following exercises sketch a proof of an important theorem on induced characters due to Artin:

THEOREM 5.24. *Let χ be a complex character of G that is rational valued. ($\chi(g) \in \mathbb{Q}$ for all $g \in G$.) Then*

$$\chi = \sum_Z \frac{a_Z}{[N(Z):Z]} {\chi_{1Z}}^G \tag{101}$$

where χ_1 is the unit character, the $a_Z \in \mathbb{Z}$, and the summation is taken over the cyclic subgroups Z of G.

3. Let χ be a rational valued character on G and let $g, h \in G$ satisfy $\langle g \rangle = \langle h \rangle$. Show that $\chi(g) = \chi(h)$. (*Hint*: If $|\langle g \rangle| = m$, then $h = g^k$ where $(k, m) = 1$. Let $\Lambda^{(m)}$ be the cyclotomic field of mth roots of unity over $\mathbb{Q}$, so $\Lambda^{(m)} = \mathbb{Q}(\varepsilon)$ where ε is a primitive mth root of unity. Show that there exists an automorphism σ of $\Lambda^{(m)}/\mathbb{Q}$ such that $\sigma\varepsilon = \varepsilon^k$ and that if $g \rightsquigarrow \operatorname{diag}\{\varepsilon_1, \varepsilon_2, \ldots, \varepsilon_n\}$, $\varepsilon_i \in \Lambda^{(m)}$, is a matrix representation, then $h = g^k \rightsquigarrow \operatorname{diag}\{\sigma\varepsilon_1, \sigma\varepsilon_2, \ldots, \sigma\varepsilon_n\}$. Hence conclude that $\chi(g) = \chi(h)$.)

4. Let χ be as in exercise 3. Define an equivalence relation $\equiv$ in G by $g \equiv h$ if $\langle g \rangle$ and $\langle h \rangle$ are conjugate in G ($\langle g \rangle = a\langle h \rangle a^{-1}$ for some $a \in G$). Let $D_1, D_2, \ldots, D_t$ be the distinct equivalence classes determined by $\equiv$. Note that these are unions of conjugacy classes and we may assume $D_1 = \{1\}$. Let $g_i \in D_i$ and let $|\langle g_i \rangle| = n_i$, $N(\langle g_i \rangle)$, the normalizer of $\langle g_i \rangle$. Show that

$$|D_i| = [G:N(\langle g_i \rangle)]\varphi(n_i) \tag{102}$$

(φ the Euler φ-function). Let Φ_i be the characteristic function of D_i, so $\Phi_i(g) = 1$ if $g \in D_i$ and $\Phi_i(g) = 0$ otherwise. Prove by induction on n_i that

$$|N(\langle g_i \rangle)|\Phi_j = \sum a_j n_j {\chi_{1\langle g_j \rangle}}^G \tag{103}$$

for $a_j \in \mathbb{Z}$ where $a_j = 0$ unless $\langle g_j \rangle$ is conjugate to a subgroup of $\langle g_i \rangle$.

5. Note that χ is an integral linear combination of the functions Φ_i. Use this and (103) to prove Artin's theorem.

5.13 BRAUER'S THEOREM ON SPLITTING FIELDS

In this section, we shall prove that if m is the exponent of a finite group G, that is, the least common multiple of the orders of its elements, then the cyclotomic field $\Lambda^{(m)}$ of mth roots of unity over $\mathbb{Q}$ is a splitting field for G. This result was conjectured by Maschke around 1900, but was not proved in complete generality until 1945 when a proof was given by Brauer, based on his theory of modular representations. Subsequently, Brauer discovered his theorem on induced characters and observed that the splitting field theorem is an easy consequence. We shall follow this approach here.

We recall that a representation ρ of G acting on V/F is called absolutely irreducible if the extension representation ρ_K is irreducible for any field K/F (p. 263). The field F is a splitting field for G if every irreducible representation of G over F is absolutely irreducible. We have proved (Theorem 5.8, p. 264) that F is a splitting field for G if and only if $F[G]$ is a direct sum of matrix algebras $M_n(F)$.

We now suppose that F is a subfield of $\mathbb{C}$. Then we can give an alternative, more intuitive definition of splitting fields. For, as we shall show, F is a splitting field for G if and only if for any representation ρ acting on a vector space $V/\mathbb{C}$, there exists a base for V such that for the corresponding matrix representation, the matrices have all of their entries in F. Since we have complete reducibility of the representations, this holds if and only if it holds for the irreducible representations. It is clear also that the condition can be formulated completely in matrix terms: Any complex matrix representation is similar to a matrix representation over F. Now suppose F is a splitting field. Then $F[G] = M_{n_1}(F) \oplus \cdots \oplus M_{n_s}(F)$ and $\mathbb{C}[G] = F[G]_{\mathbb{C}} = M_{n_1}(\mathbb{C}) \oplus \cdots \oplus M_{n_s}(\mathbb{C})$. If I_j is a minimal left ideal in $M_{n_j}(F)$, then this is an irreducible module for $F[G]$ and $\{I_j | 1 \leqslant j \leqslant s\}$ is a set of representatives for the irreducible $F[G]$-modules. The degree of the representation afforded by I_j is n_j and $|G| = \sum n_j^2$. Now $I_{j\mathbb{C}}$ is a left ideal contained in the simple component $M_{n_j}(\mathbb{C})$. Since $[I_{j\mathbb{C}} : \mathbb{C}] = n_j$, $I_{j\mathbb{C}}$ is a minimal left ideal in $M_{n_j}(\mathbb{C})$ and hence the $I_{j\mathbb{C}}$, $1 \leqslant j \leqslant s$, constitute a set of representatives of the irreducible modules for $\mathbb{C}[G]$. Since any base for I_j/F is a base for $I_{j\mathbb{C}}/\mathbb{C}$, it is clear that the irreducible complex matrix representations are similar to matrix representations over F. Conversely, suppose this condition holds for a field F. Then every irreducible complex representation has the form $\sigma_{\mathbb{C}}$ where σ is a representation of G over F. Let $\rho_1, \ldots, \rho_s$ be a set of representatives of the irreducible representations of G, n_j the degree of ρ_j, and choose a representation σ_j of G over F such that $\sigma_{j\mathbb{C}} = \rho_j$. Then the σ_j are inequivalent and irreducible, and the relation $|G| = \sum n_j^2$ implies that $F[G] = M_{n_1}(F) \oplus \cdots \oplus M_{n_s}(F)$ (see p. 259). Hence F is a splitting field for G.

Let χ and χ' be characters of inequivalent irreducible representations ρ and ρ' of G over F. If $F[G] = M_{m_1}(\Delta_1) \oplus \cdots \oplus M_{m_r}(\Delta_r)$ where the Δ_i are division algebras, then we may suppose that ρ acts on a minimal left ideal $I \subset M_{m_1}(\Delta_1)$ and ρ' acts on a minimal left ideal I' of $M_{m_2}(\Delta_2)$. When we pass to $\mathbb{C}[G] = F[G]_{\mathbb{C}}$, $M_{m_1}(\Delta_1)_{\mathbb{C}}$ and $M_{m_2}(\Delta_2)_{\mathbb{C}}$ split as direct sums of simple components of $\mathbb{C}[G]$ and these ideals in $\mathbb{C}[G]$ have no common simple components. It follows that if $\chi_1, \chi_2, \ldots, \chi_s$ are the complex irreducible characters, then $\chi = \sum m_i\chi_i$ and $\chi' = \sum m_i'\chi_i$ where the m_i and m_i' are non-negative integers and for every i, either $m_i = 0$ or $m_i' = 0$. Thus $(\chi|\chi) = \sum m_i^2 > 0$ and $(\chi|\chi') = 0$. Moreover, χ is an irreducible complex character if and only if $(\chi|\chi) = 1$, in which case, the complex irreducible representation whose character is χ is the extension $\rho_{\mathbb{C}}$. It is also clear that F is a splitting field for G if and only if for every irreducible complex character χ_i there exists a representation ρ_i of G over F such that $\chi_{\rho_i} = \chi_i$.

We can now prove

THEOREM 5.25. *If m is the exponent of G, then the cyclotomic field $\Lambda^{(m)}/\mathbb{Q}$ of the mth roots of unity is a splitting field for G.*

Proof. Let χ be an irreducible complex character of G. By Brauer's theorem on induced characters, χ is an integral linear combination of characters of the form λ^G where λ is a linear character of a subgroup H of G. Now λ is a homomorphism of H into the multiplicative group $\mathbb{C}^*$ of complex numbers and if $h \in H$, then $h^m = 1$ so $\lambda(h)^m = 1$. Hence $\lambda(h) \in \Lambda^{(m)}$. It is clear from the definition of induced representations that the representation affording λ^G has a representation by matrices with entries in $\Lambda^{(m)}$. Thus λ^G is the character of a representation of G over $\Lambda^{(m)}$. It follows that $\chi = \sum_1^r k_j\varphi_j$ where the $k_j \in \mathbb{Z}$ and $\varphi_1, \ldots, \varphi_r$ are the characters of the inequivalent irreducible representations of G over $\Lambda^{(m)}$. Then $1 = (\chi|\chi) = \sum k_j^2(\varphi_j|\varphi_j)$ and since every $(\varphi_j|\varphi_j)$ is a positive integer, all of the k_j but one are zero and the non-zero one is 1. Hence χ is a character of an irreducible representation of G over $\Lambda^{(m)}$. This implies that $\Lambda^{(m)}$ is a splitting field. □

5.14 THE SCHUR INDEX

In this section, we shall study relations between irreducible representations of a finite group G over $\mathbb{C}$ and over a subfield F of $\mathbb{C}$. We use the notations of the previous section: $\mathbb{C}[G] = M_{n_1}(\mathbb{C}) \oplus \cdots \oplus M_{n_s}(\mathbb{C})$, $F[G] = M_{m_1}(\Delta_1) \oplus \cdots \oplus M_{m_r}(\Delta_r)$ where the Δ_i are division algebras. We write also $A_i = M_{m_i}(\Delta_i)$.

Let χ be an irreducible complex character of G, ρ a representation of G over $\mathbb{C}$ affording χ. We may assume that ρ is the restriction of the regular representation of G to a minimal left ideal V of one of the simple components $M_{n_j}(\mathbb{C})$. Then V is contained in exactly one of the $A_{i\mathbb{C}}$. If $V \subset A_{i\mathbb{C}}$ then $A_i V \neq 0$, but $A_{i'} V = 0$ for every $i' \neq i$. This property is independent of the choice of the representation ρ affording χ. We shall say that A_i is the simple component of $F[G]$ *belonging to* χ. We can associate also a subfield of $\mathbb{C}/F$ with χ, namely, the subfield over F generated by the complex numbers $\chi(g)$, $g \in G$. We denote this as $F(\chi)$. We have the following

PROPOSITION 5.9. *Let χ be an irreducible complex character of the finite group G, F a subfield of $\mathbb{C}$, and let $A = A_i$ be the simple component of $F[G]$ belonging to χ. Then $F(\chi)$ is isomorphic over F to the center of A.*

Proof. As indicated, we may take the representation ρ affording χ to be the restriction of the regular representation to a minimal left ideal $V \subset A_{\mathbb{C}}$. We can regard ρ as a representation of $\mathbb{C}[G]$ and identify $F[G]$ with an F-subalgebra of $\mathbb{C}[G]$. Since ρ is irreducible, it follows from Schur's lemma that $\rho(\text{cent}\, \mathbb{C}[G]) = \mathbb{C}1$. We recall that the centers of $F[G]$ and $\mathbb{C}[G]$ are spanned by the elements $\sum_{g \in C_i} g$ where $C_1, \ldots, C_s$ are the conjugacy classes of G. If $g \in C_i$, then $\sum_{a \in G} aga^{-1}$ is a non-zero multiple of $\sum_{g \in C_i} g$. Hence the centers are spanned by the elements $c_g = \sum_{a \in G} aga^{-1}$. We have $\rho(c_g) = \gamma 1$, $\gamma \in \mathbb{C}$, and hence $\text{tr}\, \rho(c_g) = n\gamma$, where $n = \chi(1)$, the degree of ρ. On the other hand, $\text{tr}\, \rho(c_g) = \sum_{a \in G} \text{tr}\, \rho(aga^{-1}) = |G|\chi(g)$. Hence

$$\rho(c_g) = \frac{|G|}{n}\chi(g)1. \tag{104}$$

We now restrict the $\mathbb{C}$-algebra homomorphism ρ of $\mathbb{C}[G]$ to $F[G]$. This gives an F-algebra homomorphism of $F[G]$, which maps $\text{cent}\, F[G]$ onto the set of F-linear combinations of the elements $\chi(g)1$. Since ρ maps every simple component $A_i \neq A$ into 0, we have a homomorphism of $\text{cent}\, A$ onto $F(\chi)1$. Since $\text{cent}\, A$ is a field, we have an isomorphism of $\text{cent}\, A$ onto $F(\chi)$. $\square$

We shall now say that a complex character χ of G is *realizable over the subfield* F of $\mathbb{C}$ if χ is the character of a representation σ of G over F. Since a representation is determined up to equivalence by its character, it is clear that χ is realizable over F if and only if ρ is equivalent to $\sigma_{\mathbb{C}}$ for any representation ρ affording χ. Also, as we noted before, this is the case if and only if there exists a base for the space V on which ρ acts such that the entries of the matrices of the $\rho(g)$, $g \in G$, are all in F. Evidently, if χ is realizable over F, then F contains every $\chi(g)$, $g \in G$.

We shall now show that if χ is an irreducible complex character of G and F is a subfield of $\mathbb{C}$ containing all the $\chi(g)$, then there exists a positive integer d such that the character $d\chi$ (of the direct sum of d copies of the representation ρ affording χ) is realizable over F. The minimum such d is called the *Schur index of χ over F*. We prove the existence of the Schur index and give a structural description of this integer in the following

THEOREM 5.26. *Let χ be an irreducible complex character of G, F a subfield of $\mathbb{C}$ containing all of the $\chi(g)$, $g \in G$, and let $A = M_m(\Delta)$, where Δ is a division algebra, be the simple component of $F[G]$ belonging to χ. Then $[\Delta : F] = d^2$ and the character $d\chi$ is realizable over F. Moreover, if d' is any positive integer such that $d'\chi$ is realizable over F, then $d | d'$ and hence d is the Schur index of χ over F.*

Proof. Since $F = F(\chi)$, A is central simple by Proposition 5.9. Hence Δ is central, $[\Delta : F] = d^2$ (p. 222), and $[A : F] = m^2 d^2$. Any two minimal left ideals of A are isomorphic as A-modules, hence as vector spaces over F, and $A = I_1 \oplus \cdots \oplus I_m$ where I_j, $1 \leqslant j \leqslant m$, is a minimal left ideal. Hence $[I : F] = md^2$ and $[I_{\mathbb{C}} : \mathbb{C}] = md^2$. Since A is central simple, $A_{\mathbb{C}}$ is simple and hence this is one of the simple components of $\mathbb{C}[G]$. Since $[A_{\mathbb{C}} : \mathbb{C}] = m^2 d^2$, we have $A_{\mathbb{C}} = M_{md}(\mathbb{C})$ and the calculation we gave for F shows that if V is a minimal left ideal of $A_{\mathbb{C}}$, then $[V : \mathbb{C}] = md$. We have seen that the representation ρ affording χ can be taken to be the restriction of the regular representation to a minimal left ideal contained in $A_{\mathbb{C}}$. We may take this to be V. Now $I_{\mathbb{C}}$ is a left ideal in $A_{\mathbb{C}}$, and $[I_{\mathbb{C}} : \mathbb{C}] = md^2$ while $[V : \mathbb{C}] = md$. Hence $I_{\mathbb{C}}$ is a direct sum of d left ideals isomorphic to V. This implies that the direct sum of d complex irreducible representations equivalent to ρ is equivalent to $\tau_{\mathbb{C}}$ where τ is the representation of G acting on I/F. Hence $d\chi$ is realizable over F. Now suppose d' is a positive integer such that $d'\chi$ is realizable over F. Then the $\mathbb{C}[G]$-module $V^{(d')}$, a direct sum of d' copies of V, is isomorphic to a module $I'_{\mathbb{C}}$ where I' is a module for $F[G]$. Now $V^{(d')}$ is annihilated by every simple component of $F[G]$ except A and hence I' is annihilated by every simple component of $F[G]$ except A. It follows that $I' \cong I^{(h)}$ where I is a minimal left ideal of A. Then $V^{(d')} \cong I'_{\mathbb{C}} \cong (I_{\mathbb{C}})^{(h)} \cong (V^{(d)})^{(h)} \cong V^{(dh)}$. Hence $d' = dh$. Evidently this implies that d is the Schur index. $\square$

The foregoing result shows that the Schur index of χ is 1 if and only if $\Delta = F$ and $A = M_m(F)$. Hence χ is realizable over $F = F(\chi)$ if and only if $A = M_m(F)$.

Now let E be a subfield of $\mathbb{C}$ containing $F = F(\chi)$ and consider $E[G]$. Since the simple component A of $F[G]$ belonging to χ is central simple, A_E is a simple component of $E[G]$. Evidently this is the simple component of $E[G]$ belonging to χ. Hence χ is realizable over E if and only if $A_E \cong M_n(E)$, that is,

E is a splitting field over F of A (p. 220). This is the case if and only if E is a splitting field for the division algebra Δ such that $A = M_m(\Delta)$. The basic criterion for this was given in Theorem 4.7 (p. 218). According to this result, a finite dimensional extension field E/F is a splitting field for a finite dimensional central division algebra Δ/F if and only if E is isomorphic to a subalgebra E' of a matrix algebra $M_r(\Delta)$ such that the centralizer of E' in $M_r(\Delta)$ is E', and if this is the case then $[E:F] = rd$ where $[\Delta:F] = d^2$. This result and Theorem 5.26 give the following

THEOREM 5.27. *Let the notations be as in Theorem 5.26 and let E be a finite dimensional extension field of F contained in $\mathbb{C}$. Then χ is realizable over E if and only if E is isomorphic to a subalgebra E' of the matrix algebra $M_r(\Delta)$ such that $C_{M_r(\Delta)}(E') = E'$. Moreover, in this case $[E:F] = rd$ where d is the Schur index of χ over F.*

5.15 FROBENIUS GROUPS

We shall conclude this chapter by applying the theory of characters to derive an important theorem of Frobenius on finite groups. Frobenius' theorem can be viewed in two different ways: first as a theorem on transitive permutation groups and second as a theorem on abstract groups. Also, as we shall show, the result is related to the study of fixed-point-free automorphisms of finite groups. Finally, we shall consider an example of a Frobenius group whose character analysis leads to a variant of a classical proof of the quadratic reciprocity law of number theory.

First, let G be a permutation group of the set $N = \{1, 2, \dots, n\}$, $n > 1$, such that (1) G is transitive; (2) For any i, the stabilizer $\operatorname{Stab} i = \{g \in G | gi = i\} \neq 1$; and (3) No element of G except 1 fixes more than one element of $\{1, \dots, n\}$. Let $H_i = \operatorname{Stab} i$. Then condition (3) is equivalent to $H_i \cap H_j = 1$ if $i \neq j$. Also any two of the subgroups of H_i, H_j are conjugate: $H_j = {}^gH_i = gH_ig^{-1}$. For, there exists a $g \in G$ such that $gi = j$. Then $\operatorname{Stab} j = \operatorname{Stab} gi = g(\operatorname{Stab} i)g^{-1}$. It is clear also that if $g \notin H_i$, then $gi = j \neq i$ and ${}^gH_i = H_j \neq H_i$. We remark also that since $n \geqslant 2$, $H_i \neq G$.

We now introduce the following

DEFINITION 5.4. *A finite group G is called a* Frobenius group *if G contains a subgroup H such that* (i) $1 \subsetneq H \subsetneq G$, *and* (ii) *for any $g \in G - H$, $H \cap {}^gH = 1$. The subgroup H is called a* Frobenius complement *in G.*

It is clear from the foregoing remarks that if G is a permutation group satisfying conditions (1), (2), and (3) above and $n \geqslant 2$, then any one of the subgroups $H = H_i$ satisfies the conditions of Definition 5.4. Hence G is a Frobenius group with H a Frobenius complement in G. Conversely, suppose G is Frobenius with Frobenius complement H, and consider the action of G on the set G/H of left cosets of H. We know that this action is transitive. Moreover, $a(gH) = gH$ if and only if $a \in {}^gH$, so $\operatorname{Stab}(gH) = {}^gH$. Then the kernel of the action is $\bigcap_g {}^gH = 1$. Hence we can identify G with the corresponding group of permutations of G/H. It is clear that condition (2) holds and condition (3) is equivalent to $aH \neq bH \Rightarrow {}^aH \neq {}^bH$. This follows from (ii) in the definition of a Frobenius group.

Thus we see that the concept of a Frobenius group is the abstract version of the permutation group situation we considered first. We can now state

FROBENIUS' THEOREM. *Let G be a Frobenius group with Frobenius complement H. Put*

$$K = \{1\} \cup \left(G - \bigcup_{g \in G} {}^gH \right). \tag{105}$$

Then K is a normal subgroup of G, $G = KH$, and $K \cap H = 1$.

If we adopt the permutation group point of view, we see that K can be described as the union of $\{1\}$ and the set of transformations that fix no $i \in N$.

For the proof we require the following

LEMMA. *Let G be a Frobenius group with Frobenius complement H and let φ be a class function on H such that $\varphi(1) = 0$. Then $(\varphi^G)_H = \varphi$.*

Proof. By definition

$$\varphi^G(g) = \frac{1}{|H|} \sum_{a \in G} \dot{\varphi}(aga^{-1}). \tag{106}$$

Then $\varphi^G(1) = 0$ since $\varphi(1) = 0$. Now let $h \in H$, $h \neq 1$, and let $a \in G - H$. Then $aha^{-1} \notin H$ and $\dot{\varphi}(aha^{-1}) = 0$. Hence

$$\varphi^G(h) = \frac{1}{|H|} \sum_{a \in H} \dot{\varphi}(aha^{-1}) = \frac{1}{|H|} \sum_{a \in H} \varphi(aha^{-1}) = \varphi(h).$$

Thus $(\varphi^G)_H = \varphi$. $\square$

We can now give the

Proof of Frobenius' theorem. We note first that the definition of Frobenius complement H implies that distinct conjugates ${}^{g_1}H$ and ${}^{g_2}H$ have only the unit element in common and the normalizer $N_G(H) = \{g | {}^gH = H\} = H$. Hence $|\{{}^gH\}| = [G:N_G(H)] = [G:H]$. Then $|\bigcup_{g\in G} {}^gH| = [G:H]|H| - [G:H] + 1$ and so, by (105),

$$|K| = [G:H]. \tag{107}$$

Now let ψ be a complex irreducible character of H different from the unit character ψ_1 and put $\varphi = \psi - \psi(1)\psi_1$. Then φ is a generalized character of H satisfying $\varphi(1) = 0$. Hence, by the lemma, φ^G is a generalized character of G such that $(\varphi^G)_H = \varphi$. Then, by Frobenius reciprocity,

$$(\varphi^G | \varphi^G)_G = (\varphi | (\varphi^G)_H)_H = (\varphi | \varphi)_H = 1 + \psi(1)^2$$
$$(\varphi^G | \chi_1)_G = (\varphi | \psi_1)_H = -\psi(1)$$

for χ_1, the unit character of G. Now put $\psi^* = \varphi^G + \psi(1)\chi_1$. Then ψ^* is a generalized character of G and

$$\begin{aligned}(\psi^* | \psi^*)_G &= (\varphi^G | \varphi^G)_G + 2\psi(1)(\varphi^G | \chi_1)_G + \psi(1)^2 \\ &= 1 + \psi(1)^2 - 2\psi(1)^2 + \psi(1)^2 = 1.\end{aligned}$$

Also

$$(\psi^*)_H = (\varphi^G)_H + \psi(1)(\chi_1)_H = \varphi + \psi(1)\psi_1 = \psi.$$

Since ψ^* is a generalized character of G satisfying $(\psi^* | \psi^*)_G = 1$ and $\psi^*(1) = \psi(1) > 0$, ψ^* is a complex irreducible character of G. Thus for each irreducible complex character $\psi \neq \psi_1$ of H, we have defined an irreducible complex character ψ^* of G that is an extension of ψ. If ρ^* is a representation of G affording ψ^*, then $\ker \psi^* = \{g \in G | \psi^*(g) = \psi^*(1)\} = \ker \rho^*$ (p. 271). Now put

$$K^* = \bigcap_{\psi} \ker \psi^* \tag{108}$$

where the intersection is taken over the irreducible characters ψ of H, $\psi \neq \psi_1$. Evidently K^* is a normal subgroup of G. Let $k \neq 1$ be in K. Then $k \notin {}^gH$ for any $g \in G$ and $\varphi^G(k) = 0$, by definition (106). Hence $\psi^*(k) = \psi^*(1)$. Thus $K \subset K^*$. Next let $h \in H \cap K^*$. Then $\psi(h) = \psi^*(h) = \psi^*(1) = \psi(1)$. Thus h is in the kernel of every irreducible character of H. Then $\rho(h) = 1$ for the regular representation ρ of H and hence $h = 1$. Hence $H \cap K^* = 1$. Thus we have $K^* \lhd G$, $K \subset K^*$, $H \cap K^* = 1$, and $|G| = |H|[G:H] = |H||K|$. Hence HK^* is a subgroup of G and $HK^*/K^* \cong H/(H \cap K^*)$, so $|HK^*| = |H||K^*|/|H \cap K^*| = |H||K^*|$ is a factor of $|G| = |H||K|$. Hence $|K^*| \mid |K|$ and since $K^* \supset K$, $K = K^*$. We now have $K = K^* \lhd G$. $H \cap K = 1$ and $|K| = [G:H]$. Then $|KH| = |K||H| = |G|$ and hence $G = KH$. This completes the proof. □

The subgroup K is called the *Frobenius kernel* of the Frobenius group G with complement H. We note that no proof of Frobenius' theorem without characters is known.

Now let A be a group of automorphisms of a finite group G. We shall say that A is *fixed-point-free* if no $\alpha \neq 1$ in A fixes any $g \neq 1$ of G. We shall now relate the study of pairs (G, A) where G is a finite group and A is a fixed-point-free group of automorphisms $\neq 1$ of G to the study of Frobenius groups. We recall the definition of the holomorph $\text{Hol}\, G$ of a group G as the group of transformations G_L Aut G where G_L is the set of left multiplications $g_L : x \rightsquigarrow gx$ (BAI, p. 63). If $\alpha \in \text{Aut}\, G$, then $\alpha g_L = (\alpha g)_L \alpha$. Hence if $g_i \in G$, $\alpha_i \in \text{Aut}\, G$, then $(g_{1L}\alpha_1)(g_{2L}\alpha_2) = (g_1\alpha_1(g_2))_L\alpha_1\alpha_2$. Aut G and G_L are subgroups of Hol G, G_L is normal, $G_L \cap \text{Aut}\, G = 1$, and Hol $G = G_L\text{Aut}\, G$. It is clear that if A is a subgroup of Aut G, then $G_L A$ is a subgroup of the holomorph. Now suppose $A \neq 1$ and A is fixed-point-free. If $g \neq 1$ is in G, then ${}^{g_L}A = \{(g\alpha(g)^{-1})_L\alpha \mid \alpha \in A\}$. Hence ${}^{g_L}A \cap A = 1$ and ${}^{u}A \cap A = 1$ for any $u \in G_L A - A$. Thus $G_L A$ is a Frobenius group with Frobenius complement A. It is clear also that G_L is the Frobenius kernel.

Conversely, suppose that G is a Frobenius group with Frobenius complement H and let K be the Frobenius kernel. The group H acts on K by conjugation and these maps are automorphisms of K. Suppose $h \in H$ and $k \in K$ satisfies ${}^{h}k = k$. Then $kh = hk$ and ${}^{k}h = h \in H$. It follows that either $h = 1$ or $k = 1$. This implies that the homomorphism of H into Aut K sending h into the map $x \rightsquigarrow hxh^{-1}$ is a monomorphism, so H can be identified with a subgroup of Aut K. Moreover, this group of automorphisms is fixed-point-free.

A result that had been conjectured for a long time and was proved by Thompson is that the Frobenius kernel of a Frobenius group is nilpotent. The foregoing considerations show that this is equivalent to the following fact: A finite group having a group of automorphisms $\neq 1$ that is fixed-point-free is nilpotent. For a proof of this theorem, see page 138 of *Characters of Finite Groups* by W. Feit (see References).

EXAMPLE

Let B be the subgroup of $GSL_2(F), F = \mathbb{Z}/(p)$, p an odd prime, consisting of the matrices of the form

$$\begin{pmatrix} a & 0 \\ b & a^{-1} \end{pmatrix}. \tag{109}$$

Evidently, $|B| = p(p-1)$. B contains subgroups K and D where K is the set of matrices (109) with $a = 1$ and D is the set of diagonal matrices. We have $K \lhd B$, $B = DK$,

$D \cap K = 1$. Moreover,

$$\begin{pmatrix} a & 0 \\ 0 & a^{-1} \end{pmatrix}\begin{pmatrix} 1 & 0 \\ b & 1 \end{pmatrix}\begin{pmatrix} a^{-1} & 0 \\ 0 & a \end{pmatrix} = \begin{pmatrix} 1 & 0 \\ a^{-2}b & 1 \end{pmatrix} \tag{110}$$

and

$$\begin{pmatrix} 1 & 0 \\ b & 1 \end{pmatrix}\begin{pmatrix} a & 0 \\ 0 & a^{-1} \end{pmatrix}\begin{pmatrix} 1 & 0 \\ -b & 1 \end{pmatrix} = \begin{pmatrix} a & 0 \\ ab-a^{-1}b & a^{-1} \end{pmatrix}. \tag{111}$$

Put $G = B/\{1, -1\}$ and let $\bar{A}$ denote the image in G of a subset A of B under the canonical homomorphism of B onto G. We may identify K with $\bar{K}$. Since the matrix (111) is contained in D if and only if $a = \pm 1$, it follows that G is a Frobenius group with $\bar{D}$ as a complement and K as the kernel.

We wish to obtain a character table for G. Let v be a generator of the cyclic multiplicative group F^* of the field F. Then the elements of $\bar{D}$ are $\overline{\text{diag}\{v^i, v^{-i}\}}$, $1 \leqslant i \leqslant (p-1)/2$. Since elements in a complement in a Frobenius group are conjugate in the group if and only if they are conjugate in the complement, the $(p-1)/2$ elements of $\bar{D}$ determine $(p-1)/2$ distinct conjugacy classes. The class $\bar{1}$ has a single element and the classes of the elements $\overline{\text{diag}\{v^i, v^{-i}\}}$, $1 \leqslant i \leqslant (p-3)/2$, have cardinality p. Formulas (110) and (111) imply that we have two more conjugacy classes with representatives

$$\begin{pmatrix} 1 & 0 \\ 1 & 1 \end{pmatrix}, \quad \begin{pmatrix} 1 & 0 \\ v & 1 \end{pmatrix}$$

(since v is not a square in F^*). The number of elements in these classes is $(p-1)/2$. Altogether we have $(p-1)/2+2 = (p+3)/2$ conjugacy classes and hence we have this many irreducible complex characters.

Since $\bar{D}$ is cyclic of order $(p-1)/2$, it has $(p-1)/2$ linear characters. Using the homomorphism of G onto $G/K \cong \bar{D}$, we obtain $(p-1)/2$ linear characters for G, which give the following table:

	1	$\begin{pmatrix} 1 & 0 \\ 1 & 1 \end{pmatrix}$	$\begin{pmatrix} 1 & 0 \\ v & 1 \end{pmatrix}$	$\overline{\begin{pmatrix} v^i & 0 \\ 0 & v^{-i} \end{pmatrix}}$,	$1 \leqslant i \leqslant (p-3)/2$
χ_1	1	1	1	1	
χ_2	1	1	1	ω^i	$\omega = e^{4\pi i/p-1}$
.	.	.	.	.	
.	.	.	.	.	
.	.	.	.	.	
$\chi_{(p-1)/2}$	1	1	1	$\omega^{(p-3)i/2}$	

Since the conjugacy class of $\overline{\text{diag}\{v^i, v^{-i}\}}$ has cardinality p, the orthogonality relation (60) implies that the remaining entries in the columns headed by this class are 0's.

The subgroup K of G is cyclic with generator $\left(\begin{smallmatrix} 1 & 0 \\ 1 & 1 \end{smallmatrix}\right)$. Hence we have a homomorphism σ of K into $\mathbb{C}$ mapping the generator into the primitive pth root of unity $\zeta = e^{2\pi i/p}$. This defines the representation $\rho = \sigma^G$ of G. By (110),

$$\sigma\begin{pmatrix} a & 0 \\ 0 & a^{-1} \end{pmatrix}\begin{pmatrix} 1 & 0 \\ 1 & 1 \end{pmatrix}\begin{pmatrix} a^{-1} & 0 \\ 0 & a \end{pmatrix} = \zeta^{a^{-2}}$$

where the exponent is an integer. If $a \neq \pm 1$, then $\zeta^{a^{-2}} \neq \zeta$. Hence the representation of K such that

$$\begin{pmatrix} 1 & 0 \\ 1 & 1 \end{pmatrix} \to \zeta^{-a^{-2}}$$

is inequivalent to σ. It follows from Corollary 2 to Theorem 5.20 (p. 304) that $\rho = \sigma^G$ is an irreducible representation of G. If χ is its character, then the degree of ρ is $\chi(1) = (p-1)/2$. Let χ' be the remaining irreducible character. The relation

$$\begin{aligned} p(p-1)/2 = |G| &= \sum_{1}^{(p-1)/2} \chi_i(1)^2 + \chi(1)^2 + \chi'(1)^2 \\ &= (p-1)/2 + (p-1)^2/4 + \chi'(1)^2 \end{aligned}$$

gives $\chi'(1) = (p-1)/2$. Hence the last two rows of the character table have the form

χ	$(p-1)/2$	c	d	0
χ'	$(p-1)/2$	c'	d'	0

The orthogonality relations (60) for columns 1 and 2 give $(p-1)/2 + c(p-1)/2 + c'(p-1)/2 = 0$, so $c + c' = -1$. Similarly, $d + d' = -1$. The orthogonality of χ with χ_1 gives $(p-1)/2 + c(p-1)/2 + d(p-1)/2 = 0$, so $c + d = -1$. These relations imply that $c' = d, d' = c$. Hence the last two rows have the form

χ	$(p-1)/2$	c	$-1-c$	0
χ'	$(p-1)/2$	$-1-c$	c	0

and it remains to determine c. We need to distinguish two cases:

Case 1. $(-1/p) = 1$, that is, -1 is a quadratic residue mod p, or equivalently, $p \equiv 1 \pmod 4$ (BAI, p. 133). In this case $\left(\begin{smallmatrix} 1 & 0 \\ 1 & 1 \end{smallmatrix}\right)$ is conjugate to $\left(\begin{smallmatrix} 1 & 0 \\ -1 & 1 \end{smallmatrix}\right) = \left(\begin{smallmatrix} 1 & 0 \\ 1 & 1 \end{smallmatrix}\right)^{-1}$. Then c is real. Then the orthogonality relation (60) applied to the second and third columns gives $(p-1)/2 - 2c(1+c) = 0$, which implies that $c = (-1 \pm \sqrt{p})/2$.

Case 2. $(-1/p) = -1$, -1 is a non-square mod p, so $p \equiv 3 \pmod 4$. In this case, $\left(\begin{smallmatrix} 1 & 0 \\ 1 & 1 \end{smallmatrix}\right)$ is not conjugate to its inverse and hence $c \notin \mathbb{R}$. Then $\chi' = \chi^*$, so $\bar{c} = -1-c$. Then the orthogonality relation (60) for the second and third columns gives $(p-1)/2 + c^2 + (1+c)^2 = 0$, which implies that $c = (-1 \pm \sqrt{-p})/2$.

These two cases complete the character tables, although we do not know which determination of the signs gives the character χ of $\rho = \sigma^G$. We can amalgamate the two cases by writing $p^* = (-1/p)p$. Then in both cases we have

$$c = (-1 \pm \sqrt{p^*})/2. \tag{112}$$

We now apply the induced character formula (85) to the character χ of ρ. Since $\bar{D}$ is a cross section of G relative to K, we obtain

$$(-1 \pm \sqrt{p^*})/2 = \chi \begin{pmatrix} 1 & 0 \\ 1 & 1 \end{pmatrix} = \sum_{x \in S} \zeta^x, \zeta = e^{2\pi i/p} \tag{113}$$

where S is the set of positive integers less than p that are squares modulo p.

We proceed to apply formula (113) to give a proof of the law of quadratic reciprocity of elementary number theory. We introduce the Gauss sum

$$g = \sum_x \left(\frac{x}{p}\right)\zeta^x \tag{114}$$

where $1 \leqslant x \leqslant p-1$. Since $1+\sum_x \zeta^x = 0$ (by the factorization $\lambda^p - 1 = \prod_{i=0}^{p-1}(\lambda - \zeta^i)$), we obtain from (113) the formula

$$g^2 = p^*. \tag{115}$$

Now let q be an odd prime $\neq p$. We note that $(p^*/q) \equiv p^{*(q-1)/2} \pmod q$ (BAI, p. 129, exercise 13). We now work in the ring R of integers of the cyclotomic field $\mathbb{Q}(\zeta)$ and use congruences modulo qR (BAI, pp. 278–281). We have

$$\left(\frac{p^*}{q}\right) \equiv g^{q-1} \pmod{qR}. \tag{116}$$

Since for $u_i \in R$, $(\sum u_i)^q \equiv \sum u_i^q \pmod{qR}$, we have

$$\begin{aligned} g^q &= \left(\sum_x \left(\frac{x}{p}\right)\zeta^x\right)^q \equiv \sum_x \left(\frac{x}{p}\right)\zeta^{xq} \\ &= \left(\frac{q}{p}\right)\sum_x \left(\frac{q}{p}\right)\left(\frac{x}{p}\right)\zeta^{xq} \\ &= \left(\frac{q}{p}\right)\sum_x \left(\frac{xq}{p}\right)\zeta^{xq} \\ &= \left(\frac{q}{p}\right)\sum_x \left(\frac{x}{q}\right)\zeta^{x} = \left(\frac{q}{p}\right)g. \end{aligned}$$

Hence

$$\left(\frac{p^*}{q}\right)g \equiv g^q \equiv \left(\frac{q}{p}\right)g \pmod{qR}.$$

Then $(p^*/q)g^2 \equiv (q/p)g^2$ and $(p^*/q)p^* \equiv (q/p)p^*$. Since the coset of $p^* = (-1/p)p$ is a unit in R/qR, this implies that $(p^*/q) \equiv (q/p)$. Since the coset of 2 is a unit, it follows that $(p^*/q) = (q/p)$. Hence

$$\begin{aligned} \left(\frac{q}{p}\right) = \left(\frac{p^*}{q}\right) &= \left(\frac{(-1/p)p}{q}\right) = \left(\frac{(-1)^{(p-1)/2}p}{q}\right) \\ &= \left(\frac{-1}{q}\right)^{(p-1)/2}\left(\frac{p}{q}\right) \\ &= (-1)^{[(q-1)/2][(p-1)/2]}\left(\frac{p}{q}\right) \end{aligned}$$

which is the law of quadratic reciprocity whose first proof was published by Gauss in 1801.

I am indebted to D. R. Corro for pointing out this example of a Frobenius group with its application to the reciprocity law. The derivation following the formula $g^2 = p^*$ is classical (apparently first given by Jacobi in 1827). See

Sechs Beweise des Fundamentaltheorems über quadratische Reste von Carl Friederich Gauss in W. Ostwald's "Klassiker der exacten Wissenschaften," pp. 107–109.

EXERCISES

1. Show that if G is a Frobenius group with complement H, then the Frobenius kernel is the only normal subgroup of G satisfying $G = KH$, $K \cap H = 1$.

2. Show that if G is a dihedral group of order $2m$, m odd, generated by a and b such that $a^m = 1$, $b^2 = 1$, $bab^{-1} = a^{-1}$, then G is a Frobenius group.

3. A subset S of a group G is called a *T.I. set* (trivial intersection set) if $S \neq \varnothing$ and for any $g \in G$, either ${}^gS = S$ or ${}^gS \cap S \subset \{1\}$. Show that if S is T.I., then $N_G(S) = \{g \in G | S^g = S\}$ is a subgroup containing S. Show that if G is a Frobenius group with Frobenius complement H, then H is a T.I. set.

4. (Brauer-Suzuki.) Let S be a T.I. set in G and let $N = N_G(S)$. Suppose that φ and ψ are complex class functions on N such that φ and ψ are 0 outside of S and $\varphi(1) = 0$. Show that (i) $\varphi^G(g) = \varphi(g)$ for any $g \neq 1$ in S, and (ii) $(\varphi|\psi)_N = (\varphi^G|\psi^G)_G$.

5. (Brauer-Suzuki.) Let S, G, and N be as in exercise 4 and let $\psi_1, \dots, \psi_m$ be complex irreducible characters of N such that $\psi_1(1) = \cdots = \psi_m(1)$. Show that there exists $\varepsilon = \pm 1$ and complex irreducible characters $\chi_1, \dots, \chi_m$ of G such that $\chi_i - \chi_j = \varepsilon(\psi_i{}^G - \psi_j{}^G)$.

6. Let G be a Frobenius group with Frobenius complement H and Frobenius kernel K. Assume K is abelian. Use exercise 8, p. 63 of BAI to show that if ρ is an irreducible representation of G, then either (i) $K \subset \ker \rho$ or (ii) ρ is induced from a degree one representation of K. Conclude that G is an M-group if and only if H is an M-group. (This result holds without the hypothesis that K is abelian.)

   Sketch of proof: Suppose that (i) fails. Then, by Clifford's theorem, there exists an irreducible constituent λ of ρ_K such that λ is not the unit representation λ_1 of K. We claim that $T_G(\lambda) = K$. Otherwise, by Corollary 2, p. 304, there exists an $x \in G - K = \bigcup_{g \in G} {}^gH$ such that ${}^x\lambda = \lambda$. But H acts fixed-point-freely on K; so does ${}^gH$ for any g (since $K \lhd G$). Thus x induces a fixed-point-free automorphism of K, so by the exercise cited

$$K = \{xkx^{-1}k^{-1} | k \in K\}.$$

   Then

$$\begin{aligned}\lambda(xkx^{-1}k^{-1}) &= \lambda(xkx^{-1})\lambda(k^{-1})\\ &= {}^x\lambda(k)\lambda(k)^{-1}\\ &= \lambda(k)\lambda(k)^{-1}\\ &= 1.\end{aligned}$$

Hence $K \subset \ker \lambda$, contrary to $\lambda \neq \lambda_1$. Thus $T_G(\lambda) = K$ and so (ii) follows from Corollary 4 on p. 305.

7. Let F be a finite field with $q = p^m$ elements, p a prime. Define the following subgroups of $GL_2(F)$:

$$T_n = \left\{ \begin{pmatrix} a^n & 0 \\ b & 1 \end{pmatrix} \middle| a \in F^*, b \in F \right\}$$

$$D_n = \left\{ \begin{pmatrix} a^n & 0 \\ 0 & 1 \end{pmatrix} \middle| a \in F^* \right\}$$

$$K = \left\{ \begin{pmatrix} 1 & 0 \\ b & 1 \end{pmatrix} \middle| b \in F \right\}.$$

Show that if $D_n \neq 1$, then T_n is a Frobenius group with Frobenius complement D_n and Frobenius kernel K.

(*Note*: These groups have been used by W. Feit to obtain an estimate for the number of solutions in $F^{(n)}$ of an equation of the form $\sum_1^n c_i x_i^{m_i} = 0$, $c_i \in F^*$. See Feit, p. 140, in the references below.)

REFERENCES

C. W. Curtis and I. Reiner, *Representation Theory of Finite Groups and Associative Algebras*, Wiley-Interscience, New York, 1962.

W. Feit, *Characters of Finite Groups*, Benjamin, New York, 1967.

I. M. Isaacs, *Character Theory of Finite Groups*, Academic, New York, 1976.

6

Elements of Homological Algebra with Applications

Homological algebra has become an extensive area of algebra since its introduction in the mid-1940's. Its first aspect, the cohomology and homology groups of a group, was an outgrowth of a problem in topology that arose from an observation by Witold Hurewicz that the homology groups of a path-connected space whose higher homotopy groups are trivial are determined by the fundamental group π_1. This presented the problem of providing a mechanism for this dependence. Solutions of this problem were given independently and more or less simultaneously by a number of topologists: Hopf, Eilenberg and MacLane, Freudenthal, and Eckmann [see Cartan and Eilenberg (1956), p. 137, and MacLane (1963), p. 185, listed in References]. All of these solutions involved homology or cohomology groups of π_1. The next step was to define the homology and cohomology groups of an arbitrary group and to study them for their own sake. Definitions of the cohomology groups with coefficients in an arbitrary module were given by Eilenberg and MacLane in 1947. At the same time, G. Hochschild introduced cohomology groups for associative algebras. The cohomology theory of Lie algebras, which is a purely algebraic theory corresponding to the cohomology theory of Lie groups, was developed by J. L. Koszul and by Chevalley and Eilenberg. These disparate theories were pulled together by Cartan and Eilenberg (1956) in a cohesive synthesis based on a concept of derived functors from the category of modules over a ring to the category of abelian groups. The derived functors that are needed for the cohomology and homology theories are the functors Ext and Tor, which are the derived functors of the hom and tensor functors respectively.

Whereas the development of homological algebra proper dates from the period of the Second World War, several important precursors of the theory appeared earlier. The earliest was perhaps Hilbert's syzygy theorem (1890) in invariant theory, which concerned free resolutions of modules for the ring of polynomials in m indeterminates with coefficients in a field. The second cohomology group $H^2(G, \mathbb{C}^*)$ with coefficients in the multiplicative group $\mathbb{C}^*$ of non-zero complex numbers appeared in Schur's work on projective representations of groups (1904). More general second cohomology groups occurred as factor sets in Schreier's extension theory of groups (1926) and in Emmy Noether's construction of crossed product algebras (1929). The third cohomology group appeared first in a paper by O. Teichmüller (1940).

In this chapter we shall proceed first as quickly as possible to the basic definition and results on derived functors. These will be specialized to the most important instances: Ext and Tor. In the second half of the chapter we shall consider some classical instances of homology theory: cohomology of groups, cohomology of algebras with applications to a radical splitting theorem for finite dimensional algebras due to Wedderburn, homological dimension of modules and rings, and the Hilbert syzygy theorem. Later (sections 8.4 and 8.5), we shall present another application of homology theory to the Brauer group and crossed products.

6.1 ADDITIVE AND ABELIAN CATEGORIES

A substantial part of the theory of modules can be extended to a class of categories called abelian. In particular, homological algebra can be developed for abelian categories. Although we shall stick to modules in our treatment, we will find it convenient to have at hand the definitions and simplest properties of abelian categories. We shall therefore consider these in this section.

We recall that an object 0 of a category $\mathbf{C}$ is a zero object if for any object A of $\mathbf{C}$, $\hom_{\mathbf{C}}(A, 0)$ and $\hom_{\mathbf{C}}(0, A)$ are singletons. If 0 and $0'$ are zero objects, then there exists a unique isomorphism $0 \to 0'$ (exercise 3, p. 36). If A, $B \in \operatorname{ob} \mathbf{C}$, we define $0_{A,B}$ as the morphism $0_{0B}0_{A0}$ where 0_{A0} is the unique element of $\hom_{\mathbf{C}}(A, 0)$ and 0_{0B} is the unique element of $\hom_{\mathbf{C}}(0, B)$. It is easily seen that this morphism is independent of the choice of the zero object. We call $0_{A,B}$ the *zero morphism* from A to B. We shall usually drop the subscripts in indicating this element.

We can now give the following

DEFINITION 6.1. *A category* $\mathbf{C}$ *is called* additive *if it satisfies the following conditions*:

AC1. $\mathbf{C}$ *has a zero object.*

AC2. *For every pair of objects* (A, B) *in* $\mathbf{C}$, *a binary composition* $+$ *is defined on the set* $\hom_{\mathbf{C}}(A, B)$ *such that* $(\hom_{\mathbf{C}}(A, B), +, 0_{A,B})$ *is an abelian group.*

AC3. *If* $A, B, C \in \operatorname{ob} \mathbf{C}, f, f_1, f_2 \in \hom_{\mathbf{C}}(A, B)$, *and* $g, g_1, g_2 \in \hom_{\mathbf{C}}(B, C)$, *then*

$$(g_1 + g_2)f = g_1 f + g_2 f$$
$$g(f_1 + f_2) = gf_1 + gf_2.$$

AC4. *For any finite set of objects* $\{A_1, \ldots, A_n\}$ *there exists an object* A *and morphisms* $p_j : A \to A_j$, $i_j : A_j \to A$, $1 \leqslant j \leqslant n$, *such that*

$$p_j i_j = 1_{A_j}, \qquad p_k i_j = 0 \quad \text{if } j \neq k$$
$$\sum i_j p_j = 1_A. \tag{1}$$

We remark that AC2 means that we are given, as part of the definition, an abelian group structure on every $\hom_{\mathbf{C}}(A, B)$ whose zero element is the categorically defined $0_{A,B}$. AC3 states that the product fg, when defined in the category, is bi-additive. A consequence of this is that for any A, $(\hom_{\mathbf{C}}(A, A), +, \cdot, 0, 1 = 1_A)$ is a ring. We note also that AC4 implies that $(A, \{p_j\})$ is a product in $\mathbf{C}$ of the A_j, $1 \leqslant j \leqslant n$. For, suppose $B \in \operatorname{ob} \mathbf{C}$ and we are given $f_j : B \to A_j$, $1 \leqslant j \leqslant n$. Put $f = \sum i_j f_j \in \hom_{\mathbf{C}}(B, A)$. Then $p_k f = f_k$ by (1) and if $p_k f' = f_k$ for $1 \leqslant k \leqslant n$, then (1) implies that $f' = \sum i_j f_j = f$. Hence f is the only morphism from B to A such that $p_k f = f_k$, $1 \leqslant k \leqslant n$, and $(A, \{p_j\})$ is a product of the A_j. In a similar manner we see that $(A, \{i_j\})$ is a coproduct of the A_j.

It is not difficult to show that we can replace AC4 by either

AC4′. $\mathbf{C}$ is a category with a product (that is, products exist for arbitrary finite sets of objects of $\mathbf{C}$), or

AC4″. $\mathbf{C}$ is a category with a coproduct.

We have seen that AC4 $\Rightarrow$ AC4′ and AC4″ and we shall indicate in the exercises that AC1–3 and AC4′ or AC4″ imply AC4. The advantage of AC4 is that it is self-dual. It follows that the set of conditions defining an additive category is self-dual and hence if $\mathbf{C}$ is an additive category, then $\mathbf{C}^{\text{op}}$ is an additive category. This is one of the important advantages in dealing with additive categories.

If R is a ring, the categories R-**mod** and **mod**-R are additive. As in these special cases, in considering functors between additive categories, it is natural to assume that these are additive in the sense that for every pair of objects A, B, the map F of $\hom(A, B)$ into $\hom(FA, FB)$ is a group homomorphism. In this case, the proof given for modules (p. 98) shows that F preserves finite products (coproducts).

We define next some concepts that are needed to define abelian categories. Let $\mathbf{C}$ be a category with a zero (object), $f: A \to B$ in $\mathbf{C}$. Then we call $k: K \to A$ a *kernel of* f if (1) k is monic, (2) $fk = 0$, and (3) for any $g: G \to A$ such that $fg = 0$ there exists a g' such that $g = kg'$. Since k is monic, it is clear that g' is unique. Condition (2) is that

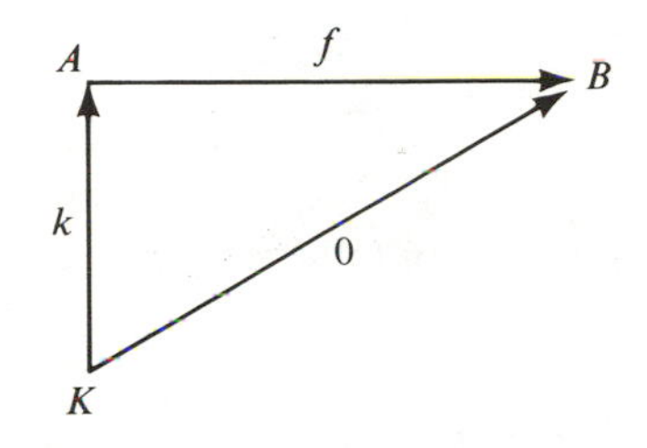

is commutative and (3) is that if the triangle in

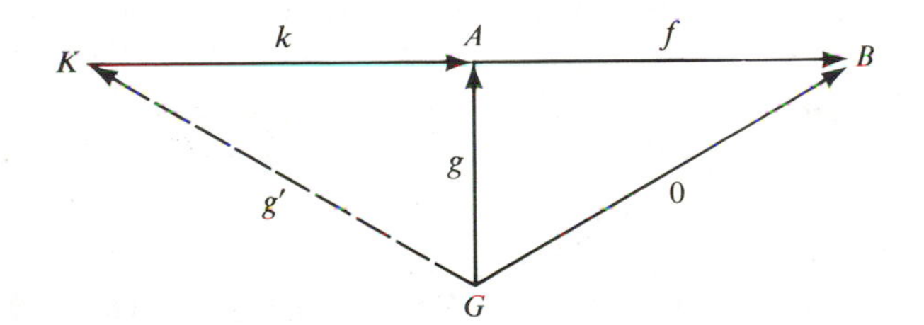

is commutative, then this can be completed by $g': G \to K$ to obtain a commutative diagram. It is clear that if k and k' are kernels of f, then there exists a unique isomorphism u such that $k' = ku$.

In a dual manner we define a *cokernel of* f as a morphism $c: B \to C$ such that (1) c is epic, (2) $cf = 0$, and (3) for any $h: B \to H$ such that $hf = 0$ there exists h' such that $h = h'c$.

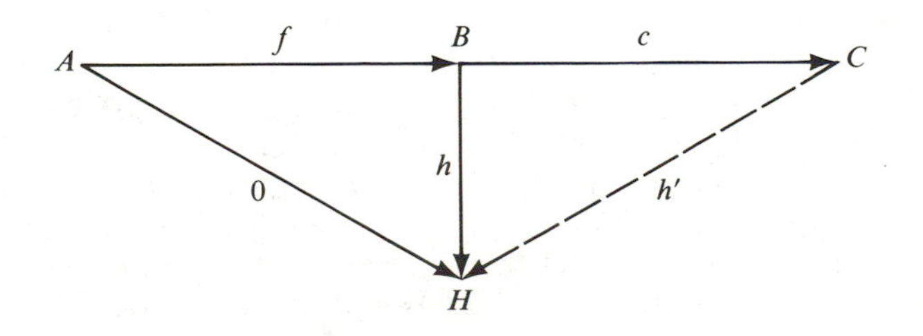

If $f: A \to B$ in R-**mod**, let $K = \ker f$ in the usual sense and let k be the injection of K in A. Then k is monic, $fk = 0$, and if g is a homomorphism of G into A such that $fg = 0$, then $gG \subset K$. Hence if we let g' be the map obtained from g by restricting the codomain to K, then $g = kg'$. Hence k is a kernel of f. Next let $C = B/fA$ and let c be the canonical homomorphism of B onto C. Then c is epic, $cf = 0$, and if $h: B \to H$ satisfies $hf = 0$, then $fA \subset \ker h$. Hence we have a unique homomorphism $h': C = B/fA \to H$ such that

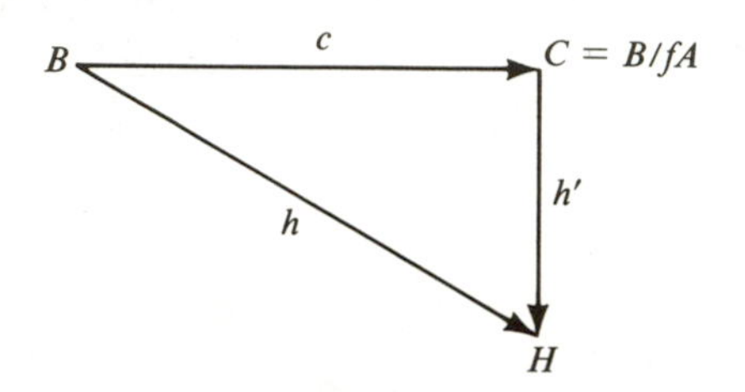

is commutative. Thus c is a cokernel of f in the category R-**mod**.

We can now give the definition of an abelian category

DEFINITION 6.2. *A category* **C** *is* abelian *if it is an additive category having the following additional properties*:

AC5. *Every morphism in* **C** *has a kernel and a cokernel.*

AC6. *Every monic is a kernel of its cokernel and every epic is a cokernel of its kernel.*

AC7. *Every morphism can be factored as $f = me$ where e is epic and m is monic.*

We have seen that if R is a ring, then the categories R-**mod** and **mod**-R are additive categories satisfying AC5. We leave it to the reader to show that AC6 and AC7 also hold for R-**mod** and **mod**-R. Thus these are abelian categories.

EXERCISES

1. Let **C** be a category with a zero. Show that for any object A in **C**, $(A, 1_A, 0)$ is a product and coproduct of A and 0.

2. Let **C** be a category, (A, p_1, p_2) be a product of A_1 and A_2 in **C**, (B, q_1, q_2) a product of B_1 and B_2 in **C**, and let $h_i: B_i \to A_i$. Show that there exists a unique $f: B \to A$ such that $h_i q_i = p_i f$. In particular, if **C** has a zero and we take $(B, q_1, q_2) = (A_1, 1_{A_1}, 0)$, then this gives a unique $i_1: A_1 \to A$ such that $p_1 i_1 = 1_{A_1}$, $p_2 i_1 = 0$. Similarly, show that we have a unique $i_2: A_2 \to A$ such that $p_1 i_2 = 0$, $p_2 i_2 = 1_{A_2}$. Show that $(i_1 p_1 + i_2 p_2) i_1 = i_1$ and $(i_1 p_1 + i_2 p_2) i_2 = i_2$. Hence conclude that $i_1 p_1 + i_2 p_2 = 1_A$. Use this to prove that the conditions AC1–AC3 and AC4′ $\Leftrightarrow$ AC4. Dualize to prove that AC1–AC3 and AC4″ $\Leftrightarrow$ AC4.

3. Show that if A and B are objects of an additive category, then $0_{A,B} = 0_{0,B}0_{A,0}$, where $0_{A,0}$ is the zero element of hom(A,0), $0_{0,B}$ is the zero element of hom(0,B), and $0_{A,B}$ is the zero element of the abelian group hom(A, B).

4. Let $(A_1 \Pi A_2, p_1, p_2)$ be a product of the objects A_1 and A_2 in the category **C**. If $f_j: B \to A_j$, denote the unique $f: B \to A_1 \Pi A_2$ such that $p_j f = f_j$ by $f_1 \Pi f_2$. Similarly, if $(A_1 \amalg A_2, i_1, i_2)$ is a coproduct and $g_j: A_j \to C$, write $g_1 \amalg g_2$ for the unique $g: A_1 \amalg A_2 \to C$ such that $gi_j = g_j$. Note that if C is additive with the i_j and p_j as in AC4, then $f_1 \Pi f_2 = i_1 f_1 + i_2 f_2$ and $g_1 \amalg g_2 = g_1 p_1 + g_2 p_2$. Hence show that if i_1, i_2, p_1, p_2 are as in AC4, so $A = A_1 \Pi A_2 = A_1 \amalg A_2$, then
$$(g_1 \amalg g_2)(f_1 \Pi f_2) = g_1 f_1 + g_2 f_2$$
(from $B \to C$). Specialize $A_1 = A_2 = 0$, $g_1 = g_2 = 1_C$ to obtain the formula
$$f_1 + f_2 = (1_C \amalg 1_C)(f_1 \Pi f_2)$$
for the addition in $\hom_{\mathbf{C}}(B, C)$.

5. Use the result of exercise 4 to show that if F is a functor between additive categories that preserves products and coproducts, then F is additive.

6.2 COMPLEXES AND HOMOLOGY

The basic concepts of homological algebra are those of a complex and homomorphisms of complexes that we shall now define.

DEFINITION 6.3. *If R is a ring, a* complex (C,d) *for R is an indexed set $C = \{C_i\}$ of R-modules indexed by $\mathbb{Z}$ together with an indexed set $d = \{d_i | i \in \mathbb{Z}\}$ of R-homomorphisms $d_i: C_i \to C_{i-1}$ such that $d_{i-1}d_i = 0$ for all i. If (C,d) and (C',d') are R-complexes, a* (chain) homomorphism *of C into C' is an indexed set $\alpha = \{\alpha_i | i \in \mathbb{Z}\}$ of homomorphisms $\alpha_i: C_i \to C'_i$ such that we have the commutativity of*

(2)

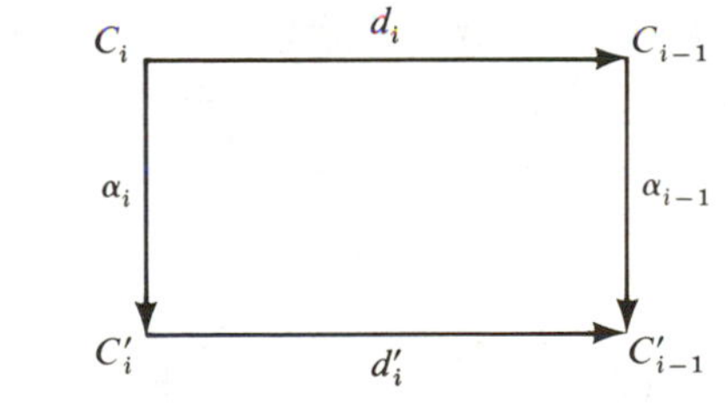

for every i. More briefly we write $\alpha d = d'\alpha$.

These definitions lead to the introduction of a category R-**comp** of complexes for the ring R. Its objects are the R-complexes (C, d), and for every pair of R-complexes (C, d), (C', d'), the set $\hom(C, C')$ is the set of chain homomorphisms of (C,d) into (C',d'). It is clear that these constitute a category, and as we proceed to show, the main features of R-**mod** carry over to R-**comp**. We note first that $\hom(C,C')$ has a natural structure of abelian group. This is obtained by defining $\alpha+\beta$ for $\alpha,\beta\in\hom(C,C')$ by $(\alpha+\beta)_i=\alpha_i+\beta_i$. The commutativity $\alpha_{i-1}d_i=d'_i\alpha_i$, $\beta_{i-1}d_i=d'_i\beta_i$ gives $(\alpha_{i-1}+\beta_{i-1})d_i=d'_i(\alpha_i+\beta_i)$, so $\alpha+\beta\in\hom(C,C')$. Since $\hom_R(C_i,C'_i)$ is an abelian group, it follows that $\hom(C,C')$ is an abelian group. It is clear also by referring to the module situation that we have the distributive laws $\gamma(\alpha+\beta)=\gamma\alpha+\gamma\beta$, $(\alpha+\beta)\delta=\alpha\delta+\beta\delta$ when these products of chain homomorphisms are defined. If (C,d) and (C',d') are complexes, we can define their *direct sum* $(C\oplus C',d\oplus d')$ by $(C+C')_i=C_i\oplus C'_i$, $d_i\oplus d'_i$ defined component-wise from $C_i\oplus C'_i$ to $C_{i-1}\oplus C'_{i-1}$ as $(x_i,x'_i)\rightsquigarrow(d_ix_i,d'_ix'_i)$. It is clear that $(d_{i-1}\oplus d'_{i-1})(d_i\oplus d'_i)=0$, so $(C\oplus C',d\oplus d')$ is indeed a complex. This has an immediate extension to direct sums of more than two complexes. Since everything can be reduced to the module situation, it is quite clear that if we endow the hom sets with the abelian group structure we defined, then the category R-**comp** becomes an abelian category.

The interesting examples of complexes will be encountered in section 4. However, it may be helpful to list some at this point, although most of these will appear to be rather special.

EXAMPLES

1. Any module M becomes a complex in which $C_i=M$, $i\in\mathbb{Z}$, and $d_i=0\colon C_i\to C_{i-1}$.

2. A *module with differentiation* is an R-module equipped with a module endomorphism δ such that $\delta^2=0$. If (M,δ) is a module with differentiation, we obtain a complex (C,d) in which $C_i=0$ for $i\leqslant 0$, $C_1=C_2=C_3=M$, $C_j=0$ for $j>3$, $d_2=d_3=\delta$, and $d_i=0$ if $i\neq 2,3$.

3. Let (M,δ) be a module with a differentiation that is *$\mathbb{Z}$-graded* in the following sense: $M=\bigoplus_{i\in\mathbb{Z}}M_i$ where the M_i are submodules and $\delta(M_i)\subset M_{i-1}$ for every i. Put $C_i=M_i$ and $d_i=\delta|M_i$. Then $C=\{C_i\}$, $d=\{d_i\}$ constitute an R-complex.

4. Any short exact sequence $0\to M'\xrightarrow{\alpha}M\xrightarrow{\beta}M''\to 0$ defines a complex in which $C_i=0$, $i\leqslant 0$, $C_1=M''$, $C_2=M$, $C_3=M'$, $C_j=0$ if $j>3$, $d_2=\beta$, $d_3=\alpha$, $d_j=0$ if $j\neq 2,3$.

We shall now define for each $i\in\mathbb{Z}$ a functor, the ith homology functor, from the category of R-complexes to the category of R-modules. Let (C,d) be a complex and let $Z_i(C)=\ker d_i$, so $Z_i(C)$ is a submodule of C_i. The elements of Z_i are called *i-cycles*. Since $d_id_{i+1}=0$, it is clear that the image $d_{i+1}C_{i+1}$ is a submodule of Z_i. We denote this as $B_i=B_i(C)$ and call its elements *i-boundaries*. The module $H_i=H_i(C)=Z_i/B_i$ is called the ith *homology module* of the complex (C,d). Evidently, $C_{i+1} \xrightarrow{d_{i+1}} C_i \xrightarrow{d_i} C_{i-1}$ is exact if and only if $H_i(C) = 0$ and hence the infinite sequence of homomorphisms

$$\cdots \leftarrow C_{i-1} \leftarrow C_i \leftarrow C_{i+1} \leftarrow \cdots$$

is exact if and only if $H_i(C) = 0$ for all i.

Now let α be a chain homomorphism of (C,d) into the complex (C',d'). The commutativity condition on (2) implies that $\alpha_i Z_i \subset Z_i' = Z_i(C')$ and $\alpha_i(B_i) \subset B_i' = B_i(C')$. Hence the map $z_i \rightsquigarrow \alpha_i z_i + B_i'$, $z_i \in Z_i$, is a homomorphism of Z_i into $H_i' = H_i(C') = Z_i'/B_i'$ sending B_i into 0. This gives the homomorphism $\tilde{\alpha}_i$ of $H_i(C)$ into $H_i(C')$ such that

$$z_i + B_i \rightsquigarrow \alpha_i z_i + B_i'. \tag{3}$$

It is trivial to check that the maps $(C,d) \rightsquigarrow H_i(C)$, $\hom(C,C') \to \hom(H_i(C), H_i(C'))$, where the latter is $\alpha \rightsquigarrow \tilde{\alpha}_i$, define a functor from R-**comp** to R-**mod**. We call this the ith *homology functor* from R-**comp** to R-**mod**. It is clear that the map $\alpha \rightsquigarrow \tilde{\alpha}_i$ is a homomorphism of abelian groups. Thus the ith homology functor is additive.

In the situations we shall encounter in the sequel, the complexes that occur will have either $C_i = 0$ for $i < 0$ or $C_i = 0$ for $i > 0$. In the first case, the complexes are called *positive* or *chain complexes* and in the second, *negative* or *cochain complexes*. In the latter case, it is usual to denote C_{-i} by C^i and d_{-i} by d^i. With this notation, a cochain complex has the appearance

$$0 \to C^0 \xrightarrow{d^0} C^1 \xrightarrow{d^1} C^2 \xrightarrow{d^2} \cdots$$

if we drop the C^{-i}, $i > 1$. It is usual in this situation to denote $\ker d^i$ by Z^i and $d^{i-1}C^{i-1}$ by B^i. The elements of these groups are respectively *i-cocycles* and *i-coboundaries* and $H^i = Z^i/B^i$ is the ith *cohomology group*. In the case of H^0, we have $H^0 = Z^0$. A chain complex has the form $0 \leftarrow C_0 \xleftarrow{d_1} C_1 \xleftarrow{d_2} C_2 \leftarrow \cdots$. In this case $H_0 = C_0/d_1C_1 = \operatorname{coker} d_1$.

EXERCISES

1. Let α be a homomorphism of the complex (C,d) into the complex (C',d'). Define $C_i'' = C_{i-1} \oplus C_i'$, $i \in \mathbb{Z}$, and if $x_{i-1} \in C_{i-1}$, $x_i' \in C_i'$, define $d_i''(x_{i-1}, x_i') = (-d_{i-1}x_{i-1}, \alpha_{i-1}x_{i-1} + d_i'x_i')$. Verify that (C'', d'') is a complex.

2. Let (C,d) be a positive complex over a field F such that $\sum \dim C_i < \infty$ (equivalently every C_i is finite dimensional and $C_n = 0$ for n sufficiently large). Let $r_i = \dim C_i$, $\rho_i = \dim H_i(C)$. Show that $\sum(-1)^i \rho_i = \sum(-1)^i r_i$.

3. (*Amitsur's complex.*) Let S be a commutative algebra over a commutative ring K. Put $S^0 = K$, $S^n = S\otimes\cdots\otimes S$, n factors, where $\otimes$ means $\otimes_K$. Note that for any n we have $n+1$ algebra isomorphisms δ^i, $1 \leqslant i \leqslant n+1$, of S^n into S^{n+1} such that

$$x_1\otimes\cdots\otimes x_n \rightsquigarrow x_1\otimes\cdots\otimes x_{i-1}\otimes 1\otimes x_i\otimes\cdots\otimes x_n.$$

For any ring R let $U(R)$ denote the multiplicative group of units of R. Then $\delta^i U(S_n) \subset U(S_{n+1})$. Define $d^n : U(S_n) \to U(S_{n+1})$, $n \geqslant -1$, by

$$d^n u = \prod_{i=1}^{n+1} (\delta^i u)^{(-1)^i}$$

(e.g., $d^2 u = (\delta^1 u)^{-1}(\delta^2 u)(\delta^3 u)^{-1}$). Note that if $i \geqslant j$, then $\delta^{i+1}\delta^j = \delta^j\delta^i$ and use this to show that $d^{n+1}d^n = 0$, the map $u \rightsquigarrow 1$. Hence conclude that $\{U(S^n), d^n | n \geqslant 0\}$ is a cochain complex.

6.3 LONG EXACT HOMOLOGY SEQUENCE

In this section we shall develop one of the most important tools of homological algebra: the long exact homology sequence arising from a short exact sequence of complexes. By a *short exact sequence of complexes* we mean a sequence of complexes and chain homomorphisms $C' \xrightarrow{\alpha} C \xrightarrow{\beta} C''$ such that $0 \to C'_i \xrightarrow{\alpha_i} C_i \xrightarrow{\beta_i} C''_i \to 0$ is exact for every $i \in \mathbb{Z}$, that is, α_i is injective, β_i is surjective, and $\ker \beta_i = \operatorname{im} \alpha_i$. We shall indicate this by saying that $0 \to C' \to C \to C'' \to 0$ is exact. We have the commutative diagram

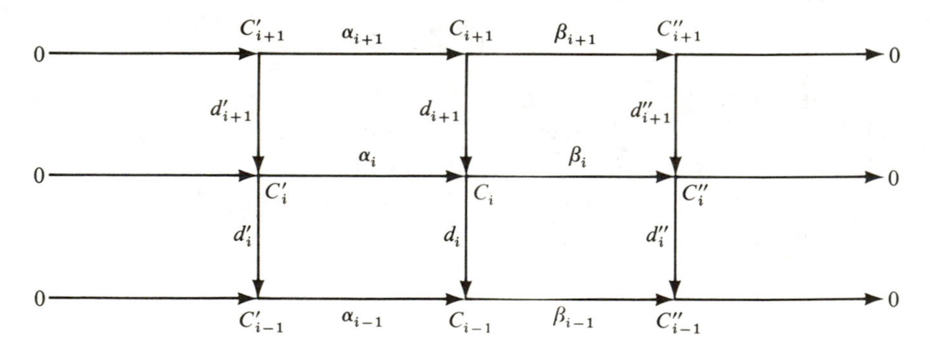

in which the rows are exact. The result we wish to prove is

THEOREM 6.1. *Let $0 \to C' \xrightarrow{\alpha} C \xrightarrow{\beta} C'' \to 0$ be an exact sequence of complexes. Then for each $i \in \mathbb{Z}$ we can define a module homomorphism $\Delta_i : H_i(C'') \to H_{i-1}(C')$ so that the infinite sequence of homology modules*

$$(4) \qquad \cdots \to H_i(C') \xrightarrow{\bar{\alpha}_i} H_i(C) \xrightarrow{\bar{\beta}_i} H_i(C'') \xrightarrow{\Delta_i} H_{i-1}(C') \xrightarrow{\bar{\alpha}_{i-1}} H_{i-1}(C) \longrightarrow \cdots$$

is exact.

Proof. First, we must define Δ_i. Let $z_i'' \in Z_i(C'')$, so $d_i''z_i'' = 0$. Since β_i is surjective, there exists a $c_i \in C_i$ such that $\beta_i c_i = z_i''$. Then $\beta_{i-1}d_i c_i = d_i''\beta_i c_i = d_i''z_i'' = 0$. Since $\ker \beta_{i-1} = \operatorname{im} \alpha_{i-1}$ and α_{i-1} is injective, there exists a unique $z_{i-1}' \in C_{i-1}'$ such that $\alpha_{i-1}z_{i-1}' = d_i c_i$. Then $\alpha_{i-2}d_{i-1}'z_{i-1}' = d_{i-1}\alpha_{i-1}z_{i-1}' = d_{i-1}d_i c_i = 0$. Since α_{i-2} is injective, $d_{i-1}'z_{i-1}' = 0$ so $z_{i-1}' \in Z_{i-1}(C')$. Our determination of z_{i-1}' can be displayed in the formula

$$z_{i-1}' \in \alpha_{i-1}^{-1}(d_i\beta_i^{-1}(z_i'')) \tag{5}$$

where $\beta_i^{-1}(\,)$ and $\alpha_{i-1}^{-1}(\,)$ denote inverse images. We had to make a choice of $c_i \in \beta_i^{-1}(z_i'')$ at the first stage. Suppose we make a different one, say, $\bar{c}_i \in \beta_i^{-1}(z_i'')$. Then $\beta_i\bar{c}_i = \beta_i c_i$ implies that $\bar{c}_i = c_i + \alpha_i c_i'$, $c_i' \in C_i'$. Then $\alpha_{i-1}(z_{i-1}' + d_i'c_i') = d_i c_i + d_i\alpha_i c_i' = d_i(c_i + \alpha_i c_i') = d_i\bar{c}_i$. Thus the replacement of c_i by $\bar{c}_i$ replaces z_{i-1}' by $z_{i-1}' + d_i'c_i'$. Hence the coset $z_{i-1}' + B_{i-1}(C')$ in $H_{i-1}(C')$ is independent of the choice of c_i and so we have a map

$$z_i'' \rightsquigarrow z_{i-1}' + B_{i-1}(C') \tag{6}$$

of $Z_i(C'')$ into $H_{i-1}(C')$. It is clear from (5) that this is a module homomorphism. Now suppose $z_i'' \in B_i(C'')$, say, $z_i'' = d_{i+1}''c_{i+1}''$, $c_{i+1}'' \in C_{i+1}''$. Then we can choose $c_{i+1} \in C_{i+1}$ so that $\beta_{i+1}c_{i+1} = c_{i+1}''$ and then $\beta_i d_{i+1}c_{i+1} = d_{i+1}''\beta_{i+1}c_{i+1} = d_{i+1}''c_{i+1}'' = z_i''$. Hence $d_{i+1}c_{i+1} \in \beta_i^{-1}(z_i'')$ and since $d_i d_{i+1} = 0$, we have $z_{i-1}' = 0$ in (5). Thus $B_i(C'')$ is in the kernel of the homomorphism (6) and

$$\Delta_i : z_i'' + B_i(C'') \rightsquigarrow z_{i-1}' + B_{i-1}(C') \tag{7}$$

is a homomorphism of $H_i(C'')$ into $H_{i-1}(C')$.

We claim that this definition of Δ_i makes (4) exact, which means that we have I: $\operatorname{im} \tilde{\alpha}_i = \ker \tilde{\beta}_i$, II: $\operatorname{im} \tilde{\beta}_i = \ker \Delta_i$, and III: $\operatorname{im} \Delta_i = \ker \tilde{\alpha}_{i-1}$.

I. It is clear that $\tilde{\beta}_i\tilde{\alpha}_i = 0$, so $\operatorname{im} \tilde{\alpha}_i \subset \ker \tilde{\beta}_i$. Suppose $z_i \in Z_i(C)$ and $\tilde{\beta}_i(z_i + B_i(C)) = 0$, so $\beta_i z_i = d_{i+1}''c_{i+1}''$, $c_{i+1}'' \in C_{i+1}''$. There exists a $c_{i+1} \in C_{i+1}$ such that $\beta_{i+1}c_{i+1} = c_{i+1}''$ and so $\beta_i(z_i - d_{i+1}c_{i+1}) = d_{i+1}''c_{i+1}'' - d_{i+1}''\beta_{i+1}c_{i+1} = 0$. Then there exists $z_i' \in C_i'$ such that $\alpha_i z_i' = z_i - d_{i+1}c_{i+1}$. Then $\alpha_{i-1}d_i'z_i' = d_i\alpha_i z_i' = d_i z_i - d_i d_{i+1}c_{i+1} = 0$. Hence $d_i'z_i' = 0$ and $z_i' \in Z_i(C')$. Now $\tilde{\alpha}_i(z_i' + B_i(C')) = \alpha_i z_i' + B_i(C) = z_i - d_{i+1}c_{i+1} + B_i(C) = z_i + B_i(C)$. Thus $z_i + B_i(C) \in \operatorname{im} \tilde{\alpha}_i$ and hence $\ker \tilde{\beta}_i \subset \operatorname{im} \tilde{\alpha}_i$ and hence $\ker \tilde{\beta}_i = \operatorname{im} \tilde{\alpha}_i$.

II. Let $z_i \in Z_i(C)$ and let $z_i'' = \beta_i z_i$. Then $\Delta_i(z_i'' + B_i(C'')) = 0$ since $z_i \in \beta_i^{-1}(z_i'')$ and $d_i z_i = 0$, so $\alpha_{i-1}0 = d_i z_i$. Thus $\Delta_i\tilde{\beta}_i(z_i + B_i(C)) = 0$ and $\operatorname{im} \tilde{\beta}_i \subset \ker \Delta_i$. Now suppose $z_i'' \in Z_i(C'')$ satisfies $\Delta_i(z_i'' + B_i(C'')) = 0$. This means that if we choose $c_i \in C_i$ so that $\beta_i c_i = z_i''$ and $z_{i-1}' \in C_{i-1}'$ so that $\alpha_{i-1}z_{i-1}' = d_i c_i$, then $z_{i-1}' = d_i'c_i'$ for some $c_i' \in C_i'$. Then $d_i c_i = \alpha_{i-1}z_{i-1}' = \alpha_{i-1}d_i'c_i' = d_i\alpha_i c_i'$ and $d_i(c_i - \alpha_i c_i') = 0$.

Also $\beta_i(c_i - \alpha_i c_i') = \beta_i c_i = z_i''$. Hence, if we put $z_i = c_i - \alpha_i c_i'$, then we shall have $z_i \in Z_i(C)$ and $\tilde{\beta}_i(z_i + B_i(C)) = \beta_i z_i + B_i(C'') = z_i'' + B_i(C'')$. Thus $\ker \Delta_i \subset \operatorname{im} \tilde{\beta}_i$ and hence $\ker \Delta_i = \operatorname{im} \tilde{\beta}_i$.

III. If $z_{i-1}' \in Z_{i-1}(C')$ and $z_{i-1}' + B_{i-1}(C') \in \operatorname{im} \Delta_i$, then we have a $z_i'' \in Z_i(C'')$ and a $c_i \in C_i$ such that $\beta_i c_i = z_i''$ and $\alpha_{i-1} z_{i-1}' = d_i c_i$. Then $\tilde{\alpha}_{i-1}(z_{i-1}' + B_{i-1}(C'))$ $= \alpha_{i-1} z_{i-1}' + B_{i-1}(C) = 0$. Hence $\operatorname{im} \Delta_i \subset \ker \tilde{\alpha}_{i-1}$. Conversely, let $z_{i-1}' + B_{i-1}(C') \in \ker \tilde{\alpha}_{i-1}$. Then $\alpha_{i-1} z_{i-1}' = d_i c_i$, $c_i \in C_i$. Put $z_i'' = \beta_i c_i$. Then $d_i'' z_i'' = d_i'' \beta_i c_i = \beta_{i-1} d_i c_i = \beta_{i-1} \alpha_{i-1} z_{i-1}' = 0$, so $z_i'' \in Z_i(C'')$. The definition of Δ_i shows that $\Delta_i(z_i'' + B_i(C'')) = z_{i-1}' + B_{i-1}(C')$. Thus $\ker \tilde{\alpha}_{i-1} \subset \operatorname{im} \Delta_i$ and hence $\ker \tilde{\alpha}_{i-1} = \operatorname{im} \Delta_i$. □

The homomorphism Δ_i that we constructed is called *the connecting homomorphism* of $H_i(C'')$ into $H_{i-1}(C')$ and (4) is the *long exact homology sequence* determined by the short exact sequence of complexes $0 \to C' \xrightarrow{\alpha} C \xrightarrow{\beta} C'' \to 0$. An important feature of the connecting homomorphism is its naturality, which we state as

THEOREM 6.2. *Suppose we have a diagram of homomorphisms of complexes*

(8)

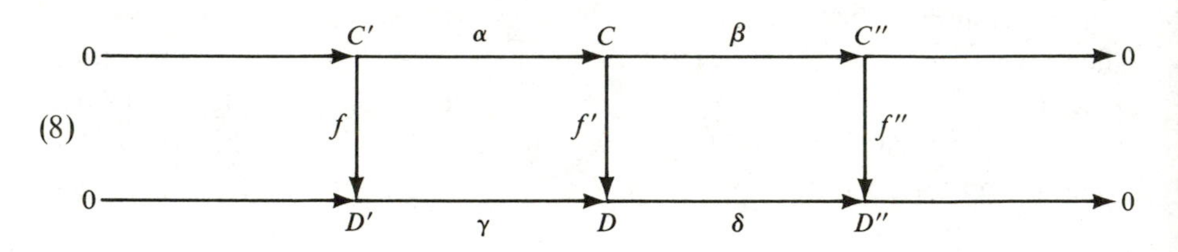

which is commutative and has exact rows. Then

(9)

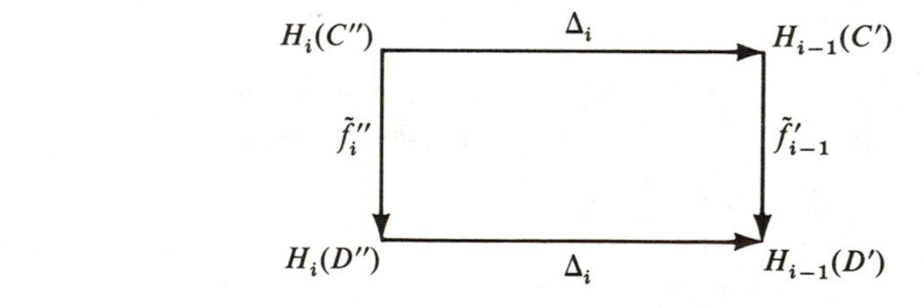

is commutative.

By the commutativity of (8) we mean of course the commutativity of

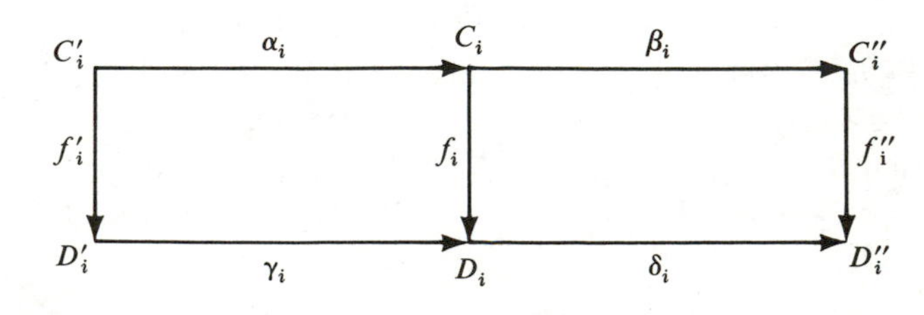

for every i. The proof of the commutativity of (9) is straightforward and is left to the reader.

EXERCISE

1. (The snake lemma.) Let

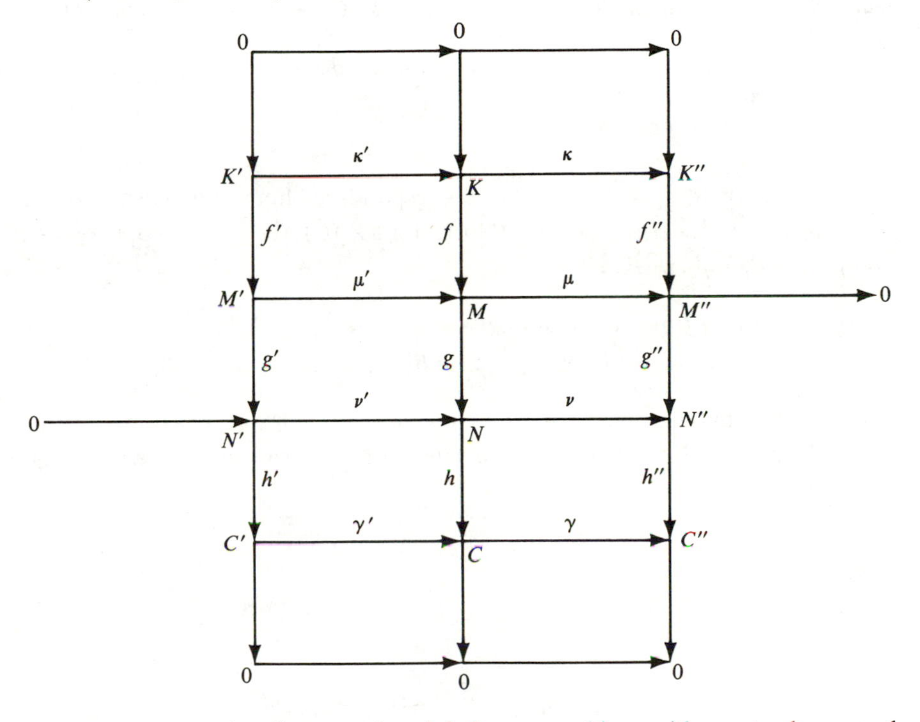

be a commutative diagram of module homomorphisms with exact columns and middle two rows exact. Let $x'' \in K''$ and let $y \in M$ satisfy $\mu y = f''x''$. Then $\nu g y = g''\mu y = g''f''x'' = 0$ and there exists a unique $z' \in N'$ such that $\nu' z' = gy$. Define $\Delta x'' = h'z'$. Show that $\Delta x''$ is independent of the choice of y and that $\Delta : K'' \to C'$ is a module homomorphism. Verify that

$$K' \xrightarrow{\kappa} K \xrightarrow{\kappa} K'' \xrightarrow{\Delta} C' \xrightarrow{\gamma'} C \xrightarrow{\gamma} C''$$

is exact. Show that if μ' is a monomorphism, then so is κ' and if ν is an epimorphism, then so is γ.

6.4 HOMOTOPY

We have seen that a chain homomorphism α of a complex (C, d) into a complex (C', d') determines a homomorphism $\bar{\alpha}_i$ of the ith homology module

$H_i(C)$ into $H_i(C')$ for every $i \in \mathbb{Z}$. There is an important relation between chain homomorphisms of (C,d) to (C',d') that automatically guarantees that the corresponding homomorphisms of the homology modules are identical. This is defined in

DEFINITION 6.4. *Let α and β be chain homomorphisms of a complex (C,d) into a complex (C',d'). Then α is said to be* homotopic *to β if there exists an indexed set $s = \{s_i\}$ of module homomorphisms $s_i : C_i \to C'_{i+1}$, $i \in \mathbb{Z}$, such that*

$$\alpha_i - \beta_i = d'_{i+1}s_i + s_{i-1}d_i.$$

We indicate homotopy by $\alpha \sim \beta$.

If $\alpha \sim \beta$, then $\tilde{\alpha}_i = \tilde{\beta}_i$ for the corresponding homomorphisms of the homology modules $H_i(C) \to H_i(C')$. For, if $z_i \in Z_i(C)$, then $\tilde{\alpha}_i : z_i + B_i \rightsquigarrow \alpha_i z_i + B'_i$ and $\tilde{\beta}_i : z_i + B_i \rightsquigarrow \beta_i z_i + B'_i$. Hence

$$\begin{aligned}\tilde{\alpha}_i(z_i + B_i) &= \alpha_i z_i + B'_i = (\beta_i + d'_{i+1}s_i + s_{i-1}d_i)z_i + B'_i \\ &= (\beta_i z_i + d'_{i+1}s_i z_i) + B'_i = \beta_i z_i + B'_i = \tilde{\beta}_i(z_i + B_i).\end{aligned}$$

It is clear that homotopy is a symmetric and reflexive relation for chain homomorphisms. It is also transitive, since if $\alpha \sim \beta$ is given by s and $\beta \sim \gamma$ is given by t, then

$$\begin{aligned}\alpha_i - \beta_i &= d'_{i+1}s_i + s_{i-1}d_i \\ \beta_i - \gamma_i &= d'_{i+1}t_i + t_{i-1}d_i.\end{aligned}$$

Hence

$$\alpha_i - \gamma_i = d'_{i+1}(s_i + t_i) + (s_{i-1} + t_{i-1})d_i.$$

Thus $s + t = \{s_i + t_i\}$ is a homotopy between α and γ.

Homotopies can also be multiplied. Suppose that $\alpha \sim \beta$ for the chain homomorphisms of $(C,d) \to (C',d')$ and $\gamma \sim \delta$ for the chain homomorphisms $(C',d') \to (C'',d'')$. Then $\gamma\alpha \sim \delta\beta$. We have, say,

$$\begin{aligned}\alpha_i - \beta_i &= d'_{i+1}s_i + s_{i-1}d_i, \\ \gamma_i - \delta_i &= d''_{i+1}t_i + t_{i-1}d'_i.\end{aligned}$$

Multiplication of the first of these by γ_i on the left and the second by β_i on the right gives

$$\begin{aligned}\gamma_i\alpha_i - \gamma_i\beta_i &= \gamma_i d'_{i+1}s_i + \gamma_i s_{i-1}d_i = d''_{i+1}\gamma_{i+1}s_i + \gamma_i s_{i-1}d_i \\ \gamma_i\beta_i - \delta_i\beta_i &= d''_{i+1}t_i\beta_i + t_{i-1}d'_i\beta_i = d''_{i+1}t_i\beta_i + t_{i-1}\beta_{i-1}d_i\end{aligned}$$

(by (2)). Hence

$$\gamma_i\alpha_i - \delta_i\beta_i = d''_{i+1}(\gamma_{i+1}s_i + t_i\beta_i) + (\gamma_i s_{i-1} + t_{i-1}\beta_{i-1})d_i.$$

Thus $\gamma\alpha \sim \delta\beta$ via $u = \{u_i\}$ where $u_i = \gamma_{i+1}s_i + t_i\beta_i$.

6.5 RESOLUTIONS

In the next section we shall achieve the first main objective of this chapter: the definition of the derived functor of an additive functor from the category of modules of a ring to the category of abelian groups. The definition is based on the concept of resolution of a module that we now consider.

DEFINITION 6.5. *Let M be an R-module. We define a* complex over M *as a positive complex $C = (C, d)$ together with a homomorphism $\varepsilon: C_0 \to M$, called an* augmentation, *such that $\varepsilon d_1 = 0$. Thus we have the sequence of homomorphisms*

$$\to C_n \xrightarrow{d_n} C_{n-1} \to \cdots \to C_1 \xrightarrow{d_1} C_0 \xrightarrow{\varepsilon} M \to 0 \tag{10}$$

where the product of any two successive homomorphisms is 0. The complex C, ε over M is called a resolution *of M if (10) is exact. This is equivalent to $H_i(C) = 0$ for $i > 0$ and $H_0(C) = C_0/d_1C_1 = C_0/\ker\varepsilon \cong M$. A complex C, ε over M is called* projective *if every C_i is projective.*

We have the following important

THEOREM 6.3. *Let C, ε be a projective complex over the module M and let C', ε' be a resolution of the module M', μ a homomorphism of M into M'. Then there exists a chain homomorphism α of the complex C into C' such that $\mu\varepsilon = \varepsilon'\alpha_0$. Moreover, any two such homomorphisms α and β are homotopic.*

Proof. The first assertion amounts to saying that there exist module homomorphisms α_i, $i > 0$, such that

(11)

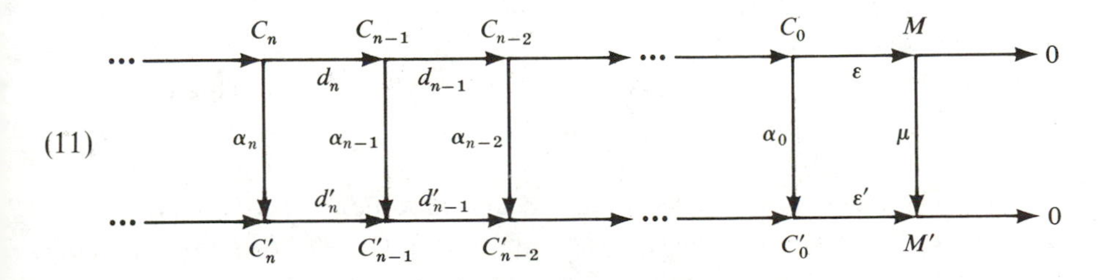

is commutative. Since C_0 is projective and $C'_0 \xrightarrow{\varepsilon'} M' \to 0$ is exact, the homomorphism $\mu\varepsilon$ of C_0 into M' can be "lifted" to a homomorphism $\alpha_0 : C_0 \to C'_0$ so that $\mu\varepsilon = \varepsilon'\alpha_0$. Now suppose we have already determined $\alpha_0, \ldots, \alpha_{n-1}$ so that the commutativity of (11) holds from C_0 to C_{n-1}. We have $d'_{n-1}\alpha_{n-1}d_n = \alpha_{n-2}d_{n-1}d_n = 0$. Hence $\alpha_{n-1}d_nC_n \subset \ker d'_{n-1} = \operatorname{im} d'_n = d'_nC'_n$. We can replace C'_{n-1} by $d'_nC'_n$ for which we have the exactness of $C'_n \to d'_nC'_n \to 0$. By the projectivity of C_n we have a homomorphism $\alpha_n : C_n \to C'_n$ such that $d'_n\alpha_n = \alpha_{n-1}d_n$. This inductive step proves the existence of α. Now let α and β satisfy the conditions. Let $\gamma = \alpha - \beta$. Then we have

$$\varepsilon'\gamma_0 = \varepsilon'\alpha_0 - \varepsilon'\beta_0 = \mu\varepsilon - \mu\varepsilon = 0,$$
$$d'_n\gamma_n = \gamma_{n-1}d_n, \qquad n \geqslant 1.$$

We have the diagram

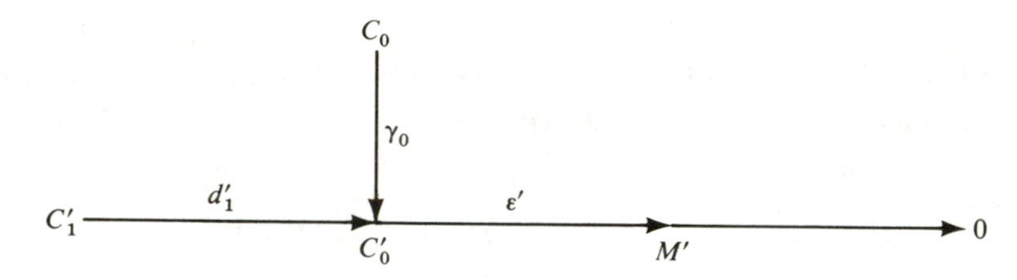

Since $\varepsilon'\gamma_0 = 0$, $\gamma_0C_0 \subset d'_1C'_1$ and we have the diagram

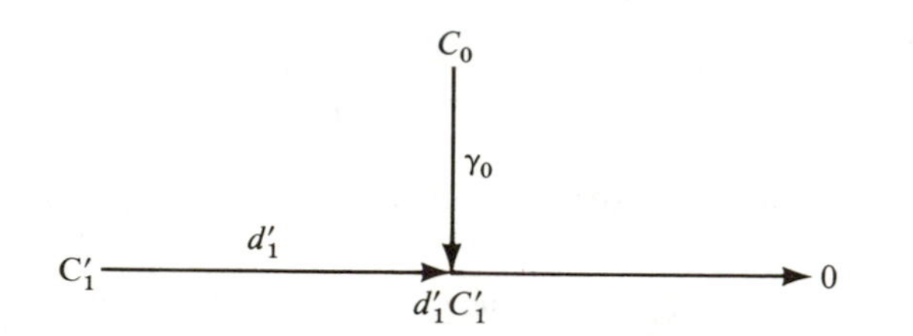

with exact row. As before, there exists a homomorphism $s_0 : C_0 \to C'_1$ such that $\gamma_0 = d'_1s_0$. Suppose we have determined $s_0, \ldots, s_{n-1}$ such that $s_i : C_i \to C'_{i+1}$ and

$$\gamma_i = d'_{i+1}s_i + s_{i-1}d_i, \qquad 0 \leqslant i \leqslant n-1.$$

Consider $\gamma_n - s_{n-1}d_n$. We have $d'_n(\gamma_n - s_{n-1}d_n) = \gamma_{n-1}d_n - d'_ns_{n-1}d_n = (\gamma_{n-1} - d'_ns_{n-1})d_n = s_{n-2}d_{n-1}d_n = 0$. It follows as before that there exists a homomorphism $s_n : C_n \to C'_{n+1}$ such that $d'_{n+1}s_n = \gamma_n - s_{n-1}d_n$. The sequence of homomorphisms $s_0, s_1, \ldots$ defines a homotopy of α to β. This completes the proof. □

The existence of a projective resolution of a module is easily established. In fact, as we shall now show, there exists a resolution (10) of M that is *free* in the sense that the modules C_i are free. First, we represent M as a homomorphic image of a free module C_0, which means that we have an exact sequence $\ker\varepsilon \xrightarrow{i} C_0 \xrightarrow{\varepsilon} M \to 0$ where C_0 is free and i is the injection of $\ker\varepsilon$. Next we obtain a free module C_1 and an epimorphism π of C_1 onto $\ker\varepsilon$. If we put $d_1 = i\pi : C_1 \to C_0$, we have the exact sequence $C_1 \xrightarrow{d_1} C_0 \xrightarrow{\varepsilon} M \to 0$. Iteration of this procedure leads to the existence of an exact sequence

$$\cdots \to C_n \xrightarrow{d_n} C_{n-1} \to \cdots \xrightarrow{d_1} C_0 \xrightarrow{\varepsilon} M \to 0$$

where the C_i are free. Then (C, d) and ε constitute a free resolution for M.

All of this can be dualized. We define a *complex under* M to be a pair D, η where D is a cochain complex and η is a homomorphism $M \to D^0$ such that $d^0\eta = 0$. Such a complex under M is called a *coresolution* of M if

$$(12) \qquad 0 \to M \xrightarrow{\eta} D^0 \xrightarrow{d^0} D^1 \xrightarrow{d^1} D^2 \to \cdots$$

is exact. We have shown in section 3.11 (p. 159) that any module M can be embedded in an injective module, that is, there exists an exact sequence $0 \to M \xrightarrow{\eta} D^0$ where D^0 is injective. This extends to $0 \to M \to D^0 \xrightarrow{\pi} \operatorname{coker}\eta$ where $\operatorname{coker}\eta = D^0/\eta M$ and π is the canonical homomorphism onto the quotient module. Next we have a monomorphism η_1 of coker η into an injective module D^1 and hence we have the exact sequence $0 \to M \xrightarrow{\eta} D^0 \xrightarrow{d^0} D^1$ where $d^0 = \eta_1\pi$. Continuing in this way, we obtain a coresolution (12) that is *injective* in the sense that every D^i is injective. The main theorem on resolutions can be dualized as follows.

THEOREM 6.4. *Let (D, η) be an injective complex under M, (D', η') a coresolution of M', λ a homomorphism of M' into M. Then there exists a homomorphism g of the complex D' into the complex D such that $\eta\lambda = g^0\eta'$. Moreover, any two such homomorphisms are homotopic.*

The diagram for the first statement is

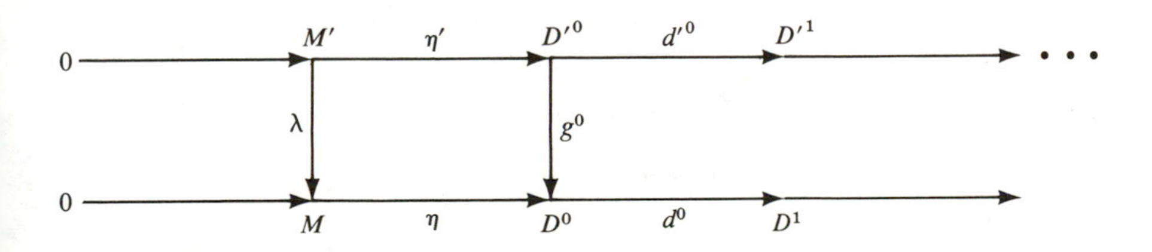

The proof of this theorem can be obtained by dualizing the argument used to prove Theorem 6.3. We leave the details to the reader.

6.6 DERIVED FUNCTORS

We are now ready to define the derived functors of an additive functor F from a category R-**mod** to the category **Ab**. Let M be an R-module and let

$$0 \leftarrow M \xleftarrow{\varepsilon} C_0 \xleftarrow{d_1} C_1 \xleftarrow{d_2} \cdots$$

be a projective resolution of M. Applying the functor F we obtain a sequence of homomorphisms of abelian groups

$$0 \longleftarrow FM \xleftarrow{F(\varepsilon)} FC_0 \xleftarrow{F(d_1)} FC_1 \xleftarrow{F(d_2)} \cdots. \tag{13}$$

Since $F(0) = 0$ for a zero homomorphism of a module into a second one and since F is multiplicative, the product of successive homomorphisms in (13) is 0 and so $FC = \{FC_i\}$, $F(d) = \{F(d_i)\}$ with the augmentation $F\varepsilon$ is a (positive) complex over FM. If F is exact, that is, preserves exactness, then (13) is exact and the homology groups $H_i(FC) = 0$ for $i \geqslant 1$. This need not be the case if F is not exact, and these homology groups in a sense measure the departure of F from exactness. At any rate, we now put

$$L_nFM = H_n(FC), \qquad n \geqslant 0. \tag{14}$$

This definition gives

$$H_0(FC) = FC_0/F(d_1)FC_1 \tag{15}$$

since we are taking the terms $FC_i = 0$ if $i < 0$.

Let M' be a second R-module and suppose we have chosen a projective resolution $0 \leftarrow M' \xleftarrow{\varepsilon'} C'_0 \xleftarrow{d'_1} C'_1 \cdots$ of M', from which we obtain the abelian groups $H_n(FC')$, $n \geqslant 0$. Let μ be a module homomorphism of M into M'. Then we have seen that we can determine a homomorphism α of the complex (C, d) into (C', d') such that $\mu\varepsilon = \varepsilon'\alpha_0$. We call α a chain homomorphism *over* the given homomorphism μ. Since F is an additive functor, we have the homomorphism $F(\alpha)$ of the complex FC into the complex FC' such that $F(\mu)F(\varepsilon) = F(\varepsilon')F(\alpha_0)$. Thus we have the commutative diagram

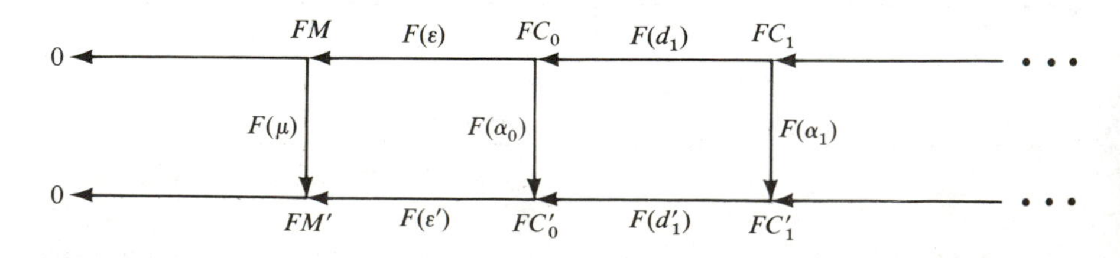

Then we have the homomorphism $\widetilde{F(\alpha_n)}$ of $H_n(FC)$ into $H_n(FC')$. This is

independent of the choice of α. For, if β is a second homomorphism of C into C' over μ, β is homotopic to α, so there exist homomorphisms $s_n: C_n \to C'_{n+1}$, $n \geqslant 0$, such that $\alpha_n - \beta_n = d'_{n+1}s_n + s_{n-1}d_n$. Since F is additive, application of F to these relations gives

$$F(\alpha_n) - F(\beta_n) = F(d'_{n+1})F(s_n) + F(s_{n-1})F(d_n).$$

Thus $F(\alpha) \sim F(\beta)$ and hence $\widetilde{F(\alpha_n)} = \widetilde{F(\beta_n)}$. We now define $L_nF(\mu) = \widetilde{F(\alpha_n)}$. Thus a homomorphism $\mu: M \to M'$ determines a homomorphism $L_nF(\mu)$: $H_n(FC) \to H_n(FC')$. We leave it to the reader to carry out the verification that L_nF defined by

$$\begin{aligned} (L_nF)M &= H_n(FC), \qquad && M \in \text{ob}\, R\text{-}\mathbf{mod} \\ (L_nF)(\mu) &= \widetilde{F(\alpha_n)}, \qquad && \mu \in \hom_R(M, M') \end{aligned}$$

is an additive functor from R-**mod** to **Ab**. This is called the nth *left derived* functor of the given functor F.

We now observe that our definitions are essentially independent of the choice of the resolutions. Let $\bar{C}$ be a second projective resolution of M. Then taking $\mu = 1$ in the foregoing argument, we obtain a unique isomorphism η_n of $H_n(FC)$ onto $H_n(F\bar{C})$. Similarly, another choice $\bar{C}'$ of projective resolution of M' yields a unique isomorphism η'_n of $H_n(FC')$ onto $H_n(F\bar{C}')$ and L_nF is replaced by $\eta'_n(L_nF)\eta_n^{-1}$.

From now on we shall assume that for every R-module M we have chosen a particular projective resolution and that this is used to determine the functors L_nF. However, we reserve the right to switch from one such resolution to another when it is convenient to do so.

We consider next a short exact sequence $0 \to M' \xrightarrow{\alpha} M \xrightarrow{\beta} M'' \to 0$ and we shall show that corresponding to such a sequence we have connecting homomorphisms

$$\Delta_n: L_nFM'' \to L_{n-1}FM', \qquad n \geqslant 1,$$

such that

$$0 \longleftarrow L_0FM'' \xleftarrow{L_0F(\beta)} L_0FM \xleftarrow{L_0F(\alpha)} L_0FM' \xleftarrow{\Delta_1} L_1FM'' \longleftarrow \cdots$$

is exact. For this purpose we require the existence of projective resolutions of short exact sequences of homomorphisms of modules. By a *projective resolution of such a sequence* $0 \to M' \to M \to M'' \to 0$ we mean projective resolutions C', ε', C, ε, C'', ε'' of M', M, and M'' respectively together with chain homomorphisms $i: C' \to C$, $p: C \to C''$ such that for each n, $0 \to C'_n \to C_n \to C''_n \to 0$ is exact and

(16)

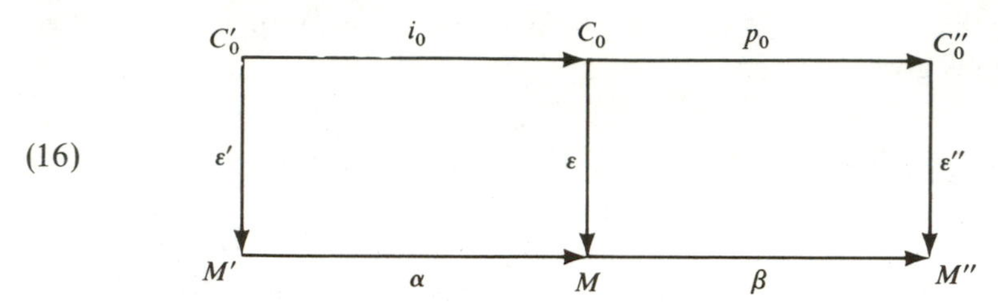

is commutative.

We shall now prove the existence of a projective resolution for any short exact sequence of modules $0 \to M' \xrightarrow{\alpha} M \xrightarrow{\beta} M'' \to 0$. We begin with projective resolutions C', ε', and C'', ε'' of M' and M'' respectively:

(17)
$$\begin{aligned} \cdots C'_2 \xrightarrow[d'_2]{} C'_1 \xrightarrow[d'_1]{} C'_0 \xrightarrow[\varepsilon']{} M' \to 0 \\ \cdots C''_2 \xrightarrow[d''_2]{} C''_1 \xrightarrow[d''_1]{} C''_0 \xrightarrow[\varepsilon'']{} M'' \to 0. \end{aligned}$$

We let $C_n = C'_n \oplus C''_n$, $n = 1, 2, 3, \ldots, i_n x'_n = (x'_n, 0)$ for $x'_n \in C'_n$, $p_n x_n = x''_n$ for $x_n = (x'_n, x''_n)$. Then $0 \to C'_n \to C'_n \oplus C''_n \to C''_n \to 0$ is exact and C_n is projective. We now define $\varepsilon x_0 = \alpha\varepsilon' x'_0 + \sigma x''_0$, $d_n x_n = (d'_n x'_n + \theta_n x''_n, d''_n x''_n)$ where $\sigma: C''_0 \to M'$, $\theta_n: C''_n \to C'_{n-1}$ are to be determined so that C, ε is a projective resolution of M which together with C', ε' and C'', ε'' constitutes a projective resolution for the exact sequence $0 \to M' \xrightarrow[\alpha]{} M \xrightarrow[\beta]{} M'' \to 0$.

We have $\varepsilon i_0 x'_0 = \varepsilon(x'_0, 0) = \alpha\varepsilon' x'_0$. Hence commutativity of the left-hand rectangle in (16) is automatic. Also $\varepsilon'' p_0 x_0 = \varepsilon'' x''_0$ and $\beta\varepsilon x_0 = \beta(\alpha\varepsilon' x'_0 + \sigma x''_0) = \beta\sigma x''_0$. Hence commutativity of the right-hand rectangle in (16) holds if and only if

(18)
$$\varepsilon'' = \beta\sigma.$$

We have $\varepsilon d_1 x_1 = \varepsilon(d'_1 x'_1 + \theta_1 x''_1, d''_1 x''_1) = \alpha\varepsilon'\theta_1 x''_1 + \sigma d''_1 x''_1$. Hence $\varepsilon d_1 = 0$ if and only if

(19)
$$\alpha\varepsilon'\theta_1 + \sigma d''_1 = 0.$$

Similarly, the condition $d_{n-1} d_n = 0$ is equivalent to

(20)
$$d'_{n-1}\theta_n + \theta_{n-1} d''_n = 0.$$

Now consider the diagram

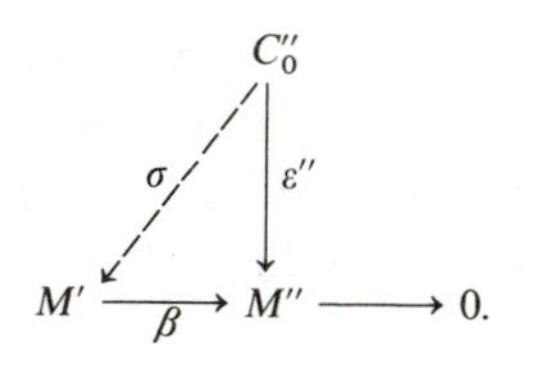

Since C_0'' is projective there exists a $\sigma: C_0'' \to M'$ such that (18) holds. Next we consider

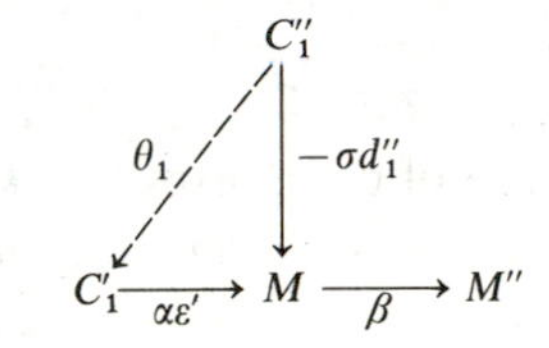

Since $\varepsilon' C_0' = M'$ the row is exact, and since C_1'' is projective and $\beta\sigma d_1'' = \varepsilon'' d_1'' = 0$, there exists a $\theta_1: C_1'' \to C_0'$ such that (19) holds (see exercise 4, p. 100). Next we consider

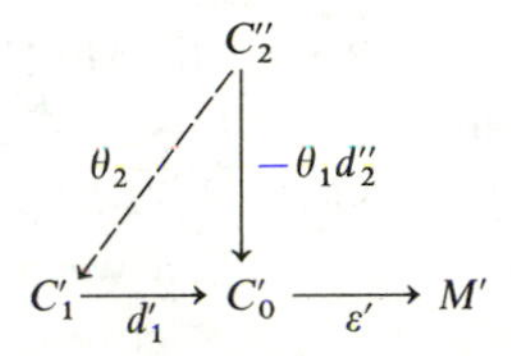

Here again the row is exact, C_2'' is projective and $\varepsilon'\theta_1 d_2'' = 0$ since $\alpha\varepsilon'\theta_1 d_2'' = -\sigma d_1'' d_2'' = 0$ and $\ker \alpha = 0$. Hence there exists $\theta_2: C_2'' \to C_1'$ such that (19) holds for $n = 2$. Finally, the same argument establishes (20) for $n > 2$ using induction and the diagram

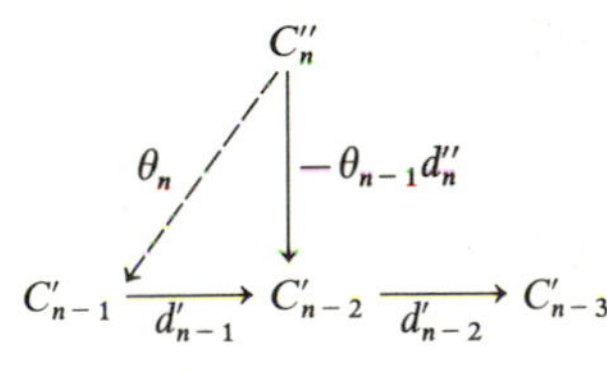

It remains to show that $\cdots C_2 \underset{d_2}{\to} C_1 \underset{d_1}{\to} C_0 \underset{\varepsilon}{\to} M \to 0$ is a resolution. For this purpose we regard this sequence of modules and homomorphisms as a complex $\bar{C}$ and similarly we regard the modules and homomorphisms in (17) as complexes $\bar{C}'$ and $\bar{C}''$. Then we have an exact sequence of complexes $0 \to \bar{C}' \to \bar{C} \to \bar{C}'' \to 0$. Since $H_i(\bar{C}') = 0 = H_i(\bar{C}'')$ it follows from (4) that $H_i(\bar{C}) = 0$. Then $\bar{C}$ provides a resolution for M which together with the resolutions for M' and M'' gives a projective resolution for $0 \to M' \to M \to M'' \to 0$.

We can now prove

THEOREM 6.5. *Let F be an additive functor from R-**mod** to **Ab**. Then for any short exact sequence of R-modules $0 \to M' \overset{\alpha}{\to} M \overset{\beta}{\to} M'' \to 0$ and any $n = 1, 2, 3, \ldots$ there exists a connecting homomorphism $\Delta_n: L_n FM'' \to L_{n-1} FM'$ such that*

$$(21) \qquad 0 \longleftarrow L_0 FM'' \xleftarrow{L_0F(\beta)} L_0 FM \xleftarrow{L_0F(\alpha)} L_0 FM' \xleftarrow{\Delta_1} L_1 FM'' \longleftarrow \cdots$$

is exact.

Proof. We construct a projective resolution C', ε', C, ε, C'', ε'', i, p for the given short exact sequence of modules. For each $n \geqslant 0$, we have the short exact sequence of projective modules $0 \to C'_n \xrightarrow{i_n} C_n \xrightarrow{p_n} C''_n \to 0$. Since C''_n is projective, this splits and consequently, $0 \to FC'_n \xrightarrow{F(i_n)} FC_n \xrightarrow{F(p_n)} FC''_n \to 0$ is split exact. Thus we have a short exact sequence of complexes $0 \to F(C') \xrightarrow{F(i)} F(C) \xrightarrow{F(p)} F(C'') \to 0$. The theorem follows by applying the long exact homology sequence to this short exact sequence of complexes. □

Everything we have done can be carried over to coresolutions, and this gives the definition and analogous results for right derived functors. We shall now sketch this, leaving the details to the reader.

Again let F be a functor from R-**mod** to **Ab**. For a given R-module M, we choose an injective coresolution $0 \to M \xrightarrow{\eta} D^0 \xrightarrow{d^0} D^1 \xrightarrow{d^1} D^2 \to \cdots$. Applying F, we obtain $0 \longrightarrow FM \xrightarrow{F(\eta)} FD^0 \xrightarrow{F(d^0)} FD^1 \xrightarrow{F(D^1)} FD^2 \longrightarrow \cdots$ and we obtain the complex $FD = \{FD^i\}$, $F(d) = \{F(d^i)\}$. Then we put $(R^nF)M = H^n(FD)$, $n \geqslant 0$. If M' is a second R-module, (D', η') a coresolution of M', then for any homomorphism $\lambda: M' \to M$ we obtain a homomorphism $R^nF(\lambda)$: $(R^nF)M' \to (R^nF)M$. This defines the *right derived functor* of the given functor F. The results we obtained for left derived functors carry over. In particular, we have an analogue of the long exact sequence given in Theorem 6.5. We omit the details.

EXERCISES

1. Show that if M is projective, then $L_0FM = FM$ and $L_nFM = 0$ for $n > 0$.
2. Show that if F is right exact, then F and L_0F are naturally equivalent.

6.7 EXT

In this section and the next we shall consider the most important instances of derived functors. We begin with the contravariant hom functor $\hom(-, N)$ defined by a fixed module N, but first we need to indicate the modifications in the foregoing procedure that are required in passing to additive contravariant functors from R-**mod** to **Ab**. Such a functor is a (covariant) functor from the opposite category R-**mod**^{op} to **Ab**, and since arrows are reversed in passing to the opposite category, the roles of injective and projective modules must be interchanged. Accordingly, to define the right derived functor of a contravariant functor G from R-**mod** to **Ab**, we begin with a projective resolution

$0 \leftarrow M \xleftarrow{\varepsilon} C_0 \xleftarrow{d_1} C_1 \leftarrow \cdots$ of the given module M. This gives rise to the sequence $0 \to GM \xrightarrow{G(\varepsilon)} GC_0 \xrightarrow{G(d_1)} GC_1 \longrightarrow \cdots$ and the cochain complex $(GC, G(d))$ where $GC = \{GC_i\}$ and $G(d) = \{G(d_i)\}$. We define $(R^nG)M = H^n(GC)$. In particular, we have $(R^0G)M = \ker(GC_0 \to GC_1)$. For any $\mu \in \hom_R(M', M)$ we obtain a homomorphism $(R^nG)(\mu): (R^nG)M \to (R^nG)M'$ and so we obtain the nth *right derived functor* R^nG of G, which is additive and contravariant. Corresponding to a short exact sequence $0 \to M' \to M \to M'' \to 0$ we have the long exact cohomology sequence

$$(22) \quad 0 \to R^0GM'' \to R^0GM \to R^0GM' \to R^1GM'' \to R^1GM \to R^1GM' \to \cdots$$

where $R^nGM' \to R^{n+1}GM''$ is given by a connecting homomorphism. The proof is almost identical with that of Theorem 6.5 and is therefore omitted.

We now let $G = \hom(-, N)$ the contravariant hom functor determined by a fixed R-module N. We recall the definition: If $M \in \text{ob}\, R\textbf{-mod}$, then $\hom(-, N)M = \hom_R(M, N)$ and if $\alpha \in \hom_R(M, M')$, $\hom(-, N)(\alpha)$ is the map α^* of $\hom_R(M', N)$ into $\hom_R(M, N)$ sending any β in the former into $\beta\alpha \in \hom_R(M, N)$. $\hom(-, N)M$ is an abelian group and α^* is a group homomorphism. Hence $\hom(-, N)$ is additive. Since $(\alpha_1\alpha_2)^* = \alpha_2^*\alpha_1^*$, the functor $\hom(-, N)$ is contravariant. We recall also that this functor is left exact, that is, if $M' \xrightarrow{\alpha} M \xrightarrow{\beta} M'' \to 0$ is exact, then $0 \to \hom(M'', N) \xrightarrow{\beta^*} \hom(M, N) \xrightarrow{\alpha^*} \hom(M', N)$ is exact (p. 105).

The nth right derived functor of $\hom(-, N)$ is denoted as $\text{Ext}^n(-, N)$; its value for the module M is $\text{Ext}^n(M, N)$ (or $\text{Ext}_R^n(M, N)$ if it is desirable to indicate R). If C, ε is a projective resolution for M, then the exactness of $C_1 \to C_0 \xrightarrow{\varepsilon} M \to 0$ implies that of $0 \to \hom(M, N) \xrightarrow{\varepsilon^*} \hom(C_0, N) \to \hom(C_1, N)$. Since $\text{Ext}^0(M, N)$ is the kernel of the homomorphism of $\hom(C_0, N)$ into $\hom(C_1, N)$ it is clear that

$$(23) \qquad \text{Ext}^0(M, N) \cong \hom(M, N)$$

under the map ε^*.

Now let $0 \to M' \to M \to M'' \to 0$ be a short exact sequence. Then we obtain the long exact sequence

$$(24) \qquad \begin{aligned} &0 \to \text{Ext}^0(M'', N) \to \text{Ext}^0(M, N) \to \text{Ext}^0(M', N) \\ &\quad \to \text{Ext}^1(M'', N) \to \text{Ext}^1(M, N) \to \cdots. \end{aligned}$$

If we use the isomorphism (23), we obtain an imbedding of the exact sequence $0 \to \hom(M'', N) \to \hom(M, N) \to \hom(M', N)$ in a long exact sequence

$$(25) \qquad \begin{aligned} &0 \to \hom(M'', N) \to \hom(M, N) \to \hom(M', N) \\ &\quad \to \text{Ext}^1(M'', N) \to \text{Ext}^1(M, N) \to \cdots. \end{aligned}$$

We can now prove

THEOREM 6.6. *The following conditions on a module M are equivalent:*

(1) *M is projective.*

(2) $\mathrm{Ext}^n(M,N) = 0$ *for all $n \geqslant 1$ and all modules N.*

(3) $\mathrm{Ext}^1(M,N) = 0$ *for all N.*

Proof. (1)$\Rightarrow$(2). If M is projective, then $0 \leftarrow M \overset{1_M}{\leftarrow} C_0 = M \leftarrow 0 \leftarrow \cdots$ is a projective resolution. The corresponding complex to calculate $\mathrm{Ext}^n(M,N)$ is $0 \to \hom(M,N) \to \hom(M,N) \to 0 \to \cdots$. Hence $\mathrm{Ext}^n(M,N) = 0$ for all $n \geqslant 1$. (2)$\Rightarrow$(3) is clear. (3)$\Rightarrow$(1). Let M be any module and let $0 \to K \overset{\eta}{\to} P \overset{\varepsilon}{\to} M \to 0$ be a short exact sequence with P projective. Then (25) and the fact that $\mathrm{Ext}^1(P,N) = 0$ yield the exactness of

$$(26) \qquad 0 \to \hom(M,N) \to \hom(P,N) \to \hom(K,N) \to \mathrm{Ext}^1(M,N) \to 0.$$

Now assume $\mathrm{Ext}^1(M,N) = 0$. Then we have the exactness of $0 \to \hom(M,N) \to \hom(P,N) \to \hom(K,N) \to 0$, which implies that the map η^* of $\hom(P,N)$ into $\hom(K,N)$ is surjective. Now take $N = K$. Then the surjectivity of η^* on $\hom(K,K)$ implies that there exists a $\zeta \in \hom(P,K)$ such that $1_K = \zeta\eta$. This implies that the short exact sequence $0 \to K \overset{\eta}{\to} P \overset{\varepsilon}{\to} M \to 0$ splits. Then M is a direct summand of a projective module and so M is projective. $\square$

The exact sequence (24) in which $0 \to K \overset{\eta}{\to} P \overset{\varepsilon}{\to} M \to 0$ is exact, M is arbitrary and P is projective gives the following formula for $\mathrm{Ext}^1(M,N)$:

$$(27) \qquad \mathrm{Ext}^1(M,N) = \operatorname{coker} \eta^* = \hom(K,N)/\mathrm{im}(\hom(P,N)).$$

We shall use this formula to relate $\mathrm{Ext}^1(M,N)$ with extensions of the module M by the module N. It is this connection that accounts for the name Ext for the functor.

If M and N are modules, we define an *extension of M by N* to be a short exact sequence

$$(28) \qquad 0 \to N \overset{\alpha}{\to} E \overset{\beta}{\to} M \to 0.$$

For brevity we refer to this as "the extension E." Two extensions E_1 and E_2 are said to be *equivalent* if there exists an isomorphism $\gamma: E_1 \to E_2$ such that

(29)

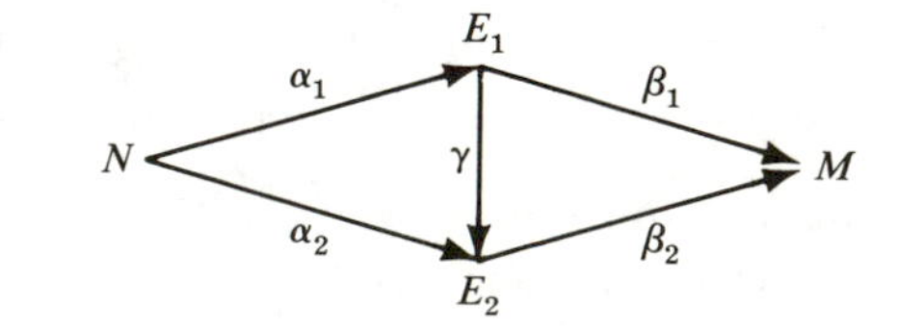

is commutative. It is easily seen that if γ is a homomorphism from the extension E_1 to E_2 making (29) commutative, then γ is necessarily an isomorphism. Equivalence of extensions is indeed an equivalence relation. It is clear also that extensions of M by N exist; for, we can construct the *split extension*

$$0 \to N \xrightarrow{i} M \oplus N \xrightarrow{p} M \to 0 \tag{30}$$

with the usual i and p.

We shall now define a bijective map of the class $E(M,N)$ of equivalence classes of extensions of M by N with the set $\mathrm{Ext}^1(M,N)$. More precisely, we shall define a bijective map of $E(M,N)$ onto $\mathrm{coker}\,\eta^*$ where $0 \to K \xrightarrow{\eta} P \xrightarrow{\varepsilon} M \to 0$ is a projective presentation of M and η^* is the corresponding map $\hom(P,N) \to \hom(K,N)\,(\eta^*(\lambda) = \lambda\eta)$. In view of the isomorphism given in (27), this will give the bijection of $E(M,N)$ with $\mathrm{Ext}^1(M,N)$.

Let $0 \to N \xrightarrow{\alpha} E \xrightarrow{\beta} M \to 0$ be an extension of M by N. Then we have the diagram

(31)

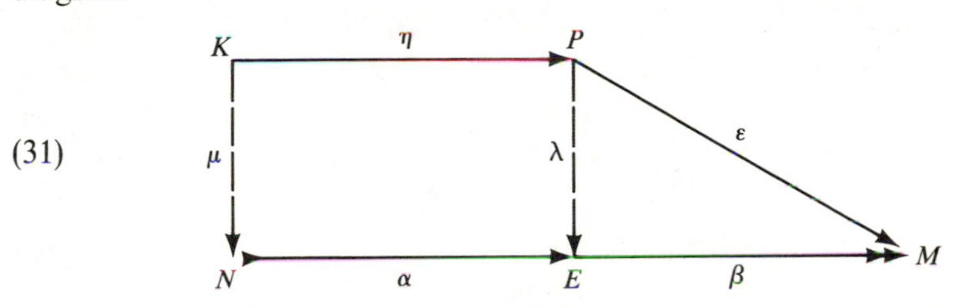

without the dotted lines. Since P is projective, there is a homomorphism $\lambda: P \to E$ making the triangle commutative. With this choice of λ there is a unique homomorphism $\mu: K \to N$ making the rectangle commutative. For, if $x \in K$, then $\beta\lambda\eta x = \varepsilon\eta x = 0$. Hence $\lambda\eta x \in \ker\beta$ and so there exists a unique $y \in N$ such that $\alpha y = \lambda\eta x$. We define μ by $x \rightsquigarrow y$. Then it is clear that μ is a homomorphism of K into N making the rectangle commutative and μ is unique. Next, let λ' be a second homomorphism of P into E such that $\beta\lambda' = \varepsilon$. Then $\beta(\lambda' - \lambda) = 0$, which implies that there exists a homomorphism $\tau: P \to N$ such that $\lambda' - \lambda = \alpha\tau$. Then $\lambda'\eta = (\lambda + \alpha\tau)\eta = \alpha(\mu + \tau\eta)$. Hence $\mu' = \mu + \tau\eta$ makes

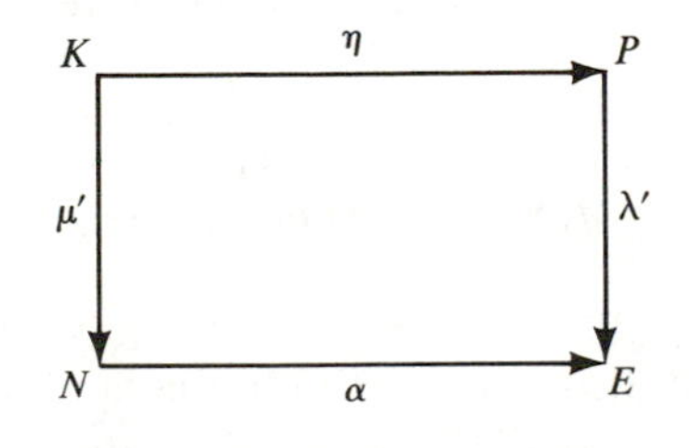

commutative. Since $\tau \in \hom(P,N)$, $\tau\eta \in \operatorname{im} \eta^*$. Thus μ and μ' determine the same element of $\operatorname{coker} \eta^*$ and we have the map sending the extension E into the element $\mu + \operatorname{im} \eta^*$ of $\operatorname{coker} \eta^*$. It is readily seen, by drawing a diagram, that the replacement of E by an equivalent extension E' yields the same element $\mu + \operatorname{im} \eta^*$. Hence we have a map of $E(M,N)$ into $\operatorname{coker} \eta^*$.

Conversely, let $\mu \in \hom(K,N)$. We form the pushout of η and μ (exercise 8, p. 37). Explicitly, we form $N \oplus P$ and let I be the submodule of elements $(-\mu(x), \eta(x))$, $x \in K$. Put $E = (N \oplus P)/I$ and let α be the homomorphism of N into E such that $\alpha(y) = (y,0) + I$. Also we have the homomorphism of $N \oplus P$ into M such that $(y,z) \rightsquigarrow \varepsilon(z), y \in N$, $z \in P$. This maps I into 0 and so defines a homomorphism β of $E = (N \oplus P)/I$ into M such that $(y,z) + I \rightsquigarrow \varepsilon(z)$. We claim that $0 \to N \xrightarrow{\alpha} E \xrightarrow{\beta} M \to 0$ is exact. First, if $\alpha(y) = (y,0) + I = 0$, then $(y,0) = (-\mu(x), \eta(x))$, $x \in K$, so $\eta(x) = 0$ and $x = 0$ and $y = 0$. Thus α is injective. Next, $\beta\alpha y = \beta((y,0) + I) = \varepsilon(0) = 0$, so $\beta\alpha = 0$. Moreover, if $\beta((y,z) + I) = 0$, then $\varepsilon(z) = 0$ so $z = \eta(x)$, $x \in K$. Then $(y,z) + I = (y + \mu(x), 0) + I = \alpha(y + \mu(x))$. Hence $\ker \beta = \operatorname{im} \alpha$. Finally, β is surjective since if $u \in M$, then $u = \varepsilon(z)$, $z \in P$, and $\beta((0,z) + I) = \varepsilon(z) = u$. If we put $\lambda(z) = (0,z) + I$, then we have the commutativity of the diagram (31). Hence the element of $\operatorname{coker} \eta^*$ associated with the equivalence class of the extension E is $\mu + \operatorname{im} \eta^*$. This shows that our map is surjective. It is also injective. For, let E be any extension such that the class of E is mapped into the coset $\mu + \operatorname{im} \eta^*$. Then we may assume that $E \rightsquigarrow \mu$ under the original map we defined, and if we form the pushout E' of μ and η, then E' is an extension such that $E' \rightsquigarrow \mu$. Now since E' is a pushout, we have a commutative diagram (29) with $E_1 = E'$ and $E_2 = E$. Then E and E' are isomorphic. Evidently, this implies injectivity.

We state this result as

THEOREM 6.7. *We have a bijective map of* $\operatorname{Ext}^1(M,N)$ *with the set* $E(M,N)$ *of equivalence classes of extensions of* M *by* N.

We shall study next the dependence of $\operatorname{Ext}^n(M,N)$ on the argument N. This will lead to the definition of a functor $\operatorname{Ext}^n(M,-)$ and a bifunctor Ext^n. Let M, N, N' be R-modules and β a homomorphism of N into N'. As before, let $C = \{C_i\}$, ε be a projective resolution of M. Then we have the diagram

$$\begin{array}{ccccccc} 0 & \longrightarrow & \hom(M,N) & \longrightarrow & \hom(C_0,N) & \longrightarrow & \cdots \\ & & \downarrow & & \downarrow & & \\ 0 & \longrightarrow & \hom(M,N') & \longrightarrow & \hom(C_0,N') & \longrightarrow & \cdots \end{array} \tag{32}$$

where the horizontal maps are as before and the vertical ones are the left multiplications β_L by β. It is clear that (32) is commutative and hence we have homomorphisms of the complex $\hom(C,N)$ into the complex $\hom(C,N')$;

consequently, for each $n \geqslant 0$ we have a homomorphism $\widetilde{\beta}^n$ of the corresponding cohomology groups. Thus we have the homomorphism $\widetilde{\beta}^n : \mathrm{Ext}^n(M,N) \to \mathrm{Ext}^n(M,N')$. It is clear that this defines a functor $\mathrm{Ext}^n(M,-)$ from R-**mod** to **Ab** that is additive and covariant.

If $\alpha \in \hom(M',M)$, we have the commutative diagram

$$\begin{array}{ccc} \hom(M,N) & \longrightarrow & \hom(M',N) \\ \downarrow & & \downarrow \\ \hom(M,N') & \longrightarrow & \hom(M',N'), \end{array}$$

which gives the commutative diagram

$$\begin{array}{ccc} \mathrm{Ext}^n(M,N) & \longrightarrow & \mathrm{Ext}^n(M',N) \\ \downarrow & & \downarrow \\ \mathrm{Ext}^n(M,N') & \longrightarrow & \mathrm{Ext}^n(M',N'). \end{array}$$

This implies as in the case of the hom functor that we can define a bifunctor Ext^n from R-**mod** to **Ab** (p. 38).

Now suppose that we have a short exact sequence $0 \to N' \to N \to N'' \to 0$. As in (32), we have the sequence of homomorphisms of these complexes: $\hom(C,N') \to \hom(C,N) \to \hom(C,N'')$. Since C_i is projective and $0 \to N' \to N \to N'' \to 0$ is exact, $0 \to \hom(C_i,N') \to \hom(C_i,N) \to \hom(C_i,N'') \to 0$ is exact for every i. Thus $0 \to \hom(C,N') \to \hom(C,N) \to \hom(C,N'') \to 0$ is exact. Hence we can apply Theorem 6.1 and the isomorphism of $\hom(M,N)$ with $\mathrm{Ext}^0(M,N)$ to obtain a second long exact sequence of Ext functors:

$$(33) \quad 0 \to \hom(M,N') \to \hom(M,N) \to \hom(M,N'') \to \mathrm{Ext}^1(M,N') \to \\ \mathrm{Ext}^1(M,N) \to \mathrm{Ext}^1(M,N'') \to \cdots.$$

We shall call the two sequences (24) and (33) the *long exact sequences in the first and second variables* respectively for Ext.

We can now prove the following analogue of the characterization of projective modules given in Theorem 6.6.

THEOREM 6.8. *The following conditions on a module N are equivalent:*

(1) *N is injective.*
(2) *$\mathrm{Ext}^n(M,N) = 0$ for all $n \geqslant 1$ and all modules M.*
(3) *$\mathrm{Ext}^1(M,N) = 0$ for all M.*

Proof. (1) $\Rightarrow$ (2). If N is injective, the exactness of $0 \leftarrow M \leftarrow C_0 \leftarrow C_1 \leftarrow \cdots$ implies that of $0 \to \hom(M,N) \to \hom(C_0,N) \to \hom(C_1,N) \to \cdots$. This implies that $\mathrm{Ext}^n(M,N) = 0$ for all $n \geqslant 1$. The implications (2) $\Rightarrow$ (3) are trivial, and (3) $\Rightarrow$ (1) can be obtained as in the proof of Theorem 6.6 by using a short exact sequence $0 \to N \to Q \to L \to 0$ where Q is injective. We leave it to the reader to complete this argument. $\square$

The functors $\text{Ext}^n(M, -)$ that we have defined by starting with the functors $\text{Ext}^n(-, N)$ can also be defined directly as the right derived functors of $\hom(M, -)$. For the moment, we denote the value of this functor for the module M as $\overline{\text{Ext}}^n(M, N)$. To obtain a determination of this group, we choose an injective coresolution $0 \to N \xrightarrow{\eta} D^0 \xrightarrow{d^1} D^1 \to \cdots$ and we obtain the cochain complex $\hom(M, D): 0 \to \hom(M, D^0) \to \hom(M, D^1) \to \cdots$. Then $\overline{\text{Ext}}^n(M, N)$ is the nth cohomology group of $\hom(M, D)$. The results we had for Ext can easily be established for $\overline{\text{Ext}}$. In particular, we can show that $\overline{\text{Ext}}^0(M, N) \cong \hom(M, N)$ and we have the two long exact sequences for $\overline{\text{Ext}}$ analogous to (24) and (33). We omit the derivations of these results. Now it can be shown that the bifunctors Ext^n and $\overline{\text{Ext}}^n$ are naturally equivalent. We shall not prove this, but shall be content to prove the following weaker result:

THEOREM 6.9. $\text{Ext}^n(M, N) \cong \overline{\text{Ext}}^n(M, N)$ *for all* n, M, *and* N.

Proof. If $n = 0$, we have $\text{Ext}^0(M, N) \cong \hom(M, N) \cong \overline{\text{Ext}}^0(M, N)$. Now let $0 \to K \to P \to M \to 0$ be a short exact sequence with P projective. Using the long exact sequence on the first variable for $\overline{\text{Ext}}$ and $\overline{\text{Ext}}^1(P, N) = 0$ we obtain the exact sequence $0 \to \hom(M, N) \to \hom(P, N) \to \hom(K, N) \to \overline{\text{Ext}}^1(M, N) \to 0$. This implies that $\overline{\text{Ext}}^1(M, N) \cong \hom(K, N)/\text{im}\hom(P, N)$. In (27) we showed that $\text{Ext}^1(M, N) \cong \hom(K, N)/\text{im}\hom(P, N)$. Hence $\text{Ext}^1(M, N) \cong \overline{\text{Ext}}^1(M, N)$. Now let $n > 1$ and assume the result for $n - 1$. We refer again to the long exact sequence on the first variable for Ext and obtain

$$0 = \text{Ext}^{n-1}(P, N) \to \text{Ext}^{n-1}(K, N) \to \text{Ext}^n(M, N) \to \text{Ext}^n(P, N) = 0$$

from which we infer that $\text{Ext}^n(M, N) \cong \text{Ext}^{n-1}(K, N)$. Hence $\text{Ext}^n(M, N) \cong \text{Ext}^{n-1}(K, N) \cong \overline{\text{Ext}}^{n-1}(K, N)$. The same argument gives $\overline{\text{Ext}}^n(M, N) \cong \overline{\text{Ext}}^{n-1}(K, N)$. Hence $\text{Ext}^n(M, N) \cong \overline{\text{Ext}}^n(M, N)$. □

EXERCISES

1. Let $R = D$, a commutative principal ideal domain. Let $M = D/(a)$, so we have the projective presentation $0 \to D \to D \to M \to 0$ where the first map is the multiplication by a and the second is the canonical homomorphism onto the factor module. Use (27) and the isomorphism of $\hom_D(D, N)$ with N mapping η into $\eta 1$ to show that $\text{Ext}^1(M, N) \cong N/aN$. Show that if $N = D/(b)$, then $\text{Ext}^1(M, N) \cong D/(a, b)$ where (a, b) is a g.c.d. of a and b. Use these results and the

fundamental structure theorem on finitely generated modules over a p.i.d. (BAI, p. 187) to obtain a formula for $\mathrm{Ext}^1(M,N)$ for any two finitely generated modules over D.

2. Give a proof of the equivalence of $\mathrm{Ext}^n(M,-)$ and $\overline{\mathrm{Ext}}^n(M,-)$.

3. Show that in the correspondence between equivalence classes of extensions of M by N and the set $\mathrm{Ext}^1(M,N)$ given in Theorem 6.7, the equivalence class of the split extension corresponds to the 0 element of $\mathrm{Ext}^1(M,N)$.

4. Let $N \overset{\alpha_i}{\rightarrowtail} E_i \overset{\beta_i}{\twoheadrightarrow} M$, $i = 1,2$, be two extensions of M by N. Form $E_1 \oplus E_2$ and let F be the submodule of $E_1 \oplus E_2$ consisting of the pairs (z_1, z_2), $z_i \in E_i$ such that $\beta_1 z_1 = \beta_2 z_2$. Let K be the subset of $E_1 \oplus E_2$ of elements of the form $(\alpha_1 y, -\alpha_2 y)$, $y \in N$. Note that K is a submodule of F. Put $E = F/K$ and define maps $\alpha : N \to E$, $\beta : E \to M$ by $\alpha y = (\alpha_1 y, 0) + K = (0, -\alpha_2 y) + K$, $\beta((z_1, z_2) + K) = \beta_1 z_1 = \beta_2 z_2$. Show that α and β are module homomorphisms and $N \overset{\alpha}{\rightarrowtail} E \overset{\beta}{\twoheadrightarrow} M$, so we have an extension of M by N. This is called the *Baer sum of the extensions* $N \overset{\alpha_i}{\rightarrowtail} E_i \overset{\beta_i}{\twoheadrightarrow} M$. Show that the element of $\mathrm{Ext}^1(M,N)$ corresponding to the Baer sum is the sum of the elements corresponding to the given extensions. Use this and exercise 3 to conclude that the set of equivalence classes of extensions of M by N form a group under the composition given by Baer sums with the zero element as the class of the split extension.

6.8 TOR

If $M \in$ **mod**-R, the category of right modules for the ring R, then $M \otimes_R$ is the functor from R-**mod** to **Ab** that maps a left R-module N into the group $M \otimes_R N$ and an element η of $\hom_R(N, N')$ into $1 \otimes \eta$. $M \otimes_R$ is additive and right exact. The second condition means that if $N' \to N \to N'' \to 0$ is exact, then $M \otimes N' \to M \otimes N \to M \otimes N'' \to 0$ is exact. The nth left derived functor of $M \otimes (= M \otimes_R)$ is denoted as $\mathrm{Tor}_n(M,-)$ (or $\mathrm{Tor}^R{}_n(M,-)$). To obtain $\mathrm{Tor}_n(M,N)$ we choose a projective resolution of $N : 0 \leftarrow N \overset{\varepsilon}{\leftarrow} C_0 \leftarrow \cdots$ and form the chain complex $M \otimes C = \{M \otimes C_i\}$. Then $\mathrm{Tor}_n(M,N)$ is the nth homology group $H_n(M \otimes C)$ of $M \otimes C$. By definition, $\mathrm{Tor}_0(M,N) = (M \otimes C_0)/\mathrm{im}(M \otimes C_1)$. Since $M \otimes$ is right exact, $M \otimes C_1 \to M \otimes C_0 \to M \otimes N \to 0$ is exact and hence $M \otimes N \cong (M \otimes C_0)/\mathrm{im}(M \otimes C_1) = \mathrm{Tor}_0(M,N)$.

The isomorphism $\mathrm{Tor}_0(M,N) \cong M \otimes N$ and the long exact sequence of homology imply that if $0 \to N' \to N \to N'' \to 0$ is exact, then

$$0 \leftarrow M \otimes N'' \leftarrow M \otimes N \leftarrow M \otimes N' \leftarrow \mathrm{Tor}_1(M, N'') \leftarrow \mathrm{Tor}_1(M,N) \leftarrow \cdots \tag{34}$$

is exact.

We recall that a right module M is flat if and only if the tensor functor $M \otimes$ from the category of left modules to the category of abelian groups is exact (p. 154). We can now give a characterization of flatness in terms of the functor Tor. The result is the following analogue of Theorem 6.6 on the functor Ext.

THEOREM 6.10. *The following conditions on a right module M are equivalent*:

(1) *M is flat.*

(2) $\operatorname{Tor}_n(M,N) = 0$ *for all $n \geqslant 1$ and all (left) modules N.*

(3) $\operatorname{Tor}_1(M,N) = 0$ *for all N.*

Proof. (1)$\Rightarrow$(2). If M is flat and $0 \leftarrow N \leftarrow C_0 \leftarrow C_1 \leftarrow \cdots$ is a projective resolution of N, then $0 \leftarrow M \otimes N \leftarrow M \otimes C_0 \leftarrow M \otimes C_1 \leftarrow \cdots$ is exact. Hence $\operatorname{Tor}_n(M,N) = 0$ for any $n \geqslant 1$. (2)$\Rightarrow$(3) is clear. (3)$\Rightarrow$(1). Let $0 \to N' \to N \to N'' \to 0$ be exact. Then the hypothesis that $\operatorname{Tor}_1(M,N') = 0$ implies that $0 \to M \otimes N' \to M \otimes N \to M \otimes N'' \to 0$ is exact. Hence M is flat. $\square$

We consider next the dependence of $\operatorname{Tor}_n(M,N)$ on M. The argument is identical with that used in considering Ext^n. Let α be a homomorphism of the right module M into the right module M' and as before let $0 \leftarrow N \leftarrow C_0 \leftarrow C_1 \leftarrow \cdots$ be a projective resolution for the left module N. Then we have the commutative diagram

$$\begin{array}{ccccccc} 0 \leftarrow & M \otimes N & \leftarrow & M \otimes C_0 & \leftarrow & M \otimes C_1 & \leftarrow \cdots \\ & \downarrow & & \downarrow & & \downarrow & \\ 0 \leftarrow & M' \otimes N & \leftarrow & M' \otimes C_0 & \leftarrow & M' \otimes C_1 & \leftarrow \cdots \end{array} \tag{35}$$

where the vertical maps are $\alpha \otimes 1_N$, $\alpha \otimes 1_{C_0}$, $\alpha \otimes 1_{C_1}$, etc. Hence we have a homomorphism of the complex $\{M \otimes C_i\}$ into the complex $\{M' \otimes C_i\}$ and a corresponding homomorphism of the homology groups $\operatorname{Tor}_n(M,N)$ into $\operatorname{Tor}_n(M',N)$. In this way we obtain a functor $\operatorname{Tor}_n(-,N)$ from **mod**-R, the category of right modules for the ring R, to the category **Ab** that is additive and covariant.

We now suppose we have a short exact sequence of right modules $0 \to M' \to M \to M'' \to 0$ and as before, let C, ε be a projective resolution for the left module N. Since the C_i are projective, $0 \to M' \otimes C_i \to M \otimes C_i \to M'' \otimes C_i \to 0$ is exact for every i. Consequently, by Theorem 6.1 and the isomorphism of $\operatorname{Tor}_0(M,N)$ with $M \otimes N$ we obtain the long exact sequence for Tor in the first variable:

$$\begin{aligned} (36) \quad & 0 \leftarrow M'' \otimes N \leftarrow M \otimes N \leftarrow M' \otimes N \leftarrow \\ & \qquad \operatorname{Tor}_1(M'',N) \leftarrow \operatorname{Tor}_1(M,N) \leftarrow \operatorname{Tor}_1(M',N) \leftarrow \cdots. \end{aligned}$$

Finally, we note that as in the case of Ext, we can define functors $\overline{\operatorname{Tor}}_n(M,N)$ using a projective resolution of the first argument M. Moreover, we can prove that $\operatorname{Tor}_n(M,N) \cong \overline{\operatorname{Tor}}_n(M,N)$. The argument is similar to that we gave for Ext and $\overline{\operatorname{Ext}}$ and is left to the reader.

EXERCISES

1. Determine $\mathrm{Tor}_1^{\mathbb{Z}}(M,N)$ if M and N are cyclic groups.
2. Show that $\mathrm{Tor}_1^{\mathbb{Z}}(M,N)$ is a torsion group for any abelian groups M and N.

6.9 COHOMOLOGY OF GROUPS

In the remainder of this chapter we shall consider some of the most important special cases of homological algebra together with their applications to classical problems, some of which provided the impetus to the development of the abstract theory.

We begin with the cohomology of groups and we shall first give the original definition of the cohomology groups of a group, which, unlike the definition of the derived functors, is quite concrete. For our purpose we require the concept of a G-module, which is closely related to a basic notion of representation theory of groups. If G is a group, we define a *G-module* A to be an abelian group (written additively) on which G acts as endomorphisms. This means that we have a map

$$(g,x) \rightsquigarrow gx \tag{37}$$

of $G \times A$ into A such that

$$\begin{aligned} g(x+y) &= gx+gy \\ (g_1g_2)x &= g_1(g_2x) \\ 1x &= x \end{aligned} \tag{38}$$

for $g, g_1, g_2 \in G$, $x, y \in A$. As in representation theory, we can transform this to a more familiar concept by introducing the group ring $\mathbb{Z}[G]$, which is the free $\mathbb{Z}$-module with G as base and in which multiplication is defined by

$$(\sum \alpha_g g)(\sum \beta_h h) = \sum \alpha_g \beta_h gh \tag{39}$$

where $\alpha_g, \beta_h \in \mathbb{Z}$. Then if A is a G-module, A becomes a $\mathbb{Z}[G]$-module if we define

$$(\sum \alpha_g g)x = \sum \alpha_g (gx). \tag{40}$$

The verification is immediate and is left to the reader. Conversely, if A is a $\mathbb{Z}[G]$-module, then A becomes a G-module if we define gx as $(1g)x$.

A special case of a G-module is obtained by taking A to be any abelian group and defining $gx = x$ for all $g \in G$, $x \in A$. This action of G is called the *trivial* action. Another example of a G-module is the *regular* G-module $A = G[\mathbb{Z}]$ in which the action is $h(\sum \alpha_g g) = \sum \alpha_g hg$.

Now let A be a G-module. For any $n = 0, 1, 2, 3, \ldots$, let $C^n(G, A)$ denote the set of functions of n variables in G into the module A. Thus if $n > 0$, then $C^n(G, A)$ is the set of maps of $\overbrace{G \times G \times \cdots \times G}^{n}$ into A and if $n = 0$, a map is just an element of A. $C^n(G, A)$ is an abelian group with the usual definitions of addition and 0: If $f, f' \in C^n(G, A)$, then

$$(f+f')(g_1, \ldots, g_n) = f(g_1, \ldots, g_n) + f'(g_1, \ldots, g_n)$$
$$0(g_1, \ldots, g_n) = 0.$$

In the case of $C^0(G, A) = A$, the group structure is that given in A.

We now define a map $\delta(=\delta_n)$ of $C^n(G, A)$ into $C^{n+1}(G, A)$. If $f \in C^n(G, A)$, then we define δf by

$$\begin{aligned} \delta f(g_1, \ldots, g_{n+1}) = {} & g_1 f(g_2, \ldots, g_{n+1}) \\ & + \sum_{i=1}^{n} (-1)^i f(g_1, \ldots, g_{i-1}, g_i g_{i+1}, \ldots, g_{n+1}) \\ & + (-1)^{n+1} f(g_1, \ldots, g_n). \end{aligned} \tag{41}$$

For $n = 0$, f is an element of A and

$$\delta f(g_1) = g_1 f - f. \tag{42}$$

For $n = 1$ we have

$$\delta f(g_1, g_2) = g_1 f(g_2) - f(g_1 g_2) + f(g_1) \tag{43}$$

and for $n = 2$ we have

$$\delta f(g_1, g_2, g_3) = g_1 f(g_2, g_3) - f(g_1 g_2, g_3) + f(g_1, g_2 g_3) - f(g_1, g_2). \tag{44}$$

It is clear that δ is a homomorphism of $C^n(G, A)$ into $C^{n+1}(G, A)$. Let $Z^n(G, A)$ denote its kernel and $B^{n+1}(G, A)$ its image in $C^{n+1}(G, A)$. It can be verified that $\delta^2 f = 0$ for every $f \in C^n(G, A)$. We shall not carry out this calculation since the result can be derived more simply as a by-product of a result that we shall consider presently. From $\delta(\delta f) = 0$ we can conclude that $Z^n(G, A) \supset B^n(G, A)$. Hence we can form the factor group $H^n(G, A) = Z^n(G, A)/B^n(G, A)$. This is called the nth *cohomology group of* G *with coefficients in* A.

The foregoing definition is concrete but a bit artificial. The special cases of $H^1(G, A)$ and $H^2(G, A)$ arose in studying certain natural questions in group theory that we shall consider in the next section. The general definition was suggested by these special cases and by the definition of cohomology groups of a simplicial complex. We shall now give another definition of $H^n(G, A)$ that is functorial in character. For this we consider $\mathbb{Z}$ as trivial G-module and we consider $\operatorname{Ext}^n(\mathbb{Z}, A)$ for a given G-module A. We obtain a particular determination of this group by choosing a projective resolution

$$0 \leftarrow \mathbb{Z} \overset{\varepsilon}{\leftarrow} C_0 \overset{d_1}{\leftarrow} C_1 \leftarrow \cdots \tag{45}$$

of $\mathbb{Z}$ as trivial $\mathbb{Z}[G]$-module. Then we obtain the cochain complex

$$0 \rightarrow \hom(C_0, A) \rightarrow \hom(C_1, A) \rightarrow \cdots \tag{46}$$

whose nth homology group is a determination of $\operatorname{Ext}^n(\mathbb{Z}, A)$.

We shall now construct the particular projective resolution (45) that will permit us to identify $\operatorname{Ext}^n(\mathbb{Z}, A)$ with the nth cohomology group $H^n(G, A)$ as we have defined it. We put

$$C_n = \overbrace{\mathbb{Z}[G] \otimes_{\mathbb{Z}} \cdots \otimes_{\mathbb{Z}} \mathbb{Z}[G]}^{n+1}. \tag{47}$$

Since $\mathbb{Z}[G]$ is a free $\mathbb{Z}$-module with G as base, C_n is a free $\mathbb{Z}$-module with base $g_0 \otimes g_1 \otimes \cdots \otimes g_n$, $g_i \in G$. We have an action of G on C_n defined by

$$g(x_0 \otimes \cdots \otimes x_n) = gx_0 \otimes x_1 \otimes \cdots \otimes x_n, \tag{48}$$

which makes C_n a $\mathbb{Z}[G]$-module. This is $\mathbb{Z}[G]$-free with base

$$(g_1, \ldots, g_n) \equiv 1 \otimes g_1 \otimes \cdots \otimes g_n, \qquad g_i \in G. \tag{49}$$

We now define a $\mathbb{Z}[G]$-homomorphism d_n of C_n into C_{n-1} by its action on the base $\{(g_1, \ldots, g_n)\}$:

$$\begin{aligned} d_n(g_1, \ldots, g_n) &= g_1(g_2, \ldots, g_n) \\ &\quad + \sum_{1}^{n-1} (-1)^i (g_1, \ldots, g_{i-1}, g_i g_{i+1}, g_{i+2}, \ldots, g_n) \\ &\quad + (-1)^n (g_1, \ldots, g_{n-1}) \end{aligned} \tag{50}$$

where it is understood that for $n = 1$ we have $d_1(g_1) = g_1 - 1 \in C_0 = \mathbb{Z}[G]$. Also we define a $\mathbb{Z}[G]$-homomorphism ε of C_0 into $\mathbb{Z}$ by $\varepsilon(1) = 1$. Then $\varepsilon(g) = \varepsilon(g1) = g1 = 1$ and $\varepsilon(\sum \alpha_g g) = \sum \alpha_g$. We proceed to show that $0 \leftarrow \mathbb{Z} \overset{\varepsilon}{\leftarrow} C_0 \overset{d_1}{\leftarrow} C_1 \leftarrow \cdots$ is a projective resolution of $\mathbb{Z}$. Since the C_i are free $\mathbb{Z}[G]$-modules, projectivity is clear. It remains to prove the exactness of the indicated sequence of $\mathbb{Z}[G]$-homomorphisms. This will be done by defining a sequence of *contracting homomorphisms*:

$$\mathbb{Z} \xrightarrow{s_{-1}} C_0 \xrightarrow{s_0} C_1 \xrightarrow{s_1} C_2 \longrightarrow \cdots.$$

By this we mean that the s_i are group homomorphisms such that

$$\varepsilon s_{-1} = 1_{\mathbb{Z}}, \qquad d_1 s_0 + s_{-1}\varepsilon = 1_{C_0}, \tag{51}$$
$$d_{n+1}s_n + s_{n-1}d_n = 1_{C_n}, \qquad n \geqslant 1.$$

We observe that $\{1\}$ is a $\mathbb{Z}$-base for $\mathbb{Z}$, G is a $\mathbb{Z}$-base for $C_0 = \mathbb{Z}[G]$, and $\{g_0(g_1,\dots,g_n) = g_0 \otimes g_1 \otimes \cdots \otimes g_n | g_i \in G\}$ is a $\mathbb{Z}$-base for C_n, $n \geqslant 1$. Hence we have unique group homomorphisms $s_{-1}: \mathbb{Z} \to C_0, s_n: C_{n-1} \to C_n$ such that

$$s_{-1}1 = 1, \qquad s_0 g_0 = (g_0), \tag{52}$$
$$s_n g_0(g_1,\dots,g_n) = (g_0, g_1,\dots,g_n), \qquad n > 0.$$

If $n > 0$, we have

$$\begin{aligned} d_{n+1}s_n g_0(g_1,\dots,g_n) &= d_{n+1}(g_0,g_1,\dots,g_n) \\ &= g_0(g_1,\dots,g_n) + \sum_{i=0}^{n-1} (-1)^{i+1}(g_0,\dots,g_ig_{i+1},\dots,g_n) \\ &\quad + (-1)^{n+1}(g_0,\dots,g_{n-1}) \end{aligned}$$

$$\begin{aligned} s_{n-1}d_n g_0(g_1,\dots,g_n) &= s_{n-1}g_0 d_n(g_1,\dots,g_n) \\ &= s_{n-1}g_0g_1(g_2,\dots,g_n) + \sum_{1}^{n-1} (-1)^i s_{n-1}g_0(g_1,\dots,g_ig_{i+1},\dots,g_n) \\ &\quad + (-1)^n s_{n-1}g_0(g_1,\dots,g_{n-1}) \\ &= (g_0g_1,g_2,\dots,g_n) + \sum_{1}^{n-1} (-1)^i (g_0,g_1,\dots,g_ig_{i+1},\dots,g_n) \\ &\quad + (-1)^n (g_0,g_1,\dots,g_{n-1}). \end{aligned}$$

Hence $d_{n+1}s_n g_0(g_1,\dots,g_n) + s_{n-1}d_n g_0(g_1,\dots,g_n) = g_0(g_1,\dots,g_n)$. This shows that the third equation in (51) holds. Similarly, one verifies the other two equations in (51).

We can now show that the $\mathbb{Z}[G]$-homomorphisms ε, d_n satisfy $\varepsilon d_1 = 0$, $d_n d_{n+1} = 0$, $n \geqslant 1$. By (49), C_1 is free with base $\{(g) = 1 \otimes g | g \in G\}$. Since $\varepsilon d_1(g) = \varepsilon(g-1) = 1 - 1 = 0$, the first equality holds. Thus if we put $\varepsilon = d_0$, then we have $d_n d_{n+1} = 0$ for $n = 0$. We note also that $s_n C_n$ for $n > 0$ contains the set of generators $\{(g_0,g_1,\dots,g_n)\}$ for C_{n+1} as $\mathbb{Z}[G]$-module. Hence it suffices to show that $d_n d_{n+1} s_n = 0$ if $n > 0$, and we may assume $d_{n-1}d_n = 0$. Then

$$d_n d_{n+1} s_n = d_n(1 - s_{n-1}d_n) = d_n - (1 - s_{n-2}d_{n-1})d_n = s_{n-2}d_{n-1}d_n = 0.$$

We can regard (45) as a $\mathbb{Z}$-complex and the sequence of maps $s_{-1}, s_0, s_1, \dots$ as a homotopy between the chain homomorphism of this complex into itself that is the identity on every C_i with the chain homomorphism that is 0 on every C_i. Then these chain homomorphisms define the same homomorphisms of the homology groups. It follows that the homology groups of the $\mathbb{Z}$-complex (45) are all 0. This means that (45) is exact and hence we have proved

THEOREM 6.11. *Let* $C_n = \overbrace{\mathbb{Z}[G]\otimes\cdots\otimes\mathbb{Z}[G]}^{n+1}$, $n \geqslant 0$, *and let* $\varepsilon: C_0 \to \mathbb{Z}$ *be the* $\mathbb{Z}[G]$*-homomorphism such that* $\varepsilon 1 = 1$, $d_n: C_n \to C_{n-1}$, *the* $\mathbb{Z}[G]$*-homomorphism such that* (50) *holds. Then* $C = \{C_n\}$, *and* ε *constitutes a free resolution for* $\mathbb{Z}$ *regarded as trivial G-module.*

To calculate $\operatorname{Ext}^n(\mathbb{Z}, A)$ for any $\mathbb{Z}[G]$-module we can use the resolution C, ε. Then $\operatorname{Ext}^n(\mathbb{Z}, A)$ is the nth homology group of the complex

$$0 \to \hom(C_0, A) \to \hom(C_1, A) \to \cdots. \tag{53}$$

Since $\{(g_1, \ldots, g_n) | g_i \in G\}$ is a $\mathbb{Z}[G]$-base for C_n, we have a bijection of $\hom(C_n, A)$ with the set $C^n(G, A)$ of functions of n variables in G into A, which associates with any $f \in \hom(C_n, A)$ its restriction to the base $\{(g_1, \ldots, g_n)\}$. The map $\hom(C_n, A) \to \hom(C_{n+1}, A)$ is right multiplication by d_{n+1}; that is, if f is a $\mathbb{Z}[G]$-homomorphism of C_n into A, then its image under $\hom(C_n, A) \to \hom(C_{n+1}, A)$ is the homomorphism $x \rightsquigarrow f(d_{n+1}x)$. If we take x to be the element $(g_1, \ldots, g_{n+1})$ of C_{n+1}, then this map is

$$\begin{aligned}
(g_1, \ldots, g_{n+1}) \rightsquigarrow f(g_1(g_2, \ldots, g_{n+1}) &+ \sum_1^n (-1)^i (g_1, \ldots, g_i g_{i+1}, \ldots, g_{n+1}) \\
&+ (-1)^{n+1}(g_1, \ldots, g_n)) \\
= g_1(f(g_2, \ldots, g_{n+1})) &+ \sum_1^n (-1)^i f(g_1, \ldots, g_i g_{i+1}, \ldots, g_{n+1}) \\
&+ (-1)^{n+1} f(g_1, \ldots, g_n) \\
= \delta f(g_1, \ldots, g_{n+1}).&
\end{aligned}$$

Thus we have the following commutative diagram

$$\begin{array}{ccc}
\hom(C_n, A) & \longrightarrow & \hom(C_{n+1}, A) \\
\downarrow & & \downarrow \\
C^n(G, S) & \xrightarrow[\delta]{} & C^{n+1}(G, A)
\end{array} \tag{54}$$

where the vertical arrows are group isomorphisms. Since the product of $\hom(C_{n-1}, A) \to \hom(C_n, A)$ and $\hom(C_n, A) \to \hom(C_{n+1}, A)$ is 0, we have $\delta^2 = 0$. Hence $0 \to C^0(G, A) \to C^1(G, A) \to C^2(G, A) \to \cdots$ is a cochain complex and this is isomorphic to the cochain complex $0 \to \hom(C_0, A) \to$

$\hom(C_1, A) \to \hom(C_2, A) \to \cdots$. It follows that these two complexes have isomorphic homology groups. We therefore have

THEOREM 6.12. *$B^n(G, A)$ is a subgroup of $Z^n(G, A)$ and $H^n(G, A) = Z^n(G, A)/B^n(G, A) \cong \mathrm{Ext}^n(\mathbb{Z}, A)$.*

We shall now switch from the original definition of the cohomology groups of G with coefficients in A to the groups $\mathrm{Ext}^n(\mathbb{Z}, A)$. From now on we use the definition $\mathrm{Ext}^n(\mathbb{Z}, A)$ for the nth cohomology group of G with coefficients in A. This definition has the advantage that it makes available the functorial results on Ext for the study of cohomology groups of a group. Also it offers considerably more flexibility since it permits us to replace the resolution of $\mathbb{Z}$ that we have used by others. Some instances of this will be given in the exercises.

We shall now look at the cohomology group $H^n(G, A)$ for $n = 0, 1, 2$. We prove first

THEOREM 6.13. *$H^0(G, A) \cong A^G$, the subgroup of A of elements x satisfying $gx = x$, $g \in G$.*

Proof. We recall that $\mathrm{Ext}^0(M, N) \cong \hom(M, N)$. Hence $H^0(G, A) \cong \hom_{\mathbb{Z}[G]}(\mathbb{Z}, A)$, the group of $\mathbb{Z}[G]$-module homomorphisms of $\mathbb{Z}$ into A. If η is such a homomorphism, η is determined by $\eta(1)$ and if $\eta(1) = x \in A$, then $x = \eta(1) = \eta(g1) = g\eta(1) = gx$, $g \in G$. Conversely, if $x \in A$ satisfies $gx = x$, $g \in G$, then the map η such that $\eta(n) = nx$ is a $\mathbb{Z}[G]$-homomorphism of $\mathbb{Z}$ into A. It follows that $\hom(\mathbb{Z}, A)$ is isomorphic (under $\eta \rightsquigarrow \eta(1)$) to A^G. $\square$

We remark that this proposition can also be proved easily by using the definitions of $Z^0(G, A)$, $B^0(G, A)$, and $Z^0(G, A)/B^0(G, A)$. We leave it to the reader to carry out such a proof.

If A is a G-module, a map f of G into A is called a *crossed homomorphism of G into A* if

$$f(gh) = gf(h) + f(g), \qquad g, h \in G. \tag{55}$$

If $x \in A$, then the map f defined by

$$f(g) = gx - x \tag{56}$$

is a crossed homomorphism of G into A since

$$gf(h) + f(g) = g(hx - x) + gx - x = ghx - x = f(gh).$$

A crossed homomorphism defined by (56) is called *principal*. It is clear that the crossed homomorphisms form an abelian group under addition of maps and that the principal ones form a subgroup. Comparison with (42) and (43) shows that the first of these groups is $Z^1(G, A)$ and the second is $B^1(G, A)$. Hence the factor group is (isomorphic to) the first cohomology group of G with coefficients in A.

We have encountered crossed homomorphisms in Galois theory in considering Speiser's equations and their additive analogue (BAI, pp. 297–299). We recall these results, the first of which constituted a generalization of Hilbert's Satz 90. Let E be a finite dimensional Galois extension field of the field F, G the Galois group of E/F, so G is finite and $|G| = [E:F]$. We have the natural actions of G on the additive group E of E and on the multiplicative group E^* of non-zero elements of E. If we consider the additive group of E as G-module, then a crossed homomorphism is a map f of G into E such that $f(gh) = f(g) + gf(h)$. Theorem 4.32 of BAI (p. 297) states that any such crossed homomorphism is principal. Thus we have the result $H^1(G, E) = 0$: The first cohomology group of the Galois group G of E/F with coefficients in E is 0. Now consider the action of G on E^*. Since the composition in E^* is multiplication, a crossed homomorphism of G into E^* is a map f of G into E^* such that

$$f(gh) = (gf(h))f(g). \tag{57}$$

These are Speiser's equations as given in (75), p. 297 of BAI. Speiser's theorem is that such an f is principal, that is, it has the form $f(g) = (gu)u^{-1}$ for some $u \in E^*$. Thus Speiser's theorem is the homological result that $H^1(G, E^*) = 1$ (using multiplicative notation).

If G is a group and A is an abelian group written multiplicatively on which G acts by automorphisms, then the group $C^2(G, A)$ is the group of functions of two variables in G to A with multiplication as composition. $Z^2(G, A)$ is the subgroup of $f \in C^2(G, A)$ such that

$$f(g,h)f(gh,k) = (gf(h,k))f(g,hk), \qquad g,h,k \in G. \tag{58}$$

This is clear from (44). Such a map is called a *factor set*. The subgroup $B^2(G, A)$ is the subgroup of maps of the form f where $f(g,h) = u(g)gu(h)u(gh)^{-1}$ where u is a map of G into A. The group $Z^2(G, A)/B^2(G, A)$ is the second cohomology group of G with coefficients in A. We shall give an interpretation of this group in the next section.

We shall conclude this section by proving the following result on cohomology groups of finite groups.

THEOREM 6.14. *If G is a finite group, A a G-module, then every element of $H^n(G, A)$, $n > 0$, has finite order a divisor of $|G|$.*

Proof. Let $f \in C^n(G, A)$ and consider the formula (41) for δf. We let g_{n+1} range over G and sum the corresponding formulas. If we denote $\sum_{g \in G} f(g_1, \ldots, g_{n-1}, g)$ by $u(g_1, \ldots, g_{n-1})$, then since $\sum_{g \in G} f(g_1, \ldots, g_{n-1}, g_n g) = u(g_1, \ldots, g_{n-1})$, the result we obtain is

$$\sum_g \delta f(g_1, \ldots, g_n, g) = g_1 u(g_2, \ldots, g_n) + \sum_{i=1}^{n-1} (-1)^i u(g_1, \ldots, g_i g_{i+1}, \ldots, g_n)$$

$$+ (-1)^n u(g_1, \ldots, g_{n-1}) + (-1)^{n+1} |G| f(g_1, \ldots, g_n)$$

$$= \delta u(g_1, \ldots, g_n) + (-1)^{n+1} |G| f(g_1, \ldots, g_n).$$

Hence if $\delta f = 0$, then $|G| f(g_1, \ldots, g_n) = \pm \delta u(g_1, \ldots, g_n) \in B^n(G, A)$. Then $|G| Z^n(G, A) \subset B^n(G, A)$, so $|G| H^n(G, A) = 0$, which proves the theorem. □

An immediate consequence of this result is the

COROLLARY. *Let G be a finite group, A a finite G-module such that $(|G|, |A|) = 1$. Then $H^n(G, A) = 0$ for every $n > 0$.*

This is clear since $|A| f = 0$ for every $f \in C^n(G, A)$.

EXERCISES

1. Let B be a right module for $\mathbb{Z}[G]$. Define the nth *homology group* of G *with coefficients in* B, $n \geqslant 0$, as $H_n(G, B) = \operatorname{Tor}_n(B, \mathbb{Z})$ where $\mathbb{Z}$ is regarded as a trivial G-module. Show that $H_0(G, B) \cong B/B_G$ where B_G is the subgroup of B generated by the elements $xg - x$, $x \in B$.

2. Let ε be the homomorphism of $\mathbb{Z}[G]$ into $\mathbb{Z}$ defined in the text and let $I = \ker \varepsilon$. Show that I is a free $\mathbb{Z}$-module with base $\{g - 1 \mid g \in G, g \neq 1\}$.

3. Let A and B be G-modules. Show that $A \otimes_{\mathbb{Z}} B$ is a G-module with the action such that $g(x \otimes y) = gx \otimes gy$.

4. Let A be a G-module and let A_t denote the abelian group A with the trivial G-action. Show that $\mathbb{Z}[G] \otimes_{\mathbb{Z}} A$ and $\mathbb{Z}[G] \otimes_{\mathbb{Z}} A_t$ are isomorphic as G-modules. (*Hint*: Show that there is a module isomorphism of $\mathbb{Z}[G] \otimes_{\mathbb{Z}} A$ onto $\mathbb{Z}[G] \otimes_{\mathbb{Z}} A_t$ such that $g \otimes x \to g \otimes g^{-1} x$, $g \in G$, $x \in A$.)

5. Use exercise 4 to prove that if A is a G-module that is $\mathbb{Z}$-free, then $\mathbb{Z}[G] \otimes_{\mathbb{Z}} A$ is $\mathbb{Z}[G]$-free.

6. Let $I = \ker \varepsilon$, $\varepsilon : \mathbb{Z}[G] \to \mathbb{Z}$, as in the text and put $I^n = I \otimes_{\mathbb{Z}} I \otimes_{\mathbb{Z}} \cdots \otimes_{\mathbb{Z}} I$, n factors. Show that the short exact sequence $0 \to I \to \mathbb{Z}[G] \xrightarrow{\varepsilon} \mathbb{Z} \to 0$, where the first map is inclusion, yields a short exact sequence $0 \to I^{n+1} \to \mathbb{Z}[G] \otimes_{\mathbb{Z}} I^n \to I^n \to 0$. Let d_n

be the homomorphism of $\mathbb{Z}[G]\otimes_{\mathbb{Z}} I^n$ into $\mathbb{Z}[G]\otimes_{\mathbb{Z}} I^{n-1}$ that is the composite of $\mathbb{Z}[G]\otimes_{\mathbb{Z}} I^n \to I^n$ and $I^n \to \mathbb{Z}[G]\otimes_{\mathbb{Z}} I^{n-1}$. Note that

$$d_n(g_0\otimes(g_1-1)\otimes\cdots\otimes(g_n-1)) = (g_1-1)\otimes\cdots\otimes(g_n-1),$$

$g_i \in G$. Show that

$$0\leftarrow\mathbb{Z}\overset{\varepsilon}{\leftarrow}\mathbb{Z}[G]\overset{d_1}{\leftarrow}\mathbb{Z}[G]\otimes I\overset{d_2}{\leftarrow}\mathbb{Z}[G]\otimes I^2\leftarrow\cdots$$

is a free resolution for $\mathbb{Z}$.

7. Let $G = \langle g\rangle$, the cyclic group of finite order m generated by the element g. Note that $\mathbb{Z}[G] \cong \mathbb{Z}[t]/(t^m-1)$, t an indeterminate. Let $D = g-1$, $N = 1+g+\cdots+g^{m-1}$, and let D', N' denote multiplication by D and N respectively in $\mathbb{Z}[G]$. Show that

$$0\leftarrow\mathbb{Z}\overset{\varepsilon}{\leftarrow}\mathbb{Z}[G]\overset{D'}{\leftarrow}\mathbb{Z}[G]\overset{N'}{\leftarrow}\mathbb{Z}[G]\overset{D'}{\leftarrow}\cdots$$

is a free resolution for $\mathbb{Z}$.

8. Use exercise 7 to show that if A is a G-module for the cyclic group of order $m < \infty$, then

$$H^{2n}(G, A) = A^G/NA, \qquad n > 0,$$
$$H^{2n+1}(G, A) = (\mathrm{Ann}_A N)/DA, \qquad n \geqslant 0,$$

where $\mathrm{Ann}_A N = \{x \in A \mid Nx = 0\}$.

6.10 EXTENSIONS OF GROUPS

By an *extension of a group G by a group A* we shall mean a short exact sequence

(59) $$1 \to A \xrightarrow{i} E \xrightarrow{p} G \to 1.$$

Thus i is injective, p is surjective, and $\ker p = iA$. Hence $iA \lhd E$. If $1 \to A \xrightarrow{i'} E' \xrightarrow{p'} G \to 1$ is a second extension of G by A, then we say that this is *equivalent* to (59) if there exists a homomorphism $h: E \to E'$ such that

(60)

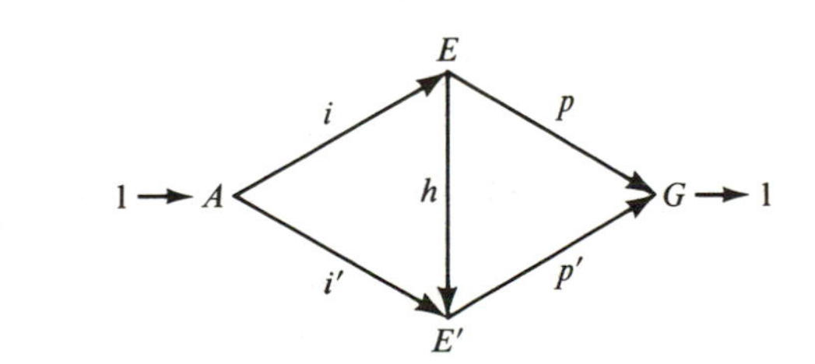

is commutative. It follows easily, as for modules (p. 348), that in this case h is an isomorphism.

We restrict our attention to extensions of a group G by an abelian group. With such an extension, we can associate an action of G on A by automorphisms and an element of the cohomology group $H^2(G, A)$ where A is regarded as G-module by the action we shall define. It will be helpful in these considerations to denote the elements of G by small Greek letters and those of A and E by small Latin letters.

We first define the action of G on A. Let $\sigma \in G$, $x \in A$. Choose an element $s \in E$ such that $ps = \sigma$ and consider the element $s(ix)s^{-1}$. Since $iA \lhd E$, $s(ix)s^{-1} \in iA$ and since i is injective, we have a unique element $y \in A$ such that $s(ix)s^{-1} = iy$. To obtain y we made a choice of an element $s \in E$ such that $ps = \sigma$. Let s' be a second element such that $ps' = \sigma$. Then $p(s's^{-1}) = 1$ and hence $s's^{-1} = ia$, $a \in A$, and $s' = (ia)s$. Since iA is abelian, we have $s'(ix)s'^{-1} = (ia)s(ix)s^{-1}(ia)^{-1} = s(ix)s^{-1}$. Thus the element y is independent of the choice of s and hence we can put $\sigma x = y$. Our definition is

$$\sigma x = y \quad \text{where } ps = \sigma, \quad s(ix)s^{-1} = iy. \tag{61}$$

It is straightforward to verify that the definition of σx gives an action of G on A by automorphisms: $(\sigma_1\sigma_2)x = \sigma_1(\sigma_2 x)$, $1x = x$, $\sigma(xy) = (\sigma x)(\sigma y)$. Except for the fact that A is written multiplicatively, we have defined a G-module structure on A. We shall now call this a G-module even though we retain the multiplicative notation in A.

Our next step is to choose for each $\sigma \in G$ an element $s_\sigma \in E$ such that $ps_\sigma = \sigma$. Thus we have a map $s: G \to E$, $\sigma \rightsquigarrow s_\sigma$ such that $ps_\sigma = \sigma$ for all σ (or $ps = 1_G$).

Let $\sigma, \tau \in G$ and consider the element $s_\sigma s_\tau s_{\sigma\tau}^{-1}$ of E. Applying p to this element gives 1. Hence there is a unique element $k_{\sigma,\tau} \in A$ such that $s_\sigma s_\tau s_{\sigma\tau}^{-1} = ik_{\sigma,\tau}$ or

$$s_\sigma s_\tau = (ik_{\sigma,\tau})s_{\sigma\tau}. \tag{62}$$

If $\rho \in G$ also, then

$$\begin{aligned} s_\rho(s_\sigma s_\tau) = s_\rho(ik_{\sigma,\tau})s_{\sigma\tau} = s_\rho(ik_{\sigma,\tau})s_\rho^{-1}s_\rho s_{\sigma\tau} &= i(\rho k_{\sigma,\tau})s_\rho s_{\sigma\tau} \\ &= i(\rho k_{\sigma,\tau})i(k_{\rho,\sigma\tau})s_{\rho\sigma\tau} = i(\rho(k_{\sigma,\tau})k_{\rho,\sigma\tau})s_{\rho\sigma\tau}. \end{aligned}$$

Similarly, $(s_\rho s_\sigma)s_\tau = i(k_{\rho,\sigma}k_{\rho\sigma,\tau})s_{\rho\sigma\tau}$. Hence, the associative law in E implies that

$$k_{\rho,\sigma}k_{\rho\sigma,\tau} = (\rho k_{\sigma,\tau})k_{\rho,\sigma\tau}. \tag{63}$$

These relations show that the map $k: G \times G \to A$ such that $(\sigma, \tau) \rightsquigarrow k_{\sigma,\tau}$ is an element of $Z^2(G, A)$ as defined in the classical definition of $H^2(G, A)$ given at the beginning of section 6.9.

We now consider the alteration in k that results from changing the map s to a second one s' satisfying $ps' = 1_G$. Then for any $\sigma \in G$, $p(s'_\sigma s_\sigma^{-1}) = (ps'_\sigma)(ps_\sigma)^{-1} = \sigma\sigma^{-1} = 1$. Hence there exists a unique $u_\sigma \in A$ such that $s'_\sigma s_\sigma^{-1} = iu_\sigma$ or $s'_\sigma = (iu_\sigma)s_\sigma$. Thus we have a map $u: \sigma \rightsquigarrow u_\sigma$ of G into A such that

(64) $$s'_\sigma = (iu_\sigma)s_\sigma, \qquad \sigma \in G.$$

Conversely, if u is any map of G into A, then s' defined by (64) satisfies $ps' = 1_G$. By (64), we have $s'_\sigma s'_\tau = (iu_\sigma)s_\sigma(iu_\tau)s_\tau = i(u_\sigma(\sigma u_\tau))s_\sigma s_\tau = i(u_\sigma(\sigma u_\tau)k_{\sigma,\tau})s_{\sigma\tau} = i(u_\sigma(\sigma u_\tau)k_{\sigma,\tau}u_{\sigma\tau}^{-1})s'_{\sigma\tau}$. Hence k is replaced by k' where

(65) $$k'_{\sigma,\tau} = u_\sigma(\sigma u_\tau)u_{\sigma\tau}^{-1}k_{\sigma,\tau}.$$

This shows that k' and k determine the same element of $H^2(G, A)$. Hence the extension determines a unique element of $H^2(G, A)$.

It follows directly from the definitions that if the extensions $1 \to A \xrightarrow{i} E \xrightarrow{p} G \to 1$ and $1 \to A \xrightarrow{i'} E' \xrightarrow{p'} G \to 1$ are equivalent, then they determine the same module action of G on A and the same element of $H^2(G, A)$. To prove the converse we shall show that the multiplication in E is determined by the action of G on A and the map k. Let $e \in E$ and put $\sigma = pe \in G$, $f = e(s_\sigma)^{-1}$. Then $pf = pe(ps_\sigma)^{-1} = \sigma\sigma^{-1} = 1$. Hence $f = ix$ for a uniquely determined element $x \in A$. Then we have the factorization

(66) $$e = (ix)s_\sigma, \qquad x \in A, \qquad \sigma \in G.$$

It is clear that the elements $x \in A$, and $\sigma \in G$ are uniquely determined by the given element e. Now let $(iy)s_\tau$, where $y \in A$ and $\tau \in G$, be a second element of E. Then

(67) $$(ix)s_\sigma(iy)s_\tau = i(x(\sigma y))s_\sigma s_\tau = i(x(\sigma y)k_{\sigma,\tau})s_{\sigma\tau}.$$

Now suppose the extensions $1 \to A \xrightarrow{i} E \xrightarrow{p} G \to 1$ and $1 \to A \xrightarrow{i'} E' \xrightarrow{p'} G \to 1$ determine the same module structure and the same element of $H^2(G, A)$. Then we can choose maps s and s' of G into E and E' respectively such that $ps = 1_G$, $p's' = 1_G$ and for any $\sigma, \tau \in G$ we have $s_\sigma s_\tau = (ik_{\sigma,\tau})s_{\sigma\tau}$, $s'_\sigma s'_\tau = (i'k_{\sigma,\tau})s'_{\sigma\tau}$. Then we have (67) and $(i'x)s'_\sigma(i'y)s'_\tau = i'(x(\sigma y)k_{\sigma,\tau})s'_{\sigma\tau}$. It follows that the map h: $(ix)(s_\sigma) \rightsquigarrow (i'x)(s'_\sigma)$ is a homomorphism of E into E' so that (60) is commutative. Hence the extensions are equivalent.

We shall now state the following basic

THEOREM 6.15. *Two extensions of G by an abelian group A are equivalent if and only if they determine the same action of G on A and the same element of $H^2(G, A)$. Let G be a group, A a G-module, and let M denote the set of extensions of G by A having a given G-module A as associated module. Then we* *have a 1–1 correspondence between the set of equivalence classes of extensions of G by A contained in M with the elements of $H^2(G, A)$.*

Proof. The first statement has been proved. To prove the second, it suffices to show that given a G-module A and a map k of $G \times G$ into A satisfying (63) there exists an extension $1 \to A \xrightarrow{i} E \xrightarrow{p} G \to 1$ whose associated action of G on A is the given one and whose element of $H^2(G, A)$ is the one determined by k. We put $E = A \times G$, the set of pairs (x, σ), $x \in A$, $\sigma \in G$, and we define a multiplication in E by

$$(68) \qquad (x, \sigma)(y, \tau) = (x(\sigma y)k_{\sigma,\tau}, \sigma\tau).$$

Then it is immediate from (63) that this multiplication is associative. If we put $\rho = \sigma = 1$ in (61), we obtain $k_{1,1}k_{1,\tau} = k_{1,\tau}k_{1,\tau}$. Hence $k_{1,\tau} = k_{1,1}$. Then

$$(k_{1,1}^{-1}, 1)(y, \tau) = (k_{1,1}^{-1}yk_{1,\tau}, \tau) = (y, \tau),$$

so $1 \equiv (k_{1,1}^{-1}, 1)$ is a left unit in E. To prove that E is a group with 1 as unit, it suffices to show that any element (x, σ) of E has a left inverse relative to 1 (BAI, p. 36, exercise 10). This follows from

$$(k_{1,1}^{-1}k_{\sigma^{-1},\sigma}^{-1}\sigma^{-1}x^{-1}, \sigma^{-1})(x, \sigma) = (k_{1,1}^{-1}k_{\sigma^{-1},\sigma}^{-1}(\sigma^{-1}x^{-1})(\sigma^{-1}x)k_{\sigma^{-1},\sigma}, 1) = (k_{1,1}^{-1}, 1).$$

Hence E is a group. Let $i : x \rightsquigarrow (xk_{1,1}^{-1}, 1)$, $p : (x, \sigma) \rightsquigarrow \sigma$. Then i is a homomorphism of A into E and p is a homomorphism of E into G. It is clear that i is injective and p is surjective. Moreover, $iA = \ker p$. Hence $1 \to A \xrightarrow{i} E \xrightarrow{p} G \to 1$ is an extension of G by A. To determine the module action of G on A determined by this extension, we note that $p(1, \sigma) = \sigma$ so we must calculate $(1, \sigma)ix(1, \sigma)^{-1} = (1, \sigma)(xk_{1,1}^{-1}, 1)(1, \sigma)^{-1} = (1, \sigma)(xk_{1,1}^{-1}, 1)$ $(k_{1,1}^{-1}k_{\sigma^{-1},\sigma}^{-1}, \sigma^{-1})$, We have seen that $k_{1,\rho} = k_{1,1}$ and if we put $\sigma = \tau = 1$ in (63), we obtain $k_{\rho,1}k_{\rho,1} = (\rho k_{1,1})k_{\rho,1}$. Hence $k_{\rho,1} = \rho k_{1,1}$. Now put $\sigma = \rho^{-1}$, $\tau = \rho$ in (63). This gives $k_{\rho,\rho^{-1}}k_{1,\rho} = (\rho k_{\rho^{-1}})k_{\rho,1}$, so $k_{\rho,\rho^{-1}}k_{1,1} = (\rho k_{\rho^{-1},\rho})k_{\rho,1}$. Thus we have

$$(69) \qquad k_{1,\rho} = k_{1,1}, \qquad k_{\rho,1} = \rho k_{1,1}, \qquad k_{\rho,\rho^{-1}}k_{1,1} = (\rho k_{\rho^{-1},\rho})k_{\rho,1}.$$

Now $(1, \sigma)(k_{1,1}^{-1}x, 1) = ((\sigma k_{1,1}^{-1})\sigma x k_{\sigma,1}, \sigma) = (\sigma x, \sigma)$ and

$$\begin{aligned}(\sigma x, \sigma)(k_{1,1}^{-1}k_{\sigma^{-1},\sigma}^{-1}, \sigma^{-1}) &= ((\sigma x)\sigma(k_{1,1})^{-1}\sigma(k_{\sigma^{-1},\sigma})^{-1}k_{\sigma,\sigma^{-1}}, 1) \\ &= ((\sigma x)k_{\sigma,1}^{-1}\sigma(k_{\sigma^{-1},\sigma})^{-1}k_{\sigma,\sigma^{-1}}, 1) = ((\sigma x)k_{1,1}^{-1}, 1)\end{aligned}$$

(by (69)). Hence the module action is the given one. Now let s be the map $\sigma \rightsquigarrow (1, \sigma)$, so $ps = 1_G$. We have $(1, \sigma)(1, \tau) = (k_{\sigma,\tau}, \sigma\tau) = (k_{1,1}^{-1}k_{\sigma,\tau}, 1)(1, \sigma\tau)$, so $s_\sigma s_\tau = (ik_{\sigma,\tau})s_{\sigma\tau}$. Hence the element of $H^2(G, A)$ associated with this extension is that determined by k. This completes the proof. □

The foregoing result in a slightly different form is classical. It was proved by Schreier, who first considered the problem of extensions as the problem of describing all of the groups that have a given normal subgroup A (not necessarily commutative) with given factor group G.

An extension (59) is said to be *split* if there exists a group homomorphism $s: G \to E$ such that $ps = 1_G$. Then $(s\sigma)(s\tau) = s\sigma\tau$ for any $\sigma, \tau \in G$ and hence $k_{\sigma,\tau}$ determined by (62) is 1. Thus the element of $H^2(G, A)$ associated with a split extension is 1. Conversely, if this is the case, then we have a map $s: G \to E$ satisfying $ps = 1_G$ for which the $k_{\sigma,\tau}$ are all 1, which means that s is a homomorphism. Thus the split extensions are those for which the associated element of $H^2(G, A)$ is 1. Evidently this implies that $H^2(G, A) = 1$ if and only if all extensions of G by A with the given G-module structure on A split. By the corollary to Theorem 6.14 (p. 361), this is the case if G and A are finite and $(|G|, |A|) = 1$.

If $A \lhd E$ and $E/A = G$ for arbitrary (not necessarily abelian) A, then we have the extension $1 \to A \xrightarrow{i} E \xrightarrow{p} G \to 1$ where i is the injection of A in E and p is the natural homomorphism of E onto G. It is readily seen from the definition that this extension splits if and only if there exists a subgroup S of E such that $E = SA$, $S \cap A = 1$. In this case E is said to be the *semi-direct product* of S and A. The result just indicated is that if A is abelian and A and G are finite with $(|A|, |G|) = 1$, then $E = SA$ and $S \cap A = 1$ for a subgroup S of E. This result, which was obtained by homological methods, can be supplemented by a little group theory to prove the following theorem due to H. Zassenhaus.

THEOREM 6.16. *Let E be a finite group, A a normal subgroup of E such that $(|A|, |E/A|) = 1$. Then E is a semi-direct product of A and a subgroup S.*

Proof. Let $|A| = m$, $|E/A| = n$. It suffices to show that there exists a subgroup S such that $|S| = n$. For, if S is such a subgroup, then $S \cap A$ is a subgroup whose order divides $|S| = n$ and $|A| = m$. Then $S \cap A = 1$. Also since $A \lhd E$, SA is a subgroup whose order is a multiple of $|S|$ and $|A|$ and so is a multiple of $mn = |E|$. Let p be a prime divisor of m and let H be a Sylow p-subgroup of A. Then H is also a Sylow p-subgroup of E. The set Syl_p of Sylow p-subgroups of E is $\{gHg^{-1} | g \in E\}$ (BAI, p. 80). Since $A \lhd E$ this is also the set of Sylow p-subgroups of A. Hence if N is the normalizer of H in E then $|\mathrm{Syl}_p| = [E:N] = [A:N \cap A]$. Thus

$$|E|/|N| = |A|/|N \cap A|$$

and hence

$$n = [E:A] = |E|/|A| = |N|/|N \cap A| = [N:N \cap A].$$

On the other hand, $|N \cap A| \,\big|\, |A|$ so $(|N \cap A|, |N/N \cap A|) = 1$. If $|N| < |E|$ we can use induction on order to conclude that N and hence E contains a subgroup of order n. Now suppose $|N| = |E|$. Then $N = E$ and $H \lhd E$. The center Z of H is non-trivial (BAI, p. 76). Since Z is a characteristic subgroup of H, $Z \lhd E$ and $A/Z \lhd E/Z$. Since $(E/Z)/(A/Z) \cong E/A$ we can apply induction to conclude that E/Z contains a subgroup L/Z of order n. Since Z is abelian and $|Z|$ is a power of p, the result we established by homology implies that L contains a subgroup S of order n. This completes the proof. □

EXERCISES

1. Let $1 \to A \xrightarrow{i} E \xrightarrow{p} G \to 1$ be an extension of G by the abelian group A and let H be the set of equivalences h of this extension: the automorphisms of E that make

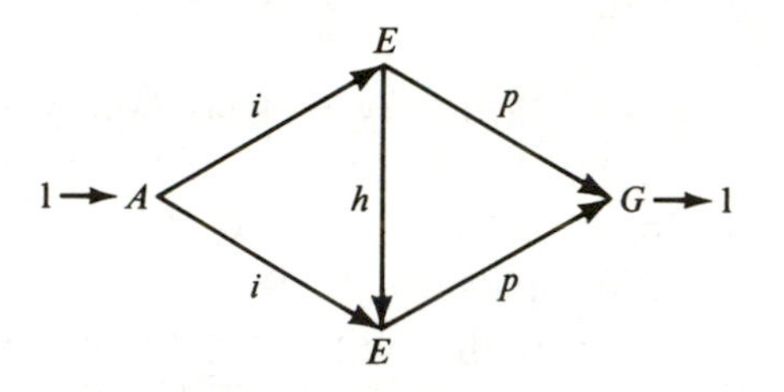

commutative. Show that H is a subgroup of $\operatorname{Aut} E$ that contains the inner automorphisms by the elements of iA. Show that the latter set is a normal subgroup J of H and $H/J \cong H^1(G, A)$.

2. (Schreier, Eilenberg and MacLane.) Let $1 \to A \xrightarrow{i} E \xrightarrow{p} G \to 1$ be an extension of G by the group A (not necessarily abelian) and let Z be the center of A. Let s be a map $\sigma \rightsquigarrow s_\sigma$ of G into E such that $ps_\sigma = \sigma$ and let $\varphi(s_\sigma)$ be the automorphism $x \rightsquigarrow y$ of A such that $s_\sigma(ix)s_\sigma^{-1} = iy$. Show that if s' is a second map of G into E satisfying $ps'_\sigma = \sigma, \sigma \in G$, then $\varphi(s'_\sigma) \in (\operatorname{Inaut} A)\varphi(s_\sigma)$ and $\varphi(s'_\sigma)|Z = \varphi(s_\sigma)|Z$. Show that $\varphi : \sigma \rightsquigarrow (\operatorname{Inaut} A)\varphi(s_\sigma)$ is a homomorphism of G into $\operatorname{Aut} A/\operatorname{Inaut} A$ and $\sigma c \equiv \varphi(s_\sigma)c, c \in Z$, defines an action of G on Z by automorphisms.

   Show that if $\sigma, \tau \in G$, then $s_\sigma s_\tau = i(k_{\sigma,\tau})s_{\sigma\tau}$ where $k_{\sigma,\tau}$ is a uniquely determined element of A. Show that if we put $\varphi(\sigma) = \varphi(s_\sigma)$, then

$$\varphi(\sigma)\varphi(\tau) = I_{k_{\sigma,\tau}}\varphi(\sigma\tau) \tag{70}$$

   where $I_{k_{\sigma,\tau}}$ is the inner automorphism $x \rightsquigarrow k_{\sigma,\tau}xk_{\sigma,\tau}^{-1}$ in A. Show that

$$k_{\rho,\sigma}k_{\rho\sigma,\tau} = (\varphi(\rho)k_{\sigma,\tau})k_{\rho,\sigma\tau} \tag{71}$$

   and

$$\varphi(1) = I_{k_{1,1}}. \tag{72}$$

   Conversely, suppose G and A are groups and k is a map of $G \times G$ into A and φ a map of G into $\operatorname{Aut} A$ such that (70)–(72) hold. Let $E = A \times G$ and define a product in E by

$$(x, \sigma)(y, \tau) = (x(\varphi(\sigma)y)k_{\sigma,\tau}, \sigma\tau) \tag{73}$$

Show that E is a group with unit $(k_{1,1}^{-1}, 1)$ and that if i is the map $x \rightsquigarrow (xk_{1,1}^{-1}, 1)$ and p is the map $(x,\sigma) \rightsquigarrow \sigma$, then $1 \to A \xrightarrow{i} E \xrightarrow{p} G \to 1$ is an extension of G by A.

Let $1 \to A \xrightarrow{i} E \xrightarrow{p} G \to 1$ and $1 \to A \xrightarrow{i'} E' \xrightarrow{p'} G \to 1$ be extensions of G by A such that the associated homomorphisms of G into Aut A/Inaut A and hence the associated actions of G on Z are the same. Show that the maps s and s' (for G to E') can be chosen so that $\varphi(s_\sigma) = \varphi'(s'_\sigma)$ (φ' defined analogously to φ). Let $k'_{\sigma,\tau}$ be defined by $s'_\sigma s'_\tau = i'(k'_{\sigma,\tau})s'_{\sigma\tau}$ and put $f(\sigma,\tau) = k'_{\sigma,\tau}k_{\sigma,\tau}^{-1}$. Show that $f(\sigma,\tau) \in Z$ and $f: (\sigma,\tau) \rightsquigarrow f(\sigma,\tau)$ is a 2-cocycle for G with values in Z (where the action of G is as defined before). Use this to establish a 1–1 correspondence between the set of equivalence classes of extensions of G by A, all having a fixed associated homomorphism of G into Aut A/Inaut A with $H^2(G,Z)$.

3. Let G be finite and let $H^2(G,\mathbb{C}^*)$ be the second cohomology group of G with coefficients in the multiplicative group $\mathbb{C}^*$ of non-zero elements of $\mathbb{C}$ where the action of G on $\mathbb{C}^*$ is trivial. The group $H^2(G,\mathbb{C}^*)$ is called the *Schur multiplier* of G. Use Theorem 6.14 to show that if $[\gamma]$ is any element of $H^2(G,\mathbb{C}^*)$, then the representative cocycle γ can be chosen to have values that are nth roots of unity, $n = |G|$. Hence conclude that $H^2(G,\mathbb{C}^*)$ is a finite group.

4. Let ρ be a projective representation of a group G. By definition, ρ is a homomorphism of G into the projective linear group $PGL(V)$ of a finite dimensional vector space V/F (exercise 4, p. 256). As in the exercise cited, for each $g \in G$, let $\mu(g)$ denote a representative of the coset $\rho(g) \in PGL(V)$, so $\mu(g_1g_2) = \gamma_{g_1,g_2}\mu(g_1)\mu(g_2)$ where $\gamma_{g_1,g_2} \in F^*$, the multiplicative group of F. Show that $\gamma: (g_1,g_2) \rightsquigarrow \gamma_{g_1,g_2}$ is a 2-cocycle of G with coefficients in F^* where the action of G on F^* is trivial. Show that if we make a second choice of representatives $\mu'(g) \in GL(V)$ for the $\rho(g), g \in G$, then the resulting 2-cocycle γ' determines the same element of $H^2(G,F^*)$ as γ. Hence show that we can associate with ρ a unique element $[\gamma]$ of $H^2(G,F^*)$. Note that if $[\gamma] = 1$, then we may take $\gamma_{g_1,g_2} = 1, g_i \in G$. Then we have $\mu(g_1g_2) = \mu(g_1)\mu(g_2)$, so ρ is essentially an ordinary representation. In this case we shall say that ρ *splits*.

5. Let the notations be as in the last exercise and let A be a subgroup of F^* containing the γ_{g_1,g_2} for a particular choice of the cocycle γ. Construct the extension E of G by A corresponding to γ as in the text. Write the elements of E as (g,a) $g \in G$, $a \in A$. Then $(g_1,a_1)(g_2,a_2) = (g_1g_2,\gamma_{g_1,g_2}a_1a_2)$. Note that $iA = \{(1,a)\}$ is contained in the center of E. Show that $\tilde{\mu}(g,a) = a\mu(g)$ defines a representation of E acting on V such that $\tilde{\mu}(g,1) = \mu(g)$.

6. Let E be an extension of G by A such that $iA \subset Z(G)$, the center of E. Let $\tilde{\mu}$ be a representation of E acting on a finite dimensional vector space over an algebraically closed field F. For $g \in G$ define $\rho(g)$ to be the element of $PGL(V)$ having representative $\tilde{\mu}(g,1) \in L(V)$. Show that ρ is a projective representation of G.

Note: The preceding exercises 3–6 give a slight indication of a rich connection between the Schur multiplier and projective representations of a finite group. This was developed in three papers by Schur. The reader may consult Huppert's *Endliche Gruppen*, Springer-Verlag, Berlin-Heidelberg-New York, 1967, pp. 628–641 for an account of Schur's theory, with references.

6.11 COHOMOLOGY OF ALGEBRAS

The definitions of homology and cohomology modules for an (associative) algebra, which are due to Hochschild, are based on the concept of a bimodule for an algebra. Let A be an algebra over a commutative ring K. We have defined a (left) module M for A (as K-algebra) to be an abelian group written additively that is both a K-module and an A-module in the usual ring sense such that $k(ax) = (ka)x = a(kx)$, $k \in K$, $a \in A$, $x \in M$ (p. 211). One has a similar definition for a right (algebra) A-module M: M is a K-module, a right A-module in the usual sense such that $k(xa) = (kx)a = x(ka)$, $k \in K$, $a \in A$, $x \in M$. Now let A and B be algebras over K. Then we define an (algebra) A-B-bimodule M to be a left A-module and a right B-module such that the K-module structures given by A and by B are the same and $(ax)b = a(xb)$, $a \in A$, $b \in B$, $x \in M$.

There is a simple device for reducing the study of A-B-bimodules to that of modules for another algebra. Let B^{op} be the opposite algebra of B and form the algebra $A \otimes B^{\mathrm{op}}$ where $\otimes$ stands for $\otimes_K$. Let M be an A-B-bimodule, x an element of M. Then we have the map of $A \times B$ into M sending $(a, b) \rightsquigarrow axb \in M$. This is K-bilinear, so we have a K-linear map of $A \otimes B$ into M such that $\sum a_i \otimes b_i \rightsquigarrow \sum a_i x b_i$. The main point of this is that for a given $\sum a_i \otimes b_i \in A \otimes B^{\mathrm{op}}$ and a given $x \in M$ we have a well-defined product $(\sum a_i \otimes b_i)x = \sum a_i x b_i \in M$. Direct verification shows that this renders M an algebra $A \otimes B^{\mathrm{op}}$-module. Conversely, if M is given as an $A \otimes B^{\mathrm{op}}$-module, then $ax = (a \otimes 1)x$, $xb = (1 \otimes b)x$, $kx = kx$ for $a \in A$, $b \in B$, $k \in K$, make M an (algebra)A-B-bimodule. It is clear from this that we can pass freely from the point of view of A-B-bimodules to that of $A \otimes B^{\mathrm{op}}$-modules and conversely.

If M and N are modules for the algebra A, a homomorphism η of M into N is a homomorphism of abelian groups satisfying $\eta(kx) = k(\eta x)$, $\eta(ax) = a(\eta x)$, $k \in K$, $a \in A$. Since $kx = (k1)x$, the first of these conditions is superfluous, so the notion coincides with that of homomorphisms of M into N in the sense of modules for the ring A. On the other hand, it is natural to endow $\hom_A(M, N)$ with a K-module structure rather than just the abelian group structure we have considered hitherto. This is done by defining $k\eta$, $k \in K$, $\eta \in \hom(M, N)$ by $(k\eta)x = k(\eta x)$. It is clear that in this way $\hom(M, N)$ becomes a K-module. Similarly, if M is a right A-module and N is a left A-module for the algebra A, then $M \otimes_A N$ is a K-module. In place of the usual functors $\hom_A(M, -)$, $\hom_A(-, N)$, $\otimes_A N$, etc., to the category of abelian groups, we now have functors to K-modules. Moreover, these are not only additive but also K-linear in the sense that the maps between the K-modules involved in the definitions are K-homomorphisms. Similar remarks apply to the derived functors. All of this is quite obvious and would be tedious to spell out in detail.

We shall therefore say nothing more about it and shall replace abelian groups by modules in what follows when it is appropriate to do so.

We now consider a single algebra A and the category of A-A-bimodules (homomorphisms = A-A-homomorphisms). Equivalently we have the category of $A \otimes_K A^{op}$-modules. We shall now write A^e for $A \otimes A^{op}$. Evidently A itself is an A-A-bimodule relative to the left and right multiplications in A. Thus A is an A^e-module in which we have

$$(74) \qquad (\sum a_i \otimes b_i)x = \sum a_i x b_i$$

for $a_i, x \in A$, $b_i \in A^{op}$. Evidently A is cyclic as A^e-module with 1 as generator. Hence we have an A^e-homomorphism

$$(75) \qquad \varepsilon : \sum a_i \otimes b_i \rightarrow (\sum a_i \otimes b_i)1 = \sum a_i b_i$$

of A^e onto A.

We are now ready to define the homology and cohomology modules of an algebra A. Let M be an A-A-bimodule ($= A^e$-module). Then we define the nth *cohomology module* $H^n(A, M)$ *of* A *with coefficients in* M as $\mathrm{Ext}^n_{A^e}(A, M)$ and the nth *homology module* $H_n(A, M)$ *of* A *with coefficients in* M as $\mathrm{Tor}^{A^e}_n(A, M)$. In both cases A is regarded as A^e-module in the manner defined above.

We now assume that A is K-free (projective would suffice). We shall define a free resolution of A as A^e-module such that the determination of $H^n(A, M)$ by this resolution can be identified with the original definition of $H^n(A, M)$ as given by Hochschild for algebras over fields. Let $X_0 = A \otimes_K A$, $X_1 = A \otimes_K A \otimes_K A$ and, in general, $X_n = A \otimes_K \otimes \cdots \otimes_K A$ with $n+2$ factors. If M and N are A-A-bimodules, then $M \otimes_K N$ is an A-A-bimodule in which $a(x \otimes y) = ax \otimes y$, $(x \otimes y)a = x \otimes ya$, $a \in A$, $x \in M$, $y \in N$ (Proposition 3.6, p. 135). It follows that X_n is an A-A-bimodule in which

$$(76) \qquad \begin{aligned} a(x_0 \otimes \cdots \otimes x_{n+1}) &= ax_0 \otimes x_1 \otimes \cdots \otimes x_{n+1} \\ (x_0 \otimes \cdots \otimes x_{n+1})a &= x_0 \otimes \cdots \otimes x_n \otimes x_{n+1}a. \end{aligned}$$

It is clear from the definitions that $X_0 = A \otimes_K A \cong A^e$, $X_1 = A \otimes_K A \otimes_K A \cong A^e \otimes A$ and for $n > 1$, $X_n \cong A^e \otimes X_{n-2}$ as A^e-modules. The isomorphism $A^e \otimes X_{n-2} \rightarrow X_n$ maps $(a \otimes b) \otimes x \rightsquigarrow a \otimes x \otimes b$, $x \in X_{n-2}$. Since A is K-free, the X_n are K-free. It follows that $X_0 \cong A^e$, $X_1 \cong A^e \otimes A$ and $X_n \cong A^e \otimes X_{n-2}$ are A^e-free.

It is clear from the usual argument with tensor products that we have a unique K-homomorphism d_n of X_n into X_{n-1} such that

$$(77) \qquad d_n(x_0 \otimes \cdots \otimes x_{n+1}) = \sum_0^n (-1)^i x_0 \otimes \cdots \otimes x_i x_{i+1} \otimes \cdots \otimes x_{n+1}.$$

It is clear from the definition of the left and right A-actions that this is an A-A-bimodule, hence, A^e-homomorphism of X_n into X_{n-1}. Together with the augmentation $\varepsilon: X_0 \to A$ we have the sequence of A^e-homomorphisms

$$0 \leftarrow A \overset{\varepsilon}{\leftarrow} X_0 \overset{d_1}{\leftarrow} X_1 \leftarrow \cdots. \tag{78}$$

We shall show that this is exact, which will prove that (78) is a free resolution of A as A^e-module. We obtain a proof of exactness in a manner similar to that given for the complex employed in the group case (p. 357). We define a contracting homomorphism

$$A \overset{s_{-1}}{\to} X_0 \overset{s_0}{\to} X_1 \to \cdots, \tag{79}$$

that is, a sequence of K-homomorphisms s_i such that

$$\varepsilon s_{-1} = 1_A, \quad d_1 s_0 + s_{-1}\varepsilon = 1_{X_0}, \quad d_{n+1}s_n + s_{n-1}d_n = 1_{X_n}, \quad n \geqslant 1. \tag{80}$$

We define s_n, $n \geqslant -1$, to be the K-homomorphism such that

$$s_n(x_0 \otimes \cdots \otimes x_{n+1}) = 1 \otimes x_0 \otimes \cdots \otimes x_{n+1}. \tag{81}$$

Then it follows directly from the definition that (80) holds. As in the group case (p. 358), this implies that (78) is exact and hence this is a free resolution of A with ε as augmentation.

For a given A-A-bimodule M we now have the cochain complex

$$0 \to \hom_{A^e}(X_0, M) \to \hom_{A^e}(X_1, M) \to \cdots \tag{82}$$

whose cohomology groups are the cohomology groups of A with coefficients in M. Now we have the sequence of isomorphisms $\hom_{A^e}(X_n, M) \cong \hom_{A^e}(A^e \otimes X_{n-2}, M) \cong \hom_{A^e}(X_{n-2} \otimes A^e, M) \cong \hom_K(X_{n-2}, \hom_{A^e}(A^e, M))$ (see p. 136) $\cong \hom_K(X_{n-2}, M)$. We can also identify $\hom_K(X_{n-2}, M)$ with the K-module of n-linear maps of $A \times \cdots \times A$, n times, into M. Such a map has $A \times \cdots \times A$ as domain and M as codomain and is a K-homomorphism of A into M if all but one of the arguments is fixed. Hence the isomorphism above becomes an isomorphism onto the K-module $C^n(A, M)$ of n-linear maps f of $A \times \cdots \times A$ into M. We now define for $f \in C^n(A, M)$, $\delta f \in C^{n+1}(A, M)$ by

$$\begin{aligned} \delta f(x_1, \ldots, x_{n+1}) &= x_1 f(x_2, \ldots, x_{n+1}) \\ &\quad + \sum_{i=1}^{n} (-1)^i f(x_1, \ldots, x_{i-1}, x_i x_{i+1}, x_{i+2}, \ldots, x_{n+1}) \\ &\quad + (-1)^{n+1} f(x_1, \ldots, x_n) x_{n+1}, \end{aligned} \tag{83}$$

$x_i \in A$. Then one can check the commutativity of

$$\begin{array}{ccc} \hom_{A^e}(X_n, M) & \rightarrow & \hom_{A^e}(X_{n+1}, M) \\ \downarrow & & \downarrow \\ C^n(A, M) & \underset{\delta}{\rightarrow} & C^{n+1}(A, M) \end{array} \tag{84}$$

where the vertical maps are the indicated isomorphisms. It follows that

$$0 \rightarrow C^0(A, M) \overset{\delta}{\rightarrow} C^1(A, M) \rightarrow \cdots \tag{85}$$

is a cochain complex isomorphic to (82). Hence it has the same cohomology groups and consequently we have the following

THEOREM 6.17. *Let A be an algebra over K that is K-free, M an A-A-bimodule, $C^n(A, M)$, $n \geqslant 0$, the K-module of n-linear maps of $A \times \cdots \times A$, n times, into M. For $f \in C^n(A, M)$ define $\delta f \in C^{n+1}(A, M)$ by* (83). *Let $Z^n(A, M) = \ker \delta$ on $C^n(A, M)$, $B^n(A, M) = \delta C^{n-1}(A, M)$. Then $B^n(A, M)$ is a submodule of $Z^n(A, M)$ and $Z^n(A, M)/B^n(A, M) \cong H^n(A, M)$, the nth cohomology module of A with coefficients in M.*

Although some of the results we shall now indicate are valid without this restriction, we continue to assume that A is K-free. Following the pattern of our discussion of the group case, we now consider $H^n(A, M)$ for $n = 0, 1, 2$, using the determination of these modules given in Theorem 6.17.

As usual, it is understood that $C^0(A, M)$ is identified with the module M. Taking $u \in M$, the definition of δu gives $(\delta u)(x) = xu - ux$, $x \in A$. Hence $Z^0(A, M)$ is the submodule of M of u such that $ux = xu$, $x \in A$. Since $C^{-1}(A, M) = 0$, we see that $H^0(A, M)$ is isomorphic to the submodule of M of u such that $ux = xu$, $x \in A$.

Next let $f \in C^1(A, M)$. Then $\delta f(x, y) = xf(y) - f(xy) + f(x)y$, $x, y \in A$ and $\delta f = 0$ if and only if f is a K-homomorphism of A into M such that

$$f(xy) = xf(y) + f(x)y. \tag{86}$$

It is natural to call such an f a *derivation* of A into the A-A-bimodule M. If $u \in M$, u determines the *inner derivation* δu such that

$$(\delta u)(x) = xu - ux \tag{87}$$

These form a submodule $\text{Inder}(A, M)$ of the module $\text{Der}(A, M)$ of derivations of A into M. The special case $n = 1$ of Theorem 6.17 gives the isomorphism

$$H^1(A, M) \cong \text{Der}(A, M)/\text{Inder}(A, M). \tag{88}$$

Now let $f \in C^2(A, M)$. Then

$$(\delta f)(x, y, z) = xf(y, z) - f(xy, z) + f(x, yz) - f(x, y)z$$

and $\delta f = 0$ if and only if

$$xf(y, z) + f(x, yz) = f(xy, z) + f(x, y)z, \tag{89}$$

$x, y, z \in A$. The set of $f \in C^2(A, M)$ satisfying this condition constitutes $Z^2(A, M)$. This contains the submodule of maps δg, $g \in C^1(A, M)$ and $(\delta g)(x, y) = xg(y) - g(xy) + g(x)y$. The quotient of $Z^2(A, M)$ by this submodule is isomorphic to $H^2(A, M)$.

The second cohomology group of an algebra in the form $Z^2(A, M)/B^2(A, M)$ made its first appearance in the literature in proofs by J. H. C. Whitehead and by Hochschild of a classical structure theorem on finite dimensional algebras over a field: the so-called Wedderburn principal theorem. We shall give a sketch of a cohomological proof of this theorem, leaving the details to be filled in by the reader in a sequence of exercises at the end of the chapter.

Let A be a finite dimensional algebra over a field F, $N = \text{rad}\, A$. Then N is a nilpotent ideal in A and $\bar{A} = A/N$ is semi-primitive. The Wedderburn principal theorem asserts that if $\bar{A}$ is *separable* in the sense that $\bar{A}_E$ is semi-primitive for every extension field E/F, then A contains a subalgebra S such that $A = S + N$ and $S \cap N = 0$, that is, $A = S \oplus N$ as vector space over F (not as algebra direct sum!).

To prove the theorem one first reduces the proof to the case in which $N^2 = 0$. This is done by introducing $B = A/N^2$ (N^2 is an ideal) whose radical is N/N^2 and $(N/N^2)^2 = 0$. If $N^2 \neq 0$, then the dimensionality $[B:F] < [A:F]$, so we may assume that the theorem holds for B. It follows easily that it holds also for A.

Now assume $N^2 = 0$. We can choose a subspace V of A such that $A = V \oplus N$. This is equivalent to choosing a linear map s of $\bar{A}$ into A such that $ps = 1_{\bar{A}}$ for p, the canonical map $x \rightsquigarrow x + N$ of A onto $\bar{A}$. Then $V = s\bar{A}$ and s is injective. Any element of A can be written in one and only one way as $s(\bar{a}) + x$, $\bar{a} \in \bar{A}$, $x \in N$. Since N is an ideal and $N^2 = 0$, we have the multiplication

$$(s(\bar{a}) + x)(s(\bar{b}) + y) = s(\bar{a})s(\bar{b}) + s(\bar{a})y + xs(\bar{b}). \tag{90}$$

If we define $\bar{a}x = s(\bar{a})x$, $x\bar{a} = xs(\bar{a})$, then N becomes an $\bar{A}$-$\bar{A}$-bimodule. Since $ps(\bar{a})s(\bar{b}) = ps(\bar{a})ps(\bar{b}) = \bar{a}\bar{b}$ and $ps(\bar{a}\bar{b}) = \bar{a}\bar{b}$, we have

$$s(\bar{a})s(\bar{b}) = s(\bar{a}\bar{b}) + f(\bar{a}, \bar{b}) \tag{91}$$

where $f(\bar{a}, \bar{b}) \in N$. The map $f \in Z^2(\bar{A}, N)$. Replacing s by the linear map $t: \bar{A} \to A$

such that $pt = 1_{\bar{A}}$ replaces f by a cohomologous cocycle. Moreover, if $f = 0$ in (91), then $S = s(\bar{A})$ is a subalgebra such that $A = S \oplus N$. Then Wedderburn's principal theorem will follow if we can prove that $H^2(\bar{A}, N) = 0$ for any separable algebra $\bar{A}$ and any $\bar{A}$-$\bar{A}$-bimodule N. A proof of this is indicated in the following exercises.

EXERCISES

1. First fill in the details of the foregoing argument: the reduction to the case $N^2 = 0$, the reduction in this case to the proof of $H^2(\bar{A}, N) = 0$.

2. Let A be a finite dimensional separable algebra over a field F. Show that there exists an extension field E/F such that $A_E = M_{n_1}(E) \oplus M_{n_2}(E) \oplus \cdots \oplus M_{n_s}(E)$. (The easiest way to do this is to use the algebraic closure $\bar{F}$ of F as defined in section 8.1. However, it can be done also without the algebraic closure.)

3. Use exercise 2 to show that $A^e = A \otimes_F A^{\mathrm{op}}$ is finite dimensional semi-simple.

4. Show that any module for a semi-simple artinian ring is projective.

5. Use exercises 3 and 4 to show that if A is finite dimensional separable, then A is a projective A^e-module. Hence conclude from Theorem 6.6, p. 347, that $H^n(A, M) = 0$ for any $n \geqslant 1$ and any M. (This completes the proof of the theorem.)

6. (A. I. Malcev.) Let $A = S \oplus N$ where $N = \mathrm{rad}\, A$ and S is a separable subalgebra of A. Let T be a separable subalgebra of A. Show that there exists a $z \in N$ such that $(1-z)T(1-z)^{-1} \subset S$.

7. Let A be an arbitrary algebra and let $J = \ker \varepsilon$ where ε is the augmentation $A^e \to A$ defined above. Show that J is the left ideal in A^e generated by the elements $a \otimes 1 - 1 \otimes a$.

8. Show that A is A^e-projective if and only if there exists an idempotent $e \in A^e$ such that $(a \otimes 1)e = (1 \otimes a)e$, $a \in A$.

6.12 HOMOLOGICAL DIMENSION

Let M be a (left) module for a ring R. There is a natural way of defining homological dimension for M in terms of projective resolutions of M. We say that M has *finite homological dimension* if M has a projective resolution C, ε for which $C_n = 0$ for all sufficiently large n. In this case the smallest integer n such that M has a projective resolution

$$\cdots 0 \to 0 \to C_n \to \cdots C_0 \xrightarrow{\varepsilon} M \to 0$$

is called the *homological dimension*, h.dim M, of M. It is clear from this definition that M is projective if and only if h.dim $M = 0$. We recall that such a module can be characterized by the property that $\mathrm{Ext}^n(M,N) = 0$ for all modules N and $n \geqslant 1$. The following result contains a generalization of this criterion.

THEOREM 6.18. *The following conditions on a module M are equivalent*:

(1) h.dim $M \leqslant n$.

(2) $\mathrm{Ext}^{n+1}(M,N) = 0$ *for all modules* N.

(3) *Given an exact sequence* $0 \to C_n \to C_{n-1} \to \cdots \to C_0 \to M \to 0$ *in which every* C_k, $k < n$, *is projective, then* C_n *is projective.*

Proof. (1)$\Rightarrow$(2). The hypothesis is that we have a projective resolution $\cdots 0 \to 0 \to C_n \to \cdots \to C_0 \to M \to 0$. Then we have the complex $0 \to \hom(C_0, N) \to \hom(C_1, N) \to \cdots \to \hom(C_n, N) \to 0 \to \cdots$. The cohomology groups of this cochain complex are the terms of the sequence $\mathrm{Ext}^0(M,N)$, $\mathrm{Ext}^1(M,N), \ldots$. Evidently we have $\mathrm{Ext}^{n+1}(M,N) = 0$ and this holds for all N.

(2)$\Rightarrow$(3). If we are given an exact sequence with the properties stated in (3), we obtain from it a sequence of homomorphisms

$$0 \to C_n \to D_n \to C_{n-1} \to D_{n-1} \to \cdots \to C_0 \to D_0 \to 0$$

where $D_k = \mathrm{im}(C_k \to C_{k-1})$ for $k > 0$, $D_0 = \mathrm{im}(C_0 \to M) = M$, $C_k \to D_k$, is obtained from $C_k \to C_{k-1}$ by restricting the codomain and $D_k \to C_{k-1}$ is an injection. Then $0 \to D_k \to C_{k-1} \to D_{k-1} \to 0$ is exact. Hence the long exact sequence for Ext in the first variable gives the exactness of

$$\mathrm{Ext}^i(C_{k-1}, N) \to \mathrm{Ext}^i(D_k, N) \to \mathrm{Ext}^{i+1}(D_{k-1}, N) \\ \to \mathrm{Ext}^{i+1}(C_{k-1}, N)$$

for $i = 1, 2, \ldots$, $1 \leqslant k \leqslant n$. Since C_{k-1} is projective, the first and last terms are 0. Thus $\mathrm{Ext}^i(D_k, N) \cong \mathrm{Ext}^{i+1}(D_{k-1}, N)$ and hence

$$\mathrm{Ext}^1(D_n, N) \cong \mathrm{Ext}^2(D_{n-1}, N) \cong \cdots \cong \mathrm{Ext}^{n+1}(D_0, N).$$

Assuming (2), we have $\mathrm{Ext}^{n+1}(D_0, N) = 0$. Hence $\mathrm{Ext}^1(D_n, N) = 0$. Since $0 \to C_n \to C_{n-1} \to \cdots$ is exact, $D_n \cong C_n$. Thus $\mathrm{Ext}^1(C_n, N) = 0$ for all N, which implies that C_n is projective.

(3)$\Rightarrow$(1). The construction of a projective resolution for M gives at the $(n-1)$-st stage an exact sequence $0 \leftarrow M \leftarrow C_0 \leftarrow \cdots \leftarrow C_{n-1}$ where all of the

C_i are projective. Let $C_n = \ker(C_{n-1} \to C_{n-2})$. Then we have the exact sequence $0 \leftarrow M \leftarrow C_0 \leftarrow \cdots \leftarrow C_n \leftarrow 0$. Assuming (3), we can conclude that C_n is projective. Then $0 \leftarrow M \leftarrow C_0 \leftarrow \cdots \leftarrow C_n \leftarrow 0 \leftarrow 0 \cdots$ is a projective resolution, which shows that $\text{h.dim}\, M \leqslant n$. $\square$

Remarks. The proof of the implication (1)$\Rightarrow$(2) shows also that if $\text{h.dim}\, M \leqslant n$, then $\text{Ext}^k(M,N) = 0$ for every $k > n$ and every module N. In a similar manner the condition implies that $\text{Tor}_k(M,N) = 0$ for all $k > n$ and all N. It is clear also that if $\text{h.dim}\, M = n$, then for any $k \leqslant n$ there exists a module N such that $\text{Ext}^k(M,N) \neq 0$.

It is clear from the fact that $\text{Ext}^n(-,N)$ is an additive functor from R-**mod** to **Ab** that $\text{Ext}^k(M' \oplus M'', N) \cong \text{Ext}^k(M',N) \oplus \text{Ext}^k(M'',N)$. An immediate consequence of this and Theorem 6.18 is that $M = M' \oplus M''$ has finite homological dimension if and only if this is the case for M' and M''. Then $\text{h.dim}\, M$ is the larger of $\text{h.dim}\, M'$ and $\text{h.dim}\, M''$. We now consider, more generally, relations between homological dimensions of terms of a short exact sequence $0 \to M' \to M \to M'' \to 0$. For any module N we have the long exact sequence

$$\cdots \to \text{Ext}^k(M'',N) \to \text{Ext}^k(M,N) \to \text{Ext}^k(M',N) \to \text{Ext}^{k+1}(M'',N) \to \cdots.$$

Suppose $\text{h.dim}\, M \leqslant n$. Then $\text{Ext}^k(M,N) = 0$ if $k > n$ and hence $\text{Ext}^k(M',N) \cong \text{Ext}^{k+1}(M'',N)$ if $k > n$. Similarly, if $\text{h.dim}\, M' \leqslant n$, then $\text{Ext}^{k+1}(M'',N) \cong \text{Ext}^{k+1}(M,N)$ for $k > n$ and if $\text{h.dim}\, M'' \leqslant n$, then $\text{Ext}^k(M,N) \cong \text{Ext}^k(M',N)$ for $k > n$. These relations imply first that if any two of the three modules M, M', M'' have finite homological dimension, then so has the third. Suppose this is the case and let $\text{h.dim}\, M = n$, $\text{h.dim}\, M' = n'$, $\text{h.dim}\, M'' = n''$. It is readily seen that the facts we have noted on the Ext's imply that we have one of the following possibilities:

I. $n \leqslant n', n''$. Then either $n = n' = n''$ or $n \leqslant n'$ and $n'' = n'+1$.
II. $n' \leqslant n'', n' < n$. Then $n = n''$.
III. $n'' \leqslant n', n'' < n$. Then $n = n'$.

From this it follows that if $n > n'$, then $n'' = n$; if $n < n'$, then $n'' = n'+1$; and if $n = n'$, then $n'' \leqslant n'+1$. We state these results as

THEOREM 6.19. *Let $0 \to M' \to M \to M'' \to 0$ be exact. Then if any two of* $\text{h.dim}\, M'$, $\text{h.dim}\, M$, $\text{h.dim}\, M''$ *are finite, so is the third. Moreover, we have* $\text{h.dim}\, M'' = \text{h.dim}\, M$ *if* $\text{h.dim}\, M' < \text{h.dim}\, M$, $\text{h.dim}\, M'' = \text{h.dim}\, M'+1$ *if* $\text{h.dim}\, M < \text{h.dim}\, M'$, *and* $\text{h.dim}\, M'' \leqslant \text{h.dim}\, M+1$ *if* $\text{h.dim}\, M = \text{h.dim}\, M'$.

The concept of homological dimension of a module leads to the definition of homological dimensions for a ring. We define the *left* (*right*) *global dimension* of a ring R as $\sup \text{h.dim}\, M$ for the left (right) modules for R. Thus the left (right) global dimension of R is 0 if and only if every left (right) R-module is projective. This is the case if and only if every short exact sequence of left (right) modules $0 \to M' \to M \to M'' \to 0$ splits (p. 150) and this happens if and only if every left (right) module for R is completely reducible. It follows that R has left (right) global dimension 0 if and only if R is semi-simple artinian (p. 208). Thus R has left global dimension 0 if and only if it has right global dimension 0. Otherwise, there is no connection between the left and right global dimensions of rings in general. We shall be interested primarily in the case of commutative rings where, of course, there is no distinction between left and right modules and hence there is only one concept of global dimension.

A (commutative) p.i.d. R that is not a field has global dimension one. For, any submodule of a free R-module is free (exercise 4, p. 155). Hence if M is any R-module, then we have an exact sequence $0 \to K \to F \to M \to 0$ in which F and K are free. Hence $\text{h.dim}\, M \leqslant 1$ for any R-module M and the global dimension of R is $\leqslant 1$. Moreover, it is not 0, since this would imply that R is semi-simple artinian and hence that R is a direct sum of a finite number of fields. Since R has no zero divisors $\neq 0$, this would imply that R is a field, contrary to assumption.

EXERCISE

1. Let M be a module over a commutative ring K, L a commutative algebra over K that is K-free. Show that $\text{h.dim}_K M = \text{h.dim}_L M_L$ ($\text{h.dim}_K M =$ homological dimension as K-module).

6.13 KOSZUL'S COMPLEX AND HILBERT'S SYZYGY THEOREM

We shall now consider homological properties of the ring $R = F[x_1, \ldots, x_m]$ of polynomials in indeterminates x_i with coefficients in a field F. Our main objective is a theorem of Hilbert that concerns graded modules for the ring R, graded in the usual way into homogeneous parts. We consider first the decomposition $R = F \oplus J$ where J is the ideal in R of polynomials with 0 constant term, or, equivalently, vanishing at 0. This decomposition permits us to define an R-module structure on F by $(a+f)b = ab$ for $a, b \in F$, $f \in J$. Note that this module is isomorphic to R/J. An important tool for the study of the homological properties of R is a certain resolution of F as R-module, which was first introduced by Koszul in a more general situation that is applicable to the study of homology of Lie algebras—see the author's *Lie Algebras*, pp. 174–185).

Koszul's complex, which provides a resolution for F, is based on the exterior algebra $E(M)$ for a free module M of rank m over $R = F[x_1, \ldots, x_m]$. We need to recall the definitions and elementary facts on exterior algebras that we obtained in Chapter 3 (p. 141). Let K be an arbitrary commutative ring, M a K-module, and $E(M)$ the exterior algebra defined by M. We recall that M is embedded in $E(M)$ and $E(M)$ is graded so that $E(M) = K \otimes M \otimes M^2 \otimes \cdots$. We recall also the basic universal property of $E(M)$, namely, $x^2 = 0$ for every $x \in M$, and if f is a K-homomorphism of M into an algebra A over K such that $f(x)^2 = 0$, then f can be extended in one and only one way to a K-algebra homomorphism of $E(M)$ into A. In particular, the map $x \rightsquigarrow -x$ in M has a unique extension to a homomorphism ι of $E(M)$ into itself and since $x \rightsquigarrow -x$ is of period two, $\iota^2 = 1_{E(M)}$.

We shall call a K-endomorphism D of $E(M)$ an *anti-derivation* if

$$D(ab) = (Da)b + \bar{a}(Db), \qquad \bar{a} = \iota a. \tag{92}$$

We require the following

LEMMA 1. *Let D be a K-homomorphism of M into $E(M)$ such that*

$$x(Dx) = (Dx)x, \qquad x \in M. \tag{93}$$

Then D can be extended in one and only one way to an anti-derivation of $E(M)$.

Proof. Consider the map

$$f: x \rightsquigarrow \begin{pmatrix} x & 0 \\ Dx & -x \end{pmatrix}$$

of M into the algebra $A = M_2(E(M))$. The condition on D implies that f is a K-homomorphism such that $f(x)^2 = 0$. Hence f has a unique extension to an algebra homomorphism of $E(M)$ into A. It is clear that the extension has the form

$$f: a \rightsquigarrow \begin{pmatrix} a & 0 \\ Da & \bar{a} \end{pmatrix}, \qquad a \in E(M)$$

where $\bar{a} = \iota a$ and D is a K-endomorphism of $E(M)$ extending the given D. The condition $f(ab) = f(a)f(b)$ implies that D is an anti-derivation. The uniqueness of the extension follows from the following readily verified facts:

1. The difference of two anti-derivations is an anti-derivation.
2. The subset on which an anti-derivation is 0 is a subalgebra.
3. M generates $E(M)$.

Now if D_1 and D_2 are anti-derivations such that $D_1|M = D_2|M = D$, then $D_1 - D_2$ is an anti-derivation such that $(D_1 - D_2)|M = 0$. Since M generates $E(M)$, we have $D_1 - D_2 = 0$ and $D_1 = D_2$. □

A particular case in which the lemma applies is that in which $D = d \in M^*$, the K-module of K-homomorphisms of M into K (the dual module of M). Since $K \subset E(M)$, d can be regarded as a K-homomorphism of M into $E(M)$. Since K is contained in the center of $E(M)$, it is clear that the condition $(dx)x = x(dx)$ of the lemma is fulfilled. Hence we have the extension d that is an anti-derivation of $E(M)$. Since $dM \subset K$ and d is an anti-derivation, we can prove by induction that $dM^i \subset M^{i-1}$, $i \geqslant 1$. We prove next that if $d \in M^*$, the anti-derivation extension d satisfies

LEMMA 2. $d^2 = 0$.

Proof. It is clear from (90) that $D1 = 0$ for any anti-derivation. Hence $dk = 0$ for $k \in K$. Then $d^2M = 0$ since $dM \subset K$. We note next that $\overline{dx} = dx = -d\bar{x}$ for $x \in M$ and if $a \in E(M)$ satisfies $\overline{da} = -d\bar{a}$, then $\overline{d(ax)} = -d(\overline{ax})$ follows from the fact that d is an anti-derivation. It follows that $\overline{da} = -d\bar{a}$ for all a. This relation implies that $d^2M^i = 0$, by induction on i. Hence $d^2 = 0$. □

We now have the chain complex

$$0 \leftarrow K \overset{d}{\leftarrow} M \overset{d}{\leftarrow} M^2 \leftarrow \cdots \tag{94}$$

determined by the element $d \in M^*$.

We shall require also a result on change of base rings for exterior algebras.

LEMMA 3. *Let M be a module over a commutative ring K, L a commutative algebra over K, and let $E(M)$ be the exterior algebra over M. Then $E(M)_L = E(M_L)$.*

Proof. Since M is a direct summand of $E(M)$, we can identify M_L with the subset of $E(M)_L$ of elements $\sum l_i \otimes x_i$, $l_i \in L$, $x_i \in M \subset E(M)$. We have $x_i^2 = 0$ and hence $x_i x_j + x_j x_i = (x_i + x_j)^2 - x_i^2 - x_j^2 = 0$ for $x_i, x_j \in M$. This implies that $(\sum l_i \otimes x_i)^2 = 0$ for every $\sum l_i \otimes x_i \in M_L$. Let $\tilde{f}$ be an L-homomorphism of M_L

into an algebra $\tilde{A}/L$ such that $\tilde{f}(\tilde{x})^2 = 0$, $\tilde{x} \in M_L$. Now $\tilde{A}$ becomes an algebra over K if we define $k\tilde{a}$, $k \in K$, $\tilde{a} \in \tilde{A}$, as $(k1)\tilde{a}$, $(k1 \in L)$. If $x \in M$, then $f: x \to \tilde{f}(1 \otimes x)$ is a K-homomorphism of M into $\tilde{A}/K$ such that $f(x)^2 = 0$. Hence this has an extension to a K-homomorphism f of $E(M)$ into $\tilde{A}/K$. Then we have the homomorphism $1 \otimes f$ of $E(M)_L$ into $(\tilde{A}/K)_L$ and we have the L-algebra homomorphism of $(\tilde{A}/K)_L$ into $\tilde{A}/L$ such that $l \otimes \tilde{a} \to l\tilde{a}$. Taking the composite we obtain an L-algebra homomorphism of $E(M)_L$ into $\tilde{A}/L$. This maps the element $1 \otimes x$, $x \in M$, into $f(x) = \tilde{f}(1 \otimes x)$. Hence it coincides with the given L-homomorphism $\tilde{f}$ on M_L. Thus we have obtained an extension of $\tilde{f}$ to an algebra homomorphism of $E(M)_L$ into $\tilde{A}/L$. Since M generates $E(M)$, M_L generates $E(M)_L$. Hence the extension is unique. We have therefore shown that $E(M)_L$ has the universal property characterizing $E(M)_L$ and so we may identify these two algebras. □

We now specialize K to $R = F[x_1, \ldots, x_m]$ where F is a field and the x_i are indeterminates. Let V be an m-dimensional vector space over F, $(y_1, \ldots, y_m)$ a base for V/F, $E(V)$ the exterior algebra determined by V. Then $E(V) = F1 \oplus V \oplus \cdots \oplus V^m$, $V^{m+1} = 0$, and V^r has the base of $\binom{m}{r}$ elements $y_{i_1} \cdots y_{i_r}$ where $i_1 < i_2 < \cdots < i_r$ (BAI, p. 415). Let $M = V_R$. Then by Lemma 3, $E(M) = R1 \oplus M \oplus \cdots \oplus M^m$ where M^r has the base $\{y_{i_1} \cdots y_{i_r}\}$ over R. Thus this module is free of rank $\binom{m}{r}$ over R. (This can also be seen directly by using the same method employed in the case V/F.) Let d be the element of $M^* = \hom_R(M, R)$ such that $dy_i = x_i$, $1 \leqslant i \leqslant m$, and let d also denote the anti-derivation in $E(M)$ extending d. Then we have the chain complex (94) and we wish to show that if ε is the canonical homomorphism of R into F obtained by evaluating a polynomial at $(0, \ldots, 0)$, then

$$0 \leftarrow F \overset{\varepsilon}{\leftarrow} R \overset{d}{\leftarrow} M \leftarrow \cdots \leftarrow M^m \leftarrow 0 \leftarrow \cdots \tag{95}$$

is a resolution of F as R-module with ε as augmentation. Since $F \subset R \subset E(M)$ and $E(M)$ is a vector space over F, we can extend ε to a linear transformation ε in $E(M)/F$ so that $\varepsilon(M^i) = 0$, $i \geqslant 1$. Then $\varepsilon E(M) = \varepsilon R = F$ and so $d\varepsilon = 0$. Also since $dE(M) \subset dM + \sum_{i \geqslant 1} M^i$ and dM is the ideal in R generated by the elements $x_i = dy_i$, we have $\varepsilon dM = 0$ and so $\varepsilon dE(M) = 0$. Thus we have

$$\varepsilon d = 0 = d\varepsilon. \tag{96}$$

Also since $\sum_{i \geqslant 1} M^i$ is an ideal in $E(M)$ and $\varepsilon | R$ is an F-algebra homomorphism of R into F, we have

$$\varepsilon(ab) = (\varepsilon a)(\varepsilon b) \tag{97}$$

for $a, b \in E(M)$. Thus ε is an F-algebra homomorphism of $E(M)$.

The proof that (95) is exact is similar to two other proofs of exactness that we have given (p. 359 and p. 373). It is based on the following

LEMMA 4. *There exists a linear transformation s in $E(M)/F$ such that*

$$sd+ds = 1-\varepsilon. \tag{98}$$

Proof. We use induction on m. If $m = 1$, then $M = Ry$ and $E(M) = R1 \oplus Ry$. Since $(1, x, x^2, \ldots)$ is a base for $R = F[x]$ over F, $(1, x, x^2, \ldots, y, xy, x^2y, \ldots)$ is a base for $E(M)/F$. We have $dx^i = 0$, $dx^iy = x^{i+1}$, $i \geqslant 0$, and $\varepsilon 1 = 1$, $\varepsilon x^{i+1} = 0$, $\varepsilon x^i y = 0$. Let s be the linear transformation in $E(M)/F$ such that $s1 = 0$, $sx^i = x^{i-1}y$, $sx^iy = 0$. Then it is readily checked that (98) holds. We note also that

$$s\varepsilon = 0 = \varepsilon s, \qquad \iota\varepsilon = \varepsilon = \varepsilon\iota, \qquad s\iota + \iota s = 0 \tag{99}$$

if ι is the automorphism defined before. Now assume the lemma for $m-1 > 0$. Let E_1 be the F-subspace of $E(M)$ spanned by the elements ${x_1}^k, {x_1}^k y_1$, $k \geqslant 0$, and E_2 the F-subspace spanned by the elements $x_2^{k_2} \cdots x_m^{k_m} y_{i_1} \cdots y_{i_r}$ where $k_j \geqslant 0$ and $2 \leqslant i_1 < i_2 < \cdots < i_r \leqslant m$. Then it is clear by looking at bases that E_1 is a subalgebra isomorphic to $E(M_1)$, $M_1 = Ry_1$, E_2 is a subalgebra isomorphic to $E(M_2)$, $M_2 = \sum_2^m Ry_j$, and we have a vector space isomorphism (not algebra isomorphism) of $E_1 \otimes_F E_2$ onto $E(M)$ such that $u \otimes v \rightsquigarrow uv$. We note also that E_1 and E_2 are stabilized by ι and by d and the induced maps are as defined in $E(M)$. Using induction we have a linear transformation s_2 in E_2 such that $s_2 d + ds_2 = 1 - \varepsilon$. Let s_1 be the linear transformation in E_1 as defined at the beginning of the proof. Since $E(M) \cong E_1 \otimes_F E_2$, there exists a unique linear transformation s in $E(M)$ such that

$$s(uv) = (s_1 u)v + (\varepsilon u)(s_2 v) \tag{100}$$

for $u \in E_1$, $v \in E_2$. Then

$$\begin{aligned}
ds(uv) &= (ds_1 u)v + (\iota s_1 u)(dv) + (d\varepsilon u)(s_2 v) \\
&\quad + (\iota\varepsilon u)(ds_2 v) \\
&= (ds_1 u)v + (\iota s_1 u)(dv) + (\varepsilon u)(ds_2 v) \\
sd(uv) &= s((du)v + (\iota u)dv) \\
&= (s_1 du)v + (\varepsilon du)(s_2 v) + (s_1 \iota u)(dv) + (\varepsilon\iota u)(s_2 dv) \\
&= (s_1 du)v + (s_1 \iota u)(dv) + (\varepsilon u)(s_2 dv).
\end{aligned}$$

Hence

$$\begin{aligned}(ds+sd)(uv) &= ((1-\varepsilon)u)v+(\varepsilon u)((1-\varepsilon)v)\\ &= uv-(\varepsilon u)v+(\varepsilon u)v-(\varepsilon u)(\varepsilon v)\\ &= (1-\varepsilon)(uv)\end{aligned}$$

since ε is an algebra homomorphism. This completes the proof. □

We can now prove that the sequence (95) is exact. First, we know that ε is surjective, so $R \to F \to 0$ is exact. We have seen also that dM is the ideal in R generated by the x_i. Since this is the ideal of polynomials vanishing at $(0,0,\ldots,0)$, we have $dM = J = \ker\varepsilon$. Hence $M \xrightarrow{d} R \to F \to 0$ is exact. It remains to show that if $z_i \in M^i$, $i \geqslant 1$, and $dz_i = 0$, then there exists a $w_{i+1} \in M^{i+1}$ such that $dw_{i+1} = z_i$. We have $z_i = 1z_i = (1-\varepsilon)z_i = (sd+ds)z_i = d(sz_i) = dw$, $w = sz_i$. Now we can write $w = \sum_0^m w_j$, $w_j \in M^j$, then $dw_j \in M^{j-1}$ so $dw = z_i$ gives $dw_{i+1} = z_i$. Thus (95) is exact.

We summarize our results in

THEOREM 6.20. *Let $R = F[x_1,\ldots,x_m]$, x_i indeterminates, F a field, and let M be the free R-module with base $(y_1,\ldots,y_m)$, $E(M)$ the exterior algebra defined by M. Let d be the anti-derivation in $E(M)$ such that $dy_i = x_i$, $1 \leqslant i \leqslant m$, ε the ring homomorphism of R into F such that $\varepsilon f = f(0,\ldots,0)$. Then $dM^i \subset M^{i-1}$ and*

$$0 \leftarrow F \xleftarrow{\varepsilon} R \xleftarrow{d} M \leftarrow \cdots \leftarrow M^m \leftarrow 0 \leftarrow \cdots$$

is a free resolution of F as R-module.

We call this resolution the *Koszul resolution* of F as R-module and the complex $M^0(=R) \leftarrow M^1 \leftarrow M^2 \leftarrow \cdots$ the *Koszul complex for R*.

Since $M^k = 0$ for $k > m$, we evidently have h.dim $F \leqslant m$. We claim that, in fact, h.dim $F = m$. This will follow from one of the remarks following Theorem 6.18, by showing that $\mathrm{Tor}_m{}^R(F,F) \neq 0$. More generally, we determine $\mathrm{Tor}_r{}^R(F,F)$ by using the Koszul resolution of the second F. Then we have the complex

$$0 \leftarrow F\otimes_R R \leftarrow F\otimes_R M \leftarrow F\otimes_R M^2 \leftarrow \cdots \tag{101}$$

whose homology groups give $\mathrm{Tor}_0{}^R(F,F)$, $\mathrm{Tor}_1{}^R(F,F),\ldots$. Now $M^r \cong R\otimes_F V^r$ where V is the m-dimensional vector space over F (as above) and $F\otimes_R M^r \cong F\otimes_R(R\otimes_F V^r) \cong (F\otimes_R R)\otimes_F V^r \cong F\otimes_F V^r \cong V^r$. If we use the base $\{y_{i_1}\cdots y_{i_r} | i_1 < i_2 < \cdots < i_r\}$ for M^r, $1 \leqslant r \leqslant m$, as before, we can follow through the chain of isomorphisms and see that the isomorphism of $F \otimes_R M^r$ onto V^r/F sends $\alpha \otimes f y_{i_1} \cdots y_{i_r}$ for $\alpha \in F$, $f \in R$ into $\alpha(\varepsilon f) y_{i_1} \cdots y_{i_r}$, $\varepsilon f = f(0, \ldots, 0)$. The definition of d gives

$$(102) \qquad d y_{i_1} \cdots y_{i_r} = \sum_1^r (-1)^{j+1} x_{i_j} y_{i_1} \cdots y_{i_{j-1}} y_{i_{j+1}} \cdots y_{i_r}.$$

Hence $(1 \otimes d)(\alpha \otimes 1 y_{i_1} \cdots y_{i_r}) \to 0$ under our isomorphism. Thus the boundary operators in (101) are all 0 and the isomorphism $F \otimes_R M^r \to V^r$ gives the F-isomorphism

$$(103) \qquad \mathrm{Tor}_r{}^R(F, F) \cong V^r.$$

In particular, $\mathrm{Tor}_m{}^R(F, F) \cong V^m \cong F$.

The result just proved that $\mathrm{h.dim}\, F = m$ implies that $R = F[x_1, \ldots, x_m]$ has global dimension $\geqslant m$. It is not difficult to supplement this result and prove that the global dimension of R is exactly m. We shall indicate this in the exercises. In the remainder of the section we shall consider a somewhat similar result on free resolutions of graded modules for R that is due to Hilbert. Of course, Hilbert had to state his theorem in a cruder form than we are able to, since the concepts that we shall use were not available to him.

We recall first the standard grading of $R = F[x_1, \ldots, x_m]$ as

$$(104) \qquad F[x_1, \ldots, x_m] = R^{(0)}(= F1) \oplus R^{(1)} \oplus R^{(2)} \oplus \cdots$$

where $R^{(i)}$ is the subspace over F of i-forms, that is, F-linear combinations of monomials of total degree i in the x's. We have $R^{(i)} R^{(j)} \subset R^{(i+j)}$.

If R is a graded ring, graded by the subgroups $R^{(i)}$ of the additive group ($R = R^{(0)} \oplus R^{(1)} \oplus \cdots$, $R^{(i)} R^{(j)} \subset R^{(i+j)}$), then an R-module M is said to be *graded* by the subgroups $M^{(i)}$, $i = 0, 1, \ldots$, of its additive group if $M = M^{(0)} \oplus M^{(1)} \oplus M^{(2)} \oplus \cdots$ and $R^{(i)} M^{(j)} \subset M^{(i+j)}$ for every i, j. The elements of $M^{(i)}$ are called *homogeneous of degree* i. A submodule N of M is called *homogeneous* if $N = \sum (N \cap M^{(i)})$, or equivalently, if $v = \sum v^{(i)} \in N$ where $v^{(i)} \in M^{(i)}$, then every $v^{(i)} \in N$. Then N is graded by the submodules $N^{(i)} = N \cap M^{(i)}$. Moreover, M/N is graded by the subgroups $(M^{(i)} + N)/N$ and $(M^{(i)} + N)/N \cong M^{(i)}/(M^{(i)} \cap N)$. If M and N are graded R-modules, a *homomorphism* η of *graded modules* (of *degree* 0) is a homomorphism in the usual sense of R-modules such that $\eta M^{(i)} \subset N^{(i)}$ for every i. Then the image $\eta(M)$ is a homogeneous submodule of N and $\ker \eta$ is a homogeneous submodule of M.

If M is a graded module for the graded ring R, we can choose a set of generators $\{u_\alpha\}$ for M such that every u_α is homogeneous (e.g., $\{u_\alpha\} = \bigcup M^{(i)}$). Let $\{e_\alpha\}$ be a set of elements indexed by the same set $I = \{\alpha\}$ as $\{u_\alpha\}$ and let L

be a free R-module with base $\{e_\alpha\}$. Then we have the homomorphism ε of L onto M such that $e_\alpha \rightsquigarrow u_\alpha$. Let $L^{(i)}$ be the set of sums of the elements of the form $a^{(j)}e_\alpha$ where $a^{(j)} \in R^{(j)}$ and $j + \deg u_\alpha = i$. Then it is readily seen that L is graded by the $L^{(i)}$ and it is clear that the epimorphism ε is a homomorphism of graded modules. It is easily seen also that if N is a graded module and $\eta: M \twoheadrightarrow N$ is a homomorphism (of graded modules), then there exists a homomorphism ζ such that

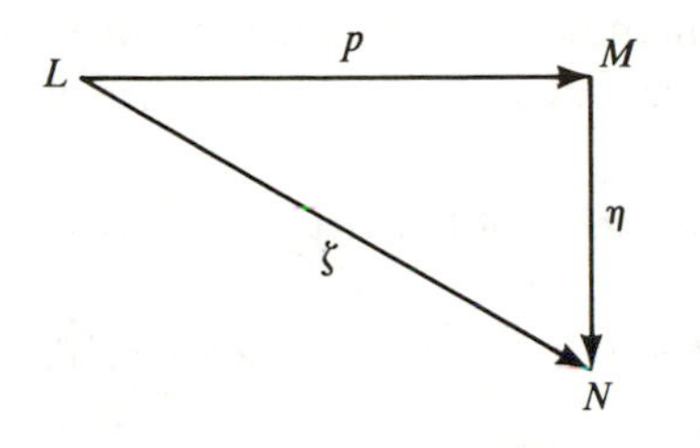

is commutative.

We now suppose that M is a graded module for $R = F[x_1, \ldots, x_m]$. Let $L = L_0$ be a free graded module for which we have an epimorphism ε of L_0 onto M and let $K_0 = \ker \varepsilon$. Then K_0 is graded and hence we can determine a free graded module L_1 with an epimorphism ε_1 of L_1 onto K_0. Combining with the injection of K_0 into L_0 we obtain $d_1: L_1 \rightarrow L_0$. Again, let $K_1 = \ker d_1$ and let L_2, ε_2 be a free graded module and epimorphism of L_2 onto K_1, d_2 the composite of this with the injection of K_1 into L_1. We continue this process. Then Hilbert's syzygy theorem states that in at most m steps we obtain a kernel K_i that is itself free, so that one does not have to continue. In a slightly different form we can state

HILBERT'S SYZYGY THEOREM. *Let M be a graded module for $R = F[x_1, \ldots, x_m]$, the ring of polynomials in m indeterminates with coefficients in a field F. Let*

$$0 \leftarrow M \overset{\varepsilon}{\leftarrow} L_0 \overset{d_1}{\leftarrow} L_1 \leftarrow \cdots \leftarrow L_{m-1} \overset{d_m}{\leftarrow} K_m$$

be an exact sequence of homomorphisms of graded modules such that every L_i is free. Then K_m is free.

The proof will be based on the result that $\text{h.dim}\, F \leqslant m$ and two further results that we shall now derive. As before, F is regarded as an R-module in which $JF = 0$, where J is the ideal of polynomials vanishing at $(0, \ldots, 0)$. Since $J = \sum_{i>0} R^{(i)}$, F becomes a graded R-module if we put $F^{(0)} = F$, $F^{(i)} = 0$ for $i \geqslant 1$. We now prove

LEMMA 5. *If M is a graded R-module and $M \otimes_R F = 0$, then $M = 0$.*

Proof. We have the exact sequence $0 \to J \to R \to F \to 0$. This gives the exact sequence $M \otimes_R J \to M \otimes_R R \to M \otimes_R F \to 0$. Since $M \otimes_R R \simeq M$ and $M \otimes_R F = 0$, by hypothesis, we have the exactness of $M \otimes_R J \to M \to 0$. The map here sends an element $\sum u_i \otimes f_i$ into $\sum f_i u_i$, so the exactness means that every element of M can be written in the form $\sum f_i u_i$ where the $u_i \in M$ and the $f_i \in J$. It follows that $M = 0$. For, otherwise, let u be a non-zero of M of minimum degree (=highest degree of the homogeneous parts). Then $u = \sum f_i u_i$, $f_i \in J$, $u_i \in M$, f_i and u_i homogeneous. Since the degrees of the f_i are >0 we have a $u_i \neq 0$ with $\deg u_i < \deg u$, contrary to the choice of u. Hence $M = 0$. □

The key result for the proof of Hilbert's theorem is

THEOREM 6.21. *If M is a graded module such that* $\mathrm{Tor}_1{}^R(M, F) = 0$, *then M is free.*

Proof. We can regard $M \otimes_R F$ as vector space over F by restricting the action of R to F. Then if $\alpha, \beta \in F$ and $u \in M$, $\alpha(u \otimes \beta) = \alpha u \otimes \beta = u \otimes \alpha\beta$. It follows that the elements of the form $u \otimes 1$, $u \in M$, u homogeneous, span $M \otimes_R F$ as vector space over F and so we can choose a base for $M \otimes_R F$ of the form $\{u_i \otimes 1 | u_i \in M\}$. Let L be a free R-module with base $\{b_i\}$. We shall prove that M is free with base $\{u_i\}$ by showing that the homomorphism η of L into M such that $\eta b_i = u_i$ is an isomorphism. Let C be the cokernel of η. Then we have the exact sequence $L \to M \xrightarrow{p} C \to 0$ and hence we have the exact sequence $L \otimes_R F \xrightarrow{\eta \otimes 1} M \otimes_R F \xrightarrow{p \otimes 1} C \otimes_R F \to 0$. The map $\eta \otimes 1$ sends $b_i \otimes 1$ into $u_i \otimes 1$ and since $\{u_i \otimes 1\}$ is a base for $M \otimes_R F$, $\eta \otimes 1$ is an isomorphism. Hence $C \otimes_R F = 0$ and so, by Lemma 5, $C = 0$. This means that η is surjective and $L \xrightarrow{\eta} M \to 0$ is exact. Let K now denote $\ker \eta$. Then we have the exact sequence $0 \to K \to L \to M \to 0$, which gives the exact sequence $\mathrm{Tor}_1{}^R(M, F) \to K \otimes_R F \to L \otimes_R F \xrightarrow{\eta \otimes 1} M \otimes_R F$. By hypothesis, $\mathrm{Tor}_1{}^R(M, F) = 0$, and we have seen that $\eta \otimes 1$ is an isomorphism. Hence $K \otimes_R F = 0$ and so, again by Lemma 5, $K = 0$. Then, $\ker \eta = 0$ and η is an isomorphism. □

Since any projective module satisfies the hypothesis $\mathrm{Tor}_1{}^R(M, F) = 0$ (Theorem 6.10, p. 354), we have the following consequence of Theorem 6.21.

COROLLARY. *Any projective R-module for $R = F[x_1, \ldots, x_m]$ that is graded free.*

We can now give the

Proof of Hilbert's syzygy theorem. The argument used in the proof of the implication (2) $\Rightarrow$ (3) in Theorem 6.18, p. 376, shows that $\mathrm{Tor}_1{}^R(K_m, F) \cong \mathrm{Tor}_{m+1}^R(M, F)$. Since h.dim $F \leqslant m$, $\mathrm{Tor}_{m+1}^R(M, F) = 0$ by one of the remarks following Theorem 6.18. Hence $\mathrm{Tor}_1{}^R(K_m, F) = 0$ and so K_m is free by Theorem 6.21. □

EXERCISES

1. Let K be a commutative ring of finite global dimension m and let $K[x]$ be the polynomial ring in an indeterminate x over K. Let M be a $K[x]$-module and $\tilde{M} = K[x] \otimes_K M$. Note that any element of $\tilde{M}$ can be written in one and only one way in the form $1 \otimes u_0 + x \otimes u_1 + x^2 \otimes u_2 + \cdots$, $u_i \in M$. Note that there is a $K[x]$-homomorphism η of $\tilde{M}$ onto M such that $f(x) \otimes u \rightsquigarrow f(x)u$ for $f(x) \in K[x]$, $u \in M$. Show that $N = \ker \eta$ is the set of elements $(x \otimes u_0 - 1 \otimes xu_0) + (x^2 \otimes u_1 - 1 \otimes x^2 u_1) + \cdots + (x^n \otimes u_{n-1} - 1 \otimes x^n u_{n-1}) + \cdots$ and that the map
$$1 \otimes u_0 + x \otimes u_1 + x^2 \otimes u_2 + \cdots \rightsquigarrow (x \otimes u_0 - 1 \otimes xu_0) + (x^2 \otimes u_1 - 1 \otimes x^2 u_1) + \cdots$$
is a $K[x]$-isomorphism, so we have an exact sequence of $K[x]$-homomorphisms $0 \to \tilde{M} \to \tilde{M} \to M \to 0$. By exercise 1, p. 278, $\mathrm{h.dim}_{K[x]} \tilde{M} = \mathrm{h.dim}_K M \leqslant m$. Hence, by Theorem 6.19, $\mathrm{h.dim}_{K[x]} M \leqslant m + 1$ and the global dimension of $K[x] \leqslant m + 1$.

2. Prove that if F is a field and x_i are indeterminates, then the global dimension of $F[x_1, \ldots, x_m]$ is m.

REFERENCES

D. Hilbert, Über die Theorie der Algebraischen Formen, *Mathematische Annalen*, vol. 36 (1890), pp. 473–534.

H. Cartan and S. Eilenberg, *Homological Algebra*, Princeton University Press, Princeton, N.J., 1956.

S. MacLane, *Homology*, Springer, New York, 1963.

P. J. Hilton and U. Stammbach, *A Course in Homological Algebra*, Springer, New York, 1970.

7

Commutative Ideal Theory: General Theory and Noetherian Rings

The ideal theory of commutative rings was initiated in Dedekind's successful "restoration" of unique factorization in the rings of algebraic integers of number fields by the introduction of ideals. Some of these, e.g., $\mathbb{Z}[\sqrt{-5}]$, are not factorial, that is, do not have unique factorization of elements into irreducible elements (see BAI, pp. 141–142). However, Dedekind showed that in these rings unique factorization does hold for ideals as products of prime ideals (definition on p. 389). A second type of ideal theory, which was introduced at the beginning of this century by E. Lasker and F. S. Macaulay, is concerned with the study of ideals in rings of polynomials in several indeterminates. This has obvious relevance for algebraic geometry. A principal result in the Lasker-Macaulay theory is a decomposition theorem with comparatively weak uniqueness properties of ideals in polynomial rings as intersections of so-called primary ideals (definition on p. 434). In 1921 Emmy Noether gave an extremely simple derivation of these results for arbitrary commutative rings satisfying the ascending chain condition for ideals. This paper, which by its effective use of conceptual methods, gave a new direction to algebra, has been one of the most influential papers on algebra published during this century.

In this chapter we shall consider the ideal theory—once called additive ideal theory—which is an outgrowth of the Lasker-Macaulay-Noether theory. In recent years the emphasis has shifted somewhat away from the use of primary decompositions to other methods, notably, localization, the use of the prime spectrum of a ring, and the study of local rings. The main motivation has continued to come from applications to algebraic geometry. However, other developments, such as the study of algebras over commutative rings, have had their influence, and of course, the subject has moved along under its own power.

We shall consider the Dedekind ideal theory in Chapter 10 after we have developed the structure theory of fields and valuation theory, which properly precede the Dedekind theory.

Throughout this chapter (and the subsequent ones) all rings are commutative unless the contrary is explicitly stated. From time to time, mainly in the exercises, applications and extension to non-commutative rings will be indicated. The first nine sections are concerned with arbitrary commutative rings. The main topics considered here are localization, the method of reducing questions on arbitrary rings to local rings via localization with respect to the complements of prime ideals, the prime spectrum of a ring, rank of projective modules, and the projective class group. The ideal theory of noetherian rings and modules is developed in sections 7.10–7.18. Included here are the important examples of noetherian rings: polynomial rings and power series rings over noetherian rings. We give also an introduction to affine algebraic geometry including the Hilbert Nullstellensatz. Primary decompositions are treated in section 7.13. After these we consider some of the basic properties of noetherian rings, notably the Krull intersection theorem, the Hilbert function of a graded module, dimension theory, and the Krull principal ideal theorem. We conclude the chapter with a section on I-adic topologies and completions.

7.1 PRIME IDEALS. NIL RADICAL

We recall that an element p of a domain D is called a prime if p is not a unit and if $p|ab$ in D implies $p|a$ or $p|b$ in D. This suggests the following

DEFINITION 7.1. *An ideal P in a (commutative) ring R is called* prime *if $P \neq R$ and if $ab \in P$ for $a, b \in R$ implies either $a \in P$ or $b \in P$.*

In other words, an ideal P is prime if and only if the complementary set $P' = R - P$ is closed under multiplication and contains 1, that is, P' is a submonoid of the multiplicative monoid of R. In congruence notation the second condition in Definition 7.1 is that if $ab \equiv 0(\mathrm{mod}\, P)$, then $a \equiv 0(\mathrm{mod}\, P)$ or $b \equiv 0(\mathrm{mod}\, P)$. The first condition is that the ring $\bar{R} = R/P \neq 0$. Hence it is clear that an ideal P is prime if and only if R/P is a domain. Since an ideal M in a commutative ring R is maximal if and only if R/M is a field and since any field is a domain, it is clear that any maximal ideal of R is prime. It is clear also that an element p is prime in the usual sense if and only if the principal ideal $(p)(= pR)$ is a prime ideal. Another thing worth noting is that if P is a prime ideal in R and A and B are ideals in R such that $AB \subset P$, then either $A \subset P$ $B \subset P$. If not, then we have $a \in A, \notin P$ and $b \in B, \notin P$. Then $ab \in AB \subset P$, contrary to the primeness of P. It is clear by induction that if P is a prime ideal and $a_1 a_2 \cdots a_n \in P$, then some $a_i \in P$ and if $A_1 A_2 \cdots A_n \subset P$ for ideals A_i, then some $A_i \subset P$.

We recall the elementary result in group theory that a group cannot be a union of two proper subgroups (exercise 14, p. 36 of BAI). This can be strengthened to the following statement: If G_1, G_2, and H are subgroups of a group G and $H \subset (G_1 \cup G_2)$, then either $H \subset G_1$ or $H \subset G_2$. The following result is a useful extension of this to prime ideals in a ring.

PROPOSITION 7.1. *Let $A, I_1, \ldots, I_n$ be ideals in a ring such that at most two of the I_j are not prime. Suppose $A \subset \bigcup_1^n I_j$. Then $A \subset I_j$ for some j.*

Proof. We use induction on n. The result is clear if $n = 1$. Hence we assume $n > 1$. Then if we have $A \subset I_1 \cup \cdots \cup \hat{I}_k \cup \cdots \cup I_n$ for some k, the result will follow by the induction hypothesis. We therefore assume $A \not\subset I_1 \cup \cdots \cup \hat{I}_k \cup \cdots \cup I_n$ for $k = 1, 2, \ldots, n$ and we shall complete the proof by showing that this leads to a contradiction. Since $A \not\subset I_1 \cup \cdots \cup \hat{I}_k \cup \cdots \cup I_n$, there exists an $a_k \in A, \notin I_1 \cup \cdots \cup \hat{I}_k \cup \cdots \cup I_n$, $k = 1, 2, \ldots, n$. Since $A \subset \bigcup I_j$, $a_k \in I_k$. If $n = 2$, it is readily seen that (as in the group theory argument) $a_1 + a_2 \in A$ but $a_1 + a_2 \notin I_1 \cup I_2$, contrary to hypothesis. If $n > 2$, then at least one of the I_j is prime. We may assume it is I_1. Then it is readily seen that

$$a_1 + a_2 a_3 \cdots a_n$$

is in A but is not in $\bigcup I_j$. Again we have a contradiction, which proves the result. □

The foregoing result is usually stated with the stronger hypothesis that every I_j is prime. The stronger form that we have proved is due to N. McCoy, who strengthened the result still further by replacing the hypothesis that A is an ideal by the condition that A is a subrng (BAI, p. 155) of the ring R. It is clear that the foregoing proof is valid in this case also. In the sequel we shall refer to Proposition 7.1 as the "prime avoidance lemma." The terminology is justified since the contrapositive form of the proposition is that if A and $I_1, \ldots, I_n$ are ideals such that $A \not\subset I_j$ for any j and at most two of the I_j are not prime, then there exists an $a \in A$ such that $a \notin \bigcup I_j$.

There is an important way of obtaining prime ideals from submonoids of the multiplicative monoid of R. This is based on

PROPOSITION 7.2. *Let S be a submonoid of the multiplicative monoid of R and let P be an ideal in R such that (1) $P \cap S = \varnothing$. (2) P is maximal with respect to property (1) in the sense that if P' is any ideal such that $P' \supsetneqq P$, then $P' \cap S \neq \varnothing$. Then P is prime.*

Proof. Let a and b be elements of R such that $a \notin P$ and $b \notin P$. Then the ideals $(a)+P$ and $(b)+P$ properly contain P and so meet S. Hence we have elements $p_1, p_2 \in P$, $x_1, x_2 \in R$, $s_1, s_2 \in S$ such that $s_1 = x_1 a + p_1$, $s_2 = x_2 b + p_2$. Then $s_1 s_2 \in S$ and

$$s_1 s_2 = x_1 x_2 ab + x_1 a p_2 + x_2 b p_1 + p_1 p_2.$$

Hence if $ab \in P$, then $s_1 s_2 \in P$, contrary to $P \cap S = \varnothing$. Thus $ab \notin P$ and we have shown that $a \notin P$, $b \notin P$ implies $ab \notin P$, so P is prime. $\square$

If S is a submonoid not containing 0, then the ideal 0 satisfies $0 \cap S = \varnothing$. Now let A be any ideal such that $A \cap S = \varnothing$ and let Λ be the set of ideals B of R such that $B \supset A$ and $B \cap S = \varnothing$. It is an immediate consequence of Zorn's lemma that Λ contains maximal elements. Such an element is an ideal $P \supset A$ satisfying the hypotheses of Proposition 7.2. Hence the following result follows from this proposition.

PROPOSITION 7.3. *Let S be a submonoid of the multiplicative monoid of R and A an ideal in R such that $S \cap A = \varnothing$. Then A can be imbedded in a prime ideal P such that $S \cap P = \varnothing$.*

Let N denote the set of nilpotent elements of R. Evidently if $z \in N$ so $z^n = 0$ for some n and if a is any element of R, then $(az)^n = a^n z^n = 0$. Hence $az \in N$. If $z_i \in N, i = 1, 2$, and n_i is an integer such that $z_i^{n_i} = 0$, then by the binomial theorem for $n = n_1 + n_2 - 1$ we have $(z_1 + z_2)^n = \sum_0^n \binom{n}{i} z_1{}^i z_2^{n-i}$. This is 0, since if $i < n_1$ then $n - i \geqslant n_2$, so $z_1{}^i z_2^{n-i} = 0$ and if $i \geqslant n_1$, then clearly $z_1{}^i z_2^{n-i} = 0$. Thus N is an ideal. We call this ideal the *nil radical* of the ring R and we shall denote it as nilrad R. If $\bar{z} = z + N$ is in the nil radical of $\bar{R} = R/N$, $N = \text{nilrad } R$, then $z^n \in N$ for some n and so $z^{mn} = 0$ for some integer mn. Then

$z \in N$ and $\bar{z} = 0$. Thus nilrad $\bar{R} = 0$. We recall that any nil ideal of a ring is contained in the (Jacobson) radical rad R (p. 192). We recall also that rad R is the intersection of the maximal left ideals of R (p. 193), so for a commutative ring rad R is the intersection of the maximal ideals of the ring. The analogous result for the nil radical is the following

THEOREM 7.1 (Krull). *The nil radical of R is the intersection of the prime ideals of R.*

Proof. Let $N = \text{nilrad}\, R$ and let $N' = \bigcap P$ where the intersection is taken over all of the prime ideals P of R. If $z \in N$, we have $z^n = 0$ for some integer n. Then $z^n \in P$ for any prime ideal P and hence $z \in P$. Hence $N \subset P$ for every prime ideal P, so $N \subset N'$. Now let $s \notin N$, so s is not nilpotent and $S = \{s^n | n = 0, 1, 2, \ldots\}$ is a submonoid of the multiplicative monoid of R satisfying $S \cap \{0\} = \varnothing$. Then by Proposition 7.3 applied to $A = 0$ there exists a prime ideal P such that $P \cap S = \varnothing$. Then $s \notin P$ and $s \notin N'$. This implies that $N' \subset N$ and so $N = N' = \bigcap P$. $\square$

If A is an ideal in R, we define the *nil radical of* A, nilrad A (sometimes denoted as $\sqrt{A}$), to be the set of elements of R that are nilpotent modulo A in the sense that there exists an integer n such that $z^n \in A$. This is just the set of elements z such that $\bar{z} = z + A$ is in the nil radical of $\bar{R} = R/A$. Thus

$$\text{nilrad}\, A = v^{-1}\ (\text{nilrad}\, \bar{R}) \tag{1}$$

where v is the canonical homomorphism of R onto $\bar{R}$. It follows from this (or it can be seen directly) that nilrad A is an ideal of R containing A. It is clear also that iteration of the process of forming the nil radical of an ideal gives nothing new: nilrad (nilrad A) = nilrad A.

We have the bijective map $B \rightsquigarrow \bar{B} = B/A$ of the set of ideals of R containing A onto the set of ideals of $\bar{R} = R/A$. Moreover, $R/B \cong \bar{R}/\bar{B}$ (BAI, p. 107). Hence B is prime in R if and only if $\bar{B}$ is prime in $\bar{R}$. Thus the set of prime ideals of $\bar{R}$ is the set of ideals $\bar{P} = P/A$ where P is a prime ideal of R containing A. Since $\bigcap(P/A) = (\bigcap P)/A$ (BAI, p. 67, exercise 2), and $\bigcap \bar{P}$ taken over the prime ideals of $\bar{R}$ is the nil radical of $\bar{R}$, we have

$$\begin{aligned}(\textstyle\bigcap P)/A = \bigcap(P/A) = \bigcap \bar{P} &= \text{nilrad}\, \bar{R} \\ &= \text{nilrad}\, R/A.\end{aligned} \tag{2}$$

Hence nilrad $A = \bigcap P$ taken over the prime ideals P of R containing A. We state this as

THEOREM 7.2. *The nil radical of an ideal A is the intersection of the prime ideals of R containing A.*

EXERCISES

1. Let A_i, $1 \leqslant i \leqslant n$, be ideals, P a prime ideal. Show that if $P \supset \bigcap_1^n A_i$, then $P \supset A_i$ for some i and if $P = \bigcap_1^n A_i$, then $P = A_i$ for some i.

2. Show that if P is a prime ideal, then $S = R - P$ is a submonoid of the multiplicative monoid of R, which is *saturated* in the sense that it contains the divisors of every $s \in S$. Show more generally that if $\{P_\alpha\}$ is a set of prime ideals, then $S = R - \bigcup P_\alpha$ is a saturated submonoid of the multiplicative monoid of R. Show that conversely any saturated submonoid of the multiplicative monoid of R has the form $R - \bigcup P_\alpha$, $\{P_\alpha\}$ a set of prime ideals of R.

3. Show that the set of zero divisors of R is a union of prime ideals.

4. (McCoy.) Show that the units of the polynomial ring $R[x]$, x an indeterminate, are the polynomials $a_0 + a_1 x + \cdots + a_n x^n$ where a_0 is a unit in R and every a_i, $i > 0$, is nilpotent. (*Hint*: Consider the case in which R is a domain first. Deduce the general case from this by using Krull's theorem.)

The next two exercises are designed to prove an important result on the radical of a polynomial ring due to S. Amitsur. In these exercises R need not be commutative.

5. Let $f(x) \in R[x]$ have 0 constant term and suppose $f(x)$ is quasi-regular with quasi-inverse $g(x)$ (p. 194). Show that the coefficients of $g(x)$ are contained in the subring generated by the coefficients of $f(x)$.

6. (Amitsur.) Show that $R[x]$ is semi-primitive if R has no nil ideals $\neq 0$. (*Hint*: Assume that $\operatorname{rad} R[x] \neq 0$ and choose an element $\neq 0$ in this ideal with the minimum number of non-zero coefficients. Show that these coefficients are contained in a commutative subring B of R. Apply exercises 4 and 5.)

7.2 LOCALIZATION OF RINGS

The tool of localization that we shall now introduce is one of the most effective ones in commutative algebra. It amounts to a generalization of the familiar construction of the field of fractions of a domain (BAI, pp. 115–119). We can view this generalization from the point of view of a universal construction that is a solution of the following problem:

Given a (commutative) ring R and a subset S of R, to construct a ring R_S and a homomorphism λ_S of R into R_S such that every $\lambda_S(s)$, $s \in S$, is invertible in R_S, and the pair (R_S, λ_S) is universal for such pairs in the sense that if η is any homomorphism of R into a ring R' such that every $\eta(s)$ is invertible, then there exists a unique homomorphism $\tilde{\eta}\colon R_S \to R'$ such that the diagram

(3)

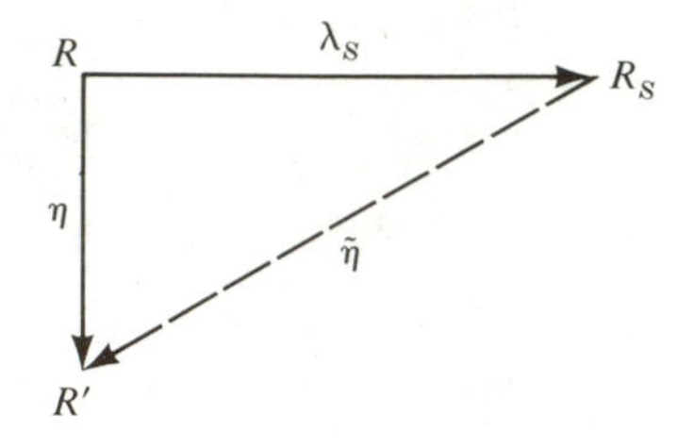

is commutative.

Before proceeding to the solution of the problem (which is easy), we shall make some remarks about the problem.

1. Since the product of two elements of a ring is invertible if and only if the elements are invertible, there is no loss in generality in assuming that S is a submonoid of the multiplicative monoid of R. The solution of the problem for an arbitrary set S can be reduced to the case in which S is a monoid by replacing the set S by the submonoid $\langle S \rangle$ generated by S. For example, if S is a singleton $\{s\}$, then we can replace it by $\langle s \rangle = \{s^n | n = 0, 1, 2, \ldots\}$.

2. The special case of the field of fractions is that in which R is a domain and $S = R^*$, the submonoid of non-zero elements of R. In this case nothing is changed if we restrict the rings R' in the statement of the problem to be fields. This is clear since the image under a homomorphism of a field either is the trivial ring consisting of one element or is a field.

3. Since a zero divisor of a ring ($\neq \{0\}$) is not invertible, we cannot expect λ_S to be injective if S contains zero divisors.

4. If the elements of S are invertible in R, then there is nothing to do: We can simply take $R_S = R$ and $\lambda_S = 1_R$, the identity map on R. Then it is clear that $\tilde{\eta} = \eta$ satisfies the condition in the problem.

5. If a solution exists, it is unique in the strong sense that if $(R_S^{(1)}, \lambda_S^{(1)})$ and $(R_S^{(2)}, \lambda_S^{(2)})$ satisfy the condition for the ring R and subset S, then there exists a unique isomorphism $\zeta\colon R_S^{(1)} \to R_S^{(2)}$ such that

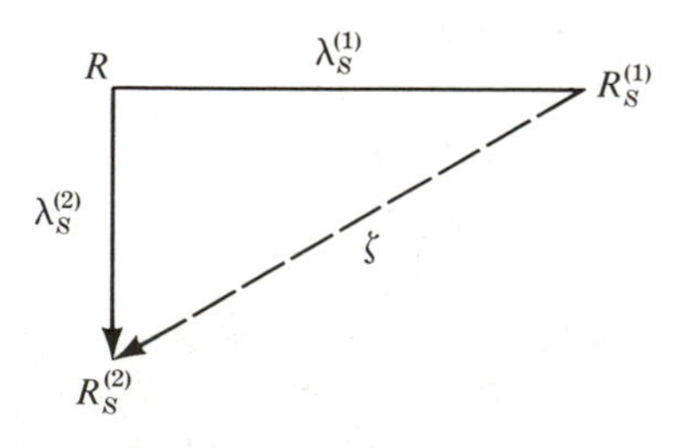

is commutative (see p. 44).

We now proceed to a construction of a pair (R_S, λ_S) for any ring R and submonoid S of its multiplicative monoid. As in the special case of a domain R and its monoid $S = R^*$ of non-zero elements, we commence with the product set $R \times S$ of pairs (a, s), $a \in R$, $s \in S$. Since the monoid S may contain zero divisors, it is necessary to modify somewhat the definition of equivalence among the pairs (a, s) that we used in the construction of the field of fractions of a domain. We define a binary relation $\sim$ on $R \times S$ by declaring that $(a_1, s_1) \sim (a_2, s_2)$ if there exists an $s \in S$ such that

$$s(s_2 a_1 - s_1 a_2) = 0 \tag{4}$$

(or $ss_2 a_1 = ss_1 a_2$). If R is a domain and S does not contain 0, then (4) can be replaced by the simpler condition $s_2 a_1 = s_1 a_2$. In the general case it is readily verified that $\sim$ is an equivalence relation. We denote the quotient set defined by this equivalence relation as R_S and we denote the equivalence class of (a, s) by a/s.

We define addition and multiplication in the set R_S by

$$a_1/s_1 + a_2/s_2 = (s_2 a_1 + s_1 a_2)/s_1 s_2 \tag{5}$$

and

$$(a_1/s_1)(a_2/s_2) = a_1 a_2/s_1 s_2. \tag{6}$$

It is a bit tedious but straightforward to check that these compositions are well-defined and that if we put $0 = 0/1$ $(=0/s$ for any $s \in S)$ and $1 = 1/1$ $(=s/s$, $s \in S)$, then $(R_S, +, \cdot, 0, 1)$ is a ring. We leave the verifications to the reader.

We now define a map λ_S (or $\lambda_S{}^R$) of R into R_S by

$$\lambda_S : a \rightsquigarrow a/1. \tag{7}$$

It is clear from (5) and (6) and the definition of 1 in R_S that λ_S is a homomorphism of R into R_S. Moreover, if $s \in S$, then $\lambda_S(s) = s/1$ has the inverse $1/s$ since $(s/1)(1/s) = s/s = 1$. Now let η be a homomorphism of R into R' such that $\eta(s)$ is invertible for every $s \in S$. It is readily verified that

$$\tilde{\eta} : a/s \rightsquigarrow \eta(a)\eta(s)^{-1} \tag{8}$$

is well-defined and this is a homomorphism of R_S into R'. Moreover, $\tilde{\eta}\lambda_S(a) = \tilde{\eta}(a/1) = \eta(a)$. Thus we have the commutativity of (3). The uniqueness of $\tilde{\eta}$ is clear also since $a/s = (a/1)(1/s) = (a/1)(s/1)^{-1}$ and hence any $\tilde{\eta}$ satisfying the commutativity condition satisfies $\tilde{\eta}(a/s) = \tilde{\eta}(\lambda_S(a)\lambda_S(s)^{-1}) = \eta(a)\eta(s)^{-1}$. Thus $\tilde{\eta}$ is the map we defined in (8) and (R_S, λ_S) is a solution of the problem we formulated at the outset. We shall call (R_S, λ_S) (or simply R_S) the *localization of R at S*.

We shall now study relations between R_S and $R_{S'}$ for S', a submonoid of the monoid S. Note first that $\lambda_S(s')$ is invertible in R_S, $s' \in S'$. Hence we have a unique homomorphism $\zeta_{S',S}: R_{S'} \to R_S$ such that $\lambda_S = \zeta_{S',S}\lambda_{S'}$. Now suppose that $\lambda_{S'}(s)$ is invertible in $R_{S'}$, $s \in S$. Then we also have a unique homomorphism $\zeta_{S,S'}: R_S \to R_{S'}$ such that $\lambda_{S'} = \zeta_{S,S'}\lambda_S$. It follows that $\zeta_{S,S'}\zeta_{S',S} = 1_{R_{S'}}$ and $\zeta_{S',S}\zeta_{S,S'} = 1_{R_S}$. Hence $\zeta_{S',S}$ and $\zeta_{S,S'}$ are isomorphisms. We shall now show that in general if $S' \subset S$, then R_S is, in fact, a localization of $R_{S'}$. We state the result in a somewhat imprecise manner as follows.

PROPOSITION 7.4. *Let S' be a submonoid of S and let $S/S' = \{s/s' | s \in S, s' \in S'\}$. Then S/S' is a submonoid of the multiplicative monoid of $R_{S'}$ and we have canonical isomorphisms of R_S and $(R_{S'})_{\lambda_{S'}(S)}$ and of $(R_{S'})_{\lambda_{S'}(S)}$ and $(R_{S'})_{S/S'}$.*

Proof. We shall obtain the first isomorphism by showing that $(R_{S'})_{\lambda_{S'}(S)}$ has the universal property of R_S. First, we have the composite homomorphism $\lambda_{\lambda_{S'}(S)}\lambda_{S'}$ of R into $(R_{S'})_{\lambda_{S'}(S)}$ obtained from the sequence of homomorphism

$$R \xrightarrow{\lambda_{S'}} R_{S'} \xrightarrow{\lambda_{\lambda_{S'}(S)}} (R_{S'})_{\lambda_{S'}(S)}.$$

Now let η be a homomorphism of R into R' such that $\eta(s)$ is invertible, $s \in S$. Since $S' \subset S$, we have a unique homomorphism $\tilde{\eta}'$ of $R_{S'}$ into R' such that $\tilde{\eta}'\lambda_{S'} = \eta$. If $s \in S$, then $\tilde{\eta}'(\lambda_{S'}(s)) = \eta(s)$ is invertible in R'. Accordingly, by the universal property of $(R_{S'})_{\lambda_{S'}(S)}$ we have a unique homomorphism $\tilde{\eta}$ of $(R_{S'})_{\lambda_{S'}(S)}$ into R' such that $\tilde{\eta}\lambda_{\lambda_{S'}(S)} = \tilde{\eta}'$. Then $\tilde{\eta}\lambda_{\lambda_{S'}(S)}\lambda_{S'} = \tilde{\eta}'\lambda_{S'} = \eta$. Moreover, $\tilde{\eta}$ is the only homomorphism of $(R_{S'})_{\lambda_{S'}(S)}$ into R' satisfying $\tilde{\eta}(\lambda_{\lambda_{S'}(S)}\lambda_{S'}) = \eta$. For, this condition implies that $(\tilde{\eta}\lambda_{\lambda_{S'}(S)})\lambda_{S'} = \tilde{\eta}'\lambda_{S'}$, which implies that $\tilde{\eta}\lambda_{\lambda_{S'}(S)} = \tilde{\eta}'$ by the universality of $(R_{S'}, \lambda_{S'})$. Then we obtain the uniqueness of $\tilde{\eta}$ by the universality of $((R_{S'})_{\lambda_{S'}(S)}, \lambda_{\lambda_{S'}(S)})$. Thus $((R_{S'})_{\lambda_{S'}(S)}, \lambda_{\lambda_{S'}(S)}\lambda_{S'})$ has the universal property of (R_S, λ_S) and so we have the required isomorphism.

The isomorphism of $(R_{S'})_{\lambda_{S'}(S)}$ and $(R_{S'})_{S/S'}$ can be seen by observing that $\lambda_{S'}(S)$ is a submonoid of S/S'. Hence we have the canonical homomorphism of $(R_{S'})_{\lambda_{S'}(S)}$ into $(R_{S'})_{S/S'}$. If $s \in S$, $s' \in S'$, then $s/s' = (s/1)(s'/1)^{-1}$ in $R_{S'}$ and $\lambda_{\lambda_{S'}(S)}(s/1)$ and $\lambda_{\lambda_{S'}(S)}(s'/1)$ are invertible in $(R_{S'})_{\lambda_{S'}(S)}$. Hence $\lambda_{\lambda_{S'}(S)}(s/s')$ is invertible. It follows from the result we obtained above that the canonical homomorphism of $(R_{S'})_{\lambda_{S'}(S)}$ into $(R_{S'})_{S/S'}$ is an isomorphism. □

EXERCISES

1. Let S and T be submonoids of the multiplicative monoid of R. Note that $ST = \{st | s \in S, t \in T\}$ is the submonoid generated by S and T. Show that $R_{ST} \cong (R_S)_{\lambda_S(T)}$.

2. Let $a, b \in R$. Show that $(R_{\langle a \rangle})_{\langle b/1 \rangle} \cong R_{\langle ab \rangle}$.
3. Let S be a submonoid of the multiplicative monoid of R. If $a, b \in S$, define $\langle a \rangle \leqslant \langle b \rangle$ if $a|b$ in S. In this case there is a unique homomorphism $\zeta_{\langle a \rangle, \langle b \rangle}$ of $R_{\langle a \rangle} \to R_{\langle b \rangle}$ such that $\lambda_{\langle b \rangle} = \zeta_{\langle a \rangle, \langle b \rangle} \lambda_{\langle a \rangle}$. Show that $R_S = \varinjlim R_{\langle a \rangle}$ (with respect to the ζ's).

7.3 LOCALIZATION OF MODULES

It is important to extend the concept of localization to R-modules. Let M be a module over R, S a submonoid of the multiplicative monoid of R. We shall construct an R_S-module M_S in a manner similar to the construction of R_S. We consider $M \times S$ the product set of pairs (x, s), $x \in M$, $s \in S$, and we introduce a relation $\sim$ in this set by $(x_1, s_1) \sim (x_2, s_2)$ if there exists an $s \in S$ such that

$$(9) \qquad s(s_2 x_1 - s_1 x_2) = 0.$$

The same calculations as in the ring case show that $\sim$ is an equivalence. Let M_S denote the quotient set and let x/s be the equivalence class of (x, s). We can make M_S into an R_S-module by defining addition by

$$(10) \qquad x_1/s_1 + x_2/s_2 = (s_2 x_1 + s_1 x_2)/s_1 s_2$$

and the action of R_S on M_S by

$$(11) \qquad (a/s)(x/t) = ax/st.$$

We can verify as in the ring case that (10) and (11) are well-defined, that $+$ and $0 = 0/s$ constitute an abelian group, and that (11) defines a module action of R_S on M_S. We shall call the R_S-module M_S the *localization of M at S* or the *S-localization of M*.

Although we are generally interested in M_S as R_S-module, we can also regard M_S as R-module by defining the action of R by $a(x/s) = (a/1)(x/s) = ax/s$. Since $a \rightsquigarrow a/1$ is a ring homomorphism, it is clear that this is a module action. We have a map λ_S (or $\lambda_S{}^M$ if it is necessary to indicate M) of M into M_S defined by $x \rightsquigarrow x/1$. This is an R-module homomorphism. The kernel of λ_S is the set of $x \in M$ for which there exists an $s \in S$ such that $sx = 0$, that is, if $\operatorname{ann} x$ denotes the annihilator ideal in R of x, then $\operatorname{ann} x \cap S \neq \varnothing$. It is clear that λ_S need not be injective. For example, if M is a $\mathbb{Z}$-module and $S = \mathbb{Z} - \{0\}$, then the torsion submodule of M is mapped into 0 by λ_S. It is clear that if S includes 0, then $M_S = 0$ and $\ker \lambda_S = M$. Another useful remark is

PROPOSITION 7.5. *If M is a finitely generated module, then $M_S = 0$ if and only if there exists an $s \in S$ such that $sM = 0$.*

Proof. If $sM = 0$, every $x/t = 0 = 0/1$ since $s(1x) = s(t0) = 0$. Conversely, suppose $M_S = 0$ and let $\{x_1, \ldots, x_n\}$ be a set of generators for M. Then $x_i/1 = 0$ implies there exists an $s_i \in S$ such that $s_i x_i = 0$. Then $sx_i = 0$ for $s = \prod_1^n s_i$; hence $sx = 0$ for any $x = \sum r_i x_i, r_i \in R$. Thus $sM = 0$. □

Let $f: M \to N$ be a homomorphism of R-modules of M into N. Then we have a corresponding R_S-homomorphism f_S of M_S into N_S defined by

$$f_S(x/t) = f(x)/t. \tag{12}$$

Again we leave the verifications to the reader. The maps $M \rightsquigarrow M_S, f \rightsquigarrow f_S$ define a functor, the *S-localization functor*, from the category of R-modules to the category of R_S-modules. We shall now show that this functor is naturally isomorphic to the functor $R_S \otimes_R$ from R-**mod** to R_S-**mod**. Thus we have

PROPOSITION 7.6. *For every R-module M we can define an R_S-isomorphism η_M of M_S onto $R_S \otimes_R M$ that is natural in M.*

Proof. We show first that there is a map η_M of M_S into $R_S \otimes_R M$ such that

$$x/s \rightsquigarrow (1/s) \otimes x. \tag{13}$$

Suppose $x_1/s_1 = x_2/s_2$, which means that we have an $s \in S$ such that $ss_2 x_1 = ss_1 x_2$. Then

$$\begin{aligned}(1/s_1)\otimes x_1 &= (ss_2/ss_2s_1)\otimes x_1 = ss_2(1/ss_2s_1)\otimes x_1\\ &= (1/ss_2s_1)\otimes ss_2x_1 = (1/ss_2s_1)\otimes ss_1x_2\\ &= ss_1(1/ss_2s_1)\otimes x_2 = (ss_1/ss_2s_1)\otimes x_2\\ &= (1/s_2)\otimes x_2.\end{aligned}$$

Hence (13) is well-defined. Direct verification shows that η_M is a group homomorphism. We note next that we have a well-defined map of the product set $R_S \times M$ into M_S such that

$$(a/s, x) \rightsquigarrow ax/s. \tag{14}$$

To check this we have to show that if $a_1/s_1 = a_2/s_2$ in R_S, then $a_1 x/s_1 = a_2 x/s_2$. Now if $a_1/s_1 = a_2/s_2$, then we have an $s \in S$ such that $sa_1 s_2 = sa_2 s_1$. Then $sa_1 s_2 x = sa_2 s_1 x$, which implies the required equality $a_1 x/s_1 = a_2 x/s_2$. Direct verification shows that (14) satisfies the condition for a balanced product of the R-modules R_S and M. Hence we have a group homomorphism η'_M of $R_S \otimes_R M$ into M_S sending $(a/s)\otimes x$ into ax/s. Following this with η_M we obtain $(1/s)\otimes ax = (a/s)\otimes x$. On the other hand, if we apply η_M to x/s, we obtain $(1/s)\otimes x$ and the application of η'_M to this gives x/s. It follows that $\eta_M\eta'_M$ is the identity map on $R_S\otimes_R M$ and $\eta'_M\eta_M$ is the identity on M_S. Hence η_M is a group isomorphism and so is $\eta'_M = \eta_M^{-1}$. Direct verification using the definitions shows that η_M and hence η_M^{-1} is an R_S-map, hence an R_S-isomorphism. It remains to show the naturality, that is, the commutativity of

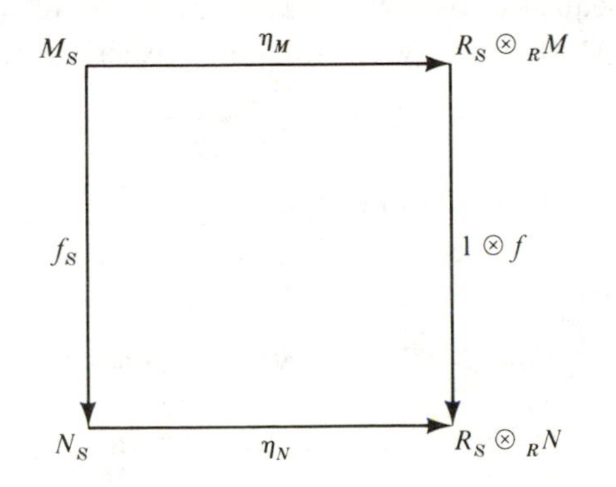

for a given R-homomorphism $f: M\to N$. This follows from the calculation

$$(1\otimes f)\eta_M(x/s) = (1\otimes f)((1/s)\otimes x) = (1/s)\otimes f(x),$$

$$\eta_N f_S(x/s) = \eta_N(f(x)/s) = (1/s)\otimes f(x),$$

which completes the proof. □

This result gives rise to a useful interplay between tensor products and localization. We recall first that the functor $N\otimes_R$ is right exact, that is, exactness of $M'\xrightarrow{f} M\xrightarrow{g} M''\to 0$ implies that of $N\otimes_R M'\xrightarrow{1\otimes f} N\otimes M\xrightarrow{1\otimes g} N\otimes M''\to 0$. Applying this with $N = R_S$ in conjunction with Proposition 7.6 shows that $M'_S\xrightarrow{f_S} M_S\xrightarrow{g_S} M''_S\to 0$ is exact. Next we prove directly that if $0\to M'\xrightarrow{f} M$ is exact, then $0\to M'_S\xrightarrow{f_S} M_S$ is exact. Suppose $f_S(x'/s) = 0$, so $f(x')/s = 0$. Then we have a $t\in S$ such that $tf(x') = 0$. Then $f(tx') = 0$ and since f is injective, $tx' = 0$ and hence $x'/s = 0$. Hence $\ker f_S = 0$ and so $0\to M'_S\xrightarrow{f_S} M_S$ is exact. We can now apply Proposition 7.6 to conclude that $0\longrightarrow R_S\otimes_R M'\xrightarrow{1\otimes f} R_S\otimes_R M$ is exact. We recall that this is the definition of flatness for R_S as R-module (p. 153). Hence we have the important

PROPOSITION 7.7. *R_S is a flat R-module.*

We remark also that we have shown that if $0\to M'\to M\to M''\to 0$ is an exact sequence of R-modules, then $0\to M'_S\to M_S\to M''_S\to 0$ is an exact sequence of R_S-modules. Thus the S-localization functor from **R-mod** to **R_S-mod** is exact.

Now let N be a submodule of M. Then the exact sequence $0 \to N \xrightarrow{\iota} M$ where ι is the injection gives the exact sequence $0 \to N_S \xrightarrow{\iota_S} M_S$ and the definition shows that ι_S is the injection of N_S. We have the exact sequence $0 \to N \xrightarrow{\iota} M \xrightarrow{\nu} M/N \to 0$ where ν is the canonical homomorphism. Hence we have the exact sequence $0 \to N_S \xrightarrow{\iota_S} M_S \xrightarrow{\nu_S} (M/N)_S \to 0$, which shows that $M_S/N_S \cong (M/N)_S$. More precisely, the foregoing exactness shows that the map

$$x/s + N_S \rightsquigarrow (x+N)/s \tag{15}$$

is an isomorphism of M_S/N_S with $(M/N)_S$.

If M and N are R-modules and f is a homomorphism $f: M \to N$, then we have the exact sequence

$$0 \to \ker f \xrightarrow{\iota} M \xrightarrow{f} N \xrightarrow{\nu} \operatorname{coker} f \to 0$$

where $\operatorname{coker} f = N/f(M)$. Hence we have the exact sequence

$$0 \to (\ker f)_S \xrightarrow{\iota_S} M_S \xrightarrow{f_S} N_S \xrightarrow{\nu_S} (\operatorname{coker} f)_S \to 0. \tag{16}$$

This implies that $(\ker f)_S$ is $\ker f_S$. Also $(\operatorname{coker} f)_S = (N/f(M))_S \cong N_S/f(M)_S = N_S/f_S(M) = \operatorname{coker} f_S$.

We recall next that we have an R_S-isomorphism of $R_S \otimes_R (M \otimes_R N)$ onto $(R_S \otimes_R M) \otimes_{R_S} (R_S \otimes N)$ sending $1 \otimes_R (x \otimes_R y)$ into $(1 \otimes_R x) \otimes_{R_S} (1 \otimes_R y)$ (exercise 13, p. 148). Hence, by Proposition 7.6, we have an R_S-isomorphism of $(M \otimes_R N)_S$ onto $M_S \otimes_{R_S} N_S$ such that

$$(x \otimes_R y)/1 \rightsquigarrow (x/1) \otimes_{R_S} (y/1). \tag{17}$$

Now let M be an R_S-module. Then M becomes an R-module by defining ax for $a \in R$ to be $(a/1)x$. It is clear that if f is a homomorphism of M as R_S-module, then f is an R-homomorphism. Now consider M_S where M is regarded as an R-module. Since $s/1$ is invertible in R_S for $s \in S$, $sx = (s/1)x = 0$ for $x \in M$ implies $x = 0$. Hence the homomorphism $\lambda_S: x \rightsquigarrow x/1$ of M into M_S is a monomorphism. Since $x/s = (1/s)(x/1)$, λ_S is surjective. Thus we can identify M as R_S-module with the S-localization M_S of M as R-module.

7.4 LOCALIZATION AT THE COMPLEMENT OF A PRIME IDEAL. LOCAL-GLOBAL RELATIONS

We recall that an ideal P in R is prime if and only if the complement, $R - P$ of P in R, is a submonoid of the multiplicative monoid of R. Of particular importance are localizations with respect to such monoids. We shall usually write M_P for M_{R-P}, f_P for f_{R-P}, etc. and call M_P the *localization of M at the prime ideal P*.

We consider first a correspondence between the ideals of R and R_P and we shall begin by considering more generally the ideals in R and in R_S for any submonoid S. Let A' be an ideal in R_S. Then

$$j(A') = \{a \in R \mid a/s \in A' \text{ for some } s \in S\} \tag{18}$$

is an ideal in R. Clearly $(j(A'))_S = A'$. Again, if we begin with an ideal A in R, then it is easily seen that $A_S = R_S$ if and only if A contains an element of S. For this reason it is natural to confine our attention to the ideals A of R that do not meet S.

We observe next that if P is a prime ideal of R such that $P \cap S = \varnothing$, then $j(P_S) = P$. For, let $a \in j(P_S)$. Then we have an element $p \in P$ and elements $s, t \in S$ such that $a/s = p/t$. Hence we have a $u \in S$ such that $uat = ups \in P$. Since $ut \in S$ and $S \cap P = \varnothing$, this implies that $a \in P$. Hence $j(P_S) \subset P$. Since the reverse inclusion $j(A_S) \supset A$ holds for any ideal A, we have $j(P_S) = P$. It is straightforward to verify two further facts: If P is a prime ideal of R such that $P \cap S = \varnothing$, then P_S is a prime ideal in R_S and if P' is a prime ideal in R_S, then $j(P')$ is prime in R that does not meet S. We leave this to the reader. Putting together these results we obtain

PROPOSITION 7.8. *The map $P \rightsquigarrow P_S$ is bijective and order-preserving from the set of prime ideals of R, which do not meet S with the set of prime ideals of R_S. The inverse map is $P' \rightsquigarrow j(P')$ (defined by* (18)*).*

We now consider the important case of localization at a prime ideal P. Since $Q \cap (R - P) = \varnothing$ means $Q \subset P$, the foregoing result specializes to

PROPOSITION 7.9. *The map $Q \rightsquigarrow Q_P$ is a bijective order-preserving map of the set of prime ideals Q contained in P with the set of prime ideals of R_P. The inverse map is $P' \rightsquigarrow j(P')$.*

It is clear that P_P contains every prime ideal of R_P. It is clear also that the elements not in P_P are units in R_P and since $R_P \neq P_P$, no element of P_P is a unit. Thus P_P is the set of non-units of R_P and hence R_P is a local ring with rad $R_P = P_P$ as its only maximal ideal. We repeat the statement of this result as

PROPOSITION 7.10. *R_P is a local ring with* rad $R_P = P_P$ *as its only maximal ideal.*

This fact accounts for the central importance of localization: It often permits a reduction of questions on commutative rings and modules over such rings to the case of local rings, since in many important instances a result will be valid for R if it holds for every R_P, P a prime ideal in R. The following result gives some basic properties of modules that hold if and only if they hold at all the localizations at prime ideals.

PROPOSITION 7.11. (1) *Let M be an R-module. If $M = 0$, then $M_S = 0$ for every localization and if $M_P = 0$ for every maximal P, then $M = 0$.* (2) *If M and N are R-modules and f is a homomorphism of M into N, then f injective (surjective) implies that f_S is injective (surjective) for the localizations at every S. On the other hand, if f_P is injective (surjective) for every maximal ideal P, then f is injective (surjective).* (3) *If M is flat, then so is every M_S and if M_P is flat for every maximal ideal P, then M is flat.*

Proof. (1) Evidently $M = 0$ implies $M_S = 0$. Now assume $M \neq 0$ and let x be a non-zero element of M. Then $\operatorname{ann} x \neq R$, so this ideal can be imbedded in a maximal ideal P of R. We claim that $x/1 \neq 0$ in M_P, so $M_P \neq 0$. This is clear since $x/1$ is the image of x under the canonical homomorphism of M into M_P and if $x/1 = 0$, then $\operatorname{ann} x \cap (R - P) \neq \varnothing$. Since $\operatorname{ann} x \subset P$, this is ruled out.

(2) Let $f: M \to N$. Then we have seen that $\ker f_S = (\ker f)_S$ and coker $f_S \cong (\operatorname{coker} f)_S$ for any submonoid S of the multiplicative monoid of R. Since a homomorphism is injective (surjective) if and only if its kernel (cokernel) is 0, (2) is an immediate consequence of (1).

(3) Suppose that M is a flat R-module and $0 \to N' \xrightarrow{f} N$ is an exact sequence of R_S-modules. Regarding N' and N as R-modules, we see that $0 \to M \otimes_R N' \to M \otimes_R N$ is exact. Then $0 \longrightarrow (M \otimes_R N')_S \xrightarrow{(1 \otimes f)_S} (M \otimes_R N)_S$ is exact, and by (16), $0 \longrightarrow M_S \otimes_{R_S} N'_S \xrightarrow{1 \otimes f_S} M_S \otimes_{R_S} N_S$ is exact. We have seen that N'_S and N_S can be identified with N' and N respectively. Hence $0 \longrightarrow M_S \otimes_{R_S} N' \xrightarrow{1 \otimes f} M_S \otimes_{R_S} N$ is exact and so M_S is R_S-flat. Now suppose that M is an R-module such that M_P is R_P-flat for every maximal ideal of P of R. Let $0 \to N' \xrightarrow{f} N$ be an exact sequence of R-modules and consider $M \otimes_R N' \xrightarrow{1 \otimes f} M \otimes_R N$. Since M_P is flat and $0 \to N'_P \xrightarrow{f_P} N_P$ is exact, $0 \longrightarrow M_P \otimes_{R_P} N'_P \xrightarrow{1 \otimes f_P} M_P \otimes_{R_P} N_P$ is exact. Then $0 \longrightarrow (M \otimes_R N')_P \xrightarrow{(1 \otimes f_P)} (M \otimes_R N)_P$ is exact. Since this holds for every maximal P, $0 \longrightarrow M \otimes_R N' \xrightarrow{1 \otimes f} M \otimes_R N$ is exact by (2). Hence M is flat. □

EXERCISES

1. Show that the nil radical of R_S is $(\operatorname{nilrad} R)_S$.

2. Show that if P is a prime ideal of R, then R_P/P_P is isomorphic to the field of fractions of the domain R/P.

3. Show that if R is a factorial domain (BAI, p. 141) and S is a submonoid of the multiplicative monoid of R not containing 0, then R_S is a factorial domain.

4. Let $\{P_i | 1 \leqslant i \leqslant n\}$ be a set of prime ideals in R and let $S = \bigcap_1^n (R - P_i)$. Show that any prime ideal of R_S has the form P_S where P is a prime ideal contained in one of the P_i.

7.5 PRIME SPECTRUM OF A COMMUTATIVE RING

Let R be a commutative ring and let $X = X(R)$ denote the set of prime ideals of R. There is a natural way of introducing a topology on the set X that permits the introduction of geometric ideas in the study of the ring R. We proceed to define this topology.

If A is any subset of R, we let $V(A)$ be the subset of X consisting of the prime ideals P containing A. Evidently $V(A) = V(I(A))$ where $I(A)$ is the ideal generated by A and since the nilradical of an ideal I is the intersection of the prime ideals containing I, it is clear that $V(I) = V(\text{nilrad}\, I)$. Also if P is a prime ideal containing $I_1 I_2$ for I_1, I_2 ideals, then either $P \supset I_1$ or $P \supset I_2$. Hence $V(I_1 I_2) = V(I_1) \cup V(I_2)$. We can now verify that the sets $V(A)$, A a subset of R, satisfy the axioms for closed sets in a topological space:

(1) $\varnothing$ and X are closed sets, since $\varnothing = V(\{1\})$ and $X = V(\{0\})$.

(2) The intersection of any set of closed sets is closed, since if $\{A_\alpha\}$ is a set of subsets of R, then

$$\bigcap_\alpha V(A_\alpha) = V(\bigcup (A_\alpha)).$$

(3) The union of two closed sets is closed, since we can take these to be $V(I_1)$ and $V(I_2)$, I_j ideals, and then $V(I_1) \cup V(I_2) = V(I_1 I_2)$.

We shall call X equipped with this topology the *prime spectrum of* R and denote it as $\operatorname{Spec} R$. The subset $X_{\max}$ of X consisting of the maximal ideals of R with the induced topology is called the *maximum spectrum*. This will be denoted as $\operatorname{Maxspec} R$. Such topologies were first introduced by M. H. Stone for Boolean rings and were considered by the present author for the primitive ideals of an arbitrary ring. In the case of commutative rings the topology is called the *Zariski topology* of X.

The open sets in $X = \operatorname{Spec} R$ are the complements $X - V(A) = X - \bigcap_{a \in A} V(\{a\}) = \bigcup_{a \in A} (X - V(\{a\}))$. We denote the set $X - V(\{a\})$ as X_a for $a \in R$. This is just the set of prime ideals P not containing a, that is, the P such that

$\bar{a} = a+P \neq 0$ in R/P. Since any open set is a union of sets X_a, these open subsets of X constitute a base for the open sets in Spec R. It is worthwhile to list the following properties of the map $a \rightsquigarrow X_a$ of R into the set of open subsets of Spec R:

(1) $X_{ab} = X_a \cap X_b$.
(2) $X_a = \varnothing$ if and only if a is nilpotent (since nilrad R is the intersection of the prime ideals of R).
(3) $X_a = X$ if and only if a is a unit in R.

If Y is a subset of X, put $\Delta_Y = \bigcap_{P \in Y} P$. This is an ideal in R and $V(\Delta_Y)$ is a closed set containing Y. On the other hand, if $V(A) \supset Y$, then $P \supset A$ for every $P \in Y$ so $\Delta_Y \supset A$ and $V(A) \supset V(\Delta_Y)$. Thus $V(\Delta_Y)$ is the closure of Y, that is, the smallest closed set in X containing Y. In particular, we see that the closure of a point P is the set of prime ideals containing P. If R is a domain, 0 is a prime ideal in R and hence the closure of 0 is the whole space X.

EXAMPLES

1. $R = \mathbb{Z}$. As we have noted, the closure of the prime ideal 0 is the whole space $X(\mathbb{Z})$. Hence Spec $\mathbb{Z}$ is not a T_1-space (a space in which points are closed sets). Now consider Maxspec $\mathbb{Z}$. The maximal ideals of $\mathbb{Z}$ are the prime ideals $(p) \neq 0$. Hence the closure of $(p) \neq 0$ is (p) and so Maxspec $\mathbb{Z}$ is a T_1-space. Let Y be an infinite set of primes $(p) \neq 0$ in $\mathbb{Z}$. Evidently $\bigcap_{(p) \in Y}(p) = 0$, so $V(\Delta_Y) = X$. Thus the closure of any infinite subset of Maxspec $\mathbb{Z}$ is the whole space. Hence the closed sets of Maxspace $\mathbb{Z}$ are the finite subsets (including $\varnothing$) and the whole space. Evidently the Hausdorff separation axiom fails in Maxspec $\mathbb{Z}$.

2. $R = F[x_1, \dots, x_r]$, F a field, x_i indeterminates. For $r = 1$, the discussion of Spec and Maxspec is similar to that of the ring $\mathbb{Z}$. For arbitrary r we remark that $F[x_1, \dots, x_r]/(x_1) \cong F[x_2, \dots, x_r]$, so the prime ideals of $F[x_1, \dots, x_r]$ containing (x_1) are in 1-1 correspondence with the prime ideals of $F[x_2, \dots, x_r]$. Hence the closure of (x_1) in Spec $F[x_1, \dots, x_r]$ is in 1-1 correspondence with Spec $F[x_2, \dots, x_r]$.

We shall now derive some of the basic properties of the prime spectrum. We prove first

PROPOSITION 7.12. Spec R *is quasi-compact.*

Proof. This means that if we have a set of open subsets O_α such that $\bigcup O_\alpha = X$, then there exists a finite subset $O_{\alpha_1}, \dots, O_{\alpha_n}$ of the O_α such that $\bigcup O_{\alpha_i} = X$. (We are following current usage that reserves "compact" for "quasi-compact Hausdorff.") Since the sets X_a form a base, it suffices to show that if $\bigcup_{a \in A} X_a = X$, then $\bigcup X_{a_i} = X$ for some finite subset $\{a_i\}$ of A. The condition $\bigcup_{a \in A} X_a = X$ gives $X - V(A) = X$ and $V(A) = \varnothing$. Then $V(I(A)) = \varnothing$ for the ideal $I(A)$ generated by A and so $I(A) = R$. Hence there exist $a_i \in A$, $x_i \in R$ such that $\sum_1^r a_i x_i = 1$. Retracing the steps, we see that $V(\{a_i\}) = \varnothing$ and $\bigcup X_{a_i} = X$. □

Let $N = \text{nilrad}\, R$, $\bar{R} = R/N$, and let v be the canonical homomorphism $a \to a+N$ of R onto $\bar{R}$. Any prime ideal P of R contains N and $v(P)$ is a prime ideal of $\bar{R}$. Moreover, every prime ideal of $\bar{R}$ has the form $v(P)$, P a prime ideal of R. We have

PROPOSITION 7.13. *The map $P \rightsquigarrow v(P)$ is a homeomorphism of* Spec R *onto* Spec $\bar{R}$.

Proof. The map is injective, since $P \supset N$ for every prime ideal P and we have seen that the map is surjective. Now if $a \in R$, $v(a) \in v(P)$ if and only if $a \in P$. This implies that if A is a subset of R, then the image of the closed set $V(A)$ in Spec R is $V(v(A))$ in Spec $\bar{R}$ and if $\bar{A}$ is a subset of $\bar{R}$, then the inverse image of the closed set $V(\bar{A})$ is $V(v^{-1}(\bar{A}))$. Hence $P \rightsquigarrow v(P)$ is a homeomorphism. □

We recall that a space X is disconnected if it contains an open and closed subset $\neq \varnothing$, $\neq X$. We shall show that Spec R is disconnected if and only if R contains an idempotent $\neq 0, 1$. This will follow from a considerably stronger result, which gives a bijection of the set of idempotents of R and the set of open and closed subsets of Spec R. To obtain this we shall need the following result on lifting of idempotents, which is of independent interest.

PROPOSITION 7.14. *Let R be a ring that is not necessarily commutative, N a nil ideal in R, and $\bar{u} = u+N$ an idempotent element of $\bar{R} = R/N$. Then there exists an idempotent e in R such that $\bar{e} = \bar{u}$. Moreover, e is unique if R is commutative.*

Proof. We have $u^2 - u = z$ where z is nilpotent, say, $z^n = 0$ and $z \in N$. Then $(u(1-u))^n = u^n v^n = 0$ where $v = 1-u$. From $u+v = 1$ we obtain

$$1 = 1^{2n-1} = (u+v)^{2n-1} = e+f$$

where e is the sum of the terms $u^i v^{2n-1-i}$ in which $n \leqslant i \leqslant 2n-1$ and f is the sum of the terms $u^i v^{2n-1-i}$ in which $0 \leqslant i \leqslant n-1$. Since $u^n v^n = 0$, any term in e annihilates any term in f. Hence $ef = 0 = fe$. Since $e+f = 1$, this gives $e^2 = e$, $f^2 = f$. Every term in e except u^{2n-1} contains the factor $uv = -z$. Hence $u^{2n-1} \equiv e \pmod{N}$. Since $u \equiv u^2 \equiv u^3 \equiv \cdots \equiv u^{2n-1} \pmod{N}$, we have $e \equiv u \pmod{N}$. This proves the first assertion.

Now assume that R is commutative. The uniqueness of e will follow if we can show that if e is an idempotent, then the only idempotent of the form $e+z$, z nilpotent, is e. The condition $(e+z)^2 = e+z$ gives $(1-2e)z = z^2$. Then $z^3 = (1-2e)z^2 = (1-2e)^2 z$ and by induction we have $(1-2e)^n z = z^{n+1}$. Since $(1-2e)^2 = 1-4e+4e = 1$, this implies that $z = 0$ and hence $e+z = e$. □

If e and f are idempotents in R, then so are $e' = 1-e$, ef, and $e \circ f = 1-(1-e)(1-f) = e+f-ef$. It is readily verified that the set E of idempotents of R is a Boolean algebra with the compositions $e \wedge f = ef$ and $e \vee f = e \circ f$ (exercise 1, p. 479 of BAI). We remark next that the open and closed subsets of a topological space X constitute a subalgebra of the Boolean algebra of subsets of X. We can now prove

THEOREM 7.3. *If e is an idempotent in R, then X_e is an open and closed subspace of* Spec R *and the map $e \rightsquigarrow X_e$ is an isomorphism of the Boolean algebra E onto the Boolean algebra of open and closed subsets of* Spec R.

Proof. Let $e = e^2 \in R$. Then $e(1-e) = 0$, so any prime ideal P of R contains one of the elements e, $1-e$ but not both. Hence $X_e \cup X_{1-e} = X$ and $X_e \cap X_{1-e} = \varnothing$, so X_e is open and closed. Now let Y be an open and closed subset of X, $Y' = X-Y$. Let $\bar{R} = R/N$, $N = \text{nilrad}\, R$, v the canonical homomorphism of R onto $\bar{R}$. We use the homeomorphism $P \rightsquigarrow v(P)$ of $X = \text{Spec}\, R$ with $\bar{X} = \text{Spec}\, \bar{R}$ to conclude that $v(Y)$ is an open and closed subset of $\bar{X}$ and $v(Y')$ is its complement. Consider $\Delta_{v(Y)}$ and $\Delta_{v(Y')}$ as defined on p. 404. These are ideals in $\bar{R}$ and $V(\Delta_{v(Y)}) = v(Y)$, $V(\Delta_{v(Y')}) = v(Y')$ since $v(Y)$ and $v(Y')$ are open and closed in $\bar{X}$. If $\bar{P}$ is a prime ideal of $\bar{R}$ containing $\Delta_{v(Y)} + \Delta_{v(Y')}$, then $\bar{P} \supset \Delta_{v(Y)}$ and $\bar{P} \supset \Delta_{v(Y')}$, so $\bar{P} \in v(Y) \cap v(Y')$. Since $v(Y) \cap v(Y') = \varnothing$, there are no such $\bar{P}$ and so $\Delta_{v(Y)} + \Delta_{v(Y')} = \bar{R}$. If $\bar{P}$ is any prime in $\bar{R}$, either $\bar{P} \supset \Delta_{v(Y)}$ or $\bar{P} \supset \Delta_{v(Y')}$, so $\bar{P} \supset \Delta_{v(Y)} \cap \Delta_{v(Y')}$. Since this holds for all $\bar{P}$ and $\bar{R}$ has no nilpotent elements $\neq 0$, $\Delta_{v(Y)} \cap \Delta_{v(Y')} = 0$. Hence $\bar{R} = \Delta_{v(Y)} \oplus \Delta_{v(Y')}$. Let $\bar{e}$ be the unit of $\Delta_{v(Y')}$ and let e be the idempotent in R such that $v(e) = \bar{e}$. Now if P is an ideal in R, the condition $P \ni e$ is equivalent to $v(P) \ni \bar{e}$, which in turn is equivalent to $v(P) \supset \Delta_{v(Y')}$ and to $v(P) \in v(Y')$ and $P \in Y'$. Hence $X_e = Y$, which shows that the map $e \rightsquigarrow X_e$ is surjective. To see that the map is injective, let e be a given idempotent in R, $\bar{e} = v(e)$. Then we have $\bar{R} = \bar{R}\bar{e} \oplus \bar{R}(\bar{1}-\bar{e})$ and $\bar{R}\bar{e}$ and $\bar{R}(\bar{1}-\bar{e})$ are ideals. Now $v(X_e)$ is the set of prime ideals of $\bar{R}$ containing $\bar{1}-\bar{e}$, hence $\bar{R}(\bar{1}-\bar{e})$, and $v(X_{1-e})$ is the set of prime ideals containing $\bar{R}\bar{e}$. We have shown that $\bar{R} = \Delta_{v(X_e)} \oplus \Delta_{v(X_{1-e})}$. Since $\Delta_{v(X_e)} \supset \bar{R}(\bar{1}-\bar{e})$ and $\Delta_{v(X_{1-e})} \supset \bar{R}\bar{e}$ and $\bar{R} = \bar{R}\bar{e} \oplus \bar{R}(\bar{1}-\bar{e})$, we have $\Delta_{v(X_e)} = \bar{R}(\bar{1}-\bar{e})$. Since $1-e$ is the only idempotent in the coset $\bar{1}-\bar{e}$, by Proposition 7.14, this implies the injectivity of $e \rightsquigarrow X_e$. If e and f are idempotents, then $X_{ef} = X_e \cap X_f$. Also $X_{1-e} = X - X_e$, $X_1 = X$, $X_0 = \varnothing$. These relations imply that $e \rightsquigarrow X_e$ is an isomorphism of Boolean algebras. $\square$

Evidently we have the following consequence of the theorem.

COROLLARY. Spec R *is connected if and only if* R *contains no idempotent* $\neq 0, 1$.

Because of this result, a commutative ring R is called *connected* if the only idempotents in R are 0 and 1.

Let f be a homomorphism of R into a second ring R'. If P' is a prime ideal in R', then $P = f^{-1}(P')$ is a prime ideal in R since if $ab \in P$, then $f(a)f(b) = f(ab) \in P'$. Thus either $f(a)$ or $f(b) \in P'$ and hence either a or $b \in P$. This permits us to define a map

$$f^*: P' \rightsquigarrow f^{-1}(P')$$

of $X' = \operatorname{Spec} R'$ into $X = \operatorname{Spec} R$. If $a \in R$, then a prime ideal P' of R' does not contain $a' = f(a)$ if and only if $P = f^*(P')$ does not contain a. Hence

$$(f^*)^{-1}(X_a) = X'_{f(a)}.$$

This implies that the inverse image of any open subset of Spec R under f^* is open in Spec R', so f^* is a continuous map of Spec R' into Spec R. It is clear that if $R' = R$ and $f = 1_R$, then $f^* = 1_{\operatorname{Spec} R}$ and if g is a homomorphism of R' into R'', then $(fg)^* = g^*f^*$. Thus the pair of maps $R \rightsquigarrow \operatorname{Spec} R$, $f \rightsquigarrow f^*$ define a contravariant functor from the category of commutative rings (homomorphisms as morphisms) into the category of topological spaces (continuous maps as morphisms).

EXERCISES

1. Let f be a homomorphism of R into R', f^* the corresponding continuous map of Spec R' into Spec R. Show that if f is surjective, then $f^*(\operatorname{Spec} R')$ is the closed subset $V(\ker f)$ of Spec R. Show also that f^* is a homeomorphism of Spec R' with the closed set $V(\ker f)$.

2. Same notations as exercise 1. Show that $f^*(\operatorname{Spec} R')$ is dense in Spec R if and only if nilrad $R \supset \ker f$. Note that this holds if f is injective.

3. Give an example of a homomorphism f of R into R' such that for some maximal ideal M' of R', $f^{-1}(M')$ is not maximal in R.

4. (D. Lazard.) Show that if P is a prime ideal in R and I is the ideal in R generated by the idempotents of R contained in P, then R/I contains no idempotents $\neq 0, 1$. Show that the set of prime ideals $P' \supset I$ is the connected component of Spec R containing P (the largest connected subset containing P). (*Hint*: Let $\bar{u}$ be an idempotent of $\bar{R} = R/I$ where $\bar{u} = u + I$. Then $u(1-u) \in I \subset P$ and we may assume $u \in P$. Show that there exists an $f \in I$ such that $f^2 = f$ and $f(u-u^2) = u-u^2$. Then $(1-f)u = g$ is an idempotent contained in P, so $g \in I$. Then $u = g + fu \in I$ and $\bar{u} = 0$.)

In exercises 5 and 6, R need not be commutative.

5. Let N be a nil ideal in R and let $\bar{u}_1, \ldots, \bar{u}_n$ be orthogonal idempotent elements of $\bar{R} = R/N$ ($\bar{u}_i^2 = \bar{u}_i$, $\bar{u}_i\bar{u}_j = 0$ if $i \neq j$). Show that there exist orthogonal idempotents e_i in R such that $\bar{e}_i = \bar{u}_i$, $1 \leqslant i \leqslant n$. Show also that if $\sum_1^n \bar{u}_i = 1$, then necessarily $\sum e_i = 1$.

6. Let R, N, $\bar{R}$ be as in exercise 5 and let $\bar{u}_{ij}$, $1 \leqslant i, j \leqslant n$, be elements of $\bar{R}$ such that $\bar{u}_{ij}\bar{u}_{kl} = \delta_{jk}\bar{u}_{il}$, $\sum \bar{u}_{ii} = 1$. Show that there exist $e_{ij} \in R$ such that $e_{ij}e_{kl} = \delta_{lk}e_{ij}$, $\sum e_{ii} = 1$, $\bar{e}_{ij} = \bar{u}_{ij}$, $1 \leqslant i, j \leqslant n$.

7.6 INTEGRAL DEPENDENCE

In BAI, pp. 278–281, we introduced the concept of R-integrality of an element of a field E for a subring R of E. The concept and elementary results derived in BAI can be extended to the general case in which E is an arbitrary commutative ring. We have the following

DEFINITION 7.2. *If E is a commutative ring and R is a subring, then an element $u \in E$ is called* R-integral *if there exists a monic polynomial $f(x) \in R[x]$, x an indeterminate, such that $f(u) = 0$.*

If $f(x) = x^n - a_{n-1}x^{n-1} - \cdots - a_0$, $a_i \in R$, then we have the relation $u^n = a_0 + a_1 u + \cdots + a_{n-1}u^{n-1}$, from which we deduce that if $M = R1 + Ru + \cdots + Ru^{n-1}$, then $uM \subset M$. Evidently M is a finitely generated R-submodule of E containing 1. Since $uM \subset M$, M is an $R[u]$-submodule of E and since $1 \in M$, M is faithful as $R[u]$-module. Hence we have the implication $1 \Rightarrow 2$ in the following.

LEMMA. *The following conditions on an element $u \in E$ are equivalent:* 1. u *is* R-integral. 2. *There exists a faithful $R[u]$-submodule of E that is finitely generated as R-module.*

Proof. Now assume 2. Then we have $u_i \in M$ such that $M = Ru_1 + Ru_2 + \cdots + Ru_n$. Then we have the relations $uu_i = \sum_{j=1}^n a_{ij}u_j$, $1 \leqslant i \leqslant n$, where the $a_{ij} \in R$.

Hence we have the relations

$$\begin{aligned}(u-a_{11})u_1-a_{12}u_2-\cdots-a_{1n}u_n&=0\\-a_{21}u_1+(u-a_{22})u_2-\cdots-a_{2n}u_n&=0\\\cdot\quad\cdot\quad\cdots\quad\cdot\quad&\cdot\\-a_{n1}u_1-a_{n2}u_2-\cdots+(u-a_{nn})u_n&=0.\end{aligned}$$

If we multiply the ith of these equations by the cofactor of the (i,j)-entry of the matrix $u1-(a_{ij})$ and add the resulting equations, we obtain the equation $f(u)u_j=0$, $1\leqslant j\leqslant n$, where $f(x)$ is the characteristic polynomial of the matrix (a_{ij}). Since the u_j generate M and M is faithful as $R[u]$-module, we have $f(u)=0$, so u is a root of the monic polynomial $f(x)$. Hence $2\Rightarrow 1$. □

We remark that the argument just used is slightly different from the one given in the field case in BAI, p. 279. If M and N are R-submodules of E that are finitely generated over R then so is $MN=\{\sum m_in_i|m_i\in M,\ n_i\in N\}$. Moreover, if $1\in M$ and $1\in N$ then $1\in MN$ and if either $uM\subset M$ or $uN\subset N$ then $uMN\subset MN$. These observations and the foregoing lemma can be used in exactly the same way as in BAI, pp. 279–280, to prove

THEOREM 7.4. *If E is a commutative ring and R is a subring, the subset R' of elements of E that are R-integral is a subring containing R. Moreover, any element of E that is R'-integral is R-integral and hence is contained in R'.*

The subring R' of E is called the *integral closure of R in E*. If $R'=E$, that is, every element of E is integral over R, then we say that E is *integral over R* (or E is an *integral extension of R*).

If A is an ideal of a ring E, then evidently $R\cap A$ is an ideal in the subring R of E. We call this the *contraction* of A to R and denote it as A^c. We also have the subring $(R+A)/A$ of E/A and the canonical isomorphism of this ring with the ring $R/A^c=R/(R\cap A)$. One usually identifies $(R+A)/A$ with R/A^c by means of the canonical isomorphism and so regards E/A as an extension of R/A^c. In this sense we have

PROPOSITION 7.15. *If E is integral over R and A is an ideal in E, then $\bar{E}=E/A$ is integral over $\bar{R}=R/A^c$.*

Proof. Let $\bar{u}=u+A$ be an element of $\bar{E}$. Then we have $a_i\in R$, $0\leqslant i\leqslant n-1$, such that $u^n=a_0+a_1u+\cdots+a_{n-1}u^{n-1}$. Then $\bar{u}^n=\bar{a}_0+\bar{a}_1\bar{u}+\cdots+\bar{a}_{n-1}\bar{u}^{n-1}$ where $\bar{a}_i=a_i+A$. Hence $\bar{u}$ is integral over $(R+A)/A$, hence over $\bar{R}=R/A^c$. □

Next suppose that S is a submonoid of the multiplicative monoid of R. Then the localization E_S contains the localization R_S as a subring. Moreover, we have

PROPOSITION 7.16. *Let R be a subring of E, R' the integral closure of R in E, S a submonoid of the multiplicative monoid of R. Then R'_S is the integral closure of R_S in E_S.*

Proof. Any element of R'_S has the form u/s, $u \in R'$, $s \in S$. We have a relation $u^n = a_0 + a_1 u + \cdots + a_{n-1}u^{n-1}$, $a_i \in R$. Hence $(u/s)^n = a_0/s^n + (a_1/s^{n-1})(u/s) + \cdots + (a_{n-1}/s)(u/s)^{n-1}$. Hence u/s is integral over R_S. Conversely, suppose that u/s is integral over R_S where $u \in E$, $s \in S$. To show that $u/s = v/t$ where $v \in R'$, $t \in S$, it suffices to show that there exists an $s' \in S$ such that $us' \in R'$. For, $u/s = us'/ss'$ has the required form. Now u/s integral over R_S implies that $u/1$ is integral over R_S, since $s/1 \in R_S$ and $u/1 = (u/s)(s/1)$. Thus we have a relation of the form $(u/1)^n = a_0/s_0 + (a_1/s_1)(u/1) + \cdots + (a_{n-1}/s_{n-1})(u/1)^{n-1}$ with $a_i \in R$, $s_i \in S$. Multiplication by $t_1^n/1$ where $t_1 = \prod_0^{n-1} s_i$ gives $(t_1 u/1)^n = a'_0/1 + (a'_1/1)(t_1 u/1) + \cdots + (a'_{n-1}/1)(t_1 u/1)^{n-1}$ where the $a'_i \in R$. Hence $[(t_1 u)^n - a'_0 - a'_1(t_1 u) - \cdots - a'_{n-1}(t_1 u)^{n-1}]/1 = 0$. Then there exists a $t_2 \in S$ such that $t_2[(t_1 u)^n - a_0 - \cdots] = 0$. Multiplication of this relation by t_2^{n-1} shows that $s'u \in R'$ for $s' = t_1 t_2$. □

We prove next

PROPOSITION 7.17. *If E is a domain that is integral over the subdomain R, then E is a field if and only if R is a field.*

Proof. Assume first that R is a field and let $u \neq 0$ be an element of E. We have a relation $u^n + a_1 u^{n-1} + \cdots + a_n = 0$ with $a_i \in R$ and since u is not a zero divisor in E, we may assume $a_n \neq 0$. Then a_n^{-1} exists in R and we have $u(u^{n-1} + a_1 u^{n-2} + \cdots + a_{n-1})(-a_n^{-1}) = 1$. Hence u is invertible and E is a field. Conversely, suppose that E is a field and let $a \neq 0$ be in R. Then a^{-1} exists in E and we have a relation $(a^{-1})^n + a_1(a^{-1})^{n-1} + \cdots + a_n = 0$ with $a_i \in R$. Multiplication by a^{n-1} gives $a^{-1} = -(a_1 + a_2 a + \cdots + a_n a^{n-1}) \in R$. Hence a is invertible in R and R is a field. □

An immediate consequence of this result and of Proposition 7.17 is the

COROLLARY 1. *Let E be a commutative ring, R a subring such that E is integral over R, and let P be a prime ideal in E. Then $P^c = P \cap R$ is maximal in R if and only if P is maximal in E.*

Proof. $\bar{E} = E/P$ is a domain and by Proposition 7.15, $\bar{E}$ is integral over $\bar{R} = R/P^c$. By Proposition 7.17, $\bar{E}$ is a field if and only if $\bar{R}$ is a field. Hence P is maximal in E if and only if P^c is maximal in R. □

We also have the following

COROLLARY 2. *Let E and R be as in Corollary* 1 *and suppose P_1 and P_2 are ideals in R such that $P_1 \supsetneqq P_2$. Then $P_1^c \supsetneqq P_2^c$.*

Proof. We have $P_1^c \supset P_2^c$. Now suppose $p = P_1^c = P_2^c$. Consider the localization E_S for $S = R - p$. By Proposition 7.16, E_S is integral over R_S. Since $P_i \cap R = p$, $P_i \cap S = \varnothing$. Then, by Proposition 7.8, $P_{1S} \supsetneqq P_{2S}$. On the other hand $P_{iS} \supset p_S$ which is the maximal ideal of the local ring R_S. Since E_S is integral over R_S it follows from Corollary 1 that P_{iS} is maximal in E_S. This contradicts $P_{1S} \supsetneqq P_{2S}$. Thus $P_1^c \supsetneqq P_2^c$. □

If P is a prime ideal in E, then it is clear that P^c is a prime ideal in R. Hence we have a map $P \rightsquigarrow P^c$ of Spec E into Spec R. This is surjective since we have the following

THEOREM 7.5 ("LYING-OVER" THEOREM). *Let E be a commutative ring, R a subring such that E is integral over R. Then any prime ideal p of R is the contraction P^c of a prime ideal P of E.*

Proof. We assume first that R is local and p is the maximal ideal of R. Let P be a maximal ideal in E. Then P^c is a maximal ideal in R, by the above corollary. Since p is the only maximal ideal in R, we have $p = P^c$.

Now let R be arbitrary and consider the localizations R_p and E_p. R_p is a local ring and E_p is integral over R_p (Proposition 7.16). Now there exists a prime ideal P' in E_p whose contraction $P'^c = P' \cap R_p = p_p$, the maximal ideal of R_p. By Proposition 7.8 (p. 401), $P' = P_p$ for a prime ideal P of E such that $P \cap (R - p) = \varnothing$ or, equivalently, $P \cap R \subset p$. Again, by Proposition 7.8, $P = j(P_p) = \{u \in E | u/s \in P_p\}$ for some $s \in R - p$. Now $p \subset j(p_p) \subset j(P_p) = P$. Hence $P \cap R = p$. □

EXERCISES

1. Let E be a commutative ring, R a subring such that (1) E is integral over R and (2) E is finitely generated as R-algebra. Show that E is finitely generated as R-module.

2. ("Going-up" theorem). Let E be a commutative ring, R a subring such that E is integral over R. Let p_1 and p_2 be prime ideals of R such that $p_1 \supset p_2$ and let P_2 be a prime ideal of E such that $P_2^c = p_2$. Show that there exists a prime ideal P_1 of E such that $P_1 \supset P_2$ and $P_1^c = p_1$.

3. Let R be a subring of a commutative ring E, R' the integral closure of R in E, I an ideal in R, $I^e = IR'$ its extension to an ideal in R'. An element $a \in E$ is *integral over* I if it satisfies an equation $f(a) = 0$ where $f(\lambda) = \lambda^m + b_1\lambda^{m-1} + \cdots + b_m$, $b_i \in I$. The subset I' of E of elements integral over I is called the *integral closure* of I in E.

   Show that $I' = \text{nilrad } I^e$ in R'. *Sketch of proof*: If $a \in E$ and $a^m + b_1a^{m-1} + \cdots + b_m = 0$ with the $b_i \in I$ then $a \in R'$ and $a^m \in I^e$. Hence $a \in \text{nilrad } I^e$ in R'. Conversely, let $a \in \text{nilrad } I^e$ in R'. Then $a \in R'$ and $a^m = \sum_i^n a_ib_i$ for some $m > 0$, $a_i \in R'$, $b_i \in I$. Let A be the R-subalgebra of E generated by the a_i. By exercise 1, A has a finite set of generators $a_1, \ldots, a_q$, $q \geq n$, as R-module. Then $a_m a_j = \sum_{k=1}^q b_{jk}a_k$, $1 \leq j \leq q$, where the $b_{jk} \in I$. As in the proof of the lemma on p. 408, this implies that a^m and hence a is integral over I.

   Remarks. (i) I' is a subring of E (in the sense of BAI, that is, I' is a subgroup of the additive group closed under multiplication). (ii) If R is integrally closed in E ($R' = R$) then $I' = \text{nilrad } I$ in R.

4. (Basic facts on contractions and extensions of ideals.) Let R be a subring of a commutative ring E. For any ideal I of E, $I^c = I \cap R$ is an ideal in R and for any ideal i of R, $i^e = iE$ is an ideal in E. Note that

$$I_1 \supset I_2 \Rightarrow I_1^c \supset I_2^c$$

$$i_1 \supset i_2 \Rightarrow i_1^e \supset i_2^e$$

$$i^{ec} \supset i, \qquad I^{ce} \subset I.$$

   Hence conclude that

$$i^{ece} = i^e, \; I^{cec} = I^c$$

   for any ideals i and I. Note that these imply that an ideal i of R is the contraction of an ideal of $E \Leftrightarrow i = i^{ec}$ and an ideal I of E is an extension of an ideal of $R \Leftrightarrow I = I^{ce}$.

7.7 INTEGRALLY CLOSED DOMAINS

An important property of domains that we shall encounter especially in the study of Dedekind domains (see Chapter 10) is given in the following

DEFINITION 7.3. *A domain D is called* integrally closed *if it is integrally closed in its field of fractions.*

It is readily seen that $\mathbb{Z}$ is integrally closed. More generally any factorial domain is integrally closed (exercise 1, below). Our main objective in this brief section is to prove a "going-down" theorem (Theorem 7.6) for integrally closed domains that will be required later (in the proof of Theorem 8.37). For the proof of this theorem we require

PROPOSITION 7.18. *Let D be an integrally closed subdomain of a domain E, F the field of fractions of D, I an ideal in D. Let $a \in E$ be integral over I. Then a is algebraic over F and its minimum polynomial $m(\lambda) = \lambda^m - b_1\lambda^{m-1} + \cdots + (-1)^m b_m$ over F has its coefficients $(-1)^j b_j \in$ nilrad I, $1 \leq j \leq m$.*

Proof. The first statement is clear. For the second let S be a splitting field over F of $m(\lambda)$ and let $m(\lambda) = (\lambda - a_1)\cdots(\lambda - a_m)$ in $S[\lambda]$ where $a_1 = a$. For any i we have an automorphism of S/F such that $a \rightsquigarrow a_i$. Since this stabilizes I it follows that every a_i is integral over I. Since b_j, $1 \leq j \leq m$, is an elementary symmetric polynomial in the a_i it follows from Remark (i) following exercise 3 of section 7.6 that b_j is integral over I. Then, by Remark (ii), $b_j \in$ nilrad I. $\square$

We shall need also the following criterion that an ideal be a contraction of a prime ideal.

PROPOSITION 7.19. *Let R be a subring of a commutative ring E, p a prime ideal of R. Then $p = P^c \equiv P \cap R$ for a prime ideal P of $E \Leftrightarrow p^{ec} \equiv Ep \cap R = p$.*

Proof. If $p = P^c$ then $p^{ec} = p$ by exercise 4 of section 7.6. Now assume this holds for a prime ideal p of R. Consider the submonoid $S = R - p$ of the multiplicative monoid of E. Since $p^{ec} = p$, $Ep \cap R = p$ and hence $Ep \cap S = \varnothing$. Since Ep is an ideal of E and $Ep \cap S = \varnothing$, the extension $(Ep)_S$ is a proper ideal of the localization E_S of E relative to the monoid S. Then $(Ep)_S$ is contained in a maximal ideal Q of E_S and $P = j(Q)$ as defined by (18) is a prime ideal of R such that $P \cap (R - p) = P \cap S = \varnothing$. Then $P \cap R \subset p$. Since $P = j(Q) \supset j((Ep)_S) \supset p$ we have $P^c = P \cap R = p$. $\square$

We can now prove

THEOREM 7.6 ("Going-down" theorem). *Let D be an integrally closed subdomain of a domain E that is integral over D. Let p_1 and p_2 be prime ideals of D such that $p_1 \supset p_2$ and suppose P_1 is a prime ideal of E such that $P_1^c = p_1$. Then there exists a prime ideal P_2 of E such that $P_2^c = p_2$ and $P_2 \subset P_1$.*

Proof. Consider the localization E_{P_1} $(= E_{E-P_1})$. It suffices to show that $p_2 E_{P_1} \cap D = p_2$. For, if this holds, then by Proposition 7.19 there exists a prime ideal

Q of E_{P_1} such that $D \cap Q = p_2$. Then $P_2 = j(Q)$ is a prime ideal of E contained in P_1 and $P_2 \cap D = j(Q) \cap D = j(Q) \cap E \cap D = Q \cap D = p_2$.

We now proceed to the proof that $p_2 E_{P_1} \cap D = p_2$. Let $a \in p_2 E_{P_1}$. Then $a = b/s$ where $b \in Ep_2$ and $s \in E - P_1$. By exercise 3 of section 7.6, b is integral over p_2 and hence by Proposition 7.18, the minimum polynomial of b over the field of fractions F of D has the form $\lambda^m - b_1\lambda^{m-1} + \cdots + (-1)^m b_m$ where $b_i \in p_2$, $1 \leq i \leq m$.

Suppose $a = b/s \in p_2 E_{P_1} \cap D$. Then $s = b/a$ and the minimum polynomial of s over F is $\lambda^m - (b_1/a)\lambda^{m-1} + \cdots + (-1)^m(b_m/a^m)$. Since s is integral over D, taking $I = D$ in Proposition 7.18, we see that every $b_i/a^i \in D$. Then $(b_i/a^i)a^i \in p_2$ and b_i/a^i and $a^i \in D$. Now suppose $a \notin p_2$. Then, since p_2 is prime in D, $b_i/a^i \in p_2$, $1 \leq i \leq m$. Thus s is integral over p_2 and hence over p_1. Then again by exercise 3, $s \in \text{nilrad}\, p_1 E \subset P_1$ contrary to $s \in E - P_1$. This shows that $a \in p_2$ so $p_2 E_{P_1} \cap D \subset p_2$. Since the reverse inequality is clear we have $p_2 E_{P_1} \cap D = p_2$. □

EXERCISES

1. Show that any factorial domain is integrally closed.
2. Let D be a domain, F its field of fractions. Show that if D is integrally closed then D_S is integrally closed for every submonoid S of the multiplicative monoid of R. On the other hand, show that if D_P is integrally closed for every maximal ideal P of D then D is integrally closed. (*Hint*: Use Proposition 7.11 on p. 402.)

7.8 RANK OF PROJECTIVE MODULES

We have shown in BAI, p. 171, that if R is a commutative ring and M is an R-module with a base of n elements, then any base has cardinality n. Hence the number n, called the *rank* of M, is an invariant. We repeat the argument in a slightly improved form. Let $\{e_i | 1 \leqslant i \leqslant n\}$ be a base for M and let $\{f_j | 1 \leqslant j \leqslant m\}$ be a set of generators. Then we have $f_j = \sum_1^n a_{ji} e_i$, $e_i = \sum_1^m b_{ik} f_k = \sum_{k,j} b_{ik} a_{kj} e_j$, which gives $\sum_{k=1}^m b_{ik} a_{kj} = \delta_{ij}$, $1 \leqslant i, j \leqslant n$. Assume $n \geqslant m$ and consider the $n \times n$ matrices

$$A = \begin{bmatrix} a_{11} & a_{12} & \cdots & a_{1n} \\ \cdot & \cdot & \cdots & \cdot \\ a_{m1} & a_{m2} & \cdots & a_{mn} \\ & 0 & & \end{bmatrix} \qquad B = \begin{bmatrix} b_{11} & \cdots & b_{1m} & \\ b_{21} & \cdots & b_{2m} & \\ \cdot & \cdots & \cdot & 0 \\ b_{n1} & \cdots & b_{nm} & \end{bmatrix}.$$

Then we have $BA = 1_n$, the $n \times n$ unit matrix. Since R is commutative, this implies $AB = 1_n$ (BAI, p. 97), which is impossible if $n > m$. Hence $n = m$. Thus we see that if M has a base of n elements, then any set of generators contains at least n elements. Hence any two bases have the same cardinality. The argument shows also that any set of n generators $f_j = \sum a_{ji}e_i$, $1 \leqslant j \leqslant n$, is a base, since the argument shows that in this case the matrix $A = (a_{ij})$ is invertible in $M_n(R)$, and this implies that the only relation of the form $\sum c_i f_i = 0$ is the one with every $c_i = 0$. We summarize these results in

PROPOSITION 7.20 *Let M be a free module over a commutative ring with base of n elements. Then* (1) *any base has cardinality n,* (2) *any set of generators contains at least n elements, and* (3) *any set of n generators is a base.*

We shall now give a method, based on localization, for extending the concept of rank to finitely generated projective modules over commutative rings R. We shall prove first that such modules are free if R is local. For this we require an important lemma for arbitrary rings known as

NAKAYAMA'S LEMMA. *Let M be a module over a ring R* (*not necessarily commutative*). *Suppose that* (1) *M is finitely generated and* (2) $(\text{rad}\, R)M = M$. *Then $M = 0$.*

Proof. Let m be the smallest integer such that M is generated by m elements $x_1, x_2, \ldots, x_m$. If $M \neq 0$, then $m > 0$. The condition $(\text{rad}\, R)M = M$ implies that $x_m = r_1 x_1 + \cdots + r_m x_m$ where the $r_i \in \text{rad}\, R$. Then we have

$$(1 - r_m)x_m = r_1 x_1 + \cdots + r_{m-1} x_{m-1}.$$

Since $r_m \in \text{rad}\, R$, $1 - r_m$ is invertible in R, and acting with $(1 - r_m)^{-1}$ on the foregoing relation shows that x_m can be expressed in terms of $x_1, \ldots, x_{m-1}$. Then $x_1, \ldots, x_{m-1}$ generate M, contrary to the choice of m. □

We recall that if B is an ideal in a ring R and M is an R-module, then BM is a submodule and the module M/BM is annihilated by B and so can be regarded in a natural way as R/B-module. In particular this holds for $B = \text{rad}\, R$. In this case we have the following consequence of Nakayama's lemma.

COROLLARY. *Let M be a finitely generated R-module. Then $x_1, \ldots, x_m \in M$ generate M if and only if the cosets $\bar{x}_1 = x_1 + (\text{rad}\, R)M, \ldots, \bar{x}_m = x_m + (\text{rad}\, R)M$ generate $M/(\text{rad}\, R)M$ as $R/\text{rad}\, R$ module.*

Proof. If $x_1, \ldots, x_m$ generate M, then evidently $\bar{x}_1, \ldots, \bar{x}_m$ generate $\bar{M} = M/(\text{rad}\, R)M$ as R-module and, equivalently, as $\bar{R} = R/\text{rad}\, R$-module. Conversely, suppose the $\bar{x}_i$ generate $\bar{M}$ as $\bar{R}$-module, hence as R-module. Then $M = (\text{rad}\ R)M + N$ where $N = \sum Rx_i$. Then $(\text{rad}\ R)(M/N) = M/N$ and M/N is finitely generated. Hence $M/N = 0$ by Nakayama's lemma, so $M = N = \sum_1^m Rx_i$. □

We can now prove

THEOREM 7.7. *If R is a (commutative) local ring, then any finitely generated projective module over R is free.*

Proof. We may assume that M is a direct summand of a free module $F = R^{(n)}: F = M \oplus N$ where M and N are submodules. Then $(\text{rad}\ R)F = (\text{rad}\ R)M + (\text{rad}\ R)N$ and $(\text{rad}\ R)M \subset M$, $(\text{rad}\ R)N \subset N$. Hence $\bar{F} = F/(\text{rad}\ R)F = \bar{M} \oplus \bar{N}$ where $\bar{M} = (M + (\text{rad}\ R)F)/(\text{rad}\ R)F$, $\bar{N} = (N + (\text{rad}\, R)F)/(\text{rad}\, R)F$. Now $\bar{R} = R/\text{rad}\, R$ is a field and evidently $\bar{F}$ is an n-dimensional vector space over $\bar{R}$, and $\bar{M}$ and $\bar{N}$ are subspaces. We can choose elements $y_1, \ldots, y_n$ so that $y_1, \ldots, y_r \in M$, $y_{r+1}, \ldots, y_n \in N$ and if $\bar{y}_i = y_i + (\text{rad}\, R)F$, then $\{\bar{y}_1, \ldots, \bar{y}_n\}$ is a base for $\bar{F}$ over $\bar{R}$. Then, by the Corollary to Nakayama's lemma, $y_1, \ldots, y_n$ generate F and, by Proposition 7.18, they form a base for F. Then $\{y_1, \ldots, y_r\}$ is a base for M so M is free. □

We now consider finitely generated projective modules over an arbitrary ring R and we prove first

PROPOSITION 7.21. *If M is a free R-module of rank n, then M_S is a free R_S-module of rank n for any submonoid S of the multiplicative monoid of R. If M is projective with n generators over R, then M_S is projective with n generators over R_S.*

Proof. To prove the first statement we recall that S-localization is a functor from R-**mod** to R_S-**mod**. Hence a relation $M \cong M_1 \oplus \cdots \oplus M_s$ for R-modules implies $M_S \cong M_{1S} \oplus \cdots \oplus M_{sS}$ for R_S-modules. Then $M \cong R \oplus \cdots \oplus R$ (n copies) implies $M_S \cong R_S \oplus \cdots \oplus R_S$ (n copies). Next suppose M is projective with n generators. This is equivalent to assuming that M is isomorphic to a direct summand of the free module $R^{(n)}$. Then M_S is R_S-projective with n generators. □

In particular, if M is projective with n generators, then for any prime ideal P, M_P is projective with n generators over the local ring R_P. Hence M_P is R_P-free of rank $n_P \leqslant n$. We call n_P the *P-rank* of M and we shall say that M *has a rank* if $n_P = n_Q$ for any two prime ideals P and Q of R. In this case the common value of the local ranks n_P is called *the rank* of the finitely generated projective module M. We shall now show that the map $P \rightsquigarrow n_P$ is a continuous map of Spec R into $\mathbb{Z}$ endowed with the discrete topology. This will imply that if Spec R is connected or, equivalently, R has no idempotents $\neq 0, 1$, then M has a rank.

The continuity we wish to establish will be an easy consequence of

PROPOSITION 7.22. *Let M be a finitely generated projective R-module, P a prime ideal in R. Then there exists $a \notin P$ such that $M_{\langle a \rangle}$ is free as $R_{\langle a \rangle}$-module.*

Proof. M_P is a free R_P-module of finite rank. If $(x_1/s_1, \ldots, x_n/s_n)$, $x_i \in M$, $s_i \in R - P$, is a base for M_P, and since the elements $s_i/1$ are units in R_P, $(x_1/1, \ldots, x_n/1)$ is also a base. Consider the R-homomorphism f of $R^{(n)}$ into M such that $(a_1, \ldots, a_n) \rightsquigarrow \sum a_i x_i$. We have the exact sequence

$$0 \to \ker f \xrightarrow{\iota} R^{(n)} \xrightarrow{f} M \xrightarrow{\nu} \operatorname{coker} f \to 0.$$

Localizing at P, we obtain the exact sequence

$$0 \to \ker f_P \xrightarrow{\iota_P} R_P^{(n)} \xrightarrow{f_P} M_P \xrightarrow{\nu_P} \operatorname{coker} f_P \to 0.$$

Since M_P is free with base $(x_1/1, \ldots, x_n/1)$, f_P is an isomorphism and hence $\ker f_P = 0$, $\operatorname{coker} f_P = 0$. Since M is finitely generated, so is its homomorphic image $\operatorname{coker} f$ and since $\operatorname{coker} f_P \simeq (\operatorname{coker} f)_P$, it follows from Proposition 7.5 that we have an element $b \notin P$ such that $b(\operatorname{coker} f) = 0$. Then $(\operatorname{coker} f)_{\langle b \rangle} = 0$ and we have the exact sequence

$$0 \longrightarrow \ker f_{\langle b \rangle} \xrightarrow{\iota_{\langle b \rangle}} R_{\langle b \rangle}^{(n)} \xrightarrow{f_{\langle b \rangle}} M_{\langle b \rangle} \longrightarrow 0.$$

Since $M_{\langle b \rangle}$ is projective, this splits, so $\ker f_{\langle b \rangle}$ is isomorphic to a homomorphic image of $R_{\langle b \rangle}^{(n)}$ and hence this is finitely generated. Since $\langle b \rangle$ is a submonoid of $R - P$, $0 = (\ker f)_P \cong \ker f_P$ is obtained by a localization of $(\ker f)_{\langle b \rangle}$ (Proposition 7.4). Since $(\ker f)_{\langle b \rangle}$ is finitely generated, this implies that there exists an element $c/1$, $c \notin P$ such that $(c/1)(\ker f)_{\langle b \rangle} = 0$. Then $\ker f_{\langle bc \rangle} = 0$. Hence if we put $a = bc$, then $a \notin P$ and we have

$$0 \longrightarrow R_{\langle a \rangle}^{(n)} \xrightarrow{f_{\langle a \rangle}} M_{\langle a \rangle} \longrightarrow 0,$$

so $M_{\langle a \rangle} \cong R_{\langle a \rangle}^{(n)}$ is free. This completes the proof. □

We now have

THEOREM 7.8. *The map* $r_M: P \rightarrow n_P(M)$ *of* Spec R *into* $\mathbb{Z}$ *(with discrete topology) is continuous.*

Proof. Let P be any point in Spec R. We have to show that there exists an open set O containing P such that $n_Q = n_P$ for all Q in O. Take $O = X_a$ where a is as in Proposition 7.20. Then $P \in X_a$ and if $Q \in X_a$, then M_Q is a localization of $M_{\langle a \rangle}$, which is a free $R_{\langle a \rangle}$-module of rank n. Then the rank of M_Q is n. Thus $n_Q = n = n_P$. $\square$

If Spec R is connected, the continuity of the map r_M implies that $n_P = n_Q$ for all $P, Q \in$ Spec R. Hence we have the

COROLLARY. *If R is connected, that is, R has no idempotents* $\neq 0, 1$, *then the rank is defined for every finitely generated projective R-module.*

Again let M be a finitely generated projective module and consider the continuous map r_M of Spec R into $\mathbb{Z}$. Since $\mathbb{Z}$ has the discrete topology, any integer n is an open and closed subset of $\mathbb{Z}$ and $r_M^{-1}(n)$ is an open and closed subset of Spec R. Moreover, Spec R is a disjoint union of these sets. Since Spec R is quasi-compact, the number of non-vacuous sets $r_M^{-1}(n)$ is finite. Thus we have positive integers $n_1, \ldots, n_s$ such that $r_M^{-1}(n_i) \neq \varnothing$ and $X = \text{Spec } R = \bigcup r_M^{-1}(n_i)$. By Theorem 7.3, the open and closed subset $r_M^{-1}(n_i)$ has the form X_{e_i} for an idempotent $e_i \in R$. Since $r_M^{-1}(n_i) \bigcap r_M^{-1}(n_j) = \varnothing$ if $i \neq j$, $e_i e_j = 0$ for $i \neq j$ and since $\bigcup r_M^{-1}(n_i) = X$, $\sum e_i = 1$. Thus the e_i are orthogonal idempotents in R with sum 1 and hence $R = R_1 \oplus \cdots \oplus R_s$ where $R_i = Re_i$ and $M = M_1 \oplus \cdots \oplus M_s$ where $M_i = e_i M = R_i M$. It is clear from the dual basis lemma (Proposition 3.11, p. 152) that M_i is a finitely generated projective R_i-module. We claim that M_i has a rank over R_i and this rank is n_i. Let P_i be a prime ideal in R_i. Then $P = P_i + \sum_{j \neq i} R_j$ is a prime ideal in R not containing e_i, so $P \in X_{e_i}$ and P contains every e_j, $j \neq i$. We have $R_P = R_{1P} \oplus \cdots \oplus R_{sP}$ and since $e_i \notin P$ and $e_i R_j = 0$ for $j \neq i$, $R_P = R_{iP}$. Similarly $M_P = M_{iP}$. Hence the rank of M_{iP} over R_{iP} is n_i. We have the homomorphism $r \rightarrow re_i$ of R into R_i in which the elements of $R - P$ are mapped into elements of $R_i - P_i$. Following this with the canonical homomorphism of R_i into R_{iP_i} gives a homomorphism of R into R_{iP_i} in which the elements of $R - P$ are mapped into invertible elements. Accordingly, we have a homomorphism of R_P into R_{iP_i} and we can use this to regard R_{iP_i} as R_P-module. Then $M_{iP_i} = R_{iP_i} \otimes_{R_P} M_{iP}$ and since M_{iP} is a free R_{iP} module of rank n_i, M_{iP_i} is a free R_{iP_i} module of rank n_i. Since this holds for every prime ideal P_i of R_i, we see that the rank of M_i over R_i is n_i. A part of what we have proved can be stated as

THEOREM 7.9. *Let M be a finitely generated projective module over a commutative ring R. Then there are only a finite number of values for the ranks $n_P(M)$ for the prime ideal P of R. If these values are $n_1,\ldots,n_s$, then $R = R_1 \oplus \cdots \oplus R_s$ where the R_i are ideals such that R_iM is finitely generated projective over R_i of rank n_i.*

EXERCISES

1. Let M be a finitely generated projective module over the commutative ring R and let P be a maximal ideal of R. Show that if M has a rank over R, then this is the dimensionality of M/PM regarded as vector space over the field R/P.
2. Let M and R be as in exercise 1 and assume that R is an algebra over R' such that R is finitely generated projective over R'. Suppose that M has a rank over R and R has a rank over R'. Show that M has a rank over R' and $\operatorname{rank} M/R' = (\operatorname{rank} M/R)(\operatorname{rank} R/R')$.
3. Show that if D is a domain, then any finitely generated projective module M over D has a rank and this coincides with the dimensionality of $F \otimes_D M$ over F where F is the field of fractions of D.

7.9 PROJECTIVE CLASS GROUP

We now make contact again with the Morita theory. We recall that if R' and R are rings, an R'-R-bimodule M is said to be invertible if there exists an R-R'-bimodule M' such that $M' \otimes_{R'} M \cong R$ as R-R-bimodule and $M \otimes_R M' \cong R'$ as R'-R'-bimodule. This is the case if and only if M is a progenerator of **mod**-R (R'-**mod**). Then $R' \cong \operatorname{End} M_R$ as rings and $M' \cong \hom(M_R, R_R)$ as R-R'-bimodule. If $R' = R$, the isomorphism classes of invertible bimodules form a group $\operatorname{Pic} R$ in which the multiplication is given by tensor products. Now suppose R is commutative. We restrict our attention to the R-R-bimodules in which the left action is the same as the right action. In effect we are dealing with (left) R-modules. The isomorphism classes of invertible modules constitute a subgroup of $\operatorname{Pic} R$ that is called the *projective class group* of the commutative ring R. The following result identifies the modules whose isomorphism classes constitute the projective class group.

THEOREM 7.10. *Let R be a commutative ring, M an R-module. Then M is invertible if and only if it is faithful finitely generated projective and rank $M = 1$.*

Proof. Suppose that M is invertible and M' is an R-module such that $M \otimes_R M' \cong R$. Then M and M' are finitely generated projective. Let P be a prime ideal in R. Then $M \otimes_R M' \cong R$ implies that $M_P \otimes_{R_P} M'_P \cong R_P$. Since these modules are free of finite rank over the local ring R_P, this relation implies that $\text{rank}\, M_P/R_P = 1$. Since this holds for all P, we see that $\text{rank}\, M = 1$. Conversely, suppose that M satisfies the stated conditions and let $M^* = \hom_R(M, R)$. Since M is faithful finitely generated projective, M is a progenerator by Theorem 3.22. Hence, as in the Morita theory, we have the isomorphism μ of $M \otimes_R M^*$ onto $\text{End}_R M$ sending $x \otimes y^*$, $x \in M$, $y^* \in M^*$, into the map $y \rightsquigarrow y^*(y)x$. Since R is commutative, any $r \in R$ determines an R-endomorphism $y \rightsquigarrow ry$. We can identify R with this set of endomorphisms and so $\text{End}_R M \supset R$. Hence to show that $M \otimes_R M^* \cong R$, it suffices to prove that $\text{End}_R M = R$. This will follow if we can show that $(\text{End}_R M)_P = R_P$ for every prime ideal P. Now $(\text{End}_R M)_P \cong R_P \otimes_R \text{End}_R M$. Since M is finitely generated projective, it is easily seen that $R_P \otimes_R \text{End}_R M \cong \text{End}_{R_P} M_P$ (see the proof of Proposition 3.14, p. 154). Now if M is a free module of rank n over a commutative ring R, then $\text{End}_R M$ is a free module of rank n^2 over R ($\cong M_n(R)$). By hypothesis, M_P has rank 1 over R_P. Hence $\text{End}_{R_P} M_P$ has rank 1 over R_P. Thus $(\text{End}_R M)_P$ has rank 1 over R_P and hence $(\text{End}_R M)_P = R_P$ for every P. This completes the proof. □

We shall not give any examples of projective class groups at this point. Later (section 10.6) we shall see that if R is an algebraic number field, then this group can be identified with a classical group of number theory.

7.10 NOETHERIAN RINGS

In the remainder of this chapter we shall be interested primarily in noetherian rings and modules. We recall that the noetherian condition that a module satisfy the ascending chain condition for submodules is equivalent to the maximum condition that every non-vacuous set of submodules contains submodules that are maximal in the set, and equivalent to the "finite basis condition" that every submodule has a finite set of generators (p. 103, exercise 1). We recall also that a (commutative) ring is noetherian if it is noetherian as a module with respect to itself, which means that every properly ascending chain of ideals in the ring terminates, that every non-vacuous set of ideals contains maximal ones, and that every ideal has a finite set of generators. Moreover, any one of these conditions implies the other two. In 1890 Hilbert based a proof of a fundamental theorem in invariant theory on the following theorem:

HILBERT'S BASIS THEOREM. *If R is a field or the ring of integers, then any ideal in the polynomial ring $R[x_1, x_2, \ldots, x_r]$, x_i indeterminates, has a finite set of generators.*

Hilbert's proof admits an immediate extension of his basis theorem to

HILBERT'S (GENERALIZED) BASIS THEOREM. *If R is a ring such that every ideal in R is finitely generated, then every ideal in $R[x_1,\dots,x_r]$ is finitely generated.*

Proof. Using induction it evidently suffices to prove the theorem for the case of one indeterminate x. We have to show that any ideal B of $R[x]$ has a finite set of generators. Let $j = 0, 1, 2, \dots$ and let $I_j = I_j(B)$ be the set of elements $b_j \in R$ such that there exists an element of the form

$$f_j = b_j x^j + a_{j-1}x^{j-1} + \cdots + a_0 \in B.$$

It is evident that I_j is an ideal in R, and since $f_j \in B$ implies that $xf_j = b_j x^{j+1} + \cdots \in B$, it is clear that $I_j \subset I_{j+1}$. Hence $I = \bigcup I_j$ is an ideal in R. This has a finite set of generators, and we may assume that these are contained in one of the ideals I_m. If $\{b_m^{(1)},\dots,b_m^{(k)}\}$ are these generators, we have polynomials $f_m^{(i)} = b_m^{(i)}x^m + g_m^{(i)} \in B$ where $\deg g_m^{(i)} < m$. Similarly, since every I_j is finitely generated for $j < m$, we have polynomials $f_j^{(i_j)} = b_j^{(i_j)}x^j + g_j^{(i_j)} \in B$ where $\deg g_j^{(i_j)} < j$ and $1 \leqslant i_j \leqslant k_j$. We claim that the finite set of polynomials

$$\{f_0^{(1)},\dots,f_0^{(k_0)},f_1^{(1)},\dots,f_1^{(k_1)},\dots,f_m^{(1)},\dots,f_m^{(k)}\}$$

generate B. Let $f \in B$. We prove by induction on $n = \deg f$ that $f = \sum_{i,j} h_{ij} f_j^{(i)}$ for suitable $h_{ij} \in R[x]$. This is clear if $f = 0$ or $\deg f = 0$, so we assume that $f = b_n x^n + f_1$ where $b_n \neq 0$ and $\deg f_1 < n$. Then $b_n \in I_n \subset I$. If $n \geqslant m$, then $b_n = \sum_i a_i b_m^{(i)}$, $a_i \in R$, and $\sum a_i x^{n-m} f_m^{(i)} \in B$ and has the same leading coefficient as f. Hence $\deg(f - \sum a_i x^{n-m} f_m^{(i)}) < n$. Since this polynomial is in B, the result follows by induction. If $n < m$, we have $b_n = \sum_i a_i b_n^{(i)}$, $a_i \in R$, and $\deg(f - \sum a_i f_n^{(i)}) < n$. Again the result follows by induction. $\square$

An alternative version of Hilbert's theorem is that if R is noetherian, then so is the polynomial ring $R[x_1,\dots,x_r]$. We also have the following stronger result.

COROLLARY. *Let R be noetherian and R' an extension ring of R, which is finitely generated (as algebra) over R. Then R' is noetherian.*

Proof. The hypothesis is that $R' = R[u_1,\dots,u_r]$ for certain $u_i \in R'$. Then R' is a homomorphic image of $R[x_1,\dots,x_r]$, x_i indeterminates. Since $R[x_1,\dots,x_r]$ is noetherian, so is R'. $\square$

There is another important class of examples of noetherian rings: rings of formal power series over noetherian rings. If R is an arbitrary ring, we can define the ring $R[[x]]$ of formal power series over R as the set of unrestricted sequences

$$a = (a_0, a_1, a_2, \ldots), \tag{19}$$

$a_i \in R$ with addition defined component-wise, $0 = (0, 0, \ldots)$, $1 = (1, 0, 0, \ldots)$, and product ab for a as above and $b = (b_0, b_1, \ldots)$ defined as $c = (c_0, c_1, \ldots)$ where

$$c_i = \sum_{j+k=i} a_j b_k. \tag{20}$$

It is easily checked that $R[[x]]$ is a ring (exercise 7, p. 127 of BAI). It is clear that R can be identified with the subring of $R[[x]]$ of elements $(a_0, 0, 0, \ldots)$ and $R[x]$ with the subring of sequences $(a_0, a_1, \ldots, a_n, 0, 0, \ldots)$, that is, the sequence having only a finite number of non-zero terms (BAI, pp. 116–118).

In dealing with formal power series, the concept of order of a series takes the place of the degree of a polynomial. If $a = (a_0, a_1, \ldots)$, we define the *order* $o(a)$ by

$$o(a) = \begin{cases} \infty & \text{if} \quad a = 0 \\ k & \text{if} \quad a_0 = \cdots = a_{k-1} = 0, \qquad a_k \neq 0. \end{cases} \tag{21}$$

Then we have

$$o(ab) \geqslant o(a) + o(b) \tag{22}$$

and

$$o(a+b) \geqslant \min(o(a), o(b)). \tag{23}$$

If R is a domain, then (22) can be strengthened to $o(ab) = o(a) + o(b)$. In any case if we define

$$|a| = 2^{-o(a)} \tag{24}$$

(with the convention that $2^{-\infty} = 0$), then we have the following properties of the map $a \rightsquigarrow |a|$ of $R[[x]]$ into $\mathbb{R}$:

(i) $|a| \geqslant 0$ and $|a| = 0$ if and only if $a = 0$.
(ii) $|a+b| \leqslant \max(|a|, |b|)$.
(iii) $|ab| \leqslant |a|\,|b|$.

The second of these implies that $|a+b| \leqslant |a| + |b|$. Hence (i) and (ii) imply that $R[[x]]$ is a metric space with distance function $d(a, b) = |a-b|$. We can therefore introduce the standard notions of convergence of sequences and series, Cauchy sequences etc.

We say that the sequence $\{a^{(i)}\}$, $a^{(i)} \in R[[x]]$, $i = 1, 2, 3, \ldots$, *converges to* $a \in R[[x]]$ if for any real $\varepsilon > 0$ there exists an integer $N = N(\varepsilon)$ such that $|a - a^{(i)}| < \varepsilon$ for all $i \geqslant N$. Then we write $\lim a^{(i)} = a$. The sequence $\{a^{(i)}\}$ is called a *Cauchy sequence* if given any $\varepsilon > 0$ there exists an integer N such that $|a^{(i)} - a^{(j)}| < \varepsilon$ for all $i, j \geqslant N$. $R[[x]]$ is complete relative to its metric in the sense that every Cauchy sequence of elements in $R[[x]]$ converges. For, let $\{a^{(i)}\}$ be a Cauchy sequence in $R[[x]]$. Then for any integer $n \geqslant 0$ we have an N such that $o(a^{(i)} - a^{(j)}) > n$ for all $i, j \geqslant N$. Let a_n be the entry in the $(n+1)$-st place of $a^{(N)}$. Then every $a^{(i)}$, $i \geqslant N$, has this element as its $(n+1)$-st entry. It is readily seen that if we take $a = (a_0, a_1, a_2, \ldots)$, then $\lim a^{(i)} = a$.

We can also define convergence of series in the usual way. We say that $a^{(1)} + a^{(2)} + \cdots = a$ if the sequence of partial sums $a^{(1)}, a^{(1)} + a^{(2)}, \ldots$ converges to a.

If we put $x = (0, 1, 0, \ldots)$, then x^i has 1 in the $(i+1)$th place, 0's elsewhere. It follows that if $a = (a_0, a_1, a_2, \ldots)$ and we identify a_j with $(a_j, 0, 0, \ldots)$, then we can write

$$(a_0, a_1, \ldots) = a_0 + a_1 x + a_2 x^2 + \ldots . \tag{25}$$

It is clear that if R is a domain, then so is $R[[x]]$. It is also easy to determine the units of $R[[x]]$, namely, $\sum a_i x^i$ is a unit if and only if a_0 is a unit in R. The condition is clearly necessary. To see that it is sufficient, we write $\sum_0^\infty a_i x^i = a_0(1 - z)$ where $z = \sum_1^\infty b_j x^j$ and $b_j = -a_0^{-1} a_j$. Then $o(z) \geqslant 1$ and $o(z^k) \geqslant k$. Hence $1 + z + z^2 + \cdots$ exists. It is readily seen that

$$(1 - z)(1 + z + z^2 + \cdots) = 1. \tag{26}$$

Hence $\sum_0^\infty a_i x^i = a_0(1 - z)$ is a unit with inverse $a_0^{-1}(\sum_0^\infty z^i)$. It is clear also that the set of power series $\sum_1^\infty b_i x^i$ of order > 0 is an ideal in $R[[x]]$. An immediate consequence of these remarks is

PROPOSITION 7.23. *If F is a field, then $F[[x]]$ is a local ring.*

Proof. The set of elements $\sum_0^\infty a_i x^i$ that are not units is the set for which $a_0 = 0$. This is an ideal. Thus the non-units constitute an ideal and hence $F[[x]]$ is local. □

It is easily seen also that if R is local, then so is $R[[x]]$. We leave the proof to the reader.

We shall prove next the following important

THEOREM 7.11. *If R is noetherian, then so is $R[[x]]$.*

Proof. The proof is quite similar to the proof of the Hilbert basis theorem. Let B be an ideal. For any $j = 0, 1, 2, \ldots$ let I_j be the set of $b_j \in R$ such that there exists an element $f_j = b_j x^j + g_j \in B$ where $o(g_j) > j$. Then I_j is an ideal in R and $I_0 \subset I_1 \subset \cdots$, so $I = \bigcup I_j$ is an ideal in R. Since R is noetherian, I has a finite set of generators and all of these are contained in I_m for some m. Let these generators be $b_m^{(1)}, \ldots, b_m^{(k)}$. It is clear from (22) and (23) that the set of elements of $R[[x]]$ of order $\geqslant m$ form an ideal. The intersection of this set with B is an ideal B_m in $R[[x]]$ containing the elements $f_m^{(i)}$, $1 \leqslant i \leqslant k$. Now let $f \in B_m$ and suppose $f = b_n x^n + g$ where $b_n \in R$, $n \geqslant m$ and $o(g) > n$. Then $b_n = \sum_i a_i b_n^{(i)}$ for $a_i \in R$ and

$$o\left(f - \sum_i a_i x^{n-m} f_m^{(i)}\right) > n.$$

Iteration of this process yields a sequence of integers $n_1 = n - m < n_2 < n_3 < \cdots$ and elements $a_{ij} \in R$, $1 \leqslant i \leqslant k$, $j = 1, 2, \ldots$ such that

$$o\left(f - \sum_{i=1}^{k} \left(\sum_{j=1}^{r} a_{ij} x^{n_j}\right) f_m^{(i)}\right) > n + n_r,$$

$r = 1, 2, \ldots$. Then $a_i = \sum_{j=1}^{\infty} a_{ij} x^{n_j}$ is well defined and $f = \sum_{i=1}^{k} a_i f_m^{(i)}$. Now consider the ideals I_j, $0 \leqslant j < m$. Choose a set of generators $\{b_j^{(1)}, \ldots, b_j^{(k_j)}\}$ for I_j and $f_j^{(i)} = b_j^{(i)} x_j + p_j^{(i)} \in B$ with $o(p_j^{(i)}) > j$. Then, as in the polynomial case,

$$\{f_0^{(1)}, \ldots, f_0^{(k_0)}, \ldots, f_m^{(1)}, \ldots, f_m^{(k)}\}$$

is a set of generators for B. □

We can iterate the process for forming formal power series to construct $R[[x, y]] = (R[[x]])[[y]]$, etc. We can also mix this construction with that of forming polynomial extensions. If we start with a noetherian R and perform these constructions a finite number of times, we obtain noetherian rings.

Another construction that preserves the noetherian property is described in the following

THEOREM 7.12. *Let R be noetherian and let S be a submonoid of the multiplicative monoid of R. Then the localization R_S is noetherian.*

Proof. Let B' be an ideal in R_S. As on p. 401 let $j(B')$ be the set of elements $b \in R$ such that $b/s \in B'$ for some $s \in S$. Then $j(B')$ is an ideal in R and $j(B')_S = B'$. Since R is noetherian, $j(B')$ has a finite set of generators $\{b_1, \ldots, b_m\}$. Then the set $\{b_1/1, \ldots, b_m/1\} \subset B'$ and generates this ideal. □

EXERCISES

1. Show that if I is an ideal in a commutative ring R such that I is not finitely generated and every ideal properly containing I is finitely generated, then I is prime. Use this to prove that if every prime ideal in R is finitely generated, then R is noetherian.

2. If R is a ring, we define the ring of *formal Laurent series* $R((x))$ over R as the set of sequences (a_i), $-\infty < i < \infty$, $a_i \in R$ such that there exists an integer n (depending on the sequence) such that $a_i = 0$ for $i < n$. Define addition and multiplication as for power series. Show that $R((x))$ is a localization of $R[[x]]$ and hence that $R((x))$ is noetherian if R is noetherian. Show that if R is a field, so is $R((x))$, and $R((x))$ contains the field $R(x)$ of rational expressions in x.

3. (Emmy Noether's finiteness theorem on invariants of a finite group.) Let $E = F[u_1, \dots, u_m]$ be a finitely generated algebra over the field F and let G be a finite group of automorphisms of E/F. Let $\operatorname{Inv} G = \{y \in E \mid sy = y, s \in G\}$. Show that $\operatorname{Inv} G$ is a finitely generated algebra over F. (*Sketch of proof.* Let $f_i(x) = \prod_{s \in G}(x - su_i) = x^n - p_{i1}x^{n-1} + p_{i2}x^{n-2} - \cdots$, where x is an indeterminate. Then $I = F[p_{11}, \dots, p_{mn}] \subset \operatorname{Inv} G$ and E is integral over I. Hence, by exercise 1, p. 411, E is a finitely generated I-module. Since I is noetherian, $\operatorname{Inv} G$ is finitely generated, say by $v_1, \dots, v_r$. Then $\operatorname{Inv} G = F[p_{11}, \dots, p_{mn}, v_1, \dots, v_r]$.)

7.11 COMMUTATIVE ARTINIAN RINGS

The study of artinian rings constitutes a major part of the structure theory of rings as developed in Chapter 4. It is interesting to see how this theory specializes in the case of commutative rings, and to consider relations between the artinian and noetherian conditions. Our first result in this direction is valid for rings that need not be commutative.

THEOREM 7.13. *If R is a ring that is left (right) artinian, then R is left (right) noetherian. Moreover, R has only a finite number of maximal ideals.*

Proof. Let $J = \operatorname{rad} R$, the Jacobson radical of R. By Theorem 4.3 (p. 202), J is nilpotent, so we have an integer n such that $J^n = 0$. We have the sequence of ideals $R \supset J \supset J^2 \supset \cdots \supset J^n = 0$. We regard R as left R-module. Then we have the sequence of R-modules $\bar{M}_i = J^i/J^{i+1}$ where $J^0 = R$ and all of these are annihilated by J, so they may be regarded as modules for the semi-primitive artinian ring $\bar{R} = R/J$. Since all modules of a semi-primitive artinian ring are completely reducible (Theorem 4.4, p. 208), this is the case for the modules $\bar{M}_i$. Moreover, any completely reducible artinian (noetherian) module is noetherian (artinian) and hence has a composition series. Accordingly, for each $\bar{M}_i$

we have a sequence of submodules

$$\bar{M}_i = \bar{M}_{i1} \supset \bar{M}_{i2} \supset \cdots \supset \bar{M}_{in_i} \supset \bar{M}_{i,n_i+1} = 0$$

such that every $\bar{M}_{ij}/\bar{M}_{i,j+1}$ is irreducible, and these can be regarded as R-modules. Corresponding to these we have a sequence of left ideals $M_{i1} = J^i \supset M_{i2} \supset \cdots \supset M_{i,n_i+1} = J^{i+1}$ such that $M_{ij}/M_{i,j+1}$ is an irreducible R-module. Putting together these sequences we obtain a composition series for R as left R-module. The existence of such a series implies that R is left noetherian as well as left artinian. The same argument applies to right artinian rings. This proves the first statement of the theorem.

To prove the second, we note that any maximal ideal I of R contains J and $\bar{I} = I/J$ is a maximal ideal of the semi-primitive artinian ring $\bar{R}$. We have $\bar{R} = \bar{R}_1 \oplus \cdots \oplus \bar{R}_s$ where the $\bar{R}_i$ are minimal ideals. It follows that the only maximal ideals of $\bar{R}$ are the ideals $\bar{M}_j = \bar{R}_1 + \cdots + \bar{R}_{j-1} + \bar{R}_{j+1} + \cdots + \bar{R}_s$ and that the only maximal ideals of R are the ideals $M_j = R_1 + \cdots + R_{j-1} + R_{j+1} + \cdots + R_s$ where R_j is the ideal in R such that $\bar{R}_j = R_j/J$. □

For commutative rings we have the following partial converse to Theorem 7.13.

THEOREM 7.14. *If R is a commutative noetherian ring that has only a finite number of prime ideals and all of these are maximal, then R is artinian.*

Proof. Since the Jacobson radical J is the intersection of the maximal ideals of R and the nil radical N is the intersection of the prime ideals of R, the hypothesis that the prime ideals are maximal implies that $J = N$. Hence J is a nil ideal and since R is noetherian, J is finitely generated. As is easily seen, this implies that J is nilpotent. Since R contains only a finite number of maximal ideals, $\bar{R} = R/J$ is a subdirect product of a finite number of fields. Then $\bar{R}$ is a direct sum of a finite number of fields (Lemma on p. 202) and hence $\bar{R}$ is artinian. As in the proof of Theorem 7.11, we have the chain of ideals $R \supset J \supset \cdots \supset J^{n-1} \supset J^n = 0$ and every J^i/J^{i+1} is a completely reducible noetherian module for the semi-primitive artinian ring $\bar{R}$. Then J^i/J^{i+1} has a composition series and R has a composition series as R-module. Then R is artinian. □

It is easy to determine the structure of commutative artinian rings. This is given in the following

THEOREM 7.15. *Let R be a commutative artinian ring. Then R can be written in one and only one way as a direct sum $R = R_1 \oplus R_2 \oplus \cdots \oplus R_s$ where the R_i are artinian and noetherian local rings and hence the maximal ideal of R_i*

is nilpotent. Conversely, if $R = R_1 \oplus \cdots \oplus R_s$ where the R_i are noetherian local rings with nilpotent maximal ideals, then R is artinian.

Proof. We base the proof of the first part on the results on modules that we obtained in connection with the Krull-Schmidt theorem (pp. 110–115) and the fact that for a commutative ring R we have the isomorphism $R \cong \mathrm{End}_R R$. We have seen that R has a composition series as R-module. Hence $R = R_1 \oplus \cdots \oplus R_s$ where the R_i are indecomposable R-modules with composition series. Hence $\mathrm{End}_R R_i$ is a local ring. Now $R_i \cong \mathrm{End}_{R_i} R_i = \mathrm{End}_R R_i$, so R_i is a local ring. Now there is only one decomposition of a ring into indecomposable ideals (p. 204). Hence the R_i are unique. The artinian property of R carries over to the R_i. Hence, by Theorem 7.13, every R_i is also noetherian. It is clear also that the maximal ideal, $\mathrm{rad}\, R_i$, of R_i is nilpotent.

To prove the converse, suppose first that R is a commutative local ring whose maximal ideal J is nilpotent. Let P be a prime ideal in R. Then $P \subset J$ and $J^n \subset P$ for some n. Since P is prime, this implies that $J = P$. Hence P is the only prime ideal of R. Since R is noetherian, R is artinian by Theorem 7.14. The general case in which $R = R_1 \oplus \cdots \oplus R_s$ follows immediately from this special case. □

If R is an artinian ring and M is a finitely generated R-module, then M has a composition series since R is also noetherian and hence M is artinian and noetherian. We can therefore define the length $l(M)$ of M as the length of any composition series for M. The use of the length provides a tool that is often useful in proving results on modules for artinian rings.

7.12 AFFINE ALGEBRAIC VARIETIES. THE HILBERT NULLSTELLENSATZ

Let F be an algebraically closed field, $F^{(n)}$ the n-dimensional vector space of n-tuples $(a_1, a_2, \ldots, a_n)$, $a_i \in F$, and $F[x_1, x_2, \ldots, x_n]$ the ring of polynomials in n indeterminates x_i over F. If S is a subset of $F[x_1, \ldots, x_n]$ we let $V(S)$ denote the set of points $(a_1, \ldots, a_n) \in F^{(n)}$ such that $f(a_1, \ldots, a_n) = 0$ for every $f \in S$ and we call $V(S)$ the *(affine algebraic) variety* defined by S. It is clear that $V(S) = V(I)$ where $I = (S)$, the ideal generated by S, and also $V(I) = V(\mathrm{nilrad}\, I)$. Hence if I_1 and I_2 are two ideals such that $\mathrm{nilrad}\, I_1 = \mathrm{nilrad}\, I_2$, then $V(I_1) = V(I_2)$. A fundamental result, due to Hilbert, is that conversely if $V(I_1) = V(I_2)$ for two ideals in $F[x_1, \ldots, x_n]$, F algebraically closed, then $\mathrm{nilrad}\, I_1 = \mathrm{nilrad}\, I_2$. There are a number of ways of proving this result. In this section we shall give a very natural proof, due to Seidenberg, which is based on Krull's theorem that the nil radical of an ideal is the intersection of the prime ideals containing it and on a general theorem of elimination theory that we proved in a completely elementary fashion in BAI, pp. 322–325. For convenience we quote the theorem on elimination of unknowns:

Let $K = \mathbb{Z}$ or $\mathbb{Z}/(p)$, p a prime, and let $A = K[t_1,\dots,t_r]$, $B = A[x_1,\dots,x_n]$ where the t's and x's are indeterminates. Suppose $F_1,\dots,F_m$, $G \in B$. Then we can determine in a finite number of steps a finite collection $\{\Gamma_1,\Gamma_2,\dots,\Gamma_s\}$ where $\Gamma_j = \{f_{j1},\dots,f_{jm_j};g_j\} \in A$ such that for any extension field F of K and any $(c_1,\dots,c_r)$, $c_i \in F$, the system of equations and inequations

$$F_1(c_1,\dots,c_r;x_1,\dots,x_n) = \cdots = F_m(c_1,\dots,c_r;x_1,\dots,x_n) = 0, \qquad (27)$$

$$G(c_1,\dots,c_r;x_1,\dots,x_n) \neq 0$$

is solvable for x's in some extension field E of F if and only if the c_i satisfy one of the systems Γ_j:

$$f_{j1}(c_1,\dots,c_r) = \cdots = f_{jn_j}(c_1,\dots,c_r) = 0, \qquad g_j(c_1,\dots,c_r) \neq 0. \qquad (28)$$

Moreover, when the conditions are satisfied, then a solution of (27) exists in some algebraic extension field E of F.

For our present purposes the important part of this result is the last statement. This implies the following theorem, which perhaps gives the real meaning of the algebraic closedness property of a field.

THEOREM 7.16. *Let F be an algebraically closed field and let $f_1,\dots,f_m$, $g \in F[x_1,\dots,x_n]$ where the x_i are indeterminates. Suppose that the system of equations and inequation*

$$f_1(x_1,\dots,x_n) = \cdots = f_m(x_1,\dots,x_n) = 0, \qquad g(x_1,\dots,x_n) \neq 0 \qquad (29)$$

has a solution in some extension field E of F. Then (29) *has a solution in F.*

Proof. The field F has one of the rings $K = \mathbb{Z}$ or $\mathbb{Z}/(p)$ as its prime ring. Now by choosing enough additional indeterminates $t_1,\dots,t_r$ we can define polynomials $F_1,\dots,F_m$, $G \in B$ as in the elimination theorem such that $F_i(c_1,\dots,c_r;\ x_1,\dots,x_n) = f_i(x_1,\dots,x_n)$, $G(c_1,\dots,c_r;\ x_1,\dots,x_n) = g(x_1,\dots,x_n)$. Then it follows from the result quoted that, if (29) has a solution in some extension field E of F, then we have a j for which (28) holds, and this in turn implies that (29) has a solution in an algebraic extension field of F. Since F is algebraically closed, the only algebraic extension field of F is F itself (BAI, p. 216, or p. 460 below). Hence (29) is solvable in F. □

Hilbert's theorem, the so-called Nullstellensatz, is an immediate consequence of this result and the characterization of the nil radical as intersection of prime ideals.

NULLSTELLENSATZ. *Let I be an ideal in the polynomial ring $F[x_1,\dots,x_n]$ in n indeterminates x_i over an algebraically closed field F and let $g \in F[x_1,\dots,x_n]$. Suppose $g(a_1,\dots,a_n) = 0$ for all $(a_1,\dots,a_n)$ on the variety $V(I)$ defined by I. Then $g \in \operatorname{nilrad} I$.*

Proof. Suppose $g \notin \operatorname{nilrad} I$. Then there exists a prime ideal $P \supset I$ such that $g \notin P$. Consider the domain $D = F[x_1,\dots,x_n]/P = F[\bar{x}_1,\dots,\bar{x}_n]$ where $\bar{x}_i = x_i + P$. Then for $f = f(x_1,\dots,x_n) \in I$, $f \in P$ and hence $f(\bar{x}_1,\dots,\bar{x}_n) = 0$. On the other hand, $g \notin P$, so $g(\bar{x}_1,\dots,\bar{x}_n) \neq 0$. Now $F[x_1,\dots,x_n]$ is noetherian, so I has a finite set of generators $f_1,\dots,f_m$. Let E be the field of fractions of D. Then E is an extension field of F containing the elements $\bar{x}_1,\dots,\bar{x}_n$ such that $f_i(\bar{x}_1,\dots,\bar{x}_n) = 0$, $1 \leqslant i \leqslant m$, $g(\bar{x}_1,\dots,\bar{x}_n) \neq 0$. Hence by Theorem 7.14, there exist $a_1,\dots,a_n \in F$ such that $f_i(a_1,\dots,a_n) = 0$, $g(a_1,\dots,a_n) \neq 0$. Since the f_i generate I, we have $f(a_1,\dots,a_n) = 0$ for all $f \in I$. This contradicts the hypothesis and proves that $g \in \operatorname{nilrad} I$. □

Evidently the Nullstellensatz implies that if I_1 and I_2 are ideals in $F[x_1,\dots,x_n]$ such that $V(I_1) = V(I_2)$, then $I_1 \subset \operatorname{nilrad} I_2$ so $\operatorname{nilrad} I_1 \subset \operatorname{nilrad}(\operatorname{nilrad} I_2) = \operatorname{nilrad} I_2$. By symmetry, $\operatorname{nilrad} I_2 \subset \operatorname{nilrad} I_1$ and so $\operatorname{nilrad} I_1 = \operatorname{nilrad} I_2$. An important special case of the Nullstellensatz is

THEOREM 7.17. *If I is a proper ideal in $F[x_1,\dots,x_n]$, F algebraically closed, then $V(I) \neq \varnothing$.*

Proof. If $V(I) = \varnothing$, then $g = 1$ satisfies the condition of the Nullstellensatz so $1 \in \operatorname{nilrad} I$ and hence $1 \in I$, contrary to the hypothesis that I is proper. □

The Nullstellensatz permits us also to determine the maximal ideals in $F[x_1,\dots,x_n]$. If $(a_1,\dots,a_n) \in F^{(n)}$, the ideal $M_{a_1,\dots,a_n} = (x_1 - a_1,\dots,x_n - a_n)$ is maximal in $F[x_1,\dots,x_n]$ since $F[x_1,\dots,x_n]/M_{a_1,\dots,a_n} \cong F$. Evidently $V(M_{a_1,\dots,a_n})$ consists of the single point $(a_1,\dots,a_n)$. If I is any proper ideal, the foregoing result and the Nullstellensatz imply that $I \subset \operatorname{nilrad} M_{a_1,\dots,a_n}$ for some point $(a_1,\dots,a_n)$. Since $M_{a_1,\dots,a_n}$ is maximal, it coincides with its nilradical, so $I \subset M_{a_1,\dots,a_n}$. Hence we see that the only maximal ideals of $F[x_1,\dots,x_n]$ are those of the form $M_{a_1,\dots,a_n}$. Moreover, since $V(M_{a_1,\dots,a_n}) = \{(a_1,\dots,a_n)\}$, we have the following result.

COROLLARY. *If F is an algebraically closed field, then the map $(a_1,\dots,a_n) \rightsquigarrow M_{a_1,\dots,a_n} = (x_1 - a_1,\dots,x_n - a_n)$ is a bijection of $F^{(n)}$ onto the set of maximal ideals of $F[x_1,\dots,x_n]$.*

We now suppose that F is any infinite field and we consider again $F^{(n)}$ and $F[x_1,\dots,x_n]$, and define the variety $V(S)$ for a subset S of $F[x_1,\dots,x_n]$ as in the algebraically closed case. Also if V_1 is a variety, we let $i(V_1)$ denote the ideal of polynomials $f \in F[x_1,\dots,x_n]$, which vanish for every $(a_1,\dots,a_n) \in V_1$. The following properties follow directly from the definition:

(1) $V(F[x_1,\dots,x_n]) = \varnothing$,
(2) $V(\varnothing) = F^{(n)}$,
(3) $V(\bigcup S_\alpha) = \bigcap V(S_\alpha)$,
(4) $V(S) = V(I)$ if $I = (S)$, the ideal generated by S,
(5) $V(I_1 I_2) = V(I_1) \cup V(I_2)$ for any ideals I_1 and I_2,
(6) $V(S_1) \supset V(S_2)$ if $S_1 \subset S_2$,
(7) $i(\varnothing) = F[x_1,\dots,x_n]$,
(8) $i(V_1) \supset i(V_2)$ if $V_1 \subset V_2$,
(9) $i(V(I_1)) \supset I_1$ for any ideal I_1,
(10) $V(i(V_1)) \supset V_1$ for any variety V_1.

Relations (6), (8), (9), and (10) have the immediate consequences that

(11) $V(i(V(I_1))) = V(I_1)$,
(12) $i(V(i(V_1))) = i(V_1)$.

We now recall an important theorem, proved in BAI on p. 136, that if F is an infinite field and $f(x_1,\dots,x_n)$ is a non-zero polynomial with coefficients in F, then there exist $a_i \in F$ such that $f(a_1,\dots,a_n) \neq 0$. Evidently this implies

(13) $i(F^{(n)}) = 0$.

Relations (1)–(5) show that the set of varieties satisfy the axioms for closed subsets of a topological space. The resulting topology is called the *Zariski topology* of the space $F^{(n)}$. The open subsets in this topology are the sets $F^{(n)} - V$, V a variety. If $V = V(S)$, then $F^{(n)} - V = \bigcup_{f \in S}(F^{(n)} - V(\{f\}))$. Thus the open sets $0_f = (F^{(n)} - V(\{f\}))$ form a base for the open subsets of $F^{(n)}$. Evidently 0_f is the set of points $(a_1,\dots,a_n)$ such that $f(a_1,\dots,a_n) \neq 0$.

We now observe that if F is algebraically closed (hence infinite), then $F^{(n)}$ with its Zariski topology is the same thing as the maximum spectrum of the ring $F[x_1,\dots,x_n]$. More precisely, we have a canonical homeomorphism between these two spaces. This is the map $(a_1,\dots,a_n) \rightsquigarrow M_{a_1,\dots,a_n}$ given in the corollary above. Since the condition that $f(a_1,\dots,a_n) = 0$ for a polynomial $f(x_1,\dots,x_n)$ is equivalent to $f \in M_{a_1,\dots,a_n}$, the map we have defined induces a bijective map of the set of closed sets in $F^{(n)}$ with the set of closed sets in Maxspec (see p. 403). Hence we have a homeomorphism.

When $F = \mathbb{R}$ or $\mathbb{C}$, the Zariski topology on $F^{(n)}$ has strikingly different properties from the usual Euclidean topology. Let us now consider some of these properties for $F^{(n)}$, F any infinite field. Since any point $(a_1,\ldots,a_n)$ is the variety defined by the ideal $M_{a_1,\ldots,a_n} = (x_1-a_1,\ldots,x_n-a_n)$, it is clear that $F^{(n)}$ is a T_1-space. However, it is not a Hausdorff space. On the contrary, $F^{(n)}$ is *irreducible* in the sense that any two non-vacuous open subsets of $F^{(n)}$ meet. In other words, any non-vacuous open subset is dense in $F^{(n)}$. Since the sets O_f constitute a base, it suffices to see that if $O_f \neq \varnothing$ and $O_g \neq \varnothing$, then $O_f \cap O_g \neq \varnothing$. The theorem on non-vanishing of polynomials implies that $O_f \neq \varnothing$ if and only if $f \neq 0$. It is clear also that $O_f \cap O_g = O_{fg}$. Since $F[x_1,\ldots,x_n]$ is a domain, it follows that $O_f \neq \varnothing$, $O_g \neq \varnothing$ imply $O_f \cap O_g \neq \varnothing$.

We note next that the space $F^{(n)}$ with the Zariski topology is *noetherian* in the sense that the ascending chain condition holds for open subsets of $F^{(n)}$. Equivalently the descending chain condition holds for varieties. Suppose that $V_1 \supset V_2 \supset \cdots$ is a descending chain of varieties. Then we have the ascending chain of ideals $i(V_1) \subset i(V_2) \subset \cdots$, so there exists an m such that $i(V_j) = i(V_{j+1})$ for all $j \geqslant m$. Since $V_j = V(i(V_j))$, we have $V_j = V_{j+1}$ for $j \geqslant m$.

Let us now look at the simplest case: $n = 1$. Here the variety $V(\{f\})$ defined by a single polynomial f is a finite set if $f \neq 0$ and is the whole space F if $f = 0$. Moreover, given any finite set $\{a_i|1 \leqslant i \leqslant r\}$ we have $V(\{f\}) = \{a_i\}$ for $f = \prod(x-a_i)$. It follows that the closed sets in the Zariski topology are the finite subsets (including the vacuous set) and the whole space. The open subsets of F are the vacuous set and the complements of finite sets. A subbase for the open sets is provided by the complements of single points, since any non-vacuous open set is a finite intersection of these sets.

It should be observed that the Zariski topology provides more open sets than one gets from the product topology obtained by regarding $F^{(n)}$ as product of n copies of F. For example, the open subset of $F^{(n)}$ defined by $x_1 + \cdots + x_n \neq 0$ cannot be obtained as a union of open subsets of the form $O_1 \times O_2 \times \cdots \times O_n$, O_i open in F.

Any polynomial $f(x_1,\ldots,x_n)$ defines a polynomial function

$$(a_1,\ldots,a_n) \rightsquigarrow f(a_1,\ldots,a_n) \tag{30}$$

of $F^{(n)}$ into F. We have considered such functions in BAI, pp. 134–138. The theorem we quoted on non-vanishing of polynomials shows that the map sending the polynomial $f(x_1,\ldots,x_n)$ into the function defined by (30) is an isomorphism of $F[x_1,\ldots,x_n]$ onto the ring of polynomial functions. More generally, consider a second space $F^{(m)}$. Then a sequence $(f_1,\ldots,f_m)$, $f_i = f_i(x_1,\ldots,x_n)$, defines a *polynomial map*

$$(a_1,\ldots,a_n) \rightsquigarrow (f_1(a_1,\ldots,a_n),\ldots,f_m(a_1,\ldots,a_n)) \tag{31}$$

of $F^{(n)}$ into $F^{(m)}$. Such maps are continuous of $F^{(n)}$ into $F^{(m)}$, both endowed with the Zariski topology. To see this we take any neighborhood O_g of $(f_1(a_1,\dots,a_n),\dots,f_m(a_1,\dots,a_n))$. Then we have a polynomial $g(x_1,\dots,x_m)$ such that $g(f_1(a_1,\dots,a_n),\dots,f_m(a_1,\dots,a_n)) \neq 0$. Then $h(x_1,\dots,x_n) = g(f_1(x_1,\dots,x_n),\dots,f_m(x_1,\dots,x_n)) \neq 0$ in $F[x_1,\dots,x_n]$ and O_h is mapped into O_g by (31). Thus (31) is continuous. We remark that since non-vacuous open subsets are dense, to prove that a polynomial map is 0 (that is, sends every element into 0), it suffices to show this for a non-vacuous open subset.

These rudimentary algebraic geometric ideas are often useful in "purely algebraic" situations. Some illustrations of the use of the Zariski topology will be given in the following exercises.

EXERCISES

In all of these we assume F algebraically closed (hence infinite). The topologies are Zariski.

1. Let $a \in M_n(F)$ and let $f_a(\lambda) = \lambda^n - \operatorname{tr}(a)\lambda^{n-1} + \cdots + (-1)^n \det a$ be the characteristic polynomial of a. Show that the maps $a \rightsquigarrow \operatorname{tr} a, \dots, a \rightsquigarrow \det a$ are polynomial functions on the n^2-dimensional vector space. Show that the set of invertible matrices is an open subset of $M_n(F)$.

2. Let $\{\rho_1,\dots,\rho_n\}$ be the characteristic roots of a (in some order) and let $g(x_1,\dots,x_n)$ be a symmetric polynomial in the x_i with coefficients in F. Show that the map $a \rightsquigarrow g(\rho_1,\dots,\rho_n)$ is a polynomial function on $M_n(F)$. Show that the set of matrices, which are similar to diagonal matrices with distinct diagonal entries, is an open subset of $M_n(F)$.

3. Let $f, g \in F[x_1,\dots,x_n]$ and suppose that $g(a_1,\dots,a_n) = 0$ for every $(a_1,\dots,a_n)$ such that $f(a_1,\dots,a_n) = 0$. Show that every prime factor of f is a factor of g.

4. If $a \in M_n(F)$, let U_a denote the linear transformation $y \rightsquigarrow aya$ of $M_n(F)$ into itself. Let $d(a) = \det U_a$. Use the Hilbert Nullstellensatz and Theorem 7.2 of BAI, p. 418, to prove that $d(a) = (\det a)^{2n}$.

5. Give an alternative proof of the result in exercise 4 by noting that it suffices to prove the relation for a in the open subset of matrices that are similar to diagonal matrices with non-zero diagonal entries. Calculate $d(a)$ and $\det a$ for a diagonal matrix.

6. Show that the nil radical of any ideal in $F[x_1,\dots,x_n]$ is the intersection of the maximal ideals containing it.

7. Let V be a variety in $F^{(n)}$. A polynomial function on V is defined to be a map $p|V$ where p is a polynomial function on $F^{(n)}$. These form an algebra $A(V)$ over F under the usual compositions of functions. Show that $A(V) \cong F[x_1,\dots,x_n]/i(V) = F[\bar{x}_1,\dots,\bar{x}_n]$, $\bar{x}_i = x_i + i(V)$. The latter algebra over F is called the *coordinate algebra*

of the variety V. V is called *irreducible* if $i(V)$ is prime. In this case the elements of the field of fractions of the coordinate algebra are called *rational functions on* V. If V' is a variety in $F^{(m)}$, a map of V into V' is called *regular* if it has the form $g|V$ where g is a polynomial map of $F^{(n)}$ into $F^{(m)}$. Show that if p' is a polynomial function on V', then $p'g$ is a polynomial function on V and the map $\eta(g): p' \rightsquigarrow p$ is an algebra homomorphism of $A(V')$ into $A(V)$. Show that $g \rightsquigarrow \eta(g)$ is a bijection of the set of regular maps of V into V' and the set of algebra homomorphisms of $A(V')$ into $A(V)$.

7.13 PRIMARY DECOMPOSITIONS

In this section we consider the classical Lasker-Noether decomposition theorems for ideals in noetherian rings as finite intersections of primary ideals. It is easy to see by using the ascending chain condition that any proper ideal I in a noetherian ring R can be written as an intersection of ideals that are indecomposable in the sense that $I \neq I_1 \cap I_2$ for any two ideals $I_j \neq I$. Moreover, in any noetherian ring, indecomposable ideals are primary in a sense that we shall define in a moment. This type of decomposition is a weak analogue of the decomposition of an element as a product of prime powers. Associated with every primary ideal is a uniquely determined prime ideal. However, primary ideals need not be prime powers and not every prime power is primary. Although the decomposition into primary ideals is not in general unique, it does have some important uniqueness properties. The establishment of these as well as the existence of the primary decomposition are the main results of the Lasker-Noether theory. We shall begin with the uniqueness questions, since these do not involve the noetherian property. The classical results can be generalized to modules and the passage to modules makes the arguments somewhat more transparent. In our discussion we shall consider the general case of modules first and then specialize these to obtain the results on ideals.

If M is a module for R and $a \in R$, then as in section 3.1, we denote the map $x \rightsquigarrow ax$, $x \in M$, by a_M. Since R is commutative, this is an endomorphism of M as R-module. We have the homomorphism $\rho_M : a \rightsquigarrow a_M$ of R into the ring of endomorphisms of M and $\ker \rho_M$ is the set of $a \in R$ such that $ax = 0$ for all x. As usual we write $\operatorname{ann}_R x$ for the ideal of elements $b \in R$ such that $bx = 0$. Evidently $\ker \rho_M = \bigcap_{x \in M} \operatorname{ann}_R x = \operatorname{ann}_R M$. If $ax = 0$ for some $x \neq 0$, then we shall call this element a of the ring R a *zero divisor* of the module M. The elements a that are not zero divisors are those for which a_M is injective. We now look at the nil radical of the ideals $\ker \rho_M$ and $\operatorname{ann}_R x$ for a particular x. By definition, the first of these is the set of elements $a \in R$ for which there exists an integer m such that $a^m \in \ker \rho_M$, that is, $a^m x = 0$ for every x or, equivalently,

${a_M}^m = (a^m)_M = 0$. Likewise $a \in \text{nilrad}(\text{ann}_R x)$ if and only if there exists an m such that $a^m x = 0$ or, equivalently, the restriction of a_M to the submodule Rx is nilpotent.

The case of primary interest is that in which $M = R/I$ where I is an ideal in R. Since we shall need to consider simultaneously such a module and the modules R/I' for I' an ideal of R containing I and since $R/I' \cong (R/I)/(I'/I)$, we need to formulate our definitions and results in terms of a module and a submodule. The style is set in the following

DEFINITION 7.4. *A submodule Q of an R-module M is called* primary *if $Q \neq M$ and for every $a \in R$ either $a_{M/Q}$ is injective or it is nilpotent.*

Evidently this means that if a is a zero divisor of M/Q, then there exists an integer m such that $a^m x \in Q$ for every $x \in M$. The set P of elements satisfying this condition is the ideal nilrad $(\text{ann}_R M/Q)$. This ideal is prime since if $a \notin P$ and $b \notin P$, then $a_{M/Q}$ and $b_{M/Q}$ are injective. Then $(ab)_{M/Q} = a_{M/Q} b_{M/Q}$ is injective and hence $ab \notin P$. We shall call P *the prime ideal associated* with the primary submodule Q. If $M = R$, the submodules are the ideals. Then the condition $a^m x \in Q$ for every $x \in R$ is equivalent to $a^m \in Q$. Thus in this case an ideal Q is primary if and only if $ab \in Q$, and $b \notin Q$ implies that there exists an m such that $a^m \in Q$. The associated prime of Q is the nilradical of this ideal.

EXAMPLES

1. Let R be a p.i.d. and $Q = (p^e) = (p)^e$, p a prime. It is readily seen that if $a \in R$ and $a \notin P = (p)$, then $a_{R/Q}$ is invertible, hence injective. On the other hand, if $a \in P$, then $a^e_{R/Q} = 0$. Hence Q is primary and P is the associated prime ideal. It is easy to see also that 0 and the ideals (p^e) are the only primary ideals in R.

2. Let $R = \mathbb{Z}$ and let M be a finite abelian group written additively and regarded as a $\mathbb{Z}$-module in the usual way. Suppose every element of M has order a power of a prime p. Then 0 is a p-primary submodule of M since if $a \in \mathbb{Z}$ and $(a, p) = 1$, then a_M is injective and if a is divisible by p, then a_M is nilpotent. The associated prime ideal of 0 is (p).

3. Let $R = F[x, y, z]/(xy - z^2)$ where F is a field and x, y, z are indeterminates. If $a \in F[x, y, z]$, let $\bar{a}$ be its image in R under the canonical homomorphism. Let $P = (\bar{x}, \bar{z})$, the ideal in R generated by $\bar{x}, \bar{z}$. Then the corresponding ideal in $F[x, y, z]$ is (x, z), since this contains $xy - z^2$. Since $F[x, y, z]/(x, z) \cong F[y]$, it follows that P is prime in R. Since it is clear from Krull's theorem that the nilradical of any power of a prime ideal is this prime, the ideal P^2 of R has P as its nilradical. However, this is not primary since $\bar{x} \notin P^2$, $\bar{y} \notin P$ but $\bar{x}\bar{y} = \bar{z}^2 \in P^2$.

We can obtain many examples of primary ideals by using the following

PROPOSITION 7.24. *If P is a maximal ideal in R, any ideal Q between P and P^e, $e \geqslant 1$, is primary with P as associated prime ideal.*

Proof. If $a \in P$, then $a^e \in P^e \subset Q$. If $a \notin P$, then we claim that $a+Q$ is a unit in R/Q. For, since P is maximal, R/P is a field. Hence we have an element $a' \in R$ and a $z \in P$ such that $aa' = 1-z$. Since $z+Q$ is nilpotent in R/Q, $(1-z)+Q$ is a unit in R/Q and hence $a+Q$ is a unit in R/Q. It is now clear that if $ab \in Q$, then $b \in Q$. Thus we have shown that if $a \in P$, then $a_{R/Q}$ is nilpotent and if $a \notin P$, then $a_{R/Q}$ is injective. Then it is clear that Q is primary and P is the associated prime. □

This result can be used to construct examples of primary ideals that are not prime powers. For instance, let $P = (x, y)$ in $F[x, y]$. This is maximal, since $F[x, y]/P \cong F$. We have $(x, y) \supsetneqq (x^2, y) \supsetneqq (x^2, xy, y^2) = P^2$. Hence (x^2, y) is a primary ideal that is not a prime power.

It is clear that if Q is a submodule of M and P is a subset of R such that 1) if $a \in P$, then there exists an integer m such that $a^m x \in Q$ for all $x \in M$ and 2) if $a \in R-P$, then $a_{M/Q}$ is injective, then Q is primary with P as associated ideal. We use this to prove

PROPOSITION 7.25. *The intersection of a finite number of primary submodules of M having the same associated prime ideal P is primary with P as associated prime ideal.*

Proof. It suffices to prove this for two submodules Q_1 and Q_2. Then $Q_1 \cap Q_2$ is a proper submodule since the Q_i are proper. Let $a \in P$. Then we have an integer m_i such that $a^{m_i} x \in Q_i$ for all $x \in M$. If we take $m = \max(m_1, m_2)$, then $a^m x \in Q = Q_1 \cap Q_2$. Next let $a \in R-P$ and let $x \in M-Q$. We may assume that $x \notin Q_1$. Then $ax \notin Q_1$ so $ax \notin Q$. Thus $a_{M/Q}$ is injective. Hence Q is primary with P as associated prime ideal. □

We have noted that for any x the set of $a \in R$ such that there exists an integer m such that $a^m x = 0$ is the ideal nilrad(ann x). Hence if Q is a submodule and $\bar{x} = x+Q$ in M/Q, then the set of a for which there exists an m such that $a^m x \in Q$ is the ideal nilrad($\text{ann}_R\, \bar{x}$). If Q is primary, we have the following important result.

PROPOSITION 7.26. *Let Q be a primary submodule of M, P the associated prime, and let $x \in M$. Let $I_x = \{a | a^m x \in Q$ for some $m\}$. Then $I_x = P$ if $x \notin Q$ and $I_x = R$ if $x \in Q$.*

Proof. Let $x \notin Q$. Evidently $I_x \supset P$, since $a_{M/Q}$ is nilpotent for every $a \in P$. On the other hand, if $a \notin P$, then $a_{M/Q}$ is injective. This precludes $a^m x \in Q$ for any m. Hence $I_x = P$. The other assertion that if $x \in Q$, then $I_x = R$ is clear. □

Now let N be a finite intersection of primary submodules Q_i of M: $N = \bigcap Q_i$. It may happen that some of these are redundant, that is, that N is an intersection of a proper subset of the Q_i. In this case we can drop some of these until we achieve an *irredundant decomposition* $N = \bigcap Q_i$, which means that for every j, $N \neq \bigcap_{i \neq j} Q_i$. This is equivalent to assuming that for every j, $Q_j \not\supset \bigcap_{i \neq j} Q_i$.

We shall prove two uniqueness theorems for decompositions of a submodule as intersection of submodules of M. Before taking this up in the general case, we illustrate the results in the special case of ideals of a ring in the following

EXAMPLE

Consider the ideal $I = (x^2, xy)$ in $F[x, y]$. Let $Q = (x)$, $Q_a = (y - ax, x^2)$ for $a \in F$. Q is prime, hence primary with associated ideal $P = Q$. Since $P_2 = (x, y)$ is maximal and $P_2 \supset Q_a \supset P_2{}^2 = (x^2, xy, y^2)$, Q_a is primary with associated prime P_2. Evidently $I \subset Q \cap Q_a$. Now let $h(x, y) \in Q_a$, so

$$h(x, y) = f(x, y)x^2 + g(x, y)(y - ax)$$

for $f(x, y), g(x, y) \in F[x, y]$. If $h(x, y) \in (x)$, we have $g(x, y) = xk(x, y)$ and then $h(x, y) \in I$. Hence $Q \cap Q_a \subset I$, so $I = Q \cap Q_a$. Since $Q \not\supset Q_a$ and $Q_a \not\supset Q$, the decomposition is irredundant. Since $Q_a \neq Q_b$ if $a \neq b$, we have distinct decompositions of I as irredundant intersections of primary ideals. It should be remarked that the associated primes P and P_2 are common to all of the decompositions and so is the primary ideal Q. This illustrates the two general uniqueness theorems, which we shall now prove.

The first of these is

THEOREM 7.18. *Let N be a submodule of M and let $N = \bigcap_1^s Q_i$ where the Q_i are primary submodules and the set is irredundant. Let $\{P_i\}$ be the set of prime ideals associated with the Q_i. Then a prime ideal $P \in \{P_i\}$ if and only if there exists an $x \in M$ such that the set $I_x(N)$ of elements $a \in R$ for which there exists an m such that $a^m x \in N$ is the ideal P. Hence the set of prime ideals $\{P_i\}$ is independent of the particular irredundant primary decomposition of N.*

Proof. Since $N = \bigcap_1^s Q_i$ and s is finite, it is clear that for any $x \in M$, $I_x(N) = \bigcap_1^s I_x(Q_i)$. By Proposition 7.26, $I_x(Q_i) = P_i$ if $x \notin Q_i$ and $I_x(Q_i) = R$ if $x \in Q_i$. Hence $I_x(N) = \bigcap P_{i_j}$ where the intersection is taken over those i_j such that $x \notin Q_{i_j}$. Now suppose that $I_x(N) = P$ a prime ideal. Then $P = \bigcap P_{i_j} \supset \prod P_{i_j}$, so $P \supset P_{i_j}$ for some i_j. Since $P_{i_j} \supset P$, we have $P = P_{i_j}$. Next consider any one of the associated primes P_i. By the irredundancy of the set of Q's we can choose an $x \notin Q_i, \in \bigcap_{j \neq i} Q_j$. Then the formula for $I_x(N)$ gives $I_x(N) = P_i$. This completes the proof. □

We shall call the set of prime ideals $\{P_i\}$ the *set of associated prime ideals of* $N = \bigcap_1^s Q_i$. These give important information on the submodule N. For example, we have

THEOREM 7.19. *Let N be a submodule of M that is an intersection of a finite number of primary submodules and let $\{P_i\}$ be the set of associated prime ideals of N. Then $\bigcup P_i$ is the set of zero divisors of M/N and $\bigcap P_i$ is the nilradical of* $\ker \rho_{M/N}$.

Proof. We may assume that $N = \bigcap Q_i$ as in the preceding theorem and P_i is associated with Q_i. (We may have $P_i = P_j$ for $i \neq j$.) The foregoing proof shows that if $a \in P_i$, we have an $x \in M - N$ such that $a^m x \in N$ for some positive integer m. If m is taken minimal, then $a^{m-1}x \notin N$ and $a(a^{m-1}x) \in N$ and so a is a zero divisor of M/N. Thus every element of $\bigcup P_i$ is a zero divisor of M/N. Conversely, let a be a zero divisor of M/N. Then we have an $x \in M, \notin N$ such that $ax \in N$. Since $x \notin N$, $x \notin Q_i$ for some i and $ax \in Q_i$. Hence a_{M/Q_i} is not injective, so $a \in P_i$. Thus $\bigcup P_i$ is the set of zero divisors of M/N. This is the first statement of the theorem. To prove the second we recall that nilrad ($\ker \rho_{M/N}$) is the set of $a \in R$ for which there exists an m such that $a^m x \in N$ for every $x \in M$. This is the intersection of the set of elements $a_i \in R$ for which there exists an m_i such that $a_i^{m_i} x \in Q_i$ for every $x \in M$. The latter set is P_i. Hence nilrad ($\ker \rho_{M/N}$) $= \bigcap P_i$. □

It is clear also from the uniqueness theorem and Proposition 7.25 that a submodule that is an intersection of primary submodules of M is primary if and only if it has only one associated prime ideal.

We now suppose that the irredundant decomposition $N = \bigcap_1^s Q_i$ into primary submodules is *normal* in the sense that distinct Q_i have distinct associated prime ideals. This can be achieved by replacing any set of the Q_i associated with the same prime by their intersection, which is primary, by Proposition 7.25. In the case of a normal decomposition we have a 1–1 correspondence between the primary submodules of the decomposition and their associated prime ideals. We shall now call a primary submodule Q_i in the normal decomposition $N = \bigcap_1^s Q_i$ *isolated* if the associated prime ideal P_i is minimal in the set of associated prime ideals of N. The second uniqueness theorem states that the isolated components persist in every normal decomposition of N as intersection of primary ideals:

THEOREM 7.20. *Let $N = \bigcap_1^s Q_i$ be a normal primary decomposition of N, P_i an associated prime that is minimal in the set of associated primes. Then we can choose an $a \in \bigcap_{j \neq i} P_j, \notin P_i$, and if a is any such element, then the isolated component Q_i corresponding to P_i can be characterized as the set of $x \in M$ such that $a^m x \in N$ for some integer m. Hence the isolated component Q_i is independent of the particular normal primary decomposition of N.*

Proof. If no a of the sort indicated exists, then $P_i \supset \bigcap_{j \neq i} P_j$ and hence $P_i \supset P_j$ for some $j \neq i$, contrary to the minimality of P_i. Now let $a \in \bigcap_{j \neq i} P_j, \notin P_i$. Then a_{M/Q_j} is nilpotent for $j \neq i$ and a_{M/Q_i} is injective. Hence if $x \in M$, there exists an integer m such that $a^m x \in Q_j$ for $j \neq i$ and hence $a^m x \in \bigcap_{j \neq i} Q_j$. On the other hand, a_{M/Q_i^m} is injective for every m, so $a^m x \in Q_i$ holds if and only if $x \in Q_i$. Since $N = \bigcap_1^s Q_k$, we see that the stated characterization of Q_i holds and its consequence is clear. □

In the example given on p. 433 the associated primes are (x) and (x, y), so (x) is minimal. The corresponding isolated component is (x) and this persists in all of the decompositions of $I = (x^2, xy)$. We remark also that in a ring R in which all non-zero primes are maximal (e.g., a p.i.d.), the primary decomposition is unique if none of the associated primes is 0.

We now assume that M is noetherian and we shall show that every submodule N of M can be expressed as a finite intersection of primary submodules. We shall call a submodule N of M (*intersection*) *indecomposable* if we cannot write N as $N_1 \cap N_2$ where $N_i \neq N$ for $i = 1, 2$. We first prove

LEMMA 1. *If M is noetherian, any submodule $N \neq M$ can be written as a finite intersection of indecomposable ones.*

Proof. If the result is false, the collection of submodules for which it is false has a maximal element N. Then N is not indecomposable, so we have submodules N_1 and N_2 such that $N_i \neq N$ and $N_1 \cap N_2 = N$. Then $N_i \supsetneq N$ and so, by the maximality of N, N_i is a finite intersection of indecomposable submodules. Then so is $N = N_1 \cap N_2$, contrary to assumption. □

The result we wish to prove on primary decompositions will be an immediate consequence of this lemma and

LEMMA 2. *If M is noetherian, then any indecomposable submodule N of M is primary.*

Proof. The proof is due to Noether and is similar to the proof of Fitting's lemma (p. 113), which it antedated. If N is not primary, we have an $a \in R$ such that $a_{M/N}$ is neither injective nor nilpotent. Consider the sequence of submodules of M/N

$$0 \subset \ker a_{M/N} \subset \ker a_{M/N}^2 \subset \cdots.$$

Since M/N is noetherian, we have an m such that $\ker a_{M/N}^m = \ker a_{M/N}^{m+1} = \cdots$. Since $a_{M/N}$ is not nilpotent, $\ker a_{M/N}^m \neq M/N$. Hence $\ker a_{M/N}^m = N_1/N$ where N_1 is a proper submodule of M containing N. Also since $a_{M/N}$ is not injective, $\ker a_{M/N}^m \neq 0$, so $N_1 \neq N$. Now consider the submodule $N_2 = a^m M + N$ and let $x \in N_1 \cap N_2$. Then $x = a^m y + z$ where $y \in M$, $z \in N$, and $a^m x \in N$. Hence $a^{2m} y \in N$ and the condition $\ker a_{M/N}^m = \ker a_{M/N}^{2m}$ gives $a^m y \in N$. Then $x = a^m y + z \in N$. Thus $N_1 \cap N_2 = N$ and $N_2 \neq N$, since $N_1 \neq M$. Hence $N = N_1 \cap N_2$ is decomposable, contrary to the hypothesis. □

Evidently the two lemmas have the following consequence.

THEOREM 7.21. *Any submodule of a noetherian module M has a decomposition as finite intersection of primary submodules of M.*

Everything we have done applies to ideals in noetherian rings. In this case we obtain that if I is a proper ideal in a noetherian ring, then $I = \bigcap Q_i$ where the Q_i are primary ideals. We may assume also that this representation of I is irredundant and the associated prime ideals P_i of the Q_i are distinct. The set $\{P_i\}$ is uniquely determined and these P_i are called the *associated primes* of I. The Q_j whose associated prime ideals P_j are minimal in the set $\{P_i\}$ are uniquely determined, that is, they persist in every normal representation of I as intersection of primary ideals.

Theorem 7.19 specializes to the following: If $\{P_i\}$ is the set of associated primes of I, then $\bigcup P_i$ is the set of zero divisors modulo I, that is, the set of elements $z \in R$ for which there exists a $y \in R - I$ such that $yz \in I$. Moreover, $\bigcap P_i$ is the nil radical of I. Evidently $\bigcap P_i = \bigcap P_j$ where $\{P_j\}$ is the set of associated primes that are minimal in $\{P_i\}$.

If I is a proper ideal in a ring R, a prime ideal P is said to be a *minimal prime over I* if $P \supset I$ and there exists no prime ideal P' such that $P \supsetneqq P' \supset I$. If R is noetherian, there are only a finite number of minimal primes over a given I: the minimal primes among the associated prime ideals of I. For, if P is any prime ideal containing I, then $P \supset \operatorname{nilrad} I = \bigcap P_j$, P_j a minimal associated prime ideal of I. Then $P \supset \prod P_j$ and $P \supset P_j$ for some j. Hence $P = P_j$ if and only if P is a minimal prime over I. Taking $I = 0$ (and $R \neq 0$) we see that a noetherian ring R contains only a finite number of prime ideals P that are minimal primes in the sense that there exists no prime ideal P' such that $P \supsetneqq P'$. These prime ideals are called the *minimal prime ideals* of R. Evidently R is a domain if and only if 0 is its only minimal prime ideal.

We also have the following result on nil radicals in the noetherian case.

THEOREM 7.22. *If I is an ideal in a noetherian ring R and N is its nil radical, then there exists an integer m such that $N^m \subset I$. In particular, if Q is a primary ideal in R and P is the associated prime, then $P^m \subset Q$ for some integer m.*

Proof. N has a finite set of generators $z_1, \ldots, z_r$, which means that N consists of the elements $\sum_1^r a_i z_i$, $a_i \in R$. We have an integer $m_i \geqslant 1$ such that $z_i^{m_i} \in I$. Put $m = \sum_1^r m_i - (r-1)$. Then any product of m of the z_i contains a factor of the form $z_i^{m_i}$ and hence is contained in I. It follows that any product of m elements of N is contained in I. Thus $N^m \subset I$. If Q is primary, the associated prime P is nilrad Q. Hence the second statement follows from the first. $\square$

EXERCISES

1. Let S be a submonoid of the multiplicative monoid of R, M an R-module, N a submodule that is an intersection of a finite number of primary submodules. Show that $M_S = N_S$ if and only if $(\bigcap P_i) \bigcap S \neq \varnothing$ where $\{P_i\}$ is the set of associated prime ideals of N. Show that if Q is a primary submodule and $Q_S \neq M_S$, then Q_S is a primary R_S-submodule of M_S. Show that if $N_S \neq M_S$ (N as before), then N_S has a decomposition as finite intersection of primary R_S-submodules.

2. Let N be a submodule of M and suppose that $N = \bigcap_1^s Q_i$ is a normal primary decomposition of N. Let $\{P_i\}$ be the set of associated primes. Call a subset T of $\{P_i\}$ *isolated* if for any $P_j \in T$, every $P_i \subset P_j$ is contained in T. Show that if T is isolated, then $\bigcap Q_j$ taken over the j such that $P_j \in T$ is independent of the given normal primary decomposition of N. This submodule is called *the isolated component of* N corresponding to T.

3. If I_1 and I_2 are ideals in a ring R, define the quotient $(I_1 : I_2) = \{a \in R \mid aI_2 \subset I_1\}$. Note that $(I_1 : I_2) \supset I_1$. Show that if R is noetherian, then $(I_1 : I_2) = I_1$ if and only if I_2 is not contained in any of the associated primes of I_1.

7.14 ARTIN-REES LEMMA. KRULL INTERSECTION THEOREM

In the remainder of this chapter we shall derive some of the main theorems of the theory of commutative noetherian rings. Some of these results were originally proved by using the primary decomposition theorems of the previous sections. Subsequently simpler proofs were devised based on a lemma that was discovered independently and about the same time by E. Artin and D. Rees. We shall begin with this result. For the proof we shall make use of a certain subring of a polynomial ring that is defined by an ideal in the coefficient ring.

Let R be a ring, $R[x]$ the polynomial ring over R in an indeterminate x, I an ideal in R. We introduce the *Rees ring* T defined by I to be the R-subalgebra of $R[x]$ generated by Ix. Thus

$$T = R + Ix + I^2x^2 + \cdots + I^nx^n + \cdots. \tag{32}$$

We now assume that R is commutative noetherian. Then I has a finite set of generators as R-module. If these are $b_1, b_2, \ldots, b_n$, then Ix is generated as R-module by $b_1x, b_2x, \ldots, b_nx$. It follows from (32) that T is generated as R-algebra by $b_1x, \ldots, b_nx$. Hence, by the Corollary to Hilbert's basis theorem (p. 418), T is noetherian.

Next let M be an R-module. Then we obtain the $R[x]$-module $R[x]\otimes_R M$ whose elements have the form $\sum x^i \otimes m_i$, $m_i \in M$. It is convenient to regard this $R[x]$-module as a set of polynomials in x with coefficients in the module M. For this purpose we introduce the set $M[x]$ of formal polynomials in x with coefficients in the module M. This set can be regarded as an $R[x]$-module in the obvious way. Using the universal property of $R[x]\otimes_R M$, we obtain a group homomorphism of $R[x]\otimes_R M$ onto $M[x]$ sending $\sum x^i \otimes m_i \rightsquigarrow \sum m_i x^i$, and using the fact that $\sum m_i x^i = 0$ implies every $m_i = 0$, we see that our homomorphism is an isomorphism. Moreover, it is clear that it is an $R[x]$-isomorphism. Hence we can identify $R[x]\otimes_R M$ with $M[x]$, and we prefer to work with the latter model. It is clear that any set of generators of M as R-module is a set of generators for $M[x]$ as $R[x]$-module.

We now consider the subset TM of $M[x]$. This is the T-submodule of $M[x]$ generated by M and it has the form

$$TM = M + (IM)x + (I^2M)x^2 + \cdots + (I^kM)x^k + \cdots. \tag{33}$$

It is clear that any set of generators for M as R-module is a set of generators for TM as T-module. Hence if M is a finitely generated R-module, then TM is a finitely generated T-module. Moreover, if R is noetherian, then T is noetherian and then TM is a noetherian T-module. We shall use this fact in the proof of the

ARTIN-REES LEMMA. *Let R be a commutative noetherian ring, I an ideal in R, M a finitely generated R-module, M_1 and M_2 submodules of M. Then there exists a positive integer k such that for every $n \geqslant k$*

(34) $$I^nM_1 \cap M_2 = I^{n-k}(I^kM_1 \cap M_2).$$

Proof. We have $I^{n-k}(I^kM_1 \cap M_2) \subset I^nM_1 \cap M_2$ for any k and $n \geqslant k$, so we have to show that there is a k such that $I^nM_1 \cap M_2 \subset I^{n-k}(I^kM_1 \cap M_2)$ for every $n \geqslant k$. For this purpose we define the ring T as before and the T-submodule TM of $M[x]$. We now define the subgroup

$$N = (M_1 \cap M_2) + (IM_1 \cap M_2)x + (I^2M_1 \cap M_2)x^2 + \cdots$$

of TM. Since $I^jM_1 \cap M_2$ is an R-submodule of I^jM and since $(Ix)(I^jM_1 \cap M_2)x^j \subset (I^{j+1}M_1 \cap M_2)x^{j+1}$, N is a T-submodule of TM. Since TM is a noetherian T-module, N has a finite set of generators, say, $u_1, \ldots, u_m$ where we can write $u_i = \sum_{j=0}^{k} n_{ij}x^j$ where $n_{ij} \in I^jM_1 \cap M_2$. Now let $n \geqslant k$ and let $u \in I^nM_1 \cap M_2$, so $ux^n \in N$. Then we have $f_i \in T$ so that $ux^n = \sum f_iu_i$. If we write $f_i = \sum f_{i_l}x^l$, $f_{i_l} \in I^l$, we obtain $ux^n = \sum f_{i_l}n_{ij}x^{j+l}$. Since $j \leqslant k$, comparing coefficients of x^n gives an expression for u as a sum of terms of the form fv where $v \in I^jM_1 \cap M_2$ with $j \leqslant k$ and $f \in I^{n-j}$. Now

$$I^{n-j}(I^jM_1 \cap M_2) = I^{n-k}I^{k-j}(I^jM_1 \cap M_2) \subset I^{n-k}(I^kM_1 \cap M_2).$$

Hence $u \in I^{n-k}(I^kM_1 \cap M_2)$. Thus $I^nM_1 \cap M_2 \subset I^{n-k}(I^kM_1 \cap M_2)$. □

Our first application of the Artin-Rees lemma is

KRULL'S INTERSECTION THEOREM. *Let R be a commutative noetherian ring, I an ideal in R, M a finitely generated R-module. Put $I^{\omega}M = \bigcap_{n=1}^{\infty} I^nM$. Then $I(I^{\omega}M) = I^{\omega}M$.*

Proof. Put $M_1 = M$, $M_2 = I^{\omega}M$. Then we have a k such that $I^nM \cap I^{\omega}M = I^{n-k}(I^kM \cap I^{\omega}M)$ for all $n \geqslant k$. Taking $n = k+1$ we obtain $I^{k+1}M \cap I^{\omega}M = I(I^kM \cap I^{\omega}M)$ and since $I^kM \cap I^{\omega}M = I^{\omega}M$ and $I^{k+1}M \cap I^{\omega}M = I^{\omega}M$, we have the required relation $I(I^{\omega}M) = I^{\omega}M$. □

In section 3.14 (p. 174) we derived a criterion for the equality $IM = M$ for I an ideal in a commutative ring R and M a finitely generated R-module: $IM = M$ if and only if $I + \text{ann}_R M = R$. In element form the latter condition is equivalent to: there exists $b \in I$ such that $(1+b)M = 0$. For, if b is such an element, then $1 = -b + (1+b)$ and $-b \in I$, $1 + b \in \text{ann}_R M$. Hence $1 \in I + \text{ann}_R M$ and since $I + \text{ann}_R M$ is an ideal, $I + \text{ann}_R M = R$. The converse is clear.

If we combine this result with the Krull intersection theorem, we see that if I is an ideal in a noetherian ring and M is a finitely generated module, then there exists a $b \in I$ such that $(1+b)I^{\omega}M = 0$. If $I = \operatorname{rad} R$, the Jacobson radical of R, then every $1+b$, $b \in I$, is invertible so we can conclude that $(\operatorname{rad} R)^{\omega}M = 0$. This follows also from Nakayama's lemma. At any rate we have

THEOREM 7.23. *Let M be a finitely generated module for a noetherian ring R and let $J = \operatorname{rad} R$. Then $\bigcap_{n=1}^{\infty} J^n M = 0$.*

We obtain an interesting specialization of the foregoing results by taking $M = R$ (which is generated by 1). Submodules of R are ideals. Then the Artin-Rees lemma states that if I, I_1, and I_2 are ideals in R, then there exists a k such that for all $n \geqslant k$

$$I^n I_1 \cap I_2 = I^{n-k}(I^k I_1 \cap I_2). \tag{35}$$

If we take $I_1 = R$ and change the notation slightly, we obtain the result that if I_1 and I_2 are ideals in a noetherian ring R, then there exists a k such that for $n \geqslant k$

$$I_1{}^n \cap I_2 = I_1^{n-k}(I_1{}^k \cap I_2). \tag{36}$$

This was the original form of the lemma given by Rees. If we take $M = R$ in Theorem 7.21, we obtain

$$\bigcap_{n=1}^{\infty} J^n = 0 \tag{37}$$

for J, the Jacobson radical of a commutative noetherian ring.

As we shall see in section 7.18, the results of this section play an important role in the study of completions of rings.

7.15 HILBERT'S POLYNOMIAL FOR A GRADED MODULE

In the same paper in which he proved the basis theorem and the theorem on syzygies, Hilbert proved another remarkable result, namely, a theorem on the dimensionality of the homogeneous components of a graded module for the polynomial ring $R = F[x_1, \ldots, x_m]$ with coefficients in a field F (see the reference on p. 387). He showed that if $M = \oplus M_n$ is graded where M_n is the homogeneous component of degree n, then there exists a polynomial $f(t)$ of degree $\leqslant m-1$ with rational coefficients such that for all sufficiently large n, $f(n) = [M_n : F]$. We shall prove this theorem in a generalized form in this section and apply it in succeeding sections to derive the main results on dimensionality in noetherian rings.

We recall that a ring R is graded by an indexed set of subgroups R_i, $i = 0, 1, 2, \ldots$, of its additive group if $R = \oplus R_i$ and $R_i R_j \subset R_{i+j}$ for all i, j. Then $R_0^{\,2} \subset R_0$ and it is easily seen that $1 \in R_0$. Hence R_0 is a subring of R. It is clear that $R^+ = \sum_{j>0} R_j$ is an ideal in R. If R is commutative, then R can be regarded as an algebra over R_0. We recall also the definition of a graded R-module M for the graded ring R (p. 384). Here we have $M = \oplus_{i \geqslant 0} M_i$ where the M_i are subgroups and $R_i M_j \subset M_{i+j}$ for all i, j. The elements of M_i are called *homogeneous of degree i*. We have the following

PROPOSITION 7.27. *If R is a graded commutative ring, then R is noetherian if and only if R_0 is noetherian and R is finitely generated as R_0-algebra. If the conditions hold and M is a finitely generated graded R-module, then every M_i is a finitely generated R_0-module.*

Proof. If R_0 is noetherian and R is a finitely generated R_0-algebra, then R is noetherian by the Corollary to Hilbert's basis theorem (p. 418). Now assume that R is noetherian. Then $R_0 \cong R/R^+$ is a homomorphic image of the noetherian ring R, hence is noetherian. The ideal R^+ of R is finitely generated as R-module by, say, $x_1, \ldots, x_m$. By replacing each x_i by its homogeneous parts, we may assume that the x_i are homogeneous elements of R^+. We claim the x_i are generators of R as R_0-algebra. It suffices to show that every homogeneous element can be written as a polynomial in the x_i with coefficients in R_0. We use induction on the degree of homogeneity. The result is clear if the degree is 0. Now suppose that u is homogeneous of degree $n > 0$. We can write $u = \sum_1^m u_i x_i$ where $u_i \in R$. Equating homogeneous parts, we may assume that u_i is homogeneous of degree $n - \deg x_i < n$, since $x_i \in R^+$. Then u_i is a polynomial in the x_j with coefficients in R_0 and hence the same is true of $u = \sum u_i x_i$.

Now suppose that R is noetherian and M is a finitely generated graded R-module. We may choose a set of generators $\{u_1, \ldots, u_r\}$ for M that are homogeneous. If the x's are chosen as before, then it is readily seen that every element of M_n is an R_0-linear combination of elements $y_i u_i$ where y_i is a monomial in the x's and $\deg y_i + \deg u_i = n$. Since the number of these elements is finite, M_n is finitely generated as R_0-module. $\square$

We now assume that (1) R is noetherian, (2) R_0 is artinian (as well as noetherian), and (3) M is a finitely generated graded R-module. Then every M_n is finitely generated as R_0-module and hence is artinian and noetherian. Hence M_n has a composition series as R_0-module and we have a uniquely determined non-negative integer $l(M_n)$, the length of the composition series for M_n. To obtain information on these lengths we introduce the *generating function* or *Poincaré series of the graded module M* as the element

$$P(M,t) = \sum_0^\infty l(M_n)t^n \tag{38}$$

of the field $\mathbb{Q}((t))$ of formal Laurent series (p. 425). We can now state the following

THEOREM 7.24 (Hilbert-Serre). *If R is generated by homogeneous elements* $x_1, \ldots, x_m$ *where* $\deg x_i = e_i > 0$, *then* $P(M,t)$ *is a rational expression of the form*

$$f(t)\Big/\prod_{i=1}^{m}(1-t^{e_i}) \tag{39}$$

where $f(t) \in \mathbb{Z}[t]$.

For the proof we require the

LEMMA. *Let* $0 \to M_0 \to M_1 \to \cdots \to M_s \to 0$ *be an exact sequence of finitely generated* R_0*-modules for an artinian ring* R_0 *and let* $l(M_i)$ *be the length of* M_i. *Then* $\sum_0^s(-1)^i l(M_i) = 0$.

Proof. The result is immediate from the definition of exactness if $s = 0$ or 1. If $s = 2$, we have the short exact sequence $0 \to M_0 \to M_1 \to M_2 \to 0$. Then $M_2 \cong M_1/M_0$ and $l(M_2) = l(M_1) - l(M_0)$, which gives $l(M_0) - l(M_1) + l(M_2) = 0$. Now suppose that $s > 2$. Then we can imbed the given sequence in a sequence of homomorphisms

$$0 \to M_0 \to \text{im}\, M_0 \to M_1 \to \text{im}\, M_1 \to \cdots \to \text{im}\, M_{s-1} \to M_s \to 0$$

where $M_i \to \text{im}\, M_i$ is obtained by restricting the codomain of $M_i \to M_{i+1}$ and where $\text{im}\, M_i \to M_{i+1}$ is an injection. Then we have the exact sequences $0 \to M_0 \to \text{im}\, M_0 \to 0$, $\quad 0 \to \text{im}\, M_{k-1} \to M_k \to \text{im}\, M_k \to 0$, $\quad 1 \leqslant k \leqslant s-1$, $0 \to \text{im}\, M_{s-1} \to M_s \to 0$, which give the relations

$$l(M_0) = l(\text{im}\, M_0)$$

$$l(M_k) = l(\text{im}\, M_{k-1}) + l(\text{im}\, M_k), \qquad 1 \leqslant k \leqslant s-1$$

$$l(M_s) = l(\text{im}\, M_{s-1}).$$

If we take the alternating sum of these equations, we obtain the required relation $\sum_0^\infty(-1)^i l(M_i) = 0$. $\square$

We can now give the

Proof of Theorem 7.24. We use induction on the number m of homogeneous generators. If $m = 0$, $R = R_0$ and M is a finitely generated R_0-module. Then $M_n = 0$ for sufficiently large n and the result holds with $f(t) = P(M,t)$.

Now assume the result holds if R has $m-1$ homogeneous generators. We act with x_m on M. This is an R-endomorphism x_m^* sending $M_n \to M_{n+e_m}$, $n = 0,1,2,\ldots$. Hence $\ker x_m^*$ and $\operatorname{im} x_m^*$ are homogeneous submodules, so we have the graded modules $K = \ker x_m^* = \oplus K_n$, $C = \operatorname{coker} x_m^* = \oplus C_n$. We have exact sequences

$$0 \to K_n \to M_n \to M_{n+e_s} \to C_{n+e_m} \to 0.$$

Hence, by the lemma,

$$l(M_n) - l(M_{n+e_m}) = l(K_n) - l(C_{n+e_m}).$$

If we multiply this relation by t^{n+e_m} and sum on $n = 0,1,2,\ldots$, we obtain

$$(40) \qquad (t^{e_m}-1)P(M,t) = P(K,t)t^{e_m} - P(C,t) + g(t)$$

where $g(t)$ is a polynomial in t. Now $x_m K = 0$ and $x_m C = 0$. Hence C and K are effectively R/Rx_m modules. Since R/Rx_m is a graded ring with $m-1$ homogeneous generators of degrees $e_1, \ldots, e_{m-1}$ and since C and K are graded modules for this ring, we have $P(k,t) = f_1(t)/\Pi_1^{m-1}(1-t^{e_i})$, $P(C,t) = f_2(t)/\Pi_1^{m-1}(1-t^{e_i})$. Substitution of these in (40) gives the theorem. □

The most important case of the foregoing result is that in which the generators can be chosen in R_1, that is, the $e_i = 1$. This is the case if $R = R_0[x_1, \ldots, x_m]$ where R_0 is artinian, the x_i are indeterminates, and the grading is the usual one in which R_n is the set of homogeneous polynomials of degree n. In fact, this case can be regarded as the most general one, since any graded ring with the $e_i = 1$ is a homomorphic image of $R_0[x_1, \ldots, x_m]$ under a graded homomorphism and any module for the image ring can be regarded as a module for $R_0[x_1, \ldots, x_m]$. For the applications it is convenient, however, to assume that R is any ring satisfying our earlier conditions and the condition that R is generated by elements of R_1. In this case, Theorem 7.24 states that $P(M,t) = f(t)/(1-t)^m$. We can use this to prove the main result on the lengths $l(M_n)$:

THEOREM 7.25 (Hilbert-Serre). *Suppose that R is generated by m homogeneous elements of degree* 1. *Then there exists a unique polynomial $\tilde{l}(t)$ of degree $\leqslant m-1$ with rational coefficients such that $l(M_n) = \tilde{l}(n)$ for sufficiently large n.*

Proof. We have $P(M,t) = \sum_0^\infty l(M_n)t^n = f(t)/(1-t)^m$ and $f(t) = (1-t)^m q(t) + r(t)$ where $\deg r(t) < m$. Then $r(t) = a_0 + a_1(1-t) + \cdots + a_{m-1}(1-t)^{m-1}$, $a_i \in \mathbb{Q}$, and hence

$$\frac{f(t)}{(1-t)^m} = q(t) + a_0 \frac{1}{(1-t)^m} + a_1 \frac{1}{(1-t)^{m-1}} + \cdots + a_{m-1} \frac{1}{1-t}. \tag{41}$$

It is readily seen, using the differentiation formula $((1-t)^{-k})' = k(1-t)^{-(k+1)}$, that

$$\frac{1}{(1-t)^k} = \frac{1}{(k-1)!} \sum_{n \geqslant 0} (n+k-1)\cdots(n+1)t^n.$$

The coefficient of t^n on the right-hand side is a polynomial in n of degree $k-1$ with rational coefficients. Hence the coefficient of t^n in $a_0/(1-t)^m + a_1/(1-t)^{m-1} + \cdots + a_{m-1}/(1-t)$ is $\tilde{l}(n)$ where $\tilde{l}$ is a polynomial of degree $\leqslant m-1$ with rational coefficients. By (37), $\tilde{l}(n) = l(n)$ if $n > \deg q(t)$. The uniqueness of $\tilde{l}$ is clear, since a polynomial $\in \mathbb{Q}(t)$ of degree $\leqslant m-1$ that vanishes for m consecutive integers must be 0. □

The polynomial $\tilde{l} = \tilde{l}(M)$ is called *Hilbert's* (*characteristic*) *polynomial* for the graded module M. Theorem 7.25 shows that $\tilde{l}$ has integral values at the non-negative integers. This does not imply that $\tilde{l} \in \mathbb{Z}[t]$. However, it is easy to determine the polynomials having this property. For any $n = 1, 2, 3, \ldots$ we define the polynomial

$$\binom{x}{n} = \frac{x(x-1)\cdots(x-n+1)}{n!} \tag{42}$$

and we put $\binom{x}{0} = 1$. Evidently the polynomials $\binom{x}{n} \in \mathbb{Q}[x]$ and since $\deg \binom{x}{n} = n$, the $\binom{x}{n}$ form a base for $\mathbb{Q}[x]/\mathbb{Q}$. As with binomial coefficients, we have

$$\Delta\binom{x}{n} \equiv \binom{x+1}{n} - \binom{x}{n} = \binom{x}{n-1}. \tag{43}$$

If x is an integer, then $\binom{x}{n}$ is a product of n consecutive integers divided by $n!$ It is readily seen that this is an integer. Hence any integral linear combination of the polynomials $1, \binom{x}{1}, \ldots, \binom{x}{n}$ has integral values for integral values of the argument. Conversely, we have

PROPOSITION 7.28. *Let $f(x)$ be a polynomial of degree n with rational coefficients such that $f(x)$ takes on integral values for some set of n consecutive integers. Then $f(x)$ is an integral linear combination of the polynomials $\binom{x}{j}$, $0 \leqslant j \leqslant n$.*

Proof. We can write $f(x) = a_0 + a_1\binom{x}{1} + \cdots + a_n\binom{x}{n}$ with rational a_i. Then $\Delta f(x) \equiv f(x+1) - f(x) = a_1 + a_2\binom{x}{1} + \cdots + a_n\binom{x}{n-1}$. Using induction on the degree we conclude that every a_i with $i > 0$ is an integer. It follows also that $a_0 \in \mathbb{Z}$. □

EXERCISES

1. Determine the lengths $l(R_n)$, $n = 0, 1, 2, \ldots$, for R_n, the set of homogeneous polynomials of degree n in $R = F[x_1, \ldots, x_m]$, F a field. Determine $P(R, t)$ as rational expression in t.

2. Let f be an integral valued function on $\mathbb{N} = \{0, 1, 2, \ldots\}$ such that $f(n+1) - f(n) = \tilde{g}(n)$ where $\tilde{g}(t) \in \mathbb{Q}[t]$ has degree $m-1$. Show that there exists a polynomial $\tilde{f}(t) \in \mathbb{Q}[t]$ of degree m such that $f(n) = \tilde{f}(n)$, $n \in \mathbb{N}$.

7.16 THE CHARACTERISTIC POLYNOMIAL OF A NOETHERIAN LOCAL RING

We consider first a general construction of a graded ring $G_I(R)$ and a graded module $G_I(M)$ from an ideal I and module M for the ring R. We put $G_I(R) = \oplus I^n/I^{n+1}$, $G_I(M) = \oplus I^nM/I^{n+1}M$, $n \geqslant 0$, where $I^0 = R$. If $\bar{a}_i \in I^i/I^{i+1}$ and $\bar{x}_j \in I^jM/I^{j+1}M$, we have $\bar{a}_i = a_i + I^{i+1}$, $\bar{x}_j = x_j + I^{j+1}M$, $x_j \in I^jM$. Then $a_ix_j + I^{i+j+1}M$ is independent of the choice of the representatives a_i, x_j. Hence if we put $\bar{a}_i\bar{x}_j = a_ix_j$, $(\sum \bar{a}_i)(\sum \bar{x}_j) = \sum \overline{a_ix_j}$, these are well defined. It is readily verified that if we take $M = R$, we obtain the structure of a ring on $G_I(R)$ with unit $1 + I$. Also $G_I(M)$ is a graded module for $G_I(R)$ with the $I^nM/I^{n+1}M$ as the homogeneous components. We shall call $G_I(R)$ and $G_I(M)$ the *graded ring* and *graded module associated with the ideal I*.

There is an alternative way of defining $G_I(R)$ and $G_I(M)$. We consider the rings T and modules TM we introduced in Section 7.14 for the proof of the Artin-Rees lemma. We defined $T = \sum_{n \geqslant 0} I^nx^n \subset R[x]$, x an indeterminate, and $TM = \sum_{n \geqslant 0}(I^nM)x^n \subset M[x] = R[x] \otimes_R M$. It is clear that T is a graded ring with nth homogeneous component $T_n = I^nx^n$ and TM is a graded module for T with nth homogeneous component $(I^nM)x^n$. We have a homomorphism of graded R-algebras of T onto $G_I(R)$ mapping bx^k, $b \in I^k$, $k \geqslant 1$, onto $b + I^{k+1}$. The kernel is IT. Hence we can identify $G_I(R)$ with T/IT. Similarly, we can identify $G_I(M)$ with TM/ITM, where $ITM = \Sigma(I^{n+1}M)x^n$.

We showed in Section 7.14 that if R is a noetherian ring, then T is noetherian and, moreover, if M is a finitely generated R-module, then TM is a finitely generated T-module. Using the foregoing isomorphisms, we obtain the following

PROPOSITION 7.29. *If R is noetherian, then $G_I(R)$ is noetherian for any ideal I and if R is noetherian and M is a finitely generated R-module, then $G_I(M)$ is a finitely generated $G_I(R)$-module.*

In the situation in which R/I is artinian we could apply the results of the last section to define a Hilbert polynomial for the graded module $G_I(M)$. For, it is clear that $G_I(R)$ is generated by its homogeneous component I/I^2 of degree one. We shall not pursue this further but instead we specialize to the case of primary interest in which R is local noetherian with maximal ideal J and $I = Q$ is a J-primary ideal (=primary ideal with J as associated prime). We recall that since J is maximal, any ideal Q such that $J^e \subset Q \subset J$ for some $e \geqslant 1$ is J-primary (Proposition 7.24, p. 435). On the other hand, if Q is J-primary, then $Q \subset J$ and $Q \supset J^e$ for some e (Theorem 7.22, p. 440). Hence the condition on Q is equivalent to $J^e \subset Q \subset J$ for some e. The condition we require that R/Q is artinian is certainly satisfied since R/Q is a local ring whose maximal ideal J/Q is nilpotent (Theorem 7.15).

We can apply the results on the Hilbert polynomial to the ring $G_Q(R)$ and module $G_Q(R)$. Suppose that $y_1, \ldots, y_m$ are elements of Q such that the cosets $y_i + Q^2$ generate Q/Q^2 as R-module. Then it is readily seen that the cosets $y_i + Q^2$ generate $Q/Q^2 + Q^2/Q^3 + \cdots$ as $G_Q(R)$-module. Hence, by the proof of Proposition 7.27, these elements generate $G_Q(R)$ as algebra over R/Q. It follows from the theorem of Hilbert-Serre that there exists a polynomial $\tilde{l}(t) \in \mathbb{Q}[t]$ of degree $\leqslant m-1$ such that $\tilde{l}(n) = l(Q^n/Q^{n+1})$ for sufficiently large n where $l(Q^n/Q^{n+1})$ is the length of Q^n/Q^{n+1} as R/Q-module.

We now switch to R/Q^n as R-module. We have the chain of submodules $R \supset Q \supset Q^2 \supset \cdots \supset Q^n$ with factors R/Q, $Q/Q^2, \ldots, Q^{n-1}/Q^n$, which can be regarded as R/Q-modules. As such they have composition series and lengths. Hence this is the case also for R/Q^n that has a length $l(R/Q^n)$. Since $l(R/Q^{n+1}) - l(R/Q^n) = l(Q^n/Q^{n+1}) = \tilde{l}(n)$ for sufficiently large n, it follows that there exists a polynomial $\chi_Q{}^R \in \mathbb{Q}[t]$ of degree $\leqslant m$ such that $\chi_Q{}^R(n) = l(R/Q^n)$ for sufficiently large n (exercise 2, p. 448). We can state this result without reference to the graded ring $G_Q(R)$ as follows:

THEOREM 7.26. *Let R be a noetherian local ring with maximal ideal J and let Q be a J-primary ideal. Suppose Q/Q^2 is generated by m elements as R-module. Then for any $n = 1, 2, 3, \ldots$, R/Q^n has a composition series as R-module* *and if $l(R/Q^n)$ is its length, then there exists a polynomial $\chi_Q{}^R \in \mathbb{Q}[t]$ of degree $\leqslant m$ such that $l(R/Q^n) = \chi_Q{}^R(n)$ for sufficiently large n.*

The polynomial $\chi_Q{}^R$ is called the *characteristic polynomial* of R *relative to* Q. The most important thing about this polynomial is its degree. For, this is independent of the choice of Q and hence is an invariant of R. To see this, let Q' be another J-primary ideal. Then there exist positive integers s and s' such that $Q'^s \subset Q$ and $Q^{s'} \subset Q'$. Then $R \supset Q^n \supset Q'^{sn}$ and $l(R/Q^n) \leqslant l(R/Q'^{sn})$. Hence $\chi_Q{}^R(n) \leqslant \chi_{Q'}^R(sn)$ for sufficiently large n. This implies that $\deg \chi_Q{}^R \leqslant \deg \chi_{Q'}^R$ and by symmetry $\deg \chi_Q{}^R = \deg \chi_{Q'}^R$. We shall denote the common degree of the $\chi_Q{}^R$ by $d = d(R)$.

7.17 KRULL DIMENSION

We shall now introduce a concept of dimension for noetherian rings that measures the size of such a ring by the length of chains of distinct prime ideals in the ring. For our purposes it is convenient to denote a finite chain of distinct prime ideals by $P_0 \supsetneqq P_1 \supsetneqq \cdots \supsetneqq P_s$ and call s the *length of the chain*. With this convention we have the following

DEFINITION 7.4. *If R is a noetherian ring, the* Krull dimension *of R, denoted as* $\dim R$, *is the supremum of the lengths of chains of distinct prime ideals in R. If no bound exists for such claims, then* $\dim R = \infty$.

We remark that the Krull dimension of a noetherian ring may be infinite. Examples of such rings have been given by M. Nagata. We observe also that in considering a chain $P = P_0 \supsetneqq P_1 \supsetneqq \cdots \supsetneqq P_s$ we can pass to the localization R with maximal ideal P. The correspondence between prime ideals of R_P and prime ideals of R contained in P gives a chain of prime ideals $P_P \supsetneqq P_{1P} \supsetneqq \cdots \supsetneqq P_{sP}$ in R. We shall therefore consider first dimension theory for noetherian local rings. If R is such a ring with maximal ideal J, then it is clear that it suffices to consider chains that begin with $P_0 = J$. The fundamental theorem on dimension in noetherian local rings is the following

THEOREM 7.27. *Let R be a noetherian local ring. Then the following three integers associated with R are equal:*

(1) *The degree $d = d(R)$ of the characteristic polynomial $\chi_Q{}^R$ of any J-primary ideal Q of R.*

(2) *The minimum number $m = m(R)$ of elements of R that generate a J-primary ideal Q of R.*
(3) $\dim R$.

(The first two numbers are evidently finite. A priori the third may not be. The assertion of the theorem is that it is, and it is the same as $d(R)$ and $m(R)$. Thus the theorem implies that the dimension of any noetherian local ring is finite.)

Proof. The theorem will be proved by showing that $d \leqslant m \leqslant \dim R \leqslant d$.

$d \leqslant m$. This is a part of Theorem 7.26.

$m \leqslant \dim R$. There is nothing to prove if $\dim R = \infty$, so we assume that $\dim R < \infty$. We use induction on $\dim R$. Now $\dim R = 0$ if and only if J is the only prime ideal in R. In this case R is artinian by Theorem 7.14. Then $J^e = 0$ for some e and 0 is J-primary. Hence $m = 0$. Thus the inequality $m \leqslant \dim R$ holds if $\dim R = 0$. Now suppose that $0 < \dim R < \infty$. Let $P_1, \ldots, P_r$ be the minimal primes of R (see p. 433). Since any prime ideal in R contains one of the P_i and every prime ideal is contained in J, $J \not\subset P_i$, $1 \leqslant i \leqslant r$. Hence, by the prime avoidance lemma (p. 390) there exists an element $x \in J, \notin \bigcup P_i$. Consider the ring $R' = R/Rx$. This is a local ring with maximal ideal J/Rx. Any chain of distinct prime ideals in R' has the form $P'_0/Rx \supsetneqq \cdots \supsetneqq P'_s/Rx$ where the P'_i are prime ideals in R containing Rx. Now P'_s contains one of the minimal primes P_i and $P'_s \neq P_i$ since $x \notin P_i$. Hence we have a chain of prime ideals $P'_0 \supsetneqq \cdots \supsetneqq P'_s \supsetneqq P_i$ for some i. This shows that $\dim R' \leqslant \dim R - 1$. On the other hand, if $y_1 + Rx, \ldots, y_{m'} + Rx$ is a set of generators for a J/Rx-primary ideal Q'/Rx in R', then $y_1, \ldots, y_{m'}$, x is a set of generators for the J-primary ideal Q' in R. Hence $m(R) \leqslant m(R') + 1$. The induction hypothesis gives $m(R') \leqslant \dim R'$. Hence $m(R) \leqslant m(R') + 1 \leqslant \dim R' + 1 \leqslant \dim R$.

$\dim R \leqslant d$. We shall prove this by induction on d. If $d = 0$, then $l(R/J^n)$ is constant for sufficiently large n. Hence $J^n = J^{n+1}$ for sufficiently large n and $J^n = 0$ by Nakayama's lemma. Hence R is artinian and J is the only prime ideal in R. (Theorems 7.14 and 7.15, pp. 426–427.) Then $\dim R = 0$ and $\dim R = 0 = d$.

Now assume $d > 0$. We have to show that if $P_0 = J \supsetneqq P_1 \supsetneqq \cdots \supsetneqq P_s$ is a chain of prime ideals in R, then $s \leqslant d$. We reduce the proof to the case of a domain by forming the domain R/P_s, which is local with maximal ideal J/P_s and has the chain of prime ideals $J/P_s \supsetneqq P_1/P_s \supsetneqq \cdots \supsetneqq P_s/P_s = 0$. We have $(J/P_s)^n = (J^n + P_s)/P_s$, so $(R/P_s)/(J/P_s)^n = (R/P_s)/((J^n + P_s)/P_s) \cong R/(J^n + P_s)$. Hence $l((R/P_s)/(J/P_s)^n) = l(R/(J^n + P_s)) \leqslant l(R/J^n)$. It follows that $d(R/P_s) \leqslant d(R)$. Hence it suffices to prove $s \leqslant d$ in the case in which R is a domain and we have the chain of prime ideals $J = P_0 \supsetneqq \cdots \supsetneqq P_{s-1} \supsetneqq 0$. Let $x \in P_{s-1}$, $x \neq 0$, and consider R/Rx, which is local with maximal ideal J/Rx and has the chain of prime ideals $J/Rx \supsetneqq \cdots \supsetneqq P_{s-1}/Rx$. We wish to show that $d(R/Rx) \leqslant d(R)-1$. As above, $l((R/Rx)/(J/Rx)^n) = l(R/(J^n+Rx))$. Now $l(R/(J^n+Rx)) = l(R/J^n) - l((J^n+Rx)/J^n)$ and $l((J^n+Rx)/J^n) = l(Rx/J^n \cap Rx)$. By the Artin-Rees lemma there exists a positive integer k such that $J^n \cap Rx = J^{n-k}(J^k \cap Rx)$ for all $n \geqslant k$. Hence

$$l((J^n+Rx)/J^n)) \geqslant l(Rx/J^{n-k}x).$$

Since $x \neq 0$ in the domain R, multiplication by x is an R-module isomorphism of R onto Rx and this maps J^{n-k} onto $J^{n-k}x$. Hence $R/J^{n-k} \cong Rx/J^{n-k}x$ as R-modules and so $l(Rx/J^{n-k}x) = l(R/J^{n-k})$. Hence

$$\begin{aligned} l((R/Rx)/(J/Rx)^n &= l(R/(J^n+Rx)) \\ &= l(R/J^n) - l(Rx/J^n \cap Rx) \\ &\leqslant l(R/J^n) - l(R/J^{n-k}). \end{aligned}$$

This implies that

$$d(R/Rx) \leqslant d(R)-1.$$

Using the induction on d we conclude that $s-1 \leqslant d(R)-1$ and so $s \leqslant d(R)$. □

We shall now apply Theorem 7.27 to arbitrary noetherian rings. If R is such a ring, we define the *height* of a prime ideal P in R as the supremum of the lengths of chain of prime ideals $P_0 = P \supsetneqq P \supsetneqq \cdots \supsetneqq P_s$. It is clear that the height, ht P, is the dimensionality of the noetherian local ring R_P. This is finite by Theorem 7.27 Evidently this has the following rather surprising consequence.

THEOREM 7.28. *Any noetherian ring satisfies the descending chain condition for prime ideals.*

Another quick consequence of the main theorem is the

GENERALIZED PRINCIPAL IDEAL THEOREM (Krull). *If I is a proper ideal in a noetherian ring and I is generated by m elements, then any minimal prime over I has height $\leqslant m$.*

Proof. Let P be a minimal prime over I. The assertion is equivalent to ht $P_P \leqslant m$ in R_P. Now P_P is a minimal prime over I_P in R_P and since P_P is the only maximal ideal in R_P, P_P is the only prime ideal of R_P containing I_P.

Hence I_P is P_P-primary. On the other hand, if $y_1, \ldots, y_m$ are generators of I, then $y_1/1, \ldots, y_m/1$ generate I_P. Hence $\operatorname{ht} P_P = \dim R_P \leqslant m$. □

A special case of the foregoing theorem is the

PRINCIPAL IDEAL THEOREM *Let y be a non-unit in a noetherian ring R. Then any minimal prime ideal over Ry has height $\leqslant 1$.*

It is clear from the definitions that the Krull dimension of a noetherian ring is the supremum of the heights of its maximal ideals or, equivalently, the supremum of the Krull dimensions of its localizations at maximal ideals.

We now consider the Krull dimension of a polynomial ring $F[x_1, \ldots, x_m]$ in m indeterminates over an algebraically closed field F. We recall that any maximal ideal M in $F[x_1, \ldots, x_m]$ is generated by m elements $x_1 - a_1$, $x_2 - a_2, \ldots, x_m - a_m$, $a_i \in F$ (p. 426). Hence $\operatorname{ht} M \leqslant m$. On the other hand, we have the chain of prime ideals

$$\sum_1^m R(x - a_i) \supsetneqq \sum_1^{m-1} R(x - a_i) \supsetneqq \cdots \supsetneqq R(x_n - a_n) \supsetneqq 0. \tag{44}$$

It follows that $\operatorname{ht} M = m$. Since this holds for any maximal ideal, it follows that $\dim F[x_1, \ldots, x_m] = m$.

We shall obtain next an important application of the generalized principal ideal theorem to systems of polynomial equations with coefficients in an algebraically closed field. This is the following

THEOREM 7.29. *Let F be an algebraically closed field and let $f_1(x_1, \ldots, x_m), \ldots, f_n(x_1, \ldots, x_m) \in R = F[x_1, \ldots, x_m]$ where the x_i are indeterminates. Assume that the system of equations*

$$f_1(x_1, \ldots, x_m) = 0, \ldots, f_n(x_1, \ldots, x_m) = 0 \tag{45}$$

has a solution in $F(m)$ and $m > n$. *Then* (45) *has an infinite number of solutions in $F^{(m)}$.*

Proof. Assume to the contrary that the system has only a finite number of solutions: $(\alpha_1^{(r)}, \ldots, \alpha_m^{(1)}), \ldots, (\alpha_1^{(r)}, \ldots, \alpha_m^{(r)})$. Let $I = \sum_1^n Rf_i$, so the variety $V(I) = S = \{(\alpha_1^{(1)}, \ldots, \alpha_m^{(1)}), \ldots, (\alpha_1^{(r)}, \ldots, \alpha_m^{(r)})\}$. Let $I_j = \sum_{i=1}^m R(x_i - \alpha_i^{(j)})$, $1 \leqslant j \leqslant r$. Then $V(I_j)$ consists of the single point $(\alpha_1^{(j)}, \ldots, \alpha_m^{(j)})$ and $S = \bigcup_1^r V(I_j) = V(\prod_1^r I_j)$ (p. 430). Hence if $I' = \prod_1^r I_j$, then $V(I') = S = V(I)$ and so, by the Hilbert Nullstellensatz, nilrad I = nilrad I'. Let P be a minimal prime ideal over I. Then $P \supset$ nilrad I = nilrad I'. Hence $P \supset I' = \prod I_j$, so $P \supset I_j$ for some j and since I_j is maximal, $P = I_j$. We have seen that the height of any maximal ideal in R is m. Hence the height of any minimal prime over $I = \sum_1^n Rf_i$ is $m > n$, contrary to the generalized principal ideal theorem. □

A useful special case of this theorem is obtained by taking the f_i to be polynomials with 0 constant terms (e.g., homogeneous polynomials of positive degree). These have the trivial solution $(0, \ldots, 0)$. The theorem shows that if there are more unknowns than equations, then the system (45) has a non-trivial solution.

If R is a noetherian local ring with maximal ideal J and $\dim R = m$, then any set of m elements $y_1, \ldots, y_m$ such that $\sum_1^m Ry_i$ is J-primary is called a *system of parameters* for R. Of particular interest for geometry are the local rings R for which J itself can be generated by $m = \dim R$ elements. Such local rings are called *regular*. The following exercises indicate some of their properties.

EXERCISES

1. Show that if R is a noetherian local ring with maximal ideal J, then $[J^2/J : R/J] \geqslant \dim R$ and equality holds if and only if R is regular.
2. Show that R is regular of Krull dimension 0 if and only if R is a field.
3. Let R be a regular local ring with $J \neq 0$. Show that $J \supsetneqq J^2$. Show that if $x \in J - J^2$, then $R' = R/Rx$ is regular with $\dim R' = \dim R - 1$.
4. Let Rp be a principal prime ideal in a commutative ring R. Show that if P is a prime ideal such that $Rp \supsetneqq P$, then $P \subset \bigcap_{n=1}^{\infty} (Rp)^n$. Use this to prove that if R is a local ring that is not a domain, then any principal prime ideal in R is minimal.
5. Prove that any regular local ring is a domain. (*Hint*: Use induction on $\dim R$. The case $\dim R = 0$ is settled by exercise 2. Hence assume that $\dim R > 0$. By exercise 3 and the induction hypothesis, R/Rx is a domain for any $x \in J - J^2$. Hence Rx is a prime ideal for any $x \in J - J^2$. Suppose that R is not a domain. Then by exercise 4, Rx is a minimal prime ideal for any $x \in J - J^2$. Let $P_1, \ldots, P_s$ be the minimal prime ideals of R. Then $J - J^2 \subset \bigcup_1^s P_i$ and $J \subset J^2 \cup \bigcup_1^s P_i$. By the prime avoidance lemma, $J \subset P_i$ for some i. Then $J = P_i$ is a minimal prime ideal in R. Hence $\dim R = 0$, contrary to hypothesis.)
6. Let M be a module for a commutative ring R. A sequence of elements $a_1, a_2, \ldots, a_n$ of R is called an *R-sequence on M* if (i) $(\sum_1^n Ra_i)M \neq M$, and (ii) For any i, $1 \leqslant i \leqslant n$, a_i is not a zero divisor of the module $M/\sum_1^{i-1} Ra_j$. If $M = R$, we call the sequence simply an *R-sequence*.

   Show that if R is a regular local ring with $\dim R = m$ and $J = \operatorname{rad} R$, then J can be generated by an R-sequence of m elements contained in $J - J^2$.

7. Let R be a noetherian local ring with maximal ideal J. Show that if J can be generated by an R-sequence, then R is regular.

7.18 I-ADIC TOPOLOGIES AND COMPLETIONS

Let I be an ideal in a ring R that for the present need not be commutative, M an R-module. We can introduce a topology in M, the *I-adic topology*, by specifying that the set of cosets $x+I^nM$, $x \in M$, $n = 1, 2, 3, \ldots$, is a base for the open sets in M. If $z \in (x+I^nM) \cap (y+I^mM)$ and $q = \max(m, n)$, then $z+I^qM \subset (x+I^nM) \cap (y+I^mM)$. Since, by definition, the open sets are the unions of the cosets $x+I^nM$, it follows that the intersection of any two open sets is open. The other axioms for open sets—namely, that the union of any set of open sets is open and that M and $\varnothing$ are open—are clearly satisfied. Hence we have a topological space. In particular, if we take $M = R$, we obtain the I-adic topology in R in which a base for the open sets is the set of cosets $x+I^n$, $x \in R$.

We recall that a group G is called a topological group if it has a topology and the map $(x, y) \rightsquigarrow xy^{-1}$ of $G \times G$ into G is continuous. A ring R is a topological ring if it has a topology such that the additive group is a topological group and multiplication is continuous. A module M for a topological ring R is a topological module if it is a topological group and the map $(a, x) \rightsquigarrow ax$ of $R \times M$ into M is continuous.

It is readily verified that R endowed with its I-adic topology is a topological ring and if M is a module endowed with the I-adic topology, then M is a topological module for R.

The I-adic topology in M is Hausdorff if and only if $I^\omega M = \bigcap I^nM = 0$. For, if this condition holds, then given distinct elements x, y in M there exists an n such that $x-y \notin I^nM$, which implies that the open neighborhoods $x+I^nM$ and $y+I^nM$ of x and y respectively are disjoint. Conversely, if $\bigcap I^nM \neq 0$, let $z \neq 0$ be in $\bigcap I^nM$. Then any open set containing 0 contains z. Hence z is in the closure of $\{0\}$ and M is not a T_1-space and hence is not Hausdorff.

From now on we shall assume that $I^\omega = 0$ and $I^\omega M = 0$, so the I-adic topologies in R and M will be Hausdorff. We recall that these conditions are satisfied if R is a commutative noetherian ring, $I = \operatorname{rad} R$, and M is a finitely generated R-module (Theorem 7.23, p. 443). We recall also that if R is a commutative noetherian ring and I is any ideal in R, then $(1+b)I^\omega = 0$ for some $b \in I$. It follows that if I is a proper ideal in a noetherian domain, then $I^\omega = 0$. An important special case is obtained by taking $R = F[x_1, \ldots, x_n]$ where F is a field and $I = \sum Rx_i$, the ideal of polynomials vanishing at $(0, \ldots, 0)$.

As in the case of rings of power series, we can introduce the concepts of convergence, Cauchy sequences, etc. It is convenient to base the definitions on the notion of the order of an element with respect to the I-adic topology. Let x be a non-zero element of M. Since $M \supset IM \supset I^2M \supset \cdots$ and $\bigcap I^nM = 0$, there exists a largest integer n such that $x \in I^nM$. We define this integer to be the *order* $o(x)$, and we complete the definition of the order function by putting $o(0) = \infty$. We use these definitions also in R. Our definitions evidently imply that $o(x) = \infty$ if and only if $x = 0$, $o(-x) = o(x)$, $o(x+y) \geqslant \min(o(x), o(y))$ and if $a \in R$, then $o(ax) \geqslant o(a) + o(x)$. Hence, if we put $|x| = 2^{-o(x)}$ and $|a| = 2^{-o(a)}$ (with the convention that $2^{-\infty} = 0$), we obtain

(i) $$|x| \geqslant 0, \qquad |x| = 0 \Leftrightarrow x = 0,$$

(ii) $$|-x| = |x|,$$

(iii) $$|x+y| \leqslant \max(|x|, |y|),$$

(iv) $$|ax| \leqslant |a||x|.$$

These hold also for R and $|\ |$ define metrics in M and R by $d(x, y) = |x-y|$. Then we have the usual properties: $d(x, y) = d(y, x)$, $d(x, y) \geqslant 0$, and $d(x, y) = 0$ if and only if $x = y$. In place of the triangle inequality $d(x, y) \leqslant d(x, z) + d(z, y)$ we have the stronger inequality

$$d(x, y) \leqslant \max(d(x, z), d(z, y)). \tag{46}$$

This has the curious consequence that all triangles are isosceles, that is, for any three elements x, y, z, two of the distances $d(x, y)$, $d(x, z)$, $d(y, z)$ are equal. For, (46) and the symmetry of $d(,)$ implies that no one of the numbers $d(x, y)$, $d(x, z)$, $d(y, z)$ can exceed the other two, hence some two are equal.

The metric $d(x, y) = |x-y|$ can be used to introduce convergence of sequences, Cauchy sequences, etc. in the usual manner. We say that the sequence of elements $\{x_n\}$ *converges to* x and we write $\lim x_n = x$ or $x_n \to x$ if for any real $\varepsilon > 0$ there exists an integer $N = N(\varepsilon)$ such that $|x - x_n| < \varepsilon$ for all $n \geqslant N$. As usual, we have that $x_n \to x$ and $y_n \to y$ imply $x_n + y_n \to x + y$ and $x_n \to x$ and $a_n \to a$ imply $a_n x_n \to ax$.

The sequence $\{x_n\}$ is called a *Cauchy sequence* if for any $\varepsilon > 0$ there exists an integer N such that $|x_m - x_n| < \varepsilon$ for all $m, n \geqslant N$. Since for $m > n$

$$\begin{aligned} |x_m - x_n| &= |(x_m - x_{m-1}) + (x_{m-1} - x_{m-2}) + \cdots + (x_{n+1} - x_n)| \\ &\leqslant \max(|x_m - x_{m-1}|, |x_{m-1} - x_{m-2}|, \ldots, |x_{n+1} - x_n|), \end{aligned}$$

$\{x_n\}$ is a Cauchy sequence if and only if for any $\varepsilon > 0$ there exists an N such that $|x_{n+1} - x_n| < \varepsilon$ for all $n \geqslant N$. We call M *complete* in the I-adic topology if every Cauchy sequence of elements of M converges.

We also have the usual definitions of convergence of series: $\sum_1^\infty x_n = x_1 + x_2 + \cdots$ converges if the sequence of partial sums $s_n = x_1 + x_2 + \cdots + x_n$ converges. If $s_n \to s$, then we write $s = x_1 + x_2 + \cdots$. We note that if M is complete, then a series $x_1 + x_2 + \cdots$ converges if and only if its nth term x_n converges to 0. For, if this condition holds, then the sequence of partial sums is Cauchy by the remark we made before. Hence $x_1 + x_2 + \cdots$ converges. Conversely if $x_1 + x_2 + \cdots$ converges, then $s_n \to s$ and $s_{n-1} \to s$, so $x_n = s_n - s_{n-1} \to 0$.

This simple criterion for convergence of series in a complete M can be used to show that if R is complete relative to the I-adic topology, then any $b \in I$ is quasi-invertible. For, since $b^n \in I^n$, $b^n \to 0$. Hence the series

$$1 + b + b^2 + \cdots$$

converges in R. Since

$$(1 + b + \cdots + b^{n-1})(1 - b) = 1 - b^n = (1 - b)(1 + b + \cdots + b^{n-1}),$$

we obtain on passing to the limit that $(1 + b + b^2 + \cdots)(1 - b) = 1 = (1 - b)(1 + b + b^2 + \cdots)$. Hence b is quasi-invertible with $\sum_1^\infty b^n$ as quasi-inverse. Thus every element of the ideal I is quasi-invertible and hence $I \subset \operatorname{rad} R$, the Jacobson radical of R. We record this result as

PROPOSITION 7.30. *If R is complete relative to the I-adic topology, then $I \subset \operatorname{rad} R$.*

If we know in addition that R/I is semi-primitive, then $I = \operatorname{rad} R$. For example, this will be the case if I is a maximal ideal in R.

Another important property of complete rings relative to I-adic topologies is the following idempotent lifting property that is closely related to Proposition 7.14. We recall that Proposition 7.14 played an important role in the study of the topology of Spec R, the space of prime ideals in R.

THEOREM 7.30. *Let R be a ring that is complete relative to the I-adic topology defined by the ideal I and let $\bar{u} = u + I$ be idempotent in $\bar{R} = R/I$. Then there exists an idempotent e in R such that $\bar{e} = \bar{u}$.*

Proof. We assume first that R is commutative. In this case we can apply Proposition 7.14, including the uniqueness part to the ring R/I^n, $n = 1, 2, \ldots$. We have the ideal I/I^n in R/I^n and I/I^n is nilpotent and $(u + I^n)^2 \equiv u + I^n (\text{mod } I/I^n)$. It follows from Proposition 7.14 that there exists a unique element

$e_n + I^n$ in R/I^n such that $(e_n + I^n)^2 = e_n + I^n$ and $e_n \equiv u (\text{mod } I)$. Then $e_n^2 \equiv e_n (\text{mod } I^n)$, $e_n \equiv u(\text{mod } I)$, and if e'_n is a second element satisfying these conditions, then $e'_n \equiv e_n(\text{mod } I^n)$. Since $I^n \supset I^{n+k}$, $k = 1, 2, \ldots$, we have $e_{n+k} \equiv e_n(\text{mod } I^n)$. Then $\{e_n\}$ is Cauchy, so $\{e_n\}$ converges to, say, e. We have $e^2 = \lim e_n^2 = \lim e_n = e$ and $u \equiv e(\text{mod } I)$.

Now let R be arbitrary. Let R_1 be the closure in R of the subring generated by u. Then R_1 is commutative and if $I_1 = I \cap R_1$, then I_1 is an ideal in R_1 such that $I_1{}^{\omega} = 0$. Moreover, since R_1 is closed in R, R_1 is complete relative to the I_1-adic topology. We have $(u + I_1)^2 = u + I_1$. The result now follows by applying the first part to the commutative ring R_1. □

An important technique in commutative ring theory is to pass from a ring R to its completion. We proceed to give constructions of the completion of a ring R and of a module M relative to their I-adic topologies (assuming $I^{\omega} = 0$ and $I^{\omega}M = 0$). Let $C(M)$ denote the set of Cauchy sequences of elements of M. $C(M)$ has a natural module structure, since the sum of two Cauchy sequences is a Cauchy sequence, and if $\{x_n\}$ is a Cauchy sequence and $a \in R$, then $\{ax_n\}$ is a Cauchy sequence. $C(M)$ contains the submodule of constant sequences $\{x\}$ and the submodule $Z(M)$ of null sequences $\{z_n\}$, which are the sequences such that $z_n \to 0$. We can form the modules $\hat{M} = C(M)/Z(M)$ and $\hat{R} = C(R)/Z(R)$. If $\{a_n\} \in C(R)$ and $\{x_n\} \in C(M)$, then $\{a_n x_n\} \in C(M)$ and this is contained in $Z(M)$ if either $\{a_n\} \in Z(R)$ or $\{x_n\} \in Z(M)$. These facts are readily verified. It follows that $Z(R)$ is an ideal in $C(R)$ and hence $\hat{R}$ has the structure of a ring. Moreover, if $\hat{a} \in \hat{R}$ and $\hat{x} \in \hat{M}$, we can define $\hat{a}\hat{x}$ by taking representatives $\{a_n\}$, $\{x_n\}$ of $\hat{a}$, $\hat{x}$ and defining $\hat{a}\hat{x} = \{a_n x_n\} + Z(M)$. In this way $\hat{M}$ becomes an $\hat{R}$-module.

Since $Z(M)$ contains no constant sequences except $\{0\}$, we can identify M with the set of cosets (with respect to $Z(M)$) of the constant sequences. In particular, we have an identification of R with a subring of $\hat{R}$ and M is an R-submodule of $\hat{M}$.

Now let $\hat{I}$ denote the subset of $\hat{R}$ of elements $\{a_n\} + Z(R)$ such that every $a_n \in I$. Then $\hat{I}$ is an ideal in $\hat{R}$ and $\hat{I}^k$ is the set of elements $\{b_n\} + Z(R)$ such that $b_n \in I^k$. It follows that $\hat{I}^{\omega} = 0$. Similarly, $\hat{I}^{\omega}\hat{M} = 0$, so we have Hausdorff topologies in $\hat{R}$ and $\hat{M}$ defined by $\hat{I}$. It is clear that the induced topologies on M and R are the I-adic topologies. Moreover, if $\{x_n\} + Z(M)$ is an element of $\hat{M}$, then $\{x_n\} + Z(M) = \lim x_n$ in $\hat{M}$. Hence M is dense in $\hat{M}$. Moreover, $\hat{M}$ is complete. For, let $\{x_n^{(k)}\} + Z(M)$, $k = 1, 2, \ldots$, be a Cauchy sequence in $\hat{M}$. For each $\{x_n^{(k)}\}$ we can choose $x^{(k)} \in M$ so that $|(\{x_n^{(k)}\} + Z(M)) - x^{(k)}| < 1/2^k$. It is easily seen that $\hat{x} = \{x^{(k)}\}$ is a Cauchy sequence and $\lim(\{x_n^{(k)}\} + Z(M)) = \hat{x}$. Thus $\hat{M}$ is complete and, similarly, $\hat{R}$ is complete relative to the $\hat{I}$-adic topologies. We call $\hat{M}$ and $\hat{R}$ the *completions* of M and R respectively.

We shall now specialize these considerations to the most interesting case in which R is a commutative noetherian ring and M is a finitely generated R-module. With the notations as before, we prove first the following

PROPOSITION 7.31. $\hat{M} = \hat{R}M$.

Proof. Let $\{x_1, \ldots, x_r\}$ be a set of generators for M as R-module and let $\hat{y}$ be any element of $\hat{M}$. Then $\hat{y} = \lim y_n$ where the $y_n \in M$. Evidently $y_{n+1} - y_n \to 0$, so if $o(y_{n+1} - y_n) = s_n$, then $s_n \to \infty$. Since $M = \sum Rx_i$, $I^m M = \sum I^m x_i$ and hence we can write $y_{n+1} - y_n = \sum_{j=1}^r a_{nj} x_j$ where $a_{nj} \in I^{s_n}$. We now write $y_i = \sum_1^r b_{ij} x_j$ where $b_{ij} \in R$. Then

$$y_n = y_1 + (y_2 - y_1) + \cdots + (y_n - y_{n-1}) = \sum_{j=1}^{r} b_{nj} x_j$$

where $b_{nj} = b_{1j} + a_{1j} + \cdots + a_{n-1\,j}$. The r sequences $\{b_{nj}\}$, $1 \leqslant j \leqslant r$, are Cauchy sequences. Hence $\lim_{n \to \infty} \{b_{nj}\} = \hat{b}_j$ in $\hat{R}$. Then $\hat{y} = \lim y_n = \sum \hat{b}_j x_j$. Thus $\hat{M} = \hat{R}M$. □

We have defined the ideal $\hat{I}$ in $\hat{R}$ as the sets of cosets $\{a_n\} + Z(R)$ where the $a_n \in I$. Evidently $I \subset \hat{I}$ and hence $\hat{R}I \subset \hat{I}$. Since $\{a_n\} + Z(R) = \lim a_n$, it is clear that I is dense in $\hat{I}$. The proof of the foregoing proposition shows that $\hat{R}I = \hat{I}$. We consider next a submodule N of M. Since R is noetherian and M is finitely generated, N is also finitely generated. Evidently $I^\omega N \subset I^\omega M = 0$. We have the I-adic topology of N and the induced topology in N as a subspace of M. We shall now prove

PROPOSITION 7.32. *If N is a submodule of the module M for the (commutative) noetherian ring R and I is an ideal in R, then the I-adic topology in N coincides with the topology induced by the I-adic topology of M.*

Proof. A base for the neighborhoods of 0 in the I-adic topology of N is the set $\{I^n N\}$ and a base for the neighborhoods of 0 in the induced topology is $\{I^n M \cap N\}$. To show that the two topologies are identical, that is, have the same open sets, it suffices to show that given any $I^n N$ there exists an $I^m M \cap N \subset I^n N$ and given any $I^n M \cap N$ there exists an $I^q N$ such that $I^q N \subset I^n M \cap N$. Since $I^n N \subset I^n M \cap N$, the second is clear. To prove the first we use the Artin-Rees lemma. According to this there exists an integer k such that for any $q \geqslant k$, $I^q M \cap N = I^{q-k}(I^k M \cap N)$. Hence if we take $q = n+k$, we obtain $I^{n+k} M \cap N = I^n(I^k M \cap N) \subset I^n N$. □

We shall prove next the important result that the completion of a noetherian ring relative to an I-adic topology is noetherian. The proof is based on the method of graded rings and modules that was introduced in section 7.16. Recall that we showed that if I is an ideal in a noetherian ring R, then the associated graded ring $G_I(R)$ is noetherian and if R is noetherian and M is a finitely generated R-module, then the graded module $G_I(M)$ is a finitely generated $G_I(R)$-module (Proposition 7.29, p. 449).

We shall need the following

PROPOSITION 7.33. *Let R be a noetherian ring, I an ideal in R such that $I^\omega = 0$; $\hat{R}$ the completion of R relative to the I-adic topology. Let $G_I(R)$ and $G_{\hat{I}}(\hat{R})$ be the graded rings of R and $\hat{R}$ associated with the ideals I and $\hat{I}$ respectively. Then $G_I(R) \cong G_{\hat{I}}(\hat{R})$.*

Proof. We have $R_i = I^i/I^{i+1}$, $\hat{R}_i = \hat{I}^i/\hat{I}^{i+1}$. It is clear from the definition of $\hat{I}$ that $I^i \cap \hat{I}^{i+1} = I^{i+1}$. By the density of R in $\hat{R}$, for any $\hat{b} \in \hat{I}^i$, there exists a $b \in R$ such that $\hat{b} - b \in \hat{I}^{i+1}$. Then $b \in I^i = \hat{I}^i \cap R$. Hence $\hat{I}^i = I^i + \hat{I}^{i+1}$. We now have $\hat{R}_i = \hat{I}^i/\hat{I}^{i+1} = (I^i + \hat{I}^{i+1})/\hat{I}^{i+1} \cong I^i/(\hat{I}^{i+1} \cap I^i) = I^i/I^{i+1} = R_i$. It follows from the definition of the multiplication in the graded rings that $G_{\hat{I}}(\hat{R}) \cong G_I(R)$. □

We require also

PROPOSITION 7.34. *Let R be complete relative to an I-adic topology and let M be an R-module such that $I^\omega M = 0$. Suppose that $G_I(M)$ is finitely generated as $G_I(R)$-module. Then M is finitely generated as R-module.*

Proof. We choose a set of generators $\{\bar{x}_1, \ldots, \bar{x}_n\}$ for $G_I(M)$ as $G_I(R)$-module. We may assume that $\bar{x}_i = x_i + I^{e_i+1}M$ and $x_i \in I^{e_i}M$, $\notin I^{e_i+1}M$. We shall show that $\{x_1, \ldots, x_n\}$ generates M as R-module. We observe first that if $\bar{u}$ is a homogeneous element of $G_I(M)$ of degree m, then we can write $\bar{u} = \sum \bar{a}_i \bar{x}_i$ where $\bar{a}_i$ is a homogeneous element of $G_I(R)$ of degree $m - e_i$. Now let u_1 be any element of M. Suppose that $o(u_1) = m_1 \geqslant 0$. Then $\bar{u}_1 = u_1 + I^{m_1+1}M \in M^{(m_1)}$ and $\bar{u}_1 \neq 0$. Hence there exist $a_{1i} \in I^{m_1 - e_i}$ such that $\bar{u}_1 = \sum \bar{a}_{1i}\bar{x}_i$. Then $u_2 = u_1 - \sum a_{1i}x_i \in I^{m_1+1}M$ and so $o(u_2) = m_2 > m_1$. We repeat the argument with u_2 and obtain $a_{2i} \in I^{m_2 - e_i}$ such that $u_2 - \sum a_{2i}x_i \in I^{m_2+1}M$. Continuing in this way we obtain a sequence of integers $m_1 < m_2 < m_3 < \cdots$, a sequence of elements $u_1, u_2, \ldots$ where $u_k \in I^{m_k}M$, and for each i a sequence $a_{1i}, a_{2i}, \ldots$ where $a_{ki} \in I^{m_k - e_i}$ such that $u_k - \sum a_{ki}x_i = u_{k+1}$. Since R is complete, the infinite series $\sum_k a_{ki}$ converge, say, $\sum_k a_{ki} = a_i$. We claim that $u_1 = \sum a_i x_i$. To see this we write

$$
\begin{aligned}
u_1 - \sum a_i x_i &= u_1 - \sum_i \left(\sum_k a_{ki} \right) x_i \\
&= u_1 - \sum_i \left(\sum_{k=1}^{q} a_{ki} \right) x_i - \sum_i \left(\sum_{l=q+1}^{\infty} a_{li} \right) x_i \\
&= u_{q+1} + \sum_i \left(\sum_{l=q+1}^{\infty} a_{li} \right) x_i \in I^q M .
\end{aligned}
$$

Since $I^\omega M = 0$, we have $u_1 - \sum a_i x_i = 0$ and $u_1 = \sum a_i x_i$. □

If M is an ideal in R, then $G_I(M)$ is an ideal in $G_I(R)$. Hence we have the following consequence of the last result.

COROLLARY. *Let R be complete with respect to an I-adic topology and suppose that $G_I(R)$ is noetherian. Then R is noetherian.*

We can now prove our main result.

THEOREM 7.31. *If R is a commutative noetherian ring, I an ideal in R such that $I^\omega = 0$, and $\hat{R}$ is the completion of R relative to the I-adic topology, then $\hat{R}$ is noetherian.*

Proof. By Proposition 7.29, $G_I(R)$ is noetherian. Hence, by Proposition 7.33, $G_I(\hat{R})$ is noetherian. Then $\hat{R}$ is noetherian by the Corollary. □

The noetherian property of rings of power series over noetherian rings (Theorem 7.12, p. 424) follows also from Theorem 7.31. Let R be noetherian and let I be the ideal in $R[x_1, \ldots, x_r]$ of polynomials $f(x_1, \ldots, x_r)$ such that $f(0, \ldots, 0) = 0$. Then $I^\omega = 0$ and so we can form the completion of $R[x_1, \ldots, x_r]$ relative to the I-adic topology. This is the ring $R[[x_1, \ldots, x_r]]$. Since $R[x_1, \ldots, x_r]$ is noetherian by the Hilbert basis theorem, so is the ring $R[[x_1, \ldots, x_r]]$.

EXERCISES

1. Let R be a ring and I an ideal in R such that $I^\omega = 0$. If $m < n$, then we have the canonical homomorphism $\theta_{m,n}$ of $R/I^n \to R/I^m$. These homomorphisms define an inverse limit $\varprojlim R/I_n$ as on p. 73. Show that this inverse limit is isomorphic to $\hat{R}$.
2. Show that if R is noetherian, then $\hat{R}$ is a flat R-module.

REFERENCES

E. Noether, Idealtheorie in Ringbereichen, *Mathematische Annalen*, vol. 83 (1921), pp. 24–66.

O. Zariski and Pierre Samuel, *Commutative Algebra*, Vol. I and II, D. van Nostrand, 1958, 1960. New printing, Springer-Verlag, New York, 1975.

M. Atiyah and I. G. MacDonald, *Introduction to Commutative Algebra*, Addison-Wesley, Reading, Mass., 1969.

P. Samuel, *Algebre Commutative*, Mimeographed notes of Ecole Normale Superieure, Paris, 1969.

I. Kaplansky, *Commutative Rings*, Allyn and Bacon, Boston, 1970.

8

Field Theory

In BAI we developed a substantial part of the theory of fields. In particular, we gave a rather thorough account of the Galois theory of equations and the relevant theory of finite dimensional extension fields. Some of the topics covered were splitting fields of a polynomial, separability, perfect fields, traces and norms, and finite fields.

We now take up again the subject of fields and we begin with a construction and proof of uniqueness up to isomorphism of the algebraic closure of a field. After this we give an alternative derivation of the main results of finite Galois theory based on a general correspondence between the subfields of a field and certain rings of endomorphisms of its additive group. We also extend the finite theory to Krull's Galois theory of infinite dimensional algebraic extension fields. An important supplement to the (finite) Galois theory is the study of crossed products, which is a classical tool for studying the Brauer group $\mathrm{Br}(F)$ associated with a field. We consider this in sections 8.4, 8.5, and 8.8. In sections 8.9 and 8.11 we study the structure of two important types of abelian extensions of a field: Kummer extensions and abelian p-extensions. These extensions can be described by certain abelian groups associated with the base field. In the case of the p-extensions, the groups are defined by rings of Witt vectors. These rings are of considerable interest aside from their application to p-extensions. The definition and elementary theory of rings of Witt vectors are discussed in section 8.10. The concept of normal closures and the structure of normal extensions are considered in section 8.7.

The first two thirds of the chapter is concerned with algebraic extension fields. After this we take up the structure of arbitrary extensions. We show that such extensions can be obtained in two stages, first, a purely transcendental one and then an algebraic one. Associated with this process are the concepts of transcendency base and transcendency degree. We consider also the more delicate analysis concerned with separating transcendency bases and a general notion of separability applicable to extensions that need not be algebraic. Derivations play an important role in these considerations. We show also that derivations can be used to develop a Galois theory for purely inseparable extension fields of exponent one.

In section 8.13 we extend the concept of transcendency degree of fields to domains. In section 8.18 we consider tensor products of fields that need not be algebraic and in section 8.19 we consider the concept of free composites of two field extensions of the same field.

8.1 ALGEBRAIC CLOSURE OF A FIELD

We recall that a field F is called algebraically closed if every monic polynomial $f(x)$ of positive degree with coefficients in F has a root in F. Since r is a root, that is, $f(r) = 0$, if and only if $x - r$ is a factor of $f(x)$ in $F[x]$, this is equivalent to the condition that the only irreducible polynomials in $F[x]$ are the linear ones and hence also to the condition that every polynomial of $F[x]$ of positive degree is a product of linear factors. Still another condition equivalent to the foregoing is that F has no proper algebraic extension field. For, if E is an extension field of F and $a \in E$ is algebraic over F, then $[F(a):F]$ is the degree of the minimum polynomial $f(x)$ of a over F, and $f(x)$ is monic and irreducible. Then $a \in F$ if and only if $\deg f(x) = 1$. Hence if E is algebraic over F and $E \supsetneq F$, then there exist irreducible monic polynomials in $F[x]$ of degree $\geqslant 2$; hence F is not algebraically closed. Conversely, if F is not algebraically closed, then there exists a monic irreducible $f(x) \in F[x]$ with $\deg f(x) \geqslant 2$. Then the field $F[x]/(f(x))$ is a proper algebraic extension of F.

We recall that if E is an extension field of the field F, then the set of elements of E that are algebraic over F constitute a subfield A of E/F (that is, a subfield of E containing F). Evidently $E = A$ if and only if E is algebraic over F. At the other extreme, if $A = F$, then F is said to be *algebraically closed in E*. In any case A is algebraically closed in E, since any element of E that is algebraic over A is algebraic over F and so is contained in A. These results were proved in BAI, pp. 270–271, as special cases of theorems on integral dependence. We shall now give alternative proofs based on the dimensionality formula

$$[E:F] = [E:K][K:F] \tag{1}$$

for E, an extension field of F, and K, an intermediate field. This was proved in BAI, p. 215, for finite dimensional extensions and the proof goes over without change in the general case. In fact, as in the finite case, one sees directly that if (u_α) is a base for E/K and (v_β) is a base for K/F, then $(u_\alpha v_\beta)$ is a base for E/F. The dimensionality formula gives the following

PROPOSITION 8.1. *$[E:F] < \infty$ if and only if $E = F(a_1, \dots, a_n)$ where a_1 is algebraic over F and every a_{i+1} is algebraic over $F(a_1, \dots, a_{i-1})$.*

The proof is clear.

This proposition gives immediately the results that were stated on algebraic elements: If a and $b \in E$ are algebraic over F, then, of course, b is algebraic over $F(a)$ so $F(a, b)$ is finite dimensional. Then the elements $a \pm b$, ab, and a^{-1} if $a \neq 0$ are algebraic. Hence the algebraic elements form a subfield A containing F. If c is algebraic over A, then evidently c is algebraic over a subfield $F(a_1, \dots, a_n)$, $a_i \in A$. Then $F(a_1, \dots, a_n, c)$ is finite dimensional over F and c is algebraic over F, so $c \in A$.

We remark also that if the a_i are algebraic, then

$$F(a_1, \dots, a_n) = F[a_1, \dots, a_n]. \tag{2}$$

This is well known if $n = 1$ (BAI, p. 214) and it follows by induction for any n.

We prove next

PROPOSITION 8.2. *If E is an algebraically closed extension field of F, then the subfield A/F of elements of E that are algebraic over F is algebraically closed.*

Proof. Let $f(x)$ be a monic polynomial of positive degree with coefficients in A. Then $f(x)$ has a root r in E and evidently r is algebraic over A. Hence $r \in A$. Hence A is algebraically closed. $\square$

This result shows that if a field F has an algebraically closed extension field, then it has one that is algebraic over F. We shall now call an extension field E/F an *algebraic closure of F* if (1) E is algebraic over F and (2) E is algebraically closed. We proceed to prove the existence and uniqueness up to isomorphism of an algebraic closure of any field F.

For countable F a straightforward argument is available to establish these results. We begin by enumerating the monic polynomials of positive degree as $f_1(x), f_2(x), f_3(x), \ldots$. Evidently this can be done. We now define inductively a sequence of extension fields beginning with $F_0 = F$ and letting F_i be a splitting field over F_{i-1} of $f_i(x)$. The construction of such splitting fields was given in BAI, p. 225. It is clear that every F_i is countable, so we can realize all of these constructions in some large set S. Then we can take $E = \bigcup F_i$ in the set. Alternatively we can define E to be a direct limit of the fields F_i. It is easily seen that E is an algebraic closure of F. We showed in BAI, p. 227, that if K_1 and K_2 are two splitting fields over F of $f(x) \in F[x]$, then there exists an isomorphism of K_1/F onto K_2/F. This can be used to prove the isomorphism theorem for algebraic closures of a countable field by a simple inductive argument.

The pattern of proof sketched above can be carried over to the general case by using transfinite induction. This is what was done by E. Steinitz, who first proved these results.

There are several alternative proofs available that are based on Zorn's lemma. We shall give one that makes use of the following result, which is of independent interest.

PROPOSITION 8.3. *If A is an algebraic extension of an infinite field F, then the cardinality $|A| = |F|$.*

Proof. A is a disjoint union of finite sets, each of which is the set of roots in A of a monic irreducible polynomial of positive degree. Then $|A|$ is the same as the cardinality of this collection of finite sets and hence the same as that of the set of corresponding polynomials. If we use the result that the product of two infinite cardinals is the larger of the two, we see that the cardinality of the set of monic polynomials of a fixed degree is $|F|$ and hence of all monic polynomials is $|F|$. Hence, $|A| = |F|$. $\square$

If F is finite, the preceding argument shows that any algebraic extension of F is either finite or countable. We can now prove the existence of algebraic closures.

THEOREM 8.1. *Any field F has an algebraic closure.*

Proof. We first imbed F in a set S in which we have a lot of elbow room. Precisely, we assume that $|S| > |F|$ if F is infinite and that S is uncountable if F is finite. We now define a set Λ whose elements are $(E, +, \cdot)$ where E is a subset of S containing F and $+, \cdot$ are binary compositions in E such that

$(E, +, \cdot)$ is an algebraic extension field of F. We partially order Λ by declaring that $(E, +, \cdot) > (E', +', \cdot')$ if E is an extension field of E'. By Zorn's lemma there exists a maximal element $(E, +, \cdot)$. Then E is an algebraic extension of F. We claim that E is algebraically closed. Otherwise we have a proper algebraic extension $E' = E(a)$ of E. Then $|E'| < |S|$, so we can define an injective map of E' into S that is the identity on E and then we can transfer the addition and multiplication on E' to its image. This gives an element of $\Lambda \gneq (E, +, \cdot)$, contrary to the maximality of $(E, +, \cdot)$. This contradiction shows that E is an algebraic closure of F. □

Next we take up the question of uniqueness of algebraic closures. It is useful to generalize the concept of a splitting field of a polynomial to apply to sets of polynomials. If $\Gamma = \{f\}$ is a set of monic polynomials with coefficients in F, then an extension field E/F is called a *splitting field over F of the set* Γ if 1) every $f \in \Gamma$ is a product of linear factors in $E[x]$ and 2) E is generated over F by the roots of the $f \in \Gamma$. It is clear that if E is a splitting field over F of Γ, then no proper subfield of E/F is a splitting field of Γ and if K is any intermediate field, then E is a splitting field over K of Γ. Since an algebraic closure E of F is algebraic, it is clear that E is a splitting field over F of the complete set of monic polynomials of positive degree in $F[x]$. The isomorphism theorem for algebraic closures will therefore be a consequence of a general result on isomorphism of splitting fields that we will now prove. Our starting point is the following result, which was proved in BAI, p. 227:

Let $\eta: a \rightsquigarrow \bar{a}$ be an isomorphism of a field F onto a field $\bar{F}$, $f(x) \in F[x]$ be monic of positive degree, $\bar{f}(x)$ the corresponding polynomial in $\bar{F}[x]$ (under the isomorphism, which is η on F and sends $x \rightsquigarrow x$), and let E and $\bar{E}$ be splitting fields over F and $\bar{F}$ of $f(x)$ and $\bar{f}(x)$ respectively. Then η can be extended to an isomorphism of E onto $\bar{E}$.

We shall now extend this to sets of polynomials:

THEOREM 8.2. *Let $\eta: a \rightsquigarrow \bar{a}$ be an isomorphism of a field F onto a field $\bar{F}$, Γ a set of monic polynomials $f(x) \in F[x]$, $\bar{\Gamma}$ the corresponding set of polynomials $\bar{f}(x) \in \bar{F}[x]$, E and $\bar{E}$ splitting fields over F and $\bar{F}$ of Γ and $\bar{\Gamma}$ respectively. Then η can be extended to an isomorphism of E onto $\bar{E}$.*

Proof. The proof is a straightforward application of Zorn's lemma. We consider the set of extensions of η to monomorphisms of subfields of E/F into $\bar{E}/\bar{F}$ and use Zorn's lemma to obtain a maximal one. This must be defined on the whole of E, since otherwise we could get a larger one by applying the result quoted to one of the polynomials $f(x) \in \Gamma$. Now if ζ is a monomorphism of E into $\bar{E}$ such that $\zeta | F = \eta$, then it is clear that $\zeta(E)$ is a splitting field over $\bar{F}$ of $\bar{\Gamma}$. Hence $\zeta(E) = \bar{E}$ and ζ is an isomorphism of E onto $\bar{E}$. □

As we have observed, this result applies in particular to algebraic closures. If we take $\bar{F} = F$ and $\eta = 1$, we obtain

THEOREM 8.3. *Any two algebraic closures of a field F are isomorphic over F.*

From now on we shall appropriate the notation $\bar{F}$ for any determination of an algebraic closure of F. If A is any algebraic extension of F, its algebraic closure $\bar{A}$ is an algebraic extension of A, hence of F, and so $\bar{A}$ is an algebraic closure of F. Consequently, we have an isomorphism of $\bar{A}/F$ onto $\bar{F}/F$. This maps A/F into a subfield of $\bar{F}/F$. Thus we see that every algebraic extension A/F can be realized as a subfield of the algebraic closure $\bar{F}/F$.

EXERCISE

1. Let E be an algebraic extension of a field F. A an algebraic closure of F. Show that E/F is isomorphic to a subfield of A/F. (*Hint*: Consider the algebraic closure $\bar{A}$ of A and note that this is an algebraic closure of F.)

8.2 THE JACOBSON-BOURBAKI CORRESPONDENCE

Let E be a field, F a subfield, and let $\mathrm{End}_F E$ be the ring of linear transformations of E regarded as a vector space over F. We have the map

$$F \rightsquigarrow \mathrm{End}_F E, \tag{3}$$

which is injective since we can recover F from $\mathrm{End}_F E$ by taking $\mathrm{End}_{\mathrm{End}_F E} E$, which is the set of multiplications $x \rightsquigarrow ax$, $a \in F$ (pp. 205–206), so applying this set to 1 gives F.

We now seek a characterization of the rings of endomorphisms $\mathrm{End}_F E$ for a fixed E that does not refer explicitly to the subfields F. We observe first that for any $b \in E$ we have the multiplication $b_E : x \rightsquigarrow bx$, $x \in E$, and this is contained in $\mathrm{End}_F E$ no matter what F we are considering. This is one of the ingredients of the characterization we shall give. The second one will be based on a topology of maps we shall now define.

Let X and Y be arbitrary sets. Then the set Y^X of maps of X into Y can be endowed with the product set (of copies of Y) topology in which Y is given the discrete topology. A base for the open sets in this topology consists of the sets $O_{f,\{x_i\}}$ where f is a map $X \to Y$, $\{x_i | 1 \leqslant i \leqslant n\}$ a finite subset of X, and $O_{f,\{x_i\}} = \{g \in Y^X | g(x_i) = f(x_i), 1 \leqslant i \leqslant n\}$. We shall call this topology the *finite topology* for Y^X and for subsets regarded as subspaces with the induced topology. If X and Y are groups, the subset $\hom(X, Y)$ of homomorphisms of X into Y is a closed subset of Y^X. For, if f is in the closure of $\hom(X, Y)$ and x_1, x_2 are arbitrary elements of X, then the open set $O_{f,\{x_1,x_2,x_1x_2\}}$ contains a $g \in \hom(X, Y)$. Since $g(x_1x_2) = g(x_1)g(x_2)$, it follows that $f(x_1x_2) = f(x_1)f(x_2)$ and since this holds for all x_1, x_2, we see that $f \in \hom(X, Y)$. In a similar fashion we see that if M and N are modules for a ring R (not necessarily commutative), then $\hom_R(X, Y)$ is a closed subset of Y^X. The density theorem for completely reducible modules (p. 197) has a simple formulation in terms of the finite topology: If M is a completely reducible R-module, then the set R_M of multiplications $a_M : x \rightsquigarrow ax$ is dense in $\mathrm{End}_{\mathrm{End}_R M} M$, that is, the closure of R_M is $\mathrm{End}_{\mathrm{End}_R M} M$. This formulation is clear from the definition of the topology. We shall now prove a result that includes a characterization of the rings $\mathrm{End}_F E$ for E a field and F a subfield.

THEOREM 8.4. *Let E be a field, Σ the set of subfields of E, Γ the set of subrings L of the ring of endomorphisms of the additive group of E such that* (1) *$L \supset E_E$, the set of multiplications a_E in E by the elements a of E, and* (2) *L is closed in the finite topology. Then the map $F \rightsquigarrow \mathrm{End}_F E$ is a bijective order-inverting map with order-inverting inverse of Σ onto Γ.*

Proof. We have seen that the indicated map is injective, and $F_1 \supset F_2$ if and only if $\mathrm{End}_{F_1} E \subset \mathrm{End}_{F_2} E$. It remains to show that it is surjective. Let L be a ring of endomorphisms of the additive group of E, and regard E as L-module in the natural way. Since E is a field, E has no ideals $\neq 0$, E, and hence E is irreducible as E-module. Since $L \supset E_E$, E is irreducible also as L-module. The inclusion $L \supset E_E$ implies that $\mathrm{End}_L E \subset \mathrm{End}_E E$. Since E is a commutative ring, $\mathrm{End}_E E = E_E$. Hence $\mathrm{End}_L E$ is a subring of E_E and so this has the form F_E, the set of multiplications in E determined by the elements of a subring F of E. By Schur's lemma, F_E is a division ring. Hence F is a subfield of E. By the density theorem, L is dense in $\mathrm{End}_{\mathrm{End}_L E} E = \mathrm{End}_F E$. Since L is closed, $L = \mathrm{End}_F E$ for the subfield F. This proves the surjectivity of $F \rightsquigarrow \mathrm{End}_F E$ onto Γ. Thus our map is bijective. $\square$

We have previously noted the inverse map from Γ to Σ. The preceding proof gives this again in slightly different and somewhat more convenient form, namely, for given $L \in \Gamma$ the corresponding subfield F is the set of $a \in E$ such that

$$a_E l = l a_E \tag{4}$$

for every $l \in L$.

The ring of endomorphisms $L = \mathrm{End}_F E$ contains E_E and hence can be regarded as E-E-bimodule by defining $al = a_E l$, $la = la_E$, $l \in L$, $a \in E$. Since la could also be interpreted as the image of a under l, we shall always write $l(a)$ when the latter is intended. For the present we shall be interested only in the left action of E on L, that is, L as (left) vector space over E.

We consider first the dimensionality $[L:E]$ (as left vector space over E). Suppose that $x_1, \ldots, x_n$ are elements of E that are linearly independent over F. Then there exist $l_i \in L = \mathrm{End}_F E$ such that

$$l_i(x_j) = \delta_{ij}, \qquad 1 \leqslant i, j \leqslant n. \tag{5}$$

Hence if $a_i \in E$, then $\sum_i a_i l_i(x_j) = a_j$, which implies that the l_i are linearly independent elements of L over E. Evidently this implies that if $[E:F] = \infty$, then $[L:E] = \infty$. On the other hand, if $[E:F] < \infty$ and $(x_1, \ldots, x_n)$ is a base, then the l_i such that (5) holds are uniquely determined. Moreover, if $l \in L$ and $l(x_j) = a_j$, then $\sum_i a_i l_i(x_j) = l(x_j)$. Since the x_j form a base for E/F and l and $\sum a_i l_i$ are linear transformations in E/F, this implies that $l = \sum a_i l_i$. Thus $(l_1, \ldots, l_n)$ is a base for L over E. We have proved the following important result.

PROPOSITION 8.4. *If F is a subfield of the field E and $L = \mathrm{End}_F E$, then $[E:F] < \infty$ if and only if $[L:E] < \infty$ and in this case we have $[E:F] = [L:E]$. Moreover, if $(x_1, \ldots, x_n)$ is a base for E/F, then the linear transformations l_i such that $l_i(x_j) = \delta_{ij}$, $1 \leqslant i \leqslant n$, form a base for L over E.*

This result shows that in the correspondence between subfields of E and rings of endomorphisms of its additive group given in Theorem 8.4, the subfields of finite codimension in E ($[E:F] < \infty$) and the subrings L such that $[L:E] < \infty$ are paired. We shall now show that in this case, condition (2) that L is closed is superfluous. This will follow from

PROPOSITION 8.5. *Any finite dimensional subspace V over E of the ring of endomorphisms of the additive group of E is closed in the finite topology.*

Proof. Let $x \in E$ and consider the map $x^*: l \rightsquigarrow l(x)$ of V into E. This is E-linear, so $x^* \in V^*$, the conjugate space of V/E. Let W be the subspace of V^*

spanned by the x^*. Since $x^*(l) = l(x) = 0$ for all x implies that $l = 0$, $W = V^*$. Hence if $[V:E] = n$, then we can choose $x_i \in E$, $1 \leqslant i \leqslant n$, such that $(x_1^*, \ldots, x_n^*)$ is a base for V^*. Then there exists a base $(l_1, \ldots, l_n)$ for V over E such that the relations $l_i(x_j) = \delta_{ij}$, $1 \leqslant i \leqslant n$, hold. Now let l be in the closure of V and let x be any element of E. Then there exist $a_i \in E$ such that $\sum_i a_i l_i(x_j) = l(x_j)$ for $1 \leqslant j \leqslant n$ and $\sum_i a_i l_i(x) = l(x)$. The first set of these equations and $l_i(x_j) = \delta_{ij}$ give $a_i = l(x_i)$, so the a_i are independent of x. Then the last equation shows that $l = \sum a_i l_i \in V$. Hence V is closed. □

Theorem 8.4 and Propositions 8.4 and 8.5 clearly imply the

JACOBSON-BOURBAKI CORRESPONDENCE. *Let F be a field, Φ the set of subfields F of E of finite codimension in E, Δ the set of subrings L of the ring of endomorphisms of the additive group of E such that* (1) *$L \supset E_E$, the set of multiplications in E by the elements of E, and* (2) *L as left vector space over E is finite dimensional. Then the map $F \rightsquigarrow \mathrm{End}_F E$ is bijective and order-inverting with order-inverting inverse of Φ onto Δ. Under this correspondence $[E:F] = [\mathrm{End}_F E : E]$ and if L is given in Δ, the corresponding subfield F of E is the set of $a \in E$ such that $a_E l = l a_E$ (or $al = la$) for all $l \in L$.*

EXERCISES

1. Let $E = (\mathbb{Z}/(p))(x)$ where p is a prime and x is an indeterminate. Let $F = (\mathbb{Z}/(p))(x^{p^e})$ where $e \geqslant 1$. Let $D_i \in \mathrm{End}_F E$ be defined by $D_i x^m = \binom{m}{i} x^{m-i}$, $1 \leqslant i \leqslant p^e - 1$. Verify that if $a \in E$, then
$$D_j a_E = \sum_{i=0}^{j} (D_{j-i} a) D_i$$
and $D_i D_j = \binom{i+j}{i} D_{i+j}$ if $i + j \leqslant p^e - 1$, $D_i D_j = 0$ if $i + j \geqslant p^e$. Prove that $(1, D_1, \ldots, D_{p^e-1})$ is a base for $\mathrm{End}_F E$ over E.

2. Let $[E:F] = n < \infty$ and let $(l_1, \ldots, l_n)$ be a base for $\mathrm{End}_F E$, $(x_1, \ldots, x_n)$ a base for E. Show that the matrix $(l_i(x_j)) \in M_n(E)$ is invertible.

3. Extend the results of this section to division rings.

8.3 FINITE GALOIS THEORY

If K and E are fields, we denote the additive group of homomorphisms of the additive group $(K, +, 0)$ into $(E, +, 0)$ by $\hom(K, E)$ and if K and E have a common subfield F, then $\hom_F(K, E)$ denotes the subset of $\hom(K, E)$ consisting of the F-linear maps of K into E. The additive group $\hom(K,E)$ has an E-K-bimodule structure obtained by defining af for $a \in E$, $f \in \hom(K,E)$ by $(af)(x) = af(x)$, $x \in K$, and fb for $b \in K$ by $(fb)(x) = f(bx)$. The subset $\hom_F(K,E)$ is a submodule of the E-K-bimodule $\hom(K,E)$. In the special case, which we considered in the last section, in which $K = E$, we have End E and $\mathrm{End}_F E$, and these are E-E-bimodules with the module compositions as left and right multiplications by the elements of E_E. The result we proved on the dimensionality of $\mathrm{End}_F E$ in the case in which $[E:F] < \infty$ can be extended to prove

PROPOSITION 8.6. *Let E and K be extension fields of the field F and assume that $[K:F] = n < \infty$. Then the dimensionality of* $\hom_F(K,E)$ *as left vector space over E is $[K:F] = n$.*

Proof. The proof is identical with that of the special case: If $(x_1, \dots, x_n)$ is a base for K/F, let l_i be the linear map of K/F into E/F such that $l_i(x_j) = \delta_{ij}$, $1 \leqslant j \leqslant n$. Then as in the special case we see that $(l_1, \dots, l_n)$ is a base for $\hom_F(K,E)$ over E. □

Let s be a ring homomorphism of K into E. The existence of such a homomorphism presupposes that the two fields have the same characteristic and then s is a monomorphism of K into E. Evidently $s \in \hom(K,E)$. Moreover, if $a, b \in K$, then $s(ab) = s(a)s(b)$. In operator form this reads

$$sa = s(a)s \tag{6}$$

for all $a \in K$. If K and E have a common subfield F, then a ring homomorphism of K/F into E/F is a ring homomorphism of K into E, which is the identity on F. Then $s \in \hom_F(K,E)$.

We recall the Dedekind independence theorem for characters: distinct characters of a group into a field are linearly independent over the field (BAI, p. 291). A consequence of this is that distinct ring homomorphisms of K into E are linearly independent over E. We use this and Proposition 8.6 to prove

PROPOSITION 8.7. *If K and E are extension fields of the field F and $[K:F] = n < \infty$, then there are at most n (ring) homomorphisms of K/F into E/F.*

Proof. If $s_1, \dots, s_m$ are distinct homomorphisms of K/F into E/F, these are elements of $\hom_F(K,E)$ that are linearly independent over E. Since the left dimensionality of $\hom_F(K,E)$ over E is n, it follows that $m \leqslant n$. □

Suppose that G is a group of automorphisms of E. Put

$$EG = \{\sum a_i s_i \mid a_i \in E, s_i \in G\}. \tag{7}$$

Then EG contains $E_E = E1$ and formula (6) shows that EG is a subring of the ring of endomorphisms, End E, of the additive group of E, since it implies that $(as)(bt) = as(b)st$ for the automorphisms s, t, and the elements $a, b \in E$. If G is finite, say, $G = \{s_1 = 1, s_2, \ldots, s_n\}$, then the s_i form a base for EG over E. Hence the dimensionality $[EG : E] = n = |G|$. Thus EG satisfies the conditions for the rings of endomorphisms in the Jacobson-Bourbaki correspondence. The corresponding subfield F is the set of $a \in E$ such that $a_E l = l a_E$ for all $l \in EG$. Since $a_E b_E = b_E a_E$, this set of conditions reduces to the finite set $a_E s_i = s_i a_E$ or $a s_i = s_i a$ for all $s_i \in G$. By (6), these conditions are equivalent to $s_i(a) = a$, $1 \leqslant i \leqslant n$. Hence the field associated with EG in the correspondence is

$$F = \text{Inv } G, \tag{8}$$

the subfield of E of elements fixed under every $s_i \in G$. We have

$$[E : F] = [EG : E] = |G| \tag{9}$$

and

$$EG = \text{End}_F E. \tag{10}$$

Now let Gal E/F be the Galois group of E/F, that is, the group of automorphisms of E fixing every element of F. Evidently Gal $E/F \supset G$. Let $t \in \text{Gal } E/F$. Then $t \in \text{End}_F E$, so by (10), $t = \sum_1^n a_i s_i$, $a_i \in E$, $s_i \in G$. By the Dedekind independence theorem $t = s_i$ for some i. Thus

$$\text{Gal } E/F = G. \tag{11}$$

We shall require also the following result on subrings of EG containing E_E.

PROPOSITION 8.8. *Let G be a finite group of automorphisms in the field E. Then the subrings of EG containing E_E are the rings EH, H a subgroup of G.*

Proof. Clearly any EH is a subring of EG containing $E_E = E1$. Conversely, let S be such a subring and let $H = S \cap G$. Clearly H is a submonoid of G. Since G is finite, H is a subgroup. It is evident that $EH \subset S$. We claim that $EH = S$. Otherwise, we have an element $a_1 s_1 + a_2 s_2 + \cdots + a_j s_j + \cdots + a_r s_r \in S$ such that no $s_j \in H$ and every $a_j \neq 0$. Evidently $r > 1$ and assuming r minimal, the proof of Dedekind's theorem gives an element with the same properties and a smaller r. This contradiction proves that $S = EH$. □

We shall say that a subfield F of E is *Galois in* E, alternatively, E is *Galois over* F if there exists a group of automorphisms G of E such that $\operatorname{Inv} G = F$. We can now state the main result on finite groups of automorphisms in a field.

THEOREM 8.5. *Let E be a field and let Λ be the set of finite groups of automorphisms of E, Ψ the set of subfields of E that are Galois in E and have finite codimension in E. Then we have the map $G \rightsquigarrow \operatorname{Inv} G$ of Λ into Ψ and the map $F \rightsquigarrow \operatorname{Gal} E/F$ of Ψ into Λ. These are inverses and are order-inverting. We have*

$$(12) \qquad |G| = [E:\operatorname{Inv} G].$$

If $F \in \Psi$ and K is a subfield of E containing F, then $K \in \Psi$. Moreover, $\operatorname{Gal} E/K$ *is a subgroup of* $\operatorname{Gal} E/F$ *and* $\operatorname{Gal} E/K$ *is normal in* $\operatorname{Gal} E/F$ *if and only if K is Galois over F, in which case*

$$(13) \qquad \operatorname{Gal} K/F \cong (\operatorname{Gal} E/F)/(\operatorname{Gal} E/K).$$

Proof. It is clear that the maps $G \rightsquigarrow \operatorname{Inv} G$, $F \rightsquigarrow \operatorname{Gal} E/F$ are order-inverting. Now $G \rightsquigarrow \operatorname{Inv} G$ is surjective, by definition of Ψ, and since $\operatorname{Gal} E/\operatorname{Inv} G = G$, it is bijective with inverse $F \rightsquigarrow \operatorname{Gal} E/F$. Also we had (9), which is the same thing as (12). Now let $F \in \Psi$ and let K be a subfield of E containing F. Then $F = \operatorname{Inv} G$ for some finite group of automorphisms G of E. Since $K \supset F$, the corresponding ring $\operatorname{End}_K E$ in the Jacobson-Bourbaki correspondence is a subring of $\operatorname{End}_F E = EG$ containing E_E. By Proposition 8 this has the form EH, for a subgroup H of G. Then $K = \operatorname{Inv} H \in \Psi$. If $s \in G$, then sHs^{-1} is a subgroup of G and $\operatorname{Inv} sHs^{-1}$ is $s(K)$. Hence H is normal in G if and only if $s(K) = K$ for every $s \in G$. Then the restriction $\bar{s} = s|K$ is an automorphism of K. The set of these is a finite group of automorphisms $\bar{G}$ of K. It is clear that $F = \operatorname{Inv} \bar{G}$. Hence if H is normal in G, then K is Galois over F. Moreover, in this case $\bar{G} = \operatorname{Gal} K/F$ and $s \rightsquigarrow \bar{s}$ is an epimorphism of G onto $\bar{G}$. Evidently the kernel is $\operatorname{Gal} E/K$. Hence we have the isomorphism (13). It remains to show that if K is Galois over F, then H is normal in G or, equivalently, $s(K) = K$ for every $s \in G$. If $s(K) \neq K$, then $s|K$ is a homomorphism of K/F into E/F, which is not an automorphism of K/F. On the other hand, the order of $\operatorname{Gal} K/F$ is $[K:F]$. This gives $[K:F]+1$ distinct homomorphisms of K/F into E/F, contrary to Proposition 8.6. This completes the proof. □

It remains to determine the structure of finite dimensional Galois extension fields. For this we use an important addendum to the isomorphism theorem for splitting fields given in BAI, p. 227, namely, that if E is a splitting field over F of a polynomial with distinct roots, then $|\operatorname{Gal} E/F| = [E:F]$. It is easily seen that the hypothesis that $f(x)$ has distinct roots can be replaced by the following: $f(x)$ is separable, that is, its irreducible factors have distinct roots. If $F' = \text{Inv}(\text{Gal } E/F)$, then $F' \supset F$ and $[E:F'] = |\text{Gal } E/F| = [E:F]$, which implies that $F = F'$ is Galois in E. Conversely, let E be finite dimensional Galois over F. Let $a \in E$ and let $\{a_1 = a, \ldots, a_r\}$ be the orbit of a under $G = \text{Gal } E/F$. Then the minimum polynomial of a over F is $g(x) = \Pi(x - a_i)$, which is a product of distinct linear factors in E. Since E is finite dimensional over F, it is generated by a finite number of elements $a, b, \ldots$. Then it is clear that E is a splitting field over F of the polynomial, which is the product of the minimum polynomials of $a, b, \ldots$. This polynomial is separable. Hence we have

THEOREM 8.6. *E is finite dimensional Galois over F if and only if E is a splitting field over F of a separable polynomial $f(x) \in F[x]$.* (Compare Theorem 4.7 on p. 238 of BAI.)

Theorems 8.5 and 8.6 give the so-called fundamental theorem of Galois theory, the pairing between the subfields of E/F and the subgroups of $G = \text{Gal } E/F$ (BAI, p. 239).

EXERCISES

1. Let E be Galois over F (not necessarily finite dimensional) and let K be a finite dimensional subfield of E/F. Show that any isomorphism of K/F into E/F can be extended to an automorphism of E.

2. Let s be a ring homomorphism of K into E, so $s \in \text{hom}(K, E)$. Show that Es is a submodule of $\text{hom}(K, E)$ as E-K-bimodule. Show that distinct ring homomorphisms of K into E give rise in this way to non-isomorphic E-K-bimodules and that any E-K-bimodule that is one-dimensional over E is isomorphic to an Es.

3. Use the theory of completely reducible modules to prove the following result: If $s_1, \ldots, s_r$ are distinct ring homomorphisms of K into E, then any E-K-submodule of $\sum_1^r Es_i$ has the form $\sum Es_{i_j}$ where $\{s_{i_j}\}$ is a subset of $\{s_i\}$.

4. Let $\bar{F}$ be the algebraic closure of F and let E and F' be finite dimensional subfields of $\bar{F}/F$ such that E is Galois over F. Let E' be the subfield of $\bar{F}/F$ generated by E and F'. Show that E' is Galois over F' and the map $g' \rightsquigarrow g'|E$ is an isomorphism of $G' = \text{Gal } E'/F'$ onto the subgroup $\text{Gal } E/(E \cap F')$ of $G = \text{Gal } E/F$ (see Lemma 4, p. 254 of BAI). Show that $E'/F' \cong E \otimes_{E \cap F'} F'$.

8.4 CROSSED PRODUCTS AND THE BRAUER GROUP

We shall now make an excursion into non-commutative algebra and give an important application of Galois theory to the study of the Brauer group of similarity classes of finite dimensional central simple algebras over a field F. We shall study a certain construction of central simple algebras called crossed products, which are defined by means of a Galois extension field E and a factor set of the Galois group into the multiplicative group E^* of non-zero elements of E. Such crossed products in the case of cyclic field extensions are called cyclic algebras. These were introduced by Dickson in 1906 and were apparently inspired by an earlier construction of infinite dimensional division algebras that had been given by Hilbert in his *Grundlagen der Geometrie* (1899). The most general crossed products were studied by Dickson and his students in the 1920's. The form of the theory that we shall present here is essentially due to Emmy Noether.

We begin with a finite dimensional Galois extension field E/F with Galois group G. We form a vector space A over E with a base $\{u_s|s\in G\}$ in 1–1 correspondence with the Galois group G. Thus the elements of A can be written in one and only one way in the form $\sum_{s\in G}\rho_s u_s$, $\rho_s\in E$. We shall introduce a multiplication in A that will render A an algebra over F. The defining relations for this multiplication are

$$u_s\rho = (s\rho)u_s, \qquad u_s u_t = k_{s,t}u_{st} \tag{14}$$

where the $k_{s,t}$ are non-zero elements of E and u_s is $1u_s$. More precisely, we define the product in A by

$$\left(\sum_{s\in G}\rho_s u_s\right)\left(\sum_{t\in G}\sigma_t u_t\right) = \sum_{s,t} k_{s,t}\rho_s(s\sigma_t)u_{st} \tag{15}$$

where the $k_{s,t}$ are non-zero elements of E. In particular, we have the product $u_s u_t = k_{s,t}u_{st}$ and hence for any $s,t,v\in G$ we have $(u_s u_t)u_v = k_{s,t}u_{st}u_v = k_{s,t}k_{st,v}u_{stv}$ and $u_s(u_t u_v) = u_s(k_{t,v}u_{tv}) = (sk_{t,v})k_{s,tv}u_{stv}$. Hence to insure associativity of the product defined by (15) we must have

$$k_{s,t}k_{st,v} = (sk_{t,v})k_{s,tv}, \qquad s,t,v\in G. \tag{16}$$

We observe that these conditions are exactly the conditions under which the map $(s,t)\rightsquigarrow k_{s,t}$ of $G\times G$ into the multiplicative group E^* of non-zero elements of E constitutes a 2-cocycle (p. 361). Here the action of G on E^* is the natural one where $s\rho$ is the image of ρ under $s\in G$. As in the case of group extensions, we call a 2-cocycle $k:(s,t)\rightsquigarrow k_{s,t}$ a *factor set* of G in E^*. From now on we assume that the $k_{s,t}$ satisfy (16), so k is a factor set. Then it is straightforward to verify that (15) is an associative product. Moreover, distributivity with respect to addition holds. The factor set conditions (16) imply that

$$k_{1,s} = k_{1,1} \quad\text{and}\quad k_{s,1} = sk_{1,1} \tag{17}$$

and these imply that

$$1 = k_{1,1}^{-1}u_1 \tag{18}$$

is the unit for A with respect to multiplication. Finally, we note that with respect to the vector space structure of A relative to the subfield F of E we have the algebra conditions

$$\alpha(ab) = (\alpha a)b = a(\alpha b) \tag{19}$$

for $\alpha \in F$, $a, b \in A$. These follow directly from (15). Hence A is an algebra over F. We call A the *crossed product of E and G with respect to the factor set k* and we write $A = (E, G, k)$ when we wish to display the ingredients in the definition.

We now investigate the structure of crossed products and we prove first

THEOREM 8.7. *$A = (E, G, k)$ is central simple over F, and $[A:F] = n^2$ where $n = [E:F]$. $E(= E1)$ is a subfield of A such that the centralizer $C_A(E) = E$.*

Proof. We have the dimensionality relation $[A:F] = [A:E][E:F] = n^2$. We note next that since $u_s u_{s^{-1}} = k_{s,s^{-1}}k_{1,1}1$ is a non-zero element of E and the same is true of $u_{s^{-1}}u_s$, u_s is invertible in A. Now let B be an ideal in A and write $\bar{a} = a + B$, $a \in A$. Suppose that $B \neq A$, so $\bar{A} = A/B \neq 0$. Then $\rho \rightarrow \bar{\rho}$ is a monomorphism of E into $\bar{A}$ and the elements $\bar{u}_s$ are invertible in $\bar{A}$. We have the relations $\bar{u}_s\bar{\rho} = \overline{s\rho}\,\bar{u}_s$. The Dedekind independence argument on shortest relations (BAI, p. 291) shows that the $\bar{u}_s$ are left linearly independent over $\bar{E} = \{\bar{\rho} | \rho \in E\}$. Hence $[\bar{A}:\bar{E}] = n$ and $[\bar{A}:\bar{E}][\bar{E}:\bar{F}] = n^2 = [A:F]$. It follows that $B = 0$ and A is simple. Let $\rho \in E$ and suppose that $\sum \rho_s u_s$ commutes with ρ. Then $\sum(\rho - s\rho)\rho_s u_s = 0$, which implies that $\rho = s\rho$ for every s such that $\rho_s \neq 0$. It follows that if $\sum \rho_s u_s$ commutes with every $\rho \in E$, then $\rho_s = 0$ for every $s \neq 1$. Then $\sum \rho_s u_s = \rho_1 u_1 \in E$. This shows that $C_A(E) = E$. Next suppose that c is in the center of A. Then $c \in E = C_A(E)$ and $cu_s = u_s c$, which gives $sc = c$ for every $s \in G$. Then $c \in F$. Hence the center is F, and A is central simple over F. □

The fact that E is a subfield of $A = (E, G, k)$ such that $C_A(E) = E$ implies that E is a splitting field for A: $A_E = E \otimes_F A \cong M_n(E)$ (Theorem 4.8, p. 221). We recall that the similarity classes $\{A\}$ of the finite dimensional central simple algebras A having a given extension field E of F as a splitting field constitute a subgroup $\mathrm{Br}(E/F)$ of the Brauer group $\mathrm{Br}(F)$ of the field F (p. 228). The result we have proved implies that any factor set k of the Galois group G into E^* determines an element $[(E, G, k)]$ of $\mathrm{Br}(E/F)$. Now consider the element

$[k] = kB^2(G, E^*)$ of $H^2(G, E^*) = Z^2(G, E^*)/B^2(G, E^*)$ (p. 361). We claim that all of the factor sets in $[k]$ determine the same element in $\mathrm{Br}(E/F)$. If $k' \in [k]$, then we have relations

$$(20) \qquad k'_{s,t} = \mu_s(s\mu_t)\mu_{st}^{-1}k_{s,t}, \qquad s, t \in G,$$

where $\mu_s \in E^*$. Let $A' = (E, G, k')$ be the crossed product defined by k' and let (u'_s) be a base for A' so that $u'_s\rho = (s\rho)u'_s$ and $u'_s u'_t = k'_{s,t}u'_{st}$. Now consider again the crossed product $A = (E, G, k)$ with base (u_s) over E as before. Put $v_s = \mu_s u_s$. Then it is clear that the v_s form a base for A over E. Moreover, we have $v_s\rho = \mu_s u_s \rho = \mu_s(s\rho)u_s = (s\rho)v_s$ and $v_s v_t = \mu_s u_s \mu_t u_t = \mu_s(s\mu_t)u_s u_t = \mu_s(s\mu_t)k_{s,t}u_{st} = \mu_s(s\mu_t)k_{s,t}\mu_{st}^{-1}v_{st} = k'_{s,t}v_{st}$. It is clear from these relations that the map $\sum \rho_s v_s \rightsquigarrow \sum \rho_s u'_s$ is an isomorphism of A onto A'. Thus $A \cong A'$ and hence $A \sim A'$, that is, $[A] = [A']$ in the Brauer group.

We can now define a map

$$(21) \qquad [k] \rightsquigarrow [(E, G, k)]$$

of $H^2(G, E^*)$ into $\mathrm{Br}(E/F)$. We shall show that this is an isomorphism. We prove first that (21) is surjective by proving

THEOREM 8.8. *Let A be a finite dimensional central simple algebra over F having the finite dimensional Galois extension field E/F as splitting field. Then there exists a factor set k such that $A \sim (E, G, k)$.*

Proof. By Wedderburn's theorem, $A \cong M_r(\Delta)$ where Δ is a central division algebra. E is a splitting field for A if and only if E is a splitting field for Δ and the condition for this is that there exists an r' such that E is a subfield of $A' = M_{r'}(\Delta)$ such that $C_{A'}(E) = E$ (Theorem 4.8, p. 221). By Theorem 4.11 $[A' : F] = [E : F][C_{A'}(E) : F] = n^2$. Evidently $A' \sim A$. We recall that by Theorem 4.9 (p. 222), any automorphism of E can be extended to an inner automorphism of A'. Accordingly, for any $s \in G$ there exists a $u_s \in A'$ such that $s\rho = u_s \rho u_s^{-1}$, $\rho \in E$. Now let $t \in G$ and consider the element $u_s u_t u_{st}^{-1}$. It is clear that this commutes with every $\rho \in E$. Hence $u_s u_t u_{st}^{-1} = k_{s,t} \in E^*$ and $u_s u_t = k_{s,t}u_{st}$. The associativity conditions $(u_s u_t)u_v = u_s(u_t u_v)$ imply as before that $k : (s, t) \rightsquigarrow k_{s,t}$ is a factor set and so we can form the crossed product (E, G, k). It is clear that $A'' = \{\sum \rho_s u_s | \rho_s \in E\}$ is a subalgebra of A'. We have a homomorphism of (E, G, k) onto this subalgebra. Since (E, G, k) is simple, this is an isomorphism. Thus $A'' \cong (E, G, k)$ and hence $[A'' : F] = n^2 = [A' : F]$. Then $A'' = A'$. Thus $A \sim A' \cong (E, G, k)$. □

We prove next the following basic multiplication formula for crossed products:

THEOREM 8.9. $(E, G, k) \otimes_F (E, G, l) \sim (E, G, kl)$.

Proof. Put $A = (E, G, k)$, $B = (E, G, l)$. In considering the tensor product $A \otimes B$ we may assume that A and B are subalgebras of an algebra AB in which every element of A commutes with every element of B, and if $x_1, \ldots, x_r \in A$ (B) are linearly independent over F, then these elements are linearly independent over B (A) in the sense that $\sum x_i y_i = 0$ for $y_i \in B$ (A) only if every $y_i = 0$. We need to distinguish between the two copies of E contained in A and B and we denote these by E_1 and E_2 respectively. We have an isomorphism $\rho \to \rho_i$ of E onto E_i and we write $s\rho_i$ for $(s\rho)_i$. We have a base (u_s) for A over E_1 and a base (v_s) for B over E_2 such that $u_s\rho_1 = (s\rho_1)u_s$, $u_s u_t = k_{s,t1}u_{st}$, $v_s\rho_2 = (s\rho_2)v_s$, $v_s v_t = l_{s,t2}v_{st}$.

Since E/F is separable, it has a primitive element: $E = F(\theta)$. Let $f(\lambda)$ be the minimum polynomial of θ over F. We have $f(\lambda) = \prod_{s \in G}(\lambda - s\theta)$ in $E[\lambda]$ and hence $f(\lambda) = \prod(\lambda - s\theta_1)$ and $f(\lambda) = \prod(\lambda - s\theta_2)$ in $E_1[\lambda]$ and $E_2[\lambda]$ respectively. Consider the Lagrange interpolation polynomial

$$f_s(\lambda) = \frac{f(\lambda)}{(\lambda - s\theta_1)f'(s\theta_1)} \tag{22}$$

$$= \prod_{t \neq s}(\lambda - t\theta_1) \Big/ \prod_{t \neq s}(s\theta_1 - t\theta_1)$$

in $E_1[\lambda]$. Since $1 - \sum_{s \in G} f_s(\lambda)$ is of degree $\leq n-1$ and this polynomial is 0 for the n distinct values $s\theta_1$, $1 - \sum f_s(\lambda) = 0$ and

$$\sum_{s \in G} f_s(\lambda) = 1. \tag{23}$$

Also if $s \neq t$, then

$$f_s(\lambda) f_t(\lambda) \equiv 0 \qquad (\operatorname{mod} f(\lambda)). \tag{24}$$

Now consider the commutative subalgebra $E_1 E_2$ of AB. Put

$$e_s = f_s(\theta_2) \in E_1 E_2. \tag{25}$$

Then $e_s \neq 0$ since $1, \theta_2, \ldots, \theta_2^{n-1}$ are linearly independent over E_1. By (23), we have $\sum_{s \in G} e_s = 1$, and by (24), $e_s e_t = 0$ if $s \neq t$. Hence multiplication of $\sum e_s = 1$

by e_t gives $e_t{}^2 = e_t$. Thus we have

$$(26) \qquad e_s{}^2 = e_s,\ e_s e_t = 0 \quad \text{if } s \neq t, \qquad \sum e_s = 1.$$

The definition of e_s gives $(\theta_2 - s\theta_1)e_s = 0$ since $(\theta_2 - s\theta_1)\prod_{t \neq s}(\theta_2 - t\theta_1) = f(\theta_2) = 0$. It follows that for $\rho_2 \in E_2$ we have

$$(27) \qquad \rho_2 e_s = (s\rho_1)e_s.$$

Then $E_1E_2 = \oplus_s E_1E_2e_s = \oplus_s E_1e_s$ and $E_1e_s \cong E$. As we have seen (p. 204), this decomposition of E_1E_2 as direct sum of simple algebras is unique.

Now consider the inner automorphisms $x \rightsquigarrow u_t x u_t^{-1}$ and $x \rightsquigarrow v_t x v_t^{-1}$ in AB. These stabilize E_1E_2 and hence permute the simple components of E_1E_2 and hence the idempotents e_s. If we apply these automorphisms to the relations (27) and take into account that the first is the identity on B and the second is the identity on A, we see that

$$\rho_2(u_t e_s u_t^{-1}) = (ts\rho_1)(u_t e_s u_t^{-1})$$
$$t\rho_2(v_t e_s v_t^{-1}) = (s\rho_1)(v_t e_s v_t^{-1}).$$

Comparison of these relations with (27) gives

$$(28) \qquad u_t e_s = e_{ts} u_t$$
$$v_t e_s = e_{st^{-1}} v_t.$$

Now put

$$(29) \qquad e_{s,t} = e_s u_s u_t^{-1}.$$

Then $e_{s,s} = e_s$ and using (28), we see that the n^2 $e_{s,t}$ constitute a set of matrix units:

$$(30) \qquad e_{s,t}e_{s',t'} = \delta_{t,s'}e_{s,t'}, \quad \sum e_{s,s} = 1.$$

Hence, by Theorem 4.6, $AB \cong M_n(C)$ where C is the centralizer in AB of the $e_{s,t}$. Equally well, $C \cong e_{1,1}ABe_{1,1}$ and we proceed to calculate this algebra. To begin with, we know that $[AB:F] = n^4$ and hence $[C:F] = n^2$ and $[e_{1,1}ABe_{1,1}:F] = n^2$. Now $e_{1,1}ABe_{1,1}$ contains $E_1e_{1,1} = E_1e_1$ and if we put

$$(31) \qquad w_s = u_s v_s e_{1,1}$$

we have, by (28), that $w_s = e_{1,1}u_s v_s$, so $w_s \in e_{1,1}ABe_{1,1}$. Now

$$w_s(\rho_1 e_{1,1}) = u_s v_s \rho_1 e_1 = s\rho_1 u_s v_s e_1 = (s\rho_1)e_{1,1}w_s$$

and

$$\begin{aligned} w_s w_t &= u_s v_s u_t v_t e_1 = u_s u_t v_s v_t e_1 = k_{s,t1} u_{st} l_{s,t2} v_{st} e_1 \\ &= k_{s,t1} l_{s,t2} e_1 w_{st} = k_{s,r1} l_{s,t1} e_1 w_{st} \\ &= (k_{s,t} l_{s,t})_1 w_{st}. \end{aligned}$$

It follows that $e_{1,1}ABe_{1,1}$ contains a subalgebra isomorphic to (E, G, kl). Comparison of dimensionalities shows that $e_{1,1}ABe_{1,1} \cong (E, G, kl)$. Hence $(E, G, k) \otimes_F (E, G, l) \sim (E, G, kl)$. □

We now see that the map $[k] \rightarrow [(E, G, k)]$ is multiplicative from the group $H^2(G, E^*)$ onto the group Br(E/F). It follows that $[1] \rightarrow [1]$, which means that $(E, G, 1) \sim 1$. To complete the proof that $[k] \rightarrow [(E, G, k)]$ is an isomorphism, we require the "only if" half of the following

THEOREM 8.10. *$(E, G, k) \sim 1$ if and only if $k \sim 1$.*

Proof. We have seen that $k \sim 1$ implies that $(E, G, k) \sim 1$. Conversely, suppose that $(E, G, k) \sim 1$. Then we have an isomorphism $a \rightarrow a'$ of $(E, G, 1)$ onto (E, G, k). The image E' of $E \subset (E, G, 1)$ is a subfield of (E, G, k) and for every $s \in G = \text{Gal}\, E/F$ we have an invertible element $v'_s \in (E, G, k)$ such that $v'_s \rho' = (s\rho)' v'_s$, $\rho \in E$, $v'_s v'_t = v'_{st}$. The isomorphism $\rho' \rightarrow \rho$ of E' into $E \subset (E, G, k)$ can be extended to an automorphism η of (E, G, k). This maps v'_s into v_s and we have $v_s \rho = (s\rho) v_s$, $v_s v_t = v_{st}$. On the other hand, for every $s \in G$ we have a u_s such that $u_s \rho = (s\rho) u_s$ and $u_s u_t = k_{s,t} u_{st}$. Then $u_s v_s^{-1}$ commutes with every $\rho \in E$ and hence $u_s v_s^{-1} = \mu_s \in E$ and $u_s = \mu_s v_s$. Then $k_{s,t} = \mu_s (s\mu_t) \mu_{st}^{-1}$, so $k \sim 1$. □

We re-state the main result we have proved as

THEOREM 8.11. *The map $[k] \rightarrow [(E, G, k)]$ is an isomorphism of $H^2(G, E^*)$ onto $Br(E/F)$.*

We have proved in Theorem 6.14 (p. 361) that if G is a finite group and A is a G-module, then $[k]^{|G|} = \{1\}$ for every $[k] \in H^n(G, A)$, $n > 0$. In particular, this holds for every $[k] \in H^2(G, E^*)$. It follows from Theorem 8.11 that $(E, G, k)^{|G|} \sim 1$ for every crossed product. There is an important improvement we can make in this result. We write $A = (E, G, k) = M_r(\Delta)$ where Δ is a division algebra. We have $[\Delta : F] = d^2$ and $[A : F] = n^2 = r^2 d^2$, so $n = rd$. The integer d is called the *index* of Δ. Since Δ is determined up to isomorphism by A, we can also call d the *index* of A. We now have

THEOREM 8.12. *If d is the index of $A = (E, G, k)$, then $A^d \sim 1$.*

We shall base the proof on matrices of semi-linear transformations. Let S be an s-semi-linear transformation in a finite dimensional vector space V over a field E: S is additive and $S(\alpha x) = (s\alpha)(Sx)$, $\alpha \in E$, $x \in V$. Let $(x_1, \ldots, x_n)$ be a base for V/E and write $Sx_i = \sum_1^n \sigma_{ji} x_j$. Then the matrix $\sigma = (\sigma_{ji})$ is called the *matrix of S relative to* $(x_1, \ldots, x_n)$. S is determined by σ and s; for, if $x = \sum \xi_i x_i$, then $Sx = \sum_{i,j} (s\xi_i)\sigma_{ji} x_j$. If T is a t-semi-linear transformation, then it is readily verified that ST is an st-semi-linear transformation whose matrix is $\sigma(s\tau)$ where $s\tau = (s\tau_{ij})$.

We can now give the

Proof of Theorem 8.12. Since $A = M_r(\Delta)$, A can be identified with $\mathrm{End}_{\Delta'} V$, the algebra of linear transformation in an r-dimensional vector space V over $\Delta' = \Delta^{\mathrm{op}}$. Since $A = (E, G, k)$ contains the subfield E, V can also be regarded as vector space over E and over F. We have $[V:F] = [V:E][E:F] = [V:E]n$ and $[V:F] = [V:\Delta'][\Delta':F] = rd^2 = nd$. Hence $[V:E] = d$. Let (u_s) be a base for $A = (E, G, k)$ over E such that $u_s \rho = (s\rho)u_s$ and $u_s u_t = k_{s,t} u_{st}$. The first of these relations shows that the element $u_s \in \mathrm{End}_{\Delta'} V$ is an s-semi-linear transformation of V/E. Let $(x_1, \ldots, x_d)$ be a base for V/E and let $M(s)$ denote the matrix of u_s relative to $(x_1, \ldots, x_d)$. Then the relation $u_s u_t = k_{s,t} u_{st}$ gives the matrix relation

$$M(s)sM(t) = k_{s,t} M(st). \tag{32}$$

Taking determinants we obtain

$$k_{s,t}^d \mu_{st} = (s\mu_t)\mu_s \tag{33}$$

where $\mu_s = \det M(s)$. Since $u_s u_s^{-1}$ is a non-zero element of E, $M(s)sM(s^{-1})$ is a non-zero scalar matrix. Hence $M(s)$ is invertible and $\mu_s \neq 0$. Then, by (33), we have $k_{s,t}^d = (s\mu_t)\mu_s \mu_{st}^{-1}$ and so $k^d \sim 1$. Hence $A^d \sim 1$ by Theorem 8.11. □

There is one further question on $\mathrm{Br}(F, E)$ and $H^2(G, E^*)$ that we need to consider. We suppose we have $E \supset L \supset F$ where L is Galois over F and $H = \mathrm{Gal}\, E/L$. Then $H \lhd G$ and $\bar{G} = \mathrm{Gal}\, L/F \cong G/H$. More precisely, we have the canonical homomorphism $s \rightsquigarrow \bar{s} = s|L$ of G onto $\bar{G}$ whose kernel is H. Hence $G/H \cong \bar{G}$ under $sH \rightsquigarrow \bar{s}$. Let A' be a finite dimensional central simple algebra over F split by L. Then A' is also split by E. By the isomorphism given in Theorem 8.11 we can associate with A' an element k' of $H^2(\bar{G}, L^*)$ and an element $k \in H^2(G, E^*)$. We shall obtain an explicit relation between $[k']$ and $[k]$, namely, we shall show that if k' is $(\bar{s}, \bar{t}) \rightsquigarrow k'_{\bar{s},\bar{t}} \in L^* \subset E^*$, then k is

$$(s, t) \rightsquigarrow k'_{\bar{s},\bar{t}}$$

where $s \rightsquigarrow \bar{s}$, $t \rightsquigarrow \bar{t}$ in the canonical homomorphism of G onto $\bar{G}$. In view of the definition of the isomorphism of $H^2(G, E^*)$ and $\mathrm{Br}(E/F)$, this is an immediate consequence of the following

THEOREM 8.13 (H. Hasse). *Let E be Galois over F with Galois group G, L a Galois subfield of E/F, $H = \mathrm{Gal}\, E/L$, $\bar{G} = \mathrm{Gal}\, L/F$. Let $A' = (L, \bar{G}, k')$ be a crossed product of L and $\bar{G}$ with the factor set k'. Then $A' \sim A = (E, G, k)$ where $k_{s,t} = k'_{\bar{s},\bar{t}}$ and $s \rightsquigarrow \bar{s}$, $t \rightsquigarrow \bar{t}$ in the canonical homomorphism of G onto $\bar{G}$.*

Proof (Artin, Nesbitt, and Thrall). We form $A = A' \otimes M_m(F)$ where $m = [E: L]$. Evidently $A \sim A'$. Since L is a subalgebra of A', we have the subalgebra $L \otimes M_m(F) \cong M_m(L)$ of A.

We now consider E as vector space over L proceed to define certain semi-linear transformations in E/L. First, we have the linear transformations $\rho_E : x \rightsquigarrow \rho x$ for $\rho \in E$. Next, we have the automorphisms $s \in G$. Since $s(\xi x) = (s\xi)(sx) = (\bar{s}\xi)(sx)$, $\xi \in L$, $x \in E$, s is an $\bar{s}$-semi-linear transformation of E/L. Moreover, we have the relation $s\rho_E = (s\rho)_E s$ (cf. (6)). We now choose a base $(x_1, \ldots, x_m)$ for E/L and consider the matrices in $M_m(L)$ relative to this base of the various semi-linear transformations we have defined. Let $\mu(\rho)$ denote the matrix of ρ_E relative to $(x_1, \ldots, x_m)$. Then $\rho \rightsquigarrow \mu(\rho)$ is a monomorphism of E, so the set of matrices $\mu(E)$ is a subfield of the algebra $M_m(L)/L$ isomorphic to E/L. Let τ_s denote the matrix of the $\bar{s}$-semi-linear transformation s relative to $(x_1, \ldots, x_m)$. Then we have the matrix relation

$$\tau_{st} = \tau_s(\bar{s}\tau_t) \tag{34}$$

and the relation $s\rho_E = (s\rho)_E s$ gives the matrix relation

$$\tau_s \bar{s}\mu(\rho) = \mu(s\rho)\tau_s. \tag{35}$$

Now let $v_{\bar{s}}$, $\bar{s} \in \bar{G}$, be a base for A' over L such that $v_{\bar{s}}\xi = (\bar{s}\xi)v_{\bar{s}}$, $\xi \in L$, and $v_{\bar{s}}v_{\bar{t}} = k'_{\bar{s},\bar{t}}v_{\bar{s}\bar{t}}$. Put $u_s = \tau_s v_{\bar{s}} \in A$ (since $\tau_s \in M_m(L) \subset A$ and $A' \subset A$). Then $u_s\mu(\rho) = \tau_s v_{\bar{s}}\mu(\rho) = \tau_s \bar{s}\mu(\rho)v_{\bar{s}} = \mu(s\rho)\tau_s v_{\bar{s}}$ (by (35)). Thus

$$u_s\mu(\rho) = \mu(s\rho)u_s. \tag{36}$$

Also $u_s u_t = \tau_s v_{\bar{s}} \tau_t v_{\bar{t}} = \tau_s(\bar{s}\tau_t)v_{\bar{s}}v_{\bar{t}} = \tau_{st}k'_{\bar{s},\bar{t}}$ (by (34)) $= k'_{s,t}\tau_{st}v_{\overline{st}} = \mu(k'_{s,t})u_{st}$. Hence

$$u_s u_t = \mu(k'_{\bar{s},\bar{t}})u_{st}. \tag{37}$$

The relations (36) and (37) imply that we have a homomorphism into A of (E, G, k) where $k_{s,t} = k'_{\bar{s},\bar{t}}$. Since (E, G, k) is simple, this is a monomorphism and since $[(E, G, k): F] = [E:F]^2$ and $[A:F] = [L:F]^2 m^2 = [L:F]^2 [E:L]^2 = [E:F]^2$, we have $A \cong (E, G, k)$. □

EXERCISES

1. Show that if L is a subfield of the Galois field E/F, then $(E, G, k)_L \cong (E, H, k')$ (over L) where $H = \text{Gal } E/L$ and k' is obtained by restriction of k to $H \times H$.
2. Let $\bar{F}$ be the algebraic closure of F and let E and F' be finite dimensional subfields of $\bar{F}/F$ such that E/F is Galois. Let E' be the subfield of $\bar{F}/F$ generated by E and F'. Let $G = \text{Gal } E/F$, $G' = \text{Gal } E'/F'$, so we have a canonical isomorphism $g \rightsquigarrow g' = g|E$ of G' with a subgroup of G (exercise 4, p. 475). Show that $(E, G, k)_{F'} \sim (E', G', k')$ where k' is obtained by restricting k to G' (identified with the subgroup of G). (*Hint*: Consider the case in which $F' \cap E = F$. Combine this with exercise 1 to obtain the general case.)

8.5 CYCLIC ALGEBRAS

The simplest type of crossed product $A = (E, G, k)$ is that in which E is a cyclic extension field of F, that is, E is Galois over F with cyclic Galois group G. Let s be a generator of G and let $u = u_s$. Then $(u_{s^i})^{-1}u^i$ centralizes E; hence, $u^i = \mu_i u_{s^i}$ where $\mu_i \in E^*$. We can replace u_{s^i} by u^i, $0 \leqslant i < n = [G:1]$. This replacement replaces the factor set k by k' where

$$k'_{s^i,s^j} = \begin{cases} 1 & \text{if } 0 \leqslant i+1 < n, \\ \gamma & \text{if } n \leqslant i+j \leqslant 2n-2. \end{cases} \tag{38}$$

The crossed product A is generated by E and u and every element of A can be written in one and only one way as

$$\rho_0 + \rho_1 u + \cdots + \rho_{n-1}u^{n-1}, \qquad \rho_i \in E. \tag{39}$$

The multiplication in A is determined by the relations

$$u\rho = (s\rho)u, \qquad u^n = \gamma. \tag{40}$$

Since u^n commutes with u and with every element of E^*, u^n is in the center F of A. Hence $\gamma \in F^*$. We shall now denote $A = (E, G, k)$ by (E, s, γ) and call this the *cyclic algebra* defined by E/F, the generator s of G and $\gamma \in F^*$.

We shall now specialize the results on $\mathrm{Br}(E/F)$ for E/F Galois to the case in which E/F is cyclic. In this case we need to consider only factor sets of the form (38). We call such a factor set a *normalized factor set defined by* γ. It is readily seen that the normalized factor set defined by γ and the normalized factor set defined by δ $(\in F^*)$ are cohomologous if and only if $\delta = \gamma N_{E/F}(\mu)$, $\mu \in E^*$. Accordingly, we are led to consider the group $F^*/N(E^*)$ where F^* is the multiplicative group of non-zero elements of F and $N(E^*)$ is the subgroup of elements of the form $N_{E/F}(\mu)$, $\mu \in E^*$. Then it is clear that we have an isomorphism of $F^*/N(E^*)$ onto $H^2(G, E^*)$ sending an element $\gamma N(E^*)$ of the first group into the class in $H^2(G, E^*)$ of the normalized factor set defined by γ. If we take into account the isomorphism given in Theorem 8.11, we obtain the following result in the cyclic case:

THEOREM 8.14. *The map $\gamma N(E^*) \rightsquigarrow [(E, s, \gamma)]$ is an isomorphism of $F^*/N(E^*)$ onto $Br(E/F)$.*

It is readily seen also that Theorem 8.13 implies the following result for E/F cyclic.

THEOREM 8.15. *Let E be a cyclic extension field of F and let L be a subfield of E/F. Let $\bar{s} = s|L$ where s is a generator of the Galois group of E/F. Then $(L, \bar{s}, \bar{\gamma}) \sim (E, s, \gamma^m)$ where $m = [E:L]$.*

We leave the proof to the reader.

EXERCISES

1. Show that $(E, s, \gamma) \sim 1$ if and only if γ is a norm in E, that is, there exists a $c \in E$ such that $\gamma = N_{E/F}(c)$.
2. Show that $(E, s, \gamma) \otimes_F (E, s, \delta) \sim (E, s, \gamma\delta)$.
3. (Wedderburn.) Prove that (E, s, γ) is a division algebra if $n = [E:F]$ is the smallest positive integer m such that $\gamma^m = N_{E/F}(c)$ for some $c \in E$.
4. Let E_0 be a cyclic extension of F_0 of dimension n. For example, we can take F_0 to be finite with q elements and $E_0 \supset F_0$ such that $[E_0 : F_0] = n$. Let s_0 be a generator of the Galois group of E_0/F_0. Let $E = E_0(t)$ be the field of rational expressions over E_0 in an indeterminate t and let s be the automorphism in E extending s_0 and fixing t. Show that if $F = F_0(t)$, then E/F is cyclic with Galois group $\langle s \rangle$. Show that (E, s, t) is a division algebra.
5. Specialize exercise 2, p. 484, to show that if the notations are as in this exercise and E/F is cyclic and $G = \langle s \rangle$, then $(E, s, \gamma)_{F'} \sim (E, s^m, \gamma)$ where $G' = \langle s^m \rangle$.

8.6 INFINITE GALOIS THEORY

In this section we shall give an extension of the subfield-subgroup correspondence of the finite Galois theory to certain infinite algebraic extensions of a field.

We observe first that if E is algebraic over F, then any homomorphism s of E into itself that fixes the elements of F is an automorphism of E/F. Since any homomorphism of a field into a non-zero ring is injective, it suffices to show that s is surjective. To see this let $a \in E$ and let $f(x)$ be the minimum polynomial of a. Let $R = \{a = a_1, \ldots, a_l\}$ be the set of roots of $f(x)$ in E. Then R is stabilized by s and since $s|R$ is injective, $s|R$ is surjective. Hence $a = sa_j$ for some j. We use this result to prove

PROPOSITION 8.9. *If E is algebraic over F, then $G = \text{Gal}\, E/F$ is a closed set in the finite topology of E^E.*

Proof. Let $s \in \overline{G}$, the closure of G in E^E. It is clear from the definition of the topology that s is a homomorphism of E into itself, fixing the elements of F. Then s is an automorphism and so $s \in G$. □

It is apparent from this result that we must restrict our attention to closed subgroups of automorphisms if we wish to obtain a 1–1 correspondence between groups of automorphisms and intermediate fields of an algebraic extension. That this is a real restriction can be seen in the following example.

EXAMPLE

Let $F_p = Z/(p)$, the field of p elements, and let $\overline{F}_p$ be the algebraic closure of F_p. Since $\overline{F}_p$ is algebraic over F_p, the map $\pi : a \rightsquigarrow a^p$ in $\overline{F}_p$ is an automorphism. We shall show that $\langle \pi \rangle$ is not closed in Gal $\overline{F}_p/F_p$. We need to recall some facts from the theory of finite fields (BAI, pp. 287–290). First, every finite field of characteristic p has cardinality p^n and for any p^n there exists a field F_{p^n} with $|F_{p^n}| = p^n$. This field is a splitting field over F_p of $x^{p^n} - x$ and is unique up to isomorphism. Moreover, every element of F_{p^n} is a root of this polynomial. The subfields of F_{p^n} are the F_{p^m} with $m|n$. Since $\overline{F}_p$ is the algebraic closure of F_p, it contains every F_{p^n}. Since $F_{p^n} \cup F_{p^m} \subset F_{p^{mn}}$, it is clear that any finite subset of $\overline{F}_p$ is contained in one of the subfields F_{p^n}. It is clear also that any automorphism of $\overline{F}_p/F_p$ stabilizes every finite subfield F_{p^n}. The Galois group of F_{p^n}/F_p consists of the powers of the map $a \rightsquigarrow a^p$. It follows that if s is any automorphism of $\overline{F}_p/F_p$, then the restriction of s to any F_{p^n} coincides with the restriction of a suitable power of π to F_{p^n}. This implies that the closure $\overline{\langle \pi \rangle}$ contains every automorphism of $\overline{F}_p/F_p$. Since the latter is closed, we have $\overline{\langle \pi \rangle} = \text{Gal}\, \overline{F}_p/F_p$. We shall now show that $\overline{\langle \pi \rangle} \neq \langle \pi \rangle$ by producing an automorphism of $\overline{F}_p/F_p$ that is not a power of π. For this purpose we choose any infinite proper subfield K of $\overline{F}_p$. For example, we can take a prime q and let K be the union of the subfields F_{p^m} for $m = q^r$, $r = 1, 2, \ldots$. Since this set of subfields is (totally) ordered, their union is a subfield and it is clear that it is an infinite subfield of $\bar{F}_p$. Moreover, it is a proper subfield, since it contains no F_{p^n} where n is not a power of q. Now let K be any proper infinite subfield of $\bar{F}_p$ and let $a \in \bar{F}_p$, $\notin K$. Let $f(x)$ be the minimum polynomial of a over K. Then $\deg f(x) > 0$ and $f(x)$ has a root $b \neq a$ in $\bar{F}_p$. Since $\bar{F}_p$ is a splitting field over $K(a)$ and over $K(b)$ of the set of polynomials $\{x^{p^n} - x/n = 1, 2, \ldots\}$, it follows from Theorem 8.2 that there exists an automorphism s of $\bar{F}_p/K$ sending $a \rightsquigarrow b$. Thus $s \neq 1$ and the subfield of s-fixed elements contains K and so is infinite. On the other hand, if $k \geqslant 1$, the set of fixed points under π^k is the finite set of solutions of $x^{p^n} = x$. It follows that $s \neq \pi^k$ for any k, so $s \notin \langle \pi \rangle$.

The infinite Galois theory of automorphisms of fields is concerned with splitting fields of separable polynomials. Let F be a field, Γ a set of separable monic polynomials with coefficients in F, and let E be a splitting field over F of Γ. We have

PROPOSITION 8.10. *Any finite subset of E is contained in a subfield L/F that is finite dimensional Galois over F.*

Proof. Let Γ' be the set of products of the polynomials in Γ. If $f \in \Gamma'$, the subfield L_f over F generated by the roots of f in E is a splitting field over F of f. By Theorem 8.6 this is finite dimensional Galois over F. Evidently $L_{fg} \supset L_f \cup L_g$. Hence $E' = \bigcup_{f \in \Gamma'} L_f$ is a subfield of E and since every $f \in \Gamma$ is a product of linear factors in $E'[x]$, we have $E' = E$. Thus E is a union of finite dimensional Galois fields over F. Evidently this implies our result. $\square$

We prove next

PROPOSITION 8.11. *Let K be a subfield of E/F that is Galois over F. Then any automorphism of E/F stabilizes K. Moreover, the map $s \rightsquigarrow s|K$ of* Gal E/F *into* Gal K/F *is surjective.*

Proof. Let $a \in K$ and let $f(x)$ be the minimum polynomial of a over F. If $s \in \text{Gal } K/F$, then $f(s(a)) = 0$, so $s(a)$ is one of the finite set of roots of $f(x)$ in E. Hence the orbit of a under Gal K/F is a finite set $\{a = a_1, \ldots, a_r\}$. It follows that $f(x) = \prod_1^r (x - a_i)$ and if $t \in \text{Gal } E/F$, then $t(a)$ is one of the a_i. Hence $t(a) \in K$. This proves the first assertion. To prove the second, we observe that E is a splitting field over K of Γ. Hence by Theorem 8.2, any automorphism of K/F can be extended to an automorphism of E/F. This proves the second assertion. $\square$

We can now prove the main theorem of infinite Galois theory.

THEOREM 8.16 (Krull). *Let E be a splitting field over F of a set Γ of monic separable polynomials and let $G = \text{Gal}\, E/F$. Let Λ be the set of closed subgroups of G, Σ the set of subfields of E/F. Then we have the map $H \rightsquigarrow \text{Inv}\, H$ of Λ into Σ and $K \rightsquigarrow \text{Gal}\, E/K$ of Σ into Λ. These are inverses and are order-inverting. A subgroup $H \in \Lambda$ is normal in G if and only if $K = \text{Inv}\, H$ is Galois over F and if this is the case, then* $\text{Gal}\, K/F \cong (\text{Gal}\, E/F)/(\text{Gal}\, E/K)$.

Proof. It is clear that if $H \in \Lambda$, then $\text{Inv}\, H \in \Sigma$ and if $K \in \Sigma$, then $\text{Gal}\, E/K$ is closed and hence is in Λ. It is clear also that the indicated maps are order-inverting. Now let $a \in E$, $\notin F$, and let L be a subfield containing a that is finite dimensional Galois over F. Then we have an automorphism of L/F that moves a. Hence, by Proposition 8.11, we have an automorphism of E/F that moves a. This shows that $\text{Inv}\, G = F$. Since we can replace F by any subfield K of E/F and G by $\text{Gal}\, E/K$, we see that $\text{Inv}(\text{Gal}\, E/K) = K$. Now let H be a closed subgroup and let $K = \text{Inv}\, H$. Let L/K be a finite dimensional Galois subfield of E/K. By Proposition 8.11 (with K replacing F and L replacing K), $\text{Gal}\, E/K$ and $H \subset \text{Gal}\, E/K$ stabilize L and the set of restrictions of the $s \in \text{Gal}\, E/K$ to L constitute $\text{Gal}\, L/K$. The set $\bar{H}$ of restrictions of the $h \in H$ to L constitute a subgroup of $\text{Gal}\, L/K$. Hence, if this subgroup is proper, by the finite Galois theory there exists a $\xi \in L$, $\notin K$ such that $\bar{h}(\xi) = \xi$, $\bar{h} \in \bar{H}$. This contradicts the definition of $K = \text{Inv}\, H$. Hence $\bar{H} = \text{Gal}\, L/K$. Thus for any $s \in \text{Gal}\, E/K$ there exists an $h \in H$ such that $s|L = t|L$. By Proposition 8.10, this implies that s is in the closure of H. Hence $s \in H$. Thus $\text{Gal}\, E/\text{Inv}\, H = H$. This proves that the two maps are inverses. Now if $K = \text{Inv}\, H$ and $s \in G$, then $s(K) = \text{Inv}\, s(H)s^{-1}$. This implies that $H \lhd G$ if and only if $s(K) = K$ for every $s \in G$. In this case K is Galois over F, since the set of automorphisms $s|K$ is a group of automorphisms in K whose set of fixed elements is F. As in the finite Galois theory, it follows that $\text{Gal}\, K/F \cong (\text{Gal}\, E/F)/(\text{Gal}\, E/K)$. Finally if K is any subfield of E/F that is Galois over F, then any automorphism of E/F stabilizes K, so $s(K) = K$ for $s \in G$. Then $H = \text{Gal}\, E/K$ is normal in G. □

In the applications of infinite Galois theory it is useful to view the topology of the Galois group in a slightly different fashion, namely, as inverse limit of finite groups. If L/F is a finite dimensional Galois subfield of E/F, then $H = \text{Gal}\, E/L$ is a normal subgroup of finite index in $G = \text{Gal}\, E/F$ since $G/H \cong \text{Gal}\, L/F$, which is a finite group. Conversely, any normal subgroup of finite index is obtained in this way. Proposition 8.10 shows that E is the union of the subfields L such that L/F is finite dimensional Galois over F. It follows that $\bigcap H = 1$ for the set H of normal subgroups of finite index in G. It is easily seen that G is the inverse limit of the finite groups G/H and that the topology in G is that of the inverse limit of finite sets. It is easily seen also from this or by using the Tychonov theorem that G is a compact set.

EXERCISES

1. Prove the last two statements made above.

2. Let F be Galois in E and assume that Gal E/F is compact. Show that E/F is algebraic and E is a splitting field over F of a set of separable polynomials. (Hence Krull's theorem is applicable to E/F.)

3. Show that Gal $\bar{F}_p/F_p$ has no elements of finite order $\neq 1$.

4. Let $E = F(t_1, t_2, \ldots)$, the field of fractions of $F[t_1, t_2, \ldots]$, the polynomial ring in an infinite number of indeterminates. Show that Gal E/F is not closed in the finite topology.

5. Show that Gal $\bar{F}_p/F_p \cong$ Gal $\bar{F}_q/F_q$ for any two primes p and q.

6. A *Steinitz number* is a formal product $N = \prod p_i^{k_i}$ over all of the primes $p_i \in Z$ where $k_i = 0, 1, 2, \ldots$ or ∞. If $M = \prod p^{l_i}$, we write $M|N$ if $l_i \leqslant k_i$ for all i. Note that the set of Steinitz numbers is a complete lattice relative to the partial order $M < N$ if $M|N$. Call the sup in the lattice the *least common multiple* (l.c.m.) of the Steinitz numbers. If E is a subfield of $\bar{F}_p$, define deg E to be the Steinitz number that is the l.c.m. of the degrees of the minimum polynomials over F_p of the elements of E. Show that this gives a 1–1 correspondence between subfields of $\bar{F}_p$ and Steinitz numbers.

7. Show that Gal $\bar{F}_p/F_p$ is uncountable.

8.7 SEPARABILITY AND NORMALITY

We shall now investigate the structure of algebraic extension fields of a given field F. Most of what we shall do becomes trivial in the characteristic 0 case but is important for fields of prime characteristic. We operate in an algebraic closure $\bar{F}$ of F and consider its subfields. This amounts to looking at all algebraic extensions E of F, since any such extension is isomorphic to a subfield of $\bar{F}/F$. It will be clear that everything we do is independent of the imbedding of E in $\bar{F}$ and we shall not call attention to this fact in our discussion. We remark also that $\bar{F}$ is an algebraic closure for any of its subfields E/F.

Let Γ be a set of monic polynomials with coefficients in F and let E be the subfield of $\bar{F}/F$ generated by the roots in $\bar{F}$ of every $f(x) \in \Gamma$. Clearly E is a splitting field over F of the set Γ. It is clear also that any homomorphism of E/F into $\bar{F}/F$ stabilizes the set of roots of every $f(x)$ and hence stabilizes E. Consequently it is an automorphism of E/F. We shall now call an algebraic extension field E/F *normal* if any irreducible polynomial in $F[x]$ having a root in E is a product of linear factors in $E[x]$. It is evident from this definition that a normal extension is a splitting field, namely, the splitting field over F of the set of minimum polynomials of its elements. The following result therefore gives an abstract characterization of splitting fields as normal extensions.

THEOREM 8.17. *If E is a splitting field over F of a set Γ of monic polynomials with coefficients in F, then E is normal over F.*

Proof. Let $f(x) \in F[x]$ be irreducible and have a root r in E. We have $f(x) = \prod_1^m (x - r_i)$ in $\bar{F}[x]$ where $r_1 = r$, and we have to show that every $r_i \in E$. Consider $E(r_i)$. This field contains $F(r_i)$ and is a splitting field over $F(r_i)$ of the set Γ. Since r_1 and r_i for any i are roots of the same irreducible polynomial in $F[x]$, we have an isomorphism of $F(r)/F$ onto $F(r_i)/F$ sending $r \rightsquigarrow r_i$. Since $E(r)$ and $E(r_i)$ are splitting fields over $F(r)$ and $F(r_i)$ of the set Γ, Theorem 8.2 shows that we have an isomorphism of $E(r)$ onto $E(r_i)$ extending the isomorphism of $F(r)$ onto $F(r_i)$. Since E is a splitting field over F of $\dot{\Gamma}$, this isomorphism stabilizes E. Since $r \in E$, we have $E(r) = E$. Hence $E(r_i) = E$ and $r_i \in E$ for every i. $\square$

If E is an arbitrary subfield of $\bar{F}/F$, we can form the splitting field N in $\bar{F}$ of the set of minimum polynomials of the elements of E. Evidently $N \supset E$ and N is normal over F. It is clear also that N is the smallest normal subfield of $\bar{F}$ containing E. We call N the *normal closure* of E.

We recall that an algebraic element is called separable if its minimum polynomial over F is separable, and an algebraic extension E/F is separable if every element of E is separable (BAI, p. 238). We remark that if $a \in E$ is separable over F, then it is separable over any intermediate field K since its minimum polynomial over K is a factor of its minimum polynomial over F. If E is Galois over F, then it is separable and normal over F. For, if $a \in E$, the orbit under Gal E/F is finite, and if this is $\{a_1 = a, a_2, \ldots, a_r\}$ then the minimum polynomial of a over F is $\prod_1^r (x - a_i)$. Since this is a product of distinct linear factors in $E[x]$, a is separable. Hence E is separable and normal over F.

Now let $S\bar{F}$ be the subset of elements of $\bar{F}$ that are separable over F. It is clear that $S\bar{F}$ is contained in the splitting field over F of all the separable monic polynomials in $F[x]$. By Theorem 8.16, the latter field is Galois over F and hence it is separable and normal over F. Then $S\bar{F}$ contains this field and hence $S\bar{F}$ coincides with the splitting field over F of all separable monic polynomials in $F[x]$. We shall call $S\bar{F}$ the *separable algebraic closure of* F.

Now let $a \in F$ be separable over $S\bar{F}$ and let $f(x)$ be its minimum polynomial over $S\bar{F}$. Then $f(x)$ has distinct roots in $\bar{F}$ and so $(f(x), f'(x)) = 1$ if $f'(x)$ is the derivative of $f(x)$ (BAI, p. 230). Now apply $G = \text{Gal}\, S\bar{F}/F$ to $f(x)$ in the obvious way. This gives a finite number of distinct irreducible polynomials $f_i(x)$, $1 \leq i \leq r$, $f_1(x) = f(x)$ in $S\bar{F}[x]$. We have $(f_i(x), f_i'(x)) = 1$ and $(f_i(x), f_j(x)) = 1$ if $i \neq j$. Then $g(x) = \prod_1^r f_i(x) \in F[x]$ and $(g(x), g'(x)) = 1$, so $g(x)$ is separable. Since $g(a) = 0$, we see that a is separable over F and so $a \in S\bar{F}$.

We can use these results to prove

THEOREM 8.18. *If E is an algebraic extension of F, then the subset of elements of E that are separable over F form a subfield SE of E/F. Any element of E that is separable over SE is contained in SE.*

Proof. The first statement is clear since $SE = S\bar{F} \cap E$. The second is immediate also since if $a \in E$ is separable over SE, then it is separable over $S\bar{F} \supset SE$. Hence $a \in S\bar{F} \cap E = SE$. □

An algebraic extension E/F is *purely inseparable* over F if $SE = F$ and an element $a \in E$ is *purely inseparable* over F if $SF(a) = F$. The second part of the last theorem shows that E is purely inseparable over SE. If the characteristic is 0, then E/F is purely inseparable if and only if $E = F$ and a is purely inseparable if and only if $a \in F$. The interesting case for these considerations is that in which the characteristic is $p \neq 0$. For the remainder of the section we shall assume that we are in this situation. We proceed to derive some useful criteria for an element of a to be separable or to be purely inseparable. We prove first

PROPOSITION 8.12. *If $a \in E$, then a is separable if and only if $F(a) = F(a^p) = F(a^{p^2}) = \cdots$.*

Proof. If a is not separable, then its minimum polynomial is of the form $f(x) = g(x^p)$ (BAI, p. 231). Then $g(x)$ is irreducible, so this is the minimum polynomial of a^p. Since $[F(a){:}F] = \deg f(x)$ and $[F(a^p){:}F] = \deg g(x)$, we have $[F(a){:}F(a^p)] = p$ and $F(a) \neq F(a^p)$. Next assume that a is separable. Then the minimum polynomial $f(x)$ of a over F has distinct roots and hence so has the minimum polynomial $h(x)$ of a over $F(a^p)$. Now a is a root of $x^p - a^p \in F(a^p)[x]$ and $x^p - a^p = (x-a)^p$. Then $h(x) | x^p - a^p$ and since $h(x)$ has distinct roots, it follows that $h(x) = x - a$. Then $a \in F(a^p)$. Taking pth powers we see that $a^p \in F(a^{p^2})$, so $a \in F(a^{p^2})$. Iteration gives the required relations $F(a) = F(a^p) = F(a^{p^2}) = \cdots$. □

We prove next the following criterion for purely inseparable elements.

PROPOSITION 8.13. *If a is purely inseparable over F, then its minimum polynomial over F has the form $x^{p^e} - b$. On the other hand, if a is a root of a polynomial of the form $x^{p^e} - b \in F[x]$, then a is purely inseparable over F.*

Proof. Let $f(x)$ be the minimum polynomial of a and let $e \geqslant 0$ be the largest integer such that $f(x)$ is a polynomial in x^{p^e}. Write $f(x) = g(x^{p^e})$, $g(x) \in F[x]$.

Then $g(x)$ is irreducible in $F[x]$ and $g(x)$ is not a polynomial in x^p. Then $g(x)$ is the minimum polynomial of $b = a^{p^e}$ and b is separable over F. Moreover, $x^{p^e} - b$ is the minimum polynomial of a over $F(b)$. Hence if a is purely inseparable over F, then $b = a^{p^e} \in F$, $F(b) = F$, and $x^{p^e} - b$ is the minimum polynomial of a over F. Next assume that a is a root of $x^{p^e} - b$, $b \in F$. Then the formula for the p^eth power of a sum implies that any $c \in F(a)$ is a root of a polynomial of the form $x^{p^e} - d = (x - c)^{p^e}$ with $d \in F$. Then the minimum polynomial of c over F has multiple roots unless it is linear. Hence if $c \in F(a)$ is separable, then $c \in F$. Thus a is purely inseparable. □

We shall now look more closely at the structure of a normal algebraic extension in the characteristic p case. Let E/F be normal and let $a \in E$. Write the minimum polynomial $f(x)$ of a over F as $g(x^{p^e})$ where $g'(x) \neq 0$. Since $g(x)$ is irreducible in $F[x]$ and has the root $a^{p^e} \in E$, $g(x) = \prod_1^r (x - b_i)$, $b_1 = a^{p^e}$, in $E[x]$. The b_i are distinct since $g'(x) \neq 0$. Now $f(x) = g(x^{p^e}) = \prod (x^{p^e} - b_i)$ is irreducible in $F[x]$ and has the root a in E. Hence $f(x) = \prod_1^{rp^e} (x - a_j)$, $a_1 = a$, in $E[x]$. It now follows that every $x^{p^e} - b_i$ has a root in E and we may assume that this is a_i. Then $x^{p^e} - b_i = (x - a_i)^{p^e}$ and $f(x) = \prod_1^r (x - a_i)^{p^e} = h(x)^{p^e}$ where $h(x) = \prod_1^r (x - a_i) \in E[x]$. The relation $f(x) = h(x)^{p^e}$ shows that the coefficients of $h(x)$ satisfy equations of the form $x^{p^e} = c \in F$. Hence these are purely inseparable elements of E. Now it is clear that the subset of purely inseparable elements is a subfield P/F and the coefficients of $h(x)$ are contained in P. Moreover, a is a root of $h(x)$ and $h(x) = \prod_1^r (x - a_i)$ with distinct a_i. Hence a is separable over P. Since a was arbitrary in E, we have proved

PROPOSITION 8.14. *If E is normal over F, then E is a separable extension of its subfield P of purely inseparable elements.*

This result is striking in that it shows that whereas any algebraic extension is built up by first making a separable extension and following this with a purely inseparable one, the order can be reversed for normal extensions. We now determine the structure of arbitrary extension fields that can be constructed in the second manner.

THEOREM 8.19. (1) *If E is an algebraic extension field of F such that E is separable over its subfield P of purely inseparable elements, then $E \cong P \otimes_F S$ where S is the maximal separable subfield of E. Conversely, if P is a purely inseparable extension of F and S is a separable algebraic extension of F, then*

$P \otimes_F S$ is an algebraic extension field of F whose subfield of separable elements is S and subfield of purely inseparable elements is P.

(2) If E is normal algebraic over F, then $E \cong P \otimes_F S$ as in part (1), with S Galois over F. Conversely, if P is purely inseparable over F and S is algebraic and Galois over F, then $P \otimes_F S$ is a normal algebraic extension field of F.

Proof. (1) Assume first that $[E:F] < \infty$. If a is any element of E and $f(x)$ is its minimum polynomial, we have seen that $f(x) = g(x^{p^e})$ where $g(x)$ is separable. Then $a^{p^e} \in S$. It follows that for any finite subset $\{a_1, \ldots, a_r\}$ in E there exists an e such that $a_i^{p^e} \in S$ for all i. Since the elements having this property form a subfield and since E is finitely generated, there exists an e such that $a^{p^e} \in S$ for every $a \in E$. Now let $(x_1, \ldots, x_n)$ be a base for E/P and write $x_i x_j = \sum_{k=1}^n c_{ijk} x_k$ where the $c_{ijk} \in P$. Then $y_i y_j = \sum d_{ijk} y_k$ for $y_i = x_i^{p^e}$, $d_{ijk} = c_{ijk}^{p^e}$. The $y_i \in S$ and the $d_{ijk} \in S \cap P = F$. The multiplication table for the y_i shows that $\sum_1^n F y_i$ is an F-subalgebra of E and $\sum_1^n P y_i$ is a P-subalgebra of E. Now let $a \in E$ and write $a = \sum_1^n a_i x_i$, $a_i \in P$. Then $a^{p^e} = \sum a_i^{p^e} y_i$. Since a is separable over P, $a \in P(a^{p^e}) = P[a^{p^e}]$. This and the formula for a^{p^e} imply that $a \in \sum P y_i$, so the y_i generate E as vector space over P. Since the number of y_i is $n = [E:P]$, these form a base for E/P. The foregoing argument shows also that if $a \in S$, then $a \in \sum F y_i$. It follows that $(y_1, \ldots, y_n)$ is a base for S/F. The existence of a set of elements that is simultaneously a base for E/P and for S/F implies that $E \cong P \otimes_F S$. This proves the first assertion when $[E:F] < \infty$. Now let $[E:F]$ be arbitrary. Let L be a finite dimensional subfield of E/F. Then $S \cap L$ is the subfield of separable elements of L and $P \cap L$ is the subfield of purely inseparable elements of L. We show next that L can be imbedded in a finite dimensional subfield L' such that L' is separable over $P \cap L'$. Let $L = F(a_1, \ldots, a_r)$ and let $f_i(x)$ be a separable polynomial with coefficients in P having a_i as root. Let $p_1, \ldots, p_s$ be the coefficients of all of the $f_i(x)$ and put $L' = F(a_1, \ldots, a_r; p_1, \ldots, p_s)$. Then $L' = (L' \cap P)(a_1, \ldots, a_r)$ and the a_i are separable over $L' \cap P$. Hence every element of L' is separable over $L' \cap P$, which means that L' satisfies the first condition. To complete the proof we need to show that any element of E is a linear combination of elements of S with coefficients in P and that elements of S that are linearly independent over F are linearly independent over P. Since any finite set of elements can be imbedded in a finite dimensional subfield L' satisfying the conditions, both of these results are clear from the first part. This completes the proof of the first statement in (1).

Conversely, assume $E = P \otimes_F S$ where P is purely inseparable and S is a separable algebraic extension of F. Let (x_α) be a base for S/F. We claim that for any $e \geqslant 0$, $(x_\alpha^{p^e})$ is also a base for S/F. The argument is similar to one we used before: Let $a \in S$ and write $a = \sum a_\alpha x_\alpha$, $a_\alpha \in F$. Then $a^{p^e} = \sum a_\alpha^{p^e} x_\alpha^{p^e} \in \sum F x_\alpha^{p^e}$, which is an algebra over F. Since $a \in F[a^{p^e}]$, $a \in \sum F x_\alpha^{p^e}$, so any a is a linear combination of the $x_\alpha^{p^e}$. To show linear independence of the $x_\alpha^{p^e}$, it is enough to take a finite subset, say, $\{x_1, \ldots, x_n\}$. Then the subfield $F(x_1, \ldots, x_n)$ is finite dimensional separable over F and has a base $(x_1, \ldots, x_n, y_1, \ldots, y_m)$. Then the elements $x_1^{p^e}, \ldots, x_n^{p^e}, y_1^{p^e}, \ldots, y_m^{p^e}$ are generators for $F(x_1, \ldots, x_n)$ as vector space over F. Since their number is the dimensionality, they are linearly independent over F. Hence $\{x_1^{p^e}, \ldots, x_n^{p^e}\}$ is a linearly independent set. Since the x_α form a base for S/F, every element of $P \otimes_F S$ can be written in one and only one way in the form $\sum b_\alpha \otimes x_\alpha$, $b_\alpha \in P$. There exists an $e \geqslant 0$ such that every $b_\alpha^{p^e} \in F$. Hence $(\sum b_\alpha \otimes x_\alpha)^{p^e} = \sum b_\alpha^{p^e} \otimes x_\alpha^{p^e} = 1 \otimes \sum b_\alpha^{p^e} x_\alpha^{p^e} \in S$ and if $\sum b_\alpha \otimes x_\alpha \neq 0$, then some $b_\alpha \neq 0$. Then $b_\alpha^{p^e} \neq 0$ and $\sum b_\alpha^{p^e} x_\alpha^{p^e} \neq 0$. Since S is a field, this element has an inverse in S and hence $\sum b_\alpha \otimes x_\alpha$ is invertible. This proves that $P \otimes_F S$ is a field. Clearly P is a purely inseparable subfield, S is a separable algebraic subfield of E, and $E = P \otimes_F S = P(S)$. Thus E is generated over P by separable algebraic elements. Hence E is separable algebraic over P. Then, by the result we proved first, $E = P' \otimes_F S'$ where P' is the subfield of purely inseparable elements and S' the subfield of separable ones. Then $P' \supset P$, $S' \supset S$ and since $E = P \otimes_F S$, it follows that $P' = P$ and $S' = S$. This completes the proof of (1).

(2) If E is normal over F, E is separable over its subfield of purely inseparable elements, by Proposition 8.14. Hence $E = S \otimes_F P$ as in (1). Let $a \in S$ have minimum polynomial $f(x)$. Then $f(x)$ is separable and $f(x) = \prod_1^r (x - a_i)$, $a_1 = a$, in $E[x]$. Since S is the set of separable elements of E, every $a_i \in S$. This implies that S is a splitting field over F of a set of separable polynomials, so S is Galois over F by Theorem 8.16. Conversely, assume that $E = S \otimes_F P$ where S is algebraic and Galois over F and P is purely inseparable. We have shown in (1) that E is an algebraic field over F. It is clear that every automorphism of S/F has a unique extension to an automorphism of E/P and this implies that E is Galois over P. Let $a \in E$ and let $f(x) \in F[x]$, $h(x) \in P[x]$ be the minimum polynomials of a over F and P respectively. Then $h(x)^{p^e} \in F[x]$ for a suitable e, so $f(x) = h(x)^{p^e}$. Since E is Galois over P, $h(x)$ is a product of linear factors in $E[x]$. Then $f(x)$ is a product of linear factors in $E[x]$. Thus E is normal over F. □

These results apply in particular to the algebraic closure $\bar{F}$ of F. We have $\bar{F} = S\bar{F} \otimes_F P\bar{F}$ where $S\bar{F}$ is the subfield of $\bar{F}$ of separable elements and $P\bar{F}$ is the subfield of purely inseparable elements. $S\bar{F}$ is Galois over F and we have called this field the separable algebraic closure of F. We recall that a field is called perfect if every polynomial with coefficients in the field is separable (BAI, p. 226). For characteristic 0 this is always the case, and for characteristic p it happens if and only if every element of the field is a pth power. It is easily seen that $P\bar{F}$ is perfect and that this is contained in any perfect extension of F. For this reason $P\bar{F}$ is called the *perfect closure* of F.

EXERCISES

1. Let E/F be algebraic and let $K = F(x_1, \ldots, x_r)$, the field of fractions of $F[x_1, \ldots, x_r]$, x_i indeterminates. Show that $E \otimes_F K \cong E(x_1, \ldots, x_r)$ and hence $E \otimes_F K$ is a field.
2. Show that $F(x) \otimes_F F(y)$ is not a field if x and y are transcendental.
3. Show that if E/F contains a purely inseparable element not in F, then $E \otimes_F E$ has a non-zero nilpotent element.
4. (J. D. Reid.) Suppose that F_0 is not perfect and that char $F_0 = p$. Let $a \in F_0$, $\notin F_0^p$ (the subfield of pth power). Let $E = F_0(x)$, x transcendental over F_0. Put $y = x^{p^2}(x^p + a)^{-1}$ and $F = F_0(y)$. Show that E is not separable over F and that $P(E/F)$, the subfield of E of purely inseparable elements over F, is F.
5. Let E and F be as in exercise 4. Show that $E \otimes_F E$ contains non-zero nilpotent elements.
6. If E is a finite dimensional extension field of F, where $S = SE$, the subfield of separable elements, put $[E:F]_s = [S:F]$ and $[E:F]_i = [E:S]$. These are called the *separability degree* and *inseparability degree* respectively of E/F. Evidently $[E:F] = [E:F]_s[E:F]_i$. Show that if K is finite dimensional separable over E, then $[K:F]_s = [K:E]_s[E:F]_s$. Show that this holds also if K is purely inseparable over F. Finally show that $[K:F]_s = [K:E]_s$ and $[K:F]_i = [K:E]_i[E:F]_i$ holds for arbitrary finite dimensional extension fields of E/F.
7. Show that an algebraic extension of a perfect field is perfect.
8. Let P be purely inseparable over F. We say that P/F *has exponent* $e \geqslant 0$ if $a^{p^e} \in F$ for every $a \in P$ but there exist a such that $a^{p^{e-1}} \notin F$. Show that P/F has an exponent if $[P:F] < \infty$ and that if P/F has exponent e, then

$$P = P^{(0)} \supsetneqq P^{(1)} \supsetneqq \cdots \supsetneqq P^{(e)} = F,$$

where $P^{(i)} = F(P^{p^i})$. Show that $P^{(i-1)}/P^{(i)}$ is purely inseparable of exponent one.
9. Let P/F be finite dimensional purely inseparable of characteristic p. Show that P has a subfield Q/F such that $P/Q = p$.
10. Let $P = F(a)$ where char $F = p$ and a is algebraic with minimum polynomial $\lambda^{p^e} - \alpha$ over F. Determine all the subfields of P/F.

8.8 SEPARABLE SPLITTING FIELDS

We shall now prove the existence of a separable splitting field for any finite dimensional central simple algebra A/F. This will enable us to upgrade the results on the Brauer group that we obtained in section 8.4 and make them apply to arbitrary finite dimensional central simple algebras. As a consequence we shall obtain the result that the Brauer group $\mathrm{Br}(F)$ for any field F is a torsion group.

The theorem we want to prove is

THEOREM 8.20. *If A is finite dimensional central simple over F, then there exists a finite dimensional separable field S/F that is a splitting field for A.*

It suffices to prove this for $A = \Delta$, a division algebra. In this case the result will follow from the Corollary to Theorem 4.7 if we can show that Δ contains a maximal subfield that is separable. To prove this we shall need a few remarks on derivations of algebras.

We note that if A is any algebra and $d \in A$, then the map $D_d : x \rightsquigarrow [dx] = dx - xd$ is a derivation. It is clear that D_d is linear and the calculation

$$\begin{aligned}[dx]y + x[dy] &= dxy - xdy + xdy - xyd \\ &= dxy - xyd = [d, xy]\end{aligned}$$

shows that $D_d(xy) = (D_d x)y + x(D_d y)$. We call D_d the *inner derivation* determined by d. Evidently we have the relation $D_d = d_L - d_R$ where d_L and d_R are the left and right multiplications determined by d. Since d_L and d_R commute, we have $D_d^{\,k} = (d_L - d_R)^k = \sum_{i=0}^{k} \binom{k}{i}(-1)^{k-i} d_L{}^i d_R^{k-i}$. If the characteristic is p, this gives $D_d^{\,p} = (d^p)_L - (d^p)_R = D_{d^p}$. We can write this out as

$$\overbrace{[d[d[\cdots[dx]\cdots]]]}^{p} = [d^p x]. \tag{41}$$

We are now ready to prove

THEOREM 8.21. *Any finite dimensional central division algebra Δ/F contains a maximal subfield S that is separable over F.*

Proof. There is nothing to prove if $\Delta = F$, so we assume $\Delta \supsetneq F$. We shall show first that Δ contains an element $a \notin F$ that is separable over F. Choose any $a \in \Delta$, $\notin F$. If $F(a)$ is not purely inseparable, then it contains a separable element not in F and we have what we want. Next suppose that $F(a)$ is purely inseparable. Then the characteristic is p and the minimum polynomial of a over F has the form $x^{p^e} - \alpha$. Put $d = a^{p^{e-1}}$. Then $d \notin F$ but $d^p \in F$. Then the inner derivation $D_d \neq 0$, but $D_d^{\,p} = D_{d^p} = 0$. Hence we can choose $a \in \Delta$ so that

$[da] = b \neq 0$, but $[db] = 0$. Now, put $c = ab^{-1}d$. Then

$$D_a c = (D_a a)b^{-1}d = d.$$

Hence $[dc] = dc - cd = d$ and $dcd^{-1} = c + 1$. We now see that the subfield $F(c)/F$ has a non-trivial automorphism, namely, the restriction to $F(c)$ of the inner automorphism $x \rightsquigarrow dxd^{-1}$. It is readily seen that if E/F is purely inseparable, then the only automorphism of E/F is the identity. Hence $F(c)$ is not purely inseparable. Then we have a separable element in $F(c)$ not in F. Thus in any case we have an $a \notin F$ that is separable over F.

We shall complete the proof by showing that any maximal separable subfield S of Δ is a maximal subfield of Δ. Otherwise, $\Delta' = C_\Delta(S) \supsetneq S$. By the double centralizer theorem (Theorem 4.10, p. 224), S is the center of Δ' so Δ' is a central division algebra over S with $[\Delta' : S] \neq 1$. By the result we proved first, we have a subfield $S(a) \neq S$ that is separable over S. This contradicts the maximality of S as separable subfield of Δ since $S(a)$ is separable over F. □

As we have seen, Theorem 8.21 implies Theorem 8.20.

Now let S be a finite dimensional separable splitting field for A and let E be the normal closure of S. Then E/F is finite dimensional Galois. Since any extension field of a splitting field is a splitting field, E/F is a splitting field for A. Let Δ be the division algebra associated with A by the Wedderburn theorem and let d be its index. Then, by Theorem 8.12, $A^d \sim 1$. Evidently this implies

THEOREM 8.22. *$Br(F)$ for any field F is a torsion group.*

The order of $\{A\}$ in Br(F) is called the *exponent* of the central simple algebra A. Since $\{A\}^d = 1$ for the index d of A, we see that the exponent e of A is a divisor of its index.

EXERCISES

1. Let p be a prime divisor of the index d of a central simple algebra A. Show that $p|e$, the exponent of A (*Hint*: Let E be a Galois splitting field of A and let $[E:F] = m$. Then $d|m$ so $p|m$. By the Galois theory and Sylow's theorem, there is a subfield K of E/F such that $[E:K] = p^l$ where p^l is the highest power of p dividing m. Let e' be the exponent of $A' = A_K$ in Br(K). Show that $e'|e$ and that e' is a power of p different from 1.)

2. Show that if Δ is a central division algebra of degree $d = p_1^{l_1} \cdots p_r^{l_r}$, $l_i > 0$, p_i distinct primes, then $\Delta \cong \Delta_1 \otimes \cdots \otimes \Delta_r$ where the degree of Δ_i is $p_i^{l_i}$. (The degree is the square root of the dimensionality. Exercise 9, p. 226, is needed in the proof.)

8.9 KUMMER EXTENSIONS

We recall that a finite dimensional extension field E/F is called abelian if E is Galois over F and the Galois group G of E/F is abelian. The same terminology can also be used for infinite dimensional extension fields. However, we shall confine our attention to the finite dimensional case. In this section and in section 8.11 we shall study two types of abelian extensions that are particularly interesting because they can be treated in a fairly elementary, purely algebraic fashion.

If G is a finite abelian group, the least common multiple e of the orders of the elements of G is called the *exponent* of G. Evidently $e \mid |G|$. In this section we shall be dealing with a base field F that for a given positive integer m contains m distinct mth roots of unity and we shall give a survey of the abelian extensions whose Galois groups have exponents $m' \mid m$. We call these extensions *Kummer m-extensions* of F.

The hypothesis that F contains m distinct mth roots of 1, or equivalently, that $x^m - 1 = \prod_1^m (x - \zeta_i)$ in $F[x]$ with distinct ζ_i implies that the characteristic of F is not a divisor of m. The set $U(m) = \{\zeta_i \mid 1 \leqslant i \leqslant m\}$ is a subgroup of the multiplicative group F^* of F and $U(m) \supset U(m')$ for every $m' \mid m$. We remark also that if G is a finite abelian group of exponent $m' \mid m$, then the character group $\hat{G}$ of G can be identified with hom$(G, U(m))$, the group of homomorphisms of G into $U(m)$. For, by definition, $\hat{G}$ is the group of homomorphisms of G into the multiplicative group of complex numbers. Hence if $\chi \in \hat{G}$ and $g \in G$, then $\chi(g)^{m'} = \chi(g^{m'}) = \chi(1) = 1$. Hence $\chi(g)$ is contained in the group of complex mth roots of 1 which can be identified with $U(m)$. Accordingly, $\hat{G}$ can be identified with hom$(G, U(m))$. We recall that $|\hat{G}| = |G|$ (p. 281). This fact has some important consequences that we shall need. We state these as a

LEMMA. *Let G be a finite abelian group, $\hat{G}$ its character group. Then*

(1) *For any $s \neq 1$ in G there exists a $\chi \in \hat{G}$ such that $\chi(s) \neq 1$.*

(2) *For any $s \in G$ let $\bar{s}$ denote the map $\chi \rightsquigarrow \chi(s)$ of $\hat{G}$ into $\mathbb{C}^*$ (or into $U(m)$, m the exponent of G). Then $\bar{s} \in \hat{\hat{G}}$ and $s \rightsquigarrow \bar{s}$ is an isomorphism of G onto $\hat{\hat{G}}$.*

(3) *A set $\{\chi_1, \chi_2, \ldots, \chi_r\}$ of characters generate $\hat{G}$ if and only if the only $s \in G$ such that $\chi_i(s) = 1$, $1 \leqslant i \leqslant r$, is $s = 1$.*

Proof. (1) Let $H = \{s \mid \chi(s) = 1 \text{ for all } \chi \in \hat{G}\}$. Then H is a subgroup and condition (1) will follow if we can show that $H = 1$. Let $\bar{G} = G/H$. Any $\chi \in \hat{G}$ defines a character $\bar{\chi}$ on $\bar{G}$ by $\bar{\chi}(gH) = \chi(g)$ and the map $\chi \rightsquigarrow \bar{\chi}$ is a monomorphism of $\hat{G}$ into $\hat{\bar{G}}$. Since $|G| = |\hat{G}|$ and $|\bar{G}| = |\hat{\bar{G}}|$, this implies that $|\bar{G}| \geqslant |G|$. Hence $|\bar{G}| = |G|$ and $H = 1$.

(2) It is clear that $\bar{s} \in \hat{\hat{G}}$ and $s \rightsquigarrow \bar{s}$ is a homomorphism. If $\bar{s} = 1$, then $\chi(s) = 1$ for all $\chi \in \hat{G}$; hence $s = 1$ by condition 1. Thus $s \rightsquigarrow \bar{s}$ is a monomorphism. Since $|G| = |\hat{G}| = |\hat{\hat{G}}|$, it follows that $s \rightsquigarrow \bar{s}$ is an isomorphism.

(3) The result established in (2) has the consequence that we can regard G as the character group of $\hat{G}$. Hence assertion (3) is equivalent to the following statement: $\{s_1, \ldots, s_r\}$ generate G if and only if the only χ in $\hat{G}$ such that $\chi(s_i) = 1$, $1 \leqslant i \leqslant r$, is $\chi = 1$. Suppose that the s_i generate G. Then $\chi(s_i) = 1$, $1 \leqslant i \leqslant r$, implies $\chi(s) = 1$ for all s and hence $\chi = 1$. On the other hand, suppose that the subgroup $H = \langle s_1, \ldots, s_r \rangle \subsetneq G$. Then $\bar{G} = G/H \neq 1$ and we have a character $\bar{\chi} \neq 1$ on $\bar{G}$. This defines a character χ on G by $\chi(g) = \bar{\chi}(gH)$, which satisfies $\chi(s_i) = 1$, $1 \leqslant i \leqslant r$, and $\chi \neq 1$. □

Now let E be an abelian extension of F whose Galois group G has exponent $m' \mid m$. Let E^* and F^* be the multiplicative groups of E and F respectively. Let $M(E^*)$ be the subset of E^* of elements whose mth powers are contained in F^* and let $N(E^*)$ be the set of mth powers of the elements of $M(E^*)$. Then $M(E^*)$ and $N(E^*)$ are subgroups of E^* and F^* respectively, $M(E^*) \supset F^*$ and $N(E^*) \supset F^{*m} = \{\alpha^m \mid \alpha \in F^*\}$.

Let $\rho \in M(E^*)$ and put

$$\chi_\rho(s) = (s\rho)\rho^{-1}, \qquad s \in G. \tag{42}$$

Since $\rho^m \in F^*$, $(s\rho)^m = \rho^m$ and $\chi_\rho(s)^m = 1$, so $\chi_\rho(s) \in U(m)$. Moreover, since $U(m) \subset F^*$,

$$\begin{aligned}\chi_\rho(st) &= (st\rho)\rho^{-1} = s((t\rho)\rho^{-1})((s\rho)\rho^{-1}) \\ &= (t\rho)\rho^{-1}(s\rho)\rho^{-1} = \chi_\rho(s)\chi_\rho(t).\end{aligned}$$

Hence $\chi_\rho : s \rightsquigarrow \chi_\rho(s)$ is a character of G. If $\rho_1, \rho_2 \in M(E^*)$, then $\chi_{\rho_1\rho_2}(s) = s(\rho_1\rho_2)(\rho_1\rho_2)^{-1} = (s\rho_1)\rho_1^{-1}(s\rho_2)\rho_2^{-1} = \chi_{\rho_1}(s)\chi_{\rho_2}(s)$. Hence $\rho \rightsquigarrow \chi_\rho$ is a homomorphism of $M(E^*)$ into $\hat{G}$. We shall now prove

THEOREM 8.23. *Let F be a field containing m distinct mth roots of 1 and let E be an abelian extension of F whose Galois group G has exponent $m' \mid m$. Let $M(E^*)$ be the subgroup of E^* of elements whose mth powers are in F^*.*

Then we have the exact sequence

$$(43) \qquad 1 \to F^* \hookrightarrow M(E^*) \xrightarrow{\lambda} \hat{G} \to 1$$

where $\hookrightarrow$ *denotes inclusion and* λ *is* $\rho \rightsquigarrow \chi_\rho$ *where* $\chi_\rho(s) = s(\rho)\rho^{-1}$, $s \in G$. *The factor group* $M(E^*)/F^* \cong G$. *We have* $E = F(M(E^*))$ *and* $E = F(\rho_1, \ldots, \rho_r)$ *for* $\rho_i \in M(E^*)$ *if and only if the cosets* $\rho_i F^*$ *generate* $M(E^*)/F^*$.

Proof. To prove the first statement we have to show that λ is surjective and its kernel is F^*. Let $\chi \in \hat{G}$. Then $\chi(st) = \chi(s)\chi(t) = (t\chi(s))\chi(t)$ and hence there exists a $\rho \in E^*$ such that $\chi(s) = (s\rho)\rho^{-1}$ (p. 361). Since $(s\rho)\rho^{-1} \in U(m)$, $s\rho^m = \rho^m$ for every $s \in G$. Hence $\rho^m \in F^*$ and $\rho \in M(E^*)$. Thus $\chi = \chi_\rho$ for $\rho \in M(E^*)$ and λ is surjective. Now suppose $\chi_\rho = 1$. Then $s\rho = \rho$, $s \in G$, so $\rho \in F^*$. Hence $\ker \lambda = F^*$.

The exactness of (43) implies that $M(E^*)/F^* \cong \hat{G}$. Since $\hat{G} \cong G$, we have $M(E^*)/F^* \cong G$ and the homomorphism λ of $M(E^*)$ onto $\hat{G}$ gives the isomorphism $\bar{\lambda}: \rho F^* \rightsquigarrow \chi_\rho$ of $M(E^*)/F^*$ onto $\hat{G}$. Accordingly, the characters $\chi_{\rho_1}, \ldots, \chi_{\rho_r}$ generate $\hat{G}$ if and only if the cosets $\rho_i F^*$ generate $M(E^*)/F^*$. Suppose this is the case and consider the field $E' = F(\rho_1, \ldots, \rho_r)$. Let $H = \text{Gal}\, E/E'$. If $t \in H$, then $t\rho_i = \rho_i$, $1 \leqslant i \leqslant r$, so $\chi_{\rho_i}(t) = 1$. This implies that $\chi(t) = 1$ for every $\chi \in \hat{G}$. Then $t = 1$ by statement (1) of the lemma. Thus $H = 1$ and hence $E = E' = F(\rho_1, \ldots, \rho_r)$. Evidently this implies that $E = F(M(E^*))$. Conversely, let $\rho_1, \ldots, \rho_r \in M(E^*)$ generate E/F and let $s \in G$. Then $s\rho_i = \rho_i$, $1 \leqslant i \leqslant r$, imply $s\rho = \rho$ for every $\rho \in E$ and $s = 1$. Thus $\chi_{\rho_i}(s) = 1$, $1 \leqslant i \leqslant r$, imply $s = 1$. Then by (3) of the lemma, the χ_{ρ_i} generate $\hat{G}$ and the cosets $\rho_i F^*$ generate $M(E^*)/F^*$. $\square$

We consider next the map $\mu: \rho \rightsquigarrow \rho^m(F^{*m})$ of $M(E^*)$ into $N(E^*)/F^{*m}$. This is an epimorphism and its kernel is the set of $\rho \in M(E^*)$ such that $\rho^m = \alpha^m$, $\alpha \in F^*$. Then $\rho\alpha^{-1} \in U(m) \subset F^*$. It follows that $\ker \mu = F^*$. Hence we have the isomorphism $\bar{\mu}: \rho F^* \rightsquigarrow \rho^m(F^{*m})$ of $M(E^*)/F^*$ onto $N(E^*)/F^{*m}$. Then, by Theorem 8.23, $N(E^*)/F^{*m} \cong G$, so this is a finite subgroup of F^*/F^{*m}. Thus we see that any Kummer extension whose Galois group has exponent $m'|m$ gives rise to a finite subgroup $N(E^*)/F^{*m}$ of F^*/F^{*m}. We remark also that if $\beta_1, \ldots, \beta_r$ are elements of $N(E^*)$ whose cosets $\beta_i F^{*m}$ generate $N(E^*)/F^{*m}$, then $E = F(\sqrt[m]{\beta_r}, \ldots, \sqrt[m]{\beta_r})$ where $\sqrt[m]{\beta}$ denotes a root of $x^m = \beta$.

We shall now show that any finite subgroup of F^*/F^{*m} can be obtained in the manner indicated from a Kummer extension, that is, if the given group is N/F^{*m}, then there exists a Kummer extension E/F such that $N(E^*) = N$. We remark that since $x^m = 1$ for any $x \in F^*/F^{*m}$, the exponent of N/F^{*m} is necessarily a divisor of m. We work in an algebraic closure $\bar{F}$ of F. If $\beta \in F$, we choose a particular element $\rho \in \bar{F}$ satisfying $\rho^m = \beta$ and write $\rho = \sqrt[m]{\beta}$. We have

THEOREM 8.24. *Let N/F^{*m} be a finite subgroup of F^*/F^{*m} and let $\beta_1, \dots, \beta_r$ be elements of N whose cosets $\beta_i F^{*m}$ generate N/F^{*m}. Then $E = F(\sqrt[m]{\beta_1}, \dots, \sqrt[m]{\beta_r})$ is a Kummer m-extension such that $N(E^*) = N$.*

Proof. We note first that E is a splitting field of

$$f(x) = (x^m - \beta_1) \cdots (x^m - \beta_r). \tag{44}$$

This is clear since $x^m - \beta_i = \prod_j (x - \zeta_j \beta_i)$ where $U(m) = \{\zeta_1, \zeta_2, \dots, \zeta_m\}$. Moreover, since $x^m - \beta_i$ has distinct roots, the polynomial $f(x)$ is separable. Then E is a splitting field over F of a separable polynomial and hence E/F is finite dimensional Galois. Let $s, t \in G = \text{Gal } E/F$. We have $s\rho_i = \zeta_{s(i)}\rho_i$ and $t\rho_i = \zeta_{t(i)}\rho_i$ for $\rho_i = \sqrt[m]{\beta_i}$. Then $st\rho_i = \zeta_{s(i)}\zeta_{t(i)}\rho_i = ts\rho_i$. Hence G is abelian. Since $s^m\rho_i = \zeta_{s(i)}^m \rho_i = \rho_i$, $s^m = 1$, and the exponent of G is $m' | m$. Then E is a Kummer m-extension of F.

It remains to show that $N(E^*) = N$. Since $\rho_i^m = \beta_i \in F^*$, the $\rho_i \in M(E^*)$. Since $E = F(\rho_1, \dots, \rho_r)$, it follows from Theorem 8.23 that the cosets $\rho_i F^*$ generate $M(E^*)/F^*$. Hence the cosets $\rho_i^m F^{*m} = \beta_i F^{*m}$ generate $N(E^*)/F^{*m}$. On the other hand, the cosets $\beta_i F^{*m}$ generate N/F^*. Hence $N = N(E^*)$. $\square$

Our results establish a 1–1 correspondence between the set of finite subgroups of F^*/F^{*m} and the set of Kummer m-extensions of F contained in $\bar{F}$. It is clear that this correspondence is order-preserving where the order is given by inclusion. Since the set of finite subgroups of F^*/F^{*m} is a lattice, we see that the set of Kummer m-extensions constitutes a lattice also and our correspondence is a lattice isomorphism.

8.10 RINGS OF WITT VECTORS

An extension field E of a field F of characteristic $p \neq 0$ is called an *abelian p-extension* of F if E is an abelian extension and $[E:F] = p^e$. Alternatively, we can define an abelian p-extension as an extension that is finite dimensional Galois with Galois group G a p-primary abelian group. We have given a construction of such extensions with $[E:F] = p$ in BAI, p. 300. These were first given by Artin and Schreier, who also constructed abelian p-extensions of p^2-dimensions, in connection with the proof of an algebraic characterization of real closed fields (see p. 674). The Artin-Schreier procedure was extended by A. A. Albert to give an inductive construction of cyclic p-extensions of p^e dimensions. Slightly later Witt gave a direct construction of all abelian p-extensions analogous to that of Kummer extensions. Witt's method was based on an ingenious definition of a ring, the ring of Witt vectors, defined by any commutative ring of characteristic p. These rings have other important applications and they are of considerable interest beyond the application we shall give to abelian p-extensions.

We start with the ring $X = \mathbb{Q}[x_i, y_j, z_k]$ in $3m$ indeterminates x_i, y_j, z_k, $0 \leqslant i,j,k \leqslant m-1$, over $\mathbb{Q}$. Consider the ring $X^{(m)}$ of m-tuples $(a_0, a_1, \ldots, a_{m-1})$, $a_i \in X$, with the component-wise addition and multiplication. Let p be a fixed prime number. We use this to define a map

$$a = (a_0, a_1, \ldots, a_{m-1}) \rightsquigarrow \varphi a = (a^{(0)}, a^{(1)}, \ldots, a^{(m-1)}) \tag{45}$$

where

$$a^{(\nu)} = a_0^{p^\nu} + pa_1^{p^{\nu-1}} + \cdots + p^\nu a_\nu, \qquad 0 \leqslant \nu \leqslant m-1. \tag{46}$$

Thus $a^{(0)} = a_0$, $a^{(1)} = a_0{}^p + pa_1, \ldots$. We introduce also the map $P: a \rightsquigarrow Pa = (a_0{}^p, a_1{}^p, \ldots, a_{m-1}^p)$. Then (46) gives

$$a^{(0)} = a_0, \qquad a^{(\nu)} = (Pa)^{(\nu-1)} + p^\nu a_\nu, \qquad \nu \geqslant 1. \tag{47}$$

Next let $A = (a^{(0)}, a^{(1)}, \ldots, a^{(m-1)})$ be arbitrary and define a map ψ by $\psi A = (a_0, a_1, \ldots, a_{m-1})$ where

$$\begin{aligned} a_0 &= a^{(0)} \\ a_\nu &= \frac{1}{p^\nu}(a^{(\nu)} - a_0^{p^\nu} - pa_1^{p^{\nu-1}} - \cdots - p^{\nu-1}a_{\nu-1}), \qquad \nu \geqslant 1. \end{aligned} \tag{48}$$

Direct verification shows that $\psi\varphi a = a$ and $\varphi\psi A = A$. Hence φ is bijective with ψ as inverse.

We shall now use the maps φ and $\psi = \varphi^{-1}$ to define a new ring structure on $X^{(m)}$. It will be convenient to denote the usual (component-wise) addition and multiplication in $X^{(m)}$ by $\oplus$ and $\odot$ respectively and write $u = (1, 1, \ldots, 1)$, the unit in $X^{(m)}$. Then we define a new addition and multiplication in $X^{(m)}$ by

$$\begin{aligned} a+b &= \varphi^{-1}(\varphi a \oplus \varphi b) \\ ab &= \varphi^{-1}(\varphi a \odot \varphi b). \end{aligned} \tag{49}$$

We have $\varphi(0, \ldots, 0) = (0, \ldots, 0)$ and $\varphi(1, 0, \ldots, 0) = (1, 1, \ldots, 1) = u$. It follows that $(X^{(m)}, +, \cdot, 0, 1)$ where $1 = (1, 0, \ldots, 0)$ is a ring and φ is an isomorphism of $(X^{(m)}, +, \cdot, 0, 1)$ onto $(X^{(m)}, \oplus, \odot, 0, u)$ (exercise 10, p. 97 of BAI). We denote the new ring as X_m and write $(a^{(0)}, a^{(1)}, \ldots, a^{(m-1)})$ for $\varphi(a_0, a_1, \ldots, a_{m-1})$, etc.

We now examine the formulas for $x+y$, xy, and $x-y$ for the "generic" vectors $x = (x_0, x_1, \ldots, x_{m-1})$, $y = (y_0, y_1, \ldots, y_{m-1})$. For example, we have

$$(x+y)_0 = x_0 + y_0, \qquad (x+y)_1 = x_1 + y_1 - \frac{1}{p}\sum_{1}^{p-1} \tbinom{p}{i} x_0{}^i y_0^{p-i},$$

$$(xy)_0 = x_0 y_0, \qquad (xy)_1 = x_0{}^p y_1 + x_1 y_0{}^p + p x_1 y_1.$$

In general, if $\circ$ denotes one of the compositions $+$, $\cdot$, or $-$, then it is clear from the definitions that $(x \circ y)_v$ is a polynomial with rational coefficients and 0 constant term in $x_0, y_0, x_1, y_1, \ldots, x_v, y_v$. Also one sees easily that

$$(x+y)_v = x_v + y_v + f_v(x_0, y_0, \ldots, x_{v-1}, y_{v-1}) \tag{50}$$

where f_v is a polynomial in the indicated indeterminates. The basic result we shall now establish is that $(x \circ y)_v$ is a polynomial in $x_0, y_0, \ldots, x_v, y_v$ with integer coefficients and 0 constant term.

Let $\mathbb{Z}[x_i, y_j] = \mathbb{Z}[x_0, y_0, \ldots, x_{m-1}, y_{m-1}]$ and write (p^μ) for $p^\mu \mathbb{Z}[x_i, y_j]$, $\mu \geqslant 0$. Then we have

LEMMA 1. *Let $\mu \geqslant 1$, $0 \leqslant k \leqslant m-1$, $a = (a_v)$, $b = (b_v)$ where $a_v, b_v \in \mathbb{Z}[x_i, y_j]$, $0 \leqslant v \leqslant m-1$, $\varphi a = (a^{(v)})$, $\varphi b = (b^{(v)})$. Then the system of congruences*

$$a_v \equiv b_v \ (p^\mu), \qquad 0 \leqslant v \leqslant k \tag{51}$$

is equivalent to

$$a^{(v)} \equiv b^{(v)} \ (p^{\mu+v}), \qquad 0 \leqslant v \leqslant k. \tag{52}$$

Proof. We have $a^{(0)} = a_0$, $b^{(0)} = b_0$, so the result is clear for $k = 0$. To prove the result by induction on k we may assume that (51) and (52) hold for $v \leqslant k-1$ and show that under these conditions $a_k \equiv b_k \ (p^\mu)$ if and only if $a^{(k)} \equiv b^{(k)} \ (p^{\mu+k})$. It is clear that $a_k \equiv b_k \ (p^\mu)$ if and only if $p^k a_k \equiv p^k b_k \ (p^{\mu+k})$. Hence by (47), it suffices to show that $(Pa)^{(k-1)} \equiv (Pb)^{(k-1)} \ (p^{\mu+k})$ holds under the induction hypothesis. We have $a_v \equiv b_v \ (p^\mu)$, $0 \leqslant v \leqslant k-1$. Since $\binom{p}{i} \equiv 0 \ (p)$, $1 \leqslant i \leqslant p-1$, this gives $a_v{}^p \equiv b_v{}^p \ (p^{\mu+1})$, $0 \leqslant v \leqslant k-1$. Hence, the induction on k applied to the vectors Pa and Pb gives $(Pa)^{(k-1)} \equiv (Pb)^{(k-1)} \ (p^{\mu+1+k-1})$, which is what is required. □

We can now prove the basic

THEOREM 8.25. *If* $x \circ y$ *denotes* $x+y$, xy, *or* $x-y$, *then* $(x \circ y)_\nu$ *is a polynomial in* $x_0, y_0, x_1, y_1, \ldots, x_\nu, y_\nu$ *with integer coefficients and* 0 *constant term.*

Proof. Since $(x \circ y)_\nu$ is a polynomial in $x_0, y_0, \ldots, x_\nu, y_\nu$ with rational coefficients and 0 constant term, it suffices to show that $(x \circ y)_\nu \in \mathbb{Z}[x_i, y_j]$. This is clear for $(x \circ y)_0$ and we assume it for $(x \circ y)_k$, $0 \leqslant k \leqslant \nu - 1$. By (47), we have

$$p^\nu (x \circ y)_\nu = (x \circ y)^{(\nu)} - (P(x \circ y))^{(\nu-1)} \tag{53}$$

and $(x \circ y)^{(\nu)} = x^{(\nu)} \dot{\pm} y^{(\nu)} \in \mathbb{Z}[x_i, y_j]$. The induction hypothesis implies that $(P(x \circ y))^{(\nu-1)} \in \mathbb{Z}[x_i, y_j]$. Hence by (53), it suffices to show that $(x \circ y)^{(\nu)} \equiv (P(x \circ y))^{(\nu-1)}\ (p^\nu)$. By (47), we have $x^{(\nu)} \equiv (Px)^{(\nu-1)}\ (p^\nu)$ and $y^{(\nu)} \equiv (Py)^{(\nu-1)}\ (p^\nu)$. Hence

$$\begin{aligned} (x \circ y)^{(\nu)} = x^{(\nu)} \dot{\pm} y^{(\nu)} &\equiv (Px)^{(\nu-1)} \dot{\pm} (Py)^{(\nu-1)} \\ &\equiv (Px \circ Py)^{(\nu-1)}\ (p^\nu). \end{aligned} \tag{54}$$

We are assuming that $(x \circ y)_k \in \mathbb{Z}[x_i, y_j]$, $0 \leqslant k \leqslant \nu - 1$. For any polynomial with integer coefficients we have $f(x_0, y_0, \ldots)^p \equiv f(x_0{}^p, y_0{}^p, \ldots)\ (p)$. It follows that $(P(x \circ y))_k \equiv ((Px) \circ (Py))_k\ (p)$, $0 \leqslant k \leqslant \nu - 1$. Hence by Lemma 1, we have

$$(P(x \circ y))^{(\nu-1)} \equiv (Px \circ Py)^{(\nu-1)}\ (p^\nu). \tag{55}$$

By (54) and (55), $(x \circ y)^{(\nu)} \equiv (P(x \circ y))^{(\nu-1)} (p^\nu)$, which is what we needed to prove. □

It is convenient to write the result we have proved as

$$\begin{aligned} (x+y)_\nu &= s_\nu(x_0, y_0, \ldots, x_\nu, y_\nu) \in \mathbb{Z}[x_i, y_j] \\ (xy)_\nu &= m_\nu(x_0, y_0, \ldots, x_\nu, y_\nu) \in \mathbb{Z}[x_i, y_j] \\ (x-y)_\nu &= d_\nu(x_0, y_0, \ldots, x_\nu, y_\nu) \in \mathbb{Z}[x_i, y_j]. \end{aligned} \tag{56}$$

Let η be an algebra endomorphism of $X/\mathbb{Q}$. Suppose that $\eta x_\nu = a_\nu$, $\eta y_\nu = b_\nu$, $0 \leqslant \nu \leqslant m-1$. Then we have $\eta x^{(\nu)} = a^{(\nu)}$, $\eta y^{(\nu)} = b^{(\nu)}$, $\eta((x+y)^{(\nu)}) = \eta x^{(\nu)} + \eta y^{(\nu)} = a^{(\nu)} + b^{(\nu)}$, and $\eta((x+y)_\nu) = (a+b)_\nu$. Hence by (56),

$$(a+b)_\nu = s_\nu(a_0, b_0, \ldots, a_\nu, b_\nu). \tag{57}$$

Similarly,

$$(ab)_\nu = m_\nu(a_0, b_0, \ldots, a_\nu, b_\nu). \tag{58}$$

(59) $$(a-b)_\nu = d_\nu(a_0, b_0, \ldots, a_\nu, b_\nu).$$

Since there exists an endomorphism of $X/\mathbb{Q}$ mapping the x_ν and y_ν into arbitrary elements of X, the foregoing formulas hold for arbitrary vectors $a = (a_0, a_1, \ldots, a_{m-1})$, $b = (b_0, b_1, \ldots, b_{m-1}) \in X_m$.

We are now ready to define the ring $W_m(A)$ of Witt vectors for an arbitrary commutative ring A of characteristic p. The set $W_m(A)$ is the set of m-tuples $(a_0, a_1, \ldots, a_{m-1})$, $a_i \in A$, with the usual definition of equality. We define addition and multiplication in $W_m(A)$ by

(60) $$(a+b)_\nu = \bar{s}_\nu(a_0, b_0, \ldots, a_\nu, b_\nu)$$
$$(ab)_\nu = \bar{m}_\nu(a_0, b_0, \ldots, a_\nu, b_\nu)$$

where the right-hand sides are the images in A of $s_\nu(x_0, y_0, \ldots, x_\nu, y_\nu)$ and $m_\nu(x_0, y_0, \ldots, x_\nu, y_\nu)$ respectively under the homomorphism of $\mathbb{Z}[x_i, y_j]$ into A such that $x_i \rightsquigarrow a_i, y_i \rightsquigarrow b_i, 0 \leqslant i \leqslant m-1$. Also we put $0 = (0, \ldots, 0)$, $1 = (1, 0, \ldots, 0)$ in $W_m(A)$. Then we have the structure $(W_m(A), +, \cdot, 0, 1)$. We shall now prove

THEOREM 8.26. *$(W_m(A), +, \cdot, 0, 1)$ is a commutative ring.*

Proof. To prove any one of the defining identities for a commutative ring—i.e., the associative laws, the commutative laws, distributive laws— we let a, b, c be any three elements of $W_m(A)$. We have the homomorphism η of $\mathbb{Z}[x_i, y_j, z_k]$ into A such that $x_\nu \rightsquigarrow a_\nu$, $y_\nu \rightsquigarrow b_\nu$, $z_\nu \rightsquigarrow c_\nu$, $0 \leqslant \nu \leqslant m-1$. Let I_m denote the subset of X_m of vectors whose components are contained in $\mathbb{Z}[x_i, y_j, z_k]$. By (57)–(59) this is a subring of X_m. Moreover, comparison of (57)–(59) with (60) shows that $(u_0, u_1, \ldots, u_{m-1}) \rightsquigarrow (\eta u_0, \eta u_1, \ldots, \eta u_{m-1})$ is a homomorphism of $(I_m, +, \cdot)$ into $(W_n(A), +, \cdot)$. This homomorphism maps $x = (x_0, x_1, \ldots, x_{m-1})$, $y = (y_0, y_1, \ldots, y_{m-1})$, $z = (z_0, z_1, \ldots, z_{m-1})$ into a, b, c respectively. Since $(xy)z = x(yz)$ in I_m, we have $(ab)c = a(bc)$. Hence the associative law of multiplication holds in $W_m(A)$. In a similar manner we can prove the other identities for multiplication and addition. The same type of argument shows that 0 is the 0-element for addition and $1 = (1, 0, \ldots, 0)$ is the unit for multiplication. To prove the existence of negatives we apply our homomorphism to $-x$. Then the image is the negative of a. It follows that $(W_m(A), +, \cdot, 0, 1)$ is a commutative ring. □

We shall call $W_m(A)$ the *ring of Witt vectors of length m over A*. It is clear that $W_1(A)$ can be identified with A since $a \rightsquigarrow (a)$ is an isomorphism. It is clear also that if η is a homomorphism of A into another commutative ring A' of characteristic p, then $(a_0, a_1, \ldots, a_{m-1}) \rightsquigarrow (\eta a_0, \eta a_1, \ldots, \eta a_{m-1})$ is a homomorphism of $W_m(A)$ into $W_m(A')$. In this way we obtain a functor W_m from the category of commutative rings of characteristic p into the category of commutative rings. In particular, if A is a subring of A', then $W_m(A)$ is a subring of $W_m(A')$.

We shall now consider some of the basic properties of the ring $W_m(A)$. For this purpose we introduce three important maps P, R, and V. We note first that since A is a commutative ring of characteristic p, we have a ring endomorphism $r \to r^p$ in A. This gives rise to the *Frobenius endomorphism* P of $W_m(A)$ where $Pa = (a_0{}^p, a_1{}^b, \ldots, a_{m-1}^p)$ for $a = (a_0, a_1, \ldots, a_{m-1})$. Next we define the *restriction map* R of $W_m(A)$ into $W_{m-1}(A)$ by

$$R(a_0, a_1, \ldots, a_{m-1}) = (a_0, a_1, \ldots, a_{m-2}) \tag{61}$$

and the *shift map* V of $W_{m-1}(A)$ into $W_m(A)$ by

$$V(a_0, a_1, \ldots, a_{m-2}) = (0, a_0, \ldots, a_{m-2}). \tag{62}$$

It is immediate that R is a ring homomorphism and we shall see that V is a homomorphism of the additive groups. We have

$$RV(a_0, a_1, \ldots, a_{m-1}) = (0, a_0, \ldots, a_{m-2}) = VR(a_0, a_1, \ldots, a_{m-1}),$$

so $VR = RV$ as maps of $W_m(A)$ into itself. It is clear also from the foregoing formula that $(VR)^m = 0$. Moreover, we have $PV = VP$ as maps of $W_{m-1}(A)$ into $W_m(A)$ and $RP = PR$ as maps of $W_m(A)$ into $W_{m-1}(A)$.

We prove next the following

LEMMA 2. *The following relations hold in Witt rings:*

$$p1 = \overbrace{1 + 1 \cdots + 1}^{p} = RV1, \tag{63}$$

$$V(a+b) = Va + Vb, \tag{64}$$

$$(Va)b = V(aPRb),\ a \in W_m(A),\ b \in W_{m+1}(A), \tag{65}$$

$$pa = RVPa. \tag{66}$$

Proof. Consider the subrings I_{m-1}, I_m, I_{m+1} of X_{m-1}, X_m, X_{m+1} respectively whose vectors have components in $\mathbb{Z}[x_i, y_j]$. We define the maps R and V for these rings in the same way as for Witt rings and we define P as before. Consider the unit $1 = (1, 0, \ldots, 0)$ of I_m. We have $\varphi 1 = (1, 1, \ldots, 1)$ and hence $(p1)^{(\nu)} = p$, $0 \leqslant \nu \leqslant m-1$. On the other hand, $RV1 = (0, 1, \ldots, 0)$, so the definition of φ gives $(RV1)^{(0)} = 0$, $(RV1)^{(\nu)} = p$, $1 \leqslant \nu \leqslant m-1$. It follows that

$(RV1)^{(v)} \equiv (p1)^{(v)}\ (p^{v+1})$, $0 \leqslant v \leqslant m-1$. By Lemma 1, this implies that $(RV1)_v \equiv (p1)_v\ (p)$. Now we have a homomorphism of I_m into $W_m(A)$, sending $x = (x_0, x_1, \ldots, x_{m-1}) \rightsquigarrow (a_0, a_1, \ldots, a_{m-1})$, $y = (y_0, y_1, \ldots, y_{m-1}) \rightsquigarrow (b_0, b_1, \ldots, b_{m-1})$. If we apply this to $RV1$ and to $p1$, we obtain formula (63) from the foregoing relations and the fact that A has characteristic p.

Next we note that $Vx = (0, x_0, \ldots, x_{m-1})$, $Vy = (0, y_0, \ldots, y_{m-1})$. Hence

$$(67) \qquad \begin{aligned} (Vx)^{(v)} &= px_0^{p^{v-1}} + p^2 x_1^{p^{v-2}} + \cdots + p^v x_{v-1} \\ &= px^{(v-1)}, \qquad 1 \leqslant v \leqslant m. \end{aligned}$$

Since $(x+y)^{(v)} = x^{(v)} + y^{(v)}$, (67) and $(Vx)^{(0)} = (Vy)^{(0)} = (V(x+y))^{(0)} = 0$ give $(V(x+y))^{(v)} = (Vx)^{(v)} + (Vy)^{(v)}$, $0 \leqslant v \leqslant m$. If we apply the homomorphism of $\mathbb{Z}[x_i, y_j]$, into A such that $x_v \rightsquigarrow a_v$, $y_v \rightsquigarrow b_v$, $0 \leqslant v \leqslant m-1$, to these relations we obtain $(V(a+b))^{(v)} = (Va)^{(v)} + (Vb)^{(v)}$, $0 \leqslant v \leqslant m$. This gives (64).

To prove (65) we shall show that

$$(68) \qquad ((Vx)y)_v \equiv V(xRPy)_v\ (p), \qquad 0 \leqslant v \leqslant m,$$

for $x = (x_0, \ldots, x_{m-1})$ and $y = (y_0, y_1, \ldots y_m)$. Put $(Vx)y = (w_0, w_1, \ldots, w_m)$, $V(xRPy) = (t_0, t_1, \ldots, t_m)$. Then we have to show that $w_v \equiv t_v(p)$, $0 \leqslant v \leqslant m$. By Lemma 1, this is equivalent to $w^{(v)} \equiv t^{(v)}\ (p^{v+1})$. This holds for $v = 0$ since $w^{(0)} = 0 = t^{(0)}$. For $v \geqslant 1$, we have, by (67), that $w^{(v)} = px^{(v-1)}y^{(v)}$ and $t^{(v)} = px^{(v-1)}(PRy)^{(v-1)}$. Since $y^{(v)} = (Py)^{(v-1)} + p^v y_v$, this gives the congruences

$$\begin{aligned} w^{(v)} = px^{(v-1)}y^{(v)} &\equiv px^{(v-1)}(Py)^{(v-1)} \\ &\equiv px^{(v-1)}(PRy)^{(v-1)} \equiv t^{(v)}\ (p^{v+1}). \end{aligned}$$

Hence (68) holds. If we apply the homomorphism of $\mathbb{Z}[x_0, \ldots, x_{m-1}, y_0, \ldots, y_m]$ to A such that $x_v \rightsquigarrow a_v$, $y_v \rightsquigarrow b_v$ to (68), we obtain the required relation (65).

If we apply R to both sides of (65), we obtain $(RVa)Rb = RV(aPRb)$. Putting $a = 1$ in this, we obtain $(RV1)(Rb) = RVPRb$. Since $RV1 = p1$ by (63) and since Rb can be taken to be any element of $W_m(A)$, this gives (66). □

We can now derive the basic properties of $W_m(A)$ that we shall need. We prove first

THEOREM 8.27. *The prime ring of $W_m(A)$ is isomorphic to $\mathbb{Z}/(p^m)$. It consists of the Witt vectors with components in the prime ring of $A\,(\cong \mathbb{Z}/(p))$.*

Proof. By (63), $p1 = RV1$. Then by (66), $p^2 1 = (RV)^2 1$. Iterating this gives $p^k 1 = (RV)^k 1$. Hence $p^{m-1} 1 = (0, 0, \ldots, 1) \neq 0$ and $p^{m} 1 = 0$. Hence the prime ring is isomorphic to $\mathbb{Z}/(p^m)$. The prime ring of A can be identified with $\mathbb{Z}/(p)$ and the subset of Witt vectors with components in $\mathbb{Z}/(p)$ is a subring of cardinality p^m. Hence it coincides with the prime ring of $W_m(A)$. $\square$

We prove next the important

THEOREM 8.28. *The map $(a_0, a_1, \ldots, a_{m-1}) \rightsquigarrow a_0$ is a homomorphism of $W_m(A)$ onto A whose kernel N is a nilpotent ideal.*

Proof. We have seen that R is a homomorphism of $W_m(A)$ into $W_{m-1}(A)$. Iteration of this gives a homomorphism $R^{m-1}: (a_0, \ldots, a_{m-1}) \rightsquigarrow (a_0)$ of $W_m(A)$ into $W_1(A)$. Since we can identify (a_0) with $a_0 \in A$, we have the homomorphism $(a_0, a_1, \ldots, a_{m-1}) \rightsquigarrow a_0$, which is clearly surjective. The kernel N of the homomorphism is the set of elements of the form $(0, a_1, \ldots, a_{m-1})$, so $N = RVW_m(A)$. We shall show that $N^m = 0$. If we apply R to (65) we obtain $(RVa)Rb = RV(aPRb)$, $a \in W_m(A)$, $b \in W_{m+1}(A)$. Since Rb can be taken to be any element c of $W_m(A)$, this gives the relation $(RVa)c = RV(aPc)$ for any $a, c \in W_m(A)$. Then $(RVa)(RVc) = RV(aPRVc) = RV(aRVPc) = RV((RVPc)a) = (RV)^2(PcPa) \in (RV)^2 W_m(A)$. Thus $N^2 \subset (RV)^2 W_m(A) = (RV)N$. Now suppose that for some $k \geqslant 2$, $N^k \subset N(RV)^{k-2}N \subset (RV)^{k-1}N$. Then if $d = RVa \in N$ and $b \in N^k$, we have $b = (RV)^k c$, $c \in W_m(A)$, since $N^k \subset (RV)^{k-1}N = (RV)^k W_m(A)$. Hence $db = (RVa)((RV)^k c) \in N(RV)^{k-1}N$, so $N^{k+1} \subset N(RV)^{k-1}N$. Moreover, if $a, c \in W_m(A)$, then

$$(RVa)((RV)^k c) = RV((RVPa)((RV)^{k-1}c)) \in RV(N(RV)^{k-2}N) \subset (RV)^k N.$$

Thus for any k we have $N^k \subset (RV)^{k-1}N = (RV)^k W_m(A)$. Since $(RV)^m = 0$, this gives $N^m = 0$. $\square$

An immediate consequence of Theorem 8.28 is the

COROLLARY. *An element $a = (a_0, a_1, \ldots, a_{m-1}) \in W_m(A)$ is a unit in $W_m(A)$ if and only if a_0 is a unit in A.*

Proof. If a is a unit, then so is a_0 since we have a homomorphism of $W_m(A)$ into A sending a into a_0. Conversely, if a_0 is a unit, then $a = a' + z$ where $z \in N$ and $a' + N$ is a unit in $W_m(A)/N$. It follows that a is a unit. $\square$

EXERCISES

1. Obtain the formulas for s_ν and m_ν as in (56) for $\nu = 0, 1, 2$.

2. Let $W_1(A), W_2(A), \ldots$ be the sequence of Witt rings defined by the commutative ring A of characteristic p. If $n \geqslant m$, we have the homomorphism R^{n-m} of $W_n(A)$ into $W_m(A)$. Show that these rings and homomorphisms define an inverse limit $W(A) = \varprojlim W_m(A)$. Show that $W(A)$ is isomorphic to the ring of Witt vectors of infinite rank defined to be the set of infinite sequences $a = (a_0, a_1, a_2, \ldots)$, $a_\nu \in A$, with the addition and multiplication given by (49), $0 = (0, 0, \ldots)$, $1 = (1, 0, 0, \ldots)$. Show that $N = \{(0, a_1, a_2, \ldots)\}$ is an ideal in $W(A)$ contained in the Jacobson radical.

3. Let A be a perfect field of characteristic p (BAI, p. 233). Show that $p^k W(A) = V^k W(A)$ where $V(a_0, a_1, a_2, \ldots) = (0, a_0, a_1, a_2, \ldots)$. Show that any element of $W(A)$ has the form $p^k a$ where a is a unit and hence show that $W(A)$ is a domain.

8.11 ABELIAN p-EXTENSIONS

We recall first the Artin-Schreier construction of cyclic extensions of dimension p of a field F of characteristic p: Let β be an element of F that is not of the form $\alpha^p - \alpha$, $\alpha \in F$, and let ρ be an element of the algebraic closure $\bar{F}$ of F such that $\rho^p - \rho = \beta$. Then $F(\rho)$ is a cyclic extension of F of dimension p and every such extension is obtained in this way. The proof makes use of the additive analogue of Hilbert's Satz 90, which can be regarded as a result on cohomology. We shall now give the extension of these results to abelian p-extensions and we consider first the extension of the cohomology theorem.

We suppose first that E is a finite dimensional Galois extension field of a field F of characteristic p and we form the ring $W_m(E)$ of Witt vectors of length $m \geqslant 1$ over E. We have the subring $W_m(F)$ of E and we have an action of the Galois group G of E/F defined by

$$s(\rho_0, \rho_1, \ldots, \rho_{m-1}) = (s\rho_0, s\rho_1, \ldots, s\rho_{m-1}). \tag{69}$$

This is an action by automorphisms and the subring of elements fixed under every $s \in G = \text{Gal}\, E/F$ is $W_m(F)$.

If $\rho = (\rho_0, \rho_1, \ldots, \rho_{m-1}) \in W_m(E)$, we define the *trace* $T(\rho) = \sum_{s \in G} s\rho$. Evidently $T(\rho) \in W_m(F)$. Since $(\rho + \sigma)_0 = \rho_0 + \sigma_0$ for $\rho = (\rho_0, \ldots, \rho_{m-1})$ and $\sigma = (\sigma_0, \ldots, \sigma_{m-1})$, $T(\rho) = (T_{E/F}(\rho_0), \ldots)$. We shall require the following

LEMMA 1. *There exists a $\rho \in W_m(E)$ such that $T(\rho)$ is a unit in $W_m(F)$.*

Proof. By the Dedekind independence property of distinct automorphisms of a field, there is a $\rho_0 \in E$ such that $T_{E/F}(\rho_0) \neq 0$. Let $\rho = (\rho_0, \ldots)$. Then $T(\rho) = (T_{E/F}(\rho_0), \ldots)$. Since $T_{E/F}(\rho_0) \neq 0$, it is a unit in F. Hence, by the Corollary to Theorem 8.28, $T(\rho)$ is a unit in $W_m(F)$. $\square$

We shall use this result to prove the following: if we regard the additive group $(W_m(E), +)$ as G-module under the action we defined, then the cohomology group $H^1(G, W_m(E)) = 0$. In explicit form the result is

THEOREM 8.29. *Let $s \rightsquigarrow \mu_s$ be a map of G into $W_m(E)$ such that $\mu_{st} = s\mu_t + \mu_s$. Then there exists a $\sigma \in W_m(E)$ such that $\mu_s = s\sigma - \sigma$, $s \in G$.*

Proof. The proof is identical with that of the special case in which $m = 1$, which was treated in Theorem 4.33 of BAI (p. 298). Choose $\rho \in W_m(E)$ so that $T(\rho)$ is a unit in $W_m(F)$ and put $\tau = T(\rho)^{-1} \sum_{t \in G} \mu_t(t\rho)$. Then

$$\begin{aligned} \tau - s\tau &= T(\rho)^{-1} \sum_t (\mu_{st}(st\rho) - s\mu_t(st\rho)) \\ &= T(\rho)^{-1} \left(\mu_s \sum_t t\rho \right) \\ &= \mu_s. \end{aligned}$$

Hence if $\sigma = -\tau$, then $\mu_s = s\sigma - \sigma$. $\square$

We recall that the Frobenius map $\rho = (\rho_0, \rho_1, \ldots, \rho_{m-1}) \rightsquigarrow P\rho = (\rho_0{}^p, \rho_1{}^p, \ldots, \rho_{m-1}^p)$ is an endomorphism of the ring $W_m(E)$. We now introduce the map $\mathscr{P}$ defined by

$$\mathscr{P}(\rho) = P\rho - \rho. \tag{70}$$

It is clear that this is an endomorphism of the additive group of $W_m(E)$ (but not of the ring $W_m(E)$). The kernel of $\mathscr{P}$ is the set of vectors ρ such that $\rho_\nu{}^p = \rho_\nu$, $0 \leqslant \nu \leqslant m-1$. This is just the set of vectors whose components are in the prime field of E. We have seen (Theorem 8.27, p. 503) that this set of vectors is the prime ring of $W_m(E)$ and can be identified with $\mathbb{Z}/(p^m)$, whose additive group is a cyclic group of order p^m. If G is any finite abelian group of exponent $p^e \leqslant p^m$, then the argument on p. 494 shows that the character group $\hat{G}$ can be identified with the group of homomorphisms of G into the cyclic additive group of the prime ring of $W_m(E)$. We shall make this identification from now on.

We now assume that the Galois group G of the extension E/F is abelian of order p^f and we choose m so that $p^m \geqslant p^e$, the exponent of G. We introduce the following subset of $W_m(E)$:

$$SW_m(E) = \{\rho \in W_m(E) \mid \mathscr{P}\rho \in W_m(F)\}. \tag{71}$$

This is a subgroup of the additive group of $W_m(E)$ containing $W_m(F)$. If $\rho \in SW_m(E)$, we define a map χ_ρ of G by $\chi_\rho(s) = s\rho - \rho$. Then $P\chi_\rho(s) = Ps\rho - P\rho =$

$sP\rho - P\rho = sP\rho - s\rho + s\rho - P\rho = P\rho - \rho + s\rho - P\rho = s\rho - \rho = \chi_\rho(s)$ or $\mathscr{P}\chi_\rho(s) = 0$. Then $\chi_\rho(s)$ is in the prime ring of $W_m(E)$. Also we have $\chi_\rho(st) = st\rho - \rho = st\rho - t\rho + t\rho - \rho = s\rho - \rho + t\rho - \rho = \chi_\rho(s) + \chi_\rho(t)$. Thus χ_ρ is a homomorphism of G into the additive group of the prime ring of $W_m(E)$ and so may be regarded as an element of the character group $\hat{G}$. Next let $\rho, \sigma \in SW_m(E)$. Then $\rho + \sigma \in SW_m(E)$ and $\chi_{\rho+\sigma}(s) = s(\rho+\sigma) - (\rho+\sigma) = s\rho - \rho + s\sigma - \sigma = \chi_\rho(s) + \chi_\sigma(s)$. Thus $\rho \rightarrow \chi_\rho$ is a homomorphism of the additive group $SW_m(E)$ into $\hat{G}$. If $\chi_\rho(s) = 0$ for all s, then $s\rho = \rho$ for all s and $\rho = \alpha \in W_m(F)$. Hence $W_m(F)$ is the kernel of $\rho \rightarrow \chi_\rho$. Finally, we note that this homomorphism is surjective. For, let $\chi \in \hat{G}$. Then $\chi(s)$ is in the prime ring and $\chi(st) = \chi(s) + \chi(t) = s\chi(t) + \chi(s)$. Then, by Theorem 8.29, there exists a $\rho \in W_m(E)$ such that $\chi(s) = s\rho - \rho$. Since $P\chi(s) = \chi(s)$, we have $s(P\rho - \rho) = P\rho - \rho$, $s \in G$. Then $\mathscr{P}\rho \in W_m(F)$ and so $\rho \in SW_m(E)$ and $\chi_\rho = \chi$. This proves the surjectivity of $\rho \rightarrow \chi_\rho$. We now have $SW_m(E)/W_m(F) \cong \hat{G} \cong G$. We have therefore proved the first two statements of the following theorem, which is a perfect analogue of Theorem 8.23 of the Kummer theory:

THEOREM 8.30. *Let F be a field of characteristic $p \neq 0$, E an abelian p-extension field of F whose Galois group G is of exponent p^e, and let $W_m(E)$ be the ring of Witt vectors over E of length m where $m \geq e$. Let $SW_m(E)$ be the additive subgroup of $W_m(E)$ of ρ such that $\mathscr{P}\rho = P\rho - \rho \in W_m(F)$. Then we have the exact sequence*

$$0 \rightarrow W_m(F) \hookrightarrow SW_m(E) \xrightarrow{\zeta} \hat{G} \rightarrow 0 \tag{72}$$

where $\hookrightarrow$ denotes inclusion and ζ is $\rho \rightarrow \chi_\rho$ where $\chi_\rho(s) = s\rho - \rho$. The factor group $SW_m(E)/W_m(F) \cong G$. The field E/F is generated by the components of the vectors $\rho \in SW_m(E)$ and

$$E = F(\rho_0^{(1)}, \ldots, \rho_{m-1}^{(1)}, \rho_0^{(2)}, \ldots, \rho_{m-1}^{(2)}, \ldots, \rho_0^{(r)}, \ldots, \rho_{m-1}^{(r)})$$

if and only if cosets $\rho^{(i)} + W_m(F)$, $\rho^{(i)} = (\rho_0^{(i)}, \ldots, \rho_{m-1}^{(i)})$, generate $SW_m(E)/W_m(F)$.

The proof of the last statement is exactly like that of the corresponding statement in Theorem 8.23. We leave it to the reader to check the details.

Following the pattern of our treatment of the Kummer theory we introduce next the set

$$QW_m(E) = \{\mathscr{P}(\rho) \mid \rho \in SW_m(E)\}. \tag{73}$$

This is a subgroup of the additive group of $W_m(F)$ containing $\mathscr{P}W_m(F)$, the subgroup of vectors $\mathscr{P}\alpha$, $\alpha \in W_m(F)$. We have the homomorphism

$$\rho \rightarrow \mathscr{P}\rho + \mathscr{P}W_m(F) \tag{74}$$

of $SW_m(E)$ onto $QW_m(E)/\mathscr{P}W_m(F)$. An element ρ is in the kernel of this homomorphism if and only if $\mathscr{P}\rho = \mathscr{P}\alpha$, $\alpha \in W_m(F)$. This is equivalent to $\mathscr{P}(\rho-\alpha) = 0$, which means $\rho-\alpha$ is in the prime ring. Thus the kernel of (74) is $W_m(F)$ and we have the isomorphism

$$QW_m(E)/\mathscr{P}W_m(F) \cong SW_m(E)/W_m(F). \tag{75}$$

Since the second of these groups is isomorphic to G, the first is also isomorphic to G. We shall show next that if Q is any subgroup of the additive group of the ring $W_m(F)$ containing $\mathscr{P}W_m(F)$ and $Q/\mathscr{P}W_m(F)$ is finite, then $Q = Q(W_m(E))$ for an abelian p-extension E/F. For this we need

LEMMA 2. *Let $\beta = (\beta_0, \beta_1, \ldots, \beta_{m-1}) \in W_m(F)$. Then there exist ρ_v, $0 \leqslant v \leqslant m-1$, in $\bar{F}$ such that $E = F(\rho_0, \rho_1, \ldots, \rho_{m-1})$ is finite dimensional separable over F and the vector $\rho = (\rho_0, \rho_1, \ldots, \rho_{m-1})$ of $W_m(E)$ satisfies $\mathscr{P}\rho = \beta$.*

Proof. If $m = 1$, we choose ρ in $\bar{F}$ so that $\rho^p - \rho = \beta$. Then $F(\rho)$ is separable, since the derivative $(x^p - x - \beta)' = -1$ and hence $x^p - x - \beta$ has distinct roots. Now suppose we have $\rho_0, \ldots, \rho_{m-2}$ in $\bar{F}$ so that $E' = F(\rho_0, \rho_1, \ldots, \rho_{m-2})$ is separable over F and $\mathscr{P}(\rho_0, \rho_1, \ldots, \rho_{m-2}) = (\beta_0, \beta_1, \ldots, \beta_{m-2})$ in $W_{m-1}(E')$. Consider the polynomial ring $E'[x]$ and the Witt ring $W_m(E'[x])$. Let $y = (\rho_0, \rho_1, \ldots, \rho_{m-2}, x)$ in this ring and form

$$\mathscr{P}y = (\rho_0{}^p, \rho_1{}^p, \ldots, \rho_{m-2}^p, x^p) - (\rho_0, \rho_1, \ldots, \rho_{m-2}, x).$$

We have $\mathscr{P}y = (\beta_0, \beta_1, \ldots, \beta_{m-2}, f(x))$, $f(x) \in E'[x]$, and since $(\beta_0, \beta_1, \ldots, \beta_{m-2}, f(x)) + (\rho_0, \rho_1, \ldots, \rho_{m-2}, x) = (\rho_0{}^p, \rho_1{}^p, \ldots, \rho_{m-2}^p, x^p)$, it follows from (50) that $x^p = f(x) + x + \gamma$ when $\gamma \in E'$. Thus $f(x) = x^p - x - \gamma$ and if we choose $\rho_{m-1} \in \bar{F}$ so that $f(\rho_{m-1}) = 0$, then the derivative argument shows that $E'(\rho_{m-1})$ is separable over E'. Hence $E = F(\rho_0, \rho_1, \ldots, \rho_{m-1})$ is separable over F. The formulas show that if $\rho = (\rho_0, \rho_1, \ldots, \rho_{m-1})$, then $\mathscr{P}\rho = (\beta_0, \beta_1, \ldots, \beta_{m-1})$. □

We can now prove

THEOREM 8.31. *Let Q be a subgroup of $(W_m(F), +)$ containing $\mathscr{P}W_m(F)$ such that $Q/\mathscr{P}W_m(F)$ is finite. Then there exists an abelian p-extension E of F such that the exponent of the Galois group of E/F is p^e, $e \leqslant m$, and $QW_m(E) = Q$.*

Proof. Let $\beta^{(1)}, \beta^{(2)}, \ldots, \beta^{(r)}$ be elements of Q such that the cosets $\beta^{(i)} + \mathscr{P}W_m(F)$ generate $Q/\mathscr{P}W_m(F)$. By Lemma 2, $\bar{F}$ contains a field E that is finite dimensional separable over F and is generated by elements $\rho_v^{(i)}$, $1 \leqslant i \leqslant r$, $0 \leqslant v \leqslant m-1$

such that $\mathscr{P}(\rho_0^{(i)},\ldots,\rho_{m-1}^{(i)}) = (\beta_0^{(i)},\ldots,\beta_{m-1}^{(i)})$ in $W_m(E)$. Let E' be the normal closure of E in $\bar{F}$ so E' is a finite dimensional Galois extension of F containing E. We form $W_m(E')$ and let the Galois group G of E'/F act on $W_m(E')$ as before. If $s \in G$ and $\rho^{(i)} = (\rho_0^{(i)},\ldots,\rho_{m-1}^{(i)})$, then $\mathscr{P}\rho^{(i)} = \beta^{(i)}$ gives $\mathscr{P}(s\rho^{(i)}) = \beta^{(i)}$. Hence $\mathscr{P}(s\rho^{(i)} - \rho^{(i)}) = 0$, so $s\rho^{(i)} - \rho^{(i)}$ is in the prime ring of $W_m(E')$. This implies that $sE \subset E$, $s \in G$. It follows that E is Galois over F and hence $E' = E$. If $s, t \in G$, then $s\rho^{(i)} = \rho^{(i)} + \gamma^{(i)}$, $t\rho^{(i)} = \rho^{(i)} + \delta^{(i)}$ where $\gamma^{(i)}, \delta^{(i)} \in W_m(F)$. Hence $ts\rho^{(i)} = \rho^{(i)} + \gamma^{(i)} + \delta^{(i)} = st\rho^{(i)}$, which implies that G is abelian. Also $s^k\rho^{(i)} = \rho^{(i)} + k\gamma^{(i)}$. Since $W_m(E)$ is of characteristic p^m, this implies that $s^{p^m} = 1$ hence G has order p^q and exponent p^e with $e \leqslant m$. Let χ_i be the character of G defined by $\rho^{(i)}$: $\chi_i(s) = s\rho^{(i)} - \rho^{(i)}$. Then it is clear that $\chi_i(s) = 1$, $1 \leqslant i \leqslant r$, implies that $s = 1$. It follows that the χ_i generate $\hat{G}$. Hence, if ρ is any element of $W_m(E)$ such that $\mathscr{P}\rho \in W_m(F)$, then $\chi_\rho = \prod \chi_i^{m_i}$. This implies that $\rho = \sum m_i\rho^{(i)} + \beta$, $\beta \in W_m(F)$, m_i integers. Then $\mathscr{P}(\rho) = \sum m_i\beta^{(i)} + \mathscr{P}(\beta) \in Q$. Since ρ is any element of $SW_m(E)$, this shows that $Q(W_m(E)) \subset Q$. The converse is clear so the proof is complete. □

The results we have obtained are analogous to the main results on Kummer extensions. They establish a 1–1 correspondence between the abelian p-extensions whose Galois groups have exponent p^e, $e \leqslant m$, with the subgroups Q of $(W_m(F), +)$ containing $\mathscr{P}W_m(F)$ as subgroup of finite index.

We shall now consider the special case of cyclic p-extensions. We observe first that the original Artin-Schreier theorem is an immediate consequence of the general theory: Any p-dimensional cyclic extension of F has the form $F(\rho)$ where $\mathscr{P}\rho = \rho^p - \rho = \beta \in F$ and β is an element such that there exists no $\alpha \in F$ satisfying $\mathscr{P}\alpha = \beta$. We shall now prove a result that if such an extension exists, then there exist cyclic extensions of F of any dimension p^m, $m \geqslant 1$. For this purpose we require

LEMMA 3. *If $\beta = (\beta_0, \beta_1, \ldots, \beta_{m-1}) \in W_m(F)$, then $p^{m-1}\beta \in \mathscr{P}W_m(F)$ if and only if $\beta_0 \in \mathscr{P}F$.*

Proof. We note first that by iteration of the formula (66) we obtain $p^{m-1}\beta = (0,\ldots,0,\beta_0^{p^{m-1}})$. Next we write

$$\begin{aligned}(0,\ldots,0,\beta_0) - (0,\ldots,0,\beta_0^{p^{m-1}}) &= [(0,\ldots,0,\beta_0) - (0,\ldots,0,\beta_0^p)] \\ &\quad + [(0,\ldots,0,\beta_0^p) - (0,\ldots,0,\beta_0^{p^2})] + \cdots \\ &\quad + [(0,\ldots,0,\beta_0^{p^{m-2}}) - (0,\ldots,0,\beta_0^{p^{m-1}})].\end{aligned}$$

Evidently the right-hand side is contained in $\mathscr{P}W_m(F)$. Hence $p^{m-1}\beta = (0,\ldots,0,\beta_0^{p^{m-1}}) \in \mathscr{P}W_m(F)$ if and only if $(0,\ldots,0,\beta_0) \in \mathscr{P}W_m(F)$. Suppose that this is the case, say, $(0,\ldots,0,\beta_0) = P\alpha - \alpha$ for $\alpha = (\alpha_0,\alpha_1,\ldots,\alpha_{m-1}) \in W_m(F)$. Applying R to this relation gives $PR\alpha - R\alpha = 0$ where $R\alpha = (\alpha_0,\ldots,\alpha_{m-2})$. Hence if $\gamma = (\alpha_0,\alpha_1,\ldots,\alpha_{m-2},0)$, then $P\gamma = \gamma$ and if $\delta = \alpha - \gamma$, then $P\delta - \delta = (0,\ldots,0,\beta_0)$. Moreover, since $R\delta = R\alpha - R\gamma = 0$, $\delta = (0,\ldots,0,\delta_{m-1})$. Then the formula (50) applied to $P\delta = \delta + (0,\ldots,0,\beta_0)$ implies that $\mathscr{P}\delta_{m-1} = \beta_0$, so $\beta_0 \in \mathscr{P}F$. Conversely, if this condition holds, then $\mathscr{P}\delta = (0,\ldots,0,\beta_0)$ for $\delta = (0,\ldots,0,\delta_{m-1})$ and hence $p^{m-1}\beta \in \mathscr{P}W_m(F)$. □

We can now prove

THEOREM 8.32. *Let F be a field of characteristic $p \neq 0$. Then there exist cyclic extensions of p^m dimensions, $m \geqslant 1$, over F if and only if there exist such extensions of p dimensions. The condition for this is $\mathscr{P}F \neq F$.*

Proof. We have seen that there exists a cyclic extension of p dimensions over F if and only if $F \neq \mathscr{P}F$. Suppose that this conditions holds and choose $\beta_0 \in F$, $\beta_0 \notin \mathscr{P}F$. Let $\beta = (\beta_0,\beta_1,\ldots,\beta_{m-1})$ where the β_i, $i > 0$, are any elements of F. By Lemma 3, $p^{m-1}\beta \notin \mathscr{P}W_m(F)$. This implies that the subgroup Q of $(W_m(F),+)$ generated by β and $\mathscr{P}W_m(F)$ has the property that $Q/\mathscr{P}W_m(F)$ is cyclic of order p^m. By Theorem 8.31, $Q = QW_m(E)$ for an abelian p-extension E/F. Moreover, we have seen that the Galois group G of E/F is isomorphic to $Q/\mathscr{P}W_m(F)$. Hence this is cyclic of order p^m and E/F is cyclic of p^m dimensions. □

EXERCISE

1. Show that if $\beta \in W_m(F)$ satisfies $p^{m-1}\beta \in \mathscr{P}W_m(F)$, then there exists a $\gamma \in W_m(F)$ such that $p\gamma \equiv \beta(\mathscr{P}W_m(F))$. Use this to prove that any cyclic extension of p^{m-1} dimensions over F can be embedded in a cyclic extension of p^m dimensions over F.

8.12 TRANSCENDENCY BASES

A finite subset $\{a_1,\ldots,a_n\}$, $n \geqslant 1$, of an extension field E/F is called *algebraically dependent over* F if the homomorphism

$$f(x_1,\ldots,x_n) \rightsquigarrow f(a_1,\ldots,a_n) \tag{76}$$

of the polynomial algebra $F[x_1,\ldots,x_n]$, x_i indeterminates, into E has a non-zero kernel. In other words, there exists a non-zero polynomial $f(x_1,\ldots,x_n)$ such that $f(a_1,\ldots,a_n) = 0$. Evidently if $\{a_1,\ldots,a_m\}$, $1 \leqslant m \leqslant n$, is algebraically dependent, then so is $\{a_1,\ldots,a_n\}$. We shall now say that an arbitrary non-vacuous subset of E is *algebraically dependent over* F if some finite subset has this property. We have the following criterion.

THEOREM 8.33. *A non-vacuous subset S of an extension field E/F is algebraically dependent over F if and only if there exists an $a \in S$ that is algebraic over $F(S-\{a\})$.*

Proof. If T is a subset of E and a is algebraic over $F(T)$, then a is algebraic over $F(U)$ for some finite subset U of T. It follows that it suffices to prove the theorem for S finite, say, $S = \{a_1, \ldots, a_n\}$. Suppose first that S is algebraically dependent. We shall prove by induction on n that there exists an a_i such that a_i is algebraic over $F(S_i)$, $S_i = S - \{a_i\}$. This is clear from the definitions if $n = 1$, so we assume that $n > 1$. We may assume also that $\{a_1, \ldots, a_{n-1}\}$ is algebraically independent (= not algebraically dependent). Then we have a polynomial $f(x_1, \ldots, x_n) \neq 0$ such that $f(a_1, \ldots, a_n) = 0$. Write $f(x_1, \ldots, x_n) = f_0(x_1, \ldots, x_{n-1})x_n{}^m + f_1(x_1, \ldots, x_{n-1})x_n^{m-1} + \cdots + f_m(x_1, \ldots, x_{n-1})$ with $f_0(x_1, \ldots, x_{n-1}) \neq 0$. Then $f_0(a_1, \ldots, a_{n-1}) \neq 0$, so $g(x) = f_0(a_1, \ldots, a_{n-1})x_n{}^m + f_1(a_1, \ldots, a_{n-1})x_n^{m-1} + \cdots + f_m(a_1, \ldots, a_{n-1})$ is non-zero polynomial in $F(a_1, \ldots, a_{n-1})[x]$ such that $g(a_n) = 0$. Hence a_n is algebraic over $F(a_1, \ldots, a_{n-1})$.

Conversely, suppose one of the a_i is algebraic over $S_i = S - \{a_i\}$. We may assume that $i = n$. Then we have elements $b_1, \ldots, b_m \in F(a_1, \ldots, a_{n-1})$ such that $g(a_n) = 0$ for $g(x) = x^m + b_1x^{m-1} + \cdots + b_m \in F(a_1, \ldots, a_{n-1})[x]$. There exist polynomials $f_0(x_1, \ldots, x_{n-1})$, $f_1(x_1, \ldots, x_{n-1}), \ldots, f_m(x_1, \ldots, x_{n-1}) \in F[x_1, \ldots, x_{n-1}]$ with $f_0(a_1, \ldots, a_{n-1}) \neq 0$ such that $b_i = f_i(a_1, \ldots, a_{n-1})f_0(a_1, \ldots, a_{n-1})^{-1}$. Then if we put

$$f(x_1, \ldots, x_n) = f_0(x_1, \ldots, x_{n-1})x^m + f_1(x_1, \ldots, x_{n-1})x^{m-1} + \cdots + f_m(x_1, \ldots, x_{n-1})$$

we shall have $f(x_1, \ldots, x_n) \neq 0$ and $f(a_1, \ldots, a_n) = 0$. Thus S is algebraically dependent over F. □

We shall now introduce a correspondence from the set E to the set $\mathscr{P}(E)$ of subsets of E, which will turn out to be a dependence relation on E in the sense defined on pp. 122–123. This is given in

DEFINITION 8.1. *An element a of E is called* algebraically dependent over F on the subset S *(which may be vacuous) if a is algebraic over $F(S)$. In this case we write $a \prec S$.*

Using this definition, Theorem 8.33 states that a non-vacuous subset S is algebraically dependent over F if and only if there exists an $a \in S$ that is algebraically dependent over F on $S-\{a\}$. We shall now show that the correspondence $\prec$ between elements of E and subsets of E satisfies the axioms for a dependence relation, that is, we have

THEOREM 8.34 *The correspondence $\prec$ of E to $\mathscr{P}(E)$ given in Definition* 8.1 *is a dependence relation.*

Proof. The axioms we have to verify are the following: (i) If $a \in S$, then $a \prec S$. (ii) If $a \prec S$, then $a \prec U$ for some finite subset U of S. (iii) If $a \prec S$ and every $b \in S$ satisfies $b \prec T$, then $a \prec T$. (iv) If $a \prec S$ and $a \not\prec S-\{b\}$ for some b in S, then $b \prec (S-\{b\}) \cup \{a\}$. Axiom (i) is clear and (ii) was noted in the proof of Theorem 8.33. To prove (iii) let A be the subfield of E of elements algebraic over $F(T)$. Then $S \subset A$ and if $a \prec S$, then a is algebraic over A. Hence $a \in A$, which means that $a \prec T$. To prove (iv) let $a \prec S$, $a \not\prec S-\{b\}$ where $b \in S$. Put $K = F(T)$ where $T = S-\{b\}$. Then a is transcendental over K and algebraic over $K(b)$. Hence, by Theorem 8.33, $\{a, b\}$ is algebraically dependent over K, so there exists a polynomial $f(x, y) \neq 0$ in indeterminates x, y with coefficients in K such that $f(a, b) = 0$. We can write $f(x, y) = a_0(x)y^m + a_1(x)y^{m-1} + \cdots + a_m(x)$ with $a_i(x) \in K[x]$ and $a_0(x) \neq 0$. Then $a_0(a) \neq 0$, so $f(a, y) \neq 0$ in $K[y]$, and $f(a, b) = 0$ shows that b is algebraic over $K(a) = F(T \cup \{a\})$. Hence $b \prec T \cup \{a\}$ as required. □

We can now apply the results that we derived on dependence relations to algebraic dependence. The concept of a base becomes that of a transcendency base, which we define in

DEFINITION 8.2. *If E is an extension field of F, a subset B of E is called a* transcendency base of E over F *if* (1) *B is algebraically independent, and* (2) *every $a \in E$ is algebraically dependent on B.*

The two results we proved in the general case now give

THEOREM 8.35. *E/F has a transcendency base and any two such bases have the same cardinality.*

It should be remarked that B may be vacuous. This is the case if and only if E is algebraic over F. The cardinality $|B|$ is called the *transcendency degree* (tr deg) of E/F. A field E is called *purely transcendental* over F if it has a transcendency base B such that $E = F(B)$.

EXERCISES

1. Let $E \supset K \supset F$. Show that tr deg $E/F =$ tr deg $E/K +$ tr deg K/F.

2. Show that if char $F \neq 3$ and $E = F(a,b)$ where a is transcendental over F and $a^3 + b^3 = 1$, then E is not purely transcendental over F.

3. Let $\mathbb{C}$ be the field of complex numbers, $\mathbb{Q}$ the rationals. Show that tr deg $\mathbb{C} = |\mathbb{C}|$. Show that if B is a transcendency base of $\mathbb{C}/\mathbb{Q}$, then any bijective map of B onto itself can be extended to an automorphism of $\mathbb{C}/\mathbb{Q}$. Hence conclude that there are as many automorphisms of $\mathbb{C}/\mathbb{Q}$ as bijective maps of $\mathbb{C}$ onto $\mathbb{C}$.

4. Show that any subfield of a finitely generated E/F is finitely generated.

5. Let $E = F(x_1, \ldots, x_m)$ where the x_i are algebraically independent. Call a rational expression $f = gh^{-1}$ *homogeneous of degree* m ($\in \mathbb{Z}$) if g is a homogeneous polynomial of degree r, h is a homogeneous polynomial of degree s, and $r - s = m$. Show that the set E_0 of homogeneous rational expressions of degree 0 is a subfield of E that is purely transcendental of transcendency degree $m - 1$ over F. Show that E is a simple transcendental extension of E_0.

8.13 TRANSCENDENCY BASES FOR DOMAINS. AFFINE ALGEBRAS

Let D be a commutative domain that is an algebra over a field F, E, the field of fractions of D, so $F \subset D \subset E$. Evidently, since $F(D)$ is a subfield of E containing D, $F(D) = E$ and hence D contains a transcendency base for E/F (see the comment (ii) on bases on p. 124). We call tr deg E/F the *transcendency degree* of D/F. This is an important concept for studying homomorphisms of domains that are algebras over the same field F. For, we have the following

THEOREM 8.36. (i) *Let D/F and D'/F be domains and suppose there exists a surjective homomorphism η of D/F onto D'/F. Then tr deg $D/F \geq$ tr deg D'/F.* (ii) *Moreover, if tr deg $D/F =$ tr deg $D'/F = m < \infty$ then η is an isomorphism.*

Proof. (i) Let B' be a transcendency base for D'/F. For each $x' \in B'$ choose an $x \in D$ such that $\eta x = x'$. Then $C = \{x\}$ is an algebraically independent subset of D. Hence, C can be augmented to a base B for E/F, E the field of fractions of D (see comment (i) on p. 124). Hence

$$\text{tr deg } D/F = |B| \geqslant |C| = |B'| = \text{tr deg } D'/F$$

(ii) Now let $B' = \{x'_1, \ldots, x'_m\}$, $C = \{x_1, \ldots, x_m\}$ where $\eta x_i = x'_i$. Since C is an algebraically independent set of cardinality tr deg E/F, $B = C$ is a transcendency base for D/F. Let a be a non-zero element of D. Then a is algebraic over

$F(x_1,\dots,x_m)$. Let $m(\lambda)=\lambda^n-\alpha_1\lambda^{n-1}+\cdots+\alpha_n$, $\alpha_i\in F(x_1,\dots,x_m)$ be the minimum polynomial of a over $F(x_1,\dots,x_m)$. Since $a\neq 0$, $\alpha_n\neq 0$. We can write $\alpha_i = g_i(x_1,\dots,x_m)g_0(x_1,\dots,x_m)^{-1}$ where $g_i(x_1,\dots,x_m)$, $g_0(x_1,\dots,x_m)\in F[x_1,\dots,x_m]$. Then we have

$$g_0(x_1,\dots,x_m)a^n+g_1(x_1,\dots,x_m)a^{n-1}+\cdots+g_n(x_1,\dots,x_m)=0$$

and hence

$$g_0(x'_1,\dots,x'_m)(\eta a)^n+\cdots+g_n(x'_1,\dots,x'_m)=0. \tag{77}$$

Since $\alpha_n\neq 0$, $g_n(x_1,\dots,x_m)\neq 0$ and since the x'_i are algebraically independent $g_n(x'_1,\dots,x'_m)=0$. Then by (77), $\eta a\neq 0$. Thus $a\neq 0\Rightarrow \eta a\neq 0$ and η is an isomorphism. □

We prove next the important

NOETHER NORMALIZATION THEOREM. *Let D be a domain which is finitely generated over a field F, say, $D=F[u_1,\dots,u_m]$. Let $\operatorname{tr\,deg} D = r\leqslant m$. Then there exists a transcendency base $\{v_i\}$ such that D is integral over $F[u_1,\dots,u_r]$.*

Proof. The result is trivial if $m=r$ so suppose $m>r$. Then the u_i are algebraically dependent. Hence there exists a non-zero polynomial

$$f(x_1,\dots,x_m)=\sum a_{j_1\dots j_m}x_1^{j_1}\dots x_m^{j_m}$$

in indeterminates x_i with coefficients in F such that $f(u_1,\dots,u_m)=0$. Let X be the set of monomials $x_1^{j_1}\dots x_m^{j_m}$ occurring in f (with non-zero coefficients). With each such monomial $x_1^{j_1}\dots x_m^{j_m}$ we associate the polynomial $j_1+j_2t+\cdots+j_mt^{m-1}\in\mathbb{Z}[t]$, t an indeterminate. The polynomials obtained in this way from the monomials in X are distinct. Since a polynomial of degree n in one indeterminate with coefficients in a field has at most n zeros in the field, it follows that there exists an integer $d\geqslant 0$ such that the integers $j_1+j_2d+\cdots+j_md^{m-1}$ obtained from the monomials in X are distinct. Now consider the polynomial $f(x_1,x_1^d+y_2,\dots,x_1^{d^{m-1}}+y_m)$ where $y_2,\dots,y_m$ are indeterminates. We have

$$\begin{aligned}f(x_1,x_1^d+y_2,\dots,x_1^{d^{m-1}}+y_m)&=\sum a_{j_1\dots j_m}x_1^{j_1}(x_1^d+y_2)^{j_2}\dots(x_1^{d^{m-1}}+y_m)^{j_m}\\&=\sum a_{j_1\dots j_m}x_1^{j_1+j_2d+\cdots+j_md^{m-1}}+g(x_1,y_2,\dots,y_m)\end{aligned}$$

where the degree of g in x_1 is less than that of $\sum a_{j_1\dots j_m}x_1^{j_1+j_2d+\cdots+j_md^{m-1}}$. Hence for a suitable $\beta\in F^*$, $\beta f(x_1,x_1^d+y_2,\dots,x_1^{d^{m-1}}+y_m)$ is monic as a polynomial in x_1 with coefficients in $F[y_2,\dots,y_m]$. If we put $w_i=u_i-u_1^{d^{i-1}}$, $2\leq i\leq m$ we have $\beta f(u_1,u_1^d+w_2,\dots,u_1^{d^{m-1}}+w_m)=0$ which implies that u_1 is integral over $D'=F[w_2,\dots,w_m]$. By induction on the number of generators, D' has a transcendency base $\{v_i\}$ such that D' is integral over $F[v_1,\ldots,v_r]$. Then D is integral over $F[v_1,\ldots,v_r]$ by the transitivity of integral dependence. □

A commutative algebra that is finitely generated over a field is called an *affine algebra*. Such an algebra is Noetherian (Corollary to the Hilbert basis theorem, p. 421). We recall that the Krull dimension of a Noetherian ring is defined to be $\operatorname{Sup} S$ for chains of prime ideals $P_0 \supsetneq P_1 \supsetneq \cdots \supsetneq P_S$ in R. We are now in a position to prove the following theorem on dimension of an affine domain.

THEOREM 8.37. *Let D be an affine domain of transcendency degree r over F. Then the Krull dimension $\dim D \geqslant r$ and $\dim D = r$ if F is algebraically closed.*

Proof. By Noether's normalization theorem we may write $D = F[u_1,\ldots,u_r, u_{r+1},\ldots,u_m]$ where the u_i. $1 \leqslant i \leqslant r$, constitute a transcendency base and the remaining u_j are integral over $F[u_1,\ldots,u_r]$. Then $F[u_1,\ldots,u_r]$ is factorial and hence is integrally closed in its field of fractions. Under these circumstances we can apply the "going-down" Theorem 7.1 to show that $\dim D = \dim F[u_1,\ldots,u_v]$. First, let $p_0 \supsetneq p_1 \supsetneq \cdots \supsetneq p_S$ be a strictly descending chain of prime ideals in $F[u_1,\ldots,u_r]$. By the lying-over Theorem 7.5, there exists a prime ideal P_0 in D such that $P_0^c = P_0 \cap F[u_1,\ldots,u_r] = p_0$. By Theorem 7.6, there exists a prime ideal P_1 in D such that $P_1^c = p_1$ and $P_0 \supset P_1$. Then $P_0 \supsetneq P_1$. Then by induction we obtain a chain of prime ideals $P_0 \supsetneq P_1 \supsetneq \cdots \supsetneq P_s$ such that $P_i \cap F[u_1,\ldots,u_r] = p_i$, $0 \leqslant i \leqslant s$. This implies that $\dim D \geqslant \dim F[u_1,\ldots,u_r]$. Next let $P_0 \supsetneq P_1 \supsetneq \cdots \supsetneq P_s$ for prime ideals P_i in D. Then, by Corollary 2 to Proposition 7.17 (p. 410), $p_0 \supsetneq p_1 \supsetneq \cdots \supsetneq p_s$, $p_i = P_1^c$ is a properly descending chain of prime ideals in $F[u_1,\ldots,u_r]$. It follows that $\dim F[u_1,\ldots,u_r] \geqslant \dim D$. Hence $\dim D = \dim F[u_1,\ldots,u_r]$. Now we have the chain of prime ideals

$$(u_1,\ldots,u_r) \supsetneq (u_1,\ldots,u_{r-1}) \supsetneq \cdots \supsetneq (u_1) \supsetneq (0)$$

in $F[u_1,\ldots,u_r]$. Hence $\dim D = \dim F[u_1,\ldots,u_r] \geqslant r = \operatorname{tr deg} D/F$. On the other hand, we have shown earlier (p. 453) that if F is algebraically closed then $\dim F[u_1,\ldots,u_r]$ for algebraically independent u_i is r. This concludes the proof. □

EXERCISE

1. Use the Noether normalization theorem to prove the Corollary to Theorem 7.15 (p. 426). {*Sketch of proof*. Let M be a maximal ideal in $F[x_1,\ldots,x_n]$, x_i indeterminates, F algebraically closed. Then $F[x_1,\ldots,x_n]/M$ is a field that is an affine

algebra $F[\bar{x}_1,\ldots,\bar{x}_n]$, $\bar{x}_i = x_i + M$. By the Noether normalization theorem and Proposition 7.17, $\operatorname{tr\,deg} F[x_1,\ldots,x_n]/M = 0$. Since F is algebraically closed, $F[x_1,\ldots,x_n]/M = F$. Then $\bar{x}_i = a_i \in F$, $1 \leqslant i \leqslant n$, and $M = (x_1 - a_1,\ldots,x_n - a_n)\}$.

8.14 LUROTH'S THEOREM

The purely transcendental extension fields E/F, especially those having a finite transcendency degree, appear to be the simplest type of extension fields. It is clear that such a field is isomorphic to the field of fractions $F(x_1,\ldots,x_n)$ of the polynomial ring $F[x_1,\ldots,x_n]$ in indeterminates $x_1,\ldots,x_n$. Even though these fields look quite innocent, as noted in BAI (p. 270), there are difficult and unsolved problems particularly on the nature of the subfields of $F(x_1,\ldots,x_n)/F$. A problem of the type mentioned in BAI, which, as far as we know remains unsolved (although it was stated as an exercise in the first edition of the author's *Lectures in Abstract Algebra* vol. III (1964), p. 160), is the following: Let the alternating group A_n operate on $F(x_1,\ldots,x_n)$ by automorphisms of this field over F so that $\pi x_i = x_{\pi(i)}$, $1 \leqslant i \leqslant n$, for $\pi \in A_n$ and let Inv A_n be the subfield of fixed points under this action. Is Inv A_n purely transcendental over F?

The one case where the situation is quite simple is that in which E has transcendency degree one. We consider this case.

Let $E = F(t)$, t transcendental, and let $u \in E$, $\notin F$. We can write $u = f(t)g(t)^{-1}$ where $f(t)$, $g(t) \in F[t]$ and $(f(t), g(t)) = 1$. If n is the larger of the degrees of $f(t)$ and $g(t)$, then we can write

$$f(t) = a_0 + a_1 t + \cdots + a_n t^n$$

$$g(t) = b_0 + b_1 t + \cdots + b_n t^n,$$

a_i, $b_i \in F$, and either a_n or $b_n \neq 0$. We have $f(t) - ug(t) = 0$, so

$$(a_n - ub_n)t^n + (a_{n-1} - ub_{n-1})t^{n-1} + \cdots + (a_0 - ub_0) = 0 \tag{78}$$

and $a_n - ub_n \neq 0$ since either $a_n \neq 0$ or $b_n \neq 0$ and $u \notin F$. Thus (78) shows that t is algebraic over $F(u)$ and $[F(t):F(u)] \leqslant n$. We shall now prove the following more precise result.

THEOREM 8.38. *Let $E = F(t)$, t transcendental over F, and let $u \in F(t)$, $\notin F$. Write $u = f(t)g(t)^{-1}$ where $(f(t), g(t)) = 1$, and let $n = \max(\deg f(t), \deg g(t))$. Then u is transcendental over F, t is algebraic over $F(u)$, and $[F(t):F(u)] = n$. Moreover, the minimum polynomial of t over $F(u)$ is a multiple in $F(u)$ of $f(x,u) = f(x) - ug(x)$.*

Proof. Put $f(x, y) = f(x) - yg(x) \in F[x, y]$, x, y indeterminates. This polynomial in x and y is of first degree in y and it has no factor $h(x)$ of positive degree since $(f(x), g(x)) = 1$. Hence it is irreducible in $F[x, y]$. Now t is algebraic over $F(u)$ so if u were algebraic over F, then t would be algebraic over F, contrary to the hypothesis. Hence u is transcendental over F. Then $F[x, u] \cong F[x, y]$ under the isomorphism over F fixing x and mapping u into y and hence $f(x, u)$ is irreducible in $F[x, u]$. It follows that $f(x, u)$ is irreducible in $F(u)[x]$ (BAI, p. 153). Since $f(t, u) = f(t) - ug(t) = 0$, it follows that $f(x, u)$ is a multiple in $F(u)$ of the minimum polynomial of t over $F(u)$. Hence $[F(t): F(u)]$ is the degree in x of $f(x, u)$. This degree is n, so the proof is complete. □

A first consequence of this theorem is that it enables us to determine the elements u that generate $F(t)$. These have the form $u = f(t)g(t)^{-1}$ where $f(t)$ and $g(t)$ have degree 1 or 0, $(f(t), g(t)) = 1$, and either $f(t)$ or $g(t) \notin F$. Then

$$u = \frac{at + b}{ct + d} \tag{79}$$

where $a, b, c, d \in F$, either $a \neq 0$ or $c \neq 0$, and $at + b$ and $ct + d$ have no common factor of positive degree. It is easily seen that this set of conditions is equivalent to the single condition

$$ad - bc \neq 0. \tag{80}$$

Now if $F(u) = F(t)$, then we have a uniquely determined automorphism of $F(t)/F$ such that $t \rightsquigarrow u$ and every automorphism is obtained in this way.

The condition (80) holds for the matrix

$$A = \begin{pmatrix} a & b \\ c & d \end{pmatrix} \tag{81}$$

if and only if A is invertible. Now consider the linear group $GL_2(F)$ of these matrices (BAI, p. 375). Any matrix $A \in GL_2(F)$ as in (81) determines a generator $u = (at + b)/(ct + d)$ of $F(t)$ and hence determines the automorphism $\eta(A)$ of $F(t)/F$ such that

$$f(t) \rightsquigarrow f(u) = f\left(\frac{at + b}{ct + d}\right). \tag{82}$$

If

$$A' = \begin{pmatrix} a' & b' \\ c' & d' \end{pmatrix}$$

$\in GL_2(F)$, then

$$\eta(A)\eta(A')t = \frac{(a'a+b'c)t+(a'b+b'd)}{(c'a+d'c)t+(c'b+d'd)} \tag{83}$$

$$= \eta(A'A)t.$$

Since any automorphism of $F(t)/F$ sends t into a generator, η is surjective. Hence by (83), η is an anti-homomorphism of $GL_2(F)$ onto Gal $F(t)/F$. The kernel consists of the matrices A as in (81) such that $(at+b)(ct+d)^{-1} = t$ or $at+b = ct^2+dt$. This gives $c = b = 0$ and $a = d$. Hence the kernel is the set of scalar matrices $a1$, $a \neq 0$. The factor group $GL_2(F)/F^*1$ is called a projective linear group and is denoted as $PGL_2(F)$. Hence Gal $F(t)/F$ is anti-isomorphic to $PGL_2(F)$ and since any group is anti-isomorphic to itself (under $g \rightsquigarrow g^{-1}$), we also have Gal $F(t)/F \cong PGL_2(F)$.

One can determine all of the subfields of E/F for $E = F(t)$, t transcendental: These have the form $F(u)$ for some u. This important result is called

LUROTH'S THEOREM. *If $E = F(t)$, t transcendental over F, then any subfield K of E/F, $K \neq F$, has the form $F(u)$, u transcendental over F.*

Proof. Let $v \in K$, $\notin F$. Then we have seen that t is algebraic over $F(v)$. Hence t is algebraic over K. Let $f(x) = x^n+k_1x^{n-1}+\cdots+k_n$ be the minimum polynomial of t over K, so the $k_i \in K$ and $n = [F(t):K]$. Since t is not algebraic over F, some $k_j \notin F$. We shall show that $K = F(u)$, $u = k_j$. We can write $u = g(t)h(t)^{-1}$ where $g(t), h(t) \in F[t]$, $(g(t), h(t)) = 1$, and $m = \max(\deg h, \deg g) > 0$. Then, as we showed in Theorem 8.38, $[E:F(u)] = m$. Since $K \supset F(u)$ and $[E:K] = n$, we evidently have $m \geqslant n$ and equality holds if and only if $K = F(u)$. Now t is a root of the polynomial $g(x)-uh(x) \in K[x]$. Hence we have a $q(x) \in K[x]$ such that

$$g(x)-uh(x) = q(x)f(x). \tag{84}$$

The coefficient k_i of $f(x)$ is in $F(t)$, so there exists a non-zero polynomial $c_0(t)$ of least degree such that $c_0(t)k_i = c_i(t) \in F[t]$ for $1 \leqslant i \leqslant n$. Then $c_0(t)f(x) = f(x,t) = c_0(t)x^n+c_1(t)x^{n-1}+\cdots+c_n(t) \in F[x,t]$, and $f(x,t)$ is primitive as a polynomial in x, that is, the $c_i(t)$ are relatively prime. The x-degree of $f(x,t)$ is n and since $k_j = g(t)h(t)^{-1}$ with $(g(t), h(t)) = 1$, the t-degree of $f(x,t)$ is $\geqslant m$. Now replace u in (84) by $g(t)h(t)^{-1}$ and the coefficients of $q(x)$ by their expressions in t. Then (84) shows that $f(x,t)$ divides $g(x)h(t)-g(t)h(x)$ in $F(t)[x]$. Since $f(x,t)$ and $g(x)h(t)-g(t)h(x) \in F[x,t]$ and $f(x,t)$ is primitive as a polynomial in x, it follows that there exists a polynomial $q(x,t) \in F[x,t]$

such that

$$g(x)h(t) - g(t)h(x) = f(x,t)q(x,t). \tag{85}$$

Since the t-degree of the left-hand side is $\leqslant m$ and that of $f(x,t)$ is $\geqslant m$, it follows that this degree is m and $q(x,t) = q(x) \in F[x]$. Then the right-hand side is primitive as a polynomial in x and so is the left-hand side. By symmetry the left-hand side is primitive as a polynomial in t also. Hence $q(x) = q \in F$. Then $f(x,t)$ has the same x-degree and t-degree so $m = n$, which implies that $K = F(u)$. $\square$

We shall now indicate some of the results that are presently known on subfields of purely transcendental extensions of transcendency degree greater than one. We use the algebraic geometric terminology in which a purely transcendental extension E/F is called a *rational* extension and a subfield of such an E/F is called *unirational*. In BAI (p. 270) we have noted some results and given some references on unirational fields of the form Inv G where G is a finite group of automorphisms of a field $F(x_1, \ldots, x_n)/F$ where the x_i are indeterminates that are permuted by G. Further results on the rationality and non-rationality of fields of the form Inv G are given in a survey article by D. J. Saltman, "Groups acting on fields: Noether's problem" in *Contemporary Mathematics* vol. 43, 1985, pp. 267–277.

An old result on subfields of rational extensions of transcendency degree two is the theorem of Castelnuovo-Zariski: if F is algebraically closed of characteristic 0 then any subfield L of a rational extension $F(x_1, x_2)$ such that $F(x_1, x_2)$ is affine over L is rational. The result does not always hold for characteristic $p \neq 0$. (See R. Hartshorne's *Algebraic Geometry*, Springer-Verlag, New York, 1977, p. 422.)

Examples of non-rational subfields of rational extensions of transcendency degree 3 over $\mathbb{C}$ are given in the following papers:

1. M. Artin and D. Mumford, "Some elementary examples of unirational varieties that are not rational," *Proc. London Math. Soc.* vol. 25, 3rd ser. (1972), pp. 75–95.
2. C. H. Clemens and P. A. Griffiths, "The intermediate jacobian of the cubic threefold," *Annals of Math.* (2) vol. 95 (1972), pp. 281–356.
3. V. A. Iskovkikh and J. Manin, "Three dimensional quartics and counterexamples to the Luroth problem," *Math. Sbornik*, vol. 86 (1971), pp. 140–166.

A. Beauville, J. L. Colliot-Thélène, J. J. Sansuc, and Sir P. Swinnerton-Dyer in "Variétés stablement rationelles non rationelle," *Annals of Math.* (2) vol. 121, pp. 283–318 have given an example of an extension $K/\mathbb{C}$ of transcendency degree three such that a purely transcendental extension of transcendency degree three over K is purely transcendental of transcendency degree six over $\mathbb{C}$ but K is not rational over $\mathbb{C}$.

EXERCISES

1. Let F_q be a field of q elements and let K be the subfield of fixed elements of $F_q(t)$, t transcendental, under Gal $F_q(t)/F_q$. Determine an element u such that $K = F_q(u)$.

2. Let $E = F[t, v]$ where t is transcendental over F and $v^2 + t^2 = 1$. Show that E is purely transcendental over F.

The following two exercises sketch proofs due to Mowaffag Hajja (to appear in *Algebras, Groups, and Geometries*) that $\operatorname{Inv} A_3$ and $\operatorname{Inv} A_4$ are rational. This improves results of Burnside published in *Messenger of Mathematics*, vol. 37 (1908), p. 165. We mention also that it has been proved recently by Takashi Maeda (to appear in the *J. of Algebra*) that $\operatorname{Inv} A_5$ for the base field $\mathbb{Q}$ is rational. The situation for $\operatorname{Inv} A_n$ with $n > 5$ is still unsettled.

3. Show that $\operatorname{Inv} A_3$ in $K = F(x_1, x_2, x_3)$ is rational.
   (*Sketch of Proof*: We distinguish three cases:
   i. F contains a primitive cube root of 1, which implies $\operatorname{char} F \neq 3$.
   ii. $\operatorname{char} F \neq 3$ but F contains no primitive cube root of 1.
   iii. $\operatorname{char} F = 3$.

   In all cases A_3 is the group of automorphisms of K/F generated by the automorphism σ such that $x_1 \rightsquigarrow x_2$, $x_2 \rightsquigarrow x_3$, $x_3 \rightsquigarrow x_1$. Also $\operatorname{Gal} K/\operatorname{Inv} A_3 = A_3$ and $[K:\operatorname{Inv} A_3] = |A_3| = 3$. In case i we put $X_j = x_1 + w^j x_2 + w^{2j} x_3$ where $w^3 = 1$, $w \neq 1$. Then $\sigma X_j = w^{-j} X_j$ and $K = F(X_1, X_2, X_3)$. Now put $Y_1 = X_1^2/X_2$, $Y_2 = X_2^2/X_1$, $Y_3 = X_3$. Then $\sigma Y_j = Y_j$ so $F(Y_1, Y_2, Y_3) \subset \operatorname{Inv} A_3$. On the other hand, $X_1^3 = Y_1^2 Y_2$ and $K = F(X_1, X_2, X_3) = F(Y_1, Y_2, Y_3, X_1)$. Hence $[K:F(Y_1, Y_2, Y_3)] \leqslant 3$. Then $\operatorname{Inv} A_3 = F(Y_1, Y_2, Y_3)$. In case ii we adjoin a primitive cube root of unity w to K to obtain $K' = K(w) = F'(x_1, x_2, x_3)$ for $F' = F(w)$. We have $w^2 + w + 1 = 0$, (w, w^2) is a base for K'/K and for F'/F and we have an automorphism τ of K'/K such that $w \rightsquigarrow w^2$. As in case i, we define $X_j = x_1 + w^j x_2 + w^{2j} x_3$, $j = 1, 2, 3$, $Y_1 = X_1^2/X_2$, $Y_2 = X_2^2/X_1$, $Y_3 = X_3 = x_1 + x_2 + x_3$. Then $\operatorname{Inv} A_3 = F'(Y_1, Y_2, Y_3)$. We have $\tau Y_1 = Y_2$, $\tau Y_2 = Y_1$, $\tau Y_3 = Y_3 \in K$. Hence if $Y_1 = wZ_1 + w^2 Z_2$ where $Z_i \in K$ then $Y_2 = w^2 Z_1 + w Z_2$ and $\sigma Y_2 = Y_2$ implies $\sigma Z_i = Z_i$ and $\operatorname{Inv} A_3 = F(Z_1, Z_2, X_3)$. In case iii we let $\Delta = \sigma - 1$ and $U_j = \Delta^j x_1$, $j = 0, 1, 2$. Then $K = F(U_0, U_1, U_2)$ and $\sigma U_0 = U_0 + U_1$, $\sigma U_1 = U_1 + U_2$, $\sigma U_2 = U_2$. Let $U = U_0 U_2 + U_1^2 - U_1 U_2$. Then $K = F(U, U_1, U_2)$, $\sigma U = U$, $\sigma U_1 = U_1 + U_2$ and $\sigma U_2 = U_2$. Hence $\sigma(U_1^3 - U_1 U_2^2) = U_1^3 - U_1 U_2^2$ and $\operatorname{Inv} A_3 \supset F(U, U_2, U_1^3 - U_1 U_2^2)$. Since $[K:F(U, U_2, U_1^3 - U_1 U_2^2)] \leqslant 3$ it follows that $\operatorname{Inv} A_3 = F(U, U_2, U_1^3 - U_1 U_2^2)$. (Cf. exercise 1, p. 271 of BAI.))

4. Show that $\text{Inv}\, A_4$ in $K = F(x_1, x_2, x_3, x_4)$ is rational.
(*Sketch of Proof*: It is clear that A_4 is generated by the automorphisms α, β, σ of K/F such that $\alpha x_1 = x_2$, $\alpha x_2 = x_1$, $\alpha x_3 = x_4$, $\alpha x_4 = x_3$; $\beta x_1 = x_3$, $\beta x_3 = x_1$, $\beta x_2 = x_4$, $\beta x_4 = x_2$, $\sigma x_1 = x_2$, $\sigma x_2 = x_3$, $\sigma x_3 = x_1$, $\sigma x_4 = x_4$. If $\text{char}\, F \neq 2$ we define

$$s = x_1 + x_2 + x_3 + x_4$$

$$z_1 = x_1 + x_2 - x_3 - x_4$$

$$z_2 = x_1 - x_2 + x_3 - x_4$$

$$z_3 = x_1 - x_2 - x_3 + x_4.$$

Then $K = F(s, z_1, z_2, z_3)$. The action of α, β, σ on (s, z_1, z_2, z_3) is given by the following table

	s	z_1	z_2	z_3
α	s	z_1	$-z_2$	$-z_3$
β	s	$-z_1$	z_2	$-z_3$
σ	s	$-z_3$	z_1	$-z_2$

Put $Y_1 = z_1 z_3/z_2$, $Y_2 = \sigma Y_1 = z_2 z_3/z_1$, $Y_3 = \sigma^2 Y_1 = z_1 z_2/z_3$. Then $F(s, Y_1, Y_2, Y_3) \subset \text{Inv}\, H$ where $H = \langle \alpha, \beta \rangle$. Since $z_1^2 = Y_1 Y_3$, $z_2^2 = Y_2 Y_3$, $F(s, Y_1, Y_2, Y_3, z_1, z_2) = K$, and $|H| = 4$, it follows as before that $\text{Inv}\, H = F(s, Y_1, Y_2, Y_3)$. Since $\sigma s_1 = s_1$ and $\sigma Y_1 = Y_2$, $\sigma Y_2 = Y_3$, $\sigma Y_3 = Y_1$, the result of exercise 3 shows that $\text{Inv}\langle \sigma \rangle$ in $F(s, Y_1, Y_2, Y_3)$ is rational over $F(s)$. It follows that $\text{Inv}\, A_4$ in K is rational over F. Now suppose $\text{char}\, F = 2$. Put $x = x_1$, $y = x_1 + x_3$, $z = x_1 + x_2$, $s = x_1 + x_2 + x_3 + x_4$, $X = xs + yz = x_1 x_4 + x_2 x_3$. Then $K = F(x, y, z, s) = F(X, y, z, s)$ and $\alpha X = X$, $\alpha y = y + s$, $\alpha z = z$, $\alpha s = s$, $\beta X = X$, $\beta y = y$, $\beta z = z + s$, $\beta s = s$. It follows that $\text{Inv}\, H$ in $K = F(X, s, y(y+s), z(z+s))$. Then $\text{Inv}\, H = F(X, s, y(y+s), z(z+s))$. If we put $X_1 = X = x_1 x_4 + x_4 x_1$, $X_2 = \sigma X = x_2 x_4 + x_3 x_1$, $X_3 = \sigma^2 X = x_3 x_4 + x_1 x_2$, then $y(y+s) = (x_1 + x_3)(x_2 + x_4) = X_1 + X_3$, $z(z+s) = (x_1 + x_2)(x_3 + x_4) = X_2 + X_1$. Hence, $\text{Inv}\, H = F(X_1, X_2, X_3, s)$. Then the rationality of $\text{Inv}\, A_4$ follows from exercise 3 since $\sigma s = s$, $\sigma X_1 = X_2$, $\sigma X_2 = X_3$, $\sigma X_3 = X_1$.)

8.15 SEPARABILITY FOR ARBITRARY EXTENSION FIELDS

In this section we shall introduce a concept of separability for arbitrary extension fields that generalizes this notion for algebraic extensions. This is based on the concept of linear disjointness, which we now define.

DEFINITION 8.3. *Let E be an extension field of F, A, and B subalgebras of E/F. Then A and B are said to be* linearly disjoint over F *if the canonical* *homomorphism of $A \otimes_F B$ into E sending $a \otimes b$ into ab, $a \in A$, $b \in B$, is a monomorphism.*

It is clear that if A and B satisfy this condition and A' and B' are subalgebras of A and B respectively, then A' and B' satisfy the condition. Let K and L be the subfields of E/F generated by A and B respectively. Then A and B are linearly disjoint over F if and only if K and L are linearly disjoint over F. To prove this it suffices to show that if $k_1, \ldots, k_m$ are F-linearly independent elements of K and $l_1, \ldots, l_n$ are F-linearly independent elements of L, then the elements $k_i l_j$, $1 \leqslant i \leqslant m$, $1 \leqslant j \leqslant n$, are linearly independent over F. This follows from the linear disjointness over F of A and B by writing $k_i = a_i a^{-1}$, $a_i, a \in A$, $1 \leqslant i \leqslant m$, $l_j = b_j b^{-1}$, $b_j, b \in B$, $1 \leqslant j \leqslant n$. Conversely, the linear disjointness of K and L over F implies that of A and B.

The following result permits establishment of linear disjointness in stages.

LEMMA 1. *Let E_1 and E_2 be subfields of E/F, K_1 a subfield of E_1/F. Then E_1 and E_2 are linearly disjoint over F if and only if the following two conditions hold:* (1) *K_1 and E_2 are linearly disjoint over F and* (2) *$K_1(E_2)$ and E_1 are linearly disjoint over K_1.*

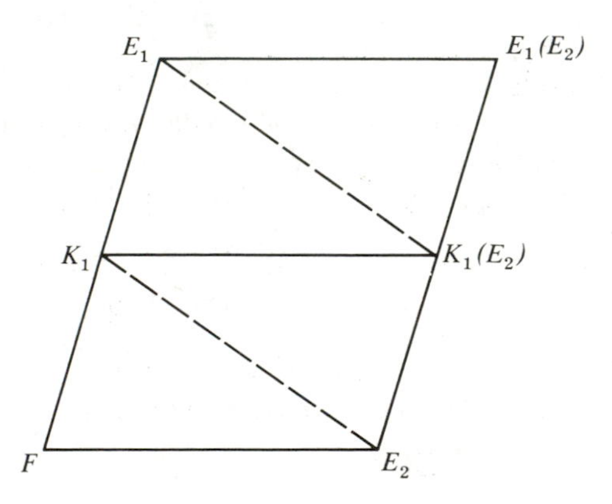

Proof. Assume the two conditions. Let (u_α) be a base for E_2/F. By (1) these elements are linearly independent over K_1 and, since they are contained in $K_1(E_2)$, they are linearly independent over E_1, by (2). Hence E_1 and E_2 are linearly disjoint over F. Conversely assume that E_1 and E_2 are linearly disjoint over F. Then (1) is clear since $K_1 \subset E_1$. Let (u_α) be a base for E_1/K_1, (v_β) a base for K_1/F, (w_γ) a base for E_2/F. Then $(u_\alpha v_\beta)$ is a base for E_1/F and since E_1 and E_2 are linearly disjoint over F, the set of elements $\{u_\alpha v_\beta w_\gamma\}$ is linearly independent. This implies that the only relations of the form $\sum d_i u_{\alpha_i} = 0$ with d_i in the subalgebra $K_1 E_2$ generated by K_1 and E_2 are the trivial ones in which every $d_i = 0$. Since $K_1(E_2)$ is the set of elements cd^{-1}, c, $d \in K_1 E_2$, it follows that (u_α) is a set of elements that is linearly independent over $K_1(E_2)$. Then (2) holds. □

We now assume that the characteristic is $p \neq 0$ and we imbed E in its algebraic closure $\bar{E}$. If $e > 0$, we denote the subset of $\bar{E}$ of elements a such that $a^{p^e} \in F$ by $F^{p^{-e}}$. This is a subfield and $F \subset F^{p^{-1}} \subset F^{p^{-2}} \subset \cdots$. Hence $F^{p^{-\infty}} \equiv \bigcup_{e \geq 1} F^{p^{-e}}$ is a subfield. Since linear disjointness is a property of finite subsets, it is clear that $F^{p^{-\infty}}$ and E are linearly disjoint over F if and only if $F^{p^{-e}}$ and E are linearly disjoint over F for every e. A result on separable algebraic extensions that we proved before can now be reformulated as

LEMMA 2. *If E/F is separable algebraic, then E and $F^{p^{-\infty}}$ are linearly disjoint over F.*

Proof. It suffices to show that if $a_1, \ldots, a_m$ are F-independent elements of E, then these are linearly independent over $F^{p^{-e}}$ for every e. This is equivalent to the following: $a_1^{p^e}, \ldots, a_m^{p^e}$ are F-independent, which is a property of separable algebraic extensions proved on p. 489. Hence E and $F^{p^{-\infty}}$ are linearly disjoint over F. □

We prove next

LEMMA 3. *If E is purely transcendental over F, then E/F and $F^{p^{-\infty}}$ are linearly disjoint over F.*

Proof. It suffices to prove the result for $E = F(x_1, \ldots, x_n)$ where the x_i are algebraically independent. Moreover, the result will follow in this case if we can show that $F[x_1, \ldots, x_n]$ and $F^{p^{-e}}$ are linearly disjoint over F for every $e > 0$. We have a base for $F[x_1, \ldots, x_n]/F$ consisting of all of the monomials $x_1^{k_1} \cdots x_n^{k_n}$, $k_i \geq 0$. The map $m \rightarrow m^{p^e}$ for the set of monomials is injective onto a subset. Hence if (m_α) is the base of monomials, then the set $\{m_\alpha^{p^e}\}$ is linearly independent over F. It follows that $\{m_\alpha\}$ is linearly independent over $F^{p^{-e}}$ and hence $F[x_1, \ldots, x_n]$ and $F^{p^{-e}}$ are linearly disjoint over F. □

An extension E/F is said to be *separably generated* over F if E has a transcendency base B such that E is separable algebraic over $F(B)$. In this case B is called a *separating transcendency base* for E/F. The example of E inseparable algebraic over F shows that E/F may not be separably generated. The example of $B = \{x^p\}$ in $F(x)$, x transcendental, shows that even if E is separably generated over F, not every transcendency base has the property that E is separable algebraic over $F(B)$.

We can now prove our main result.

THEOREM 8.39. *Let E be an extension field of a field of characteristic $p > 0$. Then the following properties of E/F are equivalent:*

(1) *Every finitely generated subfield of E/F is separably generated.*
(2) *E and $F^{p^{-\infty}}$ are linearly disjoint over F.*
(3) *E and $F^{p^{-1}}$ are linearly disjoint over F.*

Proof. (1) $\Rightarrow$ (2). To prove this we suppose that E is separable algebraic over $F(B)$, B a transcendency base of E/F. By Lemma 3, $F^{p^{-\infty}}$ and $F(B)$ are linearly disjoint over F. By Lemma 2, E and $F(B)^{p^{-\infty}}$ are linearly disjoint over $F(B)$. Since $F(B)^{p^{-\infty}} \supset F^{p^{-\infty}}(B)$ it follows that $F^{p^{-\infty}}(B)$ and E are linearly disjoint over $F(B)$. Then, by Lemma 1, E and $F^{p^{-\infty}}$ are linearly disjoint over F. Since this is a property of finite subsets, the result proved shows that (1) $\Leftrightarrow$ (2).

(2) $\Rightarrow$ (3) obviously.

(3) $\Rightarrow$ (1). Assume that E and $F^{p^{-1}}$ are linearly disjoint over F and let $K = F(a_1, \ldots, a_n)$ be a finitely generated subfield of E/F. We prove by induction on n that we can extract from the given set of generators a transcendency base $a_{i_1}, \ldots, a_{i_r}$ (where r is 0 if all the a_i are algebraic over F) such that K is separable algebraic over $F(a_{i_1}, \ldots, a_{i_r})$. The result is clear if $n = 0$, so we assume $n > 0$. The result is clear also if $a_1, \ldots, a_n$ are algebraically independent. Hence we assume that $a_1, \ldots, a_r$, $0 \leqslant r < n$, is a transcendency base for K/F. Then $a_1, \ldots, a_{r+1}$ are algebraically dependent over F, so we can choose a polynomial $f(x_1, \ldots, x_{r+1}) \neq 0 \in F[x_1, \ldots, x_{r+1}]$ of least degree such that $f(a_1, \ldots, a_{r+1}) = 0$. Then $f(x_1, \ldots, x_{r+1})$ is irreducible. We claim that f does not have the form $g(x_1{}^p, \ldots, x_{r+1}^p)$, $g \in F[x_1, \ldots, x_{r+1}]$. For $g(x_1{}^p, \ldots, x_{r+1}^p) = h(x_1, \ldots, x_{r+1})^p$ in $F^{p^{-1}}[x_1, \ldots, x_{r+1}]$ and if $f(x_1, \ldots, x_{r+1}) = g(x_1{}^p, \ldots, x_{r+1}^p)$, then $h(a_1, \ldots, a_{r+1}) = 0$. Let $m_i(x_1, \ldots, x_{r+1})$, $1 \leqslant i \leqslant u$, be the monomials occurring in h. Then the elements $m_i(a_1, \ldots, a_{r+1})$ are linearly dependent over $F^{p^{-1}}$, so by our hypothesis, these are linearly dependent over F. This gives a non-trivial polynomial relation in $a_1, \ldots, a_{r+1}$ with coefficients in F of lower degree than f, contrary to the choice of f. We have therefore shown that for some i, $1 \leqslant i \leqslant r+1$, $f(x_1, \ldots, x_{r+1})$ is not a polynomial in $x_i{}^p$ (and the other x's). Then a_i is algebraic over $F(a_1, \ldots, \hat{a}_i, \ldots, a_{r+1})$ where $\hat{a}_i$ denotes omission of a_i. It follows that $\{a_1, \ldots, \hat{a}_i, \ldots, a_{r+1}\}$ is a transcendency base for $F(a_1, \ldots, a_n)$. Then $F[a_1, \ldots, a_{i-1}, x, a_{i+1}, \ldots, a_{r+1}] \cong F[x_1, \ldots, x_{r+1}]$ in the obvious way and hence $f(a_1, \ldots, a_{i-1}, x, a_{i+1}, \ldots, a_{r+1})$ is irreducible in $F[a_1, \ldots, a_{i-1}, x, a_{i+1}, \ldots, a_{r+1}]$. Then this polynomial is irreducible in $F(a_1, \ldots, \hat{a}_i, \ldots, a_{r+1})[x]$. Since a_i is a root of $f(a_1, \ldots, a_{i-1}, x, a_{i+1}, \ldots, a_{r+1})$ and this is not a polynomial in x^p, we see that a_i is separable algebraic over $F(a_1, \ldots, \hat{a}_i, \ldots, a_{r+1})$ and hence over $L = F(a_1, \ldots, \hat{a}_i, \ldots, a_n)$. The induction hypothesis applies to L and gives us a subset

$\{a_{i_1},\ldots,a_{i_r}\}$ of $\{a_1,\ldots,\hat{a}_i,\ldots,a_n\}$ that is a separating transcendency base for L over F. Since a_i is separable algebraic over L, it follows that a_i is separable algebraic over $F(a_{i_1},\ldots,a_{i_r})$. Hence $\{a_{i_1},\ldots,a_{i_r}\}$ is a separating transcendency base for $F(a_1,\ldots,a_n)$. □

We remark that the result is applicable in particular to an algebraic extension E/F. In this case it states that if E and $F^{p^{-1}}$ are linearly disjoint over F, then E is separable and if E is separable, then E and $F^{p^{-\infty}}$ are linearly disjoint over F (which was Lemma 2). This makes it natural to extend the concept of separability for arbitrary extension fields in the following manner.

DEFINITION 8.4. *An extension field E/F is called* separable *if either the characteristic is* 0 *or the characteristic is $p \neq 0$, and the equivalent conditions of Theorem* 8.37 *hold.*

The implication (3) ⇒ (1) of Theorem 8.37 is due to MacLane. It implies an earlier result due to F. K. Schmidt, which we state as a

COROLLARY. *If F is perfect, then every extension E/F is separable.*

Proof. This is clear since F is perfect if and only if the characteristic is 0 or it is p and $F^{p^{-1}} = F$. □

The following grab-bag theorem states some properties and non-properties of separable extensions.

THEOREM 8.40. *Let E be an extension field of F, K an intermediate field. Then* (1) *If E is separable over F, then K is separable over F.* (2) *If E is separable over K and K is separable over F, then E is separable over F.* (3) *If E is separable over F, then E need not be separable over K.* (4) *If E is separable over F, it need not have a separating transcendency base over F.*

Proof. We may assume that the characteristic is $p \neq 0$. (1) This is clear since the linear disjointness of E and $F^{p^{-1}}$ over F implies that of K and $F^{p^{-1}}$ over F. (2) The hypothesis is that E and $K^{p^{-1}}$ are linearly disjoint over K and that K and $F^{p^{-1}}$ are linearly disjoint over F. Then E and $K(F^{p^{-1}})$ are linearly disjoint over K since $K(F^{p^{-1}}) \subset K^{p^{-1}}$. Hence, by Lemma 1, E and $F^{p^{-1}}$ are linearly disjoint over F and so E is separable over F. (3) Take $E = F(x)$, x transcendental, and $K = F(x^p)$. (4) Take $E = F(x, x^{p^{-1}}, x^{p^{-2}}, \ldots)$ where x is transcendental over F. Then E has transcendency degree one over F and E is not separably generated over F. □

EXERCISES

1. Let E_1/F and E_2/F be subfields of E/F such that E_1/F is algebraic and E_2/F is purely transcendental. Show that E_1 and E_2 are linearly disjoint over F.

2. Let F have characteristic $p \neq 0$. Let $E = F(a,b,c,d)$ where a,b,c are algebraically independent over F and $d^p = ab^p + c$. Show that E is not separably generated over $F(a,c)$.

3. (MacLane.) Let F be a perfect field of characteristic p, E an imperfect extension field of transcendency degree one over F. Show that E/F is separably generated.

8.16 DERIVATIONS

The concept of a derivation is an important one in the theory of fields and in other parts of algebra. We have already encountered this in several places (first in BAI, p. 434). We consider this notion now first in complete generality and then in the special case of derivations of commutative algebras and fields. In the next section we shall consider some applications of derivations to fields of characteristic p.

DEFINITION 8.5. *Let B be an algebra over a commutative ring K, A a subalgebra. A* derivation *of A into B is a K-homomorphism of A into B such that*

$$D(ab) = aD(b) + D(a)b \tag{86}$$

for $a, b \in A$. If $A = B$, then we speak of a derivation in A *(over K).*

Let $\mathrm{Der}_K(A, B)$ denote the set of derivations of A into B. Then $\mathrm{Der}_K(A,B) \subset \hom_K(A,B)$. If $D_1, D_2 \in \mathrm{Der}_K(A,B)$, then the derivation condition (86) for the D_i gives

$$(D_1 + D_2)(ab) = a(D_1 + D_2)(b) + (D_1 + D_2)(a)b.$$

Hence $D_1 + D_2 \in \mathrm{Der}_K(A,B)$. Now let $k \in K$, $D \in \mathrm{Der}_K(A,B)$. Then

$$(kD)(ab) = k(D(ab)) = k(D(a)b + aD(b)) = (kD)(a)b + a(kD(b)).$$

Hence $kD \in \mathrm{Der}_K(A,B)$ and so $\mathrm{Der}_K(A,B)$ is a K-submodule of $\hom_K(A,B)$.

Now let $B = A$ and write $\mathrm{Der}_K A$ for $\mathrm{Der}_K(A,A)$. As we shall now show, this has a considerably richer structure than that of a K-module (cf. BAI, pp. 434–435). Let $D_1, D_2 \in \mathrm{Der}_K A$, $a, b \in A$. Then

$$\begin{aligned} D_1D_2(ab) &= D_1(aD_2(b) + D_2(a)b) \\ &= aD_1D_2(b) + D_1(a)D_2(b) + D_2(a)D_1(b) + D_1D_2(a)b. \end{aligned}$$

If we interchange D_1 and D_2 in this relation and subtract we obtain

$$[D_1D_2](ab) = a[D_1D_2](b)+[D_1D_2](a)b \tag{87}$$

where we have put $[D_1D_2]$ for $D_1D_2 - D_2D_1$. This result and the fact that $\mathrm{Der}_K A$ is a K-module of $\mathrm{End}_K A$ amount to the statement that $\mathrm{Der}_K A$ is a Lie algebra of K-endomorphisms of A (BAI, p. 434).

There is still more that can be said in the special case in which K is a field of characteristic $p \neq 0$. We note first that for any K, if $D \in \mathrm{Der}_K A$, we have the Leibniz formula for D^n:

$$D^n(ab) = \sum_{i=0}^{n} \binom{n}{i} D^i(a)D^{n-i}(b), \tag{88}$$

which can be proved by induction on n. If K is a field of characteristic p, then (88) for $n = p$ becomes

$$D^p(ab) = D^p(a)b + aD^p(b). \tag{89}$$

This shows that $D^p \in \mathrm{Der}_K A$. If V is a vector space over a field K of characteristic p, then a subspace of $\mathrm{End}_K V$ that is closed under the bracket composition $[D_1D_2]$ and under pth powers is called a *p-Lie algebra* (or *restricted Lie algebra*) *of linear transformations* in V. Thus we have shown that if A is an algebra over a field of characteristic p, then $\mathrm{Der}_K A$ is a p-Lie algebra of linear transformations in A over K.

There is an important connection between derivations and homomorphisms. One obtains this by introducing the *algebra Δ of dual numbers over K*. This has the base $(1, \delta)$ over K with 1 the unit and δ an element such that $\delta^2 = 0$. If B is any algebra over K, then we can form the algebra $B \otimes_K \Delta$ and we have the map $b \to b \otimes 1$ of B into $B \otimes_K \Delta$. Since Δ is K-free, this is an algebra isomorphism and so B can be identified with its image $B \otimes 1$. We can also identify $\delta \in \Delta$ with $1 \otimes \delta$ in $B \otimes_K \Delta$. When this is done, then $B \otimes_K \Delta$ appears as the set of elements

$$b_1 + b_2\delta, \qquad b_i \in B. \tag{90}$$

This representation of an element is unique and one has the obvious K-module compositions. Moreover, if $b_i' \in B$, then

$$(b_1 + b_2\delta)(b_1' + b_2'\delta) = b_1b_1' + (b_1b_2' + b_2b_1')\delta. \tag{91}$$

Now let D be a K-homomorphism of A into B. We define a corresponding map $\alpha(D)$ of A into $B \otimes_K \Delta$ by

$$\alpha(D)\colon a \rightsquigarrow a + D(a)\delta, \tag{92}$$

which is evidently a K-homomorphism. We claim that $\alpha(D)$ is an algebra homomorphism if and only if D is a derivation. First, we have $D(1) = 0$ for any derivation since $D(1) = D(1^2) = 2D(1)$. Now

$$\begin{aligned}(\alpha(D)(a))(\alpha(D)(b)) &= (a + D(a)\delta)(b + D(b)\delta)\\ &= ab + (aD(b) + D(a)b)\delta\end{aligned}$$

and

$$\alpha(D)(ab) = ab + D(ab)\delta.$$

Thus $\alpha(D)(ab)) = (\alpha(D)(a))(\alpha(D)(b))$ if and only if D is a derivation and $\alpha(D)$ is an algebra homomorphism if and only if D is a derivation.

The homomorphisms $\alpha(D)$ have a simple characterization in terms of the map

$$\pi : b_1 + b_2\delta \rightsquigarrow b_1, \qquad b_i \in B, \tag{93}$$

of $B \otimes_K \Delta$ into B, which is a surjective K-algebra homomorphism of $B \otimes_K \Delta$ onto B. If $a \in A$, then $\alpha(D)a = a + D(a)\delta$ so $\pi\alpha(D)a = a$. Hence $\pi\alpha(D) = 1_A$. Conversely, let H be a homomorphism of A into $B \otimes_K \Delta$. For any a we write $H(a) = a_1 + a_2\delta$. This defines the maps $a \rightsquigarrow a_1, a \rightsquigarrow a_2$ of A into B, which are K-homomorphisms. The condition $\pi H = 1_A$ is equivalent to $a_1 = a$ for all a. Hence if we denote $a \rightsquigarrow a_2$ by D, then $H(a) = a + D(a)\delta$. The condition $H(ab) = H(a)H(b)$ is equivalent to: D is a derivation.

We summarize our results in

PROPOSITION 8.15. *Let A be a subalgebra of an algebra B and let D be a derivation of A into B. Then $\alpha(D): a \rightsquigarrow a + D(a)\delta$ is an algebra homomorphism of A into $B \otimes_K \Delta$ such that $\pi\alpha(D) = 1_A$. Conversely, any homomorphism H of A into $B \otimes_K \Delta$ such that $\pi H = 1_A$ has the form $\alpha(D)$, D a derivation of A into B.*

The importance of this connection between derivations and homomorphisms is that it enables us to carry over results on algebra homomorphisms to derivations. In this way we can avoid tedious calculations that would be involved in direct proofs of the results for derivations. As an illustration we prove

PROPOSITION 8.16. *Let A be a subalgebra of an algebra B, D, D_1, D_2 derivations of A into B, X a set of generators for A. Call an element $a \in A$ a* D-constant *if $Da = 0$. Then*

(1) *$D_1 = D_2$ if $D_1|X = D_2|X$.*
(2) *The set of D-constants is a subalgebra of A. Moreover, if A is a division algebra, then it is a division subalgebra.*

Proof. (1) The condition $D_1|X = D_2|X$ implies that $\alpha(D_1)|X = \alpha(D_2)|X$ for the algebra homomorphisms of A into $B \otimes_K \Delta$. Since X generates A, we have $\alpha(D_1) = \alpha(D_2)$. Hence $D_1 = D_2$.

(2) The condition that a is a D-constant is equivalent to: a is a fixed element under the homomorphism $\alpha(D)$ of $A \subset B \otimes_K \Delta$ into $B \otimes_K \Delta$. Since the set of fixed points of a homomorphism of a subalgebra A of an algebra C is a subalgebra and is a division subalgebra if A is a division algebra, the result on derivations is clear. □

We obtain next a formula for $D(a^{-1})$ for an invertible element a of A and derivation D of A into B. Since 1 is a $D =$ constant, applying D to $aa^{-1} = 1$ gives

$$D(a)a^{-1} + aD(a^{-1}) = 0.$$

Hence we have the formula

$$D(a^{-1}) = -a^{-1}D(a)a^{-1}, \tag{94}$$

which generalizes the well-known formula from calculus.

From now on we consider derivations of commutative algebras into commutative algebras, that is, we assume B commutative. Let $D \in \mathrm{Der}_K(A, B)$, $b \in B$. If we multiply the relation $D(xy) = D(x)y + xD(y)$, $x, y \in A$, by b we obtain

$$bD(xy) = bD(x)y + xbD(y).$$

This shows that bD defined by $(bD)(x) = b(D(x))$ is again a derivation. It is clear that this action of B on $\mathrm{Der}_K(A, B)$ endows $\mathrm{Der}_K(A, B)$ with the structure of a B-module.

We now consider the problem of extending a given derivation D of a subalgebra A into B to a derivation of a larger subalgebra A'. Let $\alpha(D)$ be the corresponding homomorphism of A into $B \otimes_K \Delta$ such that $\pi\alpha(D) = 1_A$. The problem of extending D to a derivation of A' amounts to that of extending $H = \alpha(D)$ to a homomorphism H' of A' into $B \otimes_K \Delta$ such that $\pi H' = 1_{A'}$. Now if H' is a homomorphism of A' into $B \otimes_K \Delta$ extending H, then $\pi H' = 1_{A'}$ will hold if and only if $\pi H'(x) = x$ holds for every x in a set of generators for A' over A. We shall use these observations in treating the extension problem.

We now suppose that $A' = A[u_1, \ldots, u_n]$, the subalgebra generated by A and a finite subset $\{u_1, \ldots, u_n\}$ of B. Let $A[x_1, \ldots, x_n]$ be the polynomial algebra over A in the indeterminates x_i and let I be the kernel of the homomorphism of $A[x_1, \ldots, x_n]$ onto A', which is the identity on A and sends $x_i \rightsquigarrow u_i$, $1 \leqslant i \leqslant n$. Suppose that we have a homomorphism s of A into a commutative algebra

C and elements v_i, $1 \leqslant i \leqslant n$, of C. Then we have the homomorphism

$$f(x_1,\ldots,x_n) \rightsquigarrow (sf)(v_1,\ldots,v_n) \tag{95}$$

of $A[x_1,\ldots,x_n]$ into C (BAI, p. 124). Here sf denotes the polynomial obtained from f by applying s to its coefficients. The homomorphism (95) induces a homomorphism

$$f(x_1,\ldots,x_n)+I \rightsquigarrow (sf)(v_1,\ldots,v_n) \tag{96}$$

of $A[x_1,\ldots,x_n]/I$ into C if and only if $(sf)(v_1,\ldots,v_n) = 0$ for every $f \in I$. Since we have the isomorphism $f(x_1,\ldots,x_n)+I \rightsquigarrow f(u_1,\ldots,u_n)$, we see that we have a homomorphism of A' into C extending s and sending $u_i \rightsquigarrow v_i$, $1 \leqslant i \leqslant n$, if and only if

$$(sf)(v_1,\ldots,v_n) = 0 \tag{97}$$

for every $f \in I$. Moreover, it is clear that it suffices to have this relation for every f in any set of generators X for the ideal I.

If $f(x_1,\ldots,x_n) \in A[x_1,\ldots,x_n]$, we write $\partial f/\partial x_i$ for the polynomial obtained from f by formal partial differentiation with respect to x_i. For example

$$\frac{\partial}{\partial x_1}(x_1^3x_2x_3+3x_1x_3^2) = 3x_1^2x_2x_3+3x_3^2.$$

If D is a derivation of A into B, then we shall write $(Df)(x_1,\ldots,x_n)$ for the polynomial in $B[x_1,\ldots,x_n]$ obtained by applying the derivation D to the coefficients of f.

We can now prove

THEOREM 8.41. *Let B be a commutative algebra, A a subalgebra, $A' = A[u_1,\ldots,u_n]$, $u_i \in B$, X a set of generators for the kernel of the homomorphism of $A[x_1,\ldots,x_n]$ onto A' such that $a \rightsquigarrow a$ for $a \in A$ and $x_i \rightsquigarrow u_i$, $1 \leqslant i \leqslant n$. Let D be a derivation of A into B. Then D can be extended to a derivation of A' into B such that $u_i \rightsquigarrow v_i$, $1 \leqslant i \leqslant n$, if and only if*

$$(Df)(u_1,\ldots,u_n)+\sum_{i=1}^{n}\frac{\partial f}{\partial x_i}(u_1,\ldots,u_n)v_i = 0 \tag{98}$$

for every $f \in X$.

Proof. The condition that D has an extension of the sort specified is that $\alpha(D)$ is extendable to a homomorphism of A' into $B \otimes_K \Delta$ sending $u_i \rightsquigarrow u_i+v_i\delta$, $1 \leqslant i \leqslant n$. This will be the case if and only if for every $f \in X$

$$(\alpha(D)f)(u_1+v_1\delta, u_2+v_2\delta,\ldots,u_n+v_n\delta) = 0. \tag{99}$$

Now let $a \in A$ and consider the monomial $ax_1^{k_1}x_2^{k_2}\dots x_n^{k_n}$, $k_i \geqslant 0$. We have

$$(a+D(a)\delta)(u_1+v_1\delta)^{k_1}\cdots(u_n+v_n\delta)^{k_n}$$
$$= au_1^{k_1}\cdots u_n^{k_n}+((Da)u_1^{k_1}\cdots u_n^{k_n}+k_1au_1^{k_1-1}u_2^{k_2}\cdots u_n^{k_n}v_1$$
$$+k_2au_1^{k_1}u_2^{k_2-1}u_3^{k_3}\cdots u_n^{k_n}v_2+\cdots+k_nau_1^{k_1}\cdots u_{n-1}^{k_{n-1}}u_n^{k_n-1}v_n)\delta.$$

Hence for any $f \in A[x_1,\dots,x_n]$ we have

$$(\alpha(D)f)(u_1+v_1\delta,u_2+v_2\delta,\dots,u_n+v_n\delta) = f(u_1,\dots,u_n)$$
$$+((Df)(u_1,\dots,u_n)$$
$$+\sum_{i=1}^{n}\frac{\partial f}{\partial x_i}(u_1,\dots,u_n)v_i)\delta.$$

Then the condition that (99) holds for all $f \in X$ is equivalent to (98). $\square$

We suppose next that S is a submonoid of the multiplicative monoid of A and we consider the localizations A_S and B_S. We can prove

THEOREM 8.42. *Let D be a derivation of A into B (commutative) and let S be a submonoid of the multiplicative monoid of A. Then there exists a unique derivation D_S of A_S into B_S such that*

(100)

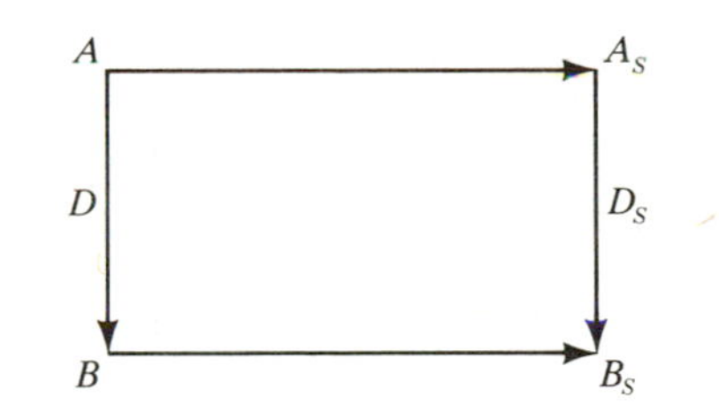

is commutative. Here the horizontal maps are the canonical homomorphisms $a \rightsquigarrow a/1$ and $b \rightsquigarrow b/1$ respectively.

Proof. We have the homomorphism $\alpha(D)$ of A into $B \otimes_K \Delta$ sending $a \rightsquigarrow a+D(a)\delta$ and the homomorphism of $B \otimes_K \Delta$ into $B_S \otimes_K \Delta$ sending $b \otimes u \rightsquigarrow b/1 \otimes u$. Hence we have the homomorphism of A into $B_S \otimes_K \Delta$ sending such that $a \rightsquigarrow a/1+(D(a)/1)\delta$. Now if $s \in S$, then $s/1+(D(s)/1)\delta$ is invertible with inverse $1/s-(D(s)/s^2)\delta$. Hence by the universal property of A_S we have a unique homomorphism $H: A_S \to B_S \otimes_K \Delta$ such that

(101)

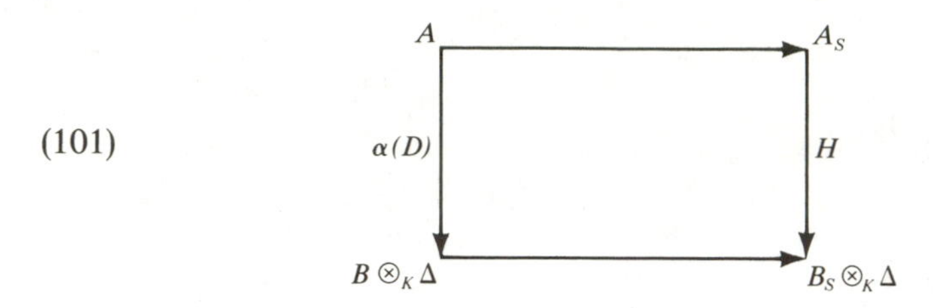

is commutative. If π_S denotes the canonical homomorphism of $B_S \otimes_K \Delta$ onto B_S, then the commutativity of (101) implies that $\pi_S H = 1_{A_S}$. Hence H has the form $\alpha(D_S)$ where D_S is a derivation of A_S into B_S. Then D_S satisfies the condition of the theorem. □

We now specialize to the case of fields. We consider an extension field E/F of a field and regard this as an algebra over F. Suppose K is a subfield of E/F and we have a derivation D of K/F into E/F and a is an element of E. If a is transcendental over K, then Theorem 8.41 shows that for any $b \in E$ there exists a derivation of $K[a]$ into E extending D and mapping a into b. Then Theorem 8.42 shows that this has a unique extension to a derivation of $K(a)$ into E. Hence if a is transcendental over K, then there exists an extension of D to a derivation of $K(a)$ sending a into b. By Proposition 8.16 (and Theorem 8.42) this is unique. Next let a be algebraic over K with minimum polynomial $f(x)$ over K. Then $K(a) = K[a]$ and Theorem 8.41 shows that D can be extended to a derivation of $K(a)$ into E sending $a \to b$ if and only if

$$(Df)(a) + f'(a)b = 0. \tag{102}$$

If a is separable, $(f(x), f'(x)) = 1$ and $f'(a) \neq 0$. Then there is only one choice we can make for b, namely,

$$b = -f'(a)^{-1}(Df)(a). \tag{103}$$

Hence in this case D can be extended in one and only one way to a derivation of $K(a)$ into E. If a is inseparable, then $f'(a) = 0$. Then (101) shows that D can be extended to a derivation of $K(a)$ if and only if the coefficients of $f(x)$ are D constants and if this is the case, the extension can be made to send a into any $b \in E$. We summarize these results in

PROPOSITION 8.17. *Let E be an extension field of F, K an intermediate field, D a derivation of K/F into E/F, a an element of E. Then*

(1) *D can be extended to a derivation of $K(a)$ into E sending a into any $b \in E$ if a is transcendental over K.*

(2) *D has a unique extension to a derivation of $K(a)$ into E if a is separable algebraic over K.*
(3) *D can be extended to a derivation of $K(a)$ into E if a is inseparable algebraic over K if and only if the coefficients of the minimum polynomial of a over K are D-constants. Moreover, if this condition is satisfied, then there exists an extension sending a into any $b \in E$.*

We now suppose that E is finitely generated over F: $E = F(a_1, \ldots, a_n)$. Let X be a set of generators for the ideal I in $F[x_1, \ldots, x_n]$, x_i indeterminates, consisting of the polynomials f such that $f(a_1, \ldots, a_n) = 0$. Then it follows from Theorems 8.41 and 8.42 that there exists a derivation D of E/F into itself such that $Da_i = b_i$, $1 \leqslant i \leqslant n$, if and only if

$$(104) \qquad \sum_{i=1}^{n} \frac{\partial f}{\partial x_i}(a_1, \ldots, a_n) b_i = 0$$

for every $f \in X$. We recall that $\mathrm{Der}_F E$ becomes a vector space over E if we define bD for $b \in E$, $D \in \mathrm{Der}_F E$ by $(bD)(x) = b(D(x))$ (p. 525). We wish to calculate the dimensionality $[\mathrm{Der}_F E : E]$ when $E = F(a_1, \ldots, a_n)$. For this purpose we introduce the n-dimensional vector space $E^{(n)}$ of n-tuples of elements of E. If $D \in \mathrm{Der}_F E$, D determines the element $(Da_1, \ldots, Da_n)$ of $E^{(n)}$ and we have the map

$$(105) \qquad \lambda_{a_1, \ldots, a_n} : D \rightsquigarrow (Da_1, \ldots, Da_n)$$

of $\mathrm{Der}_F E$ into $E^{(n)}$. Evidently this is a linear map of vector spaces over E and since a derivation is determined by its action on a set of generators, $\lambda = \lambda_{a_1, \ldots, a_n}$ is injective. Hence $[\mathrm{Der}_F E : E]$ is the dimensionality of the subspace $\lambda(\mathrm{Der}_F E)$ of $E^{(n)}$. Now let $g \in F[x_1, \ldots, x_n]$. Then g defines a map dg of $E^{(n)}$ into E by

$$(106) \qquad dg : (b_1, \ldots, b_n) \rightsquigarrow \sum_{1}^{n} \frac{\partial g}{\partial x_i}(a_1, \ldots, a_n) b_i.$$

Evidently this is linear. The result we proved before now states that $(b_1, \ldots, b_n) \in \lambda(\mathrm{Der}_F E)$ if and only if

$$(107) \qquad df(b_1, \ldots, b_n) = 0$$

for every $f \in X$. This is a system of linear equations that characterizes $\lambda(\mathrm{Der}_F E)$. This leads to a formula for $[\mathrm{Der}_F E : E]$, which we give in

THEOREM 8.43. *Let $E = F(a_1, \ldots, a_n)$ and let X be a set of generators of the ideal of polynomials in $F[x_1, \ldots, x_n]$ such that $f(a_1, \ldots, a_n) = 0$. Let dX denote* *the subspace of* $E^{(n)*}$ *spanned by the linear functions* $df, f \in X$. *Then*

$$[\mathrm{Der}_F E : E] = n - [dX : E]. \tag{108}$$

Proof. We have $[\mathrm{Der}_F E : E] = [\lambda\, \mathrm{Der}_F E : E]$, and $\lambda\, \mathrm{Der}_F E$ is the subspace of $E^{(n)}$ of elements such that $df(b_1, \ldots, b_n) = 0$ for all $f \in X$. Hence (108) follows from linear algebra. □

By the Hilbert basis theorem, we can take $X = \{f_1, \ldots, f_m\}$. Then it follows from linear algebra that $[dX : E]$ is the rank of the Jacobian matrix

$$\begin{bmatrix} \frac{\partial f_1}{\partial x_1}(a_1, \ldots, a_n) & \cdots & \frac{\partial f_1}{\partial x_n}(a_1, \ldots, a_n) \\ & \cdots & \\ \frac{\partial f_m}{\partial x_1}(a_1, \ldots, a_n) & \cdots & \frac{\partial f_m}{\partial x_n}(a_1, \ldots, a_n) \end{bmatrix} \tag{109}$$

Combining this with Theorem 8.43 we obtain the

COROLLARY. *Let* $E = F(a_1, \ldots, a_n)$ *and let* $X = \{f_1, \ldots, f_m\}$ *be a finite set of generators for the ideal of polynomials in* $F[x_1, \ldots, x_n]$ *such that* $f(a_1, \ldots, a_n) = 0$. *Then*

$$[\mathrm{Der}_F E : E] = n - \operatorname{rank} J(f_1, \ldots, f_m) \tag{110}$$

where $J(f_1, \ldots, f_m)$ *is the Jacobian matrix* (108).

We obtain next a connection between $[\mathrm{Der}_F E : E]$ and the structure of E/F. We prove first

PROPOSITION 8.18. *If* $E = E(a_1, \ldots, a_n)$, *then* $\mathrm{Der}_F E = 0$ *if and only if* E *is separable algebraic over* F.

Proof. If $a \in E$ is separable algebraic over F, then Proposition 8.17.2 applied to the derivation 0 on F shows that $D(a) = 0$ for every derivation of E/F. Hence $\mathrm{Der}_F E = 0$ if E is separable algebraic over F. Now suppose that E is not separable algebraic over F. We may assume that $\{a_1, \ldots, a_r\}$ $(r \geqslant 0)$ is a transcendency base for E/F. Let S be the subfield of elements of E that are separable algebraic over $F(a_1, \ldots, a_r)$. If $S = E$, then $r > 0$ and we have a derivation of $F(a_1, \ldots, a_r)$ into E sending a_i, $1 \leqslant i \leqslant r$, into any element we please in E. By applying Proposition 8.17.2 successively to $a_{r+1}, \ldots, a_n$, we obtain extensions of the derivation of $F(a_1, \ldots, a_r)$ to E to a derivation of E/F. Hence we can obtain a non-zero derivation of E/F. Now let $E \supsetneqq S$. Then the characteristic is $p \neq 0$ and E is purely inseparable over S. We have $0 \neq [E:S] < \infty$ and we can choose a maximal subfield K of E containing S ($K \neq E$). If $a \in E, \notin K$, then $K(a) = E$ by the maximality of K. Moreover, the minimum polynomial of a over K has the form $x^{p^e} - b$ since E is purely inseparable over S and hence over K. If $e > 1$, then $K(a^{p^{e-1}})$ is a proper subfield of E properly containing K. Hence $E = K(a)$ where $x^p - b$ is the minimum polynomial of a over K. By Proposition 8.17.3 we have a non-zero derivation of E/K. Since this is a derivation of E/F, we have $\mathrm{Der}_F E \neq 0$ in the case $E \supsetneqq S$ also. □

We can now prove the following theorem relating $[\mathrm{Der}_F E : E]$ and the structure of E/F.

THEOREM 8.44. *Let $E = F(a_1, \ldots, a_n)$. Then $[\mathrm{Der}_F E : E]$ is the smallest s such that there exists a subset $\{a_{i_1}, \ldots, a_{i_s}\}$ of $\{a_1, \ldots, a_n\}$ such that E is separable algebraic over $F(a_{i_1}, \ldots, a_{i_s})$. Moreover, $[\mathrm{Der}_F E : E]$ is the transcendency degree of E over F if and only if E is separable over F.*

Proof. We again consider the map $\lambda = \lambda_{a_1, \ldots, a_n}$ of $\mathrm{Der}_F E$ into $E^{(n)}$ defined by (105). Let $s = [\mathrm{Der}_F E : E] = [\lambda(\mathrm{Der}_F E) : E]$ and let $D_1, D_2, \ldots, D_s$ be a base for $\mathrm{Der}_F E$ over E. Then $s \leqslant n$ and $\lambda(\mathrm{Der}_F E)$ has the base $(D_1 a_1, \ldots, D_1 a_n), \ldots, (D_s a_1, \ldots, D_s a_n)$ and so the $s \times n$ matrix $(D_i a_j)$ has rank s. Hence we may suppose that the a_i are ordered so that

$$\det(D_i a_j) \neq 0 \quad \text{if } 1 \leqslant i,\ j \leqslant s. \tag{111}$$

Put $K = F(a_1, \ldots, a_s)$ and let $D \in \mathrm{Der}_K E \subset \mathrm{Der}_F E$. Then $D = \sum_1^s b_i D_i$, $b_i \in E$, and $D(a_j) = 0$ for $1 \leqslant j \leqslant s$ gives $\sum_i b_i D_i(a_j) = 0$ for $1 \leqslant j \leqslant s$. By (110), this implies that every $b_i = 0$, so $D = 0$. Hence $\mathrm{Der}_K E = 0$ and so by Proposition 8.18, E is separable algebraic over K. Conversely, let $\{a_{i_1}, \ldots, a_{i_t}\}$ be a subset of $\{a_1, \ldots, a_n\}$ such that E is separable algebraic over $F(a_{i_1}, \ldots, a_{i_t})$. By re-ordering the a's we may assume that the subset is $\{a_1, \ldots, a_t\}$. We now map $\mathrm{Der}_F E$ into $E^{(t)}$ by $D \rightsquigarrow (Da_1, \ldots, Da_t)$. The kernel of this linear map is the set of D such that $D(a_k) = 0$, $1 \leqslant k \leqslant t$, and hence it is the set of D such that $D(K) = 0$ for $K = F(a_1, \ldots, a_t)$. Now $D(K) = 0$ means that $D \in \mathrm{Der}_K(E)$ and since E is separable algebraic over K, this implies that $D = 0$. Thus the map $D \rightsquigarrow (Da_1, \ldots, Da_t)$ is injective. Hence $t \geqslant s = [\mathrm{Der}_F E : E]$. This completes the proof of the first statement. To prove the second, we note first that if $[\mathrm{Der}_F E : E] = s$, then we may assume that E is separable algebraic over $F(a_1, \ldots, a_s)$. Then $\{a_1, \ldots, a_s\}$ contains a transcendency base for E/F, so $s \geqslant r = \text{tr deg } E/F$. Moreover, if $r = s$, then $\{a_1, \ldots, s_s\}$ is a transcendency base. Hence this is a separating transcendency base and E is separable over F. Conversely, suppose that E is separable over F. Then the proof of Theorem 8.39 shows that we can choose a separating transcendency base among the

a_i, so we may assume this is $\{a_1,\dots,a_r\}$. Then E is separable algebraic over $F(a_1,\dots,a_r)$ and hence, as we showed in the first part, $r \geqslant [\mathrm{Der}_F E:E]$. Since we had $[\mathrm{Der}_F E:E] \geqslant r$, this proves that $[\mathrm{Der}_F E:E] = r$. □

EXERCISES

1. Let $E = F(a_1,\dots,a_n)$. Show that E is separable algebraic over F if and only if there exist n polynomials $f_1,\dots,f_n \in F[x_1,\dots,x_n]$ such that $f_i(a_1,\dots,a_n) = 0$ and

$$\det\left(\frac{\partial f_i}{\partial x_j}(a_1,\dots,a_n)\right) \neq 0.$$

2. Let D be a derivation in E/F, K the subfield of D-constants. Show that $a_1,\dots,a_m \in E$ are linearly dependent over K if and only if the Wronskian determinant

$$\begin{vmatrix} a_1 & a_2 & \cdots & a_m \\ Da_1 & Da_2 & \cdots & Da_m \\ \cdot & \cdot & \cdots & \cdot \\ D^{m-1}a_1 & D^{m-1}a_2 & \cdots & D^{m-1}a_m \end{vmatrix} = 0.$$

3. (C. Faith.) Let $E = F(a_1,\dots,a_n)$ and let K be a subfield of E/F. Show that $[\mathrm{Der}_F K:K] \leqslant [\mathrm{Der}_F E:E]$.

4. Let A be a subalgebra of an algebra B over a commutative ring K. Define a *higher derivation of rank m* of A into B to be a sequence of K-homomorphisms

$$D = (D_0 = 1_A, D_1,\dots,D_m) \tag{112}$$

of A into B such that

$$D_i(ab) = \sum_{j=0}^{i} D_j(a)D_{i-j}(b) \tag{113}$$

for $a,b \in A$. Let $\Delta^{(m)}$ be the algebra $K[x]/(x^{m+1})$ so $\Delta^{(m)}$ has a base $(1,\delta,\dots,\delta^m)$ where $\delta = x+(x^{m+1})$ and $\delta^{m+1} = 0$. Note that $B \otimes_K \Delta^{(m)}$ is the set of elements

$$b_0 + b_1\delta + \cdots + b_m\delta^m \tag{114}$$

where $b_i = b_i \otimes 1$, and $\delta = 1 \otimes \delta$ and that an element (113) is 0 if and only if every $b_i = 0$. Let π denote the homomorphism of $B \otimes_K \Delta^{(m)}$ into B sending $b_0 + b_1\delta + \cdots + b_m\delta^m \rightsquigarrow b_0$. Show that a sequence of maps $D = (D_0, D_1,\dots,D_m)$ of A into B is a higher derivation of rank m of A into B if and only if

$$\alpha(D): a \rightsquigarrow D_0(a) + D_1(a)\delta + \cdots + D_m(a)\delta^m$$

is a K-algebra homomorphism of A into $B \otimes_K \Delta^{(m)}$ such that $\pi\alpha(D) = 1_A$.

5. Let A and B be as in exercise 4. Define a *higher derivation of infinite rank* of A into B to be an infinite sequence of homomorphisms $D = (D_0 = 1, D_1, D_2, \dots)$ of A into B such that (113) holds. Obtain a connection between higher derivations of A into B and homomorphisms of A into $B[[x]]$, the algebra of formal power series in x with coefficients in B.

8.17 GALOIS THEORY FOR PURELY INSEPARABLE EXTENSIONS OF EXPONENT ONE

Let E/F be of characteristic $p \neq 0$ and let D be a derivation of E/F. If $a \in E$, then $D(a^p) = pa^{p-1}D(a) = 0$. Hence every element of $F(E^p)$ is a constant relative to every derivation of E/F. If $c \in F(E^p)$ and $a \in E$, then $D(ca) = cD(a)$ for $D \in \mathrm{Der}_F E$. It is natural to replace F by $F(E^p)$ in studying the derivations of E/F. We shall now do this, so with a change of notation, we may assume that $E^p \subset F$, which means that either $E = F$ or E is purely inseparable of exponent one over F (see exercise 8, p. 495). We restrict our attention also to finitely generated extensions $E = F(a_1, \ldots, a_n)$.

Let $\{a_1, \ldots, a_m\}$ be a minimal set of generators for E/F, so $m = 0$ if and only if $E = F$. Suppose that $m > 0$. Then $a_1 \notin F$ and $a_i \notin F(a_1, \ldots, a_{i-1})$ for $1 < i \leqslant m$. Since $a_i{}^p \in F$ for all i, the minimum polynomial of a_1 over F and of a_i over $F(a_1, \ldots, a_{i-1})$ for $i > 1$ has the form $x^p - b$. Hence $[F(a_1):F] = p$ and $[F(a_1, \ldots, a_i):F(a_1, \ldots, a_{i-1})] = p$, which implies that $[E:F] = p^m$. Evidently $E = F[a_1, \ldots, a_m]$ and since $a_i{}^p \in F$ for all i and $[E:F] = p^m$, the set of monomials

$$a_1{}^{k_1} a_2{}^{k_2} \cdots a_m{}^{k_m} \qquad 0 \leqslant k_i < p, \tag{115}$$

constitutes a base for E/F. It is clear also that E/F is a tensor product of the simple extensions $F(a_i)/F$.

Put $F_i = F(a_1, \ldots, \hat{a}_i, \ldots, a_m)$. Then $E = F_i(a_i)$ and the minimum polynomial of a_i over F_i has the form $x^p - b_i$, $b_i \in F$. By Proposition 8.17.3 we have a derivation D_i of E/F_i such that $D_i(a_i) = 1$. Thus we have $D_i(a_j) = \delta_{ij}$. It follows immediately that the D_i, $1 \leqslant i \leqslant m$, form a base for $\mathrm{Der}_F E$ as vector space over E and hence $[\mathrm{Der}_F E:E] = m$. Since $[E:F] = p^m$, this implies that $[\mathrm{Der}_F E:F] = mp^m$.

We now consider any field E of characteristic p, and derivations of E into itself without reference to a particular subfield of E. These are the endomorphisms D of the group $(E, +, 0)$ that satisfy the condition $D(ab) = D(a)b + aD(b)$. One deduces from this that $D(ca) = cD(a)$ if c is in the prime field P, so D can be regarded as a derivation of E/P. However, we shall simply say that D is a derivation of E into itself. Let $\mathrm{Der}\, E$ denote the set of these maps. Then $\mathrm{Der}\, E$ is a set of endomorphisms of the additive group $(E, +, 0)$ having the following closure properties: (1) $\mathrm{Der}\, E$ is a subspace of $\mathrm{End}\,(E, +, 0)$ regarded as a vector space over E by defining bL for $b \in E$, $L \in \mathrm{End}(E, +, 0)$ by $(bL)(a) = b(L(a))$, (2) if $D_1, D_2 \in \mathrm{Der}\, E$, then $[D_1, D_2] \in \mathrm{Der}\, E$, and (3) if $D \in \mathrm{Der}\, E$, then $D^p \in \mathrm{Der}\, E$. We shall now call any subset of $\mathrm{End}\,(E, +, 0)$ having these closure properties a *p-E-Lie algebra of endomorphisms of* $(E, +, 0)$. We use this terminology for want of anything better, but we should call attention to the fact that a p-E-Lie algebra need not be a Lie algebra over E in the usual sense, since the composition $[D_1, D_2]$ is not E-bilinear.

If F is a subfield of E such that $\lfloor E:F \rfloor < \infty$ and E is purely inseparable of exponent $\leqslant 1$ over F, then $\mathrm{Der}_F E$ is a p-E-Lie algebra of endomorphisms of $(E, +, 0)$. Moreover, we have seen that $[\mathrm{Der}_F E : E] < \infty$. We shall now show that every p-E-Lie algebra of derivations of E having finite dimensionality over E is obtained in this way. For, we have

THEOREM 8.45 (Jacobson). *Let E be a field of characteristic $p \neq 0$, F a subfield such that* (1) $[E:F] < \infty$ *and* (2) *E is purely inseparable of exponent $\leqslant 1$ over E. Then $\mathrm{Der}_F E$ is a p-E-Lie algebra of endomorphisms of $(E, +, 0)$ such that $p^{[\mathrm{Der}_F E : E]} = [E:F]$. Conversely, let $\mathscr{D}$ be a p-E-Lie algebra of derivations of E such that $[\mathscr{D}:E] < \infty$ and let F be the set of $\mathscr{D}$-constants of E, that is, the elements that are D-constants for every $D \in \mathscr{D}$. Then $[E:F] < \infty$ and E is purely inseparable of exponent $\leqslant 1$ over F. Moreover, $\mathscr{D} = \mathrm{Der}_F E$ and if $(D_1, \ldots, D_m)$ is a base for $\mathscr{D}$ over E, then the set of monomials*

$$D_1{}^{k_1} D_2{}^{k_2} \cdots D_m{}^{k_m}, \qquad 0 \leqslant k_i < p, \qquad (D_i^0 = 1) \tag{116}$$

form a base for $\mathrm{End}_F E$ regarded as a vector space over E.

Proof. The first statement has already been proved. To prove the second, we use the same idea we used to establish the results on finite groups of automorphisms in fields: We use the given set of endomorphisms to define a set of endomorphisms L satisfying the conditions in the Jacobson-Bourbaki correspondence. In the present case we take L to be the set of E-linear combinations of the endomorphisms given in (116). Evidently L contains $1 = D_1^0 \cdots D_m^0$, so L contains $E_E = E1$ and $[L:E] \leqslant p^m$. It remains to show that L is closed under multiplication by the D_i. We note first that if D is a derivation in E, then the condition $D(ab) = D(a)b + aD(b)$ gives the operator condition

$$Da_E = a_E D + D(a)_E \tag{117}$$

where a_E and $D(a)_E$ denote the multiplications by a and $D(a)$ respectively. Using this relation we see that $D_i(aD_1{}^{k_1} \cdots D_m{}^{k_m}) = aD_i D_1{}^{k_1} \cdots D_m{}^{k_m} + D_i(a) D_1{}^{k_1} \cdots D_m{}^{k_m}$. Hence to prove that $D_i L \subset L$, it suffices to show that $D_i D_1{}^{k_1} \cdots D_m{}^{k_m} \in L$ for all i and all k_j such that $0 \leqslant k_j < p$. We shall prove this by showing that $D_i D_1{}^{k_1} \cdots D_m{}^{k_m}$ is a linear combination with coefficients in E of the monomials $D_1{}^{k'_1} \cdots D_m{}^{k'_m}$ such that $0 \leqslant k'_j < p$ and $\sum k'_j \leqslant \sum k_j + 1$. The argument for this is very similar to one we used in the study of Clifford algebras (p. 230): We use induction on $\sum k_j$ and for a given k_j, induction on i. We have at our disposal the formulas

$$D_j^p = \sum_k b_{jk} D_k, \qquad b_{jk} \in E, \tag{118}$$

and

$$D_i D_j = D_j D_i + \sum_k d_{ijk} D_k, \qquad d_{ijk} \in E, \tag{119}$$

that follow from the conditions that $\mathscr{D}$ is closed under pth powers and under commutators. The result we want to prove is clear if $\sum k_j = 0$, so we may suppose that $\sum k_j > 0$. Then some $k \neq 0$ and we suppose that k_j is the first of the k's that is >0. Then $D_1^{k_1} \cdots D_m^{k_m} = D_j^{k_j} \cdots D_m^{k_m}$. If $i < j$, then $D_i D_j^{k_j} \cdots D_m^{k_m}$ is one of the monomials (116) for which the sum of the exponents is $\sum_l k_l + 1$. Hence the result holds in this case. The same thing is true if $i = j$ and $k_j < p-1$. Now let $i = j$, $k_j = p-1$. Then $D_i D_j^{k_j} \cdots D_m^{k_m} = D_j^p D_{j+1}^{k_{j+1}} \cdots D_m^{k_m}$ and the result follows by induction if we replace D_j^p by $\sum_k b_{jk} D_k$. Now assume that $i > j$. Then by (119),

$$D_i D_j^{k_j} \cdots D_m^{k_m} = D_j D_i D_j^{k_j - 1} \cdots D_m^{k_m} + \sum d_{ijk} D_k D_j^{k_j - 1} \cdots D_m^{k_m}$$

The result follows in this case also by applying both induction hypotheses to the right-hand side. This establishes the key result that L is closed under multiplication. Hence the Jacobson-Bourbaki correspondence is applicable to L, and this shows that if $F = \{a \mid a_E B = B a_E \text{ for } B \in L\}$, then F is a subfield such that $[E:F] = [L:E]$ and $L = \text{End}_F E$. By definition of L we have $[L:E] \leqslant p^m$ and equality holds here if and only if the monomials (116) form a base for L over E. Since $\mathscr{D}$ generates L, the conditions defining F can be replaced by $a_E D = D a_E$ for all $D \in \mathscr{D}$. By (117) this is equivalent to: a is a D-constant for every $D \in \mathscr{D}$. Hence F is the set of $\mathscr{D}$-constants and $\mathscr{D} \subset \text{Der}_F E$. Then E is purely inseparable of exponent $\leqslant 1$ over F. We have $[E:F] = [L:E] \leqslant p^m$, so $[E:F] = p^{m'}$ with $m' \leqslant m$. On the other hand, $\mathscr{D}$ contains m linearly independent derivations (over E), so $[\text{Der}_F E : E] \geqslant m$ and hence $[E:F] \geqslant p^m$. It follows that $\mathscr{D} = \text{Der}_F E$, $[\text{Der}_F E : E] = m$, and $[L:E] = [E:F] = p^m$. This completes the proof. □

EXERCISES

In these exercises we assume that E is purely inseparable of exponent $\leqslant 1$ over F of characteristic p and that E is finitely generated over F.

1. (Baer.) Show that there exists a derivation D of E/F such that F is the subfield of E of D-constants.

2. Let D be a derivation of E/F such that F is the subfield of D-constants. Show that the minimum polynomial of D as a linear transformation in E/F has the form

$$x^{p^m} + c_1 x^{p^{m-1}} + \cdots + c_m, \qquad c_i \in F$$

where $p^m = [E:F]$. Show that $(1, D, \ldots, D^{p^m-1})$ is a base for $\mathrm{End}_F E$ as vector space over E.

3. (J. Barsotti, P. Cartier.) Show that if D is a derivation in a field F of characteristic $p \neq 0$, then $D^{p-1}(a^{-1}Da) = a^{-1}D^p a - (a^{-1}Da)^p$.

4. (M. Gerstenhaber, M. Ojanguren-M.R. Sridharan.) Let E be a field of characteristic $p \neq 0$ and let V be an E subspace of Der E closed under pth powers. Show that V is a Lie subring. Note that this shows that closure under Lie products is superfluous in the statement of Theorem 8.45.

5. (Gerstenhaber, Ojanguren-Sridharan.) Extend Theorem 8.45 to obtain a 1–1 correspondence between the set of subfields F of E such that E/F is purely inseparable of exponent 1 and the set of p-E-Lie algebras of derivations of E that are closed in the finite topology.

8.18 TENSOR PRODUCTS OF FIELDS

If E/F and K/F are fields over F what can be said about the F-algebra $E \otimes_F K$? In particular, is this a field or a domain? It is easy to give examples where $E \otimes_F K$ is not a field. In fact, if $E \neq F$ then $F \otimes_F E$ is never a field. To see this we observe that by the basic property of tensor products, we have an additive group homomorphism η of $E \otimes_F E$ into E such that $a \otimes b \rightsquigarrow ab$, $a, b \in E$. Also it is clear from the definitions that this is an F-algebra homomorphism. Now let $a \in E, \notin F$. Then 1, a are F-independent in E and hence $1 \otimes 1$, $1 \otimes a$, $a \otimes 1$, and $a \otimes a$ are F-independent in $E \otimes_F E$. Hence $1 \otimes a - a \otimes 1 \neq 0$ but $\eta(1 \otimes a - a \otimes 1) = a - a = 0$. Thus $\ker \eta$ is a non-zero ideal in $E \otimes_F E$. The existence of such an ideal implies that $E \otimes_F E$ is not a field.

It is readily seen also that if x and y are indeterminates then $F(x) \otimes_F F(y)$ is a domain but not a field (see below).

Another fact worth noting is that if E is algebraic over F and $E \otimes_F K$ is a domain then this algebra is a field. This is clear if E is finite dimensional over F. For, then $E \otimes_F K$ can be regarded as a finite dimensional algebra over K ($[E \otimes_F K:K] = [E:F]$) and a finite dimensional domain is necessarily a field. The general case follows from this since any $a \in E \otimes_F K$ is contained in a subalgebra isomorphic to an algebra $E_0 \otimes_F K$ where E_0/F is finitely generated, hence finite dimensional over F. If $a \neq 0$ the corresponding element of $E_0 \otimes_F K$ is invertible. Hence a is invertible in $E \otimes_F K$.

We shall now proceed to a systematic study of tensor products of fields. In our discussion separability will mean separability in the general sense of Definition 8.4, pure inseparability of E/F will mean that E is algebraic over F and the subfield of E/F of separable elements over F coincides with F. We shall say that F is algebraically closed (separably algebraically closed) in E if every algebraic (separable algebraic) element of E is contained in F. We prove first

THEOREM 8.46. *Let E/F and K/F be extension fields of F.*

(1) *If E/F is separable and K/F is purely inseparable, then $E \otimes_F K$ is a field. On the other hand, if E/F is not separable, then there exists a purely inseparable extension K/F of exponent 1 such that $E \otimes_F K$ contains a non-zero nilpotent element.*

(2) *If E/F is separable algebraic, then $E \otimes_F K$ has no nilpotent elements for arbitrary K/F, and $E \otimes_F K$ is a field if F is separably algebraically closed in K.*

(3) *The elements of $E \otimes_F K$ are either invertible or nilpotent if either E/F is purely inseparable and K/F is arbitrary, or E/F is algebraic and F is separably algebraically closed in K.*

Proof. In (1) and in the first part of (3) we may assume the characteristic is $p \neq 0$.

(1) Assume E/F is separable and K/F is purely inseparable. The separability implies that if $a_1, \dots, a_m$ are F-independent elements of E then these elements are linearly independent over F^{1/p^e} (contained in the algebraic closure of E) for every $e = 0, 1, 2, \dots$. This implies that the elements $a_1^{p^e}, \dots, a_m^{p^e}$ are F-independent for every e. Now let K be purely inseparable over F and let $z = \sum a_i \otimes c_i \neq 0$ in $E \otimes_F K$ where $a_i \in E$, $c_i \in K$. We may assume that a_i are F-independent and we have an e such that $c_i^{p^e} \in F$, $1 \leqslant i \leqslant m$. Then $z^{p^e} = \sum a_i^{p^e} \otimes c_i^{p^e} = \sum c_i^{p^e} a_i^{p^e} \otimes 1 \neq 0$. Hence $\sum c_i^{p^e} a_i^{p^e} \neq 0$ and this element of E is invertible. Thus z^{p^e} is invertible and hence z is invertible in $E \otimes_F K$. Then $E \otimes_F K$ is a field.

Next assume E/F is not separable. Then there are F-independent elements $a_1, \dots, a_m \in E$ that are not $F^{1/p}$ independent. Hence we have $c_i^{1/p}$, $c_i \in F$ not all 0 such that $\sum c_i^{1/p} a_i = 0$. Then $\sum c_i a_i^p = 0$ but $\sum c_i a_i \neq 0$ since not every $c_i = 0$. Consider the field $K = F(c_1^{1/p}, \dots, c_m^{1/p})$. We have $K \supsetneqq F$ since $K = F$ implies the a_i are F-dependent. Now consider the element $z = \sum a_i \otimes c_i^{1/p} \in E \otimes_F K$. This is non-zero since the a_i are F-independent and not every $c_i^{1/p} = 0$ in K. On the other hand, $z^p = \sum a_i^p \otimes c_i = \sum c_i a_i^p \otimes 1 = 0$. Hence z is a non-zero nilpotent in $E \otimes_F K$.

(2) Assume E/F is separable algebraic, K/F is arbitrary. We have to show that $E \otimes_F K$ has no non-zero nilpotents and that $E \otimes_F K$ is a field if F is separably algebraically closed in K. Using the argument at the beginning of this section, we obtain a reduction to the case in which E is finitely generated, hence $[E:F] < \infty$. In this case since E is separable algebraic, $E = F[a] = F(a)$ where the minimum polynomial $m(\lambda)$ of a over F is irreducible and separable (meaning, e.g., that $(m(\lambda), m'(\lambda)) = 1$). Then $E \otimes_F K \cong K[a]$ where the minimum polynomial of a over K is $m(\lambda)$. Hence $E \otimes_F K \cong K[\lambda]/(m(\lambda))$. Since $m(\lambda)$ is separable, we have the factorization in $K[\lambda]$ of $m(\lambda)$ as $m(\lambda) = m_1(\lambda)\dots m_r(\lambda)$ where the $m_i(\lambda)$ are distinct irreducible monic polynomials. Then $E \otimes_F K \cong K[\lambda]/(m(\lambda)) \cong \bigoplus_1^r K[\lambda]/(m_i(\lambda))$ (exercise 4, p. 410 of BAI). Since $K_i[\lambda]/(m_i(\lambda))$ is a field we see that $E \otimes_F K$ is a direct sum of fields. Clearly an algebra having this structure has no non-zero nilpotent elements. This proves the first assertion of (2).

The coefficients of the $m_i(\lambda)$ are separable algebraic over F since they are elementary symmetric polynomials in some of the roots of $m(\lambda)$ and these are separable algebraic over F. Hence if F is separably algebraically closed in K then $m_i(\lambda) \in F[\lambda]$. Then $r = 1$ and $E \otimes_F K \cong K[\lambda]/(m(\lambda))$ is a field.

(3) Let E/F be purely inseparable, K/F arbitrary. Let $z = \sum_1^m a_i \otimes c_i$, $a_i \in E$, $c_i \in K$. We can choose e so that $a_i^{p^e} \in F$, $1 \leqslant i \leqslant m$. Then $z^{p^e} = \sum a_i^{p^e} \otimes c_i^{p^e} = 1 \otimes \sum_1^m a_i^{p^e} c_i^{p^e} \in 1 \otimes K$. If $z^{p^e} \neq 0$ then z^{p^e} and hence z is invertible. Otherwise, z is nilpotent.

Next let E/F be algebraic and K/F separably algebraically closed. Let S be the subfield of E/F of separable elements. Then E/S is purely inseparable. Now $E \otimes_F K \cong E \otimes_S (S \otimes_F K)$ (exercise 13 (iv), p. 148). Since S/F is separable algebraic and K/F is separable algebraically closed, $S \otimes_F K$ is a field by (2). Since E/S is purely inseparable it follows from the first part of this proof that the elements of $E \otimes_S (S \otimes_F K)$ are either nilpotent or units. Hence this holds for $E \otimes_F K$. □

We consider next tensor products of fields in which one of the factors is purely transcendental.

THEOREM 8.47. *Let E/F be purely transcendental, say, $E = F(B)$ where B is a transcendency base and let K/F be arbitrary. Then $E \otimes_F K$ is a domain and its field of fractions Q is purely transcendental over $K = 1 \otimes_F K$ with $B = B \otimes 1$ as transcendency base. Moreover, if F is algebraically closed (separably algebraically* *closed) in K then $E = F(B)$ is algebraically closed (separably algebraically closed) in $Q = K(B)$.*

Proof. For simplicity of notation we identify E and K with the corresponding subfields $E \otimes 1$ and $1 \otimes K$ of $E \otimes_F K$. These are linearly disjoint over F in $A = E \otimes_F K$. A consequence of this is that if a subset S of E is algebraically independent over F then it is algebraically independent in A over K. It suffices to see this for $S = \{s_1, \dots, s_m\}$. In this case algebraic independence over F is equivalent to the condition that the monomials $s_1^{k_1} \dots s_m^{k_m}$, $k_i \geqslant 0$ are distinct and linearly independent over F. Since this carries over on replacing F by K it follows that S is algebraically independent over K. In particular, this holds for the transcendency base B of E. Consider the subalgebra $K[B]$. If C is a finite subset of B then $K[C]$ is a domain (Theorem 2.13 of BAI, p. 128). It follows that $K[B]$ is a domain and this is a subalgebra of $A = F(B) \otimes_F K$. Let $z \in A$. Then $z = \sum a_i c_i$, $a_i \in F(B)$, $c_i \in K$. We can write $a_i = p_i q^{-1}$ where p_i, $q \in F[B]$. Then $z = pq^{-1}$ where $p = \sum p_i c_i \in K[B]$. Conversely, if $p \in K[B]$ and $q \in F[B]$, $q \neq 0$, then p and $q^{-1} \in A$ so $pq^{-1} \in A$. It follows that A is the localization $K[B]_{F[B]^*}$ of $K[B]$ with respect to the multiplicative monoid $F[B]^*$ of non-zero elements of $F[B]$. Since $K[B]$ is a domain, its localization $K[B]_{F[B]^*}$ is a domain. Moreover, this is a subalgebra of the localization $K[B]_{K[B]^*}$ which is the field of fractions Q of $K[B]$ and of A. Evidently $Q = K(B)$, the subfield of Q/K generated by B. Since B is algebraically independent over K we see that Q is purely transcendental over K with transcendency base B. This proves the first assertion of the theorem.

To prove the second assertion we shall show that if $K(B)$ contains an element that is algebraic (separable algebraic) over $F(B)$ not contained in $F(B)$, then K contains an element that is algebraic (separable algebraic) over F not contained in F. Clearly, if such an element exists it exists in $K(C)$ for some finite subset C of B. Hence it suffices to prove the result for finite B and then by induction, it is enough to prove the result for $K(x)$, x an indeterminate. Hence suppose $z \in K(x)$ is algebraic over $F(x)$ and $z \notin F(x)$. Let $\lambda^n + b_1\lambda^{n-1} + \cdots + b_n$ be the minimum polynomial of z over $F(x)$ so $n > 1$. Write $b_i = p_i q^{-1}$, p_i, $q \in F[x]$. Then $w = qz$ has minimum polynomial $\lambda^n + p_1\lambda^{n-1} + qp_2\lambda^{n-2} + \cdots + q^{n-1}p_n$. Replacing z by qz we may assume the $b_i \in F[x]$. Now write $z = rs^{-1}$ where r, $s \in K[x]$ are relatively prime. Then we have

$$-r^n = b_1 r^{n-1}s + b_2 r^{n-2}s^2 + \cdots + b_n s^n. \tag{120}$$

If $\deg s > 0$ then s has a prime factor in $K[x]$. By (120) this is also a factor of r^n, hence of r, contrary to the relative primeness of r and s. Thus s is a unit so we may assume $z \in K[x]$ so $z = z(x) = c_0 + c_1x + \cdots + c_mx^m, c_i \in K$. We claim that the c_i are algebraic over F. We have the relation $z(x)^n + b_1z(x)^{n-1} + \cdots + b_n = 0$ where $b_i = b_i(x) \in F[x]$. Hence for every $a \in F$ we have

$$z(a)^n + b_1(a)z(a)^{n-1} + \cdots + b_n(a) = 0. \tag{121}$$

Since $b_i(a) \in F$ this shows that the element $z(a)$ of K is algebraic over F. If F is infinite we choose $m + 1$ distinct elements $a_1, a_2, \ldots, a_{m+1}$ in F and write

$$\begin{aligned} z(a_1) &= c_0 + c_1a_1 + \cdots + c_ma_1^m \\ z(a_2) &= c_0 + c_1a_2 + \cdots + c_ma_2^m \\ &\ \vdots \\ z(a_m) &= c_0 + c_1a_m + \cdots + c_ma_m^m. \end{aligned} \tag{122}$$

Since the a_j are distinct the Vandermonde determinant $\det(a_k^j) \neq 0$. Hence we can solve (122) for the c's by Cramer's rule to show that every c_k is a rational expression with integer coefficients in the a_i and the $z(a_j)$. Since the a_i and $z(a_j)$ are algebraic over F it follows that every c_k is algebraic over F. If F is finite we replace F by its algebraic closure $\bar{F}$ which is infinite. Then the argument shows that every c_k is algebraic over $\bar{F}$ and since $\bar{F}$ is algebraic over F it follows again that the c_k are algebraic over F. Since $z \notin F(x)$, some $c_k \notin F$ and hence we have an element of K that is algebraic over F and is not contained in F.

Finally, we suppose that z is separable algebraic over F. Then $F(c_0, \ldots, c_m)$ contains an element not in F that is separable algebraic over F. Otherwise, this field is purely inseparable over F and hence there exists a p^f, p the characteristic such that $c_k^{p^f} \in F$, $0 \leqslant k \leqslant m$. Then $z^{p^f} = c_0^{p^f} + c_1^{p^f}x^{p^f} + \cdots + c_m^{p^f}x^{mp^f} \in F(x)$ contrary to the separability of z over F. Thus if $K(x)$ contains an element that is separable algebraic over $F(x)$ and is not in $F(x)$, then K contains an element separable algebraic over F not in F. □

In our next result we weaken the hypothesis that E/F is purely transcendental to separability. Then we have the following

THEOREM 8.48. *Let E/F be separable, K/F arbitrary. Then $E \otimes_F K$ has no non-zero nilpotent elements.*

Proof. It is clear that it suffices to prove the theorem in the case in which E/F is finitely generated. In this case E has a transcendency base B over F such that E is separable algebraic over $F(B)$ (Theorem 8.39). Then $E \otimes_F K \cong E \otimes_{F(B)} (F(B) \otimes_F K)$. By the last result $F(B) \otimes_F K$ is a domain. If Q is its field of fractions then $E \otimes_{F(B)} (F(B) \otimes_F K)$ is a subalgebra of $E \otimes_{F(B)} Q$. Since E is separable algebraic over $F(B)$, $E \otimes_{F(B)} Q$ has no nilpotent elements $\neq 0$, by

Theorem 8.46 (2). Hence $E \otimes_{F(B)} (F(B) \otimes_F K)$ has no non-zero nilpotent elements and this is true also of $E \otimes_F K$. □

Next we consider the situation in which F is separably algebraically closed in one of the factors.

THEOREM 8.49. *Let F be separably algebraically closed in E and let K/F be arbitrary. Then every zero divisor of $E \otimes_F K$ is nilpotent.*

Proof. Let B be a transcendency base for K/F. Then $E \otimes_F K \cong (E \otimes_F F(B)) \otimes_{F(B)} K$. By the last result $E \otimes_F F(B)$ is a domain and $F(B)$ is separably algebraically closed in the field of fractions Q of $E \otimes_F F(B)$. Since $F(B)$ is separably algebraically closed in Q and K is algebraic over $F(B)$, it follows from Theorem 8.46 (3) that the elements of $Q \otimes_{F(B)} K$ are either invertible or nilpotent. Now let $z \in (E \otimes_F F(B)) \otimes_{F(B)} K$ be a zero divisor in this algebra. Then z is a zero divisor in the larger algebra $Q \otimes_{F(B)} K$. Hence z is not invertible in $Q \otimes_{F(B)} K$ so z is nilpotent. Since $E \otimes_F K \cong (E \otimes_F F(B)) \otimes_{F(B)} K$ it follows that every zero divisor of $E \otimes_F K$ is nilpotent. □

We can now prove our main result on the question as to when the tensor product of two fields is a domain.

THEOREM 8.50. *Let E/F and K/F be extension fields of F. Assume* (1) *either E/F or K/F is separable and* (2) *F is separably algebraically closed in either E or K. Then $E \otimes_F K$ is a domain.*

Proof. By the last result if one of the factors has the property that F is separably algebraically closed in it then the zero divisors of $E \otimes_F K$ are nilpotent. On the other hand, by Theorem 8.48, if one of the factors is separable then $E \otimes_F K$ has no non-zero nilpotent elements. Hence $E \otimes_F K$ has no zero divisor $\neq 0$. □

A class of extension fields that is important in algebraic geometry (see Weil's *Foundations of Algebraic Geometry*, American Mathematical Society Colloquium Publication v. XXIX, 1946 and 1960) is given in the following

DEFINITION 8.6. *An extension field E/F is called* regular *if* (1) *E is separable over F and* (2) *F is algebraically closed in E.*

We remark that a separable extension field E/F contains no purely inseparable subfield. Hence we can replace condition (2) in the definition of regularity by: (2′) F is separably algebraically closed in E. The sufficiency part of the following theorem is a special case of Theorem 8.50.

THEOREM 8.51. *An extension field E/F is regular if and only if $E \otimes_F K$ is a domain for every field K/F.*

Proof. It remains to prove the necessity of the two conditions. The necessity of separability follows from Theorem 8.46 (1). Now suppose F is not algebraically closed in E. Then E contains a finite dimensional subfield $K \supsetneqq F$. Then $E \otimes_F K$ contains $K \otimes_F K$ that is not a field. Since $K \otimes K$ is finite dimensional it is not a domain. Hence we have the necessity of condition (2). □

One readily sees that if F is algebraically closed then any extension field E/F is regular. Then $E \otimes_F K$ is a domain for any K/F.

In the situation in which $E \otimes_F K$ is a domain for the extension fields E/F and K/F we shall denote the field of fractions of $E \otimes_F K$ by $E \cdot K$ (or $E \cdot_F K$).

EXERCISES

1. Show that if E/F is purely transcendental then E/F is regular.
2. Show that if E_1/F and E_2/F are regular then $E_1 \cdot E_2$ is regular.

8.19 FREE COMPOSITES OF FIELDS

Given two extension fields E/F and K/F, a natural question to ask is: What are the possible fields over F that can be generated by subfields isomorphic to E/F and K/F, respectively? To make this precise we define the *composite of E/F and K/F* as a triple (Γ, s, t) where Γ is a field over F and s and t are monomorphisms of E/F and K/F, respectively, into Γ/F such that Γ/F is generated by the subfields $s(E)$ and $t(K)$, that is, $\Gamma = F(s(E), t(K))$. The composites (Γ, s, t) and (Γ', s', t') of E/F and K/F are said to be *equivalent* if there exists an isomorphism $u: \Gamma \to \Gamma'$ such that the following two diagrams are commutative:

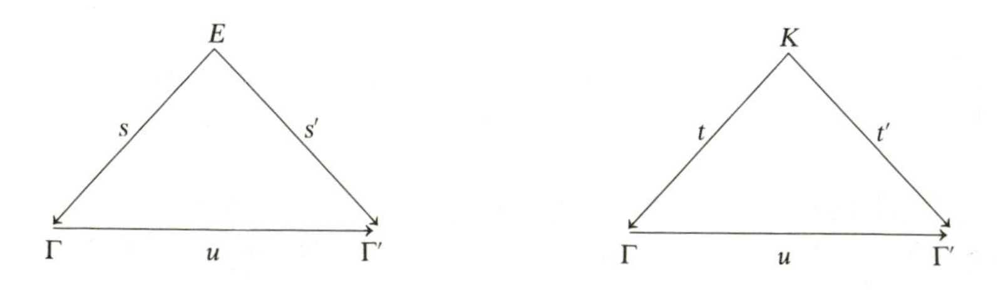

Of particular interest for algebraic geometry are the composites that are free in the sense of the following

DEFINITION 8.7. *A field composite* (Γ, s, t) *of* E/F *and* K/F *is called* free *if for any algebraically independent subsets* C *and* D *of* E/F *and* K/F, *respectively,* $s(C) \cap t(D) = \varnothing$ *and* $s(C) \cup t(D)$ *is algebraically independent in* Γ/F.

Since any algebraically independent subset can be imbedded in a transcendency base, it is clear that the condition that (Γ, s, t) is free is equivalent to the following: for every pair of transcendency bases B and B' of E/F and K/F respectively, $s(B) \cap t(B') = \varnothing$ and $s(B) \cup t(B')$ is algebraically independent. We now observe that the word "every" can be replaced by "some" in this condition; for, we have

LEMMA 1. *A composite* (Γ, s, t) *of* E/F *and* K/F *is free if and only if there exist transcendency bases* B *and* B' *of* E/F *and* K/F *respectively such that* $s(B) \cap t(B') = \varnothing$ *and* $s(B) \cup t(B')$ *is algebraically independent in* Γ/F.

Proof. The necessity of the condition is clear. To prove sufficiency, suppose we have transcendency bases B and B' for E/F and K/F such that $s(B) \cap t(B') = \varnothing$ and $s(B) \cup t(B')$ is algebraically independent over F. To prove freeness of (Γ, s, t), it obviously suffices to show that if B_1 is another transcendency base for E/F, then $s(B_1) \cap t(B') = \varnothing$ and $s(B_1) \cup t(B')$ is algebraically independent over F. Hence it suffices to show that if C is a finite algebraically independent subset of E/F and D is a finite subset of B', then $s(C) \cap t(D) = \varnothing$ and $s(C) \cup t(D)$ is algebraically independent over F. Now there exists a finite subset G of B such that C is algebraically dependent over F on G. Obviously $s(G) \cap t(D) = \varnothing$ and $s(G) \cup t(D)$ is a transcendency base for $F(s(G), s(C), t(D))/F$, so $\operatorname{tr\,deg} F(s(G), s(C), t(D))/F = |G| + |D|$. We also have $\operatorname{tr\,deg} F(s(G), s(C))/F = \operatorname{tr\,deg} F(s(G))/F = |G|$ and $\operatorname{tr\,deg} F(s(C))/F = |C|$. It follows that $\operatorname{tr\,deg} F(s(G), s(C), t(D))/F(s(C), t(D)) \leqslant |G| - |C|$ (see exercise 1, p. 517). From this and the above formula for $\operatorname{tr\,deg} F(s(G), s(C), t(D))/F$, we know that

$$\operatorname{tr\,deg} F(s(C), t(D))/F \geqslant (|G| + |D|) - (|G| - |C|) = |D| + |C|.$$

Hence $s(C) \cap t(D) = \varnothing$ and $s(C) \cup t(D)$ is algebraically independent over F. Hence (Γ, s, t) is free. $\square$

We note also that if B and B' are transcendency bases for E/F and K/F satisfying the conditions of the lemma then $s(B) \cup t(B')$ is a transcendency base for Γ/F. This is clear since the elements of $s(E)$ and $t(K)$ are algebraic over

$F(s(B), t(B'))$ and since Γ is generated by $s(E)$ and $t(K)$, it follows that Γ is algebraic over $F(s(B), t(B'))$. Hence $s(B) \cup t(B')$ is a transcendency base for Γ/F.

We can use these results to construct a free composite for any two given fields E/F, K/F. Let B and B' be transcendency bases for E/F and K/F. Suppose first that B and B' are finite, say, $B = \{\xi_1, \dots, \xi_m\}$, $B' = \{\zeta_1, \dots, \zeta_n\}$. Let Ω be the algebraic closure of the field $F(x_1, \dots, x_{m+n})$ where the x_i are indeterminates. We have monomorphisms s' and t' of $F(\xi_1, \dots, \xi_m)/F$ and $F(\zeta_1, \dots, \zeta_n)/F$, respectively, into Ω such that $s'\xi_i = x_i$, $1 \leqslant i \leqslant m$, and $t'\zeta_j = x_{m+j}$, $1 \leqslant j \leqslant n$. Since E is algebraic over $F(B)$ and Ω is algebraically closed, s' can be extended to a monomorphism s of E/F into Ω/F (exercise 1, p. 475). Similarly, t' can be extended to a monomorphism t of K/F into Ω. Then it is readily seen that if $\Gamma = F(sE, tK)$ then (Γ, s, t) is a free composite of E and K.

If either B or B' is infinite we modify the procedure used for B and B' finite as follows. We may assume $|B'| \geqslant |B|$. Then let X be a set disjoint from B and B' such that $|X| = |B'|$. We can decompose X as a disjoint union of two sets Y and Z such that $|Y| = |B|$, $|Z| = |X| = |B'|$. Let $F(X)$ be the field of fractions of the polynomial algebra $F[X]$. Let Ω be the algebraic closure of $F(X)$. Then, as before, we can define a monomorphism s of E/F into Ω/F whose restriction to B is a bijective map of B onto Y and a monomorphism t of K/F into Ω/F whose restriction to B' is a bijective map of B' onto Z. Then (Γ, s, t) for $\Gamma = F(sE, tK)$ is a free composite of E and K.

We wish to give a survey of the isomorphism classes of the free composites of two given fields E/F and K/F. First, we consider the composites of E and K that need not be free. We form $E \otimes_F K$ and let P be a prime ideal in this F-algebra. Then $(E \otimes_F K)/P$ is a domain whose field of fractions we denote as Γ_P. We have the homomorphism $s_p : a \rightsquigarrow a + P$ $(a = a \otimes 1)$ of E/F into $(E \otimes_F K)/P$ and hence into Γ_P/F. Since E is a field this is a monomorphism. Similarly, we have the monomorphism $t_P : b \rightsquigarrow b + P$ of K/F into Γ_P/F. Since $E \otimes_F K$ is generated by E and K, $(E \otimes_F K)/P$ is generated by its subalgebras $s_P(E)$, $t_P(K)$, and hence Γ_P is generated as a field by $S_P(E)$, $t_P(K)$. Thus (Γ_P, s_P, t_P) is a composite of E/F and K/F.

We note next that distinct prime ideals P and P' of $E \otimes_F K$ define inequivalent composites (Γ_P, s_P, t_P) and $(\Gamma_{P'}, s_{P'}, t_{P'})$. If these are equivalent then we have an isomorphism u of Γ_P onto $\Gamma_{P'}$ such that $s_{P'} = us_P$ and $t_{P'} = ut_P$. Then $u(a + P) = us_P a = s_{P'} a = a + P'$ for $a \in E$ and $u(b + P) = b + P'$ for $b \in K$. Then $u(\sum a_i b_i + P) = \sum a_i b_i + P'$, $a_i \in E$, $b_i \in K$. If $\sum a_i b_i \in P$ then $\sum a_i b_i + P = 0$ so $\sum a_i b_i + P' = 0$ and $\sum a_i b_i \in P'$. Thus $P \subset P'$ and by symmetry $P' \subset P$. Hence $P = P'$ contrary to our hypothesis.

Now let (Γ, s, t) be any composite of E/F and K/F. The homomorphisms s and t give rise to the homomorphism $s \otimes t$ of $E \otimes_F K$ into Γ which maps

$\sum a_i b_i$, $a_i \in E$, $b_i \in K$, into $\sum s(a_i)t(b_i)$ of Γ. Since the image is a domain the kernel is a prime ideal P. Hence we have the isomorphism u of $(E \otimes_F K)/P$ onto the subalgebra $s(E)t(K)$, which can be extended to an isomorphism u of the field of fractions Γ_P of $(E \otimes_F K)/P$ onto $\Gamma = F(s(E), t(E))$. It is clear that u is an equivalence of composites. We have therefore established a 1-1 correspondence between the set of prime ideals of $E \otimes_F K$ and the set of equivalence classes of composites of E/F and K/F.

It remains to sort out the composites (Γ_P, s_P, t_P) that are free. For this we shall prove

THEOREM 8.52. *The composite (Γ_P, s_P, t_P) is free if and only if every element of P is a zero divisor in $E \otimes_F K$.*

We shall identify E and K with their images in $E \otimes_F K$ so we regard E and K as subfields of $E \otimes_F K$ with the property that if S and S' are linearly independent subsets of E/F and K/F respectively then the map $(s, s') \rightsquigarrow ss'$ is a bijection of $S \times S'$ with SS', and SS' is a linearly independent subset of $E \otimes_F K$.

Let B and B' be transcendency bases for E/F and K/F, respectively, and let M and M' be the submonoids of the multiplicative monoids of E and K generated by B and B'. Since B and B' are algebraically independent, M and M' are linearly independent in E/F and K/F and hence MM' is linearly independent in $E \otimes_F K$. Now the subspace FMM' spanned by MM' is the subalgebra $F[B \cup B']$ and also the subalgebra $F[B]F[B']$. Thus MM' is a base for $F[B \cup B'] = F[B]F[B']$. Since $B \cup B'$ is algebraically independent, $F[B \cup B']$ is a domain. The field of fractions $F(B \cup B')$ of this domain contains the subfields $F(B)$ and $F(B')$ and it is readily seen that the subalgebra $F(B)F(B')$ generated by $F(B)$ and $F(B')$ is the set of fractions $a(cc')^{-1}$ where $a \in F[B \cup B'] = F[B]F[B']$, $c \in F[B]$ and $c' \in F[B']$.

We can now state

LEMMA 2. *The composite (Γ_P, s_P, t_P) is free if and only if $F(B)F(B') \cap P = 0$.*

Proof. By Lemma 1, (Γ_P, s_P, t_P) is free if and only if $s_P(B) \cap t_P(B') = \varnothing$ and $s_P(B) \cup t_P(B')$ is algebraically independent over F. Since $s_P(B) = \{b + P \mid b \in B\}$ and $t_P(B') = \{b' + P \mid b' \in B'\}$ these conditions hold if and only if the cosets $q + P$, $q \in MM'$, are linearly independent, where M and M' are as above. Hence (Γ_P, s_P, t_P) is free if and only if $FMM' \cap P = 0$. Since $FMM' = F[B]F[B']$ this condition is $F[B]F[B'] \cap P = 0$. Since $F(B)F(B') = \{a(cc')^{-1} \mid a \in F[B]F[B'], c \in F[B], c' \in F[B']\}$, $F[B]F[B'] \cap P = 0 \Leftrightarrow F(B)F(B') \cap P = 0$. Hence (Γ_P, s_P, t_P) is free if and only if $F(B)F(B') \cap P = 0$. $\square$

We require also

LEMMA 3. *$E \otimes_F K$ is integral over $F(B)F(B')$.*

Proof. Since B is a transcendency base for E, E is algebraic, hence, integral over $F(B)$. À fortiori E is integral over $F(B)F(B')$. Similarly, K is integral over $F(B)F(B')$. Since the subset of $E \otimes_F K$ of elements integral over $F(B)F(B')$ is a subalgebra, it follows that $E \otimes_F K = EK$ is integral over $F(B)F(B')$. □

We can now give the

Proof of Theorem 8.52. Suppose first that every element of P is a zero divisor in $E \otimes_F K$. Let $a \in P \cap F(B)F(B')$. Then a is a zero divisor in $E \otimes_F K$. We claim that a is a zero divisor in $F(B)F(B')$. To see this let $\{c_\alpha\}$ and $\{d_\beta\}$ be bases for $E/F(B)$ and $K/F(B')$ respectively, and let $\{m_\gamma\}$ and $\{n_\delta\}$ be bases for $F(B)/F$ and $F(B')/F$ respectively. Then $\{c_\alpha m_\gamma\}$ is a base for E/F and $\{d_\beta n_\delta\}$ is a base for K/F. Hence $\{c_\alpha d_\beta m_\gamma n_\delta\}$ is a base for $E \otimes_F K$ over F. Then every element of $E \otimes_F K$ can be uniquely written as a finite sum $\sum q_{\alpha\beta} c_\alpha d_\beta$ where $q_{\alpha\beta} \in F(B)F(B')$.

Now we have an element $\sum q_{\alpha\beta} c_\alpha d_\beta \neq 0$ such that $\sum a q_{\alpha\beta} c_\alpha d_\beta = a(\sum q_{\alpha\beta} c_\alpha d_\beta) = 0$. Since $a \in F(B)F(B')$ this implies that every $a q_{\alpha\beta} = 0$ and since $\sum q_{\alpha\beta} c_\alpha d_\beta \neq 0$ some $q_{\alpha\beta} \neq 0$ so a is a zero divisor in $F(B)F(B')$. Since $a \in F(B)F(B')$ and $F(B)F(B')$ is a domain this implies that $a = 0$. Hence we have proved that if every element of P is a zero divisor in $E \otimes_F K$ then $P \cap F(B)F(B') = 0$ and (Γ_P, s_P, t_P) is free by Lemma 2. Conversely, assume (Γ_P, s_P, t_P) is free so $P \cap F(B)F(B') = 0$. Let $b \in P$. Then b is integral over $F(B)F(B')$ and hence we have a relation $b^n + c_1 b^{n-1} + \cdots + c_n = 0$ for $c_i \in F(B)F(B')$ and we may assume n is minimal. Then we have $c_n = -b^n - \cdots - bc_{n-1} \in P$ since $b \in P$. Thus $c_n \in P \cap F(B)F(B')$ and hence $c_n = 0$. Then $b(b^{n-1} + c_1 b^{n-2} + \cdots + c_{n-1}) = 0$ and by the minimality of n, $b^{n-1} + c_1 b^{n-2} + \cdots + c_{n-1} \neq 0$ so b is a zero divisor. Thus freeness of (Γ_P, s_P, t_P) implies that every element of P is a zero divisor in $E \otimes_F K$. This completes the proof. □

We now consider an important special case of this theorem, namely, that in which F is separably algebraically closed in E. In this case, by Theorem 8.49, the zero divisors of $E \otimes_F K$ are nilpotent. Since the converse holds, nilrad $E \otimes_F K$ is the only prime ideal in $E \otimes_F K$ that consists of zero divisors. Hence we have

THEOREM 8.53. *If K/F is arbitrary and F is separably algebraically closed in E, then in the sense of equivalence there is only one free composite of E/F and K/F.*

This applies in particular if E is a regular extension of F. Then $E \otimes_F K$ is a domain for any K so nilrad $E \otimes_F K = 0$. Then the free composite of E/F and K/F is $E \cdot K$ which was defined to be the field of fractions of $E \otimes_F K$.

Finally, we consider a very simple case of composites that will be needed in valuation theory (see section 9.9), that in which one of the fields, say, E/F is finite dimensional. In this case all that we shall require of the foregoing discussion is the 1-1 correspondence between the set of prime ideals P of $E \otimes_F K$ and the equivalence classes of composites (Γ_P, s_P, t_P). We note also that since $E \otimes_F K$ is finite dimensional over K this ring is artinian. Hence the results we derived in section 7.11 are applicable. Using the fact that a domain that is finite dimensional over a field is a field we see that the prime ideals of $E \otimes_K F$ are maximal. By Theorem 7.13, there are only a finite number of these. Moreover, the proof of Theorem 7.13 shows how to determine the maximal ideals. We remark also that if E is algebraic over F, E/F has a vacuous transcendency base. It follows from Definition 8.7 that the composites of E/F with any K/F are free.

We now suppose E has a primitive element, say, $E = F(\theta)$. This is always true if E/F is finite dimensional separable (BAI, p. 291). We have the following

THEOREM 8.54. *Let K/F be arbitrary and let $E = F(\theta)$ where θ is algebraic over F with minimum polynomial $f(x)$ of degree n. Let $f_1(x), \dots, f_h(x)$ be the monic irreducible factors of $f(x)$ in $K[x]$. Then there are h inequivalent composites (Γ_i, s_i, t_i) of E/F and K/F where $\Gamma_i = K[x]/(f_i(x))$, s_i is the monomorphism of E/F into Γ_i/F such that $\theta \rightsquigarrow x + (f_i(x))$ and t_i is the monomorphism $b \rightsquigarrow b + (f_i(x))$ of K/F into Γ_i/F. Any composite of E/F and K/F is equivalent to one of these.*

Proof. We have the isomorphism of $E = F(\theta)$ over F onto $F[x]/(f(x))$ such that $\theta \rightsquigarrow \bar{x} = x + (f(x))$. As we have seen before p. 546), $E \otimes_F K \cong K[x]/(f(x))$. The maximal ideals of $K[x]/(f(x))$ have the form $(f_i(x))/(f(x))$ where $f_i(x)$ is a monic irreducible factor of $f(x)$ in $K[x]$. Then $\Gamma_i = E[x]/(f_i(x)) \cong (E(x)/(f(x)))/(f_i(x))/(f(x))$ is a field over F. We have the monomorphism s_i of E/F into Γ_i/F such that $\theta \rightsquigarrow x + (f_i(x))$ and the monomorphism t_i of K/F into Γ_i/F such that $b \rightsquigarrow b + (f_i(x))$, $b \in K$. Clearly $s_i(E)$ and $t_i(K)$ generate Γ_i. Hence (Γ_i, s_i, t_i) is a composite of E/F and K/F. The rest follows from what we proved before. □

EXERCISES

1. Determine the composites over $\mathbb{Q}$ of $\mathbb{R}$ with
   (1) $\mathbb{Q}(\omega)$, ω a primitive cube root of 1
   (2) $\mathbb{Q}(\sqrt{2})$

2. Determine the composites over $\mathbb{Q}$ of $\mathbb{Q}(\sqrt{-1})$ and $\mathbb{Q}(\sqrt{3})$.
3. Suppose E/F and K/F are finite dimensional Galois and $E \otimes_F K$ is a domain hence a field. Is $(E \otimes_F K)/F$ Galois?

REFERENCES

E. Steinitz, Algebraische Theorie den Körpern, *Journal für die reine und angewandete Mathematik*, vol. 137 (1909), pp. 167–309.

N. Jacobson, *Lectures in Abstract Algebra III. The Theory of Fields and Galois Theory*, 1964, New printing, Springer-Verlag, New York.

P. J. McCarthy, *Algebraic Extension of Fields*, Blaisdell, Waltham, Mass., 1966.

D. J. Winter, *The Structure of Fields*, Springer-Verlag, New York, 1974.

9

Valuation Theory

The valuation theory of fields that we consider in this chapter, like commutative ideal theory, had two sources: number theory and algebraic geometry. The number theoretic source was Hensel's discovery of p-adic numbers. Hensel introduced the ring of p-adic integers essentially as we have done in Chapter 2: as inverse limit of the rings $\mathbb{Z}/(p^n)$. Later Kürschák (1913) introduced the concept of a real valued valuation of a field and showed that Hensel's field of p-adic numbers could be viewed as the completion of $\mathbb{Q}$ relative to the p-adic valuation. The algebraic geometric source of the theory appears to have been the introduction of the concept of a place by Dedekind and Weber, for the purpose of providing a precise definition of the Riemann surface of an algebraic curve.

Valuation theory forms a solid link between number theory, algebra, and analysis. On the one hand, it permits a precise study of algebraic functions, and on the other hand, it leads to the introduction of analogues of classical analytic concepts in the study of arithmetic questions.

We begin our discussion with real valued valuations, which are now usually called absolute values. We can distinguish two types: archimedean and non-archimedean. The latter suggests a generalization to valuations into ordered abelian groups, a generalization that was introduced by Krull in 1934. In what follows we reserve the term "valuation" for Krull's valuations into ordered abelian groups. Valuations in this sense are equivalent to two other concepts: valuation rings and places. We study in some detail the problem of extension of absolute values and valuations to extension fields, especially finite dimensional extensions. In particular, we consider this problem for the field $\mathbb{Q}$.

We define the important concept of a local field that is of central interest in algebraic number theory and we give a topological characterization of these fields and of finite dimensional division algebras over local fields. Our method of treating these is an elementary one that yields at the same time a determination of the Brauer group of a local field. We consider also quadratic forms over local fields and define an invariant for these that together with the discriminant provides a complete classification of these forms.

9.1 ABSOLUTE VALUES

We assume familiarity with the properties of $\mathbb{R}$ as an ordered field such as are developed in a beginning calculus course (e.g., completeness, existence of logarithms, etc.). The concept of an absolute value on a field is a direct generalization of the absolute value defined on the field $\mathbb{C}$. This is given in

DEFINITION 9.1 *An* absolute value $|\ |$ *on a field F is a map* $a \rightsquigarrow |a|$ *of F into* $\mathbb{R}$ *such that*

(1) $|a| \geqslant 0$ *and* $|a| = 0$ *if and only if* $a = 0$.
(2) $|ab| = |a|\ |b|$.
(3) $|a+b| \leqslant |a| + |b|$ (*triangle inequality*).

As indicated earlier, the classical example is that of $\mathbb{C}$ in which $|a| = (\alpha^2+\beta^2)^{1/2}$ for $a = \alpha+\beta\sqrt{-1}$, $\alpha, \beta \in \mathbb{R}$. We shall use the notation $|\ |_\infty$ for this absolute value if it is necessary to distinguish it from other absolute values. On the subfields $\mathbb{R}$ and $\mathbb{Q}$ this reduces to the usual absolute value. We proceed to list some less familiar examples.

EXAMPLES

1. *p-adic absolute value of* $\mathbb{Q}$. Let p be a fixed prime in $\mathbb{Z}$. If $a \neq 0$ in $\mathbb{Q}$, we write $a = (b/c)p^k$ where $k \in \mathbb{Z}$ and $(b, p) = 1 = (c, p)$. The integer k is uniquely determined by a. We denote it as $v_p(a)$ and we define $v_p(0) = \infty$. Then it is easy to verify the following properties:

(i) $v_p(a) = \infty$ if and only if $a = 0$.

(ii) $v_p(ab) = v_p(a)+v_p(b)$.

(iii) $v_p(a+b) \geqslant \min(v_p(a), v_p(b))$.

Now let γ be a real number such that $0<\gamma<1$ and define a *p-adic absolute value* $|a|_p$ on $\mathbb{Q}$ by

$$|a|_p = \gamma^{v_p(a)} \tag{1}$$

(where $\gamma^\infty = 0$). Then it is clear that this is an absolute value in the sense of Definition 9.1. In fact, in place of the triangle inequality we have the stronger relation

$$|a+b| \leqslant \max(|a|, |b|) \tag{2}$$

for $|\ | = |\ |_p$.

2. Let $F(x)$ be the field of rational expressions in an indeterminate x and let $p(x)$ be a prime polynomial in $F[x]$. If $a \in F(x)$ and $a \neq 0$, we have $a = p(x)^k b(x)/c(x)$ where $k \in \mathbb{Z}$ and $(p(x), b(x)) = 1 = (p(x), c(x))$. We define $v_p(a) = k$, $v_p(0) = \infty$, and $|a|_p = \gamma^{v_p(a)}$ for some real γ, $0<\gamma<1$. Then properties 9.1.1 and 9.1.2 and equation (2) hold. Hence we have an absolute value on $F(x)$ defined by $|a|_p = \gamma^{v_p(a)}$.

An important special case of this type of absolute value is obtained by taking $F = \mathbb{C}$ and $p(x) = x-r$, $r \in \mathbb{C}$. An element $a \in \mathbb{C}(x)$ defines a rational function on the Riemann sphere $\mathbb{C} \cup \{\infty\}$ in the usual manner. If $v_p(a) = k > 0$, then this rational function has a zero of order k at r and if $v_p(a) = -k$, $k>0$, it has a pole of order k at r. Finally if $v_p(a) = 0$, then a has neither a zero nor a pole at r. Thus the value of $v_p(a)$ gives us information on the behavior of a in the neighborhood of the point r.

3. We obtain another absolute value on $F(x)$ in the following manner. If $a \neq 0$, we write $a = b(x)/c(x)$ where $b(x) = b_0+b_1x+\cdots+b_mx^m$, $c(x) = c_0+c_1x+\cdots+c_nx^n$, b_i, $c_i \in F$, $b_m \neq 0$, $c_n \neq 0$. Define $v_\infty(a) = n-m$, $v_\infty(0) = 0$, and $|a|_\infty = \gamma^{v_\infty(a)}$ where $\gamma \in \mathbb{R}$ and $0<\gamma<1$. We have

$$\begin{aligned} a &= \frac{x^m(b_0x^{-m}+b_1x^{-(m-1)}+\cdots+b_m)}{x^n(c_0x^{-n}+c_1x^{-(n-1)}+\cdots+c_n)} \\ &= \frac{b_0x^{-m}+b_1x^{-(m-1)}+\cdots+b_m}{c_0x^{-n}+c_1x^{-(n-1)}+\cdots+c_n}\,x^{-(n-m)}. \end{aligned}$$

Hence the definition of $|a|_\infty$ amounts to using the generator x^{-1} for $F(x)$ and applying the procedure in example 2 to $F[x^{-1}]$ with $p(x^{-1}) = x^{-1}$. Hence $|\ |_\infty$ is an absolute valuc.

In the special case of $F = \mathbb{C}$, $v_\infty(a)$ gives the behavior at ∞ of the rational function defined by a.

For any field F we have the *trivial absolute value* on F in which $|0| = 0$ and $|a| = 1$ if $a \neq 0$.

We now list some simple properties of absolute values that follow directly from the definition:

$$|1| = 1, \qquad |u| = 1 \quad \text{if } u^n = 1, \qquad |-a| = |a|,$$
$$|a^{-1}| = |a|^{-1} \quad \text{if } a \neq 0, \qquad \| a| - |b\|_\infty \leqslant |a-b|.$$

An absolute value on F defines a topology on F whose open sets are the unions of the spherical neighborhoods where such a neighborhood of a is defined by $\{x \mid |x-a| < r\}$ for some $r > 0$. It is easy to see that this gives a topology on F and that multiplication, addition, and subtraction are continuous functions of two variables in the topology. We can now define convergence of sequences and series in the usual way. Thus we say that $\{a_n \mid n = 1, 2, \ldots\}$ *converges* to a if for any real $\varepsilon > 0$ there exists an integer $N = N(\varepsilon)$ such that

$$|a - a_n| < \varepsilon$$

for all $n \geqslant N$. In this case we write also $\lim a_n = a$ or $a_n \to a$ ($a_n \xrightarrow[|\,|]{} a$ if it is necessary to indicate $|\ |$). The standard elementary facts on convergence of sequences in $\mathbb{C}$ carry over ($a_n \to a$ and $b_n \to b$ imply $a_n \pm b_n \to a \pm b$; $a_n b_n \to ab$; $|a_n| \to |a|$ in $\mathbb{R}$; if $a_n \to a \neq 0$, then $a_n \neq 0$ for sufficiently large k, say, $n \geqslant k$, and then $a_{n+k}^{-1} \to a^{-1}$, etc.). We define convergence of a series $\sum_1^\infty a_n$ and write $\sum a_n = s$ if $s_n \to s$ for the sequence of partial sums $s_n = a_1 + \cdots + a_n$.

It is natural to consider two absolute values $|\ |_1$ and $|\ |_2$ as equivalent if they define the same topology on F. For example, if $|\ |_p$ and $|\ |_p'$ are p-adic valuations defined by γ and γ' respectively, that is, $|a|_p = \gamma^{v_p(a)}$, $|a|_p' = \gamma'^{v_p(a)}$, then $|a|_p' = |a|_p^s$ for $s = \log \gamma'/\log \gamma > 0$. Hence any spherical neighborhood of a point defined by one of these absolute values is a spherical neighborhood defined by the other. Hence $|\ |_p$ and $|\ |_p'$ define the same topology. This is the case for any field F and any two absolute values $|\ |$ and $|\ |' = |\ |^s$ where s is a positive real number. We remark also that the topology defined by an absolute value $|\ |$ is discrete if and only if $|\ |$ is trivial: It is clear that the trivial $|\ |$ defines the discrete topology. On the other hand, if $|\ |$ is not discrete, then we have an a such that $0 < |a| < 1$. Then $a^n \xrightarrow[|\,|]{} 0$ and the set of points $\{a^n\}$ is not closed in F, so the topology is not discrete. It is now clear that the only absolute value equivalent to the discrete one $|\ |$ is $|\ |$ itself. For non-trivial absolute values we shall now show that equivalence can hold only if each absolute value is a positive power of the other. Moreover, equivalence can be assured by a simple one-sided condition. Both of these results follow from

THEOREM 9.1. *Let $|\ |_1$ and $|\ |_2$ be absolute values of a field F such that $|\ |_1$ is not trivial and $|a|_1 < 1$ for $a \in F$ implies that $|a|_2 < 1$. Then there exists a positive real number s such that $|\ |_2 = |\ |_1^s$ (that is, $|a|_2 = |a|_1^s$ for all $a \neq 0$). Hence $|\ |_1$ and $|\ |_2$ are equivalent.*

Proof (Artin). The hypothesis implies that if $|a|_1 < |b|_1$, then $|a|_2 < |b|_2$ and hence $|a|_1 > 1 = |1|_1$ implies that $|a|_2 > 1$. Since $|\ |_1$ is non-trivial, we can choose an a_0 such that $|a_0|_1 > 1$ and hence $|a_0|_2 > 1$. Now let a be any element such that $|a|_1 > 1$ and hence $|a|_2 > 1$. Let

$$t = \log |a|_1 / \log |a_0|_1.$$

Then $t > 0$ and $|a|_1 = |a_0|_1{}^t$. We claim that $|a|_2 = |a_0|_2{}^t$ also. We have $|a|_2 = |a_0|_2{}^{t'}$ with $t' > 0$ and if $t' \neq t$, then there exists either a rational number r such that $r > t$ and $r < t'$ or a rational number r such that $r < t$ and $r > t'$. In the first case, $|a_0|_2{}^r < |a|_2$ and in the second $|a_0|_2{}^r > |a|_2$. Our claim can therefore be established by showing that if r is a rational number $> t$, then $|a_0|_2{}^r > |a|_2$ and if r is a positive rational number $< t$, then $|a_0|_2{}^r < |a|_2$. Write $r = m/n$ where m and n are positive integers and suppose first that $r > t$. Then $|a|_1 < |a_0|_1^{m/n}$ and $|a^n|_1 < |a_0{}^m|_1$. Then $|a^n|_2 < |a_0{}^m|_2$ and $|a|_2 < |a_0|_2^{m/n}$. Similarly, if $r = m/n < t$, then $|a|_2 > |a_0|_2^{m/n}$. Hence we have $|a|_2 = |a_0|_2{}^t$ and

$$t = \frac{\log |a|_2}{\log |a_0|_2} = \frac{\log |a|_1}{\log |a_0|_1}$$

and

$$\frac{\log |a|_2}{\log |a|_1} = \frac{\log |a_0|_2}{\log |a_0|_1}.$$

Then $|a|_2 = |a|_1{}^s$ for $s = \log |a_0|_2 / \log |a_0|_1 > 0$ and this holds for all a such that $|a|_1 > 1$. Then if $|a|_1 < 1$, we have $|a^{-1}|_2 = |a^{-1}|_1{}^s$ and hence also $|a|_2 = |a|_1{}^s$. Thus $|a|_2 = |a|_1{}^s$ and we have seen that this implies equivalence of the absolute values. □

We have observed that the p-adic absolute value on $\mathbb{Q}$ and the absolute values on $F(x)$ defined in examples 1 and 2 satisfy the stronger triangle inequality that $|a+b| \leq \max(|a|, |b|)$. We call an absolute value *non-archimedean* if it has this property; otherwise we say that the absolute value is *archimedean*. This distinction can be settled by looking at the prime ring; for, we have

THEOREM 9.2. *An absolute value* $|\ |$ *of a field* F *is non-archimedean if and only if* $|n1| \leq 1$ *for all* $n \in \mathbb{Z}$.

Proof (Artin). If $|\ |$ is non-archimedean, then $|a_1 + \cdots + a_n| \leq \max |a_i|$. Hence $|1 + \cdots + 1| \leq |1| = 1$ and $|n1| \leq 1$ for all $n \in \mathbb{Z}$. Conversely, suppose this holds and let $a, b \in F$. Then for any positive integer n we have

$$\begin{aligned}|a+b|^n &= |a^n + \tbinom{n}{1}a^{n-1}b + \cdots + b^n| \\ &\leq |a|^n + |a^{n-1}||b| + \cdots + |b|^n \\ &\leq (n+1)\max(|a|^n, |b|^n).\end{aligned}$$

Hence $|a+b| \leqslant (n+1)^{1/n} \max(|a|, |b|)$. Since $\lim (n+1)^{1/n} = 1$ in $\mathbb{R}$, this implies $|a+b| \leqslant \max(|a|, |b|)$. □

A consequence of this criterion is

COROLLARY 1. *Any absolute value on a field of characteristic $p \neq 0$ is non-archimedean.*

Proof. If $n1$ is in the prime ring and $n1 \neq 0$, then $(n1)^{p-1} = 1$ and hence $|n1| = 1$. Also $|0| = 0$. Hence $|\ |$ is non-archimedean, by Theorem 9.2. □

The trivial absolute value is non-archimedean. It is clear also from Theorem 9.2 that if $|\ |$ is non-archimedean on a subfield of a field F, then $|\ |$ is non-archimedean on F. Hence we have

COROLLARY 2. *If $|\ |$ is trivial on a subfield, then $|\ |$ is non-archimedean.*

EXERCISES

1. Show that if $|\ |$ is an absolute value and $0 < s < 1$, then $|\ |^s$ is an absolute value. Show also that if $|\ |$ is non-archimedean, then $|\ |^s$ is an absolute value for every $s > 0$.

2. Show that if $|\ |$ is non-archimedean, then $|a+b| = |a|$ if $|a| > |b|$. Show also that if $a_1 + \cdots + a_n = 0$, then $|a_i| = |a_j|$ for some $i \neq j$.

3. Let F be a field with a non-archimedean absolute value $|\ |$ and let $f(x) = x^n + a_1 x^{n-1} + \cdots + a_n \in F[x]$. Show that if $M = \max(1, |a_i|)$ then any zero ρ of $f(x)$ in F satisfies $|\rho| \leqslant M$.

4. Let $|\ |$ be an absolute value on E and assume that $|\ |$ is trivial on a subfield F such that E/F is algebraic. Show that $|\ |$ is trivial on E.

5. Let $|\ |$ be an absolute value on F and let σ be an automorphism. Show that $|\ |_\sigma$ defined by $|a|_\sigma = |\sigma a|$ is an absolute value. Apply this to $F = \mathbb{Q}(\sqrt{2})$ and $\mathbb{Q}(\sqrt{-1})$ with $\sigma \neq 1$. In each case determine whether or not $|\ |_\sigma$ is equivalent to $|\ |$.

9.2 THE APPROXIMATION THEOREM

In this section we prove a theorem on simultaneous approximation with respect to inequivalent absolute values that is an analogue of the Chinese remainder theorem in ring theory. This will imply a strong independence property for inequivalent absolute values. The result is the following

APPROXIMATION THEOREM (Artin–Whaples). *Let* $|\ |_1, \dots, |\ |_n$ *be inequivalent non-trivial absolute values on a field* F, $a_1, \dots, a_n$ *elements of* F, ε *a positive real number. Then there exists an* $a \in F$ *such that*

$$|a - a_k|_k < \varepsilon, \qquad 1 \leqslant k \leqslant n. \tag{3}$$

We shall prove first

THEOREM 9.3. *Let* $|\ |_1, \dots, |\ |_n$ *be inequivalent non-trivial absolute values on* F. *Then, for any* k, $1 \leqslant k \leqslant n$, *there exists an* $a_k \in F$ *such that*

$$|a_k|_k > 1, \qquad |a_k|_l < 1 \quad \text{if } l \neq k. \tag{4}$$

Proof. It suffices to prove this for $k = 1$. Hence we have to show that there is an a such that $|a|_1 > 1$ and $|a|_l < 1$ for every $l > 1$. First let $n = 2$. By Theorem 9.2 there is a $b \in F$ such that $|b|_1 < 1$ and $|b|_2 \geqslant 1$ and a $c \in F$ such that $|c|_2 < 1$ and $|c|_1 \geqslant 1$. Then if we put $a = cb^{-1}$ we have $|a|_1 > 1$ and $|a|_2 < 1$. Hence the result holds for $n = 2$ and we may assume it for $n - 1 \geqslant 2$. Then we have elements b and c such that

$$|b|_1 > 1, \qquad |b|_2 < 1, \dots, |b|_{n-1} < 1,$$
$$|c|_1 > 1, \qquad |c|_n < 1.$$

We distinguish two cases.

Case I. $|b|_n \leqslant 1$. Consider $a_r = b^r c$. We have $|a_r|_k = |b|_k{}^r |c|_k$. This is > 1 if $k = 1$ and < 1 if $k > 1$ and r is sufficiently large. For any such r put $a = a_r$. Then (4) holds.

Case II. $|b|_n > 1$. Here we take $a_r = b^r/(1 + b^r)c$. If $2 \leqslant k \leqslant n - 1$, $|b|_k < 1$ so $b^r \xrightarrow[|\ |_k]{} 0$. Then $a_r \xrightarrow[|\ |_k]{} 0$ and $|a_r|_k < 1$ for sufficiently large r. Next let $k = 1$ or n. Then $|b|_k > 1$ and hence $b^r/(1 + b^r) = 1/(1 + b^{-r}) \xrightarrow[|\ |_k]{} 1$ and $a_r \xrightarrow[|\ |_k]{} c$. Since $|c|_1 > 1$ and $|c|_n < 1$, we have $|a_r|_1 > 1$ and $|a_r|_n < 1$ for sufficiently large r. Hence for a suitable $a = a_r$ we have $|a|_1 > 1$, $|a|_2 < 1, \dots, |a|_n < 1$. $\square$

We can now prove the approximation theorem. Let $|\ |_1, \dots, |\ |_n$ be inequivalent and non-trivial and let $a_1, \dots, a_n$ be elements of F. For each k, $1 \leqslant k \leqslant n$, apply Theorem 9.3 to obtain an element b_k such that $|b_k|_k > 1$ and $|b_k|_l < 1$ for all $l \neq k$. Then

$$b_k{}^r/(1 + b_k{}^r) \xrightarrow[|\ |_k]{} 1$$
$$b_k{}^r/(1 + b_k{}^r) \xrightarrow[|\ |_l]{} 0 \quad \text{if } l \neq k.$$

Hence $a_k b_k{}^r/(1+b_k{}^r) \underset{|\ |_k}{\rightarrow} a_k$ and $a_k b_k{}^r(1+b_k{}^r) \underset{|\ |_l}{\rightarrow} 0$ for $l \neq k$. Then $\sum_{k=1}^{n} a_k b_k{}^r/(1+b_k{}^r) \underset{|\ |_j}{\rightarrow} a_j$ for every j, $1 \leqslant j \leqslant n$. Hence for any $\varepsilon > 0$ we can take

$$a = \sum_{k=1}^{n} a_k b_k{}^r/(1+b_k{}^r)$$

for a sufficiently large r and we shall have the required relations $|a - a_k|_k < \varepsilon$, $1 \leqslant k \leqslant n$. □

We have seen that if $|\ |_1$ and $|\ |_2$ are equivalent, there exists a positive real s such that $|\ |_2 = |\ |_1^s$. On the other hand, it is clear from the approximation theorem that if $|\ |_1, \dots, |\ |_n$ are inequivalent and non-trivial, then there exists no $(s_1, \dots, s_n) \neq (0, \dots, 0)$, s_i real, such that

$$|a|_1^{s_1} |a|_2^{s_2} \cdots |a|_n^{s_n} = 1$$

for all $a \in F$. For we may assume $s_1 = 1$. Then, by the Approximation Theorem, there exists an a_i such that $|a_i|_1 < 1/2^i$ and $|a_i - 1|_j < 1/2^i$ for $j > 1$. Then $|a_i|_1 \rightarrow 0$ and $|a_i|_j \rightarrow 1$ and hence the foregoing relation cannot hold for all a_i.

EXERCISE

1. If p is a prime, we define the *normalized p-adic* absolute value on $\mathbb{Q}$ by $|a|_p = p^{-v_p(a)}$. Let $|\ |_\infty$ be the usual absolute value. Show that $\prod_p |a|_p = |a|_\infty^{-1}$ for all $a \in \mathbb{Q}$ in the sense that for a given a only a finite number of $|a|_p$ are $\neq 1$ and the product of these is $|a|_\infty^{-1}$. This can also be written as $\prod_q |a|_q = 1$ if q ranges over the primes $p \in \mathbb{Z}$ and ∞.

9.3 ABSOLUTE VALUES ON $\mathbb{Q}$ and $F(x)$

We shall now determine all of the absolute values on $\mathbb{Q}$ and all of the absolute values on a field $F(x)$, x transcendental, that are trivial on the subfield F. We begin with $\mathbb{Q}$ and we prove

THEOREM 9.4. *Any archimedean absolute value $|\ |$ on $\mathbb{Q}$ is equivalent to the ordinary absolute value $|\ |_\infty$.*

Proof (Artin). Let n and n' be integers > 1 and write

$$n' = a_0 + a_1 n + \cdots + a_k n^k, \qquad 0 \leqslant a_i < n.$$

Then

$$|n'| < n(1+|n|+\cdots+|n|^k) \leqslant n(k+1)\max(1, |n|^k)$$

and since $n' \geqslant n^k$, $\log n' \geqslant k \log n$, $k \leqslant \log n'/\log n$, so

$$|n'| < n\left(\frac{\log n'}{\log n}+1\right)\max(1, |n|^{\log n'/\log n}).$$

Replacing n' by n'^r for any positive integer r we obtain

$$|n'^r| < n\left(\frac{r\log n'}{\log n}+1\right)\max(1, |n|^{r\log n'/\log n}).$$

Then

$$|n'| < \left(n\frac{r\log n'}{\log n}+1\right)^{1/r}\max(1, |n|^{\log n'/\log n}).$$

For any real numbers a and b, $\lim_{r\to\infty}(ra+b)^{1/r}=1$ since $\lim_{r\to\infty} 1/r\log(ra+b)=0$. Hence the last inequality gives

$$|n'| \leqslant \max(1, |n|^{\log n'/\log n}) \tag{5}$$

for any n, n' such that n, $n' > 1$. Since $|\ |$ is archimedean, there exists an n' such that $|n'|>1$ (Theorem 9.2). Then $|n'| \leqslant |n|^{\log n'/\log n}$ so $|n|>1$ for all $n>1$. Then (5) holds for all n, $n'>1$. Hence

$$|n'|^{1/\log n'} \leqslant |n|^{1/\log n}$$

for all n, $n'>1$. By symmetry, we have

$$|n'|^{1/\log n'} = |n|^{1/\log n}.$$

Then $\log|n'|/\log n' = \log|n|/\log n = s > 0$ and so $|n| = n^s$ for all integers $n>1$. It follows that $|a| = |a|_\infty{}^s$ for all $a \neq 0$ in $\mathbb{Q}$ and hence $|\ |$ is equivalent to $|\ |_\infty$. □

We determine next the non-archimedean absolute values on $\mathbb{Q}$.

THEOREM 9.5. *Any non-trivial non-archimedean absolute value of $\mathbb{Q}$ is a p-adic absolute value for some prime p.*

Proof. We have $|n| \leqslant 1$ for every integer n. If $|n|=1$ for all $n\neq 0$ in $\mathbb{Z}$, then $|\ |$ is trivial. Hence the set P of integers b such that $|b|<1$ contains non-zero elements. Now P is an ideal in $\mathbb{Z}$, since $|b_1+b_2| \leqslant \max(|b_1|, |b_2|) < 1$

if $b_i \in P$ and $|nb| = |n|\ |b| < 1$ if $n \in \mathbb{Z}$ and $b \in P$. Also P is prime since $|n| = 1$ and $|n'| = 1$ imply that $|nn'| = 1$. Hence $P = (p)$ for some prime $p > 0$. Now put $\gamma = |p|$ so $0 < \gamma < 1$. If $r \in \mathbb{Q}$, we can write $r = p^k a/b$ where $k \in \mathbb{Z}$ and $a, b \notin (p)$. Then $|a| = 1 = |b|$ and $|r| = \gamma^k = \gamma^{v_p(p)}$. Hence $|\ |$ is the p-adic absolute value defined by γ. □

We consider next the case of $F(x)$ and we prove

THEOREM 9.6. *Let* $|\ |$ *be a non-trivial absolute value on* $F(x)$, x *transcendental, that is trivial on* F. *Then* $|\ |$ *is one of the absolute values defined in examples* 1 *and* 2 *of section* 9.1.

Proof. Since $|\ |$ is trivial on F, it is non-archimedean. We distinguish two cases.

Case I. $|x| \leqslant 1$. In this case, the fact that $|\ |$ is non-archimedean and trivial on F implies that $|b| \leqslant 1$ for every $b \in F[x]$. Since $|\ |$ is not trivial on $F(x)$ we must have a $b \in F[x]$ such that $0 < |b| < 1$. The argument proceeds as in the proof of Theorem 9.5. We let P be the subset of $b \in F[x]$ such that $|b| < 1$. This is an ideal (p) where $p = p(x)$ is a prime polynomial. Then one sees that $|\ |$ is an absolute value determined by $p(x)$ as in example 2, p. 539.

Case II. $|x| > 1$. Let $b = b_0 + b_1 x + \cdots + b_m x^m$, $b_i \in F$, $b_m \neq 0$. Then $|b_m x^m| = |x|^m > |b_i x^i|$ for $i < m$. Hence $|b| = |x|^m$ (exercise 2, p. 542). If we put $|x| = \gamma^{-1}$, $0 < \gamma < 1$, it is easy to check that $|\ |$ is an absolute value $|\ |_\infty$ as defined in example 3, p. 539. □

9.4 COMPLETION OF A FIELD

The familiar concepts of convergence, Cauchy sequence, and completeness in $\mathbb{R}$ and $\mathbb{C}$, which we have also encountered in considering topologies defined by ideals in a ring (section 7.17), carry over to fields with an absolute value. For arbitrary F and $|\ |$ we have the following definitions.

DEFINITION 9.2. *Let* F *be a field with an absolute value* $|\ |$. *A sequence* $\{a_n\}$ *of elements of* F *is called a* Cauchy sequence *if given any real* $\varepsilon > 0$, *there exists a positive integer* $N = N(\varepsilon)$ *such that*

$$|a_m - a_n| < \varepsilon \tag{6}$$

for all $m, n \geqslant N$. F *is said to be* complete *relative to* $|\ |$ *if every Cauchy sequence of elements of* F *converges* (lim a_n *exists*).

As in the case of $\mathbb{C}$, it is clear that if $a_n \to a$, then $\{a_n\}$ is a Cauchy sequence

We shall now show that any field F with an absolute value $|\ |$ has a *completion* $\hat{F}$ in the sense that

(1) $\hat{F}$ is an extension field of F and has an absolute value that is an extension of the given absolute value.
(2) $\hat{F}$ is complete relative to the absolute value.
(3) F is dense in $\hat{F}$ relative to the topology provided by the absolute value.

Since the steps of the construction of $\hat{F}$ are almost identical with familiar ones for real numbers and for metric spaces, we shall just sketch these and leave the verifications to the reader. We begin with the set $C = C(F)$ of Cauchy sequences of elements of F. This can be made into a commutative ring extension of F. If $\{a_n\}, \{b_n\} \in C$, we define $\{a_n\} + \{b_n\} = \{a_n + b_n\}$, $\{a_n\}\{b_n\} = \{a_n b_n\}$. These are contained in C. If $a \in F$, we let $\{a\}$ be the constant sequence all of whose terms are a. Then $(C, +, \cdot, \{0\}, \{1\})$ is a commutative ring containing the subring of constant sequences that is isomorphic under $a \rightsquigarrow \{a\}$ with F.

A sequence $\{a_n\}$ is called a *null* sequence if $a_n \to 0$. Let B be the set of null sequences. It is easily seen that B is an ideal in C. It is clear that $B \cap \{F\} = \{0\}$ where $\{F\}$ is the set of constant sequences $\{a\}$, $a \in F$. Hence B is a proper ideal in C. We have the

LEMMA. *B is a maximal ideal of C.*

Proof. Let B' be an ideal of C such that $B' \supsetneq B$ and let $\{a_n\} \in B'$, $\notin B$, so $\{a_n\}$ is a Cauchy sequence that is not a null sequence. Then there exists a real number $\eta > 0$ and an integer p such that $|a_n| > \eta$ for all $n \geqslant p$. Define $b_n = 1$ if $n < p$ and $b_n = a_n$ if $n \geqslant p$. Then $\{a_n\} - \{b_n\} = \{c_n\} \in B$. Also it is easy to check that $\{b_n^{-1}\} \in C$. Then $\{1\} = \{b_n^{-1}\}\{b_n\} = \{b_n^{-1}\}\{a_n\} - \{b_n^{-1}\}\{c_n\} \in B'$. Hence $B' = C$ and so B is a maximal ideal. $\square$

We now put $\hat{F} = C/B$. This is a field. The canonical homomorphism of C onto $\hat{F}$ maps $\{F\}$ onto a subfield that we can identify with F. Thus we identify an element a of F with the coset $\{a\} + B$ in C/B.

We now introduce an absolute value $|\ |$ on F. Let $\{a_n\} \in C$. Then the inequality $\| a_n| - |a_m\|_\infty \leqslant |a_n - a_m|$ shows that $\{|a_n|\}$ is a Cauchy sequence of real numbers. Hence $\lim |a_n|$ exists in $\mathbb{R}$. If $\{b_n\} \in B$, then $b_n \to 0$ and $|b_n| \to 0$. This implies that if $\{a_n'\} \in \{a_n\} + B$, then $\lim |a_n'| = \lim |a_n|$. Hence if we define

$$|\{a_n\} + B| = \lim |a_n|,$$

we obtain a map of $\hat{F}$ into $\mathbb{R}$. Since $|a| \geqslant 0$, it is clear that the values of $|\ |$ on $\hat{F}$ are non-negative. Also since $|a_n| \to 0$ implies that $a_n \to 0$, it is clear that $|\bar{a}| = 0$ for $\bar{a} \in \hat{F}$ if and only if $\bar{a} = 0$. It is straightforward to verify that $|\ |$

satisfies also conditions (2) and (3) for an absolute value (see Definition 9.1). Hence $| \ |$ is an absolute value on $\hat{F}$. It is clear also that $| \ |$ on $\hat{F}$ is an extension of $| \ |$ on F if we identify F as before with the subset of elements $\{a\}+B$ in $\hat{F}$.

We show next that F is dense in $\hat{F}$. This can be seen by showing that any $\bar{a} \in \hat{F}$ is a limit of a sequence of elements of F. In fact, if $\bar{a} = \{a_n\}+B$, then it is easy to see that $\lim a_n = \bar{a}$.

Now let $\{\bar{a}_n\}$ be a Cauchy sequence of elements of $\hat{F}$. Then there exists an $a_n \in F$ such that $|\bar{a}_n - a_n| < 1/2^n$. It follows that $\{a_n\}$ is a Cauchy sequence in F and $\lim \bar{a}_n = \bar{a}$ where $\bar{a} = \{a_n\}+B$. Thus $\hat{F}$ is complete and we have proved the existence of a completion of F.

We now consider the question of uniqueness of the completion. More generally we consider two fields $\hat{F}_i$, $i = 1, 2$, which are complete relative to valuations $| \ |_i$ and let F_i be a dense subfield of $\hat{F}_i$. Suppose we have an isomorphism s of F_1 onto F_2 that is *isometric* in the sense that $|a|_1 = |sa|_2$ for $a \in F_1$. Then s is a continuous map of F_1 into $\hat{F}_2$ and since F_1 is dense in $\hat{F}_1$, s has a unique extension to a continuous map $\bar{s}$ of $\hat{F}_1$ into $\hat{F}_2$. This is easily seen to be a homomorphism, and since $\overline{s^{-1}}$ is a homomorphism and $s^{-1}s = 1_{F_1}$ and $ss^{-1} = 1_{F_2}$ imply $\overline{s^{-1}}\bar{s} = 1_{\hat{F}_1}$, $\bar{s}\overline{s^{-1}} = 1_{\hat{F}_2}$, it follows that $\bar{s}$ is an isomorphism. It is clear also that $\bar{s}$ is unique and is isometric.

We state these results in

THEOREM 9.7. (1) *Any field F with an absolute value has a completion $\hat{F}$.* (2) *If $\hat{F}_i$, $i = 1, 2$, is complete relative to an absolute value and F_i is a dense subfield of $\hat{F}_i$, then any isometric isomorphism of F_1 onto F_2 has a unique extension to an isometric isomorphism of $\hat{F}_1$ onto $\hat{F}_2$.*

If we take $s = 1_F$, then this proves the uniqueness of the completion up to an isometric isomorphism that is the identity on F.

If we complete $\mathbb{Q}$ relative to its usual absolute value $| \ |_\infty$ we obtain classically the field $\mathbb{R}$ of real numbers. On the other hand, the completion of $\mathbb{Q}$ relative to a p-adic absolute value $|a|_p = \gamma^{v_p(a)}$, $0 < \gamma < 1$, can be identified with the field $\mathbb{Q}_p$ of p-adic numbers as defined on pp. 74–75. To see this we first consider the closure $\hat{\mathbb{Z}}$ of $\mathbb{Z}$ in the completion $\hat{\mathbb{Q}}$ relative to $| \ |_p$. An element $\alpha \in \hat{\mathbb{Z}}$ is the limit in $\hat{\mathbb{Q}}$ of a sequence of integers a_i. Such a sequence may be assumed to satisfy $a_i \equiv a_j \pmod{p^i}$, $i \leqslant j$, and so determines an element of the inverse limit $\mathbb{Z}_p = \varprojlim \mathbb{Z}/(p^i)$. It is straightforward to verify that we have an isomorphism of $\hat{\mathbb{Z}}$ with $\mathbb{Z}_p$ mapping $\lim a_i$ into the corresponding element of $\mathbb{Z}_p$. Hence we can identify these two rings. We observe next that since $\{|a|_p | a \in \mathbb{Q}\}$ is closed in $\mathbb{R}$ this set is identical with $\{|\alpha| \ | \alpha \in \hat{\mathbb{Q}}\}$. Hence given any $\beta \neq 0$ in $\hat{\mathbb{Q}}$ there exists an $e \in \mathbb{Z}$ such that $|\beta| = |p^e|$, so $\alpha = \beta p^{-e}$ satisfies $|\alpha| = 1$. Then $\alpha = \lim a_i$

where $a_i = b_i/c_i$ and $(b_i, p) = 1 = (c_i, p)$. Now there exists an $x_i \in \mathbb{Z}$ such that $x_i c_i \equiv b_i (\mathrm{mod}\ p^i)$. Then $|x_i - b_i/c_i| = |p^i|$ and so $\alpha = \lim x_i \in \mathbb{Z}_p = \hat{\mathbb{Z}}$. It follows that every element of $\hat{\mathbb{Q}}$ has the form αp^e, $\alpha \in \mathbb{Z}_p$, $e \leqslant 0$. Evidently this implies that $\hat{\mathbb{Q}}$ is the field of fractions of $\mathbb{Z}_p$ so $\mathbb{Q}$ is the field $\mathbb{Q}_p$ of p-adic numbers.

EXERCISES

In exercises 1–3, F is assumed to be complete relative to an absolute value $|\ |$.

1. Show that if $\sum a_n$ converges in F, then $\lim a_n = 0$. Show that the converse holds if $|\ |$ is non-archimedean.
2. Show that if k is a natural number then

$$v_p(k!) = \left[\frac{k}{p}\right] + \left[\frac{k}{p^2}\right] + \left[\frac{k}{p^3}\right] + \cdots$$

and hence $v_p(k!) < k(1/p - 1)$ (cf. exercise 15, p. 84 of BAI). Use this to prove that the power series

$$\exp z = \sum_0^\infty z^i/i!$$

converges for all $z \in Q_p$, $p \neq 2$, such that $|z|_p < 1$.

3. Show that $\exp(x + y) = (\exp x)(\exp y)$ if $|x|, |y| < 1$.

9.5 FINITE DIMENSIONAL EXTENSIONS OF COMPLETE FIELDS. THE ARCHIMEDEAN CASE

One of the central problems in valuation theory is: Given a field F with an absolute value $|\ |$ and a finite dimensional extension E/F, can $|\ |$ be extended to an absolute value on E? We are interested also in determining all of the extensions. We shall now consider this problem for F complete especially in the case in which the absolute value is archimedean. We prove first a uniqueness theorem.

THEOREM 9.8. *Let $|\ |$ be a non-trivial absolute value on F such that F is complete relative to $|\ |$ and let E be a finite dimensional extension field of F. Suppose that $|\ |$ can be extended to an absolute value $|\ |$ on E. Then this can be done in only one way and $|\ |$ on E is given by the formula*

$$|a| = |N_{E/F}(a)|^{1/[E:F]}. \tag{7}$$

Moreover, E is complete.

Proof. Let $\{u_1, \ldots, u_r\}$ be linearly independent elements of E, $\{a_n\}$ a sequence of elements of the form $a_n = \sum_{i=1}^{r} \alpha_{ni} u_i$ where the $\alpha_{ni} \in F$. Then $\{a_n\}$ is a Cauchy sequence if and only if the r sequences $\{\alpha_{ni}\}$, $1 \leqslant i \leqslant r$, are Cauchy sequences in F. In one direction this is clear: If every $\{\alpha_{ni}\}$ is Cauchy, then so is $\{a_n\}$. Conversely, suppose $\{a_n\}$ is Cauchy. If $r = 1$, then clearly $\{\alpha_{n1}\}$ is Cauchy. We now use induction on r. If $\{\alpha_{nr}\}$ is Cauchy, then the sequence b_n where $b_n = a_n - \alpha_{nr} u_r$ is Cauchy and since $b_n = \sum_{j=1}^{r-1} \alpha_{nj} u_j$, the result will follow by induction in this case. Now suppose $\{\alpha_{nr}\}$ is not a Cauchy sequence. Then there is a real $\varepsilon > 0$ such that for any positive integer N there exist integers $p, q \geqslant N$ such that $|\alpha_{pr} - \alpha_{qr}| > \varepsilon$. Then there exist pairs of positive integers (p_k, q_k), $p_1 < p_2 < \cdots$, $q_1 < q_2 < \cdots$ such that $|\alpha_{p_k r} - \alpha_{q_k r}| > \varepsilon$. Then $(\alpha_{p_k r} - \alpha_{q_k r})^{-1}$ exists in F and we can form the sequence $\{b_k\}$ where

$$b_k = (\alpha_{p_k r} - \alpha_{q_k r})^{-1}(a_{p_k} - a_{q_k}). \tag{8}$$

Since $|\alpha_{p_k r} - \alpha_{q_k r}|^{-1} < \varepsilon^{-1}$ and $(a_{p_k} - a_{q_k}) \to 0$, we have $b_k \to 0$. On the other hand,

$$b_k = \sum_{j=1}^{r-1} \beta_{kj} u_j + u_r, \tag{9}$$

so if $c_k = \sum_1^{r-1} \beta_{kj} u_j$, $c_k \to -u_r$ and so $\{c_k\}$ is Cauchy. Hence, by the induction hypothesis, every $\{\beta_{kj}\}$ is Cauchy and since F is complete relative to $|\ |$, $\beta_{kj} \to \beta_j \in F$. Then taking limits in (9) we obtain $\sum_{j=1}^{r-1} \beta_j u_j + u_r = 0$ contrary to the linear independence of the u_i. This proves that every $\{\alpha_{ni}\}$ is Cauchy. It is now clear also that if $a_n \to 0$, then every $\alpha_{ni} \to 0$. Also, if we take $(u_1, \ldots, u_r)$ to be a base, then we see that E is complete. It remains to prove the formula (7). If this does not hold, we have an $a = \sum_1^r \alpha_i u_i$ such that $|a^r| \neq |N_{E/F}(a)|$ ($r = [E:F]$). By replacing a by a^{-1} if necessary we may assume that $|a^r| < |N_{E/F}(a)|$. Put $b = a^r N_{E/F}(a)^{-1}$. Then $|b| < 1$ and $N(b) = N_{E/F}(b) = 1$. Since $|b| < 1$, $b^n \to 0$. Hence if we write $b^n = \sum_1^r \beta_{ni} u_i$, then $\beta_{ni} \to 0$ for every i. Since the norm map is a polynomial function from E to F, it is continuous. Hence $\beta_{ni} \to 0$ implies that $1 = N(b^n) \to 0$. This contradiction proves (7). $\square$

We shall determine next the fields that are complete relative to a real archimedean valuation. We shall see that the only possibilities are $\mathbb{R}$ and $\mathbb{C}$. We shall need the following result, which is applicable in the archimedean case since these fields have characteristic 0.

LEMMA. *Let F be a field of characteristic $\neq 2$ that is complete relative to a real valuation $|\ |$ and let E be a quadratic extension of F. Then* (7)

$$|a| = |N_{E/F}(a)|^{1/2}$$

defines a valuation of E that is an extension of $\|$.

Proof. E is Galois over F. Let $a \rightsquigarrow \bar{a}$ be the automorphism of E/F that is not the identity. Then $T(a) = a+\bar{a}$ and $N(a) = a\bar{a}$ and $a^2 - T(a)a + N(a) = 0$. If $a \in F$, then $N(a) = a^2$ and the definition shows that the map $a \rightsquigarrow |N_{E/F}(a)|^{1/2}$ is an extension of the absolute value on F. Evidently $|a|$ as defined by (7) is $\geqslant 0$ and $|a| = 0$ if and only if $N_{E/F}(a) = 0$, hence if and only if $a = 0$. The condition $|ab| = |a||b|$ follows from the multiplicative property of the norm. It remains to prove that $|a+b| \leqslant |a|+|b|$. The multiplicative property permits us to reduce this to proving that $|a+1| \leqslant |a|+1$. Since this holds in F, we may assume that $a \notin F$. Then $E = F(a)$ and the minimum polynomial of a over F is $x^2 - T(a)x + N(a) = (x-a)(x-\bar{a})$. Then $N(a+1) = (a+1)(\bar{a}+1) = N(a) + T(a) + 1$ and we have to show that $|N(a+1)| \leqslant (|N(a)|^{1/2}+1)^2$. This is equivalent to

$$|1+T(a)+N(a)| \leqslant 1+2|N(a)|^{1/2}+|N(a)|.$$

Since $|\alpha+\beta| \leqslant |\alpha|+|\beta|$ in F, this will follow if we can show that $|T(a)| \leqslant 2|N(a)|^{1/2}$ or, equivalently, $|T(a)|^2 \leqslant 4|N(a)|$. Suppose to the contrary that $|T(a)|^2 > 4|N(a)|$. Then we shall show that $a \in F$ contrary to $E = F(a)$.

We consider the equation $x^2 - \alpha x + \beta = 0$, $\alpha = T(a)$, $\beta = N(a)$, whose roots in E are a and $\bar{a}$. We shall show that this has a root in F and this will prove that $a \in F$ and will give the required contradiction. The given equation is equivalent to $x = \alpha - \beta x^{-1}$ and we have $|\alpha|^2 > 4|\beta|$. We shall obtain a solution of this as a limit of a sequence of elements $\{\gamma_n\}$ in F. We define the γ_n recursively by

$$\gamma_1 = \tfrac{1}{2}\alpha, \qquad \gamma_{n+1} = \alpha - \beta\gamma_n^{-1}.$$

To see that this makes sense we have to show that no $\gamma_n = 0$. We have $|\gamma_1| = \frac{1}{2}|\alpha| > 0$, so $\gamma_1 \neq 0$ and assuming $|\gamma_n| \geqslant \frac{1}{2}|\alpha|$ we have

$$|\gamma_{n+1}| = |\alpha - \beta\gamma_n^{-1}| \geqslant |\alpha| - |\beta||\gamma_n|^{-1} \geqslant |\alpha| - 2|\beta||\alpha|^{-1} \geqslant |\alpha| - \tfrac{1}{2}|\alpha|^2|\alpha|^{-1} = \tfrac{1}{2}|\alpha|.$$

Hence every $|\gamma_n| \geqslant \frac{1}{2}|\alpha| > 0$, so $\gamma_n \neq 0$. Now $\gamma_{n+2} - \gamma_{n+1} = \beta\gamma_{n+1}^{-1}\gamma_n^{-1}(\gamma_{n+1} - \gamma_n)$. Hence

$$|\gamma_{n+2} - \gamma_{n+1}| \leqslant \frac{4|\beta|}{|\alpha|^2}|\gamma_{n+1} - \gamma_n|,$$

so if we put $r = 4|\beta|/|\alpha|^2 < 1$ we obtain $|\gamma_{n+2} - \gamma_{n+1}| \leqslant r^n|\gamma_2 - \gamma_1|$. This inequality implies that $\{\gamma_n\}$ is a Cauchy sequence. Hence $\gamma = \lim \gamma_n$ exists in F and if we take limits in the relation $\gamma_{n+1} = \alpha - \beta\gamma_n^{-1}$, we obtain $\gamma = \alpha - \beta\gamma^{-1}$ and $\gamma^2 - \alpha\gamma + \beta = 0$. This completes the proof. □

We can now prove

OSTROWSKI'S THEOREM. *The only fields that are complete relative to an archimedean absolute value are* $\mathbb{R}$ *and* $\mathbb{C}$.

Proof (A. Ostrowski). Let F be complete relative to the archimedean absolute value $|\ |$. Then F contains $\mathbb{Q}$ and since the archimedean absolute values of $\mathbb{Q}$ are equivalent to the ordinary absolute value, the closure $\hat{\mathbb{Q}}$ of $\mathbb{Q}$ in F can be identified with $\mathbb{R}$ and the restriction of $|\ |$ to $\mathbb{R}$ is equivalent to the usual absolute value $|\ |$. If F contains an element i such that $i^2 = -1$, then F contains a subfield $\mathbb{R}(i)$ that can be identified with $\mathbb{C}$. Moreover, by Theorem 9.8, the restriction of $|\ |$ to $\mathbb{C}$ is equivalent to the usual $|\ |_\infty$. If F contains no i such that $i^2 = -1$, we can adjoin such an element to F to obtain $E = F(i)$. By the Lemma and Theorem 9.8 the absolute value $|\ |$ has a unique extension to an absolute value $|\ |$ on E and E is complete relative to $|\ |$. The foregoing considerations apply to E in place of F and if we can show that $E = \mathbb{C}$, it will follow that $F = \mathbb{R}$. Hence the proof has been reduced to showing that if F satisfies the hypotheses and, moreover, F contains $\mathbb{C}$ and the restriction of $|\ |$ to $\mathbb{C}$ is equivalent to $|\ |_\infty$, then $F = \mathbb{C}$.

We proceed to show this. Suppose that $F \supsetneqq \mathbb{C}$ and let $a \in F, \notin \mathbb{C}$. We consider the map $x \rightsquigarrow |x-a|$ of $\mathbb{C}$ into $\mathbb{R}$. Since the topology induced by $|\ |$ on $\mathbb{C}$ is the usual one, it is easily seen that this is a continuous map of $\mathbb{C}$ into $\mathbb{R}$. Let $r = \inf|\gamma - a|$ for $\gamma \in \mathbb{C}$. We claim that there exists a $\gamma_0 \in \mathbb{C}$ such that $|\gamma_0 - a| = r$. First, it is clear that if we restrict the γ so that $|\gamma - a| \leqslant r+1$, then $\inf|\gamma - a|$ for these γ is also r. Hence we consider the $\gamma \in \mathbb{C}$ such that $|\gamma - a| \leqslant r+1$. If γ_1 and γ_2 are two elements of this set, then $|\gamma_1 - \gamma_2| \leqslant 2r+2$. Hence the γ such that $|\gamma - a| \leqslant r+1$ form a closed and bounded set in $\mathbb{C}$, so by the continuity of $x \rightsquigarrow |x-a|$ we see that there is a $\gamma_0 \in \mathbb{C}$ such that $|\gamma_0 - a| = r$ and hence $|\gamma - a| \geqslant |\gamma_0 - a|$ for all $\gamma \in \mathbb{C}$. Now let $D = \{\gamma \in \mathbb{C} \mid |\gamma - a| = r\}$. This set is non-vacuous and is closed since $|x-a|$ is continuous. The argument used before shows that it is bounded. We shall now show that D is open and this will give a contradiction. The openness of D will be proved by showing that if $\gamma' \in D$, then every γ such that $|\gamma' - \gamma| < r$ is in D. If we replace a by $\gamma' - a$ and call this a, this amounts to showing that if $|a| = r$ and $|\gamma| < r$, then $|a - \gamma| = r$. To see this, let n be a positive integer and consider $a^n - \gamma^n = (a-\gamma)(a-\varepsilon\gamma)\cdots(a-\varepsilon^{n-1}\gamma)$ where ε is a primitive nth root of 1 (in $\mathbb{C}$). Then

$$\begin{aligned}|a-\gamma||a-\varepsilon\gamma|\cdots|a-\varepsilon^{n-1}\gamma| &= |a^n - \gamma^n| \\ &\leqslant |a|^n + |\gamma|^n.\end{aligned}$$

Since $|a - \varepsilon^k\gamma| \geqslant r$, this gives

$$\begin{aligned}|a-\gamma|r^{n-1} &\leqslant |a|^n\left(1 + \frac{|\gamma|^n}{|a|^n}\right) \\ &\leqslant r^n\left(1 + \frac{|\gamma|^n}{r^n}\right)\end{aligned}$$

Hence $|a-\gamma| \leqslant r(1+|\gamma|^n/r^n)$ and so if $|\gamma| < r$, then $\lim |1+(|\gamma|/r)^n| = 1$ gives $|a-\gamma| \leqslant r$. Hence $|a-\gamma| = r$. Thus D is open as well as closed and non-vacuous. Since $\mathbb{C}$ is connected, $D = \mathbb{C}$. Then for any two complex numbers γ_1 and γ_2, $|\gamma_1-\gamma_2| \leqslant |\gamma_1-a|+|\gamma_2-a| = 2r$. Since this is impossible, the proof is complete. $\square$

The extension problem for archimedean complete fields becomes trivial in view of Ostrowski's theorem. If F is such a field with the absolute value $|\ |$, then either $F = \mathbb{C}$ or $F = \mathbb{R}$. In the first case, F has no proper algebraic extension and in the second its only proper algebraic extension is $\mathbb{C}$. In both cases $|\ |$ is equivalent to $|\ |_\infty$ and if $F = \mathbb{R}$ and $E = \mathbb{C}$, the extension is given by formula (7) and E is complete. For future reference we state these results as

THEOREM 9.9. *Let F be complete relative to an archimedean absolute value $|\ |$ and let E be a finite dimensional extension field of F. Then $|\ |$ can be extended in one and only one way to an absolute value on E and this is given by* (7). *Moreover, E is complete relative to this absolute value.*

The analogous result for non-archimedean absolute values will be postponed until after we have generalized this concept to that of a valuation into an ordered abelian group. We shall consider this generalization in the next section. After that we shall prove an extension theorem for valuations and then specialize this to obtain the extension theorem for non-archimedean absolute values.

9.6 VALUATIONS

A non-archimedean absolute value $|\ |$ satisfies the condition $|a+b| \leqslant \max(|a|, |b|)$. The addition of the reals plays no role in any of the defining conditions for such an absolute value; only the multiplication and order relation for the set of non-negative reals are involved. This observation leads to a generalization of the concept of a non-archimedean absolute value to that of a valuation with values in any ordered group plus 0. This generalization has important applications to algebraic geometry; moreover, it is more natural than the original concept, since this generality is just what is needed to relate valuations to internal properties of the field. We consider first the concept of an ordered abelian group.

DEFINITION 9.3. *An* ordered abelian group *is a pair (G, H) consisting of an abelian group G and a subset H such that*

(1) *G is a disjoint union of $H \cup \{1\} \cup H^{-1}$ where $H^{-1} = \{h^{-1} | h \in H\}$.*
(2) *H is closed under the multiplication in G.*

Whenever there is no risk of confusion, we shall speak of the "ordered abelian group G" or the group G "ordered by H." We can use H to define an order relation on the set G by specifying that $g_1 > g_2$ in G if $g_1^{-1}g_2 \in H$. It is clear that this relation is transitive. Moreover, the order is total (= linear) since for any (g_1, g_2), $g_i \in G$, one and only one of the possibilities $g_1 > g_2$, $g_1 = g_2, g_2 > g_1$ (which we write also as $g_1 < g_2$) holds. Furthermore, the order is compatible with multiplication: $g_1 > g_2$ implies $gg_1 > gg_2$ for any g. Conversely, if we have an abelian group G with a total order relation $>$ compatible with multiplication, then we obtain the ordered group (G, H) in which $H = \{h|h < 1\}$. This gives an equivalent way of defining ordered abelian groups. We remark also that if (G, H) is an ordered abelian group, then so is (G, H^{-1}). Evidently, if $g_1 > g_2$ in (G, H), then $g_1 < g_2$ in (G, H^{-1}). We call (G, H^{-1}) the ordered abelian group obtained by *reversing order* in (G, H).

We have the usual elementary properties that $g_1 > g_1'$ and $g_2 > g_2'$ imply $g_1 g_2 > g_1' g_2'$ and $g_1 > g_2$ implies $g_1^{-1} < g_2^{-1}$. We note also that not every abelian group G can be ordered, that is, G may contain no subset H satisfying conditions 1 and 2. For example, we have this situation if G has torsion. For, if G is ordered, then $g^n = 1$ in G implies $g = 1$.

If G is ordered by H and G_1 is a subgroup of G, then G_1 is ordered by $H_1 = G_1 \cap H$. The order relation $>$ in G_1 is the restriction to G_1 of this relation in G. If (G, H) and (G', H') are ordered abelian groups, an *order homomorphism* of G into G' is a homomorphism of G into G' that maps H into H'. If (G_i, H_i), $1 \leqslant i \leqslant n$, is an ordered abelian group, then $H_1 \times H_2 \times \cdots \times H_n$ does not generally define an order on $G_1 \times G_2 \times \cdots \times G_n$. However, we can obtain an order in $G = G_1 \times G_2 \times \cdots \times G_n$, the *lexicographic order* (corresponding to the given indexing $i \rightsquigarrow G_i$), by taking H to be the set of elements of the form

$$(1_1, \dots, 1_{r-1}, h_r, g_{r+1}, \dots, g_n)$$

where $1 \leqslant r \leqslant n$, 1_i is the unit of G_i, $h_r \in H_r$, and the g_i are arbitrary. It is easy to check that this provides an order in G. In this order we have

$$(g_1, g_2, \dots, g_n) > (g_1', g_2', \dots, g_n')$$

if and only if for some r, $1 \leqslant r \leqslant n$, we have $g_1 = g_1', \dots, g_{r-1} = g_{r-1}'$ and $g_r > g_r'$.

We have the following familiar examples: (1) $(\mathbb{R}, +, 0)$, the additive group of reals with the usual order; (2) the multiplicative group $\mathbb{P}$ of positive reals—here the subset H defining the order is the set of positive reals < 1 and the map $x \rightsquigarrow e^x$ of $\mathbb{R}$ into $\mathbb{P}$ is an order isomorphism; and (3) $\mathbb{Z}^{(n)}$, the additive group of n-tuples of integers ordered by the lexicographic ordering.

To define general valuations of fields we need to consider ordered abelian groups with an element 0 adjoined: $V = G \cup \{0\}$, disjoint, in which we define $00 = 0$ and $g > 0$, $0g = 0 = g0$ for all $g \in G$. We can now give the following

DEFINITION 9.4. *If F is a field and V is an ordered abelian group with 0 adjoined, then we define a V-*valuation *of F to be a map φ of F into V such that*

(i) $\varphi(a) = 0 \Leftrightarrow a = 0$.
(ii) $\varphi(ab) = \varphi(a)\varphi(b)$.
(iii) $\varphi(a+b) \leqslant \max(\varphi(a), \varphi(b))$.

If we take V to be the non-negative reals, so $V = \{0\} \cup G$ where G is the set of positive real numbers, we obtain the non-archimedean absolute values defined before.

In the applications one generally encounters valuations in a form in which the ordered group is written additively and the order is reversed. Then the element 0 is replaced by an element ∞ such that $\infty + \infty = \infty$ and $\infty > a$, $\infty + a = \infty$ for all a. The definition of a valuation then has the following form.

DEFINITION 9.4′. *If F is a field and V is an additive ordered group with ∞ adjoined, then we define* an exponential V-valuation *of F to be a map v of F into V such that*

(i′) $v(a) = \infty \Leftrightarrow a = 0$.
(ii′) $v(ab) = v(a) + v(b)$.
(iii′) $v(a+b) \geqslant \min(v(a), v(b))$.

It is clear that this definition is equivalent to the first one we gave.

If φ is a valuation of F (as in Definition 9.4) and, as usual, F^* denotes the multiplicative group of non-zero elements of F, then $\varphi(F^*)$ is a subgroup of the given ordered abelian group G. We call this the *value group* of F (relative to the given φ). There is no loss in generality in replacing G by the value group $\varphi(F^*)$ if we wish to do so.

As in the special case of non-archimedean absolute values, we have $\varphi(1) = 1$ for any valuation φ since $\varphi(1)^2 = \varphi(1^2) = \varphi(1)$. Also $\varphi(-1)^2 = \varphi(1) = 1$ and since G has no elements of finite order $\neq 1$, we see that $\varphi(-1) = 1$. Then $\varphi(-a) = \varphi(-1)\varphi(a) = \varphi(a)$ and $\varphi(b^{-1}) = \varphi(b)^{-1}$. Also $\varphi(a+b) = \varphi(b)$ if $\varphi(a) < \varphi(b)$.

We shall now give some examples of valuations that are not absolute values. We base the construction on a couple of lemmas.

First, we need to extend Definition 9.4 to domains. If D is a domain and V an ordered group with 0, then we define a *V-valuation* of D as a map φ of D into V satisfying conditions (i)–(iii) of Definition 9.4. Then we have

LEMMA 1. *Let D be a domain and let φ be a V-valuation on D. Then φ has a unique extension to a V-valuation on the field of fractions F of D.*

Proof. Let $ab^{-1} = cd^{-1}$ for $a, b, c, d \in D$, $bd \neq 0$. Then $ad = bc$ and $\varphi(a)\varphi(d) = \varphi(b)\varphi(c)$. Hence $\varphi(a)\varphi(b)^{-1} = \varphi(c)\varphi(d)^{-1}$ and $ab^{-1} \rightsquigarrow \varphi(a)\varphi(b)^{-1}$ is a well-defined map of F into V. Evidently, this is an extension of the valuation φ of D. Denoting this as φ also, then it is clear that (i) and (ii) hold. To prove (iii) it suffices to show that if $a, b, d \in D$, $d \neq 0$, then $\varphi(ad^{-1} + bd^{-1}) \leqslant \max(\varphi(ad^{-1}), \varphi(bd^{-1}))$. This is equivalent to $\varphi(a+b)\varphi(d)^{-1} \leqslant \max(\varphi(a)\varphi(d)^{-1}, \varphi(b)\varphi(d)^{-1})$, which is obtained by multiplying $\varphi(a+b) \leqslant \max(\varphi(a), \varphi(b))$ by $\varphi(d)^{-1}$. The uniqueness of the extension is clear since $\varphi(ab^{-1}) = \varphi(a)\varphi(b)^{-1}$. □

We now consider a field F with an exponential V-valuation v where $V = G \cup \{\infty\}$. We form $V' = G' \cup \{\infty\}$ where $G' = \mathbb{Z} \oplus G$ ordered lexicographically: $(m, g) > (m', g')$ if either $m > m'$ or $m = m'$ and $g > g'$. We have

LEMMA 2. *The exponential V-valuation v can be extended to an exponential V'-valuation v' of $F(x)$, x an indeterminate.*

Proof. In view of Lemma 1, it suffices to prove the statement with $F(x)$ replaced by the polynomial ring $F[x]$. Let $f(x) \neq 0$ in $F[x]$. Then $f(x) = x^k(a_0 + a_1 x + \cdots + a_n x^n)$ where $k \geqslant 0$, $a_i \in F$, and $a_0 \neq 0$. Define $v'(f) = (k, v(a_0)) \in G'$. Then v' is an extension of v (if we identify V with the set of elements $(0, u)$, $u \in V$). Conditions (i′) and (ii′) of Definition 9.4′ are clear. To prove (iii′) it suffices to assume that f is as indicated and $g(x) = x^l(b_0 + b_1 x + \cdots + b_m x^m)$ where $l \geqslant 0$, $b_i \in F$, $b_0 \neq 0$. We may assume also that $v'(f) \geqslant v'(g)$. Then either $k > l$ or $k = l$ and $v(a_0) \geqslant v(b_0)$. In the first case, $v'(f+g) = (l, v(b_0)) = v'(g) = \min(v'(f), v'(g))$. In the second case, $v'(f+g) > \min(v'(f), v'(g))$ if $a_0 + b_0 = 0$ and $v'(f+g) = (k, v(a_0 + b_0)) \geqslant \min(v'(f), v'(g))$ if $a_0 + b_0 \neq 0$. Hence in both cases, (iii′) is valid. □

We can now begin with the trivial valuation on F and apply Lemma 2 to obtain an exponential valuation of $F(x_1)$, x_1 an indeterminate, for which the value group $(= v(F(x_1)^*))$ *is* $\mathbb{Z}$. Adjoining successive indeterminates we obtain an exponential valuation of $F(x_1, \ldots, x_n)$ with value group $\mathbb{Z}^{(n)}$ ordered lexicographically. We remark that $\mathbb{Z}^{(n)}$ with $n > 1$ does not satisfy Archimedes' axiom that if g_1 and g_2 are elements of $\mathbb{Z}^{(n)}$ with $g_1 > 0$, then there exists an integer n such that $ng_1 > g_2$. (For example, take $g_1 = (0, 1, \ldots)$ and $g_2 = (1, \ldots)$.) The additive group of real numbers does satisfy Archimedes' axiom. Hence $\mathbb{Z}^{(n)}$ is not order isomorphic to a subgroup of this group and the valuation we have constructed is not an absolute value if $n > 1$.

EXERCISE

1. Let (G, H) be an ordered abelian group, F a field, and let $F[G]$ be the group algebra of G over F. Show that $F[G]$ is a domain. If $f = \sum a_i g_i$, $a_i \neq 0$ in F and $g_i \in G$, define $\varphi(f) = \min g_i$ (in the ordering in G) and define $\varphi(0) = 0$. Show that φ is a valuation. Hence conclude from Lemma 1 that for any given ordered abelian group (G, H) there exists a field with a valuation whose value group is (G, H).

9.7 VALUATION RINGS AND PLACES

There are two other concepts, valuation rings and places, that are equivalent to the concept of a valuation of a field into an ordered abelian group with 0. The first of these arises in the following way. Let φ be a V-valuation of a field F and let

$$R = \{a \in F \mid \varphi(a) \leqslant 1\}. \tag{10}$$

This is a subring of F since if $a, b \in R$, then $\varphi(a - b) \leqslant \max(\varphi(a), \varphi(b))$ so $a - b \in R$ and $\varphi(ab) = \varphi(a)\varphi(b) \leqslant 1$ so $ab \in R$. Also $1 \in R$. Now suppose that $a \notin R$. Then $\varphi(a) > 1$ and $\varphi(a^{-1}) = \varphi(a)^{-1} < 1$. We therefore see that R is a valuation ring (in F) in the sense of the following

DEFINITION 9.5. *If F is a field, a* valuation ring *in F is a subring R of F such that every element of F either is contained in R or is the inverse of an element of R.*

If R is the subring of F defined by (10), then R is called *the valuation ring* of the valuation φ. We shall now show that we can turn things around: Given any valuation ring R in F, it gives rise to a valuation. To see this, let U be the set of units of R, P the set of non-units, F^* the multiplicative group of non-zero elements of F, $R^* = R \cap F^*$, $P^* = P \cap F^*$. U is a subgroup of F^*, so we can form the factor group $G' = F^*/U$. Now put

$$H' = \{bU \mid b \in P^*\} \subset G'. \tag{11}$$

Then (G', H') is an ordered abelian group: If $a \in F^*$, either $a \in R$ or $a^{-1} \in R$. If both a and $a^{-1} \in R$, then $a \in U$ and $aU = U = 1$ in G'. If $a \in R$ and $a^{-1} \notin R$, then $a \in P^*$ and $aU \in H'$. If $a \notin R$, then $a^{-1} \in P^*$ and $aU \in H'^{-1}$. It is now clear that we have condition 1 of the definition of an ordered abelian group. Now let $b_1, b_2 \in P^*$, so $b_i \neq 0$ and b_i is a non-unit of R. Then $b_1 b_2 \in P^*$ and so H' is closed under multiplication and (G, H') is an ordered abelian group. Next we adjoin an element 0 to G' to obtain $V' = G' \cup \{0\}$. We define a map

φ' of F into G' by

$$\varphi'(0) = 0, \qquad \varphi'(a) = aU \in G' \quad \text{if } a \neq 0. \tag{12}$$

We proceed to show that this is a V'-valuation of F. Conditions (i) and (ii) of Definition 9.4 are clearly satisfied. Also (iii) is clear if either $a = 0$ or $b = 0$. If $a \neq 0$ and $b \neq 0$, then either $ab^{-1} \in R^*$ or $ba^{-1} \in R^*$. We may as well assume the first. Then $\varphi'(ab^{-1}) \leqslant 1$ since $R^* = U \cup P^*$. Hence $\varphi'(a) \leqslant \varphi'(b)$. Also $ab^{-1}+1 \in R$, so $\varphi'(ab^{-1}+1) \leqslant 1$. Then

$$\varphi'(a+b) = \varphi'(b)\varphi'(ab^{-1}+1) \leqslant \varphi'(b) = \max(\varphi'(a), \varphi'(b)).$$

Thus φ' is a valuation. Moreover, since $\varphi'(a) \leqslant 1$ if and only if $a \in R$, the valuation ring associated with φ' is the given ring R. We shall call φ' the *canonical valuation* of the valuation ring R.

Now consider again an arbitrary valuation φ of F into $V = G \cup \{0\}$ where (G, H) is an ordered abelian group. Let R be the corresponding valuation ring and φ' the canonical valuation of F determined by R. We have $\varphi'(a) = aU$ if $a \neq 0$, U as before. On the other hand, we have the homomorphism $a \rightsquigarrow \varphi(a)$ of F^* into G whose kernel is U. Hence we have the monomorphism $\varphi'(a) \rightsquigarrow \varphi(a)$ of G' into G, which is an isomorphism of G' onto the value group of φ. If $bU \in H'$, then $b \in P^*$ and hence $\varphi(b) < 1$. It follows that $\varphi'(a) \rightsquigarrow \varphi(a)$ is order-preserving. We now see that the given valuation φ can be factored as $\eta\varphi'$ where φ' is the canonical valuation determined by R and η is an order isomorphism of G' onto the value group of φ.

We shall call two valuations φ_1 and φ_2 *equivalent* if there exists an order isomorphism η of the value group of φ_1 onto the value group of φ_2 such that $\varphi_2 = \eta\varphi_1$. It is clear that this implies that φ_1 and φ_2 have the same valuation rings. Moreover, we have shown that any valuation φ is equivalent to the canonical valuation φ' determined by the valuation ring R of φ. Hence any two valuations φ_1 and φ_2 are equivalent if and only if they have the same valuation ring.

Let R be a valuation ring in F, so we may assume that R is the subset of F of elements satisfying $\varphi(a) \leqslant 1$ for a valuation φ of F. Let P be the subset of R of elements b such that $\varphi(b) < 1$. Then it is clear that P is the set of non-units of R. Now P is an ideal in R since if $b_1, b_2 \in P$, then $\varphi(b_1+b_2) \leqslant \max(\varphi(b_1), \varphi(b_2)) < 1$, and if $a \in R$, then $\varphi(b_1 a) = \varphi(b_1)\varphi(a) < 1$. Hence R is a local ring, P is its maximal ideal and $\bar{R} = R/P$ is a field. This field is called the *residue field* of R or of the valuation φ. We have the canonical homomorphism $a \rightsquigarrow a+P$ of R onto the field $\bar{R}$. This leads us to introduce still another concept that is equivalent to valuation into an ordered group and valuation ring:

DEFINITION 9.6. *If F and Δ are fields, a Δ-*valued place *$\mathscr{P}$* of *F is a homomorphism of a subring R of F into Δ such that if $a \notin R$, then $a^{-1} \in R$ and $\mathscr{P}(a^{-1}) = 0$.*

Evidently the subring R specified in this definition is a valuation ring. On the other hand, if R is any valuation ring in F, then we have the canonical homomorphism of R onto the residue field. Moreover, if $a \notin R$, then $\varphi(a) > 1$ and $\varphi(a^{-1}) < 1$ so a^{-1} is a non-unit in R. Then $a^{-1} \in P$ and $a^{-1} + P = 0$. Hence $a \rightsquigarrow a + P$ is a place $\mathscr{P}'$ of F. We shall call this the *canonical place* of the valuation ring R. As in the case of valuations, if $\mathscr{P}$ is a Δ-valued place, R the associated valuation ring, and $\mathscr{P}'$ the canonical place of R, then we have a monomorphism λ of the residue field of R into the field Δ for which we have the factorization $\mathscr{P} = \lambda\mathscr{P}'$. The introduction of the concept of a place permits one to transfer the problem of extension of a valuation from a field F to an extension field to one of extending homomorphisms. We shall treat these questions in the next section.

We remark that everything we have done here can be formulated also in terms of exponential V-valuations. If v is such a valuation, then the corresponding valuation ring is

$$R = \{a \in F \mid v(a) \geqslant 0\}$$

and the maximal ideal is

$$P = \{b \in F \mid v(b) > 0\}.$$

For our purposes it will be somewhat more convenient to work with the valuations φ. However, as we have indicated, the reader will generally encounter the exponential valuations in the applications, especially to number theory.

EXERCISES

1. Let $| \ |_p$ be the p-adic absolute value of $\mathbb{Q}$ determined by γ, $0 < \gamma < 1$, as in section 9.1. Determine the value group, valuation ring and ideals of non-units for $\mathbb{Q}$ and its completion $\mathbb{Q}_p$. Show that the residue field is isomorphic to $\mathbb{Z}/(p)$.

2. Let $\mathscr{P}$ be a Δ-valued place on f with valuation ring R, $\mathscr{P}'$ a Δ'-valued place on Δ with valuation ring R'. Show that the composite $\mathscr{P}'\mathscr{P}$ is a Δ'-valued place with valuation ring $S = \{a \in R \mid \mathscr{P}(a) \in R'\}$.

3. Let D be a factorial domain, F its field of fractions, and let p be a prime in D. Show that the localization $D_{(p)}$ is a valuation ring in F.

4. Let $E = F(\xi_i, \ldots, \xi_m)$, the field of rational expressions in m indeterminates ξ_i. Show that there exists an F-valued place on E such that F is contained in the valuation ring and $\mathscr{P}(\xi_i) = a_i$ for any prescribed elements a_i of F.

9.8 EXTENSION OF HOMOMORPHISMS AND VALUATIONS

In this section we shall prove a basic theorem on extensions of a homomorphism of a subring of a field. This will enable us to prove a general theorem on extensions of a valuation of a field to any extension field. The key to everything is the following simple result.

LEMMA. *Let R be a subring of a field F, M a proper ideal in R, a a non-zero element of F, $R[a]$ and $R[a^{-1}]$ the subrings generated by R and a and a^{-1}, respectively. Then either the ideal $MR[a]$ is proper in $R[a]$ or $MR[a^{-1}]$ is proper in $R[a^{-1}]$.*

Proof. If not, we have $m_i, n_j \in M$ such that

$$1 = m_0 + m_1 a + \cdots + m_k a^k \tag{13}$$

and

$$1 = n_0 + n_1 a^{-1} + \cdots + n_l a^{-l}. \tag{14}$$

Since M is proper in R, $k + l \geqslant 2$ and we may assume $k + l$ minimal. By symmetry, we may assume also that $k \geqslant l$. Then $a^l(1 - n_0) = n_1 a^{l-1} + \cdots + n_l$ and $(1 - n_0) = (1 - n_0)m_0 + \cdots + (1 - n_0)m_k a^k$, so if we substitute the expression for $a^l(1 - n_0)$ given in the first of these relations in the last term of the second one, we obtain a relation (13) with a lower k. This contradicts the minimality of $k + l$ and proves the lemma. □

The following is the main result on extension of homomorphisms to places.

THEOREM 9.10. *Let R_0 be a subring of a field F and let $\mathscr{P}_0$ be a homomorphism of R_0 into an algebraically closed field Δ. Then $\mathscr{P}_0$ can be extended to a Δ-valued place on F.*

Proof. We consider the collection of homomorphisms of subrings of F into Δ that extend the given $\mathscr{P}_0$. By Zorn's lemma we have a maximal one defined on the subring $R \supset R_0$. We shall show that R is a valuation ring, which will prove the theorem. Let M be the kernel of $\mathscr{P}$. Since $\mathscr{P}$ maps into a field, R/M is a domain and M is a prime ideal in R. We claim that R is a local ring with M as its maximal ideal. Otherwise, we have an element $b \in R, \notin M$ such that b is not a unit. Then $\mathscr{P}(b) \neq 0$ and the subset of elements of F of the form ab^{-k}, $a \in R$, $k \geqslant 1$, is a subring R' of F properly containing R. This can be identified with the localization $R_{\langle b \rangle}$ (see p. 395). It follows that $\mathscr{P}$ can be extended to a homomorphism $\mathscr{P}'$ of R' into Δ. This contradicts the maximality of $\mathscr{P}$ and proves that R is local with M as its maximal ideal. Then R/M is a field and hence $\Gamma = \mathscr{P}(R)$ is a subfield of Δ.

If R is not a valuation ring, we have an $a \in F$ such that $a \notin R$ and $a^{-1} \notin R$. By the lemma, either $MR[a] \subsetneqq R[a]$ or $MR[a^{-1}] \subsetneqq R[a^{-1}]$. We may assume the former. Now consider the polynomial rings $R[x]$ and $\Gamma[x]$, x an indeterminate. The homomorphism $\mathscr{P}$ of R can be regarded as an epimorphism of R onto Γ. This extends to an epimorphism of $R[x]$ onto $\Gamma[x]$ mapping $x \rightsquigarrow x$. Let Z be the ideal in $R[x]$ of polynomials $f(x)$ such that $f(a) = 0$ and let Z' be its image in $\Gamma[x]$. Then Z' is an ideal in $\Gamma[x]$ and if $Z' = \Gamma[x]$, we have an $f(x) = \sum_0^n c_i x^i$, $c_i \in R$, such that $\sum_0^n c_i a^i = 0$ and $\sum \mathscr{P}(c_i)x^i = 1$. Then $\mathscr{P}(c_0) = 1$ and $\mathscr{P}(c_j) = 0$ if $j > 0$. Hence $c_0 = 1 - m_0$ and $c_j = m_j$ where $m_0, m_j \in M$. The relation $\sum_0^n c_i a^i = 0$ then implies that $1 \in MR[a]$, contrary to hypothesis. Thus $Z' \subsetneqq \Gamma[x]$ and $Z' = (g(x))$ where either $g(x) = 0$ or $g(x)$ is a monic polynomial of positive degree. In the first case we choose any $r \in \Delta$ and in the second case we choose any $r \in \Delta$ such that $g(r) = 0$. This can be done since Δ is algebraically closed. Now we have the homomorphism of $R[x]$ into Δ, which is the composite of the homomorphism of $R[x]$ into $\Gamma[x]$ with the homomorphism of $\Gamma[x]$ into Δ, which is the identity on Γ and maps x into r. The homomorphism of $R[x]$ into Δ has the form $h(x) = c_0 + c_1 x + \cdots \rightsquigarrow \mathscr{P}(c_0) + \mathscr{P}(c_1)r + \cdots$ and it maps any $f(x) \in Z$ into 0. Hence we have the homomorphism of $R[x]/Z$ into Δ such that $h(x) + Z \rightsquigarrow \mathscr{P}(c_0) + \mathscr{P}_1(c)r + \cdots$. Since Z is the set of polynomials such that $f(a) = 0$, we have the homomorphism $\mathscr{P}'$ of $R[a]$ into Δ such that $h(a) \rightsquigarrow \mathscr{P}(c_0) + \mathscr{P}(c_1)r + \cdots$. This is a proper extension of $\mathscr{P}$ contrary to the maximality of $\mathscr{P}$. Hence R is a valuation ring and the proof is complete. □

We shall now apply this result to valuations. We work first with canonical objects that are defined internally. Thus we suppose we have a field F_0 and a valuation ring R_0 in F_0. We have the canonical valuation φ'_0 and canonical place $\mathscr{P}'_0$ associated with R_0. If P_0 is the kernel of $\mathscr{P}'_0$, then P_0 is the ideal of non-units in the local ring R_0 and the residue field R_0/P_0 can be imbedded in an algebraically closed field Δ. Then $\mathscr{P}'_0$ can be regarded as a Δ-valued place of F_0. Let F be an extension field of F_0. Then R_0 is a subring of F, so by the foregoing extension theorem, $\mathscr{P}'_0$ can be extended to a Δ-valued place $\mathscr{P}$ on F. Let R be the valuation ring in F on which $\mathscr{P}$ is defined, P the ideal of non-units in R. Let U_0 be the set of units in R_0, U the set of units in R. One easily sees that $R_0 = R \cap F_0$, $P_0 = P \cap F_0$, $U_0 = U \cap F_0$. It follows that we have an order monomorphism s of the ordered group $G'_0 = F_0^*/U_0$ (ordered by the subset of elements bU_0, $b \in P_0^*$) into the ordered group $G' = F^*/U$ such that

$$a_0 U_0 \rightsquigarrow a_0 U, \qquad a_0 \in F_0^*. \tag{15}$$

Moreover, the definitions show that if φ' is the canonical valuation determined by R, then we have

$$(16) \qquad \varphi'(a_0) = s\varphi_0'(a_0), \qquad a_0 \in F_0.$$

In this sense we have an extension of the canonical valuation of F_0 to the canonical valuation on F.

An immediate consequence of this result and the relation between canonical valuations and arbitrary valuations is the following extension theorem for valuations.

THEOREM 9.11. *Let φ_0 be a valuation on F_0, F an extension field of F_0. Then there exists an ordered group G that is an extension of the value group of F_0 and a V-valuation φ of F for $V = \{0\} \cup G$ that is an extension of φ_0.*

Proof. Let R_0 be the valuation ring of φ_0, φ_0' the canonical valuation associated with R_0. Then if we apply the result just proved and the relation between φ_0 and φ_0', we see that we have a V-valued valuation φ on F, $V = \{0\} \cup G$, and an order monomorphism t of the value group of φ_0 into G such that $\varphi(a_0) = t\varphi_0(a_0)$ for every $a_0 \in F_0$. We can then identify the value group of F_0 with its image under t in G and we shall have the result stated. □

In the case of finite dimensional extension fields we have the following result.

LEMMA. *Let φ be a valuation of a field F, F_0 a subfield such that $[F:F_0] = n < \infty$. Then the value group of F is order isomorphic to a subgroup of the value group of F_0.*

Proof. For any non-zero a in F we have a relation of the form $\alpha_1 a^{n_1} + \alpha_2 a^{n_2} + \cdots + \alpha_k a^{n_k} = 0$ where the α_i are non-zero elements of F_0 and the n_i are integers such that $[F:F_0] = n \geqslant n_1 > n_2 > \cdots > n_k$. If $\varphi(\alpha_i a^{n_i}) > \varphi(\alpha_j a^{n_j})$ for all $j \neq i$, then $\varphi(\sum_1^k \alpha_j a^{n_j}) = \varphi(\alpha_i a^{n_i})$, contrary to $\sum \alpha_i a^{n_i} = 0$. Hence we have $\varphi(\alpha_i a^{n_i}) = \varphi(\alpha_j a^{n_j})$ for some pair (i, j) with $i > j$. Then $\varphi(a)^{n_i - n_j} = \varphi(\alpha_j \alpha_i^{-1})$ is in the value group G_0 of F_0. It follows that for any $a \neq 0$, $\varphi(a)^{n!} \in G_0$. Since the value group G of F has no torsion, the map $g \rightsquigarrow g^{n!}$ is an order monomorphism of G. Thus G is order isomorphic to a subgroup of G_0. □

We can use the proof of this result and Theorem 9.11 to prove the existence of extensions of non-archimedean absolute values to finite dimensional extension fields. We have

THEOREM 9.12. *Let F_0 be a field with a non-archimedean absolute value $|\ |$ and let F be a finite dimensional extension field of F_0. Then $|\ |$ can be extended to an absolute value $|\ |$ (necessarily non-archimedean) of F.*

Proof. The value group G_0 of F is a subgroup of the multiplicative group of positive reals. By Theorem 9.11 we have an extension of $\varphi_0 = |\ |$ to a valuation φ' of F whose value group G is an extension of G_0. The proof of the lemma shows that if $n = [F:F_0]$, then the map $g \rightsquigarrow g^{n!}$ is an order monomorphism of G into G_0. Then $g \rightsquigarrow (g^{n!})^{1/n!}$ is an order monomorphism of G into the positive reals, which is the identity on G_0. If we apply this to φ' we obtain the required extension $|\ |$. $\square$

An immediate consequence of this result, Theorem 9.8, and our earlier result, Theorem 9.9, on archimedean absolute values is the

COROLLARY. *Let F be complete relative to a non-trivial absolute value $|\ |$ and let E be a finite dimensional extension of F. Then the map $a \rightsquigarrow |N_{E/F}(a)|^{1/[E:F]}$ is an absolute value on E that extends $|\ |$ on F. Moreover, this is the only extension of $|\ |$ on F to an absolute value on E and E is complete relative to $|\ |$.*

EXERCISES

1. Let G be an ordered abelian group written additively. If $a \in G$, define $|a| = a$ if $a \geqslant 0$ and $|a| = -a$ if $a < 0$. Call a subgroup K of G *isolated* (or *convex*) if given any $a \in K$, K contains every b such that $|b| \leqslant a$. Show that the set of isolated subgroups is totally ordered by inclusion. The order type of this set is called the *rank* of G. (See Hausdorff's *Mengenlehre*, 3rd ed., Chapter 3.)

2. Show that an ordered group G (written additively) is of rank one if and only if it satisfies Archimedes' postulate: If $a, b \in G$ and $a > 0$, then there exists a positive integer n such that $na > b$.

3. Prove that an ordered group is of rank one if and only if it is order isomorphic to a subgroup of the ordered group of additive reals.

   (This result shows that the real valued valuations are the *rank one valuations*, that is, the V-valuations such that $V = \{0\} \cup G$ and G is of rank one.)

The next two exercises are designed to indicate a proof of Hilbert's Nullstellensatz, based on the homomorphism extension theorem. In these exercises we assume F algebraically closed.

4. Let I be a prime ideal in $F[x_1,\dots,x_n]$. Let E be the field of fractions of the domain $F[x_1,\dots,x_n]/I$ and write $\xi_i = x_i + I$, $1 \leqslant i \leqslant n$, in E. Then $E = F(\xi_1,\dots,\xi_n)$ and we may assume that $\{\xi_1,\dots,\xi_r\}$, $r \geqslant 0$, is a transcendency base for E/F. For each $i > r$ choose a polynomial

$$a_{i0}(x_1,\dots,x_r)x^{m_i} + a_{i1}(x_1,\dots,x_r)x^{m_i-1} + \cdots + a_{im_i}(x_1,\dots,x_r)$$

   in $F[x_1,\dots,x_r,x]$ such that $a_{i0}(x_1,\dots,x_r) \neq 0$ and $a_{i0}(\xi_1,\dots,\xi_r)\xi_i^{m_i} + \cdots + a_{im_i}(\xi_1,\dots\xi_r) = 0$. Let $g = g(x_1,\dots,x_n) \in F[x_1,\dots,x_n], \notin I$. Then $g(\xi_1,\dots,\xi_n) \neq 0$ and there exists a polynomial

$$b_0(x_1,\ldots,x_r)x^n + \cdots + b_n(x_1,\ldots,x_r)$$

$\in F[x_1,\ldots,x_r,x]$ such that $g(\xi_1,\ldots,\xi_n)^{-1}$ is a root of $b_0(\xi_1,\ldots,\xi_r)x^n + \cdots + b_n(\xi_1,\ldots,\xi_r)$. Show that there exists an F-algebra homomorphism η of $F[\xi_1,\ldots,\xi_r]$ into F such that $\eta b_0(\xi_1,\ldots,\xi_n)\prod a_{i0}(\xi_1,\ldots,\xi_r) \neq 0$. Use the fact that η can be extended to an F-valued place on E to show that there exist elements c_i, $1 \leqslant i \leqslant n$, such that

$$f(c_1,\ldots,c_n) = 0 \quad \text{for all } f \in I,$$
$$g(c_1,\ldots,c_n) \neq 0.$$

5. Use the result established in exercise 4 and the theorem that the nil radical of an ideal is the intersection of the prime ideals containing it to prove Hilbert's Nullstellensatz.

The next two exercises are designed to prove that if $f(x)$ and $g(x)$ are two monic irreducible separable polynomials with coefficients in a complete non-archimedean field F and $f(x)$ and $g(x)$ are sufficiently close in a sense to be defined, then the fields $F[x]/(f(x))$ and $F[x]/(g(x))$ are isomorphic. Let $\bar{F}$ be the algebraic closure of F. Observe that since any finite subset of $\bar{F}$ is contained in a finite dimensional subfield of $\bar{F}/F$, the absolute value $|\ |$ on F has a unique extension to an absolute value on $\bar{F}$.

6. Let $f(x) \in F[x]$ be monic irreducible and separable, $f(x) = \prod_1^n (x - \alpha_i)$ in $\bar{F}[x]$. Put $\delta = \min_{i \neq j} |\alpha_i - \alpha_j|$. Show that if $\beta \in \bar{F}$ and $|\beta - \alpha_1| < \delta$ then $F(\alpha_1) \subset F(\beta)$.

   (*Sketch of proof*: Consider $F(\alpha_1, \beta)/F(\beta)$. Let $\varphi(x)$ be the minimum polynomial of α_1 over $F(\beta)$. Then we may assume $\varphi(x) = \prod_1^m (x - \alpha_j)$, $m \leqslant n$. Since $\bar{F}$ is the algebraic closure of $F(\beta)$, for any j, $1 \leqslant j \leqslant m$, there exists an automorphism σ of $\bar{F}/F(\beta)$ such that $\sigma\alpha_1 = \alpha_j$. Since the extension of $|\ |$ on F to $\bar{F}$ is unique $|\sigma\gamma| = |\gamma|$ for any $\gamma \in \bar{F}$. Hence $|\beta - \alpha_j| = |\sigma(\beta - \alpha_1)| = |\beta - \alpha_1| < \delta$. If $m > 1$, we can choose $j > 1$ and obtain $|\alpha_j - \alpha_1| \geqslant \delta$ so $|\alpha_j - \alpha_1| > |\beta - \alpha_j| = |\beta - \alpha_1|$ since $\beta - \alpha_j = (\beta - \alpha_1) + (\alpha_1 - \alpha_j)$ this contradicts exercise 2 on p. 562. Hence $m = 1$, $F(\alpha_1, \beta) = F(\beta)$ and $F(\alpha_1) \subset F(\beta)$.)

7. (Krasner's lemma). Let $F, \bar{F}, f(x) = x^n + a_1x^{n-1} + \cdots + a_n$ be as in 6. Let $g(x) = x^n + b_1x^{n-1} + \cdots + b_n \in F[x]$. Show that there exists an $\varepsilon > 0$ such that if $|a_i - b_i| < \varepsilon$, $1 \leqslant i \leqslant n$, then $F[x]/(f(x)) \cong F[x]/(g(x))$; hence $g(x)$ is also irreducible and separable.

   (*Sketch of proof*: Let $M = \max(1, |a_i|)$, $\delta = \min_{i \neq j}(\alpha_i - \alpha_j)$ where the α_i are the roots of $f(x)$ in $\bar{F}$. Let $\varepsilon = \min(1, \delta^n/(M+1)^n)$. Suppose the b_i satisfy $|a_i - b_i| < \varepsilon$ and let β be a root of $g(x)$. Then $|b_i| \leqslant |a_i| + 1 \leqslant M + 1$, so by exercise 3, p. 562. $|\beta| \leqslant M + 1$ and $\prod_1^n |\beta - \alpha_i| = |f(\beta)| = |f(\beta) - g(\beta)| = |\sum (a_i - b_i)\beta^{n-i}| < \varepsilon(M+1)^n$. Hence some $|\beta - \alpha_i|$, say, $|\beta - \alpha_1| < \varepsilon^{1/n}(M+1)^{1/n} \leqslant \delta$. Then by exercise 6, $F(\alpha_1) \subset F(\beta)$. Since $[F(\alpha_1):F] = n$ and $[F(\beta):F] \leqslant n$ we have $F(\alpha_1) = F(\beta)$ and $F[x]/(f(x)) \cong F[x]/(g(x))$.)

8. (Kurschak). Prove that if F is algebraically closed with an absolute value $|\ |$, then the completion $\hat{F}$ is algebraically closed.

   (*Sketch of proof*: The case in which $|\ |$ is archimedean follows readily from Ostrowski's theorem. Now assume $|\ |$ is non-archimedean. We show first that $\hat{F}$ is perfect. This is clear if $\operatorname{char} F = 0$. If $\operatorname{char} F = p$ and $a \in \hat{F}$, let $\{a_n\}$ be a sequence of elements of F such that $a_n \to a$. Then $\{a_n^{1/p}\}$ is a Cauchy sequence so $b = \lim a_n^{1/p}$ exists in $\hat{F}$ and $b^p = a$. Thus $a^{1/p}$ exists for any $a \in \hat{F}$ and $\hat{F}$ is perfect.)

Suppose $\hat{F}$ is not algebraically closed. Then there exists a polynomial $f(x) = x^n + a_1x^{n-1} + \cdots + a_n \in \hat{F}[x]$ that is irreducible and has degree $n > 1$. This is also separable. Since F is dense in $\hat{F}$, Krasner's lemma implies the existence of a monic polynomial $g(x) \in F[x]$ that is irreducible of degree n in $\hat{F}[x]$, hence, in $F[x]$. Since F is algebraically closed this is impossible.

9.9 DETERMINATION OF THE ABSOLUTE VALUES OF A FINITE DIMENSIONAL EXTENSION FIELD

We return to the problem of extension of absolute values posed in section 9.5: Given an absolute value $|\ |$ on a field F and a finite dimensional extension field E/F, determine all of the extensions of $|\ |$ to absolute values on E. By the corollary to Theorem 9.12 we have the answer for F complete: There exists one and only one extension of the given absolute value and it is given by $|a| = |N_{E/F}(a)|^{1/n}$, $n = [E:F]$. Moreoover, E is complete.

Now suppose that F is arbitrary. Let $\hat{F}$ be the completion of F relative to $|\ |$, $|\ |$ the absolute value of $\hat{F}$ extending the given absolute value on F. Let (Γ, s, t) be a composite of $\hat{F}/F$ and E/F since $[E:F] = n < \infty$, and $\Gamma = s(\hat{F})t(E)$, $[\Gamma : s(\hat{F})] \leqslant n < \infty$. The absolute value $|\ |$ on $\hat{F}$ can be transferred to an absolute value $|\ |_s$ on $s(\hat{F})$ by putting $|s(a)|_s = |a|$ for $a \in \hat{F}$. This coincides with $|\ |$ on F and $s(\hat{F})$ is complete relative to $|\ |_s$. Since Γ is finite dimensional over $s(\hat{F})$, $|\ |_s$ has a unique extension to an absolute value $|\ |_\Gamma$ on Γ. Restricting this to $t(E)$ and defining $|b|_\Gamma = |t(b)|_\Gamma$ for $b \in E$, we obtain an absolute value on E that extends $|\ |$.

Thus we have a procedure for associating with every composite (Γ, s, t) of $\hat{F}/F$ and E/F an absolute value on E that extends $|\ |$. We shall now show that this correspondence is bijective if we identify equivalent composites. First, suppose the two composites (Γ_1, s_1, t_1) and (Γ_2, s_2, t_2) are equivalent. Then we have an isomorphism u of Γ_1 onto Γ_2 such that $us_1 = s_2$ and $ut_1 = t_2$. For $c \in \Gamma_1$, $|u(c)|_{\Gamma_2}$ defines an absolute value on Γ_1. If $a \in \hat{F}$ then $|u(s_1(a))|_{\Gamma_2} = |s_2(a)|_{\Gamma_2} = |s_2(a)|_{s_2} = |a| = |s_1(a)|_{s_1}$. Thus $|u(\ )|_{\Gamma_2}$ is an extension of $|\ |_{s_1}$. The same is true of $|\ |_{\Gamma_1}$, and since the extension of $|\ |_{s_1}$ to an absolute value of Γ_1 is unique, it follows that $|u(\ )|_{\Gamma_2} = |\ |_{\Gamma_1}$. This implies (Γ_1, s_1, t_1) and (Γ_2, s_2, t_2) provide the same absolute value on E extending $|\ |$ on F.

Conversely, suppose the absolute values on E defined by (Γ_1, s_1, t_1) and (Γ_2, s_2, t_2) are identical. Then for any $b \in E$, $|t_1(b)|_{\Gamma_1} = |t_2(b)|_{\Gamma_2}$. Next, we observe that the closure of E_i, $i = 1, 2$, with respect to the topology defined by $|\ |_{\Gamma_i}$ contains $s_i(\hat{F})$ and $t_i(E)$ and since the closure is a subalgebra it contains $\Gamma_i = s_i(\hat{F})t_i(E)$.

Thus E_i is dense in Γ_i. The map $t_1(b) \rightarrow t_2(b)$, $b \in E$, is an isometric isomorphism of $t_1(E)$ onto $t_2(E)$. Hence, by Theorem 9.2(2), this map has a unique extension to an isometric isomorphism u of Γ_1 onto Γ_2. Evidently, $ut_1 = t_2$. Moreover, since u is the identity on F and $s_i(\hat{F})$ is the completion of F relative to $|\ |_{s_i}$, the restriction of u to $s_1(\hat{F})$ is an isometric isomorphism of $s_1(\hat{F})$ onto $s_2(\hat{F})$. Also the map $s_1(a) \rightarrow s_2(a)$ is an isometric isomorphism of $s_1(\hat{F})$ onto $s_2(\hat{F})$ which is the identity on F. It follows that this coincides with the restriction of u to $s_1(\hat{F})$. Hence $us_1(a) = s_2(a)$, $a \in \hat{F}$, and u is an equivalence of (Γ_1, s_1, t_1) onto (Γ_2, s_2, t_2).

It remains to show that every extension of $|\ |$ to an absolute value on E can be obtained from a composite in the manner indicated. Let $|\ |'$ be an absolute value on E extending $|\ |$ and let $\hat{E}'$ be the completion of E relative to $|\ |'$. As usual we denote the absolute value on $\hat{E}'$ that extends $|\ |'$ on E by $|\ |'$. The field $\hat{E}' \supset \hat{F}$ and E so $\hat{E}' \supset \hat{F}E$ which is a domain finite dimensional over $\hat{F}$. Hence $\hat{F}E$ is a finite dimensional extension field of $\hat{F}$. It follows that $\hat{F}E$ is complete relative to $|\ |'$. Hence $\hat{E}' = \hat{F}E$ and we have the composite $(\hat{E}', s, t)$ where s and t are the injections of $\hat{F}$ and E. The given extension $|\ |'$ is obtained by restricting $|\ |'$ as defined on $\hat{E}'$.

We can now apply the determination we gave in section 8.19 of the composites of two fields. According to this the composites of $\hat{F}$ and E are obtained by taking $\Gamma_M = (\hat{F} \otimes_F E)/M$ where M is a maximal ideal in $\hat{F} \otimes_F E$ and letting $s: a \rightarrow a \otimes 1 + M$, $a \in \hat{F}$, $t: b \rightarrow 1 \otimes b + M$, $b \in E$. Any composite of $\hat{F}/F$ and E/F is equivalent to one of these and the composites obtained from distinct maximal ideals are inequivalent. To obtain the corresponding extension to E of $|\ |$ we define $|a \otimes 1 + M|_s = |a|$, $a \in \hat{F}$. This has a unique extension to an absolute value $|\ |_{\Gamma_M}$ on Γ_M. Then $|b|_M = |(1 \otimes b) + M|_{\Gamma_M}$, $b \in E$, defines the extension of $|\ |$ on F to E.

We can summarize our results in the following

THEOREM 9.13. *Let F be a field with an absolute value $|\ |$, E a finite dimensional extension of F, and $\hat{F}$ the completion of F. Then we have a bijection of the set of maximal ideals M of $\hat{F} \otimes_F E$ $(= E_{\hat{F}})$ with the set of absolute values on E extending $|\ |$. The absolute value $|\ |_M$ corresponding to M is given by $|b|_M = |(1 \otimes b) + M|_{\Gamma_M}$, $b \in E$ where $\Gamma_M = (\hat{F} \otimes_F E)/M$ and $|\ |_{\Gamma_M}$ is the unique extension of the absolute value $|\ |_s$ on $\hat{F} \otimes 1$ such that $|a \otimes 1 + M|_s = |a|$, $a \in \hat{F}$ to an absolute value on Γ_M. Moreover, Γ_M is isomorphic to the completion of E relative to $|\ |_M$.*

Let $\operatorname{rad} E_{\hat{F}}$ be the Jacobson radical of $E_{\hat{F}} = \hat{F} \otimes_F E$. Then $E_{\hat{F}}/\operatorname{rad} E_{\hat{F}}$ is semiprimitive artinian; hence

$$E_{\hat{F}}/\operatorname{rad} E_{\hat{F}} = \bar{I}_1 \oplus \bar{I}_2 \oplus \cdots \oplus \bar{I}_n \tag{17}$$

where the $\bar{I}_j$ are minimal ideals and are fields over $\hat{F}$ (p. 422). The maximal ideals of this algebra are the ideals $\bar{M}_j = \bar{I}_1 + \cdots + \hat{\bar{I}}_j + \cdots + \bar{I}_h$, $1 \leqslant j \leqslant h$, and $\bar{I}_j \cong (E_{\hat{F}}/\text{rad } E_{\hat{F}})/\bar{M}_j$. Since rad $E_{\hat{F}}$ is the intersection of the maximal ideals of $E_{\hat{F}}$, we have a 1–1 correspondence between the maximal ideals of $E_{\hat{F}}$ and those of $E_{\hat{F}}/\text{rad } E_{\hat{F}}$. Accordingly, we have h distinct maximal ideals $M_1, \ldots, M_h$ of $E_{\hat{F}}$ and $E_{\hat{F}}/M_j \cong (E_{\hat{F}}/\text{rad } E_{\hat{F}})/\bar{M}_j \cong \bar{I}_j$. Hence by Theorem 9.13, we have h extensions of $| \ |$ to absolute values $| \ |_1, \ldots, | \ |_h$ on E where $| \ |_j = | \ |_{M_j}$ and $\bar{I}_j$ is isomorphic to the completion $\hat{E}_j$ of E relative to $| \ |_j$. We call $n_j = [\hat{E}_j : \hat{F}]$ the *local degree* of E/F at $| \ |_j$. Then it is clear from (17) and the isomorphism $\hat{E}_j \cong \bar{I}_j$ that we have the formula

$$\sum_1^h n_j = n - [\text{rad } E_{\hat{F}} : \hat{F}] \tag{18}$$

where $n = [E:F] = [E_{\hat{F}} : \hat{F}]$. Hence we have the inequality

$$\sum_1^h n_j \leqslant [E:F] \tag{19}$$

for the local degrees.

These results assume a simple concrete form if E has a primitive element: $E = F(u)$. In this case we have

THEOREM 9.14. *Let $| \ |$ be an absolute value on F, $\hat{F}$ the corresponding completion of F, and let $E = F(u)$ where u is algebraic over F with minimum polynomial $f(x)$ over F. Let $f_1(x), \ldots, f_h(x)$ be the distinct monic irreducible factors of $f(x)$ in $\hat{F}[x]$. Then there are exactly h extensions of $| \ |$ to absolute values on E. The corresponding completions are isomorphic to the fields $\hat{F}[x]/(f_j(x))$, $1 \leqslant j \leqslant h$, and the local degree $n_j = \deg f_j(x)$.*

Proof. The field E is isomorphic to $F[x]/(f(x))$, which has the base $(1, \bar{x}, \ldots, \bar{x}^{n-1})$ where $\bar{x}^i = x^i + (f(x))$ and $n = \deg f(x)$. This is a base also for $E_{\hat{F}}/\hat{F}$, so $E_{\hat{F}} \cong \hat{F}[x]/(f(x))$. Since $\hat{F}[x]$ is a p.i.d., the maximal ideals of $\hat{F}[x]/(f(x))$ have the form $(g(x))/(f(x))$ where $g(x)$ is a monic irreducible factor of $f(x)$ in $\hat{F}[x]$. Hence $\hat{F}[x]/(f(x))$ has h maximal ideals $(f_j(x))/(f(x))$. The corresponding factor is isomorphic to $\hat{F}[x]/(f_j(x))$. The results stated follow from this and from Theorem 9.13. □

Theorem 9.14 is applicable when E/F is separable, since such an E has a primitive element (BAI, p. 291). Moreover, in this case the derivative criterion (BAI, p. 231) shows that $f(x) = f_1(x) \cdots f_h(x)$ where the $f_i(x)$ are distinct irreducible factors. Hence we have the following:

COROLLARY. *Let E be a finite dimensional separable extension of a field F with an absolute value $|\ |$. Then $[E:F] = \sum_1^h n_j$ where $n_1,\dots,n_h$ are the local degrees.*

As an example of these results, let $F = \mathbb{Q}$, $|\ |$ the usual absolute value on $\mathbb{Q}$, and let $E = \mathbb{Q}(u)$ where the minimum polynomial of u is $f(x)$. Here $\hat{F} = \mathbb{R}$ and $f(x)$ factors in $\mathbb{R}[x]$ as

$$f(x) = \prod_1^t (x-u_i) \prod_1^s g_j(x)$$

where the g_j are irreducible quadratic polynomials. There are no multiple factors. Hence the number of extensions of $|\ |$ to absolute values on E is $t+s$ of which t have local degree one and s have local degree two. The dimensionality of E/F is $t+2s$.

9.10 RAMIFICATION INDEX AND RESIDUE DEGREE. DISCRETE VALUATIONS

In this section we introduce the important concepts of ramification index and residue degree of an extension of a valuation as well as the concept of discrete valuation. Discrete valuations are the ones whose value groups are cyclic, and these are the valuations that are of primary interest in number theory and in the study of algebraic functions of one variable.

Let E be a field with a valuation φ, F a subfield. Let S be the valuation ring of φ in E, Q the maximal ideal of S, R and P the corresponding objects of F. Then $R = S \cap F$ and $P = Q \cap F$. Hence we have the monomorphism

$$a+P \rightsquigarrow a+Q \tag{20}$$

of $\bar{R} = R/P$ into $\bar{S} = S/Q$. The image is $(R+Q)/Q$. This can be identified with $\bar{R}$ by means of the isomorphism (20). In this way we can regard $\bar{R}$ as imbedded in $\bar{S}$ and we can consider the dimensionality

$$f = [\bar{S}:\bar{R}], \tag{21}$$

which is called the *residue degree of E/F relative to φ*. It is clear also that the value group $\varphi(F^*)$ is a subgroup of the value group $\varphi(E^*)$. Accordingly, we have the index

$$e = [\varphi(E^*):\varphi(F^*)]. \tag{22}$$

This is called the *ramification index* of E/F relative to φ.

We now consider non-archimedean absolute values, or equivalently, valuations whose value groups are contained in the multiplicative group of positive real numbers. Then $S = \{a \in E \mid |a| \leqslant 1\}$ and $Q = \{b \in E \mid |b| < 1\}$. We assume also that $|\ |$ is non-trivial and we show first that if $[E:F] = n < \infty$, then both the ramification index and the residue degree are finite; for we have

PROPOSITION 9.1. $ef \leqslant n = [E:F]$.

Proof. Let $u_1, \ldots, u_{f_1}$ be elements of S such that the cosets $u_i + Q$ of $\bar{S}$ are linearly independent over $\bar{R}\,(= (R+Q)/Q)$. This means that if we have elements $a_i \in R$ such that $\sum a_i u_i \in Q$, then every $a_i \in Q$. Let $b_1, \ldots, b_{e_1}$ be elements of E^* such that the cosets $|b_j|\,|F^*|$ are distinct elements of the factor group $|E^*|/|F^*|$. We proceed to show that the $e_1 f_1$ elements $u_i b_j$ are linearly independent over F. If we multiply the b_j by a suitable element of F^*, we can bring these elements into Q. Hence we may assume that the $b_j \in Q$. We shall show first that if $a_i \in F$ and $\sum a_i u_i \neq 0$, then $|\sum a_i u_i| \in |F^*|$. For if $\sum a_i u_i \neq 0$, then some $a_i \neq 0$ and we may assume that $0 \neq |a_1| \geqslant |a_i|$ for all a_i. Then $|\sum a_i u_i| = |a_1|\,|\sum a_1^{-1} a_i u_i|$ and $|\sum a_1^{-1} a_i u_i| \leqslant 1$. If $|\sum a_1^{-1} a_i u_i| < 1$, $\sum a_1^{-1} a_i u_i \in Q$ and since $|a_1^{-1} a_i| \leqslant 1$, $a_1^{-1} a_i \in R$ for every i. Then the relation $\sum a_1^{-1} a_i u_i \in Q$ contradicts the choice of the u_i. Hence $|\sum a_1^{-1} a_i u_i| = 1$ and $|\sum a_i u_i| = |a_1| \in |F^*|$.

Now suppose we have a relation $\sum_{i,j} a_{ij} u_i b_j = 0$ for $a_{ij} \in F$. If there exists a j such that $\sum_i a_{ij} u_i \neq 0$, then the relation $\sum_{i,j} a_{ij} u_i b_j = 0$ implies that we have distinct j, say, $j = 1, 2$, such that $|\sum_i a_{i1} u_i b_1| = |\sum_i a_{i2} u_i b_2| \neq 0$. Then $\sum a_{i1} u_i \neq 0$ and $\sum a_{i2} u_i \neq 0$, so $|\sum a_{i1} u_i|$ and $|\sum a_{i2} u_i| \in |F^*|$. Then $|b_1|\,|F^*| = |b_2|\,|F^*|$, contrary to the choice of the b_j. Thus the relation $\sum a_{ij} u_i b_j = 0$ implies that $\sum_i a_{ij} u_i = 0$ for every j. If we multiply the a_{ij} by a suitable non-zero element of F, we obtain a relation $\sum_i a'_{ij} u_i = 0$ with $a'_{ij} \in R$ and if some $a'_{ij} \neq 0$, we may assume one of the $a'_{ij} \notin Q$. This contradicts the choice of the u_i and proves the F-independence of the $e_1 f_1$ elements $u_i b_j$. Evidently this implies that $ef \leqslant n$. $\square$

The inequality $ef \leqslant n$ can be sharpened by passing to the completions. Consider the completion $\hat{E}$ of E, which contains the completion $\hat{F}$ of F. We observe first that E and $\hat{E}$ have the same value groups. Let $\hat{a} \in \hat{E}^*$. Since E is dense in $\hat{E}$, there exists an $a \in E$ such that $|a - \hat{a}| < |\hat{a}|$. Then $|a| = |\hat{a}|$ and hence the value group $|\hat{E}^*| = |E^*|$. Likewise $|\hat{F}^*| = |F^*|$. Hence E/F and $\hat{E}/\hat{F}$ have the same ramification index.

Now let $\hat{S}$ and $\hat{R}$ be the valuation rings of $\hat{E}$ and $\hat{F}$ respectively, $\hat{Q}$ and $\hat{P}$ the corresponding maximal ideals. We have $S = \hat{S} \cap E$, $Q = \hat{Q} \cap E$, $R = \hat{R} \cap F$, and $P = \hat{P} \cap F$. If $\hat{a} \in \hat{R}$, we have an $a \in F$ such that $|\hat{a} - a| < 1$. Then

$\hat{a}-a \in \hat{P}$, so $a \in \hat{R} \cap F = R$. Hence $\hat{R} = R+\hat{P}$ and we have the canonical isomorphism

$$(23) \qquad a+P \rightsquigarrow a+\hat{P}$$

of the residue field $\bar{R} = R/P$ onto the residue field $\hat{R}/\hat{P}$. Similarly we have the isomorphism $b+Q \rightsquigarrow b+\hat{Q}$ of S/Q onto $\hat{S}/\hat{Q}$. This is a semi-linear isomorphism of S/Q regarded as vector space over R/P onto $\hat{S}/\hat{Q}$ regarded as vector space over $\hat{R}/\hat{P}$ relative to the base field isomorphism given by (23). Hence $[S/Q:R/P] = [\hat{S}/\hat{Q}:\hat{R}/\hat{P}]$, so the residue degree as well as the ramification index is unchanged on passing to the completions. On the other hand, $[E:F] = n$ is replaced by the local degree $\hat{n} = [\hat{E}:\hat{F}]$ and this will generally be less than $[E:F]$. We have the sharper inequality $ef \leqslant \hat{n} \leqslant n$.

Considerably sharper results can be obtained if $| \ |$ is discrete. A valuation φ is called *discrete* if the value group of φ is cyclic. If this group is not trivial, then it is infinite cyclic and hence it can be realized in the multiplicative group of positive reals. Thus φ can be regarded as a non-archimedean absolute value. The p-adic absolute values on $\mathbb{Q}$ are discrete since $|p|$ is a generator of the value group. Also the absolute values of $F(x)$ that are trivial on F are discrete (p. 566). We recall also that if E is a finite dimensional extension of F, then the value group of E is isomorphic to a subgroup of the value group of F (p. 582). Hence if a valuation is discrete on F, then any extension of it to a valuation on a finite dimensional extension field is also discrete.

Suppose that $| \ |$ is discrete and non-trivial on F and let c be a generator of $|F^*|$ such that $c < 1$. Let π be an element of F^* such that $|\pi| = c$. Then $\pi \in P$. If $a \in F^*$, $|a| = |\pi|^k$ for $k \in \mathbb{Z}$, so $|a\pi^{-k}| = 1$ and $a\pi^{-k} = u$ is a unit in R. Then $a = u\pi^k$. Evidently $a \in R$ if and only if $k \geqslant 0$ and $a \in P$ if and only if $k > 0$. Hence $P = \pi R$. It is easily seen that the only ideals in R are the ideals $P^k = \pi^k R, k \geqslant 0$. Hence the valuation ring of a discrete $| \ |$ is a principal ideal domain. Moreover, this property is characteristic, since we have the following

PROPOSITION 9.2. *Let R be a local p.i.d. (= local ring that is a principal ideal domain). Then R is a valuation ring in its field F of fractions such that the corresponding valuation is discrete.*

Proof. Let rad $R = P = \pi R$. Since R is a p.i.d., R is factorial and since πR is the only maximal ideal in R, π and its associates are the only primes in R. Hence every non-zero element of R can be written in one and only one way in the form $u\pi^k$ where u is a unit and $k \geqslant 0$. Then every non-zero element of F can be written in one and only one way in the form $v\pi^l$ where v is a unit in R and $l \in \mathbb{Z}$. It follows that R is a valuation ring in F and the corresponding valuation is discrete. □

In view of Proposition 9.2 and the result preceding it, it is natural to call a commutative ring R having the properties stated in the proposition a *discrete valuation ring* (often abbreviated as DVR). As we shall see in Chapter 10, these rings play an important role in multiplicative ideal theory.

We shall now prove

PROPOSITION 9.3. *Let F be complete relative to a discrete valuation $|\ |$, E a finite dimensional extension field of F, $|\ |$ the unique extension of $|\ |$ to E. Then $ef = n = [E:F]$.*

Proof. We know that $|\ |$ is discrete on E and E is complete. Let P, R, Q, S be as before and let π and Π be chosen so that $P = \pi R$, $Q = \Pi S$. Then $|\pi|$ and $|\Pi|$ generate $|F^*|$ and $|E^*|$ respectively. Since $[|E^*|:|F^*|] = e$, we have $|\pi| = |\Pi|^e$ and hence $\pi = u\Pi^e$ where u is a unit in S. Let $u_1, \dots, u_f$ be elements of S such that the cosets $u_i + Q$ form a base for S/Q over R/P. We shall show that the ef elements

$$u_i\Pi^j, \quad 1 \leqslant i \leqslant f, \quad 0 \leqslant j \leqslant e-1, \tag{24}$$

form a base for E/F. It is clear that the cosets $\Pi^j|F^*|$, $0 \leqslant j \leqslant e-1$, are distinct elements of $|E^*|/|F^*|$. Hence the proof of Proposition 9.1 shows that the elements $u_i\Pi^j$ are linearly independent over F. It remains to show that every element of E is a linear combination of the $u_i\Pi^j$ with coefficients in F. We shall show first that every element of S can be written as $\sum_{i,j} a_{ij}u_i\Pi^j$ with $a_{ij} \in R$. Let $v \in S$. Then $|v| = |\Pi^k|$ for some $k \geqslant 0$. We can write $k = m_1 e + j_1$ where $m_1 \geqslant 0$ and $0 \leqslant j_1 \leqslant e-1$. Then $|v| = |\Pi^{m_1 e + j_1}| = |\pi^{m_1}\Pi^{j_1}|$, so $v = w\pi^{m_1}\Pi^{j_1}$ where $|w| = 1$, so $w \in S$. Then there exist $a_{i1} \in R$ such that $w - \sum a_{i1}u_i \in Q$, that is, $|w - \sum a_{i1}u_i| < 1$. Hence if we put

$$v_1 = v - \left(\sum_i a_{i1}u_i\right)\pi^{m_1}\Pi^{j_1}$$

then $v_1 = (w - \sum a_{i1}u_i)\pi^{m_1}\Pi^{j_1}$ and $|v_1| < |v| = |\Pi|^k = \Pi^{m_1 e_1 + j_1}$. Then $v = \sum b_{i1}u_i\Pi^{j_1} + v_1$ where $b_{i1} = a_{i1}\pi^{m_1} \in R$. We repeat this process with v_1 and obtain $v_1 = \sum b_{i2}u_i\Pi^{j_2} + v_2$ where $b_{i2} \in R$ and $|v_2| < |v_1|$. We substitute this in the expression for v and iterate this process to obtain

$$v = \sum_{i=1}^{f}\sum_{j=0}^{e-1} c_{ij}^{(m)} u_i\Pi^j + v_m, \qquad m = 1, 2, \dots$$

where $|v| > |v_1| > |v_2| > \cdots$ and the $c_{ij}^{(m)} \in R$. Then $v_m \to 0$ and $\sum c_{ij}^{(m)} u_i \Pi^j \to v$, Since F is complete and the ef elements $u_i \Pi^j$ are linearly independent, the proof of Theorem 9.8 shows that $\lim c_{ij}^{(m)} = c_{ij}$ exists for every i, j. Since $|c_{ij}^{(m)}| \leqslant 1$ we have $|c_{ij}| \leqslant 1$, so $c_{ij} \in R$, Hence $v = \sum c_{ij} u_i \Pi^j$ with every $c_{ij} \in R$. If v' is any element of E, we can multiply it by a non-zero element of F to obtain a v with $|v| \leqslant 1$. Then applying the result that we proved for v, we see that v' is an F-linear combination of the $u_i \Pi^j$. This completes the proof. □

We can now prove the main result on residue degrees and ramification indices.

THEOREM 9.15. *Let F be a field with a non-archimedean absolute value $|\ |$, E an extension field of F such that $[E:F] = n < \infty$, and let $|\ |_1, \ldots, |\ |_n$ be the extensions of $|\ |$ to absolute values on E. Let e_i and f_i be the ramification index and residue degree of E/F relative to $|\ |_i$. Then $\sum e_i f_i \leqslant n$ and $\sum e_i f_i = n$ if $|\ |$ is discrete and E/F is separable.*

Proof. Let $\hat{E}_i$ be the completion of E relative to $|\ |_i$ so $\hat{E}_i \supset \hat{F}$, the completion of F relative to $|\ |$. Then we have seen that if $n_i = [\hat{E}_i : \hat{F}]$, then $e_i f_i \leqslant n_i$ and $\sum n_i \leqslant n$. Hence $\sum e_i f_i \leqslant n$. We have seen also that $\sum n_i = n$ if E/F is separable and $n_i = e_i f_i$ if $|\ |$ is discrete. Hence $\sum e_i f_i = n$ if $|\ |$ is discrete and E/F is separable. □*

EXERCISES

1. Determine the value group, valuation ring and its ideal of non-units for the field $\mathbb{Q}_p$ of p-adic numbers. Show that the residue field is isomorphic to $\mathbb{Z}/(p)$. (See exercise 1, p. 560.)
2. Let R be a valuation ring in F. Show that if R is noetherian, then R is a discrete valuation ring.

9.11 HENSEL'S LEMMA

We shall prove next a reducibility criterion for polynomials with coefficients in a valuation ring, known as Hensel's lemma. In the text we treat the most

* An example in which $n \neq \sum e_i f_i$ has been given by F. K. Schmidt in "Über die Erhaltung der Kettensatz der Ideal theorie bei beliebigen endliche Körpererweiterungen," *Mathematische Zeitschrift* vol. 41 (1936), pp. 443–450.

important case of this lemma in which the valuation is discrete and the polynomials are monic. The general form of the lemma is indicated in the exercises. The proofs given are due to D. S. Rim. The key result for these considerations is the following

PROPOSITION 9.4. *Let F be complete relative to a discrete valuation $|\ |$, R the valuation ring of $|\ |$, P its maximal ideal, and $\bar{R} = R/P$. Suppose that $f(x)$ is a monic irreducible polynomial in $R[x]$. Then the image $\bar{f}(x) \in \bar{R}[x]$ is a power of an irreducible polynomial in $\bar{R}[x]$.*

Proof. Let E be a splitting field over F of $f(x)$. Then $|\ |$ has a unique extension to an absolute value $|\ |$ on E. Let S be the valuation ring determined by the absolute value on E, Q the maximal ideal of S. Let $a \in E$, $\sigma \in \mathrm{Gal}\, E/F$. Then $N_{E/F}(\sigma a) = N_{E/F}(a)$. Hence by (7), $|\sigma a| = |a|$. It follows that $\sigma(S) = S$ and $\sigma(Q) = Q$. Hence σ determines the automorphism

$$\bar{\sigma}: \bar{a} \rightsquigarrow \overline{\sigma a} \tag{25}$$

of S/Q. Evidently this is the identity on $\bar{R} = R/P$, so $\bar{\sigma} \in \mathrm{Gal}\, \bar{S}/\bar{R}$ where $\bar{S} = S/Q$. Now the given polynomial $f(x)$ is irreducible in $F[x]$. This is clear since R is a p.i.d. and hence is factorial (BAI, p. 148). We have the factorization $f(x) = \prod_1^n (x - r_i)$ in $E[x]$. Let $a_n = f(0) = \prod(-r_i)$. Then $N_{E/F}(r_i) = ((-1)^n a_n)^e$ where $e = [E:F]/n$. Since $a_n \in R$, it follows that $|r_i| \leqslant 1$ and the $r_i \in S$. Applying the canonical homomorphism of S onto $\bar{S}$ we obtain $\bar{f}(x) = \prod(x - \bar{r}_i)$. Let $\bar{r}_i$, $\bar{r}_j$ be any two of these roots. Since $f(x)$ is irreducible in $F[x]$, we have an automorphism $\sigma \in \mathrm{Gal}\, E/F$ such that $\sigma r_i = r_j$. Then $\overline{\sigma r_i} = \bar{r}_j$, which implies that $\bar{r}_i$ and $\bar{r}_j$ have the same minimum polynomial over $\bar{R}$. If this is $\bar{g}(x)$, then $\bar{f}(x)$ is a power of $\bar{g}(x)$. □

We can now prove the following version of

HENSEL'S LEMMA. *Let F be complete relative to a discrete valuation $|\ |$, R the valuation ring, P the ideal of non-units, $\bar{R} = R/P$. Suppose that $f(x)$ is a monic polynomial in $R[x]$ such that $\bar{f}(x) = \bar{\gamma}(x)\bar{\delta}(x)$ in $\bar{R}[x]$ where $\bar{\gamma}(x)$, $\bar{\delta}(x)$ are monic and $(\bar{\gamma}(x), \bar{\delta}(x)) = 1$. Then $f(x) = g(x)h(x)$ in $R[x]$ where $g(x)$ and $h(x)$ are monic and $\bar{g}(x) = \bar{\gamma}(x)$, $\bar{h}(x) = \bar{\delta}(x)$.*

Proof. We have the factorization $f(x) = \prod_1^s f_i(x)^{e_i}$ in $R[x]$ where the $f_i(x)$ are the distinct monic irreducible factors. By Proposition 9.4, $\bar{f}_i(x) = \bar{g}_i(x)^{k_i}$ where $\bar{g}_i(x)$ is monic and irreducible in $\bar{R}[x]$. Then $\bar{f}(x) = \prod_1^s \bar{g}_i(x)^{e_i k_i}$ and since $(\bar{\gamma}(x), \bar{\delta}(x)) = 1$, we may assume that $\bar{\gamma}(x) = \prod_1^r \bar{g}_j(x)^{e_j k_j}$ and $\bar{\delta}(x) =$

$\prod_{r+1}^{s} \bar{g}_l(x)^{e_l k_l}$ where $\bar{g}_j(x) \neq \bar{g}_l(x)$ for the j and l indicated. Put $g(x) = \prod_1^r f_j(x)^{e_j}$, $h(x) = \prod_{r+1}^{s} f_l(x)^{e_l}$. Then $g(x)$ and $h(x)$ satisfy the required conditions. $\square$

Hensel's lemma can often be used to conclude the existence of roots of equations $f(x) = 0$ in R from the existence of roots of $\bar{f}(x)$ in $\bar{R}$. The precise result giving this is the

COROLLARY. *Let F, R, P, $f(x)$, etc. be as in Hensel's lemma. Suppose that $\bar{f}(x)$ has $\bar{\rho}$ as a simple root in $\bar{R}[x]$. Then $f(x)$ has a root r in R such that $\bar{r} = \bar{\rho}$.*

This is clear, since we have $\bar{f}(x) = (x - \bar{\rho})\bar{\delta}(x)$ where $\bar{\delta}(\bar{\rho}) \neq 0$ so $(x - \bar{\rho}, \bar{\delta}(x)) = 1$.

Hensel's lemma can also be used in conjunction with Theorem 9.14 to determine the number of extensions and local degrees of a p-adic valuation of $\mathbb{Q}$. This is illustrated in the second exercise below.

EXERCISES

1. Show that $x^3 = 4$ has a root in $\mathbb{Q}_5$, the field of 5-adic numbers.
2. Determine the number of extensions and local degrees of the p-adic valuation of $\mathbb{Q}$ for $p = 3, 5, 11$ to the field of fifth roots of unity.
3. Show that $\mathbb{Q}_p$ has p-distinct pth roots of 1.

The next three exercises are designed to prove a general Hensel's lemma.

4. Let R be a valuation ring in the field F. Call a polynomial $f(x) \in R[x]$ primitive if some coefficient of $f(x)$ is a unit. Prove Gauss' lemma that the product of two primitive polynomials in $R[x]$ is primitive. Show that any non-zero polynomial in $F[x]$ can be written as a product $cg(x)$ where $c \in F^*$ and $g(x)$ is a primitive polynomial in $R[x]$. Show that c and $g(x)$ are determined up to a unit in R. Show that if $g(x)$ is primitive and irreducible in $R[x]$, then $g(x)$ is irreducible in $F[x]$.
5. Let F be a field, and R a valuation ring in F such that if E is any finite dimensional extension field of F, then there is a unique valuation ring S of E containing R. (Note that this holds if F is complete relative to the non-archimedean absolute value $| \ |$ and R is the valuation ring of $| \ |$.) Prove that Proposition 9.4 holds for monic polynomials in $R[x]$.
6. (Hensel's lemma.) Assume that F and R are as in exercise 5. Let $f(x) \in R[x]$ and assume $\bar{f}(x) = \bar{\gamma}(x)\bar{\delta}(x)$ in $\bar{R}[x]$ where $(\bar{\gamma}(x), \bar{\delta}(x)) = 1$ and $\deg \bar{\gamma}(x) > 0$. Then $f(x) = g(x)h(x)$ in $R[x]$ where $\bar{g}(x) = \bar{\gamma}(x)$, $\bar{h}(x) = \bar{\delta}(x)$ and $\deg g(x) = \deg \bar{\gamma}(x)$.

7. Let F be a field and let $F^{\mathbb{Q}}$ be the set of maps of $\mathbb{Q}$ into F. If $f \in F^{\mathbb{Q}}$, define $\operatorname{Supp} f = \{\alpha \in \mathbb{Q} \mid f(\alpha) \neq 0\}$. Let $P(F)$ be the subset of $F^{\mathbb{Q}}$ of f such that
   (i) $\operatorname{Supp} f \subset \mathbb{Z}n^{-1}$ for some positive integer n (depending on f),
   (ii) $\operatorname{Supp} f$ is bounded below.

   The elements of $P(F)$ can be represented as formal series $\sum_{\alpha \in \operatorname{Supp} f} f(\alpha)t^{\alpha}$ and these are called *Newton-Puiseaux series*. If $f, g \in P(F)$, define $f+g$ and fg by $(f+g)(\alpha) = f(\alpha)+g(\alpha)$, $(fg)(\alpha) = \sum_{\beta+\gamma=\alpha} f(\beta)g(\gamma)$ (which is well defined). Define 0 by $0(\alpha) = 0$ for all α and 1 by $1(\alpha) = 1$ if $\alpha = 0$ and $1(\alpha) = 0$ if $\alpha \neq 0$. Verify that $(P(F), +, \cdot, 0, 1)$ is a commutative ring. Show that any finite subset $\{f_1, \ldots, f_r\}$ of $P(F)$ is contained in a subring isomorphic to a field of Laurent series $\sum_{i \geqslant k} a_i t^i$. Hence show that $P(F)$ is a field.

   Let $v(0) = \infty$ and $v(f)$ for $f \neq 0$ be the least rational in Supp f. Verify that $v(fg) = v(f)+v(g)$, $v(f+g) \geqslant \min(v(f), v(g))$.

   Show that if λ is a positive rational number, then the map η_λ of $P(F)$ defined by $(\eta_\lambda f)(\alpha) = f(\lambda^{-1}\alpha)$ is an automorphism of $P(F)$. (Symbolically, $\eta_\lambda: \sum f(\alpha)t^{\alpha} \rightsquigarrow \sum f(\alpha)t^{\lambda\alpha}$.) We have $v(\eta_\lambda f) = \lambda v(f)$.

8. (Newton-Puiseaux.) Prove that if F is algebraically closed of characteristic 0, then $P(F)$ is algebraically closed. (*Sketch of proof*, following an exposition by S. Abhyankar: We have to show that if $g(x) \in P(F)[x]$ is monic of degree $n > 1$, then $g(x)$ has a factor of degree m, $1 \leqslant m < n$. By applying a suitable automorphism we may assume $g(x) \in F((t))[x]$ and by replacing $g(x) = x^n + f_1 x^{n-1} + \cdots + f_n$, $f_i \in F((t))$, by $g(x-(1/n)f_1)$ we may assume $f_1 = 0$. Suppose first that all of the $f_1 \in F[[t]]$ and for some f_i, $f_i(0) \neq 0$, that is, $v(f_i) \geqslant 0$. Let $\bar{g}(x) = x^n + f_2(0)x^{n-2} + \cdots + f_n(0)$. Note that $\bar{g}(x) \in F[x]$ is not a power of a linear factor and hence apply Hensel's lemma to show that if $n > 1$ then $g(x)$ has a factor of degree m, $1 \leqslant m < n$. Now assume that $g(x)$ is arbitrary of the form $x^n + f_2 x^{n-2} + \cdots + f_n$, $f_i \in F((t))$. We may assume some $f_i \neq 0$. Let $u = \inf\{v(f_i)/i \mid 2 \leqslant i \leqslant n\}$ and let r, $2 \leqslant r \leqslant n$, satisfy $v(f_r)/r = u$. Apply the automorphism η_r to the coefficients and follow this with the automorphism of $P(F)[x]$, which is the identity on the coefficients and sends $x \rightsquigarrow t^{v(f_r)}x$. Multiplying the resulting polynomial by $t^{-v(f_r)n}$ we obtain a monic polynomial satisfying the conditions considered at first. The validity of the result for this implies the result for $g(x)$.)

9.12 LOCAL FIELDS

A field F with an absolute value $|\ |$ is called a *local field* if the pair $(F, |\ |)$ satisfies the following conditions:

(1) $|\ |$ is non-archimedean discrete and non-trivial.
(2) F is complete relative to $|\ |$.
(3) The residue field of $|\ |$ is finite.

Typical examples are $\mathbb{Q}_p$ and the field $F((x))$ of formal Laurent series in one indeterminate over a finite field F. For the first, this was indicated in exercise 1, p. 561. The residue field here is $\mathbb{Z}/(p)$. The second example can be regarded as the completion of the field $F(x)$ of rational expressions in x where the absolute value is $|\ |_\infty$ as defined on p. 539. The valuation ring R of $F(x)$ is the set of rational expressions $x^k b(x)/c(x)$ such that $b(0) \neq 0$, $c(0) \neq 0$, and $k \geqslant 0$. The maximal ideal P of R is the set of these elements with $k > 0$. The map sending an element of R into its value at 0 is a homomorphism of R onto F whose kernel is P. Hence $R/P \cong F$. It follows that the residue field of the completion $\widehat{F(x)} = F((x))$ is also isomorphic to F, so $F((x))$ is a local field.

Properties (1), (2), and (3) carry over to a finite dimensional extension field; hence if F is local and E is a finite dimensional field over F, then E is local.

It is quite easy to determine all of the local fields. We prove first the

LEMMA 1. *Let F be a local field and let $|R/P| = N_P$. Then R contains N_P distinct roots $\zeta_1, \zeta_2, \ldots, \zeta_{N_P}$ of $x^{N_P} = x$ and these elements constitute a set of representatives of the cosets of P in the additive group R.*

Proof. Since N_P is the cardinality of the finite field $\bar{R} = R/P$, N_P is a power of the characteristic of $\bar{R}$ and $\bar{R}$ is a splitting field over the prime field of $x^{N_P} - x$ (see BAI, p. 287). Let ζ_0 be any element of $\bar{R}$. Then by the Corollary to Hensel's lemma (p. 593), there exists a $\zeta \in R$ such that $\zeta^{N_P} = \zeta$ and $\zeta + P = \zeta_0$. If $\zeta_0' \neq \zeta_0$ is another element of $\bar{R}$ and $\zeta' \in R$ satisfies $\zeta'^{N_P} = \zeta'$, $\zeta' + P = \zeta_0'$, then $\zeta + P \neq \zeta' + P$. Hence we can obtain N_P elements $\zeta_1, \zeta_2, \ldots, \zeta_{N_P}$ such that $\zeta_i^{N_P} = \zeta_i$ and the cosets $\zeta_i + P$ are distinct. $\square$

Put $\Lambda = \{\zeta_1, \zeta_2, \ldots, \zeta_{N_P}\}$. Let π be an element such that $\pi R = P$. Then $|\pi|$ is a generator of the value group $|F^*|$. More generally let π_k be an element such that $|\pi_k| = |\pi|^k$, $k \in \mathbb{Z}$. In particular, we can take $\pi_k = \pi^k$. Let $a \in F^*$. Then we claim that we can write

$$a = \alpha_{k_1}\pi_{k_1} + \alpha_{k_2+1}\pi_{k_2+1} + \cdots \tag{26}$$

where the $\alpha_i \in \Lambda$, $k_1 < k_2 < \ldots$, and $\alpha_k \neq 0$. Let $|a| = |\pi_{k_1}|$. Then $a\pi_{k_1}^{-1} \in R, \notin P$ so there exists an $\alpha_{k_1} \neq 0$ in Λ such that $a\pi_{k_1}^{-1} \equiv \alpha_{k_1} (\text{mod } P)$. Then $|a - \alpha_{k_1}\pi_{k_1}| < |a|$. If $a = \alpha_{k_1}\pi_{k_1}$ we have (26). Otherwise, we repeat the argument with $a - \alpha_{k_1}\pi_{k_1}$. By induction we obtain $k_1 < k_2 < \cdots$ and $\alpha_{k_1}, \alpha_{k_2}, \ldots$ non-zero in Λ such that

$$|a| > |a - \alpha_{k_1}\pi_{k_1}| > |a - \alpha_{k_1}\pi_{k_1} - \alpha_{k_2}\pi_{k_2}| > \cdots.$$

Then we obtain (26). It is clear also that the α's such that (26) holds are uniquely determined and that $a \in R$ if and only if $k \geqslant 0$.

Now let F_0 be a subfield of F such that (1) $F_0 \supset \Lambda$, (2) F_0 is closed in the topology of F, and (3) $F_0 \cap P \neq 0$. Put $R_0 = R \cap F_0$, $P_0 = P \cap F_0$. We shall call a polynomial $f(x) = x^n + b_1 x^{n-1} + \cdots + b_n \in R_0[x]$ an *Eisenstein polynomial* in $R_0[x]$ if the $b_i \in P_0$ and $b_n \notin P_0^2$. Then we have

LEMMA 2. *Let* F_0 *be a subfield of* F *such that* (1) $F_0 \supset \Lambda$, (2) F_0 *is closed in the topology of* F, *and* (3) $P_0 = F_0 \cap P \neq 0$. *Let* π *be an element of* R *such that* $P = (\pi)$. *Then* $F = F_0(\pi)$ *and* π *is algebraic over* F_0 *with minimum polynomial an Eisenstein polynomial over* $R_0 = R \cap F_0$.

Proof. It is clear that F_0 is a local subfield of F. Now let $P_0 = (\pi_0)$ (in R_0), $P = (\pi)$. Then $|\pi_0| = |\pi|^e$ for $e \geqslant 1$. If $k \in \mathbb{Z}$, we have $k = eq + r$ where $0 \leqslant r \leqslant e-1$. Hence if $\pi_k = \pi_0{}^q \pi^r$, then $|\pi_k| = |\pi^k|$. It follows as above that any $a \in F^*$ can be written as in (26) using these π_k, and that if $a \in R$, then we have $k \geqslant 0$ in (26). We can rearrange the terms of this sum and obtain

$$a = a_0 + a_1\pi + \cdots + a_{e-1}\pi^{e-1} \tag{27}$$

where each a_i has the form $\sum \alpha_q \pi_0{}^q$ and for $a \in R$ the summation is taken over the $q \geqslant 0$ so the $a_i \in R_0$. Now $|a_i \pi^i|$ has the form $|\pi|^{eq+i}$. Hence $|a_i \pi^i| \neq |a_j \pi^j|$ if $i \neq j$ and $0 \leqslant i, j \leqslant e-1$. It follows that if $\sum a_i \pi^i = 0$, then every $a_i = 0$. Thus $(1, \pi, \ldots, \pi^{e-1})$ is a base for F/F_0 and hence $F = F_0(\pi)$. Moreover, applying (27) to $a = \pi^e$ we see that π is algebraic with minimum polynomial of the form $x^e + b_1 x^{e-1} + \cdots + b_e$ where the $b_i \in R_0$. Then $N_{F/F_0}(\pi) = \pm b_e$ and $|b_e| = |N_{F/F_0}(\pi)| = |\pi^e| = |\pi_0|$ (see (7)). Hence $b_e \in P_0, \notin P_0^2$. Suppose one of the $b_i \notin P_0$. Then $\bar{f}(x) = \bar{g}_0(x) x^j$ in $(R_0/P_0)[x]$ where $j \geqslant 1$ and $x \nmid \bar{g}_0(x)$. It follows from Hensel's lemma that $f(x)$ is reducible. This contradiction shows that every $b_i \in P_0$ and hence $f(x)$ is an Eisenstein polynomial. □

We can now prove

THEOREM 9.16. *The local fields are either fields of formal Laurent series* $F_0((x))$ *where* F_0 *is a finite field or the finite algebraic extensions of fields of p-adic numbers.*

Proof. Assume first that F is of finite characteristic. Then R/P has the same characteristic. Let $\Lambda = \{\zeta_1, \zeta_2, \ldots, \zeta_{N_P}\}$, the set of elements of F such that $\zeta_i^{N_P} = \zeta_i$. Since N_P is a power of the characteristic of F, it follows that Λ is a finite subfield of F. Let $\pi \in P$ satisfy $P = (\pi)$. Then we have seen that every element of F has the form $\sum_{j \geqslant k} \alpha_j \pi^j$, $\alpha_j \in \Lambda$, and this expression is unique. It follows that $F = \Lambda((\pi))$. Now assume that F is of characteristic 0. Then F contains $\mathbb{Q}$ and the valuation is non-trivial on $\mathbb{Q}$ since R/P is finite, so $\mathbb{Z} \cap P \neq 0$. Hence F contains $\mathbb{Q}_p$ for some p. Let Λ be as before and let $F_0 = \mathbb{Q}_p(\Lambda)$. Then F_0 satisfies the conditions of Lemma 2, so $F = F_0(\pi)$ is algebraic over F_0. Then F is algebraic over $\mathbb{Q}_p$. □

We shall now analyze the structure of a finite dimensional extension field E of a local field F. Then E is also local. Let e and f be the ramification index and residue degree of the absolute value $|\ |$ of E extending the given $|\ |$ on F. Then $ef = n$. We shall show that E is built up from F in two stages: $E \supset W \supset F$ where W is *unramified* over F in the sense that the ramification index associated with W/F is 1, and E is *completely ramified* over W in the sense that the associated residue degree is 1.

Let R, P, S, Q have the usual significance (p. 568). Then $[S/Q:R/P] = f$, so

$$N_Q = |S/Q| = |R/P|^f = N_P{}^f. \tag{28}$$

Moreover, the Galois group of S/Q over R/P is cyclic with generator

$$\bar{a} \rightsquigarrow \bar{a}^{N_P}. \tag{29}$$

Let Λ_E and Λ_F denote the set of roots of $x^{N_Q} = x$ and $x^{N_P} = x$ in E and F respectively. Then $\Lambda_E \supset \Lambda_F$. Put $W = F(\Lambda_E)$. Then we have

LEMMA 3. $[W:F] = f$ *and* W/F *is unramified.*

Proof. We have the isomorphism $\zeta \rightsquigarrow \bar{\zeta} = \zeta + Q$ of the group Λ_E^* of non-zero elements of Λ_E onto the multiplicative group of the field S/Q. Hence if ζ is a primitive $(N_Q - 1)$-st root of unity in S, then $\bar{\zeta}$ is a primitive $(N_Q - 1)$-st root of unity of S/Q. We have $W = F(\zeta)$ and $S/Q = (R/P)(\bar{\zeta})$. Let $\bar{g}_0(x)$ be the minimum polynomial of $\bar{\zeta}$ over R/P, so $\deg \bar{g}_0(x) = [S/Q:R/P] = f$ and $x^{N_Q} - x = \bar{g}_0(x)\bar{h}_0(x)$ in $(R/P)[x]$. By Hensel's lemma, $x^{N_Q} - x = g(x)h(x)$ where $g(x)$ and $h(x)$ are monic in $R[x]$ and $\bar{g}(x) = \bar{g}_0(x)$, $\bar{h}(x) = \bar{h}_0(x)$. If $g(\zeta) \neq 0$, then $h(\zeta) = 0$ and $\bar{h}_0(\bar{\zeta}) = 0$, which is impossible since $\bar{g}_0(\bar{\zeta}) = 0$ and $x^{N_Q} - x$ has distinct roots. Hence $g(\zeta) = 0$. Since $\bar{g}_0(x)$ is irreducible in $(R/P)[x]$, $g(x)$ is irreducible in $R[x]$ and hence in $F[x]$. Then $g(x)$ is the minimum polynomial of ζ over F and hence $[W:F] = \deg g(x) = f$. Since $\Lambda_E \subset W$, it is clear that the residue degree of the valuation of W is f and since $[W:F] = f$, it follows that the ramification index is 1. Thus W is unramified. □

Now W is a splitting field over F of the polynomial $x^{N_Q} - x$, which has distinct roots. Hence W is Galois over F. Let $G = \mathrm{Gal}\, W/F$. As in the proof of Proposition 9.4 (p. 573) any $\sigma \in G$ determines an automorphism $\bar{\sigma}: \bar{a} \rightsquigarrow \overline{\sigma a}$, $a \in S$, contained in $\mathrm{Gal}\,(S/Q)/(R/P)$. Now σ maps Λ_E^* into itself and $\bar{\sigma}$ maps the multiplicative group of S/Q into itself. Since the map $a \rightsquigarrow \bar{a}$ of Λ_E^* is injective and $W = F(\Lambda_E)$, it follows that $\sigma \rightsquigarrow \bar{\sigma}$ is a monomorphism. Since $|G| = [W:F] = [S/Q:R/P] = |\mathrm{Gal}\,(S/Q)/(R/P)|$, it follows that $\sigma \rightsquigarrow \bar{\sigma}$ is an isomorphism of the Galois group of W/F onto the Galois group of S/Q over R/P. The latter is cyclic. Hence $\mathrm{Gal}\, W/F$ is a cyclic group.

Since W is unramified, $|W^*| = |F^*|$ and hence $[|E^*|:|W^*|] = e$. Thus e is the ramification index of E over W and this is also $[E:W]$. Hence E is completely ramified over W. By Lemma 2, if Π is chosen so that $(\Pi) = Q$, then $E = W(\Pi)$ and the minimum polynomial of Π over W is an Eisenstein polynomial. If E is unramified, then clearly $E = W$. It is easily seen also that in any case W contains every unramified subfield of E/F. We have proved the following results.

THEOREM 9.17. *Let F be a local field, E a finite dimensional extension field of F. Then E contains a unique maximal unramified subfield W. We have $[W:F] = f$, the residue degree of E over F, and W is a cyclic field over F. E is completely ramified over W and $[E:W] = e$, the ramification index. Moreover, if Π is chosen so that $Q = (\Pi)$, then $E = W(\Pi)$ and the minimum polynomial of Π over W is an Eisenstein polynomial (over $S \cap W$).*

There is one other important result in these considerations. This concerns the existence of a distinguished generator of Gal W/F for W unramified. We have such a generator $\bar{a} \rightsquigarrow \bar{a}^{N_P}$ for the Galois group of the finite field S/Q over R/P. We have a corresponding automorphism in Gal W/F. This is called the *Frobenius automorphism* of W/F. It is characterized by the property that it maps any $\zeta \in \Lambda_E$ into ζ^{N_P}.

EXERCISES

1. Let the notations be as in Theorem 9.17. Show that if the characteristic of S/Q is not a divisor of e, then Π can be chosen so that its minimum polynomial has the form $x^e - \pi$, $\pi \in P$.

2. Let E, F be as in Exercise 1, with char $S/Q \nmid e$ and assume that E/F is Galois. Show that Gal E/W is cyclic of order e. Hence show that Gal E/F is an extension of a cyclic group by a cyclic group. Show also that if E/F is an abelian extension field, then $e|(N_Q - 1)$.

3. Show that if R is a valuation ring, then any Eisenstein polynomial in $R[x]$ is irreducible.

9.13 TOTALLY DISCONNECTED LOCALLY COMPACT DIVISION RINGS

In this section we propose to show that local fields and more generally finite dimensional division algebras over local fields have a simple topological characterization: These are the non-discrete totally disconnected locally compact topological division rings. Using topological methods, we shall also determine the structure of these division rings. These results will be used in the next section to determine the Brauer group of a local field.

By a topological division ring D we mean a division ring that is a Hausdorff space in which subtraction and multiplication are continuous maps of the product space into the underlying space D and the map $x \rightsquigarrow x^{-1}$ is continuous on the subspace D^* of non-zero elements. We assume that the topology is not discrete. It is easily seen that a topological division ring is either connected or totally disconnected, which means that the only connected subsets are the points. We recall that a space is locally compact if every point has a compact neighborhood. The locally compact fields, both in the connected and totally disconnected cases, were determined by D. van Dantzig in 1931. The connected locally compact division rings were determined in 1932 by L. Pontrjagin to be one of the classical trinity: $\mathbb{H}$ (Hamilton's quaternions), $\mathbb{C}$, and $\mathbb{R}$. The totally disconnected ones were determined by the author in 1936, assuming the first countability axiom. This condition and the hypothesis that $x \rightsquigarrow x^{-1}$ is continuous were removed by Y. Otobe in 1945. In this section we begin by proving the first countability property of locally compact division rings by a simple argument given in the second edition of Pontrjagin's *Topological Groups.* After this we follow the method of our 1936 proof with some improvements. Pontrjagin's book can serve as a reference for topological definitions and results, which we shall state without proofs.*

Let D be a locally compact division ring (not discrete) and let C be a compact subset of D, W a neighborhood of 0. Since $a0 = 0$ for every a and multiplication is continuous, for any $x \in C$ there exist neighborhoods U_x of x and V_x of 0 such that $U_x V_x (= \{uv | u \in U_x, v \in V_x\}) \subset W$. Since C is compact, a finite subset $\{U_{x_1}, \dots, U_{x_n}\}$ covers C. If we put $V = \bigcap_1^n V_{x_i}$, then we have $CV \subset W$. Next let U be a compact neighborhood of 0. Then U is infinite, since D is not discrete. Let $\{b_n\}$ be an infinite sequence of distinct elements of U and put $B_k = \{b_m | m \geq k\}$. Then $B_1 \supset B_2 \supset \cdots$ and $\bigcap B_k = \varnothing$. Hence some B_k is not closed and we may assume that $B = \{b_n\}$ is not closed. Let $b \in \bar{B}$, the closure of B, $b \notin B$. Then b is a limit point of B and $b \in U$. Using a translation we may assume $b = 0$. We now claim that the set $\{Ub_n\}$ is a base for the neighborhoods of 0, that is, given any neighborhood W of 0 there exists an n such that $Ub_n \subset W$. Since U is compact, the result we proved first shows that there exists a neighborhood V of 0 such that $UV \subset W$. Since 0 is a limit point of B, there exists a b_n in $V \cap B$. Then $Ub_n \subset UV \subset W$. This proves the first

*A good bibliography of the early literature on topological algebra appears in H. Freudenthal's review of this book in *Nieuw Archief voor Wiskunde*, vol. 20 (1940), pp. 311–316.

countability axiom of D and permits us to base the topological considerations on convergence of sequences.

The concept of a Cauchy sequence in a topological abelian group is clear: $\{a_n\}$ is Cauchy if given any neighborhood U of 0 there exists an integer N such that $a_m - a_n \in U$ for all $m, n \geqslant N$. Such a sequence converges (to a limit) if a subsequence converges. From this it follows that any locally compact abelian group is complete, that is, every Cauchy sequence converges. For, if $\{a_n\}$ is Cauchy and U is a compact neighborhood of 0, then every $a_{N+p} - a_N \in U$ for N sufficiently large and $p = 1, 2, \ldots$. A subsequence of $\{a_{N+p} - a_N | p = 1, 2, \ldots\}$ converges and since this sequence is Cauchy, we have a b such that $\lim_{p\to\infty} (a_{N+p} - a_N) = b$. Then $\lim_{n\to\infty} a_n = a_N + b$.

Again let D be a locally compact division ring and let U and V be compact neighborhoods of 0 such that $V \subsetneqq U$. Then there exists a neighborhood W of 0 such that $WU \subset V$ and hence any $w \in W$ satisfies the condition

$$wU \subset V. \tag{30}$$

We shall now show that if w is any element satisfying (30), then $w^n \to 0$. Let u be a non-zero element of U. Since $w^n U \subset U$, $w^n \in Uu^{-1}$ for $n = 1, 2, 3, \ldots$, and Uu^{-1} is compact. Since the power sequence $\{w^n\}$ is contained in a compact set, to prove $w^n \to 0$ it suffices to show that 0 is the only limit point of $\{w^n\}$, that is, if $w^{n_k} \to z$ for a subsequence $\{w^{n_k}\}$ of $\{w^n\}$ then $z = 0$. Suppose not. Then there exists a subsequence $\{w^{m_k}\}$ where the m_k are differences $n_{l+1} - n_l$ such that $w^{m_k} \to 1$. Then $\lim w^{m_k} u = u$ for any $u \in U$. Since every $w^{m_k} u \in wU$ and wU is closed, $u = \lim w^{m_k} u \in wU$. Then $U \subset wU \subset V$, contrary to $V \subsetneqq U$.

We shall write $a_n \to \infty$ if no subsequence of $\{a_n\}$ converges. We shall now use the fact that W contains elements $w \neq 0$ such that $w^n \to 0$ to prove

PROPOSITION 9.5. *If D is a locally compact division ring, then a sequence of non-zero elements $\{a_n\}$ satisfies $a_n \to \infty$ if and only if $a_n^{-1} \to 0$.*

Proof. Since we are assuming continuity of $x \rightsquigarrow x^{-1}$ in D^*, $a_n \to a \neq 0$ implies that $a_n^{-1} \to a^{-1}$. Hence it suffices to show that we cannot have both $a_n \to \infty$ and $a_n^{-1} \to \infty$. Suppose this is the case and let U be a compact neighborhood of 0. By dropping some terms at the beginning, we may assume that $a_n \notin U$ and $a_n^{-1} \notin U$ for $n = 1, 2, \ldots$. Let w be an element such that $w \neq 0$ and $\lim w^n = 0$. Suppose that for a fixed j we have infinitely many n such that $a_n w^j \in U$. Since Uw^{-j} is compact, this would imply that a subsequence of $\{a_n\}$ converges, contrary to the hypothesis. Hence for every j there exists an n_j such that $a_n w^j \notin U$ for $n \geqslant n_j$. This implies that there is a subsequence $\{b_n\}$ of $\{a_n\}$ such that $b_n w^n \notin U$. Since $w^n \to 0$, for each n there exists a $k_n \geqslant n$ such that $b_n w^{k_n} \notin U$ but $b_n w^{k_n+1} \in U$. A subsequence of $\{b_n w^{k_n+1}\}$ converges, so we may assume that $b_n w^{k_n+1} \to z$. Thus $b_n w^{k_n} \to zw^{-1}$ and $zw^{-1} \neq 0$ since $b_n w^{k_n} \notin U$. Then $w^{-k_n} b_n^{-1} \to wz^{-1}$ and $b_n^{-1} = w^{k_n}(w^{-k_n} b_n^{-1}) \to 0$. This contradicts $a_n^{-1} \to \infty$. □

Let G be a locally compact totally disconnected abelian group. Then G contains a base for the neighborhoods of 0, which are compact open and closed subsets of G. (See Pontrjagin, loc. cit. p. 87.) We can now prove

PROPOSITION 9.6. *If G is a locally compact totally disconnected abelian group, then the set of compact open (hence also closed) subgroups of G form a base for the neighborhoods of 0 in G.*

Proof. Since 0 has a base of compact open and closed neighborhoods, it suffices to show that if U is such a neighborhood then there exists an open subgroup of G contained in U. Let $V = U \cap -U$ where $-U = \{-u \mid u \in U\}$ and let $H = \{h \in G \mid h + V \subset V\}$. Then H is an open subgroup of G contained in V and hence $H \subset U$. □

Next we prove the existence of compact and open subrings of a totally disconnected locally compact division ring.

PROPOSITION 9.7. *Let D be a totally disconnected locally compact division ring and let H be a compact and open subgroup of D. Then the set $R = \{a \in D \mid aH \subset H\}$ is a compact and open subring of D.*

Proof. It is clear that R is a subring and R contains an open neighborhood of 0, so R is open and hence closed. If h is a non-zero element of H, then $R \subset Hh^{-1}$. Since Hh^{-1} is compact, it follows that R is compact. □

We now let R be any compact and open subring of D and we investigate the arithmetic properties of R. We observe first that if b is a non-zero element of R, then bR and Rb are open right and left ideals of R. If I is any non-zero right (left) ideal of R, then $I = \bigcup_{b \in I} bR (\bigcup_{b \in I} Rb)$, so I is open. Then R/I is compact and discrete and hence R/I is finite. Evidently this implies that R is left and right noetherian, and if B is any ideal $\neq 0$ in R, then R/B is finite and hence is left and right artinian. We note also that the set of non-zero ideals of R is a base for the neighborhoods of 0. For if U is any compact neighborhood of 0, there exists a neighborhood V of 0 such that $RVR = \{ava' \mid a, a' \in R, v \in V\} \subset U$. If we take z to be a non-zero element of $V \cap R$, we see that RzR is a non-zero ideal of R contained in U. We can now prove

THEOREM 9.18. *Let R be a compact open subring of the totally disconnected locally compact division ring D. Then R is a local ring and if $P = \text{rad } R$, then R/P is a finite field and $P = \{b \in R | b^n \to 0\}$. Moreover, $\{P^n | n = 1, 2, \ldots\}$ is a base of the neighborhoods of* 0 *in D.*

Proof. Let $P = \{b \in R | bR \subsetneqq R\}$. The argument preceding Proposition 9.5 shows that $b^n \to 0$ for every $b \in P$. If $N = R - P$, then $N = \{a \in R | aR = R\}$ and N is a subgroup of the multiplicative group D^* of D since N is a submonoid of D^* and every element of N has a right inverse in N. It follows that if $a \in R$ and $b \in P$, then ab and $ba \in P$. Now let $b_1, b_2 \in P$ and suppose that $a = b_1 + b_2 \notin P$. Then $c_1 + c_2 = 1$ for $c_i = b_i a^{-1} \in P$. The fact that the set of ideals of R is a base for the neighborhoods of 0 and that $c_2{}^n \to 0$ implies that $1 + c_2, 1 + c_2 + c_2{}^2, \ldots$ is a Cauchy sequence. Hence $1 + c_2 + c_2{}^2 + \cdots$ exists in R and this element is the inverse of $1 - c_2 = c_1$. Then $c_1^{-1} c_1 = 1 \in P$, contrary to the definition of P. Then $b_1 + b_2 \in P$ if $b_i \in P$ and we have shown that P is an ideal in R. Since the elements of $N = R - P$ are units in R, it is clear that this set is the group of units of R, so P is the set of non-units. Since P is an ideal, R is a local ring and P is its radical. Since P is an ideal, P is open and R/P is finite. Since R/P is a division ring, R/P is a field, by Wedderburn's theorem (BAI, p. 453). We have seen that the set of ideals $\neq 0$ of R is a base of the neighborhoods of 0. Hence the last statement of the theorem will follow if we show that if B is an ideal $\neq 0$ in R, then there exists an integer n such that $P^n \subset B$. We have $P \supset B$ since P is the only maximal ideal of R. Now P/B is the radical of R/B, which is a finite ring. Since a finite ring is artinian, its radical is nilpotent. Hence $(P/B)^n = 0$ for some integer n. Then $P^n \subset B$. □

Now let R_1 be a compact and open subring of D, P_1 its radical, and let $R_2 = \{a \in D | aP_1 \subset P_1\}$. Then R_2 is a compact open subring of D containing R_1 and its radical $P_2 \supset P_1$. Continue this process to define an ascending sequence of compact subrings $R_1 \subset R_2 \subset \cdots$ with radicals $P_1 \subset P_2 \subset \cdots$. Put $R = \bigcup R_i$. This is an open subring of D. We claim that R/R_1 is finite. Otherwise, we have a set of distinct cosets $a_n + R_1$, $a_n \in R$, $n = 1, 2, 3, \ldots$. Suppose $a_n \to a$. Then $a \in R$ since R is open, hence closed. Since P_1 is a neighborhood of 0, there exists an N such that $a_n - a \in P_1$ for all $n \geqslant N$. Then $a_m - a_n \in P_1$ for $m, n \geqslant N$. Since $P_1 \subset R_1$, this implies that $a_m + R_1 = a_n + R_1$, contrary to the assumption that the cosets $a_n + R_1$ are distinct. Thus the sequence $\{a_n\}$ does not converge and since we can replace this sequence by a subsequence, we see that $a_n \to \infty$. Hence $a_n^{-1} \to 0$ and so $a_n^{-1} \in P_1$ for n sufficiently large, say, $n \geqslant N$. Since $R = \bigcup R_i$, we have an m such that $a_N \in R_m$. Since $a_N^{-1} \in P_1 \subset P_m$, we have $1 = a_N a_N^{-1} \in P_m$. This is impossible and so we have proved that R/R_1 is finite. We have $R_1/R_1 \subset R_2/R_1 \subset \cdots \subset R/R_1$ and since R/R_1 is finite we have $R = R_m$ for some m. We have therefore proved the existence of a compact open subring R of D such that if P is the radical of R, then

$$R = \{a \in D \mid aP \subset P\}. \tag{31}$$

We now prove

THEOREM 9.19. *Let R be a compact open subring of D such that* (31) *holds for P the radical of R. Then R is a valuation ring in the sense that if $a \in D$, $\notin R$, then $a^{-1} \in R$. R is a maximal compact open subring of D and is the only subring having these properties.*

Proof. If $a \notin R$, there exists a $b_1 \in P$ such that $ab_1 \notin P$. If $ab_1 \notin R$, we repeat the process. Eventually we obtain n elements $b_i \in P$ such that $ab_1 \cdots b_n \in R$, $\notin P$. Otherwise, we get an infinite sequence $\{b_k\} \subset P$ such that $ab_1 \cdots b_k \notin R$. On the other hand, $\{P^n\}$ is a base for the neighborhoods of 0, so $\lim b_1 \cdots b_k = 0$, which implies that $\lim ab_1 \cdots b_k = 0$. This contradicts $ab_1 \cdots b_k \notin R$ for all k. Thus we have an n such that $ab_1 \cdots b_n \in R$, $\notin P$. Then $ab_1 \cdots b_n = u$ is a unit in R and $a^{-1} = b_1 \cdots b_n u^{-1} \in P \subset R$. Hence R is a valuation ring in D. We have shown also that if $a \notin R$, then $a^{-1} \in P$, and since $b^n \to 0$ for every $b \in P$, $a^n \to \infty$ if $a \notin R$. If $u \in R$, $\notin P$, we have $u^n \not\to \infty$ and $u^{-n} \not\to 0$. In fact, since $R-P$ is a compact set, every sequence of powers $\{u^{n_k}\}$ contains a convergent subsequence with limit in $R-P$ and hence $\neq 0$. It is clear from these results on powers that $P = \{b \in D \mid b^n \to 0\}$, $R = \{a \in D \mid a^n \not\to \infty\}$, $D-R = \{a \in D \mid a^n \to \infty\}$. If R' is any compact open subring of D, then $R' \cap (D-R) = \varnothing$ so $R' \subset R$. This shows that R is maximal and it is the only maximal compact open subring of D. $\square$

We can use the valuation ring R to define an absolute value on D (defined as for fields). We have $\bigcap P^n = 0$. Hence given any $a \neq 0$ in R, there exists a $k \geqslant 0$ such that $a \in P^k$, $a \notin P^{k+1}$ where we put $P^0 = R$. Then we define $v(a) = k$. If $a \notin R$, then $a^{-1} \in P$ and we define $v(a) = -v(a^{-1})$. Also we define $v(0) = \infty$ and $|a| = c^{v(a)}$ where c is a fixed real number such that $0 < c < 1$. Then we have

THEOREM 9.20. *$|\ |$ is a non-archimedean absolute value on D: $|a| \geqslant 0$, $|a| = 0$ if and only if $a = 0$, $|a+b| \leqslant \max(|a|, |b|)$, $|ab| = |a||b|$.*

Proof. It suffices to prove the corresponding statements on the map v from D into $\mathbb{Z}$, namely, $v(a) = \infty$ if and only if $a = 0$, $v(a+b) \geqslant \min(v(a), v(b))$,

$v(ab) = v(a)+v(b)$. The first is clear from the definition of v. Let $a, b \in D$ satisfy $v(a) = v(b) = k \geqslant 0$. If $ab^{-1} \in P$, $a \in Pb \subset P^{k+1}$ contrary to $v(a) = k$. Hence $ab^{-1} \notin P$ and similarly $ba^{-1} \notin P$. Then $ab^{-1} \in R-P$ and $v(ab^{-1}) = 0$. If $v(a) = v(b) = k < 0$, then $v(a^{-1}) = v(b^{-1}) = -k > 0$ and again $v(ab^{-1}) = 0$. Now choose $\pi \in P, \notin P^2$. Then $\pi^k \in P^k$ for $k > 0$, so $v(\pi^k) \geqslant k$. If $v(\pi^k) = l > k$, then $\pi^k = \sum_j b_1^{(j)} \cdots b_l^{(j)}$ where $b_i^{(j)} \in P$ and every $b_i^{(j)} \notin P^2$, so $v(b_1^{(j)}) = 1 = v(\pi)$. Then $\pi^{k-1} = \sum (\pi^{-1} b_1^{(j)}) \cdots b_l^{(j)}$ and $\pi^{-1} b_1^{(j)} \in R-P$. Hence $v(\pi^{k-1}) > k-1$. This leads to the contradiction that $v(\pi) > 1$. Thus $v(\pi^k) = k$. This and our earlier result show that if $v(a) = k \neq \infty$, then there exists a $u \in R-P$ such that $a = u\pi^k$. We observe next that the characterization of the sets P, $R-P$, and $D-R$ by the properties of the power sequences of the elements in these sets shows that these sets are stabilized by inner automorphisms of D. Now let $a \neq 0$, $b \neq 0$, and let $v(a) = k$, $v(b) = l \geqslant k$. Then $a = u\pi^k$, $b = v\pi^l$ where u and v are units in R. We have $a+b = (u+v\pi^{l-k})\pi^k$ and $u+v\pi^{l-k} \in R$, so $v(a+b) \geqslant k = \min(v(a), v(b))$. We have $ab = u\pi^k v\pi^l = uv'\pi^{k+l}$ where $v' = \pi^k v \pi^{-k} \in R-P$. If $k+l > 0$, we have seen that $\pi^{k+l} \in P^{k+l}, \notin P^{k+l+1}$. Then the same relations hold for ab, so $v(ab) = k+l = v(a)+v(b)$. This is clear also if $k+l = 0$. On the other hand, if $k+l < 0$, then $(ab)^{-1} = b^{-1}a^{-1} = \pi^{-(k+l)}w$ with $w \in R-P$. Then $v((ab)^{-1}) = -(k+l)$ and again $v(ab) = k+l = v(a)+v(b)$. Hence in all cases $v(ab) = v(a)+v(b)$. We therefore have $v(a+b) \geqslant \min(v(a), v(b))$ and $v(ab) = v(a)+v(b)$ if $a \neq 0$ and $b \neq 0$. These relations are evident if either $a = 0$ or $b = 0$. Hence we have the required relations for all a, b. □

The definition of $| \ |$ shows that the spherical neighborhood of 0 defined by $|a| \leqslant c^k, k \geqslant 0$, is P^k. Hence it is clear that the topology defined by the valuation is the same as the given topology on D. It follows also that D is complete relative to $| \ |$.

Let F be a closed subfield of D such that $P_F = P \cap F \neq 0$. Then the absolute value $| \ |$ is non-trivial on F and F is complete. Moreover, if $R_F = R \cap F$, then R_F/P_F is isomorphic to a subfield of R/P. Hence R_F/P_F is finite and F is a local field. Then the results of section 9.12 are available.

Let p be the characteristic of $\bar{R} = R/P$. Then $|\bar{R}| = q = p^m$ for $m \geqslant 1$. Let ζ_0 be an element of R such that $\bar{\zeta}_0 = \zeta_0 + P$ is a primitive $(q-1)$-st root of 1. We shall now show that R contains an element ζ such that $\zeta^{q-1} = 1$ and $\bar{\zeta} = \bar{\zeta}_0$. Then ζ will be a primitive $(q-1)$-st root of unity in R. Now $\zeta_0^{q-1} - 1 \in P$ and if $\zeta_0^{q-1} = 1$, we can take $\zeta = \zeta_0$. Otherwise, let F be the closure in D of $C(\zeta_0)$ where C is the center of D. Then $P_F = P \cap F$ contains $\zeta_0^{q-1} - 1 \neq 0$, so $P_F \neq 0$, and if $R_F = F \cap R$, then $\bar{R}_F = R_F/P_F$ contains $\zeta_0 + P_F$ and this is a primitive $(q-1)$-st root of unity. The results of the previous section show that R_F contains an element ζ such that $\zeta^{q-1} = 1$ and $\zeta + P_F = \zeta_0 + P_F$. Then $\bar{\zeta} = \bar{\zeta}_0$.

We have seen that R and P are stabilized by the inner automorphisms of D. Hence any inner automorphism determines an automorphism of $\bar{R} = R/P$. In particular, if $\pi \in P - P^2$, then we have the automorphism

$$\bar{\eta}\colon \bar{a} = a + P \rightsquigarrow \overline{\pi a \pi^{-1}} \tag{32}$$

of $\bar{R} = R/P$. Since $|\bar{R}| = q = p^m$, this has the form

$$\bar{a} \rightsquigarrow \bar{a}^s, \qquad s = p^t. \tag{33}$$

On the other hand, if $u \in R - P$, then $\overline{ua} = \overline{au}$ and $\overline{uau^{-1}} = \bar{a}$. Hence the corresponding automorphism in $\bar{R}$ is the identity. Since π is determined up to a multiplier in $R - P$ by the condition that $\pi \in P - P^2$, it is clear that $\bar{\eta}$ is independent of the choice of π in $P - P^2$.

We shall now show that π can be chosen so that

$$\pi \zeta \pi^{-1} = \zeta^s. \tag{34}$$

Let $G = \langle \zeta \rangle$, the subgroup of D^* generated by ζ. Then if $\pi_0 \in P - P^2$ and $\lambda \in G$, then $\overline{\pi_0 \lambda \pi_0^{-1}} = \bar{\lambda}^s$ and hence $\lambda^{-s}\pi_0\lambda\pi_0^{-1} \equiv 1 \pmod{P}$ and $\lambda^{-s}\pi_0\lambda \equiv \pi_0 \pmod{P^2}$. Then

$$\pi = \sum_{\lambda \in G} \lambda^{-s}\pi_0\lambda \tag{35}$$

satisfies $\mu^{-s}\pi\mu = \pi$ for $\mu \in G$. Moreover, $\pi \equiv (q-1)\pi_0 \pmod{P^2}$ and since $q1 \in P, \pi \equiv -\pi_0 \pmod{P^2}$ and hence $\pi \in P - P^2$. Then $\mu^{-s}\pi\mu = \pi$ gives $\pi\mu\pi^{-1} = \mu^s$ for $\mu \in G$, so in particular we have (34).

The inner automorphism $a \rightsquigarrow \pi a \pi^{-1}$ stabilizes G and induces the automorphism $\bar{\eta}$ in $\bar{R}$. Since $\lambda \rightsquigarrow \bar{\lambda}$ is an isomorphism of G onto the multiplicative group $\bar{R}^*$, it is clear that the order of the restriction of $a \rightsquigarrow \pi a \pi^{-1}$ to G is the order r of $\bar{\eta}$. Then π^r commutes with every $\lambda \in G$ and this is the smallest power of π with this property.

The proof given on pp. 571–572 shows that every element of D has a representation as a power series in π with coefficients in $K = \{0\} \cup G$. Thus we can show that any non-zero element of D can be written in one and only one way as a series

$$a = (\lambda_0 + \lambda_1\pi + \lambda_2\pi^2 + \cdots)\pi^{-k} \tag{36}$$

where the $\lambda_i \in K$, $\lambda_0 \neq 0$, and $k \geqslant 1$. It is clear from this that an element a is in the center if and only if it commutes with π and with every $\lambda \in G$. It follows that the center C is the set of elements

$$c = (\mu_0 + \mu_1\pi^r + \mu_2\pi^{2r} + \cdots)\pi^{-kr} \tag{37}$$

where the μ_i are elements of K such that $\mu_i^s = \mu_i$. Evidently $C \cap P \neq 0$ and C is a local field. The extension field $W = C(\zeta)$ is stabilized by the inner automorphism determined by π. If σ is the induced automorphism in W/C, then C is the set of elements fixed under σ. Hence W/C is cyclic with Galois group $\langle\sigma\rangle$. It is clear also that σ has order r. Hence $[W:C] = r$. Since W is the set of elements

$$b = (\lambda_0 + \lambda_1\pi^r + \lambda_2\pi^{2r} + \cdots)\pi^{-kr} \tag{38}$$

where the $\lambda_i \in R$, it is clear that the ramification index of W over C is 1, so W is an unramified extension of C. Comparison of (36) and (38) shows that every element of D can be written in one and only one way in the form

$$w_0 + w_1\pi + \cdots + w_{r-1}\pi^{r-1} \tag{39}$$

where the $w_i \in W$. The multiplication in D is determined by that in W and the following relations for $w \in W$:

$$\pi w = \sigma(w)\pi, \qquad \pi^r \in C. \tag{40}$$

Thus D is a cyclic algebra $D = (W, \sigma, \pi^r)$ and $[D:C] = r^2$. We have therefore obtained the following structure theorem for totally disconnected locally compact division rings.

THEOREM 9.21. *Let D be a totally disconnected locally compact division ring. Then the center C of D is a local field and D is a cyclic algebra $D = (W, \sigma, \gamma)$ over C where W is unramified and γ is a generator of the maximal ideal P_C of the valuation ring of C.*

Of course, this shows that D is a finite dimensional algebra over a local field. We have seen also that a local field is either a Laurent series field over a finite field or a finite dimensional extension field of a field of p-adic numbers (Theorem 9.16, p. 597). Hence D is either a finite dimensional division algebra over a field of formal Laurent series $F_0((x))$, F_0 finite, or a finite dimensional division algebra over some p-adic field $\mathbb{Q}_p$. The first case holds if and only if the characteristic of D is a prime. We have seen also that the topology is given by the absolute value $|\ |$ defined by the unique maximal compact and open subring R of D. It is easily seen that this topology is the same as the product topology obtained by regarding D as a product of a finite number of copies of $F_0((x))$ or of $\mathbb{Q}_p$.

It is not difficult to prove the following converse of Theorem 9.21. Let D be a finite dimensional division algebra over $F_0((x))$ or $\mathbb{Q}_p$. Then we can introduce a topology in D so that D becomes a non-discrete totally disconnected locally compact division ring. We sketch the argument. Let $F = F_0((x))$ or $\mathbb{Q}_p$. We have the topology on F given by an absolute value defined as before and the valuation ring R and the maximal ideal P ($R = F_0[[x]]$ in the first case and the ring of p-adic integers in the second). Now R can be regarded as the inverse limit of the set of finite rings R/P^k (see p. 73) and its topology can be identified with the topology of the inverse limit of finite sets. Hence R is compact and totally disconnected. It follows that F is locally compact totally disconnected and not discrete. The fact that the map $x \rightsquigarrow x^{-1}$ of F^* into itself is continuous can be proved as in the case of the field $\mathbb{R}$.

Now let A be a finite dimensional algebra over F and endow A with the product topology. Then A is locally compact totally disconnected and not discrete. It is readily seen that A is a topological ring. Let $N(a)$ denote the determinant of the matrix $\rho(a)$ in a regular representation of A. Then $a \rightsquigarrow N(a)$ is a continuous map of A into F, and the set U of invertible elements of A is the open subset defined by $N(u) \neq 0$. It is easy to see that $u \rightsquigarrow u^{-1}$ is a continuous map of U into U. In the special case in which $A = D$ is a division algebra over F, $U = D^*$ and A is a topological division ring in the sense defined at the outset.

9.14 THE BRAUER GROUP OF A LOCAL FIELD

We shall now apply Theorem 9.21, the remarks following it, and the results on cyclic algebras given in section 8.5 (p. 484) to determine the Brauer group $\mathrm{Br}(F)$ of a local field F. We recall that if E is a cyclic extension of a field F, then the subgroup $\mathrm{Br}(F, E)$ of the Brauer group of F consisting of the classes of finite dimensional central simple algebras A having E as splitting field is isomorphic to $F^*/N(E^*)$ where $N(E^*)$ is the group of norms $N_{E/F}(a)$ of the non-zero elements $a \in E$. The isomorphism is implemented by choosing a generator s of $G = \mathrm{Gal}\, E/F$ and defining the cyclic algebra (E, s, γ), $\gamma \in F^*$. Then the map $\gamma(N(E^*)) \rightsquigarrow [(E, s, \gamma)]$, the similarity class of (E, s, γ), is an isomorphism of $F^*/N(E^*)$ onto $\mathrm{Br}(E/F)$. Let K be a subfield of E/F and let $\bar{s}$ be the restriction of s to K. Then $\bar{s}$ is a generator of the Galois group of K/F. The order of $\bar{s}$ is $[K:F] = r$ and we have $n = rm$ where $m = [E:K]$. Any central simple algebra split by K is split by E and we have the monomorphism of $\mathrm{Br}(K/F)$ into $\mathrm{Br}(E/F)$ sending the class of $(K, \bar{s}, \gamma)$ into that of (E, s, γ^m) (p. 485).

Now let F be a local field. We determine first the group $F^*/N(W^*)$ where W is an unramified extension of F. We shall need the following result.

LEMMA. *Let F_q be a finite field with q elements, F_{q^n} an extension field with q^n elements. Then any $a \in F_q^*$ is a norm of an element $b \in F_{q^n}^*$ that is not contained in any proper subfield of F_{q^n}.*

Proof. The automorphism $x \rightsquigarrow x^q$ generates the Galois group of F_{q^n}/F_q. Hence the norm map of $F_{q^n}^*$ is

$$x \rightsquigarrow xx^q \cdots x^{q^{n-1}} = x^{(q^n-1)/(q-1)}.$$

The kernel of this map is the set of elements such that $N_{F_{q^n}/F_q}(x) = 1$ and this has order $(q^n-1)/(q-1)$. Hence the image has order $q-1$. Since the image is contained in F_q^*, which has order $q-1$, it is clear that the norm map of $F_{q^n}^*$ is surjective on F_q^*. Moreover, for any $a \in F_q^*$ there exist $(q^n-1)/(q-1)$ elements b such that $N_{F_{q^n}/F_q}(b) = a$. On the other hand, the elements b contained in proper subfields of F_{q^n} are contained in maximal proper subfields. The cardinality of any of these is of the form q^m where m is a maximal proper divisor of n, and distinct subfields have distinct orders. It follows that the number of non-zero elements contained in proper subfields does not exceed $\sum(q^m-1)$ where the summation is taken over the maximum proper divisors m of n. Evidently this number is $< (q^n-1)/(q-1) = 1+q+\cdots+q^{n-1}$. Hence we have a $b \in F_{q^n}$ not in any proper subfield such that $N_{F_{q^n}/F_q}(b) = a$. □

The requirement that b is not contained in any proper subfield of F_{q^n} is equivalent to $F_q(b) = F_{q^n}$. This occurs if and only if the degree of the minimum polynomial of b over F_q is n. We can now prove

PROPOSITION 9.8. *Let W be an unramified extension field of the local field F, R_F the valuation ring of F, P_F its ideal of non-units. Then any element $u \in R_F - P_F$ is a norm in W.*

Proof. Let $\bar{R}_F$ be the residue field R_F/P_F and similarly let $\bar{R}_W = R_W/P_W$ where R_W is the valuation ring of W, P_W its ideal of non-units. Since W is unramified, we have $[\bar{R}_W : \bar{R}_F] = n = [W : F]$. By the lemma, if $a = \bar{u} = u + P_F$, then there exists a $b \in \bar{R}_W$ such that the minimum polynomial of b over $\bar{R}_F$ has degree n and $N_{\bar{R}_W/\bar{R}_F}(b) = a$. If $\sigma \in \mathrm{Gal}\, W/F$, then the map $\bar{\sigma}: \bar{x} \rightsquigarrow \overline{\sigma x}$, $x \in R_W$, is in $\mathrm{Gal}\, \bar{R}_W/\bar{R}_F$ and $\sigma \rightsquigarrow \bar{\sigma}$ is an isomorphism between these Galois groups. Hence for any $v \in R_W$ we have $\overline{N_{W/F}(v)} = N_{\bar{R}_W/\bar{R}_F}(\bar{v})$, so if we choose $v \in R_W$ such that $\bar{v} = b$, then $\overline{N_{W/F}(v)} = a = \bar{u}$. We can choose a monic polynomial $f(x) \in R_F[x]$ of degree n such that $\bar{f}(x)$ is the minimum polynomial of $\bar{v} = b$. Since $\deg \bar{f}(x) = n$, this is the characteristic polynomial of b (in a regular representation) and its constant term is $(-1)^n N_{\bar{R}_W/\bar{R}_F}(\bar{v}) = (-1)^n \bar{u}$. Hence we may assume that the constant term of $f(x)$ is $(-1)^n u$. Since $\bar{f}(x)$ is a separable polynomial, we can apply Hensel's lemma to conclude that there exists a $v \in R_W$ such that $f(v) = 0$. Since $\bar{f}(x)$ is irreducible in $\bar{R}_F$, $f(x)$ is irreducible in $F[x]$ and so this is the minimum polynomial over F of v. Since its degree is n, it is also the characteristic polynomial. Hence its constant term $(-1)^n u = (-1)^n N_{W/F}(v)$. Then $u = N_{W/F}(v)$ as required. □

Since W is unramified over F, we can choose a $\pi \in F$ such that $\pi \in P_W - (P_W)^2$. Then any $w \in W$ has the form $u\pi^k$ where $u \in R_W - P_W$ and $k \in \mathbb{Z}$. Then $N_{W/F}(u\pi^k) = N_{W/F}(u)\pi^{kn}$ and $N_{W/F}(u) \in R_F - P_F$. Conversely, if $v = u\pi^{kn}$ where $u \in R_F - P_F$, then $\pi^{kn} = N_{W/F}(\pi^k)$ and Proposition 9.8 shows that u is a norm. Hence v is a norm. It is clear from these results that $F^*/N(W^*)$ is a cyclic group of order n with generator $\pi N(W^*)$. Then $\mathrm{Br}(W/F)$ is a cyclic group of order n. We can obtain an isomorphism between $F^*/N(W^*)$ and $\mathrm{Br}(W/F)$ by mapping the coset $\pi^k N(W^*)$, $0 \leqslant k \leqslant n-1$, onto the class of central simple algebras over F determined by the cyclic algebra (W, σ, π^k) where σ is the Frobenius automorphism of W/F.

We can combine this result with the results of the previous section to obtain a determination of $\mathrm{Br}(F)$, namely, we have

THEOREM 9.22 (Hasse). *The Brauer group of a local field is isomorphic to the additive group of rational numbers modulo* 1 (*that is*, $\mathbb{Q}/\mathbb{Z}$).

Proof. We have seen that any finite dimensional central division algebra D over the local field F is a totally disconnected locally compact division ring. Hence Theorem 9.21 shows that D has an unramified (hence cyclic) extension field W/F as splitting field. It follows that any class $[A]$ in $\mathrm{Br}(F)$ is contained in $\mathrm{Br}(W/F)$ for some W. Then $A \sim (W, \sigma, \pi^k)$ where σ is the Frobenius automorphism and $0 \leqslant k < n = [W:F]$. Once W has been chosen, then k is uniquely determined. We now map $[A]$ into the rational number $r = k/n$. We wish to show that the rational number thus determined is independent of the choice of the splitting field. It is readily seen by using the results of section 9.12 that for any positive integer n there exists a unique (up to isomorphism) unramified extension W/F with $[W:F] = n$. Moreover, if W'/F is unramified and $[W':F] = m$, then W' is isomorphic to a subfield of W if and only if $m|n$. It follows that it suffices to show that if $W' \subset W$ is a splitting field for A, then the rational number determined by W' is the same as that determined by W. Now the restriction $\bar{\sigma}$ of the Frobenius automorphism σ of W/F is the Frobenius automorphism of W'/F. Hence $A \sim (W', \bar{\sigma}, \pi^l)$ where $0 \leqslant l < m$ and so the rational number determined by A and W' is l/m. Since $(W', \bar{\sigma}, \pi^l) \sim (W, \sigma, \pi^{ln/m}) = (W, \sigma, \pi^k)$, we have $ln/m = k$ and $k/n = l/m$. It is clear also that our map is surjective on rational numbers satisfying $0 \leqslant r < 1$. For

$r = k/n$ with $0 \leqslant k < n$ and if we take W to be the unramified extension of degree n over F, then the cyclic algebra (W, σ, π^k) is central simple with W as splitting field and this maps into $r = k/n$. If we have two central simple algebras A and B over F, we can choose an unramified extension field W that is a splitting field for both. Then $A \sim (W, \sigma, \pi^k)$ and $B \sim (W, \sigma, \pi^l)$ where $0 \leqslant k, l < n$, and $A \otimes B \sim (W, \sigma, \pi^{k+l}) \sim (W, \sigma, \pi^m)$ where $0 \leqslant m < n$ and $m/n \equiv (k+l)/n \pmod{\mathbb{Z}}$. It follows that the map $\{A\} \rightsquigarrow (k/n) + \mathbb{Z}$ is an isomorphism of $\mathrm{Br}(F)$ onto $\mathbb{Q}/\mathbb{Z}$. □

Another important consequence of our results is

THEOREM 9.23. *The exponent of a finite dimensional central simple algebra over a local field coincides with its index.*

Proof. We have to show that if D is a central division algebra over a local field F such that $[D:F] = n^2$, then the order of $\{D\}$ is n. By Theorem 9.21, $D = (W, \tau, \pi)$ where τ is a generator of the Galois group of W/F and $\pi \in P_F - P_F{}^2$. Then the exponent of D is the order of $\pi N(W^*)$. This is evidently n. □

9.15 QUADRATIC FORMS OVER LOCAL FIELDS

We shall first define an invariant, the Hasse invariant, of a non-degenerate quadratic form on a finite dimensional vector space over an arbitrary field F of characteristic $\neq 2$. In this we follow a method due to Witt that appeared in a beautiful paper of his on quadratic forms in vol. 176 (1937) of Crelle's Journal.

The definition and properties of the Hasse invariant are based on quaternion algebras and Clifford algebras. We need to recall some results on quaternion algebras and develop some formulas for tensor products of these algebras. We have denoted the quaternion algebra generated by two elements i, j satisfying the relations

$$(41) \qquad i^2 = a, \qquad j^2 = b, \qquad ij = -ji$$

where a and b are non-zero elements of F, as (a, b) (p. 232). In dealing with tensor products of quaternion algebras we abbreviate $(a, b) \otimes_F (c, d)$ to $(a, b)(c, d)$ and as usual we write $\sim$ for similarity of central simple algebras. Evidently we have

$$\text{(i)} \qquad (a, b) = (b, a).$$

It is clear also that

(ii) $$(a, b) = (as^2, bt^2)$$

for any $s \neq 0$, $t \neq 0$.

A quaternion algebra (a, b) is either a division algebra or $(a, b) \cong M_2(F)$, that is, $(a, b) \sim 1$ in the Brauer group $\mathrm{Br}(F)$. Evidently $(1, b) \sim 1$ and hence

(iii) $$(a^2, b) \sim 1.$$

The algebra (a, b) has the base $(1, i, j, k = ij)$. If $x = x_0 + x_1 i + x_2 j + x_3 k$ and

(42) $$T(x) = 2x_0, \qquad N(x) = x_0{}^2 - ax_1{}^2 - bx_2{}^2 + abx_3{}^2,$$

then $x^2 - T(x)x + N(x) = 0$. Let $(a, b)_0$ denote the subspace of elements of trace 0: $T(x) = 0$. This has the base (i, j, k) and has the quadratic norm form $N(x) = -ax_1{}^2 - bx_2{}^2 + abx_3{}^2$. It is clear that $(a, b) \sim 1$ if and only if (a, b) contains an element $z \neq 0$ such that $z^2 = 0$. This is the case if and only if $T(z) = 0 = N(z)$. Hence $(a, b) \sim 1$ if and only if the quadratic norm form on $(a, b)_0$ is a null form, that is, $-ax_1{}^2 - bx_2{}^2 + abx_3{}^2 = 0$ has a solution $\neq (0, 0, 0)$. Evidently this implies

(iv) $$(a, -a) \sim 1.$$

Since $(a, b)_0$ can be characterized as the set of elements $x \in (a, b)$ such that $x \notin F$ but $x^2 \in F$, it is clear that an isomorphism of (a, b) onto (c, d) maps $(a, b)_0$ onto $(c, d)_0$. It follows that if $(a, b) \cong (c, d)$, then the quadratic forms $-ax_1{}^2 - bx_2{}^2 + abx_3{}^2$ and $-cx_1{}^2 - dx_2{}^2 + cdx_3{}^2$ are equivalent. It is quite easy to apply the theory of composition algebras to prove the converse (see exercise 2, page 450 of BAI).

If a is a non-square, then we have the field $Z = F(\sqrt{a})$, which has the automorphism σ such that $\sigma(\sqrt{a}) = -\sqrt{a}$. Then the quaternion algebra (a, b) is the same thing as the cyclic algebra (Z, σ, b) (p. 480). Hence the multiplication formula for cyclic algebras (p. 475) gives the formula

(v) $$(a, b)(a, c) \sim (a, bc)$$

if a is a non-square. Evidently this holds also if a is a square, since in this case all three algebras are ~ 1. Since $(a, b)(a, b) \sim 1$, $(a, a).(b, b) \sim (a, a)(a, b)(b, b)(a, b) \sim (a, ab)(b, ab) \sim (ab, ab)$. Hence we have

(vi) $$(a, a)(b, b) \sim (ab, ab).$$

Iteration of this gives

(vi′) $$(a_1, a_1)(a_2, a_2) \cdots (a_r, a_r) \sim (a_1 \cdots a_r, a_1 \cdots a_r).$$

We now consider a quadratic form Q on an n-dimensional vector space V over a field F of characteristic $\neq 2$. The associated symmetric bilinear form B is defined by $B(x, y) = Q(x+y) - Q(x) - Q(y)$. Then $Q(x) = \frac{1}{2}B(x, x)$. We assume throughout that Q is non-degenerate in the sense that B is non-degenerate. We now define the *discriminant of* Q to be the discriminant of $\frac{1}{2}B$. Thus if $(v_1, \ldots, v_n)$ is an orthogonal base of V relative to Q (= relative to B), then the discriminant $d = d(Q)$ defined by this base is $\prod_1^n (\frac{1}{2}B(v_i, v_i)) = \prod_1^n Q(v_i)$.

We shall now define the Hasse invariant of Q as a certain element of the Brauer group $\mathrm{Br}(F)$. If $n = 1$, the element is the unit 1 of $\mathrm{Br}(F)$ and if $n > 1$, we define the *Hasse invariant of Q relative to an orthogonal base* $(v_1, \ldots, v_n)$ as the element of $\mathrm{Br}(F)$ determined by the tensor product

(43) $$\prod_{i<j} (Q(v_i), Q(v_j)).$$

We proceed to show that this is independent of the choice of the orthogonal base.

PROPOSITION 9.9. *Let* $(v_1, \ldots, v_n)$, $(v_1', \ldots, v_n')$ be orthogonal bases of V relative to Q. Then

(44) $$\prod_{i<j} (Q(v_i), Q(v_j)) \sim \prod_{i<j} (Q(v_i'), Q(v_j')).$$

Proof (Witt). Let U be an n-dimensional vector space equipped with a quadratic form P for which we have an orthogonal base $(u_1, \ldots, u_n)$ such that $P(u_i) = -1$, $1 \leqslant i \leqslant n$. Form $W = U \oplus V$ and define a quadratic form R on W by $R(u + v) = P(u) + Q(v)$, $u \in U, v \in V$. Then $W = U \perp V$ and the restrictions of R to U and V are P and Q respectively. We shall show that for any orthogonal base $(v_1, \ldots, v_n)$ for V we have

(45) $$\prod_{i<j} (Q(v_i), Q(v_j)) \sim C(W, R) \otimes_F (d, d)$$

where $C(W, R)$ is the Clifford algebra of R and d is a discriminant of Q. Evidently this will imply (44). Put $a_i = Q(v_i)$, $d_i = \prod_1^i a_j$, so $d_n = d$. We show first that

(46) $$C(W, R) \cong \prod_{i=1}^{n} (a_i, d_i).$$

We know that $C(W, R)$ is a central simple algebra generated by the elements u_i, v_i, $1 \leqslant i \leqslant n$, and we have the relations $u_i^2 = -1$, $v_i^2 = a_i$, $u_i u_j = -u_j u_i$, $v_i v_j = -v_j v_i$ if $i \neq j$, $u_i v_k = -v_k u_i$ for all i, k (pp. 229–230). Put

(47) $$w_1 = (v_1 u_1) \cdots (v_{n-1} u_{n-1}) v_n, \qquad w_2 = v_n u_n.$$

Since $(v_iu_i)^2 = -v_i{}^2u_i{}^2 = a_i$, $(v_iu_i)(v_ju_j) = (v_ju_j)(v_iu_i)$, $v_n(v_iu_i) = (v_iu_i)v_n$ if $i < n$ and $v_nw_2 = -w_2v_n$, we have

$$w_1{}^2 = d = d_n, \qquad w_2{}^2 = a_n, \qquad w_1w_2 = -w_2w_1. \tag{48}$$

Hence the subalgebra generated by w_1 and w_2 is (a_n, d_n) and $C(W, R) \cong (a_n, d_n) \otimes_F C'$ where C' is the centralizer in $C(W, R)$ of the subalgebra generated by w_1 and w_2 (p. 233). The elements u_i, v_i, $1 \leqslant i \leqslant n-1$, commute with w_1 and w_2 and the subalgebra generated by these elements is isomorphic to the Clifford algebra $C(W', R')$ where $W' = \sum_1^{n-1} Fu_j + \sum_1^{n-1} Fv_j$ and R' is the restriction of R to W'. Since $[C(W', R'): F] = 2^{2(n-1)}$ and $[C' : F] = 2^{2n}/4 = 2^{2(n-1)}$, we have $C' \cong C(W', R')$ and $C(W, R) \cong (a_n, d_n) \otimes C(W', R')$. The formula (46) now follows by induction on n. This and (v) and (vi′) give

$$C(W, R) \sim \prod_{i \leqslant j} (a_i, a_j) \sim (d, d) \prod_{i<j} (a_i, a_j). \tag{49}$$

Hence $\prod_{i<j} (a_i, a_j) \sim C(W, R) \otimes (d, d)$. □

In view of Proposition 9.9 it makes sense to define the *Hasse invariant* $s(Q)$ of Q to be the unit of $\mathrm{Br}(F)$ if $n = 1$ and the element of $\mathrm{Br}(F)$ defined by (43) if $n > 1$.

The Hasse invariant is either 1 or an element of order two in the Brauer group. If F is algebraically closed or is finite, then $\mathrm{Br}(F) = 1$, so in this case the Hasse invariant is trivial for any quadratic form over F. If $F = \mathbb{R}$ or a local field, then there is a unique element of order two in $\mathrm{Br}(F)$. We denote this as -1. Let Q^+ and Q^- be positive definite and negative definite quadratic forms respectively on an n-dimensional vector space over $\mathbb{R}$. Then the Hasse invariant $s(Q^\pm) = (\pm 1, \pm 1)^{n(n-1)/2} = (\pm 1)^{n(n-1)/2}$. If $n \equiv 0 \pmod 4$, then $s(Q^+) = s(Q^-)$, but these forms are inequivalent. On the other hand, we shall show that the discriminant and Hasse invariant constitute a complete set of invariants for quadratic forms over local fields: Two such forms are equivalent if and only if they have the same discriminant and the same Hasse invariant.

We develop first some results for arbitrary base fields (of characteristic $\neq 2$).

PROPOSITION 9.10. *Let $n \leqslant 3$. Then two non-degenerate quadratic forms on an n-dimensional vector space are equivalent if and only if they have the same discriminant and Hasse invariant.*

Proof. The necessity of the condition is clear and the sufficiency is clear if $n = 1$. Now let $n = 2$ and let $\mathrm{diag}\,\{a, b\}$, $\mathrm{diag}\,\{a', b'\}$ be diagonal matrices for the two quadratic forms. We are assuming that $(a, b) \sim (a', b')$ and ab and $a'b'$ differ by a square. Then $(a, b) \cong (a', b')$ and we may assume that $ab = a'b'$.

The condition $(a, b) \cong (a', b')$ implies that the quadratic forms $ax_1{}^2+bx_2{}^2-abx_3{}^2$ and $a'x_1{}^2+b'x_2{}^2-a'b'x_3{}^2$ are equivalent. Since $ab = a'b'$ we have the equivalence of $ax_1{}^2+bx_2{}^2$ and $a'x_1{}^2+b'x_2{}^2$ by Witt's cancellation theorem (BAI, p. 367). Next let $n = 3$ and assume that $Q = ax_1{}^2+bx_2{}^2+cx_3{}^2$, $Q' = a'x_1{}^2+b'x_2{}^2+c'x_3{}^2$. The hypotheses are $(a, b)(a, c)(b, c) \sim (a', b')(a', c')(b', c')$, and $d = abc$ and $d' = a'b'c'$ differ by a square, so we may assume that $d = d'$. It suffices to show that $-dQ$ and $-dQ'$, which have discriminant -1, are equivalent. A simple calculation, which we leave as an exercise, shows that $s(-dQ) \sim s(Q) \otimes (-d, -d)$. Hence $s(-dQ) \sim s(-dQ')$, so it suffices to prove the result for Q and Q' of discriminant -1. Then we may assume that $Q = ax_1{}^2+bx_2{}^2-abx_3{}^2$, $Q' = a'x_1{}^2+b'x_2{}^2-a'b'x_3{}^2$. Then $s(Q) = (a, b)(a, -ab)(b, -ab) \sim (a, b)(ab, -ab) \sim (a, b)$. Hence $(a, b) \cong (a', b')$. Since Q and Q' are the negatives of the norm forms on $(a, b)_0$ and $(a', b')_0$ respectively, it follows that Q and Q' are equivalent. □

We prove next

PROPOSITION 9.11. *Let F be a field such that every quadratic form on a five-dimensional vector space over F is a null form. Then any two non-degenerate quadratic forms on a vector space V over F are equivalent if and only if they have the same discriminant and the same Hasse invariant.*

Proof. The necessity of the condition is clear and the sufficiency holds by Proposition 9.10 if $\dim V \leqslant 3$. Hence assume $n \geqslant 4$. The hypothesis implies that any non-degenerate quadratic form P on a four-dimensional vector space U/F is universal, that is, represents every non-zero element of F. For if $a \neq 0$ we can form $U \oplus Fx$, $x \neq 0$, and define a quadratic form R on $U \oplus Fx$ by $R(u+\alpha x) = P(u) - \alpha^2 a$ for $u \in U$, $\alpha \in F$. The fact that R is a null form implies that we have a $u+\alpha x \neq 0$ such that $P(u) = \alpha^2 a$. If $\alpha = 0$ then $u \neq 0$, so P is a null form and hence P is universal. If $\alpha \neq 0$ then $P(\alpha^{-1}u) = a$. Thus P is universal. The universality of non-degenerate quadratic form on four-dimensional spaces implies that if Q is a non-degenerate quadratic form on an n-dimensional vector space V, $n \geqslant 4$, then we have an orthogonal base $(v_1, \ldots, v_n)$ with $Q(v_i) = 1$ for $i > 3$. If R denotes the restriction of Q to $Fv_1 + Fv_2 + Fv_3$, then the definitions and the formula $(1, a) \sim 1$ show that Q and R have the same discriminant and Hasse invariant. If Q' is a second non-degenerate quadratic form on an n-dimensional vector space, then we have an orthogonal base $(v'_1, \ldots, v'_n)$ with $Q(v'_i) = 1$ for $i > 3$. The conditions that $s(Q) = s(Q')$ and Q and Q' have the same discriminant imply the same conditions on the restrictions of Q and Q' to $Fv_1 + Fv_2 + Fv_3$ and $Fv'_1 + Fv'_2 + Fv'_3$. Hence these restrictions are equivalent and so Q and Q' are equivalent. □

We require one further result for general fields.

PROPOSITION 9.12. *The quadratic form* $Q = a_1x_1{}^2 + a_2x_2{}^2 + a_3x_3{}^2 + a_4x_4{}^2$ *with* $d = a_1a_2a_3a_4 \neq 0$ *is a null form if and only if* $F(\sqrt{d})$ *is a splitting field for* $(-a_3a_4, -a_2a_4)$.

Proof. Put $a = -a_3a_4$, $b = -a_2a_4$, $c = a_2a_3a_4$. Then cQ is equivalent to $dx_1{}^2 - ax_2{}^2 - bx_3{}^2 + abx_4{}^2$, and $-ax_2{}^2 - bx_3{}^2 + abx_4{}^2$ is the norm form on $(a, b)_0 = (-a_3a_4, -a_2a_4)_0$. Suppose first that $\sqrt{d} \in F$ so $F(\sqrt{d}) = F$. In this case cQ is equivalent to $x_1{}^2 - ax_2{}^2 - bx_3{}^2 + abx_4{}^2$, the norm form of (a, b), and $(a, b) \sim 1$ if and only if this norm form and hence Q is a null form. Thus the result holds in this case. Next assume $\sqrt{d} \notin F$. Then $F(\sqrt{d})$ is a splitting field of (a, b) if and only if $F(\sqrt{d})$ is a subfield of (a, b) (p. 221). The condition for this is that $(a, b)_0$ contains an element u such that $u^2 = d$. This is the case if and only if cQ and hence Q is a null form. Hence the result holds in this case also. □

We now suppose that F is a local field. Then the results of the previous section show that there is a unique element of order two in $\mathrm{Br}(F)$. This has a representative that is a cyclic algebra (W, σ, π) where W is an unramified quadratic extension of F, σ the automorphism $\neq 1$ of W/F, and π is any element of F such that $\pi \in P - P^2$ where P is the ideal of non-units in the valuation ring R of W. Since we are assuming that char $F \neq 2$, $W = F(\sqrt{a})$ and hence $(W, \sigma, \pi) = (a, \pi)$. We have

PROPOSITION 9.13. *Let F be a local field of characteristic $\neq 2$ and let A be a quaternion division algebra over F. Then any quadratic extension field E/F is a splitting field for A.*

Proof. We have $A = (W, \sigma, \pi) = (a, \pi)$. The extension field E/F is either unramified or completely ramified. In the first case $E \cong W$, so A contains a subfield isomorphic to E and hence E is a splitting field. Next assume that E is completely ramified. Then $E = F(b)$ where b is a root of a quadratic Eisenstein polynomial. By completing the square we may assume that $b^2 = \pi' \in P - P^2$. We can construct the division algebra (W', σ', π') where W' is unramified and σ' is an automorphism of period two. Then $(W', \sigma', \pi') \cong (W, \sigma, \pi) = A$, so again A contains a subfield isomorphic to E and E is a splitting field. □

The next result we shall need on quaternion algebras requires the stronger hypothesis that the residue field R/P is of characteristic $\neq 2$. This is

PROPOSITION 9.14. *If F is a local field such that* char $R/P \neq 2$, *then* $(-1,-1)/F \sim 1$.

(This does not always hold if char $R/P = 2$. For example $(-1,-1)/\mathbb{Q}_2 \nsim 1$. See exercise 5 below.)

Proof. The result is clear if -1 is a square in F. Hence we assume $W = F(\sqrt{-1}) \supsetneqq F$. Now W is unramified since the reducibility of x^2+1 in $(R/P)[x]$ implies by Hensel's lemma the existence of $\sqrt{-1}$ in F. Thus $[(R_W/P_W):(R/P)] = 2$ so the residue degree of W/F is two and hence the ramification index is 1, that is W/F is unramified. It follows that if $(-1,-1) \nsim 1$ then $(-1,-1) \cong (-1,\pi)$ where $\pi \in P_w - P_w^2$. This implies that $-\pi \in N(W^*)$. This contradicts the determination of $N(W^*)$ given on p. 610. Hence $(-1,-1) \sim 1$. □

We can now prove

PROPOSITION 9.15. *If F is a local field of characteristic $\neq 2$, then any non-degenerate quadratic form Q on a five-dimensional vector space over F is a null form.*

Proof. We may assume that $Q = \sum_1^5 a_i x_i^2$ and if we multiply Q by $\prod a_i$ we may assume that $\prod a_i$ is a square. Suppose that Q is not a null form. Then $\sum_1^4 a_j x_j^2$ is not a null form. By Proposition 9.12 $F(\sqrt{a_1a_2a_3a_4})$ is not a splitting field for $(-a_3a_4, -a_2a_4)$. Then $(-a_3a_4, -a_2a_4) \nsim 1$, so by Proposition 9.13, $a_1a_2a_3a_4$ is a square. Hence a_5 is a square. Similarly every a_i is a square and hence we may assume that $Q = \sum_1^5 x_i^2$. Then $(-1,-1) \nsim 1$ by Proposition 9.12 again. This contradicts Proposition 9.14 if the characteristic of the residue field of $F \neq 2$. Now suppose this is 2. Then F contains the field $\mathbb{Q}_2$ of 2-adic numbers and $\mathbb{Q}_2$ contains $\sqrt{-7}$. To see this we note that, by Hensel's lemma, x^2+x+2 is reducible in $\mathbb{Q}_2$. Hence $\mathbb{Q}_2$ contains $\frac{1}{2}(-1 \pm \sqrt{-7})$ and $\mathbb{Q}_2$ contains $\sqrt{-7}$. Then $1^2+1^2+1^2+2^2+(\sqrt{-7})^2 = 0$ and $\sum_1^5 x_i^2$ is a null form in $\mathbb{Q}_2$ and hence in F. This completes the proof. □

By Proposition 9.11 and 9.15 we have

THEOREM 9.24. *If F is a local field of characteristic $\neq 2$, then any two non-degenerate quadratic forms on an n-dimensional vector space V/F are equivalent if and only if they have the same discriminant and Hasse invariant.*

We show next that if $n \geqslant 3$, the two invariants are independent. The proof of Proposition 9.11 shows that it suffices to prove this for $n = 3$. Then a calculation indicated in the proof of Proposition 9.10 shows that if $Q = ax_1{}^2 + bx_2{}^2 + cx_3{}^2$, then $s(Q) \sim (-da, -db)(-d, -d)$ where $d = abc$. It is clear from this formula that for a given d, a and b can be chosen so that $s(Q) = \pm 1$. Hence a, b, c can be chosen so that the discriminant is any d and $s(Q) = \pm 1$. This result and Theorem 9.24 imply that the number of equivalence classes of non-degenerate quadratic forms over F with $n \geqslant 3$ is $2|F^*/F^{*2}|$ where F^{*2} is the subgroup of squares in F^*. It is easy to see, using an argument based on Hensel's lemma as in the proof of Proposition 9.8 on norms, that if the characteristic of the residue field is $\neq 2$, then $|F^*/F^{*2}| = 4$. Accordingly, the number of equivalence classes of non-degenerate quadratic forms for a given $n \geqslant 3$ is 8. Some information on the case $n < 3$ and the case in which the residue class has characteristic two is indicated in the exercises.

EXERCISES

1. Show that $ax_1{}^2 + bx_2{}^2$, $ab \neq 0$, is a null form if and only if $ab = -d^2$ and that $ax_1{}^2 + bx_2{}^2 + cx_3{}^2$ is a null form if and only if $s(Q) \sim (-d, -d)$, $d = abc$.

2. Let F be a finite dimensional extension of $\mathbb{Q}_2$. Show that $|F^*/F^{*2}| = 2^{[F:\mathbb{Q}_2]}$.

3. Determine the number of equivalence classes of non-degenerate quadratic forms with $n = 2$ over a local field.

4. Let F be a field such that $(a, b) \sim 1$ for every quaternion algebra over F. Show that two non-degenerate quadratic forms on an n-dimensional space over F are equivalent if and only if they have the same discriminant.

5. Show that the quadratic form $x_1^2 + x_2^2 + x_3^2$ on a three-dimensional vector space over $\mathbb{Q}_2$ is not a null form. Hence conclude that $(-1, -1)/\mathbb{Q}_2 \nsim 1$.

REFERENCES

S. Kürschák, Über Limesbildung und allgemeine Körpertheorie, *Journal für die reine und angewandete Mathematik*, vol. 142 (1913), 211–253.

E. Witt, Theorie der quadratischen Formen in beliebigen Körpern, *Journal für die reine und angewandete Mathematik*, vol. 176 (1937), 31–44.

O. F. G. Schilling, *The Theory of Valuations*, American Mathematical Society Surveys, Providence, R.I., 1950.

O. Zariski and P. Samuel, *Commutative Algebra* II, New printing, Springer, New York, 1960.

E. Artin, *Algebraic Numbers and Algebraic Functions*, Gordon and Breach, New York, 1967.

10

Dedekind Domains

In this chapter we shall study the domains in which proper non-zero ideals can be factored in one and only one way as products of prime ideals. The most notable examples are the rings of $\mathbb{Z}$-integral elements of number fields, that is, finite dimensional extension fields of the field $\mathbb{Q}$ (see BAI, pp. 278–281). These are the objects of study of algebraic number theory. Another important class of examples are the rings that occur in the study of algebraic curves. Here we begin with a field $F(x, y)$ where F is a base field (usually algebraically closed), x is transcendental over F, and y is algebraic over $F(x)$. Then the subring of elements that are integral over $F[x]$ has the factorization property stated above.

There are many equivalent ways of defining the class of domains, called Dedekind domains, in which the fundamental factorization theorem for ideals into prime ideals holds (see section 10.2). We shall take as our point of departure a definition based on the concepts of fractional ideals and of invertibility. The latter is equivalent to projectivity as a module for the given ring.

The result that the domains mentioned above are Dedekind can be deduced from a general theorem stating that if D is Dedekind with field of fractions F and E is a finite dimensional extension field of F, then the subring D' of

D-integral elements of E is a Dedekind domain. This is proved in section 10.3. In sections 10.4 and 10.5 we consider the central problem of studying the factorization in D' of extensions of prime ideals of D. This is closely related to the problem of extension of valuations from F to E that was considered in chapter 9.

Besides the study of ideal theory, we consider the structure of finitely generated modules over Dedekind domains. The special case of torsion-free modules is a classical one that was first treated by E. Steinitz (see section 10.6).

Finally, we consider the class group of a Dedekind domain as defined in section 7.9 and we show in section 10.6 that this group has a concrete realization as a classical group defined by the fractional ideals.

10.1 FRACTIONAL IDEALS. DEDEKIND DOMAINS

Let D be a domain, F its field of fractions. It is useful to extend the concept of an ideal in D to certain submodules of F given in the following

DEFINITION 10.1. *If D is a domain and F is its field of fractions, a $(D-)$* fractional ideal *I is a non-zero D-submodule of F such that there exists a non-zero a in D such that $aI \subset D$ (or, equivalently, $aI \subset I \cap D$).*

The fractional ideals contained in D are the non-zero ideals of D. These are called *integral ideals*. If I is a fractional ideal and a is a non-zero element of D such that $aI \subset D$, then aI is an integral ideal. If b is any non-zero element of F, then $Db = \{db|d \in D\}$ is a fractional ideal since it is clearly a D-submodule of F and if $b = ac^{-1}$, $a, c \in D$, then $c(Db) \subset D$. A fractional ideal of the form Db is called a *principal* ideal.

If I_1 and I_2 are fractional ideals, then so is the module sum $I_1 + I_2$ since this is a submodule, and if $a_i I_i \subset D$ for $a_i \neq 0$ in D, then $a_1 a_2(I_1 + I_2) \subset D$. The intersection $I_1 \cap I_2$ is a fractional ideal also since this is a D-submodule and $I_1 \cap I_2 \neq 0$ since if $b_i \neq 0$ is in I_i, we have an $a_i \neq 0$ in D such that $a_i b_i \in D \cap I_i$. Then $(a_1 b_1)(a_2 b_2) \neq 0$ is in $I_1 \cap I_2$. Moreover, if $a \neq 0$ satisfies $aI_1 \subset D$, then $a(I_1 \cap I_2) \subset D$. We define $I_1 I_2 = \{\sum b_{1i} b_{2i} | b_{1i} \in I_1,\ b_{2i} \in I_2\}$. Evidently this is a D-submodule of F and the argument used for $I_1 \cap I_2$ shows that $I_1 I_2$ is a fractional ideal.

We shall need to know what the homomorphisms of a fractional ideal into D look like. This information is given in

PROPOSITION 10.1. *Let f be a D-homomorphism of the fractional ideal I into D and let b be any non-zero element of I. Then f has the form*

$$a \rightsquigarrow b^{-1}f(b)a, \tag{1}$$

$a \in I$.

Proof. Choose a $d \neq 0$ in D such that $dI \subset D$. Then $db \in D$ and $da \in D$ for any $a \in I$. We have

$$daf(b) = f(dab) = dbf(a).$$

Hence $af(b) = bf(a)$ and so $f(a) = b^{-1}f(b)a$. Thus f has the form given in (1). □

If I is a fractional ideal, we define

$$I^{-1} = \{c \in F | cI \subset D\}. \tag{2}$$

It is clear that I^{-1} is a D-submodule of F. Since there exist $a \neq 0$ in D such that $aI \subset D$, it is clear that $I^{-1} \neq 0$. Moreover, if $b \neq 0$ is in $I \cap D$, then $I^{-1}b \subset D$. Hence I^{-1} is a fractional ideal. If I is the principal ideal Db, $b \neq 0$, then it is clear that $I^{-1} = Db^{-1}$. It is clear also that $I_1 \subset I_2 \Rightarrow I_1^{-1} \supset I_2^{-1}$.

If I is a fractional ideal, then $I^{-1}I \subset D$ so $I^{-1}I$ is an integral ideal. We shall now call I *invertible* if

$$I^{-1}I = D. \tag{3}$$

If $I = bD \neq 0$, then $I^{-1} = b^{-1}D$. Evidently $I^{-1}I = (b^{-1}D)(bD) = D$. Hence every principal ideal is invertible. The fractional ideals constitute a commutative monoid under multiplication with D as the unit. If I is invertible, then $I^{-1}I = D$ so I^{-1} is the inverse of I in the monoid of fractional ideals. Conversely, let I have the inverse J in this monoid. Then $JI = D$ implies that $J \subset I^{-1}$ as defined by (2). Since $I^{-1}I \subset D$, we have $I^{-1}I = D$. Since the inverse of an element of a monoid is unique, we have $I^{-1} = J$. Thus the invertible ideals constitute the group of units of the monoid of fractional ideals. We prove next

PROPOSITION 10.2. *Any invertible ideal is finitely generated.*

Proof. Let I be invertible. Then $I^{-1}I = D$ implies the existence of elements $c_i \in I^{-1}$, $b_i \in I$ such that $\sum_1^n c_ib_i = 1$. Now let $b \in I$. Then $b = b1 = \sum (bc_i)b_i$ and $a_i = bc_i \in D$. Thus $I = \sum Db_i$ and $(b_1, \dots, b_n)$ is a set of generators for I. □

Next we establish the equivalence of the conditions of invertibility and projectivity for fractional ideals:

PROPOSITION 10.3. *A fractional ideal I is invertible if and only if it is projective.*

Proof. First suppose that I is invertible. Then $1 = \sum_1^m c_i b_i$, $c_i \in I^{-1}$, $b_i \in I$. The map $f_i: a \rightsquigarrow ac_i$, $a \in I$, is in $\hom_D(I, D)$ and $a = \sum (ac_i)b_i = \sum f_i(a)b_i$. Hence I is D-projective by the "dual basis lemma" (Proposition 3.11, p. 152). Conversely, suppose that I is projective. Then we have a set of generators $\{b_\alpha\}$ of I and corresponding maps $f_\alpha \in \hom_D(I, D)$ such that for any $a \in I$, $f_\alpha(a) = 0$ for all but a finite number of α and $a = \sum f_\alpha(a) b_\alpha$. We have shown in Proposition 10.1 that if b is a non-zero element of I, then f_α has the form $a \rightsquigarrow b^{-1} f_\alpha(b)a$. Since $f_\alpha(b) = 0$ for all α except, say, $\alpha = 1, 2, \ldots, m$, we have $f_\alpha = 0$ for $\alpha \neq 1, 2, \ldots, m$ and $a = \sum_1^m b^{-1} f_i(b) a b_i$. Then $1 = \sum b^{-1} f_i(b) b_i$ and since $b^{-1} f_i(b) a \in D$ for all $a \in I$, $c_i = b^{-1} f_i(b) \in I^{-1}$. Thus $1 = \sum c_i b_i$, $c_i \in I^{-1}$, $b_i \in I$, and hence $I^{-1} I = D$. Thus I is invertible. □

We say that a fractional ideal I is a *divisor* of the fractional ideal J if there exists an integral ideal K such that $J = IK$. Since $K \subset D$ and $DI \subset I$, this implies that $J \subset I$. If I is invertible we have the important fact that the converse holds: If $I \supset J$, then I is a divisor of J. For, $D = I^{-1} I \supset I^{-1} J$, so $K = I^{-1} J$ is integral. Moreover, $J = I(I^{-1} J) = IK$.

We are interested in existence and uniqueness of factorization of ideals into prime ideals. For invertible integral ideals we have the following uniqueness property.

PROPOSITION 10.4. *Let I be integral and invertible and suppose that $I = P_1 P_2 \cdots P_m$ where the P_i are prime ideals of D. Then this is the only factorization of I as a product of prime ideals.*

Proof. We remark first that if M is a commutative monoid and a is a unit in M, then any factor of a is a unit. Applying this remark to the multiplicative monoid of fractional ideals we see that the P_i are invertible. Now let $I = Q_1 Q_2 \cdots Q_n$ where the Q_j are prime ideals. Then $P_1 \supset Q_1 \cdots Q_n$ and since P_1 is prime, we may assume that $P_1 \supset Q_1$. Since P_1 is invertible, we have $Q_1 = P_1 R_1$ where R_1 is integral. Then $R_1 \supset Q_1$. Since Q_1 is prime, either $P_1 \subset Q_1$ or $R_1 \subset Q_1$, so either $P_1 = Q_1$ or $R_1 = Q_1$. In the latter case $R_1 = P_1 R_1$ and since $I = Q_1 Q_2 \cdots Q_n = P_1 R_1 Q_2 \cdots Q_n$, R_1 is invertible. Then $D = R_1^{-1} R_1 = R_1^{-1} R_1 P_1 = P_1$, contrary to the hypothesis that P_1 is prime (hence proper). Thus $Q_1 = P_1$, so we have $P_1 P_2 \cdots P_m = P_1 Q_2 \cdots Q_n$ and multiplication by P_1^{-1} gives $D = Q_2 \cdots Q_n$ if $m = 1$ and $P_2 \cdots P_m = Q_2 \cdots Q_n$ if $m > 1$. In the first case we have $n = 1$ also, and in the second $n > 1$ and induction on m can be applied since $P_2 \cdots P_m$ is invertible. □

We shall now define the class of rings that will concern us in this chapter.

DEFINITION 10.2. *A domain D is called a* Dedekind domain *if every D-fractional ideal of F (the field of fractions of D) is invertible.*

By Proposition 10.3, this is equivalent to assuming that every fractional ideal is projective. It is clear from Proposition 10.2 that any Dedekind domain is noetherian. We have also the following important property:

PROPOSITION 10.5. *Any prime ideal $\neq 0$ in a Dedekind domain is maximal.*

Proof. Let I be a non-zero prime ideal ($=$ integral prime ideal) of the Dedekind domain D. If I is not maximal, we have an ideal J of D such that $D \supsetneqq J \supsetneqq I$. Then $I = JK$ where K is integral. Since $J \neq I$, we have $I \not\supset J$. Also $I \not\supset K$ since $K = DK \supset JK = I$, so $I \supset K$ implies $I = K$ and $K = JK$. Then $D = KK^{-1} = JKK^{-1} = J$, contrary to $D \neq J$. Thus we have $I = JK$ with $J \not\subset I$ and $K \not\subset I$. This contradicts the assumption that I is prime. □

We have the following fundamental factorization theorem for Dedekind domains:

THEOREM 10.1. *Every proper integral ideal of a Dedekind domain can be written in one and only one way as a product of prime ideals.*

Proof. In view of Proposition 10.4, all we have to do is prove that if I is a proper integral ideal in the Dedekind domain D then I is a product of prime ideals $\neq 0$. Suppose that this is not the case, so the set of proper integral ideals that are not products of prime ideals is not vacuous. Then by the noetherian property of D this set contains a maximal element I. Then I is not prime and hence I is not maximal. Then there exists an ideal I_1 in D such that $D \supsetneqq I_1 \supsetneqq I$. Then $I = I_1 I_2$ where I_2 is a proper integral ideal. We have $I_2 \supset I$, and $I_2 = I$ implies that $D = II^{-1} = I_1 I_2 I_2^{-1} = I_1$. Thus $D \supsetneqq I_i \supsetneqq I$ for $i = 1, 2$. By the maximality of I, I_i is a product of prime ideals. This gives the contradiction that $I = I_1 I_2$ is a product of prime ideals. □

An immediate consequence of this result is

COROLLARY 1. *Suppose that $I = P_1 P_2 \cdots P_m$ where the P_i are primes. Then the integral ideals $\neq D$ containing I have the form $P_{i_1} P_{i_2} \cdots P_{i_r}$ where $1 \leqslant i_1 < i_2 < \cdots < i_r \leqslant m$.*

The proof is clear.

It is clear also that we have the following consequence of Corollary 1.

COROLLARY 2. *If I is an integral ideal in a Dedekind domain D, then D/I is artinian.*

It is evident that the fractional ideals of a Dedekind domain constitute a group under multiplication with D as the unit and I^{-1} as defined by (2) as the inverse of I. The fundamental factorization theorem gives the structure of this group:

THEOREM 10.2. *The group of fractional ideals of a Dedekind domain is a direct product of the cyclic subgroups generated by the prime ideals. These are infinite.*

Proof. Let I be a fractional ideal $\neq D$ and let a be a non-zero element of D such that $aI \subset D$. Then aI is an integral ideal, so either $aI = D$ or this is a product of prime ideals. Also aD is an integral ideal, so either $aD = D$ or aD is a product of prime ideals. Since $I(aD) = aI$, $I = (aI)(aD)^{-1}$. Then it is clear that we can write $I = P_1^{k_1} P_2^{k_2} \cdots P_r^{k_r}$ where the P_i are prime ideals and the k_i are non-zero integers. It follows from Theorem 10.1 that, conversely, if the k_i are non-zero integers and the P_i are prime ideals, then $P_1^{k_1} P_2^{k_2} \cdots P_r^{k_r} \neq D$. This implies that the representation of a fractional ideal $\neq D$ as a product $P_1^{k_1} P_2^{k_2} \cdots P_r^{k_r}$ is unique. Hence the group of fractional ideals is the direct product of the cyclic subgroups generated by the primes. It is clear also that these are infinite groups. □

EXERCISES

1. Show that any finitely generated D-submodule of $F \neq 0$ is a fractional ideal.
2. Show that if D is Dedekind, then the only fractional ideal I satisfying $I^2 = I$ is $I = D$.
3. Use Theorem 10.1 to prove that any p.i.d. is factorial.
4. Give an example of a factorial domain that is not Dedekind.
5. Let $D = \mathbb{Z}[\sqrt{-3}]$, the subring of $F = \mathbb{Q}(\sqrt{-3})$ of elements $a + b\sqrt{-3}$, $a, b \in \mathbb{Z}$. Let I be the fractional D-ideal $D + Dw$ where $w = -\frac{1}{2} + \frac{1}{2}\sqrt{-3}$ (a primitive cube root of 1). Compute I^{-1} and II^{-1}. Is I invertible?
6. Let $D = \mathbb{Z}[\sqrt{-5}]$ with quotient field $F = \mathbb{Q}[\sqrt{-5}]$. Show that $I = D3 + D\sqrt{-5}$ is a projective D-module that is not free.
7. Show that any Dedekind domain that is factorial is a p.i.d.

In the remaining exercises, D is a Dedekind domain.

8. Let I_1 and I_2 be D-integral ideals and write $I_j = P_1^{e_{j1}} \cdots P_g^{e_{jg}}$ where P_i are distinct prime ideals and the $e_{jk} \geqslant 0$. Show that $I_1 + I_2 = P_1{}^{\min(e_{11},e_{21})} \cdots P_g{}^{\min(e_{1g},e_{2g})}$ and $I_1 \cap I_2 = P_1{}^{\max(e_{11},e_{21})} \cdots P_g{}^{\max(e_{1g},e_{2g})}$.

9. Show that the lattice of D-integral ideals is distributive (cf. BAI, p. 463).

10. (Chinese remainder theorem for Dedekind domains.) Let $I_1, \ldots, I_n$ be integral ideals in a Dedekind domain D, $a_1, \ldots, a_n$ elements of D. Show that the system of congruences $x \equiv a_j \pmod{I_j}$, $1 \leqslant j \leqslant n$, has a solution $x = a$ in D if and only if for any j, k, $1 \leqslant j$, $k \leqslant n$, we have $a_j \equiv a_k \pmod{I_j + I_k}$. (*Hint*: The necessity of the conditions is clear. The sufficiency is proved by induction on n. For $n = 2$ we have $a_1 - a_2 = b_1 - b_2$, $b_j \in I_j$. Then $a = a_1 - b_1 = a_2 - b_2$ satisfies $a \equiv a_j \pmod{I_j}$, $j = 1, 2$. For the inductive step one uses the distributive law $\bigcap_{j=1}^{n-1} (I_j + I_n) = \bigcap_{j=1}^{n} I_j + I_n$. See exercise 9 and p. 461f of BAI.)

11. Let I be a fractional ideal, J an integral ideal. Show that there exists an element $a \in I$ such that $I^{-1}a + J = D$. (*Hint*: Let $P_1, \ldots, P_r$ be the prime ideals dividing J. For $1 \leqslant i \leqslant r$, choose $a_i \in IP_1 \cdots P_r P_i^{-1}$, $\notin IP_1 \cdots P_r$ and put $a = \sum_1^r a_i$. Then $P_i \not\supset aI^{-1}$, $1 \leqslant i \leqslant r$, and $aI^{-1} + J = D$.)

12. Let I be a fractional ideal and a any non-zero element of I. Show that there exists a $b \in I$ such that $I = Da + Db$. (*Hint*: Apply exercise 10 to the given I and $J = aI^{-1}$.)

13. Let I be a fractional ideal, J an integral ideal. Show that I/IJ and D/J are isomorphic D-modules.

14. Let P be a prime ideal $\neq 0$ in D. Show that if $e \geqslant 1$ then $D/P^e \supset P/P^e \supset \cdots \supset P^{e-1}/P^e \supset 0$ is a composition series for D/P^e as D-module. Show that P^i/P^{i+1}, $i \geqslant 0$, is a one-dimensional vector space over the field D/P.

15. Show that if D has only a finite number of ideals $P_1, \ldots, P_s$, then D is a p.i.d. (*Hint*: Let $I = P_1^{e_1} \cdots P_s^{e_s}$ where the $e_i \geqslant 0$. For each i choose $a_i \in IP_1 \cdots P_s P_i^{-1}$, $\notin P_i^{e_i + 1}$. Put $a = \sum_1^s a_i$. Then $a \in P_i^{e_i}$, $\notin P_i^{e_i+1}$ and $I = Da$.)

10.2 CHARACTERIZATIONS OF DEDEKIND DOMAINS

We shall give a number of characterizations of Dedekind domains. The first one involves the important concept of integral closedness that we introduced in section 7.7.

It is easy to see that any factorial domain D is integrally closed (exercise 1, below). We now prove

PROPOSITION 10.6. *Any Dedekind domain is integrally closed.*

Proof. Let D be a Dedekind domain, F its field of fractions, and let u be an element of F that is D-integral. Let $f(x) = x^n + a_1 x^{n-1} + \cdots + a_n$ be a monic polynomial with coefficients in D such that $f(u) = 0$ and put $M = D1 + Du + \cdots + Du^{n-1}$. If $u = bc^{-1}$ where $b, c \in D$, then $c^{n-1}M \subset D$, so M is a fractional ideal. Evidently $M^2 = M$, which implies $M = D$ (exercise 2, p. 604). Then $u \in M = D$ and hence D is integrally closed. □

We have now established the following three properties of Dedekind domains D:

(1) D is noetherian.

(2) Every non-zero prime ideal in D is maximal.

(3) D is integrally closed.

We proceed to show that these properties characterize Dedekind domains. The proof is based on some lemmas that are of independent interest.

LEMMA 1. *Any fractional ideal for a noetherian domain D is finitely generated as D-module.*

Proof. If I is a fractional D-ideal, we have an $a \neq 0$ in D such that $I' = aI \subset D$. Then I' is an ideal in D and since D is noetherian, I' is a finitely generated D-module. Then $I = a^{-1}I'$ is finitely generated. □

LEMMA 2. *Any non-zero ideal of a noetherian domain contains a product of non-zero prime ideals.*

Proof. As in the proof of Theorem 10.1, if the result is false we have an ideal $I \neq 0$ maximal relative to the property that I does not contain a product of non-zero prime ideals. Then I is not prime and hence there exists ideals I_j, $j = 1, 2$, such that $I_j \supsetneq I$ and $I_1 I_2 \subset I$. By the maximality of I, I_j contains a product of non-zero prime ideals. Since $I \supset I_1 I_2$, this gives a contradiction. □

LEMMA 3. *Let D be an integrally closed noetherian domain, F the field of fractions, I a fractional D-ideal. Then*

$$S \equiv \{s \in F \mid sI \subset I\} = D.$$

Proof. Let $s \in S$. Since I is finitely generated and faithful as D-module and $sI \subset I$, s is D-integral by the lemma on p. 408. Since D is integrally closed, $s \in D$. Then $S \subset D$ and since $D \subset S$ follows from the definition of a fractional ideal, we have $S = D$. □

LEMMA 4. *Let D be a noetherian domain in which every non-zero prime ideal is maximal and let I be an ideal of D such that $I \subsetneq D$. Then $I^{-1} \supsetneq D$.*

Proof. Let $a \neq 0$ be in I. Then $D \supset I \supset aD$, and by Lemma 2, $aD \supset P_1 P_2 \cdots P_m$ where the P_i are prime ideals $\neq 0$. We assume that m is minimal. By the noetherian condition (or by Zorn's lemma) there exists a maximal ideal $P \supset I$. Then we have $P \supset I \supset aD \supset P_1 \cdots P_m$. Since P and the P_i are prime, the hypothesis that non-zero prime ideals are maximal implies that $P = P_i$ for some i, and we may assume that $P = P_1$. If $m = 1$ we have $aD = I$. Then $I^{-1} = a^{-1}D$ and since $I \subsetneq D$, $a^{-1} \notin D$ and $I^{-1} = a^{-1}D \supsetneq D$. Next assume that $m > 1$. Then $aD \not\supset P_2 \cdots P_m$ by the minimality of m and so we can choose a $b \in P_2 \cdots P_m, \notin aD$. Put $c = a^{-1}b$. Then $c \notin D$ and $cI \subset cP = a^{-1}bP \subset a^{-1}PP_2 \cdots P_m \subset a^{-1}(aD) = D$. Thus $c \in I^{-1}$ and $c \notin D$, so again we have $I^{-1} \supsetneq D$. □

We can now prove

THEOREM 10.3. *A domain D is Dedekind if and only if* (1) *D is noetherian.* (2) *Every non-zero prime ideal in D is maximal.* (*Equivalently, D has Krull dimension* $\leqslant 1$.) (3) *D is integrally closed.*

Proof. We have seen that (1)–(3) hold for any Dedekind domain. Now assume that D is a domain satisfying these conditions and let I be a D-fractional ideal. The fractional ideal $I^{-1}I$ is integral and $II^{-1}(II^{-1})^{-1} \subset D$, so $I^{-1}(II^{-1})^{-1} \subset I^{-1}$. It follows from Lemma 3 that $(II^{-1})^{-1} \subset D$. Since $II^{-1} \subset D$, it follows from Lemma 4 that $II^{-1} = D$. Hence every fractional ideal is invertible and D is Dedekind. □

Our next characterization of Dedekind domains will be in terms of localizations. Let D be a domain, S a submonoid of the multiplicative monoid of non-zero elements of D. We consider the subring D_S of the field of fractions F of D consisting of the elements as^{-1}, $a \in D$, $s \in S$. It is readily seen that this is isomorphic to the localization of D at S as defined in section 7.2. Hence the results on localization that were developed in sections 7.2 and 7.3 are available for the study of the rings D_S.

Let I be a fractional ideal for D and put $I_S = \{bs^{-1} | b \in I, s \in S\}$. Then I_S is a D_S-submodule of F, $I_S \neq 0$, and if a is a non-zero element of D such that $aI \subset D$, then $aI_S \subset D_S$. Since it is evident that F is the field of fractions for D_S, it follows that I_S is a D_S-fractional ideal. If I_1 and I_2 are D-fractional ideals, then

$$(I_1 + I_2)_S = I_{1S} + I_{2S}, \tag{4}$$

(5) $$(I_1I_2)_S = I_{1S}I_{2S},$$

(6) $$(I_1 \cap I_2)_S = I_{1S} \cap I_{2S}.$$

The first two of these are clear. To verify the third we note that $(I_1 \cap I_2)_S \subset I_{1S} \cap I_{2S}$ is clear. Now let $a \in I_{1S} \cap I_{2S}$, so $a = a_1 s_1^{-1} = a_2 s_2^{-1}$ where $a_i \in I_i$, $s_i \in S$. Then $b = a_1 s_2 = a_2 s_1 \in I_1 \cap I_2$ and $a = b(s_1 s_2)^{-1} \in (I_1 \cap I_2)_S$. Hence (6) holds.

We note next that if I is a finitely generated fractional D-ideal, then

(7) $$(I^{-1})_S = (I_S)^{-1}.$$

For, it follows from the definition of I^{-1} that $(I_1 + I_2)^{-1} = I_1^{-1} \cap I_2^{-1}$. It is clear also that if $b \neq 0$ in F, then $(Db)_S = D_S b$. Hence if $I = Db_1 + \cdots + Db_m$, then $I_S = D_S b_1 + \cdots + D_S b_m$ and $I_S^{-1} = \bigcap_1^m D_S b_i^{-1}$. On the other hand, $I^{-1} = \bigcap_1^m Db_i^{-1}$, and so by (6), $(I^{-1})_S = \bigcap_1^m D_S b_i^{-1} = I_S^{-1}$. Hence we have (7).

Now let I' be a fractional ideal for D_S. Then we have an element as^{-1}, $a \neq 0$ in D, $s \in S$, such that $as^{-1}I' \subset D_S$. Then $as^{-1}I'$ is a non-zero ideal contained in the localization D_S. By section 7.2 this has the form I_S for a non-zero ideal I contained in D. Then $I' = a^{-1}sI_S = ((a^{-1}s)I)_S$. Thus any fractional D_S-ideal has the form I_S for some fractional D-ideal I.

We can now prove

THEOREM 10.4. *Let D be a domain.*

(1) *If D is Dedekind, then D_S is Dedekind for every submonoid of the multiplicative monoid of non-zero elements of D.*
(2) *If D is Dedekind and P is a prime ideal $\neq 0$ in D, then D_P is a discrete valuation ring.*
(3) *If D is noetherian and D_P is a discrete valuation ring for every maximal ideal P in D, then D is Dedekind.*

(We recall that if P is prime ideal, $D_P \equiv D_S$ where S is the multiplicative monoid $D - P$.)

Proof. (1) Suppose that D is Dedekind and let I' be a D_S-fractional ideal. Then $I' = I_S$ for a fractional D-ideal I. Since D is Dedekind, D is noetherian and hence I is finitely generated. Then $(I')^{-1} = I_S^{-1} = (I^{-1})_S$ by (7) and $I'I'^{-1} = I_S(I^{-1})_S = (II^{-1})_S$ (by (5)) $= D_S$. Hence I' is invertible and D_S is Dedekind.

(2) The localization D_P at the prime ideal $P \neq 0$ is a local ring whose only maximal ideal is P_P. Since D_P is Dedekind, P_P is the only prime ideal $\neq 0$ in D_P. Hence D_P is a p.i.d. by exercise 14, p. 605. Then D_P is a discrete valuation ring by Proposition 9.2, p. 570.

(3) Now assume that D is noetherian and that D_P is a discrete valuation ring for every maximal ideal P of D. Then D_P is a p.i.d. and hence D_P is Dedekind. Let I be a fractional D-ideal. Since D is noetherian, I is finitely generated. Hence $I_P^{-1} = (I^{-1})_P$. Then $(II^{-1})_P = I_P(I^{-1})_P = I_P I_P^{-1} = D_P$ for every maximal ideal P of D. We can now apply the principal of passage from local to global: We have the injection $i: II^{-1} \hookrightarrow D$, which localizes to the injection $(II')_P \hookrightarrow D_P$. Since the latter is surjective for all P, it follows from Proposition 7.11.2 (p. 402) that $II^{-1} \hookrightarrow D$ is surjective. Thus $II^{-1} = D$ for every fractional D-ideal I. Hence D is Dedekind. □

Dedekind domains can also be characterized by the factorization property given in Theorem 10.1. In fact, we have the following stronger result:

THEOREM 10.5. *Let D be a domain with the property that every proper integral ideal of D is a product of prime ideals. Then D is Dedekind.*

Proof (Zariski-Samuel). We show first that any invertible prime ideal P of D is maximal in D. Let $a \in D$, $\notin P$. Suppose that $aD + P \neq D$. Then $aD + P = P_1 \cdots P_m$, $a^2D + P = Q_1 \cdots Q_n$ where the P_i and Q_j are prime ideals. Since $aD + P \supsetneqq P$ and $a^2D + P \supsetneqq P$ (since P is prime), we have $P_i \supsetneqq P$, $Q_j \supsetneqq P$ for all i and j. Passing to the domain $\bar{D} = D/P$ we obtain $\overline{aD} = \bar{P}_1 \cdots \bar{P}_m$ and $\overline{a^2D} = \bar{Q}_1 \cdots \bar{Q}_n$ where $x \to \bar{x}$ is the canonical homomorphism of D onto $\bar{D}$. The principal ideals $\overline{aD}$ and $\overline{a^2D}$ are invertible and the $\bar{P}_i$ and $\bar{Q}_j$ are prime. Moreover, $\bar{a}^2\bar{D} = (\overline{aD})^2$, so $\bar{Q}_1 \cdots \bar{Q}_n = \bar{P}_1{}^2 \cdots \bar{P}_m{}^2$. By Proposition 10.4, the sequence of prime ideals $\{\bar{Q}_1, \ldots, \bar{Q}_n\}$ and $\{\bar{P}_1, \bar{P}_1, \ldots, \bar{P}_m, \bar{P}_m\}$ are the same except for order. The corresponding sequences of ideals of D containing P are $\{Q_1, \ldots, Q_n\}$ and $\{P_1, P_1, \ldots, P_m P_m\}$. Hence these coincide except for order. This implies that $(aD + P)^2 = P_1{}^2 \cdots P_m{}^2 = Q_1 \cdots Q_n = a^2D + P$. Then $P \subset (aD + P)^2 = a^2D + aP + P^2 \subset aD + P^2$. Then if $p \in P$, $p = ax + y$ where $x \in D$ and $y \in P^2$. Then $ax \in P$ and since $a \notin P$, $x \in P$. Hence $P \subset aP + P^2 \subset P$, so $P = aP + P^2$. Multiplication by P^{-1} gives $D = aD + P$, contrary to hypothesis. Thus we have $aD + P = D$ for every $a \in D$, $\notin P$. This implies that P is a maximal ideal of D.

Now let P be any prime ideal $\neq 0$ in D and let b be a non-zero element of P. Then $P \supset bD$ and $bD = P_1 P_2 \cdots P_m$ where the P_i are prime and are invertible. Hence the P_i are maximal. Since P is prime, we have $P \supset P_i$ for some i. Then $P = P_i$ and so P is invertible. Since any proper integral ideal is a product of prime ideals, it follows that every integral ideal is invertible. Since any fractional ideal $I' = aI$ for I integral and $a \neq 0$, $I'(I')^{-1} = II^{-1} = D$. Hence every fractional ideal is invertible and D is Dedekind. □

The following theorem summarizes the characterizations of Dedekind domains that we have given.

THEOREM 10.6. *The following conditions on a domain are equivalent:*

(i) *D is Dedekind.*

(ii) *D is integrally closed noetherian and every non-zero prime ideal of D is maximal.*

(iii) *Every proper integral ideal of D can be written in one and only one way as a product of prime ideals.*

(iv) *D is noetherian and D_P is a discrete valuation ring for every maximal ideal P in D.*

EXERCISES

1. Prove that any factorial domain is integrally closed.

2. Prove that the following conditions characterize Dedekind domains: D is integrally closed noetherian and D/I is artinian for every integral ideal I.

3. (H. Sah.) Show that a domain D is Dedekind if and only if it has the following property:

   If I is any integral ideal of D and a is any non-zero element of I, then there exists a $b \in I$ such that $I = Da + Db$.

   (*Sketch of proof*: The proof that any Dedekind domain has this property was indicated in exercise 11, p. 605. Now suppose that D has the property. Then D is noetherian. Also it is easily seen that the property is inherited by any localization D_S. Now suppose that D is a local domain having the property and let $P \neq 0$ be its maximal ideal. Let I be any ideal $\neq 0$ in D. Then $I = IP + Db$ for some $b \in I$. It follows by Nakayama's lemma that $I = Db$. Thus D is a p.i.d. and hence a discrete valuation ring (p. 571). Now conclude the proof by using Theorem 10.4.3 (p. 628).

Exercises 4–7 were communicated to me by Tsuneo Tamagawa. The first of these is well known. In all of these exercises, D is Dedekind with F as field of fractions and x is an indeterminate.

4. If $f(x) \in D[x]$, we define the *content* $c(f)$ to be the ideal in D generated by the coefficients of f (cf. BAI, p. 151). Prove that $c(fg) = c(f)c(g)$ for $f, g \in D[x]$.

5. Let $S = \{f \in D[x] | c(f) = D\}$. Note that S is a submonoid of the multiplicative monoid. Let $D[x]_S$ denote the subring of $F(x)$ of fractions $f(x)/g(x)$ where $f(x) \in D[x]$ and $g(x) \in S$. Show that if $f(x), g(x) \in D[x]$, then $f(x)/g(x) \in D[x]_S$ if and only if $c(f) \subset c(g)$.

6. Show that $D[x]_S \cap F = D$. Let I' be an ideal in $D[x]_S$. Show that $I = I' \cap D$

is an ideal in D such that $ID[x]_S = I'$. Show that $I \rightsquigarrow I' = ID[x]_S$ is a bijective map of the set of ideals of D onto the set of ideals of $D[x]_S$.

7. Prove that $D[x]_S$ is a p.i.d.

10.3 INTEGRAL EXTENSIONS OF DEDEKIND DOMAINS

The classical examples of Dedekind domains are the rings of $\mathbb{Z}$-integers of number fields, that is, of finite dimensional extension fields of $\mathbb{Q}$ and of fields of algebraic functions of one variable. In this section we derive the general result that if D is Dedekind with field of fractions F and E is a finite dimensional extension field of F, then the subring D' of D-integral elements of E is Dedekind. This implies the Dedekind property of the classical domains and gives the fundamental theorem on the unique factorization of integral ideals as products of prime ideals in these domains.

We prove first

PROPOSITION 10.7. *Let D be a domain, F its field of fractions, E a finite dimensional extension field of F, and D' the subring of E of D-integral elements. Then any element of E has the form rb^{-1} where $r \in D'$ and $b \in D$.*

Proof. Let $u \in E$ and let $g(x) = x^m + a_1x^{m-1} + \cdots + a_m \in F[x]$ be a polynomial such that $g(u) = 0$. We can write $a_i = c_ib^{-1}$, c_i, $b \in D$. Then

$$\begin{aligned} h(x) = b^m g(b^{-1}x) &= b^m(b^{-m}x^m + a_1 b^{-(m-1)}x^{m-1} + \cdots + a_m) \\ &= x^m + ba_1x^{m-1} + b^2a_2x^{m-2} + \cdots + b^m a_m \end{aligned}$$

$\in D[x]$ and $h(bu) = 0$. Hence $r = bu \in D'$ and $u = b^{-1}r$ as required. □

Evidently this result implies that E is the field of fractions of D'. Since D' is the ring of D-integers of E, D' is integrally closed in E. Hence D' is an integrally closed domain.

We suppose next that D is integrally closed and we prove

PROPOSITION 10.8. *Let D, E, F, D' be as in Proposition 10.7 and assume that D is integrally closed. Let $r \in E$ and let $m(x)$, $f(x)$ be the minimum polynomial and characteristic polynomial respectively of r over F. Then the following conditions on r are equivalent:*

(1) *r is integral.*
(2) *$m(x) \in D[x]$.*
(3) *$f(x) \in D[x]$.*

If the conditions hold, then $T_{E/F}(r)$ and $N_{E/F}(r) \in D$.

Proof. Since $f(x)$ is a power of $m(x)$, it is clear that (2) $\Rightarrow$ (3). Also obviously (3) $\Rightarrow$ (1). Now assume (1). Let $\bar{F}$ be an algebraic closure of F containing E. We have $m(x) = \prod(x - r_i)$ in $\bar{F}[x]$ where $r_1 = r$. For any i we have an automorphism of $\bar{F}/F$ sending r into r_i. Since r is D-integral, there exists a monic polynomial $g(x) \in D[x]$ such that $g(r) = 0$. Applying an automorphism of $\bar{F}/F$ sending r into r_i, we obtain $g(r_i) = 0$. Hence every root r_i of $m(x)$ is a D-integer. Since the coefficients of $m(x)$ are symmetric polynomials in the r_i, these elements of F are D-integers and since D is integrally closed, they are contained in D. Thus $m(x) \in D[x]$ and (1) $\Rightarrow$ (2). The last assertion is clear, since $f(x) = x^n - T_{E/F}(r)x^{n-1} + \cdots + (-1)^n N_{E/F}(r)$. □

A key result for the proof of the theorem on integral closures of Dedekind domains is the following

PROPOSITION 10.9. *Let D, E, F, D' be as in Proposition 10.8 and assume that E/F is separable and D is noetherian. Then D' is a finitely generated D-module.*

Proof. Since E/F is separable, there exist $n = [E:F]$ distinct monomorphisms $\sigma_1, \ldots, \sigma_n$ of E/F into the algebraic closure $\bar{F}/F$. Moreover, if $(u_1, \ldots, u_n)$ is a base for E/F, then the matrix

$$A = \begin{bmatrix} \sigma_1(u_1) & \cdots & \sigma_1(u_n) \\ \sigma_2(u_1) & \cdots & \sigma_2(u_n) \\ \cdot & \cdots & \cdot \\ \sigma_n(u_1) & \cdots & \sigma_n(u_n) \end{bmatrix} \tag{8}$$

is invertible (BAI, p. 292). We know also that $T_{E/F}(u) = \sum_1^n \sigma_i(u)$ for $u \in E$ (BAI, p. 430). We have $({}^tA)A = (T_{E/F}(u_iu_j))$. Hence

$$d = \det(T_{E/F}(u_iu_j)) \neq 0. \tag{9}$$

We have seen in Proposition 10.7 that there exists a $b_i \neq 0$ in D such that $b_iu_i \in D'$. Hence we may assume that the $u_i \in D'$. Let $r \in D'$ and write $r = \sum_1^n a_iu_i$, $a_i \in F$. Then $ru_j \in D'$, $1 \leq j \leq n$, and $ru_j = \sum_i a_iu_iu_j$. Hence

$$T_{E/F}(ru_j) = \sum_i a_i T_{E/F}(u_iu_j) \in D. \tag{10}$$

The system of equations (10) for a_i, $1 \leq i \leq n$, can be solved by Cramer's rule. This shows that $a_i \in d^{-1}D$. Since this holds for every $r \in D'$, we see that D' is contained in the D-module $M = \sum_1^n D(d^{-1}u_i)$. Thus D' is contained in a finitely generated D-module. Since D is noetherian, M is a noetherian D-module. Hence the submodule D' is finitely generated. □

We can now prove the main result of this section.

THEOREM 10.7. *Let D be a Dedekind domain, F its field of fractions, E a finite dimensional extension field of F, D' the subring of D-integral elements of E. Then D' is Dedekind.*

Proof. We assume first that E is separable over F. In this case we shall prove the theorem by showing that D' has the three properties characterizing Dedekind domains given in Theorem 10.3, namely, (1) D' is noetherian, (2) every prime ideal $\neq 0$ in D' is maximal, and (3) D' is integrally closed. The first of these follows from Proposition 10.9, since this proposition implies that D' is a noetherian D-module. A fortiori D' is noetherian as D'-module. Now let P' be a non-zero prime ideal in D'. Let $r \neq 0$ be in P'. Then we have $f(r) = 0$ for the characteristic polynomial $f(x)$ of r. This implies that $N_{E/F}(r) \in P'$ and hence, by Proposition 10.8, $N_{E/F}(r) \in P = P' \cap D$. Since $N_{E/F}(r) \neq 0$, we see that $P \neq 0$. It is clear also that P is a prime ideal in D. Since D is Dedekind, this implies that P is a maximal ideal in D. By the Corollary to Proposition 7.17, p. 410, we conclude that P' is maximal in D'. Hence (2) holds. The remark following Proposition 10.7 shows that (3) holds. Hence D' is Dedekind in the separable case.

We assume next that char $F = p$ and that E is purely inseparable over F. Since $[E:F] < \infty$, there exists a $q = p^e$ such that $u^{p^e} \in F$ for all $u \in E$. Let $\bar{F}$ be an algebraic closure of F containing E and let $F^{1/q}(D^{1/q})$ be the subfield (subring) of $\bar{F}$ of elements v such that $v^q \in F(D)$. Then $v \rightsquigarrow v^q$ is an isomorphism of $F^{1/q}$ onto F mapping $D^{1/q}$ onto D. Since D is Dedekind, $D^{1/q}$ is Dedekind. If $u \in D'$, $u^q \in F$ and u^q is D-integral. Hence $u^q \in D$ and so $D' \subset D^{1/q}$. We shall now show that D' is Dedekind by proving that every D'-fractional ideal I of E is invertible. Let $I' = D^{1/q}I$. Then I' is a $D^{1/q}$-fractional ideal. Hence there exists a $D^{1/q}$-fractional ideal J' such that $I'J' = D^{1/q}$ and since $I' = ID^{1/q}$ and $D^{1/q}J' = J'$, we have elements $b_i \in I$, $c_i' \in J'$ such that $\sum b_i c_i' = 1$. Then $\sum b_i^q c_i'^q = 1$ and $\sum b_i c_i = 1$ where $c_i = b_i^{q-1} c_i'^q$. Now $b_i^{q-1} c_i'^q \in I'^{q-1} J'^q \subset J'$. Since $c_i'^q \in F$ and $b_i \in E$, $c_i = b_i^{q-1} c_i'^q \in E$. Thus $c_i \in J = J' \cap E$. Evidently $IJ \subset D^{1/q} \cap E$ and since the elements of $D^{1/q}$ are D-integral, $D^{1/q} \cap E \subset D'$. Hence $IJ \subset D'$, so $J \subset I^{-1}$. Since the $c_i \in J$ and $b_i \in I$ and $\sum c_i b_i = 1$, we have $II^{-1} = D'$. Thus I is invertible and D' is Dedekind.

The general case can be obtained by combining the preceding two cases in the obvious way: If E is arbitrary finite dimensional over F, let K be the maximal separable subfield of E/F. Then E is purely inseparable over K. Now

$D'' = D' \cap K$ is Dedekind by the first case and since D' is the set of D''-integral elements of E, D' is Dedekind by the second case. This completes the proof. □

The foregoing result is applicable in particular to $F = \mathbb{Q}$ and $D = \mathbb{Z}$ and to $F = K(x)$ where K is an arbitrary field, x an indeterminate, and $D = K[x]$. This is clear since these D are p.i.d. and hence are Dedekind. If E is a finite dimensional extension field of F, then the ring D' of D-integers of E is Dedekind. In the case $F = \mathbb{Q}$, $D = \mathbb{Z}$, E is a number field and in the other cases E is an algebraic function field in one variable.

EXERCISES

1. Let the notations be as in Theorem 10.7. Assume that E/F is separable and that D is a p.i.d. Show that D' is a free D-module of rank $n = [E:F]$.

Exercises 3, 4, and 5 of BAI, p. 287, are relevant for this section.

2. (Sah.) Show that if D is a Dedekind domain, then so is the ring $D((x))$ of Laurent series over D. (*Hint*: It is easily seen that $D((x))$ is a domain. If I is an ideal in $D((x))$, let $l(I)$ be the set of leading coefficients of the elements of I. Then $l(I)$ is an ideal and if $l(I) = Da_1 + \cdots + Da_m$ where a_i is the leading coefficient of $f_i \in I$, then $I = D((x))f_1 + \cdots + D((x))f_m$. Use this and exercise 3 on p. 630.

10.4 CONNECTIONS WITH VALUATION THEORY

We give first a valuation theoretic characterization of the integral closure of a subring of a field. This is

THEOREM 10.8 (Krull). *Let D be a subring of a field F and let D' be the integral closure of D in F (= the subring of D-integral elements of F). Then D' is the intersection of the valuation rings of F containing D.*

Proof. Let $u \in D'$ and let R be a valuation ring in F containing D, φ a valuation of F having R as valuation ring (e.g., the canonical valuation defined by R). Suppose that $u \notin R$. Then $\varphi(u^{-1}) < 1$. On the other hand, we have a relation $u^n + a_1u^{n-1} + \cdots + a_n = 0$, $a_i \in D$, which gives the relation

$$1 = -a_1u^{-1} - a_2u^{-2} - \cdots - a_nu^{-n}. \tag{11}$$

Since the $a_i \in D \subset R$, $\varphi(a_i) \leqslant 1$. Then $\varphi(a_iu^{-i}) < 1$. Since $\varphi(1) = 1$, we have the contradiction $1 \leqslant \max\{\varphi(a_iu^{-i})\} < 1$. Hence $u \in R$ and $D' \subset \bigcap R$ where the intersection is taken over all of the valuation rings R of F containing D.

Next, suppose that $u \notin D'$. Then u^{-1} is not a unit in the subring $D[u^{-1}]$ of F, since otherwise, its inverse $u = a_0 + a_1 u^{-1} + \cdots + a_{n-1} u^{-(n-1)}$, $a_i \in D$, and hence $u^n = a_0 u^{n-1} + a_1 u^{n-2} + \cdots + a_{n-1}$, contrary to $u \notin D'$. Since u^{-1} is not a unit in $D[u^{-1}]$, $u^{-1}D[u^{-1}]$ is a proper ideal in $D[u^{-1}]$ and this can be imbedded in a maximal ideal P_0 of $D[u^{-1}]$. Then $\Delta = D[u^{-1}]/P_0$ is a field and we have the canonical homomorphism of $D[u^{-1}]$ onto Δ, which can be regarded as a homomorphism $\mathscr{P}_0$ of $D[u^{-1}]$ into an algebraic closure $\bar{\Delta}$ of Δ. By Theorem 9.10 (p. 561), there exists a $\bar{\Delta}$-valued place $\mathscr{P}$ extending $\mathscr{P}_0$. If R is the valuation ring on which $\mathscr{P}$ is defined and P is its maximal ideal, then P is the kernel of $\mathscr{P}$. We have $R \supset D[u^{-1}] \supset D$ and $P \supset P_0$. Then $u^{-1}D[u^{-1}] \subset P_0 \subset P$, so $\mathscr{P}(u^{-1}) = 0$ and hence $u \notin R$. Thus we have shown that if $u \notin D'$, then $u \notin \bigcap R$ where R ranges over the valuation rings of F containing D. Hence $D' = \bigcap R$. $\square$

Now let D be Dedekind, F its field of fractions. Since D is integrally closed, D is the intersection of the valuation rings of F containing D. We claim that these valuation rings are the localizations D_P, for the prime ideals P of D. We have seen that every D_P is a discrete valuation ring for every prime ideal $\neq 0$ in D. This is clear also if $P = 0$, since $D_0 = F$. Now let R be a valuation ring in F containing D such that $R \neq F$. Let M be the maximal ideal of R and let $P = M \cap D$. Then P is a prime ideal in D and if $P = 0$, every non-zero element of D is invertible in R. Since F is the field of fractions of D, this implies that $F = R$, contrary to hypothesis. Thus P is an integral prime ideal in D. We have the valuation ring D_P with maximal ideal $P_P = PD_P$. If $a \in D - P$, then $a \in R - M$ and $a^{-1} \in R$. Then $R \supset D_P$. Moreover, since $M \supset P$, $M = RM \supset D_P P$. Now it is easily seen that if R_1 and R_2 are valuation rings with maximal ideals M_1 and M_2 respectively, then $R_1 \supset R_2$ and $M_1 \supset M_2$ imply $R_1 = R_2$ and $M_1 = M_2$. Hence $R = D_P$ and $M = PD_P$. Thus the valuation rings of F containing D (and $\neq F$) are just the localizations D_P, P a prime ideal $\neq 0$ in D. By Theorem 10.8, $D = \bigcap D_P$.

A prime ideal $P \neq 0$ in a Dedekind domain D determines an exponential valuation v_P for the field of fractions F of D in a manner similar to the definition of the exponential valuation v_p on $\mathbb{Q}$ defined by a prime integer p. If $a \neq 0$ in F, we write $Da = P^k P_1^{k_1} \cdots P_l^{k_l}$ where $P, P_1, \ldots, P_l$ are distinct prime ideals and k and the $k_i \in \mathbb{Z}$. Then we put $v_P(a) = k$. Moreover, we define $v_P(0) = \infty$. It is readily verified, as in the special case of $\mathbb{Q}$, that v_P is an exponential valuation of F. Hence if we take a real γ, $0 < \gamma < 1$, then $|a|_P = \gamma^{v_P(a)}$ is a non-archimedean absolute value on F. We call this a *P-adic absolute value on F*. The valuation ring of $|\ |_P$ evidently contains the localization D_P and the maximal ideal of the valuation ring contains PD_P. Hence the valuation ring of $|\ |_P$ is D_P and its maximal ideal is PD_P.

Since D_P is Dedekind and PD_P is its only prime ideal $\neq 0$, the integral ideals of D_P are the ideals $P^k D_P$, $k \geqslant 0$. Evidently, $P^k \subset P^k D_P \cap D$. On the other hand, if $a \in P^k D_P \cap D$ for $k \geqslant 1$, then $a = bc^{-1}$ where $b \in P^k$, $c \in D - P$. Then $ac \equiv 0 \pmod{P^k}$ and $c \not\equiv 0 \pmod{P}$. The latter condition implies that $c + P$ is a unit in D/P. Then $c + P^k$ is a unit in D/P^k (p. 195). It follows that $a \in P^k$. Hence $P^k D_P \cap D = P^k$ for $k \geqslant 0$.

Let $| \ |$ be a non-archimedean absolute value having D_P as valuation ring. We claim that $| \ | = | \ |_P$, a P-adic valuation determined by the prime P. To see this we observe that $PD_P = \pi D_P$ where $\pi \in PD_P$, $\notin P^2 D_P$. Then $P^k D_P = \pi^k D_P$ for $k \geqslant 0$ and any element of D_P can be written as $u\pi^k$, u a unit in D_P and $k \geqslant 0$. Then $|u\pi^k| = \gamma^k$ where $|\pi| = \gamma$, $0 < \gamma < 1$. Hence, if $a \in P^k$, $\notin P^{k+1}$, then $a \in P^k D_P$, $\notin P^{k+1} D_P$, and $|a| = \gamma^k$. On the other hand, it is clear that $Da = P^k P_1^{k_1} \cdots P_l^{k_l}$ where $P_1, \ldots, P_l$ are primes distinct from P and hence if $| \ |_P$ is the P-adic valuation determined by γ, then $|a|_P = |a|$ for $a \in D$. Then $| \ | = | \ |_P$.

We show next that the residue field D_P/P_P of $| \ |_P$ is canonically isomorphic to D/P. Since $P = P_P \cap D$, we have the monomorphism $a + P \rightsquigarrow a + P_P$ of D/P into D_P/P_P. If $a, b \in D$ and $b \notin P$, then since D/P is a field there exists a $c \in D$ such that $cb \equiv a \pmod{P}$. Then $ab^{-1} \equiv c \pmod{P_P}$, which implies that $a + P \rightsquigarrow a + P_P$ is surjective on D_P/P_P. Then this map is an isomorphism.

We summarize this collection of results in

PROPOSITION 10.10. *Let D be a Dedekind domain with field of fractions F. Then* (1) *Any valuation ring in F containing D has the form D_P where P is a prime ideal in D.* (2) *Any non-archimedean absolute value on F having D_P, $P \neq 0$, as valuation ring is a P-adic valuation.* (3) *The map $a + P \rightsquigarrow a + P_P$ is an isomorphism of D/P onto D_P/P_P.*

Now let E be a finite dimensional extension field of F and let D' be the ring of D-integers of E. Then D' is Dedekind. If P is an integral ideal of D, then the ideal in D' generated by P is $D'P$. Now $D'P \neq D'$. For, by the "Lying-over" theorem (p. 411), there exists a prime ideal P' in D' such that $P' \cap D = P$. Hence $P' \supset PD'$ and $PD' \neq D'$. It may happen that the extension ideal PD' of P is not prime. At any rate, we have a factorization

$$PD' = P_1'^{e_1} \cdots P_g'^{e_g} \tag{12}$$

where the P_i' are distinct prime ideals in D' and the $e_i > 0$.

We have treated the general problem of extension of an absolute value on a field F to a finite dimensional extension field in Chapter 9 (p. 585). We shall now treat the special case of a P-adic absolute value of the field of fractions F of a Dedekind domain in an alternative manner based on the factorization (12). We prove first

THEOREM 10.9. *Let D be a Dedekind domain, F the field of fractions of D, E a finite dimensional extension field of E, D' the subring of E of D-integral elements. Let P be an integral prime ideal in D, $|\ |_P$ a P-adic absolute value on F, (12) the factorization of PD' into prime ideals in D'. Then for each P'_i there is a unique P'_i-adic absolute value $|\ |_{P'_i}$ on E extending $|\ |_P$. Moreover, $|\ |_{P'_i}$ and $|\ |_{P'_j}$ are inequivalent if $i \neq j$ and the $|\ |_{P'_i}$ are the only extensions of $|\ |_P$ to absolute values on E.*

Proof. Let $a \neq 0$ be in F and write $Da = P^k P_1^{k_1} \cdots P_l^{k_l}$ where P and the P_j are distinct prime ideals and k and the $k_j \in \mathbb{Z}$. Then $v_P(d) = k$, and by (12),

$$\begin{aligned} D'a &= (D'P)^k (D'P_1)^{k_1} \cdots (D'P_l)^{k_l} \\ &= P_1'^{e_1 k} P_2'^{e_2 k} \cdots P_g'^{e_g k} \cdots \end{aligned} \tag{13}$$

where the terms after $P_g'^{e_g k}$ come from the factorizations of the $D'P_j$ into prime ideals in D'. Now we observe that if P_1 and P_2 are distinct prime ideals in D, then $P_1 + P_2 = D$ and hence $D'P_1 + D'P_2 = D'$. This implies that the prime ideals in D' dividing $D'P_1$ are different from those dividing $D'P_2$. It follows that $v_{P'_j}(a) = e_j k$. Hence if $|\ |_P = \gamma^{v_P()}$ for $0 < \gamma < 1$, then we shall have $|\ |_P = |\ |_{P'_j}$ on F if and only if $|\ |_{P'_j} = (\gamma^{1/e_j})^{v_{P'_j}()}$. This proves the first assertion. We have seen that the valuation ring of $|\ |_{P'_j}$ (or of $v_{P'_j}$) is $D'_{P'_j}$. Since these are distinct for distinct choices of j, it follows that $|\ |_{P'_i}$ and $|\ |_{P'_j}$ are inequivalent if $i \neq j$. Now let $|\ |'$ be any absolute value on E extending $|\ |_P$ and let R' denote the valuation ring of $|\ |'$. Then $R' \supset D_P$ and since D' is the integral closure of D in E, $R' \supset D'$ by Theorem 10.8. By Proposition 10.10.1, $R' = D'_{P'}$ for P', a prime ideal in D'. Since $R' \supset D_P$, it follows that $P' \supset P$. Hence $P' = P'_i$ for some i. By Proposition 10.10.2, $|\ |'$ is a P'_i-adic valuation. □

We recall that the ramification index of $|\ |_{P'_i}$ relative to $|\ |_P$ is the index $[|E^*|_{P'_i} : |F^*|_P]$ (p. 589). Since $|F^*|_P$ is the subgroup of $\mathbb{R}^*$ generated by γ and $|E^*|_{P'_i}$ is the subgroup generated by γ^{1/e_i}, the ramification index is e_i.

We recall also that the residue degree f_i of $|\ |_{P'_i}$ relative to $|\ |_P$ is the dimensionality $[D'_{P'_i}/P'_{iP'_i} : D_P/P_P]$ where D_P/P_P is imbedded in $D'_P/P'_{iP'_i}$ by the monomorphism $x + P_P \rightsquigarrow x + P'_{iP_i}$ (p. 589). We also have a monomorphism $a + P \rightsquigarrow a + P'_i$, $a \in D$, of D/P into D'/P'_i. Hence we can define the dimensionality $[D'/P'_i : D/P]$. Now we have the isomorphisms $\sigma : a + P \rightsquigarrow a + P_P$ and $s : x + P_i \rightsquigarrow x + P'_{iP'_i}$ of D/P onto D_P/P_P and of D'_{P_i} onto $D'_{P'_i}/P'_{iP'_i}$ respectively. We have $s((a + P)(x + P'_i)) = s(ax + P'_i) = ax + P'_{iP'_i}$ and hence $(\sigma(a + P))(s(x + P'_i)) = (a + P_P)(x + P'_{iP'_i}) = ax + P'_{iP'_i}$. Hence s is a semi-linear isomorphism of D'/P'_i over D/P onto $D'_{P'_i}/P'_{iP'_i}$ with associated field isomorphism σ. It follows that the residue degree f_i coincides with the dimensionality $[D'/P'_i : D/P]$.

The results on ramification index and residue degree can be stated as follows:

THEOREM 10.10. *Let the notations be as in Theorem* 10.9. *Let* $| \ |_P$ *be a* P*-adic absolute value on* F, $| \ |_{P'_i}$ *a* P'_i*-adic absolute value on* E *extending* $| \ |_P$. *Then* e_i *is the ramification index and* $[D'/P'_i : D/P]$ *is the residue degree of* $| \ |_{P'_i}$ *relative to* $| \ |_P$.

An immediate consequence of this result and the main theorem on residue degrees and ramification indices (Theorem 9.15, p. 592) is

THEOREM 10.11. *Let the notations be as in Theorem* 10.9. *Then* $\sum_1^g e_i f_i \leqslant n = [E:F]$ *if* $f_i = [D'/P'_i : D/P]$. *Moreover,* $\sum_1^g e_i f_i = n$ *if* E/F *is separable.*

We remark that this result has the consequence that the number g of prime ideals of D' containing $D'P$ is bounded by n and it gives considerably more information on the factorization (12).

EXERCISES

1. Let E be the cyclotomic field $\mathbb{Q}(\zeta)$ when ζ is a primitive lth root of unity, l a prime. Let $D = \mathbb{Z}$, D' be the integral closure of $\mathbb{Z}$ in E. Show that $lD' = ((\zeta-1)D')^{p-1}$ and $(\zeta-1)D'$ is a prime ideal in D'. (*Hint*: $[E:Q] = l-1$ and $f(x) = x^{l-1} + x^{l-2} + \cdots + 1$ is the minimum polynomial of ζ. We have $f(x) = \prod_1^{l-1}(x-\zeta^i)$ in $E[x]$ and every ζ^i, $1 \leqslant i \leqslant l-1$, is a primitive lth root of 1. (See BAI, p. 154.) Substituting 1 in $f(x)$ gives $l = \prod_1^{l-1}(1-\zeta^i)$. Moreover, $(1-\zeta^i)/(1-\zeta) = 1 + \zeta + \cdots + \zeta^{i-1} \in D'$. Show that $1+\zeta+\cdots+\zeta^{i-1}$ is a unit in D', so $lD' = ((\zeta-1)D')^{l-1}$. Show that l is not a unit in D' and use Theorem 10.11 to conclude that $(\zeta-1)D'$ is prime.)

The next three exercises are designed to show that in the special case of Theorem 10.11 in which E/F is Galois, the factorization (12) of $PD' = (P'_1 \cdots P'_g)^e$ and all of the residue degrees are equal. Hence $n = [E:F] = efg$.

2. Let E/F be Galois with $\text{Gal}\, E/F = G$. Show that any $\sigma \in G$ maps D' into itself. Show that if P' is a prime ideal in D', then $\sigma P'$ is a prime ideal. Show that $\bar{\sigma}: a + P' \rightsquigarrow \sigma a + \sigma P'$ is an isomorphism of D'/P' onto $D'/\sigma P'$.

3. Let P be a non-zero prime ideal of D, $PD' = P_1'^{e_1} \cdots P_g'^{e_g}$, the factorization of PD' into prime ideals in D' where $P'_i \neq P'_j$ if $i \neq j$. Show that if $\sigma \in G$, then $\{\sigma P'_1, \ldots, \sigma P'_g\} = \{P'_1, \ldots, P'_g\}$ and that G acts transitively on the set of prime ideals $\{P'_1, \ldots, P'_g\}$. (*Hint*: To prove transitivity of G on $\{P'_1, \ldots, P'_g\}$ one shows that if P' is any prime ideal of D' such that $P' \cap D = P$ then any prime ideal Q' of D' such that $Q' \cap D = P$ is one of the ideals $\sigma P'$, $\sigma \in G$. Otherwise, by Proposition 7.1, p. 390, $Q' \not\subset \bigcup_{\sigma \in G} \sigma P'$. Hence there exists an $a \in Q'$ such that $\sigma a \notin P'$ for any $\sigma \in G$. Then $N_{E/F}(a) = \prod_{\sigma \in G} \sigma a \notin P'$. This is a contradiction since $N_{E/F}(a) \in Q' \cap D = P$.)

4. Use the results of exercises 2 and 3 to show that $e_i = e_j$ for every i and j and that the fields D'/P'_i and D'/P'_j are isomorphic over D/P. Hence conclude that $[D'/P'_i : D/P] = [D'/P'_j : D/P]$. Put $e = e_i, f = f_i = [D'/P_i : D/P]$. Show that $n = efg$.

10.5 RAMIFIED PRIMES AND THE DISCRIMINANT

As in the previous section, let D be a Dedekind domain, F its field of fractions, E a finite dimensional extension field of F, and D' the subring of D-integral elements of E. Let P be a non-zero prime ideal in D and let $PD' = P_1'^{e_1} \cdots P_g'^{e_g}$ where the P_i' are distinct prime ideals in D, as in (12). We consider the ring $\bar{D}' = D'/PD'$, which we can regard also as a module over D' and by restriction as a module over D. Since P annihilates $\bar{D}'$, $\bar{D}'$ can also be regarded as a vector space over $\bar{D} = D/P$. The action of $a+P$, $a \in D$, on $x+PD'$, $x \in D'$, is $(a+P)(x+PD') = ax+PD'$. Taking into account the ring structure as well as the $\bar{D}$-vector space structure, we obtain the algebra $\bar{D}'$ over $\bar{D}$. We shall now consider the structure of this algebra. The main result on this is the following

THEOREM 10.12. *Let D be a Dedekind domain with field of fractions F and let E be a finite dimensional extension field, D' the subring of E of D-integral elements. Suppose that P is a non-zero prime ideal in D and let $PD' = P_1'^{e_1} \cdots P_g'^{e_g}$ where the P_i' are distinct prime ideals in D' and the $e_i > 0$. Then $\bar{D}' = D'/PD'$ is an algebra over the field $\bar{D} = D/P$ in which $(a+P)(x+PD') = ax+PD'$, $a \in D$, $x \in D'$. The dimensionality $[\bar{D}' : \bar{D}] = \sum_1^g e_i f_i$ where f_i is the residue degree $[D'/P_i' : \bar{D}]$. We have $[\bar{D}' : \bar{D}] \leqslant n = [E:F]$ and $[\bar{D}' : \bar{D}] = n$ if E/F is separable. $\bar{D}'$ is a direct sum of g ideals isomorphic to the algebras $D'/P_i'^{e_i}$. The radical of $\bar{D}'$ is $P_1' \cdots P_g'/PD'$ and $\bar{D}'/\mathrm{rad}\,\bar{D}'$ is a direct sum of g ideals isomorphic to the fields $\bar{D}'/P_i'$.*

Proof. Put $A_i' = P_i'^{-e_i} P_1'^{e_1} \cdots P_g'^{e_g}$. Then

$$\sum A_i' = D', \qquad A_i' \cap \sum_{j \neq i} A_j' = D'P. \tag{14}$$

This implies that

$$\bar{D}' = D'/PD' = A_1'/PD' \oplus \cdots \oplus A_g'/PD' \tag{15}$$

and $A_i'/PD' = A_i'/(A_i' \cap P_i'^{e_i}) \cong (A_i' + P_i'^{e_i})/P_i'^{e_i} = D'/P_i'^{e_i}$. Hence

$$\bar{D}' \cong D'/P_1'^{e_1} \oplus \cdots \oplus D'/P_g'^{e_g}. \tag{16}$$

We have the chain of ideals

$$D' \supset P_i' \supset P_i'^2 \supset \cdots \supset P_i'^{e_i}, \qquad 1 \leqslant i \leqslant g, \tag{17}$$

each of which is minimal over the next. Hence

$$D'/P_i'^{e_i} \supset P_i'/P_i'^{e_i} \supset \cdots \supset P_i'^{e_i-1}/P_i'^{e_i} \supset 0 \tag{18}$$

is a composition series for $D'/P_i'^{e_i}$ as D'-module. The composition factors are isomorphic to D'/P_i', $P_i'/P_i'^2, \ldots, P_i'^{e_i-1}/P_i'^{e_i}$, each of which is annihilated by P_i' and so can be regarded as a vector space over the field D'/P_i'. Since these vector spaces have no non-zero subspaces, they are one-dimensional over D'/P_i'. Hence the composition factors are isomorphic as D'-modules and hence also as D-modules and as vector spaces over $\bar{D}$. Then $[P_i'^{j-1}/P_i'^j:\bar{D}] = [D'/P_i':\bar{D}] = f_i$ for $j = 1, \ldots, e_i$. It follows from (18) that $[D'/P_i'^{e_i}:\bar{D}] = e_i f_i$. Since $D'/D'P \cong D'/P_1'^{e_1} \oplus \cdots \oplus D'/P_g'^{e_g}$, we have $[\bar{D}':\bar{D}] = \sum_1^g e_i f_i$. We have shown in Theorem 10.11 that $\sum e_i f_i \leqslant n$ and $\sum e_i f_i = n$ if E/F is separable. Hence $[\bar{D}':\bar{D}] \leqslant n$ and $= n$ if E/F is separable.

Let $e = \max\{e_i\}$. Then $(P_1' P_2' \cdots P_g')^e \subset D'P$. Hence $P_1' \cdots P_g'/D'P$ is a nilpotent ideal in $\bar{D}' = D'/D'P$ and $\bar{D}'/(P_1' \cdots P_g'/D'P) \cong D'/P_1' \cdots P_g' \cong D'/P_1' \oplus \cdots \oplus D'/P_g'$. Since every D'/P_i' is a field, it follows that $P_1' \cdots P_g'/D'P = \text{rad}\, \bar{D}'$ and $\bar{D}'/\text{rad}\, \bar{D}'$ is a direct sum of g ideals isomorphic to the fields D'/P_i'. □

We now give the following

DEFINITION 10.4. *A non-zero prime ideal P of D is said to be* unramified *in D' if $D'P = P_1' \cdots P_g'$ where the P_i' are distinct prime ideals in D' and every D'/P_i' is separable over $\bar{D} = D/P$. Otherwise, P is* ramified *in D'.*

It is clear from Theorem 10.12 that P is unramified in D' if and only if $\bar{D}' = D'/D'P$ is a direct sum of fields that are separable over $\bar{D} = D/P$. We shall show that if E/F is separable, then there are only a finite number of prime ideals P of D that are ramified in D' and in principle we shall determine these prime ideals. For this purpose we give the following

DEFINITION 10.5. *Let D, F, E, D' be as usual and assume that E/F is separable. Then the* discriminant (ideal) $d_{D'/D}$ *of D' over D is the ideal generated by all of the elements*

$$\det(T_{E/F}(u_i u_j)) \tag{19}$$

where $(u_1, \ldots, u_n)$ is a base for E/F consisting of elements $u_i \in D'$.

We have seen in the proof of Proposition 10.9 (p. 612) that $\det(T_{E/F}(u_i u_j)) \neq 0$ for every base $(u_1, \ldots, u_n)$ of a separable extension E/F. The main result on ramification is

THEOREM 10.13. *Let D be a Dedekind domain, F its field of fractions, E a finite dimensional separable extension of F, D' the subring of E of D-integral* *elements. Then a non-zero prime ideal P of D is ramified in D′ if and only if P is a divisor of the discriminant* $d_{D'/D}$.

Of course, this implies the finiteness of the set of ramified primes. To achieve the proof of Theorem 10.13 we need to obtain a trace criterion that a finite dimensional commutative algebra A over a field F be a direct sum of ideals that are separable field extensions of F.

If A is a finite dimensional algebra over a field F, we define the *trace bilinear form* $T(a,b)$ on A/F by

$$T(a,b) = T_{A/F}(ab) \tag{20}$$

where $T_{A/F}$ is the trace defined by the regular representation. Since $T_{A/F}$ is a linear function and $T_{A/F}(ab) = T_{A/F}(ba)$, $T(a,b)$ is a symmetric bilinear form (BAI, pp. 424, 426). We recall also that $T(a,b)$ is non-degenerate if and only if $\det(T(u_i,u_j)) \neq 0$ for any base $(u_1,\ldots,u_n)$ of A/F.

We require the following

LEMMA. *If A is a finite dimensional commutative algebra over a field F, then A/F is a direct sum of ideals that are separable fields over F if and only if the trace bilinear form T(a, b) is non-degenerate on A.*

Proof. Suppose first that $A = E_1 \oplus \cdots \oplus E_g$ where E_i is a separable field extension of F. Let $(u_{i1},\ldots,u_{in_i})$ be a base for E_i. Then $(u_{11},\ldots,u_{1n_1},\ldots, u_{g1},\ldots,u_{gn_g})$ is a base for A/F and

$$\det(T(u_{ik},u_{jl})) = \prod_i \det(T_i(u_{ik},u_{il})) \tag{21}$$

where T_i denotes the trace bilinear form on E_i. Since we have shown (p. 612) that T_i is non-degenerate, the foregoing formula shows that T is non-degenerate on A.

Conversely, suppose that T is non-degenerate on A. We show first that A is semi-primitive. Let $z \in \operatorname{rad} A$. Then $za = az$ is nilpotent for every $a \in A$. Since the trace of a linear transformation is the sum of its characteristic roots, the trace of a nilpotent linear transformation is 0. Hence $T(a,z) = 0$ for all a. Since T is non-degenerate, $z = 0$. Thus $\operatorname{rad} A = 0$ and A is semi-primitive. Then $A = E_1 \oplus \cdots \oplus E_g$ where E_i is a field extension of F. By (21), the trace bilinear form T_i is non-degenerate on every E_i. On the other hand, if E/F is a finite dimensional inseparable extension field of F, then the trace function is identically 0 on E. For, $E \supsetneq S$ where S is the maximal separable subfield of E. Then E is purely inseparable over S. It follows that there exists a subfield K of E containing S such that $[E:K] = p$ the characteristic. Then the minimum polynomial over K of any $a \in E$ is either of the form $x^p - \alpha$ or of the form $x - \alpha$. In either case the characteristic polynomial is of the form $x^p - \beta$ ($\beta = \alpha^p$ in the second case). Then $T_{E/K}(a) = 0$ for all a. By the transitivity of the trace (BAI, p. 426) we have $T_{E/F}(a) = T_{K/F}(T_{E/K}(a)) = 0$. It follows that the trace bilinear form on an inseparable field is identically 0. Hence every E_i is separable. □

We can now give the

Proof of Theorem 10.13. We assume first that D is a p.i.d. By Proposition 10.9, D' is finitely generated as D-module. Since there is no torsion and D is a p.i.d., D' is a free D-module. Since every element of E has the form $r^{-1}u$ where $r \in D$ and $u \in D'$, the rank of D' over $D = n = [E:F]$. Let $(v_1, \dots, v_n)$ be a D-base for D'. Then $\det(T(v_i, v_j))$ generates $d_{D'/D}$, so we have to show that $\bar{D}'$ is a direct sum of separable field extensions of $\bar{D}$ if and only if P does not contain the element $\det(T(v_i, v_j))$. Since $(v_1, \dots, v_n)$ is a D-base for D', $PD' = \sum Pv_i$ and if $\bar{v}_i = v_i + PD'$, then $(\bar{v}_1, \dots, \bar{v}_n)$ is a base for $\bar{D}'$ over $\bar{D}$. It follows that if $a \rightsquigarrow \rho(a)$ is the regular matrix representation of E/F obtained by using the base $(v_1, \dots, v_n)$, then the regular matrix representation of $\bar{D}'/\bar{D}$ using the base $(\bar{v}_1, \dots, \bar{v}_n)$ is $\bar{a} \rightsquigarrow \overline{\rho(a)}$ $(\bar{a} = a + PD')$. This implies that $\det(T(\bar{v}_i, \bar{v}_j)) = \det(T(v_i, v_j)) + P$. Hence by the Lemma, $\bar{D}'$ is a direct sum of separable fields over $\bar{D}$ if and only if $\det(T(v_i, v_j)) \notin P$. This completes the proof in the case in which D is a p.i.d.

The proof for arbitrary D can be reduced to the p.i.d. case by localizing at the prime P of D. Since we are interested in localizing D' as well as D and since P is not a prime ideal in D', we make explicit the localizing monoid as $S = D - P$. We form D_S and D'_S. By Proposition 7.16, D'_S is the integral closure of D_S in E. Since $P'_i \cap D = P$, $P'_i \cap S = \varnothing$. Hence P'_{iS} is a prime ideal in D'_S. We have $P_S D'_S = P'^{e_1}_{1S} \cdots P'^{e_g}_{gS}$ and the P'_{iS} are distinct. Since $P'_{iS} \cap D' \subset P'_{iP'_i} \cap D' = P'_i$, we have $P'_{iS} \cap D' = P'_i$. It follows as in the proof of Theorem 10.10 that we have a semi-linear map of D'_S/P'_{iS} over D_S/P_S onto D'/P'_i over D/P, which is a ring isomorphism. Hence D'_S/P'_{iS} is separable over D_S/P_S if and only if D'/P'_i is separable over D/P. Thus P is ramified in D' if and only if P_S is ramified in D'_S.

We show next that $d_{D'_S/D_S} = (d_{D'/D})_S$. Let $(u_1, \dots, u_n)$ be a base for E/F such that the $u_i \in D'$. Then the $u_i \in D'_S$ and $\det(T(u_i, u_j)) \in d_{D'_S/D_S}$. Hence $d_{D'/D} \subset d_{D'_S/D_S}$ and $(d_{D'/D})_S \subset d_{D'_S/D_S}$. Next let $(u'_1, \dots, u'_n)$ be a base for E/F such that the $u'_i \in D'_S$. Then $u'_i = u_i s^{-1}$, $u_i \in D'$, $s \in D - P$, and $\det(T(u'_i, u'_j)) = s^{-2n} \det(T(u_i, u_j)) \in (d_{D'/D})_S$. Hence $d_{D'_S/D_S} \subset (d_{D'/D})_S$, so $d_{D'_S/D_S} = (d_{D'/D})_S$. It is clear also that P is a divisor of $d_{D'/D}$ if and only if P_S is a divisor of $(d_{D'/D})_S = d_{D'_S/D_S}$. We have therefore achieved a reduction of the theorem to the rings D_S and D'_S. Since D_S is a p.i.d., the result follows from the case we considered first. □

EXAMPLE

Let m be a square-free integer, $E = \mathbb{Q}(\sqrt{m})$, and let D' be the integral closure of $\mathbb{Z}$ in E. Then D' has the base $(1, \sqrt{m})$ over $\mathbb{Z}$ if $m \equiv 2$ or $3 \pmod 4$ and D' has the base $(1, (1+\sqrt{m})/2)$ over $\mathbb{Z}$ if $m \equiv 1 \pmod 4$ (exercises 4 and 5, p. 287 of BAI). Hence in the first case, the discriminant $d_{D'/\mathbb{Z}}$ is the principal ideal generated by

$$\begin{vmatrix} 2 & 0 \\ 0 & 2m \end{vmatrix} = 4m.$$

Hence the ramified primes (which are the primes p in $\mathbb{Z}$ such that $\mathbb{Z}p$ is a square of a prime ideal in D') are 2 and the prime divisors of m.

In the second case ($m \equiv 1 \pmod 4$), $d_{D'/\mathbb{Z}}$ is the principal ideal generated by

$$\begin{vmatrix} 2 & 1 \\ 1 & \dfrac{m+1}{2} \end{vmatrix} = m.$$

Hence the ramified primes are the prime divisors of m.

EXERCISES

1. Notations as in the example. Show that if p is odd and pD unramified in D', then pD' is a prime or a product of two distinct prime ideals in D' according as the Legendre symbol $(m/p) = -1$ or $= 1$. Show that if $m \equiv 1 \pmod 4$, then $2D'$ is prime or a product of two prime ideals in D' according as $m \equiv 5$ or $m \equiv 1 \pmod 8$. (*Hint*: Use Theorems 10.9 and 9.14.)

2. Let E be the cyclotomic field $\mathbb{Q}(\zeta)$, ζ a primitive lth root of unity, l a prime, as in exercise 1, p. 618, which showed that $(\zeta - 1)D'$ is prime in D'. Show that $[D'/(\zeta-1)D' : \mathbb{Z}/l\mathbb{Z}] = 1$ and hence that $|D'/(\zeta-1)D'| = l$. Show that $D' = \mathbb{Z}[\zeta] + (\zeta-1)D'$. Use exercise 2, p. 276 of BAI, to show that $|\det T(\zeta^i, \zeta^j)| = l^{l-2}$ if $0 \leqslant i, j \leqslant l-2$. Hence show that $D' \subset l^{-(l+2)}\mathbb{Z}[\zeta]$ and $l^{(l+2)}D' \subset \mathbb{Z}[\zeta] \subset D'$. Use these results to prove that $D' = \mathbb{Z}[\zeta]$ and that $l\mathbb{Z}$ is the only prime ideal of $\mathbb{Z}$ that is ramified in D'.

3. Show that a finite dimensional commutative algebra A over a field F is a direct sum of separable fields if and only if A_E is semi-primitive for every field extension E of F. (The latter property was discussed on p. 374 where it was called separability.)

10.6 FINITELY GENERATED MODULES OVER A DEDEKIND DOMAIN

In this section we shall extend the theory of finitely generated modules over a p.i.d. that we treated in BAI (sections 3.6–3.9, pp. 179–194) to finitely generated modules over a Dedekind domain. We recall that if M is a module over a domain D, the subset T of elements $x \in M$ for which there exists an $a \neq 0$ such that $ax = 0$ is a submodule called the torsion submodule of M. M is called *torsion free* if $T = 0$. Any submodule of a torsion-free module is torsion free. Hence, since any free module is torsion free, any submodule of a free module is torsion free. It is clear also that if T is the torsion submodule of a module M, then M/T is torsion free. We shall derive next a pair of less obvious results on torsion-free modules.

PROPOSITION 10.11. *Any finitely generated torsion-free module M over a domain D is isomorphic to a submodule of a free module of finite rank.*

Proof. Let $\{u_1,\dots,u_r\}$ be a set of generators for M. We assume $M \neq 0$ and $u_1 \neq 0$. We may assume that we have an s, $1 \leqslant s \leqslant r$, such that $\{u_1,\dots,u_s\}$ are linearly independent over D in the sense that $\sum_1^s d_i u_i = 0$ holds only if every $d_i = 0$, but $\{u_1,\dots,u_s,u_{s+j}\}$ are linearly dependent over D for all j, $1 \leqslant j \leqslant r-s$. Then the submodule $\sum_1^s Du_i$ of M is free and hence the result holds if $s = r$. On the other hand, if $s < r$, then for each j there exists a non-zero $d_j \in D$ such that $d_j u_{s+j} \in \sum_1^s Du_i$. Then $du_{s+j} \in \sum_1^s Du_i$ for $d = \prod d_j \neq 0$ and so $dM \subset \sum_1^s Du_i$. Hence the D-endomorphism $x \rightsquigarrow dx$ of M maps M into the free submodule $\sum_1^s Du_i$. Since M is torsion free, this map is a monomorphism. Hence M is isomorphic to a submodule of the free module $\sum_1^s Du_i$. $\square$

We shall obtain next a generalization of the theorem that any submodule of a finitely generated free module over a p.i.d. is free (Theorem 3.7 of BAI, p. 179). This is

PROPOSITION 10.12. *Let R be a ring, which is not necessarily commutative, with the property that every left ideal of R is a projective R-module. Then any submodule M of the free module $R^{(n)}$ is a direct sum of $m \leqslant n$ submodules isomorphic to left ideals of R. Hence M is projective.*

Proof. The result is clear if $n = 1$, since the submodules of $R = R^{(1)}$ are left ideals. Hence assume that $n > 1$. Let S denote the submodule of $R^{(n)}$ of elements of the form $(x_1,\dots,x_{n-1},0)$. Evidently $S \cong R^{(n-1)}$. Now consider the R-homomorphism $p: y = (y_1,\dots,y_n) \rightsquigarrow y_n$ of M into R. The image is a left ideal I of R and the kernel is isomorphic to a submodule N of $R^{(n-1)}$. We have an exact sequence

$$0 \to N \xrightarrow{i} M \xrightarrow{p} I \to 0. \tag{22}$$

Since I is projective, this exact sequence splits (p. 150); hence $M \cong N \oplus I$.

Since $N \subset R^{(n-1)}$, we may use induction to conclude that N is isomorphic to a direct sum of $\leqslant n-1$ left ideals. Then M is isomorphic to a direct sum $m \leqslant n$ left ideals of R. Since any left ideal is projective, it follows that M is projective. □

We now assume that M is a finitely generated module over a Dedekind domain D. The defining property for D that we used is that any fractional ideal of D is invertible and hence is projective (Proposition 10.3). In particular, any integral ideal of D is projective and since 0 is trivially projective, every ideal in D is a projective module. Thus D satisfies the hypothesis of Proposition 10.12. Let T be the torsion submodule of M and put $\bar{M} = M/T$. Then $\bar{M}$ is finitely generated and torsion free. Hence, by Proposition 10.11, $\bar{M}$ is isomorphic to a submodule of a free module of finite rank. Then, by Proposition 10.12, $\bar{M}$ is projective. Since we have an exact sequence $0 \to T \to M \to \bar{M} \to 0$ and $\bar{M}$ is projective, $M \cong \bar{M} \oplus T$. Then T is a homomorphic image of M and so it, too, is finitely generated. It is clear that we have reduced the problem of classifying finitely generated D-modules into isomorphism classes to the same problem for torsion-free modules and torsion modules.

We consider first the torsion-free modules. We have seen that any such module is projective. On the other hand, if M is projective, M is a direct summand of a free module, and since D is a domain, free modules and hence projective modules over D are torsion-free. Thus M is finitely generated torsion free over D if and only if M is finitely generated projective.

In determining conditions for isomorphism of two such modules, we may assume that the modules have the form $M = I_1 \oplus \cdots \oplus I_n$ where I_j is an integral ideal. Although there is no gain in generality in doing so, it is somewhat more natural to assume that the I_j are fractional ideals. Consider the localization M_{D^*} determined by the monoid D^* of non-zero elements of D. We have $M_{D^*} \cong D_{D^*} \otimes_D M = I_{1D^*} \oplus \cdots \oplus I_{nD^*}$ and since D_{D^*} is the field F of fractions of D and $I_{jD^*} = F$, M_{D^*} is an n-dimensional vector space over F. Thus $M_{D^*} \cong F^{(n)}$. The dimensionality n is the rank of M over D as defined in section 7.7 (Corollary, p. 415). The canonical map $x \rightsquigarrow x/1$ of M into M_{D^*} is a monomorphism. It follows that we have an isomorphism of M with the D-submodule of $F^{(n)}$ consisting of the vectors $(x_1, \ldots, x_n)$ where $x_j \in I_j$. Hence we may assume that M consists of this set of vectors.

If η is an isomorphism of M onto a second finitely generated torsion-free module M', η has a unique extension to a bijective linear transformation of M_{D^*}/F onto M'_{D^*}/F. It follows that M and M' have the same rank n and we may assume also that M' is the D-submodule of $F^{(n)}$ consisting of the vectors $(x'_1, \ldots, x'_n)$ where $x'_j \in I'_j$, a fractional ideal in F, $1 \leqslant j \leqslant n$. Taking into account the form of a bijective linear transformation of $F^{(n)}$, we see that η

has the form

$$(23) \qquad (x_1, \ldots, x_n) \rightsquigarrow (x'_1, \ldots, x'_n) = (x_1, \ldots, x_n)A$$

where $A = (a_{ij}) \in GL_n(F)$, the group of $n \times n$ invertible matrices with entries in F. We have $x'_k = \sum x_j a_{jk} \in I'_k$ and if $A^{-1} = (b_{ij})$, then $x_j = \sum x'_k b_{kj} \in I_j$ if $x'_k \in I'_k$, $1 \leqslant k \leqslant n$. Conversely, if $A \in GL_n(F)$ satisfies these conditions, then (23) is an isomorphism of M onto M'.

If $n = 1$, the foregoing conditions imply that two fractional ideals I and I' are isomorphic if and only if there exists an $a \neq 0$ in F such that $I' = Ia$. We can state this result in terms of the *class group* of the Dedekind domain D, which we define to be the factor group $\mathscr{F}/\mathscr{D}$ where $\mathscr{F}$ is the multiplicative group of fractional ideals of D and $\mathscr{D}$ is the subgroup consisting of the principal ideals aD, $a \neq 0$ in F. The ideal I and $aI = (aD)I$ are said to be in the same class. Thus we have shown that I and I' are isomorphic D-modules if and only if I and I' are in the same class. It is clear that we have a 1–1 correspondence between the isomorphism classes of fractional ideals and hence of torsion-free modules of rank one with the elements of the class group $\mathscr{F}/\mathscr{D}$.

We prove next the

LEMMA. *Let I_1 and I_2 be fractional ideals. Then $M = I_1 \oplus I_2 \cong D \oplus I_1 I_2$.*

Proof. We use exercise 10 on p. 605. According to this, if $a_1 \neq 0$ in I_1 and $J = a_1 I_1^{-1}$ $(\subset D)$, then there exists an $a_2 \in I_2$ such that $a_1 I_1^{-1} + a_2 I_2^{-1} = D$. Hence we have $b_i \in I_i^{-1}$ such that $a_1 b_1 + a_2 b_2 = 1$. Then the matrix

$$(24) \qquad A = \begin{pmatrix} b_1 & -a_2 \\ b_2 & a_1 \end{pmatrix}$$

is invertible with $A^{-1} = \begin{pmatrix} a_1 & a_2 \\ -b_2 & b_1 \end{pmatrix}$. If $x_j \in I_j$, then $y_1 = x_1 b_1 + x_2 b_2 \in D$ and $y_2 = -x_1 a_2 + x_2 a_1 \in I_1 I_2$. On the other hand, if $y_1 \in D$ and $y_2 = c_1 c_2$, $c_j \in I_j$, then $x_1 = a_1 y_1 - b_2 c_1 c_2 \in I_1$ and $x_2 = a_2 y_1 + b_1 c_1 c_2 \in I_2$. Hence $(x_1, x_2) \rightsquigarrow (x_1, x_2)A$ is an isomorphism of $I_1 \oplus I_2$ onto $D \oplus I_1 I_2$. □

We can now prove the main theorem on finitely generated modules without torsion over a Dedekind domain.

THEOREM 10.14. *Any finitely generated torsion-free module M over a Dedekind domain is isomorphic to a direct sum of a finite number of integral* *ideals. Necessary and sufficient conditions for the isomorphism of* $M = I_1 \oplus \cdots \oplus I_n$ *and* $M' = I'_1 \oplus \cdots \oplus I'_m$ *where the* I_j *and* I'_k *are fractional ideals are* $m = n$ *and* $I_1 \cdots I_n$ *and* $I'_1 \cdots I'_m$ *are in the same class.*

Proof. The validity of the first statement has been noted before. Now suppose that $M \cong M'$ where M and M' are as in the statement of the theorem. Then $m = n$ and we have a matrix $A = (a_{ij}) \in GL_n(F)$ such that if $x_j \in I_j$, $1 \leqslant j \leqslant n$, then $x'_k = \sum_{j=1}^{n} x_j a_{jk} \in I'_k$. Moreover, if $A^{-1} = (b_{ij})$ and $x'_k \in I'_k$, $1 \leqslant k \leqslant n$, then $x_j = \sum x'_k b_{kj} \in I_j$. The first condition implies that $x_j a_{jk} \in I'_k$ for any $x_j \in I_j$. Thus $I_j a_{jk} \subset I'_k$, so $(I'_k)^{-1} I_j a_{jk} \subset D$ and $a_{jk} \in I_j^{-1} I'_k$. Hence if $(k_1, \dots, k_n)$ is a permutation of $1, 2, \dots, n$, then $a_{1k_1} \cdots a_{nk_n} \in (\prod I_j)^{-1}(\prod I'_k)$. This implies that $\det A \in (\prod I_j)^{-1}(\prod I'_k)$ and $(\det A)\prod I_j \subset \prod I'_j$. Similarly, $(\det A^{-1})\prod I'_j \subset \prod I_j$. Hence $\prod I'_j = (\det A)\prod I_j$ and so $\prod I_j$ and $\prod I'_j$ are in the same class. Conversely, assume that $m = n$ and that $\prod I_j$ and $\prod I'_j$ are in the same class. By the lemma, $M = I_1 \oplus I_2 \oplus \cdots \oplus I_n \cong D \oplus I_1 I_2 \oplus I_3 \oplus \cdots \oplus I_n \cong D \oplus D \oplus I_1 I_2 I_3 \oplus I_4 \oplus \cdots \oplus I_n \cong \cdots \cong D \oplus \cdots \oplus D \oplus \prod I_j$. Similarly, $M' \cong D \oplus \cdots \oplus D \oplus \prod I'_j$. Since $\prod I_j$ and $\prod I'_j$ are in the same class, these are isomorphic and hence $M \cong M'$. □

We consider next the structure of a finitely generated torsion module M over D. Our program for studying M will be to replace it by a closely related module over a p.i.d. Let $I = \text{ann}_D M$. If $M = Dx_1 + Dx_2 + \cdots + Dx_n$, then

$$I = \bigcap \text{ann}\, x_i \supset \prod \text{ann}\, x_i \neq 0.$$

Hence $I = \text{ann}_D M \neq 0$. Let $\{P_1, \dots, P_r\}$ be the prime ideal divisors of I and let

$$S = D - \bigcup_1^r P_i = \bigcap_1^r (D - P_i).$$

Then S is a submonoid of the multiplicative monoid of D and we can form the ring D_S and the D_S-module $M_S = D_S \otimes_D M$. Any prime ideal of D_S has the form P_S where P is a prime ideal such that $P \cap S = \varnothing$ (p. 401). Then $P \subset \bigcup P_i$ and hence $P \subset P_i$ for some i (Proposition 7.1, p. 390), which implies $P = P_i$. Thus D_S is a Dedekind domain with only a finite number of prime ideals, $P'_i = P_{iS}$, $1 \leqslant i \leqslant r$. Such a domain is a p.i.d. (exercise 15, p. 625). Moreover, M_S is a torsion D_S-module since $I_S \neq 0$ and $I_S M_S = 0$.

The fundamental theorem on finitely generated modules over a p.i.d. (BAI, p. 187) is applicable to the D_S-module M_S. According to this, $M_S = D_S y_1 \oplus \cdots \oplus D_S y_m$ where $\text{ann}_{D_S} y_1 \supset \text{ann}_{D_S} y_2 \supset \cdots \supset \text{ann}_{D_S} y_m$. Moreover, $\text{ann}_{D_S} y_i \supset I_S \neq 0$ and $\text{ann}_{D_S} y_i$ is a proper ideal in D_S. We shall be able to apply this result to determining the structure of M as D-module, since we can recover M from M_S by using the following

LEMMA. *Let I be a non-zero ideal in D, $P_1, \ldots, P_r$ the prime ideal divisors of I in D, $S = \bigcap(D - P_i)$. Let N be a D-module such that $IN = 0$. Then $N \cong (D/I) \otimes_D N_S$ as D-modules.*

Proof. We have $(D/I) \otimes_D N_S \cong (D/I) \otimes_D (D_S \otimes_D N) \cong ((D/I) \otimes_D D_S) \otimes_D N \cong (D/I)_S \otimes_D N$. It is readily seen that the elements $s+I$, $s \in S$, are units in D/I, which implies that $(D/I)_S \cong D/I$. Hence $(D/I)_S \otimes_D N \cong (D/I) \otimes_D N \cong N$, since $IN = 0$. □

We can now prove the following structure theorem.

THEOREM 10.15. *Let M be a finitely generated torsion module over a Dedekind domain D. Then M is isomorphic to a direct sum of cyclic modules $Dz_1 \oplus \cdots \oplus Dz_m$ such that* $\operatorname{ann} z_1 \supset \operatorname{ann} z_2 \supset \cdots \supset \operatorname{ann} z_m \supsetneq 0$. *Moreover, the ideals* $\operatorname{ann} z_i$, $1 \leqslant i \leqslant m$, *are uniquely determined.*

Proof. Let S be as before. Then $M_S = D_S y_1 \oplus \cdots \oplus D_S y_m$ where $\operatorname{ann}_{D_S} y_1 \supset \cdots \supset \operatorname{ann}_{D_S} y_m$. Then if $I'_j = \operatorname{ann}_{D_S} y_j$, $D_S y_j \cong D_S / I'_j$. Let $I_j = \{a \in D | a/s \in I'_j$ for some $s \in S\}$. Then $I_{jS} = I'_j$ (p. 401) and D/I_j is a cyclic D-module, Dz_j, such that $(D/I_j)_S \cong D_S/I_{jS} = D_S/I'_j$. Moreover, $\operatorname{ann}_D z_j = I_j$ and $I_1 \supset \cdots \supset I_m$. Now consider the module $N = Dz_1 \oplus \cdots \oplus Dz_m$. We have $N_S = (Dz_1)_S \oplus \cdots \oplus (Dz_m)_S \cong D_S y_1 \oplus \cdots \oplus D_S y_m \cong M_S$. Hence by the lemma, $M \cong N = Dz_1 \oplus \cdots \oplus Dz_m$. The uniqueness follows also by using S-localization and the corresponding uniqueness result in the p.i.d. case (BAI, p. 192). We leave the details to the reader. □

Theorems 10.14 and 10.15 and the fact that any finitely generated module over a Dedekind domain D is a direct sum of its torsion submodule and a torsion-free module constitute a direct generalization of the fundamental theorem on finitely generated modules over a p.i.d. If D is a p.i.d., then the class group $\mathscr{F}/\mathscr{D} = 1$ and the results show that M is a direct sum of copies of D and cyclic modules $Dz_1, \ldots, Dz_m$ with non-zero annihilators $I_1, \ldots, I_m$ such that $I_1 \supset \cdots \supset I_m$. This is the fundamental theorem as given in BAI, p. 187. The Dedekind domain is a p.i.d. if and only if $\mathscr{F}/\mathscr{D} = 1$. The group $\mathscr{F}/\mathscr{D}$ gives a measure of the departure of D from being a p.i.d. It is a classical group of algebraic number theory. There it is shown by transcendental methods that the class group is finite. On the other hand, it has been shown by L. Claborn (*Pacific J. of Math.*, vol. 18 (1966), 219–222) that any abelian group is the class group of a suitable Dedekind domain.

We shall now show that the class group coincides with the projective class group, which we defined in section 7.4. The latter, denoted as Pic D, is the set of isomorphism classes of faithful finitely generated projective modules of rank one with multiplication defined by the tensor product. Since D is a domain, projective D-modules are automatically faithful. Hence the elements of Pic D can be represented by the fractional ideals, and as we have seen, two such ideals are isomorphic if and only if they are in the same class. Hence we have the bijective map $[I] \rightsquigarrow I\mathscr{D}$ of Pic D onto $\mathscr{F}/\mathscr{D}$ where $[I]$ denotes the isomorphism class of the fractional ideal I. Now we have the canonical module homomorphism of $I_1 \otimes I_2$ into $I_1 I_2$ sending $a_1 \otimes a_2$ into $a_1 a_2$. Localization at all of the maximal ideals show that this is an isomorphism. Thus $I_1 \otimes I_2 \cong I_1 I_2$ and hence $[I] \rightsquigarrow I\mathscr{D}$ is a group isomorphism of Pic D with the class group.

REFERENCES

O. Zariski and P. Samuel, *Commutative Algebra* vol. I, D. van Nostrand, 1958. New printing, Springer-Verlag, New York, 1975.

S. Lang, *Algebraic Number Theory*, Addison-Wesley, Reading, Mass., 1970.

———, *Introduction to Algebraic and Abelian Functions*, Addison-Wesley, Reading, Mass., 1972.

G. J. Janusz, *Algebraic Number Fields*, Academic Press, New York, 1973.

11

Formally Real Fields

In Chapter 5 of *Basic Algebra I* we studied polynomial equations, inequations, and inequalities in a real closed field. We defined such a field to be an ordered field in which every odd degree polynomial in one indeterminate has a root and every positive element has a square root. We proved that if R is real closed, then $R(\sqrt{-1})$ is algebraically closed. The main problem we considered was that of developing an algorithm for testing the solvability of a system of polynomial equations, inequations, and inequalities in several unknowns in a real closed field. In the case of a single polynomial equation in one unknown, the classical method of J. C. F. Sturm provides a solution to this problem. We gave this method and developed a far-reaching extension of the method to the general case of systems in several unknowns. A consequence of this (also indicated in BAI) is Tarski's theorem, which states roughly that a system of polynomial equations, inequations, and inequalities that has a solution in one real closed field has a solution in every such field (see BAI, p. 340, for the precise statement).

We now resume the study of real closed fields, but we approach these from a different point of view, that of formally real fields as defined by Artin and Schreier. The defining property for such a field is that -1 is not a sum of squares in the field, or equivalently, if $\sum a_i^2 = 0$ for a_i in the field, then every $a_i = 0$. It is clear that any ordered field is formally real. On the other hand, as we shall see, every formally real field can be ordered. Hence a field is formally real if and only if it can be ordered.

The Artin–Schreier theory of formally real fields is an essential element in Artin's solution of one of the problems posed by Hilbert at the 1900 Paris International Congress of Mathematicians. This was Hilbert's seventeenth problem, which concerned positive semi-definite rational functions: Suppose $f(x_1, \ldots, x_n)$ is a rational expression in indeterminates x_i with real coefficients such that $f(a_1, \ldots, a_n) \geqslant 0$ for all $(a_1, \ldots, a_n)$ where f is defined. Then is f necessarily a sum of squares of rational expressions with real coefficients? Artin gave an affirmative answer to this question in 1927. A new method of proving this result based on model theory was developed by A. Robinson in 1955. We shall give an account of Artin's theorem. Our approach is essentially model theoretic and is based on the theorem of Tarski mentioned earlier.

Artin's theorem gives no information on the number of squares needed to express a given $f(x_1, \ldots, x_n)$. Hilbert had shown in 1893 that any positive semi-definite rational function in two variables is expressible as a sum of four squares. In 1966 in an unpublished paper, J. Ax showed that any positive semi-definite function in three variables is a sum of eight squares and he conjectured that for such functions of n variables, 2^n squares are adequate. This was proved in 1967 by A. Pfister by a completely novel and ingenious method. We shall give his proof.

We shall conclude this chapter with a beautiful characterization of real closed fields due to Artin and Schreier. It is noteworthy that the proof of this theorem initiated the study of cyclic fields of p^e dimensions over fields of characteristic p.

11.1 FORMALLY REAL FIELDS

We recall that a field F is said to be ordered if there is given a subset P (the set of positive elements) in F such that P is closed under addition and multiplication and F is the disjoint union of P, $\{0\}$, and $-P = \{-p \mid p \in P\}$ (see BAI, p. 307). Then F is totally ordered if we define $a > b$ to mean $a - b \in P$. Moreover if $a > b$, then $a + c > b + c$ for every c and $ap > bp$ for every $p \in P$. If $a \neq 0$, then $a^2 = (-a)^2 > 0$. Hence if $\sum a_i^2 = 0$ in F, then every $a_i = 0$. This is equivalent to: -1 is not a sum of squares in F. It follows that the characteristic of F is 0.

Following Artin and Schreier we now introduce the following

DEFINITION 11.1. *A field F is called* formally real *if -1 is not a sum of squares in F.*

It is clear that any field that can be ordered is formally real. The converse of this observation is a theorem of Artin and Schreier. We shall now prove this by an argument due to Serre that is based on the following

LEMMA 1. *Let P_0 be a subgroup of the multiplicative group F^* of a field F such that P_0 is closed under addition and contains all non-zero squares. Suppose that a is an element of F^* such that $-a \notin P_0$. Then $P_1 = P_0 + P_0 a = \{b + ca | b, c \in P_0\}$ is a subgroup of F^* closed under addition.*

Proof. Evidently P_1 is closed under addition and if $b_i, c_i \in P_0$ for $i = 1, 2$, then $(b_1 + c_1 a)(b_2 + c_2 a) = (b_1 b_2 + c_1 c_2 a^2) + (b_1 c_2 + b_2 c_1) a \in P_1$ since $b_1 b_2 + c_1 c_2 a^2$ and $b_1 c_2 + b_2 c_1 \in P_0$. We note next that P_1 does not contain 0, since otherwise we have $b + ca = 0$ for $b, c \in P_0$, which gives $-a = bc^{-1} \in P_0$, contrary to the hypothesis on a. Also we have

$$(b+ca)^{-1} = (b+ca)(b+ca)^{-2} = b(b+ca)^{-2} + c(b+ca)^{-2} a \in P_1$$

since $b(b+ca)^{-2}$ and $c(b+ca)^{-2} \in P_0$. Hence P_1 is a subgroup of F^*. □

Now let F be formally real and let P_0 be the set of sums $\sum a_i^2$ with every $a_i \neq 0$. Evidently P_0 is closed under addition and since $(\sum_i a_i^2)(\sum_j b_j^2) = \sum_{i,j} (a_i b_j)^2$, P_0 is closed under multiplication. Moreover, P_0 contains all of the non-zero squares and hence if $a = \sum a_i^2$, $a_i \neq 0$, then $a^{-1} = aa^{-2} \in P_0$. Thus P_0 satisfies the conditions of the lemma and so the set of subsets P' satisfying these conditions is not vacuous. We can apply Zorn's lemma to conclude that this set of subsets of F contains a maximal element P. Then it follows from the lemma that if a is any element of F^*, then either a or $-a \in P$. Hence $F = P \cup \{0\} \cup -P$ where $-P = \{-p | p \in P\}$ and since $0 \notin P$ and P is closed under addition, $P \cap -P = \varnothing$ and $0 \notin -P$. Thus P, $\{0\}$, $-P$ are disjoint and since P is closed under addition and multiplication, P gives an ordering of the field F. We therefore have the following

THEOREM 11.1. *A field F can be ordered (by a subset P) if and only if it is formally real.*

In BAI (p. 308) we defined a field R to be real closed if it is ordered and if (1) every positive element of R has a square root in R and (2) every polynomial of odd degree in one indeterminate with coefficients in R has a root in R. We showed that the ordering in a real closed field R is unique and that any automorphism of such a field is an order automorphism (Theorem 5.1, p. 308). Moreover, we proved an extension of the so-called fundamental theorem of algebra: If R is real closed, then $R(\sqrt{-1})$ is algebraically closed (Theorem 5.2, p. 309). We shall now give a characterization of real closed fields in terms of formal reality:

THEOREM 11.2. *A field R is real closed if and only if R is formally real and no proper algebraic extension of R is formally real.*

We separate off from the proof the following

LEMMA 2. *If F is formally real, then any extension field $F(r)$ is formally real if either $r = \sqrt{a}$ for $a > 0$ in F or r is algebraic over F with minimum polynomial of odd degree.*

Proof. First, let $r = \sqrt{a}$, $a > 0$. Suppose that $F(r)$ is not formally real. Then $r \notin F$ and we have $a_i, b_i \in F$ such that $-1 = \sum (a_i + b_i r)^2$. This gives $\sum a_i^2 + \sum b_i^2 a = -1$. Since $a > 0$, this is impossible.

In the second case let $f(x)$ be the minimum polynomial of r. Then the degree of $f(x)$ is odd. We shall use induction on $m = \deg f(x)$. Suppose that $F(r)$ is not formally real. Then we have polynomials $g_i(x)$ of degree $< m$ such that $\sum g_i(r)^2 = -1$. Hence we have $\sum g_i(x)^2 = -1 + f(x)g(x)$ where $g(x) \in F[x]$. Since F is formally real and the leading coefficient of $g_i(x)^2$ is a square, it follows that $\deg(-1 + f(x)g(x)) = \deg(\sum g_i(x)^2)$ is even and $< 2m$. It follows that $\deg g(x)$ is odd and $< m$. Now $g(x)$ has an irreducible factor $h(x)$ of odd degree. Let s be a root of $h(x)$ and consider $F(s)$. By the induction hypothesis, this is formally real. On the other hand, substitution of s in the relation $\sum g_i(x)^2 = -1 + f(x)g(x)$ gives the contradiction $\sum g_i(s)^2 = -1$. □

We can now give the

Proof of Theorem 11.2. Suppose that R is real closed. Then $C = R(\sqrt{-1})$ is algebraically closed and $C \supsetneq R$. Evidently C is an algebraic closure of R and so any algebraic extension of R can be regarded as a subfield of C/R. Hence if it is a proper extension it must be C, which is not formally real since it contains $\sqrt{-1}$.

Next suppose that R is formally real and no proper algebraic extension of R has this property. Let $a \in R$ be positive. Then Lemma 2 shows that

$R(\sqrt{a})$ is formally real. Hence $R(\sqrt{a}) = R$ and $\sqrt{a} \in R$. Next let $f(x)$ be a polynomial of odd degree with coefficients in R. Let $g(x)$ be an irreducible factor of $f(x)$ of odd degree and consider an extension field $R(r)$ where $g(r) = 0$. By Lemma 2, $R(r)$ is formally real. Hence $R(r) = R$. Then $r \in R$ and $f(r) = 0$. We have therefore verified the two defining properties of a real closed field. Hence R is real closed. □

We shall show next that the basic property of a real closed field R that $\sqrt{-1} \notin R$ and $C = R(\sqrt{-1})$ is algebraically closed characterizes these fields.

THEOREM 11.3. *A field R is real closed if and only if $\sqrt{-1} \notin R$ and $C = R(\sqrt{-1})$ is algebraically closed.*

Proof. It suffices to show that if R has the stated properties, then R is real closed. Suppose that R satisfies the conditions. We show first that the sum of two squares in R is a square. Let a, b be non-zero elements of R and let u be an element of C such that $u^2 = a + bi$, $i = \sqrt{-1}$. We have the automorphism $x + iy \rightsquigarrow x - iy$, $x, y \in R$, of C/R whose set of fixed points is R. Now

$$a^2 + b^2 = (a + bi)(a - bi) = u^2\overline{u^2} = (u\bar{u})^2$$

and $u\bar{u} \in R$. Thus $a^2 + b^2$ is a square in R. By induction, every sum of squares in R is a square. Since -1 is not a square in R, it is not a sum of squares and hence R is formally real. On the other hand, since C is algebraically closed, the first part of the proof of Theorem 11.2 shows that no proper algebraic extension of R is real closed. Hence R is real closed by Theorem 11.2. □

EXERCISES

1. Show that if F is formally real and the x_i are indeterminates, then $F(x_1, \ldots, x_n)$ is formally real.
2. Define an *ordering* for a domain D as for a field: a subset P of D such that P is closed under addition and multiplication and D is the disjoint union of P, $\{0\}$, and $-P$. Show that an ordering in a domain has a unique extension to its field of fractions.
3. Let F be an ordered field. Show that $F[x]$, x an indeterminate, has an ordering defined by $a_0x^n + a_1x^{n-1} + \cdots > 0$ if $a_0 > 0$.
4. Call an ordered field F *archimedean-ordered* if for any $a > 0$ in F there exists a positive integer n such that $n(= n1) > a$. Show that the field $F(x)$, x an indeterminate, ordered by using exercises 2 and 3 is not archimedean.

5. Prove that any archimedean-ordered field is order isomorphic to a subfield of $\mathbb{R}$.

6. (T. Springer.) Let Q be an anistropic quadratic form on a finite dimensional vector space V over a field of characteristic $\neq 2$. Let E be an odd dimensional extension field of F and let Q_E be the extension of Q to a quadratic form on V_E. Show that Q_E is anisotropic. (*Hint*: It suffices to assume that $E = F(\rho)$. Prove the result by induction on $[E:F]$ using an argument like that in the second part of the proof of Lemma 2.)

11.2 REAL CLOSURES

DEFINITION 11.2. *Let F be an ordered field. An extension field R of F is called a real closure of F if* (1) *R is real closed and algebraic over F and* (2) *the (unique) order in R is an extension of the given order in F.*

A central result in the Artin–Schreier theory of formally real fields is the existence and uniqueness of a real closure for any ordered field F. For the proof of this result we shall make use of Sturm's theorem (BAI, p. 312), which permits us to determine the number of roots in a real closed field R of a polynomial $f(x) \in R[x]$. Let $f(x) = x^n + a_1x^{n-1} + \cdots + a_n$ and $M = 1 + |a_1| + \cdots + |a_n|$ where $|\ |$ is defined as usual. Then every root of $f(x)$ in R lies in the interval $-M < x < M$ (BAI, exercise 4, p. 311). Define the standard sequence for $f(x)$ by

$$\begin{aligned} &f_0(x) = f(x), \qquad f_1(x) = f'(x), \qquad \text{the derivative of } f(x), \\ &f_{i-1}(x) = g_i(x)f_i(x) - f_{i+1}(x) \qquad \text{with } \deg f_{i+1} < \deg f_i \end{aligned} \tag{1}$$

for $i \geqslant 1$. Then we have an s such that $f_{s+1} = 0$, and by Sturm's theorem, the number of roots of $f(x)$ in R is $V_{-M} - V_M$ where V_a is the number of variations in sign in $f_0(a), f_1(a), \ldots, f_s(a)$. We can use this to prove the

LEMMA. *Let R_i, $i = 1, 2$, be a real closed field, F_i a subfield of R_i, $a \rightsquigarrow \bar{a}$ an order isomorphism of F_1 onto F_2 where the order in F_i is that induced from R_i. Suppose that $f(x)$ is a monic polynomial in $F_1[x]$, $\bar{f}(x)$ the corresponding polynomial in $F_2[x]$. Then $f(x)$ has the same number of roots in R_1 as $\bar{f}(x)$ has in R_2.*

Proof. If M is as above, then the first number is $V_{-M} - V_M$ and the second is $V_{-\bar{M}} - V_{\bar{M}}$ determined by the standard sequence for $\bar{f}(x)$. Since $a \rightsquigarrow \bar{a}$ is an order isomorphism of F_1 onto F_2, it is clear that these two numbers are the same. ☐

We can now prove the important

THEOREM 11.4. *Any ordered field has a real closure. If F_1 and F_2 are ordered fields with real closures R_1 and R_2 respectively, then any order isomorphism of F_1 onto F_2 has a unique extension to an isomorphism of R_1 onto R_2 and this extension preserves order.*

Proof. Let F be an ordered field, $\bar{F}$ an algebraic closure of F, and let E be the subfield of $\bar{F}$ obtained by adjoining to F the square roots of all the positive elements of F. Then E is formally real. Otherwise, E contains elements a_i such that $\sum a_i^2 = -1$. The a_i are contained in a subfield generated over F by a finite number of square roots of positive elements of F. Using induction on the first part of Lemma 2 of section 11.1 we see that this subfield is formally real contrary to $\sum a_i^2 = -1$ with a_i in the subfield. We now consider the set of formally real subfields of $\bar{F}$ containing E. This set is inductive, so by Zorn's lemma, we have a maximal subfield R in the set. We claim that R is real closed. If not, there exists a proper algebraic extension R' of R that is formally real (Theorem 11.2). Since $\bar{F}$ is an algebraic closure of R, we may assume that $R' \subset \bar{F}$ so we have $\bar{F} \supset R' \supsetneq R$. This contradicts the maximality of R. Hence R is real closed. Now suppose $a \in F$ and $a > 0$. Then $a = b^2$ for $b \in E \subset R$ and hence $a > 0$ in the order defined in R. Thus the order in R is an extension of that of F and hence R is a real closure of F.

Now let F_1 and F_2 be ordered fields, R_i a real closure of F_i, and let $\sigma\colon a \rightsquigarrow \bar{a}$ be an order isomorphism of F_1 onto F_2. We wish to show that σ can be extended to an isomorphism Σ of R_1 onto R_2. The definition of Σ is easy. Let r be an element of R_1, $g(x)$ the minimum polynomial of r over F_1 and let $r_1 < r_2 < \cdots < r_k = r < \cdots < r_m$ be the roots of $g(x)$ in R_1 arranged in increasing order. By the lemma, the polynomial $\bar{g}(x)$ has precisely m roots in R_2 and we can arrange these in increasing order as $\bar{r}_1 < \bar{r}_2 < \cdots < \bar{r}_m$. We now define Σ as the map sending r into the kth one of these roots. It is easy to see that Σ is bijective and it is clear that Σ is an extension of σ. However, it is a bit tricky to show that Σ is an isomorphism. To see this we show that if S is any finite subset of R_1, there exists a subfield E_1 of R_1/F_1 and a monomorphism η of E_1/F_1 into R_2/F_2 that extends σ and preserves the order of the elements of S, that is, if $S = \{s_1 < s_2 < \cdots < s_n\}$, then $\eta s_1 < \eta s_2 < \cdots < \eta s_n$. Let $T = S \cup \{\sqrt{s_{i+1} - s_i} \mid 1 \leqslant i \leqslant n-1\}$ and let $E_1 = F_1(T)$. Evidently, E_1 is finite dimensional over F_1 and so $E_1 = F_1(w)$. Let $f(x)$ be the minimum polynomial of w over F_1. By the lemma, $\bar{f}(x)$ has a root $\bar{w}$ in R_2, and we have a monomorphism η of E_1/F_1 into R_2/F_2 sending w into $\bar{w}$. Now $\eta(s_{i+1}) - \eta(s_i) = \eta(s_{i+1} - s_i) = \eta((\sqrt{s_{i+1} - s_i})^2) = (\eta\sqrt{s_{i+1} - s_i})^2 > 0$. Hence η preserves the order of the s_i. Now let r and s be any two elements of E_1 and apply the result just proved to the finite set S consisting of the roots of the minimum polynomials of r, s, $r+s$, and rs. Since η preserves the order of the elements of S, $\eta(r) = \Sigma(r)$, $\eta(s) = \Sigma(s)$, $\eta(r+s) = \Sigma(r+s)$, and $\eta(rs) = \Sigma(rs)$. Hence $\Sigma(r+s) = \eta(r+s) = \eta(r)+\eta(s) = \Sigma(r)+\Sigma(s)$ and similarly $\Sigma(rs) = \Sigma(r)\Sigma(s)$. Thus Σ is an isomorphism.

It remains to show that Σ is unique and is order preserving. Hence let Σ' be an isomorphism of R_1 onto R_2. Since Σ' maps squares into squares and the subsets of positive elements of the R_i are the sets of non-zero squares, it is clear that Σ' preserves order. Suppose also that Σ' is an extension of σ. Then it is clear from the definition of Σ that $\Sigma' = \Sigma$. This completes the proof. □

If R_1 and R_2 are two real closures of an ordered field F, then the identity map on F can be extended in a unique manner to an order isomorphism of R_1 onto R_2. In this sense there is a unique real closure of F.

It is easily seen that the field $\mathbb{Q}$ of rational numbers has a unique ordering. Its real closure $\mathbb{R}_0$ is called the *field of real algebraic numbers*. The field $\mathbb{C}_0 = \mathbb{R}_0(\sqrt{-1})$ is the algebraic closure of $\mathbb{Q}$. This is the *field of algebraic numbers*.

EXERCISES

1. Let F be an ordered field, E a real closed extension field whose order is an extension of the order of F. Show that E contains a real closure of F.

2. Let F be an ordered field, E an extension field such that the only relations of the form $\sum a_i b_i^2 = 0$ with $a_i > 0$ in F and $b_i \in E$ are those in which every $b_i = 0$. Show that E can be ordered so that its ordering is an extension of that of F.

11.3 TOTALLY POSITIVE ELEMENTS

An interesting question concerning fields is: what elements of a given field can be written as sums of squares? We consider this question in this section and in the next two sections; we first obtain a general criterion based on the following definition.

DEFINITION 11.3. *An element a of a field F is called* totally positive *if $a > 0$ in every ordering of F.*

It is understood that if F has no ordering, then every element of F is totally positive. Hence this is the case if F is not formally real. If F is not formally real, then $-1 = \sum a_i^2$ for $a_i \in F$, and if char $F \neq 2$, then the relation

$$a = \left(\frac{1+a}{2}\right)^2 - \left(\frac{1-a}{2}\right)^2 = \left(\frac{1+a}{2}\right)^2 + \sum a_i^2 \left(\frac{1-a}{2}\right)^2 \tag{2}$$

shows that every element of F is a sum of squares. We shall need this remark in the proof of the following criterion.

THEOREM 11.5. *Let F be a field of characteristic $\neq 2$. Then an element $a \neq 0$ in F is totally positive if and only if a is a sum of squares.*

Proof. If $0 \neq a = \sum a_i^2$, then evidently $a > 0$ in every ordering of F. Conversely, assume that $a \neq 0$ is not a sum of squares in F. Let $\bar{F}$ be an algebraic closure of F and consider the set of subfields E of $\bar{F}/F$ in which a is not a sum of squares. By Zorn's lemma, there is a maximal one; call it R. By the preceding remark, R is formally real and hence R can be ordered. We claim that $-a$ is a square in R. Otherwise, the subfield $R(\sqrt{-a})$ of $\bar{F}/F$ properly contains R, so a is a sum of squares in $R(\sqrt{-a})$. Hence we have $b_i, c_i \in R$ such that $a = \sum (b_i + c_i\sqrt{-a})^2$. This gives $\sum b_i c_i = 0$ and $a = \sum b_i^2 - a\sum c_i^2$. Hence $a(1+\sum c_i^2) = \sum b_i^2$ and $1+\sum c_i^2 \neq 0$, since R is formally real. Then if $c = 1+\sum c_i^2$,

$$a = \sum b_i^2 c^{-1} = \sum b_i^2 (1+\sum c_i^2) c^{-2}$$

so a is a sum of squares in R, contrary to the definition of R. Thus $-a = b^2$ for a $b \in R$ and hence $a = -b^2$ is negative in every ordering of R. These orderings give orderings of F and so we have orderings of F in which $a < 0$. Thus a is not totally positive. □

We shall apply this result first to determine which elements of a number field F are sums of squares. We have $F = \mathbb{Q}(r)$ where r is algebraic over $\mathbb{Q}$. If $\mathbb{R}_0$ is the field of real algebraic numbers (= the real closure of $\mathbb{Q}$), then $\mathbb{C}_0 = \mathbb{R}_0(\sqrt{-1})$ is an algebraic closure of $\mathbb{Q}$ and of F. If $n = [F:\mathbb{Q}]$, then we have n distinct monomorphisms of $F/\mathbb{Q}$ into $\mathbb{C}_0/\mathbb{Q}$. These are determined by the maps $r \rightsquigarrow r_i$, $1 \leqslant i \leqslant n$, where the r_i are the roots of the minimum polynomial $g(x)$ of r over $\mathbb{Q}$. Let $r_1, \dots, r_h$ be the r_i contained in $\mathbb{R}_0$. We call these the *real conjugates* of r and we agree to put $h = 0$ if no $r_i \in \mathbb{R}_0$. Let σ_i, $1 \leqslant i \leqslant h$, be the monomorphism of $F/\mathbb{Q}$ such that $\sigma_i r = r_i$. Each σ_i defines an ordering of F by declaring that $a > 0$ in F if $\sigma_i a > 0$ in the unique ordering of $\mathbb{R}_0$. We claim that these r orderings are distinct and that they are the only orderings of F. First, suppose that the orderings defined by σ_i and σ_j are the same. Then $\sigma_j \sigma_i^{-1}$ is an order-preserving isomorphism of the subfield $\mathbb{Q}(r_i)$ of $\mathbb{R}_0$ onto the subfield $\mathbb{Q}(r_j)$. Since $\mathbb{R}_0$ is algebraic over $\mathbb{Q}(r_i)$ and $\mathbb{Q}(r_j)$, $\mathbb{R}_0$ is the real closure of these fields. Hence, by Theorem 11.4, $\sigma_j\sigma_i^{-1}$ can be extended to an automorphism σ of $\mathbb{R}_0$. Since $\mathbb{R}_0$ is the real closure of $\mathbb{Q}$, it follows also from Theorem 11.4 that the only automorphism of $\mathbb{R}_0$ is the identity. Hence $\sigma = 1$ and $\sigma_j = \sigma_i$. Next suppose that we have an ordering of F and let R be the real closure defined by this ordering. Since R is algebraic over $\mathbb{Q}$, R is also a real closure of $\mathbb{Q}$. Hence we have an order isomorphism of $R/\mathbb{Q}$ onto $\mathbb{R}_0/\mathbb{Q}$. The restriction of this to F coincides with one of the σ_i and hence the given ordering coincides with the one defined by this σ_i.

It is clear that the σ_i that we have defined can be described as the monomorphisms of F into $\mathbb{R}_0$. Hence we have proved the following

THEOREM 11.6. *Let F be an algebraic number field, $\mathbb{R}_0$ the field of real algebraic numbers. Then we have a 1–1 correspondence between the set of orderings of F and the set of monomorphisms of F into $\mathbb{R}_0$. The ordering determined by the monomorphism σ_i is that in which $a > 0$ for $a \in F$ if $\sigma_i a > 0$ in $\mathbb{R}_0$.*

An immediate consequence of Theorems 11.5 and 11.6 is the following result due to Hilbert and E. Landau:

THEOREM 11.7. *Let F be an algebraic number field and let $\sigma_1, \dots, \sigma_h$ $(h \geqslant 0)$ be the different monomorphisms of F into the field $\mathbb{R}_0$ of real algebraic numbers. Then an element $a \in F$ is a sum of squares in F if and only if $\sigma_i a > 0$ for $1 \leqslant i \leqslant h$.*

EXERCISES

1. Let F be an ordered field, E an extension field. Show that if b is an element of E that cannot be written in the form $\sum a_i b_i^2$ for $a_i \geqslant 0$ in F and $b_i \in E$, then there exists an ordering of E extending the ordering of F in which $b < 0$.

2. Let F be an ordered field, R the real closure of F, and E a finite dimensional extension of F. Prove the following generalization of Theorem 11.6: There is a 1–1 correspondence between the set of orderings of E extending the ordering of F and the set of monomorphisms of E/F into R/F.

3. (I. Kaplansky–M. Kneser.) Let F be a field of characteristic $\neq 2$ that is not formally real. Suppose that $|F^*/F^{*2}| = n < \infty$. Show that any non-degenerate quadratic form in n variables is universal. (*Sketch of proof*: Let $a_1, a_2, \dots$ be a sequence of non-zero

elements of F and let M_k denote the set of values of the quadratic form $a_1x_1{}^2+\cdots+a_kx_k{}^2$ for $x_i \in F$. Show that there exists a $k \leqslant n$ such that $M_{k+1} = M_k$. Then $M_{k+1} = a_{k+1}F^2+M_k = M_k$ where F^2 is the set of squares of elements of F. Iteration of $M_k = a_{k+1}F^2+M_k$ gives $M_k = a_{k+1}(F^2+\cdots+F^2)+M_k$. Hence conclude that $M_k = F+M_k$ and $M_k = F$.)

11.4 HILBERT'S SEVENTEENTH PROBLEM

One of the problems in his list of twenty-three unsolved problems that Hilbert proposed in an address before the 1900 Paris International Congress of Mathematicians was the problem on positive semi-definite rational functions: Let f be a rational function of n real variables with rational coefficients that is positive semi-definite in the sense that $f(a_1,\dots,a_n) \geqslant 0$ for all real $(a_1,\dots,a_n)$ for which f is defined. Then is f necessarily a sum of squares of rational functions with rational coefficients? By a rational function of n real variables with rational coefficients we mean a map of the form $(a_1,\dots,a_n) \rightsquigarrow f(a_1,\dots,a_n)$ where $f(x_1,\dots,x_n) = g(x_1,\dots,x_n)/h(x_1,\dots,x_n)$ and g and h are polynomials in the indeterminates x_i with rational coefficients. The domain of definition of the function is a Zariski open subset defined by $h(a_1,\dots,a_n) \neq 0$ and two rational functions are regarded as equal if they yield the same values for every point of a Zariski open subset.

In 1927, making essential use of the Artin-Schreier theory of formally real fields, Artin gave an affirmative answer to Hilbert's question by proving the following stronger result:

THEOREM OF ARTIN. *Let F be a subfield of $\mathbb{R}$ that has a unique ordering and let f be a rational function with coefficients in F such that $f(a_1,\dots,a_n) \geqslant 0$ for all $a_i \in F$ for which f is defined. Then f is a sum of squares of rational functions with coefficients in F.*

Examples of fields having a unique ordering are $\mathbb{Q}$, $\mathbb{R}$, and any number field that has only one real conjugate field.

The condition that F is a subfield of $\mathbb{R}$ in Artin's theorem can be replaced by the hypothesis that F is archimedean ordered. It is easily seen that this condition is equivalent to the assumption that $F \subset \mathbb{R}$ of Artin's theorem. We shall prove a result that is somewhat stronger than Artin's in that the archimedean property of F will be replaced by a condition of density in the real closure. If F is a subfield of an ordered field E, then F is said to be *dense* in E if for any two elements $a < b$ in E there exists a $c \in F$ such that $a < c < b$. It is easily seen that $\mathbb{Q}$ is dense in this sense in $\mathbb{R}$ and this implies that any subfield of $\mathbb{R}$ is dense in $\mathbb{R}$. Hence the following theorem is a generalization of Artin's theorem.

THEOREM 11.8. *Let F be an ordered field such that* (1) *F has a unique ordering and* (2) *F is dense in its real closure. Let f be a rational function with coefficients in F such that $f(a_1,\dots,a_n) \geqslant 0$ for all $(a_1,\dots,a_n) \in F^{(n)}$ for which f is defined. Then f is a sum of squares of rational functions with coefficients in F.*

We shall give a model theoretic type of proof of Theorem 11.8 based on the following result, which was proved in BAI, p. 340.

Let R_1 and R_2 be real closed fields having a common ordered subfield F, that is, the orderings on F induced by R_1 and R_2 are identical. Suppose that we have a finite set S of polynomial equations, inequations ($f \neq 0$), and inequalities ($f > 0$) with coefficients in F. Then S has a solution in R_1 if and only if it has a solution in R_2.

We now proceed to the

Proof of Theorem 11.8. The set of rational functions of n variables with coefficients in F form a field with respect to the usual definitions of addition and multiplication. If p_i, $1 \leqslant i \leqslant n$, denotes the function such that $p_i(a_1,\dots,a_n) = a_i$, then we have an isomorphism of the field $F(x_1,\dots,x_n)$, x_i indeterminates, onto the field of rational functions such that $x_i \rightsquigarrow p_i$. Accordingly, the latter field is $F(p_1,\dots,p_n)$. Suppose that $f \in F(p_1,\dots,p_n)$ is not a sum of squares. Then by Theorem 11.5, there exists an ordering of $F(p_1,\dots,p_n)$ such that $f < 0$. Write $f = gh^{-1}$ where $g, h \in F[p_1,\dots,p_n]$. Then $gh < 0$ so if $k(x_1,\dots,x_n) = g(x_1,\dots,x_n)h(x_1,\dots,x_n)$, then $k(p_1,\dots,p_n) < 0$ and the inequality $k(x_1,\dots,x_n) < 0$ has the solution $(p_1,\dots,p_n)$ in $F(p_1,\dots,p_n)$ and a fortiori in the real closure R of $F(p_1,\dots,p_n)$. Now consider the real closure R_0 of F. Since F has a unique ordering, the orderings of F in R_0 and in R are identical. Moreover, $k(x_1,\dots,x_n) \in F[x_1,\dots,x_n]$. Hence by the result quoted, there exist $a_i \in R_0$ such that $k(a_1,\dots,a_n) < 0$ and hence such that $f(a_1,\dots,a_n) < 0$. The proof will be completed by showing that the a_i can be chosen in F.

LEMMA. *Let F be an ordered field that is dense in its real closure R and suppose that for $k(x_1,\dots,x_n) \in R[x_1,\dots,x_n]$ there exist $a_i \in R$ such that $k(a_1,\dots,a_n) < 0$. Then there exist $b_i \in F$ such that $k(b_1,\dots,b_n) < 0$.*

Proof. We use induction on n. If $n = 1$, let a' be chosen $< a$ $(= a_1)$ so that the interval $[a', a]$ contains no root of k. Then $k(x) < 0$ for all x in $[a', a]$ (BAI, p. 310) and we may choose $x = b \in F$ in $[a', a]$. Then $k(b) < 0$. Now assume the result for $n-1$ variables. Then the one-variable case shows that there exists a $b_1 \in F$ such that $k(b_1, a_2, \ldots, a_n) < 0$ and the $n-1$-variable case implies that there exist $b_2, \ldots, b_n$ such that $k(b_1, b_2, \ldots, b_n) < 0$. $\square$

Remark. It has been shown by K. McKenna that the following converse of Theorem 11.8 holds: If F is an ordered field such that any rational function f that is positive semi-definite in the sense that $f(a_1, \ldots, a_n) \geqslant 0$ for all $(a_1, \ldots, a_n)$ for which f is defined is a sum of squares, then F is uniquely ordered and is dense in its real closure.

EXERCISES

1. (J. Keisler.) Let F be ordered and let the extension field $F(x)$, x transcendental, be ordered as in exercise 4, p. 634. Show that $F(x)$ is not dense in its real closure by showing that there is no element of $F(x)$ in the interval $[\sqrt{x}, 2\sqrt{x}]$.

2. Let R be real closed and let $f(x) \in R[x]$ satisfy $f(a) \geqslant 0$ for all $a \in R$. Show that $f(x)$ is a sum of two squares in $R[x]$.

3. (C. Procesi.) Let R be a real closed field and let V be an irreducible algebraic variety defined over R, F the field of rational functions on V (see exercise 7, pp. 429–430). Let $h_1, \ldots, h_k \in F$ and let X be the set of points ρ of V such that $h_i(\rho) \geqslant 0$. Show that if $g \in F$ satisfies $g(\rho) \geqslant 0$ for all $\rho \in X$ on which g is defined, then g has the form

$$\Sigma' s_{i_1 \cdots i_j} h_{i_1} \cdots h_{i_j}$$

where Σ' indicates summation on the indices between 1 and k in strictly increasing order and the $s_{i_1 \cdots i_j}$ are sums of squares in F.

(*Sketch of proof.* The conclusion is equivalent to the following: g is a sum of squares in $F_1 = F(\sqrt{h_1}, \ldots, \sqrt{h_k})$. There is no loss in generality in assuming that g and the h_i are polynomials in the coordinate functions $p_1, \ldots, p_n$ (as in the proof of Artin's theorem), that is, we have polynomials $g(x_1, \ldots, x_n)$, $h_j(x_1, \ldots, x_n)$, $1 \leqslant j \leqslant k$, in indeterminates x_i with coefficients in R such that $g = g(p_1, \ldots, p_n)$, $h_j = h_j(p_1, \ldots, p_n)$. If g is not a sum of squares in F_1, then there exists an ordering of F_1 such that $g < 0$. Let $f_1(x_1, \ldots, x_n), \ldots, f_m(x_1, \ldots, x_n)$ be generators of the prime ideal in $R[x_1, \ldots, x_n]$ defining V. Then we have $f_1(p_1, \ldots, p_n) = 0, \ldots, f_m(p_1, \ldots, p_n) = 0$, $h_1(p_1, \ldots, p_n) \geqslant 0, \ldots, h_k(p_1, \ldots, p_n) \geqslant 0$, $g(p_1, \ldots, p_n) < 0$ in F_1 and hence in a real closure R_1 of F_1. Consequently, we have $(a_1, \ldots, a_n)$, $a_i \in R$, such that $f_1(a_1, \ldots, a_n) = 0, \ldots, f_m(a_1, \ldots, a_n) = 0$, $h_1(a_1, \ldots, a_n) \geqslant 0, \ldots, h_k(a_1, \ldots, a_n) \geqslant 0$, $g(a_1, \ldots, a_n) < 0$. This contradicts the hypothesis.)

4. (J. J. Sylvester.) Let $f(x) \in R[x]$, R real closed, x an indeterminate, and let $A = R[x]/(f(x))$. Let $T(a, b)$ be the trace bilinear form on A/R and $Q(a) = T(a, a)$ the corresponding quadratic form. Show that the number of distinct roots of $f(x)$ in R is the signature of Q (BAI, p. 359).

5. Let F be a field, $f(x)$ a monic polynomial in $F[x]$, and let $f(x) = \prod_1^n (x - r_i)$ in a splitting field E/F of $f(x)$. Put $s_j = \sum_{i=1}^n r_i^j$, so $s_j \in F$. Then the matrix

$$\begin{bmatrix} s_0 & s_1 & \cdots & s_{n-1} \\ s_1 & s_2 & \cdots & s_n \\ \cdot & \cdot & \cdots & \cdot \\ s_{n-1} & s_n & \cdots & s_{2n-2} \end{bmatrix}$$

is called the *Bézoutiant* of $f(x)$. Show that the determinant of the Bézoutiant is the discriminant of the trace form $T(a,b)$ on $A = F[x]/(f(x))$ determined by the base $\bar{1}, \bar{x}, \ldots, \bar{x}^{n-1}$ where $\bar{x}^i = x^i + (f(x))$.

6. (Sylvester.) Notations as in exercises 4 and 5. Let b_k denote the sum of the k-rowed principal (=diagonal) minors of the Bézoutiant of $f(x)$ (hence, the characteristic polynomial of the Bézoutiant is $x^n - b_1 x^{n-1} + \cdots + (-1)^n b_n$). Show that all the roots of $f(x)$ are in R if and only if every $b_i \geqslant 0$.

7. (Procesi.) Let R be a real closed field and let g be a symmetric rational function of n variables with coefficients in R such that $g(a_1, \ldots, a_n) \geqslant 0$ for all $(a_1, \ldots, a_n) \in R^{(n)}$ on which g is defined. Let b_k denote the sum of the k-rowed minors of the Bézoutiant of $f(x) = \prod_1^n (x - \rho_i)$. Show that g has the form $\sum' s_{i_1 \cdots i_j} b_{i_1} \cdots b_{i_j}$ where $\sum'$ indicates summation on the indices between 1 and k in strictly increasing order and the $s_{i_1 \cdots i_j}$ are sums of squares of symmetric functions.

11.5 PFISTER THEORY OF QUADRATIC FORMS

If an element of a field is a sum of squares, can we assert that it is a sum of n squares for a specified n? It is not difficult to see that if R is a real closed field, then any element of $R(x)$ that is a sum of squares is a sum of two squares. Hilbert showed that if a rational function of two variables over R is positive semi-definite, then it is a sum of four squares. In the next section we shall prove Pfister's theorem that if R is real closed, any element of $R(x_1, \ldots, x_n)$ that is a sum of squares is a sum of 2^n squares. We shall also sketch in the exercises of section 11.6 a proof of a theorem of Cassels that there exist elements in $R(x_1, \ldots, x_n)$ that are sums of squares but cannot be written as sums of fewer than n squares. The exact value of the number $k(n)$ such that every element in $R(x_1, \ldots, x_n)$ that is a sum of squares is a sum of $k(n)$ squares is at present unknown. The results we have indicated give the inequalities $n \leqslant k(n) \leqslant 2^n$.

The proof of Pfister's theorem is based on some results on quadratic forms that are of considerable independent interest. We devote this section to the exposition of these results.

We deal exclusively with quadratic forms Q on finite dimensional vector spaces V over a field F of characteristic $\neq 2$. As usual, we write $B(x, y) =$

$Q(x+y)-Q(x)-Q(y)$. Since the characteristic is $\neq 2$, it is preferable to replace $B(x, y)$ by $Q(x, y) = \frac{1}{2}B(x, y)$. Then $Q(x, x) = Q(x)$. We shall now indicate that V has an orthogonal base $(u_1, \ldots, u_n)$ relative to Q such that $Q(u_i) = b_i$ by writing

$$Q \sim \operatorname{diag}\{b_1, b_2, \ldots, b_n\}. \tag{3}$$

We shall also write

$$\operatorname{diag}\{b_1, \ldots, b_n\} \sim \operatorname{diag}\{c_1, \ldots, c_n\} \tag{4}$$

if the quadratic forms $\sum_1^n b_i x_i^2$ and $\sum_1^n c_i x_i^2$ are equivalent.

We now introduce some concepts and results on quadratic forms that we have not considered before. First, we consider the tensor product of quadratic forms. Let V_i, $i = 1, 2$, be a vector space over F, Q_i a quadratic form on V_i. We shall show that there is a unique quadratic form $Q_1 \otimes Q_2$ on $V_1 \otimes V_2$ such that

$$(Q_1 \otimes Q_2)(v_1 \otimes v_2) = Q_1(v_1)Q_2(v_2) \tag{5}$$

for $v_i \in V_i$. Let $(u_1^{(i)}, \ldots, u_{n_i}^{(i)})$ be a base for V_i/F. Then $(u_i^{(1)} \otimes u_j^{(2)})$ is a base for $V_1 \otimes V_2$ and we have a quadratic form Q on $V_1 \otimes V_2$ such that

$$Q(u_i^{(1)} \otimes u_j^{(2)}, u_k^{(1)} \otimes u_l^{(2)}) = Q_1(u_i^{(1)}, u_k^{(1)})Q_2(u_j^{(2)}, u_l^{(2)}). \tag{6}$$

If $v_1 = \sum a_i u_i^{(1)}$ and $v_2 = \sum b_j u_j^{(2)}$, then

$$\begin{aligned} Q(v_1 \otimes v_2) &= Q(v_1 \otimes v_2, v_1 \otimes v_2) \\ &= Q(\textstyle\sum a_i b_j u_i^{(1)} \otimes u_j^{(2)}, \sum a_k b_l u_k^{(1)} \otimes u_l^{(2)}) \\ &= \sum_{i,j,k,l} a_i a_k b_j b_l Q_1(u_i^{(1)}, u_k^{(1)}) Q_2(u_j^{(2)}, u_l^{(2)}) \\ &= (\textstyle\sum a_i a_k Q_1(u_i^{(1)}, u_k^{(1)}))(\sum b_j b_l Q_2(u_j^{(2)}, u_l^{(2)})) \\ &= Q_1(v_1)Q_2(v_2). \end{aligned}$$

Putting $Q = Q_1 \otimes Q_2$ we have (5) and since the vectors $v_1 \otimes v_2$ span $V_1 \otimes V_2$, it is clear that $Q_1 \otimes Q_2$ is unique.

If $(u_1^{(i)}, \ldots, u_{n_i}^{(i)})$ is an orthogonal base for V_i, then $(u_j^{(1)} \otimes u_k^{(2)})$ is an orthogonal base for $V_1 \otimes V_2$ and if

$$Q_i \sim \operatorname{diag}\{b_1^{(i)}, \ldots, b_{n_i}^{(i)}\},$$

then

$$
\begin{aligned}
Q_1 \otimes Q_2 \sim \operatorname{diag}\{&b_1^{(1)}b_1^{(2)}, \ldots, b_1^{(1)}b_{n_2}^{(2)}; b_2^{(1)}b_1^{(2)}, \ldots, b_2^{(1)}b_{n_2}^{(2)};\\
&\ldots; b_{n_1}^{(1)}b_1^{(2)}, \ldots, b_{n_1}^{(1)}b_{n_2}^{(2)}\}\\
= \operatorname{diag}\{&b_1^{(1)}, \ldots, b_{n_1}^{(1)}\} \otimes \operatorname{diag}\{b_1^{(2)}, \ldots, b_{n_2}^{(2)}\}
\end{aligned} \tag{7}
$$

where the tensor product of matrices is as defined on p. 250.

In a similar manner, we can define the tensor product of more than two quadratic forms and we have the following generalization of (7): If $Q_i \sim \operatorname{diag}\{b_1^{(i)}, \ldots, b_{n_i}^{(i)}\}$, then

$$Q_1 \otimes \cdots \otimes Q_r \sim \operatorname{diag}\{b_1^{(1)}, \ldots, b_{n_1}^{(1)}\} \otimes \cdots \otimes \operatorname{diag}\{b_1^{(r)}, \ldots, b_{n_r}^{(r)}\}. \tag{8}$$

It is also convenient to define the direct sum of two quadratic forms. If Q_i, $i = 1, 2$, is a quadratic form on V_i, then $Q_1 \oplus Q_2$ is defined to be the quadratic form on $V_1 \oplus V_2$ such that

$$(Q_1 \oplus Q_2)(v_1, v_2) = Q_1(v_1) + Q_2(v_2).$$

It is clear that $Q_1 \oplus Q_2$ is well-defined and if V_1 and V_2 are identified in the usual way with subspaces of $V_1 \oplus V_2$, then these subspaces are orthogonal relative to the bilinear form of $Q_1 \oplus Q_2$. If $(u_1^{(i)}, \ldots, u_{n_i}^{(i)})$ is an orthogonal base for V_i, then $(u_1^{(1)}, \ldots, u_{n_1}^{(1)}, u_1^{(2)}, \ldots, u_{n_2}^{(2)})$ is an orthogonal base for $V_1 \oplus V_2$. We have

$$Q_1 \oplus Q_2 \sim \operatorname{diag}\{b_1^{(1)}, \ldots, b_{n_1}^{(1)}, b_1^{(2)}, \ldots, b_{n_2}^{(2)}\} \tag{9}$$

if $Q_i \sim \operatorname{diag}\{b_1^{(i)}, \ldots, b_{n_i}^{(i)}\}$. If B and C are matrices, it is convenient to denote the matrix $\begin{pmatrix} B & 0 \\ 0 & C \end{pmatrix}$ by $B \oplus C$. Using this notation we can rewrite

$$Q_1 \oplus Q_2 \sim \operatorname{diag}\{b_1^{(1)}, \ldots, b_{n_1}^{(1)}\} \oplus \operatorname{diag}\{b_1^{(2)}, \ldots, b_{n_2}^{(2)}\}.$$

We now consider Pfister's results on quadratic forms that yield the theorem on sums of squares stated at the beginning of the section. Our starting point is a weak generalization of A. Hurwitz's theorem on sums of squares (see BAI, pp. 438–451). Hurwitz proved that there exist identities of the form $(\sum_1^n x_i^2)(\sum_1^n y_i^2) = \sum_1^n z_i^2$ where the z_i depend bilinearly on the x's and the y's if and only if $n = 1, 2, 4$ or 8. Pfister has shown that for any n that is a power of two, the product of any two sums of squares in a field F is a sum of squares in F. Thus at the expense of dropping the requirement that the z_i depend bilinearly on the x's and y's, we have $(\sum_1^n x_i^2)(\sum_1^n y_i^2) = \sum_1^n z_i^2$ for any n that is a power of two. More generally, we consider quadratic forms Q that are *multiplicative* in the sense that given any two vectors x and y there exists a vector z such that $Q(x)Q(y) = Q(z)$. A stronger condition on Q is given in

DEFINITION 11.4. *A quadratic form Q is said to be* strongly multiplicative *if Q is equivalent to cQ for any $c \neq 0$ represented by Q.*

This means that there exists a bijective linear transformation η of V such that $cQ(x) = Q(\eta x)$ for all x. Then if $Q(y) = c$, $Q(x)Q(y) = Q(z)$ for $z = \eta x$; hence Q strongly multiplicative implies Q multiplicative. If Q is of maximal Witt index (BAI, p. 369) on an even dimensional space, then Q can be given coordinate-wise as $\sum_1^m x_i^2 - \sum_{m+1}^{2m} x_j^2$. It is easily seen that Q is strongly multiplicative. On the other hand, the quadratic form $Q = x_1^2 - x_2^2 - x_3^2$ on a three-dimensional vector space over $\mathbb{R}$ is multiplicative but not strongly multiplicative. The fact that it is multiplicative follows from the observation that any form that is universal has this property. On the other hand, Q is not equivalent to $-Q$ by Sylvester's theorem (BAI, p. 359). For any quadratic form Q on a vector space V/F we let F_Q^* denote the set of non-zero elements of F represented by Q. It is clear that F_Q^* is closed under multiplication if and only if Q is multiplicative. Since $Q(x) = c \neq 0$ implies $Q(c^{-1}x) = c^{-1}$, it is clear that Q is multiplicative if and only if F_Q^* is a subgroup of F^*.

We proceed to derive Pfister's results. We give first a criterion for equivalence in the binary case.

LEMMA 1. $\operatorname{diag}\{b_1, b_2\} \sim \operatorname{diag}\{c_1, c_2\}$ *if and only if c_1 is represented by $b_1x_1^2 + b_2x_2^2$ and b_1b_2 and c_1c_2 differ by a square factor.*

Proof. Since b_1b_2 is a discriminant of $Q = b_1x_1^2 + b_2x_2^2$ and the b_i are represented by Q, it is clear that the conditions are necessary. Now suppose they hold. By the first condition we have a vector y such that $Q(y) = c_1$. Then $\operatorname{diag}\{b_1, b_2\} \sim \operatorname{diag}\{c_1, c\}$ and $c_1c = k^2b_1b_2$, $k \in F^*$. Also $c_1c_2 = l^2b_1b_2$. Hence $c_2 = n^2c$ and we can replace c by c_2. Thus $\operatorname{diag}\{b_1, b_2\} \sim \operatorname{diag}\{c_1, c_2\}$. □

We prove next the key lemma:

LEMMA 2. *Let Q be a strongly multiplicative quadratic form, a an element of F^*. Let $Q_a \sim$* $\operatorname{diag}\{1, a\}$. *Then $Q_a \otimes Q$ is strongly multiplicative.*

Proof. It is clear that $Q_a \otimes Q$ is equivalent to $Q \oplus aQ$. Hence, it suffices to show that the latter is strongly multiplicative. We now use the notation $\sim$ also for equivalence of quadratic forms and if $Q_1 \sim \operatorname{diag}\{a, b\}$, then we denote $Q_1 \otimes Q_2$ by $\operatorname{diag}\{a, b\} \otimes Q_2 \sim aQ_2 \oplus bQ_2$. Let k be an element of F^* represented by $Q \oplus aQ$, so $k = b + ac$ where b and c are represented by Q (possibly trivially if b or c is 0). We distinguish three cases:

Case I. $c = 0$. Then $k = b$ and $Q \sim bQ$. Hence $Q \oplus aQ \sim bQ \oplus abQ = b(Q \oplus aQ) = k(Q \oplus aQ)$.

Case II. $b = 0$. Then $k = ac$ and $k(Q \oplus aQ) = kQ \oplus kaQ = acQ \oplus a^2 cQ \sim aQ \oplus Q$ since $cQ \sim Q$ by hypothesis and $Q \sim a^2 Q$ for any $a \in F^*$. Thus $k(Q \oplus aQ) \sim Q \oplus aQ$.

Case III. $bc \neq 0$. We have $Q \oplus aQ \sim bQ \oplus acQ \sim \text{diag}\{b, ac\} \otimes Q$. Since $k = b + ac$ is represented by $bx_1{}^2 + acx_2{}^2$ and bac and $k^2 abc$ differ by a square, it follows from Lemma 1 that $\text{diag}\{b, ac\} \sim \text{diag}\{k, kabc\}$. Hence $\text{diag}\{b, ac\} \otimes Q \sim \text{diag}\{k, kabc\} \otimes Q \sim k\,\text{diag}\{1, abc\} \otimes Q \sim kQ \oplus kabcQ \sim kQ \oplus kaQ = k(Q \oplus aQ)$.

In all cases we have that $Q \oplus aQ \sim k(Q \oplus aQ)$, so $Q \oplus aQ$ is strongly multiplicative. □

It is clear that the quadratic form $Q_0 = x^2 \sim \text{diag}\{1\}$ is strongly multiplicative. Hence iterated application of Lemma 2 gives

THEOREM 11.9. *If the* $a_i \in F^*$, *then*

$$Q \sim \text{diag}\{1, a_1\} \otimes \cdots \otimes \text{diag}\{1, a_n\} \tag{10}$$

is a strongly multiplicative quadratic form. In particular, $\sum_1^{2^n} x_i{}^2 \sim \text{diag}\{1, 1\} \otimes \cdots \otimes \text{diag}\{1, 1\}$ *(n factors) is strongly multiplicative.*

We shall call the forms given in Theorem 11.9 *Pfister forms of dimension* 2^n. We prove next a type of factorization theorem for such forms.

THEOREM 11.10. *Write*

$$\text{diag}\{1, a_1\} \otimes \cdots \otimes \text{diag}\{1, a_n\} = \text{diag}\{1\} \oplus D \tag{11}$$

and let Q be a quadratic form such that $Q \sim D$. Suppose that $b_1 \neq 0$ is represented by Q. Then there exist $b_2, \ldots, b_n \in F^*$ such that

$$\text{diag}\{1, a_1\} \otimes \cdots \otimes \text{diag}\{1, a_n\} \sim \text{diag}\{1, b_1\} \otimes \cdots \otimes \text{diag}\{1, b_n\}. \tag{12}$$

Proof. By induction on n. If $n = 1$, then $b_1 = a_1 c^2$ and hence $\text{diag}\{1, a_1\} \sim \text{diag}\{1, b_1\}$. Now assume the result for n and consider

$$\text{diag}\{1, a_1\} \otimes \cdots \otimes \text{diag}\{1, a_n\} \otimes \text{diag}\{1, a\} = \text{diag}\{1\} \oplus D'. \tag{13}$$

Suppose that $b_1 \neq 0$ is represented by Q' such that $Q' \sim D'$. Then $b_1 = b_1' + ab$ where b_1' is represented by $Q \sim D$ and b is represented by $x^2 \oplus Q \sim \text{diag}\{1, a_1\} \otimes \cdots \otimes \text{diag}\{1, a_n\}$.

Case I. $b = 0$. Then $b'_1 \neq 0$ and the induction hypothesis gives elements $b_2, \ldots, b_n \in F^*$ such that (12) holds. Then

$$\begin{aligned}\operatorname{diag}\{1, a_1\} &\otimes \cdots \otimes \operatorname{diag}\{1, a_n\} \otimes \operatorname{diag}\{1, a\} \\ &\sim \operatorname{diag}\{1, b_1\} \otimes \cdots \otimes \operatorname{diag}\{1, b_n\} \otimes \operatorname{diag}\{1, a\}.\end{aligned}$$

Case II. $b'_1 = 0$. Then $b \neq 0$, $b_1 = ab$, and

$$\begin{aligned}&\operatorname{diag}\{1, a_1\} \otimes \cdots \otimes \operatorname{diag}\{1, a_n\} \otimes \operatorname{diag}\{1, a\} \\ &\quad \sim (\operatorname{diag}\{1, a_1\} \otimes \cdots \otimes \operatorname{diag}\{1, a_n\}) \oplus a(\operatorname{diag}\{1, a_1\} \otimes \cdots \otimes \operatorname{diag}\{1, a_n\}) \\ &\quad \sim (\operatorname{diag}\{1, a_1\} \otimes \cdots \otimes \operatorname{diag}\{1, a_n\}) \oplus ab(\operatorname{diag}\{1, a_1\} \\ &\qquad\qquad \otimes \cdots \otimes \operatorname{diag}\{1, a_n\}) \qquad \text{(Theorem 11.9)} \\ &\quad \sim \operatorname{diag}\{1, ab\} \otimes \operatorname{diag}\{1, a_1\} \otimes \cdots \otimes \operatorname{diag}\{1, a_n\}.\end{aligned}$$

Case III. $bb'_1 \neq 0$. The equivalence established in Case II permits us to replace a by ab. Then by Case I we have $b_2, \ldots, b_n \in F^*$ such that

$$\begin{aligned}\operatorname{diag}\{1, a_1\} &\otimes \cdots \otimes \operatorname{diag}\{1, a_n\} \otimes \operatorname{diag}\{1, ab\} \\ &\sim \operatorname{diag}\{1, b'_1\} \otimes \operatorname{diag}\{1, b_2\} \otimes \cdots \otimes \operatorname{diag}\{1, b_n\} \otimes \operatorname{diag}\{1, ab\}.\end{aligned}$$

Now

$$\begin{aligned}\operatorname{diag}\{1, ab\} \otimes \operatorname{diag}\{1, b'_1\} &\sim \operatorname{diag}\{1, ab, b'_1, abb'_1\} \\ &\sim \operatorname{diag}\{1, abb'_1\} \oplus \operatorname{diag}\{b'_1, ab\} \\ &\sim \operatorname{diag}\{1, abb'_1\} \oplus \operatorname{diag}\{b_1, b_1 abb'_1\} \qquad \text{(Lemma 1)} \\ &\sim \operatorname{diag}\{1, b_1\} \otimes \operatorname{diag}\{1, b'_1 ab\}.\end{aligned}$$

Substituting this in the first equivalence displayed, we obtain the result in this case. □

We are now ready to derive the main result, which concerns the representation of sums of squares by values of Pfister forms.

THEOREM 11.11. *Suppose that every Pfister form of dimension 2^n represents every non-zero sum of two squares in F. Then every Pfister form of dimension 2^n represents every non-zero sum of k squares in F for arbitrary k.*

Proof. By induction on k. Since any Pfister form represents 1, the case $k = 1$ is clear and the case $k = 2$ is our hypothesis. Now assume the result for $k \geqslant 2$. It suffices to show that if Q is a Pfister form of dimension 2^n and a is a sum of k squares such that $c = 1 + a \neq 0$, then c is represented by Q. This will follow if we can show that $Q \oplus -cQ$ represents 0 non-trivially. For then we shall have vectors u and v such that $Q(u) = cQ(v)$ where either $u \neq 0$ or $v \neq 0$. If either $Q(u) = 0$ or $Q(v) = 0$, then both are 0 and so Q represents 0 non-trivially. Then Q is universal and hence represents c. If $Q(u) \neq 0$ and $Q(v) \neq 0$, then these are contained in F_Q^* and hence $c = Q(u)Q(v)^{-1} \in F_Q^*$, so c is represented by Q. We now write $Q = x^2 \oplus Q'$. Since Q represents a, we have $a = a_1{}^2 + a'$ where Q' represents a'. We have $\text{diag}\{1, -c\} \otimes Q \sim Q \oplus (-cQ) \sim x^2 \oplus Q' \oplus (-cQ)$ and $Q' \oplus (-cQ)$ represents $a' - (1 + a_1{}^2 + a') = -(1 + a_1{}^2)$. If this is 0, then $c = a'$ is represented by Q. Hence we may assume that $1 + a_1{}^2 \neq 0$. Then by Theorem 11.10, $\text{diag}\{1, -c\} \otimes Q \sim \text{diag}\{1, -1 - a_1{}^2\} \otimes Q''$ where Q'' is a Pfister form of dimension 2^n. By the hypothesis, this represents $1 + a_1{}^2$. It follows that $\text{diag}\{1, -1 - a_1{}^2\} \otimes Q''$ represents 0 non-trivially. Then $Q \oplus -cQ \sim \text{diag}\{1, -1 - a_1{}^2\} \otimes Q''$ represents 0 non-trivially. This completes the proof. □

11.6 SUMS OF SQUARES IN $R(x_1, \ldots, x_n)$, R A REAL CLOSED FIELD

We now consider the field $R(x_1, \ldots, x_n)$ of rational expressions in n indeterminates $x_1, \ldots, x_n$ over a real closed field R. We wish to show that Theorem 11.11 can be applied to the field $F = R(x_1, \ldots, x_n)$. For this purpose we need to invoke a theorem that was proved by C. C. Tsen in 1936 and was rediscovered by S. Lang in 1951. To state this we require the following

DEFINITION 11.5. *A field F is called a C_i-*field *if for any positive integer d, any homogeneous polynomial f with coefficients in F of degree d in more than d^i indeterminates has a non-trivial zero in $F^{(d^i)}$.*

By a non-trivial zero we mean an element $(a_1, \ldots, a_{d^i}) \in F^{(d^i)}$ such that $(a_1, \ldots, a_{d^i}) \neq (0, \ldots, 0)$ and $f(a_1, \ldots, a_{d^i}) = 0$. The theorem we shall require is the

THEOREM OF TSEN-LANG. *If F is algebraically closed and the x's are indeterminates, then $F(x_1, \ldots, x_n)$ is a C_n-field.*

It is readily seen that any algebraically closed field is a C_0 field. Now suppose that F is not algebraically closed. Then there exists an extension field E/F such that $[E:F] = n > 1$ and n is finite. Let $(u_1, \ldots, u_n)$ be a base for E/F and let ρ be the regular matrix representation determined by this base (BAI, p. 424). Then if $u = \sum a_i u_i$, $a_i \in F$, $N_{E/F}(u) = \det(\sum a_i \rho(u_i)) \neq 0$

unless every $a_i = 0$. Let $x_1, \ldots, x_n$ be indeterminates and put $N(x_1, \ldots, x_n) = \det(\sum x_i\rho(u_i))$. This is a homogeneous polynomial of degree n and $N(a_1, \ldots, a_n) = N_{E/F}(u)$. Hence N has no zero except the trivial one $(0, \ldots, 0)$ and hence F is not a C_0-field. Thus a field is a C_0-field if and only if it is algebraically closed.

C_1-fields were introduced by Artin, who called these fields *quasi-algebraically closed*. We have indicated in a series of exercises in BAI that any finite field is a C_1-field (see exercises 2–7 on p. 137).

The proof of the Tsen-Lang theorem is an inductive one. The proof of the initial step that if F is algebraically closed then $F(x)$ is C_1 and the proof of the inductive step will make use of the following result, which is a consequence of Theorem 7.29, p. 453.

LEMMA 1. *Let F be an algebraically closed field, $f_1, \ldots, f_r$ polynomials without constant terms in n indeterminates with coefficients in F. If $n > r$, then the system of equations $f_1(x_1, \ldots, x_n) = 0, \ldots, f_r(x_1, \ldots, x_n) = 0$ has a non-trivial solution in $F^{(n)}$.*

We need also an analogous result for systems of polynomial equations in a C_i-field, which we derive below. First, we give the following

DEFINITION 11.6. *A polynomial with coefficients in F is called* normic of order i (>0) for F *if it is homogeneous of degree $d > 1$ in d^i indeterminates and has only the trivial zero in F.*

The argument used above shows that if F is not algebraically closed, then there exist normic polynomials of order 1 for F. If F is arbitrary and t is an indeterminate, then it is readily seen that $F(t)$ is not algebraically closed (for example, $\sqrt{t} \notin F(t)$). Hence there exist normic polynomials of order 1 for $F(t)$. We have also

LEMMA 2. *If there exists a normic polynomial of order i for F, then there exists a normic polynomial of order $i+1$ for $F(t)$, t an indeterminate.*

Proof. Let $N(x_1, \ldots, x_{d^i})$ be a normic polynomial of order i for F. The degree of N is d. We claim that

$$(14)\quad N(x_1, \ldots, x_{d^i}) + N(x_{d^i+1}, \ldots, x_{2d^i})t + \cdots + N(x_{(d-1)d^i+1}, \ldots, x_{d^{i+1}})t^{d-1}$$

is a normic polynomial of order $i+1$ for $F(t)$. Since this polynomial is homogeneous of degree d in d^{i+1} indeterminates, it suffices to show that it has only the trivial zero in $F(t)$. Hence suppose that $(a_1, \ldots, a_{d^{i+1}})$ is a non-trivial zero of (14). Since the polynomial is homogeneous, we may assume that the $a_i \in F[t]$ and not every a_i is divisible by t. Let a_k be the first one that is not, and suppose that $jd^i+1 \leqslant k \leqslant (j+1)d^i$. Then dividing by t^j gives the relation $N(a_{jd^i+1}, \ldots, a_{(j+1)d^i}) \equiv 0 \pmod{t}$ and hence reducing modulo t gives $N(\bar{a}_{jd^i+1}, \ldots, \bar{a}_{(j+1)d^i}) = 0$ where $\bar{a}$ is the constant term of a. Since $\bar{a}_k \neq 0$, this contradicts the hypothesis that N is normic and completes the proof. □

If $N(x_1, \ldots, x_{d^i})$ is a normic polynomial of order i and degree d, then

$$N(N(x_1, \ldots, x_{d^i}), N(x_{d^i+1}, \ldots, x_{2d^i}), \ldots, N(x_{d^{2i}-d+1}, \ldots, x_{d^{2i}})) \tag{15}$$

is homogeneous of degree d^2 in d^{2i} indeterminates. Moreover, this has only the trivial zero in F. Hence (15) is normic. Since $d > 1$, we can apply this process to obtain from a given normic polynomial normic polynomials of arbitrarily high degree. We shall use this remark in the proof of an analogue of Lemma 1 for C_i-fields:

LEMMA 3 (Artin). *Let F be a C_i-field for which there exists a normic polynomial of order i $(i > 0)$. Let $f_1, \ldots, f_r$ be homogeneous polynomials of degree d in the indeterminates $x_1, \ldots, x_n$ with coefficients in F. If $n > rd^i$, then the f's have a common non-trivial zero in F.*

Proof. Let N be a normic polynomial of order i for F. Suppose the degree of N is e. Then the number of indeterminates in N is e^i, which we can write as $e^i = rs+t$ where $s \geqslant 0$ and $0 \leqslant t < r$. We replace the first r indeterminates in N by $f_1(x_1, \ldots, x_n), \ldots, f_r(x_1, \ldots, x_n)$ respectively, the next r by $f_1(x_{n+1}, \ldots, x_{2n})$, $\ldots, f_r(x_{n+1}, \ldots, x_{2n})$, etc., and the last t indeterminates in N by 0's. This gives the polynomial

$$\begin{aligned} M = N(&f_1(x_1, \ldots, x_n), \ldots, f_r(x_1, \ldots, x_n), \\ &f_1(x_{n+1}, \ldots, x_{2n}), \ldots, f_r(x_{n+1}, \ldots, x_{2n}), \ldots, 0, \ldots, 0), \end{aligned}$$

which is homogeneous in ns indeterminates. Moreover, $\deg M = ed$. We want to have $ns > (ed)^i = d^i(rs+t)$. This will be the case if $(n-d^ir)s > d^it$. Since $n > d^ir$ and $0 \leqslant t < r$, this can be arranged by choosing e large enough—which can be done by the remark preceding the lemma. With our choice of e we can conclude from the C_i-property that M has a non-trivial zero in F. Since N is normic, it follows easily that the f_i have a common non-trivial zero in F. □

In the next lemma, for the sake of uniformity of statement, we adopt the convention that 1 is a normic polynomial of order 0. Then we have

LEMMA 4. *Let F be a C_i-field, $i \geqslant 0$, such that there exists a normic polynomial of order i for F. Then $F(t)$, t indeterminate, is a C_{i+1}-field.*

Proof. Let $f(x_1, \ldots, x_n)$ be a homogeneous polynomial of degree d with coefficients in $F(t)$ such that $n > d^{i+1}$. We have to show that f has a non-trivial zero in $F(t)$. There is no loss in generality in assuming that the coefficients of f are polynomials in t. Let r be the degree of f in t. Put

$$x_j = x_{j0} + x_{j1}t + \cdots + x_{js}t^s, \qquad 1 \leqslant j \leqslant n,$$

where the x_{jk} are indeterminates. Then

$$f(x_1, \ldots, x_n) = f_0 + f_1 t + \cdots + f_{sd+r} t^{sd+r}$$

where the f_i are homogeneous polynomials of degree d in the $n(s+1)$ x_{jk} with coefficients in F. The polynomial f will have a non-trivial zero in $F(t)$ if the f_i have a common non-trivial zero in F. By Lemmas 1 and 3 this will be the case if $n(s+1) > (sd+r+1)d^i$. Since $n > d^{i+1}$, the inequality can be satisfied by taking s sufficiently large. □

The proof of the Tsen-Lang theorem is now clear. First, Lemma 4 with $i = 0$ shows that if F is algebraically closed, then $F(x_1)$ is a C_1-field. Next, iterated application of Lemma 2 shows that there exists a normic polynomial of order i for $F(x_1, \ldots, x_i)$. Then iterated application of Lemma 4 implies that $F(x_1, \ldots, x_n)$ is a C_n-field. □

We can apply this result for the case $d = 2$ to conclude that if F is an algebraically closed field, then any quadratic form on a vector space V of 2^{n+1} dimensions over $F(x_1, \ldots, x_n)$ represents 0 non-trivially. It follows that any quadratic form on a vector space of 2^n dimensions over $F(x_1, \ldots, x_n)$ is universal. We shall make use of this result in the proof of

THEOREM 11.12. *Let R be a real closed field and let Q be a Pfister form on a 2^n-dimensional vector space over the field $R(x_1, \ldots, x_n)$. Then Q represents every non-zero sum of two squares in $R(x_1, \ldots, x_n)$.*

Proof. Let Q be a Pfister form on a 2^n-dimensional vector space V over $R(x_1, \ldots, x_n)$. We have to show that if $b = b_1{}^2 + b_2{}^2 \neq 0$, $b_i \in R(x_1, \ldots, x_n)$, then b is represented by Q. Since Q represents 1, the result is clear if $b_1 b_2 = 0$.

Hence we assume $b_1b_2 \neq 0$. Let $C = R(i)$, $i^2 = -1$, and consider the extension field $C(x_1,\ldots,x_n)$ of $R(x_1,\ldots,x_n)$ and the vector space $\tilde{V} = V_{C(x_1,\ldots,x_n)} = C(x_1,\ldots,x_n) \otimes_{R(x_1,\ldots,x_n)} V$. If (e_1, e_2) is a base for C/R, then this is a base for $C(x_1,\ldots,x_n)$ over $R(x_1,\ldots,x_n)$. Moreover, every element of $\tilde{V}$ can be written in one and only one way as $e_1u_1+e_2u_2$, $u_i \in V$ (identified with a subspace of $\tilde{V}$ in the usual way). The quadratic form Q has a unique extension to a quadratic form $\tilde{Q}$ on $\tilde{V}$. Evidently $\tilde{Q}$ is a Pfister form. Now put $q = b_1+b_2i$. Then $(1, q)$ is a base for C/R and $q^2-2b_1q+b = q^2-2b_1q+(b_1{}^2+b_2{}^2) = 0$. There exists a vector $\tilde{u} = u_1+qu_2$, $u_i \in V$, such that $\tilde{Q}(\tilde{u}) = q$. Then $Q(u_1)+2qQ(u_1,u_2)+q^2Q(u_2) = q$. Since $(1, q)$ is a base for $C(x_1,\ldots,x_n)/R(x_1,\ldots,x_n)$ and $q^2-2b_1q+b = 0$, this implies that $Q(u_1) = bQ(u_2)$. It follows that b is represented by Q. □

If we combine Artin's theorem (p. 640) with Theorems 11.11 and 11.12, we obtain the main result:

THEOREM 11.13. *Let R be a real closed field. Then any positive semi-definite rational function of n variables over R is a sum of 2^n squares.*

More generally, the same theorems show that any positive semi-definite rational function of n variables in R can be represented by a Pfister form of dimension 2^n. It is noteworthy that to prove Theorem 11.13 we had to use Pfister forms other than the sum of squares.

EXERCISES

1. (J. W. S. Cassels.) Let F be a field of characteristic $\neq 2$, x an indeterminate. Let $p(x) \in F[x]$ be a sum of n squares in $F(x)$. Show that $p(x)$ is a sum of n squares in $F[x]$.

(*Outline of proof.* The result is clear if $p = 0$. Also if $-1 = \sum_2^n a_j{}^2$ where the $a_j \in F$, then

$$p = \left(\frac{p+1}{2}\right)^2 + \sum_{j=2}^{n} \left(\frac{a_j(p-1)}{2}\right)^2.$$

Hence we may assume that $p \neq 0$ and that $Q = \sum_1^n x_i{}^2$ is anisotropic in F and hence in $F(x)$. We have polynomials $f_0(x), f_1(x), \ldots, f_n(x)$ such that $f_0(x) \neq 0$ and

$$p(x)f_0(x)^2 = f_1(x)^2 + \cdots + f_n(x)^2. \tag{16}$$

Let $f_0(x)$ be of minimum degree for such polynomials and assume that $\deg f_0(x) > 0$. Write

$$f_i(x) = f_0(x)g_i(x) + r_i(x), \qquad 0 \leqslant i \leqslant n, \tag{17}$$

where $\deg r_i(x) < \deg f_0(x)$ (we take $\deg 0 = -\infty$). Let Q' be the quadratic form such that $Q' \sim \mathrm{diag}\{-p, 1, \ldots, 1\}$ with n 1's. Then (16) gives $Q'(f) = 0$ where $f = (f_0, \ldots, f_n)$. The result holds if $Q'(g) = 0$ for $g = (g_0 = 1, g_1, \ldots, g_n)$, so assume that $Q'(g) \neq 0$. This implies that f and g are linearly independent over $F(x)$. Then $h = Q'(g, g)f - 2Q(f, g)g \neq 0$ and $Q'(h) = 0$. If $h_0 = 0$, then Q represents 0 nontrivially in $F(x)$, hence in F contrary to the hypothesis. Hence $h_0 \neq 0$. The h_i are polynomials in x and

$$h_0 = Q'(g, g)f_0 - 2Q(f, g) = \frac{1}{f_0} Q'(f_0 g - f) = \frac{1}{f_0} \sum_1^n (r_i(x))^2.$$

Then $\deg h_0 < \deg f_0$, which is a contradiction.)

2. (Cassels.) Let F, x be as in exercise 1 and let $x^2 + d$, for $d \in F$, be a sum of $n > 1$ squares in $F(x)$. Show that either -1 is a sum of $n-1$ squares in F or d is a sum of n squares in F.

3. (Cassels.) Use exercises 1 and 2 to show that if R is real closed, then $x_1{}^2 + \cdots + x_n{}^2$ is not a sum of $n-1$ squares in $R(x_1, \ldots, x_n)$.

4. (T. Motzkin.) Let R be real closed, x and y indeterminates, and let

$$p(x, y) = 1 + x^2(x^2 - 3)y^2 + x^2 y^4.$$

Verify that

$$p(x, y) = \frac{(1 - x^2 y^2)^2 + x^2(1 - y^2)^2 + x^2(1 - x^2)^2 y^2}{1 + x^2}$$

$$= \frac{(1 - x^2 - 2x^2 y^2)^2 + (x(1 - x^2)y)^2 + (x(1 - x^2)y^2)^2 + (x^2(1 - x^2)y^2)^2}{(1 + x^2)^2}$$

so $p(x, y)$ is a sum of four squares in $R(x)[y]$. Show that $p(x, y)$ is not a sum of squares in $R[x, y]$.

5. Show that any algebraic extension of a C_i-field is a C_i-field.

11.7 ARTIN-SCHREIER CHARACTERIZATION OF REAL CLOSED FIELDS

We conclude our discussion of real closed fields by establishing a beautiful characterization of these fields that is due to Artin and Schreier. We have seen that a field R is real closed if and only if $\sqrt{-1} \notin R$ and $C = R(\sqrt{-1})$ is algebraically closed (Theorem 11.3, p. 634). Artin and Schreier prove the following considerably stronger theorem:

THEOREM 11.14. *Let C be an algebraically closed field, R a proper subfield of finite codimension in C ($[C:R] < \infty$). Then R is real closed and $C = R(\sqrt{-1})$.*

We prove first some elementary lemmas on fields of characteristic $p \neq 0$. The first two of these were given in BAI, but for convenience we record again the statements and proofs.

LEMMA 1. *Let F be a field of characteristic p, a an element of F that is not a pth power. Then for any $e \geqslant 1$, the polynomial $x^{p^e}-a$ is irreducible in $F[x]$.*

Proof. If E is a splitting field of $x^{p^e}-a$, then we have the factorization $x^{p^e}-a = (x-r)^{p^e}$ in $E[x]$. Hence if $g(x)$ is a monic factor of $x^{p^e}-a$ in $F[x]$, then $g(x) = (x-r)^k$, $k = \deg g(x)$. Then $r^k \in F$ and $r^{p^e} = a \in F$. If $p^f = (p^e, k)$ there exist integers m and n such that $p^f = mp^e + nk$. Then $r^{p^f} = (r^{p^e})^m (r^k)^n \in F$. If $k < p^e$, then $f < e$ and if $b = r^{p^f}$, then $b^{p^{e-f}} = a$, contrary to the hypothesis that a is not a pth power in F. □

LEMMA 2. *If F is a field of characteristic p and $a \in F$ is not of the form $u^p - u$, $u \in F$, then $x^p - x - a$ is irreducible in $F[x]$.*

Proof. If r is a root of $x^p - x - a$ in $E[x]$, E a splitting field, then $r+1, r+2, \ldots, r+(p-1)$ are also roots of $x^p - x - a$. Hence we have the factorization

$$x^p - x - a = \prod_{i=0}^{p-1} (x - r - i)$$

in $E[x]$. If $g(x) = x^k - bx^{k-1} + \cdots$ is a factor of $x^p - x - a$, then $kr + l1 = b$ where l is an integer. Hence $k < p$ implies that $r \in F$. Since $r^p - r = a$, this contradicts the hypothesis. □

LEMMA 3. *Let F and a be as in Lemma 2 and let E be a splitting field for $x^p - x - a$. Then there exists an extension field K/E such that $[K:E] = p$.* (Compare Theorem 8.32, p. 510.)

Proof. We have $E = F(r)$ where $r^p = r + a$. We claim that the element $ar^{p-1} \in E$ is not of the form $u^p - u$, $u \in E$. For, we can write any u as $u_0 + u_1 r + \cdots + u_{p-1} r^{p-1}$, $u_i \in F$, and the condition $u^p - u = ar^{p-1}$ and the relation $r^p = r + a$ give

$$u_0{}^p + u_1{}^p(r+a) + u_2{}^p(r+a)^2 + \cdots + u_{p-1}^p (r+a)^{p-1}$$
$$- u_0 - u_1 r - \cdots - u_{p-1} r^{p-1} = ar^{p-1}.$$

Since $(1, r, \ldots, r^{p-1})$ is a base, this gives the relation $u_{p-1}^p - u_{p-1} = a$, contrary to the hypothesis on a. It now follows from Lemma 2 that $x^p - x - ar^{p-1}$

is irreducible in $E[x]$. Hence if K is the splitting field over E of this polynomial, then $[K:E] = p$. $\square$

We can now give the

Proof of Theorem 11.14. Let $C' = R(\sqrt{-1}) \subset C$. We shall show that $C' = C$. Then the result will follow from Theorem 11.3. Now C is the algebraic closure of C'; hence any algebraic extension of C' is isomorphic to a subfield of C/C' and so its dimensionality is bounded by $[C:C']$. The first conclusion we can draw from this is that C' is perfect. Otherwise, the characteristic is a prime p and C' contains an element that is not a pth power in C'. Then by Lemma 1, there exists an algebraic extension of C' that is p^e-dimensional for any $e \geqslant 1$. Since this has been ruled out, we see that C' is perfect. Then C is separable algebraic over C' and since C is algebraically closed, C is finite dimensional Galois over C'. Let $G = \text{Gal}\, C/C'$, so $|G| = [C:C']$.

Now suppose that $C \neq C'$. Then $|G| \neq 1$. Let p be a prime divisor of $|G|$. Then G contains a cyclic subgroup H of order p. If E is the subfield of H-fixed elements, then C is p-dimensional cyclic over E. If p were the characteristic, then $C = E(r)$ where the minimum polynomial of r over E has the form $x^p - x - a$ (BAI, p. 308). Then by Lemma 3, there exists a p-dimensional extension field of C. This contradicts the fact that C is algebraically closed. Hence the characteristic is not p, and since C is algebraically closed, C contains p distinct pth roots of unity. These are roots of $x^p - 1 = (x-1)(x^{p-1} + x^{p-2} + \cdots + 1)$ and since the irreducible polynomials in $E[x]$ have degree dividing $[C:E] = p$, it follows that the irreducible factors of $x^p - 1$ in $E[x]$ are linear and hence the p pth roots of unity are contained in E. Then $C = E(r)$ where the minimum polynomial of r over E is $x^p - a$, $a \in E$ (BAI, p. 308). Now consider the polynomial $x^{p^2} - a$. This factors as $\prod_{i=1}^{p^2} (x - u^i s)$ where u is a primitive p^2-root of unity and $s^{p^2} = a$. If any $u^i s \in E$, then $(u^i s)^p \in E$ and $((u^i s)^p)^p = a$, contrary to the irreducibility of $x^p - a$ in $E[x]$. It follows that the irreducible factors of $x^{p^2} - a$ in $E[x]$ are of degree p. If b is the constant term of one of these, then $b = s^p v$ where v is a power of u. Since $(s^p)^p = a$, $s^p \notin E$ and since $[C:E] = p$, $C = E(s^p) = E(bs^{-p}) = E(v)$. Since E contains all the pth roots of unity, it follows that v is a primitive p^2-root of unity.

Let P be the prime field of C and consider the subfield $P(v)$ of C. If $P = \mathbb{Q}$, we know that the cyclotomic field of p^rth roots of unity has dimensionality $\varphi(p^r)$ over $\mathbb{Q}$ (BAI, p. 272). This number goes to infinity with r. If P is of finite characteristic l, then we have seen that $l \neq p$. Then the field of p^rth roots of unity over P contains at least p^r elements, so again its dimensionality over P tends to infinity with r. Thus in both cases it follows that there exists an r such that $P(v)$ contains a primitive p^rth root of unity but no primitive p^{r+1}st root of unity. Since v is a primitive p^2-root of unity, $r \geqslant 2$. The field C contains an element w that is a primitive p^{r+1}st root of unity. We now consider the cyclotomic field $P(w)$. Let $K = \text{Gal } P(w)/P$. If P is finite, then we know that K is cyclic (BAI, p. 291). The same thing holds if $P = \mathbb{Q}$ unless $p = 2$ and $r \geqslant 2$. If K is cyclic, then it has only one subgroup of order p and hence, by the Galois correspondence, $P(w)$ contains only one subfield over which it is p-dimensional. We shall now show that $P(w)$ has two such subfields. This will imply that $p = 2$ and the characteristic is 0.

Let $h(x)$ be the minimum polynomial of w over E. Since $v \notin E$, $w \notin E$ and $C = E(w)$. Hence $\deg h(x) = p$. Moreover, $h(x)$ is a divisor of $x^{p^{r+1}} - 1 = \prod_1^{p^{r+1}} (x - w^i)$, so the coefficients of $h(x)$ are contained in the subfield $D = E \cap P(w)$. It follows that $[P(w):D] = p$. Next consider the subfield $D' = P(z)$ where $z = w^p$. The element w is a root of $x^p - z \in D'[x]$ and this polynomial is either irreducible or it has a root in D' (BAI, exercise 1 on p. 248). In the first case $[P(w):D'] = p$. In the second case, since z is a primitive p^rth root of unity, D' contains p distinct pth roots of unity and hence $x^p - z$ is a product of linear factors in $D'[x]$. Then $w \in D'$ and $D' = P(w)$. But $P(v)$ contains a primitive p^rth root of unity and hence $P(v)$ contains z, so if $D' = P(w)$, then $P(v)$ will contain a primitive p^{r+1}st root of unity w, contrary to the choice of r. Thus $[P(w):D'] = p$. Now $D' \neq D$. Otherwise, D contains a primitive p^rth root of unity and E contains a primitive p^rth root of unity. Then E contains w, contrary to the fact that $[C:E] = [E(v):E] = p$. Thus D and D' are distinct subfields of $P(w)$ of co-dimension p in $P(w)$. As we saw, this implies that the characteristic is 0 and $p = 2$. Now $C = E(v)$ and v is a primitive $2^2 = 4$th root of unity. Hence $v = \pm\sqrt{-1}$. Since E contains $\sqrt{-1}$, we contradict $[C:E] = 2$. This completes the proof. $\square$

REFERENCES

E. Artin and O. Schreier, Algebraische Konstruktion reele Körper, *Abhandlungen Mathematische Seminar Hamburgischen Universität*, vol. 5 (1927), 85–99.

E. Artin, Über die Zerlegung definiter Funktionen in Quadrate, ibid., 100–115.

T. Y. Lam, *The Algebraic Theory of Quadratic Forms*, W. A. Benjamin, Reading, Mass., 1973.

Index